CONCENTRATED ACIDS AND BASES

Name	Approximate Weight Percent	Approximate Molarity	mL of Reagent Needed to Prepare 1 L of ~ 1.0 M Solution
Acid			
Acetic	99.8	17.4	57.5
Hydrochloric	37.2	12.1	82.6
Hydrofluoric	49.0	28.9	34.6
Nitric	70.4	15.9	62.9
Perchloric	70.5	11.7	85.5
Phosphoric	85.5	14.8	67.6
Sulfuric	96.0	18.0	55.6
Base			
Ammonia[†]	28.0	14.5	69.0
Sodium hydroxide	50.5	19.4	51.5
Potassium hydroxide	45.0	11.7	85.5

[†]28.0 wt% ammonia is the same as 56.6 wt% ammonium hydroxide.

Quantitative Chemical Analysis

FOURTH EDITION

["The Experiment" by Sempé. Copyright C. Charillon, Paris.]

Quantitative Chemical Analysis

FOURTH EDITION

Daniel C. Harris

Michelson Laboratory
China Lake, California

W. H. Freeman and Company
New York

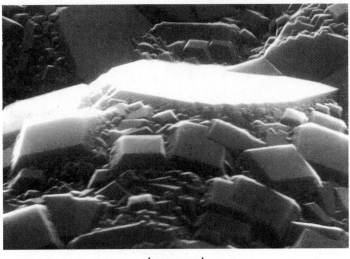

├─── 1 μm ───┤

About the Cover

The cover shows an argon plasma used to grow diamond. Hydrogen and methane added to the plasma react to form microscopic crystals of diamond on a water-cooled tungsten surface. Look carefully at the background on the cover and you will see an electron micrograph, reproduced here, showing the micrometer-size crystals grown by this process.

To help decipher the chemistry of diamond deposition, we use a spectrophotometer to study light emitted by hot molecules in the plasma. The spectrum gives information on the concentrations and temperatures of trace species, such as CH_3, CH, and C_2 involved in diamond growth. [Color photograph by D. Cornelius; diamond micrograph from C. E. Johnson; equipment supplied by W. A. Weimer and S. Reeve, Michelson Laboratory.]

. . . .

Library of Congress Cataloging-in-Publication Data

Harris, Daniel C., 1948–
    Quantitative chemical analysis/Daniel C. Harris.—4th ed.
        p.   cm.
    Includes bibliographical references and index.
    ISBN 0-7167-2508-8
    1. Chemistry, Analytic—Quantitative. I. Title.
QD101.2.H37   1995
545—dc20   94-19138
                                                         CIP

Printed in the United States of America

Fourth printing 1996, RRD

Contents

Preface

The earth is a spaceship in which each of our actions has an effect on everyone else. Analytical chemistry is an essential tool with which we monitor the environment in our attempt to preserve a habitable and hospitable planet for ourselves and future generations. With analytical chemistry we study tree rings, sediments, ice cores, and fossils to decipher nature's past actions. Analytical chemistry aids modern biology and medicine to understand how organisms function and to diagnose and treat disease. Chemical analysis is required for characterization and quality control in the manufacture of semiconductors, metals, plastics, paints, fabrics, fertilizers and nearly everything else that we use. Analytical chemistry is essential to ensure the safety of our food and water supplies. Analytical chemistry does not just belong to chemists. If you breathe the air, eat, drink, use manufactured goods, or visit the doctor when you are ill, your life depends on analytical chemistry every single day.

In writing this text my goals have been to provide a sound physical understanding of the principles of analytical chemistry and to show how these principles are applied in chemistry and related disciplines—especially in life and environmental science. I have attempted to present the subject in a rigorous, readable, and interesting manner that will appeal to students whether or not their primary interest is chemistry. I intend the material to be lucid enough for nonmajors, yet to contain the depth required by advanced undergraduates. This book grew out of an introductory analytical chemistry course that I taught mainly for nonchemistry majors at the University of California at Davis and from a course that I taught for third-year chemistry majors at Franklin and Marshall College in Lancaster, Pennsylvania.

Analytical chemistry and the tools that we use for teaching and learning it are evolving steadily. In preparing the fourth edition of this book, I sought to incorporate current developments in most application areas. Computer spreadsheets are a new feature added to the book as a tool for exploring quantitative relationships. If you are interested in almost any area of science or engineering, you will find spreadsheets to be invaluable for your future work.

The organization of the book has changed somewhat, but the philosophy is still that many chapters can be used out of sequence at the discretion of the instructor. An introduction to spectrophotometry was moved forward to Chapter 6, along with a general treatment of calibration curves, internal standards, and standard addition. Gravimetric analysis was moved to the back of the book and augmented with a discussion of modern combustion analysis. Chromatography was expanded to include capillary electrophoresis. Instructors who have used previous editions will find a new introduction to chromatography, a revised treatment of acids and bases, and my fourth different approach to computing electrochemical cell voltages. (Did I get it "right" this time?)

To help lighten the load of a very dense subject, the text is laced with interesting Boxes and Demonstrations and includes a color insert to illustrate the Demonstrations. Each chapter opens with an illustrated topic of special interest.

Learning is an active process—nobody can do it for you. The two most important ways to master this course are to work problems and to gain experience in the laboratory. This book provides Exercises and Problems at the end of each chapter. The difference between them is that complete solutions to the Exercises are found at the back of the book, but only brief answers to most Problems are at the back of the book. The Exercises are the smallest set of problems covering most of the important, and sometimes complex, topics. The Glossary at the back of the book is to help you as you read and as you review the list of terms at the end of each chapter.

The *Solutions Manual and Supplement for Quantitative Chemical Analysis* has an expanded role. In addition to giving complete solutions to the Problems, it contains supplementary problems and their solutions for each chapter. These can be an additional source of problems for students who want more practice, or they can be reproduced by instructors who wish to assign problems whose answers are not readily available. New material on nonlinear least-squares curve fitting and analysis of variance, as well as a statistics experiment, appear at the beginning of the *Solutions Manual*.

Inevitably, a book of this size and complexity is the work of many people. At W. H. Freeman and Company, John Haber and Deborah Allen played special roles in helping me develop the content and approach of this edition. Mary Louise Byrd expertly shepherded the manuscript through production, with the assistance of Lisa Douglis in laying out the pages. The book's design was created by John Hatzakis. Jodi Simpson was the copy editor who polished the manuscript, caught technical inconsistencies, and generally questioned my every word. Thank you, Jodi. At Michelson Laboratory, continual dialog with Eric Erickson, Mike Seltzer, Wayne Weimer, Scott Reeve, and Bob Green was most beneficial.

Classroom diaries from third edition users were compiled by Paul Bohn (University of Illinois, Urbana), Maurice M. Bursey (University of North Carolina), Howard Drossman (Colorado College), Ingrid Fritsch-Faules (University of Arkansas), Karen A. Henderson (University of Toronto, Scarborough), Scott Hewitt (Cal State Fullerton), Richard T. Keys (Cal State Los Angeles), and E. H. Piepmeier (Oregon State University). These diaries helped me plan the fourth edition, the manuscript of which was read and criticized by Joseph S. Alper (University of Massachusetts, Boston), Mark D. Baker (University of Guelph), E. J. Billo (Boston College), Kim Cohn (Cal State Bakersfield), Julianna V. Gilbert (University of Denver), Harry P. Hopkins (Georgia State University), John H. Nelson (University of Nevada, Reno), E. H. Piepmeier (Oregon State University), Karen B. Sentell (University of Vermont), Michael A. Sweeney (Santa Clara University), Robert Q. Thompson (Oberlin College), and David White (University of Illinois, Urbana). Ed Piepmeier was especially effective with numerous, excellent suggestions. Bob Weinberger (CE Technologies) and Howard Drossman graciously provided expert review of the new material on capillary electrophoresis. Professor A. Gustavo González (Universidad de Sevilla) made me aware of potentiometric stripping analysis. Jim Rynd (Biola University), Robert F. Mauldin (Georgia Southern University), Joseph T. Hupp (Northwestern University), Joseph E. Vitt (University of South Dakota), and Frank R. Smith (Memorial University of Newfoundland) provided valuable corrections to the third edition. My approach to many spreadsheet applications is modeled after the publications of Robert de Levie (Georgetown University), with whom I have had lively correspondence. Two students at Earlham College, Aretha Fiebig and Sean Crosson, convinced me to change the atomic weight table at the back inside cover into a periodic table.

The most special thanks go to my wife, Sally, and my son Doug, who labored with me for the last two years to create this edition. Sally typed the entire manuscript and was instrumental in every aspect of production and proofreading. She also prepared the camera-ready *Solutions Manual*. Doug, who has come of age as a scientist, checked all the solutions and made numerous suggestions for improvements.

This book is dedicated to the students who use it, who occasionally smile when they read it, who gain new insight, and who feel satisfaction after struggling to solve a problem. I have been successful if this book helps you develop critical, independent reasoning that you can apply to new problems. I truly relish your comments, criticisms, suggestions, and corrections. Please address correspondence to me at the Chemistry Division, Research Department, Michelson Laboratory, China Lake, CA 93555.

Dan Harris
November 1994

Quantitative Chemical Analysis

FOURTH EDITION

Measuring the Universe

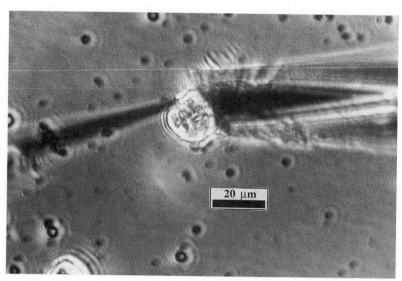

Optical micrograph showing two electrodes adjacent to an adrenal cell that releases the hormone epinephrine. The bar represents a length of 20 micrometers (20×10^{-6} meters). [Photograph courtesy R. M. Wightman, University of North Carolina.]

Every phenomenon in science is described in terms of a small set of units that enumerate quantities such as mass, length, time, temperature, and electric charge. To encompass the enormous range of measurements—from the size of a quark to the distance between galaxies—we use a set of prefixes such as *nano* (10^{-9}) and *mega* (10^6) to multiply the basic units.

The *micrograph* on this page (a photograph taken through a microscope with ×450 magnification) shows two carbon-fiber electrodes prodding an isolated cell from the adrenal gland—yours is located near your kidney. Muscular activity provokes this gland to secrete the hormone *epinephrine* (also called *adrenaline*), which in turn stimulates the release of sugar in muscle cells. Epinephrine also promotes the utilization of fat, elevates blood pressure and heart rate, and increases mental awareness. Stimulation with the well-known chemical nicotine causes the cell to release epinephrine in discrete bursts that can be measured by electrochemical reactions at the electrodes.

In this chapter we learn how to use large and small units to describe such events. You should begin to become comfortable describing the diameter of the cell as 18 micrometers (18×10^{-6} meters) and the number of molecules in a single burst as 65 attomoles (65×10^{-18} moles, or 39 million molecules), producing an electric current of 50 picoamperes (50×10^{-12} amperes). Welcome to the world of quantitative chemical analysis. May your delight be boundless!

Measurements

Chemical analysis plays an essential role not only in chemistry but in biology, medicine, environmental science, geology, oceanography, materials science, forensic science, archeology, and a multitude of other fields. Two pervasive questions that arise in all of these fields are, What chemical elements or compounds are present in this substance I am studying? and How much of the element or compound is present? Identifying the constituents of a substance to answer the first question is called **qualitative analysis,** whereas measuring how much is present is called **quantitative analysis.** This book deals mainly with quantitative analysis, for which we will study an array of powerful methods.

Bold terms should be learned and are listed in the Glossary.

1-1 Steps in a Chemical Analysis

Real analytical problems usually begin with an object that is not suitable for a laboratory experiment. The object may be a chunk of liver, a 2 000-year-old urn, a lake full of water, or a trainload of ore. The material of interest is usually **heterogeneous,** a term meaning that the composition of the material varies from one part to another. (By contrast, a **homogeneous** material has one uniform composition throughout.) Before we can perform a meaningful chemical analysis, we must obtain a small, homogeneous sample whose composition is representative of the larger object.

The *species* (chemical substance) being measured in a chemical analysis is called the **analyte.** We may need to transform the analyte into a form suitable for a particular method of analysis, and we may need to remove from the sample interfering species that would lead to false results.

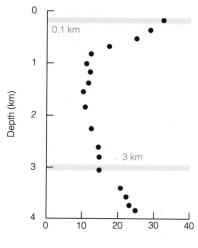

Figure 1-1 Aluminum concentration in seawater varies as a function of depth in the Atlantic Ocean, shown here for 36° 18′ north latitude and 72° 21′ west longitude. Depth is expressed in kilometers (km) and concentration in nanomoles per liter (nM), which stands for 10^{-9} moles per liter. Samples taken at depths of 0.1 and 3 km differ by a factor of 2 in aluminum concentration. [From C. I. Measures and J. M. Edmond, *Anal. Chem.* **1989**, *61*, 544. This notation refers to page 544 of volume 61 of the journal *Analytical Chemistry*, published in 1989.]

Sampling

A *lot* is the total material (the liver, the urn, the lake, etc.) from which samples are taken. A *bulk sample* (also called a *gross sample*) is taken from the lot for analysis or archiving (storing for future reference). The bulk sample must be representative of the lot, and the choice of bulk sample is critical to producing a valid analysis. From the bulk sample, a smaller *laboratory sample* is formed that must have the same composition as the bulk sample. For example, we might obtain a laboratory sample by grinding an entire solid bulk sample to a fine powder, mixing thoroughly, and keeping one bottle of powder for testing. Small test portions (called **aliquots**) of the laboratory sample are used for individual analyses.

A laborious analysis is worthless if the sample that is analyzed is not representative of the original problem. For example, Figure 1-1 shows the variation of aluminum concentration in seawater as a function of depth. If you took water from just one depth, your result would not be representative of the entire ocean. Even a shallow lake is likely to be heterogeneous, with the topmost layer in equilibrium with the atmosphere and the bottom in equilibrium with sediments. Temperature and density gradients in the lake prevent rapid mixing of the layers.

To construct a representative bulk sample from a heterogeneous material, the material is visually divided into segments. A **random sample** is collected by taking portions from the desired number of segments chosen at random. For example, if the material is divided into 1 000 segments, a computer program that generates random numbers between 1 and 1 000 may be used to select 50 segments for sampling.

For highly **segregated materials** (in which different regions clearly have different compositions), a representative **composite sample** is constructed. For example, if a solid appears to have three zones whose relative volumes are 1:6:3, then a composite sample would be made up of samples of each zone in the volume ratio 1:6:3. Sampling within each zone should be done at random. The composite sample (bulk sample) is then homogenized (perhaps by grinding), and a laboratory sample is taken. Chapter 26 provides more details of sample preparation and the statistics of sampling.

Analyzing the Sample

Most chemical analyses have the following steps (Figure 1-2):

Step 1. Obtain a representative bulk sample as described above.

Step 2. Extract from the bulk sample a smaller, homogeneous laboratory sample.

Step 3. Convert the laboratory sample into a form suitable for analysis, a process that usually involves dissolving the substance. Samples with a low concentration of analyte may need to be concentrated prior to analysis.

Step 4. Remove or *mask* species that will interfere with the chemical analysis. **Interference** is a response to a species other than the intended analyte. **Masking** refers to the transformation of an interfering species into a form that is not detected by the analytical method. For example, Ca^{2+} in natural waters is often measured with a reagent called EDTA. Dissolved aluminum interferes with this analysis, so it is masked by treating the sample with excess F^-, which reacts with Al^{3+} to form AlF_6^{3-}; the latter species does not react with EDTA.

Step 5.   Measure the concentration of the analyte in several aliquots. The purpose of *replicate measurements* is to assess the variability (uncertainty) in the analysis and to guard against a gross error in the analysis of a single aliquot. *The uncertainty of a measurement is as important as the measurement itself,* because it tells us how reliable the measurement is. (We will study uncertainty and statistics used to interpret experimental results in Chapters 3 and 4.) In some cases, different analytical methods are used with similar samples to make sure that each method is giving the same result and that the choice of analytical method is not biasing the result. (You may also wish to construct and analyze several different bulk samples to see what variations arise from your sampling procedure.)

Step 6.   Interpret your results and draw conclusions.

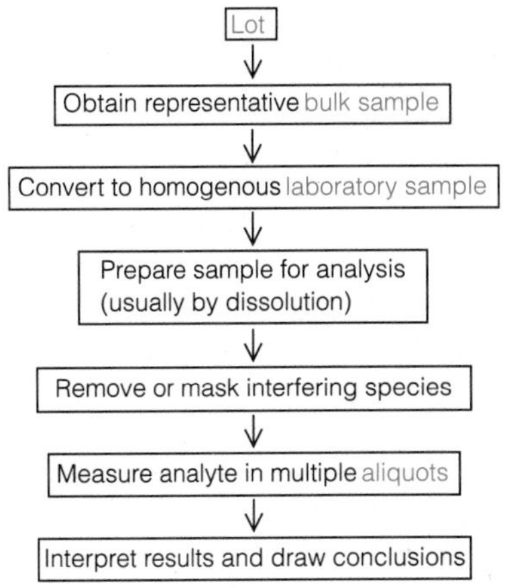

The majority of this book deals with step 5, measuring chemical concentrations in homogeneous aliquots of an unknown. We begin with a refresher on units used to express chemical and physical quantities.

Figure 1-2   Steps in a chemical analysis.

1-2 SI Units

The **SI system** of measurements derives its name from the French *Système International d'Unités*. Fundamental units (base units), from which all others are derived, are defined in Table 1-1. Standards of length, mass, and time are the *meter* (m), *kilogram* (kg), and *second* (s), respectively. Temperature is measured in *kelvins* (K), amount of substance in *moles* (mol), and electric current in *amperes* (A). Table 1-2 lists other quantities that are defined in terms of the fundamental quantities. For example, force is measured in *newtons* (N), pressure is measured in *pascals* (Pa), and energy is measured in *joules* (J), each of which can be expressed in terms of the more fundamental units of length, time, and mass.

Rather than using exponential notation, it is sometimes convenient to use the prefixes in Table 1-3 to express very large or very small quantities. For example, the pressure of the atmosphere is approximately 1.01×10^5 Pa at sea level. The number 10^5 is more than 10^3 and less than 10^6, so it is reasonable to express the pressure in multiples of 10^3 by using the prefix k (for kilo):

Pressure is force per unit area:
$1\ Pa = 1\ N/m^2$

$$\frac{1.01 \times 10^5\ \cancel{Pa}}{10^3\ \dfrac{\cancel{Pa}}{kPa}} = 1.01 \times 10^2\ kPa = 101\ kPa$$

The unit kPa is read as "kilopascals." It is a good idea to write units beside each number in a calculation and to cancel identical units in the numerator and denominator. This practice ensures that you know the units for your answer. If you intend to calculate pressure and your answer comes out with units other than pascals, then you know there is an error in your calculation.

TABLE 1-1   Fundamental SI units

Quantity	Unit	Symbol	Definition
Length	meter	m	The meter is the distance light travels in a vacuum during $\frac{1}{299\,792\,458}$ of a second. This definition fixes the speed of light at exactly 299 792 458 m/s.
Mass	kilogram	kg	One kilogram is the mass of the prototype kilogram kept at Sèvres, France. This is the only SI unit whose primary standard is not defined in terms of physical constants.
Time	second	s	The second is the duration of 9 192 631 770 periods of the radiation corresponding to the transition between two hyperfine levels of the ground state of ^{133}Cs.
Electric current	ampere	A	One ampere is the amount of constant current that will produce a force of 2×10^{-7} N/m (newtons per meter of length) when maintained in two straight, parallel conductors of infinite length and negligible cross section, separated by one meter.
Temperature[1]	kelvin	K	The thermodynamic temperature is defined such that the triple point of water (at which solid, liquid, and gaseous water are in equilibrium) is 273.16 K, and the temperature of absolute zero is 0 K.
Luminous intensity	candela	cd	One candela is the luminous intensity, in a given direction, of a source that emits monochromatic radiation of frequency 540 THz and that has a radiant intensity of $\frac{1}{683}$ W/sr in that direction. Luminous intensity refers only to radiation visible to the human eye. This quantity is not used in this book.
Amount of substance	mole	mol	One mole of substance contains as many molecules (or atoms, if the substance is a monatomic element) as there are atoms of carbon in exactly 0.012 kg of ^{12}C. The number of particles in a mole is approximately $6.022\,136\,7 \times 10^{23}$.
Plane angle	radian	rad	There are 2π radians in a circle.
Solid angle	steradian	sr	There are 4π steradians in a sphere.

TABLE 1-2 SI-derived units with special names

Quantity	Unit	Symbol	Expression in terms of other units	Expression in terms of SI base units
Frequency	hertz	Hz		$1/s$
Force	newton	N		$m \cdot kg/s^2$
Pressure	pascal	Pa	N/m^2	$kg/(m \cdot s^2)$
Energy, work, quantity of heat	joule	J	$N \cdot m$	$m^2 \cdot kg/s^2$
Power, radiant flux	watt	W	J/s	$m^2 \cdot kg/s^3$
Quantity of electricity, electric charge	coulomb	C		$s \cdot A$
Electric potential, potential difference, electromotive force	volt	V	W/A	$m^2 \cdot kg/(s^3 \cdot A)$
Capacitance	farad	F	C/V	$s^4 \cdot A^2/(m^2 \cdot kg)$
Electric resistance	ohm	Ω	V/A	$m^2 \cdot kg/(s^3 \cdot A^2)$

SI Prefixes: A Microscopic Example

The opening of this chapter showed tiny electrodes placed next to a single cell to measure epinephrine released from the cell. The hormone is contained in compartments called *vesicles* inside the cell. When the contents of a single vesicle are released from the surface of the cell (Box 1-1), the electrode responds with a burst of electric current, seen as a spike in Figure 1-3.

Now let's practice the use of SI units and prefixes. The scale bar beneath the cell shown at the beginning of this chapter indicates a length of 20 μm (20 micrometers = 20×10^{-6} m). ("Micrometers" used to be called "microns.") In comparison, the cell appears to have a diameter around 90% of the length of the bar, or 18 μm. The vesicles in Box 1-1 have a diameter of 400 nanometers (nm), which is 400×10^{-9} m = 4×10^{-7} m. In Figure 1-3, electric current is measured in picoamperes (pA; pronounced PEEK-o-amperes), which means 10^{-12} A. The burst of current labeled in the figure is 50 pA above the baseline. Because there are 6.24×10^{18} electrons per second per ampere, 50 pA amounts to

$$\text{electron flow} = (50 \times 10^{-12} \, \cancel{A}) \times (6.24 \times 10^{18} \, \frac{e^-/s}{\cancel{A}})$$
$$= 3.12 \times 10^8 \, e^-/s \qquad (1\text{-}1)$$

(The conversion between amperes and electrons is explained in Chapter 14.)

TABLE 1-3 Prefixes

Prefix	Symbol	Factor
exa	E	10^{18}
peta	P	10^{15}
tera	T	10^{12}
giga	G	10^{9}
mega	M	10^{6}
kilo	k	10^{3}
hecto	h	10^{2}
deka	da	10^{1}
deci	d	10^{-1}
centi	c	10^{-2}
milli	m	10^{-3}
micro	μ	10^{-6}
nano	n	10^{-9}
pico	p	10^{-12}
femto	f	10^{-15}
atto	a	10^{-18}
zepto	z	10^{-21}
yocto	y	10^{-24}

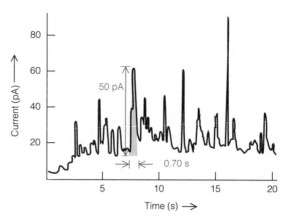

Figure 1-3   Electrode response to the release of epinephrine from the cell shown in the beginning of this chapter. Each vertical spike corresponds to the rupture of one vesicle. [K. T. Kawagoe, J. A. Jankowski, and R. M. Wightman, *Anal. Chem.* **1991,** *63,* 1589; D. J. Leszczyszyn, J. A. Jankowski, O. H. Viveros, E. J. Diliberto, Jr., J. A. Near, and R. M. Wightman, *J. Biol. Chem.* **1990,** *265,* 14736.]

EXAMPLE   **How Many Epinephrine Molecules Are in One Vesicle?**

Each burst of electric current in Figure 1-3 corresponds to a release of epinephrine from one vesicle. The electrochemical half-reaction responsible for flow of electrons to the electrode is the oxidation of epinephrine, in which each molecule gives up two electrons:

Epinephrine   Oxidized epinephrine

Consider the large burst of current whose dimensions are labeled in Figure 1-3. The area in a graph has dimensions of ordinate (y axis) × abscissa (x axis). In Figure 1-3, area has the dimensions current × time. Because current is e^-/s, current × time = $(e^-/\cancel{s}) \times \cancel{s} = e^-$. The area beneath the burst gives the number of electrons in the burst. How many molecules and how many moles of epinephrine are in this burst of current? [A **mole** (mol) is *Avogadro's number* of molecules.]

Solution   We can approximate the signal as a triangle with a height of 50 pA (= $3.12 \times 10^8 \ e^-/s$ in Equation 1-1) and a base of 0.70 s. The area of the triangle gives the number of electrons during the burst:

$$\text{area} = \tfrac{1}{2} \text{ base} \times \text{height} = \tfrac{1}{2} (0.70 \ \cancel{s}) \times \left(3.12 \times 10^8 \ \frac{e^-}{\cancel{s}}\right) = 1.1 \times 10^8 \ e^-$$

For every two electrons reaching the electrode, one molecule of epinephrine must have been released from the cell:

$$\text{molecules of epinephrine} = \frac{1.1 \times 10^8 \ \cancel{e^-}}{2 \ \cancel{e^-}/\text{molecule}} = 5.5 \times 10^7 \text{ molecules}$$

Box 1-1 Exocytosis and the Release of Epinephrine

The hormone epinephrine is contained in vesicles (compartments) of cells in the adrenal gland. When the cell is exposed to nicotine, the vesicles migrate toward the edge of the cell and fuse (merge) with the cell membrane to release the vesicle contents in a process called *exocytosis*. (The reverse process, in which cells take up molecules from their surroundings, is called *endocytosis*.)

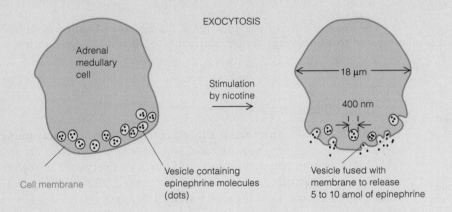

Epinephrine has the structure at the left below, which is commonly abbreviated in the form at the right.

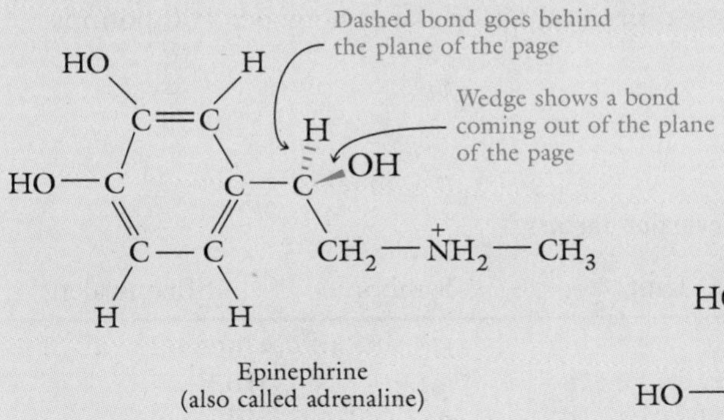

Epinephrine
(also called adrenaline)

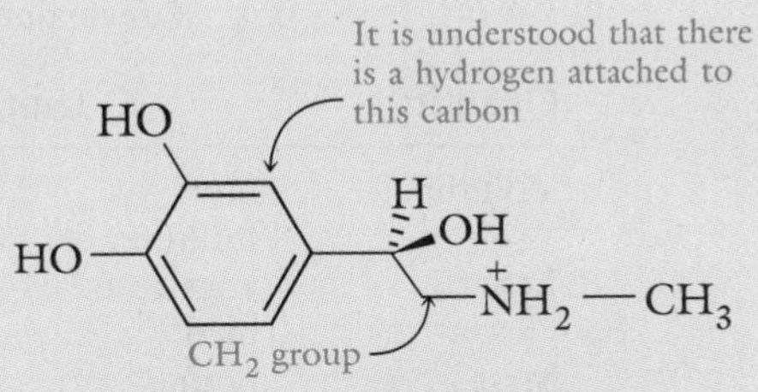

Organic chemist's shorthand drawing of
epinephrine

We use shorthand drawings to simplify the process of writing molecular structures. The rule is that each vertex in a shorthand drawing represents a carbon atom with four bonds to other atoms. Any atom not shown in the structure is assumed to be hydrogen.

For example, the shorthand shows that the carbon atom at the top right of the six-membered benzene ring forms three bonds to other carbon atoms (one single bond and one double bond), so there must be a hydrogen atom attached to this carbon atom. The carbon atom at the left side of the benzene ring forms three bonds to other carbon atoms and one bond to an oxygen atom. There is no hidden hydrogen atom attached to this carbon. In the CH_2 group adjacent to nitrogen, both hydrogen atoms are omitted in the shorthand structure.

Constants such as Avogadro's number are listed inside the front cover of this book.

Knowing that there are Avogadro's number of molecules in a mole, we can determine

$$\text{moles of epinephrine} = \frac{5.5 \times 10^7 \text{ molecules}}{6.022 \times 10^{23} \text{ molecules/mol}} = 9.1 \times 10^{-17} \text{ mol}$$

This is a tiny number. Because 10^{-17} is larger than 10^{-18}, but smaller than 10^{-15}, an appropriate prefix from Table 1-3 is atto (a), which means 10^{-18}.

$$\text{moles of epinephrine} = \frac{9.1 \times 10^{-17} \text{ mol}}{10^{-18} \text{ mol/amol}} = 91 \text{ amol}$$

The burst of current just analyzed is the largest in Figure 1-3. It has been estimated that an average vesicle contains 5 to 10 amol of epinephrine.

Conversion Between Units

One *calorie* is the energy required to heat 1 g of water from 14.5° to 15.5°C.

One *joule* is the energy expended when a force of 1 N acts over a distance of 1 m. This energy expenditure is equivalent to that required to raise 102 g (about $\frac{1}{4}$ lb) by 1 m.

1 cal = 4.184 J
1 pound (mass) ≈ 0.453 6 kg
1 mile ≈ 1.609 km

Although SI is the internationally accepted system of measurement in science, other units are encountered. Useful conversion factors are found in Table 1-4. For example, common non-SI units for energy are the *calorie* (cal) and the *Calorie* (with a capital C, which means 1 000 calories, or 1 kcal). Table 1-4 states that 1 cal is exactly 4.184 J (joules).

Your *basal metabolism* requires approximately 46 Calories per hour (h) per 100 pounds (lb) of body mass to carry out basic functions required for life, apart from doing any kind of exercise. A person walking at 2 miles per hour on a level path requires approximately 45 Calories per hour per 100 pounds of body mass beyond basal metabolism. The same person swimming at 2 miles per hour consumes 360 Calories per hour per 100 pounds.

TABLE 1-4   Conversion factors

Quantity	Unit	Symbol	SI equivalent[a]
Volume	liter	L	*10^{-3} m^3
	milliliter	mL	*10^{-6} m^3
Length	angstrom	Å	*10^{-10} m
	inch	in.	*0.025 4 m
Mass	pound	lb	*0.453 592 37 kg
Force	dyne	dyn	*10^{-5} N
Pressure	atmosphere	atm	*101 325 N/m^2
	bar	bar	*10^5 N/m^2
	torr	1 mm Hg	133.322 N/m^2
	pound/in.^{2}	psi	6 894.76 N/m^2
Energy	erg	erg	*10^{-7} J
	electron volt	eV	1.602 177 33 × 10^{-19} J
	calorie, thermochemical	cal	*4.184 J
	British thermal unit	Btu	1 055.06 J
Power	horsepower		745.700 W
Temperature	centigrade (= celsius)	°C	*K − 273.15
	fahrenheit	°F	*1.8(K − 273.15) + 32

a. An asterisk (*) indicates that the conversion is exact (by definition).

EXAMPLE Unit Conversions

Express the rate of energy use for a walking person (46 + 45 = 91 Calories per hour per 100 pounds of body mass) in kilojoules per hour per kilogram of body mass.

Solution We will convert each non-SI unit separately. First note that 91 Calories equals 91 kcal. Table 1-4 states that 1 cal = 4.184 J, or 1 kcal = 4.184 kJ, so

$$91 \ \cancel{kcal} \times 4.184 \ \frac{kJ}{\cancel{kcal}} = 381 \ kJ$$

Table 1-4 also says that 1 lb is 0.4536 kg; so 100 lb = 45.36 kg. The rate of energy consumption is therefore

$$\frac{91 \ kcal/h}{100 \ lb} = \frac{381 \ kJ/h}{45.36 \ kg} = 8.4 \ \frac{kJ/h}{kg}$$

Correct use of significant figures is discussed in Chapter 3.

You could have written this as one calculation with appropriate unit cancellation:

$$\text{rate} = \frac{91 \ \cancel{kcal}/h}{100 \ \cancel{lb}} \times 4.184 \ \frac{kJ}{\cancel{kcal}} \times \frac{1 \ \cancel{lb}}{0.4536 \ kg} = 8.4 \ \frac{kJ/h}{kg}$$

1-3 Chemical Concentrations

The minor species in a solution is called the **solute** and the major species is the **solvent.** In this text, most discussions concern **aqueous solutions,** in which the solvent is water. **Concentration** refers to how much solute is contained in a given volume or mass.

Molarity and Molality

Molarity (M) is the number of moles of a substance per liter of solution. A mole is Avogadro's number of atoms or molecules. A **liter** (L) is the volume of a cube that is 10 cm on each edge. Because 10 cm = 0.1 m, 1 L = $(0.1 \ m)^3$ = $10^{-3} \ m^3$.

$$\text{molarity (M)} = \frac{\text{moles of solute}}{\text{liters of solution}}$$

The **atomic weight** of an element is the number of grams containing Avogadro's number of atoms. The **molecular weight** (MW) of a compound is the sum of atomic weights of the atoms in the molecule. It is the number of grams containing Avogadro's number of molecules.

Atomic weights are listed inside the back cover of this book.

EXAMPLE Molarity of Salts in the Sea

(a) Typical seawater contains 2.7 g of salt (sodium chloride, NaCl) per 100 mL (= 100 × 10^{-3} L). What is the molarity of NaCl in the ocean?
(b) $MgCl_2$ has a concentration of 0.054 M in the ocean. How many grams of $MgCl_2$ are present in 25 mL of seawater?

Solution   (a) The molecular weight of NaCl is 22.99 (Na) + 35.45 (Cl) = 58.44 g/mol. The moles of salt in 2.7 g are $(2.7\ g)/(58.44\ g/mol)$ = 0.046 mol, so the molarity is

$$\text{molarity of NaCl} = \frac{\text{mol NaCl}}{\text{L seawater}} = \frac{0.046\ \text{mol}}{100 \times 10^{-3}\ \text{L}} = 0.46\ \text{M}$$

(b) The molecular weight of $MgCl_2$ is 24.30 (Mg) + 2 × 35.45 (Cl) = 95.20 g/mol, so the number of grams in 25 mL is

$$\text{grams of } MgCl_2 = 0.054\ \frac{\text{mol}}{L} \times 95.20\ \frac{g}{\text{mol}} \times (25 \times 10^{-3}\ L) = 0.13\ g$$

. . . .

Strong electrolyte: completely dissociated into ions in solution

Weak electrolyte: partially dissociated in solution

Magnesium chloride is a **strong electrolyte;** in other words, it is completely dissociated into its ions (Mg^{2+} and $2Cl^-$) in aqueous solution. The concentration of $MgCl_2$ molecules in seawater is close to zero, because there are no $MgCl_2$ molecules. Sometimes the molarity of a strong electrolyte is referred to as the **formal concentration** (F), to indicate that the substance is really converted to other species in solution. When we say that the "concentration" of $MgCl_2$ is 0.054 M in seawater, we are really referring to its formal concentration (0.054 F). The "molecular weight" of a strong electrolyte is more properly called the **formula weight** (FW), because it refers to the sum of atomic weights of the atoms in the formula, even though there are no such molecules with that formula.

A **weak electrolyte** such as acetic acid, CH_3CO_2H, is partially split into ions in solution:

Resonance structures of acetate ion

$\longleftrightarrow$ denotes resonance structures

		Formal concentration	Percent dissociated
$CH_3COH \rightleftharpoons CH_3CO^- + H^+$		0.1 F	1.3%
Acetic acid Acetate		0.01	4.1
anion		0.001	12.4

To emphasize that the charge is delocalized onto both oxygens, we can also write

Molality (m) is a designation of concentration expressing the number of moles of substance per kilogram of solvent (not total solution). Unlike molarity, molality is independent of temperature. Molarity changes with temperature because the volume of a solution usually increases when it is heated.

Percent Composition

The percentage of a component (solute) in a mixture or solution is usually expressed as a **weight percent** (wt%):

$$\text{weight percent} = \frac{\text{mass of solute}}{\text{mass of total solution or mixture}} \times 100 \qquad (1\text{-}2)$$

A common form of ethanol (CH_3CH_2OH) is 95 wt%; this expression means 95 g of ethanol per 100 g of total solution. The remainder is water. Another common percent composition is **volume percent** (vol%):

$$\text{volume percent} = \frac{\text{volume of solute}}{\text{volume of total solution}} \times 100 \qquad (1\text{-}3)$$

Although units of weight or volume should always be expressed to avoid ambiguity, weight is usually implied when units are absent.

EXAMPLE Converting Weight Percent to Molarity and Molality

Find the molarity and molality of HCl in a reagent labeled "37.0 wt% HCl, density = 1.188 g/mL." The **density** of a substance is the mass per unit volume.

Solution For *molarity*, we need to find the moles of HCl per liter of solution. The mass of a liter of solution is (1.188 g/~~mL~~)(1 000 ~~mL~~/L) = 1 188 g. The mass of HCl in 1 L is therefore

$$\text{mass of HCl per liter} = 1\,188 \frac{\text{g ~~solution~~}}{\text{L}} \times 0.370 \frac{\text{g HCl}}{\text{g ~~solution~~}}$$
$$= 439.6 \text{ g HCl/L}$$

The molecular weight of HCl is 36.46 g/mol, so the molarity is

$$\text{molarity} = \frac{\text{mol HCl}}{\text{L solution}} = \frac{439.6 \text{ g ~~HCl~~/L}}{36.46 \text{ g ~~HCl~~/mol}} = 12.1 \frac{\text{mol}}{\text{L}} = 12.1 \text{ M}$$

For *molality,* we need to find the moles of HCl per kilogram of solvent (which is H_2O). The solution is 37.0 wt% HCl, so we know that 100.0 g of solution contains 37.0 g of HCl (= 1.015 mol) and $100.0 - 37.0 = 63.0$ g of H_2O (= 0.063 0 kg). The molality is therefore

$$\text{molality} = \frac{\text{mol HCl}}{\text{kg solvent}} = \frac{1.015 \text{ mol HCl}}{0.063\,0 \text{ kg } H_2O} = 16.1 \text{ } m$$

$$\textbf{density} = \frac{\text{mass}}{\text{volume}} = \frac{\text{g}}{\text{mL}}$$

A closely related dimensionless quantity is

specific gravity =
$$\frac{\text{density of a substance}}{\text{density of water at 4°C}}$$

Because the density of water at 4°C is very close to 1 g/mL (Table 2-6), specific gravity is nearly the same as density.

Confusing abbreviations:

mol = moles

$$\text{M} = \text{molarity} = \frac{\text{mol solute}}{\text{L solution}}$$

$$m = \text{molality} = \frac{\text{mol solute}}{\text{kg solvent}}$$

Parts per Million and Parts per Billion

Sometimes composition is expressed as **parts per million** (ppm) or **parts per billion** (ppb), which mean grams of substance per million or billion grams of total solution or mixture. Because the density of a dilute aqueous solution is close to 1.00 g/mL, *we frequently equate 1 g of water with 1 mL of water*, although this equivalence is only an approximation. Therefore 1 ppm corresponds to 1 μg/mL (= 1 mg/L) and 1 ppb is 1 ng/mL (= 1 μg/L). When referring to gases, parts per million is usually expressed in units of volume rather than units of mass. For example, 8 ppm carbon monoxide in air probably means 8 μL CO per liter of air. Again, it is always best to label units to avoid confusion.

$$\text{ppm} = \frac{\text{mass of substance}}{\text{mass of sample}} \times 10^6$$

$$\text{ppb} = \frac{\text{mass of substance}}{\text{mass of sample}} \times 10^9$$

Question What does one part per thousand mean?

EXAMPLE Converting Parts per Billion to Molarity

In Figure 1-4, the concentration of $C_{29}H_{60}$ in summer rainwater is 34 ppb. What is the molarity of this alkane?

Solution A concentration of 34 ppb refers to 34 ng of $C_{29}H_{60}$ per gram of rainwater, which we equate to 34 ng/mL. Multiplying nanograms and milliliters by 1 000 gives 34 μg of $C_{29}H_{60}$ per liter of rainwater. Because the molecular weight of $C_{29}H_{60}$ is 408.8 g/mol, the molarity is

$$\text{molarity of } C_{29}H_{60} \text{ in rainwater} = \frac{34 \times 10^{-6} \text{ g/L}}{408.8 \text{ g/mol}} = 8.3 \times 10^{-8} \text{ M}$$

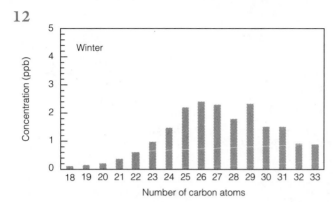

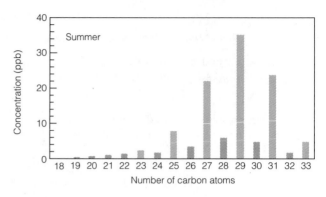

Figure 1-4 Concentrations of alkanes (straight chain hydrocarbons with the formula C_nH_{2n+2}) found in rainwater in Hannover, Germany, in the winter (left) and summer (right) in 1989 are measured in parts per billion (= μg hydrocarbon/L of rainwater). In the summer, concentrations are higher and compounds with an odd number of carbon atoms predominate. This finding implies that the pollutants in the summer are mainly derived from plants, which selectively release alkanes with an odd number of carbons, whereas the alkanes in the winter are mainly from man-made sources. [K. Levsen, S. Behnert, and H. D. Winkeler, *Fresenius J. Anal. Chem.* **1991**, *340*, 665.]

1-4 Preparing Solutions

To prepare an aqueous solution with a desired molarity of a pure solid or liquid, we weigh out the correct mass of reagent and dissolve it in the desired volume. This preparation is normally done with a volumetric flask (see Figure 2-10). If you want to prepare a solution with a particular molality of reagent, then you must mix a certain mass of reagent with the correct mass (rather than volume) of solvent.

. . . .

EXAMPLE **Preparing Solutions with Desired Molarity and Molality**

Cupric sulfate is frequently found as the pentahydrate, $CuSO_4 \cdot 5H_2O$, which means that there are 5 mol of H_2O for each mole of $CuSO_4$ in the solid crystal. The formula weight of $CuSO_4 \cdot 5H_2O$ (= $CuSO_9H_{10}$) is 249.69 g/mol. **(a)** How many grams of $CuSO_4 \cdot 5H_2O$ should be dissolved in a 250-mL volumetric flask to make a solution containing 8.00 mM Cu^{2+}? **(b)** How would you prepare a solution containing 8.00×10^{-3} *m* Cu^{2+} from anhydrous $CuSO_4$ (FW 159.61), which contains no H_2O in its formula? The term **anhydrous** means "without water."

Solution **(a)** An 8.00 mM solution contains 8.00×10^{-3} mol/L. Because 250 mL is 0.250 L, we need

$$\left(8.00 \times 10^{-3} \frac{mol}{L}\right) \times 0.250 \, L = 2.00 \times 10^{-3} \text{ mol } CuSO_4 \cdot 5H_2O$$

The correct mass of this reagent is $(2.00 \times 10^{-3} \text{ mol}) \times (249.69 \text{ g/mol}) =$ 0.499 g. The procedure is to weigh 0.499 g of solid $CuSO_4 \cdot 5H_2O$ into a 250-mL volumetric flask, add about 200 mL of distilled water and swirl to dissolve the reagent. Then dilute with distilled water up to the 250-mL mark and invert the flask 20 times to ensure complete mixing. The resulting, well-mixed solution will contain 8.00 mM Cu^{2+}.

(b) To prepare a solution containing 8.00×10^{-3} *m* (molal) Cu^{2+}, we need 8.00 mmol of $CuSO_4$ per kilogram of water. If we want approximately

250 mL of such a solution, then we would weigh $(0.250 \; \cancel{L}) \times (8.00 \; \text{mmol}/\cancel{L}) = 2.00 \; \text{mmol} = 0.319 \; \text{g CuSO}_4$ into a bottle and add exactly 250 g of water. After mixing, the concentration of Cu^{2+} should be $8.00 \times 10^{-3} \; m$. This solution is not the same as the 8.00 mM solution, because the volume of the $8.00 \times 10^{-3} \; m$ solution is not exactly 250 mL.

Dilution

Dilute solutions can be prepared from concentrated solutions. A desired volume or mass of the concentrated solution is transferred to a fresh vessel and diluted to the desired final volume or mass. (We normally deal with volumes.) Because the number of moles of reagent in V liters containing M moles per liter is the product $M \cdot V = (\text{mol}/\cancel{L}) \cdot \cancel{L}$, we can equate the number of moles in the concentrated (conc) and dilute (dil) solutions:

Dilution formula:
$$M_{conc} \cdot V_{conc} = M_{dil} \cdot V_{dil} \qquad (1\text{-}4)$$

| Moles taken from concentrated solution | Moles placed in dilute solution |

EXAMPLE Preparing 0.1 M HCl

The molarity of "concentrated" HCl purchased for laboratory use is 12.1 M. How many milliliters of this reagent should be diluted to make 1.00 L of 0.100 M HCl?

Solution The dilution formula handles this problem directly:

$$M_{conc} \cdot V_{conc} = M_{dil} \cdot V_{dil}$$

$$(12.1 \; \text{M}) \cdot (x \; \text{mL}) = (0.100 \; \text{M}) \cdot (1\,000 \; \text{mL}) \Rightarrow x = 8.26 \; \text{mL}$$

The symbol $\Rightarrow$ is read "implies that."

To make 0.100 M HCl, we would dilute 8.26 mL of concentrated HCl up to 1.00 L.

EXAMPLE A More Complicated Dilution Calculation

A solution of ammonia in water is called "ammonium hydroxide" because of the equilibrium

$$\underset{\text{Ammonia}}{NH_3} + H_2O \rightleftharpoons \underset{\text{Ammonium}}{NH_4^+} + \underset{\text{Hydroxide}}{OH^-}$$

The density of concentrated ammonium hydroxide, which contains 28.0 wt% NH_3, is 0.899 g/mL. What volume of this reagent should be diluted to make 500 mL of 0.250 M NH_3?

Solution To use Equation 1-4, we need to know the molarity of the concentrated reagent. Because the solution contains 0.899 g of solution per milliliter and there is 0.280 g of NH_3 per gram of solution (28.0 wt%), we can write

$$\text{molarity of NH}_3 = \frac{899 \, \dfrac{\text{g solution}}{\text{L}} \times 0.280 \, \dfrac{\text{g NH}_3}{\text{g solution}}}{17.03 \, \dfrac{\text{g NH}_3}{\text{mol NH}_3}} = 14.8 \text{ M}$$

Now we can find the volume of 14.8 M NH_3 required to prepare 500 mL of 0.250 M NH_3:

$$M_{conc} \cdot V_{conc} = M_{dil} \cdot V_{dil}$$

$$14.8 \, \frac{\text{mol}}{\text{L}} \times V_{conc} = 0.250 \, \frac{\text{mol}}{\text{L}} \times 0.500 \text{ L}$$

$$\Rightarrow V_{conc} = 8.45 \times 10^{-3} \text{ L} = 8.45 \text{ mL}$$

The correct procedure is to place 8.45 mL of concentrated reagent in a 500-mL volumetric flask, add about 400 mL of water, and swirl to mix. Then dilute to exactly 500 mL with water and invert the flask 20 times to mix well.

Terms to Understand†

aliquot	homogeneous	quantitative analysis
analyte	interference	random sample
anhydrous	liter	segregated material
aqueous solution	masking	SI system
atomic weight	molality	solute
composite sample	molarity	solvent
concentration	mole	specific gravity
density	molecular weight	strong electrolyte
formal concentration	parts per billion	volume percent
formula weight	parts per million	weak electrolyte
heterogeneous	qualitative analysis	weight percent

†Terms are introduced in **bold** type in the chapter and are also defined in the Glossary.

Summary

Qualitative analysis identifies substances, whereas quantitative analysis measures how much is present. Homogeneous samples are uniform, whereas heterogeneous samples vary from one part to another and require careful sampling. A random sample is appropriate if the heterogeneity is random. A composite sample is required if the material is segregated. Steps in a chemical analysis are the following: (1) Obtain a representative bulk sample. (2) Prepare a homogeneous laboratory sample from the representative sample. (3) Convert the lab sample into a form suitable for analysis, usually by dissolving the substance. (4) Remove or mask interfering species. (5) Measure the concentration of analyte in several aliquots. (6) Interpret the results.

SI base units include the meter (m), kilogram (kg), second (s), ampere (A), kelvin (K), and mole (mol). Derived quantities such as force (newton, N), pressure (pascal, Pa), and energy (joule, J) can be expressed in terms of base units. In calculations, units should be carried along with the numbers. Prefixes such as kilo- and milli- are used to denote multiples of units. Common expressions of concentration are molarity (moles of solute per liter of solution), molality (moles of solute per kilogram of solvent), formal concentration (formula units per liter), percent composition, and parts per million. You should be able to calculate quantities of reagents needed to prepare solutions; the relation $M_{conc} \cdot V_{conc} = M_{dil} \cdot V_{dil}$ is useful for this purpose.

Exercises†

A. A solution with a final volume of 500.0 mL was prepared by dissolving 25.00 mL of methanol (CH_3OH, density = 0.791 4 g/mL) in chloroform.

(a) Calculate the *molarity* of methanol in the solution.

(b) If the solution has a density of 1.454 g/mL, find the *molality* of methanol.

B. A 48.0 wt% solution of HBr in water has a density of 1.50 g/mL.

(a) Find the formal concentration of HBr.

(b) What mass of solution contains 36.0 g of HBr?

(c) What volume of solution contains 233 mmol of HBr?

(d) How much solution is required to prepare 0.250 L of 0.160 M HBr?

C. A solution contains 12.6 ppm of dissolved $Ca(NO_3)_2$ (which is a strong electrolyte dissociated into $Ca^{2+} + 2NO_3^-$). Find the concentration of NO_3^- in parts per million.

Problems§

Steps in a Chemical Analysis

1. Write the steps in a chemical analysis.

2. A heterogeneous material may be random or segregated.

(a) Explain the difference between these forms.

(b) Explain how you might construct a random sample from the random material.

(c) Explain how you might construct a composite sample from the segregated material.

Units and Conversions

3. **(a)** List the SI units of length, mass, time, electric current, temperature, and amount of substance; write the abbreviation for each.

(b) Write the units and symbols for frequency, force, pressure, energy, and power.

4. Write the names and abbreviations for each of the prefixes from 10^{-24} to 10^{18}. Which abbreviations are capitalized?

5. Write the name and number represented by each symbol. For example, for kW you should write kW = kilowatt = 10^3 watts.

(a) mW   **(c)** kΩ   **(e)** TJ   **(g)** fg

(b) pm   **(d)** μF   **(f)** ns   **(h)** dPa

6. Express the following quantities with abbreviations for units and prefixes from Tables 1-1 through 1-3.

(a) 10^{-13} joules

(b) $4.317\,28 \times 10^{-8}$ farads

(c) $2.997\,9 \times 10^{14}$ hertz

(d) 10^{-10} meters

(e) 2.1×10^{13} watts

(f) 48.3×10^{-20} moles

7. **(a)** Each adrenal cell contains about 2.5×10^4 vesicles with a total of about 150 fmol epinephrine per cell. How many attomoles (amol) of epinephrine are in each vesicle?

(b) How many molecules of epinephrine are in each vesicle?

8. How many joules per second and how many calories per hour are produced by a 100.0-horsepower engine?

9. A 120-pound woman working in an office consumes about 2.2×10^3 kcal/day, whereas the same woman climbing a mountain needs 3.4×10^3 kcal/day.

(a) Express these numbers in terms of joules per second per kilogram of body mass (= watts per kilogram).

(b) Which consumes more power (watts): the office worker or a 100-W light bulb?

10. **(a)** Refer to Table 1-4 to calculate how many meters are in 1 inch. How many inches are in 1 m?

(b) A mile contains 5 280 feet and a foot contains 12 inches. The speed of sound in the atmosphere at sea level is 345 m/s. Express the speed of sound in miles per second and miles per hour.

(c) There is a delay between lightning and thunder in a storm, because light reaches us almost instantaneously, but sound is slower. How many meters, kilometers, and miles away is a lightning bolt if the sound reaches you 3.00 s after the light?

†Detailed Solutions to Exercises are provided at the end of this book.

§Brief answers are given at the end of this book. Detailed solutions appear in the *Solutions Manual for Quantitative Chemical Analysis*. The *Solutions Manual* also contains supplementary problems (and their solutions) that may be reproduced by instructors.

11. How many joules per second (J/s) are used by a device that requires 5.00×10^3 British thermal units per hour (Btu/h)? How many watts (W) does this device use?

12. Newton's law states that force = mass × acceleration. You also know that energy = force × distance and pressure = force/area. From these relations, derive the dimensions of newtons, joules, and pascals in terms of the fundamental SI units in Table 1-1. Check your answers in Table 1-2.

13. An aqueous solution containing 20.0 wt % KI has a density of 1.168 g/mL. Find the molality (m, not M) of the KI solution.

Chemical Concentrations

14. Define the following measures of concentration.

(a) molarity

(b) molality

(c) density

(d) weight percent

(e) volume percent

(f) parts per million

(g) parts per billion

(h) formal concentration

15. Why is it more accurate to say that the concentration of a solution of acetic acid is 0.01 F rather than 0.01 M? (Despite this distinction, we will usually write 0.01 M.)

16. What is the formal concentration (expressed as mol/L = M) of NaCl when 32.0 g is dissolved in water and diluted to 0.500 L?

17. How many grams of methanol (CH_3OH, MW 32.04) are contained in 0.100 L of 1.71 M aqueous methanol (i.e., 1.71 mol CH_3OH/L solution)?

18. Any dilute aqueous solution has a density near 1.00 g/mL. Suppose the solution contains 1 ppm of solute. Express the concentration of solute in g/L, μg/L, μg/mL, and mg/L.

19. The concentration of $C_{20}H_{42}$ (MW 282.55) in winter rainwater in Figure 1-4 is 0.2 ppb. Assuming that the density of rainwater is close to 1.00 g/mL, find the molar concentration of $C_{20}H_{42}$.

20. How many grams of perchloric acid, $HClO_4$, are contained in 37.6 g of 70.5 wt % aqueous perchloric acid? How many grams of water are in the same solution?

21. The density of 70.5 wt % aqueous perchloric acid is 1.67 g/mL. Recall that grams refers to grams of *solution* (= g $HClO_4$ + g H_2O).

(a) How many grams of solution are in 1.00 L?

(b) How many grams of $HClO_4$ are in 1.00 L?

(c) How many moles of $HClO_4$ are in 1.00 L?

22. (a) The volume of a sphere of radius r is $\frac{4}{3}\pi r^3$. Find the volume of a spherical vesicle of radius 200 nm in the cell in Box 1-1. Express your answer in cubic meters (m^3) and liters, remembering that $1 \text{ L} = 10^{-3} \text{ m}^3$.

(b) Find the molar concentration of epinephrine in the vesicle if it contains 10 amol of epinephrine.

23. The concentration of sugar (glucose, $C_6H_{12}O_6$) in human blood ranges from about 80 mg/100 mL before meals up to 120 mg/100 mL after eating. Find the molarity of glucose in blood before and after eating.

24. An aqueous solution of antifreeze contains 6.067 M ethylene glycol ($HOCH_2CH_2OH$, MW 62.07) and has a density of 1.046 g/mL.

(a) Find the mass of 1.00 L of this solution and the number of grams of ethylene glycol per liter.

(b) Find the molality of ethylene glycol in this solution.

25. Protein and carbohydrates provide 4.0 Cal/g, whereas fat gives 9.0 Cal/g. (Remember that 1 Calorie, with a capital C, is really 1 kcal.) The weight percents of these components in some foods are

Food	Protein	Carbohydrate	Fat
shredded wheat	9.9%	79.9%	— %
doughnut	4.6	51.4	18.6
hamburger (cooked)	24.2	—	20.3
apple	—	12.0	—

Calculate the number of calories per gram and calories per ounce in each of these foods. (Use Table 1-4 to convert grams into ounces, remembering that there are 16 ounces in 1 pound.)

26. It is recommended that drinking water contain 1.6 ppm fluoride (F^-) for prevention of tooth decay. Consider a reservoir with a diameter of 450 m and a depth of 10 m. (The volume is $\pi r^2 h$, where r is the radius and h is the height.) How many grams of F^- should be added to give 1.6 ppm? How many grams of sodium fluoride, NaF, contain this much fluoride?

27. The *ideal gas law* states that $PV = nRT$, where P is the pressure of a gas (in atmospheres, atm), V is the volume (L), n is the number of moles, R is the gas constant [0.082 06 L·atm/(mol·K)], and T is temperature (K). The inert gases have the following volume concentrations in dry air: He, 5.24 ppm; Ne, 18.18 ppm; Ar, 0.934%; Kr, 1.14 ppm; Xe, 87 ppb.

(a) A concentration of 5.24 ppm He means 5.24 μL of He per liter of air. How many moles of He are contained in 5.24 μL at 25°C and 1 atm? This number is the molarity of He in the air.

(b) Find the molar concentrations of Ar, Kr, and Xe in air at 25°C and 1 atm.

Preparing Solutions

28. How many grams of boric acid [$B(OH)_3$, MW 61.83] should be used to make 2.00 L of 0.0500 M solution? What kind of flask is used to prepare this solution?

29. Describe how you would prepare approximately 2 L of 0.0500 m boric acid, $B(OH)_3$.

30. What is the maximum volume of 0.25 M sodium hypochlorite solution (NaOCl, laundry bleach) that can be prepared by dilution of 1.00 L of 0.80 M NaOCl?

31. How many grams of 50 wt% NaOH (FW 40.00) should be diluted to make 1.00 L of 0.10 M NaOH?

32. A bottle of concentrated aqueous sulfuric acid, labeled 98.0 wt% H_2SO_4, has a concentration of 18.0 M.

(a) How many milliliters of reagent should be diluted to make 1.00 L of 1.00 M H_2SO_4?

(b) Calculate the density of 98.0 wt% H_2SO_4.

33. What is the density of 53.4 wt% aqueous NaOH if 16.7 mL of the solution diluted to 2.00 L gives 0.169 M NaOH?

Notes and References

1. Standards for temperature measurement from 0.65 K to > 1 400 K are given by B. W. Mangum, *J. Res. Natl. Inst. Stand. Technol.* **1990**, *95*, 69. Directions for using pure metals to calibrate mercury thermometers up to 700 K are given by G. V. D. Tiers, *Anal. Chim. Acta* **1990**, *237*, 241.

Weighing Femtomoles of DNA

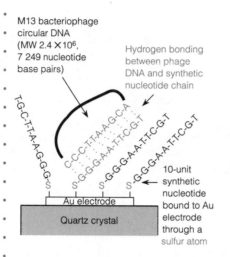

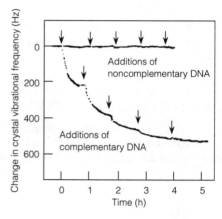

Left: Synthetic nucleotides attached through a sulfur atom to a gold electrode on the surface of a quartz crystal, which is immersed in an aqueous solution at 50°C. Each synthetic chain can bind a complementary sequence of nucleic acids in bacteriophage DNA by formation of hydrogen bonds between G and C and between A and T. *Right:* Arrows indicate additions of equal portions of DNA. Lower curve illustrates the changes in frequency that occur when DNA with a complementary sequence is added. Successive additions give smaller responses as the electrode becomes saturated with DNA. Upper curve illustrates the absence of response when noncomplementary DNA is added. [Y. Okahata, Y. Matsunobu, K. Ijiro, M. Mukae, A. Murakami, and K. Makino, *J. Am. Chem. Soc.* **1992**, *114*, 8299.]

An exquisitely sensitive detector of mass is a vibrating quartz crystal of the type that keeps time in your wristwatch and computer. When a substance becomes *adsorbed* (bound) on the crystal's surface, the vibrational frequency of the crystal changes in proportion to the adsorbed mass.

The figures show how such a crystal can detect nanogram quantities of one specific *deoxyribonucleic acid* molecule (the genetic material, DNA) from a *bacteriophage* (a virus that attacks bacteria). The crystal is coated with a synthetic chain of 10 nucleotides (designated C, G, A, and T) chosen to bind to a complementary sequence on the bacteriophage DNA. When the crystal is immersed in a solution containing many different DNA molecules, the complementary molecules bind to the crystal and lower its vibrational frequency. A frequency change of 560 Hz corresponds to 590 ng (0.25 pmol) of DNA. The smallest quantity that can be detected by this system is approximately 25 ng—or about 11 fmol.

Tools of the Trade

2

Analytical chemistry extends from simple "wet" chemical procedures to elaborate instrumental methods. In this chapter some of the basic laboratory apparatus and manipulations associated with chemical measurements are described. Throughout the remainder of this book, advice about laboratory operations will be highlighted by the icon .

2-1 Safe, Ethical Handling of Chemicals and Waste

Chemical experimentation, like driving a car or operating a household, creates hazards for you and others. *The primary safety rule is not to do something that you (or your instructor or supervisor) consider to be dangerous.* If you believe that an operation is hazardous, you should discuss it with your instructor or supervisor and not proceed until you are confident that sensible procedures and precautions are in place. If you still consider an activity to be too dangerous, don't do it.

Chemists (and automobiles and households) generate toxic wastes. Preservation of a habitable environment on our planet demands that we minimize waste production and responsibly dispose of the waste that is generated (Box 2-1). Recycling of chemicals is widely practiced in industry for economic as well as ethical reasons; it should be an important component of pollution control in your lab.[1]

Before beginning work, you should be familiar with safety precautions appropriate to your laboratory. Goggles or safety glasses with sideshields (Figure 2-1) are necessary at all times in every laboratory to protect you from flying liquids and glass, which are known to visit labs from time to time.

Figure 2-1 Goggles or safety glasses with side shields are required in every laboratory.

Box 2-1   Disposal of Chemical Waste

Many of the chemicals we commonly use are harmful to plants, animals, and people if carelessly discarded. Although many solutions can be safely poured down the drain, others put toxic waste into rivers and groundwater. For each experiment you do, your instructor needs to determine in advance how reagents will be discarded. Options include (1) pouring solutions down the drain and diluting with tap water, (2) saving the waste for disposal in an approved landfill, (3) chemically treating the waste to make it less hazardous and then pouring it down the drain or saving it for disposal at a landfill, and (4) recycling the chemical so that little waste is produced. It is critical that chemically incompatible wastes not be combined with each other and that *each waste container be properly labeled to identify the type and approximate quantity of waste.*

A few examples illustrate different approaches to managing lab waste.[2] Waste dichromate ($Cr_2O_7^{2-}$) is first reduced to Cr(III) with sodium bisulfite ($NaHSO_3$), precipitated with hydroxide to make insoluble $Cr(OH)_3$, and evaporated to a small volume of solid waste for a landfill. Waste solutions of acids can be mixed with waste solutions of base until they are nearly neutral (as determined with pH paper) and then poured down the drain. Waste iodate (IO_3^-) solutions can be treated with $NaHSO_3$ to reduce the IO_3^- to I^-. This reaction generates acid that must be neutralized with base. Then the whole solution can be safely poured down the drain. Waste Pb^{2+} solution is treated with sodium metasilicate (Na_2SiO_3) solution to precipitate insoluble $PbSiO_3$ that can be packaged for a landfill. Waste silver or gold should be chemically treated to recover these valuable metals. A toxic gas used in a fume hood is bubbled through a chemical trap to prevent its escape from the hood. Some toxic gases can be passed through a burner that converts them to harmless products.

Flame-resistant lab coats help prevent dangerous spills and flames from contacting your skin. Rubber gloves are recommended to protect your hands from splatters when pouring concentrated acids or other hazardous liquids. Food is never eaten in the lab.

Organic solvents and concentrated acids that produce harmful fumes should be handled only in a fume hood that is tested periodically for proper operation. The purpose of a fume hood is not to transfer harmful levels of pollutants to your neighbors; it is to reduce low levels of pollutants to even lower, safe levels. You should never generate large quantities of toxic fumes that are allowed to escape through the hood. A respirator is required when handling very fine powders (which you are not likely to encounter in an analytical chemistry course), which can produce a dangerous cloud of dust that could be inhaled.

Solid or liquid spills should be cleaned up immediately to prevent accidental contact by the next person who comes along. Chemical spills on your skin usually are treated first by flooding the affected area with water. In anticipation of splashes on your body or in your eyes, you must know where to find and how to operate the emergency shower and eyewash. You should also know how to operate the fire extinguisher in the lab and how to use an emergency blanket to extinguish burning clothing. A first aid kit should be available, and safety equipment and hazardous operations should be prominently labeled (Figure 2-2). You should know how and where to seek emergency medical assistance.

Proper chemical storage minimizes fire and explosion hazards and human contact with vapors. Strong acids and bases, as well as oxidizing and reducing agents, should be stored in containers that minimize the possibility

Figure 2-2   Safety equipment and hazardous operations are prominently labeled.

of accidental contact with other materials. Acid and solvent storage cabinets should be vented to a fume hood. The solvent cabinet is designed to prevent fire from getting in or out and to contain accidental spills. *All vessels should have labels that indicate what they contain.*

2-2 The Lab Notebook

The critical functions of a lab notebook are to state *what was done* and *what you observed*. The greatest error, made even by experienced scientists, is producing notebooks that are difficult to understand. Even the author of a notebook may not understand his or her own notes after a few years because of poorly labeled entries and incomplete descriptions. The habit of writing in *complete sentences* is an excellent way to avoid incomplete descriptions.

Beginning students often find it useful to write a complete description of an experiment, with formal sections dealing with purpose, methods, results, and conclusions. Arranging a notebook to accept numerical data prior to coming to the lab is an excellent way to prepare for an experiment.

The measure of scientific "truth" is the ability of different people to reproduce an experiment. Sometimes two scientists in different labs cannot reproduce each other's work, and neither has a notebook that is complete enough to allow them to understand why. Details that seem unimportant on the day of an experiment may become important months or years later. A good lab notebook will state everything that was done and will allow you or anyone else to repeat the experiment in exactly the same manner at any future date.

A notebook should contain your observations. Long after you have forgotten what happened, you should be able to rely on your notes to tell what happened and perhaps provide a key observation needed to interpret an experiment.

It is good practice to write a balanced chemical equation for every reaction you use. This helps you understand what you are doing and may point out what you do not understand about what you are doing.

Record in your notebook the names of computer disks and files where programs and data are stored. Printed graphs and program listings should state the disk file and program associated with the printed document. Printed copies of important data collected on a computer should be pasted into your notebook. The lifetime of a printed page is an order of magnitude (or more) greater than the lifetime of a magnetic disk.

The lab notebook must

1. State what was done
2. State what was observed
3. Be understandable to someone else

Without a doubt, somebody reading this book today is going to make an important discovery in the future and will seek a patent. Your lab notebook is the legal record of your discovery. For this purpose, each page in your notebook should be signed and dated. Anything of potential importance should also be signed and dated by a second person.

2-3 The Analytical Balance

An **electronic balance** is an instrument that uses an electromagnet to balance the load on the pan; the mass of the load is proportional to the current needed to balance it. Figure 2-3a shows a typical electronic analytical balance with a capacity of 100–200 g and a sensitivity of 0.01–0.1 mg. *Sensitivity* refers to the smallest increment of mass that can be measured. For smaller scale work, the microbalance in Figure 2-3b can weigh milligram quantities with a sensitivity of 0.1 μg.

The weighing operation consists of first placing a clean receiving vessel on the balance pan. The mass of the empty vessel is called the **tare.** On most

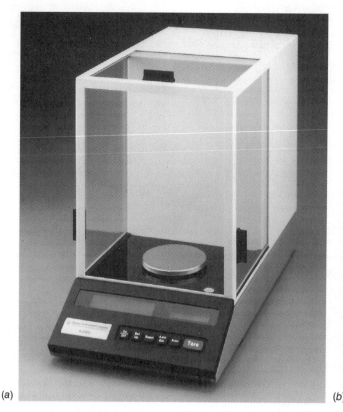

(a)

(b)

Figure 2-3 Electronic analytical balances. [Courtesy Fisher Scientific, Pittsburgh, PA.]

balances, there is a button that resets the tare reading to zero. Add the substance to be weighed to the vessel and read its new mass. If there is not an automatic tare operation, you must record the mass of the empty vessel and subtract it from that of the filled vessel. Chemicals should never be placed directly on the weighing pan. This precaution protects the balance from corrosion and allows you to recover all the chemical being weighed.

An alternate procedure, called "weighing by difference," is necessary for **hygroscopic** reagents, which rapidly absorb moisture from the air. First weigh a capped bottle containing dry reagent. Then quickly pour some reagent from the weighing bottle into a receiver. Cap the weighing bottle and weigh it again. The difference equals the mass of reagent delivered to the receiving vessel.

Principle of Operation

An object placed on a balance pushes the pan down with a force equal to mg, where m is the mass of the object and g is the acceleration of gravity. The electronic balance uses an electromagnetic restoring force to return the pan to its original position. The electric current required to generate the force is proportional to the mass of the object being weighed, which is displayed on a digital readout.

Figure 2-4 shows more detail. When a mass is placed on the pan, the null detector senses a displacement and sends an error signal to the circuit that generates a correction current. This current flows through the coil at the base of the balance pan, thereby creating a magnetic field that is repelled by the

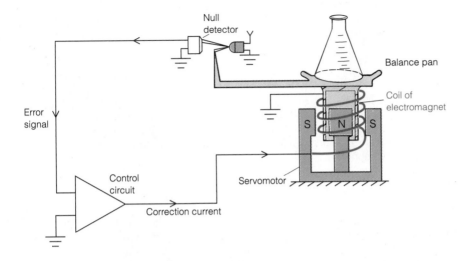

Figure 2-4 Electronic balance. Displacement of the balance pan and consequent error signal causes the control circuit to generate a correction current. The electromagnet then exerts a force restoring the pan to its initial position. N and S refer to the north and south poles of the magnet. [R. M. Schoonover, *Anal. Chem.* **1982**, *54*, 973A. See B. B. Johnson and J. D. Wells, *J. Chem. Ed.* **1986**, *63*, 86 for additional discussion.]

permanent magnet beneath the pan. As the deflection decreases, the output of the null detector decreases. The correction current needed to restore the system to its initial position is proportional to the mass on the pan.

Figure 2-5 shows the principle of operation of the older, but still common, single-pan **mechanical balance,** which uses standard masses and a balance beam to measure the mass of the material placed on the balance pan. The balance beam is suspended on a sharp central fulcrum called a *knife edge.* The mass of the pan and associated hardware hanging from the balance point (another knife edge) at the left is balanced by a counterweight at the right. An object placed on the pan pushes it down. Knobs are then adjusted to mechanically remove weights from a bar above the pan that is hidden inside the balance. The balance beam is restored to almost its original position when the masses removed from the bar are nearly equal to the mass being weighed on the pan. The slight difference from the original position is shown on an optical scale, whose reading is added to that of the knobs.

Box 2-2 describes a completely different type of microbalance that is sensitive to nanogram differences in mass.

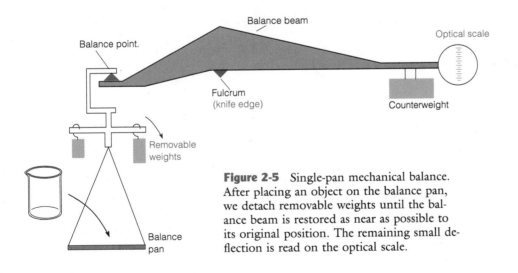

Figure 2-5 Single-pan mechanical balance. After placing an object on the balance pan, we detach removable weights until the balance beam is restored as near as possible to its original position. The remaining small deflection is read on the optical scale.

Box 2-2 The Quartz Crystal Microbalance

A vibrating quartz crystal is one of the most sensitive detectors of mass,[3] as illustrated at the beginning of this chapter. When pressure is applied to the crystal, a voltage develops between certain surfaces of the crystal. This phenomenon is called the *piezoelectric effect*. Conversely, an applied voltage causes a distortion of the crystal. Your wristwatch uses a sinusoidal voltage to set a quartz crystal into a precisely defined oscillation whose period serves as a clock.

If a thin layer of mass m is spread over area A on the quartz surface, the oscillation frequency F decreases by an amount

$$\Delta F = (2.3 \times 10^{-10})\, F^2\, \frac{m}{A}$$

where ΔF and F are given in hertz, m is measured in grams, and A is given in square meters. A 10-MHz crystal coated over 1 cm^2 of its surface with 1 µg of material will change its vibrational frequency by $\Delta F = 230$ Hz, an easily measured change.

A piezoelectric crystal coated with water-absorbing material to measure the concentration of atmospheric water vapor was sent to Mars with the Viking spacecraft that landed in 1976. Another typical application is measuring ozone (O_3) levels in the workplace.[4]

Antibodies are proteins manufactured by living organisms to bind foreign molecules (called *antigens*) and foreign cells and mark them for elimination. Antibodies are one of the primary defensive means by which your body defends itself against infection. When antibodies are chemically attached to a piezoelectric device, the quartz crystal becomes a sensitive monitor for biological antigens such as enzymes (proteins that catalyze biochemical reactions) and even whole microbes.[5]

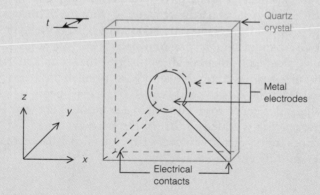

Piezoelectric quartz crystal. A sinusoidal voltage applied between the metal electrodes (along the y direction) causes the crystal to vibrate with a shearing motion along the x direction. The vibrational frequency, F, is given by $F = N/t$, where t is the thickness of the crystal (in meters) and $N = 1\,679$ Hz·m. A thickness of 0.17 mm gives a frequency of 10 MHz.

 Preventing Weighing Errors

Use a paper towel or tissue to handle the vessel you are weighing, because fingerprints will change its mass. Samples should be at ambient temperature (the temperature of the surroundings) before weighing to avoid errors due to convective air currents. A sample that has been dried in an oven usually takes about 30 minutes to cool to room temperature. During this cooling process, the sample should be placed in a desiccator. The glass doors of the balances in Figure 2-3 must be closed during weighing to prevent air currents from affecting the reading. On top-loading balances without sliding glass doors, there is usually a glass or plastic fence surrounding the pan to protect it from air currents. Sensitive balances should be placed on a heavy table, such as a marble slab, to minimize the effect of room and building vibrations on the reading. Adjustable feet and a bubble meter allow you to maintain the balance in a level position.

A mechanical balance should be in its arrested position when you load or unload the pan and in the half-arrested position when you are dialing weights. This practice protects against the application of abrupt forces that wear down the knife edges and eventually decrease the sensitivity of the balance.

Less expensive electronic balances are calibrated with a standard mass at the factory, where the force of gravity is not the same as the force of gravity in your lab. (Gravitational acceleration varies over a range of about 0.1% among different locations in the United States.) It is therefore essential that you calibrate the balance with a standard mass in your own lab. *Tolerances* (allowable deviations) for standard masses are listed in Table 2-1. More expensive balances have built-in calibration weights and an automatic calibration feature.

You can see whether magnetic materials affect your electronic balance by moving a magnet near the empty pan and looking for changes in the zero reading. Electromagnetic fields from nearby instruments may also affect readings. Dust must not be allowed to enter the gap between the coil and the permanent magnet of the servomotor.

TABLE 2-1 Tolerances (in mg) for laboratory balance weights

	Weight classification		
Denomination	S	S-1	P
Grams			
200	0.50	2.0	4.0
100	0.25	1.0	2.0
50	0.12	0.60	1.2
30	0.074	0.45	0.90
20	0.074	0.35	0.70
10	0.074	0.25	0.50
5	0.054	0.18	0.36
3	0.054	0.15	0.30
2	0.054	0.13	0.26
1	0.054	0.10	0.20
Milligrams			
500	0.025	0.080	0.16
300	0.025	0.070	0.14
200	0.025	0.060	0.12
100	0.025	0.050	0.10
50	0.014	0.042	0.085
30	0.014	0.038	0.075
20	0.014	0.035	0.070
10	0.014	0.030	0.060
5	0.014	0.028	0.055
3	0.014	0.026	0.052
2	0.014	0.025	0.050
1	0.014	0.025	0.050

Buoyancy effect:
The actual mass of an object is usually greater than the apparent mass.

Buoyancy

For accurate work, the buoyant effect of air must be considered. A sample displacing a particular volume of air appears lighter than its actual mass by an amount equal to the mass of the displaced air. The same effect applies to calibration weights in an electronic balance or removable weights in a mechanical balance. There will be a **buoyancy** correction whenever the density of the object being weighed is not equal to the density of the standard weights. If a mass m' is read from a balance, the true mass m of the object weighed in vacuum is given by[6]

Buoyancy equation:

$$m = \frac{m'\left(1 - \dfrac{d_a}{d_w}\right)}{\left(1 - \dfrac{d_a}{d}\right)}$$

(2-1)

Air density (g/L) = $(0.003\,485\,B - 0.001\,318\,V)/T$, where B is the barometric pressure (Pa), V is the vapor pressure of water in the air (Pa), and T is the air temperature (K).

where d_a = density of air (0.001 2 g/mL near 1 atm and 25°C); d_w = density of balance weights (typically 8.0 g/mL); and d = density of object being weighed. For the most accurate work, the correct air density at the particular values of temperature, pressure, and humidity should be used.

EXAMPLE   Buoyancy Correction

Find the true mass of hexane (C_6H_{14}, density = 0.660 g/mL) if the apparent mass weighed in air is 100.00 g.

Solution   Assume that the balance weights have a density of 8.0 g/mL and the density of air is 0.001 2 g/mL. Plugging these values into Equation 2-1 gives the true mass:

$$m = \frac{100.00\text{ g}\left(1 - \dfrac{0.001\,2\text{ g/mL}}{8.0\text{ g/mL}}\right)}{\left(1 - \dfrac{0.001\,2\text{ g/mL}}{0.660\text{ g/mL}}\right)} = 100.17\text{ g}$$

The error in neglecting buoyance for this low-density liquid is 0.17%.

Figure 2-6 shows the buoyancy correction (m/m') for several substances. When you weigh water with a density of 1.00 g/mL, the true mass is 1.001 1 g when the balance reads 1.000 0 g. The error of 0.11% is significant for many purposes. For NaCl with a density of 2.16 g/mL, the error is 0.04%; and for $AgNO_3$ with a density of 4.45 g/mL, the error is only 0.01%.

Figure 2-6   Buoyancy correction assuming d_a = 0.001 2 g/mL and d_w = 8.0 g/mL. The apparent mass measured in air is multiplied by the buoyancy correction to find the true mass.

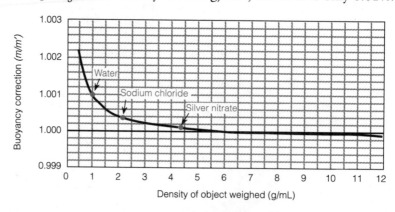

Box 2-3 Buoyancy

The *Earthwinds* balloon was designed for a 'round-the-world voyage with a three-person crew in 1993. A hot-air balloon could not carry enough propane fuel for such a trip, and a helium balloon could not carry enough ballast to cope with the daily cycle of warming and cooling that requires helium to be vented in the day and ballast to be released at night to maintain altitude. The upper helium balloon of *Earthwinds* provides lift, while the lower air balloon maintains the correct buoyancy. Air is pumped into the lower balloon during the day and pumped out of it at night.

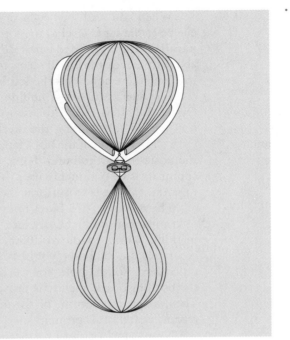

2-4 Burets

A **buret** is a precisely manufactured glass tube with graduations enabling you to measure the volume of liquid delivered. This measurement is made by reading the level before and after draining liquid from the buret. The typical glass buret in Figure 2-7a has a Teflon stopcock. A loose-fitting cap is used to keep dust out and vapors in. The graduations of Class A burets are certified to meet the tolerances in Table 2-2.

TABLE 2-2
Tolerances of Class A burets

Buret volume (mL)	Smallest graduation (mL)	Tolerance (mL)
5	0.01	±0.01
10	0.05 *or* 0.02	±0.02
25	0.1	±0.03
50	0.1	±0.05
100	0.2	±0.10

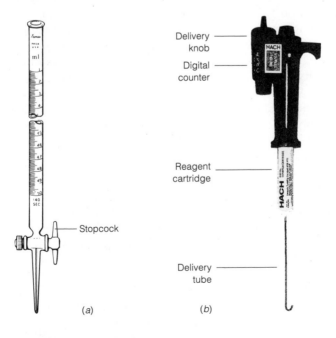

Figure 2-7 (*a*) Glass buret. [Courtesy A. H. Thomas Co., Philadelphia, PA.] (*b*) Digital titrator with plastic cartridge containing reagent solution. [Courtesy Hach Co., Loveland, CO.] A fully automatic titrator is shown in Figure 12-5.

Buret reading tips:

1. Read the bottom of the concave meniscus
2. Avoid parallax
3. Account for the thickness of the marking lines in your readings

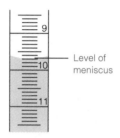

— Level of meniscus

Figure 2-8 Buret with the meniscus at 9.68 mL. You should always estimate the reading of any scale to the nearest tenth of a division. This value corresponds to 0.01 mL for the buret in this figure.

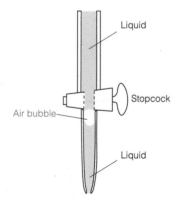

Liquid

Stopcock

Air bubble —

Liquid

Figure 2-9 The air bubble often trapped beneath the stopcock of a buret should be expelled before you use the apparatus.

When you read the height of liquid in a buret, it is important that your eye be at the same level as the top of the liquid. This practice minimizes the **parallax** error associated with reading the position of the liquid. If your eye is above this level, the liquid seems to be higher than it actually is. If your eye is too low, there appears to be less liquid than is actually present.

The surface of most liquids forms a concave **meniscus,** as shown in Figure 2-8. It is useful to use a piece of black tape on a white card as a background for locating the precise position of the meniscus. To use this card, align the top of the black tape with the bottom of the meniscus and read the position on the buret. Some solutions, especially highly colored ones, appear to have two meniscuses. In such cases, either one may be used. The important point is to perform the readings reproducibly.

The markings of a buret normally have a thickness that is *not* negligible in comparison with the distance between markings. The thickness of the markings corresponds to about 0.02 mL for a 50-mL buret. To most accurately use the buret, you should select one portion of the marking to be called zero. For example, you can say that the liquid level is *at* the mark when the bottom of the meniscus just touches the *top* of the mark on the glass. Then, when the bottom of the meniscus is at the *bottom* of the mark, the reading is 0.02 mL greater.

Near the end point of a titration, it is desirable to deliver less than one drop at a time from the buret. This practice permits a finer location of the end point, because the volume of a drop is about 0.05 mL (from a 50-mL buret). To deliver a fraction of a drop, carefully open the stopcock until part of a drop is hanging from the buret tip. (Alternatively, you can allow just a fraction of a drop to emerge from the buret by very rapidly turning the stopcock through the open position.) Then touch the inside glass wall of the receiving flask to the buret tip to transfer the liquid to the wall of the flask. Carefully tip the flask so that the main body of liquid washes over the newly added liquid, and mix the contents. Near the end of a titration, the flask should be tipped and rotated to ensure that droplets on the walls containing unreacted analyte come in contact with the bulk solution.

Liquid should drain evenly down the walls of a buret. If it does not, the buret should be cleaned with detergent and a buret brush. If this cleaning is insufficient, the buret should be soaked in peroxydisulfate–sulfuric acid cleaning solution.† Volumetric glassware should never be soaked in an alkaline cleaning solution, because glass is slowly attacked by base. The tendency of liquid to stick to the walls of a buret can be diminished by draining the buret slowly. A slowly drained buret will also provide greater reproducibility of results. *A drainage rate of no more than 20 mL/min should be used.*

One of the most common errors in using a buret is caused by failure to expel the bubble of air that often forms directly beneath the stopcock (Figure 2-9). If an air bubble is present at the start of a titration, it may become filled

. . . .

†Cleaning solution is prepared by dissolving 36 g of ammonium peroxydisulfate, $(NH_4)_2S_2O_8$, in a *loosely stoppered* 2.2-L ("one-gallon") bottle of 98 wt% sulfuric acid. More ammonium peroxydisulfate is added every few weeks, as necessary, to maintain the oxidizing capability of the solution. This solution replaces chromic acid, which is a carcinogen. [H. M. Stahr, W. Hyde, and L. Sigler, *Anal. Chem.* **1982,** *54,* 1456A.] Cleaning solution is a powerful oxidant that eats clothing and people, as well as dirt and grease, and is stored and used in a fume hood. It is loosely stoppered to prevent explosions from gas buildup. [P. S. Surdhar, *Anal. Chem.* **1992,** *64,* 310A.]

with liquid during the titration, thereby causing an error in the volume of liquid delivered from the buret. Usually the bubble can be dislodged by draining the buret for a second or two with the stopcock wide open. Sometimes a tenacious bubble must be expelled by carefully shaking the buret while draining liquid into a sink.

When you fill a buret with fresh solution, it is a good idea to rinse the buret several times with small quantities of the new solution, discarding each wash. It is not necessary to fill the whole buret with the wash solution. Simply tilt the buret to allow the whole surface to come in contact with a small volume of liquid. This same washing technique can be applied to any vessel (such as a spectrophotometer cuvet or a pipet) that must be reused without opportunity for drying.

The digital titrator in Figure 2-7b is more convenient and portable, but less accurate, than the customary glass buret in Figure 2-7a. The digital titrator is especially useful for conducting measurements in the field where samples are collected. The counter tells how much reagent from the disposable cartridge has been dispensed by rotation of the delivery knob. Prepackaged reagents are available, or you can fill an empty cartridge with your own solution. The accuracy of 1% is about 10 times poorer than the accuracy of a glass buret, but many field or quality control operations do not require greater accuracy.

Washing any glassware with a new solution is a good idea.

A syringe can be used to deliver a known *mass* of reagent (by weighing the syringe) instead of using a buret to deliver a known *volume*. The mass titration is *more accurate* than a typical volumetric titration. [W. B. Guenther, *J. Chem. Ed.* **1988,** *65,* 1097; M. A. Wade, *J. Chem. Ed.* **1988,** *65,* 467.]

2-5 Volumetric Flasks

A **volumetric flask** is calibrated to contain a particular volume of water at 20°C when the bottom of the meniscus is adjusted to the center of the line marked on the neck of the flask (Figure 2-10, Table 2-3). Most flasks bear the label "TC 20°C," which means that the flask is calibrated *to contain* the indicated volume at 20°C. (Other types of glassware may be calibrated *to deliver,* "TD," their indicated volume.) The temperature of the container and its contents is important, because both the liquid and the glass expand as they are heated.

A volumetric flask is used to prepare a solution of a known volume. Most commonly, a reagent is weighed into a volumetric flask, dissolved, and diluted to the mark. This way the mass of reagent and the final volume of the

Thermal expansion of water and glass is discussed in Section 2-9. Volumetric glassware made of Pyrex®, Kimax®, or other low-expansion glass can be safely dried in an oven heated to at least 320°C without harm, although there is rarely reason to dry glass above 150°C. [D. R. Burfield and G. Hefter, *J. Chem. Ed.* **1987,** *64,* 1054.]

TABLE 2-3
Tolerances of Class A volumetric flasks

Flask capacity (mL)	Tolerance (mL)
1	±0.02
2	±0.02
5	±0.02
10	±0.02
25	±0.03
50	±0.05
100	±0.08
200	±0.10
250	±0.12
500	±0.20
1 000	±0.30
2 000	±0.50

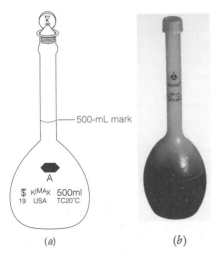

—500-mL mark

A
KIMAX 500ml
USA TC20°C

(a)

(b)

Figure 2-10 (*a*) Class A glass volumetric flask. [Courtesy A. H. Thomas Co., Philadelphia, PA.] (*b*) Class B polypropylene plastic volumetric flask for trace analysis (parts per billion concentrations) in which *analyte* (the species being measured) might be lost by *adsorption* on glass walls or contaminated by previously adsorbed species. [Courtesy Fisher Scientific, Pittsburgh, PA.] Class A flasks meet tolerances given in Table 2-3. Class B tolerances are poorer: 10 ± 0.04, 25 ± 0.06, 50 ± 0.10, 100 ± 0.16, 250 ± 0.24, 500 ± 0.40, and 1 000 ± 0.60 mL. (*Adsorption* means to be bound to the surface. *Absorption* means to go inside, as water goes into a sponge.)

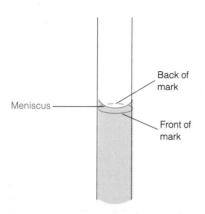

Meniscus

Back of
mark

Front of
mark

Figure 2-11   Proper position of the meniscus: at the center of the ellipse formed by the front and back of the calibration mark when viewed from above or below. Volumetric flasks and transfer pipets are calibrated to this position.

solution are both known. The solid should first be dissolved in *less* liquid than the flask is calibrated to contain. Then more liquid is added, and the solution is again mixed. The final adjustment to the mark should be done with as much well-mixed solution as possible in the flask. This practice minimizes the change in volume upon mixing pure liquid with solution already in the flask. The final drops of liquid should be added with a pipet, *not a squirt bottle,* for best control. After adjusting the solution to its final volume, the cap should be held firmly in place and the flask *inverted 20 or more times* to ensure complete mixing.

To adjust the liquid level to the *center* of the mark of a volumetric flask (or of a pipet, which is calibrated in the same way), look at the mark from *above* or *below.* The front and back of the mark will not be aligned with each other and will describe an ellipse. Adjust the liquid until the bottom of the meniscus is at the *center* of the ellipse (Figure 2-11). This places the bottom of the meniscus exactly at the center of the calibration mark, if viewed edge on.

The surface of glass is notorious for retaining traces of cations. For careful work, it is an excellent practice to soak *already thoroughly cleaned* glassware in 3–6 M HCl (in a fume hood) for >1 h, followed by many rinses with distilled water and a final soak in distilled water. H^+ replaces cations on the glass surface, thereby leaving a cleaner vessel. The HCl used in this **acid wash** procedure may be reused many times, as long as it is only used for soaking already clean glassware.

2-6 Pipets and Syringes

Pipets are used to deliver known volumes of liquid. Four common types are shown in Figure 2-12. The transfer pipet, which is the most accurate, is calibrated to deliver one fixed volume. The last drop of liquid does not drain out of the pipet; it is meant to be left in the pipet. *It should not be blown out.* The measuring pipet is calibrated to deliver a variable volume, which is determined as the difference between the volumes indicated before and after delivery. The measuring pipet in Figure 2-12 could, for example, be used to deliver 5.6 mL by starting delivery at the 1-mL mark and ending at the 6.6-mL mark. The Ostwald-Folin pipet is similar to the transfer pipet, except that the last drop *should* be blown out. When in doubt about whether the pipet is made for blowout, consult the manufacturer's catalog. Serological pipets are measuring pipets calibrated all the way to the tip. The serological pipet in Figure 2-12 will deliver 10 mL when the last drop is blown out.

In general, transfer pipets are more accurate than measuring pipets. The tolerances for Class A transfer pipets are given in Table 2-4. To deliver the calibrated volume, bring the bottom of the meniscus to the center of the calibration mark, as shown in Figure 2-11.

   Do not blow the last drop out of a transfer pipet.

Using a Transfer Pipet

Using a rubber bulb, *not your mouth,* suck liquid up past the mark. Quickly replace the bulb by putting your index finger over the end of the pipet. The liquid should still be above the mark after this maneuver. Pressing the pipet

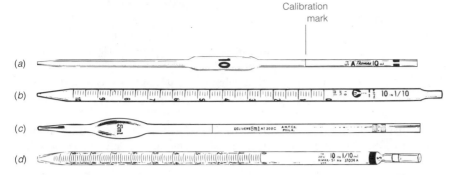

Calibration mark

(a)

(b)

(c)

(d)

Figure 2-12 Common pipets: (*a*) transfer; (*b*) measuring (Mohr); (*c*) Ostwald-Folin (blow out last drop); (*d*) serological (blow out last drop). [Courtesy A. H. Thomas Co., Philadelphia, PA.]

against the bottom of the vessel while removing the rubber bulb helps prevent liquid from draining while you get your finger in place. Wipe the excess liquid off the outside of the pipet with a clean tissue. *Touch the tip of the pipet to the side of a beaker,* and drain the liquid until the bottom of the meniscus just reaches the center of the mark. The pipet must touch the beaker during the draining so that extra liquid is not hanging from the tip of the pipet when the meniscus reaches the mark. Any liquid outside of the pipet will be drawn onto the wall of the beaker. Now transfer the pipet to the desired receiving vessel and drain it *while holding the tip against the wall of the vessel.* After the pipet stops draining, hold it against the wall for a few seconds more to be sure that everything that can drain has drained. *Do not blow out the last drop.* The pipet should be nearly vertical to ensure that the proper amount of liquid has drained. When you finish with a pipet, it should be rinsed with distilled water or placed in a pipet container to soak until it is cleaned. Solutions should never be allowed to dry inside a pipet, because removing internal deposits is very difficult.

 Delivering Small Volumes

Plastic microliter pipets, such as that in Figure 2-13a, are used to deliver volumes in the 1 to 1 000 μL (1 μL = 10^{-6} L) range. The liquid is contained

TABLE 2-4
Tolerances of Class A transfer pipets

Volume (mL)	Tolerance (mL)
0.5	±0.006
1	±0.006
2	±0.006
3	±0.01
4	±0.01
5	±0.01
10	±0.02
15	±0.03
20	±0.03
25	±0.03
50	±0.05
100	±0.08

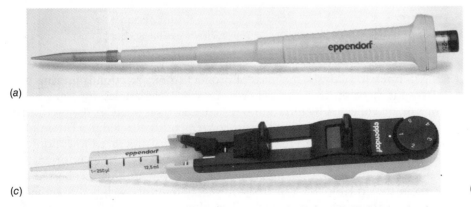

(a)

(c)

(b)

Figure 2-13 (*a*) Microliter pipet with disposable plastic tip. (*b*) Volume selection dial of microliter pipet. (*c*) Repeater® pipet that can deliver preset volumes between 10 μL and 5 mL up to 48 times at 1-s intervals without refilling. [Courtesy Brinkman Instruments, Westbury, NY.]

Needle   Barrel   Plunger

Figure 2-14 Hamilton syringe with a volume of 1 μL and divisions of 0.01 μL on the glass barrel. [Courtesy A. H. Thomas Co., Philadelphia, PA.]

Accuracy refers to the difference between the delivered volume and the desired volume. *Precision* refers to the reproducibility of replicate deliveries.

  For maximum accuracy, you can calibrate an individual pipet tip as described by B. Kratochvil and N. Motkosky, *Anal. Chem.* **1987,** *59,* 1064.

in the disposable plastic tip. The accuracy is 1–2%, and precision may be as good as 0.5%. The pipet in Figure 2-13b allows rapid delivery of repeated quantities of selected volumes from the reservoir in the syringe.

To use a microliter pipet (micropipet) like that shown in Figure 2-13a, place a fresh tip tightly on the barrel. Tips are contained in a package or dispenser so that you do not handle (and contaminate) the points with your fingers. Set the desired volume with the knob at the top of the pipet. Depress the plunger to the first stop, which corresponds to the volume that you selected. Hold the pipet *vertically,* dip it 3–5 mm into the reagent solution, and *slowly* release the plunger to suck up liquid. Withdraw the tip from the liquid by sliding it along the wall of the vessel to remove liquid from the outside of the tip. To dispense liquid, touch the micropipet tip to the wall of the receiver and gently depress the plunger to the first stop. After waiting a few seconds to allow the liquid to drain down the inside walls of the pipet tip, depress the plunger beyond the stop to squirt the last liquid out of the tip. It is a good idea to clean and wet a fresh tip by taking up and discarding two or three squirts of reagent first (using the procedure above). A used tip may be discarded, or you may rinse it well with a jet of distilled water from a squirt bottle and use it again.

Because there is air above the liquid in the tip, the volume of liquid taken into the tip depends on the angle at which the pipet is held and how far beneath the surface of reagent the tip is held during uptake. Each operator will attain slightly different precision and accuracy with this device.

The microliter *syringe,* shown in Figure 2-14, is another excellent device for dispensing tiny volumes of liquid with a precision and accuracy near 1%. Take up and discard several volumes of liquid to wash the glass walls free of contaminants and to remove air bubbles from the barrel before use. The steel needle is attacked by strong acids and will contaminate such solutions with iron.

2-7 Filtration

In **gravimetric analysis,** the mass of product from a reaction is measured to determine how much unknown was present. Precipitates for gravimetric analysis are collected by filtration, washed, and then dried. If the precipitate does not need to be ignited, it is most conveniently collected in a fritted-glass funnel (Figure 2-15). This funnel has a porous glass disk that allows liquid to pass through but retains solid particles. The filter crucible is dried and weighed before the precipitate is collected. After the product is collected and dried in the crucible, the crucible and its contents are weighed together to determine the mass of precipitate. The filter crucible is usually used with suction provided by a water aspirator, as shown in Figure 2-16. The liquid from which a substance precipitates or crystallizes is called the **mother liquor.** The liquid that passes through a filter is called the **filtrate.**

Porous glass ——

Figure 2-15 Fritted-glass Gooch filter crucible. [Courtesy A. H. Thomas Co., Philadelphia, PA.]

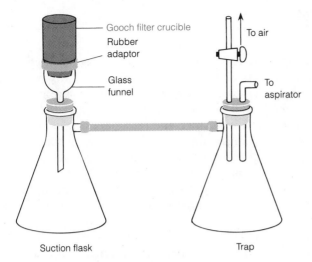

Figure 2-16 Filtration with a Gooch filter crucible. The trap prevents possible backup of tap water from the aspirator into the suction flask.

In some gravimetric procedures, a precipitate is isolated and converted to a product of known composition by **ignition** (heating to a high temperature). For example, Fe^{3+} is precipitated as a poorly defined hydrated form of $Fe(OH)_3$ and converted by ignition to Fe_2O_3 before weighing. When a gravimetric precipitate is to be ignited, it is collected in **ashless filter paper** that leaves little residue when oxidized at high temperature. The filter paper is folded in quarters, one corner torn off, and the paper placed in a conical glass funnel (Figure 2-17). The paper should fit snugly and be seated with a little distilled water. When liquid is poured into the funnel, an unbroken stream of liquid should fill the stem of the funnel. The weight of the liquid in the stem helps to speed the filtration.

For filtration, liquid containing suspended precipitate is poured down a glass rod into the filter (Figure 2-18). The rod helps prevent splattering or dripping down the side of the beaker. Particles adhering to the beaker or rod can be dislodged with a **rubber policeman** (a glass rod with a flattened piece of rubber attached to the end) and transferred with a jet of liquid from a squirt bottle containing distilled water or an appropriate wash liquid. Particles that remain in the beaker may be wiped onto a small piece of moist filter paper and the paper placed in the filter, to be ignited with the bulk of the precipitate.

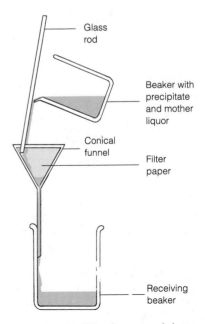

Figure 2-18 Filtering a precipitate.

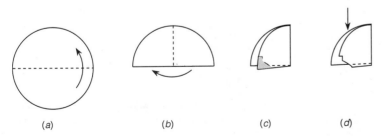

Figure 2-17 Folding filter paper for a conical funnel. (*a*) Fold the paper in half. (*b*) Then fold it in half again. (*c*) Tear off a corner, to better seat the paper in the funnel. (*d*) Open the side that was not torn when fitting the paper in the funnel.

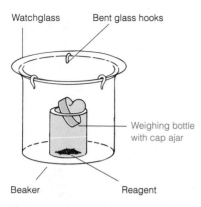

Figure 2-19 Use of watchglass as a dustcover while drying reagent or crucible in the oven.

(a)

(b)

Figure 2-20 (a) Ordinary desiccator. (b) Vacuum desiccator that can be evacuated through the sidearm at the top and then sealed by rotating the joint containing the sidearm. Drying is more efficient at low pressure. [Courtesy A. H. Thomas Co., Philadelphia, PA.]

2-8 Drying

Reagents, precipitates, and glassware are conveniently dried in an oven at 110°C. (Some reagents or precipitates require other drying temperatures.) When a filter crucible is to be brought to constant mass prior to a gravimetric analysis, the crucible should be dried for 1 h or longer and then cooled in a desiccator. The crucible is weighed and then heated again for about 30 min. When successive weighings agree to within 0.3 mg, the filter is considered to have been dried "to constant mass." A beaker and watchglass (Figure 2-19) minimize contamination by dust during drying. (Dust is a source of contamination in all experiments, so it is good practice to cover all vessels left on the benchtop whenever possible.) A microwave oven may be used instead of an electric oven for drying reagents. The microwave oven reduces each heating cycle to 5 min or less.

After a reagent or crucible has been dried at high temperature, it is cooled to room temperature in a **desiccator** (Figure 2-20), a closed chamber that contains a drying agent, or **desiccant.** The lid is greased to make an airtight seal. Desiccant is placed beneath the perforated disk near the bottom of the chamber. Data on the efficacy of common drying agents are given in Table 2-5. In addition to those listed, 98% sulfuric acid is also a common and efficient desiccant. After placing a hot object in a desiccator, leave the lid cracked open for a minute or two until the object has cooled slightly. This practice prevents the lid from popping open when the air inside becomes warm. An object being cooled prior to weighing should be given about 30 min to reach room temperature. The correct way to open a desiccator is to slide the lid sideways until it can be removed.

TABLE 2-5   Efficiencies of drying agents

Agent	Formula	Water left in atmosphere ($\mu g\ H_2O/L$)[a]
Magnesium perchlorate, anhydrous	$Mg(ClO_4)_2$	0.2
"Anhydrone"	$Mg(ClO_4)_2 \cdot 1{-}1.5H_2O$	1.5
Barium oxide	BaO	2.8
Alumina	Al_2O_3	2.9
Phosphorus pentoxide	P_4O_{10}	3.6
Lithium perchlorate, anhydrous	$LiClO_4$	13
Calcium chloride (dried at 127°C)	$CaCl_2$	67
Calcium sulfate (Drierite)	$CaSO_4$	67
Silica gel	SiO_2	70
Ascarite	$NaOH$ on asbestos	93
Sodium hydroxide	$NaOH$	513
Barium perchlorate	$Ba(ClO_4)_2$	599
Calcium oxide	CaO	656
Magnesium oxide	MgO	753
Potassium hydroxide	KOH	939

a. Moist nitrogen was passed over each desiccant, and the water remaining in the gas was condensed and weighed.

SOURCE: A. I. Vogel, *A Textbook of Quantitative Inorganic Analysis,* 3rd ed. (New York: Wiley, 1961), p. 178.

2-9 Calibration of Volumetric Glassware

When greatest accuracy is desired, it is necessary to calibrate volumetric glassware by measuring the mass of water delivered or contained in the vessel, and using the density of water to convert mass to volume. By this means you could determine, for example, that your particular 10-mL pipet delivers 10.016 mL, not 10.000 mL.

In the most careful work it is necessary to account for thermal expansion and contraction of solutions and glassware. It is therefore necessary to know the laboratory temperature at the time solutions are made and also at the time they are used. Table 2-6 shows that pure water expands about 0.02% per degree near 20°C. This change has a practical implication for the calibration of glassware and for the temperature dependence of reagent concentrations.

Small or odd-shaped vessels can be calibrated with mercury, which is easier than water to pour out of glass and is 13.6 times denser than water. Do not spill mercury. It is toxic.

. . . .

EXAMPLE **Effect of Temperature on Solution Concentration**

A dilute aqueous solution with a molarity of 0.031 46 M was prepared in the winter when the lab temperature was 17°C. What will the molarity of that solution be on a warm spring day when the temperature is 25°C?

Solution Assume that the thermal expansion of the solution is similar to the thermal expansion of pure water. Because the concentration of a solution is proportional to its density, we can write

Correction for thermal expansion:
$$\frac{c'}{d'} = \frac{c}{d} \tag{2-2}$$

where c' and d' are the concentration and density at temperature T', and c and d apply at temperature T. Using the densities in column 2 of Table 2-6, write

$$\frac{c' \text{ at } 25°C}{0.997\,05 \text{ g/mL}} = \frac{0.031\,46 \text{ M}}{0.998\,78 \text{ g/mL}} \Rightarrow c' = 0.031\,41 \text{ M}$$

The concentration has decreased by 0.16%.

The concentration of the solution decreases when the temperature increases.

. . . .

Pyrex and other borosilicate glasses, which are the most common types, expand by about 0.001 0% per degree Celsius near room temperature. Thus, if a glass container's temperature is increased 10°C, its volume will increase by about (10)(0.001 0%) = 0.010%. For all but the most accurate work, this expansion is insignificant. Soft glass expands two to three times as much as borosilicate glass.

Suppose that you want to calibrate a 25-mL pipet. You should first weigh an empty weighing bottle like the one shown in Figure 2-19. Then fill the pipet to the 25-mL mark with distilled water, drain it into the weighing bottle, and put the cap on to prevent evaporation. Weigh the bottle again to find the mass of water delivered from the pipet. Use Table 2-6 to convert the mass of water into volume of water.

TABLE 2-6   Density of water

Temperature (°C)	Density of water (g/mL)	Volume of 1 g of water (mL)	
		At temperature shown[a]	Corrected to 20°C[b]
0	0.999 842 5	—	—
4	975 0	—	—
5	966 8	—	—
10	702 6	1.001 4	1.001 5
11	608 4	1.001 5	1.001 6
12	500 4	1.001 6	1.001 7
13	380 1	1.001 7	1.001 8
14	247 4	1.001 8	1.001 9
15	102 6	1.002 0	1.002 0
16	0.998 946 0	1.002 1	1.002 1
17	777 9	1.002 3	1.002 3
18	598 6	1.002 5	1.002 5
19	408 2	1.002 7	1.002 7
20	207 1	1.002 9	1.002 9
21	0.997 995 5	1.003 1	1.003 1
22	773 5	1.003 3	1.003 3
23	541 5	1.003 5	1.003 5
24	299 5	1.003 8	1.003 8
25	047 9	1.004 0	1.004 0
26	0.996 786 7	1.004 3	1.004 2
27	516 2	1.004 6	1.004 5
28	236 5	1.004 8	1.004 7
29	0.995 947 8	1.005 1	1.005 0
30	650 2	1.005 4	1.005 3
35	0.994 034 9	—	—
37	0.993 331 6	—	—
40	0.992 218 7	—	—
100	0.958 366 5	—	—

a. Corrected for buoyancy with Equation 2-1, using 0.001 2 g/mL air density and density of weights = 8.0 g/mL.
b. Corrected for expansion of borosilicate glass (0.001 0% per degree Celsius).

EXAMPLE   Calibration of a Pipet

The empty weighing bottle weighed 10.283 g. After filling it from a 25-mL pipet, the mass was 35.225 g. If the lab temperature was 23°C, find the volume of water delivered from the pipet. What volume would be delivered if the temperature were 20°C?

Columns 3 and 4 of Table 2-6 include the effect of buoyancy.

Solution   The mass of water in the pipet is 35.225 − 10.283 = 24.942 g. From column 3 of Table 2-6, the volume of water is (24.942 g)(1.003 5 mL/g) = 25.029 mL at 23°C. This is the volume delivered by the pipet at 23°C. If the pipet and water were at 20°C instead of 23°C, the pipet would contract slightly and contain less water. However, columns 3 and 4 of Table 2-6 show that the correction from 23°C to 20°C is not significant to four decimal places. The volume delivered at 20°C would still be 25.029 mL.

Terms to Understand

acid wash	desiccant	gravimetric analysis	meniscus	rubber policeman
ashless filter paper	desiccator	hygroscopic	mother liquor	tare
buoyancy	electronic balance	ignition	parallax	volumetric flask
buret	filtrate	mechanical balance	pipet	

Summary

Safety requires you to think in advance about what you will do; do not do what seems dangerous. Know how to use safety equipment such as goggles, fume hood, lab coat, gloves, emergency shower, eyewash, and fire extinguisher. Chemicals should be stored and used in a manner that minimizes contact of solids, liquids, and vapors with people. Environmentally acceptable disposal procedures should be established in advance for every chemical that you use. Your lab notebook tells what you did and what you observed; it should be understandable to other people. It also should allow you to repeat an experiment in the same manner in the future. You should understand the principles of operation of electronic and mechanical balances and treat these as delicate equipment. Buoyancy corrections are required in accurate

work. Burets should be read in a reproducible manner and drained slowly for best results. Always interpolate between markings to obtain accuracy one decimal place beyond the graduations. Volumetric flasks are used to prepare solutions with known volume. Transfer pipets deliver fixed volumes; less accurate measuring pipets deliver variable volumes. Filtration and collection of precipitates requires careful technique, as does the drying of reagents, precipitates, and glassware in ovens and desiccators. Ignition converts a gravimetric precipitate to a known, stable composition useful for measurement by weighing. Volumetric glassware is calibrated by weighing water contained in or delivered by the vessel. In the most careful work, solution concentrations and volumes of vessels should be corrected for changes in temperature.

Exercises[†]

A. What is the true mass of water if the measured mass in the atmosphere is 5.397 4 g? When you look up the density of water, assume that the lab temperature is (a) 15°C and (b) 25°C. Take the density of air to be 0.001 2 g/mL and the density of balance weights to be 8.0 g/mL.

[†]Detailed solutions to Exercises appear at the end of this book. Numerical answers to Problems also appear at the end of the book, and full solutions to all Problems are given in the *Solutions Manual for Quantitative Chemical Analysis*. The *Solutions Manual* also contains supplementary problems (and their solutions) that may be reproduced by instructors.

B. A sample of ferric oxide (Fe_2O_3, density = 5.24 g/mL) obtained from ignition of a gravimetric precipitate weighed 0.296 1 g in the atmosphere. What is the true mass in vacuum?

C. A solution of potassium permanganate ($KMnO_4$) was found by titration to be 0.051 38 M at 24°C. What was the molarity when the lab temperature dropped to 16°C?

D. Water was drained from a buret between the 0.12 and 15.78 mL marks. The apparent volume delivered was $15.78 - 0.12 = 15.66$ mL. Measured in the air at 22°C, the mass of water delivered was 15.569 g. What was the true volume?

Problems

Safety and Lab Notebook

1. What is the primary safety rule and what is your implied responsibility to make it work?

2. After the safety features and procedures in your laboratory have been explained to you, make a list of them.

3. In Box 2-1, why is Pb^{2+} converted to $PbSiO_3$ before disposal in an approved landfill?

4. State three essential attributes of a lab notebook.

Analytical Balance

5. Explain the principles of operation of electronic and mechanical balances.

6. Explain why the buoyancy correction is zero in Figure 2-6 when the density of the object being weighed is 8.0 g/mL.

7. Pentane (C_5H_{12}) is a liquid with a density of 0.626 g/mL near 25°C. Use Equation 2-1 to find the true mass of pentane when the mass weighed in air is 14.82 g. Assume that the air density is 0.0012 g/mL and the balance weight density is 8.0 g/mL.

8. Referring to Figure 2-6, predict which compound below will have the smallest percentage of buoyancy correction and which will have the greatest.

Substance	Density (g/mL)	Substance	Density (g/mL)
acetic acid	1.05	mercury	13.6
carbon tetrachloride	1.59	lead dioxide	9.4
sulfur	2.07	lead	11.4
lithium	0.53	iridium	22.4

9. Use the equation in Box 2-2 to calculate the change in frequency for an 8.1-MHz quartz crystal when 200 ng of DNA binds to electrodes with a total area of 16×10^{-6} m^2.

10. (a) Use the ideal gas law (see Chapter 1, Problem 27) to calculate the density (g/mL) of helium at 20°C and 1.00 atm.

(b) Find the true mass of sodium metal (density = 0.97 g/mL) weighed in a glove box with a helium atmosphere, if the apparent mass was 0.823 g and the balance weight density is 8.0 g/mL.

11. (a) What is the vapor pressure of water in the air at 20°C if the humidity is 42%? The vapor pressure of water at 20°C at equilibrium is 2330 Pa. (*Hint:* Humidity refers to the fraction of the equilibrium water vapor pressure present in the air.)

(b) Find the air density (g/mL, not g/L) under the conditions of (a) if the barometric pressure is 94.0 kPa.

(c) What is the true mass of water in (b) if the balance indicates that 1.0000 g is present (balance weight density = 8.0 g/mL)?

Glassware and Thermal Expansion

12. What do the symbols "TD" and "TC" mean on volumetric glassware?

13. Describe the procedure for preparing 250.0 mL of 0.1500 M K_2SO_4 with a volumetric flask.

14. When would it be preferable to use a plastic volumetric flask instead of a more accurate glass flask?

15. Describe the procedure for transferring 5.00 mL of liquid using a transfer pipet.

16. Which is more accurate, a transfer pipet or a measuring pipet?

17. What would you do differently when delivering 1.00 mL from a 1-mL serological pipet instead of a 1-mL measuring pipet?

18. What is the purpose of the trap in Figure 2-16? What is the purpose of the watchglass in Figure 2-19?

19. Which drying agent is more efficient, Drierite or phosphorus pentoxide?

20. An empty 10-mL volumetric flask weighs 10.2634 g. When filled to the mark with distilled water and weighed again in the air at 20°C, the mass is 20.2144 g. What is the true volume of the flask at 20°C?

21. By what percentage does a dilute aqueous solution expand when heated from 15° to 25°C? If a 0.5000 M solution is prepared at 15°C, what would its molarity be at 25°C?

22. The true volume of a 50-mL volumetric flask is 50.037 mL at 20°C. What mass of water measured (a) in vacuum and (b) in air at 20°C would be contained in the flask?

23. You want to prepare 500.0 mL of 1.000 M KNO_3 at 20°C, but the lab (and water) temperature is 24°C at the time of preparation. How many grams of solid KNO_3 (density = 2.109 g/mL) should be dissolved in a volume of 500.0 mL at 24°C so that the concentration will be 1.000 M at 20°C? What apparent mass of KNO_3 weighed in air is required?

Notes and References

1. Handling and disposing of chemicals is described in *Prudent Practices for Handling Hazardous Chemicals in Laboratories* (1983) and *Prudent Practices for Disposal of Chemicals from Laboratories* (1983), both available from National Academy Press, 2101 Constitution Avenue N.W., Washington, DC 20418. See also G. Lunn and E. B. Sansone, *Destruction of Hazardous Chemicals in the Laboratory* (New York: Wiley, 1990).

2. M.-A. Armour, *J. Chem. Ed.* **1988**, *65*, A64; W. A. Walton, *J. Chem. Ed.* **1987**, *64*, A69.

3. J. Hlavay and G. G. Guilbault, *Anal. Chem.* **1977**, *49*, 1890; Z. Lin, C. M. Yip, I. S. Joseph, and M. D. Ward, *Anal. Chem.* **1993**, *65*, 1546. A quartz crystal for use in aqueous solutions *without* electrode contacts has been described [T. Nomura, F. Tanaka, T. Yamada, and H. Ito, *Anal. Chim. Acta* **1991**, *243*, 273]. Piezoelectric crystals can also be used to measure the effect of mass adsorbed on the crystal surface with *surface acoustic waves* [D. S. Ballantine, Jr., and H. Wohltjen, *Anal. Chem.* **1989**, *61*, 704A] and *flexural plate waves* [J. W. Grate, S. W. Wenzel, and R. M. White, *Anal. Chem.* **1991**, *63*, 1552].

4. H. M. Fog and B. Rietz, *Anal. Chem.* **1985**, *57*, 2634.

5. R. C. Ebersole and M. D. Ward, *J. Am. Chem. Soc.* **1988**, *110*, 8623; S. Borman, *Anal. Chem.* **1987**, *59*, 1161A; H. Muramatsu, K. Kajiwara, E. Tamiyua, and I. Karube, *Anal. Chim. Acta* **1986**, *188*, 257.

6. R. Batting and A. G. Williamson, *J. Chem. Ed.* **1984**, *61*, 51; J. E. Lewis and L. A. Woolf, *J. Chem. Ed.* **1971**, *48*, 639; F. F. Cantwell, B. Kratochvil, and W. E. Harris, *Anal. Chem.* **1978**, *50*, 1010.

The results are back from the lab: John Smith is pregnant.

[Courtesy 3M Company, St. Paul, MN.]

Experimental Error

Not all laboratory errors are so monumental as that depicted on the opposite page, but there is error associated with every measurement. There is no way to measure the "true value" of anything. The best we can do in a chemical analysis is to carefully apply a technique that experience tells us is reliable. Repetition of one type of measurement several times tells us the reproducibility (*precision*) of the measurement. Measuring the same quantity by different methods gives us confidence of nearness to the "truth" (*accuracy*), if the results agree with one another. In this chapter we will learn to determine the uncertainty in a calculated result by examining the uncertainties in measured quantities. We will also learn to use an appropriate number of digits to reflect the precision of a result. You should always be cognizant of the uncertainty associated with a result and, therefore, of how reliable that result may be.

3-1 Significant Figures

The number of **significant figures** is the minimum number of digits needed to write a given value in scientific notation without loss of accuracy. The number 142.7 has four significant figures, because it can be written as 1.427×10^2. If you wrote 1.4270×10^2, you would be implying that you know the value of the digit after 7, which is not the case for the number 142.7. The number 1.4270×10^2 therefore has five significant figures.

The number 6.302×10^{-6} has four significant figures, because all four digits are necessary. You could write the same number as 0.000 006 302, which also has just *four* significant figures. The zeros to the left of the 6 are

Definition of significant figures.

merely holding decimal places. The number 92 500 is ambiguous with regard to the number of significant figures. It could mean any of the following:

$$9.25 \times 10^4 \qquad \text{3 significant figures}$$
$$9.250 \times 10^4 \qquad \text{4 significant figures}$$
$$9.2500 \times 10^4 \qquad \text{5 significant figures}$$

It is preferable to write one of the three numbers above, instead of 92 500, to indicate how many figures are actually known.

Zeros are significant when they occur (1) in the middle of a number or (2) at the end of a number on the right-hand side of a decimal point.

The last (farthest to the right) significant figure in a measured quantity always has some associated uncertainty. The minimum uncertainty would be ±1 in the last digit. The scale of a Spectronic 20 spectrophotometer is drawn in Figure 3-1. The needle shown in the figure appears to be at an absorbance value of 0.234. We say that there are three significant figures because the numbers 2 and 3 are completely certain and the number 4 is an estimate. The value could be read 0.233 or 0.235 by different people. The percent transmittance is near 58.3. Because the transmittance scale is smaller than the absorbance scale at this point, there is probably more uncertainty in the last digit of transmittance. A reasonable estimate of the uncertainty might be 58.3 ± 0.2. There are three significant figures in the number 58.3.

In general, when reading the scale of any apparatus, you should interpolate between the markings. It is usually possible to estimate to the nearest tenth of the distance between two marks. Thus on a 50-mL buret, which is graduated to 0.1 mL, you should read the level to the nearest 0.01 mL. When using a ruler calibrated in millimeters, estimate distances to the nearest tenth of a millimeter.

Significant zeros below are *bold*:

106   0.010**6**   0.106   0.106**0**

Interpolation: Estimate all readings to the nearest tenth of the distance between scale divisions.

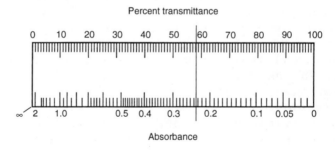

Percent transmittance

Figure 3-1   Scale of a Bausch and Lomb Spectronic 20 spectrophotometer. Percent transmittance is a linear scale and absorbance is a logarithmic scale.

3-2 Significant Figures in Arithmetic

This section deals with determining the number of digits to retain in the answer after you have performed arithmetic operations with your data. Rounding should only be done on the *final answer* (not intermediate results), to avoid accumulating round-off errors.

Addition and Subtraction

Often, the numbers to be added or subtracted have equal numbers of digits. In this case the answer should go to the *same decimal place* as in the individual numbers. For example,

$$
\begin{array}{r}
1.362 \times 10^{-4} \\
+ \; 3.111 \times 10^{-4} \\
\hline
4.473 \times 10^{-4}
\end{array}
$$

The number of significant figures in the answer may exceed or be less than that in the original data.

$$
\begin{array}{r}
5.345 \\
+ \; 6.728 \\
\hline
12.073
\end{array}
\qquad
\begin{array}{r}
7.26 \times 10^{14} \\
-6.69 \times 10^{14} \\
\hline
0.57 \times 10^{14}
\end{array}
$$

If the numbers being added do not have the same number of significant figures, we are usually limited by the least certain one. For example, in calculating the formula weight of KrF_2, the answer is known only to the second decimal place, because we are limited by our knowledge of the atomic weight of Kr.

$$
\begin{array}{rl}
18.998\,403\,2 & (F) \\
+ \; 18.998\,403\,2 & (F) \\
+ \; 83.80 & (Kr) \\
\hline
121.796\,806\,4 &
\end{array}
$$

Not significant

The number $121.796\,806\,4$ should be rounded to 121.80 as the final answer.

When rounding off, look at *all* the digits *beyond* the last place desired. In the example above, the digits 6 806 4 lie beyond the last significant decimal place. Because this number is more than half way to the next higher digit, we round the 9 up to 10 (i.e., we round up to 121.80 instead of down to 121.79). If the insignificant figures were less than half way, we would round down. For example, 121.794 8 is correctly rounded to 121.79.

In the special case where the number is exactly halfway, round to the nearest *even* digit. Thus, 43.550 00 is rounded to 43.6, if we can only have three significant figures. If we are retaining only three figures, 1.425×10^{-9} becomes 1.42×10^{-9}. The number $1.425\,01 \times 10^{-9}$ would become 1.43×10^{-9}, because 501 is more than halfway to the next digit. The rationale for rounding to the nearest even digit is that it avoids systematically increasing or decreasing results through successive round-off errors. When this rounding rule is applied consistently, half the round-offs will be up and half down.

In adding or subtracting numbers expressed in scientific notation, all numbers should first be converted to the same exponent. For example, to do the following addition we could write

$$
\begin{array}{r}
1.632 \times 10^5 \\
+ \; 4.107 \times 10^3 \\
+ \; 0.984 \times 10^6 \\
\hline
\end{array}
\quad \Rightarrow \quad
\begin{array}{r}
1.632 \quad\; \times 10^5 \\
+ \; 0.041\,07 \times 10^5 \\
+ \; 9.84 \quad\;\; \times 10^5 \\
\hline
11.51 \quad\; \times 10^5
\end{array}
$$

The sum $11.513\,07 \times 10^5$ is rounded to 11.51×10^5 because the number 9.84×10^5 limits us to two decimal places when all numbers are expressed as multiples of 10^5.

Rules for rounding off numbers.

Addition and subtraction: Express all numbers with the same exponent and align all numbers with respect to the decimal point. Round off the answer according to the number of decimal places in the number with the fewest decimal places.

Challenge Show that the answer would still have four significant figures even if all numbers were expressed as multiples of 10^4 instead of 10^5.

Multiplication and Division

In multiplication and division, we are normally limited to the number of digits contained in the number with the fewest significant figures. For example,

$$
\begin{array}{cccc}
3.26 \times 10^{-5} & 4.3179 \times 10^{12} & & 34.60 \\
\underline{\times\ 1.78} & \underline{\times\ 3.6\quad\ \times 10^{-19}} & & \underline{\div\ \ 2.46287} \\
5.80 \times 10^{-5} & 1.6\quad\ \ \times 10^{-6} & & 14.05
\end{array}
$$

The power of 10 has no influence on the number of figures that should be retained.

Logarithms and Antilogarithms

The base 10 **logarithm** of n is the number a, whose value is such that $n = 10^a$:

A refresher on logarithms is found in Appendix A.

Logarithm of n: $\boxed{n = 10^a \Leftrightarrow \log n = a}$

The number n is said to be the **antilogarithm** of a. A logarithm is composed of a **mantissa** and a **character.**

$$\log 339 = \underset{\underset{\text{Character}\ \ \text{Mantissa}}{|\quad |}}{2.530}$$

Number of digits in *mantissa* of log x = number of significant figures in x:

$$\log \underbrace{(5.403 \times 10^{-8})}_{4\ \text{digits}} = -7.\underbrace{2674}_{4\ \text{digits}}$$

The number 339 can be written 3.39×10^2. *The number of digits in the mantissa of log 339 should equal the number of significant figures in 339.* The logarithm of 339 is properly expressed as 2.530. The *character,* 2, corresponds to the exponent in 3.39×10^2.

To see that the third decimal place is the last significant place, consider the following results:

$$10^{2.531} = 340\ (339.6)$$
$$10^{2.530} = 339\ (338.8)$$
$$10^{2.529} = 338\ (338.1)$$

The numbers in parentheses are the results prior to rounding to three figures. Changing the exponent by one digit in the third decimal place changes the answer by one digit in the last (third) place of 339.

In converting a logarithm to its antilogarithm, *the number of significant figures in the antilogarithm should equal the number of digits in the mantissa.* Thus

Number of digits in antilog x (= 10^x) = number of figures in *mantissa* of x:

$$\underbrace{10^{6.142}}_{3\ \text{digits}} = \underbrace{1.39}_{3\ \text{digits}} \times 10^6$$

$$\text{antilog}\ (-3.\underbrace{42}_{}) = 10^{-3.\underbrace{42}_{}} = \underbrace{3.8}_{} \times 10^{-4}$$
$$\underset{2\ \text{digits}\qquad 2\ \text{digits}\ \ 2\ \text{digits}}{}$$

The following examples show the proper use of significant figures for logs and antilogs.

$$\log 0.001237 = -2.9076 \qquad \text{antilog}\ 4.37 = 2.3 \times 10^4$$
$$\log 1237 = 3.0924 \qquad\qquad 10^{4.37} = 2.3 \times 10^4$$
$$\log 3.2 = 0.51 \qquad\qquad 10^{-2.600} = 2.51 \times 10^{-3}$$

3-3 Significant Figures and Graphs

If you draw a graph on a computer, consider whether the graph is intended to display qualitative behavior of data (Figure 3-2) or precise values. If someone will use the graph (such as the calibration curve in Figure 3-3) to read points, it should at least have tic marks on both sides of the horizontal and vertical scales. Better still is a fine grid superimposed on the graph.

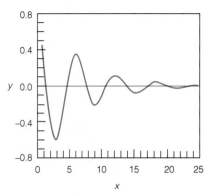

Figure 3-2 Example of a graph intended to show the qualitative behavior of the function $y = e^{-x/6} \cos x$. You are not expected to be able to read coordinates accurately on this graph.

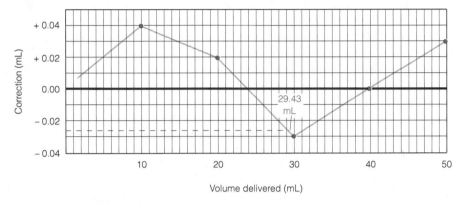

Figure 3-3 Calibration curve for a 50-mL buret. The grid allows you to read coordinates accurately. The procedure used to obtain this curve is given in Chapter 27.

The rulings on a sheet of graph paper should be compatible with the number of significant figures of the coordinates. The graph in Figure 3-4a has reasonable rulings on which to graph the points (0.53, 0.65) and (1.08, 1.47). There are rulings corresponding to every 0.1 unit, and it is easy to estimate the position of the 0.01 unit. The graph in Figure 3-4b is the same size but does not have rulings fine enough for estimating the position of the 0.01 unit.

In general, a graph must be at least as accurate as the data being plotted. Therefore, it is a good practice to use the most finely ruled graph paper available. Paper ruled with 10 lines per centimeter or 20 lines per inch is usually best. Plan the coordinates so that the data are spread over as much of the sheet of paper as possible.

The accuracy of a graph should be consistent with the accuracy of the data being plotted.

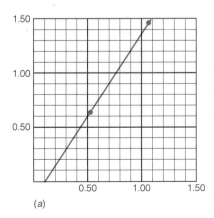

(a)

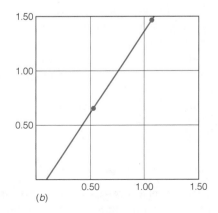

(b)

Figure 3-4 Graphs demonstrating choice of rulings in relation to significant figures in the data. Graph (b) does not have fine enough divisions to plot data that are accurate to the hundredths place.

3-4 Types of Error

Experimental error can be classified as either *systematic* or *random*.

Systematic Error

Systematic error is a consistent error that can be detected and corrected. Box 3-1 describes Standard Reference Materials designed to reduce systematic errors.

A **systematic error,** also called a **determinate error,** can in principle be discovered and corrected. One example would be using a pH meter that has been standardized incorrectly. Suppose you think that the pH of the buffer used to standardize the meter is 7.00, but it is really 7.08. If the meter is otherwise working properly, all of your pH readings will be 0.08 pH unit too low. When you read a pH of 5.60, the actual pH of the sample is 5.68. This systematic error could be discovered by using another buffer of known pH to test the meter.

Ways to detect systematic error:

1. Analyze samples of known composition. Your method should reproduce the known answer. (See Box 15-2 for an example.)

2. Analyze "blank" samples containing none of the analyte being sought. If you observe a nonzero result, your method responds to more than you intended.

3. Use different analytical methods to measure the same quantity. If the results do not agree, there is an error associated with one (or more) of the methods.

4. Samples of the same material can be analyzed by different people in different laboratories (using the same or different methods). Disagreements beyond the expected random error indicate systematic errors.

Another example of systematic error involves the use of an uncalibrated buret. The manufacturer's tolerance for a Class A 50-mL buret is ±0.05 mL. When you believe that you have delivered 29.43 mL, the actual volume could be 29.40 mL and still be within the tolerance limits. One way to correct for this type of error is by constructing an experimental calibration curve, such as that shown in Figure 3-3. In this procedure, distilled water is delivered from the buret into a flask and weighed. By knowing the density of water, you can determine the volume of water from its mass. The graph of correction factors assumes that the level of liquid is always close to the zero mark at the start of the titration. Using Figure 3-3, you would apply a correction factor of −0.03 mL to the measured value of 29.43 mL to reach the correct value of 29.40 mL.

The systematic error associated with using the buret whose calibration is shown in Figure 3-3 is a positive error in some regions and a negative error in others. The key feature of systematic error is that, with care and cleverness, you can detect and correct it.

Random Error

Random error is due to the limitations of physical measurement and cannot be eliminated. A better experiment may reduce the magnitude of the random error but cannot eliminate it entirely.

Random error is also called **indeterminate error.** It arises from natural limitations on our ability to make physical measurements. As the name implies, random error is sometimes positive and sometimes negative. It is always present, cannot be corrected, and is the ultimate limitation on the determination of a quantity. One type of random error is that associated with reading a scale. Different people reading the absorbance or transmittance in Figure 3-1 would report a range of values reflecting their subjective interpolations between the markings. One person reading the same instrument several times probably would also report several different readings. Another type of indeterminate error may result from electrical noise in an instrument. This sort of instability is normally random. Positive and negative fluctuations occur with approximately equal frequency and cannot be completely eliminated.

Precision and Accuracy

Precision: reproducibility
Accuracy: nearness to the "truth"

Precision is a measure of the reproducibility of a result. **Accuracy** refers to how close a measured value is to the "true" value.

Box 3-1 Standard Reference Materials

Inaccurate laboratory measurements can mean wrong medical diagnosis and treatment, lost production time, wasted energy and materials, manufacturing rejects, and product liability problems. To minimize errors in laboratory measurements, the U.S. National Institute of Standards and Technology distributes more than 1000 *standard reference materials,* such as metals, chemicals, rubber, plastics, engineering materials, radioactive substances, and environmental and clinical standards that can be used to test the accuracy of analytical procedures used in different laboratories.†

For example, in treating patients with epilepsy, physicians depend on laboratory tests to ascertain that the blood serum concentrations of anticonvulsant drugs are in the correct range. Low levels lead to seizures and high levels to drug toxicity. Tests of identical serum specimens revealed unacceptably large differences in results among different laboratories. The National Institute of Standards and Technology was asked to develop a standard reference material containing known levels of antiepilepsy drugs in serum. The reference material would allow different laboratories to detect and correct errors in their assay procedures.

Before introduction of this reference material, called SRM 900, five different laboratories analyzing identical samples reported a range of results with relative errors of 40–110% of the expected value. The graph below shows that after SRM 900 distribution the error in the analyses was reduced to 20–40%.

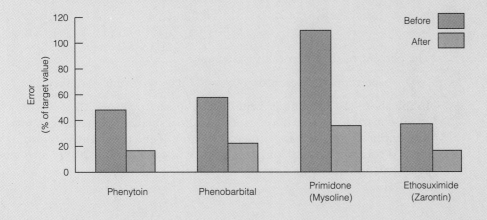

†A catalog is available from Standard Reference Materials Program, Room 204, Building 202, National Institute of Standards and Technology, Gaithersburg, MD 20899.

The result of an experiment may be very reproducible, but wrong. For example, if you made an error in preparing a solution for a titration, the solution would not have the desired concentration. You might then do a series of highly reproducible titrations but report an incorrect result because the concentration of solution was not what you intended. In such a case we would say that the precision of the result is excellent but the accuracy is poor. Conversely, it is possible to make a series of poorly reproducible measurements clustered around the correct value. In this case, the precision is low but the accuracy is high. An ideal procedure provides both precision and accuracy.

Accuracy is defined as nearness to the "true" value. The word *true* is in quotes because somebody had to *measure* the "true" value, and there is error associated with *every* measurement. The "true" value is best obtained by an experienced worker using a well-tested procedure. It is desirable to test the result by using different procedures, because, even though each method might be precise, systematic error could lead to poor agreement between methods. Good agreement among several methods affords us some confidence, but never proof, that the results are "true."

Absolute and Relative Uncertainty

Absolute uncertainty is an expression of the margin of uncertainty associated with a measurement. If the estimated uncertainty in reading a perfectly calibrated buret is ±0.02 mL, we call the quantity ±0.02 mL the absolute uncertainty associated with the reading. Absolute uncertainty has the same units as the measurement.

Relative uncertainty is an expression comparing the size of the absolute uncertainty to the size of its associated measurement. The relative uncertainty of a buret reading of 12.35 ± 0.02 mL is a dimensionless quotient.

Relative uncertainty:
$$\text{relative uncertainty} = \frac{\text{absolute uncertainty}}{\text{magnitude of measurement}}$$

$$= \frac{0.02 \text{ mL}}{12.35 \text{ mL}} = 0.002$$

The percent relative uncertainty is simply

$$\text{percent relative uncertainty} = 100 \times \text{relative uncertainty}$$

In the above example the percent relative uncertainty is 0.2%.

A constant absolute uncertainty leads to a smaller relative uncertainty as the magnitude of the measurement increases. If the uncertainty in reading a buret is constant at ±0.02 mL, the relative uncertainty is 0.2% for a volume of 10 mL and 0.1% for a volume of 20 mL.

3-5 Propagation of Uncertainty

It is usually possible to estimate or measure the random error associated with a particular measurement, such as the length of an object or the temperature of a solution. The uncertainty could be based on your estimate of how well you can read an instrument or on your experience with a particular method. When possible, uncertainty is usually expressed as the *standard deviation* or as a *confidence interval;* these parameters are based on a series of replicate measurements. The discussion that follows applies only to random error; it is assumed that any systematic error has been detected and corrected.

Standard deviation and confidence interval are defined and discussed in Chapter 4.

In most experiments it is necessary to perform arithmetic operations on several numbers, each of which has its associated random error. The most likely uncertainty in the result is not simply the sum of the individual errors, because some of these are likely to be positive and some negative. We expect some amount of cancellation of errors.

Addition and Subtraction

Suppose you wish to perform the following arithmetic, in which the experimental uncertainties are given in parentheses.

$$
\begin{aligned}
1.76\ (\pm0.03) &\leftarrow e_1 \\
+1.89\ (\pm0.02) &\leftarrow e_2 \\
-0.59\ (\pm0.02) &\leftarrow e_3 \\
\hline
3.06\ (\pm e_4) &
\end{aligned}
\tag{3-1}
$$

The arithmetic answer is 3.06; but what is the uncertainty associated with this result?

We start by calling the three uncertainties e_1, e_2, and e_3. For addition and subtraction, the uncertainty in the answer is obtained by manipulating the *absolute uncertainties* of the individual terms.

Uncertainty in addition or subtraction:
$$
e_4 = \sqrt{e_1^2 + e_2^2 + e_3^2}
\tag{3-2}
$$

> For addition and subtraction, use absolute uncertainty.

For the sum in Equation 3-1, we can write

$$
e_4 = \sqrt{(0.03)^2 + (0.02)^2 + (0.02)^2} = 0.04_1
$$

The absolute uncertainty associated with the sum is ±0.04, and we can write the answer as 3.06 ± 0.04. Although there is only one significant figure in the uncertainty, we wrote it initially as 0.04_1, with the first insignificant figure subscripted. The reason for retaining one or more insignificant figures is to avoid introducing round-off errors into later calculations through the number 0.04_1. The insignificant figure was subscripted as a reminder of where the last significant figure should be at the conclusion of the calculations.

If you wish to express the percent relative uncertainty in the sum of Equation 3-1, you can write

$$
\text{percent relative uncertainty} = \frac{0.04_1}{3.06} \times 100 = 1._3\%
$$

The uncertainty, 0.04_1, is $1._3\%$ of the result, 3.06. The subscript 3 in $1._3\%$ is not significant. It is sensible to drop the insignificant figures now and express the final result as

$$
3.06\ (\pm0.04) \qquad \text{(absolute uncertainty)}
$$

or

$$
3.06\ (\pm1\%) \qquad \text{(relative uncertainty)}
$$

> For addition and subtraction, use absolute uncertainty. Relative uncertainty can be found at the end of the calculation.

Multiplication and Division

For multiplication and division, first convert all uncertainties to percent relative uncertainties (or relative uncertainties). Then calculate the error of the product or quotient as follows:

Uncertainty in multiplication or division:
$$
\%e_4 = \sqrt{(\%e_1)^2 + (\%e_2)^2 + (\%e_3)^2}
\tag{3-3}
$$

> For multiplication and division, use relative uncertainty.

Some advice: Retain one or more extra insignificant figures until you have finished your entire calculation. Then round to the correct number of digits. When storing intermediate results in a calculator, keep all of the digits without rounding.

For example, consider the following operations:

$$\frac{1.76\ (\pm 0.03) \times 1.89\ (\pm 0.02)}{0.59\ (\pm 0.02)} = 5.6_4 \pm\ ?$$

First convert all the absolute uncertainties to percent relative uncertainties.

$$\frac{1.76\ (\pm 1._7\%) \times 1.89\ (\pm 1._1\%)}{0.59\ (\pm 3._4\%)} = 5.6_4 \pm?$$

Then find the relative uncertainty of the answer by using Equation 3-3.

$$\%e_4 = \sqrt{(1._7)^2 + (1._1)^2 + (3._4)^2} = 4._0\%$$

The answer is $5.6_4\ (\pm 4._0\%)$.

To convert the relative uncertainty to absolute uncertainty, find $4._0\%$ of the answer.

$$4._0\% \times 5.6_4 = 0.04_0 \times 5.6_4 = 0.2_3$$

The answer is $5.6_4\ (\pm 0.2_3)$. Finally, drop all the figures that are not significant. The result can be expressed as

$$5.6\ (\pm 0.2) \qquad \text{(absolute uncertainty)}$$

For multiplication and division, use relative uncertainty. Absolute uncertainty can be found at the end of the calculation.

$$5.6\ (\pm 4\%) \qquad \text{(relative uncertainty)}$$

There are only two significant figures because we are limited by the denominator, 0.59, in the original problem.

Mixed Operations

Now consider the following mixed operations:

$$\frac{[1.76\ (\pm 0.03) - 0.59\ (\pm 0.02)]}{1.89\ (\pm 0.02)} = 0.619_0 \pm\ ?$$

First work out the difference in the numerator, using absolute uncertainties.

$$1.76\ (\pm 0.03) - 0.59\ (\pm 0.02) = 1.17 \pm 0.03_6$$

because $\sqrt{(0.03)^2 + (0.02)^2} = 0.03_6$.

Then convert to relative uncertainties:

$$\frac{1.17\ (\pm 0.03_6)}{1.89\ (\pm 0.02)} = \frac{1.17\ (\pm 3._1\%)}{1.89\ (\pm 1._1\%)} = 0.619_0\ (\pm 3._3\%)$$

because $\sqrt{(3._1)^2 + (1._1)^2} = 3._3$.

The relative uncertainty in the result is $3._3\%$. The absolute uncertainty is $0.03_3 \times 0.619_0 = 0.02_0$. The final answer can be written as

$$0.619\ (\pm 0.02_0) \qquad \text{(absolute uncertainty)}$$

or

$$0.619\ (3._3\%) \qquad \text{(relative uncertainty)}$$

The result of a calculation ought to be written in a manner consistent with the uncertainty in the result.

Because the uncertainty spans the last *two* places of the result, it would also be reasonable to write the result as

$$0.62\ (\pm 0.02)$$

or

$$0.62\ (\pm 3\%)$$

The treatment of mixed operations is more complicated if a quantity appears more than once in an expression, as in $(a - b)/(a - c)$. In such cases, we must use the method described in Appendix C.

The *Real* Rule on Significant Figures

The number of figures used to express a calculated result should be consistent with the uncertainty in that result. For example, the quotient

$$\frac{0.002\,364\ (\pm0.000\,003)}{0.025\,00\ (\pm0.000\,05)} = 0.094\,6\ (\pm0.000\,2)$$

is properly expressed with *three* significant figures, even though the original data have four figures. *The first uncertain figure of the answer is the last significant figure.* The quotient

The real rule: The first uncertain figure is the last significant figure.

$$\frac{0.002\,664\ (\pm0.000\,003)}{0.025\,00\ (\pm0.000\,05)} = 0.106\,6\ (\pm0.000\,2)$$

is expressed with *four* figures because the uncertainty occurs in the fourth place. The quotient

$$\frac{0.821\ (\pm0.002)}{0.803\ (\pm0.002)} = 1.022\ (\pm0.004)$$

is expressed with *four* figures even though the dividend and divisor each have *three* figures.

Exponents and Logarithms

For the function $y = x^a$, the percent relative uncertainty in y ($\%e_y$) is equal to a times the percent relative uncertainty in x ($\%e_x$).

To calculate a power or root on your calculator, use the y^x button. For example, to find a cube root ($y^{1/3}$), just raise y to the $0.333\,333\,333\ldots$ power with the y^x button.

Uncertainty for powers and roots:
$$y = x^a \Rightarrow \%e_y = a\ \%e_x \tag{3-4}$$

For example, if $y = \sqrt{x} = x^{1/2}$, a 2% uncertainty in x will yield a $\left(\frac{1}{2}\right)(2\%) = 1\%$ uncertainty in y. If $y = x^3$, a 2% uncertainty in x leads to a $(3)(2\%) = 6\%$ uncertainty in y.

If y is the base 10 logarithm of x, then the absolute uncertainty in y (e_y) is proportional to the relative uncertainty in x (e_x/x).

Uncertainty for logarithm:
$$y = \log x \Rightarrow e_y = \frac{1}{\ln 10}\frac{e_x}{x} \approx 0.434\,29\frac{e_x}{x} \tag{3-5}$$

You should not work with percent relative uncertainty ($100 \times e_x/x$) in calculations with logs and antilogs, because one side of Equation 3-5 has relative uncertainty and the other has absolute uncertainty.

Use relative uncertainty (e_x/x), not percent relative uncertainty [$100 \times e_x/x$], in calculations involving $\log x$, $\ln x$, 10^x, and e^x.

The **natural logarithm** (ln) of x is the number y, whose value is such that $x = e^y$, where e ($= 2.718\,28\ldots$) is called the base of the natural logarithm. The absolute uncertainty in y is equal to the relative uncertainty in x.

Uncertainty for natural logarithm:
$$y = \ln x \Rightarrow e_y = \frac{e_x}{x} \tag{3-6}$$

TABLE 3-1   Summary of rules for propagation of uncertainty

Function	Uncertainty	Function[a]	Uncertainty[b]
$y = x_1 + x_2$	$e_y = \sqrt{e_{x_1}^2 + e_{x_2}^2}$	$y = x^a$	$\%e_y = a\,\%e_x$
$y = x_1 - x_2$	$e_y = \sqrt{e_{x_1}^2 + e_{x_2}^2}$	$y = \log x$	$e_y = \dfrac{1}{\ln 10}\dfrac{e_x}{x} \approx 0.434\,29\,\dfrac{e_x}{x}$
$y = x_1 \cdot x_2$	$\%e_y = \sqrt{\%e_{x_1}^2 + \%e_{x_2}^2}$	$y = \ln x$	$e_y = \dfrac{e_x}{x}$
$y = \dfrac{x_1}{x_2}$	$\%e_y = \sqrt{\%e_{x_1}^2 + \%e_{x_2}^2}$	$y = 10^x$	$\dfrac{e_y}{y} = (\ln 10)\,e_x \approx 2.3026\,e_x$
		$y = e^x$	$\dfrac{e_y}{y} = e_x$

a. x represents a variable and a represents a constant that has no uncertainty.
b. e_x/x is the relative error in x and $\%e_x$ is $100 \times e_x/x$.

Now consider $y = $ antilog x, which is the same as saying $y = 10^x$. In this case the relative uncertainty in y is proportional to the absolute uncertainty in x.

Uncertainty for 10^x: $$y = 10^x \Rightarrow \frac{e_y}{y} = (\ln 10)e_x \approx 2.3026\,e_x \qquad (3\text{-}7)$$

If $y = e^x$, the relative uncertainty in y equals the absolute uncertainty in x.

Uncertainty for e^x: $$y = e^x \Rightarrow \frac{e_y}{y} = e_x \qquad (3\text{-}8)$$

Appendix C gives a general rule for propagation of uncertainty for any function.

Table 3-1 summarizes rules for propagation of uncertainty. You need not memorize the rules for exponents, logs, and antilogs, but you should be able to use them.

EXAMPLE   Uncertainty in H⁺ Concentration

Consider the function pH $= -\log [H^+]$, where $[H^+]$ is the concentration of H^+ in moles per liter. For pH $= 5.21 \pm 0.03$, find the concentration of $[H^+]$ and its uncertainty.

Solution   First, we need to solve the equation pH $= -\log [H^+]$ for $[H^+]$. If we have the equality $a = b$, then it is true that $10^a = 10^b$. If pH $= -\log [H^+]$, then $\log [H^+] = -$pH and $10^{\log[H^+]}\,(=[H^+]) = 10^{-pH}$. We therefore need to find the uncertainty in the equation

$$[H^+] = 10^{-pH} = 10^{-(5.21 \pm 0.03)}$$

In Table 3-1, the relevant function is $y = 10^x$, in which $y = [H^+]$ and $x = -(5.21 \pm 0.03)$. For $y = 10^x$, the table tells us that $e_y/y = 2.3026\,e_x$.

$$\frac{e_y}{y} = 2.3026\,e_x = (2.3026)(0.03) = 0.0691 \qquad (3\text{-}9)$$

The relative uncertainty in y $(= e_y/y)$ is 0.0691. Inserting the value $y = 10^{-5.21} = 6.17 \times 10^{-6}$ into Equation 3-9 gives the answer

$$\frac{e_y}{y} = \frac{e_y}{6.17 \times 10^{-6}} = 0.0691 \Rightarrow e_y = 4.26 \times 10^{-7}$$

The concentration of H^+ is 6.17 (± 0.426) $\times 10^{-6}$ = 6.2 (± 0.4) $\times 10^{-6}$ M. An uncertainty of 0.03 in pH gives an uncertainty of 6.9% in H^+. Notice that extra digits were retained in the intermediate results and were not rounded off until the final answer.

Terms to Understand

absolute uncertainty	indeterminate error	random error
accuracy	logarithm	relative uncertainty
antilogarithm	mantissa	significant figure
character	natural logarithm	systematic error
determinate error	precision	

Summary

The number of significant figures in a value is the minimum number required to write the value in scientific notation. The first uncertain digit is the last significant figure. In addition and subtraction, the last significant figure is determined by the number with the fewest decimal places (when all exponents are equal). In multiplication and division, the number of figures is usually limited by the factor with the smallest number of digits. The number of figures in the mantissa of the logarithm of a quantity should equal the number of significant figures in the quantity. Random (indeterminate) error affects the precision (reproducibility) of a result, whereas systematic

(determinate) error affects the accuracy (nearness to the "true" value). Systematic error can be discovered and eliminated by a clever person, but some random error is always present. Propagation of uncertainty in addition and subtraction requires absolute uncertainties ($e_3 = \sqrt{e_1^2 + e_2^2}$), whereas multiplication and division utilize relative uncertainties ($\%e_3 = \sqrt{\%e_1^2 + \%e_2^2}$). Other rules for propagation of error are found in Table 3-1, which you need not memorize but should be able to use. Always retain more digits than necessary during a calculation and round off to the appropriate number of digits at the end.

Exercises

A. Write each answer with a reasonable number of figures. Find the absolute uncertainty and percent relative uncertainty for each answer.

(a) [12.41 (± 0.09) $\div$ 4.16 (± 0.01)] $\times$ 7.0682 (± 0.0004) = ?

(b) 3.26 (± 0.10) $\times$ 8.47 (± 0.05) $-$ 0.18 (± 0.06) = ?

(c) 6.843 (± 0.008) $\times$ 10^4 $\div$ [2.09 (± 0.04) $-$ 1.63 (± 0.01)] = ?

(d) $\sqrt{3.24 \pm 0.08}$

(e) $(3.24 \pm 0.08)^4$

(f) $\log(3.24 \pm 0.08)$

(g) $10^{3.24 \pm 0.08}$

B. Suppose that you have a bottle of aqueous solution labeled "53.4 (± 0.4) wt% NaOH—density = 1.52 (± 0.01) g/mL."

(a) How many milliliters of 53.4% NaOH are needed to prepare 2.000 L of 0.169 M NaOH?

(b) If the uncertainty in delivering the NaOH is ± 0.10 mL, calculate the absolute uncertainty in the molarity (0.169 M). You may assume negligible uncertainty in the molecular weight of NaOH and in the final volume, 2.000 L.

C. Consider a solution containing 37.0 (± 0.5) wt% HCl in water. The density of the solution is 1.18 (± 0.01) g/mL. To deliver 0.05000 ($\pm 2\%$) mol of HCl requires 4.18 ($\pm x$) mL of solution. Find x.

Caution: In this problem you have been given the uncertainty in the *answer* to a calculation. You need to find the uncertainty in one factor along the way in the calculation. Be sure to propagate uncertainties in the right direction. For example, if $a = b \cdot c$, then $\%e_a^2 = \%e_b^2 + \%e_c^2$. If you are given $\%e_a$, the value of $\%e_c$ must be *smaller* than $\%e_a$, and $\%e_c^2 = \%e_a^2 - \%e_b^2$.

Problems

Significant Figures

1. Indicate how many significant figures there are in the following numbers.

(a) 1.903 0   **(b)** 0.039 10   **(c)** 1.40 × 10⁴

2. Round each number as indicated.

(a) 1.236 7 to 4 significant figures

(b) 1.238 4 to 4 significant figures

(c) 0.135 2 to 3 significant figures

(d) 2.051 to 2 significant figures

(e) 2.005 0 to 3 significant figures

3. Round each number to three significant figures.

(a) 0.216 74   **(b)** 0.216 5   **(c)** 0.216 500 3

4. *Vernier scale.* The figure at the bottom of this page shows a scale found on instruments such as a micrometer caliper used for accurately measuring dimensions of objects. The lower scale slides along the upper scale and is used to interpolate between the markings on the upper scale. In diagram (a), the reading (at the left-hand zero of the lower scale) is between 1.4 and 1.5 on the upper scale. To find the exact reading, observe which mark on the lower scale is aligned with a mark on the upper scale. Because the 6 on the lower scale is aligned with the upper scale, the correct reading is 1.46. Write the correct readings in diagrams (b) and (c) and indicate how many significant figures are in each reading.

5. Write each answer with the correct number of digits.

(a) 1.021 + 2.69 = 3.711

(b) 12.3 − 1.63 = 10.67

(c) 4.34 × 9.2 = 39.928

(d) 0.060 2 ÷ (2.113 × 10⁴) = 2.849 03 × 10⁻⁶

(e) $\log(4.218 \times 10^{12}) = ?$

(f) $\text{antilog}(-3.22) = ?$

(g) $10^{2.384} = ?$

6. Using the correct number of significant figures, find the formula weight of **(a)** $BaCl_2$ and **(b)** $C_{31}H_{32}O_8N_2$.

7. Write each answer with the correct number of significant figures.

(a) 1.0 + 2.1 + 3.4 + 5.8 = 12.300 0

(b) 106.9 − 31.4 = 75.500 0

(c) $107.868 - (2.113 \times 10^2) + (5.623 \times 10^3) = 5519.568$

(d) $(26.14/37.62) \times 4.38 = 3.043\,413$

(e) $(26.14/37.62 \times 10^8) \times (4.38 \times 10^{-2}) = 3.043\,413 \times 10^{-10}$

(f) $(26.14/3.38) + 4.2 = 11.933\,7$

(g) $\log(3.98 \times 10^4) = 4.599\,9$

(h) $10^{-6.31} = 4.897\,79 \times 10^{-7}$

Types of Error

8. Why do we use quotation marks around the word *true* in the statement that accuracy refers to how close a measured value is to the "true" value?

9. Explain the difference between systematic error and random error.

10. Is John Smith really pregnant?

11. Cheryl, Cynthia, Carmen, and Chastity shot these targets (see next page) at Girl Scout camp. Match each target with the proper description.

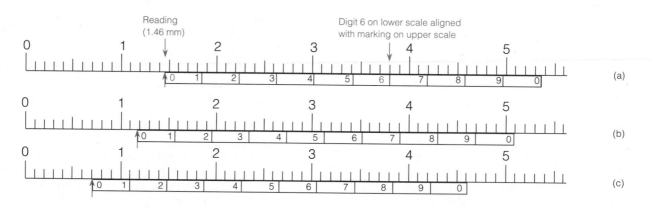

Reading
(1.46 mm)

Digit 6 on lower scale aligned
with marking on upper scale

(a)

(b)

(c)

Cheryl   Cynthia   Carmen   Chastity

(a) accurate and precise

(b) accurate but not precise

(c) precise but not accurate

(d) neither precise nor accurate

12. Rewrite the number $3.123\,56\ (\pm 0.167\,89\%)$ in the forms (a) number ($\pm$ absolute uncertainty) and (b) number ($\pm$ percent relative uncertainty) with an appropriate number of digits.

Propagation of Uncertainty

13. Find the absolute and percent relative uncertainty and express each answer with a reasonable number of significant figures.

(a) $6.2\ (\pm 0.2) - 4.1\ (\pm 0.1) = ?$

(b) $9.43\ (\pm 0.05) \times 0.016\ (\pm 0.001) = ?$

(c) $[6.2\ (\pm 0.2) - 4.1\ (\pm 0.1)] \div 9.43\ (\pm 0.05) = ?$

(d) $9.43\ (\pm 0.05) \times \{[6.2\ (\pm 0.2) \times 10^{-3}] + [4.1\ (\pm 0.1) \times 10^{-3}]\} = ?$

14. Find the absolute and percent relative uncertainty and express each answer with a reasonable number of significant figures.

(a) $9.23\ (\pm 0.03) + 4.21\ (\pm 0.02) - 3.26\ (\pm 0.06) = ?$

(b) $91.3\ (\pm 1.0) \times 40.3\ (\pm 0.2)/21.1\ (\pm 0.2) = ?$

(c) $[4.97\ (\pm 0.05) - 1.86\ (\pm 0.01)]/21.1\ (\pm 0.2) = ?$

(d) $2.016\,4\ (\pm 0.000\,8) + 1.233\ (\pm 0.002) + 4.61\ (\pm 0.01) = ?$

(e) $2.016\,4\ (\pm 0.000\,8) \times 10^3 + 1.233\ (\pm 0.002) \times 10^2 + 4.61\ (\pm 0.01) \times 10^1 = ?$

(f) $[3.14\ (\pm 0.05)]^{1/3} = ?$

(g) $\log[3.14\ (\pm 0.05)] = ?$

15. Verify the following calculations.

(a) $\sqrt{3.141\,5\ (\pm 0.001\,1)} = 1.772\,4_3(\pm 0.000\,3_1)$

(b) $\log[3.141\,5\ (\pm 0.001\,1)] = 0.497\,1_4(\pm 0.000\,1_5)$

(c) $\text{antilog}[3.141\,5\ (\pm 0.001\,1)] = 1.385_2\ (\pm 0.003_5) \times 10^3$

(d) $\ln[3.141\,5\ (\pm 0.001\,1)] = 1.144\,7_0(\pm 0.000\,3_5)$

(e) $\log\left(\dfrac{\sqrt{0.104\ (\pm 0.006)}}{0.051\,1\ (\pm 0.000\,9)}\right) = 0.80_0(\pm 0.01_5)$

16. (a) Show that the formula weight of NaCl is $58.442\,5\ (\pm 0.000\,9)$ g/mol.

(b) To prepare a solution of NaCl, you weight out $2.634\ (\pm 0.002)$ g and dissolve it in a volumetric flask whose volume is $100.00\ (\pm 0.08)$ mL. Express the molarity of the resulting solution, along with its uncertainty, with an appropriate number of digits.

17. The constant $\hbar$ (read "h bar") is defined as $h/2\pi$, where h is Planck's constant $[6.626\,075\,5\ (\pm 0.000\,004\,0) \times 10^{-34}$ J·s$]$. Calculate the value and absolute uncertainty of $\hbar$. The number 2 is an integer (infinitely accurate) and π is also an exact number. The first 10 digits of π are $3.141\,592\,653$.

18. Consider a buoyancy correction made with Equation 2-1. What is the true mass in vacuum of water weighed at 24°C in the air if the apparent mass is $1.034\,6 \pm 0.000\,2$ g? Assume that the density of air is $0.001\,2 \pm 0.000\,1$ g/mL and the density of the balance weights is 8.0 ± 0.5 g/mL. The uncertainty in the density of water in Table 2-6 is negligible in comparison to the uncertainty in the density of air.

Is My Red Blood Cell Count High Today?

Red blood cells (erythrocytes, Er) tangled in fibrin threads (Fi) in a blood clot. Stacks of erythrocytes in a clot are called a rouleaux formation (Ro). [From R. H. Kardon *Tissues and Organs* (San Francisco: W. H. Freeman, 1978), p. 39.]

All measurements contain experimental error, so it is never possible to be completely certain of a result. Nevertheless, we often seek the answers to questions such as "Is my red blood cell count today higher than usual?" If today's count is twice as high as usual, it is probably truly higher than normal. But what if the "high" count is not excessively above "normal" counts?

Count on "normal" days	Today's count
$\left. \begin{array}{l} 5.1 \\ 5.3 \\ 4.8 \\ 5.4 \\ 5.2 \end{array} \right\} \times 10^6$ cells/μL	5.6×10^6 cells/μL

The number 5.6 is higher than the five normal values, but the random variation in normal values might lead us to expect that 5.6 will be observed on some "normal" days.

We will never be able to answer the question with complete certainty, but the study of statistics allows us to say that today's value is expected to be observed on 1 out of 20 normal days. It is still up to you to decide what to do with this information.

Statistics

As illustrated on the facing page, experimental measurements carry with them some variability, so no conclusion can be drawn with certainty. Statistics gives us tools to accept conclusions that have a high probability of being correct and to reject conclusions that do not.[1] We begin with a discussion of the Gaussian distribution of data points expected in a properly designed experiment in which the probability of obtaining a "high" result is equal to the probability of obtaining a "low" result.

4-1 Gaussian Distribution

We say that the variation in experimental data is *normally distributed* when replicate measurements exhibit the bell-shaped distribution illustrated in Figure 4-1. In this hypothetical case, a manufacturer tested the lifetimes of 4 768 electric light bulbs. The bar graph shows the number of bulbs with a lifetime in each 20-hour interval. The smooth curve is the **Gaussian distribution** that best fits the data. Any finite set of data will vary somewhat from the Gaussian curve. As the number of data points increases, the line connecting them should more closely approach the Gaussian curve.

Mean Value and Standard Deviation

The data for light bulb lifetimes, and the corresponding Gaussian curve, are characterized by two parameters. The arithmetic **mean, $\bar{x}$,** also called the

The mean gives the center of the distribution. The standard deviation measures the width of the distribution.

average, is the sum of the measured values divided by *n*, the number of measurements:

Mean:
$$\bar{x} = \frac{\sum_i x_i}{n}$$
(4-1)

where each x_i is the lifetime of an individual bulb. The Greek capital sigma, Σ, means summation: $\sum_i x_i = x_1 + x_2 + x_3 + \quad + x_n$. In Figure 4-1 the mean value is indicated by the arrow at 845.2 h.

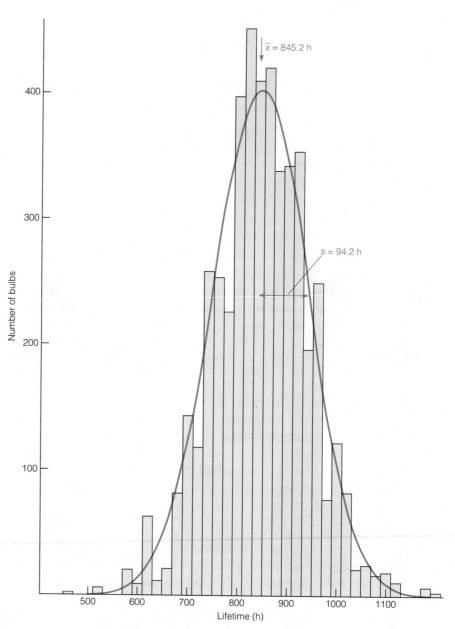

Figure 4-1 Bar graph and Gaussian curve describing the lifetime of a hypothetical set of electric light bulbs. The smooth curve has the same mean, standard deviation, and area as the bar graph. Any finite set of data, however, will differ from the bell-shaped curve.

The **standard deviation,** s, measures how closely the data are clustered about the mean. *The smaller the standard deviation, the more closely the data are clustered about the mean* (Figure 4-2).

An experiment that produces a small standard deviation is more *precise* than one that produces a large standard deviation. Greater precision does not necessarily imply greater *accuracy,* which means nearness to the "truth."

Standard deviation:
$$s = \sqrt{\frac{\sum_i (x_i - \bar{x})^2}{n - 1}}$$

(4-2)

For the data in Figure 4-1, $s = 94.2$ h. A set of light bulbs having a small standard deviation in lifetime is more uniformly manufactured than a set with a large standard deviation.

For an *infinite* set of data, the mean is designated by the lowercase Greek letter mu, μ (the population mean), and the standard deviation is written as a lowercase Greek sigma, σ (the population standard deviation). We can never measure μ and σ, but the values of $\bar{x}$ and s approach μ and σ as the number of measurements increases.

The **degrees of freedom** of the system are given by the quantity $n - 1$ in Equation 4-2. The square of the standard deviation is called the **variance.**

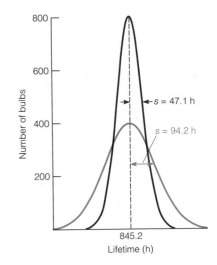

Figure 4-2 Gaussian curves for two sets of light bulbs, one having a standard deviation half as great as the other. The number of bulbs described by each curve is the same.

· ·

EXAMPLE   **Mean and Standard Deviation**

Suppose that four measurements are made: 821, 783, 834, and 855. Find the average and the standard deviation.

Solution   The average is

$$\bar{x} = \frac{821 + 783 + 834 + 855}{4} = 823._2$$

To avoid accumulating round-off errors, retain one more digit for the average and the standard deviation than was present in the original data. The standard deviation is

$$s = \sqrt{\frac{(821 - 823.2)^2 + (783 - 823.2)^2 + (834 - 823.2)^2 + (855 - 823.2)^2}{(4 - 1)}} = 30._3$$

The average and the standard deviation should both end at the *same decimal place*. For $\bar{x} = 823._2$, we will write $s = 30._3$.

· ·

Your calculator probably has a standard deviation function on it. Learn to use it and see that you get $s = 30.2696 \ldots$ in the example above. If you get $26.2 \ldots$, your calculator is using n (instead of $n - 1$) in the denominator of Equation 4-2. You can correct for this by multiplying your answer by $\sqrt{n/(n - 1)}$.

Standard Deviation and Probability

The formula for a normalized Gaussian curve is

Gaussian curve:
$$y = \frac{1}{\sigma\sqrt{2\pi}} e^{-(x - \mu)^2/2\sigma^2}$$

(4-3)

where e (= 2.718 28 . . .) is the base of the natural logarithm. (We will define the word "normalized" shortly.) To describe the curve for a finite set of data, we approximate μ by $\bar{x}$ and σ by s. A graph of Equation 4-3 is shown in Figure 4-3, in which the values $\sigma = 1$ and $\mu = 0$ are used for simplicity. The most probable value of x is μ, and the curve is symmetric about $x = \mu$. The probability of observing a particular value of x is proportional to the ordinate (y value). Thus in Figure 4-3 the maximum probability for any measurement occurs at $x = \mu = 0$. The probability of measuring the value $x = 1$ is $0.242/0.399 = 0.607$ times the probability of measuring the value $x = 0$.

In dealing with a Gaussian curve, it is especially useful to express deviations from the mean value in multiples of the standard deviation. That is, we transform x into z, given by

When $z = +1$, x is one standard deviation above the mean. When $z = -2$, x is two standard deviations below the mean.

$$z = \frac{x - \mu}{\sigma} \approx \frac{x - \bar{x}}{s} \qquad (4\text{-}4)$$

Table 4-1 gives ordinate and area values for the Gaussian curve in Figure 4-3.

The probability of measuring z in a certain *range* is proportional to the *area* of that range. For example, the probability of observing z between -2 and -1 is 0.136. This probability corresponds to the shaded area in Figure 4-3. The area under each portion of the Gaussian curve is given in Table 4-1. Because the sum of the probabilities of all the measurements must be unity, the area under the whole curve from $z = -\infty$ to $z = +\infty$ must be unity. The number $1/(\sigma\sqrt{2\pi})$ in Equation 4-3 is called the *normalization factor*. It guarantees that the area under the entire curve is unity. A Gaussian curve whose area is unity is called a *normal error curve*.

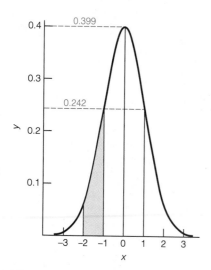

Figure 4-3 A Gaussian curve in which $\mu = 0$ and $\sigma = 1$. A Gaussian curve whose area is unity is called a normal error curve. In this case, the ordinate, x, is equal to z, defined as $z = (x - \mu)/\sigma$.

TABLE 4-1   Ordinate and area for the normal (Gaussian) error curve,

$$y = \frac{1}{\sqrt{2\pi}} e^{-z^2/2}$$

$|z|$[a]	y	Area[b]	$|z|$	y	Area	$|z|$	y	Area
0.0	0.3989	0.0000	1.4	0.1497	0.4192	2.8	0.0079	0.4974
0.1	0.3970	0.0398	1.5	0.1295	0.4332	2.9	0.0060	0.4981
0.2	0.3910	0.0793	1.6	0.1109	0.4452	3.0	0.0044	0.498650
0.3	0.3814	0.1179	1.7	0.0941	0.4554	3.1	0.0033	0.499032
0.4	0.3683	0.1554	1.8	0.0790	0.4641	3.2	0.0024	0.499313
0.5	0.3521	0.1915	1.9	0.0656	0.4713	3.3	0.0017	0.499517
0.6	0.3332	0.2258	2.0	0.0540	0.4773	3.4	0.0012	0.499663
0.7	0.3123	0.2580	2.1	0.0440	0.4821	3.5	0.0009	0.499767
0.8	0.2897	0.2881	2.2	0.0355	0.4861	3.6	0.0006	0.499841
0.9	0.2661	0.3159	2.3	0.0283	0.4893	3.7	0.0004	0.499904
1.0	0.2420	0.3413	2.4	0.0224	0.4918	3.8	0.0003	0.499928
1.1	0.2179	0.3643	2.5	0.0175	0.4938	3.9	0.0002	0.499952
1.2	0.1942	0.3849	2.6	0.0136	0.4953	4.0	0.0001	0.499968
1.3	0.1714	0.4032	2.7	0.0104	0.4965			

a. $z = (x - \mu)/\sigma$.
b. The area refers to the area between $z = 0$ and $z =$ the value in the table. Thus the area from $z = 0$ to $z = 1.4$ is 0.4192. The area from $z = -0.7$ to $z = 0$ is the same as from $z = 0$ to $z = 0.7$. The area from $z = -0.5$ to $z = +0.3$ is $(0.1915 + 0.1179) = 0.3094$. The total area between $z = -\infty$ and $z = +\infty$ is unity. A more complete table can be found in any edition of the *CRC Standard Math Tables* (Boca Raton, FL: CRC Press).

EXAMPLE   Gaussian Curve

Suppose the manufacturer offers to replace free of charge any bulb that burns out in less than 600 h. What fraction of bulbs should she keep available as replacements?

Solution   To answer this question, we express the desired interval in multiples of the standard deviation. Then we find the area of the interval by using Table 4-1. Because $\bar{x} = 845.2$ and $s = 94.2$, $z = (600 - 845.2)/94.2 = -2.60$. The area under the curve between the mean value and $z = -2.60$ is given as $0.495\,3$ in Table 4-1. Because the entire area from $-\infty$ to the mean value is $0.500\,0$, the area from $-\infty$ to -2.60 must be $0.004\,7$. In other words, the area to the left of 600 hours in Figure 4-1 is only 0.47% of the entire area under the normal error curve. Only 0.47% of the bulbs are expected to fail in less than 600 hours. If the manufacturer sells 1 million bulbs a year, she should make $1\,004\,700$ bulbs so that she will have enough bulbs to meet the replacement demand.

EXAMPLE   Interpolating Area Beneath the Gaussian Curve

What fraction of bulbs is expected to have a lifetime between 900 and 1 000 h?

Solution   To answer this question, *we need to find the fraction of the area of the Gaussian curve between x = 900 and x = 1 000 h.* When $x = 900$, $z = (900 - 845.2)/94.2 = 0.582$. When $x = 1\,000$, $z = (1\,000 - 845.2)/94.2 = 1.643$. Figure 4-4 shows that we can find the area between 900 and 1 000 h by first finding the area from $\bar{x}$ to 1 000 h and then subtracting the area from $\bar{x}$ to 900 h.

Table 4-1 can be used to find the area from $\bar{x}$ to 900 hours ($z = 0.582$) as follows. The area between the mean and $z = 0.5$ is $0.191\,5$. The area between the mean and $z = 0.6$ is $0.225\,8$. The area between $z = 0.5$ and $z = 0.6$ is therefore $0.225\,8 - 0.191\,5 = 0.034\,3$. Because $z = 0.582$ lies 82% of the way from $z = 0.5$ to 0.6, we can estimate that the interval $z = 0.5$ to 0.582 contains 82% of the area of the interval $z = 0.5$ to 0.6. This procedure is called *linear interpolation.*

$$\begin{pmatrix} \text{area between} \\ z = 0.500 \text{ and} \\ z = 0.582 \end{pmatrix} = \underbrace{\left(\frac{0.582 - 0.500}{0.600 - 0.500}\right)}_{\substack{\text{Fraction of} \\ \text{interval between} \\ z = 0.5 \text{ and } z = 0.6}} \underbrace{(0.225\,8 - 0.191\,5)}_{\substack{\text{Area between} \\ z = 0.5 \text{ and } z = 0.6}} = 0.028\,1$$

$$\begin{aligned} \frac{\text{area between } 0}{\text{and } 0.582} &= \begin{pmatrix} \text{area between } 0 \\ \text{and } 0.5 \end{pmatrix} + \begin{pmatrix} \text{area between } 0.5 \\ \text{and } 0.582 \end{pmatrix} \\ &= 0.191\,5 + 0.028\,1 = 0.219\,6 \end{aligned}$$

To find the area between $\bar{x}$ and 1 000 h ($z = 1.643$), we need to interpolate between $z = 1.6$ and $z = 1.7$:

$$\text{area between 0 and } 1.643 = 0.449\,6$$

The area between 900 and 1 000 h is $0.449\,6 - 0.219\,6 = 0.230\,0$. That is, 23.00% of the bulbs are expected to have a lifetime between 900 and 1 000 h.

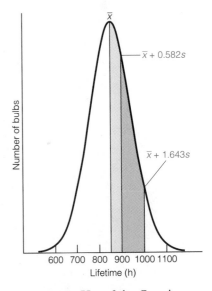

Figure 4-4   Use of the Gaussian curve to find the fraction of bulbs with a lifetime between 900 and 1 000 hours. We find the area between $\bar{x}$ and 1 000 h and subtract the lightly shaded area between $\bar{x}$ and 900 h.

You can find the area more accurately if you integrate Equation 4-3 with a computer or scientific calculator. The integrated area between 900 and 1 000 h is 0.230 2, which is not very different from our interpolated value.

. . . .

Range	Percentage of measurements
$\mu \pm 1\sigma$	68.3
$\mu \pm 2\sigma$	95.5
$\mu \pm 3\sigma$	99.7

The significance of the standard deviation is that it measures the width of the Gaussian error curve. The larger the value of σ, the broader the curve. In a Gaussian curve such as Figure 4-3, 68.3% of the area is in the range from $\mu - 1\sigma$ to $\mu + 1\sigma$. That is, more than two-thirds of the measurements are expected to lie within one standard deviation of the mean. Also, 95.5% of the area lies within $\mu \pm 2\sigma$, and 99.7% of the area lies within $\mu \pm 3\sigma$ (Table 4-1).

Suppose that you use two different experimental methods to measure the percentage of sulfur in coal: Method A has a standard deviation of 0.4%, and method B has a standard deviation of 1.1%. You can expect that approximately two-thirds of a large number of measurements from method A will lie within 0.4% of the mean. For method B, two-thirds will lie within 1.1% of the mean.

Standard Deviation of the Mean

The more times you measure a quantity, the more confident you can be that the average value of your measurements is close to the true population mean, μ. In fact, the uncertainty decreases in proportion to $1/\sqrt{n}$, where n is the number of measurements. If s is the standard deviation of n measurements, the quantity $s/\sqrt{n}$ is called the *standard deviation of the mean,* or the standard error of the mean.

$$\text{standard deviation of the mean} = \frac{s}{\sqrt{n}} \tag{4-5}$$

This expression tells us that we can decrease the uncertainty of the mean by a factor of 2 $(= \sqrt{4})$ if we make four times as many measurements and by a factor of 3.16 $(= \sqrt{10})$ if we make 10 times as many measurements.

4-2 Student's *t*

Student's *t* is a statistical tool used most frequently to express confidence intervals and for comparing results from different experiments. It is the tool you could use to evaluate the probability that your red blood cell count will be found in a certain range on "normal" days.

"Student" was the pseudonym of W. S. Gossett, whose employer, the Guinness Breweries of Ireland, restricted publication for proprietary reasons. Because of the importance of Gossett's work, he was allowed to publish it (*Biometrika* **1908,** *6,* 1), but under an assumed name.

Confidence Intervals

From a limited number of measurements, it is impossible to find the true population mean, μ, or the true standard deviation, σ. What we can determine are $\bar{x}$ and s, the sample mean and the sample standard deviation. The **confidence interval** is an expression stating that the true mean, μ, is

TABLE 4-2   Values of Student's *t*

Degrees of freedom	Confidence level (%)						
	50	90	95	98	99	99.5	99.9
1	1.000	6.314	12.706	31.821	63.657	127.32	636.619
2	0.816	2.920	4.303	6.965	9.925	14.089	31.598
3	0.765	2.353	3.182	4.541	5.841	7.453	12.924
4	0.741	2.132	2.776	3.747	4.604	5.598	8.610
5	0.727	2.015	2.571	3.365	4.032	4.773	6.869
6	0.718	1.943	2.447	3.143	3.707	4.317	5.959
7	0.711	1.895	2.365	2.998	3.500	3.832	5.041
8	0.706	1.860	2.306	2.896	3.355	3.690	4.781
9	0.703	1.833	2.262	2.821	3.250	3.581	4.587
10	0.700	1.812	2.228	2.764	3.169	3.581	4.587
15	0.691	1.753	2.131	2.602	2.947	3.252	4.073
20	0.687	1.725	2.086	2.528	2.845	3.153	3.850
25	0.684	1.708	2.068	2.485	2.787	3.078	3.725
30	0.683	1.697	2.042	2.457	2.750	3.030	3.646
40	0.681	1.684	2.021	2.423	2.704	2.971	3.551
60	0.679	1.671	2.000	2.390	2.660	2.915	3.460
120	0.677	1.658	1.980	2.358	2.617	2.860	3.373
∞	0.674	1.645	1.960	2.326	2.576	2.807	3.291

Note: In calculating confidence intervals, σ may be substituted for s in Equation 4-6 if you have a great deal of experience with a particular method and have therefore determined its "true" population standard deviation. If σ is used instead of s, the value of t to use in Equation 4-6 comes from the bottom row of Table 4-2.

likely to lie within a certain distance from the measured mean, $\bar{x}$. The confidence interval of μ is given by

Confidence interval:
$$\mu = \bar{x} \pm \frac{ts}{\sqrt{n}}$$
(4-6)

where s is the measured standard deviation, n is the number of observations, and t is Student's t, taken from Table 4-2.

EXAMPLE   Calculating Confidence Intervals

Suppose that the carbohydrate content of a glycoprotein (a protein with sugars attached to it) is determined to be 12.6, 11.9, 13.0, 12.7, and 12.5 g of carbohydrate per 100 g of protein in replicate analyses. Find the 50% and 90% confidence intervals for the carbohydrate content.

Solution   First calculate $\bar{x}$ (= 12.5_4) and s (= 0.4_0) for the five measurements. To calculate the 50% confidence interval, look up t in Table 4-2 under 50% and across from 4 degrees of freedom. (Recall that degrees of freedom = $n - 1$.) The value of t is 0.741, so the 50% confidence interval is

$$\mu = \bar{x} \pm \frac{ts}{\sqrt{n}} = 12.5_4 \pm \frac{(0.741)(0.4_0)}{\sqrt{5}} = 12.5_4 \pm 0.1_3$$

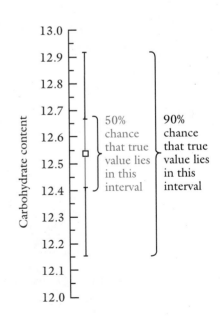

The 90% confidence interval is

$$\mu = \bar{x} \pm \frac{ts}{\sqrt{n}} = 12.5_4 \pm \frac{(2.132)(0.4_0)}{\sqrt{5}} = 12.5_4 \pm 0.3_8$$

These calculations mean that there is a 50% chance that the true mean, μ, lies within the range $12.5_4 \pm 0.1_3$ (12.4_1 to 12.6_7). There is a 90% chance that μ lies within the range $12.5_4 \pm 0.3_8$ (12.1_6 to 12.9_2).

Confidence Intervals as Estimates of Experimental Uncertainty

In Chapter 3 we learned about rules for propagation of uncertainty in calculations. For example, suppose you were dividing a mass by a volume to find density; you learned that the uncertainty in density is based on the uncertainties in mass and volume. The most common estimates of uncertainty in experimentally measured quantities are the standard deviation and the confidence interval.

Suppose you measure the volume of a vessel five times and observe values of 6.375, 6.372, 6.374, 6.377, and 6.375 mL. The average is $\bar{x} = 6.374_6$ mL and the standard deviation is $s = 0.001_8$ mL. You could choose a confidence interval (such as 90%) for the estimate of uncertainty. Using Equation 4-6 with four degrees of freedom, you find that the 90% confidence interval is $\pm ts/\sqrt{n} = \pm(2.132)(0.001_8)/\sqrt{5} = \pm 0.001_7$. By this criterion, the volume to use for the density calculation is $6.374_6 \pm 0.001_7$ mL.

We can reduce uncertainty by making more measurements. If we make 21 measurements and have the same mean and standard deviation, the 90% confidence interval is reduced to $\pm ts/\sqrt{n} = \pm(1.725)(0.001_8)/\sqrt{21} = \pm 0.000\,7$. The volume for the calculation would be $6.374\,6 \pm 0.000\,7$ mL. By making enough measurements, we can reduce the uncertainty in the mean volume to the fourth decimal place, even though the standard deviation is in the third decimal place.

Frequently, we report the standard deviation of n measurements as the estimated uncertainty. For five measurements, we would report a volume of $6.374_6 \pm 0.001_8$ mL. When reporting results in this manner, always indicate the total number of measurements, so that proper confidence intervals can be calculated. The significance of $6.374_6 \pm 0.001_8$ derived from 25 measurements is different from the significance of $6.374_6 \pm 0.001_8$ derived from five measurements.

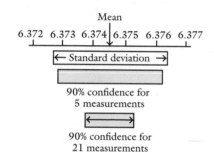

Mean

6.372 6.373 6.374 6.375 6.376 6.377

← Standard deviation →

90% confidence for
5 measurements

90% confidence for
21 measurements

The Meaning of a Confidence Interval

The experiment in Figure 4-5 illustrates the meaning of confidence intervals. A computer chose numbers at random from a Gaussian population with a population mean (μ) of 10 000 and a population standard deviation (σ) of 1 000 in Equation 4-3. In trial 1, four numbers were chosen, and their mean and standard deviation were calculated with Equations 4-1 and 4-2. The 50% confidence interval was then calculated with Equation 4-6, using $t = 0.765$ from Table 4-2 (50% confidence, 3 degrees of freedom). This trial is plotted as the first point at the left in Figure 4-5a; the square is centered at the mean value of 9 526, and the error bar extends from the lower limit to the upper limit of the 50% confidence interval (± 290). The experiment was repeated 100 times (the computer drew a total of 400 numbers) to produce all the

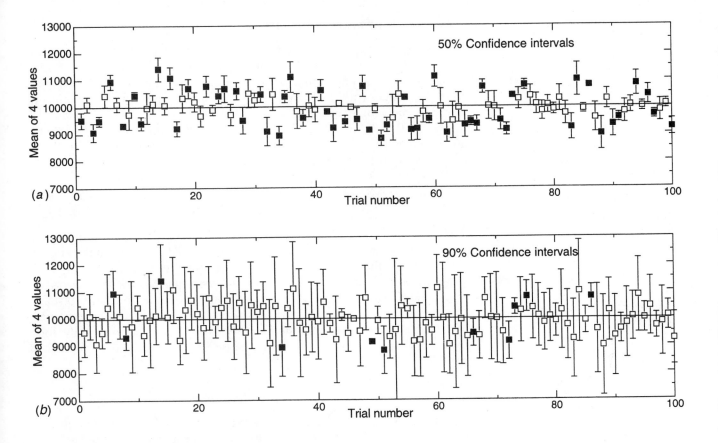

Figure 4-5 Two different confidence intervals for the same set of random data: (a) 50%; (b) 90%.

points in Figure 4-5a. Each set of four numbers has a different mean and confidence interval.

The 50% confidence interval is defined such that if we repeated this experiment an infinite number of times, 50% of the error bars in Figure 4-5a would include the true population mean of 10 000. In fact, I did the experiment 100 times, and 45 of the error bars in Figure 4-5a pass through the horizontal line at 10 000.

Figure 4-5b shows the same experiment with the same set of random numbers, but this time the 90% confidence interval was calculated. For an infinite number of experiments, we would expect 90% of the confidence intervals to include the population mean of 10 000. In fact, 89 of the 100 error bars in Figure 4-5b cross the horizontal line at 10 000.

Comparison of Means with Student's *t*

We use Student's *t* in three ways to compare one set of measurements with another to decide whether they are "the same" or "different" from each other. We arbitrarily adopt the 95% confidence level as a conservative standard: We will say that *two results do not differ from each other unless there is >95% chance that our conclusion is correct.*

Case 1: Comparing a Measured Result to a "Known" Value

In this case we claim to know that a sample contains 0.031 9% Ni, either on the basis of a very reliable measurement or on the basis of the way the sample was prepared from known materials. We are testing a new analytical method to see if it agrees with the known value. We use the 95% confidence interval to make the judgment. The measured values with the new method are 0.032 9, 0.032 2, 0.033 0, and 0.032 3% Ni, giving a mean of $\bar{x} = 0.032\,6_0$ and standard deviation of $s = 0.000\,4_1$. The 95% confidence interval is calculated with Equation 4-6.

$$95\% \text{ confidence interval:} \quad \bar{x} \pm \frac{(3.182)(0.000\,4_1)}{\sqrt{4}} = 0.032\,6_0 \pm 0.000\,6_5$$
$$= 0.031\,9_5 \text{ to } 0.033\,2_5$$

The known value of 0.031 9 lies outside the 95% confidence limit; therefore, we conclude that the new method produces a value different from the known value, because the chance that they are "the same" is <5%.

Statistical tests do not relieve us of the ultimate subjective decision to accept or reject a conclusion. The tests only provide guidance in the form of probabilities. In the foregoing example, we tentatively concluded that values obtained with the new method are systematically high. However, with only four values, it would be wise to run the analysis several more times to attain higher confidence in our conclusion.

Statistical tests only give us probabilities. They do not relieve us of the responsibility of interpreting our results.

Case 2: Comparing Replicate Measurements

We can use statistics to decide whether two sets of replicate measurements give "the same" or "different" results, within a stated confidence level. An example comes from the work of Lord Rayleigh (John W. Strutt), who is remembered today for landmark investigations of light scattering, blackbody radiation, and elastic waves in solids. His Nobel Prize in 1904 was received for the discovery of the inert gas argon, a discovery that came about when he noticed a small discrepancy between two sets of measurements of the density of nitrogen gas.

In Rayleigh's time, it was known that dry air was composed of about $\frac{1}{5}$ oxygen and $\frac{4}{5}$ nitrogen. Rayleigh removed the oxygen from air by reaction with red hot copper (to make solid CuO) and measured the density of the remaining gas by collecting it in a fixed volume at constant temperature and pressure. He also prepared the same volume of nitrogen by chemical decomposition of nitrous oxide (N_2O), nitric oxide (NO), or ammonium nitrite ($NH_4^+NO_2^-$). Table 4-3 and Figure 4-6 show the mass of gas collected in each experiment. The average mass collected from air (2.310 11 g) is 0.46% greater than the average mass of the same volume of gas from chemical sources (2.299 47 g).

If Rayleigh's measurements had not been performed with care, this difference might have been attributed to experimental error. Instead, Rayleigh understood that the discrepancy was outside his margin of error, and he postulated that nitrogen from the air was mixed with a heavier gas, which turned out to be argon.

Let's see how to use the *t* test to decide whether nitrogen isolated from air is "significantly" heavier than nitrogen isolated from chemical sources. In this case we have two sets of measurements, each with its own uncertainty and no "known," "true" value. We assume that the population standard deviation (σ) for each method is essentially the same.

TABLE 4-3   Masses of nitrogen-rich gas isolated by Lord Rayleigh

From air (g)	From chemical decomposition (g)
2.310 17	2.301 43
2.309 86	2.298 90
2.310 10	2.298 16
2.310 01	2.301 82
2.310 24	2.298 69
2.310 10	2.299 40
2.310 28	2.298 49
—	2.298 89
Average	
2.310 11	2.299 47
Standard deviation	
0.000 14$_3$	0.001 38

SOURCE: R. D. Larsen, *J. Chem. Ed.* **1990**, *67*, 925.

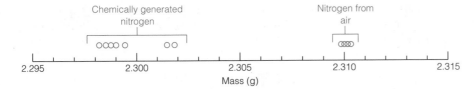

Figure 4-6 Lord Rayleigh's measurements of the mass of constant volumes of nitrogen-rich gas (at constant temperature and pressure) isolated from air or generated by decomposition of nitrogen compounds. Rayleigh recognized that the difference between the two clusters was outside of his experimental error and deduced that a heavier component, which turned out to be argon, was present in nitrogen isolated from air.

For two sets of data consisting of n_1 and n_2 measurements (with averages $\bar{x}_1$ and $\bar{x}_2$), we calculate a value of t by using the formula

$$t = \frac{\bar{x}_1 - \bar{x}_2}{s_{pooled}} \sqrt{\frac{n_1 n_2}{n_1 + n_2}} \qquad (4\text{-}7)$$

where

$$s_{pooled} = \sqrt{\frac{\displaystyle\sum_{set\ 1} (x_i - \bar{x}_1)^2 + \sum_{set\ 2} (x_j - \bar{x}_2)^2}{n_1 + n_2 - 2}} = \sqrt{\frac{s_1^2 (n_1 - 1) + s_2^2 (n_2 - 1)}{n_1 + n_2 - 2}} \qquad (4\text{-}8)$$

Here s_{pooled} is a *pooled* standard deviation making use of both sets of data. The value of t from Equation 4-7 is to be compared with the value of t in Table 4-2 for $n_1 + n_2 - 2$ degrees of freedom. *If the calculated t is greater than the tabulated t at the 95% confidence level, the two results are considered to be different.*

If $t_{calculated} > t_{table}$ (95%), the difference is significant.

. . . .

EXAMPLE **Is Lord Rayleigh's N_2 from Air Denser than N_2 from Chemicals?**

The average mass of nitrogen from air in Table 4-3 is $\bar{x}_1 = 2.310\,11$ g, with a standard deviation of $s_1 = 0.000\,14_3$ (for $n_1 = 7$ measurements). The average mass from chemical sources is $\bar{x}_2 = 2.299\,47$ g, with a standard deviation of $s_2 = 0.001\,38$ (for $n_2 = 8$ measurements).

Solution To answer the question, we calculate s_{pooled} with Equation 4-8,

$$s_{pooled} = \sqrt{\frac{0.000\,14_3{}^2 (7 - 1) + 0.001\,38^2 (8 - 1)}{7 + 8 - 2}} = 0.001\,02$$

and t with Equation 4-7:

$$t = \frac{2.310\,11 - 2.299\,47}{0.001\,02} \sqrt{\frac{7 \cdot 8}{7 + 8}} = 20.2$$

For $7 + 8 - 2 = 13$ degrees of freedom in Table 4-2, t lies between 2.228 and 2.131 at the 95% confidence level. The observed value of $t = 20.2$ is greater than the tabulated t, so the difference is significant. In fact, the tabulated value of t for 99.9% confidence is about 4.3. The difference is significant beyond the 99.9% confidence level. Our eyes were not lying to us in Figure 4-6: N_2 from the air is undoubtedly denser than N_2 from chemical sources. This observation led Rayleigh to the discovery of argon as a heavy constituent of air.

. . . .

Equations 4-7 and 4-8 are based on the assumption that the population standard deviation is the same for both sets of measurements. If this is not true, then we should use the equations

$$t = \frac{\bar{x}_1 - \bar{x}_2}{\sqrt{s_1^2/n_1 + s_2^2/n_2}} \tag{4-7'}$$

and

$$\text{degrees of freedom} = \left\{ \frac{(s_1^2/n_1 + s_2^2/n_2)^2}{\left(\dfrac{(s_1^2/n_1)^2}{n_1 + 1} + \dfrac{(s_2^2/n_2)^2}{n_2 + 1}\right)} \right\} - 2 \tag{4-8'}$$

rounded to the nearest integer. For Rayleigh's data in Figure 4-6, we suspect that the population standard deviation from air is smaller than that from chemical sources. Using Equations 4-7' and 4-8', we find $t = 21.7$ and degrees of freedom $= 7.22 \approx 7$. This value of t still far exceeds the values in Table 4-2 for 7 degrees of freedom at 95 or 99.9% confidence.

Case 3: Comparing Individual Differences

This case applies when we use two different methods to make single measurements on several different samples. No measurement has been duplicated. Consider the cholesterol content of six sets of human blood plasma measured by two different techniques in Table 4-4. Each sample has a different cholesterol content. Method B gives a lower result than method A in five out of the six samples. Is method B systematically different from method A?

To answer this question, we perform a t test on the individual *differences* between results for each sample:

$$t = \frac{\bar{d}}{s_\text{d}} \sqrt{n} \tag{4-9}$$

where

$$s_\text{d} = \sqrt{\frac{\sum(d_i - \bar{d})^2}{n - 1}} \tag{4-10}$$

TABLE 4-4   Comparison of two methods for measuring cholesterol

Plasma sample	Cholesterol content (g/L)		Difference (d_i)
	Method A	Method B	
1	1.46	1.42	0.04
2	2.22	2.38	−0.16
3	2.84	2.67	0.17
4	1.97	1.80	0.17
5	1.13	1.09	0.04
6	2.35	2.25	0.10
			$\bar{d} = +0.06_0$

Box 4-1 Analytical Chemistry and the Law

As a person who will either derive or use analytical results, you should be aware of this warning published in a report entitled "Principles of Environmental Analysis":[2]

Analytical chemists must always emphasize to the public that *the single most important characteristic of any result obtained from one or more analytical measurements is an adequate statement of its uncertainty interval.* Lawyers usually attempt to dispense with uncertainty and try to obtain unequivocal statements; therefore, an uncertainty interval must be clearly defined in cases involving litigation and/or enforcement proceedings. Otherwise, a value of 1.001 without a specified uncertainty, for example, may be viewed as legally exceeding a permissible level of 1.

Some legal limits make no scientific sense. The Delaney Amendment to the U.S. Federal Food, Drug, and Cosmetic Act of 1958 states that "no additive [in processed food] shall be deemed to be safe if it is found to induce cancer when ingested by man or animal . . . " This means that no detectable level of any carcinogenic (cancer-causing) pesticide may remain in processed foods, even if the level is far below that which can be shown to cause cancer. (Interestingly, the law does not apply to fresh food, which may contain pesticides.) The law was passed at a time when the sensitivity of analytical procedures was relatively poor, so the detection limit was relatively high. As analytical procedures became more sensitive, the ability to detect chemical residues decreased by 10^3–10^6. A concentration that may have been acceptable 30 years ago is now 10^6 times above the legal limit, regardless of whether there is any evidence that such a low level is harmful. The U.S. Supreme Court upheld this law as recently as 1993.

The quantity $\bar{d}$ is the average difference between methods A and B, and n is the number of pairs of data (six in this case). For the results in Table 4-4, the standard deviation, s_d, of the differences is calculated to be

$$s_d = \sqrt{\frac{(0.04 - \bar{d})^2 + (-0.16 - \bar{d})^2 + (0.17 - \bar{d})^2 + (0.17 - \bar{d})^2 + (0.04 - \bar{d})^2 + (0.10 - \bar{d})^2}{6 - 1}} = 0.12_2$$

(using the value $\bar{d} = 0.06_0$). Putting this value into Equation 4-9 gives

$$t = \frac{0.06_0}{0.12_2} \sqrt{6} = 1.20$$

The calculated value of t (= 1.20) is below the value of 2.571 listed in Table 4-2 for 95% confidence and 5 degrees of freedom. The two techniques are *not* significantly different at the 95% confidence level.

You may appreciate Box 4-1 at this time.

4-3 Control Charts

A **control chart** is a visual representation of confidence intervals for a Gaussian distribution. In quality control monitoring, it quickly warns us when a measurement strays dangerously far from an intended *target value*.

As an example, consider a vitamin C manufacturing line in which the intended average number of milligrams of vitamin C per tablet, designated μ,

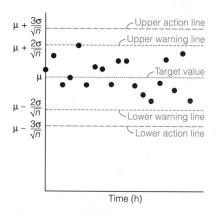

$\mu + \dfrac{3\sigma}{\sqrt{n}}$ — Upper action line

$\mu + \dfrac{2\sigma}{\sqrt{n}}$ — Upper warning line

μ — Target value

$\mu - \dfrac{2\sigma}{\sqrt{n}}$ — Lower warning line

$\mu - \dfrac{3\sigma}{\sqrt{n}}$ — Lower action line

Time (h)

Figure 4-7 Control chart for process monitoring. Consecutive measurements at a warning line—or a single observation outside the action line—indicate that the process is out of control.

Control charts tell us whether a process is remaining within expected limits and when there is a systematic drift away from the target value.

is the target value. Many analyses over a long time tell us the population standard deviation, σ, associated with the manufacturing process. We must monitor the process to see that it does not stray far from the target value.

For quality control, 25 tablets are removed from the manufacturing line each hour and analyzed. If the mean vitamin C content of the 25 tablets is $\bar{x}$, the 95% confidence interval is calculated with Equation 4-6:

$$95\% \text{ confidence: } \bar{x} \pm \frac{t\sigma}{\sqrt{n}} = \bar{x} \pm \frac{1.960\sigma}{\sqrt{25}}$$

in which t is taken from the bottom line of Table 4-2 because the true population standard deviation, σ, is known from a large number of measurements.

The value of t for 95% confidence at the bottom of Table 4-2 is equivalent to the number of standard deviations in the Gaussian distribution (Table 4-1) containing 95% of the area under the curve. We know from Table 4-1 that 95.5% of the area is contained within $\pm 2\sigma$ and 99.7% is contained within $\pm 3\sigma$.

$$95.5\% \text{ confidence: } \quad \bar{x} \pm \frac{2\sigma}{\sqrt{n}} \qquad 99.7\% \text{ confidence: } \quad \bar{x} \pm \frac{3\sigma}{\sqrt{n}}$$

In the control chart in Figure 4-7, each point shows the average vitamin C content of 25 tablets measured each hour. The $\pm 2\sigma/\sqrt{n}$ limits are designated *warning lines* and the $\pm 3\sigma/\sqrt{n}$ limits are designated *action lines*. We expect that about 1 measurement in 20 (4.5%) will be at the warning lines but that only 3 measurements in 1 000 (0.3%) will be at the action lines. It is very unlikely that we would observe two consecutive measurements at the warning line (probability $= 0.045 \times 0.045 = 0.002\,0$), so the process should be shut down and examined whenever two consecutive warning levels are found. A single observation at the action level might be cause for shutting down the process for troubleshooting.

4-4 Q Test for Bad Data

Sometimes one datum appears to be inconsistent with the remaining data. When this happens, you are faced with the decision of whether to retain or discard the questionable point.

Real data from a spectrophotometric analysis are given in Table 4-5. In this analysis a color is developed in proportion to the amount of protein in

TABLE 4-5   **Spectrophotometer readings for protein analysis by the Lowry method**

Sample (μg)	Absorbance of three independent samples			Range	Average with all data
0	0.099	0.099	0.100	0.001	0.099_3
5	0.185	0.187	0.188	0.003	0.188_7
10	0.282	0.272	0.272	0.010	0.275_3
15	0.392	0.345	0.347	0.047	0.361_2
20	0.425	0.425	0.430	0.005	0.426_7
25	0.483	0.488	0.496	0.013	0.489_0

the sample. Scanning across the three absorbance values in each row, we note that the number 0.392 seems out of line: It is inconsistent with the other values for 15 μg, and the range of values for the 15-μg samples is much bigger than the range for the other samples. The linear relation between the average values of absorbance up to the 20-μg sample also indicates that the value 0.392 is in error (Figure 4-8).

It is reasonable to ask whether all three absorbances for the 25-μg samples are low for some unknown reason, because this point falls below the straight line in Figure 4-8. The answer to this question can be discovered only by repetition of the experiment. Many repetitions of this particular analysis show that the 25-μg point is consistently below the straight line.

If you have more repetitions (at least four), you can use the **Q test** to help decide whether to retain or discard a questionable datum. Consider the five results 12.53, 12.56, 12.47, 12.67, and 12.48. Is 12.67 a "bad point"? To apply the Q test, arrange the data in order of increasing value and calculate Q, defined as

$$Q = \frac{\text{gap}}{\text{range}} \tag{4-11}$$

Gap = 0.11

12.47 12.48 12.53 12.56   (12.67) — Questionable value (too high?)

Range = 0.20

The *range* is the total spread of the data. The *gap* is the difference between the questionable point and the nearest value.

If Q (observed) > Q (tabulated), the questionable point should be discarded. For the numbers above, Q = 0.11/0.20 = 0.55. Referring to Table 4-6, we see that the critical value of Q at the 90% confidence limit is 0.64. *Because the observed Q is smaller than the tabulated Q, the questionable point should be retained.* That is, there is more than a 10% chance that the value 12.67 is a member of the same population as the other four numbers.

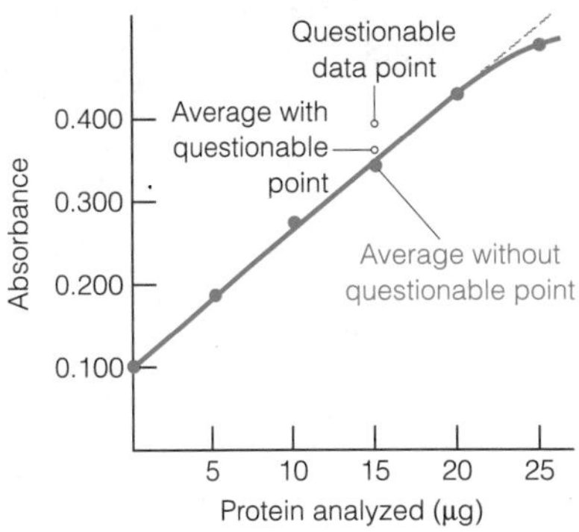

Figure 4-8 *Calibration curve* showing the average absorbance values in Table 4-5 versus micrograms of protein analyzed. Note that the *blank value* for 0 μg of protein is an experimental point on the curve.

If Q (observed) > Q (tabulated), discard the questionable point.

4-5 Finding the "Best" Straight Line

Often we seek to draw the "best" straight line through a set of data points, as in Figure 4-8. This section provides a very important recipe for finding that "best" line.

Method of Least Squares

The method of least squares assumes that the errors in the y values are substantially greater than the errors in the x values.[3] This condition is often true and applies to Figure 4-8, because the variation in absorbance is greater than the uncertainty in the concentrations of our standards. A second assumption is that the uncertainties (standard deviations) in all of the y values are similar.

Suppose we seek to draw the best straight line through the points in Figure 4-9 by minimizing the vertical deviations between the points and the line. We minimize only the vertical deviations because we assume that uncertainties in the y values are much greater than uncertainties in the x values.

TABLE 4-6   Values of Q for rejection of data

Q (90% confidence)[a]	Number of observations
0.76	4
0.64	5
0.56	6
0.51	7
0.47	8
0.44	9
0.41	10

a. Q = gap/range. If Q (observed) > Q (tabulated), the value in question can be rejected with 90% confidence.

SOURCE: R. B. Dean and W. J. Dixon, *Anal. Chem.* **1951**, *23*, 636; see also D. R. Rorabacher, *Anal. Chem.* **1991**, *63*, 139.

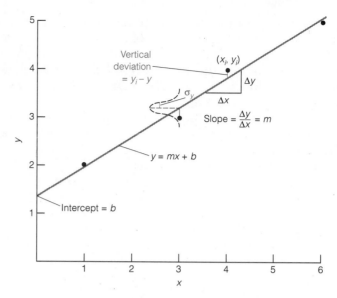

Figure 4-9   Least-squares curve fitting. The points $(1, 2)$ and $(6, 5)$ do not fall exactly on the solid line, but they are too close to the line to show their deviations. The Gaussian curve drawn over the point $(3, 3)$ is a schematic indication of the fact that each value of y_i is normally distributed about the straight line. That is, the most probable value of y will fall on the line, but there is a finite probability of measuring y some distance from the line.

Let the equation of the line be

The equation of a straight line is discussed in Appendix B.

$$y = mx + b \qquad (4\text{-}12)$$

in which m is the **slope** and b is the **y-intercept.** The vertical deviation for the point (x_i, y_i) will be given by $y_i - y$, where y is the ordinate of the straight line when $x = x_i$.

$$\text{vertical deviation} = d_i = y_i - y = y_i - (mx_i + b) \qquad (4\text{-}13)$$

Some of the deviations are positive and some are negative. Because we wish to minimize the magnitude of the deviations irrespective of their signs, we can square all the deviations so that we are dealing only with positive numbers:

$$d_i^2 = (y_i - y)^2 = (y_i - mx_i - b)^2$$

Because we seek to minimize the squares of the deviations, this is called the *method of least squares.* It can be shown that minimizing the squares of the deviations (rather than simply their magnitudes) corresponds to assuming that the set of y values is the most probable set.

Deriving the values of m and b that minimize the sum of the squares of the vertical deviations involves some calculus, which we omit. We will express the final solution for slope and intercept in terms of *determinants,* which summarize certain arithmetic operations. The **determinant** $\begin{vmatrix} e & f \\ g & h \end{vmatrix}$ represents the value $eh - fg$. So, for example,

$$\begin{vmatrix} 6 & 5 \\ 4 & 3 \end{vmatrix} = 6 \times 3 - 5 \times 4 = -2$$

TABLE 4-7 Calculations for least-squares analysis

x_i	y_i	$x_i y_i$	x_i^2	$d_i\ (= y_i - mx_i - b)$	d_i^2
1	2	2	1	0.038 47	0.001 479 9
3	3	9	9	−0.192 29	0.036 975
4	4	16	16	0.192 33	0.036 991
6	5	30	36	−0.038 43	0.001 476 9
$\Sigma x_i = 14$	$\Sigma y_i = 14$	$\Sigma(x_i y_i) = 57$	$\Sigma(x_i^2) = 62$		$\Sigma(d_i^2) = 0.076\,923$

The slope and the intercept of the "best" straight line are found to be

Least-squares "best" line

$$\begin{cases} \text{slope:} \quad m = \begin{vmatrix} \Sigma(x_i y_i) & \Sigma x_i \\ \Sigma y_i & n \end{vmatrix} \div D & (4\text{-}14) \\[2em] \text{intercept:} \quad b = \begin{vmatrix} \Sigma(x_i^2) & \Sigma(x_i y_i) \\ \Sigma x_i & \Sigma y_i \end{vmatrix} \div D & (4\text{-}15) \end{cases}$$

where the number D is given by

$$D = \begin{vmatrix} \Sigma(x_i^2) & \Sigma x_i \\ \Sigma x_i & n \end{vmatrix} \qquad (4\text{-}16)$$

Translation of least-squares equations:

$$m = \frac{n\,\Sigma(x_i y_i) - \Sigma x_i\, \Sigma y_i}{n\,\Sigma(x_i^2) - \left(\Sigma x_i\right)^2}$$

$$b = \frac{\Sigma(x_i^2)\, \Sigma y_i - \Sigma(x_i y_i)\, \Sigma x_i}{n\,\Sigma(x_i^2) - \left(\Sigma x_i\right)^2}$$

and n is the number of points.

Now we apply these equations to the points in Figure 4-9 to find the slope and intercept of the best straight line through the four points. The work is set out in Table 4-7. Noting that $n = 4$ and putting the various sums into the determinants in Equations 4-14, 4-15, and 4-16 give

$$m = \begin{vmatrix} 57 & 14 \\ 14 & 4 \end{vmatrix} \div \begin{vmatrix} 62 & 14 \\ 14 & 4 \end{vmatrix} = \frac{32}{52} = 0.615\,38$$

$$b = \begin{vmatrix} 62 & 57 \\ 14 & 14 \end{vmatrix} \div \begin{vmatrix} 62 & 14 \\ 14 & 4 \end{vmatrix} = \frac{70}{52} = 1.346\,15$$

The equation of the best straight line through the points in Figure 4-9 is therefore

$$y = 0.615\,38x + 1.346\,15$$

We tackle the question of how many significant figures should be associated with m and b in the next section.

How Reliable Are Least-Squares Parameters?

To estimate the uncertainties (expressed as standard deviations) in the slope and intercept, an uncertainty analysis must be performed on Equations 4-14 and 4-15. Because the uncertainties in m and b are related to the uncertainty in measuring each value of y, we first estimate the standard deviation that describes the population of y values. This standard deviation, σ_y, characterizes the little Gaussian curve inscribed in Figure 4-9.

We estimate σ_y, the population standard deviation of all y values, by calculating s_y, the standard deviation, for the four measured values of y. The deviation of each value of y_i from the center of its Gaussian curve is just $d_i =$

If you know $\bar{x}$ and $n - 1$ of the individual values, you can calculate the nth value. Therefore, the problem has just $n - 1$ degrees of freedom once $\bar{x}$ is known.

$y_i - y = y_i - (mx_i + b)$ (Equation 4-13). The standard deviation of these vertical deviations is given by an equation analogous to Equation 4-2:

$$\sigma_y \approx s_y = \sqrt{\frac{\sum(d_i - \bar{d})^2}{(\text{degrees of freedom})}} \tag{4-17}$$

Because the average deviation, $\bar{d}$, is 0 for the best straight line, the numerator of Equation 4-17 reduces to $\sum(d_i^2)$.

The degrees of freedom is the number of independent pieces of information available. In Equation 4-2 we set the degrees of freedom equal to $n - 1$. The rationale for so doing is that we began with n degrees of freedom, but one degree of freedom is "lost" in determining the average value ($\bar{x}$). That is, only $n - 1$ pieces of information are available in addition to the average value. If you know $n - 1$ values and you also know the average value, then the nth value is fixed and you can calculate it.

How does this apply to Equation 4-17? We began with n points. But two degrees of freedom were "used up" in determining the slope and the intercept of the best line. Therefore, $n - 2$ degrees of freedom remain. Equation 4-17 should be written

$$\sigma_y \approx s_y = \sqrt{\frac{\sum(d_i^2)}{n - 2}} \tag{4-18}$$

where d_i is given by Equation 4-13.

The uncertainty analysis for Equations 4-14 and 4-15 uses the method of Appendix C and leads to the following results:

Standard deviation of slope and intercept

$$\sigma_m^2 = \frac{\sigma_y^2 n}{D} \tag{4-19}$$

$$\sigma_b^2 = \frac{\sigma_y^2 \sum(x_i^2)}{D} \tag{4-20}$$

where σ_m is our estimate of the standard deviation of the slope, σ_b is our estimate of the standard deviation of the intercept, σ_y is given by Equation 4-18, and D is given by Equation 4-16.

Now we can finally address the question of significant figures for the slope and the intercept of the line in Figure 4-9. In Table 4-7 we see that $\sum(d_i^2) = 0.076\,923$. Putting this number into Equation 4-18 gives

$$\sigma_y^2 \approx s_y^2 = \frac{0.076\,923}{4 - 2} = 0.038\,462$$

At last we can plug numbers into Equations 4-19 and 4-20 to find

$$\sigma_m^2 = \frac{\sigma_y^2 n}{D} = \frac{(0.038\,462)(4)}{52} = 0.002\,958\,6$$

$$\sigma_b^2 = \frac{\sigma_y^2 \sum(x_i^2)}{D} = \frac{(0.038\,462)(62)}{52} = 0.045\,859$$

or

$$\sigma_m = 0.054\,39 \quad \text{and} \quad \sigma_b = 0.214\,15$$

Combining the results for m, σ_m, b, and σ_b, we write

$$\text{slope:} \quad \frac{0.615\,38}{\pm 0.054\,39} = 0.62 \pm 0.05 \text{ or } 0.61_5 \pm 0.05_4 \qquad (4\text{-}21)$$

$$\text{intercept:} \quad \frac{1.346\,15}{\pm 0.214\,15} = 1.3 \pm 0.2 \text{ or } 1.3_5 \pm 0.2_1 \qquad (4\text{-}22)$$

The first digit of the uncertainty is the last significant figure.

where the uncertainties represent one standard deviation. *The first decimal place of the standard deviation is the last significant figure of the slope or intercept.*

If we wanted to express the uncertainty as a confidence interval instead of one standard deviation, we would multiply the uncertainties in Equations 4-21 and 4-22 by the appropriate value of Student's t from Table 4-2 for $n - 2$ degrees of freedom. For example, the 95% confidence interval for the slope is $\pm t\sigma_m = \pm(4.303)(0.054) = \pm 0.23$. In this book we will express most uncertainties as one standard deviation.

A Practical Example

What good is all of this? One real application of a least-squares analysis involves the determination of protein concentration by using the absorbance values in Table 4-5. To obtain the numbers in this table, known protein standards were analyzed. Each standard leads to a certain absorbance measured with a spectrophotometer. Figure 4-8 shows that the absorbances for standards containing 0 to 20 μg of protein fall on a straight line. These standards provide 14 absorbance values in Table 4-5—if we omit the value 0.392. Using these 14 values, we calculate the least-squares parameters for Figure 4-8 to be

$$m = 0.016\,3_0 \qquad \sigma_m = 0.000\,2_2$$
$$b = 0.104_0 \qquad \sigma_b = 0.002_6$$
$$\sigma_y = 0.005_9$$

Now suppose that the absorbance of an unknown sample is 0.246. How many micrograms of protein does it contain, and what uncertainty is associated with the answer? Note that in Figure 4-8 the y axis is absorbance and the x axis is micrograms of protein. Solving for concentration gives

$$x = \frac{y\,(\pm\sigma_y) - b\,(\pm\sigma_b)}{m\,(\pm\sigma_m)} \qquad (4\text{-}23)$$

Follow the rules for propagation of uncertainty for subtraction and division.

$$x = \frac{0.246\,(\pm 0.005_9) - 0.104_0\,(\pm 0.002_6)}{0.016\,3_0\,(\pm 0.000\,2_2)}$$

The uncertainties in the numerator are combined, using the rule for subtraction with absolute uncertainties ($e = \sqrt{0.005_9{}^2 + 0.002_6{}^2} = 0.006_4$):

$$x = \frac{0.142_0\,(\pm 0.006_4)}{0.016\,3_0\,(\pm 0.000\,2_2)}$$

The uncertainty of the quotient is computed in a similar manner, using relative uncertainties instead of absolute uncertainties:

$$x = \frac{0.142_0\,(\pm 4._5\%)}{0.016\,3_0\,(\pm 1._3\%)}$$

$$x = 8.71\,(\pm 4.7\%) = 8.7\,(\pm 0.4) \text{ μg of protein}$$

The unknown has 8.7 ($\pm$0.4) µg of protein. The estimate of uncertainty is the standard deviation associated with the number 8.7.

The calculation of uncertainty in Equation 4-23 neglects the fact that the calculated slope and intercept are not independent of each other. A full treatment gives the result

$$(\text{uncertainty in } x)^2 = \frac{\sigma_y^2}{m^2}\left[1 + \left(\frac{y-b}{m}\right)^2\left(\frac{n}{D}\right) + \frac{\sum(x_i^2)}{D} - 2\left(\frac{y-b}{m}\right)\left(\frac{\sum x_i}{D}\right)\right] \quad (4\text{-}24)$$

where D is given by Equation 4-16. If we measure several values of y and take their average, the uncertainty in x will be reduced. In this case, we change the first term within the brackets from 1 to $1/k$, where k is the number of y values that have been averaged. For the example in Equation 4-23, one value of y (0.246) has been measured, and application of Equation 4-24 gives an uncertainty in x of $\pm 0.3_7$ µg instead of the $\pm 0.4_1$ µg calculated with Equation 4-23.

Whereas Equation 4-24 is somewhat painful to use by hand, it is easy to incorporate into a spreadsheet, which is introduced in the next section. Most of the problems in this text were worked with Equation 4-23 for historical reasons. We recommend using Equation 4-24 in a spreadsheet as the routine method of error analysis for least-squares problems.

Your brain is smarter than your calculator!

A final caution is in order. Before using your calculator or computer to find the least-squares straight line automatically, look at a graph of your data. This reality check gives you an opportunity to reject bad data or to decide that a straight line is not an appropriate function. Mindless use of a computer program may not do justice to your hard-earned data. *The operation in which a human being evaluates his or her data should never be sacrificed.*

Spreadsheets

A computer spreadsheet is a simple but powerful tool for manipulating quantitative information.[4] In this book, spreadsheets will help you explore phenomena such as the shapes of titration curves and the concentrations of species at equilibrium, which depend on otherwise repugnant doses of arithmetic. Spreadsheets readily allow you to conduct "what if" experiments in which you investigate the effect of a stronger acid, a higher temperature, a different ionic strength, and so on. Spreadsheets allow you to plot your results or to transfer the numbers to a graphing program. *Examining graphs is critical to understanding quantitative relationships.*

Together, spreadsheets and graphing programs allow us to explore complex mathematical relationships with little pain.

Any spreadsheet program is suitable for the exercises in this book. Our specific instructions apply to Microsoft Works® or Microsoft Excel® operating on an Apple Macintosh computer. You will need directions for the particular computer and software available to you. Although this book can be used with no loss of continuity if you skip the spreadsheet exercises, they will enrich your understanding of analytical chemistry and give you a tool that will be valuable to you outside this course.

Getting Started

Consider an example in which we plot the function $y = x^{1.5}$ over the range $0 \le x \le 2$, with an interval of 0.2 between values of x. The blank spreadsheet shown in Figure 4-10a has columns labeled A, B, C, etc., and rows numbered

Columns

Rows

	A	B
1		
2		Cell B2
3		
4		
5		
6		
7		
8		
9		
10		
11		
12	Cell A12	

(a)

	A	B
1	x	
2	0	
3	0.2	
4	0.4	
5	0.6	
6	0.8	
7	1	
8	1.2	
9	1.4	
10	1.6	
11	1.8	
12	2	

(b)

Figure 4-10 Spreadsheet for calculating values of $y = x^{1.5}$. The formula for cell B2 is "=A2 ^ 1.5".

	A	B
1	x	y
2	0	0
3	0.2	
4	0.4	
5	0.6	
6	0.8	
7	1	
8	1.2	
9	1.4	
10	1.6	
11	1.8	
12	2	

(c)

	A	B
1	x	y
2	0	0
3	0.2	0.0894427
4	0.4	0.2529822
5	0.6	0.4647580
6	0.8	0.7155417
7	1	1
8	1.2	1.3145341
9	1.4	1.6565023
10	1.6	2.0238577
11	1.8	2.4149534
12	2	2.8284271

(d)

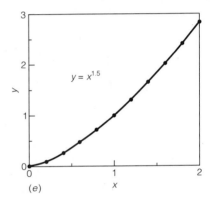

(e)

1, 2, 3, . . . The box in column B, row 2, is called *cell* B2. Here are the steps we follow:

Step 1. Label cell A1 with the letter "x". Starting in cell A2 and working down, type in the desired values of x (0, 0.2, 0.4, . . . , 2), as shown in Figure 4-10b.

Step 2. Label cell B1 "y". We want cell B2 to contain the value $y = (0)^{1.5}$, cell B3 to contain the value $y = (0.2)^{1.5}$, and so on. To do this, select cell B2 and enter the formula =A2 ^ 1.5. This entry tells the spreadsheet to take the value in cell A2, raise it to the power 1.5, and write the result in cell B2 (Figure 4-10c, $0^{1.5} = 0$). (In some software, formulas begin with + or @ instead of =.)

Step 3. Now select cells B2 through B12 and give a FILL DOWN command, which repeats the calculation in cell B2 for each cell down to B12 (Figure 4-10d). In cell B8, for example, the spreadsheet evaluates the number in cell A8 (1.2) raised to the power 1.5. (In some software, COPY is used instead of FILL DOWN.)

Step 4. Graph your results with the computer (Figure 4-10e). This may be done by your spreadsheet, or you may need to copy data from the spreadsheet to a graphing program. In either case, you will need specific instructions for your system.

Three kinds of entries:
  labels
  numbers
  formulas

In this simple example, we made three types of entries. The *labels* "x" and "y" in cells A1 and B1 tell us what is in the spreadsheet. We then typed *numbers* into cells A2–A12. In cell B2 we entered a *formula* (=A2 ^ 1.5) that necessarily begins with an equal sign.

Arithmetic Operations and Functions

Addition, subtraction, multiplication, division, and exponentiation have the symbols +, −, *, /, and ^ . *Functions,* such as Exp($\cdot$), are built into the spreadsheet. Exp($\cdot$) raises e to the power in parentheses. Other functions, such as Ln($\cdot$), Log($\cdot$), Sin($\cdot$), and Cos($\cdot$), are also available. In some spreadsheets, functions are written in all capital letters.

Order of operations:

1. Exponentiation
2. Multiplication and division (in order from left to right)
3. Addition and subtraction (in order from left to right)

Operations within parentheses are evaluated first, from the innermost set.

The order of arithmetic operations in formulas is ^ first, followed by * and / (evaluated in order from left to right as they appear), finally followed by + and − (also evaluated from left to right). Make liberal use of parentheses to be sure that the computer does what you intend. The contents of parentheses are evaluated first, before carrying out operations outside the parentheses. Here are some examples:

$$2*4 \wedge 2/3-5*6 \equiv (2*(4 \wedge 2)/3)-(5*6) = \frac{2*4^2}{3} - (5*6) = \frac{32}{3} - 30 \approx -19.33$$

$$(2*4) \wedge 2/(3-5)*6 \equiv \frac{(2*4)^2}{(3-5)} * 6 = \frac{64*6}{-2} = -192$$

$$2*4 \wedge (2/3-5*6) \equiv 2*4^{[(2/3)-30]} \approx 2*4^{-29.333} \approx 4.37 \times 10^{-18}$$

When in doubt about how an expression that you are writing will be evaluated by the computer, use parentheses to force it to do what you intend.

Plotting a Gaussian Curve

Now let's look at a slightly more complicated example that introduces standards we will use to make our spreadsheets more readable. We will compute coordinates of the smooth Gaussian curve in Figure 4-1, which has the same mean and standard deviation as the light bulbs in the bar graph.

Step 1. *Set up the formula.* The formula for a *normalized* Gaussian curve (one whose area is unity) was given by Equation 4-3:

$$y = \frac{1}{\sigma\sqrt{2\pi}} e^{-(x-\mu)^2/2\sigma^2} \qquad (4\text{-}3)$$

Because the curve in Figure 4-1 applies to 4768 light bulbs, we suppose that the correct equation is

$$y = \frac{4768}{s\sqrt{2\pi}} e^{-(x-\bar{x})^2/2s^2} \qquad (4\text{-}25)$$

where s is the standard deviation for the 4768 light bulbs and $\bar{x}$ is the mean lifetime.

Step 2. *Set up a list of constants.* It is convenient to have a list of constants that we do not have to type over and over again. We list these in column A, as shown in Figure 4-11. In cell A1, we type the label "mean =". In cell A2, we type the number 845.2, which is the mean lifetime in

	A	B	C	D
1	mean =	x (hours)	y (bulbs)	20*Column C
2	845.2	500	0.024	0.490
3	standard dev.	525	0.063	1.251
4	94.2	550	0.149	2.977
5	total bulbs =	575	0.330	6.601
6	4768	600	0.682	13.644
7	sqrt(2 pi) =	625	1.314	26.283
8	2.506628	650	2.359	47.185
9		675	3.947	78.949
10		700	6.156	123.112
11		725	8.946	178.923
12		750	12.118	242.350
13		775	15.297	305.936
14		800	17.997	359.940
15		825	19.734	394.676
16		845.2	20.193	403.855
17		850	20.167	403.331
18				
19	Formula for cell C2 =			
20	($A$6/($A$4*$A$8))*Exp(−(B2−$A$2)^2/(2*$A$4^2))			

Figure 4-11 A spreadsheet for calculating points on the smooth Gaussian curve in Figure 4-1. Note that we place constants in column A, label each column at the top, and write formulas at the bottom for future reference.

hours. Similarly, we type labels and values for the standard deviation, number of bulbs, and the value of $\sqrt{2\pi}$ (computed on a calculator).

Step 3. *Enter values of the independent variable (x).* We put the label "*x* (hours)" in cell B1 and then type in values from 500 to 850 h in cells B2–B17 for this abbreviated example.

Step 4. *Enter the formula for calculating the dependent variable (y).* Column C, labeled "*y* (bulbs)" in cell C1, will contain ordinate (*y*) values for the Gaussian curve. The formula in cell C2 is the spreadsheet translation of Equation 4-25:

$$\frac{4768}{s\sqrt{2\pi}}e^{-(x-\bar{x})^2/2s^2} = \frac{\text{cell A6}}{\text{cell A4}*\text{cell A8}} e^{-(x-\text{cell A2})^2/[2*(\text{cell A4})^2]}$$

Formula in cell C2:

$$=(\$A\$6/(\$A\$4*\$A\$8))*\text{Exp}(-(B2-\$A\$2)\hat{\ }2/(2*\$A\$4\hat{\ }2)) \tag{4-26}$$

In other words, we first replaced the mean, standard deviation, number of bulbs, and $\sqrt{2\pi}$ by the contents of cells A2, A4, A6, and A8, respectively. The dollar signs signify *absolute references* for constants used throughout the spreadsheet. No matter which value of *y* we are calculating, we are going to want to use the mean in cell $A$2 and the standard deviation in cell $A$4.

All of the cells in Equation 4-26 are written as absolute references, except B2, which is written without dollar signs. B2 without dollar signs is called a *relative reference*. It tells the spreadsheet working on cell C2 to move one column to the left and find a value in cell B2. When we FILL DOWN to compute the rest of column C, cell C6 will move over one column to cell B6 to find a value for the calculation. That is, values are located *relative* to the cell in which the computation is occurring. In contrast, the absolute reference $A$4 tells the spreadsheet to get the value in cell $A$4—no matter where the computation is occurring.

Role of a human

Step 5. *Sanity check.* Now we have a problem. The maximum value of the Gaussian curve at 845.2 h is 20.193 in cell C16. Figure 4-1 goes up to a value near 400. Inspection of the bar graph in Figure 4-1 shows that the bars are plotted for groups of 20 hours. Was a factor of 20 needed in the calculations? Yes. The correct formula for the Gaussian curve must be

$$y = \frac{4\,768*20}{s\sqrt{2\pi}}\,e^{-(x - \bar{x})^2/2s^2}$$

The beauty of a spreadsheet is that this is an easy change to make. We simply define column D to be 20 times column C by typing the formula "=20*C2" in cell D2. After a FILL DOWN operation, the values in cells D2 to D17 look pretty good, with a maximum value of 403.855.

Role of a computer

Step 6. *Plot the graph.* Use your spreadsheet or your favorite graphing program to plot the values of y (column D) versus the values of x (column B). If you use a separate graphing program, be sure to copy and paste the data from the spreadsheet into the graphing program by computer, not by hand.

Your role in this process is to set up the problem and carry out sanity checks. Let the computer do everything else and your work will be easier and contain fewer mistakes.

Documentation and Readability

If your spreadsheet cannot be read by another person without your help, it needs better documentation. (The same is true of your lab notebook!)

If you look at your spreadsheet next week, you will probably not know what formulas were used. Therefore, we added two lines of text (labels) in cells A19 and A20 of Figure 4-11. In cell A19, we wrote "Formula for cell C2 = ", and in cell A20, we wrote "($A$6/($A$4*$A$8))*Exp(−(B2−$A$2) ^ 2/ (2*$A$4 ^ 2))". Such *documentation* is an excellent practice for every spreadsheet.

For additional readability in Figure 4-11, we set columns C and D to display just 3 decimal places, even though the computer retains many more for its calculations. It does not throw away the digits that are not displayed. You should learn to control the way numbers are displayed in your spreadsheet.

A Spreadsheet for Least Squares

You should construct your own version of the least-squares spreadsheet, which will probably be your most used spreadsheet.

Figure 4-12 translates the least-squares computations of Table 4-7 into a spreadsheet. Values of x and y are entered in columns B and C, and the total number of points ($n = 4$) is entered in cell A8. The products xy and x^2 are computed in columns D and E. The sums of columns B through G appear on line 7. For example, the sum of xy values in cells D2 through D5 is computed with the statement "= Sum(D2:D5)". The least-squares parameters D, m, and b are computed in cells A10, A12, and A14 by using Equations 4-14

	A	B	C	D	E	F	G
1	n	x	y	xy	x^2	d	d^2
2	1	1	2	2	1	0.0385	0.0015
3	2	3	3	9	9	−0.1923	0.0370
4	3	4	4	16	16	0.1923	0.0370
5	4	6	5	30	36	−0.0385	0.0015
6		-------------Column Sums [Example: D7 = Sum(D2:D5)]-------------					
7	n=	14	14	57	62	0.0000	0.0770
8	4						
9	D=	sigma(y)=		A10: D = A8*E7–B7*B7			
10	52.0000	0.1962		A12: m = (D7*A8–B7*C7)/A10			
11	m=	sigma(m)=		A14: b = (E7*C7–D7*B7)/A10			
12	0.6154	0.0544		B10: sigma(y) = Sqrt(G7/(A8–2))			
13	b=	sigma(b)=		B12: sigma(m) = B10*Sqrt(A8/A10)			
14	1.3462	0.2143		B14: sigma(b) = B10*Sqrt(E7/A10)			
15				F2: d = C2–$A$12*B2–$A$14			
16							
17	Propagation of uncertainty calculation with Equation 4-24:						
18							
19	Measured y =		Derived x =		Uncertainty in x =		
20	2.72		2.2325			0.37368	
21							
22	Derived x in cell C20 = (A20–A14)/A12						
23	Uncertainty in x in cell F20 =						
24		(B10/A12)*Sqrt(1+C20*C20*A8/A10+E7/A10–2*C20*B7/A10)					

Figure 4-12   A spreadsheet for least-squares calculations.

through 4-16. The deviations, d, in column F make use of the slope and intercept. Column G contains the squares of the deviations. Standard deviations σ_y, σ_m, and σ_b are then computed in cells B10, B12, and B14, using Equations 4-18 through 4-20. On the right side of the spreadsheet, we document the formulas used for each cell.

At the bottom of the spreadsheet, we use Equation 4-24 to evaluate propagation of uncertainty. A measured value of y is entered in cell A20 and derived values of x and its uncertainty are computed in cells C20 and F20. This spreadsheet tells us that for a measured value of $y = 2.72$ in Figure 4-9, the value of x is $2.2_3 \pm 0.3_7$.

Terms to Understand

average	Gaussian distribution	standard deviation
confidence interval	intercept	Student's t
control chart	mean	t test
degrees of freedom	Q test	variance
determinant	slope	

Summary

The results of many measurements of an experimental quantity follow a Gaussian distribution. The measured mean, $\bar{x}$, approaches the true mean, μ, as the number of measurements becomes very large. The broader the distribution, the greater is σ, the standard deviation. For a limited number of measurements, an estimate of the standard deviation is given by $s = \sqrt{[\Sigma(x_i - x)^2]/(n - 1)}$. About two-thirds of all measurements lie within $\pm 1\sigma$ and 95% lie within $\pm 2\sigma$. The probability of observing a value within a certain interval is proportional to the area of that interval, given in Table 4-1.

After you select a confidence level, Student's t is used to find confidence intervals ($\mu = \bar{x} \pm ts/\sqrt{n}$) and to compare mean values measured by different methods. A control chart is used to monitor the performance of a process to see whether it remains within expected limits. The Q test helps you to decide whether or not a questionable datum should be discarded. It is best to repeat the measurement several times to increase the probability that your decision is correct.

The method of least squares is used to find the slope and the intercept of the best straight line through a series of points. The standard deviations of slope and intercept are used in analyzing the error associated with an experimental measurement.

Exercises

A. For the numbers 116.0, 97.9, 114.2, 106.8, and 108.3, find the mean, standard deviation, range, and 90% confidence interval for the mean. Using the Q test, decide whether the number 97.9 should be discarded.

B. Suppose that the mileage at which 10 000 sets of automobile brakes had been 80% worn through was recorded. The average was 62 700, and the standard deviation was 10 400 miles.

(a) What fraction of brakes is expected to be 80% worn in less than 45 800 miles?

(b) What fraction is expected to be 80% worn at a mileage between 60 000 and 70 000 miles?

C. It is found from a reliable assay that the ATP (adenosine triphosphate) content of a certain type of cell is 111 μmol/100 mL. You have developed a new assay, which gave the following values for replicate analyses: 117, 119, 111, 115, 120 μmol/100 mL. The average value is $116._4$. Can you be 95% confident that your method produces a result different from the "known" value?

D. The Ca content of a powdered mineral sample was analyzed five times by each of two methods, with similar standard deviations. Are the mean values significantly different at the 95% confidence level?

Method 1:

　　　0.0271　0.0282　0.0279　0.0271　0.0275

Method 2:

　　　0.0271　0.0268　0.0263　0.0274　0.0269

E. A common procedure for protein determination is the dye-binding assay of Bradford.[5] In this method, a dye binds to the protein and, as a consequence, the color of the dye changes from brown to blue. The amount of blue color is proportional to the amount of protein present.

Protein (μg):　0.00　9.36　18.72　28.08　37.44

Absorbance
at 595 nm:　　0.466　0.676　0.883　1.086　1.280

(a) Using the method of least squares, determine the equation of the best straight line through these points. Use the standard deviation of the slope and intercept to express the equation in the form $y = [m(\pm\sigma_m)]x + [b(\pm\sigma_b)]$ with a reasonable number of significant figures.

(b) Make a graph showing the experimental data and the calculated straight line.

(c) An unknown protein sample gave an absorbance of 0.973. Calculate the number of micrograms of protein in the unknown, and estimate its uncertainty.

Problems

Gaussian Distribution

1. What is the relation between the standard deviation of a procedure and the precision of the procedure? What is the relation between standard deviation and accuracy?

2. Use Table 4-1 to state what fraction of a Gaussian population lies within the following intervals.

(a) $\mu \pm \sigma$　　　　**(b)** $\mu \pm 2\sigma$　　　　**(c)** μ to $+\sigma$
(d) μ to $+0.5\sigma$　　**(e)** $-\sigma$ to -0.5σ

3. The ratio of the number of atoms of the isotopes ^{69}Ga and ^{71}Ga in samples from different sources was measured in an effort to understand differences in reported values of the atomic weight of gallium.[6] Results for eight samples were as follows.

Sample	^{69}Ga/^{71}Ga	Sample	^{69}Ga/^{71}Ga
1	1.526 60	5	1.528 94
2	1.529 74	6	1.528 04
3	1.525 92	7	1.526 85
4	1.527 31	8	1.527 93

Find the (a) mean, (b) standard deviation, and (c) variance.

4. (a) Calculate the fraction of bulbs in Figure 4-1 expected to have a lifetime greater than 1 000 h.

(b) Calculate the fraction expected to have a lifetime between 800 and 900 h.

5. Write the equation of the smooth Gaussian curve in Figure 4-1. (Because the curve represents the results of 4 768 measurements, and each bar on the graph corresponds to a 20-h interval, you must use a factor of 4 768 × 20 in the numerator of Equation 4-3.) Use the equation to calculate the value of y when $x = 1 000$ h and see if your calculated value agrees with the value on the graph.

6. Consider a Gaussian distribution with a population mean of 14.49_6 and a population standard deviation of 0.10_7. Find the fraction of measurements expected between 14.55 and 14.60 if many measurements are made.

7. Set up a spreadsheet like the one in Figure 4-11 to calculate the coordinates of the Gaussian curve in Figure 4-1 from 500 to 1 200 h. Use a computer to graph your results.

8. Repeat Problem 7, but use the values 50, 100, and 150 for the standard deviation. Superimpose all three curves on a single graph.

Student's t

9. What is the meaning of a confidence interval?

10. What fraction of the vertical bars in Figure 4-5a is expected to include the population mean (10 000) if many experiments are carried out?

11. List the three different cases that we studied for comparison of means, and write the equations used in each case.

12. The percentage of an additive in gasoline was measured six times with the following results: 0.13, 0.12, 0.16, 0.17, 0.20, 0.11%. Find the 90% and 99% confidence intervals for the percentage of the additive.

13. Sample 8 of Problem 3 was analyzed seven times, with $\bar{x} = 1.527 93$ and $s = 0.000 07$.

(a) Find the 99% confidence interval for sample 8.

(b) How many of the first seven samples in Problem 3 lie outside the 99% confidence limit for sample 8? Is the variation among samples due to irreproducibility in the analysis or real differences among samples?

14. The Ti content (wt%) of five different ore samples (each with a different Ti content) was measured by each of two methods.

Sample	% Composition by method 1	% Composition by method 2
A	0.013 4	0.013 5
B	0.014 4	0.015 6
C	0.012 6	0.013 7
D	0.012 5	0.013 7
E	0.013 7	0.013 6

Do the two analytical techniques give results that are significantly different at the 95% confidence level?

15. The Ti content (wt%) of two different ore samples was measured several times by the same method. Are the mean values significantly different at the 95% confidence level?

Sample 1: 0.013 4 0.013 8 0.012 8 0.013 3 0.013 7
Sample 2: 0.013 5 0.014 2 0.013 7 0.014 1 0.014 3

16. If you measure a quantity four times and the standard deviation is 1.0% of the average, can you be 90% confident that the true value is within 1.2% of the measured average?

17. Students measured the concentration of HCl in a solution by titrations using different indicators to find the end point.[7] Is the difference between indicators 1 and 2 significant at the 95% confidence level? Answer the same question for indicators 2 and 3.

Indicator	Mean HCl concentration (M) (± standard deviation)	Number of measurements
1. Bromothymol blue	0.095 65 ± 0.002 25	28
2. Methyl red	0.086 86 ± 0.000 98	18
3. Bromocresol green	0.086 41 ± 0.001 13	29

18. The calcium content of a person's urine was determined on two different days:

Day	Average Ca (mg/L)	Number of measurements
1	238	4
2	255	5

The analytical method applied to many samples yields a standard deviation known to be 14 mg/L. Are the two average values significantly different at the 95% confidence level?

19. One definition of the *detection limit* of an analytical method is the minimum concentration of a substance that can be reported with 99% confidence to be greater than the blank. Here is a procedure for measuring the detection limit:[8]

1. After estimating the detection limit from previous experience with the method, prepare a sample whose concentration is one to five times the detection limit.

2. Measure the concentration of replicate samples n times ($n \geq 7$).

3. Analyze n blank samples (containing no analyte).

4. Subtract the average blank from each sample to obtain n measurements.

5. Compute the standard deviation (s) of the n measurements.

6. To find the detection limit, multiply s by the value of Student's t in Table 4-2 corresponding to $n - 1$ degrees of freedom and 98% (not 99%) confidence:

$$\text{detection limit} = t \cdot s$$

(using Student's t for 98% confidence and $n - 1$ degrees of freedom).

The accompanying diagram shows that there is a 1% chance that the mean value is below this detection limit. Eight measurements of the detection limit of a compound by a chromatographic method gave values of 4.7, 5.4, 6.2, 6.0, 4.6, 5.6, 5.2, and 5.8 ppm. Eight blanks gave values of 0.6, 1.2, 2.2, 0.5, 1.6, 1.8, 1.7, and 1.1 ppm. Find the detection limit.

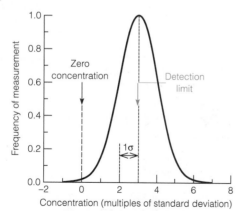

Detection limit: 1% of the area of the Gaussian curve lies to the left of zero concentration. We use the value of Student's t for 98% confidence, because 1% of the area lies to the left of 0 and 1% lies to the right of 6 on this graph.

Control Charts and Q Test

20. How is a control chart used? What are three indications that a process is going out of control?

21. Using the Q test, decide whether the value 216 should be rejected from the set of results 192, 216, 202, 195, and 204.

Finding the "Best" Straight Line

22. Suppose that you carry out an analytical procedure to generate a calibration curve such as that in Figure 4-8. Then you analyze an unknown and find an absorbance that gives a negative concentration for the analyte (the species being analyzed). What does this mean?

23. Set up a spreadsheet as in Figure 4-12 to solve linear least-squares problems, including calculating the standard deviations σ_y, σ_m, and σ_b. Use your spreadsheet to reproduce the results for the points in Table 4-7.

24. Use a graphics program to plot the results of the previous problem. Display vertical error bars ($\pm\sigma_y$) at each point. Have the program draw the least-squares straight line through the data. (*Note:* Most graphics packages can draw the least-squares line automatically. Use this feature routinely to examine the quality of your data.)

25. A straight line is drawn through the points (3.0, -3.87×10^4), (10.0, -12.99×10^4), (20.0, -25.93×10^4), (30.0, -38.89×10^4), and (40.0, -51.96×10^4), using the method of least squares. The results are $m = -1.29872 \times 10^4$, $b = 256.695$, $\sigma_m = 13.190$, $\sigma_b = 323.57$, and $\sigma_y = 392.9$. Express the slope and intercept and their uncertainties with the correct significant figures.

26. Consider the case of $y = -40.00 \times 10^4$ in the context of Problem 25. Calculate the value of x, along with its one-standard-deviation uncertainty and 95% confidence interval.

27. Consider the least-squares problem illustrated in Figure 4-9. Suppose that a single new measurement produces a y value of 2.58.

(a) Use the method of Equation 4-23 to calculate the corresponding x value and its uncertainty.

(b) Use the method of Equation 4-24 to calculate the uncertainty in x.

(c) Suppose you measure y four times and the average is 2.58. Use Equation 4-24 to calculate the uncertainty based on four measurements, not one.

28. Consider a suspension of living cells with a concentration $C_t = 4.13 (\pm 0.09) \times 10^{-13}$ mol of cells per liter. Each cell has n binding sites for the hormone H. The dissociation constant, K, is the equilibrium constant for the reaction

$$CH \overset{K}{\rightleftharpoons} C + H$$

where C is a cell, H is the hormone, and CH is hormone attached to the cell. To determine the number of binding sites per cell, we vary the concentration of cells and measure the concentrations of free hormone, H_f, and bound hormone, H_b. A graph of H_b/H_f versus H_b, called a *Scatchard plot,* is then drawn. It can be shown that

$$\frac{H_b}{H_f} = -\left(\frac{1}{K}\right)H_b + \frac{nC_t}{K}$$

That is, a graph of H_b/H_f versus H_b should give a straight line with a slope of $-1/K$ and an intercept of nC_t/K. For the data below, find n and its uncertainty, with a reasonable number of significant figures. Do not assign significant figures until the *end* of the calculation.

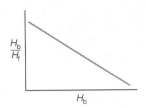

intercept: $b = 3.1386\ (\pm0.2899) \times 10^{-3}$
slope: $m = -1.3645\ (\pm0.1328) \times 10^{6}\ \text{M}^{-1}$

29. *Low-temperature limits in the lab.* The lowest temperature achieved for systems of nuclear spins are listed below.

Year (= x)	Temperature (K)	Log [temperature] (= y)
1979	5×10^{-8}	−7.3
1983	2×10^{-8}	−7.7
1989	2×10^{-9}	−8.7
1990	8×10^{-10}	−9.1

(a) Make a graph of y (= log [temperature]) versus x (= year) and find the slope and intercept and their standard deviations.

(b) Use the slope and intercept to extrapolate the line to predict the temperature expected in the year 2010. (Do not consider uncertainties.) Is there any reason to consider this a valid prediction? (*Extrapolation* means to go beyond the measured data. *Interpolation* means to go between measured data.)

Notes and References

1. An excellent source of additional reading on statistics is J. C. Miller and J. N. Miller, *Statistics for Analytical Chemistry* (New York: Wiley, 1984).

2. L. H. Keith, W. Crummett, J. Deegan, Jr., R. A. Libby, J. K. Taylor, and G. Wentler, *Anal. Chem.* **1983,** *55,* 2210. The murky interface between analytical chemistry and the law is discussed by W. E. Harris, *Anal. Chem.* **1992,** *64,* 665A; and J. G. Grasselli, *Anal. Chem.* **1992,** *64,* 677A.

3. An outstanding, readable article showing how to apply a simple least-squares procedure to many complicated nonlinear problems is by R. de Levie, *J. Chem. Ed.* **1986,** *63,* 10. The case in which both the x and y coordinates have substantial uncertainty is treated by P. J. Ogren and J. R. Norton, *J. Chem. Ed.* **1992,** *69,* A131; D. York, *Can. J. Phys.* **1966,** *44,* 1079; and J. A. Irvin and T. I. Quickenden, *J. Chem. Ed.* **1983,** *60,* 711. A general approach to nonlinear least-squares curve fitting is given by W. E. Wentworth, *J. Chem. Ed.* **1965,** *42,* 96,

162. The statistical significance of least-square parameters is discussed by M. D. Pattengill and D. E. Sands, *J. Chem. Ed.* **1979,** *56,* 244; and E. Heilbronner, *J. Chem. Ed.* **1979,** *56,* 240.

4. H. Freiser, *Concepts and Calculations in Analytical Chemistry: A Spreadsheet Approach* (Boca Raton, FL: CRC Press, 1992); R. de Levie, *A Spreadsheet Workbook for Quantitative Chemical Analysis* (New York: McGraw-Hill, 1992).

5. M. Bradford, *Anal. Biochem.* **1976,** *72,* 248.

6. J. W. Gramlich and L. A. Machlan, *Anal. Chem.* **1985,** *57,* 1788.

7. D. T. Harvey, *J. Chem. Ed.* **1991,** *68,* 329. This is a good first experiment for statistics and acid-base chemistry.

8. C. L. Grant, A. D. Hewitt, and T. F. Jenkins, *Am. Lab.,* February 1991, p. 15.

Chemical Equilibrium in the Environment

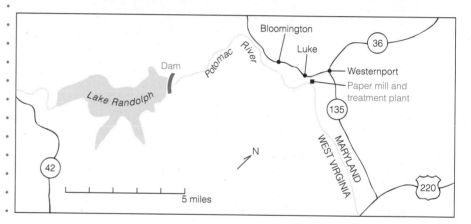

Paper mill on the Potomac River near Westernport, Maryland, neutralizes acid mine drainage in the water. Upstream of the mill, the river is acidic and lifeless; below the mill, the river teems with life. [Photo courtesy C. Dalpra, Potomac River Basin Commission.]

Part of the North Branch of the Potomac River runs crystal clear through the scenic Appalachian Mountains, but it is lifeless—a victim of acid drainage from abandoned coal mines. As the river passes a paper mill and wastewater treatment plant near Westernport, Maryland, the pH rises from an acidic, lethal value of 4.5 to a neutral value of 7.2, at which fish and plants thrive. This happy "accident" comes about because calcium carbonate exiting the paper mill equilibrates with massive quantities of carbon dioxide from bacterial respiration at the sewage treatment plant. The resulting soluble bicarbonate neutralizes the acidic river and restores life downstream of the plant.[1]

$$CaCO_3(s) + CO_2(aq) + H_2O(l) \rightleftharpoons Ca^{2+}(aq) + 2HCO_3^-(aq)$$

Calcium carbonate   Dissolved calcium bicarbonate

$$HCO_3^-(aq) + H^+(aq) \xrightarrow{\text{neutralization}} CO_2(g) + H_2O(l)$$

Chemical Equilibrium

5

Equilibria govern many diverse phenomena such as the unfolding of protein chains in human cells, the action of acid rain on minerals, and simple aqueous reactions used in analytical chemistry. In this chapter we study equilibria involving the solubility of ionic compounds, complex formation, and acid-base reactions. Chemical equilibrium provides a foundation not only for chemical analysis but also for other areas of science, such as biochemistry, geology, and oceanography.

5-1 The Equilibrium Constant

For the reaction

$$a\text{A} + b\text{B} \rightleftharpoons c\text{C} + d\text{D} \tag{5-1}$$

we write the **equilibrium constant**, K, in the form

Equilibrium constant:
$$K = \frac{[\text{C}]^c[\text{D}]^d}{[\text{A}]^a[\text{B}]^b} \tag{5-2}$$

The equilibrium constant is more correctly expressed as a ratio of *activities* rather than of concentrations. We reserve a discussion of activity for Chapter 8.

where the superscript letters denote stoichiometric coefficients and each capital letter stands for a chemical species. The symbol [A] stands for the concentration of A relative to its standard state (defined below). By definition, *a reaction is favored whenever $K > 1$.*

In the thermodynamic derivation of the equilibrium constant, each quantity in Equation 5-2 is expressed as the *ratio* of the concentration of each species to its concentration in its **standard state.** For solutes, the standard state is 1 M. For gases, the standard state is 1 atm; and for solids and liquids, the standard states are the pure solid or liquid. It is understood (but rarely

Equilibrium constants are dimensionless.

written) that the term [A] in Equation 5-2 really means $[A]/(1\ M)$ if A is a solute. If D is a gas, [D] really means (pressure of D in atmospheres)$/(1\ atm)$. To emphasize that [D] means pressure of D, we usually write P_D in place of [D]. The terms of Equation 5-2 are actually dimensionless; therefore all equilibrium constants are dimensionless.

For the ratios $[A]/(1\ M)$ and $[D]/(1\ atm)$ to be dimensionless, [A] *must* be expressed in moles per liter (M), and [D] *must* be expressed in atmospheres. If C were a pure liquid or solid, the ratio [C]/(concentration of C in its standard state) would be unity (1) because the standard state is the pure liquid or solid. If [C] is a solvent, the concentration is so close to that of pure liquid C that the value of [C] is still essentially 1.

The take-home lesson is this: In evaluating an equilibrium constant,

1. The concentration of solutes should be expressed as moles per liter.

2. The concentrations of gases should be expressed in atmospheres.

3. The concentrations of pure solids, pure liquids, and solvents are omitted because they are unity.

Equilibrium constants are dimensionless, but in specifying concentrations you must use units of molarity (M) for solutes and atmospheres (atm) for gases.

These conventions are arbitrary, but you must use them if you wish to use tabulated values of equilibrium constants, standard reduction potentials, and free energies.

Manipulating Equilibrium Constants

Consider the reaction

Throughout this text, you should assume that all species in chemical equations are in aqueous solution, unless otherwise specified.

$$HA \rightleftharpoons H^+ + A^- \qquad K_1 = \frac{[H^+][A^-]}{[HA]}$$

If the direction of a reaction is reversed, the new value of K is simply the reciprocal of the original value of K.

Equilibrium constant for reverse reaction:
$$H^+ + A^- \rightleftharpoons HA \qquad K_1' = \frac{[HA]}{[H^+][A^-]} = 1/K_1$$

If two reactions are added, the new value of K is the product of the two individual values:

$$
\begin{array}{ll}
HA \rightleftharpoons \cancel{H^+} + A^- & K_1 \\
\cancel{H^+} + C \rightleftharpoons CH^+ & K_2 \\
\hline
HA + C \rightleftharpoons CH^+ + A^- & K_3
\end{array}
$$

If a reaction is reversed, then $K' = 1/K$. If two reactions are added, then $K_3 = K_1K_2$.

Equilibrium constant for sum of reactions:
$$K_3 = K_1 K_2 = \frac{[\cancel{H^+}][A^-]}{[HA]} \cdot \frac{[CH^+]}{[\cancel{H^+}][C]} = \frac{[A^-][CH^+]}{[HA][C]}$$

If n reactions are added, the overall equilibrium constant is the product of all n individual equilibrium constants.

EXAMPLE   Combining Equilibrium Constants

The equilibrium constant for the reaction

$$H_2O \rightleftharpoons H^+ + OH^-$$

is called K_w ($= [H^+][OH^-]$) and has the value 1.0×10^{-14} at 25°C. Given that

$$NH_3(aq) + H_2O \rightleftharpoons NH_4^+ + OH^- \qquad K_{NH_3} = 1.8 \times 10^{-5}$$

find the equilibrium constant for the reaction

$$NH_4^+ \rightleftharpoons NH_3(aq) + H^+$$

Solution The third reaction can be obtained by reversing the second reaction and adding it to the first reaction:

$$
\begin{array}{ll}
\cancel{H_2O} \rightleftharpoons H^+ + \cancel{OH^-} & K_1 = K_w \\
NH_4^+ + \cancel{OH^-} \rightleftharpoons NH_3(aq) + \cancel{H_2O} & K_2 = 1/K_{NH_3} \\
\hline
NH_4^+ \rightleftharpoons H^+ + NH_3(aq) & K_3 = K_w \cdot \dfrac{1}{K_{NH_3}} = 5.6 \times 10^{-10}
\end{array}
$$

5-2 Equilibrium and Thermodynamics

The equilibrium constant is directly related to the thermodynamics of a chemical reaction. The heat released by the reaction (*enthalpy*) and the degree of disorder of reactants and products (*entropy*) independently contribute to the degree to which the reaction is favored or disfavored.

Enthalpy

The **enthalpy change,** ΔH, for a reaction is the heat absorbed when the reaction takes place under constant applied pressure. The *standard enthalpy change,* $\Delta H°$, refers to the heat absorbed when all reactants and products are in their standard states.† In this reaction, $\Delta H° = -75.15$ kJ/mol at 25°C:

$$HCl(g) \rightleftharpoons H^+(aq) + Cl^-(aq) \tag{5-3}$$

$\Delta H = (+)$
Heat is absorbed
Endothermic

$\Delta H = (-)$
Heat is liberated
Exothermic

By convention, the negative sign of $\Delta H°$ indicates that heat is given off by the products when Reaction 5-3 occurs; consequently, the solution becomes warmer when the reaction occurs. For other reactions, ΔH is positive, which means that heat is absorbed by reactants as they react. Thus, the solution gets colder during the reaction. A reaction for which ΔH is positive is said to be **endothermic.** Whenever ΔH is negative, the reaction is **exothermic.**

Entropy

The **entropy,** S, of a substance is a measure of its "disorder," which we will not attempt to define in a quantitative way. The greater the disorder, the greater the entropy. In general, a gas is more disordered (has higher entropy) than a liquid, which is more disordered than a solid. Ions in aqueous solution are normally more disordered than in their solid salt. For example, for Reaction 5-4, we find $\Delta S° = +76$ J/(K·mol) at 25°C.

$\Delta S = (+)$
Products more disordered than reactants

$\Delta S = (-)$
Products less disordered than reactants

$$KCl(s) \rightleftharpoons K^+(aq) + Cl^-(aq) \tag{5-4}$$

†The precise definition of the standard state contains subtleties beyond the scope of this text. For Reaction 5-3, the standard state of H^+ or Cl^- is the hypothetical state in which each ion is present at a concentration of 1 M but behaves as if it were in an infinitely dilute solution. That is, the standard concentration is 1 M, but the standard physical behavior is what would be observed in a very dilute solution in which each ion is unaffected by surrounding ions.

This value indicates that a mole of $K^+(aq)$ plus a mole of $Cl^-(aq)$ is more disordered than a mole of $KCl(s)$ separated from the solvent water. For Reaction 5-3, $\Delta S° = -131.5$ J/(K·mol) at 25°C. The aqueous ions are less disordered than gaseous HCl separated from the solvent water.

Free Energy

Systems at constant temperature and pressure, which are common laboratory conditions, have a tendency toward lower enthalpy and increased entropy. A chemical reaction is driven toward the formation of products by a *negative* value of ΔH (heat given off), or a *positive* value of ΔS (entropy increases), or both. When ΔH is negative and ΔS is positive, the reaction is clearly favored. When ΔH is positive and ΔS is negative, the reaction is clearly disfavored.

When ΔH and ΔS are both positive or both negative, what decides whether a reaction will be favored? The change in **Gibbs free energy**, ΔG, is the arbiter between opposing tendencies of ΔH and ΔS. At constant temperature,

Free energy:
$$\Delta G = \Delta H - T\Delta S \tag{5-5}$$

The effects of entropy and enthalpy are combined in Equation 5-5. A *reaction is favored if ΔG is negative.*

For the dissociation of HCl (Reaction 5-3), $\Delta H°$ favors the reaction and $\Delta S°$ disfavors it. To find the net result, we must evaluate $\Delta G°$:

$$\begin{aligned}
\Delta G° &= \Delta H° - T\Delta S° \\
&= (-75.15 \times 10^3 \text{ J}) - (298.15 \text{ K})(-131.5 \text{ J/K}) \\
&= -35.94 \text{ kJ/mol}
\end{aligned}$$

Note that 25.00°C = 298.15 K.

$\Delta G°$ is negative, and the reaction is favored under standard conditions. The favorable enthalpy change is greater than the unfavorable entropy change in this case.

The relation between $\Delta G°$ and the equilibrium constant for a reaction is

Free energy and equilibrium:
$$K = e^{-\Delta G°/RT} \tag{5-6}$$

where R is the gas constant [= 8.314510 J/(K·mol)] and T is temperature in kelvins. For Reaction 5-3, we find

$$\begin{aligned}
K &= e^{-(-35.94 \times 10^3 \text{ J/mol})/[8.314510 \text{ J/(K·mol)}](298.15 \text{ K})} \\
&= 1.98 \times 10^6
\end{aligned}$$

$\Delta G = (+)$
Reaction is disfavored

$\Delta G = (-)$
Reaction is favored

Because the equilibrium constant is large, $HCl(g)$ is very soluble in water and is nearly completely ionized to H^+ and Cl^- when it dissolves.

To summarize, ΔG takes into account the tendency for heat to be liberated (ΔH negative) and disorder to increase (ΔS positive). A reaction is favored ($K > 1$) if $\Delta G°$ is negative. A reaction is disfavored ($K < 1$) if $\Delta G°$ is positive.

Le Châtelier's Principle

Suppose that a system at equilibrium is subjected to a change that disturbs the system. **Le Châtelier's principle** states that the direction in which the system proceeds back to equilibrium is such that the change is partially offset.[2]

To see what this prediction means, let's see what happens when we attempt to change the concentration of one species in the reaction:

$$BrO_3^- + 2Cr^{3+} + 4H_2O \rightleftharpoons Br^- + Cr_2O_7^{2-} + 8H^+ \qquad (5\text{-}7)$$

for which the equilibrium constant is given by

$$K = \frac{[Br^-][Cr_2O_7^{2-}][H^+]^8}{[BrO_3^-][Cr^{3+}]^2} = 1 \times 10^{11} \quad \text{at } 25°C$$

Notice that H_2O is omitted from K, because it is the solvent.

In one particular equilibrium state of this system, the following concentrations exist:

$$[H^+] = 5.0 \text{ M} \qquad [Cr_2O_7^{2-}] = 0.10 \text{ M} \qquad [Cr^{3+}] = 0.003\,0 \text{ M}$$

$$[Br^-] = 1.0 \text{ M} \qquad [BrO_3^-] = 0.043 \text{ M}$$

Suppose that the equilibrium is disturbed by adding dichromate to the solution to increase the concentration of $[Cr_2O_7^{2-}]$ from 0.10 to 0.20 M. In what direction will the reaction proceed to reach equilibrium?

According to the principle of Le Châtelier, the reaction should go back to the left to partially offset the increase in dichromate, which appears on the right side of Reaction 5-7. We can verify this algebraically by setting up the **reaction quotient,** Q, which has the same form as the equilibrium constant. The only difference is that Q is evaluated with whatever concentrations happen to exist, even though the solution is not at equilibrium. When the system reaches equilibrium, $Q = K$. For Reaction 5-7,

The reaction quotient has the same form as the equilibrium constant, but the concentrations are generally not the equilibrium concentrations.

$$Q = \frac{(1.0)(0.20)(5.0)^8}{(0.043)(0.003\,0)^2} = 2 \times 10^{11} > K$$

Because $Q > K$, the reaction must go to the left to decrease the numerator and increase the denominator, until $Q = K$.

In general,

If $Q < K$, then the reaction must proceed to the right to reach equilibrium. If $Q > K$, then the reaction must proceed to the left to reach equilibrium.

1. If a reaction is at equilibrium and products are added (or reactants are removed), the reaction goes to the left.

2. If a reaction is at equilibrium and reactants are added (or products are removed), the reaction goes to the right.

When the temperature of a system is changed, so is the equilibrium constant. Equations 5-5 and 5-6 can be combined to predict the effect of temperature on K:

$$K = e^{-\Delta G°/RT} = e^{-(\Delta H° - T\Delta S°)/RT}$$
$$= e^{(-\Delta H°/RT + \Delta S°/R)} = e^{-\Delta H°/RT} \cdot e^{\Delta S°/R} \qquad (5\text{-}8)$$

The term $e^{\Delta S°/R}$ is independent of T. The term $e^{-\Delta H°/RT}$ increases with increasing temperature if $\Delta H°$ is positive. The term $e^{-\Delta H°/RT}$ decreases with increasing temperature if $\Delta H°$ is negative. Equation 5-8 therefore tells us that

1. The equilibrium constant of an endothermic reaction increases if the temperature is raised.

2. The equilibrium constant of an exothermic reaction decreases if the temperature is raised.

Heat can be treated as if it were a reactant in an endothermic reaction and a product in an exothermic reaction.

These statements can be understood in terms of Le Châtelier's principle as follows. Consider an endothermic reaction:

$$\text{heat} + \text{reactants} \rightleftharpoons \text{products}$$

If the temperature is raised, then heat is added to the system. The reaction proceeds to the right to partially offset this change.[3]

In dealing with equilibrium problems, we are making *thermodynamic* predictions, not *kinetic* predictions. We are calculating what must happen for a system to reach equilibrium, but not how long it will take. Some reactions are over in an instant; others will not reach equilibrium in a million years. For example, a stick of dynamite remains unchanged indefinitely, until a spark sets off the spontaneous, explosive decomposition. The size of an equilibrium constant tells us nothing about the rate of reaction. A large equilibrium constant does not imply that a reaction is fast.

5-3 Solubility Product

Why should anyone care about solubility? Our immediate payoff will be to understand the basis for gravimetric analysis, precipitation titrations, and electrochemical reference cells. The fish below Westernport in the Potomac River, in the discussion at the beginning of this chapter, are alive only because CO_2 increases the solubility of $CaCO_3(s)$, thereby bringing bicarbonate ion into the river where it can neutralize acid mine drainage.

The equilibrium constant omits the pure solid because its concentration is unchanged.

The **solubility product** is the equilibrium constant for the reaction in which a solid salt dissolves, to give its constituent ions in solution. The concentration of the solid is omitted from the equilibrium constant because the solid is in its standard state. A table of solubility products can be found in Appendix F.

As a typical example, consider the dissolution of mercurous chloride (Hg_2Cl_2) in water. The reaction is

$$Hg_2Cl_2(s) \rightleftharpoons Hg_2^{2+} + 2Cl^- \tag{5-9}$$

for which the solubility product, K_{sp}, is

$$K_{sp} \rightleftharpoons [Hg_2^{2+}][Cl^-]^2 = 1.2 \times 10^{-18}$$

Mercurous ion, Hg_2^{2+}, is a *dimer* (pronounced DIE·mer), which means that it consists of two identical units bound together:

250 pm	[Hg—Hg]$^{2+}$
	+1 oxidation state of mercury

Anions such as OH^-, S^{2-}, and CN^- stabilize Hg(II), thereby causing Hg(I) to disproportionate. **Disproportionation** is a process in which an element in an intermediate oxidation state gives products in higher and lower oxidation states:

$$Hg_2^{2+} + 2CN^- \rightarrow$$
$$Hg(CN)_2(aq) + Hg(l)$$

Although Hg_2^{2+} is, by far, the most common form of Hg(I), complexes of monatomic Hg^+ ion have been characterized.[4]

A solution containing excess, undissolved solid is said to be **saturated** with that solid. The solution contains all the solid capable of being dissolved under the prevailing conditions. What will be the concentration of Hg_2^{2+} in a solution saturated with Hg_2Cl_2?

In Reaction 5-9 we see that two Cl^- ions are produced for each Hg_2^{2+} ion. If the concentration of Hg_2^{2+} is x M, the concentration of dissolved Cl^- must be $2x$ M. We can display this relationship neatly in a concentration table:

	$Hg_2Cl_2(s) \rightleftharpoons Hg_2^{2+} + 2Cl^-$		
Initial concentration:	solid	0	0
Final concentration:	solid	x	$2x$

Putting these values of concentration into the solubility product gives

To find the cube root of a number with a calculator, raise the number to the 0.333 333 33· · · power.

$$[Hg_2^{2+}][Cl^-]^2 = (x)(2x)^2 = 1.2 \times 10^{-18}$$
$$4x^3 = 1.2 \times 10^{-18}$$
$$x = 6.7 \times 10^{-7} \text{ M}$$

The concentration of Hg_2^{2+} ion is calculated to be 6.7×10^{-7} M, and the concentration of Cl^- is $(2)(6.7 \times 10^{-7}) = 13.4 \times 10^{-7}$ M

The physical meaning of the solubility product is this: If an aqueous solution is left in contact with excess solid Hg_2Cl_2, the solid will dissolve until the condition $[Hg_2^{2+}][Cl^-]^2 = K_{sp}$ is satisfied. Thereafter, the amount of undissolved solid remains constant. Unless excess solid remains, there is no guarantee that $[Hg_2^{2+}][Cl^-]^2 = K_{sp}$. If Hg_2^{2+} and Cl^- are mixed together (with appropriate counterions) such that the product $[Hg_2^{2+}][Cl^-]^2$ exceeds K_{sp}, then Hg_2Cl_2 will precipitate.

The solubility product does not tell the entire story on the solubility of slightly soluble salts. The concentration of *undissociated* species may be as great as that of dissociated ions. That is, the salt MX(*s*) can give MX(*aq*) as well as $M^+(aq)$ and $X^-(aq)$. In the case of $CaSO_4$, for example, the total concentration of dissolved material is 15–19 mM.[5] Using activity coefficients as described in Problem 9-16, we estimate that the concentration of dissociated Ca^{2+} should be 9 mM. This value suggests that about half of the dissolved $CaSO_4$ is undissociated.

5-4 The Common Ion Effect

Now we vary the problem a little by adding a second source of Cl^-. What will be the concentration of Hg_2^{2+} in a solution containing 0.030 M NaCl saturated with Hg_2Cl_2? The concentration table now looks like this:

$$Hg_2Cl_2(s) \rightleftharpoons Hg_2^{2+} + 2Cl^-$$

	$Hg_2Cl_2(s)$	Hg_2^{2+}	$2Cl^-$
Initial concentration:	solid	0	0.030
Final concentration:	solid	x	$2x + 0.030$

The initial concentration of Cl^- is due to the dissolved NaCl (which is completely dissociated into Na^+ and Cl^-). The final concentration of Cl^- is due to contributions from NaCl and Hg_2Cl_2.

The proper solubility equation is

$$[Hg_2^{2+}][Cl^-]^2 = (x)(2x + 0.030)^2 = K_{sp} \qquad (5\text{-}10)$$

But think about the size of x. In the previous example, $x = 6.7 \times 10^{-7}$ M, which is pretty small compared with 0.030 M. In the present example, we anticipate that x will be even smaller than 6.7×10^{-7}, by Le Châtelier's principle. Addition of a product (Cl^- in this case) to Reaction 5-9 displaces the reaction toward the left. In the presence of a second source of Cl^-, there will be less dissolved Hg_2^{2+} than in its absence.

This application of Le Châtelier's principle is called the **common ion effect.** *A salt will be less soluble if one of its constituent ions is already present in the solution.*

Getting back to Equation 5-10, we expect that $2x \ll 0.030$. As an approximation, we will ignore $2x$ in comparison to 0.030. The equation simplifies to

$$(x)(0.030)^2 = K_{sp} = 1.2 \times 10^{-18}$$
$$x = 1.3 \times 10^{-15}$$

Because $2x = 2.6 \times 10^{-15}$ is much smaller than 0.030, it was clearly justified to neglect $2x$ to solve the problem. The answer also illustrates the common

Common ion effect: A salt is less soluble if one of its ions is already present in the solution.

It is important to confirm at the end of the calculation that the approximation $2x \ll 0.030$ is valid. Box 5-1 discusses approximations.

Demonstration 5-1 Common Ion Effect[6]

Fill two large test tubes about one-third full with saturated aqueous KCl containing no excess solid. The solubility of KCl is approximately 3.7 M, so the solubility product (ignoring activity effects introduced later) is

$$K_{sp} \approx [K^+][Cl^-] = (3.7)(3.7) = 13.7$$

Now add an equal volume of 6 M HCl to one test tube and 12 M HCl to the other. Even though a common ion, Cl^-, is added in each case, KCl precipitates only in one tube.

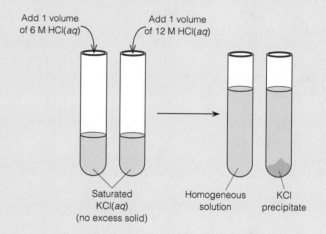

Add 1 volume of 6 M HCl(aq) Add 1 volume of 12 M HCl(aq)

Saturated KCl(aq) (no excess solid)   Homogeneous solution   KCl precipitate

To understand your observations, calculate the concentrations of K^+ and Cl^- in each tube after HCl addition. Then evaluate the reaction quotient, $Q = [K^+][Cl^-]$, for each tube. Explain your observations.

ion effect. In the absence of Cl^-, the solubility of Hg_2^{2+} was 6.7×10^{-7} M. In the presence of 0.030 M Cl^-, the solubility of Hg_2^{2+} is reduced to 1.3×10^{-15} M.

What is the maximum Cl^- concentration at equilibrium in a solution in which the concentration of Hg_2^{2+} is somehow *fixed* at 1.0×10^{-9} M? Our concentration table looks like this now:

	$Hg_2Cl_2(s) \rightleftharpoons$	Hg_2^{2+}	$+ 2Cl^-$
Initial concentration:	0	1.0×10^{-9}	0
Final concentration:	solid	1.0×10^{-9}	x

$[Hg_2^{2+}]$ is not x in this example, so there is no reason to set $[Cl^-] = 2x$. The problem is solved by plugging each concentration into the solubility product:

$$[Hg_2^{2+}][Cl^-]^2 = K_{sp}$$
$$(1.0 \times 10^{-9})(x)^2 = 1.2 \times 10^{-18}$$
$$x = [Cl^-] = 3.5 \times 10^{-5} \text{ M}$$

Box 5-1   The Logic of Approximations

Many real problems in chemistry (and other sciences) are difficult to solve without judicious numerical approximations. For example, rather than solving the equation

$$(x)(2x + 0.030)^2 = 1.2 \times 10^{-18}$$

we used the approximation that $2x \ll 0.030$, and therefore solved the much simpler equation

$$(x)(0.030)^2 = 1.2 \times 10^{-18}$$

But how can we be sure that our solution fits the original problem?

Whenever you use an approximation, you are assuming that the approximation is true. *If the assumption is true, then it will not create a contradiction. If the assumption is false, then it will lead to a contradiction.* You can actually test an assumption by using it and seeing whether you are right or wrong when you are done.

You may object to this reasoning, feeling "How can the truth of an assumption be tested by using the assumption?" Suppose you wish to test the statement "Gail can swim 100 meters." To see if the statement is true, you can assume that it is true. If Gail can swim 100 meters, then you could dump her in the middle of a lake with a radius of 100 meters and expect her to swim to shore. If she comes ashore alive, then your assumption was correct and no contradiction is created. If she does not make it to shore, then there is a contradiction. Your assumption must have been wrong. There are only two pos-

Gail's lake

sibilities: Either the assumption is correct and using it is correct, or the assumption is wrong and using it is wrong. (A third possibility in this case is that there are sharks in the lake.)

Example 1. $(x)(3x + 0.01)^3 = 10^{-12}$
$(x)(0.01)^3 = 10^{-12}$ (assuming $3x \ll 0.01$)
$x = 10^{-12}/(0.01)^3 = 10^{-6}$
No contradiction: $3x = 3 \times 10^{-6} \ll 0.01$.
The assumption is true.

Example 2. $(x)(3x + 0.01)^3 = 10^{-8}$
$(x)(0.01)^3 = 10^{-8}$ (assuming $3x \ll 0.01$)
$x = 10^{-8}/(0.01)^3 = 0.01$
A contradiction: $3x = 0.03 > 0.01$.
The assumption is false.

In Example 2 the assumption leads to a contradiction, so the assumption cannot be correct. When this happens, you must solve the quartic equation $x(3x + 0.01)^3 = 10^{-8}$.

You can try to solve the quartic equation exactly, but approximate methods are usually easier. For the equation $x(3x + 0.01)^3 = 10^{-8}$, one quick procedure is trial-and-error guessing. The first guess, $x = 0.01$, comes from the (incorrect) assumption that $3x \ll 0.01$ made in Example 2.

Guess	$x(3x + 0.01)^3$
$x = 0.01$	6.4×10^{-7}
$x = 0.005$	7.8×10^{-8}
$x = 0.002$	8.2×10^{-9}
$x = 0.0024$	1.22×10^{-8}
$x = 0.0022$	1.006×10^{-8}
$x = 0.00218$	9.86×10^{-9}
$x = 0.00219$	9.96×10^{-9} ← Best value of x with 3 figures

(continued)

(continued)

Example 3. $(x)(3x + 0.01)^3 = 10^{-9}$
$(x)(0.01)^3 = 10^{-9}$ (assuming $3x \ll 0.01$)
$x = 10^{-9}/(0.01)^3 = 10^{-3}$

In this case $3x = 0.003$. This is less than 0.01, but not a great deal less. Whether or not the assumption is adequate depends on your purposes. If you need an answer accurate to within a factor of 2, the approximation is acceptable. If you need an answer accurate to 1%, the approximation is not adequate. The correct answer is 6.059×10^{-4}.

Most spreadsheets allow you to use a *circular definition* to solve Example 3, which can be rearranged to the form $x = 10^{-9}/(3x + 0.01)^3$. In a blank spreadsheet, enter the formula = 1E−9/(3*A1+0.01) ^ 3 in cell A1. The spreadsheet may query you before dealing with your unorthodox request, but it will probably proceed to change the value in cell A1 several times until reaching 6.059×10^{-4}, which satisfies the equality.

5-5 Separation by Precipitation

A useful application of the solubility product is the separation of one substance from another by precipitating one from solution.[7] As an example, consider a solution containing plumbous (Pb^{2+}) and mercurous (Hg_2^{2+}) ions, each at a concentration of 0.010 M. Each forms an insoluble iodide, but the mercurous iodide is considerably less soluble, as indicated by the smaller value of K_{sp}.

$$PbI_2(s) \rightleftharpoons Pb^{2+} + 2I^- \qquad K_{sp} = 7.9 \times 10^{-9}$$

$$Hg_2I_2(s) \rightleftharpoons Hg_2^{2+} + 2I^- \qquad K_{sp} = 1.1 \times 10^{-28}$$

Is it possible to *completely* separate the Pb^{2+} and Hg_2^{2+} by selectively precipitating the latter with iodide?

To answer this, we must first define complete separation. "Complete" can mean anything we choose. Let us ask whether 99.99% of the Hg_2^{2+} can be precipitated without causing Pb^{2+} to precipitate. That is, we seek to reduce the Hg_2^{2+} concentration to 0.01% of its original value without precipitating Pb^{2+}. The concentration of Hg_2^{2+} in the solution after precipitation will be 0.01% of 0.010 M, or 1.0×10^{-6} M. The minimum concentration of I^- needed to effect this precipitation is calculated as follows:

Complete separation means whatever you want it to. You must define what *you* mean by complete.

	$Hg_2I_2(s) \rightleftharpoons$	Hg_2^{2+}	$+ 2I^-$
Initial concentration:	0	0.010	0
Final concentration:	solid	1.0×10^{-6}	x

$$\underbrace{\phantom{1.0 \times 10^{-6}}}_{0.01\% \text{ of } 0.010 \text{ M}}$$

$$[Hg_2^{2+}][I^-]^2 = K_{sp}$$
$$(1.0 \times 10^{-6})(x)^2 = 1.1 \times 10^{-28}$$
$$x = [I^-] = 1.0 \times 10^{-11} \text{ M}$$

The quantity x is the concentration of I^- needed to reduce $[Hg_2^{2+}]$ to 10^{-6} M.

Will this concentration of I^- cause 0.010 M Pb^{2+} to precipitate? We can answer this by seeing whether the solubility product of PbI_2 is exceeded.

$$Q = [Pb^{2+}][I^-]^2 = (0.010)(1.0 \times 10^{-11})^2$$
$$= 1.0 \times 10^{-24} < K_{sp} \quad \text{(for } PbI_2)$$

The reaction quotient, Q, is 1.0×10^{-24}, which is less than K_{sp} (= 7.9×10^{-9}) for PbI_2. Therefore, the Pb^{2+} will not precipitate, and the "complete" separation of Pb^{2+} and Hg_2^{2+} is feasible. We predict that adding I^- to a solution of Pb^{2+} and Hg_2^{2+} will precipitate virtually all of the mercurous ion before any plumbous ion precipitates.

Life should be so easy! What we have just done is to make a (valid) thermodynamic prediction. If the system comes to equilibrium, we can achieve the desired separation. However, occasionally one substance *coprecipitates* with the other. For example, some Pb^{2+} might become attached to the surface of the Hg_2I_2 crystal, or might even occupy sites within the crystal. (In **coprecipitation,** a substance whose solubility is not exceeded precipitates along with another substance whose solubility is exceeded.) The foregoing calculation says that the separation is worth trying. *However, only an experiment can show whether the separation will actually work.*

Question If you want to know whether a small amount of Pb^{2+} coprecipitates with Hg_2I_2, should you measure the Pb^{2+} concentration in the mother liquor (the solution) or the Pb^{2+} concentration in the precipitate? Which measurement is more sensitive? By "sensitive," we mean responsive to a small amount of coprecipitation.
(Answer: Measure Pb^{2+} in precipitate.)

5-6 Complex Formation

If anion X^- precipitates metal M^+, it is often observed that a high concentration of X^- causes solid MX to redissolve. We explain this by postulating the formation of **complex ions,** such as MX_2^-, which consist of two or more simple ions bonded to each other.

Lewis Acids and Bases

In complex ions such as PbI^+, PbI_3^-, and PbI_4^{2-}, iodide is said to be the **ligand** of Pb^{2+}. A ligand is any atom or group of atoms attached to the species of interest. We say that Pb^{2+} acts as a *Lewis acid* and I^- acts as a *Lewis base* in these complexes. A **Lewis acid** accepts a pair of electrons from a **Lewis base** when the two form a bond:

$$^{++}Pb\,\square + [\,:\ddot{\underset{..}{I}}:]^- \rightarrow [Pb-\ddot{\underset{..}{I}}:]^+$$

Room to accept electrons   A pair of electrons to be donated

Lewis acid + Lewis base $\rightleftharpoons$ adduct
Electron pair acceptor   Electron pair donor

The product of a reaction between a Lewis acid and a Lewis base is often called an *adduct*. The bond between a Lewis acid and a Lewis base is called a *dative* or *coordinate covalent* bond.

Effects of Complex Ion Formation on Solubility

If Pb^{2+} and I^- reacted only to form solid PbI_2, then the solubility of Pb^{2+} would always be very low in the presence of excess I^-:

$$PbI_2(s) \overset{K_{sp}}{\rightleftharpoons} Pb^{2+} + 2I^- \qquad K_{sp} = [Pb^{2+}][I^-]^2 = 7.9 \times 10^{-9} \quad (5\text{-}11)$$

The notation for these equilibrium constants is discussed in Box 5-2.

The experimental observation, however, is that high concentrations of I^- cause solid PbI_2 to dissolve. We explain this by the formation of a series of complex ions between Pb^{2+} and I^-:

$$Pb^{2+} + I^- \overset{K_1}{\rightleftharpoons} PbI^+ \qquad K_1 = [PbI^+]/[Pb^{2+}][I^-] = 1.0 \times 10^2 \qquad (5\text{-}12)$$

$$Pb^{2+} + 2I^- \overset{\beta_2}{\rightleftharpoons} PbI_2(aq) \quad \beta_2 = [PbI_2(aq)]/[Pb^{2+}][I^-]^2 = 1.4 \times 10^3 \quad (5\text{-}13)$$

$$Pb^{2+} + 3I^- \overset{\beta_3}{\rightleftharpoons} PbI_3^- \qquad \beta_3 = [PbI_3^-]/[Pb^{2+}][I^-]^3 = 8.3 \times 10^3 \qquad (5\text{-}14)$$

$$Pb^{2+} + 4I^- \overset{\beta_4}{\rightleftharpoons} PbI_4^{2-} \qquad \beta_4 = [PbI_4^{2-}]/[Pb^{2+}][I^-]^4 = 3.0 \times 10^4 \qquad (5\text{-}15)$$

The species $PbI_2(aq)$ in Reaction 5-13 is *dissolved* PbI_2, containing two iodine atoms bound to a lead atom. Reaction 5-13 is *not* the reverse of Reaction 5-11, which refers to solid PbI_2.

At low I^- concentrations, the solubility of lead is governed by the solubility of $PbI_2(s)$. However, at high I^- concentrations, Reactions 5-12 through 5-15 are driven to the right (Le Châtelier's principle), and the total concentration of dissolved lead is considerably greater than that of Pb^{2+} alone (Figure 5-1).

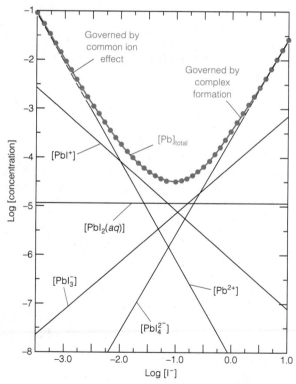

Figure 5-1   Simultaneous equilibria of complex ions. Concentrations of dissolved lead species are shown as a function of the concentration of free iodide. To the left of the minimum, $[Pb]_{total}$ is governed by the solubility product for $PbI_2(s)$. As $[I^-]$ is increased, $[Pb]_{total}$ decreases because of the common ion effect. At high values of $[I^-]$, $PbI_2(s)$ redissolves because of its reaction with I^- to form soluble complex ions, such as PbI_4^{2-}. Note logarithmic scales.

Box 5-2 Notation for Formation Constants

Formation constants are the equilibrium constants for complex ion formation. The **stepwise formation constants,** designated K_i, are defined as follows:

$$M + X \overset{K_1}{\underset{}{\rightleftharpoons}} MX \qquad K_1 = [MX]/[M][X]$$

$$MX + X \overset{K_2}{\underset{}{\rightleftharpoons}} MX_2 \qquad K_2 = [MX_2]/[MX][X]$$

$$MX_{n-1} + X \overset{K_n}{\underset{}{\rightleftharpoons}} MX_n \qquad K_n = [MX_n]/[MX_{n-1}][X]$$

The **overall,** or **cumulative, formation constants** are denoted β_i:

$$M + 2X \overset{\beta_2}{\underset{}{\rightleftharpoons}} MX_2 \qquad \beta_2 = [MX_2]/[M][X]^2$$

$$M + nX \overset{\beta_n}{\underset{}{\rightleftharpoons}} MX_n \qquad \beta_n = [MX_n]/[M][X]^n$$

A useful relationship is that $\beta_n = K_1 K_2 \cdots K_n$.

A most useful characteristic of chemical equilibrium is that *all equilibrium conditions are satisfied simultaneously* (if all reactions are fast enough to attain equilibrium). If we somehow know the concentration of I^-, we can calculate the concentration of Pb^{2+} by substituting this value into the equilibrium constant expression given with Reaction 5-11, regardless of whether there are 0, 4, or 1006 other reactions involving Pb^{2+}. *The concentration of Pb^{2+} that satisfies any one of the equilibria must satisfy all of the equilibria. There can be only one concentration of Pb^{2+} in the solution.*

. . . .

EXAMPLE Effect of I^- on the Solubility of Pb^{2+}

Find the concentrations of the species PbI^+, $PbI_2(aq)$, PbI_3^-, and PbI_4^{2-} in a solution saturated with $PbI_2(s)$ and containing dissolved I^- with a concentration of **(a)** 0.0010 M and **(b)** 1.0 M.

Solution **(a)** From the K_{sp} expression for Reaction 5-11, we calculate

$$[Pb^{2+}] = K_{sp}/[I^-]^2 = 7.9 \times 10^{-3} \text{ M}$$

if $[I^-] = 0.0010$ M. From Reactions 5-12 through 5-15, we then calculate the concentrations of the other lead-containing species:

$$[PbI^+] = K_1[Pb^{2+}][I^-] = (1.0 \times 10^2)(7.9 \times 10^{-3})(1.0 \times 10^{-3})$$
$$= 7.9 \times 10^{-4} \text{ M}$$
$$[PbI_2(aq)] = \beta_2[Pb^{2+}][I^-]^2 = 1.1 \times 10^{-5} \text{ M}$$
$$[PbI_3^-] = \beta_3[Pb^{2+}][I^-]^3 = 6.6 \times 10^{-8} \text{ M}$$
$$[PbI_4^{2-}] = \beta_4[Pb^{2+}][I^-]^4 = 2.4 \times 10^{-10} \text{ M}$$

(b) If, instead, we take $[I^-] = 1.0$ M, then the concentrations are calculated to be

$$[Pb^{2+}] = 7.9 \times 10^{-9} \text{ M} \qquad [PbI_3^-] = 6.6 \times 10^{-5} \text{ M}$$
$$[PbI^+] = 7.9 \times 10^{-7} \text{ M} \qquad [PbI_4^{2-}] = 2.4 \times 10^{-4} \text{ M}$$
$$[PbI_2(aq)] = 1.1 \times 10^{-5} \text{ M}$$

. . . .

The total concentration of dissolved lead in the previous example is

$$[Pb]_{total} = [Pb^{2+}] + [PbI^+] + [PbI_2(aq)] + [PbI_3^-] + [PbI_4^{2-}]$$

When $[I^-] = 10^{-3}$ M, $[Pb]_{total} = 8.7 \times 10^{-3}$ M, of which 91% is Pb^{2+}. As $[I^-]$ increases, $[Pb]_{total}$ decreases by the common ion effect operating in Reaction 5-11. However, at sufficiently high $[I^-]$, complex formation takes over and $[Pb]_{total}$ increases (Figure 5-1). When $[I^-] = 1.0$ M, $[Pb]_{total} = 3.2 \times 10^{-4}$ M, of which 76% is PbI_4^{2-}.

Spreadsheets for Complex Formation

Data for Figure 5-1 can be generated almost painlessly with a spreadsheet. Figure 5-2 shows a spreadsheet in which the equilibrium constants are entered in column A. Column B contains the independent variable, $\log [I^-]$. For this abbreviated example, we entered integers from -3 to $+1$. To generate Figure 5-1, more points are needed. As should be your practice, the formulas used to generate the rest of the table are listed at the bottom.

The concentration $[I^-]$ in cell C2 is computed with the expression $= 10 \wedge B2$. This expression means "raise 10 to the power $\log[I^-]$ found in cell B2." As another example, the concentration $[PbI_4^{2-}]$ in cell H2 is computed with the expression $=\$A\$10*D2*C2 \wedge 4$, which is equivalent to $\beta_4[Pb^{2+}][I^-]^4$. The absolute cell address $A$10 is chosen for the equilibrium constant β_4 because we want that value every time β_4 is required. Logarithms of the various concentrations are calculated in the lower half of the table. The function for the base 10 logarithm is $=\log10(\cdot)$.

	A	B	C	D	E	F	G	H	I
1	Ksp =	Log[I–]	[I–]	[Pb]	[PbI]	[PbI2]	[PbI3]	[PbI4]	Pb(total)
2	7.9E–09	–3	0.001	0.0079	0.00079	1.11E–5	6.56E–8	2.3E–10	0.00870
3	K1 =	–2	0.01	7.9E–05	7.9E–05	1.11E–5	6.56E–7	2.37E–8	1.70E–4
4	100	–1	0.1	7.9E–07	7.9E–06	1.11E–5	6.56E–6	2.37E–6	2.87E–5
5	B2 =	0	1	7.9E–09	7.9E–07	1.11E–5	6.56E–5	2.37E–4	3.14E–4
6	1400	1	10	7.9E–11	7.9E–08	1.11E–5	6.56E–4	0.0237	0.02436
7	B3 =								
8	8300			log[Pb]	log[PbI]	log[PbI2]	log[PbI3]	log[PbI4]	log(Total)
9	B4 =			–2.1024	–3.1024	–4.9562	–7.1833	–9.6253	–2.0604
10	30000			–4.1024	–4.1024	–4.9562	–6.1833	–7.6253	–3.7702
11				–6.1024	–5.1024	–4.9562	–5.1833	–5.6253	–4.5425
12	Cell C2 = 10^B2			–8.1024	–6.1024	–4.9562	–4.1833	–3.6253	–3.5025
13	Cell D2 = $A$2/C2^2			–10.102	–7.1024	–4.9562	–3.1833	–1.6253	–1.6132
14	Cell E2 = $A$4*D2*C2								
15	Cell F2 = $A$6*D2*C2^2		Cell H2 = $A$10*D2*C2^4					Cell D9 = log10(D2)	
16	Cell G2 = $A$8*D2*C2^3		Cell I2 = D2+E2+F2+G2+H2						

Figure 5-2   Spreadsheet for calculating concentrations for Figure 5-1.

Tabulated Equilibrium Constants Are Usually Not "Constant"

If you look up the equilibrium constant of a chemical reaction in two different books, there is an excellent chance that the values will be different (sometimes by a factor of 10 or more).[8] This discrepancy occurs because the constant may have been determined under different conditions by different investigators, perhaps using different techniques.

One of the most common sources of variation in the reported value of K is the ionic composition of the solution. It is important to note whether K is reported for a particular, stated ionic composition or whether the value has been extrapolated to zero ionic strength. If you need to use an equilibrium constant for your own work, you should choose a value of K measured under conditions as close as possible to those you will employ. If the value of K is critical to your experiment, it is best to measure K yourself under the precise conditions of your experiment.

5-7 Protic Acids and Bases

Reactions of acids and bases pervade every branch of science having anything to do with chemistry. In aqueous chemistry, an **acid** is most usefully defined as a substance that increases the concentration of H_3O^+ (**hydronium ion**) when added to water. Conversely, a **base** decreases the concentration of H_3O^+ in aqueous solution. As we shall see shortly, a decrease in H_3O^+ concentration necessarily requires an increase in OH^- concentration. Therefore, a base is a substance that increases the concentration of OH^- in aqueous solution.

The word *protic* refers to chemistry involving transfer of H^+ from one molecule to another. The species H^+ is also called a *proton* because it is what remains when a hydrogen atom loses its electron. Hydronium ion, H_3O^+, is a combination of H^+ with H_2O. Although H_3O^+ is a more accurate representation than H^+ for the hydrogen ion in aqueous solution, we use H_3O^+ and H^+ interchangeably in this book.

Brønsted-Lowry Acids and Bases

Brønsted and Lowry classified *acids* as *proton donors* and *bases* as *proton acceptors*. These definitions include the definition given above. For example, HCl is an acid (a proton donor), and it increases the concentration of H_3O^+ in water:

$$HCl + H_2O \rightleftharpoons H_3O^+ + Cl^-$$

The Brønsted-Lowry definition does not require that H_3O^+ be formed. This definition can therefore be extended to nonaqueous solvents and even to the gas phase:

$$\underset{\substack{\text{Hydrochloric acid} \\ \text{(acid)}}}{HCl(g)} + \underset{\substack{\text{Ammonia} \\ \text{(base)}}}{NH_3(g)} \rightleftharpoons \underset{\substack{\text{Ammonium chloride} \\ \text{(salt)}}}{NH_4^+Cl^-(s)}$$

For the remainder of this text, when we speak of acids and bases, we are speaking of Brønsted-Lowry acids and bases.

Brønsted-Lowry acid: proton donor

Brønsted-Lowry base: proton acceptor

Salts

Any ionic solid, such as ammonium chloride, is called a **salt.** In a formal sense, a salt can be thought of as the product of an acid-base reaction. When an acid and a base react stoichiometrically, they are said to **neutralize** each other. Most salts are *strong electrolytes.* This designation means that they dissociate completely into their component ions when dissolved in water. Thus, ammonium chloride gives NH_4^+ and Cl^- in aqueous solution:

$$NH_4^+Cl^-(s) \rightarrow NH_4^+(aq) + Cl^-(aq)$$

Conjugate Acids and Bases

The products of any reaction between an acid and base may also be classified as acids and bases:

Conjugate acids and bases are related by the gain or loss of one proton.

Acetate is a base because it can accept a proton to make acetic acid. The methylammonium ion is an acid because it can donate a proton and become methylamine. Acetic acid and the acetate ion are said to be a **conjugate acid-base pair.** Methylamine and the methylammonium ion are likewise conjugate. *Conjugate acids and bases are related to each other by the gain or loss of one H^+.*

The Nature of H^+ and OH^-

We will write H^+ when we really mean H_3O^+.

It is certain that the proton does not exist by itself in water. The simplest formula found in some crystalline salts is H_3O^+. For example, crystals of the compound perchloric acid monohydrate contain tetrahedral perchlorate ions and pyramidal hydronium (also called *hydroxonium*) ions:

A solid wedge is a bond coming out of the plane of the page and a dashed wedge is a bond to an atom behind the page.

The formula $HClO_4 \cdot H_2O$ is a way of specifying the composition of the substance when we are ignorant of its structure. A more accurate formula would be $H_3O^+ClO_4^-$.

Bond enthalpy refers to the heat needed to break a chemical bond in the gas phase.

The average dimensions of the H_3O^+ cation that occurs in many crystals are shown in Figure 5-3. The bond enthalpy of the OH bond of H_3O^+ is 544 kJ/mol, about 84 kJ/mol greater than the OH bond enthalpy in H_2O. In

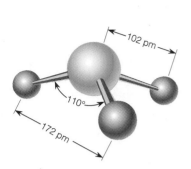

Figure 5-3 Typical structure of hydronium ion, H_3O^+, found in many crystals.

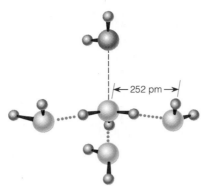

Figure 5-4 Environment of H_3O^+ in dilute aqueous solution. Three molecules of water are bound by strong hydrogen bonds (dotted lines), and one molecule (at the top) is held by much weaker ion-dipole interactions (dashed line).

aqueous solution the H_3O^+ cation is tightly associated with three molecules of water through exceptionally strong hydrogen bonds (Figure 5-4). The $O-H\cdots O$ hydrogen-bonded distance of 252 pm (picometers, 10^{-12} m) may be compared with an $O-H\cdots O$ distance of 283 pm between hydrogen-bonded water molecules. The $H_5O_2^+$ cation is another simple species in which a hydrogen ion is shared by two water molecules.[9]

$$H-\underset{\displaystyle H}{\overset{\displaystyle H}{}}\quad\underset{\displaystyle H}{\overset{\displaystyle H}{}}$$

$$\overset{H}{\underset{H}{\diagdown}} O\cdots H\cdots O \overset{H}{\underset{H}{\diagup}}$$
$$\leftarrow 243 \text{ pm} \rightarrow$$

In the gas phase, H_3O^+ may be found inside a dodecahedral shell of 20 water molecules, as shown in Figure 5-5.

Less is known of the structure of OH^- in solution, but the ion $H_3O_2^-$ ($OH^-\cdot H_2O$) has been observed by X-ray crystallography.[10] The central $O\cdots H\cdots O$ linkage contains the shortest hydrogen bond involving H_2O that has ever been observed.

$$\overset{}{\underset{H}{\diagup}} O\cdots H\cdots O \overset{H}{\diagup}$$
$$\leftarrow 229 \text{ pm} \rightarrow$$

$H_3O_2^-$ has also been observed as a ligand, forming metal—$O\cdots H\cdots O$—metal bridges in several compounds. The $O\cdots H\cdots O$ distances range from 244 to 252 pm.[11]

We will ordinarily write H^+ in most chemical equations, although we really mean H_3O^+. To emphasize the chemistry of water, we will write H_3O^+. For example, water can be either an acid or a base. Water is an acid with respect to methoxide:

$$H-\overset{..}{\underset{..}{O}}-H + CH_3-\overset{..}{\underset{..}{O}}{}^- \rightleftharpoons H-\overset{..}{\underset{..}{O}}{}^- + CH_3-\overset{..}{\underset{..}{O}}-H$$

Water　　　　Methoxide　　　Hydroxide　　　Methanol

But with respect to hydrogen bromide, water is a base:

$$H_2O \quad + \quad HBr \quad \rightleftharpoons \quad H_3O^+ \quad + \quad Br^-$$

Water   Hydrogen bromide   Hydronium ion   Bromide

Autoprotolysis

Water undergoes self-ionization, called **autoprotolysis,** in which it acts as both an acid and a base:

$$H_2O + H_2O \rightleftharpoons H_3O^+ + OH^- \tag{5-16}$$

or

$$H_2O \rightleftharpoons H^+ + OH^- \tag{5-17}$$

Reactions 5-16 and 5-17 mean the same thing.

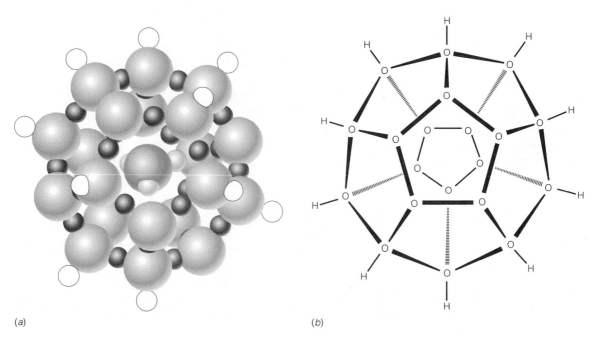

(a)   (b)

Figure 5-5 $H_3O^+(H_2O)_{20}$ cluster found in the gas phase. (*a*) Hydronium ion (H_3O^+) is located in the center of the dodecahedral cage of 20 water molecules held together by hydrogen bonds involving 30 of the 40 hydrogen atoms [*Chem. Eng. News,* 8 April 1991)]. (*b*) Diagram showing 20 oxygen atoms at the vertices of a dodecahedron, which has 12 pentagonal faces. Hydrogen atoms not involved in hydrogen bonding between oxygen atoms within the dodecahedron are shown. The structure is thought to be *fluxional,* with hydrogen bonds rapidly breaking and reforming between different water molecules in the dodecahedron. [S. Wei, Z. Shi, and A. W. Castleman, Jr., *J. Chem. Phys.* **1991,** *94,* 3268.]

Challenge   Make a copy of Figure 5-5b. Draw in 30 hydrogen-bonded hydrogen atoms between the oxygen atoms to show how this structure is held together.

Protic solvents have a reactive H^+, and all protic solvents undergo autoprotolysis. An example is acetic acid:

$$2CH_3COH \rightleftharpoons CH_3C^+\!\!\!\begin{smallmatrix}OH\\OH\end{smallmatrix} + CH_3C-O^- \quad \text{(in acetic acid)} \quad (5\text{-}18)$$

The extent of these reactions of water or of acetic acid is very small. The *autoprotolysis constants* (equilibrium constants) for Reactions 5-17 and 5-18 are 1.0×10^{-14} and 3.5×10^{-15}, respectively, at 25°C.

Examples of protic solvents (acidic proton **bold**):

$$H_2O \qquad CH_3CH_2OH$$
Water   Ethanol

Examples of **aprotic** solvents (no acidic protons):

$$CH_3CH_2OCH_2CH_3 \qquad CH_3CN$$
Diethyl ether   Acetonitrile

5-8 pH

The autoprotolysis constant for H_2O is given the special symbol K_w, where the subscript w denotes water.

Autoprotolysis of water: $\quad H_2O \overset{K_w}{\rightleftharpoons} H^+ + OH^- \quad (5\text{-}19)$

The equilibrium constant K_w is

$$K_w = [H^+][OH^-] \quad (5\text{-}20)$$

As is true of all equilibrium constants, K_w varies with temperature (Table 5-1). The value of K_w at 25.00°C is 1.01×10^{-14}.

Recall that H_2O (the solvent) is omitted from the equilibrium constant. The value $K_w = 1.0 \times 10^{-14}$ at 25°C will be accurate enough for our purposes in this text.

TABLE 5-1 Temperature dependence of K_w[a]

Temperature (°C)	K_w	$pK_w = -\log K_w$
0	1.14×10^{-15}	14.944
5	1.85×10^{-15}	14.734
10	2.92×10^{-15}	14.535
15	4.51×10^{-15}	14.346
20	6.81×10^{-15}	14.167
24	1.00×10^{-14}	14.000
25	1.01×10^{-14}	13.996
30	1.47×10^{-14}	13.833
35	2.09×10^{-14}	13.680
40	2.92×10^{-14}	13.535
45	4.02×10^{-14}	13.396
50	5.47×10^{-14}	13.262
100	5.45×10^{-13}	12.264
150	2.31×10^{-12}	11.637
200	5.15×10^{-12}	11.288
250	6.43×10^{-12}	11.192
300	3.93×10^{-12}	11.406
350	5.07×10^{-13}	12.295

a. Concentrations in the product $[H^+][OH^-]$ in this table are expressed in molality rather than in molarity.

SOURCE: H. S. Harned and B. B. Owen, *The Physical Chemistry of Electrolytic Solutions,* 3rd ed. (New York: Chapman and Hall, 1958). This book is the most authoritative reference in its field.

pH ≈ −log[H⁺]

Take the log of both sides of Equation 5-20 to derive Equation 5-22:

$K_w = [H^+][OH^-]$
$\log K_w = \log[H^+] + \log[OH^-]$
$-\log K_w = pH + pOH$
$= 14.00 \text{ at } 25°C$

pH is usually measured with a *glass electrode*, whose operation is described in Chapter 15.

E X A M P L E   Concentration of H⁺ and OH⁻ in Pure Water at 25°C

Calculate the concentration of H⁺ and OH⁻ in pure water at 25°C.

Solution   The stoichiometry of Reaction 5-19 tells us that H⁺ and OH⁻ are produced in a 1:1 mole ratio. Their concentrations must be equal. Calling each concentration *x*, we can write

$$K_w = 1.0 \times 10^{-14} = [H^+][OH^-] = [x][x] \Rightarrow x = 1.0 \times 10^{-7} \text{ M}$$

The concentrations of H⁺ and OH⁻ are both 1.0×10^{-7} M.

E X A M P L E   Concentration of OH⁻ When [H⁺] Is Known

What is the concentration of OH⁻ if $[H^+] = 1.0 \times 10^{-3}$ M?

Solution   Putting $[H^+] = 1.0 \times 10^{-3}$ M into Equation 5-20 gives

$$K_w = 1.0 \times 10^{-14} = (1.0 \times 10^{-3})[OH^-] \Rightarrow [OH^-] = 1.0 \times 10^{-11} \text{ M}$$

A concentration $[H^+] = 1.0 \times 10^{-3}$ M gives $[OH^-] = 1.0 \times 10^{-11}$ M. *As the concentration of* H⁺ *increases, the concentration of* OH⁻ *necessarily decreases, and vice versa.* A concentration of $[OH^-] = 1.0 \times 10^{-3}$ M gives $[H^+] = 1.0 \times 10^{-11}$ M.

An approximate definition of **pH** is the negative logarithm of the H⁺ concentration.

Approximate definition of pH: $pH \approx -\log [H^+]$ (5-21)

In Chapter 8 we will define pH more accurately in terms of *activities,* but for most purposes Equation 5-21 is a good working definition of pH. In pure water at 25°C with $[H^+] = 1.0 \times 10^{-7}$ M, the pH is $-\log(1.0 \times 10^{-7}) = 7.00$. If the concentration of OH⁻ is 1.0×10^{-3} M, then $[H^+] = 1.0 \times 10^{-11}$ M and the pH is 11.00.

A useful relation between the concentrations of H⁺ and OH⁻ is

$$pH + pOH = -\log K_w = 14.00 \quad \text{at } 25°C \qquad (5\text{-}22)$$

where $pOH = -\log[OH^-]$, just as $pH = -\log[H^+]$. This is a fancy way of saying that if pH = 3.58, then pOH = 14.00 − 3.58 = 10.42, or $[OH^-] = 10^{-10.42} = 3.8 \times 10^{-11}$ M.

A solution is **acidic** if $[H^+] > [OH^-]$. A solution is **basic** if $[H^+] < [OH^-]$. At 25°C, an acidic solution has a pH below 7 and a basic solution has a pH above 7.

$$pH \quad \underset{\longleftarrow \text{Acidic} \longrightarrow \quad \text{Neutral} \quad \longleftarrow \text{Basic} \longrightarrow}{\overset{-1 \; 0 \; 1 \; 2 \; 3 \; 4 \; 5 \; 6 \; 7 \; 8 \; 9 \; 10 \; 11 \; 12 \; 13 \; 14 \; 15}{\rule{8cm}{0.4pt}}}$$

pH values for various common substances are shown in Figure 5-6.

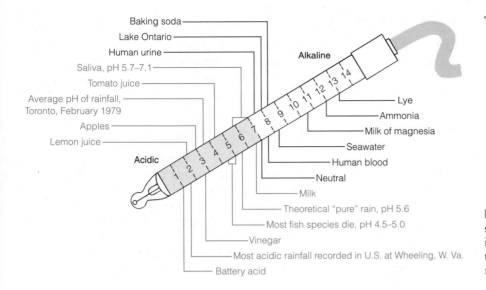

Figure 5-6 pH of various common substances. The most acidic rainfall in the United States is more acidic than lemon juice. (See Box 15-2 for more on acid rain.)

Although the pHs of most solutions fall in the range 0–14, these are not the limits of pH. A pH of −1.00, for example, means −log[H$^+$] = −1.00, or [H$^+$] = 10 M. This concentration is easily attained in a concentrated solution of a strong acid such as HCl.

Is There Such a Thing as Pure Water?

In most labs, the answer is "no." Pure water at 25°C should have a pH of 7.00. Distilled water from the tap in most labs is acidic because it contains CO_2 from the atmosphere. The CO_2 is an acid by virtue of the reaction

$$CO_2 + H_2O \rightleftharpoons \underset{\text{Bicarbonate}}{HCO_3^-} + H^+$$

Carbon dioxide can be largely removed by first boiling the water and then protecting it from the atmosphere. Alternatively, the water can be distilled under an inert atmosphere, such as pure N_2.

About 100 years ago, careful measurements of the conductivity of water were made by Friedrich Kohlrausch and his students. In removing ionic impurities, they found it necessary to distill the water *42 consecutive times* under vacuum to reduce conductivity to a limiting value.

5-9 Strengths of Acids and Bases

Acids and bases are commonly classified as strong or weak, depending on whether they react "completely" or only "partly" to produce H$^+$ or OH$^-$. Because there is a continuous range of possibilities for "partial" reaction, there is no sharp distinction between weak and strong. However, some compounds react so completely that they are easily considered strong acids or bases, and, by convention, everything else is termed weak.

Strong Acids and Bases

The most common strong acids and bases are listed in Table 5-2. By definition, a strong acid or base is completely dissociated in aqueous solution. That is, the equilibrium constants for the following reactions are very large.

$$HCl(aq) \rightleftharpoons H^+ + Cl^-$$
$$KOH(aq) \rightleftharpoons K^+ + OH^-$$

Virtually no undissociated HCl or KOH exists in aqueous solution. Demonstration 5-2 shows one consequence of the strong acid behavior of HCl.

Notice that, even though the hydrogen halides HCl, HBr, and HI are strong acids, HF is *not* a strong acid. Box 5-3 gives an explanation of this unexpected observation. For most practical purposes, the hydroxides of the alkaline earth metals (Mg^{2+}, Ca^{2+}, Sr^{2+}, and Ba^{2+}) can be considered to be strong bases, although they are far less soluble than alkali metal hydroxides and also have some tendency to form MOH^+ complexes (Table 5-3).

Weak Acids and Bases

All weak acids, HA, react with water by donating a proton to H_2O:

Dissociation of weak acid: $HA + H_2O \overset{K_a}{\rightleftharpoons} H_3O + A^-$ (5-23)

which means exactly the same as

Dissociation of weak acid: $HA \overset{K_a}{\rightleftharpoons} H^+ + A^-$ (5-24)

The equilibrium constant is called K_a, the **acid dissociation constant.**

Acid dissociation constant: $K_a = \dfrac{[H^+][A^-]}{[HA]}$ (5-25)

By definition, a weak acid is one that is only partially dissociated in water. This definition means that K_a is "small" for a weak acid.

Weak bases, B, react with water by abstracting a proton from H_2O:

Base hydrolysis: $B + H_2O \overset{K_b}{\rightleftharpoons} BH^+ + OH^-$ (5-26)

The equilibrium constant K_b is called the **base hydrolysis constant** or, mistakenly, the **base "dissociation" constant.**

Base hydrolysis constant: $K_b = \dfrac{[BH^+][OH^-]}{[B]}$ (5-27)

By definition, a weak base is one for which K_b is "small."

Common Classes of Weak Acids and Bases

Acetic acid is a typical weak acid.

$$\underset{\substack{\text{Acetic} \\ \text{acid (HA)}}}{CH_3\overset{\displaystyle O}{\overset{\|}{C}}OH} \rightleftharpoons \underset{\text{Acetate (A}^-)}{CH_3\overset{\displaystyle O}{\overset{\|}{C}}O^-} + H^+ \qquad K_a = 1.75 \times 10^{-5} \qquad (5\text{-}28)$$

TABLE 5-2   Common strong acids and bases

Formula	Name
Acids	
HCl	Hydrochloric acid (hydrogen chloride)
HBr	Hydrogen bromide
HI	Hydrogen iodide
$H_2SO_4{}^a$	Sulfuric acid
HNO_3	Nitric acid
$HClO_4$	Perchloric acid
Bases	
LiOH	Lithium hydroxide
NaOH	Sodium hydroxide
KOH	Potassium hydroxide
RbOH	Rubidium hydroxide
CsOH	Cesium hydroxide
R_4NOH^b	Quaternary ammonium hydroxide

a. For H_2SO_4, only the first proton ionization is complete. Dissociation of the second proton has an equilibrium constant of 1.0×10^{-2}.
b. This is a general formula for any hydroxide salt of an ammonium cation containing four organic groups. An example is tetrabutylammonium hydroxide: $(CH_3CH_2CH_2CH_2)_4N^+OH^-$.

TABLE 5-3   Equilibria of alkaline earth metal hydroxides

$$M(OH)_2(s) \rightleftharpoons M^{2+} + 2OH^-$$

$$K_{sp} = [M^{2+}][OH^-]^2$$

$$M^{2+} + OH^- \rightleftharpoons MOH^+$$

$$K_1 = [MOH^+]/[M^{2+}][OH^-]$$

Metal	$\log K_{sp}$	$\log K_1$
Mg^{2+}	−11.15	2.58
Ca^{2+}	−5.19	1.30
Sr^{2+}	—	0.82
Ba^{2+}	—	0.64

Note: 25°C and ionic strength = 0.

Demonstration 5-2 The HCl Fountain

The complete dissociation of HCl into H^+ and Cl^- makes HCl(g) extremely soluble in water.

$$HCl(g) \rightleftharpoons HCl(aq) \tag{a}$$

$$HCl(aq) \rightleftharpoons H^+(aq) + Cl^-(aq) \tag{b}$$

$$\text{net reaction:} \quad HCl(g) \rightleftharpoons H^+(aq) + Cl^-(aq) \tag{c}$$

Because the equilibrium of Reaction b lies far to the right, it pulls Reaction a to the right as well.

Challenge The standard free energy change ($\Delta G°$) for Reaction c is -36.0 kJ/mol. Show that the equilibrium constant is 2.0×10^6.

The extreme solubility of HCl(g) in water is the basis for the HCl fountain, assembled as shown below. In Figure a, an inverted 250-mL round-bottom flask containing air is set up with its inlet tube leading to a source of HCl(g) and its outlet tube directed into an inverted bottle of water. As HCl is admitted to the flask, air is displaced. When the bottle is filled with air, the flask is filled mostly with HCl(g).

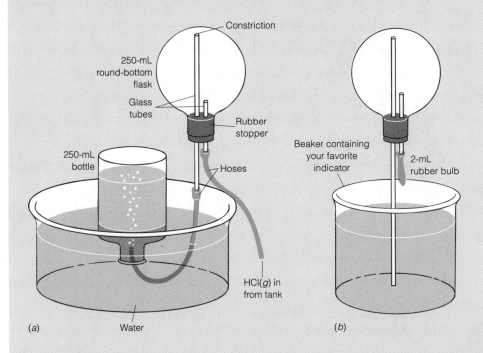

(a)

(b)

The hoses are disconnected and replaced with a beaker of indicator and a rubber bulb (Figure b). For an indicator, we use slightly alkaline methyl purple, which is green above pH 5.4 and purple below pH 4.8. When ~ 1 mL of water is squirted from the rubber bulb into the flask, a vacuum is created and indicator solution is drawn up into the flask, making a fascinating fountain (Color Plate 1).

Question Why is a vacuum created when water is squirted into the flask, and why does the indicator change color when it enters the flask?

Box 5-3 The Strange Behavior of Hydrofluoric Acid[12]

The hydrogen halides HCl, HBr, and HI are all strong acids, which means that the reactions

$$HX(aq) + H_2O \rightarrow H_3O^+ + X^-$$

(X = Cl, Br, I) all go to completion. Why, then, does HF behave as a weak acid? The answer is odd. First, HF does completely give up its proton to H_2O:

$$HF(aq) \quad \rightarrow \quad \underset{\substack{\text{Hydronium} \\ \text{ion}}}{H_3O^+} \quad + \quad \underset{\substack{\text{Fluoride} \\ \text{ion}}}{F^-}$$

But fluoride forms the strongest hydrogen bond of any ion. The hydronium ion remains tightly associated with F^- through a hydrogen bond. We call such an association an **ion pair.**

$$\underset{\text{An ion pair}}{H_3O^+ + F^- \rightleftharpoons F^- \cdots H_3O^+}$$

Ion pairs are common in nonaqueous solvents, which cannot promote ion dissociation as well as water. But the hydronium–fluoride ion pair is unusual for aqueous solutions.

Thus, HF does not behave as a strong acid because the F^- and H_3O^+ ions remain associated with each other. Dissolving one mole of the strong acid HCl in water creates one mole of free H_3O^+. Dissolving one mole of the "weak" acid HF in water creates very little free H_3O^+.

Hydrofluoric acid is not unique in its propensity to form ion pairs. Many moderately strong acids, such as those below, are thought to exist predominantly as ion pairs in aqueous solution ($HA + H_2O \rightleftharpoons A^- \cdots H_3O^+$).[13]

$$CF_3CO_2H$$

Trifluoroacetic acid
$K_a = 0.31$

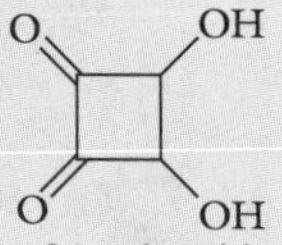

Squaric acid
$K_a = 0.29$

Acetic acid is representative of the carboxylic acids, which have the general structure

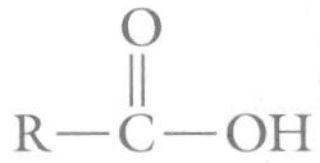

$$\underset{\text{A carboxylic acid}}{R-\overset{\overset{\textstyle O}{\|}}{C}-OH}$$

where R is an organic substituent. *Most* **carboxylic acids** *are weak acids, and most* **carboxylate anions** *are weak bases.*

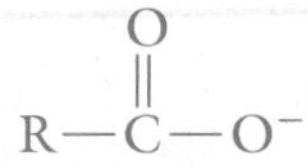

$$\underset{\text{A carboxylate anion}}{R-\overset{\overset{\textstyle O}{\|}}{C}-O^-}$$

Carboxylic acids (RCO_2H) and ammonium ions (R_3NH^+) are weak acids. Carboxylate anions (RCO_2^-) and amines (R_3N) are weak bases.

Methylamine is a typical weak base.

$$\underset{\text{Methylamine (B)}}{CH_3\ddot{N}H_2} + H_2O \rightleftharpoons \underset{\substack{\text{Methylammonium} \\ \text{ion (BH}^+\text{)}}}{CH_3\overset{+}{N}H_3} + OH^- \qquad K_b = 4.4 \times 10^{-4}$$

(5-29)

Methylamine is a representative amine, a nitrogen-containing compound.

$R\ddot{N}H_2$	a primary amine	RNH_3^+
$R_2\ddot{N}H$	a secondary amine	$R_2NH_2^+$ } ammonium ions
$R_3\ddot{N}$	a tertiary amine	R_3NH^+

Amines *are weak bases, and* **ammonium ions** *are weak acids.* The "parent" of all amines is ammonia, NH_3. When a base such as methylamine reacts with water, the product formed is the conjugate acid. That is, the methylammonium ion produced in Reaction 5-29 is a weak acid:

$$\underset{BH^+}{CH_3\overset{+}{\ddot{N}}H_3} \overset{K_a}{\rightleftharpoons} \underset{B}{CH_3\ddot{N}H_2} + H^+ \qquad K_a = 2.3 \times 10^{-11} \qquad (5\text{-}30)$$

The methylammonium ion is the conjugate acid of methylamine.

> Although we will usually write a **base** as **B** and an **acid** as **HA**, it is important to realize that BH^+ is also an **acid** and A^- is also a **base.**

You should learn to recognize whether a compound will have acidic or basic properties. The salt methylammonium chloride, for example, will dissociate completely in aqueous solution to give the methylammonium cation and the chloride anion:

$$\underset{\substack{\text{Methylammonium} \\ \text{chloride}}}{CH_3\overset{+}{\ddot{N}}H_3Cl^-(s)} \rightarrow CH_3\overset{+}{\ddot{N}}H_3(aq) + Cl^-(aq) \qquad (5\text{-}31)$$

The methylammonium ion, being the conjugate acid of methylamine, is a weak acid (Reaction 5-30). The chloride ion is neither an acid nor a base. It is the conjugate base of HCl, a strong acid. This means that Cl^- *has virtually no tendency to associate with* H^+, or else HCl would not be a strong acid. We predict (correctly) that a solution of methylammonium chloride will be acidic, because the methylammonium ion is an acid and Cl^- is not a base.

> Methylammonium chloride is a weak acid because
> 1. It dissociates into $CH_3NH_3^+$ and Cl^-.
> 2. $CH_3NH_3^+$ is a weak acid, being conjugate to CH_3NH_2, a weak base.
> 3. Cl^- has no basic properties. It is conjugate to HCl, a strong acid. That is, HCl dissociates completely.

> **Challenge** Phenol (C_6H_5OH) is a weak acid. Explain why a solution of the ionic compound potassium phenolate ($C_6H_5O^-K^+$) will be basic.

Polyprotic Acids and Bases

Polyprotic acids and bases are compounds that can donate or accept more than one proton. For example, oxalic acid is diprotic and phosphate is tribasic:

$$K_{a1} = 5.60 \times 10^{-2} \qquad (5\text{-}32)$$

$$K_{a2} = 5.42 \times 10^{-5} \qquad (5\text{-}33)$$

$$K_{b1} = 1.4 \times 10^{-2} \qquad (5\text{-}34)$$

$$K_{b2} = 1.59 \times 10^{-7} \qquad (5\text{-}35)$$

> *Notation for acid and base equilibrium constants:* K_{a1} refers to the acidic species with the most protons and K_{b1} refers to the basic species with the least protons. The subscript a in acid dissociation constants will usually be omitted.

$$\text{(structure)} + H_2O \rightleftharpoons \text{(structure)} + OH^- \qquad K_{b3} = 1.42 \times 10^{-12}$$
$$\text{(5-36)}$$

Phosphoric
acid

The standard notation for successive acid dissociation constants of a poly-protic acid is K_1, K_2, K_3, and so on, with the subscript a usually omitted. We retain or omit the subscript as dictated by clarity. For successive base hydrolysis constants, we retain the subscript b. The examples above illustrate that K_{a1} (or K_1) *refers to the acidic species with the most protons, and* K_{b1} *refers to the basic species with the least number of protons.* Carbonic acid, an interesting diprotic carboxylic acid derived from CO_2, is described in Box 5-4.

Relation Between K_a and K_b

A most important relationship exists between K_a and K_b of a conjugate acid-base pair in aqueous solution. We can derive this result with the acid HA and its conjugate base A^-.

$$HA \rightleftharpoons H^+ + A^- \qquad K_a = \frac{[H^+][A^-]}{[HA]}$$

$$\underline{A^- + H_2O \rightleftharpoons HA + OH^- \qquad K_b = \frac{[HA][OH^-]}{[A^-]}}$$

$$H_2O \rightleftharpoons H^+ + OH^- \qquad K_w = K_a \cdot K_b$$

$$= \frac{[H^+][A^-]}{[HA]} \cdot \frac{[HA][OH^-]}{[A^-]}$$

When the reactions above are added, their equilibrium constants must be multiplied, thereby obtaining a most useful result:

$K_a \cdot K_b = K_w$ for a conjugate acid-base pair in aqueous solution.

Relation between K_a and K_b
for a conjugate pair:

$$\boxed{K_a \cdot K_b = K_w} \qquad \text{(5-37)}$$

Equation 5-37 applies to any acid and its conjugate base in aqueous solution.

EXAMPLE Finding K_b for the Conjugate Base

The value of K_a for acetic acid is 1.75×10^{-5} (Reaction 5-28). Find K_b for the acetate ion.

Solution This one is trivial:

$$K_b = \frac{K_w}{K_a} = \frac{1.0 \times 10^{-14}}{1.75 \times 10^{-5}} = 5.7 \times 10^{-10}$$

Box 5-4   Carbonic Acid[14]

Carbonic acid is the acidic form of dissolved carbon dioxide.

$$CO_2(g) \rightleftharpoons CO_2(aq)$$

$$K = \frac{[CO_2(aq)]}{P_{CO_2}} = 0.0344$$

$$CO_2(aq) + H_2O \rightleftharpoons \underset{\substack{| \\ HO \quad OH \\ \text{Carbonic acid}}}{\overset{\substack{O \\ \| }}{C}}$$

$$K = \frac{[H_2CO_3]}{[CO_2(aq)]} \approx 1.3 \times 10^{-3}$$

$$\underset{\text{Bicarbonate}}{H_2CO_3 \rightleftharpoons HCO_3^- + H^+}$$

$$K_{a1} = 4.45 \times 10^{-7}$$

$$\underset{\text{Carbonate}}{HCO_3^- \rightleftharpoons CO_3^{2-} + H^+}$$

$$K_{a2}$$

Its behavior as a diprotic acid appears anomalous at first, because the value of K_{a1} is about 10^2 to 10^4 times smaller than K_a for other carboxylic acids.

CH_3CO_2H Acetic acid	$K_a = 1.75 \times 10^{-5}$	HCO_2H Formic acid	$K_a = 1.80 \times 10^{-4}$
$N{\equiv}CCH_2CO_2H$ Cyanoacetic acid	$K_a = 3.37 \times 10^{-3}$	$HOCH_2CO_2H$ Glycolic acid	$K_a = 1.48 \times 10^{-4}$

The reason for this seeming anomaly is not that H_2CO_3 is unusual but, rather, that the value commonly given for K_{a1} applies to the equation

$$\text{all dissolved } CO_2 \rightleftharpoons HCO_3^- + H^+$$
$$(= CO_2(aq) + H_2CO_3)$$

$$K_{a1} = \frac{[HCO_3^-][H^+]}{[CO_2(aq) + H_2CO_3]} = 4.45 \times 10^{-7}$$

Only about 0.2% of dissolved CO_2 is in the form H_2CO_3. When the true value of $[H_2CO_3]$ is used instead of the value $[H_2CO_3 + CO_2(aq)]$, the value of the equilibrium constant becomes

$$K_{a1} = \frac{[HCO_3^-][H^+]}{[H_2CO_3]} = 2 \times 10^{-4}$$

The hydration of CO_2 (reaction of CO_2 with H_2O) and dehydration of H_2CO_3 are surprisingly slow reactions, which can be demonstrated readily in a classroom.[14] Living cells utilize the enzyme *carbonic anhydrase* to speed the rate at which H_2CO_3 and CO_2 equilibrate in order to process this key metabolite. The active site of the enzyme provides an environment just right for the reaction of CO_2 with OH^-, lowering the *activation energy* (the energy barrier for the reaction) from 50 down to 26 kJ/mol,[15] and increasing the rate of reaction by more than a factor of 10^6.

EXAMPLE   Finding K_a for the Conjugate Acid

The value of K_b for methylamine is 4.4×10^{-4} (Reaction 5-29). Find K_a for the methylammonium ion.

Solution   Once again, $K_a = \dfrac{K_w}{K_b} = 2.3 \times 10^{-11}$

For a diprotic acid, we can derive results relating each of two acids and their conjugate bases:

$$H_2A \rightleftharpoons H^+ + HA^- \quad K_{a1} \qquad HA^- \rightleftharpoons H^+ + A^{2-} \quad K_{a2}$$
$$\underline{HA^- + H_2O \rightleftharpoons H_2A + OH^- \quad K_{b2}} \qquad \underline{A^{2-} + H_2O \rightleftharpoons HA^- + OH^- \quad K_{b1}}$$
$$H_2O \rightleftharpoons H^+ + OH^- \quad K_w \qquad H_2O \rightleftharpoons H^+ + OH^- \quad K_w$$

The final results are

General relation between K_a and K_b:

$$K_{a1} \cdot K_{b2} = K_w \tag{5-38}$$

$$K_{a2} \cdot K_{b1} = K_w \tag{5-39}$$

Challenge Derive the following results for a triprotic acid:

$$K_{a1} \cdot K_{b3} = K_w \tag{5-40}$$

$$K_{a2} \cdot K_{b2} = K_w \tag{5-41}$$

$$K_{a3} \cdot K_{b1} = K_w \tag{5-42}$$

Terms to Understand

acid	common ion effect	Lewis acid
acid dissociation constant (K_a)	complex ion	Lewis base
acid solution	conjugate acid-base pair	ligand
amine	coprecipitation	neutralization
ammonium ion	cumulative formation constant	overall formation constant
aprotic solvent	disproportionation	pH
autoprotolysis	endothermic	polyprotic acids and bases
base	enthalpy change	protic solvent
base "dissociation" constant (K_b)	entropy	reaction quotient
base hydrolysis constant (K_b)	equilibrium constant	salt
basic solution	exothermic	saturated solution
Brønsted-Lowry acid	Gibbs free energy	solubility product
Brønsted-Lowry base	hydronium ion	standard state
carboxylate anion	ion pair	stepwise formation constant
carboxylic acid	Le Châtelier's principle	

Summary

For the reaction $aA + bB \rightleftharpoons cC + dD$, the equilibrium constant is $K = [C]^c[D]^d / [A]^a[B]^b$. Solute concentrations should be expressed in moles per liter; gas concentrations should be in atmospheres; and the concentrations of pure solids, liquids, and solvents are omitted. If the direction of a reaction is changed, $K' = 1/K$. If two reactions are added, $K_3 = K_1 K_2$. The value of the equilibrium constant can be calculated from the free energy change for a chemical reaction: $K = e^{-\Delta G°/RT}$. The equation $\Delta G = \Delta H - T\Delta S$ summarizes the observations that a reaction is favored if it liberates heat (exothermic, negative ΔH) or increases disorder (positive ΔS). Le Châtelier's principle predicts the effect on a chemical reaction when reactants or products are added, or temperature is changed. The reaction quotient, Q, is used to tell how a system must change to reach equilibrium.

The solubility product, which is the equilibrium constant for the dissolution of a solid salt, is used to calculate the solubility of a salt in aqueous solution. If one of the ions of that salt is already present in the solution, the solubility of the salt is decreased (the common ion effect). You should be able to evaluate the feasibility of selectively precipitating one ion from a solution containing other ions. At high concentration of ligand, a precipitated metal ion may redissolve because of the formation of soluble complex ions. In a metal ion complex, the metal is a Lewis acid (electron pair acceptor) and the ligand is a Lewis base (electron pair donor).

Setting up and solving equilibrium problems are important skills, as are the use and testing of numerical approximations.

Brønsted-Lowry acids are proton donors, and Brønsted-Lowry bases are proton acceptors. An acid increases the concentration of H_3O^+ in aqueous solution, and a base increases the concentration of OH^-. An acid-base pair related through the gain or loss of a single proton is described as conjugate. When a proton is transferred from one molecule to another molecule of a protic solvent, the reaction is called autoprotolysis.

The definition of pH is $pH = -\log[H^+]$ (which will be modified to include activity later). K_a refers to the equilibrium constant for the dissociation of an acid: $HA + H_2O \rightleftharpoons H_3O^+ + A^-$. K_b is the base hydrolysis constant for the reaction $B + H_2O \rightleftharpoons BH^+ + OH^-$. When either K_a or K_b is large, the acid or base is said to be strong; otherwise the acid or base is weak. Common strong acids and bases are listed in Table 5-2, which you should memorize. The most common weak acids are carboxylic acids (RCO_2H), and the most common weak bases are amines ($R_3N:$). Carboxylate anions (RCO_2^-) are weak bases, and ammonium ions (R_3NH^+) are weak acids. For a conjugate acid-base pair in water, $K_a \cdot K_b = K_w$. For polyprotic acids, we denote the successive acid dissociation constants as $K_{a1}, K_{a2}, K_{a3}, \cdots$, or just $K_1, K_2, K_3, \cdots$. For polybasic species, we denote successive hydrolysis constants $K_{b1}, K_{b2}, K_{b3}, \cdots$. For a polyprotic system with n protons, the relation between the ith K_a and the $(n-i)$th K_b is $K_{ai} \cdot K_{b(n-1)} = K_w$.

Exercises

A. Consider the following equilibria, in which all ions are aqueous:

(1) $Ag^+ + Cl^- \rightleftharpoons AgCl(aq)$ $K = 2.0 \times 10^3$

(2) $AgCl(aq) + Cl^- \rightleftharpoons AgCl_2^-$ $K = 9.3 \times 10^1$

(3) $AgCl(s) \rightleftharpoons Ag^+ + Cl^-$ $K = 1.8 \times 10^{-10}$

(a) Calculate the numerical value of the equilibrium constant for the reaction $AgCl(s) \rightleftharpoons AgCl(aq)$.

(b) Calculate the concentration of $AgCl(aq)$ in equilibrium with excess undissolved solid AgCl.

(c) Find the numerical value of K for the reaction $AgCl_2^- \rightleftharpoons AgCl(s) + Cl^-$.

B. Reaction 5-7 is allowed to come to equilibrium in a solution initially containing 0.0100 M BrO_3^-, 0.0100 M Cr^{3+}, and 1.00 M H^+.

(a) Set up an equation analogous to Equation 5-10 to find the concentrations at equilibrium.

(b) Noting that $K = 1 \times 10^{11}$ for Reaction 5-7, it is reasonable to guess that the reaction will go nearly all the way "to completion." That is, we expect both the concentration of Br^- and $Cr_2O_7^{2-}$ to be close to 0.00500 M at equilibrium. (Why?) Using these concentrations for Br^- and $Cr_2O_7^{2-}$, find the concentrations of BrO_3^- and Cr^{3+} at equilibrium. Note that Cr^{3+} is the limiting reagent in this problem. The reaction uses up the Cr^{3+} before consuming all of the BrO_3^-.

C. How many grams of lanthanum iodate, $La(IO_3)_3$, which dissociates into $La^{3+} + 3IO_3^-$, will dissolve in 250.0 mL of (a) water; (b) 0.050 M $LiIO_3$?

D. Which will be more soluble (moles of metal dissolved per liter of solution)?

(a) $Ba(IO_3)_2$ ($K_{sp} = 1.5 \times 10^{-9}$)   or $Ca(IO_3)_2$ ($K_{sp} = 7.1 \times 10^{-7}$)

(b) $TlIO_3$ ($K_{sp} = 3.1 \times 10^{-6}$)   or $Sr(IO_3)_2$ ($K_{sp} = 3.3 \times 10^{-7}$)

E. Fe(III) can be precipitated from acidic solution by addition of OH^- to form $Fe(OH)_3(s)$. At what concentration of OH^- will the concentration of Fe(III) be reduced to 1.0×10^{-10} M? If Fe(II) is used instead, what concentration of OH^- is necessary to reduce the Fe(II) concentration to 1.0×10^{-10} M?

F. It is desired to perform a 99% complete separation of 0.010 M Ca^{2+} from 0.010 M Ce^{3+} by precipitation with oxalate ($C_2O_4^{2-}$). Given the solubility products below, decide whether this is feasible.

$$CaC_2O_4 \qquad K_{sp} = 1.3 \times 10^{-8}$$

$$Ce_2(C_2O_4)_3 \qquad K_{sp} = 3 \times 10^{-29}$$

G. For a solution of Ni^{2+} and ethylenediamine, the following equilibrium constants apply at 20°C:

$$Ni^{2+} + H_2NCH_2CH_2NH_2 \rightleftharpoons Ni(en)^{2+} \quad \log K_1 = 7.52$$

Ethylenediamine
(abbreviated en)

$$Ni(en)^{2+} + en \rightleftharpoons Ni(en)_2^{2+} \qquad \log K_2 = 6.32$$

$$Ni(en)_2^{2+} + en \rightleftharpoons Ni(en)_3^{2+} \qquad \log K_3 = 4.49$$

Calculate the concentration of free Ni^{2+} in a solution prepared by mixing 0.100 mol of en plus 1.00 mL of 0.0100 M Ni^{2+} and diluting to 1.00 L with dilute base (which keeps all of the en in its unprotonated form). [*Hint:* Assume that nearly all of the Ni is in the form $Ni(en)_3^{2+}$ (i.e., $[Ni(en)_3^{2+}] = 1.00 \times 10^{-5}$ M). You will need to calculate the concentrations of $Ni(en)^{2+}$ and $Ni(en)_2^{2+}$ to verify that the assumption does not lead to a contradiction.]

H. If each of the following is dissolved in water, will the solution be acidic, basic, or neutral?

(a) Na^+Br^-

(b) $Na^+CH_3CO_2^-$

(c) $NH_4^+Cl^-$

(d) K_3PO_4

(e) $(CH_3)_4N^+Cl^-$

(f) $(CH_3)_4N^+$ —CO_2^-

I. Succinic acid dissociates as follows:

$$HOCCH_2CH_2CO^- + H^+ \qquad K_1 = 6.2 \times 10^{-5}$$

$$^-OCCH_2CH_2CO^- + H^+ \qquad K_2 = 2.3 \times 10^{-6}$$

Calculate K_{b1} and K_{b2} for the following reactions:

J. Histidine is a triprotic amino acid:

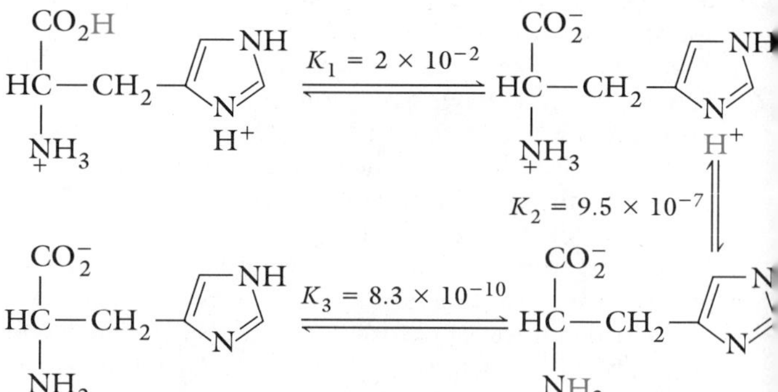

$K_1 = 2 \times 10^{-2}$

$K_2 = 9.5 \times 10^{-7}$

$K_3 = 8.3 \times 10^{-10}$

What is the value of the equilibrium constant for the reaction?

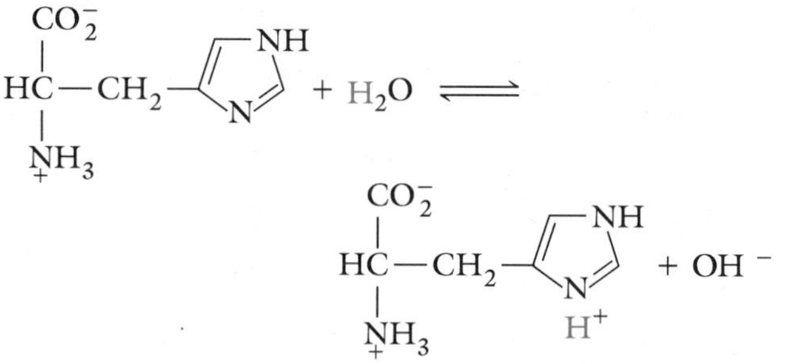

K. (a) Using the values of K_w in Table 5-1, calculate the pH of distilled water at 0°, 20°, and 40°C.

(b) The equilibrium constant for the reaction $D_2O \rightleftharpoons D^+ + OD^-$ is $K = [D^+][OD^-] = 1.35 \times 10^{-15}$ at 25°C (when concentrations are expressed in moles per liter). In this equation, D stands for deuterium, which is the isotope ^2H. What is the pD ($= -\log [D^+]$ for neutral D_2O?

Problems

Equilibrium and Thermodynamics

1. To evaluate the equilibrium constant in Equation 5-2, we are required to express the concentrations of solutes in moles per liter, gases in atmospheres, and omit solids, liquids, and solvents. Explain why.

2. Why do we say that the equilibrium constant for the reaction $H_2O \rightleftharpoons H^+ + OH^-$ (or any other reaction) is dimensionless?

3. Predictions about the direction of a reaction based on Gibbs free energy or Le Châtelier's principle are said to be *thermodynamic,* not *kinetic.* Explain what this means.

4. Write the expression for the equilibrium constant for each of the following reactions. Write the pressure of a gaseous molecule, X, as P_X.

(a) $3Ag^+(aq) + PO_4^{3-}(aq) \rightleftharpoons Ag_3PO_4(s)$

(b) $C_6H_6(l) + \frac{15}{2} O_2(g) \rightleftharpoons 3H_2O(l) + 6CO_2(g)$

5. For the reaction $2A(g) + B(aq) + 3C(l) \rightleftharpoons D(s) + 3E(g)$, the concentrations at equilibrium are found to be

A: 2.8×10^3 Pa D: 16.5 M

B: 1.2×10^{-2} M E: 3.6×10^4 torr

C: 12.8 M

Find the numerical value of the equilibrium constant that would appear in a conventional table of equilibrium constants.

6. From the equations

$$HOCl \rightleftharpoons H^+ + OCl^- \qquad\qquad K = 3.0 \times 10^{-8}$$

$$HOCl + OBr^- \rightleftharpoons HOBr + OCl^- \qquad K = 15$$

find the value of K for the reaction $HOBr \rightleftharpoons H^+ + OBr^-$. All species are aqueous.

7. (a) A favorable entropy change occurs when ΔS is positive. Does the order of the system increase or decrease when ΔS is positive?

(b) A favorable enthalpy change occurs when ΔH is negative. Does the system absorb heat or give off heat when ΔH is negative?

(c) Write the relation between ΔG, ΔH, and ΔS. Use the results of (a) and (b) to state whether ΔG must be positive or negative for a spontaneous change.

8. For the reaction $HCO_3^- \rightleftharpoons H^+ + CO_3^{2-}$, $\Delta G° = +59.0$ kJ/mol at 298.15 K. Find the value of K for the reaction.

9. The formation of tetrafluoroethylene from its elements is highly exothermic:

$$2F_2(g) \quad + \quad 2C(s) \quad \rightleftharpoons \quad F_2C{=}CF_2(g)$$
$$\text{Fluorine} \qquad \text{Graphite} \qquad \text{Tetrafluoroethylene}$$

(a) If a mixture of F_2, graphite, and C_2F_4 is at equilibrium in a closed container, will the reaction go to the right or to the left if F_2 is added?

(b) Rare bacteria from the planet Telfon eat C_2F_4 and make Telfon for their cell walls. Will the reaction go to the right or to the left when these bacteria are added?

```
         F   F   F   F   F   F
         |   |   |   |   |   |
 · · · —C — C — C — C — C — C · · · Teflon
         |   |   |   |   |   |
         F   F   F   F   F   F
```

(c) Will the reaction go right or left if solid graphite is added? (Neglect any effect of increased pressure due to the decreased volume in the vessel when solid is added.)

(d) Will the reaction go right or left if the container is crushed to one-eighth its original volume?

(e) Does the equilibrium constant become larger or smaller if the container is heated?

10. When $BaCl_2 \cdot H_2O(s)$ is dried in an oven, it loses gaseous water:

$$BaCl_2 \cdot H_2O(s) \rightleftharpoons BaCl_2(s) + H_2O(g)$$
$$\Delta H° = 63.11 \text{ kJ/mol at } 25°C$$
$$\Delta S° = +148 \text{ J/(K·mol) at } 25°C$$

(a) Calculate the vapor pressure of gaseous H_2O above $BaCl_2 \cdot H_2O$ at 298 K.

(b) Assuming that $\Delta H°$ and $\Delta S°$ are not temperature dependent (a poor assumption), estimate the temperature at which the vapor pressure of $H_2O(g)$ above $BaCl_2 \cdot H_2O(s)$ will be 1 atm.

11. The equilibrium constant for the reaction of ammonia with water has the following values in the range 5–10°C:

$$NH_3(aq) + H_2O \rightleftharpoons NH_4^+ + OH^-$$
$$K = 1.479 \times 10^{-5} \text{ at } 5°C$$
$$K = 1.570 \times 10^{-5} \text{ at } 10°C$$

(a) Assuming that $\Delta H°$ and $\Delta S°$ are constant in the interval 5°–10°C, use Equation 5-8 to find the value of $\Delta H°$ for the reaction in this temperature range.

(b) Describe how Equation 5-8 could be used to make a linear graph to determine $\Delta H°$, if $\Delta H°$ and $\Delta S°$ were constant over some range of temperature.

12. For the reaction $H_2(g) + Br_2(g) \rightleftharpoons 2HBr(g)$, $K = 7.2 \times 10^{-4}$ at 1 362 K and $\Delta H°$ is positive. A vessel is charged with 48.0 Pa HBr, 1 370 Pa H_2, and 3 310 Pa Br_2 at 1 362 K.

(a) Will the reaction proceed to the left or to right to reach equilibrium?

(b) Calculate the pressure (in pascals) of each species in the vessel at equilibrium.

(c) If the mixture at equilibrium is compressed to half its original volume, will the reaction proceed to the left or right to reestablish equilibrium?

(d) If the mixture at equilibrium is heated from 1 362 to 1 407 K, will HBr be formed or consumed in order to reestablish equilibrium?

Solubility Product

13. Use the solubility product to calculate the solubility of CuBr (FW 143.45) in water expressed as (a) moles per liter and (b) grams per 100 mL.

14. Use the solubility product to calculate the solubility of $Ag_4Fe(CN)_6$ (FW 643.43) ($\rightarrow 4Ag^+ + Fe(CN)_6^{4-}$) in water expressed as (a) moles per liter; (b) grams per 100 mL; (c) ppb Ag^+ ($\approx$ ng Ag^+/mL).

15. Ag^+ at a concentration of 10–100 ppb (ng/mL) is an effective disinfectant for swimming pool water. However, for human health reasons, the concentration should not exceed a few hundred parts per billion. One way to maintain an appropriate concentration of Ag^+ is to add a slightly soluble silver salt to the pool. For each salt below, calculate the concentration of Ag^+ (in parts per billion) that would exist at equilibrium.

AgCl:　$K_{sp} = 1.8 \times 10^{-10}$

AgBr:　$K_{sp} = 5.0 \times 10^{-13}$

AgI:　$K_{sp} = 8.3 \times 10^{-17}$

16. For the reaction below, $K = 6.9 \times 10^{-18}$.

$$Cu(OH)_{1.5}(SO_4)_{0.25}(s) \rightleftharpoons Cu^{2+} + \tfrac{3}{2}OH^- + \tfrac{1}{4}SO_4^{2-}$$

What will be the concentration of Cu^{2+} in a solution saturated with $Cu(OH)_{1.5}(SO_4)_{0.25}$?

17. The solubility product for $BaF_2(s)$ is given by $[Ba^{2+}][F^-]^2 = K_{sp}$. This equation is plotted in the figure. At the point A, $[Ba^{2+}][F^-]^2 > K_{sp}$ and the solution is said to be *supersaturated*. It contains more solute than should be present at equilibrium. Suppose that a supersaturated solution with composition A on the graph is prepared by some means. Is the following statement true or false? "The composition of the solution at equilibrium will be given by the point B, which is connected to A by a line of slope = 2."

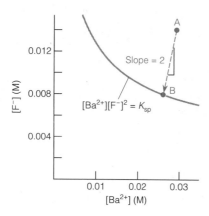

Common Ion Effect

18. Find the solubility (grams per liter) of $CaSO_4$ (FW 136.14) in **(a)** distilled water; **(b)** 0.50 M $CaCl_2$.

19. Express the solubility of $AgIO_3$ in 10.0 mM KIO_3 as a fraction of its solubility in pure water.

20. **(a)** Calculate the solubility (milligrams per liter) of $Zn_2Fe(CN)_6$ in distilled water. This salt dissociates as follows:

$$Zn_2Fe(CN)_6(s) \rightleftharpoons 2Zn^{2+} + \underset{\text{Ferrocyanide}}{Fe(CN)_6^{4-}}$$

$$K_{sp} = 2.1 \times 10^{-16}$$

(b) Calculate the molar concentration of ferrocyanide in a solution of 0.040 M $ZnSO_4$ saturated with $Zn_2Fe(CN)_6$. $ZnSO_4$ will be completely dissociated into Zn^{2+} and SO_4^{2-}.

(c) What molar concentration of $K_4Fe(CN)_6$ should be added to a suspension of solid $Zn_2Fe(CN)_6$ in water to give $[Zn^{2+}] = 5.0 \times 10^{-7}$ M?

21. Let's consider how to use *trial-and-error for solving complicated equations*. In Box 5-2 we tried to solve the equation $x(3x + 0.01)^3 - 10^{-8} = 0$. The solutions are called the *roots* of the equation. For a polynomial whose highest power is x^n, there will be n roots, some of which may be imaginary (involving $\sqrt{-1}$) and some of which may be equal to each other. If we are solving for a concentration, we only care about positive real roots, because a concentration cannot be negative.

Set up a spreadsheet in which column A is x and column B is $x(3x + 0.01)^3 - 10^{-8}$. Vary x from -5 to $+5$ and you will see that the function $x(3x + 0.01)^3 - 10^{-8}$ has roots near zero. Use systematic guessing to vary the last decimal place of x until the value of $x(3x + 0.01)^3 - 10^{-8}$ is closest to 0. Then investigate the next decimal place to the right and refine your answer. Find one positive root and one negative root, each with four significant digits.

22. Your spreadsheet may allow a *circular definition*, in which a cell is defined in terms of itself. This is a very helpful feature for solving equations. For example, to solve the equation $x = \sqrt{4x + 1}$, you can type the function "=Sqrt(4*A1+1)" in cell A1. If your spreadsheet allows this, it will find a value of x that satisfies the condition $x = \sqrt{4x + 1}$ (to some number of decimal places). Try this out to see whether your software permits this. (Answer: 4.236 067 98.)

Separation by Precipitation

23. Consideration of solubility products leads us to expect that cation A^{3+} can be 99.999 9% separated from cation B^{2+} by addition of anion X^-. When the separation is tried, we find 0.2% contamination of the precipitated AX_3 with B^{2+}. Explain what may be happening.

24. What concentration of carbonate must be added to 0.10 M Zn^{2+} to precipitate 99.9% of the Zn^{2+}?

25. If a solution containing 0.10 M Cl^-, Br^-, I^-, and CrO_4^{2-} is treated with Ag^+, in what order will the anions precipitate?

26. A given solution contains 0.050 0 M Ca^{2+} and 0.030 0 M Ag^+. Can 99% of either ion be precipitated by addition of sulfate, without precipitating the other metal ion? What will be the concentration of Ca^{2+} when Ag_2SO_4 begins to precipitate?

Complex Formation

27. Explain why the total solubility of lead species in Figure 5-1 first decreases and then increases as the concentration of I^- increases. Give an example of the chemistry that occurs in each of the two domains.

28. Identify the Lewis acids in the following reactions:

(a) $BF_3 + NH_3 \rightleftharpoons F_3B^- \!-\! \overset{+}{N}H_3$

(b) $F^- + AsF_5 \rightleftharpoons AsF_6^-$

29. The cumulative formation constant for $SnCl_2(aq)$ in 1.0 M $NaNO_3$ is $\beta_2 = 12$. Find the concentration of $SnCl_2(aq)$ for a solution in which the concentrations of Sn^{2+} and Cl^- are both somehow fixed at 0.20 M.

30. Given the equilibria below, calculate the concentrations of each zinc-containing species in a solution saturated with $Zn(OH)_2(s)$ and containing $[OH^-]$ at a fixed concentration of 3.2×10^{-7} M.

$$Zn(OH)_2(s): \quad K_{sp} = 3.0 \times 10^{-16}$$
$$Zn(OH)^+: \quad \beta_1 = 2.5 \times 10^4$$
$$Zn(OH)_3^-: \quad \beta_3 = 7.2 \times 10^{15}$$
$$Zn(OH)_4^{2-}: \quad \beta_4 = 2.8 \times 10^{15}$$

31. Although KOH, RbOH, and CsOH show no evidence of association between metal and hydroxide in aqueous solution, Li^+ and Na^+ do form complexes with OH^-:

$$Li^+ + OH^- \rightleftharpoons LiOH(aq) \quad K_1 = \frac{[LiOH(aq)]}{[Li^+][OH^-]} = 0.83$$

$$Na^+ + OH^- \rightleftharpoons NaOH(aq) \quad K_1 = 0.20$$

Calculate the fraction of sodium in the form $NaOH(aq)$ in 1 F NaOH.

32. Consider the following equilibria:[16]

$$AgI(s) \rightleftharpoons Ag^+ + I^- \quad K_{sp} = 4.5 \times 10^{-17}$$
$$Ag^+ + I^- \rightleftharpoons AgI(aq) \quad \beta_1 = 1.3 \times 10^8$$
$$Ag^+ + 2I^- \rightleftharpoons AgI_2^- \quad \beta_2 = 9.0 \times 10^{10}$$
$$Ag^+ + 3I^- \rightleftharpoons AgI_3^{2-} \quad \beta_3 = 5.6 \times 10^{13}$$
$$Ag^+ + 4I^- \rightleftharpoons AgI_4^{3-} \quad \beta_4 = 2.5 \times 10^{14}$$
$$2Ag^+ + 6I^- \rightleftharpoons Ag_2I_6^{4-} \quad K_{26} = 7.6 \times 10^{29}$$
$$3Ag^+ + 8I^- \rightleftharpoons Ag_3I_8^{5-} \quad K_{38} = 2.3 \times 10^{46}$$

Prepare a spreadsheet in which the concentration of I^- varies from $\log[I^-] = -8$ to $\log[I^-] = 0$ in 0.5-log unit increments. Use the spreadsheet to calculate the concentrations of all the aqueous silver species, as well as the total concentration of dissolved silver ($[Ag]_{total}$), and prepare a graph analogous to Figure 5-1. In calculating total dissolved silver, note that 1 mol of $Ag_2I_6^{4-}$ has 2 mol of Ag and that 1 mol of $Ag_3I_8^{5-}$ has 3 mol of Ag. Here are suggested headings for your spreadsheet columns: $\log[I^-]$, $[I^-]$, $[Ag^+]$, $[AgI(aq)]$, $[AgI_2^-]$, $[AgI_3^{2-}]$, $[AgI_4^{3-}]$, $[Ag_2I_6^{4-}]$, $[Ag_3I_8^{5-}]$, $[Ag]_{total}$, $\log[Ag^+]$, $\log[AgI(aq)]$,

$\log[AgI_2^-]$, $\log[AgI_3^{2-}]$, $\log[AgI_4^{3-}]$, $\log[Ag_2I_6^{4-}]$, $\log[Ag_3I_8^{5-}]$, $\log[Ag]_{total}$.

Acids and Bases

33. Distinguish Lewis acids and bases from Brønsted-Lowry acids and bases. Give an example of each.

34. Fill in the blanks below:

(a) The product of a reaction between a Lewis acid and a Lewis base is called _____.

(b) The bond between a Lewis acid and a Lewis base is called _____ or _____.

(c) Brønsted-Lowry acids and bases related by gain or loss of one proton are said to be _____.

(d) A solution is *acidic* if _____. A solution is *basic* if _____.

35. Why is the pH of distilled water usually <7? How can you prevent this from happening?

36. Gaseous SO_2 is created by combustion of sulfur-containing fuels, especially coal. Explain how SO_2 in the atmosphere makes acidic rain.

37. Use electron dot structures to show why tetramethylammonium hydroxide, $(CH_3)_4N^+OH^-$, is an ionic compound. That is, show why hydroxide is not covalently bound to the rest of the molecule.

38. Identify the Brønsted-Lowry acids among the reactants in the following reactions:

(a) $KCN + HI \rightleftharpoons HCN + KI$

(b) $PO_4^{3-} + H_2O \rightleftharpoons HPO_4^{2-} + OH^-$

39. Write the autoprotolysis reaction of H_2SO_4.

40. Identify the conjugate acid-base pairs in the following reactions:

(a) $H_3\overset{+}{N}CH_2CH_2\overset{+}{N}H_3 + H_2O \rightleftharpoons$
$H_3\overset{+}{N}CH_2CH_2NH_2 + H_3O^+$

(b)

Benzoic acid   Pyridine

Benzoate   Pyridinium

pH

41. Calculate the concentration of H^+ and the pH of the following solutions:

(a) 0.010 M HNO_3 **(b)** 0.035 M KOH

(c) 0.030 M HCl　　(e) 0.010 M $[(CH_3)_4N^+]OH^-$

(d) 3.0 M HCl

Tetramethylammonium
hydroxide

42. Use Table 5-1 to calculate the pH of pure water at (a) 0°C; (b) 25°C; (c) 100°C.

43. The equilibrium constant for the reaction $H_2O \rightleftharpoons H^+ + OH^-$ is 1.0×10^{-14} at 25°C. What is the value of K for the reaction $4H_2O \rightleftharpoons 4H^+ + 4OH^-$?

44. An acidic solution containing 0.010 M La^{3+} is treated with NaOH until $La(OH)_3$ precipitates. At what pH does this occur?

45. Consider the values of K_w in Table 5-1. Use Le Châtelier's principle to decide whether the autoprotolysis of water is endothermic or exothermic at (a) 25°C; (b) 100°C; (c) 300°C.

Strengths of Acids and Bases

46. Make a list of the common strong acids and strong bases. Memorize this list.

47. Write the formulas and names for two classes of weak acids and two classes of weak bases.

48. Write the K_a reaction for trichloroacetic acid, Cl_3CCO_2H, and for the anilinium ion, $-NH_3^+$.

49. Write the K_b reactions for pyridine and for sodium 2-mercaptoethanol.

Pyridine　　　　　HOCH$_2$CH$_2$$\ddot{S}$:$^-Na^+$

Sodium 2-mercaptoethanol

50. Write the K_a and K_b reactions of $NaHCO_3$.

51. Write the stepwise acid-base reactions for the following ions in water. Write the correct symbol (e.g., K_{b1}) for the equilibrium constant for each reaction.

(a)　$H_3\overset{+}{N}CH_2CH_2\overset{+}{N}H_3$

Ethylenediammonium ion

(b)　$^-OCCH_2CO^-$ (with O=C, C=O)

Malonate ion

52. Which is a stronger acid, (a) or (b)?

(a)　Cl_2HCCOH (with C=O)

Dichloroacetic acid
$K_a = 5.0 \times 10^{-2}$

(b)　ClH_2CCOH (with C=O)

Chloroacetic acid
$K_a = 1.36 \times 10^{-3}$

Which is a stronger base, (c) or (d)?

(c)　H_2NNH_2

Hydrazine
$K_b = 3.0 \times 10^{-6}$

(d)　H_2NCNH_2 (with C=O)

Urea
$K_b = 1.5 \times 10^{-14}$

53. Write the K_b reaction of CN^-. Given that the K_a value for HCN is 6.2×10^{-10}, calculate K_b for CN^-.

54. Write the K_{a2} reaction of phosphoric acid (H_3PO_4) and the K_{b2} reaction of disodium oxalate ($Na_2C_2O_4$).

55. From the K_b values for phosphate in Equations 5-34 through 5-36, calculate the three K_a values of phosphoric acid.

56. From the equilibrium constants below, calculate the equilibrium constant for the reaction $HO_2CCO_2H \rightleftharpoons 2H^+ + C_2O_4^{2-}$. All species in this problem are aqueous.

$$HOCCOH \rightleftharpoons H^+ + HOCCO^-$$

Oxalic acid

$$K_1 = 5.6 \times 10^{-2}$$

$$HOCCO^- \rightleftharpoons H^+ + {}^-OCCO^-$$

Oxalate

$$K_2 = 5.4 \times 10^{-5}$$

57. (a) Using only K_{sp} from Table 5-3, calculate how many moles of $Ca(OH)_2$ will dissolve in 1.00 L of water.

(b) How will the solubility calculated in (a) be affected by the K_1 reaction in Table 5-3?

58. The planet Aragonose (which is made mostly of the mineral aragonite, whose composition is $CaCO_3$) has an atmosphere containing methane and carbon dioxide, each at a pressure of 0.10 atm. The oceans are saturated with aragonite and have a concentration of H^+ equal to 1.8×10^{-7} M. Given the following equilibria, calculate how many grams of calcium are contained in 2.00 L of Aragonose seawater.

$$CaCO_3(s, \text{aragonite}) \rightleftharpoons Ca^{2+}(aq) + CO_3^{2-}(aq)$$

$$K_{sp} = 6.0 \times 10^{-9}$$

$$CO_2(g) \rightleftharpoons CO_2(aq)$$

$$K_{CO_2} = 3.4 \times 10^{-2}$$

$$CO_2(aq) + H_2O(l) \rightleftharpoons HCO_3^-(aq) + H^+(aq)$$

$$K_1 = 4.4 \times 10^{-7}$$

$$HCO_3^-(aq) \rightleftharpoons H^+(aq) + CO_3^{2-}(aq)$$

$$K_2 = 4.7 \times 10^{-11}$$

Remain calm! Reverse the first reaction, add all the reactions together, and see what cancels.

Notes and References

1. D. P. Sheer and D. C. Harris, *J. Water Pollution Control Federation* **1982,** *54,* 1441.

2. More than you ever wanted to know about Le Châtelier's principle can be found in F. G. Helfferich, *J. Chem. Ed.* **1985,** *62,* 305; R. Fernández-Prini, *J. Chem. Ed.* **1982,** *59,* 550; and L. K. Brice, *J. Chem. Ed.* **1983,** *60,* 387.

3. The solubility of the vast majority of ionic compounds increases with temperature, despite the fact that the standard heat of solution ($\Delta H°$) is negative for about half of them. Discussions of this seeming contradiction can be found in G. M. Bodner, *J. Chem. Ed.* **1980,** *57,* 117; and R. S. Treptow, *J. Chem. Ed.* **1984,** *61,* 499.

4. A. Renuka, *J. Chem. Ed.* **1993,** *70,* 871.

5. A. K. Sawyer, *J. Chem. Ed.* **1983,** *60,* 416.

6. E. Koubek, *J. Chem. Ed.* **1993,** *70,* 155.

7. A nice classroom demonstration of selective precipitation by addition of Pb^{2+} to a solution containing carbonate and iodide is described by T. P. Chirpich, *J. Chem. Ed.* **1988,** *65,* 359.

8. Some of the most useful compilations of equilibrium constants are found in L. G. Sillén and A. E. Martell, *Stability Constants of Metal-Ion Complexes,* Chemical Society (London) Special Publications No. 17 and 25, 1964 and 1971; and A. E. Martell and R. M. Smith, *Critical Stability Constants* (New York: Plenum Press, 1974–1989), a multivolume collection. Books dealing with the experimental measurement of equilibrium constants include A. Martell and R. Motekaitis, *Determination and Use of Stability Constants* (New York: VCH Publishers, 1992); K. A. Conners, *Binding Constants: The Measurement of Molecular Complex Stability* (New York: Wiley, 1987); M. Meloun, *Computation of Solution Equilibria* (New York: Wiley, 1988); and D. J. Leggett, Ed., *Computational Methods for the Determination of Formation Constants* (New York: Plenum Press, 1985).

9. F. A. Cotton, C. K. Fair, G. E. Lewis, G. N. Mott, K. K. Ross, A. J. Schultz, and J. M. Williams, *J. Am. Chem. Soc.* **1984,** *106,* 5319.

10. K. Abu-Dari, K. N. Raymond, and D. P. Freyberg, *J. Am. Chem. Soc.* **1979,** *101,* 3688.

11. A. Bino and D. Gibson, *Inorg. Chem.* **1984,** *23,* 109.

12. P. A. Giguère, *J. Chem. Ed.* **1979,** *56,* 571; P. A. Giguère and S. Turrell, *J. Chem. Soc.* **1980,** *102,* 5473.

13. R. I. Gelb, L. M. Schwartz, and D. A. Laufer, *J. Chem. Soc.* **1981,** *103,* 5664.

14. M. Kern, *J. Chem. Ed.* **1960,** *37,* 14; H. S. Harned and R. Davis, Jr., *J. Am. Chem. Soc.* **1943,** *65,* 2030.

15. J. Åqvist, M. Fothergill, and A. Warshel, *J. Am. Chem. Soc.* **1993,** *115,* 631; D. N. Silverman and S. Lindskog, *Acc. Chem. Res.* **1988,** *21,* 30.

16. The equilibrium constants apply in 4 M $NaClO_4$ at 25°C. The structures of some of these complexes, and such unusual species as Ag_3I^{2+}, are discussed by G. B. Kauffman, M. Karabassi, and G. Bergerhoff, *J. Chem. Ed.* **1984,** *61,* 729.

The Ozone Hole[1]

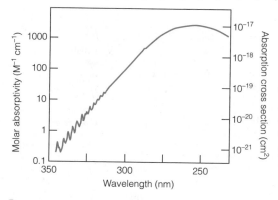

Spectrum of ozone showing maximum absorption of ultraviolet radiation at a wavelength near 260 nm. [Adapted from R. P. Wayne, *Chemistry of Atmospheres* (Oxford: Clarendon Press, 1991).]

Ozone, formed at altitudes of 20 to 40 km by the action of solar ultraviolet radiation ($h\nu$) on O_2, absorbs ultraviolet radiation that causes sunburns and skin cancers.

$$O_2 \xrightarrow{h\nu} 2O \qquad O + O_2 \rightarrow \underset{\text{Ozone}}{O_3}$$

In 1985 the British Antarctic Survey reported that ozone in the Antarctic stratosphere had decreased by 50% in early spring, relative to levels observed in the preceding 20 years. Observations have since shown that this "ozone hole" occurs only in early spring and is gradually becoming worse.

The current explanation of this phenomenon begins with chlorofluorocarbons such as Freon-12® (CCl_2F_2) from refrigerators. These compounds diffuse to the stratosphere, where they catalyze ozone decomposition:

(1) $CCl_2F_2 \xrightarrow{h\nu} CCIF_2 + Cl$ Photochemical Cl formation

(2) $Cl + O_3 \rightarrow ClO + O_2$

(3) $O_3 \xrightarrow{h\nu} O + O_2$

(4) $O + ClO \rightarrow Cl + O_2$

Catalytic O_3 destruction
Net reaction of (2)–(4):
$2O_3 \rightarrow 3O_2$

A single Cl atom in this chain reaction can destroy $>10^5$ molecules of O_3. The chain is terminated when Cl or ClO reacts with hydrocarbons or NO_2 to form HCl or $ClONO_2$.

Stratospheric clouds formed during the Antarctic winter catalyze the reaction of HCl with $ClONO_2$ to form Cl_2, which is subsequently split by sunlight into Cl atoms to initiate O_3 destruction:

$$HCl + ClONO_2 \xrightarrow[\text{polar clouds}]{\text{surface of}} Cl_2 + HNO_3 \qquad Cl_2 \xrightarrow{h\nu} 2Cl$$

Polar stratospheric clouds require winter cold to form. It is only when the sun is rising in September and October, and the clouds are still present, that conditions are right for ozone destruction.

To protect life from ultraviolet radiation, international treaties now ban or phase out chlorofluorocarbons, and there is an effort to find safe substitutes. However, ozone levels may not return to historic values until late in the next century. A pending issue is the widespread use of the agricultural fumigant bromomethane (CH_3Br), which is also a potent ozone destroyer.

A First Look at Spectrophotometry

6

Spectrophotometry is any procedure that uses light to measure chemical concentrations. In this chapter we discuss how absorption of electromagnetic radiation is used in quantitative analysis. This foundation should allow you to use spectrophotometry in the laboratory and to appreciate its application to other topics throughout this book. We will also learn to use *standard curves, standard addition,* and *internal standards,* which are calibration procedures common to all areas of analytical chemistry. In Chapters 19 through 21 more detailed discussions of spectrophotometry are presented.

6-1 Properties of Light

It is convenient to describe light in terms of both particles and waves. Light waves consist of perpendicular, oscillating electric and magnetic fields. For simplicity, a *plane-polarized* wave is shown in Figure 6-1. In this figure the electric field is confined to the *xy* plane, and the magnetic field is confined to the *xz* plane. The **wavelength,** λ, is the crest-to-crest distance between waves.

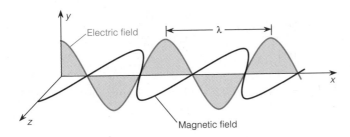

Figure 6-1 Plane-polarized electromagnetic radiation of wavelength λ, propagating along the *x* axis. The electric field of plane-polarized light is confined to a single plane. Ordinary, unpolarized light has electric field components in all planes.

Following the discovery of the Antarctic ozone "hole" in 1985, atmospheric chemist Susan Solomon led the first expedition in 1986 specifically intended to make chemical measurements of the Antarctic atmosphere by using balloons and ground-based spectroscopy. The expedition discovered that ozone depletion occurred after polar sunrise and that the concentration of chemically active chlorine in the stratosphere was ~ 100 times greater than had been predicted from gas-phase chemistry. Solomon's group identified chlorine as the culprit in ozone destruction and polar stratospheric clouds as the catalytic surface for the release of so much chlorine.

The **frequency,** ν, is the number of complete oscillations that the wave makes each second. The unit of frequency is reciprocal seconds, s^{-1}. One oscillation per second is also called 1 **hertz** (Hz). A frequency of 10^6 s^{-1} is therefore said to be 10^6 Hz, or 1 *megahertz* (MHz).

The relation between frequency and wavelength is

Relation between frequency and wavelength: $\nu\lambda = c$ (6-1)

where c is the speed of light (2.99792458×10^8 m/s in vacuum). In a medium other than vacuum, the speed of light is c/n, where n is the **refractive index** of that medium. For visible wavelengths in most substances, $n > 1$, so visible light travels more slowly through matter than through vacuum. When light moves between two media with different refractive indexes, the frequency remains constant but the wavelength changes.

With regard to energy, it is more convenient to think of light as particles called **photons.** Each photon carries the energy, E, which is given by

Relation between energy and frequency: $E = h\nu$ (6-2)

where h is **Planck's constant** ($= 6.6260755 \times 10^{-34}$ J·s).

Equation 6-2 states that energy is proportional to frequency. Combining Equations 6-1 and 6-2, we can write

$$E = \frac{hc}{\lambda} = hc\tilde{\nu} \tag{6-3}$$

where $\tilde{\nu}$ ($= 1/\lambda$) is called the **wavenumber.** We see that energy is inversely proportional to wavelength and directly proportional to wavenumber. Red light, with a longer wavelength than blue light, is thus less energetic than blue light. The SI unit for wavenumber is reciprocal meters, m^{-1}. However, the most common unit of wavenumber in the literature is cm^{-1}, read "reciprocal centimeters" or "wavenumber."

The major regions of the **electromagnetic spectrum** are labeled in Figure 6-2. The various names reflect the history of physical science. There are no discontinuities in the properties of radiation from one region of the spectrum to another. Visible light, which is the kind our eyes detect, represents only a very small fraction of the electromagnetic spectrum.

· · · · · · · · · · · · · · · · · · · ·

EXAMPLE   Photon Energies

By how many kilojoules per mole is the energy of a molecule increased when it absorbs (a) visible light with a wavelength of 500 nm or (b) infrared radiation with a wavenumber of 1251 cm^{-1}?

Solution   For the visible light, the energy increase is

$$\Delta E = h\nu = h\frac{c}{\lambda}$$

$$= (6.6261 \times 10^{-34} \text{ J·s}) \frac{(2.9979 \times 10^8 \text{ m/s})}{(500 \text{ nm})(10^{-9} \text{ m/nm})}$$

$$= 3.97 \times 10^{-19} \text{ J/molecule}$$

$$(3.97 \times 10^{-19} \text{ J/molecule})(6.022 \times 10^{23} \text{ molecules/mol}) = 239 \text{ kJ/mol}$$

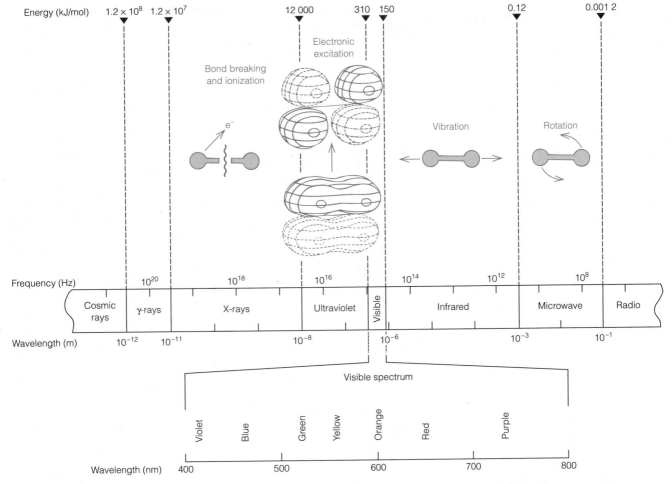

Figure 6-2 Electromagnetic spectrum showing representative molecular processes that occur when light in each region is absorbed. The visible spectrum spans the wavelength range 380–780 nanometers (1 nm = 10^{-9} m).

For the infrared light, the energy increase is

$$\Delta E = h\nu = h\frac{c}{\lambda} = hc\tilde{\nu} \quad \left(\text{recall that } \tilde{\nu} = \frac{1}{\lambda}\right)$$

$$= (6.626\,1 \times 10^{-34} \text{ J·s})(2.997\,9 \times 10^{8} \text{ m/s})(1\,251 \text{ cm}^{-1})(100 \text{ cm/m})$$

$$= 2.485 \times 10^{-20} \text{ J/molecule} = 14.97 \text{ kJ/mol}$$

6-2 Absorption of Light

When a molecule absorbs a photon, the energy of the molecule is increased. We say that the molecule is promoted to an **excited state** (Figure 6-3). If a molecule emits a photon, the energy of the molecule is lowered. The lowest energy state of a molecule is called the **ground state.** Figure 6-2 indicates

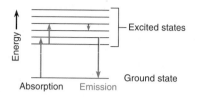

Figure 6-3 Absorption of light increases the energy of a molecule. Emission of light decreases its energy.

that microwave radiation stimulates rotational motion of molecules when it is absorbed. Similarly, infrared radiation stimulates vibrations of molecules, and visible light and ultraviolet radiation cause electrons to be promoted to higher energy orbitals. X-rays and short-wavelength ultraviolet radiation break chemical bonds and ionize molecules. Medical X-rays damage the human body and should be minimized.

When light is absorbed by a sample, the **radiant power** of the beam of light is decreased. Radiant power, P, refers to the energy per second per unit area of the light beam. A rudimentary spectrophotometric experiment is illustrated in Figure 6-4. Light is passed through a **monochromator** (a prism, a grating, or even a filter) to select one wavelength. (*Monochromatic* light consists of a single color wavelength.) Light of this wavelength, with radiant power P_0, strikes a sample of length b. The radiant power of the beam emerging from the other side of the sample is P. Some of the light may be absorbed by the sample, so $P \le P_0$.

The **transmittance**, T, is defined as the fraction of the original light that passes through the sample.

Transmittance:
$$T = \frac{P}{P_0} \qquad (6\text{-}4)$$

Therefore, T has the range 0 to 1. The *percent transmittance* is simply $100 \cdot T$ and ranges between 0 and 100%. A more useful quantity is the **absorbance**, sometimes called optical density, defined as

Relation between transmittance and absorbance:

P/P_0	$\% \, T$	A
1	100	0
0.1	10	1
0.01	1	2

Absorbance:
$$A = \log_{10}\left(\frac{P_0}{P}\right) = -\log T \qquad (6\text{-}5)$$

When no light is absorbed, $P = P_0$ and $A = 0$. If 90% of the light is absorbed, then 10% is transmitted and $P = P_0/10$. This ratio gives $A = 1$. If only 1% of the light is transmitted, $A = 2$.

The reason absorbance is so important is that it is directly proportional to the concentration, c, of light-absorbing species in the sample:

Box 6-1 explains why absorbance, not transmittance, is directly proportional to concentration.

Beer's law:
$$A = \epsilon b c \qquad (6\text{-}6)$$

Equation 6-6, which lies at the heart of spectrophotometry as applied to analytical chemistry, is called the Beer-Lambert law, or simply **Beer's law.** The absorbance, A, is dimensionless, but some authors write "absorbance units" after absorbance. The concentration of the sample, c, is usually given in units of moles per liter (M). The pathlength, b, is commonly expressed in centimeters. The quantity ϵ (epsilon) is·called the **molar absorptivity** (or *extinction coefficient* in the older literature) and has the units $M^{-1} \, cm^{-1}$ (because the product $\epsilon b c$ must be dimensionless). Molar absorptivity is the characteristic of a substance that tells how much light is absorbed at a particular wavelength.

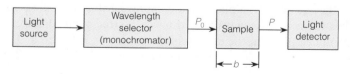

Figure 6-4 Schematic diagram of a single-beam spectrophotometric experiment.

Box 6-1 Why Is There a Logarithmic Relation Between Transmittance and Concentration?

Beer's law, Equation 6-6, states that the *absorbance* of a sample is directly proportional to the concentration of the absorbing species. The fraction of light passing through the sample (the *transmittance*) is related logarithmically, not linearly, to the sample concentration. Why should this be?

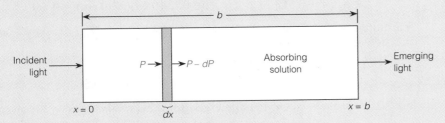

Imagine light of radiant power P passing through an infinitesimally thin layer of solution whose thickness is dx. The decrease in power (dP) is proportional to the incident power (P), to the concentration of absorbing species (c), and to the thickness of the section (dx):

$$dP = -\beta Pc\, dx$$

where β is a constant of proportionality and the minus sign indicates a decrease in P as x increases. The rationale for saying that the decrease in power is proportional to the incident power may be understood from a numerical example. If 1 photon out of 1 000 incident photons is absorbed in a thin layer of solution, we would expect that 2 photons out of 2 000 incident photons would be absorbed. The decrease in photons (power) is proportional to the incident flux of photons (power).

The equation above can be rearranged and integrated quite simply:

$$-\frac{dP}{P} = \beta c\, dx \;\Rightarrow\; -\int_{P_0}^{P} \frac{dP}{P} = \beta c \int_{0}^{b} dx$$

The limits of integration are $P = P_0$ at $x = 0$ and $P = P$ at $x = b$.

$$-\ln P - (-\ln P_0) = \beta cb \;\Rightarrow\; \ln\left(\frac{P_0}{P}\right) = \beta cb$$

Finally, converting ln to log, using the relation $\ln z = (\ln 10)(\log z)$, gives

$$\log\left(\frac{P_0}{P}\right) = \underbrace{\left(\frac{\beta}{\ln 10}\right)}_{\text{Constant} \equiv \epsilon} cb$$

or

$$A = \epsilon cb$$

which is Beer's law!

The logarithmic relation of P_0/P and concentration arises from the fact that, in each infinitesimal portion of the total volume, *the decrease in power is proportional to the power incident upon that section.* As light travels through the sample, the drop in power in each succeeding layer decreases, because the magnitude of the incident power that reaches each layer is decreasing. Molar absorptivity ranges from 0 (when the probability for photon absorption approaches 0) to approximately 10^5 M^{-1} cm^{-1} (when the probability for photon absorption approaches unity).

EXAMPLE Absorbance, Transmittance, and Beer's Law

Find the absorbance and transmittance of a 0.002 40 M solution of a substance with a molar absorptivity of 313 M^{-1} cm^{-1} in a cell with a 2.00-cm pathlength.

Solution Equation 6-6 gives us the absorbance:

$$A = \epsilon bc = (313 \ M^{-1} \ cm^{-1})(0.002\,40 \ M)(2.00 \ cm) = 1.50$$

Transmittance is obtained from Equation 6-5 by raising 10 to the power on each side of the equation:

$$\log T = -A$$
$$T = 10^{\log T} = 10^{-A} = 10^{-1.50} = 0.0316$$

Just 3.16% of the incident light emerges from this solution.

If $x = y$, $10^x = 10^y$.

Equation 6-6 could be written

$$A_\lambda = \epsilon_\lambda bc$$

because the values of A and ϵ depend on the wavelength of light. The quantity ϵ is simply a coefficient of proportionality between absorbance and the product bc. The larger the value of ϵ, the greater A is. An **absorption spectrum** is a graph showing how A (or ϵ) varies with wavelength. Demonstration 6-1 illustrates the meaning of an absorption spectrum and shows several examples of spectra.

The color of a solution is the complement of the color of the light that it absorbs.

The part of a molecule responsible for light absorption is called a **chromophore.** Any substance that absorbs visible light will appear colored when white light is transmitted through it or reflected from it. The substance absorbs certain wavelengths of the white light, and our eyes detect the wavelengths that are not absorbed. A rough guide to colors is given in Table 6-1. The observed color is called the *complement* of the absorbed color. As an example, bromophenol blue has a visible absorbance maximum at 591 nm, and its observed color is blue.

TABLE 6-1 Colors of visible light

Wavelength of maximum absorption (nm)	Color absorbed	Color observed
380–420	Violet	Green-yellow
420–440	Violet-blue	Yellow
440–470	Blue	Orange
470–500	Blue-green	Red
500–520	Green	Purple
520–550	Yellow-green	Violet
550–580	Yellow	Violet-blue
580–620	Orange	Blue
620–680	Red	Blue-green
680–780	Purple	Green

The spectrum of visible light can be projected on a screen in a darkened room in the following manner:[2] Four layers of plastic diffraction grating† are mounted on a cardboard frame having a square hole large enough to cover the lens of an overhead projector. This assembly is taped over the projector lens facing the screen. An opaque cardboard surface with two 1 × 3 cm slits is placed on the working surface of the projector.

When the lamp is turned on, the white image of each slit is projected on the center of the screen. A visible spectrum appears on either side of each image. By placing a beaker of colored solution over one slit, you can see its color projected on the screen where the white image previously appeared. The spectrum beside the colored image loses its intensity in regions where the colored species absorbs light.

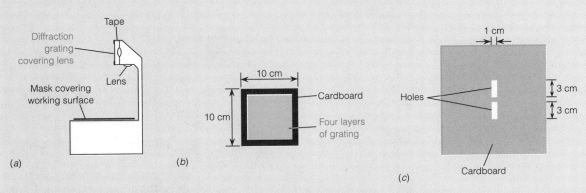

(*a*) Overhead projector. (*b*) Diffraction grating mounted on cardboard. (*c*) Mask for working surface.

Shown in Color Plate 2a are the spectrum of white light and the absorption spectra of three different colored solutions. You can see that potassium dichromate, which appears orange or yellow, absorbs blue wavelengths. Bromophenol blue absorbs orange wavelengths and appears blue to our eyes. The absorption of phenolphthalein is located near the center of the visible spectrum. For comparison, the spectra of these three solutions recorded with a spectrophotometer are shown in Color Plate 2b.

This same setup can be used to demonstrate fluorescence and the properties of colors.[2]

. . . .
†An $8\frac{1}{2}$ × 11 inch sheet of plastic diffraction grating is available from Edmund Scientific Co., 5975 Edscorp Building, Barrington, NJ 08007, catalog no. 40,267.

When Beer's Law Fails

Beer's law states that absorbance is proportional to the concentration of the absorbing species. It works very well for dilute solutions (≤0.01 M) of most substances.

As a solution becomes more concentrated, solute molecules begin to influence each other as a result of their proximity. When one solute molecule interacts with another, the properties of each (including absorption of light) are likely to change. The apparent result is that a graph of absorbance versus concentration is no longer a straight line. In the extreme case, at very high concentration, the solute *becomes* the solvent. Clearly, we cannot expect the electrical properties of a molecule to be the same in different solvents. Sometimes nonabsorbing solutes in a solution interact with the absorbing species and alter the apparent absorptivity.

> Beer's law works for dilute solutions in which the absorbing species is not participating in a concentration-dependent equilibrium.

Box 6-2 Not Just Fluorescent Lamps

A fluorescent lamp is a glass tube filled with mercury vapor; the inner walls are coated with a *phosphor* (luminescent substance) consisting of a calcium halophosphate ($Ca_5(PO_4)_3F_{1-x}Cl_x$) doped with Mn^{2+} and Sb^{3+}. The mercury atoms, promoted to an excited state by electric current passing through the lamp, emit mostly ultraviolet radiation at 254 and 185 nm. This radiation is absorbed by the Sb^{3+} dopant, and some of the energy is passed on to Mn^{2+}. The Sb^{3+} emits blue light, and the Mn^{2+} emits yellow light, with the combined emission appearing white. The emission spectrum is shown below. Fluorescent lamps are important energy-saving devices because they are more efficient than incandescent lamps in the conversion of electricity to light.

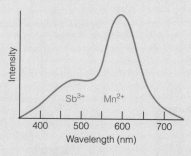

Emission spectrum of the phosphor used in fluorescent lamps. [J. A. De-Luca, *J. Chem. Ed.* **1980**, *57*, 541.]

A little known occurrence of photoemission is that from most white fabrics. Just for fun, turn on an ultraviolet lamp in a darkened room containing several people (*but do not look directly at the lamp*). You will discover a surprising amount of emission from white fabrics (shirts, pants, shoelaces, and lots more) that have been treated with fluorescent compounds to enhance their whiteness. You may also be surprised to see fluorescence from teeth and from recently bruised areas of skin that show no surface damage.

The physical interaction of two solutes is just one example of a chemical equilibrium (the association of two solutes) affecting the apparent absorbance. An even simpler example is that of a weak electrolyte, such as a weak acid. In concentrated solution, the predominant form of the acid will be the undissociated form, HA. As the solution is diluted, more dissociation occurs. If the absorptivity of A^- is not the same as that of HA, the solution will appear not to obey Beer's law as it is diluted.

Molecules Emit Light Too

A molecule can dissipate the energy of an absorbed photon by emitting a photon or by creating heat throughout the medium. Short-lived emission is called *fluorescence* and long-lived emission is *phosphorescence*. We return to a more technical distinction between these terms in Chapter 19. In general, any emission of light is called *luminescence*. Box 6-2 describes some interesting examples of luminescence.

Alternatively, after a photon is absorbed, a bond may break and *photochemistry* may occur (as in the reaction $O_2 \xrightarrow{h\nu} 2O$ in the upper atmosphere). That is, the energy of the excited molecule could overcome the activation energy of a chemical reaction. In some chemical reactions (not necessarily stimulated by light), part of the energy released is in the form of light. Light emitted during a chemical reaction is called *chemiluminescence*.[3]

6-3 The Spectrophotometer

The minimum requirements for a spectrophotometer (a device to measure absorbance of light) were shown in Figure 6-4. Light from a continuous source is passed through a monochromator, which selects a narrow band of wavelengths from the incident beam. This **"monochromatic" light** travels through a sample of pathlength b, and the radiant power of the emergent light is measured. The sample is usually contained in a cell called a **cuvet** that has flat quartz faces. Quartz is used because it transmits visible and ultraviolet light. Most types of glass and plastic absorb ultraviolet radiation, so glass or clear plastic can be used only for measurements at visible wavelengths.

The device in Figure 6-4 is called a single-beam instrument, because it has only one beam of light. We do not measure the incident radiant power, P_0, directly. Rather, the radiant power of light passing through a *reference cuvet* containing pure solvent is *defined* as P_0. This cuvet is then removed and replaced by an identical one containing sample. The radiant power of light striking the detector is then taken as P, permitting T or A to be determined. The reference cuvet, containing pure solvent, compensates for reflection, scattering, or absorption of light by the cuvet and solvent. The radiant power of light striking the detector would not be the same if the reference cuvet were removed from the beam.

A single-beam spectrophotometer is inconvenient because two different samples must be placed alternately in the beam. It is somewhat inaccurate because both the output of the source and the response of the detector fluctuate over time. If there is a change in either one between the time of the measurement of the reference solution and that of the sample solution, the apparent absorbance will be in error. A single-beam instrument is poorly suited to continuous measurements of absorbance, like those in a kinetics experiment, because both the source intensity and the detector response drift.

"Monochromatic" means one wavelength. Although it is impossible to produce truly monochromatic light, the better the monochromator, the narrower is the range of wavelengths in the emerging beam.

Double-Beam Strategy

In a *double-beam spectrophotometer* (Figure 6-5), light passes alternately through the sample and the reference cuvets. This is accomplished by a motor

The double-beam instrument makes essentially continuous measurements of the light emerging from the sample and the reference cells.

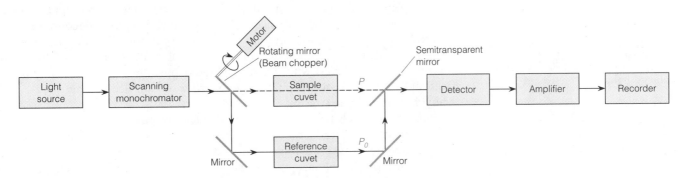

Figure 6-5 Schematic diagram of a double-beam scanning spectrophotometer. The incident beam is passed alternately through the sample and reference cuvets by the rotating beam chopper.

that rotates a mirror into and out of the light path. When the *chopper* is not diverting the beam, the light passes through the sample, and the detector measures the radiant power that we call P in Equation 6-4. When the chopper diverts the beam through the reference cuvet, the detector measures P_0. The beam is chopped several times per second, and the circuitry automatically compares P and P_0 to obtain transmittance and absorbance. This procedure provides automatic correction for changes of the source intensity and detector response with time and wavelength, because the power emerging from the two samples is compared so frequently. Most research-quality spectrophotometers also provide for automatic wavelength scanning and continuous recording of the absorbance.

 In recording an absorbance spectrum, it is a routine procedure to first record the baseline spectrum with reference solutions (pure solvent or a reagent blank) in both cuvets. The absorbance of the reference is then subtracted from the absorbance of the sample to obtain the true absorbance of the sample at each wavelength.

Color Plate 3 shows the dispersion of visible light by a grating.

A typical double-beam ultraviolet-visible spectrophotometer is shown in Figure 6-6. In this instrument, visible light comes from a quartz-halogen lamp (like that in an automobile headlight), and the ultraviolet source is a deuterium arc lamp in which an electric spark causes D_2 gas to dissociate and emit ultraviolet light in the range 200–400 nm. Only one lamp is used at a time. Grating 1 is used to select a narrow band of wavelengths to enter the monochromator, which selects an even narrower band to pass through the sample. After being chopped and passing through the sample and reference cells, the signal is detected by a *photomultiplier tube* that creates an electric current proportional to the radiant power reaching the detector.

 Precautions

Most spectrophotometers exhibit their minimum relative uncertainty at intermediate levels of absorbance ($A \approx 0.4$–0.9). If too little light gets through the sample (high absorbance), the intensity is hard to measure. If too much light gets through (low absorbance), then it is hard to distinguish the difference between the transmittances of the sample and the reference cuvets. It is therefore desirable to adjust the concentration of the sample so that its absorbance falls in the intermediate range.

Samples should be dust free because small particles scatter light and increase the apparent "absorbance" of the sample. Keeping all containers covered will lower the concentration of dust in solutions. Filtering the final solution through a very fine filter may be necessary in critical work. Cuvets should only be handled with a tissue to avoid putting fingerprints on the faces and must be kept scrupulously clean to avoid surface contamination, which also leads to scattering.

Keep your fingers off the clear faces of the cuvet! Fingerprints scatter and absorb light.

Slight mismatch between sample and reference cuvets, over which you have little control, leads to systematic errors in spectrophotometry. For reproducible readings, it is important to position a cuvet in the spectrophotometer as reproducibly as possible. Slight misplacement of the cuvet in its holder, or turning a flat cuvet around by 180°, or rotation of a circular cuvet, all lead to random errors in absorbance measurements.

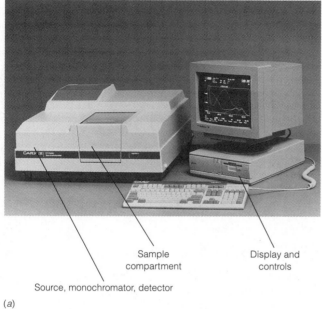

Sample
compartment

Display and
controls

Source, monochromator, detector

(a)

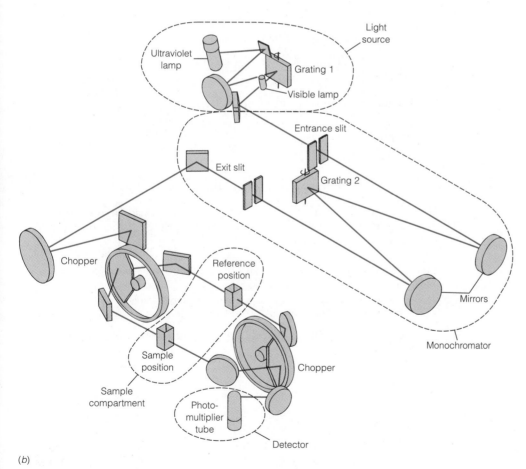

Light
source

Ultraviolet
lamp

Grating 1

Visible lamp

Entrance slit

Exit slit

Grating 2

Chopper

Reference
position

Mirrors

Monochromator

Sample
position

Chopper

Sample
compartment

Photo-
multiplier
tube

Detector

(b)

Figure 6-6 (*a*) Varian Cary 3E Ultraviolet-Visible Spectrophotometer.
(*b*) Schematic diagram of optical train. [Courtesy Varian Australia Pty Ltd.,
Victoria, Australia.]

6-4 Beer's Law in Chemical Analysis

Spectrophotometric analyses employing visible radiation are called *colorimetric* analyses.

For a compound to be analyzed by spectrophotometry, it must absorb light, and this absorption should be distinguishable from that due to other species in the sample. Because most compounds absorb ultraviolet radiation, ultraviolet absorbance tends to be inconclusive, and analysis is usually restricted to the visible spectrum. If there are no interfering species, however, ultraviolet absorbance can be used. Solutions of proteins are normally assayed in the ultraviolet region at 280 nm because the aromatic groups present in virtually every protein have an absorbance maximum at 280 nm. In this section, we introduce the use of Beer's law with a simple example and then discuss the measurement of iron in blood serum.

EXAMPLE Measuring Benzene in Hexane

(a) Pure hexane has negligible ultraviolet absorbance above a wavelength of 200 nm. A solution prepared by dissolving 25.8 mg of benzene (C_6H_6, MW 78.114) in hexane and diluting to 250.0 mL had an absorption peak at 256 nm with an absorbance of 0.266 in a 1.000-cm cell. Find the molar absorptivity of benzene at this wavelength.

Solution   The concentration of benzene is

$$[C_6H_6] = \frac{(0.025\,8 \text{ g})/(78.114 \text{ g/mol})}{0.250\,0 \text{ L}} = 1.32_1 \times 10^{-3} \text{ M}$$

We find the molar absorptivity from Beer's law:

$$\text{molar absorptivity} = \epsilon = \frac{A}{bc}$$

$$= \frac{(0.266)}{(1.000 \text{ cm})(1.32_1 \times 10^{-3} \text{ M})} = 201._3 \text{ M}^{-1} \text{ cm}^{-1}$$

(b) A sample of hexane contaminated with benzene had an absorbance of 0.070 at 256 nm in a cell with a 5.000-cm pathlength. Find the concentration of benzene in milligrams per liter.

Solution   We use the molar absorptivity from (a) in Beer's law:

$$[C_6H_6] = \frac{A}{\epsilon b} = \frac{0.070}{(201._3 \text{ M}^{-1} \text{ cm}^{-1})(5.000 \text{ cm})} = 6.9_5 \times 10^{-5} \text{ M}$$

$$[C_6H_6] = (6.9_5 \times 10^{-5} \text{ mol/L})(78.114 \times 10^3 \text{ mg/mol}) = 5.4 \text{ mg/L}$$

Serum Iron

Iron for biosynthesis is transported through the bloodstream, not as a free ion, but attached to the protein transferrin. The procedure described below measures the iron content of transferrin.[4] This analysis requires only about 1 μg of Fe to provide an accuracy of 2–5%. Human blood usually contains about 45 vol% cells and 55% plasma (liquid). If blood is collected without an anticoagulant, the blood clots; the liquid that remains is called *serum*. Serum normally contains about 1 μg of Fe/mL attached to transferrin.

To measure the serum iron content, three steps are needed:

Step 1. Fe^{3+} in transferrin is reduced to Fe^{2+} and thereby released from the protein. Commonly employed reducing agents are hydroxylamine hydrochloride ($NH_3OH^+Cl^-$), thioglycolic acid, or ascorbic acid.

$$2Fe^{3+} + 2HSCH_2CO_2H \rightarrow$$
$$\text{Thioglycolic acid}$$

$$2Fe^{2+} + HO_2CCH_2S{-}SCH_2CO_2H + 2H^+$$

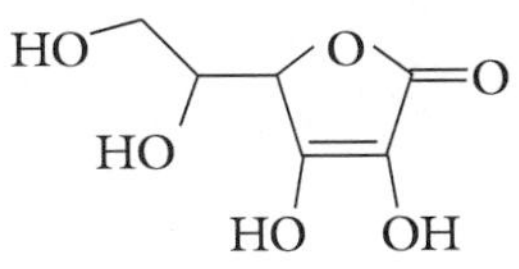

Step 2. Trichloroacetic acid (Cl_3CCO_2H) is added to precipitate all the proteins, leaving Fe^{2+} in solution. The proteins are removed by centrifugation. If protein were left in the solution, it would partially precipitate in the final solution. Light scattering by particles of precipitate would then be mistaken for absorbance.

$$\text{protein}(aq) \xrightarrow{CCl_3CO_2H} \text{protein}(s)$$

Step 3. A measured volume of supernatant liquid from Step 2 is transferred to a fresh vessel and treated with excess ferrozine to form a purple complex, whose absorbance is measured (Figure 6-7). A buffer is also added to keep the pH in a range in which the formation of the ferrozine-iron complex is complete.

Supernate refers to the liquid remaining above the solid that collects at the bottom of a tube during centrifugation.

$$Fe^{2+} + 3\text{ferrozine}^{2-} \rightarrow (\text{ferrozine})_3Fe^{4-}$$
$$\text{(purple complex)}$$

In most spectrophotometric analyses, it is important to prepare a **reagent blank** containing all reagents, but with analyte replaced by distilled water. Any absorbance of the blank is due to the color of uncomplexed ferrozine plus the color caused by the iron impurities in the reagents and glassware. *The absorbance of the blank is subtracted from all other absorbances before any calculations are done.*

The blank should contain all sources of absorbance other than the analyte.

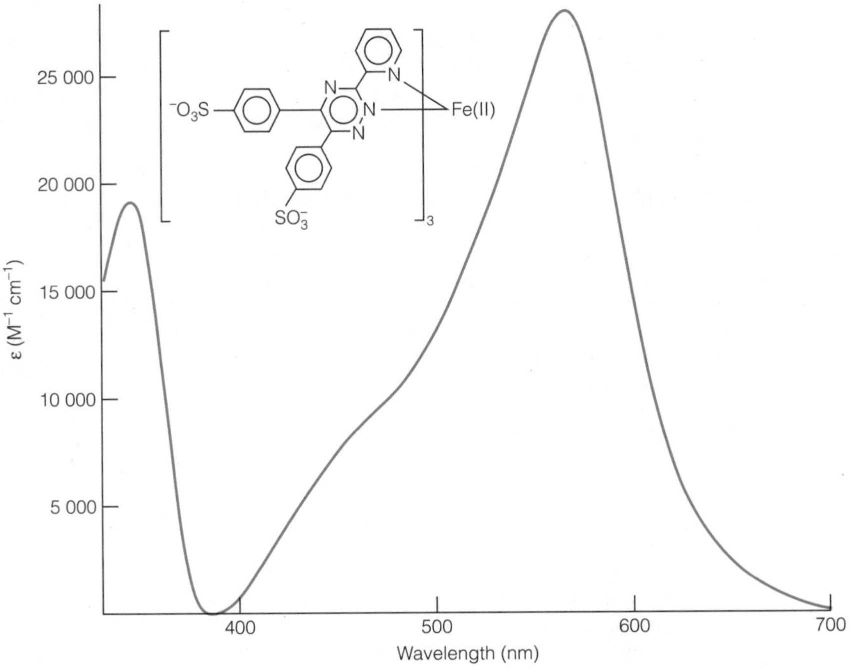

Figure 6-7 Visible absorption spectrum of the complex (ferrozine)$_3$Fe(II) used in the colorimetric analysis of iron.

It is also important to use a series of iron standards to establish a *calibration curve* such as that in Figure 6-8, where Beer's law is clearly obeyed. The standards should be prepared by the same procedure used to prepare the unknowns. The absorbance of the unknown should fall within the region covered by the standards, so that there is no question about the validity of the calibration curve.

If all samples and standards are prepared in the same way and with identical volumes, then the quantity of iron in the unknown can be read directly from the calibration curve. For example, if the unknown has an absorbance of 0.357 (after subtracting the absorbance of the blank), Figure 6-8 tells us that the sample contains 3.59 μg of iron.

In the serum iron determination just described, the values obtained would be about 10% high due to the reaction of serum copper with the ferrozine. This interference can be eliminated if neocuproine or thiourea is added.[5] These reagents form strong complexes with copper, thereby **masking** it.

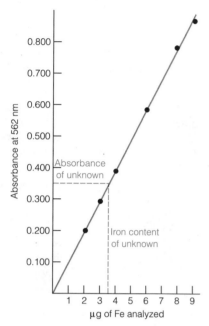

Figure 6-8   Calibration curve showing the validity of Beer's law for the (ferrozine)$_3$Fe(II) complex used in the serum iron determination. Each sample was diluted to a final volume of 5.00 mL. Therefore, 1.00 μg of iron is equivalent to a concentration of 3.58×10^{-6} M.

EXAMPLE   Serum Iron Analysis

Serum iron and standard iron solutions were analyzed as follows:

1.   To 1.00 mL of sample are added 2.00 mL of reducing agent and 2.00 mL of acid to reduce and release Fe from transferrin.

2.   The serum proteins are precipitated with 1.00 mL of 30 wt % trichloroacetic acid. The volume change of the solution is negligible when the protein precipitates and can be said to remain 1.00 + 2.00 + 2.00 + 1.00 = 6.00 mL (assuming no changes in volume due to mixing). The mixture is centrifuged to remove protein.

3.   A 4.00-mL aliquot of solution is transferred to a fresh test tube and treated with 1.00 mL of solution containing ferrozine and buffer. The absorbance of this solution is measured after a 10-min waiting period.

4.   For an absorbance standard, a volume of 1.00 mL containing 3.00 μg of Fe is diluted to 6.00 mL with other reagents. Then 4.00 mL [containing $(4.00/6.00) \times 3.00\ \mu g = 2.00\ \mu g$ of Fe] is transferred to a new vessel and diluted with 1.00 mL of reagent.

The following data were obtained:

Sample	A (at 562 nm in 1.000-cm cell)
Blank	0.038
3.00-μg Fe standard	0.239
Serum sample	0.129

Assuming that Beer's law has been shown to be valid in control experiments, use the data above to find the concentration of Fe in the serum. Also, calculate the molar absorptivity of $(ferrozine)_3Fe^{4-}$.

Solution Because the sample and standard were prepared in an identical manner, their Fe ratio must be equal to their absorbance ratio (corrected for the blank absorbance).

$$\frac{\text{Fe in sample}}{\text{Fe in standard}} = \frac{\text{corrected absorbance of sample}}{\text{corrected absorbance of standard}}$$

$$= \frac{0.129 - 0.038}{0.239 - 0.038} = 0.453$$

The standard contained 3.00 μg of Fe, so the sample must have contained $(0.453)(3.00 \text{ μg}) = 1.359 \text{ μg}$ of Fe. The concentration of Fe in the serum is

$$[\text{Fe}] = \text{moles of Fe/liters of serum}$$

$$= \left(\frac{1.359 \times 10^{-6} \text{ g Fe}}{55.847 \text{ g Fe/mol Fe}}\right) \Big/ (1.00 \times 10^{-3} \text{ L}) = 2.43 \times 10^{-5} \text{ M}$$

To find ε for $(ferrozine)_3Fe^{4-}$, we use the absorbance of the standard. Recall that the known solution used in the measurement of absorbance contained 2.00 μg of Fe in 5.00 mL (Step 4). The final concentration of Fe in the diluted standard is

$$[\text{Fe}] = \left(\frac{2.00 \times 10^{-6} \text{ g Fe}}{55.847 \text{ g Fe/mol Fe}}\right) \Big/ (5.00 \times 10^{-3} \text{ L}) = 7.16 \times 10^{-6} \text{ M}$$

All Fe is converted to $(ferrozine)_3Fe^{4-}$. The molar absorptivity is

$$\epsilon = \frac{A}{bc} = \frac{0.239 - 0.038}{(1.00 \text{ cm})(7.16 \times 10^{-6} \text{ M})} = 2.81 \times 10^4 \text{ M}^{-1} \text{ cm}^{-1}$$

Primary Standards

The validity of any analysis ultimately depends on measuring the response of the procedure to a **primary standard** whose composition is known. Primary standards are normally pure enough to be weighed and used directly. They should be 99.9% pure (or better), should not decompose under ordinary storage, and should be stable when dried (by heating or vacuum). Pure iron wire (with a shiny, rust-free surface) dissolved in acid is used to prepare the most accurate iron standards. Some common iron compounds such as ferrous ammonium sulfate $(Fe(NH_4)_2(SO_4)_2 \cdot 6H_2O)$ and ferrous ethylene-diammonium sulfate $(Fe(H_3NCH_2CH_2NH_3)(SO_4)_2 \cdot 4H_2O)$ are suitable for less accurate work.

6-5 Calibration Curves

In a **calibration curve,** spectrophotometric absorbance (or some other response of an analytical instrument) is plotted as a function of the known quantity of analyte derived from a primary standard. Figure 6-8 was an example of a calibration curve for iron, and Figure 4-8 was an example of a calibration curve for protein. We adopt the following general procedure for constructing a calibration curve, using the protein analysis as an example:

TABLE 6-2   Spectrophotometer data used to construct calibration curve

Amount of protein (μg)	Absorbance of independent samples			Corrected absorbance		
0	0.099	0.099	0.100	-0.000_3	-0.000_3	0.000_7
5	0.185	0.187	0.188	0.085_7	0.087_7	0.088_7
10	0.282	0.272	0.272	0.182_7	0.172_7	0.172_7
15	0.345	0.347	—	0.245_7	0.247_7	—
20	0.425	0.425	0.430	0.325_7	0.325_7	0.330_7
25	0.483	0.488	0.496	0.383_7	0.388_7	0.396_7

Solutions with known concentrations are said to be *standard solutions*.

Step 1. Prepare known samples of analyte, covering a convenient range of concentration, and measure the response of the analytical procedure to these standards. This procedure generates the data in the left half of Table 6-2.

Nonzero absorbance of the blank may arise from the color of starting reagents, reactions of impurities, and reactions of major species other than analyte. Blank values may vary from one set of reagents to another, but corrected absorbance should not.

Step 2. Subtract the average absorbance (0.099_3) of the three *blank* samples from each measured absorbance to obtain *corrected absorbance*. The blank measures the response of the procedure when no protein is present.

Step 3. Make a graph of corrected absorbance versus quantity of protein analyzed (Figure 6-9). Use the method of least squares described in Section 4-5 to find the best straight line through the linear portion of the data, up to and including 20 μg of protein. Find the slope and intercept and their uncertainties with Equations 4-14, 4-15, 4-19, and 4-20.

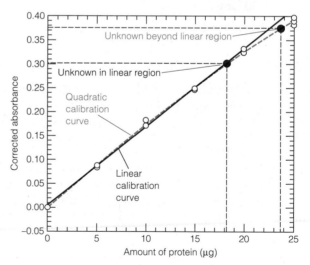

Figure 6-9 Calibration curve showing corrected absorbance versus amount of protein in spectrophotometric protein analysis and based on data from Table 6-2. The equation of the solid straight line fitting the 14 data points (open circles) for 0 to 20 μg, derived by the method of least squares described in Section 4-5, is $y = 0.0163_0 (\pm0.0002_2)x + 0.004_7 (\pm0.002_6)$. The standard deviation of y is $s_y = 0.005_9$. The equation of the dashed quadratic curve that fits all 17 data points, determined by the nonlinear least-squares program in the *Solutions Manual and Supplement for Quantitative Chemical Analysis*, is $y = -1.1_7 (\pm0.2_1) \times 10^{-4}x^2 + 0.0185_8 (\pm0.0004_6)x - 0.0007 (\pm0.0010)$, with $s_y = 0.004_6$.

Step 4. When an unknown solution is analyzed at a future time, a blank is run
at the same time. That blank absorbance value is subtracted from the
unknown absorbance to obtain the corrected absorbance.

- - - - - - - - - - - - - - - - - -

EXAMPLE   Using a Linear Calibration Curve

An unknown protein sample subjected to the analysis in Figure 6-9 gave an
absorbance of 0.406, and a blank carried through the same procedure had an
absorbance of 0.104. How many micrograms of protein are in the unknown?

Solution   The corrected absorbance is $0.406 - 0.104 = 0.302$, which lies on
the linear portion of the calibration curve (lower unknown point in Figure
6-9). The equation of the straight line in Figure 6-9 is

$$\text{absorbance} = 0.016\,3_0 \times (\mu\text{g of protein}) + 0.004_7$$

The quantity of unknown protein is therefore

$$\mu\text{g of protein} = \frac{\text{absorbance} - 0.004_7}{0.016\,3_0} = \frac{0.302 - 0.004_7}{0.016\,3_0} = 18.2_4 \ \mu\text{g}$$

Propagation of uncertainty   If you want to estimate the reliability of the
result, insert the uncertainties from the caption of Figure 6-9 into the calculations above and use the rules for propagation of uncertainty from Chapter 3.
For the uncertainty in absorbance, we use the standard deviation of y ($s_y =
0.005_9$) as an estimate:

$$\mu\text{g of protein} = \frac{0.302 \ (\pm 0.005_9) - 0.004_7 \ (\pm 0.002_6)}{0.016\,3_0 \ (\pm 0.000\,2_2)} = 18.2 \ (\pm 0.4) \ \mu\text{g}$$

A better way to estimate the uncertainty is with Equation 4-24.

- - - - - - - - - - - - - - - - - - - -

We prefer calibration procedures with a **linear response,** in which the
corrected analytical signal (= signal from sample − signal from blank) is
proportional to the quantity of analyte. Although we try to work in the *linear
range,* it is possible to obtain valid results beyond the linear region (>20 μg)
in Figure 6-9. The dashed curve that goes up to 25 μg of protein comes from
a least-squares fit of the data to the equation $y = ax^2 + bx + c$.

In general, it is not reliable to extrapolate any calibration curve, linear or
nonlinear, beyond the measured range of standards. You should always
measure standards in the entire concentration range of interest.

Box 6-3 explains how to use a nonlinear calibration curve.

6-6 Standard Addition and Internal Standards

The validity of any analytical result ultimately depends on measuring the
response of the analytical procedure to known standards. Although calibration curves are the most common way to measure such responses, *standard
additions* and *internal standards* are also used. Calibration curves, standard
addition, and internal standards apply to most kinds of analyses, not just
spectrophotometric methods.

Box 6-3 Using a Nonlinear Calibration Curve

Consider an unknown whose corrected absorbance of 0.375 lies beyond the limit of the linear calibration line in Figure 6-9. We can fit all the data points in Figure 6-9 with the quadratic equation

$$y = -1.17 \times 10^{-4}x^2 + 0.018\,58x - 0.000\,7 \qquad \text{(a)}$$

whose coefficients were found with the nonlinear least-squares program given in the *Solutions Manual and Supplement for Quantitative Chemical Analysis* or with the spreadsheet in Supplementary Problem 12 of Chapter 18 in the *Solutions Manual*.

To find the quantity of protein in the unknown, substitute the measured absorbance into Equation a:

$$0.375 = -1.17 \times 10^{-4}x^2 + 0.018\,58x - 0.000\,7$$

This can be rearranged to

$$1.17 \times 10^{-4}x^2 - 0.018\,58x + 0.375\,7 = 0$$

which has the standard form of a quadratic equation

$$ax^2 + bx + c = 0$$

whose two possible solutions are

$$x = \frac{-b + \sqrt{b^2 - 4ac}}{2a} \qquad x = \frac{-b - \sqrt{b^2 - 4ac}}{2a}$$

Substituting $a = 1.17 \times 10^{-4}$, $b = -0.018\,58$, and $c = 0.375\,7$ into these equations gives

$$x = 135\ \mu g \quad \text{and} \quad x = 23.8\ \mu g$$

The graph in Figure 6-9 tells us that the correct choice is 23.8 µg, not 135 µg.

Estimating uncertainty How can we estimate the uncertainty in the quantity of protein if the absorbance of the unknown is 0.375 ± 0.006? A simple way to do this is to substitute the upper and lower limits (0.381 and 0.369) back into Equation a:

upper limit: $0.381 = -1.17 \times 10^{-4}x^2 + 0.018\,58x - 0.000\,7 \Rightarrow x = 24.2\ \mu g$

lower limit: $0.369 = -1.17 \times 10^{-4}x^2 + 0.018\,58x - 0.000\,7 \Rightarrow x = 23.3\ \mu g$

A reasonable way to express the answer is 23.8 ± 0.5 µg.

Standard Addition[6]

In **standard addition,** known quantities of analyte are added to the unknown, and the increased signal lets us deduce how much analyte was in the original unknown. This method requires a linear response to analyte.

For example, consider atomic emission analysis in which the intensity of light given off by a solution aspirated (sucked) into a flame is proportional to the concentration of a certain element in the solution. A sample with unknown initial concentration of analyte $[X]_i$ gives an emission intensity I_X. Then a known concentration of standard, S, is added to the sample and an emission intensity I_{S+X} is observed for this second solution. Addition of standard to the unknown changes the concentration of the original analyte because of dilution. We call the diluted concentration of the analyte $[X]_f$, where f stands for "final." We designate the concentration of standard in the second solution as $[S]_f$. (Bear in mind that the chemical species X and S are the same.)

Because emission is directly proportional to analyte concentration, we can say that

$$\frac{\text{concentration of analyte in initial solution}}{\text{concentration of analyte plus standard in final solution}}$$

$$= \frac{\text{signal from initial solution}}{\text{signal from final solution}}$$

Standard addition equation: $$\frac{[X]_i}{[S]_f + [X]_f} = \frac{I_X}{I_{S+X}} \qquad (6\text{-}7)$$

By expressing the diluted concentration of analyte, $[X]_f$, in terms of the initial concentration of analyte, $[X]_i$, we can solve for $[X]_i$, because everything else in Equation 6-7 is known.

Derivation of Equation 6-7:

$I_X = k[X]_i$ where k is a constant of proportionality

$I_{S+X} = k([S]_f + [X]_f)$ where k is the same constant

Dividing one equation by the other gives

$$\frac{I_X}{I_{S+X}} = \frac{k[X]_i}{k([S]_f + [X]_f)} = \frac{[X]_i}{[S]_f + [X]_f}$$

EXAMPLE Standard Addition

A blood serum sample containing sodium ion gives a signal of 4.27 mV on a light intensity meter in an atomic emission experiment. The Na^+ concentration in the serum is then increased by 0.104 M by a standard addition, without significantly diluting the sample. This "spiked" serum sample gives a signal of 7.98 mV in atomic emission. Find the original concentration of Na^+ in the serum.

Solution Plugging this information into Equation 6-7 allows us to solve for $[Na^+]$ in the original serum:

$$\frac{[Na^+]_i}{[0.104 \text{ M}] + [Na^+]_i} = \frac{4.27 \text{ mV}}{7.98 \text{ mV}} \Rightarrow [Na^+]_i = 0.120 \text{ M}$$

If the standard addition dilutes the unknown, then the concentration of X in the denominator of Equation 6-7 should be the *diluted* concentration. For example, if 5.00 mL of 2.08 M NaCl is added to 95.0 mL of unknown, then the unknown has been diluted to 95.0% of its original concentration. The moles of added NaCl are $(0.005\,00 \text{ L})(2.08 \text{ M}) = 0.010\,4$ mol and the concentration of added NaCl is $0.010\,4$ mol$/0.100$ L $= 0.104$ M. Equation 6-7 is therefore

$$\frac{[Na^+]_i}{[0.104 \text{ M}] + 0.950[Na^+]_i} = \frac{4.27 \text{ mV}}{7.98 \text{ mV}} \Rightarrow [Na^+]_i = 0.113 \text{ M}$$

An alternative procedure is to make a series of standard additions and to plot the results as shown in Figure 6-10. The *x* axis is the concentration of added analyte *after* it has been mixed with sample. The *x* intercept of the extrapolated line is equal to the concentration of unknown *after* it has been diluted to the final volume. In Figure 6-10, this concentration is near 0.042 M. The most useful range of standard additions should increase the analytical signal to between 1.5 and 3 times its original value.

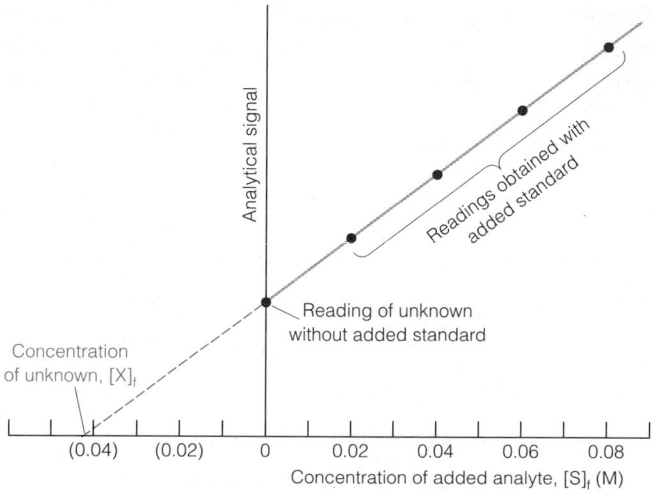

Figure 6-10   Graphic treatment of standard additions.

In standard addition, the matrix is constant for all samples.

Use of dilution factor $\left(= \dfrac{V_i}{V_f}\right)$:

final concentration

$= \dfrac{V_i}{V_f} \times$ initial concentration.

Standard addition is especially appropriate when the sample composition is unknown or complex and affects the analytical signal. The **matrix** is everything in the unknown, other than analyte. In atomic emission spectroscopy, for example, components of the matrix may react with analyte atoms to make molecules that do not emit light at the intended wavelength. The matrix may have unknown constituents that you could not incorporate into standard solutions to make a calibration curve. In general, we may define a **matrix effect** as a change in the analytical signal caused by anything in the sample other than analyte. When we add a small volume of concentrated standard, we do not change the concentration of the matrix very much. The assumption in using a standard addition is that the matrix has the same effect on added analyte as it has on the original analyte in the unknown.

In many calculations, it is necessary to find the concentration of a sample after it has been diluted from initial volume V_i to final volume V_f. In the preceding example, 2.08 M NaCl was diluted from $V_i = 5.00$ mL to $V_f = 5.00 + 95.0 = 100.0$ mL. The NaCl concentration is therefore diluted from 2.08 M to $(V_i/V_f)(2.08$ M$) = 0.104$ M. The quantity (V_i/V_f) is called a **dilution factor.** From now on, we will generally use dilution factors instead of the more laborious calculation of moles and liters.

Internal Standards

An **internal standard** is a known amount of a compound, different from analyte, that is added to the unknown. The signal due to analyte (X) is compared with that of the internal standard (S). A known mixture of standard and analyte is prepared beforehand to measure the relative response of the analytical method to the two species. The key relationship is

Internal standard equation:

$$\frac{\text{concentration ratio (X/S) in unknown}}{\text{concentration ratio in standard mixture}} = \frac{\text{signal ratio (X/S) in unknown}}{\text{signal ratio in standard mixture}}$$

(6-8)

To use Equation 6-8, you must be sure that the analytical method gives linear responses to both the analyte and the standard.

Internal standards are desirable whenever unknown losses of sample are likely to occur during handling or analysis. If a known quantity of standard is added to the unknown prior to any handling losses, the ratio of standard and analyte will remain constant thereafter, because the same fraction of each is lost in any operation.

Consider the *chromatogram* in Figure 6-11. In this experiment, a mixture of unknown X and standard S in an appropriate solvent is applied to a column packed with a solid material that adsorbs S more tightly than X. Washing the column with solvent eventually rinses each compound down through a detector, in which an electrochemical reaction senses the compounds. The area under each peak in the chromatogram is proportional to the quantity of compound passing through the detector. When an analytical signal is proportional to analyte concentration (signal = $k[X]$), the constant of proportionality, k, is sometimes called the *response factor*.

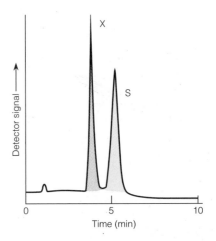

Figure 6-11 Chromatogram illustrating use of internal standard. A known amount of compound S is added to the unknown containing compound X. By measuring the relative areas of the signals from both compounds and by previously measuring the relative response of the detector to each compound, we can tell how much of compound X is in the unknown.

. . . .

EXAMPLE Using an Internal Standard

In a preliminary experiment, a solution containing 0.0837 M X and 0.0666 M S gave peak areas of 423 for X and 347 for S. (Areas are measured in arbitrary units.) To analyze the unknown, 10.0 mL of 0.146 M S was added to 10.0 mL of unknown, and the mixture was diluted to 25.0 mL in a volumetric flask. This mixture gave the chromatogram in Figure 6-11, for which the area of peak X was 553 and the area of peak S was 582. Find the concentration of X in the unknown.

Solution For the standard mixture, we can write

$$\text{standard mixture:} \quad \frac{\text{area of X}}{\text{area of S}} = \frac{423}{347} \quad \text{when} \quad \frac{[X]}{[S]} = \frac{0.0837 \text{ M}}{0.0666 \text{ M}}$$

In the mixture of unknown plus standard, the concentration of S is

$$[S] = \underbrace{(0.146 \text{ M})}_{\substack{\text{Initial} \\ \text{concentration}}} \underbrace{\left(\frac{10.0}{25.0}\right)}_{\substack{\text{Dilution} \\ \text{factor}}} = 0.0584 \text{ M}$$

This mixture gave the chromatogram in Figure 6-11, for which

$$\text{unknown mixture:} \quad \frac{\text{area of X}}{\text{area of S}} = \frac{553}{582} \quad \text{when} \quad \frac{[X]}{[S]} = \frac{\text{unknown}}{0.0584 \text{ M}}$$

We now set up the proportion in Equation 6-8 to find the concentration of X in the mixture:

$$\frac{\text{concentration ratio (X/S) in unknown}}{\text{concentration ratio in standard mixture}} = \frac{\text{signal ratio (X/S) in unknown}}{\text{signal ratio in standard mixture}}$$

$$\frac{[X] / 0.0584 \text{ M}}{0.0837 \text{ M} / 0.0666 \text{ M}} = \frac{553/582}{423/347}$$

$$\Rightarrow [X] = 0.0572 \text{ M in mixture with S}$$

Because X was diluted from 10.0 to 25.0 mL when the mixture with S was prepared, the original concentration of X in the unknown was (25.0/10.0)(0.0572 M) = 0.143 M.

. . . .

Internal standards are useful when unknown losses of sample are likely to occur during handling. Concentration ratios and signal ratios do not change, even if sample volume changes.

Terms to Understand

absorbance	ground state	photon
absorption spectrum	hertz	primary standard
Beer's law	internal standard	radiant power
calibration curve	linear response	reagent blank
chromophore	masking	refractive index
cuvet	matrix	spectrophotometry
dilution factor	matrix effect	standard addition
electromagnetic spectrum	molar absorptivity	transmittance
excited state	monochromatic light	wavelength
frequency	monochromator	wavenumber

Summary

Light can be thought of as waves whose wavelength (λ) and frequency (ν) have the important relation $\lambda\nu = c$, where c is the speed of light. Alternatively, light may be viewed as consisting of photons whose energy (E) is given by $E = h\nu = hc/\lambda = hc\tilde{\nu}$, where h is Planck's constant and $\tilde{\nu}$ ($=1/\lambda$) is the wavenumber. Absorption of light is commonly measured by absorbance (A) or transmittance (T), defined as $A = \log(P_0/P)$ and $T = P/P_0$, where P_0 is the incident radiant power and P is the exiting radiant power. Absorption spectroscopy is useful in quantitative analysis because absorbance is proportional to the concentration of the absorbing species in dilute solution (Beer's law): $A = \epsilon bc$. In this equation, b is pathlength, c is concentration, and ϵ is the molar absorptivity (a constant of proportionality). If a sample does not obey Beer's law, it is likely that a chemical reaction is altering the concentration of the chromophore when the sample concentration is changed. A molecule that absorbs light is promoted to an excited state from which it may emit light (luminesce), create heat, or carry out photochemistry. Emission of light during a chemical reaction is called chemiluminescence.

The basic components of a spectrophotometer include a radiation source, a monochromator, a sample cell, and a detector. In double-beam spectrophotometry, light is passed alternately through the sample and the reference cuvets by a rotating beam chopper, and the light beams emerging from each are continually compared to measure absorbance. To minimize errors in spectrophotometry, samples should be free of particles and cuvets must be clean and positioned reproducibly in the sample holder. Measurements should be made at a wavelength of maximum absorbance. Instrument errors tend to be minimized if the absorbance falls in the range $A \approx 0.4\text{-}0.9$.

To construct a calibration curve from a series of standard solutions, first subtract the average blank value from each measured point. (Interfering species should be removed or masked prior to analysis.) Most often, a calibration curve is linear over some useful range and is fitted by the method of least squares. The uncertainties in slope and intercept and standard deviation of y are important parameters derived from such a fit and are required for error estimates in subsequent analyses. Nonlinear calibration curves may also be used, by fitting the data to nonlinear functions.

Standard additions and internal standards require a linear response for the analytical procedure. In standard addition, known quantities of analyte are added to the unknown, and the increased signal allows us to deduce how much analyte was originally present. Because signal is proportional to concentration, we can say $[X]_i/([S]_f + [X]_f) = I_X/I_{S+X}$, where $[X]_i$ is the initial concentration of unknown, ($[S]_f + [X]_f$) is the final total concentration of standard plus unknown, I_X is the signal from the original analyte, and I_{S+X} is the signal after standard addition. When a series of standard additions is made, a graph like that shown in Figure 6-10 is used to extrapolate back to the abscissa to find the original concentration of analyte. Standard addition is appropriate when the sample matrix is unknown or complex and affects the analytical signal.

An internal standard is a known amount of a compound, other than analyte, that is added to the unknown. Comparison of the signals from the analyte and standard tells us how much analyte is present. A previous calibration is necessary to find the relative response of the analysis to each component. The relevant equation is [(concentration ratio in unknown)/(concentration ratio in standard mixture)] = [(signal ratio in unknown)/(signal ratio in standard mixture)]. You should use dilution factors to streamline all volumetric calculations.

Exercises

A. (a) What value of absorbance corresponds to 45.0% T?

(b) If a 0.0100 M solution exhibits 45.0% T at some wavelength, what will be the percent transmittance for a 0.0200 M solution of the same substance?

B. (a) A 3.96×10^{-4} M solution of compound A exhibited an absorbance of 0.624 at 238 nm in a 1.000-cm cuvet; a blank solution containing only solvent had an absorbance of 0.029 at the same wavelength. Find the molar absorptivity of compound A.

(b) The absorbance of an unknown solution of compound A in the same solvent and cuvet was 0.375 at 238 nm. Find the concentration of A in the unknown.

(c) A concentrated solution of compound A in the same solvent was diluted from an initial volume of 2.00 mL to a final volume of 25.00 mL and then had an absorbance of 0.733. What is the concentration of A in the concentrated solution?

C. Ammonia can be determined spectrophotometrically by reaction with phenol in the presence of hypochlorite (OCl^-):

Phenol
(colorless)

Ammonia
(colorless)

(blue) ($\lambda_{max} = 625$ nm)

A 4.37-mg sample of protein was chemically digested to convert its nitrogen to ammonia and then diluted to 100.0 mL. Then 10.0 mL of the solution was placed in a 50-mL volumetric flask and treated with 5 mL of phenol solution plus 2 mL of sodium hypochlorite solution. The sample was diluted to 50.0 mL, and the absorbance at 625 nm was measured in a 1.00-cm cuvet after 30 min. For reference, a standard solution was prepared from 0.0100 g of NH_4Cl (FW 53.49) dissolved in 1.00 L of water. Then 10.0 mL of this standard was placed in a 50-mL volumetric flask and analyzed in the same manner as the unknown. A reagent blank was prepared by using distilled water in place of unknown.

Sample	Absorbance at 625 nm
Blank	0.140
Reference	0.308
Unknown	0.592

(a) Calculate the molar absorptivity of the blue product.

(b) Calculate the weight percent of nitrogen in the protein.

D. Cu^+ reacts with neocuproine to form the colored complex $(neocuproine)_2Cu^+$, with an absorption maximum at 454 nm (Section 6-4). Neocuproine is particularly useful because it reacts with few other metals. The copper complex is soluble in 3-methyl-1-butanol (isoamyl alcohol), an organic solvent that does not dissolve appreciably in water. In other words, when isoamyl alcohol is added to water, a two-layered mixture results, with the denser water layer at the bottom. If $(neocuproine)_2Cu^+$ is present, virtually all of it goes into the organic phase. For the purpose of this problem, assume that the isoamyl alcohol does not dissolve in the water at all and that all of the colored complex will be in the organic phase. Suppose that the following procedure is carried out (see the figure at the top of page 146):

1. A rock containing copper is pulverized, and all metals are extracted from it with strong acid. The acidic solution is neutralized with base and made up to 250.0 mL in flask A.

2. Next 10.00 mL of the solution is transferred to flask B and treated with 10.00 mL of a reducing agent to reduce all Cu to Cu^+. Then 10.00 mL of buffer is added to bring the pH to a value suitable for complex formation with neocuproine.

3. After that, 15.00 mL of this solution is withdrawn and placed in flask C. To the flask is added 10.00 mL of an aqueous solution containing neocuproine and 20.00 mL of isoamyl alcohol. After shaking well and allowing the phases to separate, all $(neocuproine)_2Cu^+$ is in the organic phase.

4. A few milliliters of the upper layer are withdrawn, and the absorbance at 454 nm is measured in a 1.00-cm cell. A blank carried through the same procedure gave an absorbance of 0.056.

(a) Suppose that the rock contained 1.00 mg of Cu. What will be the concentration of Cu (moles per liter) in the isoamyl alcohol phase?

(b) If the molar absorptivity of $(neocuproine)_2Cu^+$ is 7.90×10^3 M^{-1} cm^{-1}, what will be the observed absorbance? Remember that a blank carried through the same procedure gave an absorbance of 0.056.

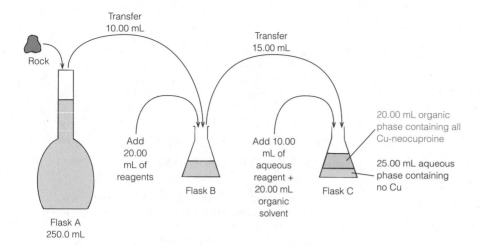

(c) A rock is analyzed and found to give a final absorbance of 0.874 (uncorrected for the blank). How many milligrams of Cu are in the rock?

E. *Calibration curve.* The figure in the right column shows an infrared absorption peak for solutions containing 10–50 vol% acetone, $(CH_3)_2C{=}O$, in water. Absorbance is plotted as a function of *wavenumber,* which is proportional to the frequency of light. The shape of the spectrum is somewhat different for each mixture, which shifts the *baseline* (dotted line) and position of the absorption peak. The baseline absorbance is the blank absorbance that must be subtracted from each peak absorbance to obtain corrected absorbance.

(a) Using a millimeter ruler, and measuring to the nearest 0.1 mm, find how many millimeters correspond to 0.8 absorbance units (from 0.4 to 1.2) on the ordinate (*y* axis) of the spectrum.

(b) Measure the corrected height of the peak corresponding to 50 vol% acetone, using the distance from the baseline marked "50%." By comparing your measurement in **(b)** to that in **(a)**, find the corrected absorbance of the 50 vol% acetone peak. For example, if 0.8 absorbance units = 58.4 mm and the peak height is 63.7 mm, then the corrected absorbance is found from the proportion

$$\frac{\text{corrected peak height (mm)}}{\text{length for 0.8 absorbance units (mm)}}$$

$$= \frac{\text{corrected peak absorbance}}{0.8 \text{ absorbance units}}$$

$$\frac{63.7 \text{ mm}}{58.4 \text{ mm}} = \frac{\text{corrected peak absorbance}}{0.8 \text{ absorbance units}}$$

$$\Rightarrow \text{corrected peak absorbance} = 0.873$$

(c) The baselines for 10 and 50 vol% acetone are shown in the figure. Draw baselines for the 20, 30, and 40 vol% acetone spectra. Use the procedure in **(b)** to find the corrected absorbance for the 10–40 vol% solutions.

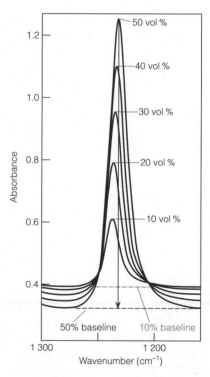

Infrared absorption spectra of 10–50 vol% acetone in water. Vertical arrow shows corrected absorbance for 50 vol% acetone, which is obtained by subtracting baseline absorbance from peak absorbance. [Spectra from A. Afran, *Am. Lab.,* February 1993, p. 40MMM.]

(d) Make a calibration curve showing corrected absorbance versus volume percent of acetone. (This curve does not go through the origin.)

(e) Find the equation of the best straight line through the data, using the method of least squares in Section 4-5.

(f) Find the volume percent of acetone in a solution whose corrected absorbance is 0.611.

(g) *Propagation of uncertainty.* Find the uncertainty in slope and intercept and the standard deviation of y (s_y) for the least-squares line in (e). Following the procedure at the end of Section 4-5, compute the uncertainty in the answer to (f). Consider the uncertainty in the unknown absorbance (0.611) to be $\pm s_y$.

F. *Standard addition.* An unknown sample of Ni^{2+} gave a current of 2.36 μA in an electrochemical analysis. When 0.500 mL of solution containing 0.0287 M Ni^{2+} was added to 25.0 mL of unknown, the current increased to 3.79 μA.

(a) Denoting the initial, unknown concentration as $[Ni^{2+}]_i$, write an expression for the final concentration, $[Ni^{2+}]_f$, after 25.0 mL of unknown was mixed with 0.500 mL of standard. Use the dilution factor for this calculation.

(b) In a similar manner, write the final concentration of added standard Ni^{2+}, designated as $[S]_f$.

(c) Use Equation 6-7 to find $[Ni^{2+}]_i$ in the unknown.

G. *Internal standard.* A solution was prepared by mixing 5.00 mL of unknown (element X) with 2.00 mL of solution containing 4.13 μg of standard (element S) per milliliter and diluting to 10.0 mL. The measured signal ratio in an atomic absorption experiment was

$$\frac{\text{signal due to X}}{\text{signal due to S}} = 0.808$$

In a separate experiment, it was found that for equal concentrations of X and S, the signal due to X was 1.31 times more intense than the signal due to S. Find the concentration of X in the unknown.

Problems

Properties of Light

1. Fill in the blanks.

(a) If you double the frequency of electromagnetic radiation, you _____ the energy.

(b) If you double the wavelength, you _____ the energy.

(c) If you double the wavenumber, you _____ the energy.

2. (a) How much energy (in kilojoules) is carried by 1 mol of photons of red light with $\lambda = 650$ nm?

(b) How many kilojoules are carried by 1 mol of photons of violet light with $\lambda = 400$ nm?

3. Calculate the frequency (in hertz), wavenumber (in cm^{-1}), and energy (in joules per photon and joules per mole of photons) of visible light with a wavelength of 562 nm.

4. Which molecular processes correspond to the energies of microwave, infrared, visible, and ultraviolet photons?

5. The characteristic orange light produced by sodium in a flame is due to an intense emission called the sodium "D" line. This "line" is actually a doublet, with wavelengths (measured in vacuum) of 589.15788 and 589.75537 nm. The index of refraction of air at a wavelength near 589 nm is 1.0002926. Calculate the frequency, wavelength, and wavenumber of each component of the "D" line, measured in air.

Absorption of Light and the Spectrophotometer

6. Explain the difference between transmittance, absorbance, and molar absorptivity. Which one is proportional to concentration?

7. What is an absorption spectrum?

8. Why does a compound whose visible absorption maximum is at 480 nm (blue-green) appear to be red?

9. What is the difference between luminescence and chemiluminescence?

10. What is the difference between a single-beam and a double-beam spectrophotometer? What are the advantages of the double-beam instrument?

11. Why is it most accurate to measure the absorbances in the range $A = 0.4$–0.9?

12. The absorbance of a 2.31×10^{-5} M solution of a compound is 0.822 at a wavelength of 266 nm in a 1.00-cm cell. Calculate the molar absorptivity at 266 nm.

13. What color would you expect to observe for a solution of $Fe(\text{ferrozine})_3^{4-}$, which has a visible absorbance maximum at 562 nm?

14. When I was a boy, Uncle Wilbur let me watch as he analyzed the iron content of runoff from his banana ranch. A 25.0-mL sample was acidified with nitric acid and treated with excess KSCN to form a red complex. (KSCN itself is colorless.) The solution was then diluted to 100.0 mL and put in a variable-pathlength cell. For comparison, a 10.0-mL reference sample of 6.80×10^{-4} M Fe^{3+} was treated with HNO_3 and KSCN and diluted to 50.0 mL. The reference was placed in a cell with a 1.00-cm light path. The runoff sample exhibited the same absorbance as the reference when the pathlength of the runoff cell was 2.48 cm. What was the concentration of iron in Uncle Wilbur's runoff?

15. The *absorption cross section* on the ordinate of the ozone absorption spectrum at the front of this chapter is defined by the relationship

$$\text{transmittance } T = e^{-n\sigma b}$$

where n is the number of absorbing molecules per cubic centimeter, σ is the absorption cross section (cm^2), and b is the pathlength (cm). The total ozone in the atmosphere is approximately 8×10^{18} molecules above each square centimeter of the Earth's surface (from the surface up to the top of the atmosphere). If this were compressed into a 1-cm-thick layer, the concentration would be 8×10^{18} molecules/cm^3.

(a) Using the ozone spectrum at the beginning of the chapter, estimate the transmittance and absorbance of this 1-cm^3 sample at 325 and 300 nm.

(b) Sunburns are caused by radiation in the 295- to 310-nm region. At the center of this region, the transmittance of atmospheric ozone is 0.14. Calculate the absorption cross section for $T = 0.14$, $n = 8 \times 10^{18}$ molecules/cm^3, and $b = 1$ cm. By what percentage does the transmittance increase if the ozone concentration decreases by 1% to 7.92×10^{18} molecules/cm^3?

(c) Atmospheric ozone is measured in *Dobson units,* of which one unit is equivalent to 2.69×10^{16} molecules of O_3 above each square centimeter of the earth's surface. (One Dobson unit is defined as the thickness [in hundredths of a millimeter] that the ozone column would occupy if it were compressed to 1 atm pressure at 0°C.) The graph below shows variations in ozone concentration as a function of latitude and season. Using an absorption cross section of 2.5×10^{-19} cm^2, calculate the transmittance in the winter and in the summer at 30°–50° N latitude, at which the ozone varies between 290 and 350 Dobson units. By what percentage is the ultraviolet transmittance greater in winter than in summer?

Beer's Law in Chemical Analysis

16. What is the purpose of neocuproine in the serum iron analysis?

17. A compound with a molecular weight of 292.16 was dissolved in a 5-mL volumetric flask. A 1.00-mL aliquot was withdrawn, placed in a 10-mL volumetric flask, and diluted to the mark. The absorbance measured at 340 nm was 0.427 in a 1.000-cm cuvet. The molar absorptivity for this compound at 340 nm is $\epsilon_{340} = 6130$ M^{-1} cm^{-1}.

(a) Calculate the concentration of compound in the cuvet.

(b) What was the concentration of compound in the 5-mL flask?

(c) How many milligrams of compound were used to make the 5-mL solution?

18. If a sample for spectrophotometric analysis is placed in a 10-cm cell, the absorbance will be 10 times greater than the absorbance in a 1-cm cell. Will the absorbance of the reagent-blank solution also be increased by a factor of 10?

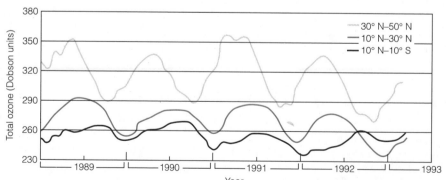

Variation in atmospheric ozone at different latitudes. [From P. S. Zurer, *Chem. Eng. News,* 24 May 1993, p. 8. Reprinted with permission. Copyright 1993 American Chemical Society.]

19. Nitrite ion, NO_2^-, is used as a preservative for bacon and other foods. It has been the center of controversy because it is potentially carcinogenic. A spectrophotometric determination of NO_2^- makes use of the following reactions:

$$HO_3S-\langle\ \rangle-NH_2 + NO_2^- + 2H^+ \longrightarrow$$

Sulfanilic acid

$$HO_3S-\langle\ \rangle-\overset{+}{N}\equiv N + 2H_2O$$

$$HO_3S-\langle\ \rangle-\overset{+}{N}\equiv N + \langle\langle\ \rangle\rangle-NH_2 \longrightarrow$$

1-Aminonaphthalene

$$HO_3S-\langle\ \rangle-N=N-\langle\langle\ \rangle\rangle-NH_2 + H^+$$

(colored product)
(λ_{max} = 520 mm)

An abbreviated procedure for the determination is given below:

1. To 50.0 mL of unknown solution containing nitrite is added 1.00 mL of sulfanilic acid solution.

2. After 10 min, 2.00 mL of 1-aminonaphthalene solution and 1.00 mL of buffer are added.

3. After 15 min, the absorbance is read at 520 nm in a 5.00-cm cell.

The following solutions were analyzed:

A. 50.0 mL of food extract known to contain no nitrite (that is, a negligible amount); final absorbance = 0.153.

B. 50.0 mL of food extract suspected of containing nitrite; final absorbance = 0.622.

C. Same as B, but with 10.0 μL of 7.50×10^{-3} M $NaNO_2$ added to the 50.0-mL sample; final absorbance = 0.967.

(a) Calculate the molar absorptivity, ϵ, of the colored product. Remember that a 5.00-cm cell was used.

(b) How many micrograms of NO_2^- were present in 50.0 mL of food extract?

20. *Kinetic method for trace determination of iron.* Very sensitive spectrophotometric analyses are based on the ability of traces of analyte to catalyze a reaction of a species present at much higher concentration than the analyte. An example is the Fe^{3+} catalysis of oxidation of N,N-dimethyl-p-phenylenediamine (DPD) by H_2O_2.

$$H_2\ddot{N}-\langle\ \rangle-\ddot{N}(CH_3)_2 \xrightarrow[Fe^{3+}]{H_2O_2}$$

DPD

$$H_2\ddot{N}-\langle\ \rangle-\overset{\cdot+}{N}(CH_3)_2$$

DPD^+

The reactant DPD has no visible absorbance, but the product has an absorbance maximum at 514 nm. The analytical procedure follows: To a 25-mL volumetric flask is added a measured volume of unknown containing Fe^{3+}. After addition of 1 mL of acetate buffer (pH 5.7, treated to remove Fe^{3+} impurity) and 1 mL of 3.5% H_2O_2, the solution is diluted to the mark. The solution is kept at $25.0 \pm 0.1°C$. At time $t = 0$, 1.00 mL of 0.024 M DPD (also treated to remove Fe^{3+}) is added. After rapid mixing, a portion of the solution is transferred to a spectrophotometer cell kept at 25.0°C. The absorbance at 514 nm is recorded at $t = 10.00$ min. The concentrations of all reactants are essentially constant throughout the 10-min period. The appearance of product is a linear function of time and Fe^{3+} concentration, as shown below.

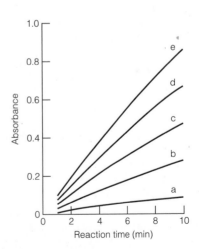

Absorbance versus time for different Fe^{3+} concentrations (ng/mL) in 26.00-mL reaction solution: (a) 0; (b) 0.40; (c) 0.80; (d) 1.20; (e) 1.60. [From K. Hirayama and N. Unohara, *Anal. Chem.* **1988**, *60*, 2573.]

(a) From the graph above, find the constants k and b in the equation $A_t = k[Fe^{3+}] + b$, where A_t is the absorbance at time t, $[Fe^{3+}]$ is the iron concentration in ng/mL, and b is the reagent blank value for a solution with no deliberate addition of Fe^{3+}. Use $t = 10$ min.

(b) The analytical procedure was carried out starting with 5.00 mL of unknown. The absorbance at 10 min was 0.515. Find the concentration of Fe^{3+} in the unknown, expressed as ng/mL and mol/L.

Calibration Curves

21. Explain why the validity of an analytical result ultimately depends on knowing the composition of some primary standard.

22. Using the calibration curve in Figure 6-9, find the quantity of unknown protein that gives a measured absorbance of 0.264 when a blank has an absorbance of 0.095.

23. *Propagation of uncertainty with a calibration curve.* Consider the linear calibration curve in Figure 6-9. Suppose that you find absorbance values of 0.265, 0.269, 0.272, and 0.258 for four identical samples of unknown and absorbances of 0.099, 0.091, 0.101, and 0.097 for four blanks.

(a) Find the average and standard deviation of the absorbance of the unknown and the blank.

(b) Find the corrected absorbance (and its uncertainty) by subtracting the average blank from the average absorbance of the unknown. Consider the standard deviations to be the uncertainties in each quantity.

(c) Substitute for y the corrected absorbance (and its uncertainty) from (b) into the equation of the calibration line: $y = 0.0163_0 (\pm 0.000 2_2)x + 0.004_7 (\pm 0.002_6)$. Solve for x ($=$ micrograms of protein) and its uncertainty.

(d) Suppose that instead of measuring four unknowns and four blanks, you just measured one unknown and one blank and obtained a corrected absorbance of 0.169. Because you do not know the uncertainty associated with this corrected absorbance, use the standard deviation of y ($s_y = 0.005_9$, computed with Equation 4-18) as an estimate of its uncertainty. Find the micrograms of protein and its uncertainty.

24. *Propagation of uncertainty with a calibration curve.* The mass spectrometric signal measured for standard concentrations of methane in hydrogen is given below:

CH_4 (vol%):

| 0 | 0.062 | 0.122 | 0.245 | 0.486 | 0.971 | 1.921 |

Signal (mV):

| 9.1 | 47.5 | 95.6 | 193.8 | 387.5 | 812.5 | 1 671.9 |

(a) Construct a calibration curve. First, subtract the blank value (9.1) from all other values. Then use the method of least squares to find the slope and intercept and their uncertainties.

(b) Replicate measurements of an unknown gave signals of 152.1, 154.9, 153.9, and 155.1 mV, and measurements of a blank gave 8.2, 9.4, 10.6, and 7.8 mV. Find the average and standard deviation of the unknown and the blank.

(c) Find the corrected signal (and its uncertainty) by subtracting the average blank from the average signal for the unknown. Consider the standard deviations to be the uncertainties in each quantity.

(d) Use the answers to (a) and (c) to find the concentration (and its uncertainty) of the unknown.

25. The figure at the top of page 151 gives replicate measurements of As(III) concentration by an electrochemical method.

(a) Using a millimeter ruler, measure each peak height to the nearest 0.1 mm. Noting the length that corresponds to 200 nA in the figure, make a table showing the observed current (nA) for each concentration (μM) of As(III). The blanks appear to be near zero, so we disregard them in this problem.

(b) Construct a calibration curve with 24 points (A–D) and find the slope and intercept and their uncertainties and the standard deviation of y (s_y), using the method of least squares.

(c) Calculate the concentration (and its uncertainty) of As(III) in an unknown that gave a current of 501 ($\pm s_y$) nA.

26. *Nonlinear calibration curve.* Following the procedure in Box 6-3, find how many micrograms of protein are contained in a sample with a corrected absorbance of 0.350 in Figure 6-9.

27. *Logarithmic calibration curve.* Calibration data spanning 5 orders of magnitude for an electrochemical determination of p-nitrophenol are given below:

p-Nitrophenol	Current
0.010 (μg/mL)	0.215 (nA)
0.029 9	0.846
0.117	2.65
0.311	7.41
1.02	20.8
3.00	66.7
10.4	224
31.2	621
107	2 020
310	5 260

Data from L. R. Taylor, *Am. Lab.*, February 1993, p. 44, Figure 4.

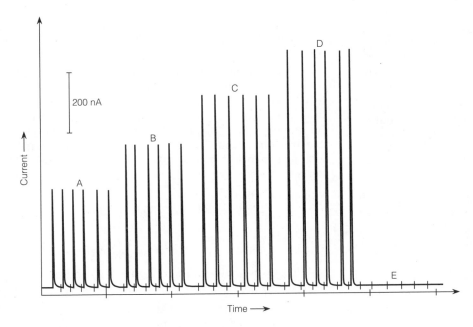

For Problem 25: Electrochemical analysis of As(III). Replicate samples correspond to (A) 20 μM; (B) 30 μM; (C) 40 μM; (D) 50 μM As(III); and (E) blanks. [From I. G. R. Gutz, O. L. Angnes, and J. J. Pedrotti, *Anal. Chem.* **1993**, *65*, 500.]

If you try to plot these data on a linear graph extending from 0 to 310 μg/mL and from 0 to 5 260 nA, most of the points will be bunched up near the origin. To handle data with such a large range, a logarithmic plot is helpful.

(a) Make a graph of log(current) versus log(concentration). Over what range is the log-log calibration linear?

(b) Find the equation of the calibration curve in the form log(current) = m × log(concentration) + b.

28. *Disposable tube detectors for water analysis.* The capillary tube shown below is filled with particles of porous glass impregnated with a colorimetric reagent that reacts with a desired analyte. For example, to measure Al^{3+} in a wastewater stream, the glass contains the reagent alizarin red S, which forms a red complex with Al^{3+}. When the tube is dipped into water, liquid is taken up by capillary action. The liquid moves from left to right in the diagram below, and the front of color formation lags behind the front of liquid travel.

If a fixed volume of liquid is passed through the tube, *the length of the colored region is proportional to the concentration of Al^{3+} in the water.* Explain why this is so. (Similar sensors are used to detect hazardous gases such as phosphine, PH_3, in air.)

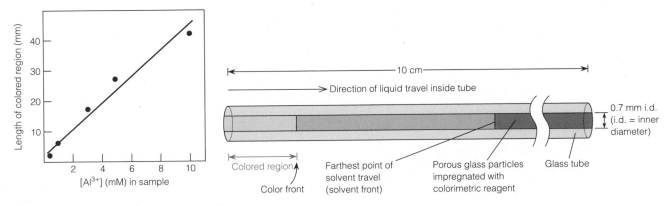

Calibration graph for Al^{3+}, assayed with a tube containing alizarin red S. [From I. Kuselman, B. I. Kuyavskaya, and O. Lev, *Anal. Chim. Acta* **1992**, *256*, 65.]

Standard Addition and Internal Standards

29. State when standard additions and internal standards, instead of a calibration curve, are desirable, and why.

30. Why is it desirable in the method of standard addition to add a small volume of concentrated standard rather than a large volume of dilute standard? Consider matrix effects in your answer.

31. *Standard addition.* An unknown sample of Cu^{2+} gave an absorbance of 0.262 in an atomic absorption analysis. Then 1.00 mL of solution containing 100.0 ppm (= $\mu g/mL$) Cu^{2+} was mixed with 95.0 mL of unknown, and the mixture was diluted to 100.0 mL in a volumetric flask. The absorbance of the new solution was 0.500.

(a) Denoting the initial, unknown concentration as $[Cu^{2+}]_i$, write an expression for the final concentration, $[Cu^{2+}]_f$, after dilution. Units of concentration in this problem are parts per million.

(b) In a similar manner, write the final concentration of added standard Cu^{2+}, designated as $[S]_f$.

(c) Use Equation 6-7 to find $[Cu^{2+}]_i$ in the unknown.

32. *Internal standard.* As described in the section on internal standards, a solution containing 3.47 mM X and 1.72 mM S gave peak areas of 3 473 and 10 222, respectively, in a chromatographic analysis. Then 1.00 mL of 8.47 mM S was added to 5.00 mL of unknown X, and the mixed solution was diluted to 10.0 mL. This solution gave peak areas of 5 428 and 4 431 for X and S, respectively.

(a) Find the concentration of S (mM) in the 10.0 mL of mixed solution.

(b) Find the concentration of X (mM) in the 10.0 mL of mixed solution.

(c) Find the concentration of X in the original unknown.

33. *Propagation of error in standard addition.* The equation of the straight line in a standard addition graph such as Figure 6-10 is signal (± 0.09) (μA) = $[0.371_3$ ($\pm 0.005_4$) ($\mu A/ppm$)] $\times$ concentration (ppm) + 4.2_2 ($\pm 0.1_3$) (μA). Find the concentration of unknown, and

its uncertainty. Does the standard deviation of y (± 0.09) enter into your calculation?

34. *Standard addition.* An assay for substance X is based on its ability to catalyze a reaction that produces radioactive Y. The quantity of Y produced in a fixed time is proportional to the concentration of X in the solution. An unknown containing X in a complex, unknown matrix with an initial volume of 50.0 mL was treated with increments of standard 0.531 M X and the following results were obtained:

Volume of added X (μL):

0	100.0	200.0	300.0	400.0

Counts/min of radioactive Y:

1 084	1 844	2 473	3 266	4 010

(a) Prepare a graph similar to Figure 6-10 and find the concentration of X in the original unknown.

(b) Using the method of least squares, find the uncertainties in the slope and intercept of the straight line in (a) and calculate the uncertainty in the answer to (a).

35. *Internal standard.* Chloroform can be used as an internal standard in the determination of the pesticide DDT in a polarographic analysis in which each compound is reduced at an electrode surface. A mixture containing 0.500 mM chloroform and 0.800 mM DDT gave polarographic currents of 15.3 μA for chloroform and 10.1 μA for DDT. (This analysis is carried out in a mixed solvent of 3:1 [vol/vol] dioxane and water to dissolve the nonpolar organic compounds and carry out electrochemistry. Tetramethylammonium bromide, $[CH_3]_4N^+Br^-$, is used as an electrolyte to give the solution electrical conductivity.) An unknown solution (10.0 mL) containing DDT was placed in a 100-mL volumetric flask and 10.2 μL of chloroform (MW 119.39, density = 1.484 g/mL) was added. After diluting to the mark with 3:1 (vol/vol) dioxane/water containing tetramethylammonium bromide, polarographic signals of 29.4 and 8.7 μA were observed for the chloroform and DDT, respectively. Find the concentration of DDT in the unknown.

. . . .

Notes and References

1. O. B. Toon and R. P. Turco, "Polar Stratospheric Clouds and Ozone Depletion," *Scientific American,* June 1991; R. S. Stolarski, "The Antarctic Ozone Hole," *Scientific American,* January 1988.

2. D. H. Alman and F. W. Billmeyer, Jr., *J. Chem. Ed.* **1976,** *53,* 166. For other approaches to spectroscopy in lecture halls, see F. H. Juergens, *J. Chem. Ed.* **1988,** *65,* 266, 1006; and S. Solomon, C. Hur, A. Lee, and K. Smith, *J. Chem. Ed.* **1994,** *71,* 250.

3. For a report on chemiluminescence in chemical analysis, see M. L. Grayeski, *Anal. Chem.* **1987,** *59,* 1243A.

4. D. C. Harris, *J. Chem. Ed.* **1978,** *55,* 539.

5. J. R. Duffy and J. Gaudin, *Clin. Biochem.* **1977,** *10,* 122.

6. For a general discussion of standard addition methods, see M. Bader, *J. Chem. Ed.* **1980,** *57,* 703.

The Earliest
Known Buret

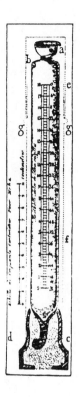

The buret, invented by F. Descroizilles in the early 1800s, was used in the same manner as a graduated cylinder is used today. The stopcock was introduced in 1846. This buret and its progeny have terrorized generations of analytical chemistry students.[1]

Illustration from E. R. Madsen, *The Development of Titrimetric Analysis 'till 1806* (Copenhagen: E. E. C. Gad Publishers, 1958.)

Volumetric Analysis

7

In **volumetric analysis,** the volume of reagent needed to react with analyte is measured. In this chapter we discuss general principles that apply to any volumetric procedure, and then we focus on precipitation titrations, about which you already know more than you might realize. We also introduce spectrophotometric titrations, which are especially useful in biochemistry.

7-1 Principles of Volumetric Analysis

In a **titration,** increments of the reagent solution—the **titrant**—are added to the analyte until their reaction is complete. Titrant is usually delivered from a buret, as shown in Figure 7-1.

The principal requirements for a titration reaction are that it have a large equilibrium constant and proceed rapidly. That is, each increment of titrant should be completely and quickly consumed by analyte until the analyte is used up. The most common titrations are based on acid-base, oxidation-reduction, complex formation, or precipitation reactions.

Methods of determining when the analyte has been consumed include (1) detecting a sudden change in the voltage or current between a pair of electrodes (Figure 7-9), (2) observing an indicator color change (Color Plate 4), and (3) monitoring a spectrophotometric absorbance change (Figure 7-5). An **indicator** is a compound with a physical property (usually color) that changes abruptly near the equivalence point. The change is caused by the disappearance of analyte or appearance of excess titrant.

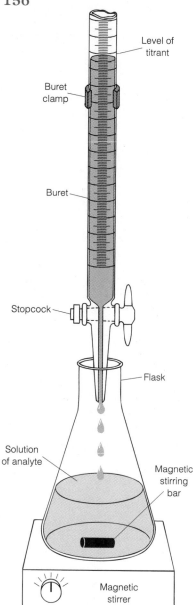

Figure 7-1 Typical setup for a titration. The analyte is contained in the flask, and the titrant in the buret. The stirring bar is a magnet coated with Teflon, which is inert to almost all solutions. The bar is spun by a rotating magnet inside the stirring motor.

The **equivalence point** occurs when the quantity of titrant added is the exact amount necessary for stoichiometric reaction with the analyte. For example, oxalic acid reacts with permanganate in hot acidic solution:

$$
\underset{\substack{\text{Analyte}\\\text{Oxalic acid}\\\text{(colorless)}}}{5\,HO-\overset{\overset{\text{O}}{\|}}{C}-\overset{\overset{\text{O}}{\|}}{C}-OH} + \underset{\substack{\text{Titrant}\\\text{Permanganate}\\\text{(purple)}}}{2\,MnO_4^-} + 6H^+ \longrightarrow 10CO_2 + 2Mn^{2+} + 8H_2O \underset{\substack{\text{(colorless)}\ \ \text{(colorless)}}}{} \tag{7-1}
$$

If the unknown contains 5.000 mmol of oxalic acid, the equivalence point is reached when 2.000 mmol of MnO_4^- has been added, because Reaction 7-1 requires 2 mol of permanganate to react with 5 mol of oxalic acid.

The equivalence point is the ideal result we seek in a titration. What we actually measure is the **end point,** which is marked by a sudden change in a physical property of the solution. In Reaction 7-1, a convenient end point is the abrupt appearance of the purple color of permanganate in the flask. Up to the equivalence point, all of the added permanganate is consumed by the oxalic acid, and the titration solution remains colorless. After the equivalence point, unreacted MnO_4^- ion builds up until there is enough to see. The *first trace* of purple color marks the end point. The better your eyes, the closer will be your measured end point to the true equivalence point. Here, the end point cannot exactly equal the equivalence point because extra MnO_4^-, beyond that needed to react with oxalic acid, is required to show the purple color.

The difference between the end point and the equivalence point is an inescapable **titration error.** By choosing an appropriate physical property, in which a change is easily observed (such as indicator color, optical absorbance of a reactant or product, or pH), it is possible for the end point to be very close to the equivalence point. It is also usually possible to estimate the titration error with a **blank titration,** in which the same procedure is carried out without analyte. For example, a solution containing no oxalic acid could be titrated with MnO_4^- to see how much is needed to form an observable purple color. This volume of MnO_4^- is then subtracted from the volume observed in the analytical titration.

The validity of an analytical result depends on knowing the amount of one of the reactants used. The concentration of titrant is known if the titrant was prepared by dissolving a weighed amount of pure reagent in a known volume of solution. In such a case, we call the reagent a **primary standard,** because it is pure enough to be weighed and used directly. A primary standard should be 99.9% pure, or better. It should not decompose under ordinary storage, and it should be stable when dried by heating or vacuum, because drying is required to remove traces of adsorbed water from the atmosphere. Primary standards for many elements are given in Table 26-8. Box 3-1 described Standard Reference Materials that are available from the U.S. National Institute of Standards and Technology and allow different laboratories to test the accuracy of their procedures.

In most cases, the titrant is not available as a primary standard. Instead, a solution having the approximate desired concentration is used to titrate a weighed, primary standard. By this procedure, called **standardization,** we determine the precise concentration of the solution to be used in the analysis. We then say that the solution is a **standard solution.** In all cases, the validity of the analytical result ultimately depends on knowing the composition of some primary standard.

In a **direct titration,** titrant is added to the analyte until the reaction is complete. Occasionally it is more convenient to perform a **back titration,** in which a known *excess* of one standard reagent is added to the analyte. Then a second standard reagent is used to titrate the excess of the first reagent. Back titrations are useful when the end point of the back titration is clearer than the end point of the direct titration, or when an excess of the first reagent is required for complete reaction with the analyte.

To appreciate the difference between direct and back titrations, consider first the addition of permanganate titrant to oxalic acid analyte in Reaction 7-1; this reaction is a direct titration. For a back titration, an *excess* (but known amount) of permanganate could be added to consume the oxalic acid. Then the excess permanganate could be back-titrated with standard Fe^{2+} to measure how much permanganate had been consumed by the oxalic acid.

7-2 Volumetric Procedures and Calculations

In this section we study examples that illustrate stoichiometry calculations in volumetric analysis. The key step is to *relate the moles of titrant to the moles of analyte.* We also introduce the Kjeldahl titration as a representative, widely used volumetric procedure.

. . . .

EXAMPLE **Standardization of Titrant Followed by Analysis of Unknown**

The calcium content of urine can be determined by the following procedure:

1. Ca^{2+} is precipitated as calcium oxalate in basic solution:

$$Ca^{2+} + C_2O_4^{2-} \rightarrow Ca(C_2O_4)\cdot H_2O(s)$$

$$\text{Oxalate} \qquad \text{Calcium oxalate}$$

2. After the precipitate is washed with ice-cold water to remove free oxalate, the solid is dissolved in acid, which gives Ca^{2+} and $H_2C_2O_4$ in solution.

3. The dissolved oxalic acid is heated to 60°C and titrated with standardized potassium permanganate until the purple end point is observed (Reaction 7-1).

Standardization Suppose that 0.356 2 g of $Na_2C_2O_4$ is dissolved in a 250.0-mL volumetric flask. If 10.00 mL of this solution requires 48.36 mL of $KMnO_4$ solution for titration, find the molarity of the permanganate solution.

Solution The concentration of the oxalate solution is

$$\frac{0.356\,2 \text{ g } Na_2C_2O_4/(134.00 \text{ g } Na_2C_2O_4/\text{mol})}{0.250\,0 \text{ L}} = 0.010\,63 \text{ M}$$

The moles of $C_2O_4^{2-}$ in 10.00 mL are $(0.010\,63 \text{ mol/L})(0.010\,00 \text{ L}) = 1.063 \times 10^{-4}$ mol $= 0.106\,3$ mmol. Reaction 7-1 requires 2 mol of permanganate to react with 5 mol of oxalate, so the moles of MnO_4^- delivered must have been

$$\text{moles of } MnO_4^- = \frac{2}{5} \text{ (moles of } C_2O_4^{2-}) = 0.042\,53 \text{ mmol}$$

Reaction 7-1 requires $2MnO_4^-$ for $5C_2O_4^{2-}$.

Note that mmol/mL is the same as mol/L = M.

The concentration of MnO_4^- in the titrant is therefore

$$\text{molarity of } MnO_4^- = \frac{0.042\,53 \text{ mmol}}{48.36 \text{ mL}} = 8.795 \times 10^{-4} \text{ M}$$

Analysis of Unknown   Suppose that the calcium in a 5.00-mL urine sample was precipitated by the procedure above, redissolved, and required 16.17 mL of the standard MnO_4^- solution. Find the concentration of Ca^{2+} in the urine.

Solution   In 16.17 mL of MnO_4^- there are $(0.016\,17 \text{ L})(8.795 \times 10^{-4}$ mol/L$) = 1.422 \times 10^{-5}$ mol of MnO_4^-. This quantity will react with

Reaction 7-1 requires $5C_2O_4^{2-}$ for $2MnO_4^-$.

$$\text{moles of } C_2O_4^{2-} = \frac{5}{2} (\text{moles of } MnO_4^-)$$

$$= 3.555 \times 10^{-5} \text{ mol} = 0.035\,55 \text{ mmol}$$

Because there is one oxalate ion for each calcium ion in $Ca(C_2O_4)\cdot H_2O$, there must have been 0.035 55 mmol of Ca^{2+} in 5.00 mL of urine:

$$[Ca^{2+}] = \frac{0.035\,55 \text{ mmol}}{5.00 \text{ mL}} = 0.007\,11 \text{ M}$$

EXAMPLE   **Titration of a Mixture**

A solid mixture weighing 1.372 g containing only sodium carbonate and sodium bicarbonate required 29.11 mL of 0.734 4 M HCl for complete titration:

$$Na_2CO_3 + 2HCl \rightarrow 2NaCl(aq) + H_2O + CO_2$$
$$\text{FW 105.99}$$

$$NaHCO_3 + HCl \rightarrow NaCl(aq) + H_2O + CO_2$$
$$\text{FW 84.01}$$

Find the mass of each component in the mixture.

Solution   Let's denote the grams of Na_2CO_3 by x and grams of $NaHCO_3$ by $1.372 - x$. The moles of each component must be

$$\text{mol } Na_2CO_3 = \frac{x \text{ g}}{105.99 \text{ g/mol}} \qquad \text{mol } NaHCO_3 = \frac{(1.372 - x) \text{ g}}{84.01 \text{ g/mol}}$$

We know that the total moles of HCl used was $(0.029\,11 \text{ L})(0.734\,4 \text{ M}) = 0.021\,38$ mol. From the stoichiometry of the two reactions, we can say that

$$2 (\text{mol } Na_2CO_3) + \text{mol } NaHCO_3 = 0.021\,38$$

$$2 \left(\frac{x}{105.99} \right) + \frac{1.372 - x}{84.01} = 0.021\,38 \Rightarrow x = 0.724 \text{ g}$$

The mixture contains 0.724 g Na_2CO_3 and $1.372 - 0.724 = 0.648$ g $NaHCO_3$.

Kjeldahl Nitrogen Analysis

Developed in 1883, the **Kjeldahl nitrogen analysis** remains one of the most accurate and widely used methods for determining nitrogen in substances

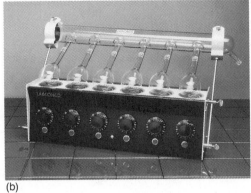

(a)   (b)

Figure 7-2 (*a*) Kjeldahl digestion flask with long neck to minimize loss by spattering. (*b*) Six-port manifold for multiple samples provides for exhaust of fumes. [Courtesy Fisher Scientific, Pittsburgh, PA.]

such as protein, milk, cereal, and flour. The solid is first *digested* (decomposed and dissolved) in boiling sulfuric acid, which converts nitrogen to ammonium ion, NH_4^+, and oxidizes other elements present:

Kjeldahl digestion: organic C, H, N $\xrightarrow[\text{H}_2\text{SO}_4]{\text{Boiling}}$ $NH_4^+ + CO_2 + H_2O$ (7-2)

Mercury, copper, and selenium compounds catalyze the digestion process. To speed the rate of reaction, the boiling point of concentrated (98 wt%) sulfuric acid (338°C) is raised by adding K_2SO_4. Digestion is carried out in a long-neck *Kjeldahl flask* (Figure 7-2) that prevents loss of sample by spattering. [A convenient alternative to the Kjeldahl flask is to use H_2SO_4 and H_2O_2 in a microwave bomb (a pressurized vessel), like that shown in Figure 26-6.]

After digestion is complete, the solution containing NH_4^+ is made basic, and the liberated NH_3 is distilled (with a large excess of steam) into a receiver containing a known amount of HCl (Figure 7-3). Excess, unreacted HCl is titrated with standard NaOH to determine how much HCl was consumed by NH_3.

neutralization of NH_4^+: $NH_4^+ + OH^- \rightarrow NH_3(g) + H_2O$ (7-3)

distillation of NH_3 into standard HCl: $NH_3 + H^+ \rightarrow NH_4^+$ (7-4)

titration of unreacted HCl with NaOH: $H^+ + OH^- \rightarrow H_2O$ (7-5)

Each atom of nitrogen in the starting material is converted into one NH_4^+ ion.

· · · · · · · · · · · · · · · ·

EXAMPLE Kjeldahl Analysis

A typical protein contains 16.2 wt% nitrogen. A 0.500-mL aliquot of protein solution was digested, and the liberated NH_3 was distilled into 10.00 mL of 0.02140 M HCl. The unreacted HCl required 3.26 mL of 0.0198 M NaOH for complete titration. Find the concentration of protein (mg protein/mL) in the original sample.

Solution The original number of moles of HCl in the receiver was (10.00 mL)(0.02140 mmol/mL) = 0.2140 mmol. The NaOH required for titration of unreacted HCl in Reaction 7-5 was (3.26 mL)(0.0198 mmol/mL) =

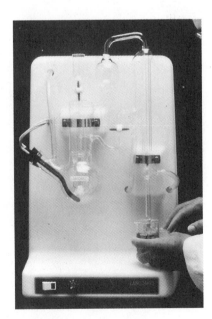

Figure 7-3 Kjeldahl distillation unit employs electric immersion heater in flask at left to carry out distillation in 5 min. Beaker at right collects liberated NH_3 in standard HCl. [Courtesy Fisher Scientific, Pittsburgh, PA.] Simpler apparatus is shown in Experiment 27-6.

0.064 5 mmol. The difference, $0.2140 - 0.0645 = 0.1495$ mmol, must equal the quantity of NH_3 produced in Reaction 7-3 and distilled into the HCl.

Because 1 mol of nitrogen in the protein gives rise to 1 mol of NH_3, there must have been 0.149 5 mmol of nitrogen in the protein, corresponding to

$$(0.149\,5\ \text{mmol})\left(14.006\,74\ \frac{\text{mg N}}{\text{mmol}}\right) = 2.093\ \text{mg N}$$

If the protein contains 16.2 wt% N, there must be

$$\frac{2.093\ \text{mg N}}{0.162\ \text{mg N/mg protein}} = 12.9\ \text{mg protein}$$

$$\Rightarrow \frac{12.9\ \text{mg protein}}{0.500\ \text{mL}} = 25.8\ \frac{\text{mg protein}}{\text{mL}}$$

7-3 Spectrophotometric Titrations

Absorption of light is one of many physical properties whose change may be used to monitor the progress of a titration. For example, a solution of the iron-transport protein, transferrin (Figure 7-4), can be titrated with iron to measure the transferrin content. Transferrin without iron, called apotransferrin, is colorless. Each protein molecule with a molecular weight of

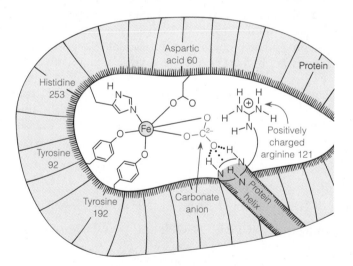

Figure 7-4   Each of the two iron-binding sites of transferrin is located at the base of a cleft in the protein, shown schematically in this diagram. The Fe^{3+} ion binds to one nitrogen atom from the amino acid histidine and three oxygen atoms from tyrosine and aspartic acid. The fifth and sixth ligand sites of the metal are occupied by oxygen atoms from a carbonate anion (CO_3^{2-}), which is anchored in place by electrostatic interaction with the positively charged amino acid arginine and by hydrogen bonding to part of the protein helix. When transferrin is taken up by a cell, it is brought to a compartment whose pH is lowered to 5.5. H^+ then reacts with the carbonate ligand to make HCO_3^- and H_2CO_3, thereby releasing Fe^{3+} from the protein. [Adapted from E. N. Baker, B. F. Anderson, H. M. Baker, M. Haridas, G. E. Norris, S. V. Rumball, and C. A. Smith, *Pure Appl. Chem.* **1990,** *62,* 1067.]

81 000 binds two Fe^{3+} ions. When the iron binds to the protein, a red color with an absorbance maximum at 465 nm develops. The appearance of the red color may be used to follow the course of a titration of an unknown amount of apotransferrin with a standard solution of Fe^{3+}.

$$\text{apotransferrin} + 2Fe^{3+} \rightarrow (Fe^{3+})_2\text{transferrin} \qquad (7\text{-}6)$$
$$\text{(colorless)} \qquad\qquad\qquad\qquad \text{(red)}$$

Figure 7-5 shows the results of a titration of 2.000 mL of a solution of apotransferrin with 1.79×10^{-3} M ferric nitrilotriacetate solution. As iron is added to the protein, the red color develops and the absorbance increases. When the protein is saturated with iron, no further color can form, and the curve levels off. The extrapolated intersection of the two straight portions of the titration curve at 203 μL in Figure 7-5 is taken as the end point. The absorbance continues to rise slowly after the equivalence point because the ferric nitrilotriacetate complex has some absorbance at 465 nm.

In constructing the graph in Figure 7-5, the effect of dilution should be considered, because the volume is different at each point. Each point plotted on the graph represents the absorbance that would be observed *if the solution had not been diluted from its original volume of 2.000 mL.*

$$\text{corrected absorbance} = \left(\frac{\text{total volume}}{\text{initial volume}}\right)(\text{observed absorbance}) \qquad (7\text{-}7)$$

· ·

EXAMPLE   Correcting Absorbance for the Effect of Dilution

The absorbance measured after adding 125 μL (= 0.125 mL) of ferric nitrilotriacetate to 2.000 mL of apotransferrin was 0.260. Calculate the corrected absorbance that should be plotted in Figure 7-5.

Solution   The total volume was 2.000 + 0.125 = 2.125 mL. If the volume had been 2.000 mL, the absorbance would have been greater than 0.260 by a factor of 2.125/2.000.

$$\text{corrected absorbance} = \left(\frac{2.125 \text{ mL}}{2.000 \text{ mL}}\right)(0.260) = 0.276$$

The absorbance plotted in Figure 7-5 is 0.276.

· ·

Turbidimetry and Nephelometry

Some titrations, such as the reaction of Ba^{2+} with SO_4^{2-} to give $BaSO_4(s)$, produce a fine, suspended precipitate instead of a soluble product. The reaction mixture becomes progressively cloudier until the equivalence point is reached. A cloudy solution, which is said to be *turbid,* scatters rather than absorbs light. After the equivalence point, the **turbidity** (light scattering) ceases to change, because no more precipitate forms. Because scattering is dependent on particle size, the reaction must be carried out in a reproducible manner to obtain precise results. A glycerol-alcohol mixture can be used to help stabilize the particles and prevent rapid settling of solid.

In *turbidimetry,* the transmittance of light through a suspension of precipitate is measured during the titration. The transmittance decreases until the equivalence point is reached and then ceases to change. In

Ferric nitrilotriacetate is used because Fe^{3+} forms a hydroxide that precipitates in neutral solution. Nitrilotriacetate binds Fe^{3+} through four ligand atoms shown in **bold** type:

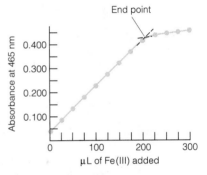

$$^{-}O_2C-N\begin{array}{c} -CO_2^- \\ \\ -CO_2^- \end{array}$$

Nitrilotriacetate anion

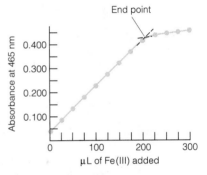

Figure 7-5 Spectrophotometric titration of transferrin with ferric nitrilotriacetate. Absorbance is corrected as if no dilution had taken place. The initial absorbance of the solution, before iron is added, is due to a colored impurity.

The signal (scattered light intensity) in nephelometry is increased when the intensity of the light source is increased. The signal (transmittance) in turbidimetry is not affected by the intensity of the light source.

nephelometry, light scattered at 90° to the incident beam by the turbid solution is measured. Scattering increases until the equivalence point is reached and then ceases to change.

In turbidimetry we measure the *fraction* of light scattered (the transmittance), which is independent of the intensity of the light source. In nephelometry we measure the *absolute intensity* of the scattered light, which increases if the source intensity is increased. Therefore, the sensitivity of nephelometry can be raised by increasing the power of the light source or by using a more sensitive detector.

Turbidimetric and nephelometric titrations are not very precise because end point detection depends on particle size, which is not very reproducible. However, the sensitivity is excellent; sulfate can be analyzed at the parts-per-million level by formation of $BaSO_4$.

7-4 The Precipitation Titration Curve

We now focus our attention on the details of precipitation titrations as an illustration of principles that underlie all kinds of titrations. We first study how the concentrations of analyte and titrant vary during a titration and then derive simple, approximate equations that can be used with a calculator to graph titration curves. Ultimately, we introduce more powerful equations that are suitable for use with spreadsheets. It is important first to learn to use the approximate equations to gain insight into the events occurring at each stage of a titration.

The titration curve is a graph showing how the concentration of one of the reactants varies as titrant is added. Rather than plotting concentration directly, it is more useful to plot the p function:

In Chapter 8 we define the p function correctly in terms of activities instead of concentrations. For now, we'll use pX = $-\log[X]$.

p function:
$$pX = -\log_{10}[X] \tag{7-8}$$

where [X] is the concentration of X.

Consider the titration of 25.00 mL of 0.100 0 M I^- with 0.050 00 M Ag^+

$$I^- + Ag^+ \rightarrow AgI(s) \tag{7-9}$$

and suppose that we are monitoring the Ag^+ concentration with an electrode. Reaction 7-9 is the reverse of the dissolution of $AgI(s)$, whose solubility product is rather small:

$$AgI(s) \rightleftharpoons Ag^+ + I^- \qquad K_{sp} = [Ag^+][I^-] = 8.3 \times 10^{-17} \tag{7-10}$$

Because the equilibrium constant for the titration reaction 7-9 is large ($K = 1/K_{sp} = 1.2 \times 10^{16}$), the equilibrium lies far to the right. It is reasonable to say that each aliquot of Ag^+ reacts completely with I^-, leaving only a very tiny amount of Ag^+ in solution. At the equivalence point, there will be a sudden increase in the Ag^+ concentration because all the I^- has been consumed and we are now adding Ag^+ directly to the solution.

What volume of Ag^+ titrant solution is needed to reach the equivalence point? To calculate this volume, which we designate V_e, first note that 1 mol of Ag^+ is required for each mole of I^-.

V_e = volume of titrant at equivalence point

$$\text{mol } I^- = \text{mol } Ag^+$$

$$(0.025\,00 \text{ L})(0.100\,0 \text{ mol } I^-/L) = (V_e)(0.050\,00 \text{ mol } Ag^+/L)$$

$$V_e = 0.050\,00 \text{ L} = 50.00 \text{ mL}$$

An easy way to find V_e in this particular example is to note that the concentration of Ag^+ is half the concentration of I^-. Therefore, the volume of Ag^+ required will be twice the volume of I^- to be titrated, and 25.00 mL of I^- will require 50.00 mL of Ag^+.

The titration curve has three distinct regions. The calculations are different, depending on whether we are before, at, or after the equivalence point. Let's consider each region separately.

Before the Equivalence Point

Consider the point at which the volume of Ag^+ added is 10.00 mL. Because there are more moles of I^- than Ag^+ at this point, virtually all the Ag^+ is "used up" to make $AgI(s)$. We want to find the very small concentration of Ag^+ remaining in solution after reaction with I^-. One way to do this is to imagine that Reaction 7-9 has gone to completion and that some AgI redissolves (Reaction 7-10). The solubility of Ag^+ will be determined by the concentration of free I^- remaining in the solution:

$$[Ag^+] = \frac{K_{sp}}{[I^-]} \tag{7-11}$$

When $V < V_e$, the concentration of unreacted I^- regulates the solubility of AgI.

The free I^- is overwhelmingly due to the I^- that has not been precipitated by 10.00 mL of Ag^+. By comparison, the I^- due to dissolution of $AgI(s)$ is negligible.

So let's find the concentration of unprecipitated I^-. The moles of I^- remaining in solution will be

moles of I^- = original moles of I^- − moles of Ag^+ added

$$= (0.025\,00\text{ L})(0.100\text{ mol/L}) - (0.010\,00\text{ L})(0.050\,00\text{ mol/L})$$

$$= 0.002\,000\text{ mol } I^-$$

Because the volume is 0.035 00 L (25.00 mL + 10.00 mL), the concentration is

$$[I^-] = \frac{0.002\,000\text{ mol } I^-}{0.035\,00\text{ L}} = 0.057\,14\text{ M} \tag{7-12}$$

The concentration of Ag^+ in equilibrium with this much I^- is

$$[Ag^+] = \frac{K_{sp}}{[I^-]} = \frac{8.3 \times 10^{-17}}{0.057\,14} = 1.4_5 \times 10^{-15}\text{ M} \tag{7-13}$$

Finally, the p function we seek is

$$pAg^+ = -\log[Ag^+] = 14.84 \tag{7-14}$$

There are two significant figures in the concentration of Ag^+ because there are two significant figures in K_{sp}. The two figures in $[Ag^+]$ translate into two figures in the *mantissa* of the p function, which is correctly written as 14.84.

The step-by-step calculation just outlined is a safe, but tedious, way to find the concentration of I^-. We will now examine a streamlined procedure that is well worth learning. Bear in mind that $V_e = 50.00$ mL. When 10.00 mL of Ag^+ has been added, the reaction is one-fifth complete because 10.00 mL out of the 50.00 mL of Ag^+ needed for complete reaction has been added. Therefore, four-fifths of the I^- remains unreacted. If there were no dilution, the concentration of I^- would be four-fifths of its original value. However, the original volume of 25.00 mL has been increased to 35.00 mL.

$\log (1.4_5 \times 10^{-15}) = 14.84$

Two significant figures / Two digits in mantissa

Significant figures in logarithms are discussed in Section 3-2.

If no I^- had been consumed, the concentration would be the original value of $[I^-]$ times $(25.00/35.00)$. Accounting for both the reaction and the dilution, we can write

Streamlined calculation well worth using.

$$[I^-] = \left(\frac{4.000}{5.000}\right) \underbrace{(0.1000 \text{ M})}_{} \left(\frac{25.00}{35.00}\right) = 0.057\,14 \text{ M}$$

Original volume of I^- solution

Total volume of solution

$\underbrace{}_{\substack{\text{Fraction} \\ \text{remaining}}}$ $\underbrace{}_{\substack{\text{Original} \\ \text{concentration}}}$ $\underbrace{}_{\substack{\text{Dilution} \\ \text{factor}}}$

This is the same result found in Equation 7-12.

EXAMPLE Using the Streamlined Calculation

To reinforce the streamlined procedure, let's calculate pAg^+ when V_{Ag^+}, the volume added from the buret, is 49.00 mL.

Solution Because $V_e = 50.00$ mL, the fraction of I^- reacted is $49.00/50.00$, and the fraction remaining is $1.00/50.00$. The total volume is $25.00 + 49.00 = 74.00$ mL.

$$[I^-] = \underbrace{\left(\frac{1.00}{50.00}\right)}_{\substack{\text{Fraction} \\ \text{remaining}}} \underbrace{(0.1000 \text{ M})}_{\substack{\text{Original} \\ \text{concentration}}} \underbrace{\left(\frac{25.00}{74.00}\right)}_{\substack{\text{Dilution} \\ \text{factor}}} = 6.76 \times 10^{-4} \text{ M}$$

$$[Ag^+] = K_{sp}/[I^-] = 1.2_3 \times 10^{-13} \text{ M}$$

$$pAg^+ = -\log[Ag^+] = 12.91$$

The concentration of Ag^+ is negligible compared with the concentration of unreacted I^-, even though the titration is 98% complete.

At the Equivalence Point

Now we have added exactly enough Ag^+ to react with all the I^-. We can imagine that all the AgI precipitates and some redissolves to give equal concentrations of Ag^+ and I^-. The value of pAg^+ is found by setting $[Ag^+] = [I^-] = x$ in the solubility product:

When $V = V_e$, the concentration of Ag^+ is determined by the solubility of pure AgI.

$$[Ag^+][I^-] = K_{sp}$$

$$(x)(x) = 8.3 \times 10^{-17} \Rightarrow x = 9.1 \times 10^{-9} \text{ M}$$

$$pAg^+ = -\log x = 8.04$$

This value of pAg^+ is independent of the original concentration or volumes.

After the Equivalence Point

Now the concentration of Ag^+ is determined almost completely by the amount of Ag^+ added *after* the equivalence point. Virtually all the Ag^+

added *before* the equivalence point has precipitated as AgI. Suppose that $V_{Ag^+} = 52.00$ mL. The amount added past the equivalence point is 2.00 mL. The calculation proceeds as follows:

$$\text{mol Ag}^+ = (0.00200 \text{ L})(0.05000 \text{ mol Ag}^+/\text{L}) = 0.000100 \text{ mol}$$

$$[\text{Ag}^+] = (0.000100 \text{ mol})/(0.07700 \text{ L}) = 1.30 \times 10^{-3} \text{ M}$$

Total volume = 77.00 mL

$$\text{pAg}^+ = 2.89$$

We could justifiably use three significant figures for the mantissa of pAg^+, because there are now three significant figures in the value of $[\text{Ag}^+]$. But, to be consistent with our earlier results, we will retain only two figures. For consistency, we will generally express p functions in this text with two decimal places.

A somewhat streamlined calculation can save time. The concentration of Ag^+ in the buret is 0.05000 M, and 2.00 mL of this solution is being diluted to $(25.00 + 52.00) = 77.00$ mL. Hence $[\text{Ag}^+]$ is given by

Volume of excess Ag^+

$$[\text{Ag}^+] = 0.05000 \text{ M} \left(\frac{2.00}{77.00}\right) = 1.30 \times 10^{-3} \text{ M}$$

Original concentration of Ag^+ — Dilution factor — Total volume of solution

When $V > V_e$, the concentration of Ag^+ is determined by the excess Ag^+ added from the buret.

The streamlined calculation.

The Shape of the Titration Curve

Figure 7-6 shows the complete titration curve and also illustrates the effect of reactant concentrations on the titration. The equivalence point is the steepest

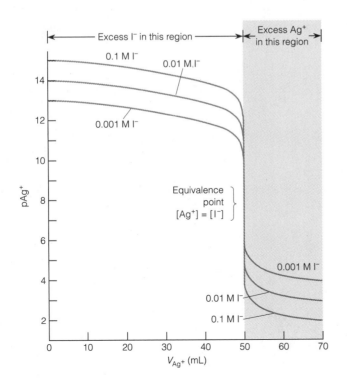

Figure 7-6 Titration curves showing the effect of diluting the reactants. *Outer curve:* 25.00 mL of 0.1000 M I^- titrated with 0.05000 M Ag^+. *Middle curve:* 25.00 mL of 0.01000 M I^- titrated with 0.005000 M Ag^+. *Inner curve:* 25.00 mL of 0.001000 M I^- titrated with 0.0005000 M Ag^+.

point of the curve. It is the point of maximum slope (a negative slope in this case) and is therefore an inflection point (at which the second derivative, by definition, is zero):

$$\text{steepest slope:} \quad \frac{dy}{dx} \quad \text{reaches its greatest value}$$

$$\text{inflection point:} \quad \frac{d^2y}{dx^2} = 0$$

A *complexometric* titration involves complex ion formation between titrant and analyte.

In titrations involving 1:1 stoichiometry of reactants, the equivalence point is the steepest point of the titration curve. This is true of acid-base, complexometric, and redox titrations as well. For stoichiometries other than 1:1, such as $2Ag^+ + CrO_4^{2-} \rightarrow Ag_2CrO_4(s)$, the curve is not symmetric near the equivalence point. The equivalence point is not at the center of the steepest section of the curve, and it is not an inflection point. In practice, conditions are chosen such that titration curves are steep enough for the steepest point to be a good estimate of the equivalence point, regardless of the stoichiometry.

At the equivalence point, the titration curve is steepest for the least soluble precipitate.

Figure 7-7 illustrates how K_{sp} affects the titration of halide ions. The least soluble product, AgI, gives the sharpest change at the equivalence point. However, even for AgCl, the curve is steep enough to locate the equivalence point with little uncertainty. The larger the equilibrium constant for any titration reaction, the more pronounced will be the change in concentration near the equivalence point.

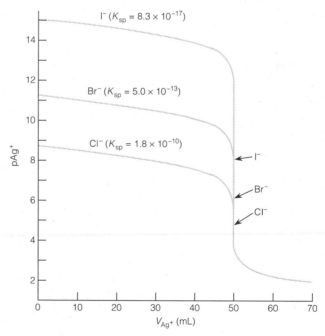

Figure 7-7 Titration curves showing the effect of K_{sp}. Each curve is calculated for 25.00 mL of 0.100 0 M halide titrated with 0.050 00 M Ag^+. Equivalence points are marked by arrows.

EXAMPLE **Calculating Concentrations During a Precipitation Titration**

A solution containing 25.00 mL of 0.041 32 M $Hg_2(NO_3)_2$ was titrated with 0.057 89 M KIO_3.

$$Hg_2^{2+} + 2IO_3^- \rightarrow Hg_2(IO_3)_2(s)$$
$$\text{Iodate}$$

The solubility product for $Hg_2(IO_3)_2$ is 1.3×10^{-18}. Calculate the concentration of Hg_2^{2+} ion in the solution (a) after addition of 34.00 mL of KIO_3; (b) after addition of 36.00 mL of KIO_3; and (c) at the equivalence point.

Solution The reaction requires 2 moles of IO_3^- per mole of Hg_2^{2+}. The volume of iodate needed to reach the equivalence point is found as follows:

moles of IO_3^- = 2(moles of Hg_2^{2+}) — Notice which side of the equation has the 2

$$(V_e)(0.057\,89\ M) = 2(25.00\ mL)(0.041\,32\ M) \Rightarrow V_e = 35.69\ mL$$

(a) When $V = 34.00$ mL, the precipitation of Hg_2^{2+} is not yet complete.

Original volume of Hg_2^{2+}

$$[Hg_2^{2+}] = \left(\frac{35.69 - 34.00}{35.69}\right)(0.041\,32\ M)\left(\frac{25.00}{25.00 + 34.00}\right) = 8.29 \times 10^{-4}\ M$$

Fraction remaining | Original concentration of Hg_2^{2+} | Dilution factor | Total volume of solution

(b) When $V = 36.00$ mL, the precipitation is complete. We have gone $(36.00 - 35.69) = 0.31$ mL *past* the equivalence point. The concentration of excess IO_3^- is

Volume of excess IO_3^-

$$[IO_3^-] = (0.057\,89\ M)\left(\frac{0.31}{25.00 + 36.00}\right) = 2.9 \times 10^{-4}\ M$$

Original concentration of IO_3^- | Dilution factor | Total volume of solution

The concentration of Hg_2^{2+} in equilibrium with solid $Hg_2(IO_3)_2$ plus this much IO_3^- is

$$[Hg_2^{2+}] = \frac{K_{sp}}{[IO_3^-]^2} = \frac{1.3 \times 10^{-18}}{(2.9 \times 10^{-4})^2} = 1.5 \times 10^{-11}\ M$$

(c) At the equivalence point, we have exactly enough IO_3^- to react with all the Hg_2^{2+}. We can write

$$Hg_2(IO_3)_2(s) \rightleftharpoons \underset{x}{Hg_2^{2+}} + \underset{2x}{2IO_3^-}$$

$$(x)(2x)^2 = K_{sp} \Rightarrow x = [Hg_2^{2+}] = 6.9 \times 10^{-7}\ M$$

7-5 Titration of a Mixture

When a mixture is titrated, the product with the smaller K_{sp} precipitates first, if the stoichiometry of the different possible precipitates is the same.

The precipitation of I^- and Cl^- with Ag^+ produces two distinct breaks in the titration curve. The first corresponds to the reaction of I^- and the second to the reaction of Cl^-.

Before Cl^- precipitates, the calculations for AgI precipitation are just as they were in Section 7-4.

If a mixture of two ions is titrated, the less soluble precipitate will be formed first. If the two solubility products are sufficiently different, the first precipitation will be nearly complete before the second commences.

Consider the titration with $AgNO_3$ of a solution containing KI and KCl. Because $K_{sp}(AgI) << K_{sp}(AgCl)$, the first Ag^+ added will precipitate AgI. Further addition of Ag^+ continues to precipitate I^- with no effect on Cl^-. When precipitation of I^- is almost complete, the concentration of Ag^+ abruptly increases. Then, when the concentration of Ag^+ is high enough, AgCl begins to precipitate and $[Ag^+]$ levels off again. Finally, when the Cl^- is consumed, another abrupt change in $[Ag^+]$ occurs. Qualitatively, we expect to see two breaks in the titration curve. The first corresponds to the AgI equivalence point, and the second to the AgCl equivalence point.

Figure 7-8 shows an experimental curve for this titration. The apparatus used to measure the curve is shown in Figure 7-9, and the theory of how this system measures Ag^+ concentration is discussed in Chapter 15.

The I^- end point is taken as the intersection of the steep and nearly horizontal curves shown in the inset of Figure 7-8. The reason for using the intersection is that the precipitation of I^- is not quite complete when Cl^- begins to precipitate. Therefore, the end of the steep portion (the intersection) is a better approximation of the equivalence point than is the middle of the steep section. The Cl^- end point is taken as the midpoint of the second steep section, at 47.41 mL. The moles of Cl^- in the sample correspond to the moles of Ag^+ delivered between the first and second end points. That is, it requires 23.85 mL of Ag^+ to precipitate I^-, and $(47.41 - 23.85) = 23.56$ mL of Ag^+ to precipitate Cl^-.

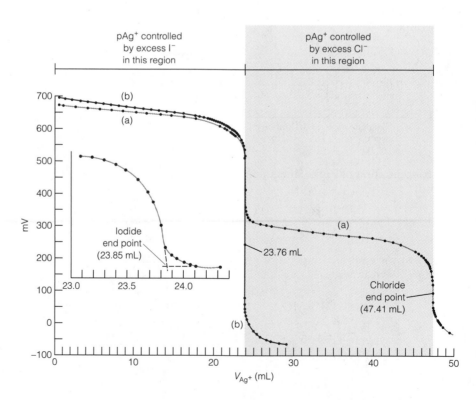

Figure 7-8 Experimental titration curves. (*a*) Titration curve for 40.00 mL of 0.050 2 M KI plus 0.050 0 M KCl titrated with 0.084 5 M $AgNO_3$. The inset is an expanded view of the region near the first equivalence point. (*b*) Titration curve for 20.00 mL of 0.100 4 M I^- titrated with 0.084 5 M Ag^+.

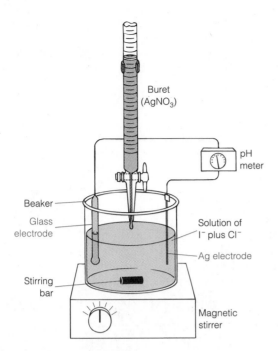

Figure 7-9   Apparatus for measuring the titration curves in Figure 7-8. The silver electrode responds to changes in the Ag^+ concentration, and the glass electrode provides a constant reference potential in this experiment. The measured voltage changes by approximately 59 mV for each factor-of-10 change in $[Ag^+]$. All solutions, including $AgNO_3$, were maintained at pH 2.0 by using 0.010 M sulfate buffer prepared from H_2SO_4 and KOH.

Comparison of the I^-/Cl^- and pure I^- titration curves in Figure 7-8 shows that the I^- end point is 0.38% too high in the I^-/Cl^- titration. We expect the first end point at 23.76 mL, but it is observed at 23.85 mL. There are two factors contributing to this high value. One is random experimental error, which is always present. This is as likely to be positive as negative. However, the end point in some titrations, especially Br^-/Cl^- titrations, is found to be systematically 0–3% high, depending on conditions. This discrepancy has been attributed to a small amount of *coprecipitation* of AgCl with AgBr. **Coprecipitation** means that even though the solubility product of AgCl has not been exceeded, a little Cl^- becomes attached to the solid AgBr as it precipitates and carries down a corresponding amount of Ag^+. A high concentration of nitrate anion reduces the extent of coprecipitation.

The second end point in Figure 7-8 corresponds to the total precipitation of both halides. It is observed at the expected value of V_{Ag^+}. The concentration of Cl^-, found from the *difference* between the two end points, will be slightly low in Figure 7-8, because the first end point is slightly high.

 7-6 **Calculating Titration Curves with a Spreadsheet**

By now you should understand the chemistry that occurs at different stages of a titration, and you should know how to calculate the shape of a titration curve by hand. However, spreadsheets are more powerful and more versatile than hand calculations, and are less prone to error. In this section we describe how to use a spreadsheet to compute titration curves, and we recommend that you learn to use this tool. If a spreadsheet is not available, you can skip this section with no loss in continuity.

The Equations

Consider the addition of V_M liters of cation M^+ (whose initial concentration is C_M^0) to V_X^0 liters of solution containing anion X^- with a concentration C_X^0.

$$M^+ + X^- \overset{K_{sp}}{\rightleftharpoons} MX(s) \tag{7-15}$$

$$\underset{\substack{\text{Titrant} \\ C_M^0, V_M}}{} \quad \underset{\substack{\text{Analyte} \\ C_X^0, V_X^0}}{}$$

We know that the total moles of added M $(= C_M^0 \cdot V_M)$ must equal the moles of M^+ in solution $(= [M^+](V_M + V_X^0))$ plus the moles of precipitated $MX(s)$. (This equality is called a *mass balance,* even though it is really a *mole balance.*) In a similar manner, we can write a mass balance for X.

The mass balance states that the sum of moles of all forms of a species in a mixture equals the total moles of that species delivered to the solution. The mass balance is discussed further in Chapter 9.

mass balance for M: $\quad C_M^0 \cdot V_M = [M^+](V_M + V_X^0) + \text{mol } MX(s) \tag{7-16}$

mass balance for X: $\quad C_X^0 \cdot V_X^0 = [X^-](V_M + V_X^0) + \text{mol } MX(s) \tag{7-17}$

Equating mol $MX(s)$ from Equation 7-16 with mol $MX(s)$ from Equation 7-17 gives

$$C_M^0 \cdot V_M - [M^+](V_M + V_X^0) = C_X^0 \cdot V_X^0 - [X^-](V_M + V_X^0)$$

which can be rearranged to

Precipitation of X^- with M^+: $\quad V_M = V_X^0 \left(\dfrac{C_X^0 + [M^+] - [X^-]}{C_M^0 - [M^+] + [X^-]} \right) \tag{7-18}$

Equation 7-18 relates the volume of added M^+ to the concentrations of $[M^+]$ and $[X^-]$ and to the constants V_X^0, C_X^0, and C_M^0. To use Equation 7-18 in a spreadsheet, we enter values of pM and compute corresponding values of V_M, as shown in Figure 7-10 for the iodide titration of Figure 7-7. Column C of Figure 7-10 is calculated with the formula $[M^+] = 10^{-pM}$ and column D is given by $[X^-] = K_{sp}/[M^+]$. Column E is calculated from Equation 7-18. The first input value of pM (15.08) corresponds to a small addition of Ag^+. You can start wherever you like. If your initial value of pM is before the true starting point, then the value of V_M in column E will be negative. In practice, you will want more points than we have shown so that you can make an accurate graph of the titration curve.

For the titration

$$xM^{m+} + mX^{x-} \rightleftharpoons M_xX_m(s) \qquad K_{sp} = [M^{m+}]^x[X^{x-}]^m \tag{7-19}$$

the corresponding result is

Precipitation of X^{x-} with M^{m+}: $V_M = V_X^0 \left(\dfrac{xC_X^0 + m[M^{m+}] - x[X^{x-}]}{mC_M^0 - m[M^{m+}] + x[X^{x-}]} \right) \tag{7-20}$

where $[X^{x-}]$ is given by Equation 7-19 as $(K_{sp}/[M^{m+}]^x)^{1/m}$.

To show off our great power, let's derive an equation for the shape of the titration curve for a mixture (initial volume $= V^0$) of the anions X^- (initial concentration $= C_X^0$) and Y^- (initial concentration $= C_Y^0$) titrated with M^+ (initial concentration $= C_M^0$, volume added $= V_M$) to precipitate $MX(s)$ and

	A	B	C	D	E
1	Ksp(AgI) =	pAg	[Ag+]	[I–]	Vm
2	8.3E-17	15.08	8.32E-16	9.98E-02	0.035
3	Vo =	15	1.00E-15	8.30E-02	3.195
4	25	14	1.00E-14	8.30E-03	39.322
5	Co(I) =	12	1.00E-12	8.30E-05	49.876
6	0.1	10	1.00E-10	8.30E-07	49.999
7	Co(Ag) =	8	1.00E-08	8.30E-09	50.000
8	0.05	6	1.00E-06	8.30E-11	50.001
9		4	1.00E-04	8.30E-13	50.150
10		3	1.00E-03	8.30E-14	51.531
11		2	1.00E-02	8.30E-15	68.750
12	C2 = 10^-B2				
13	D2 = $A$2/C2				
14	E2 = $A$4*($A$6+C2-D2)/($A$8-C2+D2)				

Figure 7-10 Spreadsheet for titration of 25 mL of 0.1 M I^- with 0.05 M Ag^+.

MY(s), whose solubility products are K_{sp}^X and K_{sp}^Y. The three mass balances are

mass balance for M: $C_M^0 \cdot V_M = [M^+](V_M + V^0)$

$$+ \text{ mol } MX(s) + \text{mol } MY(s) \qquad (7\text{-}21)$$

mass balance for X: $C_X^0 \cdot V^0 = [X^-](V_M + V^0) + \text{mol } MX(s) \qquad (7\text{-}22)$

mass balance for Y: $C_Y^0 \cdot V^0 = [Y^-](V_M + V^0) + \text{mol } MY(s) \qquad (7\text{-}23)$

Equation 7-22 can be solved for mol MX(s), and Equation 7-23 can be solved for mol MY(s). When these two expressions are substituted into Equation 7-21, we can solve for V_M to find

Precipitation of X^- + Y^- with M^+: $$V_M = V^0 \left(\frac{C_X^0 + C_Y^0 + [M^+] - [X^-] - [Y^-]}{C_M^0 - [M^+] + [X^-] + [Y^-]} \right) \qquad (7\text{-}24)$$

The IF Statement

Equation 7-24 allows us to compute the titration curve for a mixture, but we must be aware of some tricky subtleties. For the first part of the titration of a mixture of I^- and Cl^-, only I^- is precipitating. In this region, we can say $[I^-] = K_{sp}^{AgI}/[Ag^+]$. The value of $[Cl^-]$ is just the initial concentration corrected for dilution ($[Cl^-] = C_{Cl}^0 \cdot \{V^0/(V_M + V^0)\}$), because precipitation of AgCl has not commenced. Precipitation of Cl^- begins shortly before the first equivalence point, when the product $[Ag^+][Cl^-]$ exceeds K_{sp}^{AgCl}.

The value of $[Cl^-]$ in Equation 7-24 must therefore be computed with an IF statement:

$$\text{IF } [Ag^+] < K_{sp}^{AgCl}/[Cl^-]$$

$$\text{then } [Cl^-] = C_{Cl}^0 \cdot \{V^0/(V_M + V^0)\}$$

$$\text{otherwise } [Cl^-] = K_{sp}^{AgCl}/[Ag^+]$$

	A	B	C	D	E	F	G
1	Ksp(AgCl)=	pAg	[Ag+]	Cl– Diluted Value	[Cl–]	[I–]	Vm
2	1.8E-10	14.5	3.16E-15	4.11E-02	4.11E-02	2.62E-02	8.651
3	Ksp(AgI)=	14	1.00E-14	3.44E-02	3.44E-02	8.30E-03	18.060
4	8.3E-17	13	1.00E-13	3.17E-02	3.17E-02	8.30E-04	23.143
5	Vo=	12	1.00E-12	3.14E-02	3.14E-02	8.30E-05	23.701
6	40	11	1.00E-11	3.14E-02	3.14E-02	8.30E-06	23.757
7	Co(Cl)=	10	1.00E-10	3.14E-02	3.14E-02	8.30E-07	23.763
8	0.05	9	1.00E-09	3.14E-02	3.14E-02	8.30E-08	23.763
9	Co(I)=	8	1.00E-08	2.77E-02	1.80E-02	8.30E-09	32.078
10	0.0502	7	1.00E-07	2.34E-02	1.80E-03	8.30E-10	45.608
11	Co(Ag)=	6	1.00E-06	2.29E-02	1.80E-04	8.30E-11	47.247
12	0.0845	5	1.00E-05	2.29E-02	1.80E-05	8.30E-12	47.424
13		4	1.00E-04	2.28E-02	1.80E-06	8.30E-13	47.534
14		3	1.00E-03	2.26E-02	1.80E-07	8.30E-14	48.479
15		2	1.00E-02	2.02E-02	1.80E-08	8.30E-15	59.168
16							
17	C2 = 10^-B2						
18	D2 = $A$8*($A$6/($A$6+G2))						
19	E2 = IF((D2<($A$2/C2)),D2,$A$2/C2)						
20	F2 = $A$4/C2						
21	G2 = $A$6*($A$8+$A$10+C2-E2-F2)/($A$12-C2+E2+F2)						

Figure 7-11 Spreadsheet for titration of 40.00 mL of 0.0502 M I⁻ plus 0.0500 M Cl⁻ with 0.0845 M Ag⁺. This corresponds to curve (a) in Figure 7-8.

You will need to verify the format for an IF statement with your spreadsheet. A common syntax for testing a logical condition is

IF(logic_statement, value_if_true, value_if_false)

This expression means that if logic_statement is true, value_if_true is returned. Otherwise, value_if_false is returned. For example, IF(C2>$A$6, 0.05, 0) returns a value of 0.05 if C2>$A$6 and a value of 0 if C2≤$A$6. For the titration of I⁻ and Cl⁻, [Cl⁻] is calculated with the statement

$$\text{IF } ([Ag^+] < K_{sp}^{AgCl}/[Cl^-], C_{Cl}^0 \cdot \{V^0/(V_M + V^0)\}, K_{sp}^{AgCl}/[Ag^+])$$

A sample spreadsheet for the titration of I⁻ and Cl⁻ is given in Figure 7-11. In column D we calculate the concentration of Cl⁻ on the basis of the dilution from the initial volume V^0 to the final volume $V_M + V^0$. In column E we calculate the concentration of Cl⁻ with the IF statement that checks to see whether or not AgCl has precipitated.

There is a very significant subtlety in this spreadsheet. Column D requires the value in column G. But column G requires the value in column E, which uses the value in column D. This is called a *circular definition*. Most spreadsheets allow you to do this and may question you to see if it is really what you meant to do. The first time the value in cell D2 is computed, there

The spreadsheet in Figure 7-11 uses an IF statement in column E. Columns D, E, and G are related by circular definitions.

is no value in cell G2, so G2 = 0 is used in cell D2. Then cells E2 and G2 are computed with the value in cell D2. You will need to follow instructions for your spreadsheet to carry out several iterations to reach self-consistent values in columns D through G.

7-7 End-Point Detection

Three techniques are commonly employed to detect the end point in precipitation titrations:

1. Potentiometric methods: using electrodes, as shown in Figure 7-9 (these techniques are discussed in Chapter 15)

2. Indicator methods: discussed in this section

3. Light-scattering methods: turbidimetry and nephelometry

In the remainder of this section we discuss three types of indicator methods applied to the titration of Cl^- with Ag^+. Titrations with Ag^+ are called **argentometric titrations.** The three indicator methods are

1. **Mohr titration:** formation of a colored precipitate at the end point

2. **Volhard titration:** formation of a soluble, colored complex at the end point

3. **Fajans titration:** adsorption of a colored indicator on the precipitate at the end point.

Mohr Titration[2]

In the Mohr titration, Cl^- is titrated with Ag^+ in the presence of CrO_4^{2-} (chromate, dissolved as Na_2CrO_4).

$$\text{titration reaction:} \quad Ag^+ + Cl^- \rightarrow \underset{\text{(white)}}{AgCl(s)}$$

$$\text{end-point reaction:} \quad 2Ag^+ + CrO_4^{2-} \rightarrow \underset{\text{(red)}}{Ag_2CrO_4(s)}$$

The AgCl precipitates before Ag_2CrO_4. The color of AgCl is white; dissolved CrO_4^{2-} is yellow; and Ag_2CrO_4 is red. The end point is indicated by the first appearance of red Ag_2CrO_4. Reasonable control of CrO_4^{2-} concentration is required for Ag_2CrO_4 precipitation to occur at the desired point in the titration, and the pH should be in the range 4–10.5.[3]

Because some excess Ag_2CrO_4 is necessary for visual detection, the color is not seen until after the true equivalence point. We can correct for this titration error in two ways. One is by means of a blank titration with no chloride present. The volume of Ag^+ needed to form a detectable red color is then subtracted from V_{Ag^+} in the Cl^- titration. Alternatively, we can standardize the $AgNO_3$ by the Mohr method, using a standard NaCl solution and titrant volumes similar to those for the titration of unknown. The Mohr method is useful for Cl^- and Br^-, but not for I^- or SCN^- (thiocyanate).

If the pH is too low, the concentration of CrO_4^{2-} is reduced by the equilibria $2CrO_4^{2-} + 2H^+ \rightleftharpoons 2HCrO_4^- \rightleftharpoons Cr_2O_7^{2-} + H_2O$. If the pH is too high, AgOH(s) may precipitate.

(a)

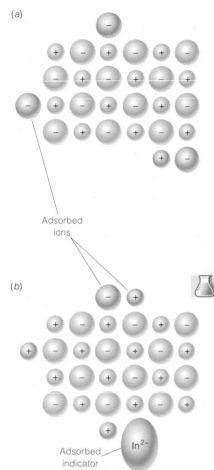

Adsorbed
ions

(b)

Adsorbed
indicator

In^{2-}

Figure 7-12 Ions from a solution are adsorbed on the surface of a growing crystallite. (*a*) A crystal growing in the presence of excess lattice anions (anions that belong in the crystal) will have a slight negative charge because anions are predominantly adsorbed. (*b*) A crystal growing in the presence of excess lattice cations will have a slight positive charge and can therefore adsorb a negative indicator ion. Anions and cations in the solution that do not belong in the crystal lattice are less likely to be adsorbed than are ions belonging to the lattice. These diagrams omit other ions in solution. Overall, each solution plus its growing crystallites must have zero total charge.

Volhard Titration

The Volhard titration is actually a procedure for the titration of Ag^+. To determine Cl^-, a back titration is necessary. First, the Cl^- is precipitated by a known, excess quantity of standard $AgNO_3$.

$$Ag^+ + Cl^- \rightarrow AgCl(s)$$

The AgCl is isolated, and the excess Ag^+ is titrated with standard KSCN in the presence of Fe^{3+}.

$$Ag^+ + SCN^- \rightarrow AgSCN(s)$$

When all the Ag^+ has been consumed, the SCN^- reacts with Fe^{3+} to form a red complex.

$$Fe^{3+} + SCN^- \rightarrow FeSCN^{2+}$$
$$\text{(red)}$$

The appearance of the red color signals the end point. Knowing how much SCN^- was required for the back titration tells us how much Ag^+ was left over from the reaction with Cl^-. Because the total amount of Ag^+ is known, the amount consumed by Cl^- can then be calculated.

In the analysis of Cl^- by the Volhard method, the end point slowly fades because AgCl is more soluble than AgSCN. The AgCl slowly dissolves and is replaced by AgSCN. To prevent this secondary reaction from happening, two techniques are commonly used. One is to filter off the AgCl and titrate only the Ag^+ in the filtrate. An easier procedure is to shake a few milliliters of nitrobenzene, $C_6H_5NO_2$, with the precipitated AgCl prior to the back titration. Nitrobenzene coats the AgCl and effectively isolates it from attack by SCN^-. Br^- and I^-, whose silver salts are *less* soluble than AgSCN, may be titrated by the Volhard method without isolating the silver halide precipitate.

Fajans Titration

The Fajans titration uses an **adsorption indicator.** To see how this works, we must consider the electric charge of a precipitate. When Ag^+ is added to Cl^-, there will be excess Cl^- ions in solution prior to the equivalence point. Some Cl^- is selectively adsorbed on the AgCl surface, imparting a negative charge to the crystal surface (Figure 7-12a). After the equivalence point, there is excess Ag^+ in solution. Adsorption of the Ag^+ cations on the crystal surface creates a positive charge on the particles of precipitate (Figure 7-12b). The abrupt change from negative charge to positive charge occurs at the equivalence point.

Common adsorption indicators are anionic dyes, which are attracted to the positively charged particles produced immediately after the equivalence point. The adsorption of the negatively charged dye on the positively charged surface changes the color of the dye by interactions that are not well understood. The color change signals the end point in the titration. Because the indicator reacts with the precipitate surface, it is desirable to have as much surface area as possible. This means performing the titration under conditions that tend to keep the particles as small as possible, because small particles have more surface area than an equal volume of large particles. Low

Demonstration 7-1 **Fajans Titration**

The Fajans titration of Cl^- with Ag^+ convincingly demonstrates the utility of indicator end points in precipitation titrations. Dissolve 0.5 g of NaCl plus 0.15 g of dextrin in 400 mL of water. The purpose of the dextrin is to retard coagulation of the AgCl precipitate. Add 1 mL of dichlorofluorescein indicator solution containing 1 mg/mL of dichlorofluorescein in 95% aqueous ethanol or 1 mg/mL of the sodium salt in water. Titrate the NaCl solution with a solution containing 2 g of $AgNO_3$ in 30 mL of water. About 20 mL are required to reach the end point.

Color Plate 4a shows the yellow color of the indicator in the NaCl solution prior to the titration. Color Plate 4b shows the milky-white appearance of the AgCl suspension during titration, before the end point is reached. The pink suspension in Color Plate 4c appears at the end point, when the anionic indicator becomes adsorbed to the cationic particles of precipitate.

electrolyte concentration helps to prevent coagulation of the precipitate and maintain small particle size.

The indicator most commonly used for AgCl is dichlorofluorescein.

Dichlorofluorescein   Tetrabromofluorescein
(eosin)

This dye has a greenish yellow color in solution but turns pink when it is adsorbed on AgCl (Demonstration 7-1). Because the indicator is a weak acid and must be present in its anionic form, the pH of the reaction must be controlled. The dye eosin is useful in the titration of Br^-, I^-, and SCN^-. It gives a sharper end point than dichlorofluorescein and is more sensitive (i.e., less halide is required for titration). It cannot be used for AgCl because the eosin anion is more strongly bound to AgCl than is Cl^- ion. Eosin will bind to the AgCl crystallites even before the particles become positively charged.

In all argentometric titrations, but especially with adsorption indicators, strong light (such as daylight through a window) should be avoided. Light causes decomposition of the silver salts, and adsorbed indicators are especially light sensitive.

Various applications of precipitation titrations are listed in Table 7-1. Whereas the Mohr and Volhard methods are specifically applicable to argentometric titrations, the Fajans method is used for a wider variety of reactions. Because the Volhard titration is carried out in acidic solution (typically 0.2 M HNO_3), it avoids certain interferences that would affect other titrations. Silver salts of anions such as CO_3^{2-}, $C_2O_4^{2-}$, and AsO_4^{3-} are soluble in acidic solution, so these anions do not interfere with the analysis.

TABLE 7-1   **Applications of precipitation titrations**

Species analyzed	Notes
	MOHR METHOD
Cl^-, Br^-	Ag_2CrO_4 end point is used.
	VOLHARD METHOD
Br^-, I^-, SCN^-, CNO^-, AsO_4^{3-}	Precipitate removal is unnecessary.
Cl^-, PO_4^{3-}, CN^-, $C_2O_4^{2-}$, CO_3^{2-}, S^{2-}, CrO_4^{2-}	Precipitate removal required.
BH_4^-	Back titration of Ag^+ left after reaction with BH_4^-: $$BH_4^- + 8Ag^+ + 8OH^- \rightarrow 8Ag(s) + H_2BO_3^- + 5H_2O$$
K^+	K^+ is first precipitated with a known excess of $(C_6H_5)_4B^-$. Remaining $(C_6H_5)_4B^-$ is precipitated with a known excess of Ag^+. Unreacted Ag^+ is then titrated with SCN^-.
	FAJANS METHOD
Cl^-, Br^-, I^-, SCN^-, $Fe(CN)_6^{4-}$	Titration with Ag^+. Detection with such dyes as fluorescein, dichlorofluorescein, eosin, bromophenol blue.
F^-	Titration with $Th(NO_3)_4$ to produce ThF_4. End point detected with alizarin red S.
Zn^{2+}	Titration with $K_4Fe(CN)_6$ to produce $K_2Zn_3[Fe(CN)_6]_2$. End-point detection with diphenylamine.
SO_4^{2-}	Titration with $Ba(OH)_2$ in 50 vol% aqueous methanol using alizarin red S as indicator.
Hg_2^{2+}	Titration with NaCl to produce Hg_2Cl_2. End point detected with bromophenol blue.
PO_4^{3-}, $C_2O_4^{2-}$	Titration with $Pb(CH_3CO_2)_2$ to give $Pb_3(PO_4)_2$ or PbC_2O_4. End point detected with dibromofluorescein (PO_4^{3-}) or fluorescein ($C_2O_4^{2-}$).

Terms to Understand

adsorption indicator	equivalence point	standardization
argentometric titration	Fajans titration	titrant
back titration	indicator	titration
blank titration	Kjeldahl nitrogen analysis	titration error
coprecipitation	Mohr titration	turbidity
direct titration	primary standard	Volhard titration
end point	standard solution	volumetric analysis

Summary

The volume of reagent (titrant) required for stoichiometric reaction of analyte is measured in volumetric analysis. The stoichiometric point of the reaction is called the equivalence point. What we measure by an abrupt change in a physical property (such as the color of an indicator or the potential of an electrode) is the end point. The small difference between the end point and the equivalence point is an unavoidable titration error. This error can be reduced by subtracting the results of a blank titration, in which the same procedure is carried out in the absence of analyte, or by standardizing the titrant, using the same reaction and a similar volume as that used for analyte.

The validity of an analytical result depends on knowing the amount of a primary standard. A solution with an approximately desired concentration can be standardized by titrating a primary standard. In a direct titration, titrant is added to analyte until the reaction is complete. In a back titration, a known excess of reagent is added to analyte, and the excess is titrated with a second standard reagent. The key to all calculations of volumetric analysis is to relate the known moles of titrant to the unknown moles of analyte.

In Kjeldahl analysis, a nitrogen-containing organic compound is digested with a catalyst in boiling H_2SO_4. Nitrogen is converted to ammonium ion, which is converted to ammonia with base and distilled into standard HCl. The excess, unreacted HCl tells us how much nitrogen was present in the original analyte.

In a spectrophotometric titration, the absorbance of a solution is monitored as titrant is added. For many reactions, there is an abrupt change in absorbance when the equivalence point is reached. Light scattering can be used to monitor the course of precipitation titrations. In turbidimetry, decreased transmittance of the solution is observed while precipitate is formed. In nephelometry, increased optical scatter perpendicular to the incident light is observed as the precipitate forms. The most general indicator method for precipitation titrations, the Fajans titration, is based on the adsorption of a charged indicator by the charged surface of the precipitate after the equivalence point. The Volhard titration, used to measure Ag^+, is based on the reaction of Fe^{3+} with SCN^- after the precipitation of AgSCN is complete. The Mohr titration utilizes the precipitation of colored Ag_2CrO_4 after the argentometric titration of Cl^- or Br^-.

The concentrations of reactants and products during a precipitation titration are calculated in three ways. Before the end point, there is excess analyte. The concentration of titrant can be found from the solubility product of the precipitate and the known concentration of excess analyte. At the equivalence point, the concentrations of both reactants are governed by the dissociation equilibrium of product. After the equivalence point, the concentration of analyte can be determined from the solubility product of precipitate and the known concentration of excess titrant.

Exercises

A. Ascorbic acid (vitamin C) reacts with I_3^- according to the equation

Ascorbic acid $+ I_3^- + H_2O \longrightarrow$

Dehydroascorbic acid $+ 3I^- + 2H^+$

Starch is used as an indicator in the reaction. The end point is marked by the appearance of a deep blue starch-iodine complex when the first fraction of a drop of unreacted I_3^- remains in the solution.

(a) If 29.41 mL of I_3^- solution is required to react with 0.1970 g of pure ascorbic acid, what is the molarity of the I_3^- solution?

(b) A vitamin C tablet containing ascorbic acid plus an inert binder was ground to a powder, and 0.4242 g was titrated by 31.63 mL of I_3^-. Find the weight percent of ascorbic acid in the tablet.

B. A solution of NaOH was standardized by titration of a known quantity of the primary standard, potassium hydrogen phthalate:

Potassium hydrogen phthalate
FW 204.233

$+ \text{NaOH} \longrightarrow$

$+ H_2O$

The NaOH was then used to find the concentration of an unknown solution of H_2SO_4:

$$H_2SO_4 + 2NaOH \rightarrow Na_2SO_4 + H_2O$$

(a) Titration of 0.824 g of potassium hydrogen phthalate required 38.314 g of NaOH solution to reach the end point detected by phenolphthalein indicator. Find the concentration of NaOH (mol NaOH/kg solution).

(b) A 10.00-mL aliquot of H_2SO_4 solution required 57.911 g of NaOH solution to reach the phenolphthalein end point. Find the molarity of H_2SO_4.

C. A solid sample weighing 0.237 6 g contained only malonic acid (MW 104.06) and aniline hydrochloride (MW 129.59). It required 34.02 mL of 0.087 71 M NaOH to neutralize the sample. Find the weight percent of each component in the solid mixture. The reactions are

$$CH_2(CO_2H)_2 + 2OH^- \longrightarrow CH_2(CO_2^-)_2 + H_2O$$

  Malonic acid   Malonate

$$NH_3^+Cl^- + OH^- \longrightarrow$$

Aniline hydrochloride

$$NH_2 + H_2O + Cl^-$$

Aniline

D. The metal chelator semi-xylenol orange is yellow at pH 5.9 but turns red ($\lambda_{max} = 490$ nm) when it reacts with Pb^{2+}. A 2.025-mL sample of semi-xylenol orange at pH 5.9 was titrated with 7.515×10^{-4} M $Pb(NO_3)_2$, with the results shown below.

Total μL Pb^{2+} added	Absorbance at 490 nm in 1-cm cell
0.0	0.227
6.0	0.256
12.0	0.286
18.0	0.316
24.0	0.345
30.0	0.370
36.0	0.399
42.0	0.425
48.0	0.445
54.0	0.448
60.0	0.449
70.0	0.450
80.0	0.447

Make a graph of A versus microliters of Pb^{2+} added. Be sure to correct the absorbances in the table for the effect of dilution. That is, the corrected absorbance is what would be observed if the volume were not changed from its initial value of 2.025 mL. Assuming that the reaction of semi-xylenol orange with Pb^{2+} has a 1:1 stoichiometry, find the molarity of semi-xylenol orange in the original solution.

E. Suppose that 50.0 mL of 0.080 0 M KSCN is titrated with 0.040 0 M Cu^+. The solubility product of CuSCN is 4.8×10^{-15}. At each of the following volumes of titrant, calculate pCu^+, and construct a graph of pCu^+ versus milliliters of Cu^+ added: 0.10, 10.0, 25.0, 50.0, 75.0, 95.0, 99.0, 100.0, 100.1, 101.0, 110.0 mL.

F. Construct a graph of pAg^+ versus milliliters of Ag^+ for the titration of 40.00 mL of solution containing 0.050 00 M Br^- and 0.050 00 M Cl^-. The titrant is 0.084 54 M $AgNO_3$. Calculate pAg^+ at the following volumes: 2.00, 10.00, 22.00, 23.00, 24.00, 30.00, 40.00 mL, second equivalence point, 50.00 mL.

G. Consider the titration of 50.00 (± 0.05) mL of a mixture of I^- and SCN^- with 0.068 3 ($\pm 0.000 1$) M Ag^+. The first equivalence point is observed at 12.6 (± 0.4) mL, and the second occurs at 27.7 (± 0.3) mL.

(a) Find the molarity and the uncertainty in molarity of thiocyanate in the original mixture.

(b) Suppose that the uncertainties listed above are all the same, except that the uncertainty of the first equivalence point (12.6 $\pm$? mL) is variable. What is the maximum uncertainty (milliliters) of the first equivalence point if the uncertainty in SCN^- molarity is to be $\leq 4.0\%$?

Problems

Volumetric Procedures and Calculations

1. Explain the following statement: "The validity of an analytical result ultimately depends on knowing the composition of some primary standard."

2. Distinguish between the terms *end point* and *equivalence point*.

3. How does a blank titration reduce titration error?

4. What is the difference between a direct titration and a back titration?

5. How many milliliters of 0.100 M KI are needed to react with 40.0 mL of 0.040 0 M $Hg_2(NO_3)_2$ if the reaction is $Hg_2^{2+} + 2I^- \rightarrow Hg_2I_2(s)$?

6. For Reaction 7-1, how many milliliters of 0.165 0 M $KMnO_4$ are needed to react with 108.0 mL of 0.165 0 M oxalic acid? How many milliliters of 0.165 0 M oxalic acid are required to react with 108.0 mL of 0.165 0 M $KMnO_4$?

7. Ammonia reacts with hypobromite, OBr^-, according to the equation

$$2NH_3 + 3OBr^- \rightarrow N_2 + 3Br^- + 3H_2O$$

What is the molarity of a hypobromite solution if 1.00 mL of the OBr^- solution reacts with 1.69 mg of NH_3?

8. The Kjeldahl procedure was used to analyze 256 μL of a solution containing 37.9 mg protein/mL. The liberated NH_3 was collected in 5.00 mL of 0.033 6 M HCl, and the remaining acid required 6.34 mL of 0.0100 M NaOH for complete titration. What is the weight percent of nitrogen in the protein?

9. Arsenic (III) oxide (As_2O_3) is available in pure form and is a useful (and poisonous) primary standard for standardizing many oxidizing agents, such as MnO_4^-. The As_2O_3 is first dissolved in base and then titrated with MnO_4^- in acidic solution. A small amount of iodide (I^-) or iodate (IO_3^-) is used to catalyze the reaction between H_3AsO_3 and MnO_4^-. The reactions are

$$As_2O_3 + 4OH^- \rightleftharpoons 2HAsO_3^{2-} + H_2O$$

$$HAsO_3^{2-} + 2H^+ \rightleftharpoons H_3AsO_3$$

$$5H_3AsO_3 + 2MnO_4^- + 6H^+ \rightarrow$$
$$5H_3AsO_4 + 2Mn^{2+} + 3H_2O$$

(a) A 3.214-g aliquot of $KMnO_4$ (FW 158.034) was dissolved in 1.000 L of water, heated to cause any reactions with impurities to occur, cooled, and filtered. What is the theoretical molarity of this solution if no MnO_4^- was consumed by impurities?

(b) What mass of As_2O_3 (FW 197.84) would be just sufficient to react with 25.00 mL of the $KMnO_4$ solution in (a)?

(c) It was found that 0.146 8 g of As_2O_3 required 29.98 mL of $KMnO_4$ solution for the faint color of unreacted MnO_4^- to appear. In a blank titration, 0.03 mL of MnO_4^- was required to produce enough color to be seen. Calculate the molarity of the permanganate solution.

10. A 0.238 6-g sample contained only NaCl and KBr. It was dissolved in water and required 48.40 mL of 0.048 37 M $AgNO_3$ for complete titration of both halides [giving $AgCl(s)$ and $AgBr(s)$]. Calculate the weight percent of Br in the solid sample.

11. A solid mixture weighing 0.054 85 g contained only ferrous ammonium sulfate and ferrous chloride. The sample was dissolved in 1 M H_2SO_4, and the Fe^{2+} required 13.39 mL of 0.012 34 M Ce^{4+} for complete oxidation to Fe^{3+} ($Ce^{4+} + Fe^{2+} \rightarrow Ce^{3+} + Fe^{3+}$). Calculate the weight percent of Cl in the original sample.

$FeSO_4 \cdot (NH_4)_2SO_4 \cdot 6H_2O$	$FeCl_2 \cdot 6H_2O$
Ferrous ammonium sulfate	Ferrous chloride
FW 392.13	FW 234.84

12. A cyanide solution with a volume of 12.73 mL was treated with 25.00 mL of Ni^{2+} solution (containing excess Ni^{2+}) to convert the cyanide to tetracyano-nickelate(II):

$$4CN^- + Ni^{2+} \rightarrow Ni(CN)_4^{2-}$$

The excess Ni^{2+} was then titrated with 10.15 mL of 0.013 07 M ethylenediaminetetraacetic acid (EDTA). One mole of this reagent reacts with 1 mol of Ni^{2+}:

$$Ni^{2+} + EDTA^{4-} \rightarrow Ni(EDTA)^{2-}$$

$Ni(CN)_4^{2-}$ does not react with EDTA. If 39.35 mL of EDTA was required to react with 30.10 mL of the original Ni^{2+} solution, calculate the molarity of CN^- in the 12.73-mL cyanide sample.

Spectrophotometric Titrations

13. Why does the slope of the absorbance versus volume graph in Figure 7-5 change abruptly at the equivalence point?

14. Explain why nephelometry can detect much smaller quantities of precipitate than can turbidimetry.

15. A 2.00-mL solution of apotransferrin was titrated as illustrated in Figure 7-5. It required 163 μL of 1.43 mM ferric nitrilotriacetate to reach the end point.

(a) How many moles of Fe(III) (= ferric nitrilotriacetate) were required to reach the end point?

(b) Each apotransferrin molecule binds two ferric ions. Find the concentration of apotransferrin in the 2.00-mL solution.

16. The iron-binding site of transferrin in Figure 7-4 can accommodate certain other metal ions besides Fe^{3+} and certain other anions besides CO_3^{2-}. Data are given below for the titration of transferrin (3.57 mg in 2.00 mL) with 6.64 mM Ga^{3+} solution in the presence of the anion oxalate, $C_2O_4^{2-}$, and in the absence of a suitable

Titration in presence of $C_2O_4^{2-}$		Titration in absence of anion	
Total μL Ga^{3+} added	Absorbance at 241 nm	Total μL Ga^{3+} added	Absorbance at at 241 nm
0.0	0.044	0.0	0.000
2.0	0.143	2.0	0.007
4.0	0.222	6.0	0.012
6.0	0.306	10.0	0.019
8.0	0.381	14.0	0.024
10.0	0.452	18.0	0.030
12.0	0.508	22.0	0.035
14.0	0.541	26.0	0.037
16.0	0.558		
18.0	0.562		
21.0	0.569		
24.0	0.576		

anion. Prepare a graph similar to Figure 7-5, showing both sets of data. Indicate the theoretical equivalence point for the binding of one and two Ga^{3+} ions per molecule of protein and the observed end point. How many Ga^{3+} ions are bound to transferrin in the presence and in the absence of oxalate?

17. Consider the titration of sulfate ion by Ba^{2+} to precipitate $BaSO_4$. Sketch how the transmittance of the sulfate solution would change during a turbidimetric titration and how scatter would change during a nephelometric titration.

Shape of a Precipitation Curve

18. Describe the chemistry that occurs in each of the following regions in curve (a) in Figure 7-8: (i) before the first equivalence point; (ii) at the first equivalence point; (iii) between the first and second equivalence points; (iv) at the second equivalence point; and (v) past the second equivalence point. For each region except (ii), write the equation that you would use to calculate $[Ag^+]$.

19. Consider the titration of 25.00 mL of 0.082 30 M KI with 0.051 10 M $AgNO_3$. Calculate pAg^+ at the following volumes of added $AgNO_3$:

(a) 39.00 mL   (b) V_e (c) 44.30 mL

20. A solution of volume 25.00 mL containing 0.031 1 M $Na_2C_2O_4$ was titrated with 0.025 7 M $La(ClO_4)_3$ to precipitate lanthanum oxalate:

$$2La^{3+} + 3C_2O_4^{2-} \rightarrow La_2(C_2O_4)_3(s)$$

$\underset{\text{Lanthanum}}{}$ $\underset{\text{Oxalate}}{}$

(a) What volume of $La(ClO_4)_3$ is required to reach the equivalence point?

(b) Find pLa^{3+} when 10.00 mL of $La(ClO_4)_3$ has been added.

(c) Find pLa^{3+} at the equivalence point.

(d) Find pLa^{3+} when 25.00 mL of $La(ClO_4)_3$ has been added.

21. A solution containing 10.00 mL of 0.100 M LiF was titrated with 0.0100 M $Th(NO_3)_4$ to precipitate ThF_4.

(a) What volume of $Th(NO_3)_4$ is needed to reach the equivalence point?

(b) Calculate pTh^{4+} when 1.00 mL of $Th(NO_3)_4$ has been added.

22. A 0.0100 M $Na_3[Co(CN)_6]$ solution with a volume of 50.0 mL was treated with 0.0100 M $Hg_2(NO_3)_2$ solution to precipitate $(Hg_2)_3[Co(CN)_6]_2$. Find $pCo(CN)_6$ when 90.0 mL of $Hg_2(NO_3)_2$ was added.

Titration of a Mixture

23. Calculate the value of pAg^+ at each of the following points in titration (a) in Figure 7-8:

(a) 10.00 mL   (d) second equivalence point

(b) 20.00 mL   (e) 50.00 mL

(c) 30.00 mL

24. A mixture having a volume of 10.00 mL and containing 0.1000 M Ag^+ and 0.1000 M Hg_2^{2+} was titrated with 0.1000 M KCN to precipitate $Hg_2(CN)_2$ and AgCN.

(a) Calculate pCN^- at each of the following volumes of added KCN: 5.00, 10.00, 15.00, 19.90, 20.10, 25.00, 30.00, 35.00 mL.

(b) Should any AgCN be precipitated at 19.90 mL?

25. One procedure[4] for determining the halogen content of organic compounds makes use of the argentometric titration in Figure 7-9. To 50 mL of anhydrous ether is added a carefully weighed sample (10–100 mg) of unknown, plus 2 mL of sodium dispersion and 1 mL of methanol. (Sodium dispersion is finely divided solid sodium suspended in oil. With methanol, it makes sodium methoxide, $CH_3O^-Na^+$, which attacks the organic compound, liberating halides.) Excess sodium is destroyed by slow addition of 2-propanol, after which 100 mL of water is added. (Sodium should not be treated directly with water, because the H_2 produced can explode in the presence of O_2: $2Na + 2H_2O \rightarrow 2NaOH + H_2$.) This procedure gives a two-phase mixture, with an ether layer floating on top of the aqueous layer that contains the halide salts. The aqueous layer is adjusted to pH 4 and titrated with Ag^+, using the electrodes shown in Figure 7-9. How much 0.025 70 M $AgNO_3$ solution will be required to reach each equivalence point when 82.67 mg of 1-bromo-4-chlorobutane ($BrCH_2CH_2CH_2CH_2Cl$; MW 171.464) is analyzed?

Using Spreadsheets

26. Derive Equation 7-20.

27. Derive Equation 7-24.

28. Derive an expression analogous to Equation 7-18 for the titration of M^+ (concentration = C_M^0, volume = V_M^0) with X^- (titrant concentration = V_X^0). Your equation should allow you to compute the volume of titrant (V_X) as a function of $[X^-]$.

29. Use Equation 7-18 to reproduce the curves in Figure 7-7. Plot your results on a single graph.

30. Prepare a graph with Equation 7-24 for the titration of 50.00 mL of 0.050 00 M Br^- + 0.050 00 M Cl^- with 0.1000 M Ag^+. Use your spreadsheet to find the fraction of each anion precipitated at the first equivalence point. (Answer: 99.723% of Br^- and 0.277% of Cl^- are precipitated.)

31. Derive an expression analogous to Equation 7-24 for titration of a mixture of three anions. Compute the shape of the titration curve for 50.00 mL of 0.050 00 M I^- + 0.050 00 M Br^- + 0.050 00 M Cl^- with 0.100 0 M Ag^+.

End-Point Detection

32. Why does the surface charge of a precipitate change sign at the equivalence point?

33. Examine the procedure in Table 7-1 for the Fajans titration of Zn^{2+}. Do you expect the charge on the precipitate to be positive or negative after the equivalence point?

34. Describe how to analyze a solution of NaI by using the Volhard titration.

35. Why is nitrobenzene used in the Volhard titration of chloride?

36. Why is a blank titration useful in the Mohr titration?

37. A 30.00-mL solution containing an unknown amount of I^- was treated with 50.00 mL of 0.365 0 M $AgNO_3$. The precipitated AgI was filtered off, and the filtrate (plus Fe^{3+}) was titrated with 0.287 0 M KSCN. When 37.60 mL had been added, the solution turned red. How many milligrams of I^- were present in the original solution?

38. Suppose that 20.00 mL of 0.100 0 M Br^- is titrated with 0.080 0 M $AgNO_3$ and the end point detected by the Mohr method. The equivalence point occurs at 25.00 mL. If the titration error is to be $\leq 0.1\%$, then the Ag_2CrO_4 should precipitate between 24.975 and 25.025 mL. Calculate the upper and lower limits for the CrO_4^{2-} concentration in the titration solution at the equivalence point if the precipitation of Ag_2CrO_4 is to commence between 24.975 and 25.025 mL.

Notes and References

1. For more history, see C. Duval, *J. Chem. Ed.* **1951**, *28*, 508; A. Johansson, *Anal. Chim. Acta* **1988**, *206*, 97; J. T. Stock, *Anal. Chem.* **1993**, *65*, 344A; and F. Szabadvary, *History of Analytical Chemistry* (New York: Pergamon, 1966), p. 236.

2. A classroom demonstration of the Mohr titration is described by R. J. Stolzberg, *J. Chem. Ed.* **1988**, *65*, 621.

3. It is especially important to run indicator blanks for Mohr titrations in the pH range 4 to 6, because different people see the end point differently. The upper pH limit is 10.5 in the absence of ammonia. Above pH 9 in the presence of ammonia, AgCl may not precipitate because of the stability of $Ag(NH_3)_2^+$. See A. C. Finlayson, *J. Chem. Ed.* **1992**, *69*, 559.

4. M. L. Ware, M. D. Argentine, and G. W. Rice, *Anal. Chem.* **1988**, *60*, 383.

Hydrated Radii

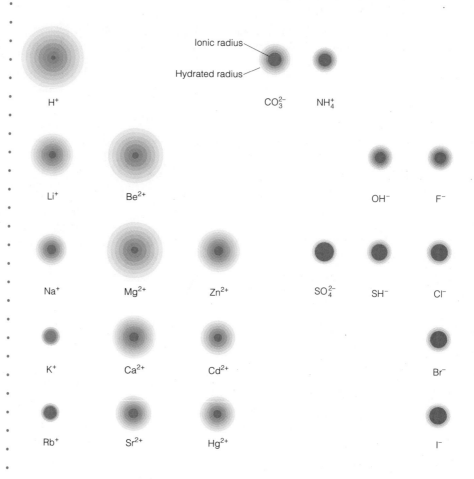

Ionic radius

Hydrated radius

H⁺   CO₃²⁻   NH₄⁺

Li⁺   Be²⁺   OH⁻   F⁻

Na⁺   Mg²⁺   Zn²⁺   SO₄²⁻   SH⁻   Cl⁻

K⁺   Ca²⁺   Cd²⁺   Br⁻

Rb⁺   Sr²⁺   Hg²⁺   I⁻

An ion in aqueous solution is surrounded by a tightly held sheath of water molecules whose electric dipoles are attracted to the electrically charged ion. The greater the charge of the ion, the more it attracts solvent molecules (e.g., $Mg^{2+} > Na^+$). The larger the radius of the naked ion, the more diffuse is its electric charge and the less it attracts solvent molecules (e.g., $K^+ < Na^+$). The chart compares radii of naked ions with those of hydrated ions. Activity coefficients, which we study in this chapter, are related to the hydrated radius, which is the effective size of an ion in aqueous solution.

Activity

In Chapter 5 we wrote the equilibrium constant for the reaction $aA + bB \rightleftharpoons cC + dD$ in the form

$$K = \frac{[C]^c[D]^d}{[A]^a[B]^b} \qquad (8\text{-}1)$$

This statement is not strictly correct, because $[C]^c[D]^d/[A]^a[B]^b$ is not constant under all conditions. In this chapter we discuss why the concentrations should be replaced by *activities* and how activities are used.

8-1 The Effect of Ionic Strength on Solubility of Salts

Consider a saturated solution of $Hg_2(IO_3)_2$ in distilled water. On the basis of the solubility product, we expect the concentration of mercurous ion to be 6.9×10^{-7} M:

$$Hg_2(IO_3)_2(s) \rightleftharpoons Hg_2^{2+} + 2IO_3^- \qquad K_{sp} = 1.3 \times 10^{-18} \qquad (8\text{-}2)$$

$$K_{sp} = [Hg_2^{2+}][IO_3^-]^2 = x(2x)^2$$

$$x = [Hg_2^{2+}] = 6.9 \times 10^{-7} \text{ M}$$

This concentration is indeed observed when $Hg_2(IO_3)_2$ is dissolved in distilled water.

However, a seemingly strange effect is observed when a salt such as KNO_3 is added to the solution. Neither K^+ nor NO_3^- reacts with either Hg_2^{2+} or IO_3^-. Yet, when 0.050 M KNO_3 is added to the saturated solution of

Demonstration 8-1 **Effect of Ionic Strength on Ion Dissociation**[1]

This experiment demonstrates the effect of ionic strength on the dissociation of the red ferric thiocyanate complex:

$$Fe(SCN)^{2+} \rightleftharpoons Fe^{3+} + SCN^-$$

(red) (pale yellow) (colorless)

Prepare a solution of 1 mM $FeCl_3$ by dissolving 0.27 g of $FeCl_3 \cdot 6H_2O$ in 1 L of water containing 3 drops of 15 M (concentrated) HNO_3. The purpose of the acid is to slow the precipitation of Fe^{3+}, which occurs in a few days and necessitates the preparation of fresh solution for this demonstration.

To demonstrate the effect of ionic strength on the dissociation reaction, mix 300 mL of the 1 mM $FeCl_3$ solution with 300 mL of 1.5 mM NH_4SCN or KSCN. Divide the pale red solution into two equal portions and add 12 g of KNO_3 to one of them to increase the ionic strength to 0.4 M. As the KNO_3 dissolves, the red $Fe(SCN)^{2+}$ complex dissociates and the color fades noticeably (Color Plate 5).

Addition of a few crystals of NH_4SCN or KSCN to either solution drives the reaction toward formation of $Fe(SCN)^{2+}$, thereby intensifying the red color. This reaction demonstrates Le Châtelier's principle—adding a product creates more reactant.

Addition of an "inert" salt increases the solubility of an ionic compound.

$Hg_2(IO_3)_2$, more solid dissolves until the concentration of Hg_2^{2+} has increased by about 50% (to 1.0×10^{-6} M).

It turns out to be a general observation that adding any "inert" salt (such as KNO_3) to any sparingly soluble salt [such as $Hg_2(IO_3)_2$] increases the solubility of the sparingly soluble salt. By "inert," we mean a salt whose ions do not react with the compound of interest. Whenever we add salts to a solution, we say that the *ionic strength* of the solution is increased. The precise definition of ionic strength will be given shortly.

The Explanation

How can we explain the increased solubility induced by the addition of ions to the solution? Consider one particular Hg_2^{2+} ion and one particular IO_3^- ion in the solution. The IO_3^- ion is surrounded by cations (K^+, Hg_2^{2+}) and anions (NO_3^-, IO_3^-) in its region of solution. However, on the average, there will be more cations than anions around any one chosen anion because cations are attracted to the anion, but anions are repelled. These interactions give rise to a region of net positive charge around any particular anion. We call this region the **ionic atmosphere** (Figure 8-1). Ions are continually diffusing into and out of the ionic atmosphere. The net charge in the atmosphere, averaged over time, is less than the charge of the anion at the center. A similar phenomenon accounts for a region of negative charge surrounding any cation in solution.

The ionic atmosphere attenuates (decreases) the attraction between ions in solution. The cation plus its negative atmosphere has less positive charge than the cation alone. The anion plus its ionic atmosphere has less negative charge than the anion alone. The net attraction between the cation-plus-atmosphere and the anion-plus-atmosphere is smaller than it would be between pure cation and anion in the absence of ionic atmospheres. *The greater the ionic strength of a solution, the higher the charge in the ionic atmosphere. The consequences of this effect are that each ion-plus-atmosphere*

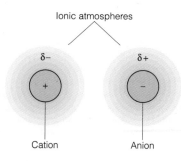

Ionic atmospheres

δ− δ+

Cation Anion

Figure 8-1 An ionic atmosphere, shown here as a spherical cloud of charge δ+ or δ−, surrounds ions in solution. The charge of the atmosphere is less than the charge of the central ion. The greater the ionic strength of the solution, the greater is the charge in each ionic atmosphere.

contains less charge and there is less attraction between any particular cation and anion.

Increasing the ionic strength of a solution therefore reduces the attraction between any particular Hg_2^{2+} ion and any IO_3^- ion, relative to their attraction in distilled water. The effect is to reduce their tendency to come together, thereby increasing the solubility of $Hg_2(IO_3)_2$.

Increasing ionic strength promotes the dissociation of compounds into ions. Thus, each of the following reactions is driven to the right if the ionic strength is raised from, say, 0.01 to 0.1 M:

$$Hg_2(IO_3)_2(s) \rightleftharpoons Hg_2^{2+} + 2IO_3^-$$

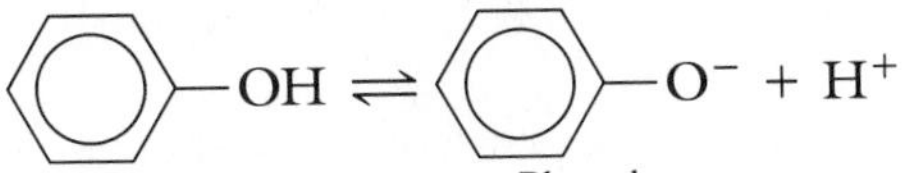

Phenol   Phenolate

$$Fe(SCN)^{2+} \rightleftharpoons Fe^{3+} + SCN^-$$
Thiocyanate

> Ion dissociation is increased by increasing the ionic strength. Demonstration 8-1 shows this effect.

What Do We Mean by "Ionic Strength"?

Ionic strength, μ, is a measure of the total concentration of ions in solution. The more highly charged an ion, the more it is counted:

Ionic strength:
$$\mu = \frac{1}{2}(c_1 z_1^2 + c_2 z_2^2 + \cdots) = \frac{1}{2}\sum_i c_i z_i^2 \qquad (8\text{-}3)$$

where c_i is the concentration of the *i*th species and z_i is its charge. The sum extends over *all* ions in solution.

EXAMPLE Calculation of Ionic Strength

Find the ionic strength of **(a)** 0.10 M $NaNO_3$; **(b)** 0.10 M Na_2SO_4; and **(c)** 0.020 M KBr plus 0.030 M $ZnSO_4$.

Solution

(a) $\mu = \frac{1}{2}\{[Na^+]\cdot(+1)^2 + [NO_3^-]\cdot(-1)^2\} = \frac{1}{2}\{[0.10]\cdot1 + [0.10]\cdot1\} = 0.10$ M

(b) $\mu = \frac{1}{2}\{[Na^+]\cdot(+1)^2 + [SO_4^{2-}]\cdot(-2)^2\} = \frac{1}{2}\{(0.20)\cdot1 + (0.10)\cdot4\} = 0.30$ M

Note that $[Na^+] = 0.20$ M because there are two moles of Na^+ per mole of Na_2SO_4.

(c) $\mu = \frac{1}{2}\{[K^+]\cdot(+1)^2 + [Br^-]\cdot(-1)^2 + [Zn^{2+}]\cdot(+2)^2 + [SO_4^{2-}]\cdot(-2)^2\}$

$= \frac{1}{2}\{(0.020)\cdot1 + (0.020)\cdot1 + (0.030)\cdot4 + (0.030)\cdot4\} = 0.14$ M

$NaNO_3$ is called a 1:1 electrolyte because the cation and anion both have a charge of 1. For 1:1 electrolytes, the ionic strength equals the molarity. For any other stoichiometry (such as the 2:1 electrolyte, Na_2SO_4), the ionic strength is greater than the molarity.

Electrolyte	Molarity	Ionic strength
1:1	M	M
2:1	M	3M
3:1	M	6M
2:2	M	4M

8-2   Activity Coefficients

The equilibrium "constant" in Equation 8-1 does not predict that there should be any effect of ionic strength on ionic reactions. To account for the effect of ionic strength, the concentrations should be replaced by **activities:**

Activity of C:
$$\mathcal{A}_C = [C]\gamma_C$$

(8-4)

↑ Activity of C   ↑ Concentration of C   Activity coefficient of C

Do not confuse the terms activity *and* activity coefficient.

where C is any species. The activity of each species is its concentration multiplied by its **activity coefficient.**

The correct form of the equilibrium constant is

This is the "real" equilibrium constant.

General form of equilibrium constant:
$$K = \frac{\mathcal{A}_C^c \, \mathcal{A}_D^d}{\mathcal{A}_A^a \, \mathcal{A}_B^b} = \frac{[C]^c\gamma_C^c \, [D]^d\gamma_D^d}{[A]^a\gamma_A^a \, [B]^b\gamma_B^b}$$

(8-5)

Equation 8-5 allows for the effect of ionic strength on a chemical equilibrium because the activity coefficients depend on ionic strength.

For Reaction 8-2, the equilibrium constant is

$$K_{sp} = \mathcal{A}_{Hg_2^{2+}} \, \mathcal{A}_{IO_3^-}^2 = [Hg_2^{2+}]\gamma_{Hg_2^{2+}}[IO_3^-]^2\gamma_{IO_3^-}^2$$

If the concentrations of Hg_2^{2+} and IO_3^- are to increase with increasing ionic strength, the values of the activity coefficients must decrease with increasing ionic strength.

At low ionic strength, activity coefficients approach unity, and the thermodynamic equilibrium constant (Equation 8-5) approaches the "concentration" equilibrium constant (Equation 8-1). One way to measure a thermodynamic equilibrium constant is to measure the concentration ratio (Equation 8-1) at successively lower ionic strengths and then extrapolate to zero ionic strength. Very commonly, tabulated equilibrium constants are not true thermodynamic constants but just the concentration ratio measured under a particular set of conditions.

Activity Coefficients of Ions

Detailed treatment of the ionic atmosphere model leads to the **extended Debye-Hückel equation,** relating activity coefficients to ionic strength:

Extended Debye-Hückel equation:
$$\log \gamma = \frac{-0.51z^2 \sqrt{\mu}}{1 + (\alpha\sqrt{\mu}/305)}$$
(at 25°C)   (8-6)

In Equation 8-6, γ is the activity coefficient of an ion of charge $\pm z$ and size α (picometers, pm; 1 pm = 10^{-12} m) in an aqueous medium of ionic strength μ. The equation works fairly well for $\mu \leq 0.1$ M.

The value of α is the effective **hydrated radius** of the ion and its tightly bound sheath of water molecules. Small, highly charged ions bind solvent molecules more tightly and have *larger* hydrated radii than do larger or less highly charged ions. F^-, for example, has a hydrated radius greater than that of I^- (Figure 8-2). Each anion attracts solvent molecules mainly by electrostatic interaction between the negative ion and the positive pole of the H_2O dipole:

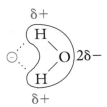

Table 8-1 lists the sizes and activity coefficients of many common ions. All ions of the same size (α) and charge appear in the same row and have the same activity coefficients. For example, Ba^{2+} and succinate ion [$^-O_2CCH_2CH_2CO_2^-$, listed as $(CH_2CO_2)_2$] each have a hydrated radius of 500 pm and are listed among the charge = ±2 ions. In an aqueous solution with an ionic strength of 0.001 M, both of these ions have an activity coefficient of 0.868.

Over the range of ionic strengths from 0 to 0.1 M, the effect of each variable on activity coefficients is as follows:

1. As ionic strength increases, the activity coefficient decreases (Figure 8-3). For all ions, γ approaches unity as μ approaches zero.

2. As the charge of the ion increases, the departure of its activity coefficient from unity increases. Activity corrections are much more important for an ion with a charge of ±3 than for one with a charge of ±1 (Figure 8-3). Note that the activity coefficients in Table 8-1 depend on the magnitude of the charge but not on its sign.

3. The smaller the hydrated radius of the ion, the more important activity effects become.

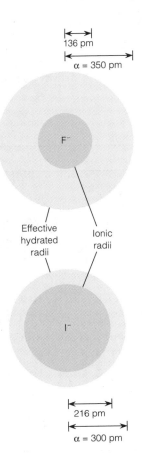

Figure 8-2 Ionic and hydrated radii of fluoride and iodide. The *smaller* F^- ion binds water molecules more tightly than does I^-, so F^- has the *larger* hydrated radius.

- -

EXAMPLE Using Table 8-1

Find the activity coefficient of Hg_2^{2+} in a solution of 0.033 M $Hg_2(NO_3)_2$.

Solution The ionic strength is

$$\mu = \tfrac{1}{2}\,([Hg_2^{2+}]\cdot 2^2 + [NO_3^-]\cdot(-1)^2)$$

$$= \tfrac{1}{2}\,([0.033]\cdot 4 + [0.066]\cdot 1) = 0.10\ \text{M}$$

In Table 8-1, Hg_2^{2+} is listed under the charge ±2 and has a size of 400 pm. Thus $\gamma = 0.355$ when $\mu = 0.1$ M.

- -

How to Interpolate

Estimating a number that lies *between* two values in a table is called *interpolation*. (Estimating a number that lies *beyond* values in a table is called *extrapolation*.) In *linear interpolation,* we assume that values between two

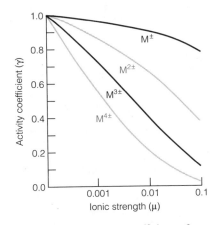

Figure 8-3 Activity coefficients for differently charged ions with a constant hydrated radius of 500 pm. At zero ionic strength, $\gamma = 1$. The greater the charge of the ion, the more rapidly γ decreases as ionic strength increases. Note that the abscissa is logarithmic.

TABLE 8-1 Activity coefficients for aqueous solutions at 25°C

Ion	Ion size (α, pm)	Ionic strength (μ, M)				
		0.001	0.005	0.01	0.05	0.1
Charge = ±1						
H^+	900	0.967	0.933	0.914	0.86	0.83
$(C_6H_5)_2CHCO_2^-$, $(C_3H_7)_4N^+$	800	0.966	0.931	0.912	0.85	0.82
$(O_2N)_3C_6H_2O^-$, $(C_3H_7)_3NH^+$, $CH_3OC_6H_4CO_2^-$	700	0.965	0.930	0.909	0.845	0.81
Li^+, $C_6H_5CO_2^-$, $HOC_6H_4CO_2^-$, $ClC_6H_4CO_2^-$, $C_6H_5CH_2CO_2^-$, $CH_2{=}CHCH_2CO_2^-$, $(CH_3)_2CHCH_2CO_2^-$, $(CH_3CH_2)_4N^+$, $(C_3H_7)_2NH_2^+$	600	0.965	0.929	0.907	0.835	0.80
$Cl_2CHCO_2^-$, $Cl_3CCO_2^-$, $(CH_3CH_2)_3NH^+$, $(C_3H_7)NH_3^+$	500	0.964	0.928	0.904	0.83	0.79
Na^+, $CdCl^+$, ClO_2^-, IO_3^-, HCO_3^-, $H_2PO_4^-$, HSO_3^-, $H_2AsO_4^-$, $Co(NH_3)_4(NO_2)_2^+$, $CH_3CO_2^-$, $ClCH_2CO_2^-$, $(CH_3)_4N^+$, $(CH_3CH_2)_2NH_2^+$, $H_2NCH_2CO_2^-$	450	0.964	0.928	0.902	0.82	0.775
$^+H_3NCH_2CO_2H$, $(CH_3)_3NH^+$, $CH_3CH_2NH_3^+$	400	0.964	0.927	0.901	0.815	0.77
OH^-, F^-, SCN^-, OCN^-, HS^-, ClO_3^-, ClO_4^-, BrO_3^-, IO_4^-, MnO_4^-, HCO_2^-, $H_2citrate^-$, $CH_3NH_3^+$, $(CH_3)_2NH_2^+$	350	0.964	0.926	0.900	0.81	0.76
K^+, Cl^-, Br^-, I^-, CN^-, NO_2^-, NO_3^-	300	0.964	0.925	0.899	0.805	0.755
Rb^+, Cs^+, NH_4^+, Tl^+, Ag^+	250	0.964	0.924	0.898	0.80	0.75
Charge = ± 2						
Mg^{2+}, Be^{2+}	800	0.872	0.755	0.69	0.52	0.45
$CH_2(CH_2CH_2CO_2^-)_2$, $(CH_2CH_2CH_2CO_2^-)_2$	700	0.872	0.755	0.685	0.50	0.425
Ca^{2+}, Cu^{2+}, Zn^{2+}, Sn^{2+}, Mn^{2+}, Fe^{2+}, Ni^{2+}, Co^{2+}, $C_6H_4(CO_2^-)_2$, $H_2C(CH_2CO_2^-)_2$, $(CH_2CH_2CO_2^-)_2$	600	0.870	0.749	0.675	0.485	0.405
Sr^{2+}, Ba^{2+}, Cd^{2+}, Hg^{2+}, S^{2-}, $S_2O_4^{2-}$, WO_4^{2-}, $H_2C(CO_2^-)_2$, $(CH_2CO_2^-)_2$, $(CHOHCO_2^-)_2$	500	0.868	0.744	0.67	0.465	0.38
Pb^{2+}, CO_3^{2-}, SO_3^{2-}, MoO_4^{2-}, $Co(NH_3)_5Cl^{2+}$, $Fe(CN)_5NO^{2-}$, $C_2O_4^{2-}$, $Hcitrate^{2-}$	450	0.867	0.742	0.665	0.455	0.37
Hg_2^{2+}, SO_4^{2-}, $S_2O_3^{2-}$, $S_2O_6^{2-}$, $S_2O_8^{2-}$, SeO_4^{2-}, CrO_4^{2-}, HPO_4^{2-}	400	0.867	0.740	0.660	0.445	0.355
Charge = ±3						
Al^{3+}, Fe^{3+}, Cr^{3+}, Sc^{3+}, Y^{3+}, In^{3+}, lanthanides[a]	900	0.738	0.54	0.445	0.245	0.18
$citrate^{3-}$	500	0.728	0.51	0.405	0.18	0.115
PO_4^{3-}, $Fe(CN)_6^{3-}$, $Cr(NH_3)_6^{3+}$, $Co(NH_3)_6^{3+}$, $Co(NH_3)_5H_2O^{3+}$	400	0.725	0.505	0.395	0.16	0.095
Charge = ±4						
Th^{4+}, Zr^{4+}, Ce^{4+}, Sn^{4+}	1100	0.588	0.35	0.255	0.10	0.065
$Fe(CN)_6^{4-}$	500	0.57	0.31	0.20	0.048	0.021

a. Lanthanides are elements 57–71 in the periodic table.
SOURCE: J. Kielland, *J. Am. Chem. Soc.* **1937**, *59*, 1675.

entries of a table lie on a straight line. For example, consider a table in which y = 0.67 when x = 10 and y = 0.83 when x = 20. What is the value of y when x = 16?

	x value	
	10	20
y value	0.67	0.83

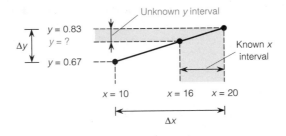

To interpolate a value of y, we can set up a proportion:

$$\frac{\text{unknown } y \text{ interval}}{\Delta y} = \frac{\text{known } x \text{ interval}}{\Delta x}$$

$$\frac{0.83 - y}{0.83 - 0.67} = \frac{20 - 16}{20 - 10}$$

$$\Rightarrow y = 0.76_6$$

For x = 16, our estimate of y is 0.76_6.

. . . .

EXAMPLE Interpolating Activity Coefficients

Calculate the activity coefficient of H^+ when μ = 0.025 M.

Solution H^+ is the first entry in Table 8-1.

$$H^+: \quad \mu = 0.01 \quad 0.025 \quad 0.05$$
$$\gamma = 0.914 \quad ? \quad 0.86$$

The linear interpolation is set up as follows:

$$\frac{\text{unknown } \gamma \text{ interval}}{\Delta \gamma} = \frac{\text{known } \mu \text{ interval}}{\Delta \mu}$$

$$\frac{0.86 - \gamma}{0.86 - 0.914} = \frac{0.05 - 0.025}{0.05 - 0.01}$$

$$\Rightarrow \gamma = 0.89_4$$

Another Solution A more correct and slightly more tedious calculation uses Equation 8-6, with the hydrated radius α = 900 pm listed for H^+ in Table 8-1:

$$\log \gamma_{H^+} = \frac{(-0.51)(1^2) \sqrt{0.025}}{1 + (900\sqrt{0.025}/305)} = -0.054_{98}$$

$$\gamma_{H^+} = 10^{-0.054_{98}} = 0.88_1$$

The difference between this calculated value and the interpolated value is less than 2%. Equation 8-6 is easy to implement in a spreadsheet.

. . . .

For neutral species, $\mathcal{A}_C \approx [C]$.

Activity Coefficients of Nonionic Compounds

Neutral molecules, such as benzene and acetic acid, are not surrounded by an ionic atmosphere, because they have no charge. To a good approximation, their activity coefficients are unity when the ionic strength is less than 0.1 M. For all problems in this text, we set $\gamma = 1$ for neutral molecules. That is, *the activity of a neutral molecule will be assumed to be equal to its concentration.*

For gaseous reactants such as H_2, the activity is written

$$\mathcal{A}_{H_2} = P_{H_2}\gamma_{H_2}$$

For gases, $\mathcal{A} \approx P$ (atm).

where P_{H_2} is pressure in atmospheres. The activity of a gas is called its **fugacity,** and the activity coefficient is called the *fugacity coefficient.* Deviation of gas behavior from the ideal gas law results in deviation of the fugacity coefficient from unity. For most gases at or below 1 atm, $\gamma \approx 1$. Therefore, for all gases, *we assume that $\mathcal{A} = P$ (atm).*

Mean Activity Coefficients

Prediction of $\gamma_\pm$ for $La(NO_3)_3$ from the individual coefficients for La^{3+} and NO_3^- in Table 8-1 at $\mu = 0.1$ M:
$$\gamma_\pm = [(0.18)^1(0.755)^3]^{1/4} = 0.53$$
Observed value = 0.59

Theories dealing with activity coefficients derive coefficients for individual ions. In experiments, ions are available only in pairs, so only the **mean activity coefficient,** $\gamma_\pm$, can be derived from various measurements. For a salt with the stoichiometry (cation)$_m$(anion)$_n$, the mean activity coefficient is related to the individual coefficients by the equation

$$\gamma_\pm^{m+n} = \gamma_+^m\gamma_-^n \qquad (8\text{-}7)$$

where γ_+ is the activity coefficient of the cation and γ_- is the activity coefficient of the anion.

High Ionic Strengths

At high ionic strength, γ increases with increasing μ.

The extended Debye-Hückel equation 8-6 predicts that the activity coefficient, γ, will decrease as ionic strength, μ, increases. In fact, above an ionic strength of approximately 1 M, most activity coefficients increase, as shown for sulfuric acid in Figure 8-4. We should not be too surprised that the activity coefficients in, say, 4 M sulfuric are not the same as those in dilute aqueous solution, because the 4 M solution contains 32 wt% H_2SO_4. The "solvent" is no longer just H_2O but, rather, a mixture of H_2O and H_2SO_4.

The simple ionic atmosphere model does not account for behavior at high ionic strength. One successful model is based on the change in dielectric constant of the solution in the immediate vicinity of each ion. Hereafter we limit our attention to dilute aqueous solutions in which Equation 8-6 applies.

8-3 Using Activity Coefficients

This section presents three examples illustrating the proper use of activity coefficients in equilibrium calculations. The general prescription is trivial: *Write each equilibrium constant with activities in place of concentrations. Use the ionic strength of the solution to find the correct values of activity coefficients.*

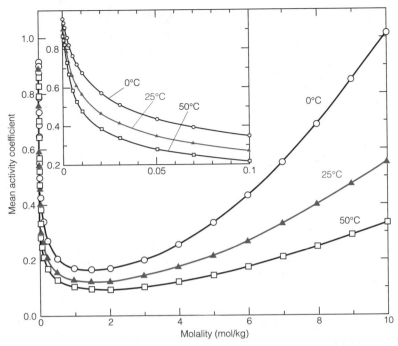

Figure 8-4 Mean activity coefficient of sulfuric acid as a function of ionic strength and temperature. Activity coefficients generally increase at sufficiently high ionic strength (>1 M). Inset shows behavior at low ionic strength where the extended Debye-Hückel equation is valid. [Data from H. S. Harned and W. J. Hamer, *J. Am. Chem. Soc.* **1935**, *57*, 27. An authoritative source on electrolyte solutions is H. S. Harned and B. B. Owen, *The Physical Chemistry of Electrolyte Solutions* (New York: Reinhold, 1958).]

A Simple Solubility Problem

In the first example, we will calculate the concentration of Ca^{2+} in a 0.0125 M solution of $MgSO_4$ saturated with CaF_2. The relevant equilibrium is

$$CaF_2(s) \rightleftharpoons Ca^{2+} + 2F^- \qquad K_{sp} = 3.9 \times 10^{-11}$$

The equilibrium expression is set up with the aid of a little table:

	$CaF_2(s) \rightleftharpoons Ca^{2+} + 2F^-$		
Initial concentration:	solid	0	0
Final concentration:	solid	x	$2x$

$$
\begin{aligned}
K_{sp} &= \mathcal{A}_{Ca^{2+}} \, \mathcal{A}_{F^-}^2 = [Ca^{2+}]\gamma_{Ca^{2+}}[F^-]^2\gamma_{F^-}^2 \\
&= [x]\gamma_{Ca^{2+}}[2x]^2\gamma_{F^-}^2
\end{aligned}
\tag{8-8}
$$

To find the values of γ to use in Equation 8-8, we need to calculate the ionic strength. The ionic strength is due to the dissolved $MgSO_4$ *and* the dissolved CaF_2. However, K_{sp} for CaF_2 is quite small, so we start by guessing that the contribution of CaF_2 to the ionic strength will be negligible. For 0.0125 M $MgSO_4$, we calculate

$$\mu = \tfrac{1}{2}\,([0.0125]\cdot 2^2 + [0.0125]\cdot(-2)^2) = 0.0500 \text{ M}$$

Using $\mu = 0.0500$ M, we look at Table 8-1 to find $\gamma_{Ca^{2+}} = 0.485$ and $\gamma_{F^-} = 0.81$. Substituting into Equation 8-8 gives

$$3.9 \times 10^{-11} = [x](0.485)[2x]^2(0.81)^2$$

$$x = [Ca^{2+}] = 3.1 \times 10^{-4} \text{ M}$$

Our assumption was correct; the contribution of CaF_2 to the ionic strength is negligible.

Note that both $[F^-]$ and γ_{F^-} are squared.

Question What is the value of μ, including the contribution of CaF_2?

The Common Ion Effect

As the next example, let's find the concentration of Ca^{2+} in a 0.050 M NaF solution saturated with CaF_2. The ionic strength is again 0.050 M, this time owing to the presence of NaF.

	$CaF_2(s)$	$\rightleftharpoons$	Ca^{2+}	+	$2F^-$
Initial concentration:	solid		0		0.050
Final concentration:	solid		x		$2x + 0.050$

Assuming that $2x \ll 0.050$, we solve the problem as follows:

$$K_{sp} = [Ca^{2+}]\gamma_{Ca^{2+}}[F^-]^2\gamma_{F^-}^2$$

$$3.9 \times 10^{-11} = (x)(0.485)(0.050)^2(0.81)^2$$

$$x = [Ca^{2+}] = 4.9 \times 10^{-8}$$

Question Why is the solubility of CaF_2 lower in this solution than in $MgSO_4$ solution?

A Problem That Requires an Iterative Solution

As a final example, we calculate the solubility of LiF in distilled water:

$$LiF(s) \rightleftharpoons Li^+ + F^- \qquad K_{sp} = 1.7 \times 10^{-3}$$

The ionic strength is determined by the concentration of LiF dissolved in the solution. But we do not know that concentration. As a first approximation, we can calculate the concentrations of Li^+ and F^- by neglecting activity coefficients. Calling x_1 our first approximation for $[Li^+]$ (and $[F^-]$), we can write

$$K_{sp} \approx [Li^+][F^-] = x_1^2 \Rightarrow x_1 = [Li^+] = [F^-] = 0.041 \text{ M}$$

As a second approximation, we assume that $\mu = 0.041$ M, which is the result of our first approximation. Interpolating in Table 8-1 gives $\gamma_{Li^+} = 0.851$ and $\gamma_{F^-} = 0.830$ for $\mu = 0.041$ M. Putting these values into the expression for K_{sp} gives

$$K_{sp} = [Li^+]\gamma_{Li^+}[F^-]\gamma_{F^-}$$

$$= [x_2](0.851)[x_2](0.830) \Rightarrow x_2 = 0.049 \text{ M}$$

This is a method of *successive approximations*. Each cycle of calculations is called one *iteration*.

As a third approximation, we assume that $\mu = 0.049$ M. Using this new ionic strength to find new values of activity coefficients in Table 8-1 gives

$$K_{sp} = [x_3](0.837)[x_3](0.812) \Rightarrow x_3 = 0.050 \text{ M}$$

Using $\mu = 0.050$ M gives a fourth approximation:

$$K_{sp} = [x_4](0.835)[x_4](0.81) \Rightarrow x_4 = 0.050 \text{ M}$$

The fourth answer is the same as the third answer. We have reached a self-consistent result, which must therefore be correct.

As LiF dissolves, it increases the ionic strength and increases its own solubility.

8-4 pH Revisited

The definition of pH given in Chapter 5, pH $\approx -\log [H^+]$, is not correct. The real definition is

$$pH = -\log \mathcal{A}_{H^+} = -\log [H^+]\gamma_{H^+} \qquad (8\text{-}9)$$

The real definition of pH!

When we measure pH with a pH meter, we are measuring the negative logarithm of the hydrogen ion *activity*, not its concentration.

. . . .

EXAMPLE pH of Pure Water at 25°C

Let's calculate the pH of pure water by using activity coefficients correctly.

Solution The relevant equilibrium is

$$H_2O \overset{K_w}{\rightleftharpoons} H^+ + OH^- \qquad (8\text{-}10)$$

$$K_w = \mathcal{A}_{H^+}\mathcal{A}_{OH^-} = [H^+]\gamma_{H^+}[OH^-]\gamma_{OH^-} \qquad (8\text{-}11)$$

The stoichiometry of the reaction tells us that H^+ and OH^- are produced in a 1:1 mole ratio. Their concentrations must therefore be equal. Calling each concentration x, we can write

$$K_w = 1.0 \times 10^{-14} = (x)\gamma_{H^+}(x)\gamma_{OH^-}$$

But the ionic strength of pure water is so small that it is reasonable to expect that $\gamma_{H^+} = \gamma_{OH^-} = 1$. Using these values in the preceding equation gives

$$1.0 \times 10^{-14} = x^2 \Rightarrow x = 1.0 \times 10^{-7} \text{ M}$$

The concentrations of H^+ and OH^- are both 1.0×10^{-7} M. Their activities are also both 1.0×10^{-7} because each activity coefficient is very close to 1.00. The pH is

$$pH = -\log [H^+]\gamma_{H^+} = -\log(1.0 \times 10^{-7})(1.00) = 7.00$$

. . . .

. . . .

EXAMPLE pH of Water Containing a Salt

Now let's calculate the pH of water containing 0.10 M KCl at 25°C.

Solution Reaction 8-10 tells us that $[H^+] = [OH^-]$. However, the two values of γ in Equation 8-11 are not equal. The ionic strength of 0.10 M KCl is 0.10 M. According to Table 8-1, the activity coefficients of H^+ and OH^- are 0.83

and 0.76, respectively, when the ionic strength is 0.10 M. Putting these values into Equation 8-11 gives

$$K_w = [H^+]\gamma_{H^+}[OH^-]\gamma_{OH^-}$$

$$1.0 \times 10^{-14} = (x)(0.83)(x)(0.76)$$

$$x = 1.26 \times 10^{-7} \text{ M}$$

The concentrations of H^+ and OH^- are equal and are both greater than 1.0×10^{-7} M. The activities of H^+ and OH^- are not equal in this solution:

$$\mathcal{A}_{H^+} = [H^+]\gamma_{H^+} = (1.26 \times 10^{-7})(0.83) = 1.05 \times 10^{-7}$$

$$\mathcal{A}_{OH^-} = [OH^-]\gamma_{OH^-} = (1.26 \times 10^{-7})(0.76) = 0.96 \times 10^{-7}$$

Finally, we calculate the pH:

$$\text{pH} = -\log \mathcal{A}_{H^+} = -\log (1.05 \times 10^{-7}) = 6.98$$

These two examples show that the pH of water changes from 7.00 to 6.98 when we add 0.10 M KCl. KCl is not an acid or a base. The small change in pH arises because KCl affects the activities of H^+ and OH^-. The pH change of 0.02 units lies at the current limit of accuracy of pH measurements and is hardly important. However, the *concentration* of H^+ in 0.10 M KCl (1.26×10^{-7} M) is 26% greater than the concentration of H^+ in pure water (1.00×10^{-7} M).

The change in pH of water when a salt is added is an example of a *matrix effect*. The **matrix** is the medium containing whatever we are interested in. In this case, we are interested in H^+ in a matrix of 0.10 M KCl. The matrix changes the activity of H^+, even though there is no direct chemical reaction between H^+ and KCl. We will mention other instances in later chapters where the matrix can have significant effects on chemical analysis.

Terms to Understand

activity
activity coefficient
extended Debye-Hückel equation
fugacity

hydrated radius
ionic atmosphere
ionic strength

matrix
mean activity coefficient
pH

Summary

The true thermodynamic equilibrium constant for the reaction $a\text{A} + b\text{B} \rightleftharpoons c\text{C} + d\text{D}$ is $K = \mathcal{A}_C^c \mathcal{A}_D^d / (\mathcal{A}_A^a \mathcal{A}_B^b)$, where $\mathcal{A}_i$ is the activity of the ith species. The activity is the product of the concentration (c) and the activity coefficient (γ): $\mathcal{A}_i = c_i\gamma_i$. For nonionic compounds and gases, we assume that $\gamma_i = 1$. For ionic species, the activity coefficient depends on the ionic strength, defined as $\mu = \frac{1}{2}\sum_i c_i z_i^2$. The activity coefficient decreases as the ionic strength increases, at least for low ionic strengths.

The extent of dissociation of ionic compounds increases with ionic strength because the ionic atmosphere of each ion diminishes the attraction of ions for one another. You should be able to perform equilibrium calculations, using activities instead of concentrations. You should be able to estimate activity coefficients by interpolation in Table 8-1. pH is defined in terms of the activity of H^+, not just the concentration: $\text{pH} = -\log \mathcal{A}_{H^+} = -\log [H^+]\gamma_{H^+}$.

Exercises

A. Calculate the ionic strength of

(a) 0.02 M KBr

(b) 0.02 M Cs_2CrO_4

(c) 0.02 M $MgCl_2$ plus 0.03 M $AlCl_3$

B. Find the activity (not the activity coefficient) of the $(C_3H_7)_4N^+$ (tetrapropylammonium) ion in a solution containing 0.0050 M $(C_3H_7)_4N^+Br^-$ plus 0.0050 M $(CH_3)_4N^+Cl^-$.

C. Using activities, find the solubility of AgSCN (moles of Ag^+/L) in (a) 0.060 M KNO_3 and (b) 0.060 M KSCN.

D. Using activities, find the concentration of OH^- in a solution of 0.025 M $CaCl_2$ saturated with $Mn(OH)_2$.

E. Calculate the mean activity coefficient for $MgCl_2$ at a concentration of 0.020 M from the data in Table 8-1.

F. Using activities correctly, calculate the pH and concentration of H^+ in pure water containing 0.050 M LiBr at 25°C.

G. A 40.0-mL solution of 0.0400 M $Hg_2(NO_3)_2$ was titrated with 60.0 mL of 0.100 M KI to precipitate Hg_2I_2 ($K_{sp} = 4.5 \times 10^{-29}$).

(a) What volume of KI is needed to reach the equivalence point?

(b) Calculate the ionic strength of the solution when 60.0 mL of KI has been added.

(c) Without ignoring activity, calculate pHg_2^{2+} ($= -\log \mathscr{A}_{Hg_2^{2+}}$) when 60.0 mL of KI has been added.

Problems

Ionic Strength

1. Explain why the solubility of an ionic compound increases as the ionic strength of the solution increases (at least up to ~ 0.5 M).

2. Which statements are true? In the ionic strength range 0–0.1 M, activity coefficients decrease with

(a) increasing ionic strength

(b) increasing ionic charge

(c) decreasing hydrated radius

3. Calculate the ionic strength of

(a) 0.0087 M KOH

(b) 0.0002 M $La(IO_3)_3$

(c) 0.02 M $CuSO_4$

(d) 0.01 M $CuCl_2$ + 0.05 M $(NH_4)_2SO_4$ + 0.02 M $Gd(NO_3)_3$

Activity Coefficients

4. Explain the following observations:

(a) Mg^{2+} has a greater hydrated radius than Ba^{2+}.

(b) Hydrated radii decrease in the order $Sn^{4+} > In^{3+} > Cd^{2+} > Rb^+$.

(c) H^+ has a hydrated radius of 900 pm, whereas that of OH^- is just 350 pm.

5. Find the activity coefficient of each ion at the indicated ionic strength:

(a) SO_4^{2-} ($\mu = 0.01$ M) (b) Sc^{3+} ($\mu = 0.005$ M)

(c) Eu^{3+} ($\mu = 0.1$ M)

(d) $(CH_3CH_2)_3NH^+$ ($\mu = 0.05$ M)

6. Use interpolation in Table 8-1 to find the activity coefficient of H^+ when $\mu = 0.030$ M.

7. Calculate the activity coefficient of Zn^{2+} when $\mu = 0.083$ M by

(a) using Equation 8-6

(b) using linear interpolation with Table 8-1

8. The observed mean activity coefficient ($\gamma_\pm$) for HCl at a concentration of 0.005 M is 0.93. What is the value of ($\gamma_\pm$) calculated from Table 8-1?

9. The equilibrium constant for dissolution in water of a nonionic compound, such as diethyl ether $(CH_3CH_2OCH_2CH_3)$, can be written

$$ether(l) \rightleftharpoons ether(aq) \qquad K = [ether(aq)]\gamma_{ether}$$

At low ionic strength, $\gamma \approx 1$ for all nonionic compounds. At high ionic strength, ether and most other neutral molecules can be *salted out* of aqueous solution. That is, when a high concentration (typically >1 M) of an ionic compound (such as NaCl) is added to an aqueous solution, neutral molecules usually become *less* soluble. Does the activity coefficient, γ_{ether}, increase or decrease at high ionic strength?

10. The temperature-dependent form of the extended Debye-Hückel equation 8-6 is

$$\log \gamma = \frac{(-1.825 \times 10^6)(\epsilon T)^{-3/2}z^2\sqrt{\mu}}{1 + \alpha\sqrt{\mu}/(2.00\sqrt{\epsilon T})}$$

where ϵ is the (dimensionless) dielectric constant† of water, T is the temperature in kelvins, z is the charge of the ion of interest, μ is the ionic strength of the solution (mol/L), and α is the size of the ion in picometers. The dependence of ϵ on temperature is given by

$$\epsilon = 79.755e^{(-4.6 \times 10^{-3})(T - 293.15)}$$

Calculate the activity coefficient of SO_4^{2-} at 50.00°C when $\mu = 0.100$ M. Compare your value with the one in Table 8-1.

11. An empirical equation describing activity coefficients at ionic strengths higher than 0.1 M is

$$\log \gamma = \frac{-A\sqrt{\mu}}{1 + B\sqrt{\mu}} + C\mu$$

where γ is the activity coefficient, μ is the ionic strength, and A, B, and C are constants characteristic of a particular solution. An appropriate measure of single ion activity coefficients can be made with ion-selective electrodes. One set of measurements for Cu^{2+} gives $A = 3.23$ (± 0.32), $B = 2.57$ (± 0.32), and $C = 0.198$ (± 0.012).[2]

(a) Ignoring the uncertainties in the parameters A, B, and C, calculate the activity coefficient of Cu^{2+} at the following ionic strengths: 0.001, 0.01, 0.1, 0.5, 1.0, 1.5, 2.0, and 3.0 M. Make a graph of γ versus μ and compare your values of γ with those in Table 8-1.

(b) *Propagation of uncertainty.* Find the uncertainty in γ when $\mu = 0.1$ M.

12. *Extended Debye-Hückel equation.* Use Equation 8-6 to calculate the activity coefficient (γ) as a function of ionic strength (μ) for the values $\mu = 0.0001, 0.0003, 0.001, 0.003, 0.01, 0.03$, and 0.1 M.

(a) For an ionic charge of ± 1 and a size $\alpha = 400$ pm, make a table of γ ($= 10 \wedge [\log \gamma]$) for each value of μ.

(b) Do the same for ionic charges of ± 2, ± 3, and ± 4.

(c) Use a graphics program to plot γ versus $\log \mu$. Your graph will be similar to Figure 8-3.

- - - - - - - - - - - - - - - -

†The *dielectric constant* of a solvent is a measure of how well that solvent can separate oppositely charged ions. The force of attraction (in newtons) between two ions of charge q_1 and q_2 (in coulombs) separated by a distance r (in meters) is

$$\text{force} = -(8.988 \times 10^9)\frac{q_1 q_2}{\epsilon r^2}$$

where ϵ is the (dimensionless) dielectric constant. The larger the value of ϵ, the smaller the attraction between ions. Water, with $\epsilon \approx 80$, separates ions very well. The value of ϵ for benzene is 2, and the value of ϵ for vacuum is 1.

Using Activity Coefficients

13. Calculate the solubility of Hg_2Br_2 (expressed as moles of Hg_2^{2+} per liter) in

(a) 0.000 33 M $Mg(NO_3)_2$

(b) 0.003 3 M $Mg(NO_3)_2$

(c) 0.033 M $Mg(NO_3)_2$

14. Find the concentration of Ba^{2+} in a 0.033 3 M $Mg(IO_3)_2$ solution saturated with $Ba(IO_3)_2$.

15. Calculate the concentration of Pb^{2+} in a saturated solution of PbF_2 in water.

16. (a) Using K_{sp} for $CaSO_4$, calculate the concentration of dissolved Ca^{2+} in a saturated aqueous solution of $CaSO_4$.

(b) At the end of Section 5-3, it was stated that the observed total concentration of dissolved calcium is 15–19 mM. Explain.

17. *Finding solubility by iteration.* Use a spreadsheet for the iterative computation of the solubility of LiF in distilled water, as described in Section 8-3. The solubility product is given by $K_{sp} = 0.0017 = [Li^+]\gamma_{Li^+}[F^-]\gamma_{F^-}$. The problem is that you do not know the ionic strength, so you cannot find the activity coefficients. As in Section 8-3, initially set the ionic strength to zero and use successive cycles of calculation (iterations) to home in on the correct solution.

(a) Set up the spreadsheet shown on page 197, in which ionic strength in cell B2 is initially given the value 0. The formulas for activity coefficients in cells A8 and A10 are from the extended Debye-Hückel equation. The formula in cell C2 comes from solving the solubility equation $[Li^+] = [F^-] = (K_{sp}/\gamma_{Li^+}\gamma_{F^-})^{1/2}$. With a value of 0 in cell B2, your spreadsheet should compute values of 1 in cells A8 and A10 and the solubility 0.041 23 M in cell C2.

(b) Because the ionic strength of a 1:1 electrolyte is equal to the concentration, copy the value 0.041 23 into cell B2. This will give new values of activity coefficients in cells A8 and A10 and a new solubility in cell C2. Copy the new concentration from cell C2 into cell B2 and repeat this procedure as many times as you like until you have a constant answer to as many decimal places as you like.

(c) Your spreadsheet may allow circular references, which simplifies the calculations. Instead of inputting a numerical value for ionic strength in cell B2, use the formula "=C2" in cell B2. This command tells the spreadsheet to use the solubility from cell C2 as the ionic strength in cell B2. Try this out and see what happens.

	A	B	C
1	Size (pm) of Li+ =	Ionic strength	x = [Li+] = [F−]
2	600	0	0.0412310562
3	Size (pm) of F− =		
4	350		
5	Ksp =		
6	0.0017		
7	Activity coeff (Li+) =		
8	1		
9	Activity coeff (F−) =		
10	1		
11			
12	Formulas used:		
13	A8 = 10^((−0.51)*Sqrt(B2)/(1+(A2*Sqrt(B2)/305)))		
14	A10 = 10^((−0.51)*Sqrt(B2)/(1+(A4*Sqrt(B2)/305)))		
15	C2 = Sqrt(A6/(A8*A10))		

18. *More solubility by iteration.* Use the procedure from the previous exercise to find the solubility of $Ca(OH)_2$ in distilled water. The relevant equilibrium is $Ca(OH)_2(s) \rightleftharpoons Ca^{2+} + 2OH^-$, for which $K_{sp} = 6.5 \times 10^{-6} = [Ca^{2+}]\gamma_{Ca^{2+}}[OH^-]^2\gamma_{OH^-}^2$. The relation between the ionic strength and $[Ca^{2+}]$ is $\mu = 3[Ca^{2+}]$, because we are dealing with a 2:1 electrolyte.

19. A 0.0100 M $Na_3[Co(CN)_6]$ solution with a volume of 50.0 mL was treated with 0.0100 M $Hg_2(NO_3)_2$ solution to precipitate $(Hg_2)_3[Co(CN)_6]_2$. Using activity coefficients, find $pCo(CN)_6$ when 90.0 mL of $Hg_2(NO_3)_2$ was added.

pH Revisited

20. Find the activity coefficient of H^+ in a solution containing 0.010 M HCl plus 0.040 M $KClO_4$. What is the pH of the solution?

21. Using activities correctly, calculate the pH of a solution containing 0.010 M NaOH plus 0.0040 M $Ca(NO_3)_2$. What would be the pH if you neglected activities?

Notes and References

1. D. R. Driscol, *J. Chem. Ed.* **1979**, *56*, 603. An experiment (or demonstration) dealing with the effect of ionic strength on the acid dissociation reaction of bromocresol green can be found in J. A. Bell (Ed.), *Chemical Principles in Practice* (Reading, MA: Addison-Wesley, 1967, pp. 105–108; and in R. W. Ramette, *J. Chem. Ed.* **1963**, *40*, 252.

2. I. Uemasu and Y. Umezawa, *Anal. Chem.* **1983**, *55*, 386.

Erosion of Marble by Acid Rain

Erosion of carbonate stone. The column at the left is exposed only internally, but the column at the right shows the effects of acid rain across its entire surface. [Courtesy P. A. Baedecker, U.S. Geological Survey.]

Limestone and marble are two important building materials whose main constituent is calcite, the common crystalline form of calcium carbonate. This mineral is not very soluble in neutral or basic solution ($K_{sp} = 4.5 \times 10^{-9}$), but it dissolves in acidic solution by virtue of two *coupled equilibria,* in which the product of one reaction is consumed in the next reaction:

$$CaCO_3(s) \rightleftharpoons Ca^{2+} + CO_3^{2-}$$

Calcite   Carbonate

$$CO_3^{2-} + H^+ \rightleftharpoons HCO_3^-$$

Bicarbonate

Carbonate produced in the first reaction is protonated to form bicarbonate in the second reaction. Le Châtelier's principle tells us that if we remove a product of the first reaction, we will draw the reaction to the right, making calcite more soluble.

Erosion caused by acidic rainfall (Box 15-2) is an increasingly important problem. In this chapter we introduce the tools to understand the interplay between coupled equilibria in chemical systems.

Systematic Treatment
of Equilibrium

The study of chemical equilibrium provides a basis for most techniques in analytical chemistry, such as titrations and chromatography, and for the application of chemistry to other disciplines, such as geology and biology. Many chemical systems are exceedingly complex because of the large number of reactions and species involved. In these complex systems, the product of one reaction can be a reactant in another, and the balance between competing reactions can be affected by factors such as pH. The devastating effect of acid rain on sphinxes and marble statues is a case in which the normally low solubility of minerals in water is increased by acidity.

The *systematic treatment of equilibrium* described in this chapter gives us tools to deal with all types of chemical equilibria, regardless of their complexity. We have already used elements of the systematic treatment to derive spreadsheet equations for precipitation titrations in Chapter 7. After setting up general equations, we will often introduce specific conditions or judicious approximations that allow us to simplify many calculations and to understand the underlying chemistry. However, now that spreadsheets are so easy to use, there is less reason to learn approximate methods. When you have mastered the systematic treatment of equilibrium, you should be able to use a spreadsheet to explore the quantitative behavior of complex systems.

The systematic procedure is to write as many algebraic equations as there are unknowns (species) in the problem. The equations are generated by writing all the chemical equilibrium conditions plus two more: the balances of charge and of mass. Although there is only one charge balance for a given solution, there may be several different mass balances.

Solutions must have zero total charge.

9-1 Charge Balance

The **charge balance** is an algebraic statement of electroneutrality of the solution. *The sum of the positive charges in solution equals the sum of the negative charges in solution.*

Consider a sulfate ion with concentration 0.0167 M. Because the charge on SO_4^{2-} is -2, the charge contributed by sulfate is $(-2)(0.0167) = -0.0334$ M. In general, an ion with a charge of $\pm n$ and a concentration of [A] will contribute $\pm n$[A] to the charge of the solution. The charge balance equates the magnitude of the total positive charge to the magnitude of the total negative charge.

Suppose that a solution contains the following ionic species: H^+, OH^-, K^+, $H_2PO_4^-$, HPO_4^{2-}, and PO_4^{3-}. The charge balance is written

$$[H^+] + [K^+] = [OH^-] + [H_2PO_4^-] + 2[HPO_4^{2-}] + 3[PO_4^{3-}] \quad (9\text{-}1)$$

The coefficient of each term in the charge balance equals the magnitude of the charge on each ion.

This statement says that the total charge contributed by H^+ and K^+ equals the magnitude of the charge contributed by all of the anions on the right side of the equation. *The coefficient in front of each species always equals the magnitude of the charge on the ion.* This statement is true because a mole of, say, PO_4^{3-} contributes three moles of negative charge. If $[PO_4^{3-}] = 0.01$ M, the negative charge is $3[PO_4^{3-}] = 3(0.01) = 0.03$ M.

Equation 9-1 appears unbalanced to many people. "The right side of the equation has much more charge than the left side!" you might think. But you would be wrong.

For example, consider a solution prepared by weighing out 0.0250 mol of KH_2PO_4 plus 0.0300 mol of KOH and diluting to 1.00 L. The concentrations of the species at equilibrium are calculated to be

$$[H^+] = 3.9 \times 10^{-12} \text{ M} \qquad [H_2PO_4^-] = 1.4 \times 10^{-6} \text{ M}$$
$$[K^+] = 0.0550 \text{ M} \qquad [HPO_4^{2-}] = 0.0226 \text{ M}$$
$$[OH^-] = 0.0026 \text{ M} \qquad [PO_4^{3-}] = 0.0024 \text{ M}$$

This calculation, which you should be able to do when you have finished studying acids and bases, takes into account the reaction of OH^- with $H_2PO_4^-$ to produce HPO_4^{2-} and PO_4^{3-}.

Are the charges balanced? Yes, indeed. Plugging into Equation 9-1, we find

$$[H^+] + [K^+] = [OH^-] + [H_2PO_4^-] + 2[HPO_4^{2-}] + 3[PO_4^{3-}]$$
$$3.9 \times 10^{-12} + 0.0550 = 0.0026 + 1.4 \times 10^{-6} + 2(0.0226) + 3(0.0024)$$
$$0.0550 = 0.0550$$

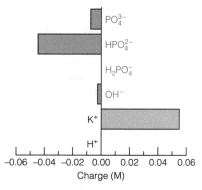

Figure 9-1 Charge contributed by each ion in 1 L of solution containing 0.0250 mol KH_2PO_4 plus 0.0300 mol KOH. The sum of positive charges equals the sum of negative charges.

The total positive charge (to three figures) is 0.0550 M, and the total negative charge is also 0.0550 M (Figure 9-1). The charges must be balanced in every solution. Otherwise your beaker with excess positive charge would glide across the lab bench and smash into another beaker with excess negative charge. (The force between beakers of "charged solutions" could be very large. See Problem 6.)

The general form of the charge balance for any solution is

Σ [positive charges] = Σ [negative charges]. *Activity coefficients do not appear in the charge balance.* The charge contributed by 0.1 M H^+ is *exactly* 0.1 M. Think about this.

Charge balance: $n_1[C_1] + n_2[C_2] + \cdots = m_1[A_1] + m_2[A_2] + \cdots \quad (9\text{-}2)$

where $[C_i]$ = concentration of the ith cation; n_i = charge of the ith cation; $[A_i]$ = concentration of the ith anion; and m_i = magnitude of the charge of the ith anion.

EXAMPLE Writing a Charge Balance

Write the charge balance for a solution containing H_2O, H^+, OH^-, ClO_4^-, $Fe(CN)_6^{3-}$, CN^-, Fe^{3+}, Mg^{2+}, CH_3OH, HCN, NH_3, and NH_4^+.

Solution Neutral species (H_2O, CH_3OH, HCN, and NH_3) do not appear in the charge balance. Therefore, the correct equation is

$$[H^+] + 3[Fe^{3+}] + 2[Mg^{2+}] + [NH_4^+]$$
$$= [OH^-] + [ClO_4^-] + 3[Fe(CN)_6^{3-}] + [CN^-]$$

9-2 Mass Balance

The **mass balance**, also called the *material balance*, is a statement of the conservation of matter. The mass balance states that *the sum of the amounts of all species in a solution containing a particular atom (or group of atoms) must equal the amount of that atom (or group) delivered to the solution.* It is easier to see this relation through particular examples than by a general statement.

The mass balance is a statement of the conservation of matter. It really refers to conservation of atoms, not to mass.

Suppose that a solution is prepared by dissolving 0.050 mol of acetic acid in water to give a total volume of 1.00 L. The acetic acid will partially dissociate into acetate:

$$CH_3CO_2H \rightleftharpoons CH_3CO_2^- + H^+$$
Acetic acid   Acetate

The mass balance states that the sum of the amounts of dissociated and undissociated acetic acid in the solution must equal the amount of acetic acid put into the solution.

Mass balance for acetic acid in water:
$$0.050 \text{ M} = [CH_3CO_2H] + [CH_3CO_2^-]$$
What we put into   Undissociated   Dissociated
the solution   product   product

When a compound dissociates in several ways, the mass balance must include all the groups that result. Phosphoric acid (H_3PO_4), for example, can dissociate into $H_2PO_4^-$, HPO_4^{2-}, and PO_4^{3-}. The mass balance for a solution prepared by dissolving 0.025 0 mol of H_3PO_4 in 1.00 L is

$$0.025\,0 \text{ M} = [H_3PO_4] + [H_2PO_4^-] + [HPO_4^{2-}] + [PO_4^{3-}]$$

Activity coefficients do not appear in the mass balance. The concentration of each species gives an exact count of the number of atoms of that species.

EXAMPLE Mass Balance When the Total Concentration Is Known

Write the mass balances for K^+ and for phosphate in a solution prepared by mixing 0.025 0 mol KH_2PO_4 plus 0.030 0 mol KOH and diluting to 1.00 L.

Solution The total concentration of K^+ is 0.025 0 M + 0.030 0 M, so one trivial mass balance is

$$[K^+] = 0.055\,0 \text{ M}$$

The total concentration of *all forms* of phosphate is 0.025 0 M. The mass balance for phosphate is

$$[H_3PO_4] + [H_2PO_4^-] + [HPO_4^{2-}] + [PO_4^{3-}] = 0.025\,0 \text{ M}$$

Now consider a solution prepared by dissolving $La(IO_3)_3$ in water.

$$La(IO_3)_3(s) \overset{K_{sp}}{\rightleftharpoons} La^{3+} + 3IO_3^-$$
$$\text{Iodate}$$

We do not know how much La^{3+} or IO_3^- is dissolved, but we do know that there must be three iodate ions for each lanthanum ion dissolved. The mass balance is

$$[IO_3^-] = 3[La^{3+}]$$

We have already used this sort of relation in solubility problems. Perhaps the following table will jog your memory:

	$La(IO_3)_3(s) \rightleftharpoons La^{3+} + 3IO_3^-$		
Initial concentration:	solid	0	0
Final concentration:	solid	x	$3x$

When we set $[IO_3^-] = 3x$, we are saying that $[IO_3^-] = 3[La^{3+}]$.

EXAMPLE   Mass Balance When the Total Concentration Is Unknown

Write the mass balance for a saturated solution of the slightly soluble salt Ag_3PO_4, which produces PO_4^{3-} and $3Ag^+$ when it dissolves.

Solution   If the phosphate in solution remained as PO_4^{3-}, we could write

$$[Ag^+] = 3[PO_4^{3-}]$$

because three silver ions are produced for each phosphate ion. However, because phosphate reacts with water to give HPO_4^{2-}, $H_2PO_4^-$, and H_3PO_4, the correct mass balance is

Atoms of Ag = 3 (atoms of P)

$$[Ag^+] = 3\{[PO_4^{3-}] + [HPO_4^{2-}] + [H_2PO_4^-] + [H_3PO_4]\}$$

That is, the number of atoms of Ag^+ must equal three times the total number of atoms of phosphorus, regardless of how many species contain phosphorus atoms.

9-3   Systematic Approach to Equilibrium Problems

Now that we have considered the charge and mass balances, we are ready for the systematic treatment of equilibrium. The general prescription follows these steps:

Step 1.   Write all the *pertinent reactions*.

Step 2.   Write the *charge balance*.

Step 3.   Write the *mass balance*.

Step 4.   Write the *equilibrium constant* for each chemical reaction. This step is the only one in which activity coefficients enter.

Step 5.   *Count the equations and unknowns.* At this point you should have as many equations as unknowns (chemical species). If not, you must either write more equilibria or fix some concentrations at known values.

Step 6.   By hook or by crook, *solve* for all the unknowns.

Steps 1 and 6 are usually the heart of the problem. Knowing (or guessing) what chemical equilibria exist in a given solution requires a fair degree of chemical intuition. In this text you will usually be given some help with Step 1. Unless we know all the relevant equilibria, it is not possible to calculate the composition of a solution correctly. Because we do not know all the chemical reactions, we undoubtedly oversimplify many equilibrium problems.

Step 6 is a mathematical problem, not a chemical problem. With n equations involving n unknowns, the problem can always be solved, at least in principle. In the simplest cases, you can do this by hand, but for most problems a spreadsheet is employed or approximations are made. How the systematic treatment of equilibrium works is best understood by studying some examples.

A Simple Example: The Ionization of Water

The dissociation of water into H^+ and OH^- occurs in every aqueous solution:

$$H_2O \overset{K_w}{\rightleftharpoons} H^+ + OH^- \qquad K_w = 1.0 \times 10^{-14} \quad \text{at } 25°C \qquad (9\text{-}3)$$

Let us apply the systematic treatment of equilibrium to find the concentrations of H^+ and OH^- in pure water.

Step 1.   Pertinent reactions. The only one is Reaction 9-3.

Step 2.   Charge balance. The only ions are H^+ and OH^-, so the charge balance is

$$[H^+] = [OH^-] \qquad (9\text{-}4)$$

Step 3.   Mass balance. In Reaction 9-3 we see that one H^+ ion is generated each time one OH^- ion is made. The mass balance is simply

$$[H^+] = [OH^-]$$

which is the same as the charge balance for this particularly trivial system.

Step 4.   Equilibrium constants. The only one is

$$K_w = [H^+]\gamma_{H^+} [OH^-]\gamma_{OH^-} = 1.0 \times 10^{-14} \qquad (9\text{-}5)$$

This is the only step in which activity coefficients are introduced into the problem.

Step 5.   Count the equations and unknowns. We have two equations, 9-4 and 9-5, and two unknowns, $[H^+]$ and $[OH^-]$.

It requires n equations to solve for n unknowns.

Step 6.   Solve.

Recall that γ approaches 1 as μ approaches 0.

Now we must stop and decide what to do about the activity coefficients. Later, it will be our custom to ignore them unless the calculation requires an accurate result. In the present problem, we anticipate that the ionic strength of pure water will be very low (because the only ions are small amounts of H^+ and OH^-). Therefore, it is reasonable to suppose that γ_{H^+} and γ_{OH^-} are both unity, because $\mu \approx 0$.

Putting the equality $[H^+] = [OH^-]$ into Equation 9-5 gives

$$[H^+]\gamma_{H^+}[OH^-]\gamma_{OH^-} = 1.0 \times 10^{-14}$$

$$[H^+] \cdot 1 \cdot [H^+] \cdot 1 = 1.0 \times 10^{-14}$$

$$[H^+] = 1.0 \times 10^{-7}\ M$$

The ionic strength is 10^{-7} M. The assumption that $\gamma_{H^+} = \gamma_{OH^-} = 1$ is good.

Because $[H^+] = [OH^-]$, $[OH^-] = 1.0 \times 10^{-7}$ M also. We have solved the problem. As a reminder, the pH is given by

$$pH = -\log \mathscr{A}_{H^+} = -\log[H^+]\gamma_{H^+}$$

This is the correct definition of pH. Whenever we neglect activity coefficients, we will write $pH = -\log[H^+]$.

$$= -\log(1.0 \times 10^{-7})(1) = 7.00$$

We Will Usually Omit Activity Coefficients

Although it is proper to write all equilibrium constants in terms of activities, the algebraic complexity of manipulating activity coefficients may obscure the chemistry of a problem. In all later chapters, we omit activity coefficients unless there is a particular point to be made. There will be occasional problems in which activity is used, so that you do not forget about it entirely. However, there is no loss in continuity if the activity problems are skipped.

Another Simple Example: The Solubility of Hg_2Cl_2

Let's apply the systematic treatment of equilibrium to calculate the concentration of Hg_2^{2+} in a saturated aqueous solution of Hg_2Cl_2.

Step 1.   Pertinent reactions. The two reactions that come to mind are

$$Hg_2Cl_2(s) \overset{K_{sp}}{\rightleftharpoons} Hg_2^{2+} + 2Cl^- \tag{9-6}$$

$$H_2O \overset{K_w}{\rightleftharpoons} H^+ + OH^- \tag{9-7}$$

Reaction 9-7 occurs in every aqueous solution.

Step 2.   Charge balance. Equating positive and negative charges gives

Multiply $[Hg_2^{2+}]$ by 2 because 1 mol of this ion has 2 mol of charge.

$$[H^+] + 2[Hg_2^{2+}] = [Cl^-] + [OH^-] \tag{9-8}$$

Step 3.   Mass balance. There are two mass balances in this system. One is the trivial statement that $[H^+] = [OH^-]$, because both arise only from the ionization of water. If there were any other reactions involving H^+ or OH^-, we could not automatically say $[H^+] = [OH^-]$. A slightly more interesting mass balance is

Two Cl^- ions are produced for each Hg_2^{2+} ion.

$$[Cl^-] = 2[Hg_2^{2+}] \tag{9-9}$$

Step 4.   Equilibrium constants. There are two:

$$K_{sp} = [Hg_2^{2+}][Cl^-]^2 = 1.2 \times 10^{-18} \tag{9-10}$$

$$K_w = [H^+][OH^-] = 1.0 \times 10^{-14} \tag{9-11}$$

We have ignored the activity coefficients in these equations.

Step 5. Count the equations and unknowns. There are four equations (9-8 to 9-11) and four unknowns: $[H^+]$, $[OH^-]$, $[Hg_2^{2+}]$, and $[Cl^-]$.

Step 6. Solve. Because we have not written any chemical reactions between the ions produced by H_2O and the ions produced by Hg_2Cl_2, there are really two separate problems. One is the trivial problem of water ionization, which we have already solved:

$$[H^+][OH^-] = 1.0 \times 10^{-14}$$

A slightly more interesting problem is the Hg_2Cl_2 equilibrium. Noting that $[Cl^-] = 2[Hg_2^{2+}]$, we can write

$$[Hg_2^{2+}][Cl^-]^2 = [Hg_2^{2+}](2[Hg_2^{2+}])^2 = K_{sp}$$

$$[Hg_2^{2+}] = (K_{sp}/4)^{1/3} = 6.7 \times 10^{-7} \text{ M}$$

This result is exactly what you would have found by writing a table of concentrations and solving by the methods of Chapter 5.

Satisfy yourself that the calculated concentrations fulfill the charge balance condition.

9-4 The Dependence of Solubility on pH

We are now in a position to study some cases that are not trivial. In this section we present two examples in which there are *coupled equilibria*. That is, the product of one reaction is a reactant in the next reaction.

Solubility of CaF₂

Let's set up the equations needed to find the solubility of CaF_2 in water. There are three pertinent reactions. First, $CaF_2(s)$ dissolves:

$$CaF_2(s) \rightleftharpoons Ca^{2+} + 2F^- \qquad K_{sp} = 3.9 \times 10^{-11} \qquad (9\text{-}12)$$

The fluoride ion can then react with water to give $HF(aq)$:

$$F^- + H_2O \rightleftharpoons HF + OH^- \qquad K_b = 1.5 \times 10^{-11} \qquad (9\text{-}13)$$

The equilibrium constant is called K_b because F^- is functioning as a base when it removes H^+ from H_2O.

Finally, for every aqueous solution, we can write

$$H_2O \overset{K_w}{\rightleftharpoons} H^+ + OH^- \qquad (9\text{-}14)$$

If Reaction 9-13 occurs, then the solubility of CaF_2 is greater than that predicted by the solubility product, because F^- produced in Reaction 9-12 is consumed in Reaction 9-13. According to Le Châtelier's principle, Reaction 9-12 will be driven to the right. Reactions 9-12 and 9-13 are *coupled equilibria*. The systematic treatment of equilibrium allows us to find the net effect of all three reactions.

Step 1. Pertinent reactions. The three are 9-12 through 9-14.

Step 2. Charge balance.

$$[H^+] + 2[Ca^{2+}] = [OH^-] + [F^-] \qquad (9\text{-}15)$$

Step 3. Mass balance. If all fluoride remained in the form F^-, we could write $[F^-] = 2[Ca^{2+}]$ from the stoichiometry of Reaction 9-12. But some F^-

reacts to give HF. The total moles of fluorine atoms is equal to the sum of F^- plus HF, and the mass balance is

$$\underbrace{[F^-] + [HF]}_{\substack{\text{Total concentration} \\ \text{of fluorine}}} = 2[Ca^{2+}] \tag{9-16}$$

Step 4.   Equilibrium constants.

$$K_{sp} = [Ca^{2+}][F^-]^2 = 3.9 \times 10^{-11} \tag{9-17}$$

$$K_b = \frac{[HF][OH^-]}{[F^-]} = 1.5 \times 10^{-11} \tag{9-18}$$

$$K_w = [H^+][OH^-] = 1.0 \times 10^{-14} \tag{9-19}$$

Step 5.   Count the equations and unknowns. There are five equations (9-15 through 9-19) and five unknowns: $[H^+]$, $[OH^-]$, $[Ca^{2+}]$, $[F^-]$, $[HF]$.

If you insist on solving the problem without fixing the pH, see Box 9-1.

The pH could be fixed at a desired value by adding a buffer, which is discussed in Chapter 10.

Step 6.   Solve. This is no simple matter for these five equations. Instead, let us ask a simpler question: What will be the concentrations of $[Ca^{2+}]$, $[F^-]$, and $[HF]$ if the pH is somehow *fixed* at the value 3.00? To fix the pH at 3.00 means that $[H^+] = 1.0 \times 10^{-3}$ M.

Once we know the value of $[H^+]$, there is a straightforward procedure for solving the equations. Setting $[H^+] = 1.0 \times 10^{-3}$ M in Equation 9-19 gives

$$[OH^-] = \frac{K_w}{[H^+]} = 1.0 \times 10^{-11}$$

Putting this value of $[OH^-]$ into Equation 9-18 gives

$$\frac{[HF]}{[F^-]} = \frac{K_b}{[OH^-]} = 1.5$$

$$[HF] = 1.5\,[F^-]$$

Substituting this expression for $[HF]$ into the mass balance (Equation 9-16) gives

$$[F^-] + [HF] = 2[Ca^{2+}]$$

$$[F^-] + 1.5[F^-] = 2[Ca^{2+}]$$

$$[F^-] = 0.80[Ca^{2+}]$$

Finally, we use this value of $[F^-]$ in the solubility product (Equation 9-17):

$$[Ca^{2+}][F^-]^2 = K_{sp}$$

$$[Ca^{2+}](0.80[Ca^{2+}])^2 = K_{sp}$$

$$[Ca^{2+}] = 3.9 \times 10^{-4} \text{ M}$$

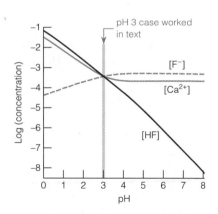

Figure 9-2   pH dependence of the concentration of Ca^{2+}, F^-, and HF in a saturated solution of CaF_2. As the pH is lowered, H^+ reacts with F^- to make HF, and the concentration of Ca^{2+} increases. Note the logarithmic scale.

Challenge   Use the concentration of Ca^{2+} that we just calculated to show that the concentrations of $[F^-]$ and $[HF]$ are 3.1×10^{-4} M and 4.7×10^{-4} M, respectively.

You should realize that the charge balance equation (9-15) is no longer valid if the pH is fixed by external means. To adjust the pH, an ionic compound must necessarily have been added to the solution. Equation 9-15 is therefore incomplete, because it omits those ions. However, we did not use Equation 9-15 to solve the problem because we omitted pH as a variable when we gave it a numerical value.

Fixing the pH invalidates the original charge balance. There exists a new charge balance, but we do not know enough to write an equation expressing this fact.

If we had selected a pH other than 3.00, we would have found a different set of concentrations because of the coupling of Reactions 9-12 and 9-13. Figure 9-2 shows this pH dependence of the concentrations of Ca^{2+}, F^-, and HF. At high pH, there is very little HF, so $[F^-] \approx 2[Ca^{2+}]$. At low pH, there is very little F^-, so $[HF] \approx 2[Ca^{2+}]$. The concentration of Ca^{2+} increases at low pH, because Reaction 9-12 is drawn to the right by the reaction of F^- with H^+ to make HF.

High pH means there is a low concentration of H^+. Low pH means that there is a high concentration of H^+.

In general, salts of basic ions are more soluble at low pH, because the basic ions react with H^+. Basic ions include F^-, OH^-, S^{2-}, CO_3^{2-}, $C_2O_4^{2-}$, and PO_4^{3-}. Figure 9-3 shows that marble, which is largely $CaCO_3$, dissolves more readily as the acid content of rain increases. Much of the acid in rain comes from SO_2 emissions generated by the combustion of fuels containing sulfur, and from nitrogen oxides produced by all types of combustion. SO_2, for example, reacts in the air to make sulfuric acid that finds its way back to the ground in rainfall.

$$SO_2 + H_2O \rightarrow H_2SO_3 \xrightarrow{\text{oxidation}} H_2SO_4$$

Monuments all over the world are degraded by acid rain. Box 9-2 describes another relation between pH and solubility—that between pH and tooth decay.

Solubility of HgS

The mineral cinnabar, consisting of red HgS, is the pigment for the color vermilion. When HgS dissolves in water, some reactions that may occur are

$$HgS(s) \rightleftharpoons Hg^{2+} + S^{2-} \qquad K_{sp} = 5 \times 10^{-54} \qquad (9\text{-}20)$$

$$S^{2-} + H_2O \rightleftharpoons HS^- + OH^- \qquad K_{b1} = 0.80 \qquad (9\text{-}21)$$

$$HS^- + H_2O \rightleftharpoons H_2S(aq) + OH^- \qquad K_{b2} = 1.1 \times 10^{-7} \qquad (9\text{-}22)$$

(We are ignoring the equilibrium $H_2S(aq) \rightleftharpoons H_2S(g)$.)

$$H_2O \rightleftharpoons H^+ + OH^- \qquad K_w = 1.0 \times 10^{-14} \qquad (9\text{-}23)$$

Because S^{2-} is a strong base, it reacts with H_2O to give HS^-, thereby drawing Reaction 9-20 to the right and increasing the solubility of HgS. Now let's set up the equations to find the composition of a solution of HgS.

Step 1. Pertinent reactions. These are 9-20 through 9-23. It is worth restating that if there are any other significant reactions, the calculated composition will be wrong. We are necessarily limited by our knowledge of the chemistry of the system.

Step 2. Charge balance. If the pH is not adjusted by external means, the charge balance is

$$2[Hg^{2+}] + [H^+] = 2[S^{2-}] + [HS^-] + [OH^-] \qquad (9\text{-}24)$$

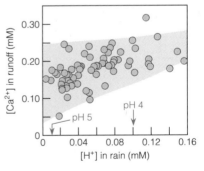

Figure 9-3 Measured calcium concentration in acid rain runoff from marble stone (which is largely $CaCO_3$) roughly increases as the H^+ concentration of the rain increases. [Data from P. A. Baedecker and M. M. Reddy, *J. Chem. Ed.* **1993**, *70*, 104. This article describes a student laboratory experiment on erosion of stone.]

| Box 9-1 | All Right, Dan, How Would You *Really* Solve the CaF$_2$ Problem? |

Suppose that we did not simplify the CaF$_2$ problem by specifying a fixed pH. How can we find the composition of the system if it simply contains CaF$_2$ dissolved in water? I would do it with a spreadsheet, without a doubt!

First, we set up a spreadsheet with the concentrations of H$^+$, OH$^-$, Ca^{2+}, and F$^-$. The strategy is to search for a pH that satisfies the charge balance, Equation 9-15. In contrast to fixing the pH by adding a magic potion to the solution, we are not adding anything except CaF$_2$, so Equation 9-15 is rigorously correct.

	A	B	C	D	E	F	G
1	Ksp=	pH	[H+]	[OH–]	[Ca2+]	[F–]	[H+]+2[Ca2+]
2	3.9E–11						–[OH–]–[F–]
3	Kb=	6	1.0E–06	1.0E–08	2.1E–04	4.3E–04	1.63058E–6
4	1.5E–11	7	1.0E–07	1.0E–07	2.1E–04	4.3E–04	6.40867E–8
5	Kw=	8	1.0E–08	1.0E–06	2.1E–04	4.3E–04	–9.8359E–7
6	1.E–14	7.1	7.9E–08	1.3E–07	2.1E–04	4.3E–04	4.44666E–9
7		7.2	6.3E–08	1.6E–07	2.1E–04	4.3E–04	–5.4957E–8
8		7.15	7.1E–08	1.4E–07	2.1E–04	4.3E–04	–2.5089E–8
9		7.11	7.8E–08	1.3E–07	2.1E–04	4.3E–04	–1.4526E–9
10		7.108	7.8E–08	1.3E–07	2.1E–04	4.3E–04	–2.727E–10
11		7.107	7.8E–08	1.3E–07	2.1E–04	4.3E–04	3.1712E–10
12		7.1075	7.8E–08	1.3E–07	2.1E–04	4.3E–04	2.2174E–11
13							
14	Formulas:		E3=(0.25*$A$2*(1+$A$4/D3)^2)^0.333333333				
15	C3=10^–B3		F3=2*E3/(1+$A$4/D3)				
16	D3=$A$6/C3		G3=C3+2*E3–D3–F3				

Step 3. Mass balance. For each atom of Hg in solution, there is one atom of S. If all of the S remained as S^{2-}, the mass balance would be [Hg^{2+}] = [S^{2-}]. But S^{2-} reacts with water to give HS$^-$ and H$_2$S. The mass balance is therefore

$$[Hg^{2+}] = [S^{2-}] + [HS^-] + [H_2S] \qquad (9\text{-}25)$$

Step 4. Equilibrium constants.

$$K_{sp} = [Hg^{2+}][S^{2-}] = 5 \times 10^{-54} \qquad (9\text{-}26)$$

$$K_{b1} = \frac{[HS^-][OH^-]}{[S^{2-}]} = 0.80 \qquad (9\text{-}27)$$

In the spreadsheet, column B contains *guessed* values of pH. Column C contains $[H^+]$ calculated with the formula $[H^+] = 10^{-pH}$. Column D gives $[OH^-]$ calculated from $[OH^-] = K_w/[H^+]$. To find $[F^-]$ and $[Ca^{2+}]$, we do a little algebra. Rearranging Equation 9-18 gives

$$[HF] = \frac{K_b[F^-]}{[OH^-]} \qquad (a)$$

Substituting expression (a) for [HF] into Equation 9-16 and solving for $[F^-]$ gives

$$[F^-] + [HF] = [F^-] + \frac{K_b[F^-]}{[OH^-]} = 2[Ca^{2+}]$$

$$\Rightarrow [F^-] = \frac{2[Ca^{2+}]}{1 + \dfrac{K_b}{[OH^-]}} \qquad (b)$$

Substituting expression (b) for $[F^-]$ into Equation 9-17 gives an expression for $[Ca^{2+}]$:

$$[Ca^{2+}][F^-]^2 = [Ca^{2+}]\left(\frac{2[Ca^{2+}]}{1 + \dfrac{K_b}{[OH^-]}}\right)^2 = K_{sp}$$

$$\Rightarrow [Ca^{2+}] = \left(\frac{K_{sp}}{4}\left(1 + \frac{K_b}{[OH^-]}\right)^2\right)^{1/3} \qquad (c)$$

In the spreadsheet, column E is calculated with expression (c) and column F is computed with expression (b).

Column G is the net charge in the solution. *We will vary the pH in column B until column G has a value close to zero, at which point we must have found the correct pH.*

How close is "close to zero"? The spreadsheet gives an example. Guessing pH values of 6 and 7 in the spreadsheet gives net positive charges in column G, whereas using a pH of 8 gives a negative charge. Therefore, the net charge passes through zero between pH 7 and pH 8. Successive guesses show that it also passes through zero between pH 7.1 and pH 7.2. We continue to guess more decimal places for pH to bracket the net charge in column G around zero. We could continue to some arbitrary precision, but in practice we usually only care to know pH to two or three decimal places. The succession of guesses in the spreadsheet gives us a pH of 7.108 that is correct to three decimal places.

Is this pH reasonable? (Always ask yourself this question at the end of any problem.) F^- is a pretty weak base ($K_b = 1.5 \times 10^{-11}$), and there will not be very much F^- dissolved (because $K_{sp} = 3.9 \times 10^{-11}$). Therefore, we do expect the pH to be in the neighborhood of 7.

$$K_{b2} = \frac{[H_2S][OH^-]}{[HS^-]} = 1.1 \times 10^{-7} \qquad (9\text{-}28)$$

$$K_w = [H^+][OH^-] = 1.0 \times 10^{-14} \qquad (9\text{-}29)$$

Step 5. Count the equations and unknowns. There are six equations (9-24 through 9-29) and six unknowns: $[Hg^{2+}]$, $[S^{2-}]$, $[HS^-]$, $[H_2S]$, $[H^+]$, and $[OH^-]$.

Step 6. Solve.

In the absence of a spreadsheet, we can simplify matters by assuming that the pH is *fixed* by some external means. This assumption invalidates the

Box 9-2   pH and Tooth Decay

The enamel covering of teeth contains the mineral hydroxyapatite, a calcium hydroxyphosphate. This slightly soluble mineral dissolves in acid, because both PO_4^{3-} and OH^- react with H^+:

$$Ca_{10}(PO_4)_6(OH)_2 + 14H^+ \rightleftharpoons 10Ca^{2+} + 6H_2PO_4^- + 2H_2O$$

Hydroxyapatite

Decay-causing bacteria that adhere to teeth produce lactic acid by metabolizing sugar.

$$\begin{array}{c} OH \\ | \\ CH_3CHCO_2H \end{array} \qquad \text{Lactic acid}$$

Lactic acid lowers the pH at the surface of the tooth to less than 5. At any pH below about 5.5, hydroxyapatite dissolves and tooth decay occurs, as shown in the electron micrographs. Fluoride inhibits tooth decay because it forms fluorapatite, $Ca_{10}(PO_4)_6F_2$, which is less soluble and more acid resistant than hydroxyapatite.

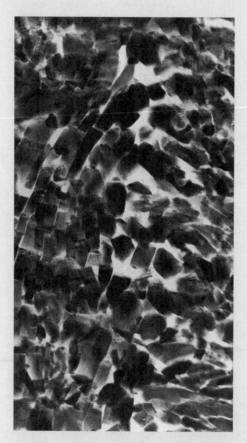 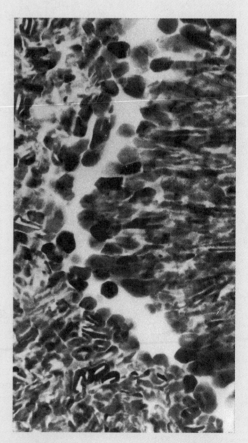

Transmission electron micrographs of *(left)* normal human tooth enamel, showing crystals of hydroxyapatite; *(right)* decayed enamel, showing regions where the mineral has been dissolved by acid. [Courtesy D. B. Scott, J. W. Simmelink, and V. K. Nygaard, Case Western Reserve University, School of Dentistry. Originally published in *J. Dent. Res.* **1974**, *53*, 165.]

charge balance (9-24), but we do not need it because there is one less unknown when the pH is known. Alternatively, with a spreadsheet, we can find the pH of the saturated solution by the method outlined in Box 9-1.

From a pH that is either known or guessed, we can always say that $[OH^-] = K_w/[H^+]$. From Equation 9-27, we can therefore write

$$[HS^-] = \frac{K_{b1}[S^{2-}]}{[OH^-]} \tag{9-30}$$

Substituting Equation 9-30 into Equation 9-28 gives

$$[H_2S] = \frac{K_{b2}[HS^-]}{[OH^-]} = \frac{K_{b1}K_{b2}[S^{2-}]}{[OH^-]^2} \tag{9-31}$$

Now we can use these values of $[H_2S]$ and $[HS^-]$ in the mass balance (Equation 9-25) to write

$$[Hg^{2+}] = [S^{2-}] + [HS^-] + [H_2S]$$

$$[Hg^{2+}] = [S^{2-}] + \frac{K_{b1}}{[OH^-]}[S^{2-}] + \frac{K_{b1}K_{b2}}{[OH^-]^2}[S^{2-}]$$

$$[Hg^{2+}] = [S^{2-}]\left(1 + \frac{K_{b1}}{[OH^-]} + \frac{K_{b1}K_{b2}}{[OH^-]^2}\right) \tag{9-32}$$

Putting this relation between $[Hg^{2+}]$ and $[S^{2-}]$ into the solubility product solves the problem:

$$K_{sp} = [Hg^{2+}][S^{2-}]$$

$$K_{sp} = [Hg^{2+}]\frac{[Hg^{2+}]}{\left(1 + \frac{K_{b1}}{[OH^-]} + \frac{K_{b1}K_{b2}}{[OH^-]^2}\right)}$$

Finally solving for $[Hg^{2+}]$ gives

$$[Hg^{2+}] = \left(K_{sp}\left(1 + \frac{K_{b1}}{[OH^-]} + \frac{K_{b1}K_{b2}}{[OH^-]^2}\right)\right)^{1/2} \tag{9-33}$$

Now let's evaluate this expression for a case that is not too outrageous for a calculator. Suppose that we somehow fix the pH at 8.00 by adding a buffer. A pH of 8.00 means that $[OH^-] = K_w/[H^+] = 1.0 \times 10^{-6}$ M. Putting this value of $[OH^-]$ into Equation 9-33 (which was derived without benefit of the charge balance), we find $[Hg^{2+}] = 2.1 \times 10^{-24}$ M. Using a spreadsheet like the one in Box 9-1, we find that the pH of a saturated solution of HgS in distilled water is 7.00, because only a very tiny amount of HgS dissolves.

Challenge Show that the other concentrations at pH 8.00 are

$$[S^{2-}] = 2.4 \times 10^{-30} \text{ M}$$

$$[HS^-] = 1.9 \times 10^{-24} \text{ M}$$

$$[H_2S] = 2.1 \times 10^{-25} \text{ M}$$

The spreadsheet also allows us to construct a graph showing the effect of an externally imposed pH on the composition of a saturated solution of HgS (Figure 9-4). Below pH $\approx$ 6, H_2S is the predominant form of sulfur in solution, so $[Hg^{2+}] \approx [H_2S]$. Above pH $\approx$ 8, HS^- is the predominant form, and $[Hg^{2+}] \approx [HS^-]$. The solubility of HgS increases as the pH is lowered because S^{2-} reacts with H^+, drawing Reaction 9-20 to the right.

The charge balance is invalid because we are adding new ions to adjust the pH.

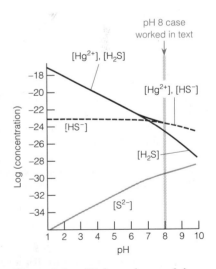

Figure 9-4 pH dependence of the concentrations of species in a saturated solution of HgS. As the pH is lowered, H^+ reacts with S^{2-} to make HS^- and H_2S, and the concentration of Hg^{2+} increases. Note the logarithmic ordinate.

Recapitulation

We have been finding the solubility of salts of the type MA, where M is a metal and A is a basic anion. The anion can react with water to give HA, H_2A, $\cdots$, and so on. A general procedure for finding the composition of the system when the pH is fixed is the following:

1. Set up all the equations, using the systematic treatment of equilibrium.

2. Write expressions for each protonated form of the anion (H_nA) in terms of A.

3. Substitute the values of H_nA from Step 2 into the mass balance to derive a relation between M and A.

4. Put the relation between M and A into the solubility product to solve for the concentrations of M and A and other species.

Armed with a spreadsheet, and using the method in Box 9-1, it is not hard to find the composition of a system whose pH is not arbitrarily adjusted. The spreadsheet also allows us to construct diagrams such as Figure 9-4, in which we can appreciate the behavior of the concentrations over a range of pH.

In all equilibrium problems, we are ultimately limited by how much of the system's chemistry is understood. Unless we know all of the relevant equilibria, it is not possible to correctly calculate the composition of a solution. From ignorance of some of the chemical reactions, we undoubtedly oversimplify many equilibrium problems. An excellent book that will teach you more about equilibrium calculations is *Unified Equilibrium Calculations* by W. B. Guenther (New York: Wiley, 1991).

Terms to Understand

charge balance   mass balance

Summary

In the systematic treatment of equilibrium, we write down all the pertinent equilibrium expressions, as well as the charge and mass balances. The charge balance states that the sum of all positive charges in solution equals the sum of all negative charges. The mass balance states that the sum of the moles of all forms of an element in solution must equal the moles of that element delivered to the solution. We make certain that we have as many equations as unknowns and then set about to solve for the concentration of each species, using algebra, insight, approximations, magic, or anything else. A spreadsheet lets us solve complicated problems with relatively little use of magic. Application of this procedure to salts of basic anions shows that their solubilities increase at low pH because the anion is protonated in acidic solution.

Exercises

A. Write the charge balance for a solution prepared by dissolving CaF_2 in H_2O. Consider that the CaF_2 may give Ca^{2+}, F^-, and CaF^+

B. (a) Write the mass balance for a solution of $CaCl_2$ in water, where the aqueous species are Ca^{2+} and Cl^-.

(b) Write a mass balance where the aqueous species are Ca^{2+}, Cl^-, and $CaCl^+$.

C. (a) Write the mass balance for a saturated solution of CaF_2 in water in which the reactions are

$$CaF_2(s) \rightleftharpoons Ca^{2+} + 2F^-$$

$$F^- + H^+ \rightleftharpoons HF(aq)$$

(b) Write a mass balance for CaF_2 in water where, in addition to the previous reactions, the following reaction occurs:

$$HF(aq) + F^- \rightleftharpoons HF_2^-$$

D. Write a mass balance for an aqueous solution of $Ca_3(PO_4)_2$, where the aqueous species are Ca^{2+}, PO_4^{3-}, HPO_4^{2-}, $H_2PO_4^-$, and H_3PO_4.

E. **(a)** Find the concentrations of Ag^+, CN^-, and HCN in a saturated solution of AgCN whose pH is somehow fixed at 9.00. Consider the following equilibria:

$$AgCN(s) \rightleftharpoons Ag^+ + CN^-$$
$$K_{sp} = 2.2 \times 10^{-16}$$

$$CN^- + H_2O \rightleftharpoons HCN(aq) + OH^-$$
$$K_b = 1.6 \times 10^{-5}$$

(b) *Activity problem.* Use activity coefficients to answer (a). Assume that the ionic strength is fixed at 0.10 M by addition of an inert salt. When using activities, the statement that the pH is 9.00 means that $-\log[H^+]\gamma_{H^+} = 9.00$.

F. Calculate the solubility of ZnC_2O_4 (g/L) in a solution held at pH 3.00. Consider the equilibria

$$ZnC_2O_4(s) \rightleftharpoons Zn^{2+} + \underset{\text{Oxalate}}{C_2O_4^{2-}}$$
$$K_{sp} = 7.5 \times 10^{-9}$$

$$C_2O_4^{2-} + H_2O \rightleftharpoons HC_2O_4^- + OH^-$$
$$K_{b1} = 1.8 \times 10^{-10}$$

$$HC_2O_4^- + H_2O \rightleftharpoons H_2C_2O_4 + OH^-$$
$$K_{b2} = 1.8 \times 10^{-13}$$

Problems

Charge Balance

1. State in words the meaning of the charge balance equation.

2. Write a charge balance for a solution containing H^+, OH^-, Ca^{2+}, HCO_3^-, CO_3^{2-}, $Ca(HCO_3)^+$, $Ca(OH)^+$, K^+, and ClO_4^-.

3. Write a charge balance for a solution of H_2SO_4 in water if the H_2SO_4 ionizes to HSO_4^- and SO_4^{2-}.

4. Write the charge balance for an aqueous solution of arsenic acid, H_3AsO_4, in which the acid can dissociate to $H_2AsO_4^-$, $HAsO_4^{2-}$, and AsO_4^{3-}. Look up the structure of arsenic acid in Appendix G and write the structure of $HAsO_4^{2-}$.

5. **(a)** Suppose that $MgBr_2$ dissolves to give Mg^{2+} and Br^-. Write a charge balance equation for this aqueous solution.

(b) What is the charge balance if, in addition to Mg^{2+} and Br^-, $MgBr^+$ is formed?

6. This problem demonstrates what would happen if charge balance did not exist in a solution. The force (newtons, N) between two charges q_1 and q_2 (coulombs, C) is given by

$$\text{force} = -(8.988 \times 10^9)\frac{q_1 q_2}{r^2}$$

where r is the distance (meters) between the two charges. What is the force (in newtons and pounds) between two beakers separated by 1.5 m if one beaker contains 250 mL of solution with 1.0×10^{-6} M excess negative charge and the other has 250 mL of 1.0×10^{-6} M excess positive charge? Note that there are 9.648×10^4 coulombs per mole of charge and 0.2248 pounds per newton.

Mass Balance

7. State in words the meaning of the mass balance equation.

8. Suppose that $MgBr_2$ dissolves to give Mg^{2+} and Br^-.

(a) Write the mass balance equation for Mg^{2+} for a 0.20 M solution of $MgBr_2$ in water.

(b) Write a mass balance equation for Br^- for a 0.20 M solution of $MgBr_2$ in water.

Now suppose that in addition to Mg^{2+} and Br^-, $MgBr^+$ can be formed.

(c) Write a mass balance equation for Mg^{2+} for a 0.20 M solution of $MgBr_2$ in water.

(d) Write a mass balance equation for Br^- for a 0.20 M solution of $MgBr_2$ in water.

9. For a 0.1 M aqueous solution of sodium acetate, $Na^+CH_3CO_2^-$, one mass balance is simply $[Na^+] = 0.1$ M. Write a mass balance involving acetate.

10. Consider the dissolution of the compound X_2Y_3, which might give the following species: $X_2Y_2^{2+}$, X_2Y^{4+}, $X_2Y_3(aq)$, and Y^{2-}. Use the mass balance for this solution to find an expression for $[Y^{2-}]$ in terms of the other concentrations. Simplify your answer as much as possible.

Systematic Treatment of Equilibrium

11. Why are activity coefficients excluded from the charge and mass balances?

12. Use the systematic treatment of equilibrium to calculate the concentration of Zn^{2+} in a saturated aqueous solution of $Zn_2[Fe(CN)_6]$ which dissociates into Zn^{2+} and $Fe(CN)_6^{4-}$ (ferrocyanide).

13. Calculate the concentration of each ion in a solution of 4.0×10^{-8} M $Mg(OH)_2$, which is completely dissociated to Mg^{2+} and OH^-.

14. **(a)** Calculate the ratio $[Pb^{2+}]/[Sr^{2+}]$ in a solution of distilled water saturated with PbF_2 *and* SrF_2.

(b) Calculate the concentrations of Pb^{2+}, Sr^{2+}, and F^- in the solution above.

15. Consider an aqueous system in which the following equilibria can occur:

$$M^{2+} + X^- \rightleftharpoons MX^+ \qquad K_1$$

$$MX^+ + X^- \rightleftharpoons MX_2(aq) \qquad K_2$$

Derive an equation giving the concentration of M^{2+} when 0.10 mol of MX_2 is dissolved in 1.00 L. Your equation should contain $[M^{2+}]$, K_1, and K_2 as the only variables.

16. *Ion pairing.* In a solution of zinc sulfate, some fraction of the cations and anions are associated by electrostatic attraction as $Zn^{2+}SO_4^{2-}$, which behaves as a single species in the solution. The tightly bound cation and anion are called an *ion pair*. Consider a solution in which the formal concentration of $ZnSO_4$ is 0.010 M. In the absence of ion pairing, $[Zn^{2+}] = [SO_4^{2-}] = 0.010$ M. In fact, ion pairing is significant:

$$Zn^{2+} + SO_4^{2-} \rightleftharpoons Zn^{2+}SO_4^{2-} \qquad \text{(ion pair)}$$

$$K = \frac{[Zn^{2+}SO_4^{2-}]\gamma_{Zn^{2+}SO_4^{2-}}}{[Zn^{2+}]\gamma_{Zn^{2+}}[SO_4^{2-}]\gamma_{SO_4^{2-}}} = 200$$

Because the species $Zn^{2+}SO_4^{2-}$ is neutral, its activity coefficient can be taken as unity. The mass balance for the solution is

$$0.010 \text{ M} = [Zn^{2+}] + [Zn^{2+}SO_4^{2-}]$$

and we also know that $[Zn^{2+}] = [SO_4^{2-}]$, even though we do not know the value of either concentration.

(a) Neglecting activity coefficients, use the ion pair equilibrium to calculate the concentration of Zn^{2+} in the solution.

(b) Use the answer from **(a)** to compute the ionic strength of the solution and the activity coefficients of Zn^{2+} and SO_4^{2-}. Then repeat the calculation of **(a)**, using activity coefficients.

(c) Repeat the procedure in **(b)** two more times to find a good estimate of $[Zn^{2+}]$. What percentage of the salt is ion paired?

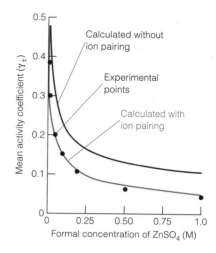

(d) With the ionic strength from **(c)**, calculate the mean activity coefficient, $\gamma_\pm$, of zinc sulfate by using Equation 8-7. In the figure above the upper curve was calculated by neglecting ion pairing. The lower curve was calculated with ion pairing, and the dots are experimental values of $\gamma_\pm$. [Data from S. O. Russo and G. I. H. Hanania, *J. Chem. Ed.* **1989,** *66,* 148, which includes other good examples of ion pairing.]

Dependence of Solubility on pH

17. Why does the solubility of a salt of a basic anion increase with decreasing pH? Considering minerals such as galena (PbS), cerussite ($PbCO_3$), kaolinite ($Al_2(OH)_4Si_2O_5$), and bauxite (AlOOH), explain how acid rain mobilizes trace quantities of potentially toxic metallic elements from relatively inert forms into the environment, where the metals can be taken up by plants and animals.

18. Use the procedure in Section 9-4 to calculate the concentrations of Ca^{2+}, F^-, and HF in a saturated aqueous solution of CaF_2 held at pH 2.00.

19. A certain metal salt of acrylic acid has the formula $M(H_2C=CHCO_2)_2$. Find the concentration of M^{2+} in a saturated aqueous solution of this salt in which $[OH^-]$ is maintained at the value 1.8×10^{-10} M. The equilibria are

$$M(CH_2=CHCO_2)_2(s) \rightleftharpoons M^{2+} + 2H_2C=CHCO_2^-$$
$$K_{sp} = 6.3 \times 10^{-14}$$

$$H_2C=CHCO_2^- + H_2O \rightleftharpoons H_2C=CHCO_2H + OH^-$$
$$\text{Acrylic acid}$$
$$K_b = 1.8 \times 10^{-10}$$

20. Consider a saturated aqueous solution of the slightly soluble $R_3NH^+Br^-$, where R is an organic group.

$R_3NH^+Br^-(s) \rightleftharpoons R_3NH^+ + Br^- \quad K_{sp} = 4.0 \times 10^{-8}$

$R_3NH^+ \rightleftharpoons R_3N + H^+ \qquad K_a = 2.3 \times 10^{-9}$

Calculate the solubility (moles per liter) of $R_3NH^+Br^-$ in a solution maintained at pH 9.50.

21. **(a)** Use the systematic treatment of equilibrium to find how many moles of PbO will dissolve in a 1.00-L solution in which the pH is fixed at 10.50. Consider the equilibrium involving Pb^{2+} to be

$$PbO(s) + H_2O \rightleftharpoons Pb^{2+} + 2OH^-$$
$$K = 5.0 \times 10^{-16}$$

(b) Answer the same question asked in **(a)**, but also consider the reaction

$$Pb^{2+} + H_2O \rightleftharpoons PbOH^+ + H^+$$
$$K_a = 1.3 \times 10^{-18}$$

(c) *Activity problem.* Answer **(a)** by using activity coefficients, assuming the ionic strength is fixed at 0.050 M.

22. Calculate the molarity of Ag^+ in a saturated aqueous solution of Ag_3PO_4 at pH 6.00 if the equilibria are

$$Ag_3PO_4(s) \rightleftharpoons 3Ag^+ + PO_4^{3-}$$
$$K_{sp} = 2.8 \times 10^{-18}$$

$$PO_4^{3-} + H_2O \rightleftharpoons HPO_4^{2-} + OH^-$$
$$K_{b1} = 2.3 \times 10^{-2}$$

$$HPO_4^{2-} + H_2O \rightleftharpoons H_2PO_4^- + OH^-$$
$$K_{b2} = 1.6 \times 10^{-7}$$

$$H_2PO_4^- + H_2O \rightleftharpoons H_3PO_4 + OH^-$$
$$K_{b3} = 1.4 \times 10^{-12}$$

23. When ammonium sulfate dissolves, both the anion and cation have acid-base reactions:

$$(NH_4)_2SO_4(s) \rightleftharpoons 2NH_4^+ + SO_4^{2-}$$
$$K_{sp} = 276$$

$$NH_4^+ \rightleftharpoons NH_3(aq) + H^+$$
$$K_a = 5.70 \times 10^{-10}$$

$$SO_4^{2-} + H_2O \rightleftharpoons HSO_4^- + OH^-$$
$$K_b = 9.80 \times 10^{-13}$$

(a) Write a charge balance for this system.

(b) Write a mass balance for this system.

(c) Find the concentration of $NH_3(aq)$ if the pH is somehow fixed at 9.25.

24. Consider the following simultaneous equilibria:

$$FeG^+ + G^- \rightleftharpoons FeG_2(aq) \qquad K_2 = 3.2 \times 10^3$$

$$G^- + H_2O \rightleftharpoons HG + OH^- \qquad K_b = 6.0 \times 10^{-5}$$

where G is the amino acid glycine, $^+H_3NCH_2CO_2^-$. Suppose that 0.05000 mol of FeG_2 is dissolved in 1.00 L.

(a) Write the charge balance for the solution.

(b) Write two independent mass balances for the solution.

(c) *Using activity coefficients,* find the concentration of FeG^+ if the pH is fixed at 8.50 and the ionic strength is 0.10 M. For FeG^+, use $\gamma = 0.79$; and for G^-, use $\gamma = 0.78$.

25. Use the equations in Box 9-1 to reconstruct Figure 9-2. Your spreadsheet should have columns for pH (varied from 0 to 14 in increments of 0.5), $[H^+]$, $[OH^-]$, $[Ca^{2+}]$, $[F^-]$, and $[HF]$. Note the logarithmic ordinate in Figure 9-2.

26. Use Equations 9-26, 9-30, 9-31, and 9-33 to reconstruct Figure 9-4. Your spreadsheet should have columns for pH (varied from 0 to 14 in increments of 0.5), $[H^+]$, $[OH^-]$, $[Hg^{2+}]$, $[S^{2-}]$, $[HS^-]$, and $[H_2S]$.

27. Use a spreadsheet like the one in Box 9-1 to calculate the pH of a saturated aqueous solution of zinc oxalate, ZnC_2O_4, whose chemistry is given in Exercise F.

28. Construct a graph analogous to Figure 9-4 for the system Ag_3PO_4 in Problem 22. Find the pH of a saturated aqueous solution of Ag_3PO_4.

Measuring pH Inside Single Cells

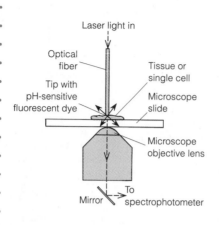

Micrograph shows light emitted from the tip of an optical fiber inserted into a rat embryo. The emission spectrum indicates that the pH of fluid surrounding the fiber is 7.50 ± 0.05 in 10 embryos. Prior to the development of this microscopic method, 1 000 embryos had to be homogenized for a single measurement. [Courtesy R. Kopelman and W. Tan, University of Michigan.]

An exquisitely wrought optical fiber with a pH-sensitive, fluorescent dye bound to its tip can be used to measure the pH inside embryos and even single cells.[1] The fiber is inserted into the specimen and laser light is directed down the fiber. Dye molecules at the tip of the fiber absorb the laser light and then emit fluorescence, the spectrum of which depends on the pH of the surrounding medium. pH is one of the most important factors controlling the rate and thermodynamics of every biological process.

Monoprotic Acid-Base Equilibria

Understanding the behavior of acids and bases is essential to virtually every application of chemistry. It would be difficult to have a meaningful discussion of any process, from protein folding to the weathering of rocks, without understanding acids and bases. If your major field of interest is *not* chemistry, Chapters 10 through 12 are probably the most important ones in this book for you. In this chapter we examine acid-base equilibria in gory detail, and we discuss how buffers operate. We extend our studies in Chapter 11 to diprotic acids and bases (compounds with two acidic protons) and generalize to polyprotic systems. Nearly every biological macromolecule is polyprotic. In Chapter 12, we apply our knowledge of equilibrium to acid-base titrations, which are most useful in analytical chemistry. You should already be familiar with the basic concepts of acids and bases, which were presented in Sections 5-7 through 5-9.

10-1 Strong Acids and Bases

What could be easier than calculating the pH of 0.10 M HBr? Because HBr is a **strong acid,** the reaction

$$HBr + H_2O \rightarrow H_3O^+ + Br^-$$

goes to completion and the concentration of H_3O^+ is 0.10 M. It will be our custom to write H^+ instead of H_3O^+, and we can say

$$pH = -\log[H^+] = -\log(0.10) = 1.00$$

Table 5-2 gave a list of strong acids and bases that you must memorize.

EXAMPLE   **Activity Coefficient in a Strong-Acid Calculation**

Let's repeat the calculation of the pH of 0.10 M HBr properly, using activity coefficients.

Solution   The ionic strength of 0.10 M HBr is 0.10 M, at which the activity coefficient of H^+ is 0.83 (Table 8-1). The pH is given by

$$pH = -\log[H^+]\gamma_{H^+} = -\log(0.10)(0.83) = 1.08$$

The activity correction is not very large, and hereafter we generally neglect activity coefficients in our calculations.

How do we calculate the pH of 0.10 M KOH? KOH is a **strong base** (that is, a base that dissociates completely), so $[OH^-] = 0.10$ M. Using $K_w = [H^+][OH^-]$, we write

> If you know $[OH^-]$, you can always find $[H^+]$, because $[H^+] = K_w/[OH^-]$.

$$[H^+] = \frac{K_w}{[OH^-]} = \frac{1.0 \times 10^{-14}}{0.10} = 1.0 \times 10^{-13}$$

$$pH = -\log[H^+] = 13.00$$

Finding the pH of other concentrations of KOH is pretty trivial:

$[OH^-]$ (M)	$[H^+]$ (M)	pH
$10^{-3.00}$	$10^{-11.00}$	11.00
$10^{-4.00}$	$10^{-10.00}$	10.00
$10^{-5.00}$	$10^{-9.00}$	9.00

A generally useful relation is that

> pH + pOH = 14.00 at 25°C

Relation between pH and pOH:

$$pH + pOH = -\log K_w = 14.00 \text{ at } 25°C \qquad (10\text{-}1)$$

The Dilemma

What is the pH of 1.0×10^{-8} M KOH? Applying our usual reasoning, we calculate

$$[H^+] = K_w/(1.0 \times 10^{-8}) = 1.0 \times 10^{-6} \Rightarrow pH = 6.00$$

But how can the base KOH produce an acidic solution (pH < 7) when dissolved in pure water? It's impossible.

The Cure

Clearly, there is something wrong with our calculation. In particular, we have not considered the contribution of OH^- from the ionization of water. In pure water, $[OH^-] = 1.0 \times 10^{-7}$ M, which is greater than the amount of KOH added to the solution.

To handle this problem, we resort to the systematic treatment of equilibrium, using charge and mass balances, as well as all relevant equilibria.

The species in solution are K^+, OH^-, and H^+. All of the K^+ and some of the OH^- come from KOH. The charge balance is

$$[K^+] + [H^+] = [OH^-] \qquad (10-2)$$

and the mass balance is a rather trivial statement in this case:

$$[K^+] = 1.0 \times 10^{-8} \text{ M}$$

The only equilibrium to consider is

$$[H^+][OH^-] = K_w$$

There are three equations and three unknowns ($[H^+]$, $[OH^-]$, $[K^+]$), so we have enough information to solve the problem.

Because we are seeking the pH, let's set $[H^+] = x$. Using the values $[K^+] = 1.0 \times 10^{-8}$ M and $[H^+] = x$, and substituting into Equation 10-2, we find

$$[OH^-] = [K^+] + [H^+] = 1.0 \times 10^{-8} + x$$

Using this value of $[OH^-]$ in the K_w equilibrium enables us to solve the problem:

$$[H^+][OH^-] = K_w$$

$$(x)(1.0 \times 10^{-8} + x) = 1.0 \times 10^{-14}$$

$$x^2 + (1.0 \times 10^{-8})x - (1.0 \times 10^{-14}) = 0$$

$$x = \frac{-1.0 \times 10^{-8} \pm \sqrt{(1.0 \times 10^{-8})^2 - 4(1)(-1.0 \times 10^{-14})}}{2(1)}$$

$$= 9.6 \times 10^{-8} \quad \text{or} \quad -1.1 \times 10^{-7} \text{ M}$$

Solution of a quadratic equation:
$$ax^2 + bx + c = 0$$
$$x = \frac{-b \pm \sqrt{b^2 - 4ac}}{2a}$$

Retain all digits in your calculator because b^2 is sometimes nearly equal to $4ac$. If you round off before computing $b^2 - 4ac$, your answer may be garbage.

Rejecting the negative solution (because a concentration cannot be negative), we conclude that

$$[H^+] = 9.6 \times 10^{-8} \text{ M}$$

$$pH = -\log[H^+] = 7.02$$

This pH is eminently reasonable, because 10^{-8} M KOH should be very slightly basic.

Figure 10-1 shows the pH calculated for different concentrations of a strong base or a strong acid dissolved in water. We can think of these curves in terms of three regions:

1. When the concentration is "high" ($\geq 10^{-6}$ M), the pH has that value we would calculate by just considering the concentration of added H^+ or OH^-. That is, the pH of $10^{-5.00}$ M KOH *is* 9.00.

2. When the concentration is "low" ($\leq 10^{-8}$ M), the pH is 7.00. We have not added enough acid or base to significantly affect the pH of the water itself.

3. At intermediate concentrations ($\sim 10^{-6} - 10^{-8}$ M), the effects of water ionization and the added acid or base are comparable. Only in this region is it necessary to do a systematic equilibrium calculation.

Region 1 is the only practical case. Unless you were to protect 10^{-7} M KOH from the air, the pH would be overwhelmingly governed by dissolved CO_2, not the KOH. To obtain a pH near 7, you should use a buffer, not a strong acid or base.

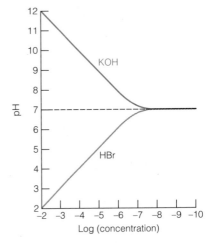

Figure 10-1 Graph showing the calculated pH as a function of the concentration of a strong acid or strong base dissolved in water. Only in the concentration range 10^{-6} to 10^{-8} M would we need to use the systematic treatment of equilibrium.

Any acid or base suppresses water ionization, as predicted by Le Châtelier's principle.

Water Almost Never Produces 10^{-7} M H^+ and 10^{-7} M OH^-

The common misconception that dissociation of water always produces 10^{-7} M H^+ and 10^{-7} M OH^- is true *only* in pure water with no added acid or base. In a 10^{-4} M solution of HBr, for example, the pH is 4. The concentration of OH^- is

$$[OH^-] = K_w/[H^+] = 10^{-10} \text{ M}$$

But the only source of $[OH^-]$ is the dissociation of water. If water produces only 10^{-10} M OH^-, it must also be producing only 10^{-10} M H^+, because it makes one H^+ for every OH^-. In a 10^{-4} M HBr solution, water dissociation produces only 10^{-10} M OH^- and 10^{-10} M H^+.

Question   What concentrations of H^+ and OH^- are produced by H_2O dissociation in 10^{-2} M NaOH?

10-2 Weak Acids and Bases

Let's review the meaning of the **acid dissociation constant K_a** for the acid HA:

Of course, you know that K_a is really $\mathcal{A}_{H^+}\mathcal{A}_{A^-}/\mathcal{A}_{HA}$.

Weak-acid equilibrium:
$$HA \overset{K_a}{\rightleftharpoons} H^+ + A^- \qquad K_a = \frac{[H^+][A^-]}{[HA]} \qquad (10\text{-}3)$$

A **weak acid** is one that is not completely dissociated. That is, Reaction 10-3 does not go to completion. For a base, B, the **base "dissociation" constant** or **base hydrolysis constant**, K_b, is defined by the reaction

Hydrolysis refers to a reaction with water.

Weak-base equilibrium:
$$B + H_2O \overset{K_b}{\rightleftharpoons} BH^+ + OH^- \qquad K_b = \frac{[BH^+][OH^-]}{[B]}$$
$$(10\text{-}4)$$

A **weak base** is one for which Reaction 10-4 does not go to completion.
  pK refers to the negative logarithm of an equilibrium constant:

$$pK_w = -\log K_w = -\log[H^+][OH^-]$$

$$pK_a = -\log K_a = -\log\frac{[A^-][H^+]}{[HA]}$$

$$pK_b = -\log K_b = -\log\frac{[BH^+][OH^-]}{[B]}$$

As K_a increases, pK_a decreases. The smaller pK_a is, the stronger the acid is.

As a K value gets bigger, its p function decreases, and vice versa. Comparing formic and benzoic acids, we see that formic acid is stronger, with a bigger dissociation constant and a smaller pK_a, than benzoic acid.

$$\underset{\text{Formic acid}}{\overset{\displaystyle O}{\overset{\|}{HCOH}}} \quad \rightleftharpoons \quad H^+ + \underset{\text{Formate}}{HCO_2^-} \qquad\qquad \begin{matrix} K_a = 1.80 \times 10^{-4} \\ \mathbf{pK_a = 3.745} \end{matrix}$$

$$\underset{\text{Benzoic acid}}{\bigcirc\!\!-\overset{\displaystyle O}{\overset{\|}{C}OH}} \quad \rightleftharpoons \quad H^+ + \underset{\text{Benzoate}}{\bigcirc\!\!-CO_2^-} \qquad \begin{matrix} K_a = 6.28 \times 10^{-5} \\ \mathbf{pK_a = 4.202} \end{matrix}$$

The acid HA and its corresponding base, A^-, are said to be a **conjugate acid-base pair,** because they are related by the gain or loss of a proton. Similarly, B and BH^+ are a conjugate pair. An important relationship between K_a and K_b for a conjugate acid-base pair is

HA and A^- are a *conjugate acid-base pair.* B and BH^+ are also conjugate.

Relation between K_a and K_b for conjugate pair: $\quad K_a \cdot K_b = K_w \qquad (10\text{-}5)$

Weak Is Conjugate to Weak

The conjugate base of a weak acid is a weak base. The conjugate acid of a weak base is a weak acid. Let's examine these statements. Consider a weak acid, HA, with $K_a = 10^{-4}$. The conjugate base, A^-, has $K_b = K_w/K_a = 10^{-10}$. That is, if HA is a weak acid, A^- is a weak base. If the K_a value were 10^{-5}, then the K_b value would be 10^{-9}. As HA becomes a weaker acid, A^- becomes a stronger base (but never a strong base). Conversely, the greater the acid strength of HA, the less the base strength of A^-. However, if either A^- or HA is weak, so is its conjugate. If HA is strong (such as HCl), its conjugate base (Cl^-) is *so* weak that it is not a base at all in water.

The conjugate base of a weak acid is a weak base. The conjugate acid of a weak base is a weak acid. *Weak is conjugate to weak.*

Using Appendix G

A table of acid dissociation constants appears in Appendix G. Each compound is shown in its *fully protonated form.* Methylamine, for example, is shown as $CH_3NH_3^+$, which is really the methylammonium ion. The value of K_a (2.3×10^{-11}) given for methylamine is actually K_a for the methylammonium ion. To find K_b for methylamine, we write $K_b = K_w/K_a = 1.0 \times 10^{-14}/2.3 \times 10^{-11} = 4.3 \times 10^{-4}$.

For polyprotic acids and bases, several K_a values are given. Pyridoxal phosphate is given in its fully protonated form as follows:[2]

pK_a	K_a
1.4 (POH)	0.04
3.44 (OH)	3.6×10^{-4}
6.01 (POH)	9.8×10^{-7}
8.45 (NH)	3.5×10^{-9}

pK_1 (1.4) is for dissociation of one of the phosphate protons, and pK_2 (3.44) is for the hydroxyl proton. The third most acidic proton is the other phosphate proton, for which $pK_3 = 6.01$, and the NH^+ group is the least acidic ($pK_4 = 8.45$).

10-3 Weak-Acid Equilibria

Let's compare the ionization of *ortho* and *para* hydroxybenzoic acids:

o-Hydroxybenzoic acid
(salicylic acid)

$pK_a = 2.97$

p-Hydroxybenzoic acid

$pK_a = 4.58$

The acetyl $\left(CH_3\overset{\displaystyle O}{\overset{\|}{C}} - \right)$ derivative of *o*-hydroxybenzoic acid is the active ingredient in aspirin.

Acetylsalicylic acid

It is thought that the reason the *o*-hydroxy acid is more than an order of magnitude stronger than the *p*-hydroxy acid is that the conjugate base of the *o*-hydroxy acid is stabilized by strong intramolecular hydrogen bonding.

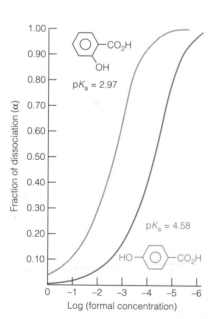

Bonding between the hydroxyl and carboxyl groups is not possible in the *para* isomer because the two functional groups are too far apart. Since the *ortho* isomer is the stronger acid, we expect that its solution will have a lower pH than an equimolar solution of the *para* isomer.

A Typical Weak-Acid Problem

Formal concentration is the total number of moles of a compound dissolved in a liter. The formal concentration of a weak acid refers to the total amount of HA placed in the solution, regardless of the fact that some has changed into A^-.

The problem is to find the pH of a solution of the weak acid HA, given the formal concentration of HA and the value of K_a. Let's call the formal concentration F and use the systematic treatment of equilibrium:

charge balance: $[H^+] = [A^-] + [OH^-]$ (10-6)

mass balance: $F = [A^-] + [HA]$ (10-7)

equilibria: $HA \rightleftharpoons H^+ + A^-$ $K_a = \dfrac{[H^+][A^-]}{[HA]}$ (10-8)

$H_2O \rightleftharpoons H^+ + OH^-$ $K_w = [H^+][OH^-]$

There are four equations and four unknowns ($[A^-]$, $[HA]$, $[H^+]$, $[OH^-]$), so the problem is solved if we can just do the necessary algebra.

But it's not so easy to solve these simultaneous equations. If you combine them, you will discover that a cubic equation results. At this point the chemist steps in and cries, "Wait! There is no reason to solve a cubic equation. We can make an excellent, simplifying approximation. (Besides, I have trouble solving cubic equations.)"

In a solution of any respectable weak acid, the concentration of H^+ due to acid dissociation will be much greater than the concentration due to water dissociation. When HA dissociates, it produces A^-. When H_2O dissociates, it produces OH^-. If the acid dissociation is much greater than the water dissociation, we can say $[A^-] \gg [OH^-]$, and Equation 10-6 reduces to

$$[H^+] \approx [A^-]   (10-9)$$

Using Equations 10-7, 10-8, and 10-9, we can set up a solution. First set $[H^+] = x$. Equation 10-9 says that $[A^-] = x$, also. Equation 10-7 says that $[HA] = F - [A^-] = F - x$. Putting these values into Equation 10-8 gives

$$K_a = \frac{[H^+][A^-]}{[HA]} = \frac{(x)(x)}{F - x}$$

Setting $F = 0.0500$ M and $K_a = 1.07 \times 10^{-3}$ for *o*-hydroxybenzoic acid, the equation is readily solved, because it is just a quadratic equation.

$$\frac{x^2}{F - x} = 1.07 \times 10^{-3}$$

Figure 10-2 The fraction of dissociation of a weak electrolyte increases as the electrolyte is diluted. The stronger acid is more dissociated than the weaker acid at all concentrations.

$$x^2 + (1.07 \times 10^{-3})x - 5.35 \times 10^{-5} = 0$$

$$x = 6.80 \times 10^{-3} \quad \text{(negative root rejected)}$$

$$[H^+] = [A^-] = x = 6.80 \times 10^{-3} \text{ M}$$

$$[HA] = F - x = 0.043\,2 \text{ M}$$

$$pH = -\log x = 2.17$$

Was the approximation $[H^+] \approx [A^-]$ justified? The calculated pH is 2.17, which means that $[OH^-] = K_w/[H^+] = 1.47 \times 10^{-12}$ M.

For uniformity, we will usually express pH values to the 0.01 decimal place, regardless of what is justified by significant figures.

$[A^-]$ (from HA dissociation) $= 6.80 \times 10^{-3}$ M
    implies that $[H^+]$ from HA dissociation $= 6.80 \times 10^{-3}$ M

$[OH^-]$ (from H_2O dissociation) $= 1.47 \times 10^{-12}$ M
    implies that $[H^+]$ from H_2O dissociation $= 1.47 \times 10^{-12}$ M

The assumption that H^+ is derived mainly from HA is excellent.

In a solution of a weak acid, H^+ is derived almost entirely from the weak acid, not from H_2O dissociation.

Fraction of Dissociation

The **fraction of dissociation, α,** is defined as the fraction of the acid in the form A^-:

Fraction of dissociation:
$$\alpha = \frac{[A^-]}{[A^-] + [HA]} = \frac{x}{x + (F - x)} = \frac{x}{F} \qquad (10\text{-}10)$$

α is the fraction of HA that has dissociated: $\alpha = \dfrac{[A^-]}{[A^-] + [HA]}$

For 0.050 0 M *o*-hydroxybenzoic acid, we find

$$\alpha = \frac{6.80 \times 10^{-3} \text{ M}}{0.050\,0 \text{ M}} = 0.136$$

That is, the acid is 13.6% dissociated at a formal concentration of 0.050 0 M.

The variation of α with formal concentration is shown in Figure 10-2. All **weak electrolytes** (compounds that are only partially dissociated) dissociate more as they are diluted. We see in Figure 10-2 that *o*-hydroxybenzoic acid is more dissociated than *p*-hydroxybenzoic acid at the same formal concentration. This is reasonable, because the *ortho* isomer is a stronger acid than the *para* isomer. Demonstration 10-1 illustrates electrical conductivity of weak electrolytes.

The Essence of a Weak-Acid Problem

When faced with the problem of finding the pH of a weak acid, you should immediately realize that $[H^+] = [A^-] = x$ and proceed to set up and solve the equation

Equation for weak acids:
$$\frac{[H^+][A^-]}{[HA]} = \frac{x^2}{F - x} = K_a \qquad (10\text{-}11)$$

The way to do it.

where F is the formal concentration of HA. The approximation $[H^+] = [A^-]$ would be poor only if the acid were too dilute or too weak, neither of which constitutes a practical problem.

Demonstration 10-1   Conductivity of Weak Electrolytes

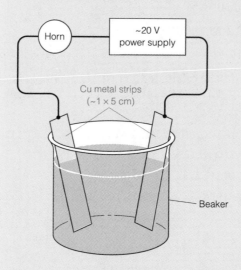

Apparatus for demonstrating conductivity of electrolyte solutions.

The relative conductivity of strong and weak acids is directly related to their different degrees of dissociation in aqueous solution. To demonstrate conductivity, we use a Radio Shack piezo alerting buzzer, but any kind of buzzer or light bulb could easily be substituted for the electric horn. The voltage required will depend on the buzzer or light chosen.

When a conducting solution is placed in the beaker, the horn sounds. First show that distilled water and sucrose solution are nonconductive. Solutions of the strong electrolytes NaCl or HCl are conductive. Compare strong and weak electrolytes by demonstrating that 1 mM HCl gives a loud sound, whereas 1 mM acetic acid gives little or no sound. With 10 mM acetic acid, the strength of the sound varies noticeably as the electrodes are moved away from each other in the beaker.[3]

EXAMPLE   A Weak-Acid Problem

Find the pH of 0.100 M trimethylammonium chloride.

$$\left[\begin{array}{c} H \\ | \\ N \\ H_3C \diagup \diagdown CH_3 \\ H_3C \end{array} \right]^+ \qquad Cl^-$$ Trimethylammonium chloride

Solution   We must first realize that salts of this type are *completely dissociated* to give $(CH_3)_3NH^+$ and Cl^-. We then recognize that the trimethylammonium ion is a weak acid, being the conjugate acid of trimethylamine, $(CH_3)_3N$, a typical weak organic base. Cl^- has no basic or acidic properties and should be ignored. Looking in Appendix G, we find the trimethylammonium ion listed under the name trimethylamine, but drawn as the trimethylammonium ion. The value of pK_a is 9.800, so

$$K_a = 10^{-pK_a} = 1.58 \times 10^{-10}$$

From here, everything is downhill.

$$(CH_3)_3NH^+ \underset{}{\overset{K_a}{\rightleftharpoons}} (CH_3)_3N + H^+$$
$$ F - x x x$$

$$\frac{x^2}{0.100 - x} = 1.58 \times 10^{-10}$$

$$x = 3.97 \times 10^{-6} \text{ M} \Rightarrow \text{pH} = 5.40$$

Do It with a Spreadsheet

The systematic treatment of a weak-acid problem is based on a charge balance (Equation 10-6), a mass balance (Equation 10-7), and two equilibria (Equation 10-8 plus the K_w equation). The eminently reasonable approximation $[H^+] \approx [A^-]$ reduces the problem to a quadratic equation. However, if the acid dissociation constant, K_a, or the formal concentration of weak acid, F, is sufficiently small, this approximation breaks down. With a spreadsheet, we can treat all four equations without approximations.

We begin with the K_a equation:

$$K_a = \frac{[H^+][A^-]}{[HA]}$$

We substitute $[A^-] = [H^+] - [OH^-]$ from the charge balance, and $[HA] = F - [A^-] = F - [H^+] + [OH^-]$ from the mass and charge balances:

$$K_a = \frac{[H^+]([H^+] - [OH^-])}{F - [H^+] + [OH^-]}$$

Charge balance:
$$[H^+] = [OH^-] + [A^-]$$

Mass balance:
$$F = [HA] + [A^-]$$

A little algebra now takes us to a usable equation for a spreadsheet:

$$K_a (F - [H^+] + [OH^-]) = [H^+] ([H^+] - [OH^-])$$

$$K_aF - K_a[H^+] + K_a\underbrace{[OH^-]}_{K_w/[H^+]} - [H^+]^2 + \underbrace{[H^+][OH^-]}_{K_w} = 0$$

$$\underbrace{1 \cdot [H^+]^2}_{a} + \underbrace{K_a[H^+]}_{b} \underbrace{- (K_w + K_aF + K_aK_w/[H^+])}_{c} = 0$$

If we multiply all terms by $[H^+]$ to remove $[H^+]$ from the denominator of the last term, we would have a cubic equation that is painful to solve. However, it is mathematically legitimate to consider the entire term in parentheses as one entity in the quadratic equation

$$a[H^+]^2 + b[H^+] + c = 0$$

where $a = 1$, $b = K_a$, and $c = -(K_w + K_aF + K_aK_w/[H^+])$. The solution to this one is easy:

Weak acid: $$[H^+] = \frac{-K_a + \sqrt{K_a^2 + 4(K_w + K_aF + K_aK_w/[H^+])}}{2} \quad (10\text{-}12)$$

Unabridged weak-acid equation for a spreadsheet.

	A	B	C	D	E
1	Kw=	Ka	H+	pH	A−
2	1.E−14	1.00E−02	9.991E−06	5.000	9.990E−06
3	F=	1.00E−04	9.162E−06	5.038	9.161E−06
4	0.00001	1.00E−06	2.704E−06	5.568	2.700E−06
5		1.00E−08	3.272E−07	6.485	2.966E−07
6		1.00E−10	1.049E−07	6.979	9.526E−09
7					
8	C2 = 0.5*(−B2+Sqrt(B2*B2+4*($A$2+B2*$A$4+B2*$A$2/C2)))				
9	D2 = −Log10(C2)				
10	E2 = C2−$A$2/C2				

Figure 10-3 The spreadsheet for a weak-acid problem does not rely on the approximation $[A^-] \approx [H^+]$. For K_a values of 10^{-8} and 10^{-10}, $[H^+] \neq [A^-]$.

We rejected the negative root of the quadratic equation (with a minus sign in front of the radical) because $[H^+]$ must be positive.

Now we use the nifty capabilities of a spreadsheet to solve Equation 10-12. In Figure 10-3 we placed the value of K_w in cell A2 and the arbitrarily selected formal concentration of weak acid, $F = 10^{-5}$ M, in cell A4. A range of values of K_a is entered in column B. Equation 10-12 is written in cell C2 to calculate $[H^+]$, and the pH $(= -\log[H^+])$ is found in column D. The concentration $[A^-]$ in column E is found from the charge balance, $[A^-] = [H^+] - [OH^-]$.

The key feature of this spreadsheet is the definition of $[H^+]$ in terms of itself in cell C2. That is, $[H^+]$ appears on both sides of Equation 10-12. (This is called a *circular definition.*) Most spreadsheets will accept this seemingly ridiculous notion and merrily solve the equation by successive approximations. Your computer may require prompting to carry out the iterative solution.

Figure 10-4 shows results for different concentrations of weak acids. At the lowest concentrations, the stronger acids (pK_a = 2 to 4) are fully

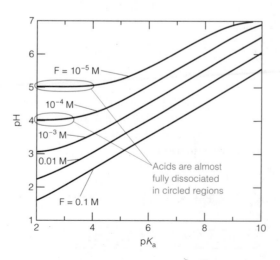

Figure 10-4 pH of weak acids calculated as a function of pK_a and formal concentration, using the spreadsheet approach in Figure 10-3.

dissociated, so 10^{-5} F HA gives a pH of 5. At higher concentrations, full dissociation does not occur, and the pH of, say, 10^{-3} F HA is not 3. As the acid becomes more dilute or weaker, the pH approaches 7. For very dilute solutions of very weak acids, such as 10^{-5} F HA with $pK_a = 8$, the approximation $[H^+] \approx [A^-]$ has broken down. The spreadsheet uses the full equations, so you never have to worry about whether an approximation is true.

10-4 Weak-Base Equilibria

The treatment of weak bases is almost the same as that of weak acids.

$$B + H_2O \overset{K_b}{\rightleftharpoons} BH^+ + OH^- \qquad K_b = \frac{[BH^+][OH^-]}{[B]}$$

We suppose that nearly all of the OH^- comes from the reaction of $B + H_2O$, and little comes from dissociation of H_2O. Setting $[OH^-] = x$, we must also set $[BH^+] = x$, because one BH^+ is produced for each OH^-. If the formal concentration of base $(= [B] + [BH^+])$ is called F, we can write

$$[B] = F - [BH^+] = F - x$$

Plugging these values into the K_b equilibrium expression, we get

Equation for weak base:
$$\frac{[BH^+][OH^-]}{[B]} = \frac{x^2}{F - x} = K_b \qquad (10\text{-}13)$$

which looks a lot like a weak-acid problem, except that now $x = [OH^-]$.

A weak-base problem has the same algebra as a weak-acid problem, except $K = K_b$ and $x = [OH^-]$. See Problem 24.

A Typical Weak-Base Problem

Let's work one problem using the weak base cocaine as an example.

Cocaine

If the formal concentration is 0.037 2 M, the problem is formulated as follows:

$$B + H_2O \rightleftharpoons BH^+ + OH^-$$
$$\underset{0.037\,2\,-\,x}{} \qquad \underset{x}{} \qquad \underset{x}{}$$

$$\frac{x^2}{0.037\,2 - x} = 2.6 \times 10^{-6} \Rightarrow x = 3.1_0 \times 10^{-4}$$

Question Considering the final answer, was it justified to neglect water dissociation as a source of OH^-? What concentration of OH^- is produced by H_2O dissociation in this solution?

Because $x = [OH^-]$, we can write

$$[H^+] = K_w/[OH^-] = 1.0 \times 10^{-14}/3.1_0 \times 10^{-4} = 3.2_2 \times 10^{-11}$$

$$pH = -\log[H^+] = 10.49$$

This is a reasonable pH for a weak base.

What fraction of cocaine has reacted with water in this solution? We can formulate α for a base, called the **fraction of association,** as the fraction that has reacted with water:

For a base, α is the fraction that has reacted with water:

$$\alpha = \frac{[BH^+]}{[B] + [BH^+]}$$

Fraction of association:

$$\alpha = \frac{[BH^+]}{[BH^+] + [B]} = \frac{x}{F} = 0.008\,3 \qquad (10\text{-}14)$$

Only 0.83% of the base has reacted.

Conjugate Acids and Bases—Revisited

HA and A^- are a conjugate acid-base pair. So are BH^+ and B.

Earlier we noted that **the conjugate base of a weak acid is a weak base,** and **the conjugate acid of a weak base is a weak acid.** We also derived an exceedingly important relation between the equilibrium constants for a conjugate acid-base pair:

$$K_a \cdot K_b = K_w$$

In aqueous solution,

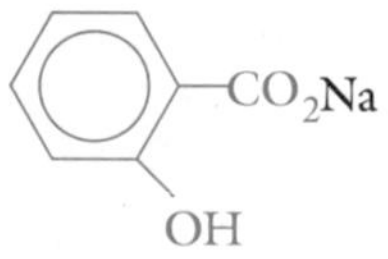

gives

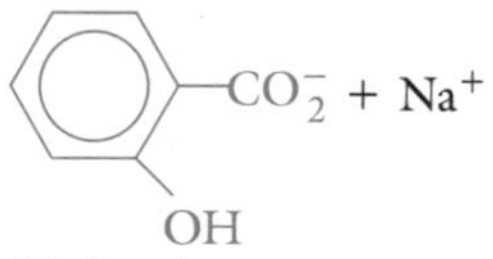

o-Hydroxybenzoate

In Section 10-3 we considered *o*- and *p*-hydroxybenzoic acids, designated HA. Now consider their conjugate bases. For example, the salt sodium *o*-hydroxybenzoate will dissolve to give the Na^+ cation (which has no acid-base chemistry) and the *o*-hydroxybenzoate anion, which is a weak base.

The acid-base chemistry is the reaction of *o*-hydroxybenzoate with water:

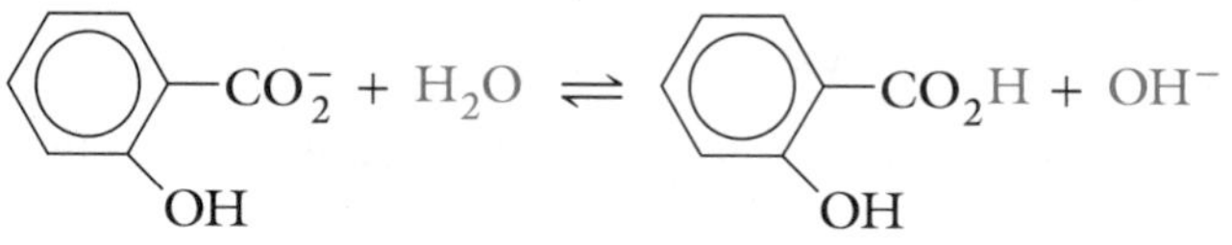

$$\frac{x^2}{F - x} = K_b$$

From the value of K_a for each isomer, we can calculate K_b for the conjugate base.

Isomer of hydroxybenzoic acid	K_a	K_b
ortho	1.07×10^{-3}	9.35×10^{-12}
para	2.63×10^{-5}	3.80×10^{-10}

Using each value of K_b, and letting F = 0.050 0 M, we find

pH of 0.050 0 M *o*-hydroxybenzoate = 7.83

pH of 0.050 0 M *p*-hydroxybenzoate = 8.64

These are reasonable pH values for solutions of weak bases. Furthermore, as expected, the conjugate base of the stronger acid is the weaker base.

EXAMPLE A Weak-Base Problem

Find the pH of 0.10 M ammonia.

Solution When ammonia is dissolved in water, its reaction is

$$NH_3 + H_2O \overset{K_b}{\rightleftharpoons} \underset{\substack{\text{Ammonium} \\ \text{ion}}}{NH_4^+} + OH^-$$

$$\underset{\substack{\text{Ammonia} \\ F-x}}{} \qquad \underset{x}{} \quad \underset{x}{}$$

In Appendix G we find the ammonium ion, NH_4^+, listed next to ammonia. pK_a for ammonium ion is 9.244. Therefore, K_b for NH_3 is

$$K_b = \frac{K_w}{K_a} = \frac{10^{-14.00}}{10^{-9.244}} = 1.75 \times 10^{-5}$$

To find the pH of 0.10 M NH_3, we set up and solve the equation

$$\frac{[NH_4^+][OH^-]}{[NH_3]} = \frac{x^2}{0.10 - x} = K_b = 1.75 \times 10^{-5}$$

$$x = [OH^-] = 1.3_1 \times 10^{-3} \text{ M}$$

$$[H^+] = \frac{K_w}{[OH^-]} = 7.6_1 \times 10^{-12} \text{ M}$$

$$pH = -\log[H^+] = 11.12$$

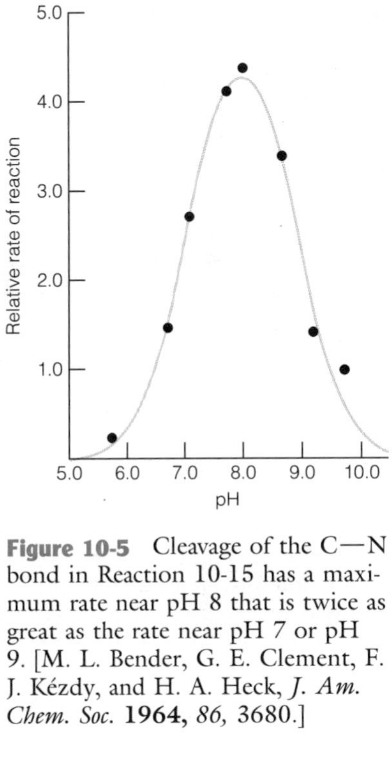

Figure 10-5 Cleavage of the C—N bond in Reaction 10-15 has a maximum rate near pH 8 that is twice as great as the rate near pH 7 or pH 9. [M. L. Bender, G. E. Clement, F. J. Kézdy, and H. A. Heck, *J. Am. Chem. Soc.* **1964,** *86,* 3680.]

10-5 Buffers

A *buffered solution resists changes in pH when acids or bases are added or when dilution occurs.* The **buffer** consists of a mixture of an acid and its conjugate base. The importance of buffers in all areas of science is immense. Biochemists are particularly concerned with buffers because the proper functioning of any biological system is critically dependent upon pH. Figure 10-5 shows how the rate of the enzyme-catalyzed Reaction 10-15 varies with pH.

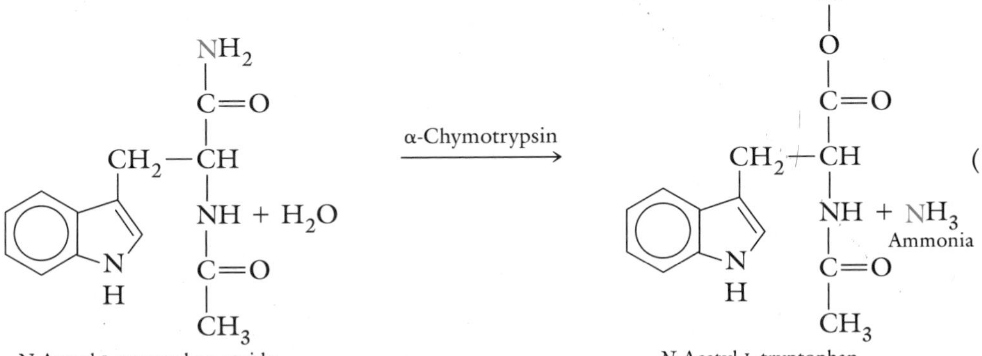

N-Acetyl-L-tryptophan amide → *N*-Acetyl-L-tryptophan (10-15)

In the absence of the enzyme α-chymotrypsin, the rate of this reaction is negligible. For any organism to survive, it must control the pH of each subcellular compartment so that each of its enzyme-catalyzed reactions can proceed at the proper rate.

When you mix a weak acid with its conjugate base, you get what you mix!

Mixing a Weak Acid and Its Conjugate Base

The purpose of this section is to show that *if you mix A moles of a weak acid with B moles of its conjugate base, the moles of acid remain close to A and the moles of base remain close to B.* Very little reaction occurs to change either concentration.

To understand why this should be so, look at the K_a and K_b reactions in terms of Le Châtelier's principle. Consider an acid with $pK_a = 4.00$ and its conjugate base with $pK_b = 10.00$. We will calculate the fraction of acid that dissociates in a 0.100 M solution of HA.

$$\underset{0.100-x \quad x \qquad x}{HA \rightleftharpoons H^+ + A^-} \qquad pK_a = 4.00$$

$$\frac{x^2}{F-x} = K_a \Rightarrow x = 3.11 \times 10^{-3}$$

$$\text{fraction of dissociation} = \alpha = \frac{x}{F} = 0.031\,1$$

The acid is only 3.11% dissociated under these conditions. In a solution containing 0.100 mol of A^- dissolved in 1.00 L, the extent of reaction of A^- with water is even smaller.

$$\underset{0.100-x \qquad\quad x \quad\; x}{A^- + H_2O \rightleftharpoons HA + OH^-} \qquad pK_b = 10.00$$

$$\frac{x^2}{F-x} = K_b \Rightarrow x = 3.16 \times 10^{-6}$$

$$\text{fraction of association} = \alpha = \frac{x}{F} = 3.16 \times 10^{-5}$$

HA *dissociates very little, and adding extra* A^- *to the solution will make the HA dissociate even less. Similarly,* A^- *does not react very much with water, and adding extra HA makes* A^- *react even less.* If 0.050 mol of A^- plus 0.036 mol of HA are added to water, there will be close to 0.050 mol of A^- and close to 0.036 mol of HA in the solution at equilibrium.

This approximation breaks down for dilute solutions or at extremes of pH. We will test the validity of the approximation at the end of this chapter.

Henderson-Hasselbalch Equation

The central equation dealing with buffers is the **Henderson-Hasselbalch equation,** which is merely a rearranged form of the K_a equilibrium expression.

$\log xy = \log x + \log y$

$$K_a = \frac{[H^+][A^-]}{[HA]}$$

$$\log K_a = \log\frac{[H^+][A^-]}{[HA]} = \log[H^+] + \log\frac{[A^-]}{[HA]}$$

$$\underbrace{-\log[H^+]}_{pH} = \underbrace{-\log K_a}_{pK_a} + \log\frac{[A^-]}{[HA]}$$

Henderson-Hasselbalch equation: $\qquad \boxed{pH = pK_a + \log\frac{[A^-]}{[HA]}} \qquad (10\text{-}16)$

The Henderson-Hasselbalch equation tells us the pH of a solution, provided we know the ratio of the concentrations of conjugate acid and base, as well as pK_a for the acid. If a solution is prepared from the weak base B and its conjugate acid, the analogous equation is

$$pH = pK_a + \log\frac{[B]}{[BH^+]} \quad\quad\text{pK_a applies to \textit{this} acid} \quad\quad (10\text{-}17)$$

where pK_a is the acid dissociation constant of the weak acid BH^+. The important features of Equations 10-16 and 10-17 are that the base (A^- or B) appears in the numerator of both equations, and the equilibrium constant is K_a of the acid in the denominator.

Challenge Show that if activities are not neglected, the correct form of the Henderson-Hasselbalch equation is

$$pH = pK_a + \log\frac{[A^-]\gamma_{A^-}}{[HA]\gamma_{HA}} \quad\quad (10\text{-}18)$$

Properties of the Henderson-Hasselbalch Equation

In Equation 10-16 you can see that if $[A^-] = [HA]$, then $pH = pK_a$.

$$pH = pK_a + \log\frac{[A^-]}{[HA]} = pK_a + \log 1 = pK_a$$

Regardless of how complex a solution may be, whenever $pH = pK_a$, $[A^-]$ must equal $[HA]$. This relation is true because *all equilibria must be satisfied simultaneously in any solution at equilibrium.* If there are 10 different acids and bases in the solution, the 10 forms of Equation 10-16 must all give the same pH, because **there can be only one concentration of H^+ in a solution.**

Another feature of the Henderson-Hasselbalch equation is that for every power-of-10 change in the ratio $[A^-]/[HA]$, the pH changes by one unit (Table 10-1). As the base (A^-) increases, the pH goes up. As the acid (HA) increases, the pH goes down. For any conjugate acid-base pair, you can say, for example, that if $pH = pK_a - 1$, there must be 10 times as much HA as A^-. Therefore ten-elevenths is in the form HA and one-eleventh is in the form A^-.

The names of Henderson and Hasselbalch appear to be associated with Equation 10-16 because they recognized that the concentrations $[A^-]$ and $[HA]$ can be set equal to their formal concentrations and were among the first people to apply the equation to practical problems.

When $[A^-] = [HA]$, $pH = pK_a$.

TABLE 10-1 Change of pH with change of $[A^-]/[HA]$

$[A^-]/[HA]$	pH
100:1	$pK_a + 2$
10:1	$pK_a + 1$
1:1	pK_a
1:10	$pK_a - 1$
1:100	$pK_a - 2$

EXAMPLE Using the Henderson-Hasselbalch Equation

Sodium hypochlorite (NaOCl, the active ingredient of almost all bleaches) was dissolved in a solution buffered to pH 6.20. Find the ratio $[OCl^-]/[HOCl]$ in this solution.

Solution In Appendix G we find that $pK_a = 7.53$ for hypochlorous acid, HOCl. The pH is known, so the ratio $[OCl^-]/[HOCl]$ can be calculated from the Henderson-Hasselbalch equation.

$$HOCl \rightleftharpoons H^+ + OCl^-$$

$$pH = pK_a + \log\frac{[OCl^-]}{[HOCl]}$$

$$6.20 = 7.53 + \log \frac{[\text{OCl}^-]}{[\text{HOCl}]}$$

$$-1.33 = \log \frac{[\text{OCl}^-]}{[\text{HOCl}]}$$

$$10^{-1.33} = 10^{\log ([\text{OCl}^-]/[\text{HOCl}])} = \frac{[\text{OCl}^-]}{[\text{HOCl}]}$$

$$0.047 = \frac{[\text{OCl}^-]}{[\text{HOCl}]}$$

Finding the ratio $[\text{OCl}^-]/[\text{HOCl}]$ requires knowing only the pH and the pK_a. We do not need to know what else is in the solution, nor how much NaOCl was added, nor the volume of the solution.

A Buffer in Action

For illustration, we choose a widely used buffer called "tris," which is short for tris(hydroxymethyl)aminomethane.

In Appendix G we find pK_a for the conjugate acid of tris to be 8.075. An example of a salt containing the BH^+ cation is tris hydrochloride, which is BH^+Cl^-. When BH^+Cl^- is dissolved in water, it dissociates completely to BH^+ and Cl^-.

EXAMPLE   A Buffer Solution

Find the pH of a solution prepared by dissolving 12.43 g of tris (FW 121.136) plus 4.67 g of tris hydrochloride (FW 157.597) in 1.00 L of water.

Solution   The concentrations of B and BH^+ added to the solution are

$$[\text{B}] = \frac{12.43 \text{ g/L}}{121.136 \text{ g/mol}} = 0.1026 \text{ M}$$

$$[\text{BH}^+] = \frac{4.67 \text{ g/L}}{157.597 \text{ g/mol}} = 0.0296 \text{ M}$$

Assuming that what we mixed stays in the same form, we can simply plug these concentrations into the Henderson-Hasselbalch equation to find the pH:

$$\text{pH} = pK_a + \log \frac{[\text{B}]}{[\text{BH}^+]} = 8.075 + \log \frac{0.1026}{0.0296} = 8.61$$

Notice that *the volume of solution is irrelevant to finding the pH,* because volume cancels in the numerator and denominator of the log term:

$$pH = pK_a + \log \frac{\text{moles of B}/\cancel{\text{L of solution}}}{\text{moles of BH}^+/\cancel{\text{L of solution}}}$$

$$= pK_a + \log \frac{\text{moles of B}}{\text{moles of BH}^+}$$

The pH of a buffer is nearly independent of volume.

EXAMPLE Effect of Adding Acid to a Buffer

If we add 12.0 mL of 1.00 M HCl to the solution used in the previous example, what will be the new pH?

Solution The key to this problem is to realize that *when a strong acid is added to a weak base, both react completely to give* BH^+ (see Box 10-1). In the present example we are adding 12.0 mL of 1.00 M HCl, which contains (0.0120 L) (1.00 mol/L) = 0.0120 mol of H^+. This much H^+ will consume 0.0120 mol of B to create 0.0120 mol of BH^+. This is shown conveniently in a little table:

	B Tris	+	H^+ from HCl	→	BH^+
Initial moles:	0.1026		0.0120		0.0296
Final moles:	0.0906		—		0.0416
	$(0.1026 - 0.0120)$				$(0.0296 + 0.0120)$

The table contains enough information for us to calculate the pH.

$$pH = pK_a + \log \frac{\text{moles of B}}{\text{moles of BH}^+}$$

$$= 8.075 + \log \frac{0.0906}{0.0416} = 8.41$$

The volume of the solution is irrelevant.

Question Does the pH change in the right direction when HCl is added?

The preceding example illustrates that *the pH of a buffer does not change very much when a limited amount of a strong acid or base is added.* Addition of 12.0 mL of 1.00 M HCl changed the pH from 8.61 to 8.41. Addition of 12.0 mL of 1.00 M HCl to 1.00 L of unbuffered solution would have lowered the pH to 1.93.

But *why* does a buffer resist changes in pH? It does so because the strong acid or base is consumed by B or BH^+. If you add HCl to tris, B is converted to BH^+. If you add NaOH, BH^+ is converted to B. As long as you don't use up the B or BH^+ by adding too much HCl or NaOH, the log term of the

A buffer resists changes in pH · · ·

· · · because the buffer consumes the added acid or base.

Henderson-Hasselbalch equation does not change very much and the pH does not change very much. Demonstration 10-2 illustrates what happens when the buffer does get used up. The buffer has its maximum capacity to resist changes of pH when pH = pK_a. We will return to this point later.

EXAMPLE   Calculating How to Prepare a Buffer Solution

How many milliliters of 0.500 M NaOH should be added to 10.0 g of tris hydrochloride to give a pH of 7.60 in a final volume of 250 mL?

Solution   The number of moles of tris hydrochloride in 10.0 g is (10.0 g)/(157.597 g/mol) = 0.063 5. We can make a table to help solve the problem.

Reaction with OH⁻:	BH^+	+	OH^-	→	B
Initial moles:	0.063 5		x		—
Final moles:	0.063 5 − x		—		x

The Henderson-Hasselbalch equation allows us to find x, because we know pH and pK_a.

$$pH = pK_a + \log \frac{\text{mol B}}{\text{mol BH}^+}$$

$$7.60 = 8.075 + \log \frac{x}{0.063\,5 - x}$$

$$-0.475 = \log \frac{x}{0.063\,5 - x}$$

$$10^{-0.475} = \frac{x}{0.063\,5 - x} \Rightarrow x = 0.015\,9 \text{ mol}$$

This many moles of NaOH is contained in

$$\frac{0.015\,9 \text{ mol}}{0.500 \text{ mol/L}} = 0.031\,8 \text{ L} = 31.8 \text{ mL}$$

Reasons why a calculation would be wrong:

1. You might have ignored activity coefficients.
2. The temperature might not be just right.
3. The approximations that [HA] = F_{HA} and [A⁻] = F_{A^-} could be in error.
4. The pK_a reported for tris in your favorite table is probably not what you would measure in your lab.
5. You will probably make an arithmetic error anyway.

 Preparing a Buffer in Real Life!

If you really wanted to prepare a tris buffer of pH 7.60, you would *not* do it by calculating what to mix. Suppose that you wish to prepare 1.00 L of buffer containing 0.100 M tris at a pH of 7.60. You have available solid tris hydrochloride and approximately 1 M NaOH. Here's how to do it:

1. Weigh out 0.100 mol of tris hydrochloride and dissolve it in a beaker containing about 800 mL of water.

2. Place a pH electrode in the solution and monitor the pH.

3. Add NaOH solution until the pH is exactly 7.60.

Strong Plus Weak Reacts Completely

A strong acid reacts with a weak base essentially "completely" because the equilibrium constant is large.

$$B + H^+ \rightleftharpoons BH^+ \qquad K = \frac{1}{K_a} \quad \text{(for } BH^+\text{)}$$

Weak base · Strong acid

If B is tris(hydroxymethyl)aminomethane, then the equilibrium constant for reaction with HCl is

$$K = \frac{1}{K_a} = \frac{1}{10^{-8.075}} = 1.2 \times 10^8$$

A strong base reacts "completely" with a weak acid because the equilibrium constant is, again, very large.

$$OH^- + HA \rightleftharpoons A^- + H_2O \qquad K = \frac{1}{K_b} \quad \text{(for } A^-\text{)}$$

Strong base · Weak acid

If HA is acetic acid, then the equilibrium constant for reaction with NaOH is

$$K = \frac{1}{K_b} = \frac{K_a \text{ for HA}}{K_w} = 1.7 \times 10^9$$

The reaction of a strong acid with a strong base is even more complete than a strong + weak reaction:

$$H^+ + OH^- \rightleftharpoons H_2O \qquad K = \frac{1}{K_w} = 10^{14}$$

Strong acid · Strong base

If you mix a strong acid, a strong base, a weak acid, and a weak base, the strong acid and base will neutralize each other until one is used up. The remaining strong acid or base will then react with the weak base or weak acid.

4. Transfer the solution to a volumetric flask and wash the beaker a few times. Add the washings to the volumetric flask.

5. Dilute to the mark and mix.

You do not simply add the calculated quantity of NaOH because it would not give exactly the desired pH. The reason for using 800 mL of water in the first step is so that the volume will be reasonably close to the final volume during pH adjustment. Otherwise, the pH will change slightly when the sample is diluted to its final volume and the ionic strength changes.

Buffer Capacity

The **buffer capacity,** β, is a measure of how well a solution resists changes in pH when strong acid or base is added. Buffer capacity is defined as

Buffer capacity:
$$\beta = \frac{dC_b}{d\text{pH}} = -\frac{dC_a}{d\text{pH}} \tag{10-19}$$

Demonstration 10-2 **How Buffers Work**

A buffer resists changes in pH because the added acid or base is consumed by the buffer. As the buffer is used up it becomes less resistant to changes in pH.

In this demonstration, a mixture containing approximately a 10:1 mole ratio of HSO_3^-: SO_3^{2-} is prepared. Because pK_a for HSO_3^- is 7.2, the pH should be approximately

$$pH = pK_a + \log \frac{[SO_3^{2-}]}{[HSO_3^-]} = 7.2 + \log \frac{1}{10} = 6.2$$

When formaldehyde is added, the net reaction is the consumption of HSO_3^-, but not of SO_3^{2-}.

$$H_2C=O + HSO_3^- \rightarrow H_2C{\overset{O^-}{\underset{SO_3H}{<}}} \rightarrow H_2C{\overset{OH}{\underset{SO_3^-}{<}}} \quad (A)$$
Formaldehyde   Bisulfite

$$H_2C=O + SO_3^{2-} \rightarrow H_2C{\overset{O^-}{\underset{SO_3^-}{<}}}$$
Sulfite
$$(B)$$

$$H_2C{\overset{O^-}{\underset{SO_3^-}{<}}} + HSO_3^- \rightarrow H_2C{\overset{OH}{\underset{SO_3^-}{<}}} + SO_3^{2-}$$

(In sequence A, bisulfite is consumed directly. In sequence B, the net reaction is destruction of HSO_3^-, with no change in the SO_3^{2-} concentration.)

We can prepare a table showing how the pH should change as the HSO_3^- reacts.

Percentage of reaction completed	$[SO_3^{2-}]:[HSO_3^-]$	Calculated pH
0	1:10	6.2
90	1:1	7.2
99	1:0.1	8.2
99.9	1:0.01	9.2
99.99	1:0.001	10.2

You can see that through 90% completion the pH should rise by just 1 unit. In the next 9% of the reaction, the pH will rise by another unit. At the end of the reaction, the change in pH should be very abrupt.

where C_a and C_b are the number of moles of strong acid and strong base per liter needed to produce a unit change in pH. The larger the value of β, the more resistant the solution is to pH change.

Figure 10-6a shows C_b versus pH for a solution containing 0.100 F HA with $pK_a = 5.00$. The ordinate (C_b) is the formal concentration of strong base needed to be mixed with 0.100 F HA to give the indicated pH. For example, a solution containing 0.050 F OH$^-$ plus 0.100 F HA would have a pH of 5.00 (neglecting activities).

The lower curve in Figure 10-6, which is the derivative of the upper curve, shows the buffer capacity for the same system. The most notable feature of buffer capacity is that it reaches a maximum when pH = pK_a. That is, *a buffer is most effective in resisting changes in pH when pH = pK_a* (i.e., when [HA] = [A$^-$]).

In the formaldehyde clock reaction, formaldehyde is added to a solution containing HSO_3^-, SO_3^{2-}, and phenolphthalein indicator. Phenolphthalein is colorless below a pH of 8.5 and red above this pH. What is observed is that the solution remains colorless for more than a minute. Suddenly the pH shoots up and the liquid turns pink. Monitoring the pH with a glass electrode gave the results shown in the figure.

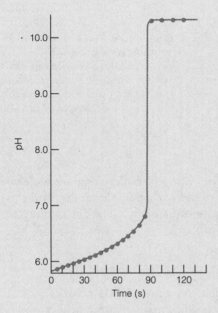

Graph of pH versus time in the formaldehyde clock reaction.

Procedure All solutions should be fresh. Prepare a solution of formaldehyde by diluting 9 mL of 37 wt% formaldehyde to 100 mL. Dissolve 1.5 g of $NaHSO_3$ and 0.18 g of Na_2SO_3 in 400 mL of water, and add ~1 mL of phenolphthalein indicator solution (Table 12-4). Add 23 mL of the formaldehyde solution to the well-stirred buffer solution to initiate the clock reaction. The time of reaction can be adjusted by changing the temperature, concentrations, or volume.

In choosing a buffer for an experiment, you should *seek one whose pK$_a$ is as close as possible to the desired pH. The useful pH range of a buffer is usually considered to be pK$_a$ ± 1 pH unit.* Outside this range, there is not enough of either the weak acid or the weak base to react with added base or acid. Clearly, the buffer capacity can be increased by increasing the concentration of the buffer.

Choose a buffer whose pK$_a$ is close to the desired pH.

The buffer capacity curve in Figure 10-6b continues upward at high pH (and at low pH, which is not shown) simply because there is a high concentration of OH^-. Addition of a small amount of acid or base to a large amount of OH^- (or H^+) will not have a very great effect on pH. A solution of high pH is buffered by the H_2O/OH^- conjugate acid-conjugate base pair. A solution of low pH is buffered by the H_3O^+/H_2O conjugate acid-conjugate base pair.

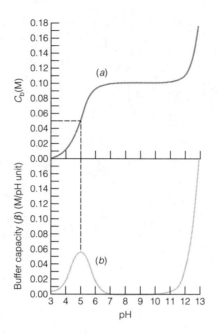

Figure 10-6 (a) C_b versus pH for a solution containing 0.100 F HA with $pK_a = 5.00$. (b) Buffer capacity versus pH for the same system reaches a maximum when $pH = pK_a$. The lower curve is the derivative of the upper curve.

Table 10-2 lists pK_a values for common buffers that are widely used in biochemistry. The measurement of pH with glass electrodes, and the buffers used by the U.S. National Institute of Standards and Technology to define the pH scale, are described in Chapter 15.

Buffer pH Depends on Ionic Strength and Temperature

Changing ionic strength changes pH.

The *correct* Henderson-Hasselbalch equation, 10-18, includes activity coefficients. Failure to include activity coefficients is the principal reason why our calculated pH values will not be in perfect agreement with measured pH values. The activity coefficients also predict that the pH of a buffer will vary with ionic strength. Adding an inert salt (such as NaCl) or altering the buffer volume will change the pH. This effect can be significant for highly charged species, such as citrate or phosphate. When a 0.5 M stock solution of phosphate buffer at pH 6.6 is diluted to 0.05 M, the pH rises to 6.9, which is a rather significant effect.

Changing temperature changes pH.

Most buffers exhibit a noticeable dependence of pK_a on temperature. Tris has an exceptionally large dependence, approximately -0.031 pK_a unit per degree, near room temperature. A solution of tris made up to pH 8.08 at 25°C will have pH ≈ 8.7 at 4° and pH ≈ 7.7 at 37°C.

When What You Mix Is Not What You Get

What you mix is not what you get in dilute solutions or at extremes of pH.

In a dilute solution, or at extremes of pH, the concentrations of HA and A^- in solution are not equal to their formal concentrations. This can be seen as follows. Suppose we mix F_{HA} moles of HA and F_{A^-} moles of the salt Na^+A^-. The mass and charge balances are

$$\text{mass balance:} \quad F_{HA} + F_{A^-} = [HA] + [A^-]$$

$$\text{charge balance:} \quad [Na^+] + [H^+] = [OH^-] + [A^-]$$

Substituting the equality $F_{A^-} = [Na^+]$ into the charge balance, and doing a little algebra, leads to the pair of equations

$$[HA] = F_{HA} - [H^+] + [OH^-] \qquad (10\text{-}20)$$

$$[A^-] = F_{A^-} + [H^+] - [OH^-] \qquad (10\text{-}21)$$

So far we have assumed that $[HA] \approx F_{HA}$ and $[A^-] \approx F_{A^-}$, and we used these values in the Henderson-Hasselbalch equation. A more rigorous procedure is to use Equations 10-20 and 10-21. We see that if F_{HA} or F_{A^-} is small, or if $[H^+]$ or $[OH^-]$ is large, then the approximations $[HA] \approx F_{HA}$ and $[A^-] \approx F_{A^-}$ are not good. In acidic solutions, $[H^+] \gg [OH^-]$, so $[OH^-]$ can be ignored in Equations 10-20 and 10-21. In basic solutions, $[H^+]$ can be neglected.

EXAMPLE A Dilute Buffer Prepared from a Moderately Strong Acid

What will be the pH if 0.0100 mol of HA (with $pK_a = 2.00$) and 0.0100 mol of A^- are dissolved in water to make 1.00 L of solution?

Solution   Because the solution will be acidic (pH $\approx pK_a = 2.00$), we can neglect $[OH^-]$ in Equations 10-20 and 10-21. Setting $[H^+] = x$ in Equations 10-20 and 10-21, we use the K_a equation to find $[H^+]$.

$$HA \rightleftharpoons H^+ + A^-$$
$$0.0100 - x \quad x \quad 0.0100 + x$$

$$K_a = \frac{[H^+][A^-]}{[HA]} = \frac{(x)(0.0100 + x)}{(0.0100 - x)} = 10^{-2.00} \Rightarrow x = 0.00414 \text{ M}$$

$$pH = -\log[H^+] = 2.38$$

The concentrations of HA and A^- are not what we mixed:

$$[HA] = F_{HA} - [H^+] = 0.00586 \text{ M}$$

$$[A^-] = F_{A^-} + [H^+] = 0.0141 \text{ M}$$

In this example, HA is too strong and the concentrations are too low for HA and A^- to be equal to their formal concentrations.

The Henderson-Hasselbalch equation (with activity coefficients) is *always* true, because it is just a rearrangement of the K_a equilibrium expression. Approximations that are not always true are the statements $[HA] \approx F_{HA}$ and $[A^-] \approx F_{A^-}$.

In summary, a buffer consists of a mixture of a weak acid and its conjugate base. The buffer is most useful when pH $\approx pK_a$. Over a reasonable range of concentration, the pH of a buffer is nearly independent of concentration. A buffer resists changes in pH because it reacts with added acids or bases. If too much acid or base is added, the buffer will be consumed and no longer resists changes in pH.

Problem 41 shows how to use Equations 10-20 and 10-21 to deal with buffers.

TABLE 10-2   Structures and pK_a values for commonly used buffers

Name	Structure[a]	pK_a (~25°C)	Formula weight
Phosphoric acid	H_3PO_4	2.15 (pK_1)	97.995
Citric acid	$HO_2CCH_2\overset{\overset{\textstyle OH}{\textstyle \vert}}{C}CH_2CO_2H$, with $\overset{\vert}{CO_2H}$	3.13 (pK_1)	192.125
Formic acid	HCO_2H	3.74	46.026
Succinic acid	$HO_2CCH_2CH_2CO_2H$	4.21 (pK_1)	118.089
Citric acid	$H_2(citrate)^-$	4.76 (pK_2)	192.125
Acetic acid	CH_3CO_2H	4.76	60.053
Succinic acid	$H(succinate)^-$	5.64 (pK_2)	118.089
2-(N-Morpholino)ethanesulfonic acid (MES)	$O\overset{+}{\underset{}{N}}HCH_2CH_2SO_3^-$	6.15	195.240
Cacodylic acid	$(CH_3)_2AsO_2H$	6.19	153.033
Citric acid	$H(citrate)^{2-}$	6.40 (pK_3)	192.125
N-2-Acetamidoiminodiacetic acid (ADA)	$H_2NCCH_2\overset{+}{N}H$ with $CH_2CO_2^-$ and CH_2CO_2H	6.60	190.156
1,3-Bis[tris(hydroxymethyl)methylamino]propane (BIS-TRIS propane)	$(HOCH_2)_3C\overset{+}{N}H_2(CH_2)_3NHC(CH_2OH)_3$	6.80	283.345
Piperazine-N,N'-bis(2-ethanesulfonic acid) PIPES	$^-O_3SCH_2CH_2\overset{+}{N}H\quad H\overset{+}{N}CH_2CH_2SO_3^-$	6.80	302.373
N-2-Acetamido-2-aminoethanesulfonic acid (ACES)	$H_2NCCH_2\overset{+}{N}H_2CH_2CH_2SO_3^-$	6.90	182.200
3-(N-Morpholino)-2-hydroxypropanesulfonic acid (MOPSO)	$O\overset{+}{\underset{}{N}}HCH_2\overset{\overset{\textstyle OH}{\textstyle \vert}}{C}HCH_2SO_3^-$	6.95[b]	225.265
Imidazole hydrochloride	imidazole Cl^-	6.99	104.539

a. The protonated form of each molecule is shown. Acidic hydrogen atoms are shown in **bold** type. Several buffers in this table are widely used in biomedical research because of their relatively weak binding of metal ions and physiologic inertness (C. L. Bering, *J. Chem. Ed.* **1987,** *64,* 803). However, the buffers ADA, BICINE, ACES, and TES have greater metal-binding ability than formerly thought (R. Nakon and C. R. Krishnamoorthy, *Science* **1983,** *221,* 749). Lutidine buffers for the pH range 3 to 8 with limited metal-binding power have been described by U. Bips, H. Elias, M. Hauröder, G. Kleinhans, S. Pfeifer, and K. J. Wannowius, *Inorg. Chem.* **1983,** *22,* 3862.
b. Temperature and ionic strength dependence of HEPES and MOPSO are given by D. Feng, W. F. Koch, and Y. C. Wu, *Anal. Chem.* **1989,** *61,* 1400; and by Y. C. Wu, P. A. Berezansky, D. Feng, and W. F. Koch, *Anal. Chem.* **1993,** *65,* 1084.

TABLE 10-2 (continued)

Name	Structure[a]	pK_a (~25°C)	Formula weight
3-(N-Morpholino)propanesulfonic acid (MOPS)	$O\overset{+}{\diagdown}NHCH_2CH_2CH_2SO_3^-$	7.20	209.266
Phosphoric acid	$H_2PO_4^-$	7.20 (pK_2)	97.995
N-Tris(hydroxymethyl)methyl-2-amino-ethanesulfonic acid (TES)	$(HOCH_2)_3C\overset{+}{N}H_2CH_2CH_2SO_3^-$	7.50	229.254
N-2-Hydroxyethylpiperazine-N'-2-ethanesulfonic acid (HEPES)	$HOCH_2CH_2N\diagdown\diagup\overset{+}{N}HCH_2CH_2SO_3^-$	7.56[b]	238.308
N-2-Hydroxyethylpiperazine-N'-3-propanesulfonic acid (HEPPS)	$HOCH_2CH_2N\diagdown\diagup\overset{+}{N}HCH_2CH_2CH_2SO_3^-$	8.00	252.335
N-Tris(hydroxymethyl)methylglycine (TRICINE)	$(HOCH_2)_3C\overset{+}{N}H_2CH_2CO_2^-$	8.15	179.173
Glycine amide hydrochloride	$H_3\overset{+}{N}CH_2\overset{O}{\overset{\|}{C}}NH_2 \ Cl^-$	8.20	110.543
Tris(hydroxymethyl)aminomethane hydrochloride (TRIS hydrochloride)	$(HOCH_2)_3C\overset{+}{N}H_3 \ Cl^-$	8.08	157.597
N,N-Bis(2-hydroxyethyl)glycine (BICINE)	$(HOCH_2CH_2)_2\overset{+}{N}HCH_2CO_2^-$	8.35	163.174
Glycylglycine	$H_3\overset{+}{N}CH_2\overset{O}{\overset{\|}{C}}NHCH_2CO_2^-$	8.40	132.119
Boric acid	$B(OH)_3$	9.24 (pK_1)	61.833
Cyclohexylaminoethanesulfonic acid (CHES)	$\diagup\diagdown\!\!-\overset{+}{N}H_2CH_2CH_2SO_3^-$	9.50	207.294
3-(Cyclohexylamino)propanesulfonic acid (CAPS)	$\diagup\diagdown\!\!-\overset{+}{N}H_2CH_2CH_2CH_2SO_3^-$	10.40	221.321
Phosphoric acid	HPO_4^{2-}	12.15 (pK_3)	97.995
Boric acid	$OB(OH)_2^-$	12.74 (pK_2)	61.833

Terms to Understand

acid dissociation constant, K_a
base "dissociation" constant, K_b
base hydrolysis constant, K_b
buffer
buffer capacity

conjugate acid-base pair
fraction of association, α (of a base)
fraction of dissociation, α (of an acid)
Henderson-Hasselbalch equation
pK

strong acid
strong base
weak acid
weak base
weak electrolyte

Summary

Strong *acids or bases.* For practical concentrations ($\gtrsim 10^{-6}$ M), pH or pOH can be found by inspection. When the concentration is near 10^{-7} M, we use the systematic treatment of equilibrium to calculate pH. At still lower concentrations, the pH is 7.00, set by autoprotolysis of the solvent.

Weak acids. For the reaction $HA \rightleftharpoons H^+ + A^-$, we set up and solve the equation $K_a = x^2/(F - x)$, where $[H^+] = [A^-] = x$, and $[HA] = F - x$. The fraction of dissociation is given by $\alpha = [A^-]/([HA] + [A^-]) = x/F$. The term pK_a is defined as $pK_a = -\log K_a$.

Weak bases. For the reaction $B + H_2O \rightleftharpoons BH^+ + OH^-$, we set up and solve the equation $K_b = x^2/(F - x)$, where $[OH^-] = [BH^+] = x$, and $[B] = F - x$. The conjugate acid of a weak base is a weak acid, and the conjugate base of a weak acid is a weak base. For a conjugate acid-base pair, $K_a \cdot K_b = K_w$.

Buffers. A buffer is a mixture of a weak acid and its conjugate base. It resists changes in pH because it reacts

with added acid or base. The pH is given by the Henderson-Hasselbalch equation.

$$pH = pK_a + \log \frac{[A^-]}{[HA]}$$

where pK_a applies to the species in the denominator. The concentrations of HA and A^- are essentially unchanged from those used to prepare the solution. The pH of a buffer is nearly independent of dilution, but the buffer capacity increases as the concentration of buffer increases. The maximum buffer capacity is found at pH = pK_a, and the useful range of a buffer is approximately pH = pK_a ± 1.

The conjugate base of a weak acid is a weak base. The weaker the acid, the stronger the base. However, if one member of a conjugate pair is weak, so is its conjugate. The relation between K_a for an acid and K_b for its conjugate base in aqueous solution is $K_a \cdot K_b = K_w$. When a strong acid (or base) is added to a weak base (or acid), they react nearly completely.

Exercises

A. Using activity coefficients correctly, find the pH of 1.0×10^{-2} M NaOH.

B. Calculate the pH of

(a) 1.0×10^{-8} M HBr

(b) 1.0×10^{-8} M H_2SO_4 (the H_2SO_4 dissociates completely to $2H^+$ plus SO_4^{2-} at this low concentration)

C. What is the pH of a solution prepared by dissolving 1.23 g of 2-nitrophenol (MW 139.110) in 0.250 L?

D. The pH of 0.010 M o-cresol is 6.05. Find the pK_a for this weak acid.

o-Cresol

E. Calculate the limiting value of the fraction of dissociation (α) of a weak acid ($pK_a = 5.00$) as the concentra-

tion of HA approaches zero. Repeat the same calculation for $pK_a = 9.00$.

F. Find the pH of 0.050 M sodium butanoate (the sodium salt of butanoic acid, also called butyric acid).

G. The pH of 0.10 M ethylamine is 11.80.

(a) Without referring to Appendix G, find K_b for ethylamine.

(b) Using the results of **(a)**, calculate the pH of 0.10 M ethylammonium chloride.

H. Which of the following bases would be most suitable for preparing a buffer of pH 9.00?

(a) NH_3 (ammonia, $K_b = 1.75 \times 10^{-5}$)

(b) $C_6H_5NH_2$ (aniline, $K_b = 3.99 \times 10^{-10}$)

(c) H_2NNH_2 (hydrazine, $K_b = 3.0 \times 10^{-6}$)

(d) C_5H_5N (pyridine, $K_b = 1.69 \times 10^{-9}$)

I. A solution contains 63 different conjugate acid-base pairs. Among them is acrylic acid and acrylate ion, with the ratio [acrylate]/[acrylic acid] = 0.75. What is the pH of the solution?

$$H_2C=CHCO_2H \qquad pK_a = 4.25$$
Acrylic acid

J. **(a)** Find the pH of a solution prepared by dissolving 1.00 g of glycine amide hydrochloride (Table 10-2) plus 1.00 g of glycine amide in 0.100 L.

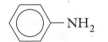

Glycine amide
$C_2H_6N_2O$
MW 74.083

(b) How many grams of glycine amide should be added to 1.00 g of glycine amide hydrochloride to give 100 mL of solution with pH 8.00?

(c) What would be the pH if the solution in **(a)** is mixed with 5.00 mL of 0.100 M HCl?

(d) What would be the pH if the solution in **(c)** is mixed with 10.00 mL of 0.100 M NaOH?

K. A solution with an ionic strength of 0.10 M containing 0.0100 M phenylhydrazine has a pH of 8.13. Using activity coefficients correctly, find pK_a for the phenylhydrazinium ion found in phenylhydrazine hydrochloride. Assume that $\gamma_{BH^+} = 0.80$.

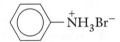

Phenylhydrazine
B

Phenylhydrazine hydrochloride
BH^+Cl^-

Problems

Strong Acids and Bases

1. Why doesn't water dissociate to produce 10^{-7} M H$^+$ and 10^{-7} M OH$^-$ when some HBr is added?

2. Calculate the pH of the following solutions.

(a) 1.0×10^{-3} M HBr

(b) 1.0×10^{-2} M KOH

3. Calculate the pH of 5.0×10^{-8} M HClO$_4$. What fraction of the total H$^+$ in this solution is derived from dissociation of water?

4. **(a)** The measured pH of 0.100 M HCl at 25°C is 1.092. From this information, calculate the activity coefficient of H$^+$ and compare your answer to that in Table 8-1.

(b) The measured pH of 0.0100 M HCl + 0.0900 M KCl at 25°C is 2.102. From this information, calculate the activity coefficient of H$^+$ in this solution.

(c) The ionic strength of both solutions above is the same. What can you conclude about the dependence of activity coefficients on the particular ions in a solution?

Weak-Acid Equilibria

5. Write the chemical reaction whose equilibrium constant is

(a) K_a for benzoic acid

(b) K_b for benzoate ion

(c) K_b for aniline

(d) K_a for anilinium ion

Benzoic acid

Potassium benzoate

Aniline

Anilinium bromide

6. Find the pH and fraction of dissociation (α) of a 0.100 M solution of the weak acid HA with $K_a = 1.00 \times 10^{-5}$.

7. BH$^+$ClO$_4^-$ is a salt formed from the base B ($K_b = 1.00 \times 10^{-4}$) and perchloric acid. It dissociates into BH$^+$, a weak acid, and ClO$_4^-$, which is neither an acid nor a base. Find the pH of 0.100 M BH$^+$ClO$_4^-$.

8. Find the pH and concentrations of $(CH_3)_3N$ and $(CH_3)_3NH^+$ in a 0.060 M solution of trimethyl-ammonium chloride.

9. *When is a weak acid weak and when is a weak acid strong?* Show that the weak acid HA will be 92% dissociated when dissolved in water if the formal concentration is one-tenth of K_a (F = K_a/10). Show that the fraction of dissociation is 27% when F = $10K_a$. At what formal concentration will the acid be 99% dissociated? Compare your answer with the left-hand curve in Figure 10-2.

10. A 0.0450 M solution of benzoic acid has a pH of 2.78. Calculate pK_a for this acid.

11. A 0.0450 M solution of HA is 0.60% dissociated. Calculate pK_a for this acid.

12. Barbituric acid dissociates as follows:

Barbituric acid
HA   A⁻

(a) Calculate the pH and fraction of dissociation of $10^{-2.00}$ M barbituric acid.

(b) Calculate the pH and fraction of dissociation of $10^{-10.00}$ M barbituric acid.

13. Using activity coefficients correctly, calculate the fraction of dissociation, α, of 50.0 mM hydroxybenzene (phenol) in 0.050 M LiBr. Assume that the size of $C_6H_5O^-$ is 600 pm.

14. Cr^{3+} (and most other metal ions with charge ≥2) is acidic by virtue of the hydrolysis reaction

$$Cr^{3+} + H_2O \rightleftharpoons Cr(OH)^{2+} + H^+ \qquad K_{a1} = 10^{-3.80}$$

[Further reactions produce $Cr(OH)_2^+$, $Cr(OH)_3$, and $Cr(OH)_4^-$.] Considering only the K_{a1} reaction, find the pH of 0.010 M $Cr(ClO_4)_3$. What fraction of chromium is in the form $Cr(OH)^{2+}$?

15. The temperature dependence of pK_a for acetic acid is given in the following table. Is the dissociation of this acid endothermic or exothermic **(a)** at 5°C; **(b)** at 45°C?

Temperature (°C)	pK_a	Temperature (°C)	pK_a
0	4.781	30	4.757
5	4.770	35	4.762
10	4.762	40	4.769
15	4.758	45	4.777
20	4.756	50	4.787
25	4.756		

16. *pH of a weak acid.* Create a spreadsheet like the one in Figure 10-3 to compute and plot the graph in Figure 10-4. For each of the five formal concentrations of HA (10^{-5} to 10^{-1} F), at what pK_a do the concentrations of $[H^+]$ and $[A^-]$ differ by more than 10%?

Weak-Base Equilibria

17. Find the pH and fraction of association (α) of a 0.100 M solution of the weak base B with $K_b = 1.00 \times 10^{-5}$.

18. Find the pH and concentrations of $(CH_3)_3N$ and $(CH_3)_3NH^+$ in a 0.060 M solution of trimethylamine.

19. Find the pH of 0.050 M NaCN.

20. Calculate the fraction of association (α) for 1.00×10^{-1}, 1.00×10^{-2}, and 1.00×10^{-12} M sodium acetate. Does α increase or decrease with dilution?

21. For a 0.10 M solution of a base that has pH = 9.28, find K_b for the base.

22. For a 0.10 M solution of a base that is 2.0% hydrolyzed ($\alpha = 0.020$), find K_b for the base.

23. Show that the limiting fraction of association of a weak base in water, as the concentration of the base approaches zero, is $\alpha = 10^7 K_b/(1 + 10^7 K_b)$.

24. *pH of a weak base.* Derive the equation below (analogous to Equation 10-12) for the concentration of OH^- from the reaction $B + H_2O \rightleftharpoons BH^+ + OH^-$.

Weak base: $[OH^-] =$

$$\frac{-K_b + \sqrt{K_b^2 + 4(K_w + K_bF + K_bK_w/[OH^-])}}{2}$$

Use this equation in a spreadsheet to find the pH of 10^{-3} M B with $K_b = 10^{-5}$ and of 10^{-5} M B with $K_b = 10^{-9}$.

Buffers

25. Describe the operations you would perform to prepare exactly 100 mL of 0.200 M acetate buffer, pH 5.00, starting with pure liquid acetic acid and solutions containing ~3 M HCl and ~3 M NaOH.

26. Why is the pH of a buffer nearly independent of concentration?

27. Why does buffer capacity increase as the concentration of buffer increases?

28. Why is the buffer capacity maximum when pH = pK_a?

29. Explain the following statement: The Henderson-Hasselbalch equation (with activity coefficients) is *always* true; what may not be correct are the values of $[A^-]$ and $[HA]$ that we choose to use in the equation.

30. Which of the following acids would be most suitable for preparing a buffer of pH 3.10?

(a) hydrogen peroxide   **(c)** cyanoacetic acid

(b) propanoic acid   **(d)** 4-aminobenzenesulfonic acid

31. A buffer was prepared by dissolving 0.100 mol of the weak acid HA ($K_a = 1.00 \times 10^{-5}$) plus 0.050 mol of its conjugate base Na^+A^- in 1.00 L. Find the pH.

32. Write the Henderson-Hasselbalch equation for a solution of formic acid. Calculate the quotient $[HCO_2^-]/[HCO_2H]$ at **(a)** pH 3.000; **(b)** pH 3.745; **(c)** pH 4.000.

33. Given that pK_b for nitrite ion (NO_2^-) is 10.85, find the quotient $[HNO_2]/[NO_2^-]$ in a solution of sodium nitrite at (a) pH 2.00; (b) pH 10.00.

34. (a) Describe the correct procedures for preparing 0.250 L of 0.0500 M HEPES (Table 10-2), pH 7.45.

(b) Would you need NaOH or HCl to bring the pH to 7.45?

35. How many milliliters of 0.246 M HNO_3 should be added to 213 mL of 0.00666 M ethylamine to give a pH of 10.52?

36. (a) Write the chemical reactions whose equilibrium constants are K_b and K_a for imidazole and imidazole hydrochloride, respectively.

(b) Calculate the pH of a solution prepared by mixing 1.00 g of imidazole with 1.00 g of imidazole hydrochloride and diluting to 100.0 mL.

(c) Calculate the pH of the solution if 2.30 mL of 1.07 M $HClO_4$ is added to the solution.

(d) How many milliliters of 1.07 M $HClO_4$ should be added to 1.00 g of imidazole to give a pH of 6.993?

37. Calculate the pH of a solution prepared by mixing 0.0800 mol of chloroacetic acid plus 0.0400 mol of sodium chloroacetate in 1.00 L of water.

(a) First do the calculation assuming that the concentrations of HA and A^- equal their formal concentrations.

(b) Then do the calculation using the real values of [HA] and [A^-] in the solution.

(c) Using first your head, and then the Henderson-Hasselbalch equation, find the pH of a solution prepared by dissolving all of the following in one beaker containing a total volume of 1.00 L: 0.180

mol of $ClCH_2CO_2H$, 0.020 mol of $ClCH_2CO_2Na$, 0.080 mol HNO_3, and 0.080 mol $Ca(OH)_2$. Assume that $Ca(OH)_2$ dissociates completely.

38. Calculate how many milliliters of 0.626 M KOH should be added to 5.00 g of HEPES (Table 10-2) to give a pH of 7.40.

39. Use Equations 10-20 and 10-21 to find the concentrations of HA and A^- in a solution prepared by mixing 0.00200 mol of acetic acid plus 0.00400 mol of sodium acetate in 1.00 L of water.

40. Calculate the pH of a solution prepared by mixing 0.0100 mol of the base B ($K_b = 10^{-2.00}$) with 0.0200 mol of BH^+Br^- and diluting to 1.00 L. First calculate the pH assuming that [B] = 0.0100 and [BH^+] = 0.0200 M. Compare this answer with the pH calculated without making such an assumption. Which calculation is more correct?

41. *Buffers.* Substitute Equations 10-20 and 10-21 into the K_a equilibrium constant to show that $K_a = [H^+](F_{A^-} + [H^+] - [OH^-])/(F_{HA} - [H^+] + [OH^-])$. Solve for [$H^+$] in the same manner in which Equation 10-12 was derived to show that

buffer: $$[H^+] = -\frac{1}{2}(F_{A^-} + K_a) + \frac{1}{2}\sqrt{(F_{A^-} + K_a)^2 + 4(K_a F_{HA} + K_w + K_a K_w/[H^+])}$$

Create a spreadsheet like the one in Figure 10-3 to find the pH when

(a) $F_{A^-} = F_{HA} = 0.001$ M; p$K_a = 6$

(b) $F_{A^-} = F_{HA} = 10^{-6}$ M; p$K_a = 6$

(c) $F_{A^-} = F_{HA} = 0.01$ M; p$K_a = 1$

(d) $F_{A^-} = F_{HA} = 0.01$ M; p$K_a = 12$

Notes and References

1. W. Tan, S.-Y. Shi, S. Smith, D. Birnbaum, and R. Kopelman, *Science* **1992,** *258,* 778; W. Tan, S.-Y. Shi, and R. Kopelman, *Anal. Chem.* **1992,** *64,* 2985.

2. Acid dissociation constants do not tell us which protons dissociate in each step. Assignments for pyridoxal phosphate come from nuclear magnetic resonance spectroscopy [B. Szpoganicz and A. E. Martell, *J. Am. Chem. Soc.* **1984,** *106,* 5513].

3. Inexpensive apparatus for a quantitative conductivity

experiment is described by T. R. Rettich, *J. Chem. Ed.* **1989,** *66,* 168; and D. A. Katz and C. Willis, *J. Chem. Ed.* **1994,** *71,* 330.

4. F. B. Dutton and G. Gordon in H. N. Alyea and F. B. Dutton, Eds., *Tested Demonstrations in Chemistry,* 6th ed. (Easton, PA: Journal of Chemical Education, 1965), p. 147; R. L. Barrett, *J. Chem. Ed.* **1955,** *32,* 78. A kinetics experiment and more detailed discussions of the mechanism of this reaction are given by M. G. Burnett, *J. Chem. Ed.* **1982,** *59,* 160; and P. Warneck, *J. Chem. Ed.* **1989,** *66,* 334.

Proteins Are Polyprotic Acids and Bases

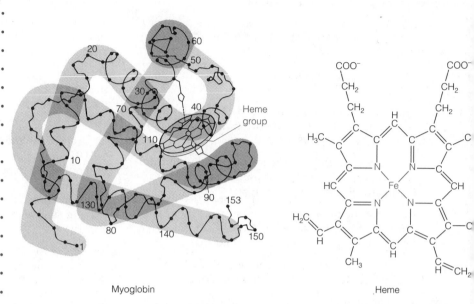

Myoglobin

Heme

Structure at the left shows the folded polypeptide backbone of myoglobin. Substituents are deleted for clarity. The heme group, whose structure is shown at the right, contains an iron atom that binds O_2, CO, and other small molecules. [From M. F. Perutz, "The Hemoglobin Molecule." Copyright © 1964 by Scientific American, Inc. All rights reserved.]

Proteins perform biological functions such as structural support, catalysis of chemical reactions, immune response to foreign substances, transport of molecules across membranes, and control of genetic expression. The three-dimensional structure and function of a protein is determined by the sequence of *amino acids* from which the protein is made. Box 11-1 shows how amino acids are connected to make a *polypeptide*. Myoglobin, which is shown here, folds into several helical (spiral) regions that control access of oxygen and other small molecules to the heme group, whose function is to store O_2 in muscle cells. Of the 153 amino acids in sperm-whale myoglobin, 35 have basic side groups and 23 are acidic.

Polyprotic Acid-Base Equilibria

11

Now that we are experts in monoprotic acids and bases, it is time to extend our capabilities to **polyprotic** systems—those which can donate or accept more than one proton. After studying **diprotic** systems (with two acidic hydrogens), the extension to three or more acidic sites will be straightforward. Then, having grappled with the nuts and bolts of acids and bases, we can step back and take a qualitative look at the big picture, to think about which species are dominant at any given pH. Finally, we develop a set of equations giving the fraction of a polyprotic acid or base in each protonated form as a function of pH. You will encounter these equations whenever equilibria involving acids and bases arise.

11-1 Diprotic Acids and Bases

Amino acids are the building blocks of proteins (Box 11-1). The general structure of an amino acid is

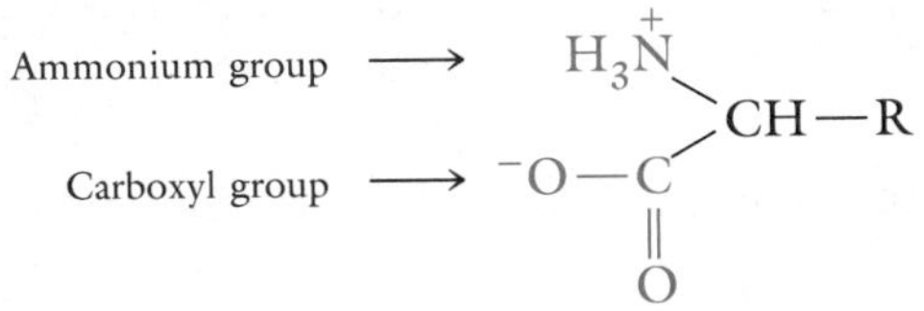

$$\text{Ammonium group} \longrightarrow \quad \underset{\displaystyle \overset{\displaystyle ^-O-\underset{\underset{O}{\|}}{C}}{}}{\overset{+}{H_3N}}\diagdown \overset{\displaystyle }{\underset{\displaystyle }{CH-R}}$$

where R is a different group for each compound. The carboxyl group, drawn here in its ionized (basic) form, is a stronger acid than the ammonium group.

Box 11-1   Proteins Are Composed of Amino Acids

Proteins are polymers made of amino acid subunits:

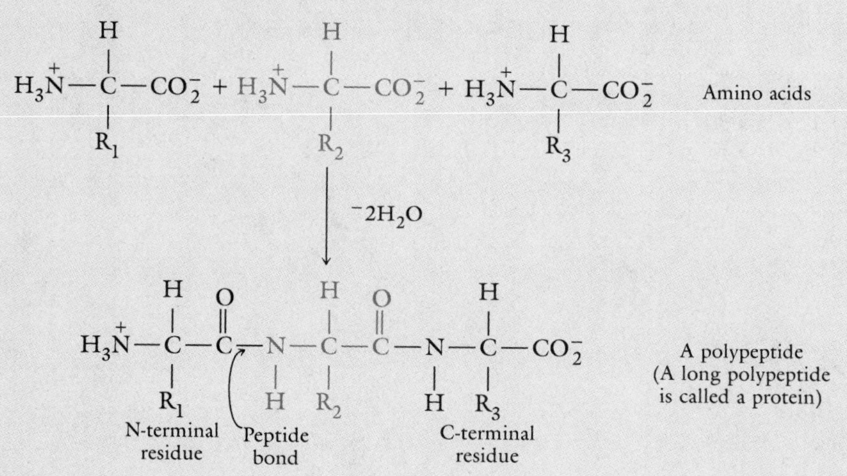

$$\underset{R_1}{\overset{H}{H_3\overset{+}{N}-C-CO_2^-}} + \underset{R_2}{\overset{H}{H_3\overset{+}{N}-C-CO_2^-}} + \underset{R_3}{\overset{H}{H_3\overset{+}{N}-C-CO_2^-}} \qquad \text{Amino acids}$$

$\downarrow ^-2H_2O$

A polypeptide
(A long polypeptide
is called a protein)

N-terminal residue   Peptide bond   C-terminal residue

The structure and function of the protein arise from the sequence of substituents (R groups) on the amino acids. Of the 20 common amino acids, three have basic substituents and four have acidic substituents.

A *zwitterion* is a molecule with positive and negative charges.

Therefore, the nonionized form rearranges spontaneously to the **zwitterion**:

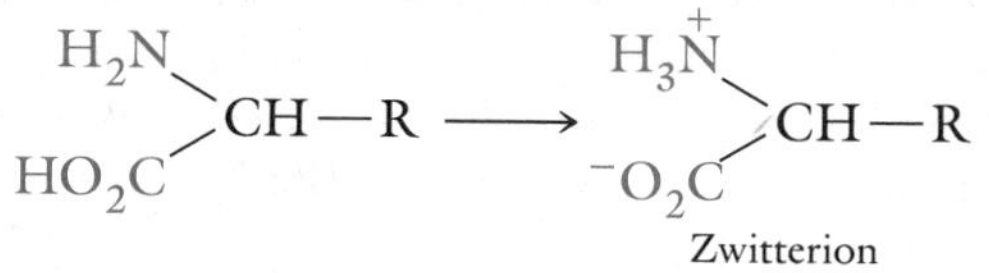

Zwitterion

At low pH, both the ammonium group and the carboxyl group are protonated. At high pH, neither is protonated. Acid dissociation constants of amino acids are listed in Table 11-1, where each compound is drawn in its fully protonated form.

In our discussion, we will take as a specific example the amino acid leucine, designated HL.

The side chain in leucine is an isobutyl group: $(CH_3)_2CHCH_2-$.

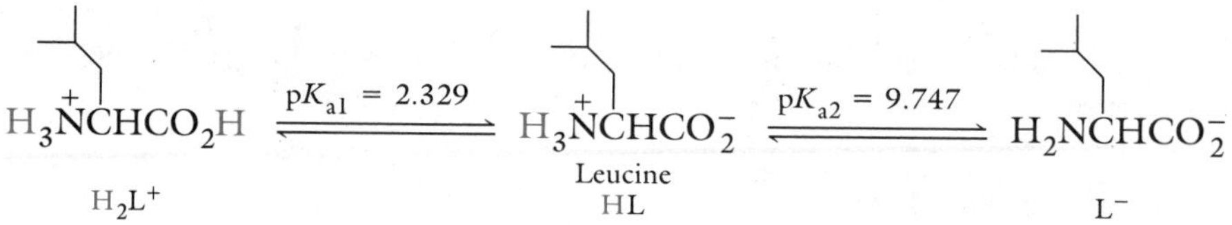

$$H_3\overset{+}{N}CHCO_2H \underset{}{\overset{pK_{a1}=2.329}{\rightleftharpoons}} H_3\overset{+}{N}CHCO_2^- \underset{}{\overset{pK_{a2}=9.747}{\rightleftharpoons}} H_2NCHCO_2^-$$

H_2L^+ Leucine L^-
                HL

The equilibrium constants refer to the following reactions:

We customarily omit the subscript a in K_{a1} and K_{a2}. We always write the subscript b in K_{b1} and K_{b2}.

Diprotic acid:

$$H_2L^+ \rightleftharpoons HL + H^+ \qquad K_{a1} \equiv K_1 \qquad (11\text{-}1)$$

$$HL \rightleftharpoons L^- + H^+ \qquad K_{a2} \equiv K_2 \qquad (11\text{-}2)$$

Diprotic base:

$$L^- + H_2O \rightleftharpoons HL + OH^- \qquad K_{b1} \qquad (11\text{-}3)$$

$$HL + H_2O \rightleftharpoons H_2L^+ + OH^- \qquad K_{b2} \qquad (11\text{-}4)$$

Recall that the relations between the acid and base equilibrium constants are

Relations between K_a's and K_b's:

$$K_{a1} \cdot K_{b2} = K_w \qquad (11\text{-}5)$$

$$K_{a2} \cdot K_{b1} = K_w \qquad (11\text{-}6)$$

We now set out to calculate the pH and composition of individual solutions of 0.0500 M H_2L^+, 0.0500 M HL, and 0.0500 M L^-. Our methods are general. They do not depend on the charge type of the acids and bases. That is, we would use the same procedure to find the pH of the diprotic H_2A, where A is anything, or H_2L^+, where HL is leucine.

The Acidic Form, H_2L^+

Easy stuff.

A salt such as leucine hydrochloride contains the protonated species, H_2L^+, which can dissociate twice, as indicated in Reactions 11-1 and 11-2. Because $K_1 = 4.69 \times 10^{-3}$, H_2L^+ is a weak acid. HL is an even weaker acid, because $K_2 = 1.79 \times 10^{-10}$. It appears that the H_2L^+ will dissociate only partly, and the resulting HL will hardly dissociate at all. For this reason, we make the (superb) approximation that a solution of H_2L^+ behaves as a monoprotic acid, with $K_a = K_1$.

With this approximation, the calculation of the pH of 0.0500 M H_2L^+ is a trivial matter.

$$\underset{\substack{H_2L^+ \\ 0.0500 - x}}{H_3\overset{+}{N}CHCO_2H} \rightleftharpoons \underset{\substack{HL \\ x}}{H_3\overset{+}{N}CHCO_2^-} + \underset{x}{H^+}$$

H_2L^+ can be treated as monoprotic with $K_a = K_{a1}$.

$$K_a = K_1 = 4.69 \times 10^{-3}$$

$$\frac{x^2}{F - x} = K_a \Rightarrow x = 1.31 \times 10^{-2} \text{ M}$$

$$[HL] = x = 1.31 \times 10^{-2} \text{ M}$$

$$[H^+] = x = 1.31 \times 10^{-2} \text{ M} \Rightarrow pH = 1.88$$

$$[H_2L^+] = F - x = 3.69 \times 10^{-2} \text{ M}$$

What is the concentration of L^- in the solution? We have already assumed that it is very small, but it cannot be zero. We can calculate $[L^-]$ from the K_{a2} equation, with the concentrations of HL and H^+ computed above.

$$K_{a2} = \frac{[H^+][L^-]}{[HL]} \Rightarrow [L^-] = \frac{K_{a2}[HL]}{[H^+]} \qquad (11\text{-}7)$$

$$[L^-] = \frac{(1.79 \times 10^{-10})(1.31 \times 10^{-2})}{(1.31 \times 10^{-2})} = 1.79 \times 10^{-10} \text{ M} \ (= K_{a2})$$

The approximation $[H^+] \approx [HL]$ reduces Equation 11-7 to $[L^-] = K_{a2} = 1.79 \times 10^{-10}$ M.

Our approximation is confirmed by this last result. The concentration of L^- is about eight orders of magnitude smaller than that of HL. As a source of protons, the dissociation of HL is indeed negligible relative to the dissociation of H_2L^+. For most diprotic acids, K_1 is sufficiently larger than K_2 for this approximation to be valid. Even if K_2 were just 10 times less than K_1,

TABLE 11-1   **Acid dissociation constants of amino acids**

Amino acid	Structure[a]	Carboxylic acid[b]	Ammonium group[b]	Substituent[b]
Alanine		$pK_a = 2.348$	$pK_a = 9.867$	
Arginine		$pK_a = 1.823$	$pK_a = 8.991$	$(pK_a = 12.48)$
Asparagine		$pK_a = 2.14^c$	$pK_a = 8.72^c$	
Aspartic acid		$pK_a = 1.990$	$pK_a = 10.002$	$pK_a = 3.900$
Cysteine		$(pK_a = 1.71)$	$pK_a = 10.77$	$pK_a = 8.36$
Glutamic acid		$pK_a = 2.23$	$pK_a = 9.95$	$pK_a = 4.42$
Glutamine		$pK_a = 2.17^c$	$pK_a = 9.01^c$	
Glycine		$pK_a = 2.350$	$pK_a = 9.778$	
Histidine		$pK_a = 1.7^c$	$pK_a = 9.08^c$	$pK_a = 6.02^c$
Isoleucine		$pK_a = 2.319$	$pK_a = 9.754$	

a. The acidic protons are shown in **bold** type. Each amino acid is written in its fully protonated form.
b. pK_a values refer to 25°C and zero ionic strength unless marked by c. Values considered to be uncertain are enclosed in parentheses.

TABLE 11-1 (continued)

Amino acid	Structure[a]	Carboxylic acid[b]	Ammonium group[b]	Substituent[b]
Leucine	NH_3^+ $CH-CH_2CH(CH_3)_2$ CO_2H	$pK_a = 2.329$	$pK_a = 9.747$	
Lysine	NH_3^+ $CH-CH_2CH_2CH_2CH_2NH_3^+$ CO_2H	$pK_a = 2.04^c$	$pK_a = 9.08^c$	$pK_a = 10.69^c$
Methionine	NH_3^+ $CH-CH_2CH_2SCH_3$ CO_2H	$pK_a = 2.20^c$	$pK_a = 9.05^c$	
Phenylalanine	NH_3^+ $CH-CH_2$⟨○⟩ CO_2H	$pK_a = 2.20$	$pK_a = 9.31$	
Proline	H_2N^+ (ring) HO_2C	$pK_a = 1.952$	$pK_a = 10.640$	
Serine	NH_3^+ $CH-CH_2OH$ CO_2H	$pK_a = 2.187$	$pK_a = 9.209$	
Threonine	NH_3^+ CH_3 $CH-CH$ CO_2H OH	$pK_a = 2.088$	$pK_a = 9.100$	
Tryptophan	NH_3^+ $CH-CH_2$ (indole) CO_2H	$pK_a = 2.35^c$	$pK_a = 9.33^c$	
Tyrosine	NH_3^+ $CH-CH_2$⟨○⟩$-OH$ CO_2H	$pK_a = 2.17^c$	$pK_a = 9.19$	$pK_a = 10.47$
Valine	NH_3^+ $CH-CH(CH_3)_2$ CO_2H	$pK_a = 2.286$	$pK_a = 9.718$	

c. For these entries the ionic strength is 0.1 M, and the constant refers to a product of concentrations instead of activities.
SOURCE: A. E. Martell and R. M. Smith, *Critical Stability Constants,* Vol. 1 (New York: Plenum Press, 1974).

the value of $[H^+]$ calculated by ignoring the second ionization would be in error by only 4%. The error in pH would be only 0.01 pH unit. In summary, *a solution of a diprotic acid behaves like a solution of a monoprotic acid, with* $K_a = K_{a1}$.

More easy stuff.

The Basic Form, L^-

The fully basic species, L^-, would be found in a salt such as sodium leucinate, which could be prepared by treating leucine with an equimolar quantity of NaOH. Dissolving sodium leucinate in water gives a solution of L^-, the fully basic species. The K_b values for this dibasic anion are

$$L^- + H_2O \rightleftharpoons HL + OH^- \qquad K_{b1} = K_w/K_{a2} = 5.59 \times 10^{-5}$$

$$HL + H_2O \rightleftharpoons H_2L^+ + OH^- \qquad K_{b2} = K_w/K_{a1} = 2.13 \times 10^{-12}$$

Hydrolysis is the reaction of anything with water. Specifically, the reaction $L^- + H_2O \rightleftharpoons HL + OH^-$ is called hydrolysis.

K_{b1} tells us that L^- will not **hydrolyze** (react with water) very much to give HL. Furthermore, K_{b2} tells us that the resulting HL is such a weak base that hardly any further reaction to make H_2L^+ will occur.

We therefore treat L^- as a monobasic species, with $K_b = K_{b1}$. The results of this (fantastic) approximation can be outlined as follows.

L^- can be treated as monobasic with $K_b = K_{b1}$.

$$\underset{\substack{L^- \\ 0.0500 - x}}{H_2NCHCO_2^-} + H_2O \rightleftharpoons \underset{\substack{HL \\ x}}{H_3\overset{+}{N}CHCO_2^-} + \underset{x}{OH^-}$$

$$K_b = K_{b1} = \frac{K_w}{K_{a2}} = 5.59 \times 10^{-5}$$

$$\frac{x^2}{F - x} = 5.59 \times 10^{-5} \Rightarrow x = 1.64 \times 10^{-3} \text{ M}$$

$$[HL] = x = 1.64 \times 10^{-3} \text{ M}$$

$$[H^+] = K_w/x = 6.08 \times 10^{-12} \text{ M} \Rightarrow pH = 11.22$$

$$[L^-] = F - x = 4.84 \times 10^{-2} \text{ M}$$

The concentration of H_2L^+ can be found from the K_{b2} (or K_{a1}) equilibrium.

$$K_{b2} = \frac{[H_2L^+][OH^-]}{[HL]} = \frac{[H_2L^+]x}{x} = [H_2L^+]$$

We find that $[H_2L^+] = K_{b2} = 2.13 \times 10^{-12}$ M, and the approximation that $[H_2L^+]$ is insignificant relative to $[HL]$ is well justified. In summary, if there is any reasonable separation between K_{a1} and K_{a2} (and, therefore, between K_{b1} and K_{b2}), *the fully basic form of a diprotic acid can be treated as monobasic, with* $K_b = K_{b1}$.

A tougher problem.

The Intermediate Form, HL

A solution prepared from leucine, HL, is more complicated than one prepared from either H_2L^+ or L^-, because HL is both an acid and a base.

HL is both an acid and a base.

$$HL \rightleftharpoons H^+ + L^- \qquad K_a = K_{a2} = 1.79 \times 10^{-10} \qquad (11\text{-}8)$$

$$HL + H_2O \rightleftharpoons H_2L^+ + OH^- \qquad K_b = K_{b2} = 2.13 \times 10^{-12} \qquad (11\text{-}9)$$

A molecule that can both donate and accept a proton is said to be **amphiprotic.** The acid dissociation reaction (11-8) has a larger equilibrium constant than the base hydrolysis reaction (11-9), so we expect that a solution of leucine will be acidic.

However, we cannot simply ignore Reaction 11-9, even if K_a and K_b differ by several orders of magnitude. Both reactions proceed to nearly equal extent, because H^+ produced in Reaction 11-8 reacts with OH^- from Reaction 11-9, thereby driving Reaction 11-9 to the right.

To treat this case correctly, we resort to the systematic treatment of equilibrium. The procedure is applied to leucine, whose intermediate form (HL) has no net charge. However, the results apply to the intermediate form of *any* diprotic acid, regardless of its charge.

Our problem deals with 0.0500 M leucine, in which both Reactions 11-8 and 11-9 can happen. The charge balance is

$$[H^+] + [H_2L^+] = [L^-] + [OH^-]$$

which can be rearranged to

$$[H_2L^+] - [L^-] + [H^+] - [OH^-] = 0$$

Using the acid dissociation equilibria, we replace $[H_2L^+]$ with $[HL][H^+]/K_1$, and $[L^-]$ with $[HL]K_2/[H^+]$. Also, we can always write $[OH^-] = K_w/[H^+]$. Putting these expressions into the charge balance gives

$$\frac{[HL][H^+]}{K_1} - \frac{[HL]K_2}{[H^+]} + [H^+] - \frac{K_w}{[H^+]} = 0$$

which can be solved for $[H^+]$. First we multiply all terms by $[H^+]$:

$$\frac{[HL][H^+]^2}{K_1} - [HL]K_2 + [H^+]^2 - K_w = 0$$

Then we factor out $[H^+]^2$ and rearrange:

$$[H^+]^2\left(\frac{[HL]}{K_1} + 1\right) = K_2[HL] + K_w$$

$$[H^+]^2 = \frac{K_2[HL] + K_w}{\frac{[HL]}{K_1} + 1}$$

Multiplying the numerator and denominator by K_1 and taking the square root of both sides give

$$[H^+] = \sqrt{\frac{K_1K_2[HL] + K_1K_w}{K_1 + [HL]}} \tag{11-10}$$

Up to this point we have made no approximations, except to neglect activity coefficients. We solved for $[H^+]$ in terms of known constants plus the single unknown, [HL]. Where do we proceed from here?

Luckily, a chemist gallops down from the mountains on her white stallion to provide the missing insight: "The major species will be HL, because it is both a weak acid and a weak base. Neither Reaction 11-8 nor Reaction 11-9 goes very far. For the concentration of HL in Equation 11-10, you can simply substitute the value 0.0500 M."

The missing insight!

K_1 and K_2 in this equation are both *acid* dissociation constants (K_{a1} and K_{a2}).

Taking the chemist's advice, we write Equation 11-10 in its most useful form.

Intermediate form of diprotic acid: $$[H^+] \approx \sqrt{\frac{K_1 K_2 F + K_1 K_w}{K_1 + F}} \qquad (11\text{-}11)$$

where F is the formal concentration of HL ($= 0.0500$ M in the present case).

At long last we can calculate the pH of 0.0500 M leucine:

$$[H^+] = \sqrt{\frac{(4.69 \times 10^{-3})(1.79 \times 10^{-10})(0.0500) + (4.69 \times 10^{-3})(1.0 \times 10^{-14})}{4.69 \times 10^{-3} + 0.0500}}$$

$$= 8.76 \times 10^{-7} \text{ M} \Rightarrow \text{pH} = 6.06$$

The concentrations of H_2L^+ and L^- can be found from the K_1 and K_2 equilibria, using $[H^+] = 8.76 \times 10^{-7}$ M and $[HL] = 0.0500$ M.

$$[H_2L^+] = \frac{[H^+][HL]}{K_1} = \frac{(8.76 \times 10^{-7})(0.0500)}{4.69 \times 10^{-3}} = 9.34 \times 10^{-6} \text{ M}$$

$$[L^-] = \frac{K_2[HL]}{[H^+]} = \frac{(1.79 \times 10^{-10})(0.0500)}{8.76 \times 10^{-7}} = 1.02 \times 10^{-5} \text{ M}$$

If $[H_2L^+] + [L^-]$ is not much less than $[HL]$ and if you wish to refine your values of $[H_2L^+]$ and $[L^-]$, try the method in Box 11-2.

Was the approximation $[HL] \approx 0.0500$ M a good one? It certainly was, because $[H_2L^+]$ ($= 9.34 \times 10^{-6}$ M) and $[L^-]$ ($= 1.02 \times 10^{-5}$ M) are small in comparison with $[HL]$ (≈ 0.0500 M). Nearly all the leucine remained in the form HL. Note also that $[H_2L^+]$ is nearly equal to $[L^-]$; this result confirms that Reactions 11-8 and 11-9 proceed equally, even though K_a is 84 times bigger than K_b for leucine.

Usually Equation 11-11 is a fair-to-excellent approximation. It applies to the intermediate form of any diprotic acid, regardless of its charge type.

An even simpler form of Equation 11-11 results from two conditions that usually exist. First, if $K_2F \gg K_w$, the second term in the numerator of Equation 11-11 can be dropped.

$$[H^+] \approx \sqrt{\frac{K_1 K_2 F + \cancel{K_1 K_w}}{K_1 + F}}$$

Then, if $K_1 \ll F$, the first term in the denominator can also be neglected.

$$[H^+] \approx \sqrt{\frac{K_1 K_2 F}{\cancel{K_1} + F}}$$

Canceling F in the numerator and denominator gives

$$[H^+] \approx \sqrt{K_1 K_2} \qquad \text{or}$$

Recall that $\log(x^{1/2}) = \frac{1}{2}\log x$ and $\log xy = \log x + \log y$.

$$\log[H^+] \approx \frac{1}{2}(\log K_1 + \log K_2)$$

$$-\log[H^+] \approx -\frac{1}{2}(\log K_1 + \log K_2)$$

The pH of the intermediate form of a diprotic acid is close to midway between the two pK_a values and is almost independent of concentration.

Intermediate form of diprotic acid: $$\text{pH} \approx \frac{pK_1 + pK_2}{2} \qquad (11\text{-}12)$$

Equation 11-12 is a good one to keep in your head. It gives a pH of 6.04 for leucine, compared with pH $= 6.06$ from Equation 11-11. Equation 11-12 says that *the pH of the intermediate form of a diprotic acid is close to midway between pK_1 and pK_2, regardless of the formal concentration.*

EXAMPLE   **pH of the Intermediate Form of a Diprotic Acid**

Potassium hydrogen phthalate, KHP, is a salt of the intermediate form of phthalic acid. Calculate the pH of 0.10 M and of 0.010 M KHP.

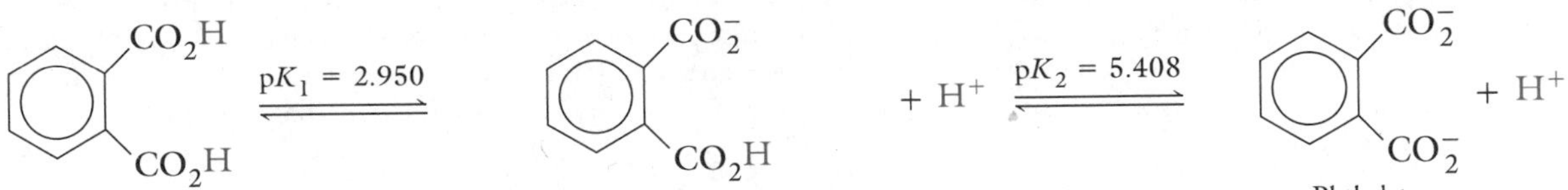

Phthalic acid
H_2P

$pK_1 = 2.950$

Monohydrogen phthalate
HP^-
Potassium hydrogen phthalate = K^+HP^-

$+ H^+$ $pK_2 = 5.408$

Phthalate
P^{2-}

$+ H^+$

Solution   The pH of potassium hydrogen phthalate is estimated as $(pK_1 + pK_2)/2 = 4.18$, regardless of concentration, using Equation 11-12. With Equation 11-11, we calculate pH = 4.18 for 0.10 M K^+HP^- and pH = 4.20 for 0.010 M K^+HP^-.

Advice: When faced with the intermediate form of a diprotic acid, use Equation 11-11 to calculate the pH. The answer should be close to $(pK_1 + pK_2)/2$.
   A summary of results for leucine is given below.

Solution	pH	$[H^+]$ (M)	$[H_2L^+]$ (M)	$[HL]$ (M)	$[L^-]$ (M)
0.0500 M H_2A	1.88	1.31×10^{-2}	3.69×10^{-2}	1.31×10^{-2}	1.79×10^{-10}
0.0500 M HA^-	6.06	8.76×10^{-7}	9.34×10^{-6}	5.00×10^{-2}	1.02×10^{-5}
0.0500 M A^{2-}	11.22	6.08×10^{-12}	2.13×10^{-12}	1.64×10^{-3}	4.84×10^{-2}

Summary of Diprotic Acid Calculations

The way to calculate the pH and composition of solutions prepared from different forms of a diprotic acid (H_2A, HA^-, or A^{2-}) is outlined below.

SOLUTION OF H_2A

1. Treat H_2A as a monoprotic acid with $K_a = K_1$, to find $[H^+]$, $[HA^-]$, and $[H_2A]$.

$$H_2A \underset{F-x}{\overset{K_1}{\rightleftharpoons}} \underset{x}{H^+} + \underset{x}{HA^-}$$

$$\frac{x^2}{F-x} = K_1$$

2. Use the K_2 equilibrium to solve for $[A^{2-}]$.

$$[A^{2-}] = \frac{K_2[\cancel{HA^-}]}{[\cancel{H^+}]} = K_2$$

Box 11-2 **Successive Approximations**

The method of *successive approximations* is a good way to deal with difficult equations that do not have simple solutions. Consider a case in which the concentration of the intermediate (amphiprotic) species of a diprotic acid is not very close to F, the formal concentration of the solution. This situation occurs when K_1 and K_2 are not very far apart, and F is small. Consider a solution of 1.00×10^{-3} M HM^-, where HM^- is the intermediate form of malic acid.

$$K_1 = 4.0 \times 10^{-4} \qquad pK_1 = 3.40 \qquad\qquad K_2 = 8.9 \times 10^{-6} \qquad pK_2 = 5.05$$

Malic acid H_2M $\qquad\qquad HM^- \qquad\qquad M^{2-}$

As a first approximation, we assume that $[HM^-] \approx 1.00 \times 10^{-3}$ M. Plugging this value into Equation 11-11, we calculate first approximations for $[H^+]$, $[H_2M]$, and $[M^{2-}]$.

$$[H^+]_1 = \sqrt{\frac{K_1 K_2 (0.001\,00) + K_1 K_w}{K_1 + (0.001\,00)}} = 5.04 \times 10^{-5} \text{ M}$$

$$\Rightarrow [H_2M]_1 = 1.26 \times 10^{-4} \text{ M and } [M^{2-}]_1 = 1.77 \times 10^{-4} \text{ M}$$

Clearly, $[H_2M]$ and $[M^{2-}]$ are not negligible relative to $F = 1.00 \times 10^{-3}$ M, so we need to revise our estimate of $[HM^-]$. The mass balance gives us a second approximation:

$$[HM^-]_2 = F - [H_2M]_1 - [M^{2-}]_1$$
$$= 0.001\,00 - 0.000\,126 - 0.000\,177 = 0.000\,697 \text{ M}$$

Using the value $[HM^-]_2 = 0.000\,697$ in Equation 11-10 gives

$$[H^+]_2 = \sqrt{\frac{K_1 K_2 (0.000\,697) + K_1 K_w}{K_1 + (0.000\,697)}} = 4.76 \times 10^{-5} \text{ M}$$

$$\Rightarrow [H_2M]_2 = 8.29 \times 10^{-5} \text{ M and } [M^{2-}]_2 = 1.30 \times 10^{-4} \text{ M}$$

The values of $[H_2M]_2$ and $[M^{2-}]_2$ can be used to calculate a third approximation for $[HM^-]$:

$$[HM^-]_3 = F - [H_2M]_2 - [M^{2-}]_2 = 0.000\,787 \text{ M}$$

Plugging $[HM^-]_3$ into Equation 11-10 gives

$$[H^+]_3 = 4.86 \times 10^{-5}$$

and the procedure can be repeated once more to get

$$[H^+]_4 = 4.83 \times 10^{-5}$$

SOLUTION OF HA^-

1. Use the approximation $[HA^-] \approx F$ and find the pH with Equation 11-11.

$$[H^+] = \sqrt{\frac{K_1 K_2 F + K_1 K_w}{K_1 + F}}$$

The pH should be close to $(pK_1 + pK_2)/2$.

We are homing in on an estimate of $[H^+]$ in which the uncertainty is already less than 1%. This is more accuracy than is justified by the accuracy of the equilibrium constants, K_1 and K_2. The fourth approximation for $[H^+]$ gives pH = 4.32, a value that we can compare with pH = 4.30 from the first approximation and pH = 4.23 from the formula pH $\approx$ (pK_1 + pK_2)/2. Considering the uncertainty in pH measurements, all this calculation was hardly worth the effort. However, the concentration of $[HM^-]_5$ is 0.000 768 M, which is 23% less than the original estimate ($[HM^-]_1 \approx$ F = 0.001 00 M).

Successive approximations can be carried out by hand, but the process is more easily and reliably performed with a spreadsheet, as shown here. Column B uses the same formulas as column C, except that $[HM^-]$ in column B is equated to the formal concentration, F, in cell $A$6. Once column C has been set up, it can be copied into further columns for each approximation.

	A	B	C	D	E
1	K1 =	1st approx.	2nd approx.	3rd approx.	4th approx.
2	0.0004	[HM–]	[HM–]	[HM–]	[HM–]
3	K2 =	1.000E–03	6.974E–04	7.866E–04	7.604E–04
4	0.0000089	[H+]	[H+]	[H+]	[H+]
5	F =	5.043E–05	4.757E–05	4.858E–05	4.830E–05
6	0.001	[H2M]	[H2M]	[H2M]	[H2M]
7	Kw =	1.261E–04	8.293E–05	9.553E–05	9.181E–05
8	1.E–14	[M2–]	[M2–]	[M2–]	[M2–]
9		1.765E–04	1.305E–04	1.441E–04	1.401E–04
10		pH	pH	pH	pH
11		4.297	4.323	4.314	4.316
12					
13	B3 = $A$6				
14	C3 = $A$6–B7–B9				
15	C5 = Sqrt(($A$2*$A$4*C3+$A$2*$A$8)/($A$2+C3))				
16	C7 = C5*C3/$A$2				
17	C9 = $A$4*C3/C5				
18	C11 = –Log10(C5)				

2. Using $[H^+]$ from Step 1 and $[HA^-] \approx$ F, solve for H_2A and A^{2-}, using the K_1 and K_2 equilibria.

$$[H_2A] = \frac{[HA^-][H^+]}{K_1}$$

$$[A^{2-}] = \frac{K_2[HA^-]}{[H^+]}$$

SOLUTION OF A^{2-}

1. Treat A^{2-} as monobasic, with $K_b = K_{b1} = K_w/K_{a2}$ to find [A^{2-}], [HA$^-$], and [H$^+$].

$$A^{2-} + H_2O \overset{K_{b1}}{\rightleftharpoons} HA^- + OH^-$$
$$\underset{F-x}{} \qquad \underset{x}{} \quad \underset{x}{}$$

$$\frac{x^2}{F-x} = K_{b1} = \frac{K_w}{K_{a2}}$$

$$[H^+] = \frac{K_w}{[OH^-]} = \frac{K_w}{x}$$

2. Use the K_1 equilibrium to solve for [H$_2$A].

$$[H_2A] = \frac{[HA^-][H^+]}{K_{a1}} = \frac{[HA^-](K_w/[OH^-])}{K_{a1}} = K_{b2}$$

11-2 Diprotic Buffers

A buffer made from a diprotic (or polyprotic) acid is treated in the same way as a buffer made from a monoprotic acid. For the acid H$_2$A, we can write the two Henderson-Hasselbalch equations below, both of which are *always* true. If we happen to know [H$_2$A] and [HA$^-$], then we will use the pK_1 equation. If we know [HA$^-$] and [A^{2-}], we will use the pK_2 equation.

$$pH = pK_1 + \log \frac{[HA^-]}{[H_2A]}$$

$$pH = pK_2 + \log \frac{[A^{2-}]}{[HA^-]}$$

All Henderson-Hasselbalch equations (with activity coefficients) are always true for a solution at equilibrium.

- -

EXAMPLE   A Diprotic Buffer System

Find the pH of a solution prepared by dissolving 1.00 g of potassium hydrogen phthalate and 1.20 g of disodium phthalate in 50.0 mL of water.

Solution   The structures of monohydrogen phthalate and phthalate were shown in the previous example. The formula weights are KHP = C$_8$H$_5$O$_4$K = 204.223 and Na$_2$P = C$_8$H$_4$O$_4$Na$_2$ = 210.097. The pH is given by

$$pH = pK_2 + \log \frac{[P^{2-}]}{[HP^-]} = 5.408 + \log \frac{(1.20\ g)/(210.097\ g/mol)}{(1.00\ g)/(204.223\ g/mol)} = 5.47$$

We used pK_2 because K_2 is the acid-dissociation constant of HP$^-$, which appears in the denominator of the Henderson-Hasselbalch equation. Notice that the volume of solution was not used in answering the question.

- -

- -

EXAMPLE   Preparing a Buffer in a Diprotic System

How many milliliters of 0.800 M KOH should be added to 3.38 g of oxalic acid to give a pH of 4.40 when diluted to 500 mL?

$$\begin{matrix} O & O \\ \parallel & \parallel \\ HOC & COH \end{matrix}$$

Oxalic acid
(H_2Ox)

Formula weight = 90.036

$pK_1 = 1.252$

$pK_2 = 4.266$

Solution We know that a 1:1 mole ratio of HOx^-:Ox^{2-} would have pH = $pK_2 = 4.266$. If the pH is to be 4.40, there must be more Ox^{2-} than HOx^- present. We must add enough base to convert all of the H_2Ox to HOx^-, plus enough additional base to convert the right amount of HOx^- to Ox^{2-}.

$$H_2Ox + OH^- \longrightarrow HOx^- + H_2O$$

$$pH \approx \frac{pK_1 + pK_2}{2} = 2.76$$

$$HOx^- + OH^- \longrightarrow Ox^{2-} + H_2O$$

A 1:1 mixture would have
pH = pK_2 = 4.27

In 3.38 g of H_2Ox, there are $0.037\,5_4$ mol. The volume of 0.800 M KOH needed to react with this much H_2Ox to make HOx^- is $(0.037\,5_4 \text{ mol})/(0.800 \text{ M}) = 46.9_2$ mL.

To produce a pH of 4.40 requires

	HOx^-	+	OH^-	$\rightarrow$	Ox^{2-}
Initial moles:	$0.037\,5_4$		x		—
Final moles:	$0.037\,5_4 - x$		—		x

$$pH = pK_2 + \log\frac{[Ox^{2-}]}{[HOx^-]}$$

$$4.40 = 4.266 + \log\frac{x}{0.037\,5_4 - x} \Rightarrow x = 0.021\,6_4 \text{ mol}$$

The volume of KOH needed to deliver $0.021\,6_4$ mol is $(0.021\,6_4 \text{ mol})/(0.800 \text{ M}) = 27.0_5$ mL. The total volume of KOH needed to bring the pH up to 4.40 is $46.9_2 + 27.0_5 = 73.9_7$ mL.

· ·

11-3 Polyprotic Acids and Bases

The treatment of diprotic acids and bases can be extended to polyprotic systems. By way of review, let's write the pertinent equilibria for a triprotic system.

$$H_3A \rightleftharpoons H_2A^- + H^+ \qquad K_{a1} = K_1$$

$$H_2A^- \rightleftharpoons HA^{2-} + H^+ \qquad K_{a2} = K_2$$

$$HA^{2-} \rightleftharpoons A^{3-} + H^+ \qquad K_{a3} = K_3$$

$$A^{3-} + H_2O \rightleftharpoons HA^{2-} + OH^- \qquad K_{b1} = \frac{K_w}{K_{a3}}$$

$$HA^{2-} + H_2O \rightleftharpoons H_2A^- + OH^- \qquad K_{b2} = \frac{K_w}{K_{a2}}$$

$$H_2A^- + H_2O \rightleftharpoons H_3A + OH^- \qquad K_{b3} = \frac{K_w}{K_{a1}}$$

The acid-base equilibria for a triprotic system:

$$K_{bi} = \frac{K_w}{K_{a(3-i)}}$$

The K values in Equations 11-13 and 11-14 are K_a values for the triprotic acid.

Triprotic systems are treated as follows:

1. H_3A is treated as a monoprotic weak acid, with $K_a = K_1$.

2. H_2A^- is treated as the intermediate form of a diprotic acid.

$$[H^+] \approx \sqrt{\frac{K_1 K_2 F + K_1 K_w}{K_1 + F}} \qquad (11\text{-}13)$$

3. HA^{2-} is also treated as the intermediate form of a diprotic acid. However, HA^{2-} is "surrounded" by H_2A^- and A^{3-}, so the equilibrium constants to use are K_2 and K_3, instead of K_1 and K_2.

$$[H^+] \approx \sqrt{\frac{K_2 K_3 F + K_2 K_w}{K_2 + F}} \qquad (11\text{-}14)$$

4. A^{3-} is treated as monobasic, with $K_b = K_{b1} = K_w/K_{a3}$.

· ·

EXAMPLE   A Triprotic System

Find the pH of 0.10 M H_3His^{2+}, 0.10 M H_2His^+, 0.10 M HHis, and 0.10 M His^-, where His stands for the amino acid histidine.

Solution

0.10 M H_3His^{2+}: Treating this as a monoprotic acid, we write

$$\underset{F-x}{H_3His^{2+}} \rightleftharpoons \underset{x}{H_2His^+} + \underset{x}{H^+}$$

$$\frac{x^2}{F - x} = K_1 = 2 \times 10^{-2} \Rightarrow x = 3._{58} \times 10^{-2} \text{ M} \Rightarrow pH = 1.45$$

0.10 M H_2His^+: Using Equation 11-13, we write

$$[H^+] = \sqrt{\frac{(2 \times 10^{-2})(9.5 \times 10^{-7})(0.10) + (2 \times 10^{-2})(1.0 \times 10^{-14})}{2 \times 10^{-2} + 0.10}}$$

$$= 1.2_6 \times 10^{-4} \text{ M} \Rightarrow pH = 3.90$$

Note that $(pK_1 + pK_2)/2 = 3.86$.

0.10 M HHis: Using Equation 11-14, we write

$$[H^+] = \sqrt{\frac{(9.5 \times 10^{-7})(8.3 \times 10^{-10})(0.10) + (9.5 \times 10^{-7})(1.0 \times 10^{-14})}{9.5 \times 10^{-7} + 0.10}}$$

$$= 2.8_1 \times 10^{-8} \text{ M} \Rightarrow pH = 7.55$$

Note that $(pK_2 + pK_3)/2 = 7.55$

0.10 M His⁻: Treating this as monobasic, we can write

$$\underset{\text{F} - x}{\text{His}^-} + H_2O \rightleftharpoons \underset{x}{\text{HHis}} + \underset{x}{\text{OH}^-}$$

$$\frac{x^2}{\text{F} - x} = K_{b1} = \frac{K_w}{K_{a3}} = 1.2 \times 10^{-5} \Rightarrow x = 1.0_9 \times 10^{-3} \text{ M}$$

$$pH = -\log\left(\frac{K_w}{x}\right) = 11.04$$

We have reduced acid-base problems to just three types. When you encounter an acid or base, decide whether you are dealing with an acidic, basic, or intermediate form. Then do the appropriate arithmetic to answer the question at hand.

Three forms of acids and bases:
  acidic
  basic
  intermediate (amphiprotic)

11-4   Which Is the Principal Species?

We are often faced with the problem of identifying which species of acid, base, or intermediate is predominant under given conditions. A simple example is this: What is the principal form of benzoic acid in an aqueous solution at pH 8?

Benzoic acid
$pK_a = 4.20$

The pK_a for benzoic acid is 4.20. This means that at pH 4.20, there is a 1:1 mixture of benzoic acid (HA) and benzoate ion (A⁻). At $pH = pK_a + 1$ (= 5.20), the quotient [A⁻]/[HA] is 10:1. At $pH = pK_a + 2$ (= 6.20), the quotient [A⁻]/[HA] is 100:1. As the pH increases, the quotient [A⁻]/[HA] increases still further.

For a monoprotic system, the basic species, A⁻, is the predominant form when pH > pK_a. The acidic species, HA, is the predominant form when pH < pK_a. The predominant form of benzoic acid at pH 8 is the benzoate anion, $C_6H_5CO_2^-$.

At $pH = pK_a$, [A⁻] = [HA] because
$$pH = pK_a + \log\frac{[A^-]}{[HA]}.$$

pH	Major species
<pK_a	HA
>pK_a	A⁻

EXAMPLE   Principal Species—Which One and How Much?

What is the predominant form of ammonia in a solution at pH 7.0? Approximately what fraction is in this form?

pH	Major species
$pH < pK_1$	H_2A
$pK_1 < pH < pK_2$	HA^-
$pH > pK_2$	A^{2-}

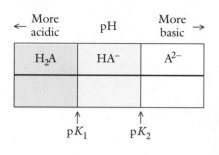

← More acidic	pH	More basic →
H_2A	HA^-	A^{2-}

↑ pK_1 ↑ pK_2

Solution In Appendix G we find $pK_a = 9.24$ for the ammonium ion (NH_4^+, the conjugate acid of ammonia, NH_3). At pH = 9.24, $[NH_4^+] = [NH_3]$. Below pH 9.24, NH_4^+ will be the predominant form. Because pH = 7.0 is about 2 pH units below pK_a, the quotient $[NH_4^+]/[NH_3]$ will be around 100:1. More than 99% is in the form NH_4^+.

For polyprotic systems, the reasoning is the same, but there are several values of pK_a. Consider oxalic acid, H_2Ox, with $pK_1 = 1.25$ and $pK_2 = 4.27$. At pH = pK_1, $[H_2Ox] = [HOx^-]$. At pH = pK_2, $[HOx^-] = [Ox^{2-}]$. The chart at the left shows the major species in each pH region.

EXAMPLE   Principal Species in a Polyprotic System

The amino acid arginine has the following forms:

H_3Arg^{2+} $\xrightarrow{pK_1 = 1.82}$ H_2Arg^+ $\xrightarrow{pK_2 = 8.99}$

$\xrightarrow{pK_3 = 12.48}$ Arg$^-$

HArg
Arginine

Note that the α-ammonium group (the one next to the carboxyl group at the left) is more acidic than the substituent (at the right). This information comes from Appendix G. What is the principal form of arginine at pH 10.0? Approximately what fraction is in this form? What is the second most abundant form at this pH?

Solution We know that at pH = pK_2 = 8.99, $[H_2Arg^+] = [HArg]$. At pH = pK_3 = 12.48, $[HArg] = [Arg^-]$. At pH = 10.0, the major species is HArg. Because pH 10.0 is about one pH unit higher than pK_2, we can say that $[HArg]/[H_2Arg^+] \approx 10:1$. About 90% of arginine is in the form HArg. The second most important species is H_2Arg^+, which makes up about 10% of the arginine.

EXAMPLE   More on Polyprotic Systems

In the pH range 1.82 to 8.99, H_2Arg^+ is the principal form of arginine. Which is the second most prominent species at pH 6.0?   At pH 5.0?

Solution We know that the pH of the pure intermediate (amphiprotic) species, H_2Arg^+, is given by Equation 11-12:

$$pH \text{ of } H_2Arg^+ \approx \frac{1}{2}(pK_1 + pK_2) = 5.40$$

Above pH 5.40 (and below pH = pK_2), we expect that HArg, the conjugate base of H_2Arg^+, will be the second most important species. Below pH 5.40 (and above pH = pK_1), we anticipate that H_3Arg^{2+} will be the second most important species.

Figure 11-1 summarizes how we think of a triprotic system. We determine the principal species by comparing the pH of the solution with the various pK_a values.

Go back and read the Example "Preparing a Buffer in a Diprotic System" on page 258. See if it makes more sense now.

11-5 Fractional Composition Equations

We now derive equations that give the fraction of each species of acid or base at a given pH. These will be useful for acid-base and EDTA titrations, as well as for electrochemical equilibria.

Monoprotic Systems

Our goal is to find an expression for the fraction of an acid in each form (HA and A^-), and we can do this by combining the equilibrium constant with the mass balance. Consider an acid with formal concentration F:

$$HA \overset{K_a}{\rightleftharpoons} H^+ + A^- \qquad K_a = \frac{[H^+][A^-]}{[HA]}$$

$$\text{mass balance:} \quad F = [HA] + [A^-]$$

Rearranging the mass balance gives $[A^-] = F - [HA]$, which can be plugged into the K_a equilibrium to give

$$K_a = \frac{[H^+](F - [HA])}{[HA]}$$

or, with a little algebra,

$$[HA] = \frac{[H^+]F}{[H^+] + K_a} \qquad (11\text{-}15)$$

The *fraction* of molecules in the form HA is called α_{HA}.

$$\alpha_{HA} = \frac{[HA]}{[HA] + [A^-]} = \frac{[HA]}{F} \qquad (11\text{-}16)$$

α_{HA} = fraction of species in the form HA

α_{A^-} = fraction of species in the form A^-

$$\boxed{\alpha_{HA} + \alpha_{A^-} = 1}$$

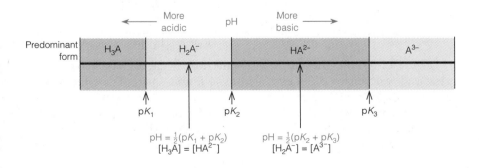

Figure 11-1 The predominant molecular form of a triprotic system (H_3A) in the various pH intervals.

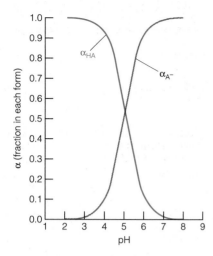

Figure 11-2 Fractional composition diagram of a monoprotic system with $pK_a = 5.00$. Below pH 5, HA is the dominant form, whereas above pH 5, A^- dominates.

Express $[HA^-]$ in terms of $[H_2A]$ from the previous equation to obtain the final result.

α_{H_2A} = fraction of species in the form H_2A

α_{HA^-} = fraction of species in the form HA^-

$\alpha_{A^{2-}}$ = fraction of species in the form A^{2-}

$$\alpha_{H_2A} + \alpha_{HA^-} + \alpha_{A^{2-}} = 1$$

The general form of α for the polyprotic acid H_nA is

$$\alpha_{H_nA} = \frac{[H^+]^n}{D}$$

$$\alpha_{H_{n-1}A} = \frac{K_1[H^+]^{n-1}}{D}$$

$$\alpha_{H_{n-j}A} = \frac{K_1K_2\cdots K_j[H^+]^{n-j}}{D}$$

where $D = [H^+]^n + K_1[H^+]^{n-1} + K_1K_2[H^+]^{n-2} + \cdots + K_1K_2K_3\cdots K_n$.

Dividing Equation 11-15 by F gives

Fraction in the form HA: $\quad \alpha_{HA} = \dfrac{[HA]}{F} = \dfrac{[H^+]}{[H^+] + K_a}$ (11-17)

In a similar manner, the fraction in the form A^-, designated α_{A^-} (what we called the *fraction of dissociation*, α, previously), can be obtained:

Fraction in the form A^-: $\quad \alpha_{A^-} = \dfrac{[A^-]}{F} = \dfrac{K_a}{[H^+] + K_a}$ (11-18)

Figure 11-2 shows α_{HA} and α_{A^-} for a system with $pK_a = 5.00$. At low pH, almost all of the acid is in the form HA. At high pH, almost everything is in the form A^-.

Diprotic Systems

Now we will derive the fractional composition equations for a diprotic system. The derivation follows the same pattern used for the monoprotic system.

$$H_2A \overset{K_1}{\rightleftharpoons} H^+ + HA^-$$

$$HA^- \overset{K_2}{\rightleftharpoons} H^+ + A^{2-}$$

$$K_1 = \frac{[H^+][HA^-]}{[H_2A]} \Rightarrow [HA^-] = [H_2A]\frac{K_1}{[H^+]}$$

$$K_2 = \frac{[H^+][A^{2-}]}{[HA^-]} \Rightarrow [A^{2-}] = [HA^-]\frac{K_2}{[H^+]} = [H_2A]\frac{K_1K_2}{[H^+]^2}$$

mass balance: $\quad F = [H_2A] + [HA^-] + [A^{2-}]$

$$F = [H_2A] + \frac{K_1}{[H^+]}[H_2A] + \frac{K_1K_2}{[H^+]^2}[H_2A]$$

$$F = [H_2A]\left(1 + \frac{K_1}{[H^+]} + \frac{K_1K_2}{[H^+]^2}\right)$$

For a diprotic system, we designate the fraction in the form H_2A as α_{H_2A}, the fraction in the form HA^- as α_{HA^-}, and the fraction in the form A^{2-} as $\alpha_{A^{2-}}$. From the definition of α_{H_2A}, we can write

Fraction in the form H_2A: $\alpha_{H_2A} = \dfrac{[H_2A]}{F} = \dfrac{[H^+]^2}{[H^+]^2 + [H^+]K_1 + K_1K_2}$ (11-19)

In a similar manner, we can derive the following equations:

Fraction in the form HA^-: $\alpha_{HA^-} = \dfrac{[HA^-]}{F} = \dfrac{K_1[H^+]}{[H^+]^2 + [H^+]K_1 + K_1K_2}$ (11-20)

Fraction in the form A^{2-}: $\alpha_{A^{2-}} = \dfrac{[A^{2-}]}{F} = \dfrac{K_1K_2}{[H^+]^2 + [H^+]K_1 + K_1K_2}$ (11-21)

Figure 11-3 shows the curves for the fractions α_{H_2A}, α_{HA^-}, and $\alpha_{A^{2-}}$ for fumaric acid, whose two pK_a values are only 1.5 units apart. The value of α_{HA^-} rises only to 0.72 because the two pK values are so close together. There is a substantial amount of both H_2A and A^{2-} in the region $pK_1 < pH < pK_2$.

Equations 11-19 through 11-21 apply equally well to B, BH^+, and BH_2^{2+} obtained by dissolving the base B in water. The fraction α_{H_2A} applies to the

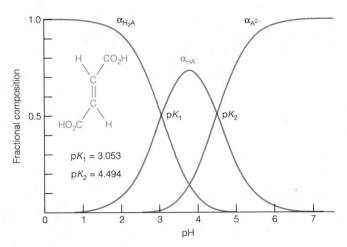

Figure 11-3 Fractional composition diagram for fumaric acid (*trans*-butenedioic acid). At low pH, H_2A is the dominant form. At intermediate pH, HA^- is dominant, and at high pH, A^{2-} dominates. Because pK_1 and pK_2 are not separated very much, the fraction of HA^- never gets very close to unity.

acidic form BH_2^{2+}. Similarly, α_{HA^-} applies to BH^+ and $\alpha_{A^{2-}}$ applies to B. The constants K_1 and K_2 are the acid dissociation constants of BH_2^{2+} ($K_1 = K_w/K_{b2}$ and $K_2 = K_w/K_{b1}$).

11-6 Isoelectric and Isoionic pH

Biochemists often refer to the isoelectric or isoionic pH of polyprotic molecules, such as proteins. These terms can be understood in terms of a simple diprotic system, such as the amino acid alanine.

$$\underset{\substack{\text{Alanine cation}\\ H_2A^+}}{\overset{\overset{\displaystyle CH_3}{\displaystyle |}}{H_3\overset{+}{N}CHCO_2H}} \;\rightleftharpoons\; \underset{\substack{\text{Neutral zwitterion}\\ HA}}{\overset{\overset{\displaystyle CH_3}{\displaystyle |}}{H_3\overset{+}{N}CHCO_2^-}} + H^+ \qquad pK_1 = 2.35$$

$$\overset{\overset{\displaystyle CH_3}{\displaystyle |}}{H_3\overset{+}{N}CHCO_2^-} \;\rightleftharpoons\; \underset{\substack{\text{Alanine anion}\\ A^-}}{\overset{\overset{\displaystyle CH_3}{\displaystyle |}}{H_2NCHCO_2^-}} + H^+ \qquad pK_2 = 9.87$$

The **isoionic point** (or isoionic pH) is the pH obtained when the pure, neutral polyprotic acid HA (the neutral zwitterion) is dissolved in water. The only ions in such a solution are H_2A^+, A^-, H^+, and OH^-. Most of the alanine is in the form HA, and the concentrations of H_2A^+ and A^- are *not* equal to each other.

The **isoelectric point** (or isoelectric pH) is the pH at which the *average* charge of the polyprotic acid is zero. Most of the molecules are in the uncharged form HA and the concentrations of H_2A^+ and A^- *are* equal to each other. It is important to realize that there is always some H_2A^+ and some A^- in equilibrium with HA.

Isoionic pH is the pH of the pure, neutral, polyprotic acid.

Isoelectric pH is the pH at which the average charge of the polyprotic acid is zero.

Alanine is the intermediate form of a diprotic acid, so we use Equation 11-11 to find the pH.

When pure alanine is dissolved in water, the pH of the solution, by definition, is the isoionic pH. Because alanine (HA) is the intermediate form of a diprotic acid (H_2A^+), the pH is given by

Isoionic point:
$$[H^+] = \sqrt{\frac{K_1K_2F + K_1K_w}{K_1 + F}} \qquad (11\text{-}22)$$

where F is the formal concentration of alanine. For 0.10 M alanine, the isoionic pH is

$$[H^+] = \sqrt{\frac{K_1K_2(0.10) + K_1K_w}{K_1 + (0.10)}} = 7.6 \times 10^{-7} \text{ M} \Rightarrow \text{pH} = 6.12$$

The isoelectric point is the pH at which the concentrations of H_2A^+ and A^- are equal, and, therefore, the average charge of alanine is zero. We calculate the isoelectric pH by first writing expressions for the concentrations of the cation and anion.

$$[H_2A^+] = \frac{[HA][H^+]}{K_1} \qquad [A^-] = \frac{K_2[HA]}{[H^+]}$$

Setting $[H_2A^+] = [A^-]$, we find

$$\frac{[HA][H^+]}{K_1} = \frac{K_2[HA]}{[H^+]} \Rightarrow [H^+] = \sqrt{K_1K_2}$$

which gives

The isoelectric point is midway between the two pK_a values "surrounding" the neutral, intermediate species.

Isoelectric point:
$$\text{pH} = \frac{pK_1 + pK_2}{2} \qquad (11\text{-}23)$$

For a diprotic amino acid, the isoelectric pH is halfway between the two pK_a values. The isoelectric pH of alanine is $\frac{1}{2}(2.35 + 9.87) = 6.11$.

The isoelectric and isoionic points for a polyprotic acid have similar values. At the isoelectric pH, the average charge of the molecule is zero; thus $[H_2A^+] = [A^-]$ and pH $= \frac{1}{2}(pK_1 + pK_2)$. At the isoionic point, the pH is given by Equation 11-22, and $[H_2A^+]$ is not exactly equal to $[A^-]$. However, the isoionic pH is quite close to the isoelectric pH.

For a protein, the *isoionic* pH is the pH of a solution containing the pure protein with no other ions except H^+ and OH^-. Proteins are usually isolated in a charged form together with various counterions (such as Na^+, NH_4^+, or Cl^-). When such a protein is subjected to intensive *dialysis* (Demonstration 26-1) against pure water, the pH in the protein compartment approaches the isoionic point if the counterions are free to pass through the semipermeable dialysis membrane that retains the protein. The *isoelectric* point is the pH at which the protein has no net charge. This property forms the basis for a sensitive technique used in the separation of proteins, as described in Box 11-3.

11-7 Microequilibrium Constants

2,3-Diphosphoglycerate reduces the affinity of hemoglobin for O_2. One method by which humans adapt to high altitudes is to decrease the concentration of 2,3-diphosphoglycerate in the blood and thereby increase the affinity of hemoglobin for the scarce O_2 in the air.

When two or more sites in a molecule have similar acid dissociation constants, there is an equilibrium of protons among those sites. Consider the molecule 2,3-diphosphoglycerate, which regulates the ability of your hemoglobin to bind O_2 in red blood cells. The species designated $H_aH_bA^{3-}$ at the left on the next page is a diprotic acid that can lose a proton from phosphate at site a or site b.

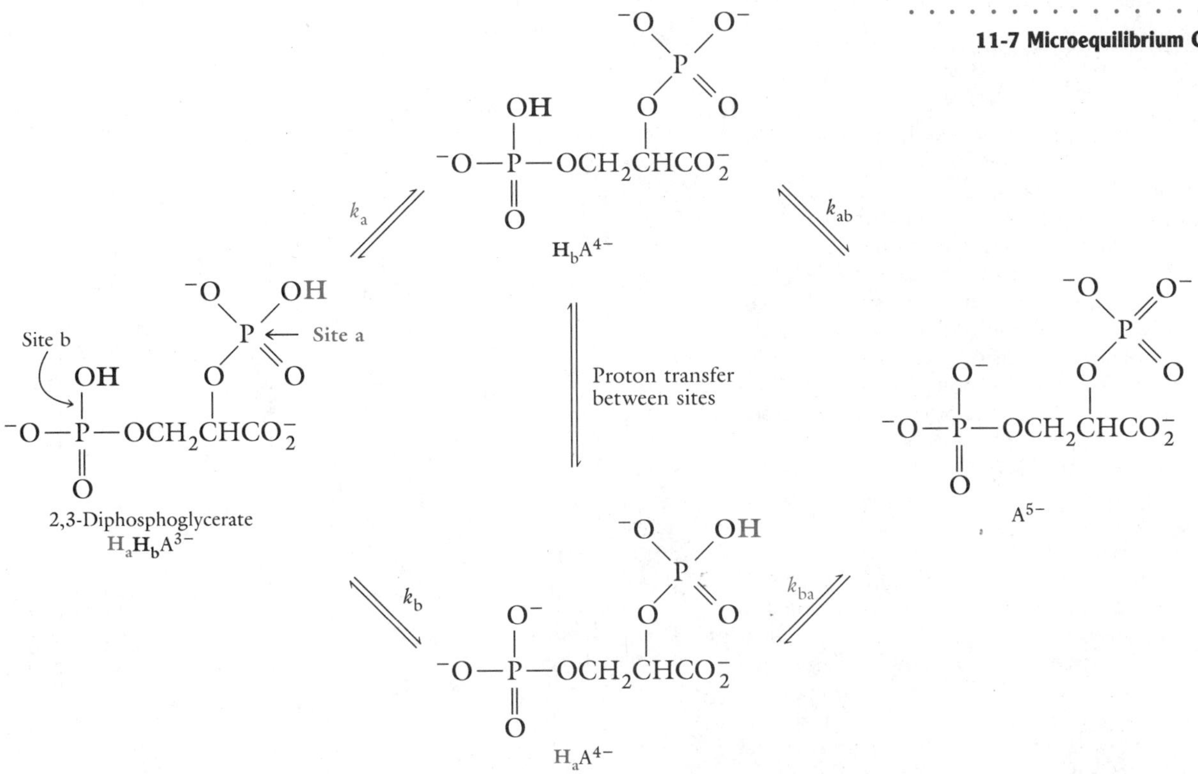

$$\overset{-O}{\underset{O}{\overset{|}{P}}}\overset{O^-}{\underset{}{}}$$

OH   O   O

$^-O-P-OCH_2CHCO_2^-$

O

H_bA^{4-}

k_a k_{ab}

Site b $\overset{-O}{\underset{O}{P}}\overset{OH}{}$ ← Site a

OH   O   O

$^-O-P-OCH_2CHCO_2^-$

O

2,3-Diphosphoglycerate
$H_aH_bA^{3-}$

Proton transfer
between sites

$\overset{-O}{\underset{}{P}}\overset{O^-}{\underset{O}{}}$

O^- O   O

$^-O-P-OCH_2CHCO_2^-$

O

A^{5-}

k_b k_{ba}

$\overset{-O}{\underset{}{P}}\overset{OH}{}$

O^- O   O

$^-O-P-OCH_2CHCO_2^-$

O

H_aA^{4-}

A **microequilibrium constant** describes the reaction of a chemically distinct site in a molecule. Each site has a unique constant associated with it.

$$\text{loss of } H_a \text{ from } H_aH_bA^{3-}: \quad k_a = \frac{[H_bA^{4-}][H^+]}{[H_aH_bA^{3-}]} = 2.9 \times 10^{-7}$$

$$\text{loss of } H_b \text{ from } H_aH_bA^{3-}: \quad k_b = \frac{[H_aA^{4-}][H^+]}{[H_aH_bA^{3-}]} = 1.5 \times 10^{-7}$$

$$\text{loss of } H_b \text{ from } H_bA^{4-}: \quad k_{ab} = \frac{[A^{5-}][H^+]}{[H_bA^{4-}]} = 7.5 \times 10^{-8}$$

$$\text{loss of } H_a \text{ from } H_aA^{4-}: \quad k_{ba} = \frac{[A^{5-}][H^+]}{[H_aA^{4-}]} = 1.5 \times 10^{-7}$$

By contrast, ordinary equilibrium constants describe the gain or loss of protons without regard to which sites in the molecule participate in the chemistry. The microequilibrium constants are related to the conventional equilibrium constants as follows:

$$K_1 = \frac{([H_aA^{4-}] + [H_bA^{4-}])[H^+]}{[H_aH_bA^{3-}]} = k_a + k_b \qquad (11\text{-}24)$$

$$K_2 = \frac{[A^{5-}][H^+]}{[H_aA^{4-}] + [H_bA^{4-}]} = \frac{k_{ab}k_{ba}}{k_{ab} + k_{ba}} \qquad (11\text{-}25)$$

Microequilibrium constants may be measured by *nuclear magnetic resonance spectroscopy*, in which the spectroscopic signals for molecules protonated at site a can be distinguished from those protonated at site b.[1] The results of one such experiment are displayed in Figure 11-4, in which

Box 11-3 Isoelectric Focusing

At its *isoelectric point,* the average charge of all forms of a protein is zero. It will therefore not migrate in an electric field at its isoelectric pH. This effect is the basis of a very sensitive technique of protein separation called **isoelectric focusing.** A mixture of proteins is subjected to a strong electric field in a medium specifically designed to have a pH gradient. Positively charged molecules move toward the negative pole and negatively charged molecules move toward the positive pole. The proteins migrate in one direction or the other until they reach the point where the pH is the same as their isoelectric pH. At this point they have no net charge and no longer move. Thus, each protein in the mixture is focused in one small region at its isoelectric pH.

An example of isoelectric focusing is shown in the accompanying figure. In this experiment a mixture of seven proteins (and, apparently, a host of impurities) was applied to a polyacrylamide gel containing a mixture of polyprotic compounds called *ampholytes.* Each end of the gel was placed in contact with a conducting solution and several hundred volts were applied across the length of the gel. The ampholytes migrated until they formed a stable pH gradient (ranging from about pH 3 at one end of the gel to pH 10 at the other).[2] The proteins migrated until they reached regions with their isoelectric pH, at which point they had no net charge and ceased migrating. If a molecule diffuses out of its isoelectric region, it becomes charged and immediately migrates back to its isoelectric zone in the gel. When the proteins finished migrating, the field was removed. The proteins were precipitated in place on the gel and stained with a dye to make their positions visible.

The stained gel is shown at the bottom of the figure. A spectrophotometer scan of the dye peaks is shown on the graph, and a profile of measured pH is also plotted. Each dark band of stained protein gives an absorbance peak. The instrument that measures absorbance as a function of position along the gel is called a *densitometer.*

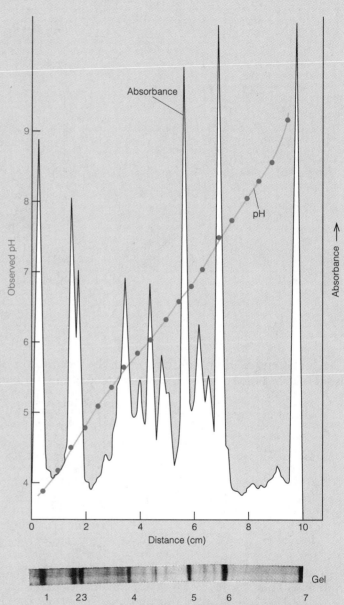

Isoelectric focusing of a mixture of proteins: (1) soybean trypsin inhibitor; (2) β-lactoglobulin A; (3) β-lactoglobulin B; (4) ovotransferrin; (5) horse myoglobin; (6) whale myoglobin; and (7) cytochrome *c.* [Courtesy BioRad Laboratories, Richmond, CA.]

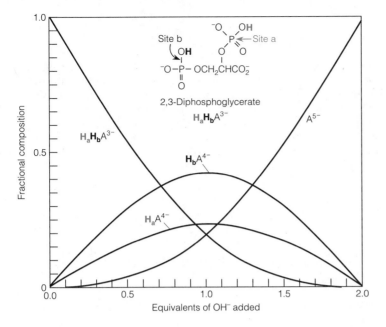

Figure 11-4 Fractional composition diagram for treatment of 2,3-diphosphoglycerate with base. Equilibrium among the species is described by microequilibrium constants. [From California State Science Fair project of Douglas Harris, 1992.]

$H_aH_bA^{3-}$ was titrated with OH^-. Protons from both sites are lost simultaneously, but because k_a is about twice as great at k_b, there is always a greater concentration of H_bA^{4-} than of H_aA^{4-} throughout the reaction.

Terms to Understand

amino acid	isoelectric focusing	microequilibrium constants
amphiprotic	isoelectric point	polyprotic acids and bases
diprotic acids and bases	isoionic point	zwitterion
hydrolysis		

Summary

Diprotic acids and bases fall into three categories:

1. The fully acidic form, H_2A, behaves as a monoprotic acid, $H_2A \rightleftharpoons H^+ + HA^-$, for which we solve the equation $K_{a1} = x^2/(F - x)$, where $[H^+] = [HA^-] = x$, and $[H_2A] = F - x$. After finding $[HA^-]$ and $[H^+]$, the value of $[A^{2-}]$ can be found from the K_{a2} equilibrium condition.
2. The fully basic form, A^{2-}, behaves as a base, $A^{2-} + H_2O \rightleftharpoons HA^- + OH^-$, for which we solve the equation $K_{b1} = x^2/(F - x)$, where $[OH^-] = [HA^-] = x$, and $[A^{2-}] = F - x$. After finding these concentrations, $[H_2A]$ can be found from the K_{a1} or K_{b2} equilibrium conditions.
3. The intermediate (amphiprotic) form, HA^-, is both an acid and a base. Its pH is given by

$$[H^+] = \sqrt{\frac{K_1K_2F + K_1K_w}{K_1 + F}}$$

where K_1 and K_2 are acid dissociation constants for H_2A, and F is the formal concentration of the intermediate. In most cases this equation reduces to the form pH $\approx \frac{1}{2}(pK_1 + pK_2)$, in which pH is independent of concentration.

In triprotic systems, there are two intermediate forms. The pH of each is found with an equation analogous to that for the intermediate form of a diprotic system. Triprotic systems also have one fully acidic and one fully basic form; these can be treated as monoprotic for the purpose of calculating pH. For polyprotic buffers, we simply write the appropriate Henderson-Hasselbalch equation connecting the two principal species in the system. The pK in this equation is the one that applies to the acid in the denominator of the log term.

The principal species of a monoprotic or polyprotic system is found by comparing the pH with the various pK_a values. For pH < pK_1, the fully protonated species,

H_nA, is the predominant form. For $pK_1 < pH < pK_2$, the form $H_{n-1}A^-$ is favored; and at each successive pK value, the next deprotonated species becomes the principal species. Finally, at pH values higher than the highest pK, the fully basic form (A^{n-}) is dominant. The fractional composition of a solution is expressed by α, given in Equations 11-17 and 11-18 for a monoprotic system and Equations 11-19 through 11-21 for a diprotic system.

The isoelectric pH of a polyprotic compound is that pH at which the average charge of all species is zero. For a diprotic amino acid whose amphiprotic form is neutral, the isoelectric pH is given by $pH = \frac{1}{2}(pK_1 + pK_2)$. The isoionic pH of a polyprotic species is the pH that would exist in a solution containing only the ions derived from the neutral polyprotic species and from H_2O.

Microequilibrium constants apply to gain or loss of protons from chemically distinct sites in a polyprotic system. In contrast, ordinary equilibrium constants describe the net gain or loss of protons, without regard to which sites are involved.

Exercises

A. Find the pH and the concentrations of H_2SO_3, HSO_3^-, and SO_3^{2-} in each of the following solutions:

(a) 0.050 M H_2SO_3

(b) 0.050 M $NaHSO_3$

(c) 0.050 M Na_2SO_3

B. (a) How many grams of $NaHCO_3$ (FW 84.007) must be added to 4.00 g of K_2CO_3 (FW 138.206) to give a pH of 10.80 in 500 mL of water?

(b) What will be the pH if 100 mL of 0.100 M HCl are added to the solution in (a)?

(c) How many milliliters of 0.320 M HNO_3 should be added to 4.00 g of K_2CO_3 to give a pH of 10.00 in 250 mL?

C. How many milliliters of 0.800 M KOH should be added to 3.38 g of oxalic acid ($H_2C_2O_4$, FW 90.035) to give a pH of 2.40 when diluted to 500 mL?

D. Calculate the pH of a 0.010 M solution of each amino acid in the form drawn below.

(a) 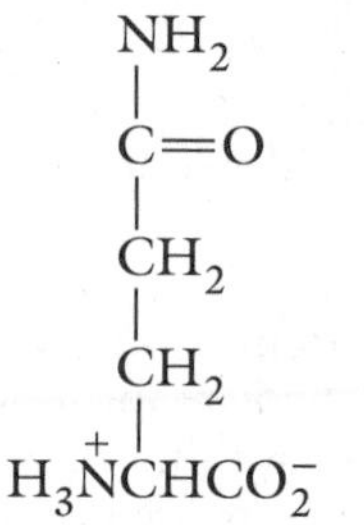 Glutamine

(b) 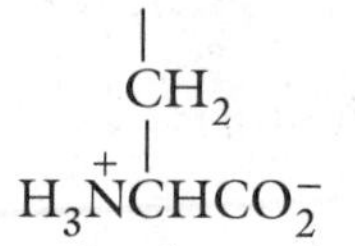 Cysteine⁻

(c) 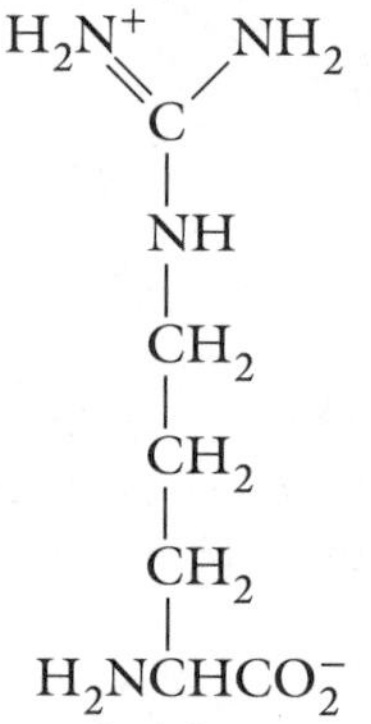 Arginine

E. (a) Draw the structure of the predominant form (principal species) of 1,3-dihydroxybenzene at pH 9.00 and at pH 11.00.

(b) What is the second most prominent species at each pH?

(c) Calculate the percentage in the major form at each pH.

F. Draw the structures of the predominant forms of glutamic acid and tyrosine at pH 9.0 and pH 10.0. What is the second most abundant species at each pH?

G. Calculate the isoionic pH of 0.010 M lysine.

H. Neutral lysine can be written HL. The other forms of lysine are H_3L^{2+}, H_2L^+, and L^-. The isoelectric point is the pH at which the *average* charge of lysine is zero. Therefore, at the isoelectric point $2[H_3L^{2+}] + [H_2L^+] = [L^-]$. Use this condition to calculate the isoelectric pH of lysine.

Problems

Diprotic Acids and Bases

1. Consider HA^-, the intermediate form of a diprotic acid. K_a for this species is 10^{-4} and K_b is 10^{-8}. Nonetheless, both the K_a and K_b reactions proceed to nearly the same extent when NaHA is dissolved in water. Explain.

2. Write the general structure of an amino acid. Why do some amino acids in Table 11-1 have two pK values and others three?

3. Write the chemical reactions whose equilibrium constants are K_{b1} and K_{b2} for the amino acid proline. Find the values of K_{b1} and K_{b2}.

4. Consider the diprotic acid H_2A with $K_1 = 1.00 \times 10^{-4}$ and $K_2 = 1.00 \times 10^{-8}$. Find the pH and concentrations of H_2A, HA^-, and A^{2-} in each of the following solutions:

(a) 0.100 M H_2A

(b) 0.100 M NaHA

(c) 0.100 M Na_2A

5. We will abbreviate malonic acid, $CH_2(CO_2H)_2$, as H_2M. Find the pH and concentrations of H_2M, HM^-, and M^{2-} in each of the following solutions:

(a) 0.100 M H_2M

(b) 0.100 M NaHM

(c) 0.100 M Na_2M

6. Calculate the pH of 0.300 M piperazine. Calculate the concentration of each form of piperazine in this solution.

7. Use the method of Box 11-2 to calculate the concentrations of H^+, H_2A, HA^-, and A^{2-} in 0.001 00 M monosodium oxalate, NaHA.

8. In this problem we will calculate the pH of the intermediate form of a diprotic acid correctly, taking activities into account.

(a) Derive Equation 11-11 for a solution of potassium hydrogen phthalate (K^+HP^- in the example following Equation 11-12). Do not neglect activity coefficients in this derivation.

(b) Calculate the pH of 0.050 M KHP, using the results in (a). Assume that the sizes of both HP^- and P^{2-} are 600 pm.

9. *Water in equilibrium with atmospheric CO_2 has a pH of 5.7.* CO_2 dissolves in water to give "carbonic acid" (which is really mostly dissolved CO_2, as described in Box 5-4).

$$CO_2(g) \rightleftharpoons CO_2(aq) \qquad K = 10^{-1.5}$$

(The equilibrium constant is called the *Henry's law constant* for carbon dioxide, because Henry's law states that the solubility of a gas in a liquid is proportional to the pressure of the gas.) The acid dissociation constants listed for "carbonic acid" in Appendix G apply to $CO_2(aq)$. Given that P_{CO_2} in the atmosphere is $10^{-3.5}$ atm, find the pH of water in equilibrium with the atmosphere.

10. *Thermodynamics and propagation of uncertainty.* The bisulfite ion exists in the following equilibrium forms:

$$HSO_3^- \qquad\qquad SO_3H^-$$

The temperature dependence of the equilibrium constant at an ionic strength of 1.0 M is $\ln K = -3.23 \ (\pm 0.53) + 1.44 \ (\pm 0.15) \times 10^3 \times (1/T)$, where T is temperature in kelvins. Because $\ln K$ must be dimensionless, the number -3.23 is dimensionless and the number 1.44×10^3 has the unit of kelvins.

(a) Using Equation 5-8, calculate the enthalpy change, $\Delta H°$, and the entropy change, $\Delta S°$, for the isomerization reaction. Include uncertainties in your answers.

(b) Calculate the quotient $[SO_3H^-]/[HSO_3^-]$ at 298 K, including estimated uncertainty.

11. *Intermediate form of diprotic acid.* Use the method described in Box 11-2 to find the pH and concentration of HA^- in a 0.01 F solution of the amphiprotic salt Na^+HA^- derived from the diprotic acid H_2A with $pK_1 = 4$ and

(a) $pK_2 = 8$ (b) $pK_2 = 5$ (c) $pK_2 = 4.5$

Case (c) is a tricky one that is going to require you to help the spreadsheet with some intelligent guessing of $[HA^-]$.

Diprotic Buffers

12. How many grams of Na_2CO_3 (FW 105.99) should be mixed with 5.00 g of $NaHCO_3$ (FW 84.01) to produce 100 mL of buffer with pH 10.00?

13. How many milliliters of 0.202 M NaOH should be added to 25.0 mL of 0.0233 M salicylic acid (2-hydroxybenzoic acid, FW 138.12) to adjust the pH to 3.50?

14. Describe how you would prepare exactly 100 mL of 0.100 M picolinate buffer, pH 5.50. Possible starting materials are pure picolinic acid (pyridine-2-carboxylic acid, FW 123.11), 1.0 M HCl, and 1.0 M NaOH. Approximately how many milliliters of the HCl or NaOH will be required?

15. How many grams of Na_2SO_4 (FW 142.04) should be added to how many grams of sulfuric acid (FW 98.08) to give 1.00 L of buffer with pH 2.10 and a total sulfur ($= SO_4^{2-} + HSO_4^- + H_2SO_4$) concentration of 0.200 M?

Polyprotic Acids and Bases

16. Phosphate, present to an extent of 0.01 M, is one of the main buffers in blood plasma, whose pH is 7.45. Would phosphate be as useful if the plasma pH were 8.5?

17. Starting with the fully protonated species, write the stepwise acid dissociation reactions of the amino acids glutamic acid and tyrosine. Be sure to remove the protons in the correct order. Which species are the neutral molecules that we call glutamic acid and tyrosine?

18. **(a)** Calculate the quotient $[H_3PO_4]/[H_2PO_4^-]$ in 0.050 0 M KH_2PO_4.

(b) Find the same quotient for 0.050 0 M K_2HPO_4.

19. Find the pH and the concentration of each species of lysine in a solution of 0.010 0 M lysine·HCl, lysine monohydrochloride.

20. **(a)** Using activity coefficients correctly, calculate the pH of a solution containing a 2.00:1.00 mole ratio of $HC^{2-}:C^{3-}$, where H_3C is citric acid. Assume that the ionic strength is 0.010 M.

(b) What will be the pH if the ionic strength is raised to 0.10 M and the mole ratio $HC^{2-}:C^{3-}$ is kept constant?

Which Is the Principal Species?

21. The acid HA has $pK_a = 7.00$.

(a) Which is the principal species, HA or A^-, at pH 6.00?

(b) Which is the principal species at pH 8.00?

(c) What is the quotient $[A^-]/[HA]$ at pH 7.00? At pH 6.00?

22. The diprotic acid H_2A has $pK_1 = 4.00$ and $pK_2 = 8.00$.

(a) At what pH is $[H_2A] = [HA^-]$?

(b) At what pH is $[HA^-] = [A^{2-}]$?

(c) Which is the principal species, H_2A, HA^-, or A^{2-} at pH 2.00?

(d) Which is the principal species at pH 6.00?

(e) Which is the principal species at pH 10.00?

23. The base B has $pK_b = 5.00$.

(a) What is the value of pK_a for the acid BH^+?

(b) At what pH is $[BH^+] = [B]$?

(c) Which is the principal species, B or BH^+, at pH 7.00?

(d) What is the quotient $[B]/[BH^+]$ at pH 12.00?

24. Draw the structure of the predominant form of pyridoxal-5-phosphate at pH 7.00.

Fractional Composition Equations

25. The acid HA has $pK_a = 4.00$. Use Equations 11-17 and 11-18 to find the fraction in the form HA and the fraction in the form A^- at pH = 5.00. Does your answer agree with what you expect for the quotient $[A^-]/[HA]$ at pH 5.00?

26. A dibasic compound, B, has $pK_{b1} = 4.00$ and $pK_{b2} = 6.00$. Find the fraction in the form BH_2^{2+} at pH 7.00, using Equation 11-19. Note that K_1 and K_2 in Equation 11-19 are acid dissociation constants for BH_2^{2+} ($K_1 = K_w/K_{b2}$ and $K_2 = K_w/K_{b1}$).

27. What fraction of ethane-1,2-dithiol is in each form (H_2A, HA^-, A^{2-}) at pH 8.00? At pH 10.00?

28. Calculate α_{H_2A}, α_{HA^-}, and $\alpha_{A^{2-}}$ for *cis*-butenedioic acid at pH 1.00, 1.91, 6.00, 6.33, and 10.00.

29. **(a)** Derive equations for α_{H_3A}, $\alpha_{H_2A^-}$, $\alpha_{HA^{2-}}$, and $\alpha_{A^{3-}}$ for a triprotic system.

(b) Calculate the values of these fractions for phosphoric acid at pH 7.00.

30. A solution containing acetic acid, oxalic acid, ammonia, and pyridine has a pH of 9.00. What fraction of ammonia is not protonated?

31. A solution was prepared by using 10.0 mL of 0.100 M cacodylic acid and 10.0 mL of 0.080 0 M NaOH. To this mixture was added 1.00 mL of 1.27×10^{-6} M morphine. Calling morphine B, calculate the fraction of morphine present in the form BH^+.

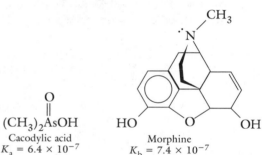

$(CH_3)_2AsOH$
Cacodylic acid
$K_a = 6.4 \times 10^{-7}$

Morphine
$K_b = 7.4 \times 10^{-7}$

32. *Fractional composition in a diprotic system.* Create a spreadsheet that uses Equations 11-19 through 11-21 to compute the three curves in Figure 11-3. Use graphics software to plot these three curves in a beautifully labeled figure.

33. *Fractional composition in a triprotic system.* For a triprotic system, the fractional composition equations are

$$\alpha_{H_3A} = \frac{[H^+]^3}{D} \qquad \alpha_{HA^{2-}} = \frac{K_1K_2[H^+]}{D}$$

$$\alpha_{H_2A^-} = \frac{K_1[H^+]^2}{D} \qquad \alpha_{A^{3-}} = \frac{K_1K_2K_3}{D}$$

where $D = [H^+]^3 + K_1[H^+]^2 + K_1K_2[H^+] + K_1K_2K_3$. Use these equations to create a fractional composition diagram analogous to Figure 11-3 for the amino acid tyrosine. What is the fraction of each species at pH 10.00?

34. *Fractional composition in a tetraprotic system.* Prepare a fractional composition diagram analogous to Figure 11-3 for the tetraprotic system derived from hydrolysis of Cr^{3+}.

$$Cr^{3+} + H_2O \rightleftharpoons Cr(OH)^{2+} + H^+$$
$$K_{a1} = 10^{-3.80}$$

$$Cr(OH)^{2+} + H_2O \rightleftharpoons Cr(OH)_2^+ + H^+$$
$$K_{a2} = 10^{-6.40}$$

$$Cr(OH)_2^+ + H_2O \rightleftharpoons Cr(OH)_3\ (aq) + H^+$$
$$K_{a3} = 10^{-6.40}$$

$$Cr(OH)_3\ (aq) + H_2O \rightleftharpoons Cr(OH)_4^- + H^+$$
$$K_{a4} = 10^{-11.40}$$

(Yes, the values of K_{a2} and K_{a3} are equal.)

(a) Use these equilibrium constants to prepare a fractional composition diagram for this tetraprotic system.

(b) You should do this next part with your head and your calculator, not your spreadsheet: The solubility of $Cr(OH)_3$ is given by

$$Cr(OH)_3(s) \rightleftharpoons Cr(OH)_3\ (aq) \qquad K_s = 10^{-6.84}$$

What concentration of $Cr(OH)_3$ (aq) is in equilibrium with solid $Cr(OH)_3(s)$? What concentration of $Cr(OH)^{2+}$ is in equilibrium with $Cr(OH)_3(s)$ if the solution pH is adjusted to 4.00?

Isoelectric and Isoionic pH

35. What is the difference between the isoelectric and isoionic pH of a protein with many different acidic and basic substituents?

36. What is wrong with the following statement? At its isoelectric point, the charge on all molecules of a particular protein is zero.

37. Calculate the isoelectric and isoionic pH of 0.010 M threonine.

Notes and References

1. A student experiment using nuclear magnetic resonance to find microequilibrium constants for *N,N*-dimethyl-1,3-propanediamine $[(CH_3)_2NCH_2CH_2CH_2-NH_2]$ is described by O. F. Onasch, H. M. Schwartz, D. A. Aikens, and S. C. Bunce, *J. Chem. Ed.* **1991,** *68,* 791.

2. pH gradient formation and the structures of gradient-forming buffers are described by P. G. Righetti, E. Gianazza, C. Gelfi, M. Chiari, and P. K. Sinha, *Anal. Chem.* **1989,** *61,* 1602.

Acid-Base Titration of a Protein

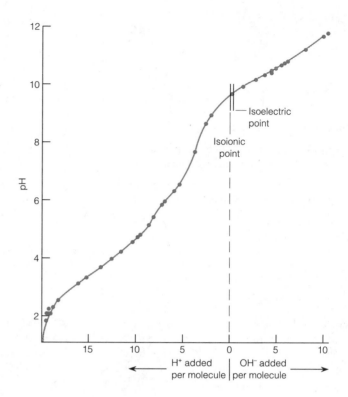

Acid-base titration of the enzyme ribonuclease. The abscissa represents moles of acid or base added per mole of enzyme. The isoionic point is the pH of the pure protein with no ions present except H^+ and OH^- The isoelectric point is the pH at which the average charge on the protein is zero. [From C. T. Tanford and J. D. Hauenstein, *J. Am. Chem. Soc.* **1956,** *78,* 5287.]

The enzyme ribonuclease is a protein with 124 amino acids. Its function is to cleave ribonucleic acid (RNA) into small fragments. An aqueous solution containing just the pure protein, with no other ions present except H^+ and OH^- derived from the protein and water, is said to be *isoionic.* From this point near pH 9.6 shown in the graph, the protein can be titrated with either acid or base. Of the 124 amino acids of the neutral enzyme, 16 can be protonated by acid and 20 can lose protons to added base. From the shape of the titration curve, it is possible to deduce the approximate pK_a for each titratable group. This information, in turn, provides insight into the immediate environment of that amino acid in its place in the protein. For ribonuclease, it was found that three tyrosine residues exhibit "normal" values of pK_a (~9.95) and three others have $pK_a > 12$. The interpretation is that three tyrosine groups are accessible to solvent and OH^-, and three are buried inside the protein where they cannot be easily titrated. The solid line in the figure is calculated from the values of pK_a deduced for all the titratable groups.

Acid-Base Titrations

In this chapter we discuss the principles of acid-base titrations, which are used throughout the realm of chemical analysis. Most often, we are simply interested in how much of an acidic or a basic substance is present in a sample. However, by analyzing a titration curve, we can deduce the quantities of acidic and basic components in a mixture and what their pK values are. For example, the titration curve for ribonuclease on the facing page can be analyzed to reveal that there are three tyrosine amino acids in a normal aqueous environment and three that are buried inside the protein, shielded from solvent. In this chapter, we will learn to predict the shapes of titration curves. We will also learn how acid-base indicators work and how to select one for a particular titration. A powerful spreadsheet approach to computing titration curves at the end of the chapter will allow you to deal with virtually any acid-base problem.

12-1 Titration of Strong Acid with Strong Base

For each type of titration studied in this chapter, *our goal is to construct a graph showing how the pH changes as titrant is added.* If you can do this, then you understand what is happening during the titration, and you will be able to interpret an experimental titration curve.

The first step in each case is to write the chemical reaction between titrant and analyte. We then use that reaction to calculate the composition and pH after each addition of titrant. As a simple example, let's focus our attention on the titration of 50.00 mL of 0.020 00 M KOH with 0.100 0 M HBr. The chemical reaction between titrant and analyte is merely

$$H^+ + OH^- \rightarrow H_2O$$

First write the reaction between *titrant* and *analyte*.

The titration reaction.

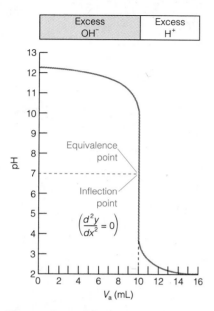

Figure 12-1 Calculated titration curve showing how pH changes as 0.100 0 M HBr is added to 50.00 mL of 0.020 00 M KOH. The equivalence point is also an inflection point.

Because the equilibrium constant for this reaction is $1/K_w = 10^{14}$, it is fair to say that it "goes to completion." *Any amount of H^+ added will consume a stoichiometric amount of OH^-.*

One useful beginning is to calculate the volume of HBr (V_e) needed to reach the equivalence point:

$$(V_e(\text{mL}))(0.100\,0\text{ M}) = (50.00\text{ mL})(0.020\,00\text{ M}) \Rightarrow V_e = 10.00\text{ mL}$$

$\underbrace{}$ mmol of HBr at equivalence point $\underbrace{}$ mmol of OH^- being titrated

It is helpful to bear in mind that when 10.00 mL of HBr have been added, the titration is complete. Prior to this point there will be excess, unreacted OH^- present. After V_e, there will be excess H^+ in the solution.

In the titration of any strong base with any strong acid, there are three regions of the titration curve that represent different kinds of calculations:

1. Before reaching the equivalence point, the pH is determined by excess OH^- in the solution.

2. At the equivalence point, the H^+ is just sufficient to react with all of the OH^- to make H_2O. The pH is determined by the dissociation of water.

3. After the equivalence point, pH is determined by the excess H^+ in the solution.

We will do one sample calculation for each region. Complete results are shown in Table 12-1 and Figure 12-1.

TABLE 12-1   Calculation of the titration curve for 50.00 mL of 0.020 00 M KOH treated with 0.100 0 M HBr

	mL HBr added (V_a)	Concentration of unreacted OH^- (M)	Concentration of excess H^+ (M)	pH
	0.00	0.020 0		12.30
	1.00	0.017 6		12.24
	2.00	0.015 4		12.18
	3.00	0.013 2		12.12
	4.00	0.011 1		12.04
	5.00	0.009 09		11.95
Region 1	6.00	0.007 14		11.85
(excess OH^-)	7.00	0.005 26		11.72
	8.00	0.003 45		11.53
	9.00	0.001 69		11.22
	9.50	0.000 840		10.92
	9.90	0.000 167		10.22
	9.99	0.000 016 6		9.22
Region 2	10.00	—	—	7.00
	10.01		0.000 016 7	4.78
	10.10		0.000 166	3.78
	10.50		0.000 826	3.08
	11.00		0.001 64	2.79
Region 3	12.00		0.003 23	2.49
(excess H^+)	13.00		0.004 76	2.32
	14.00		0.006 25	2.20
	15.00		0.007 69	2.11
	16.00		0.009 09	2.04

Region 1: Before the Equivalence Point

When 3.00 mL of HBr have been added, the reaction is three-tenths complete (recall that 10.00 mL of HBr are required to reach the equivalence point). The fraction of OH^- left unreacted is seven-tenths. The concentration of OH^- remaining in the flask is

$$[OH^-] = \underbrace{\left(\frac{10.00 - 3.00}{10.00}\right)}_{\substack{\text{Fraction} \\ \text{of } OH^- \\ \text{remaining}}} \underbrace{(0.020\,00\text{ M})}_{\substack{\text{Initial} \\ \text{concentration} \\ \text{of } OH^-}} \underbrace{\left(\frac{\overset{\substack{\text{Initial volume} \\ \text{of } OH^-}}{}}{50.00 + 3.00}\right)}_{\substack{\text{Dilution factor} \\ \text{of solution}}} = 0.013\,2\text{ M}$$

$$\text{(12-1)}$$

$$[H^+] = \frac{K_w}{[OH^-]} = \frac{1.0 \times 10^{-14}}{0.013\,2} = 7.5_8 \times 10^{-13}\text{ M} \Rightarrow pH = 12.12$$

Equation 12-1 is an example of the streamlined calculation that was introduced in Section 7-4 in connection with precipitation titrations. This equation tells us that the concentration of OH^- is equal to a certain fraction of the initial concentration, with a correction for dilution. The dilution factor equals the initial volume of analyte divided by the total volume of solution.

In Table 12-1 the volume of acid added is designated V_a. pH is expressed to the 0.01 decimal place, regardless of what is justified by significant figures. We do this for the sake of consistency and also because 0.01 is near the limit of accuracy in pH measurements.

Challenge Using a setup similar to Equation 12-1, calculate $[OH^-]$ when 6.00 mL of HBr have been added. Check your pH against the value in Table 12-1.

Region 2: At the Equivalence Point

Region 2 is the equivalence point, where enough H^+ has been added to react with all the OH^-. We could prepare the same solution by dissolving KBr in water. The pH is determined by the dissociation of water:

$$H_2O \rightleftharpoons \underset{x}{H^+} + \underset{x}{OH^-}$$

$$K_w = x^2 \Rightarrow x = 1.00 \times 10^{-7}\text{ M} \Rightarrow pH = 7.00$$

The pH at the equivalence point in the titration of any strong base (or acid) with strong acid (or base) will be 7.00 at 25°C.

As we will soon discover, *the pH is not 7.00 at the equivalence point in the titration of weak acids or bases.* The pH is 7.00 only if the titrant and analyte are both strong.

At the equivalence point, pH = 7.00, but *only* in a strong acid-strong base reaction.

Region 3: After the Equivalence Point

Beyond the equivalence point, we are adding excess HBr to the solution. The concentration of excess H^+ at, say, 10.50 mL is given by

$$[H^+] = \underbrace{(0.100\,0\text{ M})}_{\substack{\text{Initial} \\ \text{concentration} \\ \text{of } H^+}} \underbrace{\left(\frac{\overset{\substack{\text{Volume of} \\ \text{excess } H^+}}{0.50}}{50.00 + 10.50}\right)}_{\substack{\text{Dilution} \\ \text{factor}}} = 8.26 \times 10^{-4}\text{ M}$$

After the equivalence point, there is excess H^+.

$$pH = -\log[H^+] = 3.08$$

At $V_a = 10.50$ mL, there is an excess of just $V_a - V_e = 10.50 - 10.00 = 0.50$ mL of HBr. That is the reason why 0.50 appears in the dilution factor.

The Titration Curve

The complete titration curve is shown in Figure 12-1. Characteristic of all analytically useful titrations is a sudden change in pH near the equivalence point. The equivalence point is where the slope $(d\text{pH}/dV_a)$ is greatest (and the second derivative is zero, which makes it an *inflection point*). To repeat an important statement, the pH at the equivalence point is 7.00 *only* in a strong acid-strong base titration. If one or both of the reactants are weak, the equivalence point pH is *not* 7.00—and might be quite far from 7.00.

12-2 Titration of Weak Acid with Strong Base

The titration of a weak acid with a strong base allows us to put all our knowledge of acid-base chemistry to work. The example we will examine is the titration of 50.00 mL of 0.020 00 M MES with 0.100 0 M NaOH. MES is an abbreviation for 2-(*N*-morpholino)ethanesulfonic acid, which is a weak acid with $pK_a = 6.15$.

The *titration reaction* is

Always start by writing the titration reaction.

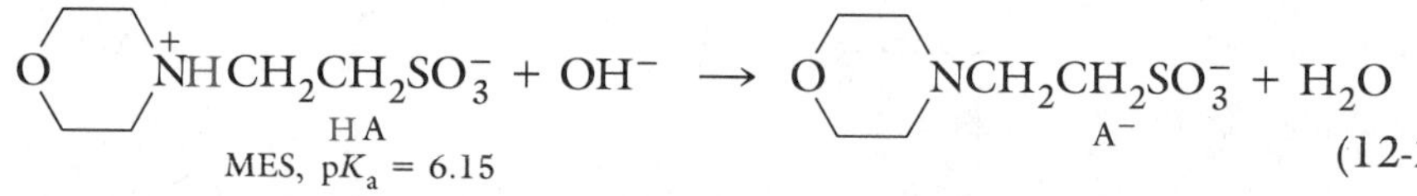

$$O \overset{+}{\diagup}\diagdown NHCH_2CH_2SO_3^- + OH^- \longrightarrow O \diagup\diagdown NCH_2CH_2SO_3^- + H_2O$$
$$\underset{\substack{HA \\ \text{MES, } pK_a = 6.15}}{} \qquad\qquad\qquad \underset{A^-}{} \qquad (12\text{-}2)$$

Reaction 12-2 is the reverse of the K_b reaction for the base A^-. The equilibrium constant for Reaction 12-2 is $K = 1/K_b = 1/(K_w/K_a \text{ (for HA)}) = 7.1 \times 10^7$. The equilibrium constant is so large that we can say that the reaction goes "to completion" after each addition of OH^-. As we saw in Box 10-1, *strong plus weak react completely.*

strong + weak → complete reaction

It is very helpful first to calculate the volume of base, V_b, needed to reach the equivalence point.

$$\underbrace{(V_b(\text{mL}))(0.100\,0\text{ M})}_{\text{mmol of base}} = \underbrace{(50.00\text{ mL})(0.020\,00\text{ M})}_{\text{mmol of HA}} \Rightarrow V_b = 10.00\text{ mL}$$

The titration calculations for this problem are of four types:

1. Before any base is added, the solution contains just HA in water. This is a weak-acid problem in which the pH is determined by the equilibrium

$$HA \overset{K_a}{\rightleftharpoons} H^+ + A^-$$

2. From the first addition of NaOH until immediately before the equivalence point, there is a mixture of unreacted HA plus the A^- produced by Reaction 12-2. *Aha! A buffer!* We can use the Henderson-Hasselbalch equation to find the pH.

3. At the equivalence point, "all" of the HA has been converted to A^-. The problem is the same as if the solution had been made by merely dissolving

A^- in water. We have a weak-base problem in which the pH is determined by the reaction

$$A^- + H_2O \overset{K_b}{\rightleftharpoons} HA + OH^-$$

4. Beyond the equivalence point, excess NaOH is being added to a solution of A^-. To a good approximation, the pH is determined by the strong base. We calculate the pH as if we had simply added excess NaOH to water. We will neglect the very small effect from having A^- present as well.

Region 1: Before Base Is Added

Before adding any base, we have a solution of $0.020\,00$ M HA with $pK_a = 6.15$. This is simply a weak-acid problem.

The initial solution contains just the *weak acid* HA.

$$\underset{F - x \quad x \quad x}{HA \rightleftharpoons H^+ + A^-} \qquad K_a = 10^{-6.15}$$

$$\frac{x^2}{0.020\,00 - x} = K_a \Rightarrow x = 1.19 \times 10^{-4} \Rightarrow pH = 3.93$$

Region 2: Before the Equivalence Point

Once we begin to add OH^-, a mixture of HA and A^- is created by the titration reaction (12-2). This mixture is a buffer whose pH can be calculated with the Henderson-Hasselbalch equation (10-16) once we know the quotient $[A^-]/[HA]$.

Before the equivalence point, there is a mixture of HA and A^-, which is a *buffer*.

Suppose we wish to calculate the quotient $[A^-]/[HA]$ when 3.00 mL of OH^- have been added. Because $V_e = 10.00$ mL, we have added enough base to react with three-tenths of the HA. We can make a table showing the relative concentrations before and after the reaction:

We only need *relative* concentrations because the pH of a buffer depends on the quotient $[A^-]/[HA]$.

Titration reaction:		$HA + OH^- \rightarrow A^- + H_2O$		
Relative initial quantities ($HA \equiv 1$):	1	$\frac{3}{10}$	—	—
Relative final quantities:	$\frac{7}{10}$	—	$\frac{3}{10}$	—

Once we know the *quotient* $[A^-]/[HA]$ in any solution, we know its pH:

$$pH = pK_a + \log \frac{[A^-]}{[HA]} = 6.15 + \log \frac{3/10}{7/10} = 5.78$$

The point at which the volume of titrant is $\frac{1}{2}V_e$ is a special one in any titration.

Titration reaction:		$HA + OH^- \rightarrow A^- + H_2O$		
Relative initial quantities:	1	$\frac{1}{2}$	—	—
Relative final quantities:	$\frac{1}{2}$	—	$\frac{1}{2}$	—

$$pH = pK_a + \log \frac{1/2}{1/2} = pK_a$$

Box 12-1 The Answer to a Nagging Question

Consider the titration of 100 mL of a 1.00 M solution of the weak acid HA with 1.00 M NaOH. The equivalence point occurs at $V_b = 100$ mL. At any point between $V_b = 0$ and $V_b = 100$ mL, part of the HA will be converted to A^- and part will be left as HA. Suppose 10 mL of NaOH had been added; we would then calculate the pH as follows:

Titration reaction:	$HA + OH^- \rightarrow A^- + H_2O$			
Initial mmol:	100	10	—	—
Final mmol:	90	—	10	—

$$pH = pK_a + \log \frac{[A^-]}{[HA]} = pK_a + \log \frac{10}{90}$$

Upon reflection, you might think "Wait a minute! The initial solution contains some A^- in equilibrium with the HA. There must be less than 100 mmol of HA and more than 0 mmol of A^- at the start. Doesn't this make our answer wrong?"

It does not, and the reason is easy to see with numbers. Suppose that the HA dissociates to give 1 mmol of A^- before any NaOH is added.

$$HA \overset{K_a}{\rightleftharpoons} H^+ + A^- \qquad (A)$$

Initial solution: 99 mmol 1 mmol 1 mmol

The solution contains 1 mmol of H^+, 1 mmol of A^-, and 99 mmol of HA.

When 10 mmol of OH^- is added, 1 mmol reacts with the strong acid, H^+, and 9 mmol is left to react with the weak acid, HA.

$$H^+ + OH^- \rightarrow H_2O$$

1 mmol 1 mmol 1 mmol

$$\qquad (B)$$

$$HA + OH^- \rightarrow A^-$$

9 mmol 9 mmol 9 mmol

The resulting solution contains 1 mmol of A^- from Reaction A and 9 mmol of A^- from Reaction B. The total A^- is 10 mmol. The total HA is 90 mmol. These are exactly the same numbers that we calculated by ignoring the initial dissociation of HA.

Moral: You can treat a mixture of the weak acid HA plus OH^- as if no dissociation of HA had occurred before adding the NaOH.

When $V_b = \frac{1}{2}V_e$, pH $= pK_a$. This is a landmark in any titration.

When the volume of titrant is $\frac{1}{2}V_e$, pH $= pK_a$ for the acid HA (neglecting activity coefficients). If you have an experimental titration curve, you can find the approximate value of pK_a by reading the pH when $V_b = \frac{1}{2}V_e$, where V_b is the volume of added base. (To find the true value of pK_a requires activity coefficients.)

In considering the amount of OH^- needed to react with HA, you need not worry about the small amount of HA that dissociates to A^- in the absence of OH^-. The reason for this is explained in Box 12-1.

Advice. As soon as you recognize a mixture of HA and A^- in any solution, *you have a buffer!* You can calculate the pH if you can find the quotient $[A^-]/[HA]$.

$$pH = pK_a + \log\frac{[A^-]}{[HA]}$$

Learn to recognize buffers! They lurk in every corner of acid-base chemistry.

Region 3: At the Equivalence Point

At the equivalence point, the quantity of NaOH is exactly enough to consume the HA.

At the equivalence point, HA has been converted to A^-, a *weak base*.

Titration reaction:	$HA + OH^- \rightarrow A^- + H_2O$			
Initial relative quantities:	1	1	—	—
Final relative quantities:	—	—	1	—

The resulting solution contains "just" A^-. We could have prepared the same solution by dissolving the salt Na^+A^- in distilled water. *A solution of* Na^+A^- *is merely a solution of a weak base.*

To compute the pH of a weak base, we write the reaction of the weak base with water:

$$\underset{F-x}{A^-} + H_2O \rightleftharpoons \underset{x}{HA} + \underset{x}{OH^-} \qquad K_b = \frac{K_w}{K_a}$$

The only tricky point is that the formal concentration of A^- is no longer 0.020 00 M, which was the initial concentration of HA. The A^- has been diluted by NaOH from the buret:

$$F' = \underbrace{(0.020\,00\ M)}_{\substack{\text{Initial} \\ \text{concentration} \\ \text{of HA}}} \underbrace{\left(\frac{\overset{\text{Initial volume of HA}}{50.00}}{\underset{\substack{\text{Total volume} \\ \text{of solution}}}{50.00 + 10.00}}\right)}_{\substack{\text{Dilution} \\ \text{factor}}} = 0.016\,7\ M$$

With this value of F', we can solve the problem:

$$\frac{x^2}{F'-x} = K_b = \frac{K_w}{K_a} = 1.43 \times 10^{-8} \Rightarrow x = 1.54 \times 10^{-5}\ M$$

$$pH = -\log[H^+] = -\log\frac{K_w}{x} = 9.18$$

The pH at the equivalence point in this titration is 9.18. **It is not 7.00.** The equivalence-point pH will *always* be above 7 for the titration of a weak acid, because the acid is converted to its conjugate base at the equivalence point.

The pH is higher than 7 at the equivalence point in the titration of a weak acid with a strong base.

Region 4: After the Equivalence Point

Now we are adding NaOH to a solution of A^-. The base NaOH is so much stronger than base A^- that it is a fair approximation to say that the pH is determined by the concentration of excess OH^- in the solution.

Here we assume the pH is governed by the excess OH^-.

Let's calculate the pH when $V_b = 10.10$ mL. This is just 0.10 mL past V_e. The concentration of excess OH^- is

$$[OH^-] = \underbrace{(0.100\,0\ M)}_{\substack{\text{Initial} \\ \text{concentration} \\ \text{of } OH^-}} \overbrace{\left(\underbrace{\frac{0.10}{50.00 + 10.10}}_{\substack{\text{Dilution} \\ \text{factor}}}\right)}^{\substack{\text{Volume of} \\ \text{excess } OH^-}} = 1.66 \times 10^{-4}\ M$$

Total volume of solution

$$pH = -\log\frac{K_w}{[OH^-]} = 10.22$$

Challenge   Compare the concentration of OH^- from excess titrant at $V_b = 10.10$ mL to the concentration of OH^- from hydrolysis of A^-. Satisfy yourself that it is fair to neglect the contribution of A^- to the pH after the equivalence point.

Landmarks in a titration:

At $V_b = V_e$, curve is steepest.

At $V_b = \frac{1}{2}V_e$, pH = pK_a and the slope is minimal.

The Titration Curve

A summary of the calculations for the titration of MES with NaOH is shown in Table 12-2. The calculated titration in Figure 12-2 has two easily identified points. One is the equivalence point, which is the steepest part of the curve. The other landmark is the point where $V_b = \frac{1}{2}V_e$ and pH = pK_a. This latter point is also an inflection point, having the minimum slope.

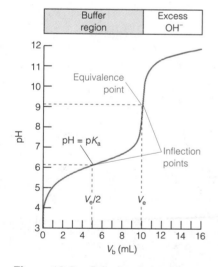

Figure 12-2   Calculated titration curve for the reaction of 50.00 mL of 0.020 00 M MES with 0.100 0 M NaOH. Landmarks occur at half of the equivalence volume (pH = K_a) and at the equivalence point, which is the steepest part of the curve.

TABLE 12-2   Calculation of the titration curve for 50.00 mL of 0.020 00 M MES treated with 0.100 0 M NaOH

	mL base added (V_b)	pH
Region 1	0.00	3.93
(weak acid)	0.50	4.87
	1.00	5.20
	2.00	5.55
	3.00	5.78
	4.00	5.97
	5.00	6.15
Region 2	6.00	6.33
(buffer)	7.00	6.52
	8.00	6.75
	9.00	7.10
	9.50	7.43
	9.90	8.15
Region 3	10.00	9.18
(weak base)	10.10	10.22
	10.50	10.91
	11.00	11.21
	12.00	11.50
Region 4	13.00	11.67
(excess OH^-)	14.00	11.79
	15.00	11.88
	16.00	11.95

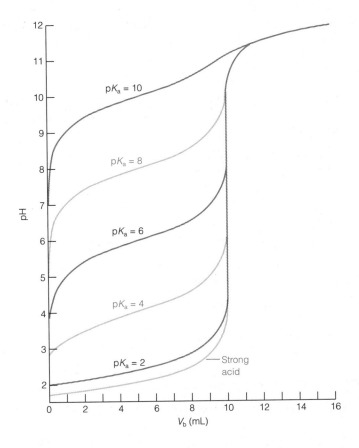

Figure 12-3 Calculated curves showing the titration of 50.0 mL of 0.020 0 M HA with 0.100 M NaOH. As the acid becomes weaker, the change in slope at the equivalence point becomes less abrupt.

The *buffer capacity* measures the ability to resist changes in pH.

If you look back at Figure 10-6, you will note that the maximum *buffer capacity* occurs when pH = pK_a. That is, the solution is most resistant to pH changes when pH = pK_a (and $V_b = \frac{1}{2}V_e$); the slope ($d\text{pH}/dV_b$) is therefore at its minimum.

Figure 12-3 shows how the titration curve depends on the acid dissociation constant of HA. As K_a decreases (pK_a increases), the inflection near the equivalence point decreases, until the equivalence point becomes too shallow to detect. A similar phenomenon occurs as the concentrations of analyte and titrant decrease (Problem 59). *It is not practical to titrate an acid or base when its strength is too weak or its concentration too dilute.*

12-3 Titration of Weak Base with Strong Acid

The titration of a weak base with a strong acid is just the reverse of the titration of a weak acid with a strong base. The *titration reaction* is

$$B + H^+ \rightarrow BH^+$$

Because the reactants are a weak base and a strong acid, the reaction goes essentially to completion after each addition of acid. There are four distinct regions of the titration curve:

1. Before acid is added, the solution contains just the weak base, B, in water. The pH is determined by the K_b reaction:

$$\underset{\text{F}-x}{\text{B}} + \text{H}_2\text{O} \overset{K_b}{\rightleftharpoons} \underset{x}{\text{BH}^+} + \underset{x}{\text{OH}^-}$$

When $V_a = 0$, we have a *weak-base* problem.

When $0 < V_a < V_e$, we have a *buffer*.

2. Between the initial point and the equivalence point, there is a mixture of B and BH⁺—*Aha! A buffer!* The pH is computed by using

$$pH = pK_a \text{ (for BH}^+) + \log\frac{[B]}{[BH^+]}$$

In adding acid (increasing V_a), we reach the special point where $V_a = \frac{1}{2}V_e$ and pH = pK_a (for BH⁺). As before, pK_a (and therefore pK_b) can be determined easily from the titration curve.

3. At the equivalence point, B has been converted into BH⁺, a weak acid. The pH is calculated by considering the acid dissociation reaction of BH⁺.

$$BH^+ \rightleftharpoons B + H^+ \qquad K_a = \frac{K_w}{K_b}$$
$$ F-x \quad x \quad x$$

When $V_a = V_e$, the solution contains the *weak acid* BH⁺.

The formal concentration of BH⁺, F′, is not the same as the original formal concentration of B, because some dilution has occurred. Because the solution contains BH⁺ at the equivalence point, it is acidic. *The pH at the equivalence point must be below 7.*

For $V_a > V_e$, there is excess *strong acid*.

4. After the equivalence point, there is excess H⁺ in the solution. We treat this problem by considering only the concentration of excess H⁺ and neglecting the contribution of weak acid, BH⁺.

EXAMPLE   Titration of Pyridine with HCl

Consider the titration of 25.00 mL of 0.083 64 M pyridine with 0.106 7 M HCl.

$$K_b = 1.69 \times 10^{-9}$$

Pyridine

The titration reaction is

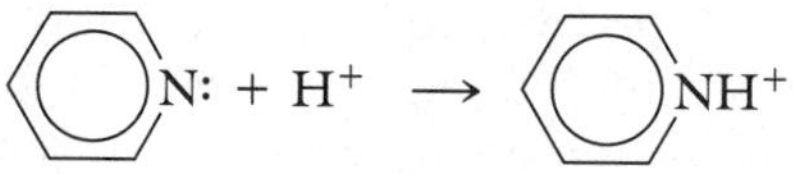

and the equivalence point occurs at 19.60 mL:

$$\underbrace{(V_e(\text{mL}))(0.106\,7\text{ M})}_{\text{mmol of HCl}} = \underbrace{(25.00\text{ mL})(0.083\,64\text{ M})}_{\text{mmol of pyridine}} \Rightarrow V_e = 19.60\text{ mL}$$

Find the pH when V_a = 4.63 mL.

Solution   Part of the pyridine has been neutralized, so there is a mixture of pyridine and pyridinium ion—*Aha! A buffer!* The fraction of pyridine that has been titrated is 4.63/19.60 = 0.236, because it takes 19.60 mL to titrate the whole sample. The fraction of pyridine remaining is (19.60 − 4.63)/19.60 = 0.764. The pH is

$$pH = pK_a \left(= -\log\frac{K_w}{K_b}\right) + \log\frac{[B]}{[BH^+]}$$

$$= 5.23 + \log\frac{0.764}{0.236} = 5.74$$

12-4 Titrations in Diprotic Systems

The principles developed for titrations of monoprotic acids and bases are readily extended to titrations of polyprotic acids and bases. These principles are demonstrated in this section by a few sample calculations.

A Typical Case

The upper curve in Figure 12-4 is calculated for the titration of 10.0 mL of 0.100 M base (B) with 0.100 M HCl. The base is dibasic, with $pK_{b1} = 4.00$ and $pK_{b2} = 9.00$. The titration curve has reasonably sharp breaks at both equivalence points, corresponding to the reactions

$$B + H^+ \rightarrow BH^+$$

$$BH^+ + H^+ \rightarrow BH_2^{2+}$$

The volume at the first equivalence point is 10.00 mL because

$$\underbrace{(V_e(\text{mL}))(0.100 \text{ M})}_{\text{mmol of HCl}} = \underbrace{(10.00 \text{ mL})(0.100 \text{ 0 M})}_{\text{mmol of B}}$$

$$V_e = 10.00 \text{ mL}$$

The volume at the second equivalence point must be $2V_e$, because the second reaction requires exactly the same number of moles of HCl as the first reaction.

The calculation of pH at each point along the curve is very similar to that for the corresponding point in the titration of a monobasic compound. Let's examine points A through E in Figure 12-4.

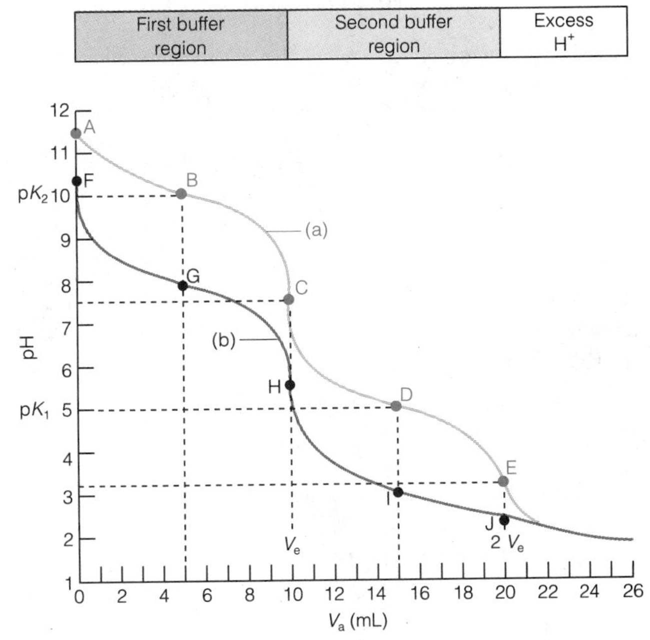

$V_{e2} = 2V_{e1}$, always.

Figure 12-4 (a) Titration curve for the reaction of 10.0 mL of 0.100 M base ($pK_{b1} = 4.00$, $pK_{b2} = 9.00$) with 0.100 M HCl. The two equivalence points are C and E. Points B and D are the half-neutralization points, whose pH values equal pK_{a2} and pK_{a1}, respectively. (b) Titration curve for the reaction of 10.0 mL of 0.100 M nicotine ($pK_{b1} = 6.15$, $pK_{b2} = 10.85$) with 0.100 M HCl. There is no sharp break at the second equivalence point, J, because the pH is too low.

Recall that the fully basic form of a dibasic compound can be treated as if it were monobasic. (The K_{b2} reaction can be neglected.)

Point A

Before any acid is added, the solution contains just B, a weak base, whose pH is governed by the reaction

$$\underset{0.100 - x}{B} + H_2O \overset{K_{b1}}{\rightleftharpoons} \underset{x}{BH^+} + \underset{x}{OH^-}$$

$$\frac{x^2}{0.100 - x} = 1.00 \times 10^{-4} \Rightarrow x = 3.11 \times 10^{-3}$$

$$[H^+] = \frac{K_w}{x} \Rightarrow pH = 11.49$$

Point B

At any point between A (the initial point) and C (the first equivalence point), we have a buffer containing B and BH^+. Point B is halfway to the equivalence point, so $[B] = [BH^+]$. The pH is calculated by using the Henderson-Hasselbalch equation *for the weak acid*, BH^+, whose acid dissociation constant is K_{a2} for BH_2^{2+}. The value of K_{a2} is

Of course, you remember that

$$K_{b1} = \frac{K_w}{K_{a2}}$$

$$K_{b2} = \frac{K_w}{K_{a1}}$$

$$K_{a2} = \frac{K_w}{K_{b1}} = 10^{-10.00}$$

To calculate the pH at point B, we write

$$pH = pK_{a2} + \log\frac{[B]}{[BH^+]} = 10.00 + \log 1 = 10.00$$

So the pH at point B is just pK_{a2}.

To calculate the quotient $[B]/[BH^+]$ at any point in the buffer region, just find what fraction of the way from point A to point C the titration has progressed. For example, if $V_a = 1.5$ mL, then

$$\frac{[B]}{[BH^+]} = \frac{8.5}{1.5}$$

because 10.0 mL are required to reach the equivalence point and we have added just 1.5 mL. The pH at $V_a = 1.5$ mL is given by

$$pH = 10.00 + \log\frac{8.5}{1.5} = 10.75$$

Point C

BH^+ is the *intermediate form* of a diprotic acid.

At the first equivalence point, B *has been converted to* BH^+, *the intermediate form of the diprotic acid*, BH_2^{2+}. BH^+ *is both an acid and a base.* As described in Section 11-1, the pH is given by

$$pH \approx \frac{1}{2}(pK_1 + pK_2)$$

$$[H^+] \approx \sqrt{\frac{K_1 K_2 F + K_1 K_w}{K_1 + F}} \tag{12-3}$$

where K_1 and K_2 are the acid dissociation constants of BH_2^{2+}.

The formal concentration of BH^+ is calculated by considering the dilution of the original solution of B.

$$F = (0.100 \text{ M}) \underbrace{\left(\frac{10.0}{20.0}\right)}_{} = 0.0500 \text{ M}$$

Initial volume of B

Initial concentration of B · Dilution factor · Total volume of solution

Plugging all the numbers into Equation 12-3 gives

$$[H^+] = \sqrt{\frac{(10^{-5})(10^{-10})(0.0500) + (10^{-5})(10^{-14})}{10^{-5} + 0.0500}} = 3.16 \times 10^{-8}$$

$$pH = 7.50$$

Note that in this example $pH = (pK_{a1} + pK_{a2})/2$.

Point C in Figure 12-4 shows where the intermediate form of a diprotic acid lies on a titration curve. This is the least-buffered point on the whole curve, because the pH changes most rapidly if small amounts of acid or base are added. There is a common misconception that the intermediate form of a diprotic acid behaves as a buffer when, in fact, it is the worst choice for a buffer.

Point D

At any point between C and E, we can consider the solution to be a buffer containing BH^+ (the base) and BH_2^{2+} (the acid). When $V_a = 15.0$ mL, $[BH^+] = [BH_2^{2+}]$ and

$$pH = pK_{a1} + \log\frac{[BH^+]}{[BH_2^{2+}]} = 5.00 + \log 1 = 5.00$$

Challenge Show that if V_a were 17.2 mL, the ratio in the log term would be

$$\frac{[BH^+]}{[BH_2^{2+}]} = \frac{20.0 - 17.2}{17.2 - 10.0} = \frac{2.8}{7.2}$$

Point E

Point E is the second equivalence point, at which the solution is formally the same as one prepared by dissolving BH_2Cl_2 in water. The formal concentration of BH_2^{2+} is

$$F = (0.100 \text{ M})\left(\frac{10.0}{30.0}\right) = 0.0333 \text{ M}$$

Original volume of B

Total volume of solution

The pH is determined by the acid dissociation reaction of BH_2^{2+}.

$$\underset{F-x}{BH_2^{2+}} \rightleftharpoons \underset{x}{BH^+} + \underset{x}{H^+} \qquad K_{a1} = \frac{K_w}{K_{b2}}$$

At the second equivalence point, we have made BH_2^{2+}, which can be treated as a monoprotic weak acid.

$$\frac{x^2}{0.0333 - x} = 1.0 \times 10^{-5} \Rightarrow x = 5.72 \times 10^{-4} \Rightarrow pH = 3.24$$

Beyond the second equivalence point ($V_a > 20.0$ mL), the pH of the solution can be calculated from the volume of strong acid added to the solution. For example, at $V_a = 25.00$ mL, there is an excess of 5.00 mL of 0.100 M HCl in a total volume of $10.00 + 25.00 = 35.00$ mL. The pH is found by writing

$$[H^+] = (0.100 \text{ M})\left(\frac{5.00}{35.00}\right) = 1.43 \times 10^{-2} \text{ M} \Rightarrow \text{pH} = 1.85$$

Blurred End Points

When the pH is too low or too high or when pK_a values are too close together, end points are obscured.

Titrations of many diprotic acids or bases have two clear end points, as shown in the upper curve in Figure 12-4. Some titrations, however, may not show both end points, as illustrated by the lower curve of the same figure. This latter curve is calculated for the titration of 10.0 mL of 0.100 M nicotine ($pK_{b1} = 6.15$, $pK_{b2} = 10.85$) with 0.100 M HCl. The two reactions are

Nicotine (B)   BH$^+$ BH$_2^{2+}$

Both reactions occur, but there is almost no perceptible break at the second equivalence point, because BH_2^{2+} is too strong an acid (or, equivalently, BH^+ is too weak a base). As the acidic end of the titration is approached (pH $\lesssim 3$), the approximation that all the HCl reacts with BH^+ to give BH_2^{2+} is not true. To calculate the pH between points I and J requires the systematic treatment of equilibrium. Later in this chapter, we will describe how to calculate the whole curve with a spreadsheet.

In the titration of ribonuclease at the beginning of this chapter, there is a continuous change in pH, with no clear breaks. The reason is that 29 groups are titrated in the pH interval shown. The 29 end points are so close together that a nearly uniform rise results. The curve can be analyzed to find the many pK_a values, but this analysis requires a computer, and the individual pK_a values will not be determined very precisely.

12-5 Finding the End Point with a pH Electrode

Box 12-2 illustrates an important application of acid-base titrations in environmental analysis.

Titrations are most commonly performed either to find out how much analyte is present or to measure equilibrium constants of the analyte. We can obtain the information necessary for both purposes by monitoring the pH of the solution as the titration is performed.

Figure 12-5 shows an *autotitrator* in which the entire operation is performed automatically. Titrant from the plastic bottle at the rear is dispensed in small increments by a syringe pump while the pH is measured by electrodes immersed in the beaker of analyte on the stirrer. (We will learn how these electrodes work in Chapter 15.) The instrument waits for the pH to stabilize after each addition, before adding the next increment. pH is displayed on the screen, and the end point is computed automatically by finding the maximum slope in the titration curve.

Box 12-2   Alkalinity and Acidity

Alkalinity is defined as the capacity of natural water to react with H^+ to reach pH 4.5, which is the second equivalence point in the titration of carbonate (CO_3^{2-}) with H^+. To a good approximation, alkalinity is determined by OH^-, CO_3^{2-}, and HCO_3^-:

$$\text{alkalinity} \approx [OH^-] + 2[CO_3^{2-}] + [HCO_3^-]$$

When water whose pH is greater than 4.5 is titrated with acid to pH 4.5 (measured with a pH meter), all OH^-, CO_3^{2-}, and HCO_3^- will have reacted. Other basic species also react, but OH^-, CO_3^{2-}, and HCO_3^- account for most of the alkalinity in most water samples. Alkalinity is normally expressed as millimoles of H^+ needed to bring 1 L of water to pH 4.5.

Alkalinity and *hardness* (dissolved Ca^{2+} and Mg^{2+}, Box 13-2) are important characteristics of irrigation water. Alkalinity in excess of the $Ca^{2+} + Mg^{2+}$ content is called "residual sodium carbonate." Water with a residual sodium carbonate content equivalent to ≥ 2.5 mmol H^+/L is not suitable for irrigation. Residual sodium carbonate between 1.25 and 2.5 mmol H^+/L is marginal, whereas a content ≤ 1.25 mmol H^+/L is suitable for irrigation.

Acidity of natural waters refers to the total acid content that can be titrated to pH 8.3 with NaOH. This pH is the second equivalence point for titration of carbonic acid (H_2CO_3) with OH^-. Almost all weak acids that might be present in the water will also be titrated in this procedure. Acidity is expressed as mmol of OH^- needed to bring 1 L of water to pH 8.3.

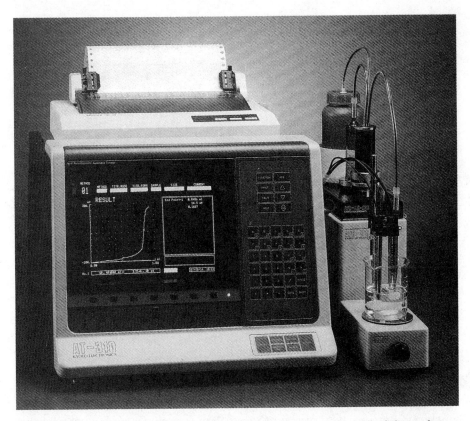

Figure 12-5   Autotitrator delivers titrant from the bottle at the back right to the beaker of analyte on the stirring motor at the front right. Electrodes immersed in the analyte solution monitor pH (or concentrations of specific ions) and display results on the screen during the titration. [Courtesy Kyoto Electronics Manufacturing USA, Springfield, NJ.]

Figure 12-6 (a) Experimental points in the titration of 1.430 mg of xylenol orange, a hexaprotic acid, dissolved in 1.000 mL of aqueous 0.10 M $NaNO_3$. The titrant was 0.065 92 M NaOH. (b) The first derivative, $\Delta pH/\Delta V$, of the titration curve. (c) The second derivative, $\Delta(\Delta pH/\Delta V)/\Delta V$, which is the derivative of the middle curve. Derivatives for the first end point are calculated in Table 12-3. End points are taken as maxima in the derivative curve and zero crossings of the second derivative.

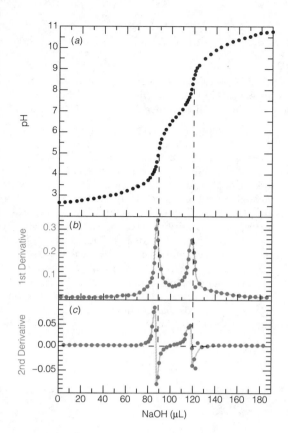

The upper part of Figure 12-6 shows experimental results for the manual titration of a hexaprotic weak acid, H_6A, with NaOH. Because the compound is difficult to purify, only a tiny amount was available for titration. Just 1.430 mg was dissolved in 1.00 mL of water and titrated with microliter quantities of 0.065 92 M NaOH, delivered with a Hamilton syringe.

The curve in Figure 12-6 shows two clear breaks, near 90 and 120 μL, which correspond to titration of the *third* and *fourth* protons of H_6A.

$$H_4A^{2-} + OH^- \rightarrow H_3A^{3-} + H_2O \qquad (\sim 90 \ \mu L \text{ equivalence point})$$

$$H_3A^{3-} + OH^- \rightarrow H_2A^{4-} + H_2O \qquad (\sim 120 \ \mu L \text{ equivalence point})$$

The first two and last two equivalence points give unrecognizable end points, because they occur at pH values that are too low or too high.

Using Derivatives to Find the End Point

The end point has maximum slope and the second derivative is zero.

The end point is taken as the point where the slope ($d pH/dV$) of the titration curve is greatest. The slope (first derivative) displayed in the middle of Figure 12-6 was calculated as shown in Table 12-3. The first two columns of this table give experimental volumes and pH measurements. (The pH meter was precise to three digits, even though accuracy ends in the second decimal place.) To compute the first derivative, each pair of volumes is averaged and the quantity $\Delta pH/\Delta V$ is calculated. Here ΔpH is the change in pH between consecutive readings and ΔV is the change in volume between

TABLE 12-3 Computation of first and second derivatives for a titration curve

μL NaOH	pH	First derivative μL	$\dfrac{\Delta pH}{\Delta \mu L}$	Second derivative μL	$\dfrac{\Delta(\Delta pH/\Delta \mu L)}{\Delta \mu L}$
85.0	4.245				
		85.5	0.155		
86.0	4.400			86.0	0.0710
		86.5	0.226		
87.0	4.626			87.0	0.0810
		87.5	0.307		
88.0	4.933			88.0	0.0330
		88.5	0.340		
89.0	5.273			89.0	−0.0830
		89.5	0.257		
90.0	5.530			90.0	−0.0680
		90.5	0.189		
91.0	5.719			91.25	−0.0390
		92.0	0.130		
93.0	5.980				

consecutive additions. The last two columns of Table 12-3 and the lower portion of Figure 12-6 give the second derivative, computed in an analogous manner. The end point corresponds to the volume at which the second derivative is zero. A graph on the scale of Figure 12-7 allows us to make good estimates of the end-point volumes.

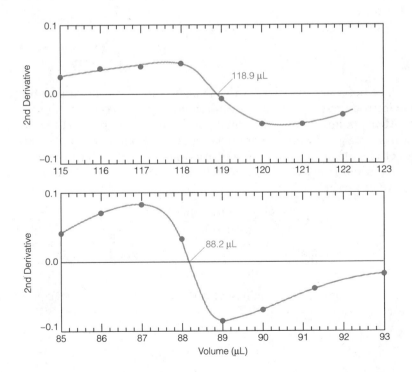

Figure 12-7 Enlargement of the end-point regions in the second derivative curve shown in Figure 12-6c.

E X A M P L E   Computing Derivatives of a Titration Curve

Let's see how the first and second derivatives in Table 12-3 are calculated.

Solution   The first number in the third column, 85.5, is the average of the first two volumes (85.0 and 86.0) in the first column. The derivative $\Delta pH / \Delta V$ is calculated from the first two pH values and the first two volumes:

$$\frac{\Delta pH}{\Delta V} = \frac{4.400 - 4.245}{86.0 - 85.0} = 0.155$$

The coordinates ($x = 85.5$, $y = 0.155$) are one point in the graph of the first derivative in Figure 12-6.

The second derivative is computed from the first derivative in an analogous manner. The first volume in the fifth column of Table 12-3 is 86.0, which is the average of 85.5 and 86.5. The second derivative is found as follows:

$$\frac{\Delta(\Delta pH / \Delta V)}{\Delta V} = \frac{0.226 - 0.155}{86.5 - 85.5} = 0.071\ 0$$

 The coordinates ($x = 86.0$, $y = 0.071\ 0$) are plotted in the second derivative graph at the bottom of Figure 12-6. These calculations are tedious by hand, but they are trivial with a spreadsheet.

Using a Gran Plot to Find the End Point[1]

One problem with using derivatives to find the end point is that titration data are most difficult to obtain right near the end point, because buffering is minimal and electrode response is sluggish. A **Gran plot** is a graphical method that allows us to use data from before the end point (typically $0.8\ V_e$ to V_e) to locate the end point.

Consider the titration of a weak acid, HA:

$$HA \rightleftharpoons H^+ + A^- \qquad K_a = \frac{[H^+]\gamma_{H^+}[A^-]\gamma_{A^-}}{[HA]\gamma_{HA}}$$

It will be necessary to include activity coefficients in this discussion because a pH electrode responds to hydrogen ion *activity*, not concentration.

> Strong plus weak react completely.

At any point between the initial point and the end point of the titration, it is usually a good approximation to say that each mole of NaOH converts 1 mol of HA into 1 mol of A^-. If we have titrated V_a mL of HA (whose formal concentration is F_a) with V_b mL of NaOH (whose formal concentration is F_b), we can write

$$[A^-] = \frac{\text{moles of } OH^- \text{ delivered}}{\text{total volume}} = \frac{V_b F_b}{V_b + V_a}$$

$$[HA] = \frac{\text{original moles of } HA - \text{moles of } OH^-}{\text{total volume}} = \frac{V_a F_a - V_b F_b}{V_a + V_b}$$

Substitution of these values of $[A^-]$ and $[HA]$ into the equilibrium constant gives

$$K_a = \frac{[H^+]\gamma_{H^+} V_b F_b \gamma_{A^-}}{(V_a F_a - V_b F_b)\gamma_{HA}}$$

which can be rearranged to

$$\underbrace{V_b[H^+]\gamma_{H^+}}_{10^{-pH}} = \frac{\gamma_{HA}}{\gamma_{A^-}}K_a\left(\frac{V_aF_a - V_bF_b}{F_b}\right) \qquad (12\text{-}4)$$

$$\mathcal{A}_{H^+} = [H^+]\gamma_{H^+} = 10^{-pH}$$

The term on the left is $V_b \cdot 10^{-pH}$, because $[H^+]\gamma_{H^+} = 10^{-pH}$. The term in parentheses on the right is

$$\frac{V_aF_a - V_bF_b}{F_b} = \frac{V_aF_a}{F_b} - V_b = V_e - V_b$$

$$V_aF_a = V_eF_b \Rightarrow V_e = \frac{V_aF_a}{F_b}$$

Equation 12-4 can, therefore, be written in the form

Gran plot equation: $\quad V_b \cdot 10^{-pH} = \dfrac{\gamma_{HA}}{\gamma_{A^-}}K_a(V_e - V_b) \qquad (12\text{-}5)$

A graph of $V_b \cdot 10^{-pH}$ versus V_b is called a *Gran plot*. If γ_{HA}/γ_{A^-} is constant, the graph is a straight line with a slope of $-K_a\gamma_{HA}/\gamma_{A^-}$ and an intercept of V_e on the abscissa (the x axis). A Gran plot for the titration in Figure 12-6 is shown in Figure 12-8. Any units can be used for V_b, but the same units should be used on both axes. In Figure 12-8, V_b was expressed in microliters on both axes.

The beauty of a Gran plot is that it enables us to use data taken before the end point to find the end point. The slope of the Gran plot enables us to find K_a. Although we derived the Gran function for a monoprotic acid, the same plot ($V_b \cdot 10^{-pH}$ versus V_b) applies to polyprotic acids (such as H_6A in Figure 12-6).

The Gran function, $V_b \cdot 10^{-pH}$, does not actually go to zero, because 10^{-pH} is never zero. The curve must be extrapolated to find V_e. The reason the function does not reach zero is that we have used the approximation that every mole of OH^- generates 1 mol of A^-, which is not true as V_b approaches V_e. As a practical matter, only the linear portion of the Gran plot is used.

Another source of curvature in the Gran plot is the changing ionic strength, which causes γ_{HA}/γ_{A^-} to vary. In Figure 12-6 this variation was avoided by maintaining nearly constant ionic strength with $NaNO_3$. Even without added salt, the last 10–20% of data before V_e gives a fairly straight line because the value of γ_{HA}/γ_{A^-} does not change very much.

Gran plot:

Plot $V_b \cdot 10^{-pH}$ versus V_b.

x intercept $= V_e$

slope $= -K_a\gamma_{HA}/\gamma_{A^-}$

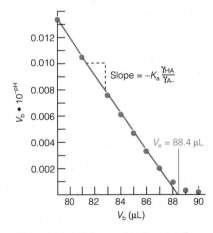

Figure 12-8 Gran plot for the first equivalence point of Figure 12-6. This plot gives an estimate of V_e that differs from that in Figure 12-7 by 0.2 μL (88.4 versus 88.2 μL). The last 10–20% of volume prior to V_e is normally used for a Gran plot.

Challenge Show that when a weak base, B, is titrated with a strong acid, the appropriate Gran function is

$$V_a \cdot 10^{+pH} = \left(\frac{1}{K_a} \cdot \frac{\gamma_B}{\gamma_{BH^+}}\right)(V_e - V_a) \qquad (12\text{-}6)$$

where V_a is the volume of strong acid added and K_a is the acid dissociation constant of BH^+. A graph of $V_a \cdot 10^{+pH}$ versus V_a should be a straight line with a slope of $-\gamma_B/\gamma_{BH^+}K_a$ and an intercept of V_e on the V_a axis.

12-6 Finding the End Point with Indicators

An acid-base **indicator** is itself an acid or base whose various protonated species have different colors. An example is thymol blue.

Red (R)   Yellow (Y⁻)   Blue (B²⁻)

One of the most common indicators is phenolphthalein, usually used for its colorless ↔ pink transition at pH 8.0–9.6.

Colorless phenophthalein

$2H^+ \| 2OH^-$

$O + 2H_2O$

Pink phenolphthalein

Thymol blue has two useful color changes. Below pH 1.7, the predominant species is red; between pH 1.7 and pH 8.9, the predominant species is yellow; and above pH 8.9, the predominant species is blue. For simplicity, we designate the three species R, Y⁻, and B²⁻, respectively. The sequence of color changes for thymol blue is shown in Color Plate 6.

The equilibrium between R and Y⁻ can be written

$$R \rightleftharpoons Y^- + H^+ \qquad K_1 = \frac{[Y^-][H^+]}{[R]}$$

$$pH = pK_1 + \log\frac{[Y^-]}{[R]} \qquad (12\text{-}7)$$

At pH = 1.7 (= pK_1), there will be a 1:1 mixture of the yellow and red species, which will appear orange. As a very crude rule of thumb, we can say that the solution will appear red when $[Y^-]/[R] \lesssim 1/10$ and yellow when $[Y^-]/[R] \gtrsim 10/1$. From Equation 12-7, we can see that the solution will be red when pH ≈ pK_1 − 1 and yellow when pH ≈ pK_1 + 1. In tables of indicator colors, thymol blue is listed as red below pH 1.2 and yellow above pH 2.8. By comparison, the pH values predicted by our rule of thumb are 0.7 and 2.7. Between pH 1.2 and pH 2.8, the indicator exhibits various shades of orange. Whereas most indicators have a single color change, thymol blue undergoes another transition, from yellow to blue, between pH 8.0 and pH 9.6. In this range, various shades of green are seen.

Acid-base indicator color changes form the basis of Demonstration 12-1. Box 12-3 shows how optical absorption by indicators allows us to measure pH.

Choosing an Indicator

Choose an indicator whose color change comes as close as possible to the theoretical pH of the equivalence point.

A titration curve for which pH = 5.54 at the equivalence point is shown in Figure 12-9. An indicator with a color change near this pH would be useful in determining the end point of the titration. You can see in Figure 12-9 that the pH drops steeply (from 7 to 4) over a small volume interval. Therefore, any indicator with a color change in this pH interval would provide a fair approximation to the equivalence point. The closer the point of color change is to pH 5.54, the more accurate will be the end point. The difference

Demonstration 12-1 Indicators and the Acidity of CO_2

This one is just plain fun.[2] Fill two 1-L graduated cylinders with 900 mL of water and place a magnetic stirring bar in each. Add 10 mL of 1 M NH_3 to each. Then put 2 mL of phenolphthalein indicator solution in one and 2 mL of bromothymol blue indicator solution in the other. Both indicators will have the color of their basic species.

Drop a few chunks of Dry Ice (solid CO_2) into each cylinder. As the CO_2 bubbles through each cylinder, the solutions become more acidic. First the pink phenolphthalein color disappears. After some time, the pH drops just low enough for bromothymol blue to change from blue to its green intermediate color. The pH does not go low enough to turn bromothymol blue to its yellow color.

Add about 20 mL of 6 M HCl to *the bottom* of each cylinder, using a length of Tygon tubing attached to a funnel. Then stir each solution for a few seconds on a magnetic stirrer. Explain what happens. The sequence of events is shown in Color Plate 7.

between the observed end point (color change) and the true equivalence point is called the **indicator error.**

If you dump half a bottle of indicator into your reaction, you will introduce a different indicator error. Because indicators are acids or bases, they react with analyte or titrant. We use indicators under the assumption that the moles of indicator are negligible with respect to the moles of analyte. Never use more than a few drops of dilute indicator solution.

A list of common indicators is given in Table 12-4. Many of the indicators in the table would provide a useful end point for the titration in Figure 12-9. For example, if bromocresol purple were used, we would use the purple-to-yellow color change as the end point. The last trace of purple should disappear near pH 5.2, which is quite close to the true equivalence point in Figure 12-9. If bromocresol green were used as the indicator, a color change from blue to green (= yellow + blue) would mark the end point (Box 12-4).

In general, *we seek an indicator whose transition range overlaps the steepest part of the titration curve as closely as possible.* The steepness of the titration curve near the equivalence point in Figure 12-9 ensures that the indicator error caused by the noncoincidence of the end point and equivalence point will not be large. For example, if the indicator end point were at pH 6.4 (instead of 5.54), the error in V_e would be only 0.25% in this particular case. You can estimate the indicator error by calculating what volume of titrant is required to attain pH 6.4 instead of pH 5.54.

In strong acid, the colorless form of phenolphthalein turns orange-red. In strong base, the red species loses its color. [G. Wittke, *J. Chem. Ed.* **1983,** *60,* 239.]

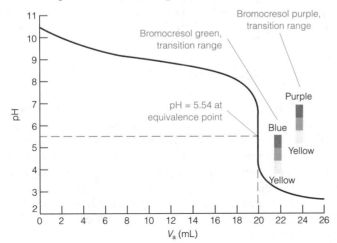

Orange-red
(formed in 65–98% H_2SO_4)

Colorless
(formed above pH 11)

Figure 12-9 Calculated titration curve for the reaction of 100 mL of 0.0100 M base (pK_b = 5.00) with 0.0500 M HCl.

TABLE 12-4   **Common indicators**

Indicator	Transition range (pH)	Acid color	Base color	Preparation
Methyl violet	0.0–1.6	Yellow	Violet	0.05 wt% in H_2O
Cresol red	0.2–1.8	Red	Yellow	0.1 g in 26.2 mL 0.01 M NaOH. Then add ~225 mL H_2O.
Thymol blue	1.2–2.8	Red	Yellow	0.1 g in 21.5 mL 0.01 M NaOH. Then add ~225 mL H_2O.
Cresol purple	1.2–2.8	Red	Yellow	0.1 g in 26.2 mL 0.01 M NaOH. Then add ~225 mL H_2O.
Erythrosine, disodium	2.2–3.6	Orange	Red	0.1 wt% in H_2O
Methyl orange	3.1–4.4	Red	Yellow	0.01 wt% in H_2O
Congo red	3.0–5.0	Violet	Red	0.1 wt% in H_2O
Ethyl orange	3.4–4.8	Red	Yellow	0.1 wt% in H_2O
Bromocresol green	3.8–5.4	Yellow	Blue	0.1 g in 14.3 mL 0.01 M NaOH. Then add ~225 mL H_2O.
Methyl red	4.8–6.0	Red	Yellow	0.02 g in 60 mL ethanol. Then add 40 mL H_2O.
Chlorophenol red	4.8–6.4	Yellow	Red	0.1 g in 23.6 mL 0.01 M NaOH. Then add ~225 mL H_2O.
Bromocresol purple	5.2–6.8	Yellow	Purple	0.1 g in 18.5 mL 0.01 M NaOH. Then add ~225 mL H_2O.
p-Nitrophenol	5.6–7.6	Colorless	Yellow	0.1 wt% in H_2O
Litmus	5.0–8.0	Red	Blue	0.1 wt% in H_2O
Bromothymol blue	6.0–7.6	Yellow	Blue	0.1 g in 16.0 mL 0.01 M NaOH. Then add ~225 mL H_2O.
Phenol red	6.4–8.0	Yellow	Red	0.1 g in 28.2 mL 0.01 M NaOH. Then add ~225 mL H_2O.
Neutral red	6.8–8.0	Red	Yellow	0.01 g in 50 mL ethanol. Then add 50 mL H_2O.
Cresol red	7.2–8.8	Yellow	Red	See above.
α-Naphtholphthalein	7.3–8.7	Pink	Green	0.1 g in 50 mL ethanol. Then add 50 mL H_2O.
Cresol purple	7.6–9.2	Yellow	Purple	See above.
Thymol blue	8.0–9.6	Yellow	Blue	See above.
Phenolphthalein	8.0–9.6	Colorless	Red	0.05 g in 50 mL ethanol. Then add 50 mL H_2O.
Thymolphthalein	8.3–10.5	Colorless	Blue	0.04 g in 50 mL ethanol. Then add 50 mL H_2O.
Alizarin yellow	10.1–12.0	Yellow	Orange-red	0.01 wt% in H_2O
Nitramine	10.8–13.0	Colorless	Orange-brown	0.1 g in 70 mL ethanol. Then add 30 mL H_2O.
Tropaeolin O	11.1–12.7	Yellow	Orange	0.1 wt% in H_2O

 Practical Notes

Acids and bases listed in Table 12-5 can be obtained in pure enough form to be used as *primary standards*.[3] Note that NaOH and KOH are not primary standards, because the reagent-grade materials contain carbonate (from reaction with atmospheric CO_2) and adsorbed water. Solutions of NaOH and KOH must be standardized against a primary standard. Potassium hydrogen phthalate is among the most convenient compounds for this purpose. Dilute solutions of NaOH for titrations are prepared by diluting a stock solution of 50 wt % aqueous NaOH. Sodium carbonate is relatively insoluble in this stock solution and settles to the bottom.

TABLE 12-5 Primary standards

Compound	Formula weight	Notes

ACIDS

Potassium hydrogen phthalate — 204.233 — The pure commercial material is dried at 105°C and used to standardize base. A phenolphthalein end point is satisfactory.

Compound	Formula weight	Notes
HCl Hydrochloric acid	36.461	HCl and water distill as an *azeotrope* (a mixture) whose composition ($\sim$6 M) depends on pressure. The composition is tabulated as a function of the pressure during distillation. See Problem 51 for more information.
$KH(IO_3)_2$ Potassium hydrogen iodate	389.912	This is a strong acid, so any indicator with an end point between $\sim$5 and $\sim$9 is adequate.

BASES

Compound	Formula weight	Notes
$H_2NC(CH_2OH)_3$ Tris(hydroxymethyl)aminomethane (also called tris or tham)	121.136	The pure commercial material is dried at 100–103°C and titrated with strong acid. The end point is in the range pH 4.5–5.

$$H_2NC(CH_2OH)_3 + H^+ \rightarrow H_3\overset{+}{N}C(CH_2OH)_3$$

Compound	Formula weight	Notes
HgO Mercuric oxide	216.59	Pure HgO is dissolved in a large excess of I^- or Br^-, whereupon $2OH^-$ are liberated:

$$HgO + 4I^- + H_2O \rightarrow HgI_4^{2-} + 2OH^-$$

The base is titrated using an indicator end point.

Compound	Formula weight	Notes
Na_2CO_3 Sodium carbonate	105.989	Primary standard grade Na_2CO_3 is commercially available. Alternatively, recrystallized $NaHCO_3$ can be heated for 1 h at 260–270°C to produce pure Na_2CO_3. Sodium carbonate is titrated with acid to an end point of pH 4–5. Just before the end point, the solution is boiled to expel CO_2.
$Na_2B_4O_7 \cdot 10H_2O$ Borax	381.367	The recrystallized material is dried in a chamber containing an aqueous solution saturated with NaCl and sucrose. This procedure gives the decahydrate in pure form. The standard is titrated with acid to a methyl red end point.

$$\text{“}B_4O_7 \cdot 10H_2O^{2-}\text{”} + 2H^+ \rightarrow 4B(OH)_3 + 5H_2O$$

Alkaline solutions (e.g., 0.1 M NaOH) must be protected from the atmosphere; otherwise they absorb CO_2:

$$OH^- + CO_2 \rightarrow HCO_3^-$$

NaOH and KOH must be standardized with primary standards.

CO_2 changes the concentration of strong base over a period of time and decreases the extent of reaction near the end point in the titration of weak acids. If the solutions are kept in tightly capped polyethylene bottles, they can be used for about a week with little change.

Strongly basic solutions attack glass and are best stored in plastic containers. Such solutions should not be kept in a buret longer than necessary. Boiling 0.01 M NaOH in a flask for 1 h decreases the molarity by 10%, owing to the reaction of OH^- with glass.

Box 12-3 **What Does a Negative pH Mean?**

In the 1930s Louis Hammett and his students measured the strengths of very weak acids and bases, using a weak reference base (B), such as p-nitroaniline ($pK_a = 0.99$), whose base strength could be measured in aqueous solution.

Suppose that some p-nitroaniline and a second base, C, are dissolved in a strong acid, such as 2 M HCl. The pK_a of CH^+ can be measured relative to that of BH^+ by first writing a Henderson-Hasselbalch equation for each acid:

$$pH = pK_a \text{ (for } BH^+) + \log \frac{[B]\gamma_B}{[BH^+]\gamma_{BH^+}}$$

$$pH = pK_a \text{ (for } CH^+) + \log \frac{[C]\gamma_C}{[CH^+]\gamma_{CH^+}}$$

Setting the two equations equal (because there is only one pH) gives

$$\underbrace{pK_a \text{ (for } CH^+) - pK_a \text{ (for } BH^+)}_{\Delta pK_a} = \log \frac{[B][CH^+]}{[C][BH^+]} + \log \frac{\gamma_B\gamma_{CH^+}}{\gamma_C\gamma_{BH^+}}$$

In solvents of high dielectric constant, the second term on the right, above, is close to zero because the ratio of activity coefficients is close to unity. Neglecting this last term gives an operationally useful result:

$$\Delta pK_a \approx \log \frac{[B][CH^+]}{[C][BH^+]}$$

That is, if you have a way to find the concentrations of B, BH^+, C, and CH^+ and if you know pK_a for BH^+, then you can find pK_a for CH^+.

Concentrations can be measured spectrophotometrically or by nuclear magnetic resonance,[4] so pK_a for CH^+ can be determined. Then, using CH^+ as the reference, the pK_a for another compound, DH^+, can be measured. This procedure can be extended to measure the strengths of successively weaker bases, far too weak to be protonated in water.

The acidity of a solvent that protonates the weak base, B, is called the **Hammett acidity function:**

Hammett acidity function: $\qquad\qquad H_0 = pK_a \text{ (for } BH^+) + \log \frac{[B]}{[BH^+]}$

12-8 Titrations in Nonaqueous Solvents

In acid-base chemistry there are three common reasons why we might choose a nonaqueous solvent:

1. The reactants or products might be insoluble in water.
2. The reactants or products might react with water.
3. The analyte is too weak an acid or base to be titrated in water.

For dilute aqueous solutions, H_0 approaches pH. For concentrated acids, H_0 is a measure of the acid strength.

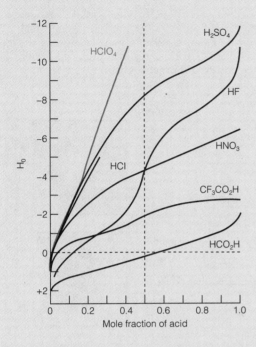

Hammett acidity function, H_0, for aqueous solutions of acids. [Data from R. A. Cox and K. Yates, *Can. J. Chem.* **1983**, *61*, 2225, which reviews acidity functions.]

When we refer to *negative* pH values, we are usually referring to H_0 values. For example, as measured by its ability to protonate very weak bases, 8 M $HClO_4$ has a "pH" close to -4. The figure shows why $HClO_4$ is considered to be a stronger acid than other mineral acids. The values of H_0 for several powerfully acidic solvents are given below.

Acid	Name	H_0
H_2SO_4 (100%)	sulfuric acid	-11.93
$H_2SO_4 \cdot SO_3$	fuming sulfuric acid (oleum)	-14.14
HSO_3F	fluorosulfuric acid	-15.07
$HSO_3F + 10\%$ SbF_5	"super acid"	-18.94
$HSO_3F + 7\%$ $SbF_5 \cdot 3SO_3$	—	-19.35

The pH inside microscopic vesicles (compartments) within living cells can be estimated by infusing an appropriate indicator and then measuring the spectrum of the indicator inside the vesicle. The "pH" of acidic solid catalysts can be measured by observing the color of Hammett indicators adsorbed on the catalyst surface.

In order for us to observe a sharp break at the equivalence point in a titration, the equilibrium constant for the titration reaction must be large. If the analyte or the titrant is too weak, the equilibrium constant is not large enough. Figure 12-3 showed that an analyte acid with $pK_a \gtrsim 8$ does not give a distinct potentiometric end point when titrated with OH^- in water. However, a sharper end point might be observed if that same acid were titrated in a nonaqueous solvent[5] with a base that is stronger than OH^-.

Box 12-4 World Record Small Titration

The titration of 29 fmol (f = femto = 10^{-15}) of HNO_3 in a 1.9-pL (p = pico = 10^{-12}) water drop under a layer of hexane in a Petri dish was carried out with 2% precision using KOH delivered by diffusion from the 1-μm-diameter tip of a glass capillary tube. A mixture of the indicators bromothymol blue and bromocresol purple, with a distinct yellow-to-purple transition at pH 6.7, allowed the end point to be observed through a microscope by a video camera.

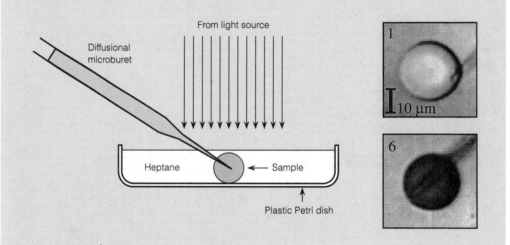

Titration of femtoliter samples. The frames show a larger 8.7-pL droplet before and after the end point, with the microburet making contact at the right side. [From M. Gratzl and C. Yi, *Anal. Chem.* **1993**, *65*, 2085. Photo courtesy M. Gratzl, Case Western Reserve University.]

The Leveling Effect

The strongest acid that can exist in water is H_3O^+ and the strongest base is OH^-. If an acid stronger than H_3O^+ is dissolved in water, it protonates H_2O to make H_3O^+. If a base stronger than OH^- is dissolved in water, it deprotonates H_2O to make OH^-. Because of this **leveling effect,** $HClO_4$ and HCl behave as if they had the same acid strength; both are *leveled* to H_3O^+:

$$HClO_4 + H_2O \rightarrow H_3O^+ + ClO_4^-$$

$$HCl + H_2O \rightarrow H_3O^+ + Cl^-$$

In acetic acid solvent, which is less basic than H_2O, $HClO_4$ and HCl are not leveled to the same strength:

In acetic acid solution, $HClO_4$ is a stronger acid than HCl; but in aqueous solution, they are leveled to the strength of H_3O^+.

$$HClO_4 + CH_3CO_2H \rightleftharpoons CH_3CO_2H_2^+ + ClO_4^- \qquad K = 1.3 \times 10^{-5}$$
$$\text{Acetic acid}$$
$$\text{solvent}$$

$$HCl + CH_3CO_2H \rightleftharpoons CH_3CO_2H_2^+ + Cl^- \qquad K = 2.8 \times 10^{-9}$$

The equilibrium constants show that $HClO_4$ is a stronger acid than is HCl in acetic acid solvent.

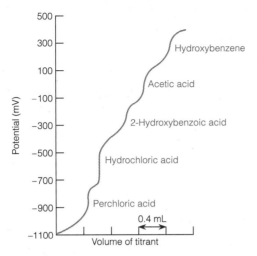

Figure 12-10 Titration of a mixture of acids with tetrabutylammonium hydroxide in methyl isobutyl ketone solvent shows that the order of acid strength is $HClO_4 > HCl > $ 2-hydroxybenzoic acid $>$ acetic acid $>$ hydroxybenzene. Measurements were made with a glass electrode and a platinum reference electrode. The ordinate is proportional to pH, with increasing pH as the potential becomes more positive. [D. B. Bruss and G. E. A. Wyld, *Anal. Chem.* **1957,** *29,* 232.]

Figure 12-10 shows a titration curve for a mixture of five acids titrated with 0.2 M tetrabutylammonium hydroxide in methyl isobutyl ketone solvent. This solvent is not protonated to a great extent by any of the acids. We see that perchloric acid is a stronger acid than HCl in this solvent as well.

Now consider a base that is too weak to give a distinct end point when titrated with a strong acid in water.

titration with $HClO_4$ in H_2O: $B + H_3O^+ \rightleftharpoons BH^+ + H_2O$

The reason why the end point cannot be recognized is that the equilibrium constant for the titration reaction is not large enough. If a stronger acid than H_3O^+ were available, the titration reaction might have a large enough equilibrium constant to give a distinct end point. If the same base were dissolved in acetic acid and titrated with $HClO_4$ in acetic acid, a clear end point might be observed. The reaction

titration with $HClO_4$ in CH_3CO_2H: $B + HClO_4 \rightleftharpoons \underbrace{BH^+ClO_4^-}_{\text{An ion pair}}$

might have a large equilibrium constant, because $HClO_4$ is a much stronger acid than H_3O^+. (The product in this reaction is written as an ion pair because acetic acid has too low a dielectric constant to allow ions to separate extensively.)

Chemistry in Micelles

An increasing amount of synthetic and analytical chemistry of nonpolar organic molecules is conducted in aqueous solution with *micelles,* instead of using organic solvents. The **micelle** in Figure 12-11 is an aggregate of molecules with ionic head groups and long, nonpolar tails. Such molecules

Question Where do you think the end point for the acid $H_3O^+ClO_4^-$ would come in Figure 12-10?

A base too weak to be titrated by H_3O^+ in water might be titrated by $HClO_4$ in acetic acid solvent.

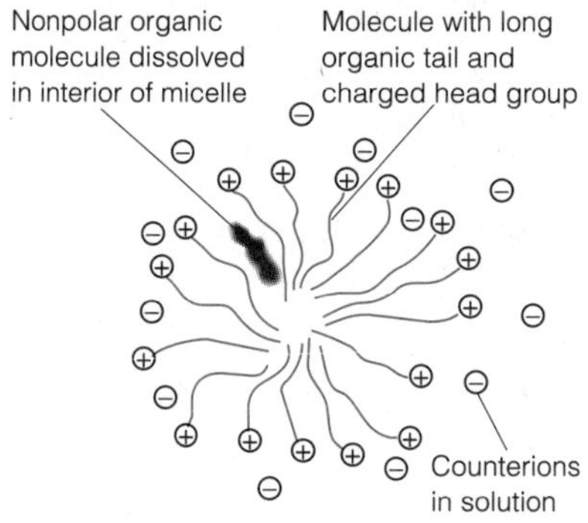

Figure 12-11 Structure of a micelle formed when ionic molecules with long, nonpolar tails aggregate in aqueous solution. The interior of the micelle resembles a nonpolar organic solvent, whereas the exterior charged groups interact strongly with water.

are called **surfactants,** an example of which is cetyltrimethylammonium bromide:

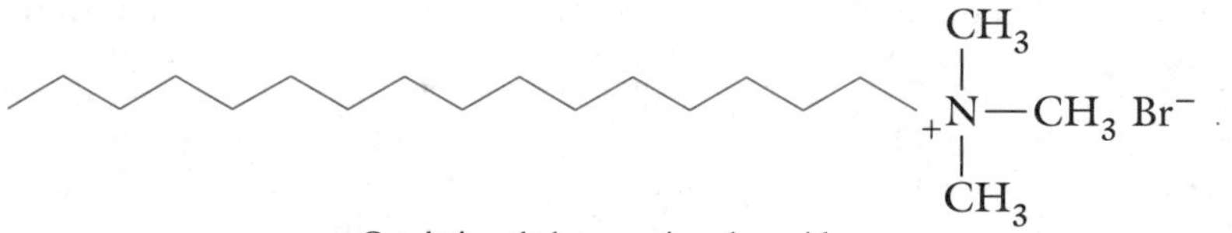

Cetyltrimethylammonium bromide
$C_{16}H_{33}N(CH_3)_3{}^+Br^-$

At very low concentration, the dissolved surfactant molecules are not associated. When the concentration of surfactant exceeds the *critical micelle concentration,* spontaneous aggregation into micelles begins to occur. Isolated surfactant molecules exist in equilibrium with micelles. The nonpolar organic solute in Figure 12-11 finds a friendly environment inside the micelle, which resembles a pocket of nonpolar organic solvent.

Now consider the titration of the very weak acid lidocaine hydrochloride with KOH in aqueous solution. This acid is so weak that it gives almost no perceptible break at the equivalence point of the titration curve, which looks similar to the topmost curve of Figure 12-3.

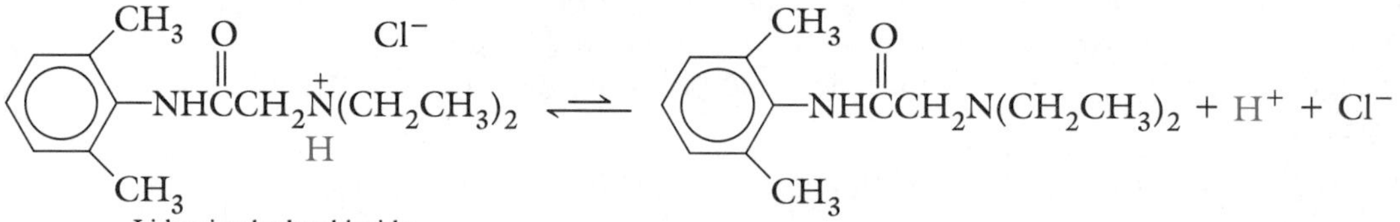

Lidocaine hydrochloride
$R_3NH^+Cl^-$

R_3N

When surfactant is added to a solution of R_3NH^+, it promotes dissociation to $R_3N + H^+$ by dissolving the nonpolar R_3N in the micelles. That is, surfactant increases K_a for this weak acid. When such a solution is titrated with KOH, it gives the upper curve in Figure 12-12. The break at the equivalence point is not sharp, but it is recognizable. Adding toluene to the aqueous micelle solution gives a two-phase mixture in which the end point is even sharper, as shown in the lower curve of Figure 12-12.

You should consider adapting analytical procedures to replace organic solvents by aqueous surfactants. This practice reduces the volume of hazardous waste that must be disposed of.

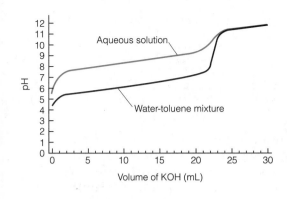

Figure 12-12   Titration of lidocaine hydrochloride in 1.25 mM cetyltrimethylammonium bromide, which forms micelles. [S. A. Tucker, V. L. Amszi, and W. E. Acree, Jr., *J. Chem. Ed.* **1993,** *70,* 80.]

12-9   Calculating Titration Curves with Spreadsheets

The beginning of this chapter was critical for developing your understanding of the chemistry that occurs during titrations. However, the approximations we used are of limited value when concentrations are too dilute or equilibrium constants are not of the right magnitude or K_a values are too closely spaced, like those in a protein. In this section we develop equations that allow us to deal with titrations in a general manner, using spreadsheets.[6] You should understand the principles well enough to derive what you need to treat specific problems as they arise.

Titrating a Weak Acid with a Strong Base

Consider the titration of a volume V_a of acid HA (initial concentration C_a) with a volume V_b of NaOH of concentration C_b. The charge balance for this solution is

charge balance: $[H^+] + [Na^+] = [A^-] + [OH^-]$

and the concentration of Na^+ is just

$$[Na^+] = \frac{C_b V_b}{V_a + V_b}$$

because we have diluted $C_b V_b$ moles of NaOH to a total volume of $V_a + V_b$. Similarly, the formal concentration of the weak acid is

$$F_{HA} = [HA] + [A^-] = \frac{C_a V_a}{V_a + V_b}$$

because we have diluted $C_a V_a$ moles of HA to a total volume of $V_a + V_b$.

Now we finally get to use the fractional composition equations from Chapter 11. Equation 11-18 told us that

$$[A^-] = \alpha_{A^-} \cdot F_{HA} = \frac{\alpha_{A^-} \cdot C_a V_a}{V_a + V_b} \qquad (12\text{-}8)$$

where $\alpha_{A^-} = K_a/([H^+] + K_a)$ and K_a is the acid dissociation constant of HA. Substituting for $[Na^+]$ and $[A^-]$ in the charge balance gives

$$[H^+] + \frac{C_b V_b}{V_a + V_b} = \frac{\alpha_{A^-} \cdot C_a V_a}{V_a + V_b} + [OH^-]$$

α_{A^-} = fraction of acid in the form A^-:

$$\alpha_{A^-} = \frac{[A^-]}{F_{HA}}$$

$\phi = C_bV_b/C_aV_a$ is the fraction of the way to the equivalence point:

ϕ	Volume of base
0.5	$V_b = \frac{1}{2}V_e$
1	$V_b = V_e$
2	$V_b = 2V_e$

2-(N-morpholino)ethanesulfonic acid
MES, $pK_a = 6.15$

The spreadsheet in Figure 12-13 is excellent for finding the pH of a weak acid. Just search for the pH at which $V_b = 0$.

which you might be clever enough to rearrange, with some algebra, to the form

Fraction of titration for weak acid by strong base:

$$\phi = \frac{C_bV_b}{C_aV_a} = \frac{\alpha_{A^-} - \dfrac{[H^+] - [OH^-]}{C_a}}{1 + \dfrac{[H^+] - [OH^-]}{C_b}} \qquad (12\text{-}9)$$

At last! Equation 12-9 is really useful. It relates the volume of titrant (V_b) to the pH and a bunch of constants. The quantity ϕ, which is the quotient C_bV_b/C_aV_a, gives the fraction of the way to the equivalence point, V_e. When $\phi = 1$, the volume of base added, V_b, is equal to V_e. Equation 12-9 works backward from the way you are accustomed to thinking, because you need to put in pH (on the right) to get out volume (on the left).

Let's set up a spreadsheet to use Equation 12-9 to calculate the titration curve for 50.00 mL of the weak acid 0.020 00 M MES with 0.100 0 M NaOH, which was shown in Figure 12-2 and Table 12-2. The equivalence volume is $V_e = 10.00$ mL. We use the quantities in Equation 12-9 as follows:

$C_b = 0.1$ $[H^+] = 10^{-pH}$

$C_a = 0.02$ $[OH^-] = K_w/[H^+]$

$V_a = 50$

$K_a = 7.0_8 \times 10^{-7}$ $\alpha_{A^-} = \dfrac{K_a}{[H^+] + K_a}$

$K_w = 10^{-14}$

$\boxed{\text{pH is the input}}$ $\boxed{V_b = \dfrac{\phi C_a V_a}{C_b}}$ is the output

The input to the spreadsheet in Figure 12-13 is pH in column B and the output is V_b in column G. From the pH, the values of $[H^+]$, $[OH^-]$, and α_{A^-} are computed in columns C, D, and E. Equation 12-9 is used in column F to find the fraction of titration, ϕ. From this value, we calculate the volume of titrant, V_b, in column G.

How do we know what pH values to put in? Trial-and-error allows us to find the starting pH, by putting in a pH and seeing if V_b is positive or negative. In a few tries, it is easy to home in on the pH at which $V_b = 0$. In Figure 12-13 we see that a pH of 3.00 is too low, because ϕ and V are both negative. Input values of pH are spaced as closely as you like, so that you can generate a smooth titration curve. To save space, we only show a few points in Figure 12-13, including the midpoint (pH 6.15 $\Rightarrow V_b = 5.00$ mL) and the end point (pH 9.18 $\Rightarrow V_b = 10.00$ mL). This spreadsheet reproduces Table 12-2 without approximations except neglecting activity coefficients. It gives correct results even when the approximations used in Table 12-2 fail.

Titrating a Weak Acid with a Weak Base

Now consider the titration of V_a mL of acid HA (initial concentration C_a) with V_b mL of base B whose concentration is C_b. Let the acid dissociation constant of HA be K_a and the acid dissociation constant of BH^+ be K_{BH^+}. The charge balance is

charge balance: $[H^+] + [BH^+] = [A^-] + [OH^-]$

As before, we can say that $[A^-] = \alpha_{A^-} \cdot F_{HA}$, where $\alpha_{A^-} = K_a/([H^+] + K_a)$ and $F_{HA} = C_aV_a/(V_a + V_b)$.

	A	B	C	D	E	F	G
1	Cb =	pH	[H+]	[OH−]	Alpha[A−]	Phi	Vb (mL)
2	0.1	3.00	1.00E-03	1.00E-11	7.075E-04	-4.880E-02	-0.488
3	Ca =	3.93	1.17E-04	8.51E-11	5.990E-03	1.153E-04	0.001
4	0.02	4.00	1.00E-04	1.00E-10	7.030E-03	2.028E-03	0.020
5	Va =	5.00	1.00E-05	1.00E-09	6.612E-02	6.561E-02	0.656
6	50	6.15	7.08E-07	1.41E-08	5.000E-01	5.000E-01	5.000
7	Ka =	7.00	1.00E-07	1.00E-07	8.762E-01	8.762E-01	8.762
8	7.08E-7	8.00	1.00E-08	1.00E-06	9.861E-01	9.861E-01	9.861
9	Kw =	9.18	6.61E-10	1.51E-05	9.991E-01	1.000E+00	10.000
10	1.E-14	10.00	1.00E-10	1.00E-04	9.999E-01	1.006E+00	10.059
11		11.00	1.00E-11	1.00E-03	1.000E+00	1.061E+00	10.606
12		12.00	1.00E-12	1.00E-02	1.000E+00	1.667E+00	16.667
13							
14	C2 = 10^-B2						
15	D2 = $A$10/C2						
16	E2 = $A$8/(C2+$A$8)						
17	F2 = (E2-(C2-D2)/$A$4)/(1+(C2-D2)/$A$2) [Equation 12-9]						
18	G2 = F2*$A$4*$A$6/$A$2						

Figure 12-13 Spreadsheet that uses Equation 12-9 to calculate the titration curve for 50 mL of the weak acid, 0.02 M MES (pK_a = 6.15), treated with 0.1 M NaOH. We provide pH as input, and the spreadsheet tells us what volume of base is required to generate that pH.

We can write an analogous expression for $[BH^+]$, which is a monoprotic weak acid. If the acid were HA, we would use Equation 11-17 to say

$$[HA] = \alpha_{HA}F_{HA} \qquad \alpha_{HA} = \frac{[H^+]}{[H^+] + K_a}$$

where K_a applies to the acid HA. For the weak acid BH^+, we write

$$[BH^+] = \alpha_{BH^+} \cdot F_B \qquad \alpha_{BH^+} = \frac{[H^+]}{[H^+] + K_{BH^+}}$$

α_{HA} is the fraction of acid in the form HA:

$$\alpha_{HA} = \frac{[HA]}{F_{HA}}$$

α_{BH^+} is the fraction of base in the form BH^+:

$$\alpha_{BH^+} = \frac{[BH^+]}{F_B}$$

where the formal concentration of base is $F_B = C_b V_b/(V_a + V_b)$.

Substituting for $[BH^+]$ and $[A^-]$ in the charge balance gives

$$[H^+] + \frac{\alpha_{BH^+} \cdot C_b V_b}{V_a + V_b} = \frac{\alpha_{A^-} \cdot C_a V_a}{V_a + V_b} + [OH^-]$$

which can be rearranged to the useful result

Fraction of titration for weak acid by weak base:
$$\phi = \frac{C_b V_b}{C_a V_a} = \frac{\alpha_{A^-} - \dfrac{[H^+] - [OH^-]}{C_a}}{\alpha_{BH^+} + \dfrac{[H^+] - [OH^-]}{C_b}} \qquad (12\text{-}10)$$

Equation 12-10 for a weak base looks just like Equation 12-9 for a strong base, except that α_{BH^+} replaces 1 in the denominator. Table 12-6 gives many useful results.

TABLE 12-6   **Titration equations for spreadsheets**

Calculation of ϕ

Titrating strong acid with strong base

$$\phi = \frac{C_b V_b}{C_a V_a} = \frac{1 - \dfrac{[H^+] - [OH^-]}{C_a}}{1 + \dfrac{[H^+] - [OH^-]}{C_b}}$$

Titrating weak acid (HA) with strong base

$$\phi = \frac{C_b V_b}{C_a V_a} = \frac{\alpha_{A^-} - \dfrac{[H^+] - [OH^-]}{C_a}}{1 + \dfrac{[H^+] - [OH^-]}{C_b}}$$

Titrating strong base with strong acid

$$\phi = \frac{C_a V_a}{C_b V_b} = \frac{1 + \dfrac{[H^+] - [OH^-]}{C_b}}{1 - \dfrac{[H^+] - [OH^-]}{C_a}}$$

Titrating weak base (B) with strong acid

$$\phi = \frac{C_a V_a}{C_b V_b} = \frac{\alpha_{BH^+} + \dfrac{[H^+] - [OH^-]}{C_b}}{1 - \dfrac{[H^+] - [OH^-]}{C_a}}$$

Titrating weak acid (HA) with weak base (B)

$$\phi = \frac{C_b V_b}{C_a V_a} = \frac{\alpha_{A^-} - \dfrac{[H^+] - [OH^-]}{C_a}}{\alpha_{BH^+} + \dfrac{[H^+] - [OH^-]}{C_b}}$$

Titrating weak base (B) with weak acid (HA)

$$\phi = \frac{C_a V_a}{C_b V_b} = \frac{\alpha_{BH^+} + \dfrac{[H^+] - [OH^-]}{C_b}}{\alpha_{A^-} - \dfrac{[H^+] - [OH^-]}{C_a}}$$

Titrating H_2A with strong base ($\rightarrow \rightarrow A^{2-}$)

$$\phi = \frac{C_b V_b}{C_a V_a} = \frac{\alpha_{HA^-} + 2\alpha_{A^{2-}} - \dfrac{[H^+] - [OH^-]}{C_a}}{1 + \dfrac{[H^+] - [OH^-]}{C_b}}$$

Titrating H_3A with strong base ($\rightarrow \rightarrow \rightarrow A^{3-}$)

$$\phi = \frac{C_b V_b}{C_a V_a} = \frac{\alpha_{H_2A^-} + 2\alpha_{HA^{2-}} + 3\alpha_{A^{3-}} - \dfrac{[H^+] - [OH^-]}{C_a}}{1 + \dfrac{[H^+] - [OH^-]}{C_b}}$$

Titrating dibasic B with strong acid ($\rightarrow \rightarrow BH_2^{2+}$)

$$\phi = \frac{C_a V_a}{C_b V_b} = \frac{\alpha_{BH^+} + 2\alpha_{BH_2^{2+}} + \dfrac{[H^+] - [OH^-]}{C_b}}{1 - \dfrac{[H^+] - [OH^-]}{C_a}}$$

Titrating tribasic B with strong acid ($\rightarrow \rightarrow \rightarrow BH_3^{3+}$)

$$\phi = \frac{C_a V_a}{C_b V_b} = \frac{\alpha_{BH^+} + 2\alpha_{BH_2^{2+}} + 3\alpha_{BH_3^{3+}} + \dfrac{[H^+] - [OH^-]}{C_b}}{1 - \dfrac{[H^+] - [OH^-]}{C_a}}$$

Symbols

ϕ = fraction of the way to the first equivalence point
C_a = initial concentration of acid
C_b = initial concentration of base

α = fraction of dissociation of acid or fraction of association of base
V_a = volume of acid
V_b = volume of base

Terms to Understand

Gran plot	indicator error	micelle
Hammett acidity function	leveling effect	surfactant
indicator		

TABLE 12-6 (continued)

Calculation of α

Monoprotic systems

$$\alpha_{HA} = \frac{[H^+]}{[H^+] + K_a} \qquad \alpha_{A^-} = \frac{K_a}{[H^+] + K_a}$$

$$\alpha_{BH^+} = \frac{[H^+]}{[H^+] + K_{BH^+}} \qquad \alpha_B = \frac{K_{BH^+}}{[H^+] + K_{BH^+}}$$

Symbols

K_a = acid dissociation constant of HA
K_{BH^+} = acid dissociation constant of BH^+ ($= K_w/K_b$)

Diprotic systems

$$\alpha_{H_2A} = \frac{[H^+]^2}{[H^+]^2 + [H^+]K_1 + K_1K_2} \qquad \alpha_{HA^-} = \frac{[H^+]K_1}{[H^+]^2 + [H^+]K_1 + K_1K_2} \qquad \alpha_{A^{2-}} = \frac{K_1K_2}{[H^+]^2 + [H^+]K_1 + K_1K_2}$$

$$\alpha_{BH_2^{2+}} = \frac{[H^+]^2}{[H^+]^2 + [H^+]K_1 + K_1K_2} \qquad \alpha_{BH^+} = \frac{[H^+]K_1}{[H^+]^2 + [H^+]K_1 + K_1K_2} \qquad \alpha_B = \frac{K_1K_2}{[H^+]^2 + [H^+]K_1 + K_1K_2}$$

Symbols

K_1 and K_2 for the acid are the acid dissociation constants of H_2A and HA^-, respectively.
K_1 and K_2 for the base refer to the acid dissociation constants of BH_2^{2+} and BH^+, respectively:
   $K_1 = K_w/K_{b2}$; $K_2 = K_w/K_{b1}$

Triprotic systems

$$\alpha_{H_3A} = \frac{[H^+]^3}{[H^+]^3 + [H^+]^2K_1 + [H^+]K_1K_2 + K_1K_2K_3} \qquad \alpha_{H_2A^-} = \frac{[H^+]^2K_1}{[H^+]^3 + [H^+]^2K_1 + [H^+]K_1K_2 + K_1K_2K_3}$$

$$\alpha_{HA^{2-}} = \frac{[H^+]K_1K_2}{[H^+]^3 + [H^+]^2K_1 + [H^+]K_1K_2 + K_1K_2K_3} \qquad \alpha_{A^{3-}} = \frac{K_1K_2K_3}{[H^+]^3 + [H^+]^2K_1 + [H^+]K_1K_2 + K_1K_2K_3}$$

Summary

In the titration of a strong acid with a strong base (or vice versa), the titration reaction is $H^+ + OH^- \rightarrow H_2O$, and the pH is determined by the concentration of excess unreacted analyte or titrant. The pH at the equivalence point is 7.00.

The titration curve for a weak acid treated with a strong base ($HA + OH^- \rightarrow A^- + H_2O$) can be divided into four regions:
1. Before any base is added, the pH is determined by the acid dissociation reaction of the weak acid ($HA \rightleftharpoons H^+ + A^-$).
2. Between the initial point and the equivalence point, there is a buffer made from A^- (which is equivalent to the moles of added base) and the excess unreacted HA.

$pH = pK_a + \log([A^-]/[HA])$. At the special point when $V_b = \frac{1}{2}V_e$, $pH = pK_a$ (neglecting activity coefficients).
3. At the equivalence point, the weak acid has been converted to its conjugate base, A^-, whose pH is governed by hydrolysis ($A^- + H_2O \rightleftharpoons HA + OH^-$). The pH is necessarily above 7.00.
4. After the equivalence point, the pH is determined by the concentration of excess strong base.

The titration curve for the reaction of a weak base with a strong acid also has four regions. At the start, the pH is governed by hydrolysis of base, B. Between the initial and equivalence points, there is a buffer consisting of B plus BH^+. When $V_a = \frac{1}{2}V_e$, $pH = pK_a$ (for BH^+). At the equivalence point, B has been converted to its conjugate acid, BH^+, whose pH must be below 7.00. Beyond the equivalence point, the pH is governed by the concentration of excess strong acid titrant.

The titration of a diprotic acid, H_2A, with a strong base has six regions:

1. The initial pH is determined by the dissociation of H_2A, which behaves as a monoprotic weak acid ($H_2A \rightleftharpoons H^+ + HA^-$).

2. Between the initial point and the first equivalence point, the solution is buffered by H_2A plus HA^-, whose pH is given by pH = pK_{a1} + log ($[HA^-]/[H_2A]$). When $V_b = \frac{1}{2}V_e$, pH = pK_{a1}.

3. At the first equivalence point, H_2A has been converted to HA^-:

$$[H^+] = \sqrt{\frac{K_1K_2F' + K_1K_w}{K_1 + F'}} \Rightarrow pH \approx \frac{1}{2}(pK_{a1} + pK_{a2})$$

where F', the formal concentration of HA^-, contains a correction for dilution of starting material.

4. Between the first and second equivalence points, there is a buffer consisting of HA^- and A^{2-}, whose pH is given by pH = pK_{a2} + log ($[A^{2-}]/[HA^-]$). When $V_b = \frac{3}{2}V_e$, pH = pK_{a2}.

5. At the second equivalence point, HA^- has been converted to its conjugate base, A^{2-}, whose pH is

governed by hydrolysis ($A^{2-} + H_2O \rightleftharpoons HA^- + OH^-$).

6. Beyond the second equivalence point, the pH is governed by the concentration of excess strong base titrant.

If the reactants are too dilute, or if the equilibrium constant for the titration reaction is not large enough, no sharp end point will be observed. Such titration curves can be simulated with the spreadsheet equations developed at the end of the chapter. In these derivations, we write a charge balance for the titration reaction and substitute for each of the concentrations, using fractional composition equations.

The end point of a titration can be found by computing the finite derivative ($\Delta pH/\Delta V$) or second derivative ($\Delta(\Delta pH/\Delta V)/\Delta V$) of a graph of pH versus volume of titrant. At the end point, the first derivative is maximum and the second derivative is zero. Alternatively, a Gran plot, such as a graph of $V_b \cdot 10^{-pH}$ versus V_b, allows us to use more reliable data from before the equivalence point to find the equivalence point. When choosing an indicator, select one whose color transition range matches the pH of the equivalence point of the titration as closely as possible. Acids or bases too weak to be titrated in H_2O may be titrated in a nonaqueous solvent in which the titrant is not leveled to the strength of H_3O^+ or OH^-.

Exercises

A. Calculate the pH at each of the following points in the titration of 50.00 mL of 0.0100 M NaOH with 0.100 M HCl. Volume of acid added: 0.00, 1.00, 2.00, 3.00, 4.00, 4.50, 4.90, 4.99, 5.00, 5.01, 5.10, 5.50, 6.00, 8.00, and 10.00 mL. Make a graph of pH versus volume of HCl added.

B. Calculate the pH at the following points for the titration of 50.0 mL of 0.0500 M formic acid with 0.0500 M KOH: V_b = 0.0, 10.0, 20.0, 25.0, 30.0, 40.0, 45.0, 48.0, 49.0, 49.5, 50.0, 50.5, 51.0, 52.0, 55.0, and 60.0 mL. Draw a graph of pH versus V_b.

C. Calculate the pH at the following points for the titration of 100.0 mL of 0.100 M cocaine (Section 10-4, $K_b = 2.6 \times 10^{-6}$) with 0.200 M HNO_3: V_a = 0.0, 10.0, 20.0, 25.0, 30.0, 40.0, 49.0, 49.9, 50.0, 50.1, 51.0, and 60.0 mL. Draw a graph of pH versus V_a.

D. Consider the titration of 50.0 mL of 0.0500 M malonic acid with 0.100 M NaOH. Calculate the pH at each point listed and sketch the titration curve: V_b = 0.0, 8.0, 12.5, 19.3, 25.0, 37.5, 50.0, and 56.3 mL.

E. Write the chemical reactions (including structures of reactants and products) that occur when the amino acid, histidine, is titrated with perchloric acid. (Histidine is a molecule with no net charge.) A solution containing 25.0 mL of 0.0500 M histidine was titrated with 0.0500 M

$HClO_4$. Calculate the pH at the following values of V_a: 0, 4.0, 12.5, 25.0, 26.0, and 50.0 mL.

F. Select indicators from Table 12-4 that would be useful for the titration in Figures 12-1, 12-2, and the pK_a = 8 curve in Figure 12-3. Select a different indicator for each titration and state what color change you would use as the end point.

G. *Spectrophotometry with indicators.* Acid-base indicators are themselves acids or bases. Consider an indicator, HIn, which dissociates according to the equation

$$HIn \overset{K_a}{\rightleftharpoons} H^+ + In^-$$

Suppose that the molar absorptivity, ϵ, is 2080 $M^{-1} cm^{-1}$ for HIn and 14200 $M^{-1} cm^{-1}$ for In^-, at a wavelength of 440 nm.

(a) Write the expression giving the absorbance at 440 nm of a solution containing HIn at a concentration [HIn] and In^- at a concentration $[In^-]$. Assume the cell pathlength is 1.00 cm. Note that absorbance is additive. The total absorbance is the sum of absorbances of all components.

(b) A solution containing the indicator at a formal concentration of 1.84×10^{-4} M is adjusted to pH

6.23 and found to exhibit an absorbance of 0.868 at 440 nm. Calculate pK_a for this indicator.

H. When 100.0 mL of a weak acid was titrated with 0.093 81 M NaOH, 27.63 mL was required to reach the equivalence point. The pH at the equivalence point was 10.99. What was the pH when only 19.47 mL of NaOH had been added?

I. A 0.100 M solution of the weak acid HA was titrated with 0.100 M NaOH. The pH measured when $V_b = \frac{1}{2}V_e$ was 4.62. Using activity coefficients correctly, calculate pK_a. The size of the A^- anion is 450 pm.

J. The table below lists the data points around the second apparent end point in Figure 12-6.

(a) Prepare a table analogous to Table 12-3, showing the first and second derivatives. Plot both derivatives versus V_b and locate the end point from each plot.

(b) Prepare a Gran plot analogous to Figure 12-8. Use least squares to find the best straight line and find the end point. You will have to use your judgment as to which points lie on the "straight" line.

K. *Indicator error.* Consider the titration in Figure 12-2 in which the equivalence point pH in Table 12-2 is 9.18 at a volume of 10.00 mL.

(a) Suppose you used the yellow-to-blue transition of thymol blue indicator to find the end point. According to Table 12-4, the last trace of green disappears near pH 9.6. What volume of base is required to reach pH 9.6? The difference between this volume and 10 mL is the indicator error.

(b) If you used cresol red with a color change at pH 8.8, what would be the indicator error?

V_b (μL)	pH	V_b (μL)	pH	V_b (μL)	pH	V_b (μL)	pH
107	6.921	114	7.457	117	7.878	120	8.591
110	7.117	115	7.569	118	8.090	121	8.794
113	7.359	116	7.705	119	8.343	122	8.952

Problems

Titration of Strong Acid with Strong Base

1. Distinguish the terms *end point* and *equivalence point*.

2. Consider the titration of 100.0 mL of 0.100 M NaOH with 1.00 M HBr. Find the pH at the following volumes of acid added and make a graph of pH versus V_a: $V_a = 0, 1, 5, 9, 9.9, 10, 10.1,$ and 12 mL.

Titration of Weak Acid with Strong Base

3. Why is it not practical to titrate an acid or base that is too weak or too dilute?

4. A weak acid HA ($pK_a = 5.00$) was titrated with 1.00 M KOH. The acid solution had a volume of 100.0 mL and a molarity of 0.100 M. Find the pH at the following volumes of base added and make a graph of pH versus V_b: $V_b = 0, 1, 5, 9, 9.9, 10, 10.1,$ and 12 mL.

5. *Titration of HA with NaOH.* At what fraction of V_e does pH = $pK_a - 1$? At what fraction of V_e does pH = $pK_a + 1$? Use these two points, plus $V_b = 0, \frac{1}{2}V_e, V_e,$ and $1.2 V_e$ to sketch the titration curve for the reaction of 100 mL of 0.100 M anilinium bromide ("aminobenzene · HBr") with 0.100 M NaOH.

6. What is the pH at the equivalence point when 0.100 M hydroxyacetic acid is titrated with 0.050 0 M KOH?

7. Find the equilibrium constant for the reaction of MES (Table 10-2) with NaOH.

8. When 22.63 mL of aqueous NaOH was added to 1.214 g of cyclohexylaminoethanesulfonic acid (FW 207.29, Table 10-2) dissolved in 41.37 mL of water, the pH was 9.24. Calculate the molarity of the NaOH.

9. *Use activity coefficients correctly* to calculate the pH after 10.0 mL of 0.100 M trimethylammonium bromide was titrated with 4.0 mL of 0.100 M NaOH.

Titration of Weak Base with Strong Acid

10. Why is the equivalence-point pH necessarily below 7 when a weak base is titrated with strong acid?

11. A 100.0-mL aliquot of 0.100 M weak base B ($pK_b = 5.00$) was titrated with 1.00 M HClO$_4$. Find the pH at the following volumes of acid added and make a graph of pH versus V_a: $V_a = 0, 1, 5, 9, 9.9, 10, 10.1,$ and 12 mL.

12. At what point in the titration of a weak base with a strong acid is the maximum buffer capacity reached? This is the point at which a given small addition of acid causes the least pH change.

13. What is the equilibrium constant for the reaction between benzylamine and HCl?

14. A solution containing 50.0 mL of 0.031 9 M benzylamine was titrated with 0.050 0 M HCl. Calculate the pH at the following volumes of added acid: $V_a = 0,$ 12.0, $\frac{1}{2}V_e$, 30.0, V_e, and 35.0 mL.

15. Calculate the pH of a solution made by mixing 50.00 mL of 0.100 M NaCN with

(a) 4.20 mL of 0.438 M $HClO_4$

(b) 11.82 mL of 0.438 M $HClO_4$

(c) What is the pH at the equivalence point with 0.438 M $HClO_4$?

Titrations in Diprotic Systems

16. At the opening of this chapter is the titration curve for an enzyme. Is the molecule positively charged, negatively charged, or neutral at its isoionic point? Explain how you know.

17. The base Na^+A^-, whose anion is dibasic, was titrated with HCl to give the lower curve in Figure 12-4. Is point H, the first equivalence point, the isoelectric point or the isoionic point?

18. The figure below compares the titration of a monoprotic weak acid with a monoprotic weak base with the titration of a diprotic acid with strong base.

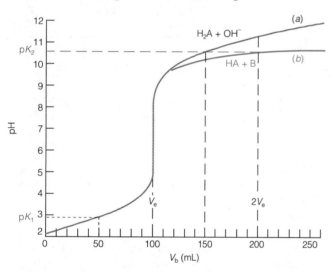

(a) Titration of 100 mL of 0.50 M H_2A (pK_1 = 2.86, pK_2 = 10.64) with 0.050 M NaOH (b) Titration of 100 mL of the weak acid HA (0.50 M, pK_a = 2.86) with the weak base B (0.050 M, pK_b = 3.36).

(a) Write the reaction between the weak acid and weak base and show that the equilibrium constant is $10^{7.78}$. This large value means that the reaction goes "to completion" after each addition of reagent.

(b) Why does pK_2 intersect the upper curve at $\frac{3}{2}V_e$ and the lower curve at $2V_e$? On the lower curve, "pK_2" is pK_a for the acid, BH^+.

19. The dibasic compound B (pK_{b1} = 4.00, pK_{b2} = 8.00) was titrated with 1.00 M HCl. The initial solution of B was 0.100 M and had a volume of 100.0 mL. Find the pH at V_a = 0, 1, 5, 9, 10, 11, 15, 19, 20, and 22 mL. Sketch the titration curve.

20. A 100.0-mL aliquot of 0.100 M diprotic acid H_2A (pK_1 = 4.00, pK_2 = 8.00) was titrated with 1.00 M NaOH. Find the pH at the following volumes of base added and make a graph of pH versus V_b: V_b = 0, 1, 5, 9, 10, 11, 15, 19, 20, and 22 mL.

21. Calculate the pH at 10.0-mL intervals (from 0 to 100 mL) in the titration of 40.0 mL of 0.100 M piperazine with 0.100 M HCl. Graph pH versus V_a.

22. Calculate the pH when 25.0 mL of 0.020 0 M 2-aminophenol has been titrated with 10.9 mL of 0.015 0 M $HClO_4$.

23. Consider the titration of 50.0 mL of 0.100 M sodium glycinate with 0.100 M HCl.

(a) Calculate the pH at the second equivalence point.

(b) Show that our approximate method of calculations gives incorrect (physically unreasonable) values of pH at V_a = 90.0 and V_a = 101.0 mL.

24. A solution containing 0.100 M glutamic acid (the molecule with no net charge) was titrated with 0.025 0 M RbOH.

(a) Draw the structures of reactants and products.

(b) Calculate the pH at the first equivalence point.

25. Find the pH of the solution when 0.010 0 M tyrosine is titrated to the equivalence point with 0.004 00 M $HClO_4$.

26. This problem deals with the amino acid cysteine, which we will abbreviate H_2C.

(a) A 0.030 0 M solution was prepared by dissolving dipotassium cysteine, K_2C, in water. Then 40.0 mL of this solution was titrated with 0.060 0 M $HClO_4$. Calculate the pH at the first equivalence point.

(b) Calculate the quotient $[C^{2-}]/[HC^-]$ in a solution of 0.050 0 M cysteinium bromide (the salt $H_3C^+Br^-$).

27. How many grams of dipotassium oxalate (FW 166.22) must be added to 20.0 mL of 0.800 M $HClO_4$ to give pH 4.40 when the solution is diluted to 500 mL?

28. When 5.00 mL of 0.103 2 M NaOH was added to 0.112 3 g of alanine (MW 89.094) in 100.0 mL of 0.10 M KNO_3, the measured pH was 9.57. *Using activity coefficients correctly,* find pK_2 for alanine. Consider the ionic strength of the solution to be 0.10 M and consider each ionic form of alanine to have an activity coefficient of 0.77.

Finding the End Point with a pH Electrode

29. What is a Gran plot used for?

30. Data for the titration of 100.00 mL of a weak acid by NaOH are given below. Find the end point by preparing a Gran plot, using the last 10% of the volume prior to V_e.

mL NaOH	pH	mL NaOH	pH
0.00	4.14	21.61	6.27
1.31	4.30	21.77	6.32
2.34	4.44	21.93	6.37
3.91	4.61	22.10	6.42
5.93	4.79	22.27	6.48
7.90	4.95	22.37	6.53
11.35	5.19	22.48	6.58
13.46	5.35	22.57	6.63
15.50	5.50	22.70	6.70
16.92	5.63	22.76	6.74
18.00	5.71	22.80	6.78
18.35	5.77	22.85	6.82
18.95	5.82	22.91	6.86
19.43	5.89	22.97	6.92
19.93	5.95	23.01	6.98
20.48	6.04	23.11	7.11
20.75	6.09	23.17	7.20
21.01	6.14	23.21	7.30
21.10	6.15	23.30	7.49
21.13	6.16	23.32	7.74
21.20	6.17	23.40	8.30
21.30	6.19	23.46	9.21
21.41	6.22	23.55	9.86
21.51	6.25		

31. Prepare a second derivative graph to find the end point from the titration data below.

mL NaOH	pH	mL NaOH	pH
10.679	7.643	10.729	5.402
10.696	7.447	10.733	4.993
10.713	7.091	10.738	4.761
10.721	6.700	10.750	4.444
10.725	6.222	10.765	4.227

Finding the End Point with Indicators

32. Explain the origin of the rule of thumb that indicator color changes occur at $pK_{HIn} \pm 1$.

33. Write the formula of a compound with a negative pK_a.

34. Consider the titration in Figure 12-2, for which the pH at the equivalence point is calculated to be 9.18. If thymol blue is used as an indicator, what color will be observed through most of the titration prior to the equivalence point? At the equivalence point? After the equivalence point?

35. What color do you expect to observe for cresol purple indicator (Table 12-4) at the following pH values?
(a) 1.0 (b) 2.0 (c) 3.0

36. Cresol red has *two* transition ranges listed in Table 12-4. What color would you expect it to be at the following pH values?
(a) 0 (b) 1 (c) 6 (d) 9

37. Would the indicator bromocresol green, with a transition range of pH 3.8–5.4, ever be useful in the titration of a weak acid with a strong base?

38. (a) What is the pH at the equivalence point when 0.030 0 M NaF is titrated with 0.060 0 M $HClO_4$?

(b) Why would an indicator end point probably not be useful in this titration?

39. A titration curve for $NaCO_3$ titrated with HCl is shown below. Suppose that *both* phenolphthalein and bromocresol green are present in the titration solution. State what colors you expect to observe at the following volumes of added HCl:
(a) 2 ml (b) 10 mL (c) 19 mL

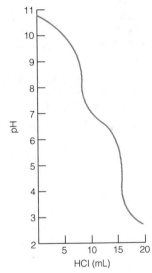

HCl (mL)

40. In the Kjeldahl nitrogen determination (Reactions 7-3 through 7-5), the final product is a solution of NH_4^+ ion in HCl solution. It is necessary to titrate the HCl without titrating the NH_4^+ ion.

(a) Calculate the pH of pure 0.010 M NH_4Cl.

(b) Select an indicator that would allow you to titrate HCl but not NH_4^+.

41. A 10.231-g sample of window cleaner containing NH_3 was diluted with 39.466 g of H_2O. Then 4.373 g of solution was titrated with 14.22 mL of 0.106 3 M HCl to reach a bromocresol green end point. Find the weight percent of NH_3 (MW 17.031) in the cleaner.

42. Spectrophotometric properties of a particular indicator are given below:

$$HIn \quad \overset{pK_a = 7.95}{\rightleftharpoons} \quad In^- + H^+$$

$\lambda_{max} = 395$ nm
$\epsilon_{395} = 1.80 \times 10^4$ M^{-1} cm^{-1}
$\epsilon_{604} = 0$

$\lambda_{max} = 604$ nm
$\epsilon_{604} = 4.97 \times 10^4$ M^{-1} cm^{-1}

where ϵ is the molar absorptivity. A solution with a volume of 20.0 mL containing 1.40×10^{-5} M indicator plus 0.0500 M benzene-1,2,3-tricarboxylic acid was treated with 20.0 mL of aqueous KOH. The resulting solution had an absorbance at 604 nm of 0.118 in a 1.00-cm cell. Calculate the molarity of the KOH solution.

43. A certain acid-base indicator exists in three colored forms:

$$H_2In \quad \overset{pK_1 = 1.00}{\rightleftharpoons} \quad HIn^- \quad \overset{pK_2 = 7.95}{\rightleftharpoons} \quad In^{2-}$$

$\lambda_{max} = 520$ nm
$\epsilon_{520} = 5.00 \times 10^4$
Red
$\epsilon_{435} = 1.67 \times 10^4$
$\epsilon_{572} = 2.03 \times 10^4$

$\lambda_{max} = 435$ nm
$\epsilon_{435} = 1.80 \times 10^4$
Yellow
$\epsilon_{520} = 2.13 \times 10^3$
$\epsilon_{572} = 2.00 \times 10^2$

$\lambda_{max} = 572$ nm
$\epsilon_{572} = 4.97 \times 10^4$
Red
$\epsilon_{520} = 2.50 \times 10^4$
$\epsilon_{435} = 1.15 \times 10^4$

The units of molar absorptivity (ϵ) are M^{-1} cm^{-1}. A solution containing 10.0 mL of 5.00×10^{-4} M indicator was mixed with 90.0 mL of 0.1 M phosphate buffer (pH 7.50). Calculate the absorbance of this solution at 435 nm in a 1.00-cm cell.

44. A solution was prepared by mixing 25.00 mL of 0.0800 M aniline, 25.00 mL of 0.0600 M sulfanilic acid, and 1.00 mL of 1.23×10^{-4} M HIn and then diluting to 100.0 mL. (HIn stands for protonated indicator.)

Anilinium ion
$pK_a = 4.601$

Sulfanilic acid
$pK_a = 3.232$

$$HIn \quad \rightleftharpoons H^+ + In^-$$

$\epsilon_{325} = 2.45 \times 10^4$
$\epsilon_{550} = 2.26 \times 10^4$

$\epsilon_{325} = 4.39 \times 10^3$ M^{-1} cm^{-1}
$\epsilon_{550} = 1.53 \times 10^4$

The absorbance measured at 550 nm in a *5.00-cm* cell was 0.110. Find pK_a for HIn.

Practical Notes

45. Give the name and formula of a primary standard used to standardize **(a)** HCl and **(b)** NaOH.

46. Why is it more accurate to use a primary standard with a high equivalent mass (the mass required to provide or consume 1 mol of H$^+$) than one with a low equivalent mass?

47. Explain how to use potassium hydrogen phthalate to standardize a solution of NaOH.

48. A solution was prepared from 1.023 g of the primary standard tris (Table 12-5) plus 99.367 g of water; 4.963 g of the solution was titrated with 5.262 g

of aqueous HNO$_3$ to reach the methyl red end point. Calculate the concentration of the HNO$_3$ (expressed as mol HNO$_3$/kg solution).

49. A solution was prepared by dissolving 0.1947 g of HgO (Table 12-5) in 20 mL of water containing 4 g of KBr. Titration with HCl required 17.98 mL to reach a phenolphthalein end point. Calculate the molarity of the HCl.

50. How many grams of potassium hydrogen phthalate should be weighed into a flask to standardize ~0.05 M NaOH if you wish to use ~30 mL of base for the titration?

51. Constant-boiling aqueous HCl is a primary standard for acid-base titrations. When ~20 wt% HCl (FW 36.461) is distilled, the composition of the distillate varies in a regular manner with the barometric pressure:

P (torr)	HCl[a] (g/100 g solution)
770	20.196
760	20.220
750	20.244
740	20.268
730	20.292

a. From C. W. Foulk and M. Hollingsworth, *J. Am. Chem. Soc.* **1923,** *45,* 1223. (Corrected for current atomic weights.)

Suppose that constant-boiling HCl was collected at a pressure of 746 torr.

(a) Make a graph of the data in the table to find the weight percent of HCl collected at 746 torr.

(b) What mass of distillate (weighed in air, using weights whose density is 8.0 g/mL) should be dissolved in 1.0000 L to give 0.10000 M HCl? The density of distillate over the whole range in the table is close to 1.096 g/mL. You will need this density to change the mass measured in vacuum to mass measured in air. See Section 2-3 for buoyancy corrections.

52. What is meant by the leveling effect?

53. Considering the pK_a values below,[7] explain why dilute sodium methoxide (NaOCH$_3$) and sodium ethoxide (NaOCH$_2$CH$_3$) are leveled to the same base strength in aqueous solution. Write the chemical reactions that occur when these bases are added to water.

CH$_3$OH
$pK_a = 15.54$

CH$_3$CH$_2$OH
$pK_a = 16.0$

HOH
$pK_a = 15.74$ (for $K_a = [H^+][OH^-]/[H_2O]$)

54. The base B is too weak to titrate in water.

(a) Which solvent, pyridine or acetic acid, would be more suitable for titration of B with HClO$_4$? Why?

(b) Which solvent would be more suitable for the titration of a very weak acid with tetrabutylammonium hydroxide? Why?

55. Explain why sodium amide ($NaNH_2$) and phenyl lithium (C_6H_5Li) are leveled to the same base strength in aqueous solution. Write the chemical reactions that occur when these are added to water.

56. Why does surfactant improve the sharpness of the end point in the titration in Figure 12-12?

Calculating Titration Curves with Spreadsheets

57. Derive the following equation analogous to those in Table 12-6 for the titration of potassium hydrogen phthalate (K^+HP^-) with NaOH:

$$\phi = \frac{C_b V_b}{C_a V_a} = \frac{\alpha_{HP^-} + 2\alpha_{P^{2-}} - 1 - \dfrac{[H^+] - [OH^-]}{C_a}}{1 + \dfrac{[H^+] - [OH^-]}{C_b}}$$

58. *Effect of pK_a in the titration of weak acid with strong base.* Use Equation 12-9 with a spreadsheet such as the one shown in Figure 12-13 to compute and plot the family of curves in Figure 12-3. For a strong acid, choose a large K_a, such as $K_a = 10^2$ or $pK_a = -2$.

59. *Effect of concentration in the titration of weak acid with strong base.* Use your spreadsheet from the previous problem to prepare a family of titration curves for $pK_a = 6$, with the following combinations of concentrations: (a) $C_a = 20$ mM, $C_b = 100$ mM; (b) $C_a = 2$ mM, $C_b = 10$ mM; (c) $C_a = 0.2$ mM, $C_b = 1$ mM.

60. *Effect of pK_b in the titration of weak base with strong acid.* Using Table 12-6, prepare a spreadsheet to compute and plot a family of curves analogous to Figure 12-3 for the titration of 50.0 mL of 0.020 0 M B ($pK_b = -2.00, 2.00, 5.00, 8.00,$ and 10.00) with 0.100 M HCl. (The value $pK_b = -2.00$ represents a strong base.) In the expression for α_{BH^+}, $K_{BH^+} = K_w/K_b$.

61. *Titrating weak acid with weak base.*

(a) Use a spreadsheet to prepare a family of graphs for the titration of 50.0 mL of 0.020 0 M HA ($pK_a = 4.00$) with 0.100 M B ($pK_b = 3.00, 6.00,$ and 9.00).

(b) Write the acid-base reaction that occurs when acetic acid and sodium benzoate (the salt of benzoic acid) are mixed and find the equilibrium constant for the reaction. Find the pH of a solution prepared by mixing 212 mL of 0.200 M acetic acid with 325 mL of 0.050 0 M sodium benzoate.

62. *Titrating diprotic acid with strong base.* Prepare a family of graphs for the titration of 50.0 mL of 0.020 0 M H_2A with 0.100 M NaOH. Consider the following cases: (a) $pK_1 = 4.00, pK_2 = 8.00$; (b) $pK_1 = 4.00, pK_2 = 6.00$; (c) $pK_1 = 4.00, pK_2 = 5.00$.

63. *Titrating nicotine with strong acid.* Prepare a spreadsheet to reproduce the lower curve in Figure 12-4.

64. *Titrating triprotic acid with strong base.* Graph the titration of 50.0 mL of 0.020 0 M histidine·2HCl with 0.100 M NaOH. Treat histidine·2HCl with the triprotic acid equation in Table 12-6.

65. *A tetraprotic system.* Write an equation for the titration of a tetrabasic base with strong acid ($B + H^+ \rightarrow \rightarrow \rightarrow \rightarrow BH_4^{4+}$). You can do this by inspection from Table 12-6 or you can derive it from the charge balance for the titration reaction. Use a spreadsheet to graph the titration of 50.0 mL of 0.020 0 M sodium pyrophosphate ($Na_4P_2O_7$) with 0.100 M $HClO_4$.

Notes and References

1. For more on Gran plots, see G. Gran, *Anal. Chim. Acta* **1988**, *206,* 111; F. J. C. Rossotti and H. Rossotti, *J. Chem. Ed.* **1965**, *42,* 375; L. M. Schwartz, *J. Chem. Ed.* **1987** *64,* 947; and L. M. Schwartz, *J. Chem. Ed.* **1992**, *69,* 879.

2. Fascinating indicator demonstrations using universal indicator (a mixed indicator with many color changes) are described in J. T. Riley, *J. Chem. Ed.* **1977**, *54,* 29.

3. Instructions for purifying and using primary standards can be found in L. Meites, *Handbook of Analytical Chemistry* (New York: McGraw-Hill, 1963), pp. 3-32–3-35; J. Bassett, R. C. Denney, G. H. Jeffery, and J. Mendham, *Vogel's Textbook of Quantitative Inorganic Analysis,* 4th ed. (Essex: Longman, 1978), pp. 296–306;

I. M. Kolthoff and V. A. Stenger, *Volumetric Analysis,* Vol. 2 (New York: Wiley, 1947).

4. D. Fărcaşiu and A. Ghenciu, *J. Am. Chem. Soc.* **1993**, *115,* 10901.

5. J. S. Fritz, *Acid-Base Titrations in Nonaqueous Solvents* (Boston: Allyn and Bacon, 1973); J. Kucharsky and L. Safarik, *Titrations in Non-Aqueous Solvents* (New York: Elsevier, 1963); W. Huber, *Titrations in Nonaqueous Solvents* (New York: Academic Press, 1967); I. Gyenes, *Titration in Non-Aqueous Media* (Princeton, NJ: Van Nostrand, 1967).

6. R. de Levie, *J. Chem. Ed.* **1993**, *70,* 209.

7. P. Ballinger and F. A. Long, *J. Am. Chem. Soc.* **1960**, *82,* 795.

A Chelating Ligand Captures Its Prey

In a nerve impulse the *hydrophilic* ("water-loving") ions, K$^+$ and Na$^+$, must cross the *hydrophobic* ("water-hating") cell membrane. Carrier molecules with polar interiors and nonpolar exteriors perform this function.

A class of antibiotics called *ionophores,* including nonactin, gramicidin, and nigericin, alters the permeability of bacterial cells to metal ions and thus disrupts their metabolism. Ionophores are *chelating* (pronounced KEE-late-ing) ligands, which means that they bind metal ions through more than one ligand atom. One molecule of nonactin locks onto a K$^+$ ion through eight oxygen atoms. The sequence below shows how the conformation of nonactin changes as it wraps itself around a K$^+$ ion.

Structure of nonactin with ligand atoms highlighted

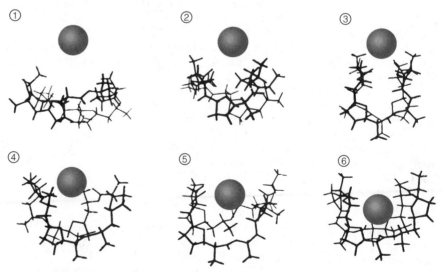

[Illustration from T. J. Marrone and K. M. Merz, Jr., *J. Am. Chem. Soc.* **1992,** *114,* 7542.]

EDTA Titrations

E_{DTA} is a merciful abbreviation for *ethylenediaminetetraacetic acid*, a compound that forms strong 1:1 complexes with most metal ions. In this chapter we see how complex formation is used in analytical chemistry.

13-1 Metal-Chelate Complexes

Metal ions are **Lewis acids,** accepting electron pairs from electron-donating ligands that are **Lewis bases.** Cyanide is termed a **monodentate** ligand because it binds to a metal ion through only one atom (the carbon atom). Most transition metal ions bind six ligand atoms. A ligand that attaches to a metal ion through more than one ligand atom is said to be **multidentate.** A multidentate ligand such as EDTA is called a **chelating ligand.**

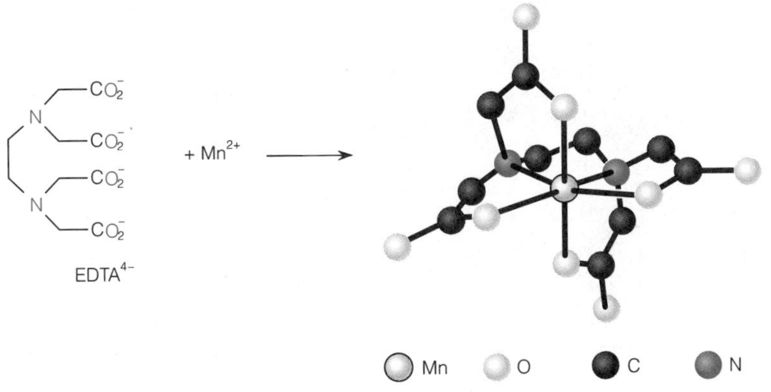

$$Ag^+ + 2:\bar{C} \equiv N: \rightleftharpoons$$

Lewis acid (electron-pair acceptor)   Lewis base (electron-pair donor)

$$[N \equiv C - Ag - C \equiv N]^-$$

Complex ion

EDTA^{4-}

+ Mn^{2+} ⟶

○ Mn   ○ O   ● C   ● N

Figure 13-1 EDTA forms strong 1:1 complexes with most metal ions, binding through four oxygen and two nitrogen atoms. The six-coordinate geometry of $Mn^{2+}-$EDTA found in the compound $KMnEDTA \cdot 2H_2O$ was deduced from X-ray crystallography. [J. Stein, J. P. Fackler, Jr., G. J. McClune, J. A. Fee, and L. T. Chan, *Inorg. Chem.* **1979**, *18,* 3511.]

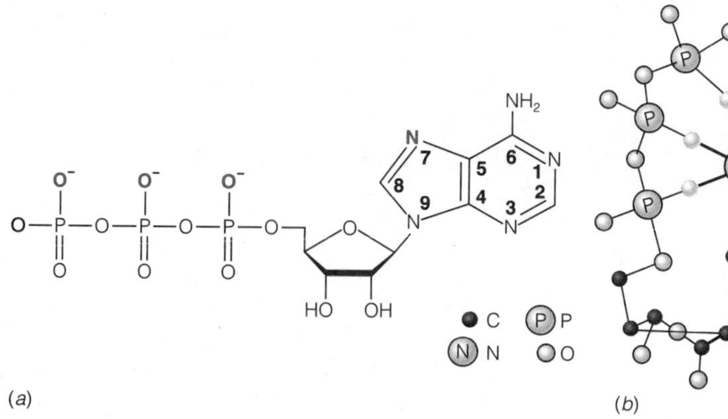

Figure 13-2 (*a*) Structure of adenosine triphosphate (ATP), with ligand atoms shown in **bold** type. (*b*) Possible structure of a metal-ATP complex, with four bonds to ATP and two bonds to H_2O ligands. There is controversy over whether N_7 is bound directly to the metal or whether a molecule of water forms a hydrogen-bonded bridge between N_7 and the metal ion.

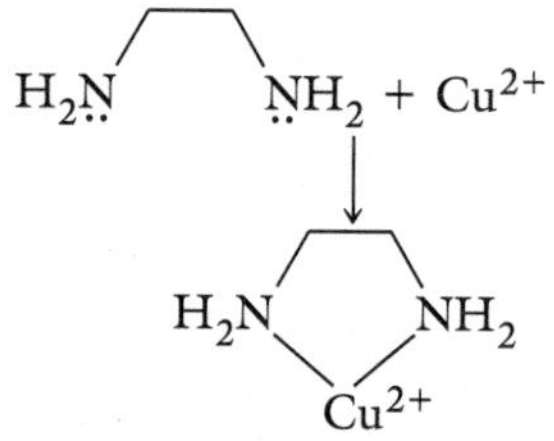

(a)

(b)

Bidentate bonding

One simple chelating ligand is ethylenediamine ($H_2NCH_2CH_2NH_2$, also called 1,2-diaminoethane), whose binding to a metal ion is shown in the margin. We say that ethylenediamine is *bidentate* because it binds to the metal through two ligand atoms.

An important *tetradentate* ligand is adenosine triphosphate (ATP), which binds to divalent metal ions (such as Mg^{2+}, Mn^{2+}, Co^{2+}, and Ni^{2+}) through four of their six coordination positions (Figure 13-2). The fifth and sixth positions are occupied by water molecules. The biologically active form of ATP is generally the Mg^{2+} complex.

The aminocarboxylic acids in Figure 13-3 are common synthetic chelating agents. The nitrogen atoms and carboxylate oxygen atoms are the potential ligand atoms in these molecules (Figures 13-4 and 13-5). When these atoms bind to a metal ion, the ligand atoms lose their protons.

A titration based on complex formation is called a **complexometric titration.** Ligands other than NTA in Figure 13-3 form strong 1:1 complexes with all metal ions, except univalent ions such as Li^+, Na^+, and K^+. *The stoichiometry is 1:1 regardless of the charge on the ion.* The equilibrium constant for the reaction of a metal with a ligand is called the **formation constant**, K_f, also termed the **stability constant.**

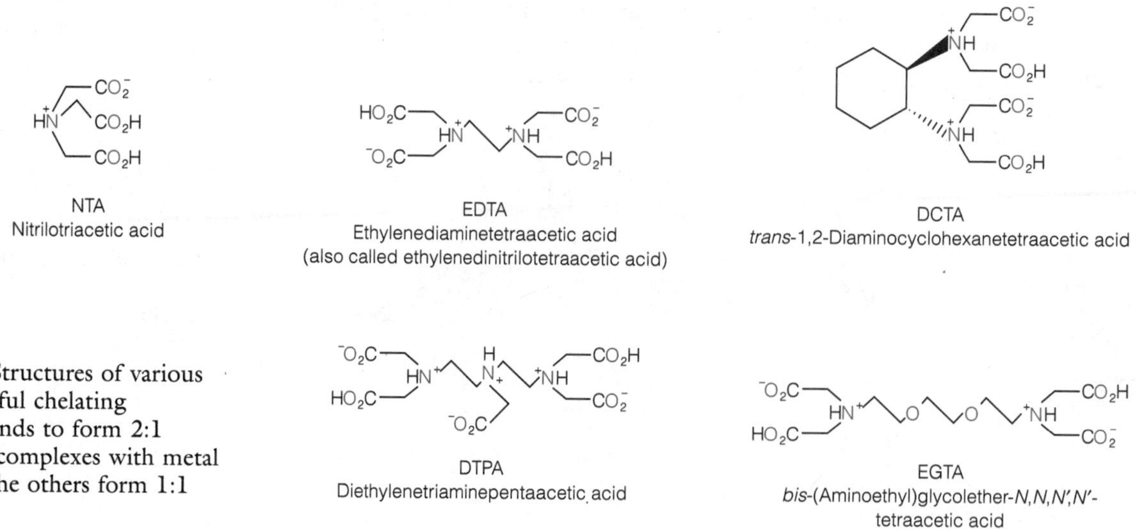

NTA
Nitrilotriacetic acid

EDTA
Ethylenediaminetetraacetic acid
(also called ethylenedinitrilotetraacetic acid)

DCTA
trans-1,2-Diaminocyclohexanetetraacetic acid

Figure 13-3 Structures of various analytically useful chelating agents. NTA tends to form 2:1 (ligand:metal) complexes with metal ions, whereas the others form 1:1 complexes.

DTPA
Diethylenetriaminepentaacetic acid

EGTA
bis-(Aminoethyl)glycolether-*N,N,N',N'*-tetraacetic acid

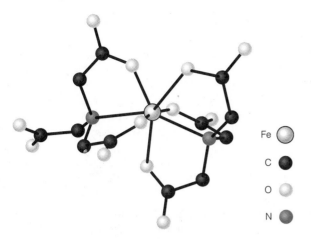

Figure 13-4 Structure of $Fe(NTA)_2^{3-}$ in the salt $Na_3[Fe(NTA)_2]\cdot 5H_2O$. The ligand at the right binds to Fe through three oxygen atoms and one nitrogen atom. The other ligand uses two oxygen atoms and one nitrogen atom. Its third carboxyl group is uncoordinated. The Fe atom is seven coordinate. [W. Clegg, A. K. Powell, and M. J. Ware, *Acta Crystallogr.* **1984,** *C40,* 1822.]

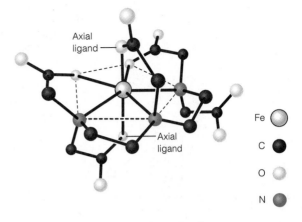

Figure 13-5 Structure of $Fe(DTPA)^{2-}$ found in the salt $Na_2[Fe(DTPA)]\cdot 2H_2O$. The seven-coordinate pentagonal bipyramidal coordination environment of the iron atom features three nitrogen and two oxygen ligands in the equatorial plane (dashed lines) and two axial oxygen ligands. The axial Fe—O bond lengths are 11 to 19 pm shorter than the more crowded equatorial Fe—O bonds. One carboxyl group of the ligand is uncoordinated. [D. C. Finnen, A. A. Pinkerton, W. R. Dunham, R. H. Sands, and M. O. Funk, Jr., *Inorg. Chem.* **1991,** *30,* 3960.]

The Chelate Effect

The **chelate effect** is the ability of multidentate ligands to form more stable metal complexes than those formed by similar monodentate ligands. For example, the reaction of Cd^{2+} with two molecules of ethylenediamine is more favorable than its reaction with four molecules of methylamine:

Chelate effect: A multidentate ligand forms more stable complexes than a similar monodentate ligand.

$$Cd^{2+} + 2H_2\ddot{N}CH_2CH_2\ddot{N}H_2 \rightleftharpoons \left[\begin{array}{c} NH_2 \\ NH_2 \end{array} Cd \begin{array}{c} H_2N \\ H_2N \end{array} \right]^{2+}$$

Ethylenediamine

$$K = 2 \times 10^{10} \qquad (13\text{-}1)$$

$$Cd^{2+} + 4CH_3\ddot{N}H_2 \rightleftharpoons \left[\begin{array}{c} CH_3NH_2 \\ CH_3NH_2 \end{array} Cd \begin{array}{c} H_2NCH_3 \\ H_2NCH_3 \end{array} \right]^{2+}$$

Methylamine

$$K = 3 \times 10^6 \qquad (13\text{-}2)$$

At pH 12 in the presence of 2 M ethylenediamine and 4 M methylamine, the quotient $[Cd(ethylenediamine)_2^{2+}]/[Cd(methylamine)_4^{2+}]$ is 200.

The chelate effect can be understood from thermodynamics. The two tendencies that drive a chemical reaction are decreasing enthalpy (negative ΔH, liberation of heat) and increasing entropy (positive ΔS, more disorder).

A reaction is favorable if $\Delta G < 0$

$$\Delta G = \Delta H - T\Delta S$$

The reaction is favored by negative ΔH and positive ΔS.

In Reactions 13-1 and 13-2, four Cd—N bonds are formed, and ΔH is about the same for both reactions.

However, Reaction 13-1 represents the coming together of *three* molecules (Cd^{2+} + 2 ethylenediamine), whereas in Reaction 13-2 *five* molecules (Cd^{2+} + 4 methylamine) are involved. More order is created in Reaction 13-2 than in Reaction 13-1. If the enthalpy change (ΔH) of each reaction is about the same, the entropy change (ΔS) will favor Reaction 13-1 over 13-2.[1] An important medical use of the chelate effect is discussed in Box 13-1.

13-2 EDTA

One mole of EDTA reacts with *one* mole of metal ion.

EDTA is, by far, the most widely used chelator in analytical chemistry. By direct titration or through an indirect sequence of reactions, virtually every element of the periodic table can be analyzed with EDTA.

Acid-Base Properties

EDTA is a hexaprotic system, designated H_6Y^{2+}. The highlighted, acidic hydrogen atoms, whose pK_a values are listed here,[2] are the ones that are lost upon metal-complex formation.

$$\begin{array}{ll}
pK_1 = & 0.0 \\
pK_2 = & 1.5 \\
pK_3 = & 2.0 \\
pK_4 = & 2.66 \\
pK_5 = & 6.16 \\
pK_6 = & 10.24
\end{array}$$

HO$_2$CCH$_2$ \ $^+$ / CH$_2$CO$_2$H
HNCH$_2$CH$_2$NH
HO$_2$CCH$_2$ / $^+$ \ CH$_2$CO$_2$H
H_6Y^{2+}

The first four pK values apply to carboxyl protons, and the last two are for the ammonium protons. The neutral acid is tetraprotic, with the formula H_4Y. A commonly used reagent is the disodium salt, $Na_2H_2Y \cdot 2H_2O$.[3]

The fraction of EDTA in each of its protonated forms is plotted in Figure 13-6. As in Section 11-5, we can define α for each species as the fraction of

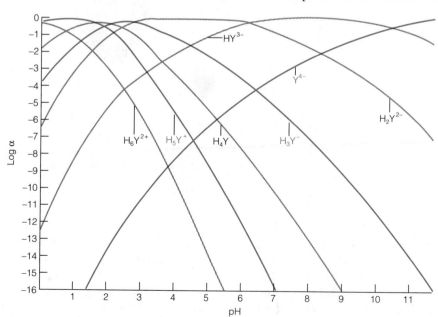

Figure 13-6 Fractional composition diagram for EDTA. Note that the ordinate is logarithmic.

Box 13-1 Chelation Therapy and Thalassemia

Oxygen is carried in the human circulatory system by the iron-containing protein hemoglobin, which consists of two pairs of subunits, designated α and β. β-Thalassemia major is a genetic disease in which the β subunits of hemoglobin are not synthesized in adequate quantities. Children afflicted with this disease can survive only with frequent transfusions of normal red blood cells. The problem with this treatment is that the patient accumulates 4–8 g of iron per year from the hemoglobin in the transfused cells. The body has no mechanism for excreting such large quantities of iron, and most patients die by age 20 from the toxic effects of iron overload.

To enhance iron excretion, intensive chelation therapy is used. The most successful drug so far is *desferrioxamine B,* whose iron complex (ferrioxamine B) is shown here, along with a graph illustrating its life-prolonging effects. The formation constant for ferrioxamine B is $10^{30.6}$.

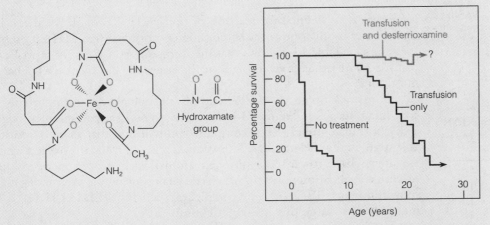

Structure of the iron complex ferrioxamine B (left) and graph showing success of transfusions and transfusions plus chelation therapy (right). [Data from P. S. Dobbin and R. C. Hider, *Chem. Brit.* **1990**, *26*, 565.]

Used in conjunction with ascorbic acid—vitamin C, a reducing agent that reduces Fe^{3+} to the more soluble Fe^{2+}—desferrioxamine clears several grams of iron per year from an overloaded patient. The ferrioxamine complex is excreted in the urine.

Clinical trials demonstrate that desferrioxamine reduces the incidence of heart and liver disease in thalassemia patients and maintains approximately correct iron balance. However, desferrioxamine is expensive and must be taken by continuous injection. It is not absorbed through the intestine. Many potent iron chelators have been tested to find an effective one that can be taken orally. In the long term, bone marrow transplants or gene therapy might be effective cures for the disease.

EDTA in that form. For example, $\alpha_{Y^{4-}}$ is defined as

Fraction of EDTA in the form Y^{4-}:

$$\alpha_{Y^{4-}} = \frac{[Y^{4-}]}{[H_6Y^{2+}] + [H_5Y^+] + [H_4Y] + [H_3Y^-] + [H_2Y^{2-}] + [HY^{3-}] + [Y^{4-}]}$$

$$\alpha_{Y^{4-}} = \frac{[Y^{4-}]}{[\text{EDTA}]} \tag{13-3}$$

where [EDTA] is the total concentration of all *free* EDTA species in the solution. By "free" we mean EDTA not complexed to metal ions. Following the derivation in Section 11-5, it can be shown that $\alpha_{Y^{4-}}$ is given by

$$\alpha_{Y^{4-}} = \frac{K_1 K_2 K_3 K_4 K_5 K_6}{\begin{array}{c}([H^+]^6 + [H^+]^5 K_1 + [H^+]^4 K_1 K_2 + [H^+]^3 K_1 K_2 K_3 + [H^+]^2 K_1 K_2 K_3 K_4 \\ + [H^+] K_1 K_2 K_3 K_4 K_5 + K_1 K_2 K_3 K_4 K_5 K_6)\end{array}} \quad (13\text{-}4)$$

Table 13-1 gives values for $\alpha_{Y^{4-}}$ as a function of pH.

EDTA Complexes

The formation constant K_f of a metal-EDTA complex is the equilibrium constant for the reaction

Formation constant: $\quad M^{n+} + Y^{4-} \rightleftharpoons MY^{n-4} \qquad K_f = \dfrac{[MY^{n-4}]}{[M^{n+}][Y^{4-}]} \quad (13\text{-}5)$

Note that K_f is defined for reaction of the species Y^{4-} with the metal ion. This species is only one of the seven different forms of free EDTA present in the solution. Table 13-2 shows that the formation constants for most EDTA complexes are quite large and tend to be larger for more positively charged metal ions.

In many complexes, EDTA engulfs the metal ion, forming the six-coordinate species in Figure 13-1. If you try to build a space-filling model of a six-coordinate metal-EDTA complex, you will find that there is considerable strain in the chelate rings. This strain is relieved when the oxygen ligands are drawn back toward the nitrogen atoms. Such distortion opens up a seventh coordination position, which can be occupied by a water molecule, as shown in Figure 13-7. In some complexes, such as $Ca(EDTA)(H_2O)_2^{2-}$, the metal ion is so large that it is eight coordinate.[4]

The formation constant can still be formulated as in Equation 13-5, even if there are water molecules attached to the product. This is true because solvent (H_2O) is omitted from the reaction quotient.

Conditional Formation Constant

The formation constant in Equation 13-5 describes the reaction between Y^{4-} and a metal ion. As you can see in Figure 13-6, most of the EDTA is not Y^{4-} below pH 10.24. The species HY^{3-}, H_2Y^{2-}, and so on, predominate at lower pH. It is convenient to express the fraction of free EDTA in the form Y^{4-} by rearranging Equation 13-3 to give

$$[Y^{4-}] = \alpha_{Y^{4-}}[EDTA]$$

where [EDTA] refers to the total concentration of all EDTA species not bound to metal ion.

The equilibrium constant for Reaction 13-5 can now be rewritten as

$$K_f = \frac{[MY^{n-4}]}{[M^{n+}][Y^{4-}]} = \frac{[MY^{n-4}]}{[M^{n+}]\alpha_{Y^{4-}}[EDTA]}$$

If the pH is fixed by a buffer, then $\alpha_{Y^{4-}}$ is a constant that can be combined with K_f:

Conditional formation constant: $\quad K_f' = \alpha_{Y^{4-}} K_f = \dfrac{[MY^{n-4}]}{[M^{n+}][EDTA]} \quad (13\text{-}6)$

Equation 13-5 does not imply that Y^{4-} is the only species that reacts with M^{n+}. It simply says that the equilibrium constant is expressed in terms of the concentration of Y^{4-}.

T A B L E 1 3 - 1 **Values of $\alpha_{Y^{4-}}$ for EDTA at 20°C and μ = 0.10 M**

pH	$\alpha_{Y^{4-}}$
0	1.3×10^{-23}
1	1.9×10^{-18}
2	3.3×10^{-14}
3	2.6×10^{-11}
4	3.8×10^{-9}
5	3.7×10^{-7}
6	2.3×10^{-5}
7	5.0×10^{-4}
8	5.6×10^{-3}
9	5.4×10^{-2}
10	0.36
11	0.85
12	0.98
13	1.00
14	1.00

Only some of the free EDTA is in the form Y^{4-}.

TABLE 13-2 Formation constants for metal-EDTA complexes

Ion	$\log K_f$	Ion	$\log K_f$	Ion	$\log K_f$
Li^+	2.79	Mn^{3+}	25.3 (25°C)	Ce^{3+}	15.98
Na^+	1.66	Fe^{3+}	25.1	Pr^{3+}	16.40
K^+	0.8	Co^{3+}	41.4 (25°C)	Nd^{3+}	16.61
Be^{2+}	9.2	Zr^{4+}	29.5	Pm^{3+}	17.0
Mg^{2+}	8.79	Hf^{4+}	29.5 ($\mu = 0.2$)	Sm^{3+}	17.14
Ca^{2+}	10.69	VO^{2+}	18.8	Eu^{3+}	17.35
Sr^{2+}	8.73	VO_2^+	15.55	Gd^{3+}	17.37
Ba^{2+}	7.86	Ag^+	7.32	Tb^{3+}	17.93
Ra^{2+}	7.1	Tl^+	6.54	Dy^{3+}	18.30
Sc^{3+}	23.1	Pd^{2+}	18.5 (25°C,	Ho^{3+}	18.62
Y^{3+}	18.09		$\mu = 0.2$)	Er^{3+}	18.85
La^{3+}	15.50	Zn^{2+}	16.50	Tm^{3+}	19.32
V^{2+}	12.7	Cd^{2+}	16.46	Yb^{3+}	19.51
Cr^{2+}	13.6	Hg^{2+}	21.7	Lu^{3+}	19.83
Mn^{2+}	13.87	Sn^{2+}	18.3 ($\mu = 0$)	Am^{3+}	17.8 (25°C)
Fe^{2+}	14.32	Pb^{2+}	18.04	Cm^{3+}	18.1 (25°C)
Co^{2+}	16.31	Al^{3+}	16.3	Bk^{3+}	18.5 (25°C)
Ni^{2+}	18.62	Ga^{3+}	20.3	Cf^{3+}	18.7 (25°C)
Cu^{2+}	18.80	In^{3+}	25.0	Th^{4+}	23.2
Ti^{3+}	21.3 (25°C)	Tl^{3+}	37.8 ($\mu = 1.0$)	U^{4+}	25.8
V^{3+}	26.0	Bi^{3+}	27.8	Np^{4+}	24.6 (25°C, $\mu = 1.0$)
Cr^{3+}	23.4				

Note: The stability constant is the equilibrium constant for the reaction $M^{n+} + Y^{4-} \rightleftharpoons MY^{n-4}$. Values in table apply at 20°C, and ionic strength 0.1 M, unless otherwise noted.
SOURCE: A. E. Martell and R. M. Smith, *Critical Stability Constants,* Vol. 1 (New York: Plenum Press, 1974), pp. 204–211.

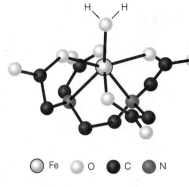

Figure 13-7 Seven-coordinate geometry of $Fe(EDTA)(H_2O)^-$. Other metal ions that form seven-coordinate EDTA complexes include Fe^{2+}, Mg^{2+}, Cd^{2+}, Co^{2+}, Mn^{2+}, Ru^{3+}, Cr^{3+}, Co^{3+}, V^{3+}, Ti^{3+}, In^{3+}, Sn^{4+}, Os^{4+}, and Ti^{4+}. Some of these same ions also form six-coordinate EDTA complexes. Eight-coordinate complexes are formed by Ca^{2+}, Er^{3+}, Yb^{3+}, and Zr^{4+}. [T. Mizuta, J. Wang, and K. Miyoshi, *Bull. Chem. Soc. Japan* **1993,** *66,* 2547.]

The number $K_f' = \alpha_{Y^{4-}} K_f$ is called the **conditional formation constant** or the *effective formation constant.* It describes the formation of MY^{n-4} at any particular pH.

The conditional formation constant is useful because it allows us to look at EDTA complex formation as if the uncomplexed EDTA were all in one form:

$$M^{n+} + EDTA \rightleftharpoons MY^{n-4} \qquad K_f' = \alpha_{Y^{4-}} K_f$$

At any given pH, we can find $\alpha_{Y^{4-}}$ and evaluate K_f'.

With the conditional formation constant, we can treat EDTA complex formation as if all the free EDTA were in one form.

EXAMPLE Using the Conditional Formation Constant

The formation constant in Table 13-2 for FeY^- is $10^{25.1} = 1.3 \times 10^{25}$. Calculate the concentration of free Fe^{3+} in solutions of 0.10 M FeY^- at pH 8.00 and at pH 2.00.

Solution The complex formation reaction is

$$Fe^{3+} + EDTA \rightleftharpoons FeY^- \qquad K_f' = \alpha_{Y^{4-}} K_f$$

where EDTA on the left side of the equation refers to all forms of unbound EDTA ($= Y^{4-}$, HY^{3-}, H_2Y^{2-}, H_3Y^-, etc.). Using values of $\alpha_{Y^{4-}}$ from Table 13-1, we find

at pH 8.00: $K_f' = (5.6 \times 10^{-3})(1.3 \times 10^{25}) = 7.3 \times 10^{22}$
at pH 2.00: $K_f' = (3.3 \times 10^{-14})(1.3 \times 10^{25}) = 4.3 \times 10^{11}$

Because dissociation of FeY^- must produce equal quantities of Fe^{3+} and EDTA, we can write

	Fe^{3+} +	EDTA ⇌	FeY^-
Initial concentration (M):	0	0	0.10
Final concentration (M):	x	x	$0.10 - x$

$$\frac{0.10 - x}{x^2} = K_f' = 7.3 \times 10^{22} \quad \text{at pH 8.00}$$
$$= 4.3 \times 10^{11} \quad \text{at pH 2.00}$$

Solving for x ($= [Fe^{3+}]$), we find $[Fe^{3+}] = 1.2 \times 10^{-12}$ M at pH 8.00 and 4.8×10^{-7} M at pH 2.00. *Using the conditional formation constant at a fixed pH, we treat the dissociated EDTA as if it were a single species.*

· · · · · · · · · · · · · · · · · · · ·

You can see from the example that a metal-EDTA complex becomes less stable at lower pH. For a titration reaction to be effective, it must go "to completion." Figure 13-8 shows how pH affects the titration of Ca^{2+} with EDTA. Below pH ≈ 8, the break at the end point is not sharp enough to allow accurate determination, because the conditional formation constant for CaY^{2-} is too small.

Control of pH can be used to select which metals will be titrated by EDTA and which will not.

Figure 13-9 gives the minimum pH needed for titration of many metal ions. The figure provides a strategy for the selective titration of one ion in the presence of another. For example, a solution containing both Fe^{3+} and Ca^{2+} could be titrated with EDTA at pH 4. At this pH, Fe^{3+} is titrated without interference from the Ca^{2+} ion.

· · · · · · · · · · · · · · · · · · · ·

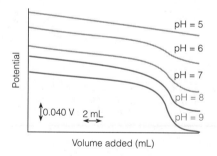

Figure 13-8 Titration of Ca^{2+} with EDTA as a function of pH. As the pH is lowered, the end point becomes less distinct. The potential was measured with mercury and calomel electrodes, as described in Exercise B in Chapter 15. [C. N. Reilley and R. W. Schmid, *Anal. Chem.* **1958,** *30,* 947.]

EXAMPLE **What Is the Minimum K_f' for "Complete" Titration?**

Suppose we define "complete" to mean 99.9% complexation. What value of K_f' is required for complete reaction at the equivalence point in a titration?

Solution Let the formal concentration of MY^{n-4} be F at the equivalence point. For 99.9% complexation, $[M^{n+}] = [\text{EDTA}] = 10^{-3}$ F.

$$M^{n+} + \text{EDTA} \rightleftharpoons MY^{n-4}$$
$$10^{-3}\,\text{F} \quad 10^{-3}\,\text{F} \quad \text{F} - 10^{-3}\,\text{F}$$

$$K_f' = \frac{[MY^{n-4}]}{[M^{n+}][\text{EDTA}]} = \frac{\text{F} - 10^{-3}\,\text{F}}{(10^{-3}\,\text{F})(10^{-3}\,\text{F})} \approx \frac{\text{F}}{10^{-6}\,\text{F}^2} = \frac{10^6}{\text{F}}$$

For $\text{F} = 10^{-2}$ M, K_f' would have to be $10^6/10^{-2} = 10^8$ for 99.9% complete reaction. The points in Figure 13-9 are based on $K_f' = 10^8$, which corresponds to 99.9% complete reaction of 10^{-2} M metal ion at the equivalence point.

· · · · · · · · · · · · · · · · · · · ·

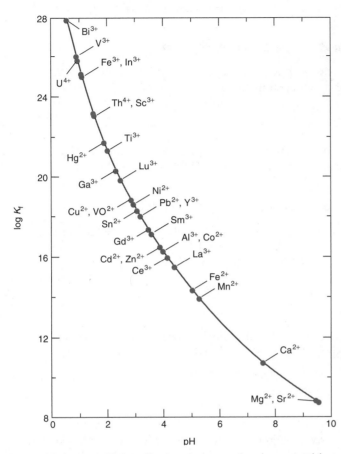

Figure 13-9 Minimum pH for effective titration of various metal ions by EDTA. The minimum pH was specified arbitrarily as the pH at which the conditional formation constant for each metal-EDTA complex is 10^8.

13-3 EDTA Titration Curves

In this section we calculate the concentration of free metal ion during the course of the titration of metal with EDTA. This titration is analogous to that of a strong acid by a weak base. The metal ion plays the role of H^+, and EDTA is the base. The titration reaction is

$$M^{n+} + EDTA \rightleftharpoons MY^{n-4} \qquad K'_f = \alpha_{Y^{4-}} K_f \qquad (13\text{-}7)$$

K'_f is the effective formation constant at the fixed pH of the solution.

If K'_f is large, we can consider the reaction to be complete at each point in the titration.

The titration curve has three natural regions (Figure 13-10).

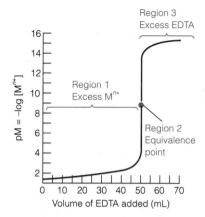

Figure 13-10 Three regions in an EDTA titration illustrated for reaction of 50.0 mL of 0.0500 M M^{n+} with 0.0500 M EDTA, assuming $K_f' = 1.15 \times 10^{16}$. The concentration of free M^{n+} decreases as the titration proceeds.

Region 1: Before the Equivalence Point

In this region there is excess M^{n+} left in solution after the EDTA has been consumed. The concentration of free metal ion is equal to the concentration of excess, unreacted M^{n+}. The dissociation of MY^{n-4} is negligible.

Region 2: At the Equivalence Point

There is exactly as much EDTA as metal in the solution. We can treat the solution as if it had been made by dissolving pure MY^{n-4}. Some free M^{n+} is generated by the slight dissociation of MY^{n-4}:

$$MY^{n-4} \rightleftharpoons M^{n+} + EDTA$$

In this reaction, EDTA refers to the total concentration of free EDTA in all of its forms. At the equivalence point, $[M^{n+}] = [EDTA]$.

Region 3: After the Equivalence Point

Now there is excess EDTA, and virtually all the metal ion is in the form MY^{n-4}. The concentration of free EDTA can be equated to the concentration of excess EDTA added after the equivalence point.

Titration Calculations

Let's calculate the shape of the titration curve for the reaction of 50.0 mL of 0.0500 M Mg^{2+} (buffered to pH 10.00) with 0.0500 M EDTA:

$$Mg^{2+} + EDTA \rightarrow MgY^{2-}$$

The value of α_{Y4-} comes from Table 13-1.

$$K_f' = \alpha_{Y4-} K_f = (0.36)(6.2 \times 10^8) = 2.2 \times 10^8$$

The equivalence volume will be 50.0 mL. Because K_f' is large, it is reasonable to say that the reaction goes to completion with each addition of titrant. We seek to make a graph in which pMg^{2+} $(= -\log[Mg^{2+}])$ is plotted versus milliliters of added EDTA.

Region 1: Before the Equivalence Point

Before the equivalence point, there is excess unreacted M^{n+}.

Consider the addition of 5.0 mL of EDTA. Because the equivalence point is 50.0 mL, one-tenth of the Mg^{2+} will be consumed and nine-tenths remains.

$$[Mg^{2+}] = \underbrace{\left(\frac{50.0 - 5.0}{50.0}\right)}_{\substack{\text{Fraction} \\ \text{remaining} \\ (= 9/10)}} \underbrace{(0.0500)}_{\substack{\text{Original} \\ \text{concentration} \\ \text{of } Mg^{2+}}} \underbrace{\left(\frac{50.0}{55.0}\right)}_{\substack{\text{Dilution} \\ \text{factor}}}$$

Initial volume of Mg^{2+}

Total volume of solution

$$= 0.0409 \text{ M} \Rightarrow pMg^{2+} = -\log[Mg^{2+}] = 1.39$$

In a similar manner, we could calculate pMg^{2+} for any volume of EDTA less than 50.0 mL.

Region 2: At the Equivalence Point

Virtually all the metal is in the form MgY^{2-}. Assuming negligible dissociation, the concentration of MgY^{2-} is equal to the original concentration of Mg^{2+}, with a correction for dilution.

At the equivalence point, the major species is MY^{n-4}, in equilibrium with small, equal amounts of free M^{n+} and EDTA.

$$[MgY^{2-}] = (0.050\,0\ M)\left(\frac{50.0}{100.0}\right) = 0.025\,0\ M$$

where 50.0 is the Initial volume of Mg^{2+}, $0.050\,0\ M$ is the Original concentration of Mg^{2+}, $\frac{50.0}{100.0}$ is the Dilution factor, and 100.0 is the Total volume of solution.

The concentration of free Mg^{2+} is small and unknown. We can write

$$Mg^{2+} + EDTA \rightleftharpoons MgY^{2-}$$

	Mg^{2+}	EDTA	MgY^{2-}
Initial concentration (M):	—	—	0.025 0
Final concentration (M):	x	x	0.025 0 − x

$$\frac{[MgY^{2-}]}{[Mg^{2+}][EDTA]} = K_f' = 2.2 \times 10^8$$

$$\frac{0.025\,0 - x}{x^2} = 2.2 \times 10^8 \Rightarrow x = 1.06 \times 10^{-5}\ M$$

$$pMg^{2+} = -\log x = 4.97$$

[EDTA] refers to the total concentration of all forms of EDTA not bound to metal.

Region 3: After the Equivalence Point

In this region virtually all the metal is in the form MgY^{2-}, and there is excess, unreacted EDTA. The concentrations of MgY^{2-} and excess EDTA are easily calculated. For example, at 51.0 mL, there is 1.0 mL of excess EDTA.

After the equivalence point, virtually all the metal is present as MY^{n-4}. There is a known excess of EDTA present. A small amount of free M^{n+} exists in equilibrium with the MY^{n-4} and EDTA.

$$[EDTA] = (0.050\,0)\left(\frac{1.0}{101.0}\right) = 4.95 \times 10^{-4}\ M$$

where 1.0 is the Volume of excess EDTA, $0.050\,0$ is the Original concentration of EDTA, $\frac{1.0}{101.0}$ is the Dilution factor, and 101.0 is the Total volume of solution.

$$[MgY^{2-}] = (0.050\,0)\left(\frac{50.0}{101.0}\right) = 2.48 \times 10^{-2}\ M$$

where 50.0 is the Original volume of Mg^{2+}, $0.050\,0$ is the Original concentration of Mg^{2+}, $\frac{50.0}{101.0}$ is the Dilution factor, and 101.0 is the Total volume of solution.

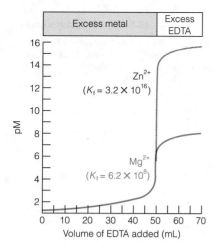

Figure 13-11 Theoretical titration curves for the reaction of 50.0 mL of 0.0500 M metal ion with 0.0500 M EDTA at pH 10.00.

The lower the pH, the less distinct is the end point.

The concentration of Mg^{2+} is governed by

$$\frac{[MgY^{2-}]}{[Mg^{2+}][EDTA]} = K_f' = 2.2 \times 10^8$$

$$\frac{[2.48 \times 10^{-2}]}{[Mg^{2+}](4.95 \times 10^{-4})} = 2.2 \times 10^8$$

$$[Mg^{2+}] = 2.2 \times 10^{-7} \text{ M} \Rightarrow pMg^{2+} = 6.65$$

The same sort of calculation can be used for any volume past the equivalence point.

The Titration Curve

The calculated titration curve in Figure 13-11 shows a distinct break at the equivalence point, where the slope is greatest. For comparison, the titration curve for 0.0500 M Zn^{2+} is also shown in Figure 13-11. The change in pZn^{2+} at the equivalence point is even greater than the change in pMg^{2+}, because the formation constant for ZnY^{2-} is greater than that of MgY^{2-}.

The completeness of reaction (and hence the sharpness of the equivalence point) is determined by the conditional formation constant, $\alpha_{Y^{4-}}K_f$, which is pH dependent. Because $\alpha_{Y^{4-}}$ decreases drastically as the pH is lowered, the pH is an important variable determining whether a titration is feasible. The end point is more distinct at high pH. However, the pH must not be so high that metal hydroxide precipitates. The effect of pH on the titration of Ca^{2+} was shown in Figure 13-8.

13-4 Doing It with a Spreadsheet

Let's see how to reproduce the EDTA titration curves in Figure 13-11 using one equation that applies to the entire titration. Because the reactions are carried out at fixed pH, the equilibria and mass balances are sufficient to solve for all unknowns.

Consider the titration of metal ion M (concentration = C_m, initial volume = V_m) with a solution of ligand L (concentration = C_1, volume added = V_1) to form a 1:1 complex:

$$M + L \rightleftharpoons ML \qquad K_f = \frac{[ML]}{[M][L]} \Rightarrow [ML] = K_f[M][L] \qquad (13-8)$$

The mass balances for metal and ligand are

$$\text{mass balance for M: } [M] + [ML] = \frac{C_m V_m}{V_m + V_1}$$

$$\text{mass balance for L: } [L] + [ML] = \frac{C_1 V_1}{V_m + V_1}$$

Substituting the product $K_f[M][L]$ (from Equation 13-8) for [ML] in the mass balances gives

$$[M](1 + K_f[L]) = \frac{C_m V_m}{V_m + V_1} \qquad (13-9)$$

total metal in all species

$$= \frac{\text{initial moles of metal}}{\text{total volume}}$$

$$= \frac{(C_m V_m)}{(V_m + V_1)}$$

total ligand in all species

$$= \frac{\text{moles of ligand added}}{\text{total volume}}$$

$$= \frac{(C_1 V_1)}{(V_m + V_1)}$$

$$[L](1 + K_f[M]) = \frac{C_1 V_1}{V_m + V_1} \Rightarrow [L] = \frac{\dfrac{C_1 V_1}{V_m + V_1}}{1 + K_f[M]} \quad (13\text{-}10)$$

Now we substitute the last expression for [L] from Equation 13-10 back into Equation 13-9

$$[M]\left(1 + K_f \frac{\dfrac{C_1 V_1}{V_m + V_1}}{1 + K_f[M]}\right) = \frac{C_m V_m}{V_m + V_1}$$

and do about five lines of algebra to solve for the fraction of titration, ϕ:

Spreadsheet equation for titration of M with L:

$$\phi = \frac{C_1 V_1}{C_m V_m} = \frac{1 + K_f[M] - \dfrac{[M] + K_f[M]^2}{C_m}}{K_f[M] + \dfrac{[M] + K_f[M]^2}{C_1}} \quad (13\text{-}11)$$

Replace K_f by K_f' if L = EDTA.

As in acid-base titrations in Table 12-6, ϕ is the fraction of the way to the equivalence point. When $\phi = 1$, $V_1 = V_e$. When $\phi = \frac{1}{2}$, $V_1 = \frac{1}{2} V_e$, etc.

For a titration with EDTA, you can follow the derivation through and find that the formation constant, K_f, should be replaced in Equation 13-11 by the effective formation constant, K_f', which applies at the fixed pH of the titration. Figure 13-12 shows a spreadsheet in which Equation 13-11 is used to calculate points along the lower curve in Figure 13-11. As in acid-base titrations, your input in column B is pM and the output in column E is volume of titrant. To find the initial point, we vary pM until V_1 is close to

	A	B	C	D	E
1	Cm =	pM	[M]	Phi	V(ligand)
2	0.05	1.301	5.00E-02	0.000	−0.002
3	Vm =	1.3882	4.09E-02	0.100	5.001
4	50	2.00	1.00E-02	0.667	33.333
5	C(ligand) =	3.00	1.00E-03	0.961	48.039
6	0.05	4.00	1.00E-04	0.996	49.803
7	Kf′ =	4.972	1.07E-05	1.000	50.000
8	2.2E+08	5.00	1.00E-05	1.000	50.003
9		6.00	1.00E-06	1.005	50.225
10		6.6437	2.27E-07	1.020	51.000
11		7.00	1.00E-07	1.045	52.273
12					
13	C2 = 10^−B2				
14	D2 = (1+$A$8*C2-(C2+C2*C2*$A$8)/$A$2)/				
15		(C2*$A$8+(C2+C2*C2*$A$8)/$A$6)			
16	E2 = D2*$A$2*$A$4/$A$6				

Figure 13-12 Spreadsheet for the titration of 50.0 mL of 0.0500 M Mg^{2+} with 0.0500 M EDTA at pH 10.00. This spreadsheet reproduces the calculations of Section 13-3 in a much less painful manner.

zero. Similarly, pM was varied to find the points corresponding to volumes of 5.00, 50.00, and 51.00 mL used in the preceding section.

If you reverse the process and titrate ligand with metal ion, the fraction of the way to the equivalence point is the inverse of the fraction in Equation 13-11:

Replace K_f by K_f' if L = EDTA.

Spreadsheet equation for titration of L with M:

$$\phi = \frac{C_m V_m}{C_l V_l} = \frac{K_f[M] + \dfrac{[M] + K_f[M]^2}{C_l}}{1 + K_f[M] - \dfrac{[M] + K_f[M]^2}{C_m}} \quad (13\text{-}12)$$

13-5 Auxiliary Complexing Agents

The titration curve for Zn^{2+} in Figure 13-11 is not realistic because the pH is high enough to precipitate $Zn(OH)_2$ ($K_{sp} = 3.0 \times 10^{-16}$) before EDTA is added. The titration of Zn^{2+} is usually carried out in ammonia buffer, which not only fixes the pH, but serves to complex the metal ion and keep it in solution. The ammonia is called an **auxiliary complexing agent,** because it complexes the metal ion until EDTA is introduced. Let's see how this works.

Metal-Ligand Equilibria

A Supplementary Problem for Chapter 13 in the *Solutions Manual* gives a nonlinear least-squares exercise in which spectrophotometric data are used to measure a formation constant.

Consider a metal ion that forms two complexes with the auxiliary complexing ligand L:

$$M + L \rightleftharpoons ML \qquad \beta_1 = \frac{[ML]}{[M][L]} \quad (13\text{-}13)$$

$$M + 2L \rightleftharpoons ML_2 \qquad \beta_2 = \frac{[ML_2]}{[M][L]^2} \quad (13\text{-}14)$$

Overall (β) and stepwise (K) formation constants were distinguished in Box 5-2. The relation between β_n and the stepwise formation constants is $\beta_n = K_1 K_2 K_3 \cdots K_n$.

The equilibrium constants, β_i, are called *overall* or **cumulative formation constants.** The fraction of metal ion in the uncomplexed state, M, can be expressed as

$$\alpha_M = \frac{[M]}{C_M} \quad (13\text{-}15)$$

where C_M refers to the total concentration of all forms of M (= M, ML, and ML_2 in this case).

Now let's find a useful expression for α_M. The mass balance for metal is simply

$$C_M = [M] + [ML] + [ML_2]$$

Equations 13-13 and 13-14 allow us to say $[ML] = \beta_1[M][L]$ and $[ML_2] = \beta_2[M][L]^2$. Therefore,

$$C_M = [M] + \beta_1[M][L] + \beta_2[M][L]^2$$
$$= [M]\{1 + \beta_1[L] + \beta_2[L]^2\}$$

If the metal forms more than two complexes, Equation 13-16 takes the form

$$\alpha_M = \frac{1}{1 + \beta_1[L] + \beta_2[L]^2 + \cdots + \beta_n[L]^n}$$

Substituting this last result into Equation 13-15 gives the desired result:

Fraction of free metal ion:

$$\alpha_M = \frac{[M]}{[M]\{1 + \beta_1[L] + \beta_2[L]^2\}} = \frac{1}{1 + \beta_1[L] + \beta_2[L]^2} \quad (13\text{-}16)$$

EXAMPLE Ammonia Complexes of Zinc

In a solution containing Zn^{2+} and NH_3, the complexes $Zn(NH_3)^{2+}$, $Zn(NH_3)_2^{2+}$, $Zn(NH_3)_3^{2+}$, and $Zn(NH_3)_4^{2+}$ are all present. If the concentration of free NH_3 is 0.10 M, find the fraction of zinc in the form Zn^{2+}. This question means that the concentration of *unprotonated* NH_3 is 0.10 M. (At any pH, there will also be some concentration of NH_4^+, in equilibrium with NH_3.) The question also implies that no other complexing agents are present.

Solution   Appendix I gives four stepwise formation constants (K_1, K_2, K_3, and K_4) for the complexes $Zn(NH_3)^{2+}$, $Zn(NH_3)_2^{2+}$, $Zn(NH_3)_3^{2+}$, and $Zn(NH_3)_4^{2+}$. The overall formation constants are related to the stepwise formation constants as follows:

$$\beta_1 = K_1 = 10^{2.18}$$

$$\beta_2 = K_1K_2 = 10^{2.18}10^{2.25} = 10^{4.43}$$

$$\beta_3 = K_1K_2K_3 = 10^{2.18}10^{2.25}10^{2.31} = 10^{6.74}$$

$$\beta_4 = K_1K_2K_3K_4 = 10^{2.18}10^{2.25}10^{2.31}10^{1.96} = 10^{8.70}$$

The constants β_1 through β_4 refer to four equilibria analogous to Reactions 13-13 and 13-14. The appropriate form of Equation 13-16 is

$$\alpha_{Zn^{2+}} = \frac{1}{1 + \beta_1[L] + \beta_2[L]^2 + \beta_3[L]^3 + \beta_4[L]^4} \qquad (13\text{-}17)$$

Putting in [L] = 0.10 M and the four values of β_i gives $\alpha_{Zn^{2+}} = 1.8 \times 10^{-5}$, which means that very little zinc is in the form Zn^{2+} in the presence of 0.10 M NH_3.

EDTA Titration with an Auxiliary Complexing Agent

Now consider a titration of Zn^{2+} by EDTA in the presence of NH_3. The extension of Equation 13-7 requires a new, effective formation constant that accounts for the fact that only some of the EDTA is in the form Y^{4-} and only some of the zinc is in the form Zn^{2+}:

$$K_f'' = \alpha_{Zn^{2+}}\alpha_{Y^{4-}}K_f \qquad (13\text{-}18)$$

In this expression, $\alpha_{Zn^{2+}}$ is given by Equation 13-17 and $\alpha_{Y^{4-}}$ by Equation 13-4. For particular values of pH and NH_3 concentration, we can calculate a value of K_f'' and proceed with titration calculations as in Section 13-3, substituting K_f'' for K_f'. An assumption here is that EDTA is a much stronger complexing agent than ammonia, and essentially all of the EDTA added at any point in the titration is bound to Zn^{2+}, until the Zn^{2+} is consumed.

K_f'' is the effective formation constant at a given fixed pH and given fixed concentration of auxiliary complexing agent.

EXAMPLE   EDTA Titration in the Presence of Ammonia

Consider the titration of 50.0 mL of 1.00×10^{-3} M Zn^{2+} with 1.00×10^{-3} M EDTA at pH 10.00 in the presence of 0.10 M NH_3. (This is the concentration of NH_3. There is also NH_4^+ in the solution.) The equivalence point is at 50.0 mL. Find pZn^{2+} after addition of 20.0, 50.0, and 60.0 mL of EDTA.

Solution   In Equation 13-17 we found that $\alpha_{Zn^{2+}} = 1.8 \times 10^{-5}$. Table 13-1 tells us that $\alpha_{Y^{4-}} = 0.36$. Therefore, the effective formation constant is

$$K''_f = \alpha_{Zn^{2+}}\alpha_{Y^{4-}}K_f = (1.8 \times 10^{-5})(0.36)(10^{16.50}) = 2.0_5 \times 10^{11}$$

(a) *Before the equivalence point—20.0 mL:* Because the equivalence point is 50.0 mL, the fraction of Zn^{2+} remaining is 30.0/50.0. The dilution factor is 50.0/70.0. Therefore, the concentration of zinc not bound to EDTA is

$$C_{Zn^{2+}} = \left(\frac{30.0}{50.0}\right)(1.00 \times 10^{-3})\left(\frac{50.0}{70.0}\right) = 4.3 \times 10^{-4} \text{ M}$$

However, nearly all of the zinc not bound to EDTA is bound to NH_3. The concentration of free Zn^{2+} is

$[Zn^{2+}] = \alpha_{Zn^{2+}}C_{Zn^{2+}}$. This follows from Equation 13-15.

$$[Zn^{2+}] = \alpha_{Zn^{2+}}C_{Zn^{2+}} = (1.8 \times 10^{-5})(4.3 \times 10^{-4}) = 7.7 \times 10^{-9} \text{ M}$$

$$\Rightarrow pZn^{2+} = -\log[Zn^{2+}] = 8.11$$

(b) *At the equivalence point—50.0 mL:* At the equivalence point, the dilution factor is 50.0/100.0, so $[ZnY^{2-}] = (50.0/100.0)(1.00 \times 10^{-3}) = 5.00 \times 10^{-4}$ M. As in Section 13-3, we can write

	$C_{Zn^{2+}}$ + EDTA		$\rightleftharpoons$	ZnY^{2-}
Initial concentration (M):	0	0		5.00×10^{-4}
Final concentration (M):	x	x		$5.00 \times 10^{-4} - x$

$$K''_f = 2.05 \times 10^{11} = \frac{[ZnY^{2-}]}{[C_{Zn^{2+}}][EDTA]} = \frac{5.00 \times 10^{-4} - x}{x^2}$$

$$\Rightarrow x = C_{Zn^{2+}} = 4.9 \times 10^{-8} \text{ M}$$

$$[Zn^{2+}] = \alpha_{Zn^{2+}}C_{Zn^{2+}} = (1.8 \times 10^{-5})(4.9 \times 10^{-8}) = 8.9 \times 10^{-13} \text{ M}$$

$$\Rightarrow pZn^{2+} = -\log[Zn^{2+}] = 12.05$$

(c) *After the equivalence point—60.0 mL:* Now we are past the equivalence point, so almost all of the zinc is in the form ZnY^{2-}. With a dilution factor of 50.0/110.0 for zinc, we can write

$$[ZnY^{2-}] = \left(\frac{50.0}{110.0}\right)(1.00 \times 10^{-3}) = 4.5 \times 10^{-4} \text{ M}$$

We also know the concentration of excess EDTA, whose dilution factor is 10.0/110.0:

$$[EDTA] = \left(\frac{10.0}{110.0}\right)(1.00 \times 10^{-3}) = 9.1 \times 10^{-5} \text{ M}$$

Once we know $[ZnY^{2-}]$ and $[EDTA]$, we can use the equilibrium constant to find $[Zn^{2+}]$:

$$\frac{[ZnY^{2-}]}{[Zn^{2+}][EDTA]} = \alpha_{Y^{4-}}K_f = K'_f = (0.36)(10^{16.50}) = 1.1 \times 10^{16}$$

$$\frac{[4.5 \times 10^{-4}]}{[Zn^{2+}][9.1 \times 10^{-5}]} = 1.1 \times 10^{16}$$

$$\Rightarrow [Zn^{2+}] = 4.3 \times 10^{-16} \Rightarrow pZn^{2+} = 15.36$$

Note that past the equivalence point the problem did not depend on the presence of NH_3, because we knew the concentrations of both $[ZnY^{2-}]$ and $[EDTA]$.

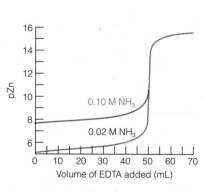

Figure 13-13 Titration curves for the reaction of 50.0 mL of 1.00×10^{-3} M Zn^{2+} with 1.00×10^{-3} M EDTA at pH 10.00 in the presence of two different concentrations of NH_3.

Figure 13-13 compares the calculated titration curves for Zn^{2+} in the presence of different concentrations of auxiliary complexing agent. The greater the concentration of NH_3, the smaller the change of pZn^{2+} near the equivalence point. When an auxiliary ligand is used, its amount must be kept below the level that would obliterate the end point of the titration. Color Plate 8 shows the appearance of a Cu^{2+}-ammonia solution during an EDTA titration.

13-6 Metal Ion Indicators

The most common technique to detect the end point in EDTA titrations is to use a metal ion indicator. Alternatives include a mercury electrode (Figure 13-8 and Exercise B in Chapter 15) and an ion-selective electrode (Section 15-6). A pH electrode will follow the course of the titration in unbuffered solution, because H_2Y^{2-} releases $2H^+$ when it forms a metal complex.

A **metal ion indicator** is a compound whose color changes when it binds to a metal ion. Several common indicators are shown in Table 13-3. *For an indicator to be useful, it must bind metal less strongly than EDTA does.*

A typical analysis is illustrated by the titration of Mg^{2+} with EDTA, using Eriochrome black T as the indicator.

$$\underset{\text{(red)}}{MgIn} + \underset{\text{(colorless)}}{EDTA} \rightarrow \underset{\text{(colorless)}}{MgEDTA} + \underset{\text{(blue)}}{In} \qquad (13\text{-}19)$$

At the start of the experiment, a small amount of indicator (In) is added to the colorless solution of Mg^{2+} to form a small amount of red complex. As EDTA is added, it reacts first with free, colorless Mg^{2+}. When free Mg^{2+} is used up, the last EDTA added before the equivalence point displaces indicator from the red MgIn complex. The change from the red of MgIn to the blue of unbound In signals the end point of the titration (Demonstration 13-1).

Most metal ion indicators are also acid-base indicators, with pK_a values listed in Table 13-3. Because the color of free indicator is pH dependent, most indicators can be used only in certain pH ranges. For example, xylenol orange (pronounced ZY-leen-ol) changes from yellow to red when it binds to a metal ion at pH 5.5. This is an easy color change to observe. At pH 7.5, the change is from violet to red and rather difficult to see. A spectrophotometer can be used to measure color change, but it is more convenient if we can see it.

Solutions of azo indicators (compounds with —N=N— bonds) deteriorate rapidly and should probably be prepared each week. Murexide solution should be prepared daily.

For an indicator to be useful in the titration of a metal with EDTA, the indicator must give up its metal ion to the EDTA. If a metal does not freely dissociate from an indicator, the metal is said to **block** the indicator. Eriochrome black T is blocked by Cu^{2+}, Ni^{2+}, Co^{2+}, Cr^{3+}, Fe^{3+}, and Al^{3+}. It cannot be used for the direct titration of any of these metals. It can be used for a back titration, however. For example, excess standard EDTA can be added to Cu^{2+}. Then indicator is added and the excess EDTA is back-titrated with Mg^{2+}.

End-point detection methods:

1. Metal ion indicators
2. Mercury electrode
3. Glass (pH) electrode
4. Ion-selective electrode

The indicator must release its metal to EDTA.

Question What will be the color change when the back titration is performed?

TABLE 13-3   Common metal ion indicators

Name	Structure	pK_a	Color of free indicator	Color of metal ion complex
Eriochrome black T	(H_2In^-)	$pK_2 = 6.3$ $pK_3 = 11.6$	H_2In^- red HIn^{2-} blue In^{3-} orange	Wine red
Calmagite	(H_2In^-)	$pK_2 = 8.1$ $pK_3 = 12.4$	H_2In^- red HIn^{2-} blue In^{3-} orange	Wine red
Murexide	(H_4In^-)	$pK_2 = 9.2$ $pK_3 = 10.9$	H_4In^- red-violet H_3In^{2-} violet H_2In^{3-} blue	Yellow (with Co^{2+}, Ni^{2+}, Cu^{2+}); red with Ca^{2+}
Xylenol orange	(H_3In^{3-})	$pK_2 = 2.32$ $pK_3 = 2.85$ $pK_4 = 6.70$ $pK_5 = 10.47$ $pK_6 = 12.23$	H_5In^- yellow H_4In^{2-} yellow H_3In^{3-} yellow H_2In^{4-} violet HIn^{5-} violet In^{6-} violet	Red
Pyridylazo-naphthol (PAN)	(HIn)	$pK_a = 12.3$	HIn orange-red In^- pink	Red
Pyrocatechol violet	(H_3In^-)	$pK_1 = 0.2$ $pK_2 = 7.8$ $pK_3 = 9.8$ $pK_4 = 11.7$	H_4In red H_3In^- yellow H_2In^{2-} violet HIn^{3-} red-purple	Blue

---

Demonstration 13-1 Metal Ion Indicator Color Changes

This demonstration illustrates the color change associated with Reaction 13-19 and shows how a second dye can be added to a solution to produce a more easily detected color change.

Stock solutions

Eriochrome black T: dissolve 0.1 g of the solid in 7.5 mL of triethanolamine plus 2.5 mL of absolute ethanol

Methyl red: dissolve 0.02 g in 60 mL of ethanol; then add 40 mL of water

Buffer: add 142 mL of concentrated (14.5 M) aqueous ammonia to 17.5 g of ammonium chloride and dilute to 250 mL with water

$MgCl_2$: 0.05 M

EDTA: 0.05 M $Na_2H_2EDTA \cdot 2H_2O$

Prepare a solution containing 25 mL of $MgCl_2$, 5 mL of buffer, and 300 mL of water. Add 6 drops of Eriochrome black T indicator and titrate with EDTA. Note the color change from wine red to pale blue at the end point (Color Plate 9a).

For some observers, the change of indicator color is not as sharp as desired. The colors can be affected by adding an "inert" dye whose color alters the appearance of the solution before and after the titration. Adding 3 mL of methyl red (or many other yellow dyes) produces an orange color prior to the end point and a green color after it. This sequence of colors is shown in Color Plate 9b.

13-7 EDTA Titration Techniques

Because so many elements can be analyzed by titration with EDTA, there is an extensive literature dealing with many variations of the basic procedure.[5] In this section we discuss several important techniques.

Direct Titration

In a **direct titration,** analyte is titrated with standard EDTA. The analyte is buffered to an appropriate pH at which the conditional formation constant for the metal-EDTA complex is large and the free indicator has a distinctly different color from the metal-indicator complex.

> The larger the effective formation constant, the more abrupt is the change in metal ion concentration at the end point.

An *auxiliary complexing agent,* such as ammonia, tartrate, citrate, or triethanolamine, may be employed to prevent the metal ion from precipitating in the absence of EDTA. For example, the direct titration of Pb^{2+} is carried out in ammonia buffer at pH 10 in the presence of tartrate, which complexes the metal ion and does not allow $Pb(OH)_2$ to precipitate. The lead-tartrate complex must be less stable than the lead-EDTA complex, or the titration would not be feasible.

Back Titration

In a **back titration,** a known excess of EDTA is added to the analyte. The excess EDTA is then titrated with a standard solution of a second metal ion.

A back titration is necessary if the analyte precipitates in the absence of EDTA, if it reacts too slowly with EDTA under titration conditions, or if it blocks the indicator. The metal ion used in the back titration must not displace the analyte metal ion from its EDTA complex.

EXAMPLE   A Back Titration

Ni^{2+} can be analyzed by a back titration using standard Zn^{2+} at pH 5.5 with xylenol orange indicator. A solution containing 25.00 mL of Ni^{2+} in dilute HCl is treated with 25.00 mL of 0.052 83 M Na_2EDTA. The solution is neutralized with NaOH, and the pH is adjusted to 5.5 with acetate buffer. The solution turns yellow when a few drops of indicator are added. Titration with 0.022 99 M Zn^{2+} requires 17.61 mL to reach the red end point. What is the molarity of Ni^{2+} in the unknown?

Solution   The unknown was treated with 25.00 mL of 0.052 83 M EDTA, which contains (25.00 mL)(0.052 83 M) = 1.320 8 mmol of EDTA. Back titration required (17.61 mL)(0.022 99 M) = 0.404 9 mmol of Zn^{2+}. Because 1 mol of EDTA reacts with 1 mol of any metal ion, there must have been

$$1.320\,8 \text{ mmol EDTA} - 0.404\,9 \text{ mmol Zn}^{2+} = 0.915\,9 \text{ mmol Ni}^{2+}$$

The concentration of Ni^{2+} is 0.915 9 mmol/25.00 mL = 0.036 64 M.

An EDTA back titration can be used to prevent precipitation of analyte. For example, Al^{3+} precipitates as $Al(OH)_3$ at pH 7 in the absence of EDTA. An acidic solution of Al^{3+} can be treated with excess EDTA, adjusted to pH 7–8 with sodium acetate, and boiled to ensure complete complexation of the ion. The $Al^{3+}-EDTA$ complex is stable at this pH. The solution is then cooled; Eriochrome black T indicator is added; and back titration with standard Zn^{2+} is performed.

Displacement Titration

For metal ions that do not have a satisfactory indicator, a **displacement titration** may be feasible. In this procedure the analyte usually is treated with excess $Mg(EDTA)^{2-}$ to displace Mg^{2+}, which is later titrated with standard EDTA.

$$M^{n+} + MgY^{2-} \rightarrow MY^{n-4} + Mg^{2+} \tag{13-20}$$

Hg^{2+} is determined in this manner. The formation constant of $Hg(EDTA)^{2-}$ must be greater than the formation constant of $Mg(EDTA)^{2-}$, or else the displacement of Mg^{2+} from $Mg(EDTA)^{2-}$ does not occur.

There is no suitable indicator for Ag^+. However, Ag^+ will displace Ni^{2+} from the tetracyanonickelate(II) ion:

$$2Ag^+ + Ni(CN)_4^{2-} \rightarrow 2Ag(CN)_2^- + Ni^{2+}$$

The liberated Ni^{2+} can then be titrated with EDTA to find out how much Ag^+ was added.

Challenge Calculate the equilibrium constant for Reaction 13-20 if $M^{n+} = Hg^{2+}$. Why is $Mg(EDTA)^{2-}$ used for a displacement titration?

Indirect Titration

Anions that form precipitates with certain metal ions can be analyzed with EDTA by **indirect titration.** For example, sulfate can be analyzed by precipitation with excess Ba^{2+} at pH 1. The $BaSO_4$ precipitate is filtered, washed, and then boiled with excess EDTA at pH 10 to bring Ba^{2+} back into solution as $Ba(EDTA)^{2-}$. The excess EDTA is back-titrated with Mg^{2+}.

Alternatively, an anion can be precipitated with excess metal ion. The precipitate is filtered and washed, and the excess metal ion in the filtrate is titrated with EDTA. Anions such as CO_3^{2-}, CrO_4^{2-}, S^{2-}, and SO_4^{2-} can be determined by indirect titration with EDTA.[6]

Masking

A **masking agent** is a reagent that protects some component of the analyte from reaction with EDTA. For example, Al^{3+} in a mixture of Mg^{2+} and Al^{3+} can be titrated by first masking the Al^{3+} with F^-, thereby leaving only the Mg^{2+} to react with EDTA.

Masking is used to prevent one element from interfering in the analysis of another element. Masking is not restricted to EDTA titrations. See Box 13-2 for an important application of masking.

Box 13-2   Water Hardness

Hardness refers to the total concentration of alkaline earth ions in water. Because the concentrations of Ca^{2+} and Mg^{2+} are usually much greater than the concentrations of other alkaline earth ions, hardness can be equated to $[Ca^{2+}] + [Mg^{2+}]$. Hardness is commonly expressed as the equivalent number of milligrams of $CaCO_3$ per liter. Thus, if $[Ca^{2+}] + [Mg^{2+}] = 1$ mM, we would say that the hardness is 100 mg $CaCO_3$ per liter because 100 mg $CaCO_3$ = 1 mmol $CaCO_3$. Water whose hardness is less than 60 mg $CaCO_3$ per liter is considered to be "soft." *Individual hardness* refers to the individual concentration of each alkaline earth ion.

Hard water reacts with soap to form insoluble curds.

$$Ca^{2+} + 2RSO_3^- \rightarrow \underset{\text{Precipitate}}{Ca(RSO_3)_2(s)} \qquad (A)$$
$$\underset{\text{Soap}}{}$$

Enough soap to consume the Ca^{2+} and Mg^{2+} must be used before the soap will be useful for cleaning. It is not believed that hard water is unhealthy. Hardness is beneficial in irrigation water because the alkaline earth ions tend to *flocculate* (cause to aggregate) colloidal particles in soil and thereby increase the permeability of the soil to water.

To measure total hardness, the sample is treated with ascorbic acid (or hydroxylamine) to reduce Fe^{3+} to Fe^{2+} and with cyanide to mask Fe^{2+}, Cu^+, and several other minor metal ions. Titration with EDTA at pH 10 in ammonia buffer then gives the total concentrations of Ca^{2+} and Mg^{2+}. The concentration of Ca^{2+} may be determined separately if the titration is carried out at pH 13 without ammonia. At this pH, $Mg(OH)_2$ precipitates and is inaccessible to the EDTA. Interference by many metal ions can be reduced with the right choice of indicators.[7]

Insoluble carbonates are converted to soluble bicarbonates by excess carbon dioxide:

$$CaCO_3(s) + CO_2 + H_2O \rightarrow Ca(HCO_3)_2 \ (aq) \qquad (B)$$

Heating converts bicarbonate to carbonate (driving off CO_2) and causes $CaCO_3$ to precipitate. The reverse of Reaction B forms a solid scale that clogs boiler pipes. The fraction of hardness due to $Ca(HCO_3)_2(aq)$ is called *temporary hardness* because this calcium is lost (by precipitation of $CaCO_3$) upon heating. Hardness arising from other salts (mainly dissolved $CaSO_4$) is called *permanent hardness,* because it is not removed by heating.

N(CH$_2$CH$_2$OH)$_3$
Triethanolamine

$$\underset{\text{2,3-Dimercaptopropanol}}{\overset{\overset{\displaystyle SH}{|}}{HOCH_2CHCH_2SH}}$$

Cyanide is a common masking agent that forms complexes with Cd^{2+}, Zn^{2+}, Hg^{2+}, Co^{2+}, Cu^+, Ag^+, Ni^{2+}, Pd^{2+}, Pt^{2+}, Fe^{2+}, and Fe^{3+}, but not with Mg^{2+}, Ca^{2+}, Mn^{2+}, or Pb^{2+}. When cyanide is added to a solution containing Cd^{2+} and Pb^{2+}, only the Pb^{2+} can react with EDTA. (*Caution:* Cyanide forms toxic gaseous HCN below pH 11. Cyanide solutions should be strongly basic and are best handled in a hood.) Fluoride can mask Al^{3+}, Fe^{3+}, Ti^{4+}, and Be^{2+}. (*Caution:* HF formed by F^- in acidic solution is extremely hazardous and should not contact skin and eyes. It may not be immediately painful, but the affected area should be flooded with water and then treated with calcium gluconate gel that you have on hand *before* the accident. First aiders must wear rubber gloves to protect themselves.) Triethanolamine masks Al^{3+}, Fe^{3+}, and Mn^{2+}; and 2,3-dimercaptopropanol masks Bi^{3+}, Cd^{2+}, Cu^{2+}, Hg^{2+}, and Pb^{2+}.

Demasking releases metal ion from a masking agent. Cyanide complexes can be demasked with formaldehyde:

$$M(CN)_m^{n-m} + mH_2CO + mH^+ \rightarrow mH_2C \underset{CN}{\overset{OH}{\diagdown}} + M^{n+}$$

Thiourea reduces Cu^{2+} to Cu^+ and masks Cu^+. Cu^{2+} can be liberated from the Cu^+-thiourea complex with H_2O_2. The selectivity afforded by masking, demasking, and pH control allows individual components of complex mixtures of metal ions to be analyzed by EDTA titration.

Terms to Understand

auxiliary complexing agent	cumulative formation constant	Lewis base
back titration	demasking	masking agent
blocking	direct titration	metal ion indicator
chelate effect	displacement titration	monodentate
chelating ligand	formation constant	multidentate
complexometric titration	indirect titration	stability constant
conditional formation constant	Lewis acid	

Summary

In a complexometric titration the analyte and titrant form a complex ion, and the equilibrium constant is called the formation constant, K_f. Chelating (multidentate) ligands form more stable complexes than do monodentate ligands, because the entropy of complex formation favors the binding of one large ligand rather than many small ligands. Synthetic aminocarboxylic acids such as EDTA have large metal-binding constants and are widely used in analytical chemistry.

Formation constants for EDTA are defined in terms of $[Y^{4-}]$, even though there are six protonated forms of EDTA. Because the fraction $(\alpha_{Y^{4-}})$ of free EDTA in the form Y^{4-} depends on pH, we define an effective (or conditional) formation constant as $K_f' = \alpha_{Y^{4-}}K_f = [MY^{n-4}]/[M^{n+}][EDTA]$. This constant describes the hypothetical reaction $M^{n+} + EDTA \rightleftharpoons MY^{n-4}$, where

EDTA refers to all forms of EDTA not bound to metal ion. Titration calculations fall into three categories: When excess unreacted M^{n+} is present, pM is calculated directly from $pM = -\log[M^{n+}]$. When excess EDTA is present, we know both $[MY^{n-4}]$ and $[EDTA]$, so $[M^{n+}]$ can be calculated from the conditional formation constant. At the equivalence point, the condition $[M^{n+}] = [EDTA]$ allows us to solve for $[M^{n+}]$. A single spreadsheet equation applies in all three regions of the titration curve.

EDTA titration curves become sharper as the formation constant increases and as the pH is raised. Auxiliary complexing agents, which compete with EDTA for the metal ion and thereby limit the sharpness of the titration curve, are sometimes necessary to keep the metal in solution. Calculations for a solution containing EDTA and an auxiliary complexing agent utilize the conditional formation constant $K''_f = \alpha_M \alpha_{Y^{4-}} K_f$, where α_M is the fraction of free metal ion not complexed by the auxiliary ligand.

For end-point detection, we commonly use metal ion indicators, a glass electrode, an ion-selective electrode, or a mercury electrode. When a direct titration is not suitable, because the analyte is unstable, reacts slowly

with EDTA, or has no suitable indicator, a back titration of excess EDTA or a displacement titration of $Mg(EDTA)^{2-}$ may be feasible. Masking prevents interference by unwanted species. Indirect EDTA titration procedures are available for the analysis of many anions or other species that do not react directly with the reagent.

Exercises

A. The potassium ion in a 250.0 ($\pm$0.1)-mL water sample was precipitated with sodium tetraphenylborate:

$$K^+ + (C_6H_5)_4B^- \rightarrow KB(C_6H_5)_4(s)$$

The precipitate was filtered, washed, and dissolved in an organic solvent. Treatment of the organic solution with an excess of Hg(II)-EDTA then gave the following reaction:

$$4HgY^{2-} + (C_6H_5)_4B^- + 4H_2O \rightarrow$$

$$H_3BO_3 + 4C_6H_5Hg^+ + 4HY^{3-} + OH^-$$

The liberated EDTA was titrated with 28.73 ($\pm$0.03) mL of 0.043 7 ($\pm$0.000 1) M Zn^{2+}. Find the concentration (and absolute uncertainty) of K^+ in the original sample.

B. A 25.00-mL sample of unknown containing Fe^{3+} and Cu^{2+} required 16.06 mL of 0.050 83 M EDTA for complete titration. A 50.00-mL sample of the unknown was treated with NH_4F to protect the Fe^{3+}. Then the Cu^{2+} was reduced and masked by addition of thiourea. Upon addition of 25.00 mL of 0.050 83 M EDTA, the Fe^{3+} was liberated from its fluoride complex and formed an EDTA complex. The excess EDTA required 19.77 mL of 0.018 83 M Pb^{2+} to reach an end point using xylenol orange. Find the concentration of Cu^{2+} in the unknown.

C. Calculate pGa^{3+} (to the 0.01 decimal place) at each of the following points in the titration of 50.0 mL of 0.040 0 M EDTA with 0.080 0 M $Ga(NO_3)_3$ at pH 4.00. Make a graph of pGa^{3+} versus volume of titrant.

(a) 0.1 mL	(d) 15.0 mL	(g) 25.0 mL
(b) 5.0 mL	(e) 20.0 mL	(h) 26.0 mL
(c) 10.0 mL	(f) 24.0 mL	(i) 30.0 mL

D. Calculate the concentration of H_2Y^{2-} at the equivalence point in Exercise C.

E. Suppose that 0.010 0 M Fe^{3+} is titrated with 0.005 00 M EDTA at pH 2.00.

(a) What is the concentration of free Fe^{3+} at the equivalence point?

(b) What is the quotient $[H_3Y^-]/[H_2Y^{2-}]$ in the solution when the titration is just 63.7% of the way to the equivalence point?

F. A solution containing 20.0 mL of 1.00×10^{-3} M Co^{2+} in the presence of 0.10 M $C_2O_4^{2-}$ at pH 9.00 was titrated with 1.00×10^{-2} M EDTA. Using formation constants from Table 13-2 and Appendix I, calculate pCo^{2+} for the following volumes of added EDTA: 0, 1.00, 2.00, and 3.00 mL. Consider the concentration of $C_2O_4^{2-}$ to be fixed at 0.10 M. Sketch a graph of Co^{2+} versus mL of added EDTA.

G. Iminodiacetic acid, abbreviated H_2X in this problem, forms 2:1 complexes with many metal ions:

$$H_2\overset{+}{N}\underset{CH_2CO_2H}{\overset{CH_2CO_2H}{\big\langle}} = H_3X^+$$

$$\alpha_{X^{2-}} = \frac{[X^{2-}]}{[H_3X^+] + [H_2X] + [HX^-] + [X^{2-}]}$$

$$Cu^{2+} + 2X^{2-} \rightleftharpoons CuX_2^{2-} \qquad K = 3.5 \times 10^{16}$$

A solution of volume 25.0 mL containing 0.120 M iminodiacetic acid buffered to pH 7.00 was titrated with 25.0 mL of 0.050 0 M Cu^{2+}. Given that $\alpha_{X^{2-}} = 4.6 \times 10^{-3}$ at pH 7.00, calculate the concentration of Cu^{2+} in the resulting solution.

Problems

EDTA

1. What is the chelate effect and why does it occur?

2. State (in words) what $\alpha_{Y^{4-}}$ means. Calculate $\alpha_{Y^{4-}}$ for EDTA at **(a)** pH 3.50; **(b)** pH 10.50.

3. (a) Find the conditional formation constant for $Mg(EDTA)^{2-}$ at pH 9.00.

(b) Find the concentration of free Mg^{2+} in 0.050 M $Na_2[Mg(EDTA)]$ at pH 9.00.

4. *Metal ion buffers.* By analogy to a hydrogen ion buffer, a metal ion buffer tends to maintain a particular metal ion concentration in solution. A mixture of the acid HA and its conjugate base A^- is a hydrogen ion buffer that maintains a pH defined by the equation $K_a = [A^-][H^+]/[HA]$. A mixture of CaY^{2-} and Y^{4-} serves as a Ca^{2+} buffer governed by the equation $1/K_f' = [EDTA][Ca^{2+}]/[CaY^{2-}]$. How many grams of $Na_2EDTA \cdot 2H_2O$ (FW 372.23) should be mixed with 1.95 g of $Ca(NO_3)_2 \cdot 2H_2O$ (FW 200.12) in a 500-mL volumetric flask to give a buffer with $pCa^{2+} = 9.00$ at pH 9.00?

EDTA Titration Curves

5. The ion M^{n+} (100.0 mL of 0.0500 M metal ion buffered to pH 9.00) was titrated with 0.0500 M EDTA.

(a) What is the equivalence volume, V_e, in milliliters?

(b) Calculate the concentration of free metal ion at $V = \frac{1}{2}V_e$.

(c) What fraction ($\alpha_{Y^{4-}}$) of free EDTA is in the form Y^{4-} at pH 9.00?

(d) The formation constant (K_f) is $10^{12.00}$. Calculate the value of the conditional formation constant $K_f'(= \alpha_{Y^{4-}} K_f)$.

(e) Calculate the concentration of free metal ion at $V = V_e$.

(f) What is the concentration of free metal ion at $V = 1.100 V_e$?

6. Calculate pCo^{2+} at each of the following points in the titration of 25.00 mL of 0.02026 M Co^{2+} by 0.03855 M EDTA at pH 6.00:

(a) 12.00 mL   (b) V_e (c) 14.00 mL

7. Consider the titration of 25.0 mL of 0.02000 M $MnSO_4$ with 0.01000 M EDTA in a solution buffered to pH 8.00. Calculate pMn^{2+} at the following volumes of added EDTA and sketch the titration curve:

(a) 0 mL   (d) 49.0 mL   (g) 50.1 mL
(b) 20.0 mL   (e) 49.9 mL   (h) 55.0 mL
(c) 40.0 mL   (f) 50.0 mL   (i) 60.0 mL

8. Using the same volumes as in Problem 7, calculate pCa^{2+} for the titration of 25.00 mL of 0.02000 M EDTA with 0.01000 M $CaSO_4$ at pH 10.00.

9. Calculate the molarity of HY^{3-} in a solution prepared by mixing 10.00 mL of 0.0100 M $VOSO_4$, 9.90 mL of 0.0100 M EDTA, and 10.0 mL of buffer with a pH of 4.00.

10. [image] *Titration of metal ion with EDTA.* Use Equation 13-11 to compute curves (pM versus mL of EDTA added) for the titration of 10.00 mL of 1.00 mM M^{2+} (= Cd^{2+} or Hg^{2+}) with 10.0 mM EDTA at pH 5.00. Plot both curves on one graph.

11. [image] *Effect of pH on the EDTA titration.* Use Equation 13-11 to compute curves (pCa^{2+} versus mL of EDTA added) for the titration of 10.00 mL of 1.00 mM Ca^{2+} with 1.00 mM EDTA at pH 5.00, 6.00, 7.00, 8.00, and 9.00. Plot all curves on one graph and compare your results to Figure 13-8.

12. [image] *Titration of EDTA with metal ion.* Use Equation 13-12 to reproduce the results of Exercise C.

Auxiliary Complexing Agents

13. State the purpose of an auxiliary complexing agent and give an example of its use.

14. According to Appendix I, Cu^{2+} forms two complexes with acetate:

$$Cu^{2+} + CH_3CO_2^- \rightleftharpoons Cu(CH_3CO_2)^+ \qquad K_1(= \beta_1)$$

$$Cu(CH_3CO_2)^+ + CH_3CO_2^- \rightleftharpoons Cu(CH_3CO_2)_2(aq) \quad K_2$$

(a) Find the value of β_2 for the reaction

$$Cu^{2+} + CH_3CO_2^- \rightleftharpoons Cu(CH_3CO_2)_2(aq)$$

$$\beta_2 = K_1 K_2$$

(b) Consider 1.00 L of solution prepared by mixing 1.00×10^{-4} mol $Cu(ClO_4)_2$ and 0.100 mol CH_3CO_2Na. Use Equation 13-16 to find the fraction of copper in the form Cu^{2+}.

15. Calculate pCu^{2+} at each of the following points in the titration of 50.00 mL of 0.00100 M Cu^{2+} with 0.00100 M EDTA at pH 11.00 in a solution whose NH_3 concentration is somehow *fixed* at 0.100 M:

(a) 0 mL   (c) 45.00 mL   (e) 55.00 mL
(b) 1.00 mL   (d) 50.00 mL

16. Consider the derivation of the fraction α_M in Equation 13-16.

(a) Derive the following expressions for the fractions α_{ML} and α_{ML_2}:

$$\alpha_{ML} = \frac{\beta_1[L]}{1 + \beta_1[L] + \beta_2[L]^2}$$

$$\alpha_{ML_2} = \frac{\beta_2[L]^2}{1 + \beta_1[L] + \beta_2[L]^2}$$

(b) Calculate the values of α_{ML} and α_{ML_2} for the conditions in Problem 14.

17. *Microequilibrium constants for binding of metal to a protein.* The iron-transport protein, transferrin, has two distinguishable metal-binding sites, designated a and b. The microequilibrium formation constants for each site are defined as follows:

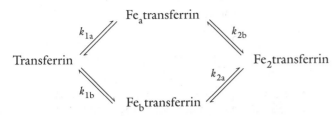

For example, the formation constant k_{1a} refers to the reaction Fe^{3+} + transferrin $\rightleftharpoons$ Fe_atransferrin, in which the metal ion binds to site a:

$$k_{1a} = \frac{[Fe_a\text{transferrin}]}{[Fe^{3+}][\text{transferrin}]}$$

(a) Write the chemical reactions corresponding to the conventional macroscopic formation constants, K_1 and K_2.

(b) Show that $K_1 = k_{1a} + k_{1b}$ and $K_2^{-1} = k_{2a}^{-1} + k_{2b}^{-1}$.

(c) Show that $k_{1a}k_{2b} = k_{1b}k_{2a}$. This means that if you know any three of the microequilibrium constants, you automatically know the fourth one.

(d) *A challenge to your sanity.* Based on the equilibrium constants below, *find the equilibrium fraction of each of the four species in the diagram above in circulating blood that is 40% saturated with iron* (i.e., Fe/transferrin = 0.80, because each protein can bind 2Fe).

Effective formation constants for blood plasma at pH 7.4

$k_{1a} = 6.0 \times 10^{22}$	$k_{2a} = 2.4 \times 10^{22}$
$k_{1b} = 1.0 \times 10^{22}$	$k_{2b} = 4.2 \times 10^{21}$
$K_1 = 7.0 \times 10^{22}$	$K_2 = 3.6 \times 10^{21}$

The binding constants are so large that you may assume that there is negligible free Fe^{3+}. To get started, let's use the abbreviations [T] = [transferrin], [FeT] = $[Fe_aT] + [Fe_bT]$, and $[Fe_2T]$ = $[Fe_2$transferrin]. Now we can write

mass balance for protein:

$$[T] + [FeT] + [Fe_2T] \equiv 1 \qquad (a)$$

mass balance for iron:

$$\frac{[FeT] + 2[Fe_2T]}{[T] + [FeT] + [Fe_2T]} = [FeT] + 2[Fe_2T] = 0.8 \quad (b)$$

combined equilibria:

$$\frac{K_1}{K_2} = 19.44 = \frac{[FeT]^2}{[T][Fe_2T]} \qquad (c)$$

Now you have three equations with three unknowns and should be able to tackle this problem.

18. *Spreadsheet equation for auxiliary complexing agent.* Consider the titration of metal M (concentration = C_m, initial volume = V_m) with a solution of EDTA (concentration = C_{EDTA}, volume added = V_{EDTA}) in the presence of an auxiliary complexing ligand (such as ammonia). Follow the general procedure of the derivation in Section 13-4 to show that the master equation for the titration is

$$\phi = \frac{C_{EDTA}V_{EDTA}}{C_m V_m}$$

$$= \frac{1 + K_f''[M]_{tot} - \dfrac{[M]_{tot} + K_f''[M]_{tot}^2}{C_m}}{K_f''[M]_{tot} + \dfrac{[M]_{tot} + K_f''[M]_{tot}^2}{C_{EDTA}}}$$

where K_f'' is the effective formation constant in the presence of auxiliary complexing agent at the fixed pH of the titration (Equation 13-18) and $[M]_{tot}$ is the total concentration of metal not bound to EDTA. $[M]_{tot}$ is the same as C_M in Equation 13-15. (We changed the symbol to avoid confusing C_M and C_m.) The equation above is the same as Equation 13-11, with [M] replaced by $[M]_{tot}$ and K_f replaced by K_f''.

19. *Auxiliary complexing agent.* Use the equation derived in Problem 18 for this exercise.

(a) Prepare a spreadsheet to reproduce the three points (20, 50, and 60 mL) calculated in the EDTA titration of zinc in the presence of ammonia in the example in Section 13-5.

(b) Use your spreadsheet to plot the curve for the titration of 50.00 mL of 0.0500 M Ni^{2+} by 0.100 M EDTA at pH 11.00 in the presence of 0.100 M oxalate as auxiliary ligand.

20. *A spreadsheet for formation of the complexes ML and ML_2.* Consider the titration of metal M (concentration = C_m, initial volume = V_m) with a solution of ligand L (concentration = C_l, volume added = V_l), which can form 1:1 and 2:1 complexes:

$$M + L \rightleftharpoons ML \qquad \beta_1 = \frac{[ML]}{[M][L]}$$

$$M + 2L \rightleftharpoons ML_2 \qquad \beta_2 = \frac{[ML_2]}{[M][L]^2}$$

Let α_M be the fraction of metal in the form M, α_{ML} be the fraction in the form ML, and α_{ML_2} be the fraction in the form ML_2. Following the derivation in Section 13-5, you could show that these fractions are given by

$$\alpha_M = \frac{1}{1 + \beta_1[L] + \beta_2[L]^2}$$

$$\alpha_{ML} = \frac{\beta_1[L]}{1 + \beta_1[L] + \beta_2[L]^2}$$

$$\alpha_{ML_2} = \frac{\beta_2[L]^2}{1 + \beta_1[L] + \beta_2[L]^2}$$

This means that the concentrations of ML and ML_2 are

$$[ML] = \alpha_{ML}\frac{C_m V_m}{V_m + V_l} \qquad [ML_2] = \alpha_{ML_2}\frac{C_m V_m}{V_m + V_l}$$

because $C_m V_m/(V_m + V_l)$ is the total concentration of all metal in the solution. The mass balance for ligand is

$$[L] + [ML] + 2[ML_2] = \frac{C_l V_l}{V_m + V_l}$$

By substituting expressions for [ML] and $[ML_2]$ into the mass balance, show that the master equation for a titration of metal by ligand is

$$\phi = \frac{C_l V_l}{C_m V_m} = \frac{\alpha_{ML} + 2\alpha_{ML_2} + ([L]/C_m)}{1 - ([L]/C_l)}$$

21. *Titration of M with L to form ML and ML_2.* Use the equation derived in Problem 20, where M is Cu^{2+} and L is acetate. Consider the addition of 0.500 M acetate to 10.00 mL of 0.0500 M Cu^{2+} at pH 7.00 (so that all ligand is present as $CH_3CO_2^-$, not CH_3CO_2H). Formation constants for $Cu(CH_3CO_2)^+$ and $Cu(CH_3CO_2)_2$ are given in Appendix I. Construct a spreadsheet in which the input is pL and the output is (a) [L], (b) V_l, (c) [M], (d) [ML], and (e) $[ML_2]$. Prepare a graph showing the concentrations of L, M, ML, and ML_2 as V_l ranges from 0 to 3 mL.

Metal Ion Indicators

22. Explain why the change from red to blue in Reaction 13-19 occurs suddenly at the equivalence point instead of gradually throughout the entire titration.

23. List four methods for detecting the end point of an EDTA titration.

24. Calcium ion was titrated with EDTA at pH 11, using calmagite as indicator (Table 13-3). Which is the principal species of calmagite at pH 11? What color was observed before the equivalence point? after the equivalence point?

25. Pyrocatechol violet (Table 13-3) is to be used as a metal ion indicator in an EDTA titration. The procedure is as follows:

1. Add a known excess of EDTA to the unknown metal ion.
2. Adjust the pH with a suitable buffer.
3. Back-titrate the excess chelate with standard Al^{3+}.

From the following available buffers, select the best buffer, and then state what color change will be observed at the end point. Explain your answer.

(a) pH 6–7; (b) pH 7–8; (c) pH 8–9; (d) pH 9–10

EDTA Titration Techniques

26. Give three circumstances in which an EDTA back titration might be necessary.

27. Describe what is done in a displacement titration and give an example.

28. Give an example of the use of a masking agent.

29. What is meant by water hardness? Explain the difference between temporary and permanent hardness.

30. How many milliliters of 0.0500 M EDTA are required to react with 50.0 mL of 0.0100 M Ca^{2+}? with 50.0 mL of 0.0100 M Al^{3+}?

31. A 50.0-mL sample containing Ni^{2+} was treated with 25.0 mL of 0.0500 M EDTA to complex all the Ni^{2+} and leave excess EDTA in solution. The excess EDTA was then back-titrated, requiring 5.00 mL of 0.0500 M Zn^{2+}. What was the concentration of Ni^{2+} in the original solution?

32. A 50.0-mL aliquot of solution containing 0.450 g of $MgSO_4$ (FW 120.37) in 0.500 L required 37.6 mL of EDTA solution for titration. How many milligrams of $CaCO_3$ (FW 100.09) will react with 1.00 mL of this EDTA solution?

33. A 1.000-mL sample of unknown containing Co^{2+} and Ni^{2+} was treated with 25.00 mL of 0.03872 M EDTA. Back titration with 0.02127 M Zn^{2+} at pH 5 required 23.54 mL to reach the xylenol orange end point. A 2.000-mL sample of unknown was passed through an ion-exchange column that retards Co^{2+} more than Ni^{2+}. The Ni^{2+} that passed through the column was treated with 25.00 mL of 0.03872 M EDTA and required 25.63 mL of 0.02127 M Zn^{2+} for back titration. The Co^{2+} emerged from the column later. It, too, was treated with 25.00 mL of 0.03872 M EDTA. How many milliliters of 0.02127 M Zn^{2+} will be required for back titration?

34. A 50.0-mL solution containing Ni^{2+} and Zn^{2+} was treated with 25.0 mL of 0.0452 M EDTA to bind all the metal. The excess unreacted EDTA required 12.4 mL of 0.0123 M Mg^{2+} for complete reaction. An excess of the reagent 2,3-dimercapto-1-propanol was then added to displace the EDTA from zinc. Another 29.2 mL of Mg^{2+} was required for reaction with the liberated EDTA. Calculate the molarity of Ni^{2+} and Zn^{2+} in the original solution.

35. Sulfide ion was determined by indirect titration with EDTA. To a solution containing 25.00 mL of 0.043 32 M $Cu(ClO_4)_2$ plus 15 mL of 1 M acetate buffer (pH 4.5) was added 25.00 mL of unknown sulfide solution with vigorous stirring. The CuS precipitate was filtered and washed with hot water. Then ammonia was added to the filtrate (which contains excess Cu^{2+}) until the blue color of $Cu(NH_3)_4^{2+}$ was observed. Titration with 0.039 27 M EDTA required 12.11 mL to reach the murexide end point. Calculate the molarity of sulfide in the unknown.

36. *Indirect EDTA determination of cesium.* Cesium ion does not form a strong EDTA complex, but it can be analyzed by adding a known excess of $NaBiI_4$ in cold concentrated acetic acid containing excess NaI. Solid $Cs_3Bi_2I_9$ is precipitated, filtered, and removed. The excess yellow BiI_4^- is then titrated with EDTA. The end point occurs when the yellow color disappears. (Sodium thiosulfate is used in the reaction to prevent the liberated I^- from being oxidized to yellow aqueous I_2 by O_2 from the air.) The precipitation is fairly selective for Cs^+. The ions Li^+, Na^+, K^+, and low concentrations of Rb^+ do not interfere, although Tl^+ does. Suppose that 25.00 mL of unknown containing Cs^+ was treated with 25.00 mL of 0.086 40 M $NaBiI_4$ and the unreacted BiI_4^- required 14.24 mL of 0.043 7 M EDTA for complete titration. Find the concentration of Cs^+ in the unknown.

Notes and References

1. Discussion of the chelate effect and the role of factors other than entropy are found in articles by J. J. R. Fraústo da Silva, *J. Chem. Ed.* **1983,** *60,* 390; and C.-S. Chung, *J. Chem. Ed.* **1984,** *61,* 1062.

2. pK_1 applies at 25°C, $\mu = 1.0$ M. The other values shown apply at 20°C, $\mu = 0.1$ M. [A. E. Martell and R. M. Smith, *Critical Stability Constants,* Vol. 1 (New York: Plenum Press, 1974), p. 204.]

3. H_4Y can be dried at 140°C for 2 h and used as a primary standard. It can be dissolved by adding NaOH solution from a plastic container. NaOH solution from a glass bottle should not be used because it contains alkaline earth metals leached from the glass. Reagent-grade $Na_2H_2Y \cdot 2H_2O$ contains ~0.3 wt% excess water. It may be used in this form with suitable correction for the mass of excess water, or dried to the composition $Na_2H_2Y \cdot 2H_2O$ at 80°C.

4. R. L. Barnett and V. A. Uchtman, *Inorg. Chem.* **1979,** *18,* 2674.

5. G. Schwarzenbach and H. Flaschka, *Complexometric Titrations,* H. M. N. H. Irving, trans. (London: Methuen, 1969); H. A. Flaschka, *EDTA Titrations* (New York: Pergamon Press, 1959); and C. N. Reilley, A. J. Barnard, Jr., and R. Puschel in L. Meites, Ed., *Handbook of Analytical Chemistry* (New York: McGraw-Hill, 1963), pp. 3-76–3-234.

6. Indirect EDTA titration of monovalent *cations* is described by I. M. Yurist, M. M. Talmud, and P. M. Zaitsev, *J. Anal. Chem. U.S.S.R.* **1987,** *42,* 911.

7. The indicator 2-[(4-phenylthioacetic acid)azo]-1,8-dihydroxynaphthalene-3, 6-disulfonic acid eliminates interference by numerous metal ions without the need for masking agents. [D. P. S. Rathore, P. K. Bhargava, M. Kumar, and R. K. Talra, *Anal. Chim. Acta* **1993,** *281,* 173.]

An Atomic-Scale Galvanic Cell

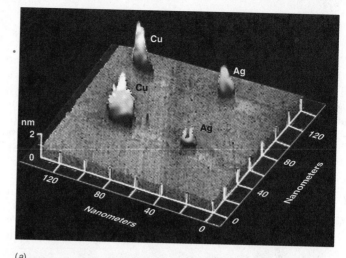

(a)

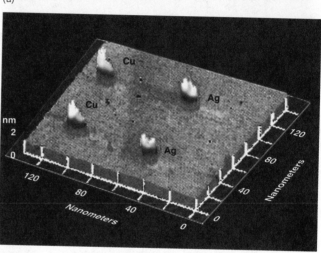

(b)

[Courtesy R. M. Penner, University of California, Irvine.] From W. Li, J. A. Virtanen, and R. M. Penner, *J. Phys. Chem.* **1992**, *96*, 6529.

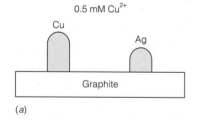

(a)

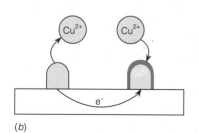

(b)

Galvanic cells, such as the battery in your calculator, produce electricity through chemical reactions. One of the smallest galvanic cells ever observed is shown in the scanning tunneling electron microscope images here. In (a), there are two mounds of silver at the right and two mounds of copper at the left, electrochemically deposited on a graphite surface. The mounds are 50–80 atoms in diameter and 7–20 atoms high. Image (b) shows that after 45 min in 0.5 mM $CuSO_4$ solution, some of the copper has dissolved and a 2-atom-thick layer of copper has plated out onto the silver.

As copper metal dissolves, electrons left behind pass through the graphite into the silver. (See the figure in the margin.) There they react with Cu^{2+} ions from the solution, and a layer of copper grows on the silver. This process is thermodynamically favorable only for the first one or two atomic layers of copper, and then the process ceases. Even though this world's smallest battery does no useful work, larger batteries driven by spontaneous chemical reactions help each of us every day.

Fundamentals of Electrochemistry

A major branch of analytical chemistry uses electrical measurements for analytical purposes. In addition to measurements on macroscopic systems, electrochemistry is now applied to microscopic and nanoscopic problems, such as measurement of neurotransmitters from a single cell at the front of Chapter 1. Another electroanalytical device, the scanning tunneling electron microscope, rasters a tiny electrode over a surface and uses electric current measurements to discern atomic-scale features of the surface. Such an instrument created the images on the facing page. Electrochemistry is used not only for chemical measurements but also for making tiny chemical structures, such as the copper and silver pillars shown on the facing page and layers of material in some electronic chips.

In this chapter we review fundamental concepts of electricity and electrochemical cells. The principles we develop will lay a foundation for the discussions of potentiometric measurements, electrogravimetric and coulo-metric analysis, polarography, and amperometric methods in the next four chapters.

14-1 Basic Concepts

A **redox reaction** involves transfer of electrons from one species to another. A species is said to be **oxidized** when it *loses electrons*. It is **reduced** when it *gains electrons*. An **oxidizing agent,** also called an **oxidant,** takes electrons from another substance and becomes reduced. A **reducing agent,** also called

Oxidation: loss of electrons
Reduction: gain of electrons

Oxidizing agent: takes electrons
Reducing agent: gives electrons

a **reductant,** gives electrons to another substance and is oxidized in the process. In the reaction

$$\underset{\substack{\text{Oxidizing} \\ \text{agent}}}{Fe^{3+}} \quad + \quad \underset{\substack{\text{Reducing} \\ \text{agent}}}{V^{2+}} \quad \rightarrow \quad Fe^{2+} \quad + \quad V^{3+} \tag{14-1}$$

Fe^{3+} is the oxidizing agent because it takes an electron from V^{2+}. V^{2+} is the reducing agent because it gives an electron to Fe^{3+}. Fe^{3+} is reduced, and V^{2+} is oxidized as the reaction proceeds from left to right. A review of oxidation numbers and balancing redox equations appears in Appendix D.

Chemistry and Electricity

The *quantity* of electrons that flow from a reaction is proportional to the quantity of analyte that reacts.

When the electrons involved in a redox reaction can be made to flow through an electric circuit, we can learn something about the reaction by measuring the voltage and current in the circuit. In Reaction 14-1, one electron must be transferred to oxidize one atom of V^{2+} and to reduce one atom of Fe^{3+}. If we know how many moles of electrons are transferred from V^{2+} to Fe^{3+}, then we know how many moles of product have been formed.

The voltage produced by a cell in which Reaction 14-1 takes place is related to the free energy change when electrons flow from V^{2+} to Fe^{3+}. The free energy change would be different if electrons were flowing, for example, from Sn^{2+} to Fe^{3+}. In techniques such as polarography, the voltage can be used to identify the reactants. Voltage is also related to the concentrations of reactants and products, as we shall see when we study the Nernst equation.

The electric force (voltage) is related to the identity and concentrations of reactants and products.

Electric Charge

Electric charge (q) is measured in **coulombs** (C). The magnitude of the charge of a single electron is 1.602×10^{-19} C, so 1 mol of electrons has a charge of 9.649×10^4 C, which is called the **Faraday constant** (F).

Faraday constant (F): 96 485.309 C/mol

Relation between charge and moles:

$$\underset{\text{Coulombs}}{q} \quad = \quad \underset{\text{Moles}}{n} \quad \cdot \quad \underset{\frac{\text{Coulombs}}{\text{Mole}}}{F} \tag{14-2}$$

- -

EXAMPLE   **Relating Coulombs to Quantity of Reaction**

If 5.585 g of Fe^{3+} was reduced in Reaction 14-1, how many coulombs of charge must have been transferred from V^{2+} to Fe^{3+}?

Solution   First, we find that 5.585 g of Fe^{3+} equals 0.100 0 mol of Fe^{3+}. Because each Fe^{3+} ion requires one electron in Reaction 14-1, 0.100 0 mol of electrons must have been transferred. Using the Faraday constant, we find that 0.100 0 mol of electrons corresponds to

$$(0.100\,0 \text{ mol e}^-) \left(9.649 \times 10^4 \, \frac{C}{\text{mol e}^-} \right) = 9.649 \times 10^3 \text{ C}$$

- -

Electric Current

The quantity of charge flowing each second through a circuit is called the **current**. The unit of current is the **ampere**, abbreviated A. A current of one ampere represents a charge of one coulomb per second flowing past a point in a circuit.

$1 \text{ A} = 1 \text{ C/s}$

· ·

EXAMPLE **Relating Current to Rate of Reaction**

Suppose that electrons are forced into a platinum wire immersed in a solution containing Sn^{4+} (Figure 14-1), which is reduced to Sn^{2+} at a constant rate of 4.24 mmol/h. How much current flows into the solution?

Solution *Two* electrons are required to reduce *one* Sn^{4+} ion:

$$Sn^{4+} + 2e^- \rightarrow Sn^{2+}$$

If Sn^{4+} is reacting at a rate of 4.24 mmol/h, electrons flow at a rate of $2(4.24) = 8.48$ mmol/h, which corresponds to

$$\frac{8.48 \text{ mmol/h}}{3\,600 \text{ s/h}} = 2.356 \times 10^{-3} \text{ mmol/s} = 2.356 \times 10^{-6} \text{ mol/s}$$

To find the current, we convert moles of electrons per second to coulombs per second:

$$\text{current} = \frac{\text{coulombs}}{\text{second}} = \frac{\text{moles}}{\text{second}} \cdot \frac{\text{coulombs}}{\text{mole}}$$

$$= \left(2.356 \times 10^{-6} \frac{\text{mol}}{\text{s}}\right)\left(9.649 \times 10^4 \frac{\text{C}}{\text{mol}}\right)$$

$$= 0.227 \text{ C/s} = 0.227 \text{ A}$$

A current of 0.227 A can also be expressed as 227 mA.

· ·

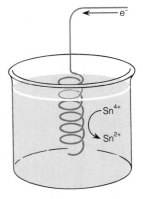

Figure 14-1 Electrons flowing into a coil of Pt wire at which Sn^{4+} ions in solution are reduced to Sn^{2+} ions. This process could not happen by itself because there is not a complete circuit. If Sn^{4+} is to be reduced at this Pt electrode, some other species must be oxidized at some other place.

In the foregoing example, we encountered a Pt **electrode,** which conducts electrons into or out of the chemical species involved in a redox reaction. Platinum is commonly used as an *inert* electrode; it does not participate in the redox chemistry except as a conductor of electrons.

Voltage, Work, and Free Energy

The difference in **electric potential** (E) between two points is a measure of the work that is needed (or can be done) when an electric charge moves from one point to the other. Potential difference is measured in **volts** (V). Work has the dimensions of energy, whose units are joules (J).

When a charge, q, moves through a potential difference, E, the work done is

It costs energy to move like charges toward each other. Energy is released when opposite charges move toward each other.

Relation between work and voltage:

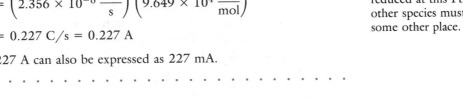

$$\underset{\text{Joules}}{\text{work}} = \underset{\text{Volts}}{E} \cdot \underset{\text{Coulombs}}{q} \qquad (14\text{-}3)$$

One **joule** of energy is gained or lost when one *coulomb* of charge is moved between points whose potentials differ by one volt. Equation 14-3 tells us that the dimensions of volts are joules per coulomb ($1 \text{ V} = 1 \text{ J/C}$).

EXAMPLE   Electrical Work

How much work is required to move 2.36 mmol of electrons through a potential difference of 1.05 V?

Solution   To use Equation 14-3, we must convert moles of electrons to coulombs of charge. The relation is simply

$$q = nF = (2.36 \times 10^{-3} \text{ mol}) (9.649 \times 10^4 \text{ C/mol}) = 2.277 \times 10^2 \text{ C}$$

The work required is

$$\text{work} = E \cdot q = (1.05 \text{ V}) (2.277 \times 10^2 \text{ C}) = 239 \text{ J}$$

The greater the potential difference between two points, the stronger will be the "push" on a charged particle traveling between those points. A 12-V battery will "push" electrons through a circuit eight times harder than a 1.5-V dry cell.

The free energy change, ΔG, for a chemical reaction conducted reversibly at constant temperature and pressure equals the maximum possible electrical work that can be done by the reaction on its surroundings.

You might review Chapter 5 for a brief discussion of ΔG.

$$\text{work done on surroundings} = -\Delta G \tag{14-4}$$

The negative sign in Equation 14-4 indicates that the free energy of a system decreases when the work is done on the surroundings.

Combining Equations 14-2, 14-3, and 14-4 produces a relationship of utmost importance to chemistry:

$$\Delta G = -\text{work} = -E \cdot q$$

Relation between free energy and electric potential difference:

$$\Delta G = -nFE \tag{14-5}$$

Equation 14-5 relates the free energy change of a chemical reaction to the electrical potential difference (i.e., the voltage) that can be generated by the reaction.

Ohm's Law

Ohm's law states that the current, I, flowing through a circuit is directly proportional to the potential difference (voltage) across the circuit and inversely proportional to the **resistance, R,** of the circuit.

The greater the voltage, the more current will flow. The greater the resistance, the less current will flow.

Ohm's law:

$$I = \frac{E}{R} \tag{14-6}$$

The units of resistance are **ohms,** assigned the Greek symbol Ω (omega). A current of one ampere will flow through a circuit with a potential difference of one volt if the resistance of the circuit is one ohm.

Power

Power, *P,* is the work done per unit time. The SI unit of power is J/s, better known as the **watt** (W).

$$P = \frac{\text{work}}{s} = \frac{E \cdot q}{s} = E \cdot \frac{q}{s} \qquad (14\text{-}7)$$

Because q/s is the current, *I,* we can write

$$P = E \cdot I \qquad (14\text{-}8)$$

A cell capable of delivering one ampere at a potential difference of one volt has a power output of one watt.

power (watts) = work per second
$$P = E \cdot I = (IR) \cdot I = I^2R$$

. . . .

EXAMPLE   Using Ohm's Law

A schematic diagram of a very simple circuit is shown in Figure 14-2. The battery generates a potential difference of 3.0 V, and the resistor has a resistance of 100 Ω. We assume that the resistance of the wire connecting the battery and the resistor is negligible. How much current and how much power are delivered by the battery in this circuit?

Solution   The current flowing through this circuit is

$$I = \frac{E}{R} = \frac{3.0 \text{ V}}{100 \text{ }\Omega} = 0.030 \text{ A} = 30 \text{ mA}$$

The power produced by the battery must be

$$P = E \cdot I = (3.0 \text{ V})(0.030 \text{ A}) = 90 \text{ mW}$$

. . . .

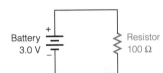

Figure 14-2   A simple electric circuit with a battery and a resistor.

What happens to the energy needed to push electrons through the circuit in Figure 14-2? Ideally, the only place at which energy is lost is in the resistor. *The energy appears as heat in the resistor.* The power (90 mW) equals the rate at which heat is produced in the resistor.

14-2 Galvanic Cells

A **galvanic cell** (also called a *voltaic cell)* is one that uses a *spontaneous* chemical reaction to generate electricity. To accomplish this, one reagent must be oxidized and another must be reduced. The two cannot be in contact, or electrons would simply flow directly from the reducing agent to the oxidizing agent. Instead, the oxidizing and reducing agents are physically separated, and electrons are forced to flow through an external circuit in order to go from one reactant to the other.

A galvanic cell uses a spontaneous chemical reaction to generate electricity.

A Cell in Action

Figure 14-3 shows a galvanic cell containing two electrodes suspended in an aqueous solution of $CdCl_2$. One electrode is a strip of cadmium metal; the

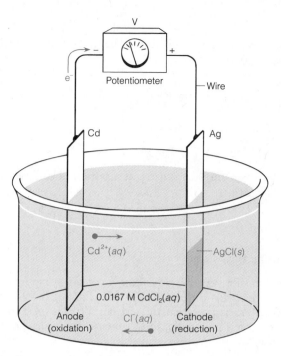

Figure 14-3 A simple galvanic cell. The **potentiometer** is a device for measuring voltage. Its two terminals (connectors) are labeled "+" and "−" and are called the positive and negative terminals. When electrons flow into the negative terminal, the voltage is positive.

other is a piece of metallic silver coated with solid AgCl. The chemical reactions occurring in this cell are

reduction: $2AgCl(s) + 2e^- \rightleftharpoons 2Ag(s) + 2Cl^-(aq)$

oxidation: $Cd(s) \rightleftharpoons Cd^{2+}(aq) + 2e^-$

net reaction: $Cd(s) + 2AgCl(s) \rightleftharpoons Cd^{2+}(aq) + 2Ag(s) + 2Cl^-(aq)$

The net reaction is composed of a reduction reaction and an oxidation reaction, each of which is called a **half-reaction.** The two half-reactions are written with equal numbers of electrons so that their sum includes no free electrons.

Oxidation of Cd metal—to produce Cd^{2+} (aq)—provides electrons that flow through the circuit to the Ag electrode in Figure 14-3. At the surface of the Ag electrode, Ag^+ (from AgCl) is reduced to $Ag(s)$. The chloride from AgCl goes into solution. The free energy change for the net reaction, −150 kJ per mole of Cd, provides the driving force that pushes electrons through the circuit.

Recall that ΔG is *negative* for a spontaneous reaction.

EXAMPLE Voltage Produced by a Chemical Reaction

Calculate the voltage that would be measured by the potentiometer in Figure 14-3.

Solution Because $\Delta G = -150$ kJ/mol of Cd, we can use Equation 14-5 (where n is the number of moles of electrons transferred in the balanced net reaction) to write

$$E = -\frac{\Delta G}{nF} = -\frac{-150 \times 10^3 \text{ J}}{(2 \text{ mol}) \left(9.649 \times 10^4 \dfrac{\text{C}}{\text{mol}}\right)}$$

$$= +0.777 \text{ J/C} = +0.777 \text{ V}$$

Reminder: 1 J/C = 1 volt

A spontaneous chemical reaction (negative ΔG) produces a *positive voltage*.

Chemists define the electrode at which *reduction* occurs as the **cathode**. The **anode** is the electrode at which *oxidation* occurs. In Figure 14-3, Ag is the cathode because reduction takes place at its surface ($2AgCl + 2e^- \rightarrow 2Ag + 2Cl^-$) and Cd is the anode because it is oxidized ($Cd \rightarrow Cd^{2+} + 2e^-$).

cathode $\leftrightarrow$ reduction
anode $\leftrightarrow$ oxidation

Salt Bridge

Consider the cell in Figure 14-4, in which the reactions are intended to be

cathode: $2Ag^+(aq) + 2e^- \rightleftharpoons 2Ag(s)$

anode: $Cd(s) \rightleftharpoons Cd^{2+}(aq) + 2e^-$

net reaction: $Cd(s) + 2Ag^+(aq) \rightleftharpoons Cd^{2+}(aq) + 2Ag(s)$ (14-9)

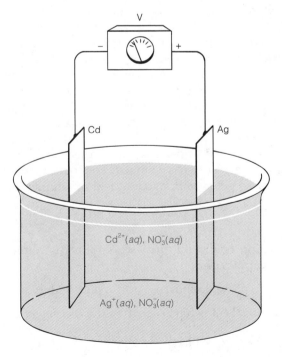

Figure 14-4 A cell that will not work. The solution contains $Cd(NO_3)_2$ and $AgNO_3$.

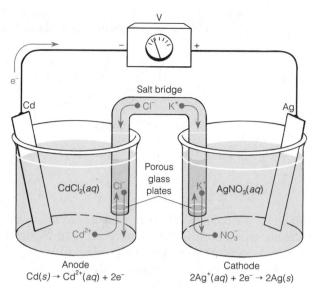

Figure 14-5   A cell that works—thanks to the salt bridge!

The cell in Figure 14-4 is *short-circuited.*

The purpose of a salt bridge is to maintain electroneutrality (no charge buildup) throughout the cell. See Demonstration 14-1.

Preparing a salt bridge.

Boundary between separate phases: |

The net reaction is spontaneous, but little current flows through the circuit because Ag^+ ions are not forced to be reduced at the Ag electrode. The Ag^+ ions in solution can react directly at the $Cd(s)$ surface, giving the same net reaction with no flow of electrons through the external circuit.

We can separate the reactants into two *half-cells,* if we connect the two halves with a **salt bridge,** as shown in Figure 14-5. The salt bridge consists of a U-shaped tube filled with a gel containing KCl (or any other electrolyte not involved in the cell reaction). The ends of the bridge are covered with porous glass disks that allow ions to diffuse but minimize mixing of the solutions inside and outside the bridge. When the galvanic cell is operating, K^+ from the bridge migrates into the cathode compartment and NO_3^- migrates from the cathode into the bridge. Ion migration exactly offsets the charge buildup that would otherwise occur as electrons flow into the silver electrode. In the other half-cell, Cd^{2+} migrates into the bridge while Cl^- migrates into the anode compartment to compensate for the buildup of positive charge that would otherwise occur.

A salt bridge is prepared by heating 3 g of agar with 30 g of KCl in 100 mL of water until a clear solution is obtained. The solution is poured into the U-tube and allowed to gel. The bridge is stored in saturated aqueous KCl.

Line Notation

A shorthand notation employing just two symbols is commonly used to describe electrochemical cells:

$$| \text{ phase boundary} \qquad || \text{ salt bridge}$$

The cell in Figure 14-3 is represented by the *line diagram*

$$Cd(s) \mid CdCl_2(aq) \mid AgCl(s) \mid Ag(s)$$

Each phase boundary is indicated by a vertical line. The electrodes are shown at the extreme left- and right-hand sides of the diagram. The cell in Figure

A salt bridge is an ionic medium with a *semipermeable* barrier on each end. Small molecules and ions can cross a semipermeable barrier, but large molecules such as gelled agar cannot. You can demonstrate a "proper" salt bridge by filling a U-tube with agar and KCl (as described in the text section entitled "Salt Bridge") and constructing the cell shown here. The pH meter is a potentiometer whose negative terminal is the reference electrode socket.

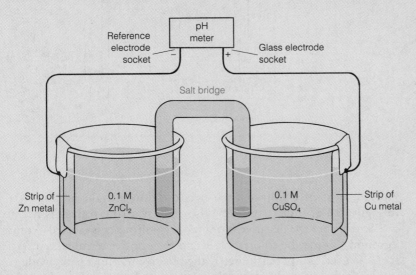

Setup for the galvanic cell

You should be able to write the two half-reactions for this cell and use the Nernst equation (14-15) to calculate the theoretical voltage. Measure the voltage with a conventional salt bridge. Then replace the salt bridge with one made of filter paper freshly soaked in NaCl solution and measure the voltage again. Finally, replace the filter-paper salt bridge with two fingers of the same hand and measure the voltage again. The human body is really just a bag of salt housed in a semipermeable membrane. The small differences in voltage observed when the salt bridge is replaced can be attributed to the junction potential discussed in Section 15-3.

Challenge One hundred and eighty students at Virginia Polytechnic Institute and State University made a salt bridge by holding hands.[1] Their resistance was lowered from $10^6 \ \Omega$ per student to $10^4 \ \Omega$ per student by wetting everyone's hands. Can your class beat this record?

14-5 can be written as follows:

$$Cd(s) \mid CdCl_2(aq) \parallel AgNO_3(aq) \mid Ag(s)$$

Salt bridge: $\parallel$

The composition of the salt bridge is not specified in cell notation.

14-3 Standard Potentials

The voltage measured in the experiment in Figure 14-5 is the difference in electric potential between the Ag electrode on the right and the Cd electrode on the left. Voltage tells us how much work can be done by electrons flowing from one side to the other (Equation 14-3). The potentiometer (voltmeter)

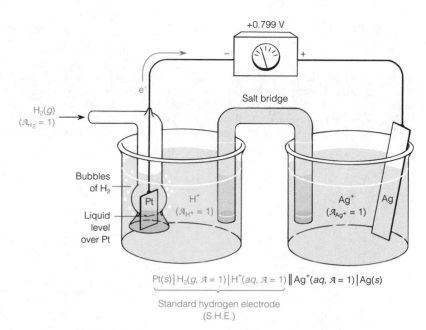

Figure 14-6 Cell used to measure the standard potential of the reaction $Ag^+ + e^- \rightleftharpoons Ag(s)$. This cell is hypothetical because it is usually not possible to adjust the activity of a species to 1.

$$Pt(s)\,|\,H_2(g,\,\mathcal{A}=1)\,|\,H^+(aq,\,\mathcal{A}=1)\,\|\,Ag^+(aq,\,\mathcal{A}=1)\,|\,Ag(s)$$

Standard hydrogen electrode
(S.H.E.)

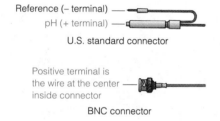

Reference (– terminal)
pH (+ terminal)

U.S. standard connector

Positive terminal is
the wire at the center
inside connector

BNC connector

indicates a positive voltage when electrons flow into the negative terminal, as in Figure 14-5. If electrons flow the other way, the voltage is negative.

Sometimes the negative terminal of a voltmeter is labeled "common" or "ground." It may be colored black and the positive terminal red. When a pH meter is used as a potentiometer, the positive terminal is the wide receptacle to which the glass pH electrode is connected. The negative terminal is the narrow receptacle to which the reference electrode is connected. In a BNC connection, the center wire is the positive input and the outer connection is the negative input. (See the illustration in the margin.)

To predict the voltage that will be observed when different half-cells are connected to each other, the **standard reduction potential** ($E°$) for each half-cell is measured by an experiment shown in an idealized form in Figure 14-6. The half-reaction of interest in this diagram is

$$Ag^+ + e^- \rightleftharpoons Ag(s) \tag{14-10}$$

which occurs in the half-cell at the right connected to the *positive* terminal of the potentiometer. The term *standard* means that the activities of all species are unity. For Reaction 14-10, this means that $\mathcal{A}_{Ag^+} = 1$ and, by definition, the activity of $Ag(s)$ is also unity.

The left half-cell, connected to the *negative* terminal of the potentiometer, is called the **standard hydrogen electrode** (S.H.E.). It consists of a catalytic Pt surface in contact with an acidic solution in which $\mathcal{A}_{H^+} = 1$. A stream of $H_2(g)$ bubbled through the electrode saturates the solution with $H_2(aq)$. The activity of $H_2(g)$ is unity if the pressure of $H_2(g)$ is 1 atm. The reaction that comes to equilibrium at the surface of the Pt electrode is

$$H^+(aq,\,\mathcal{A}=1) + e^- \rightleftharpoons \tfrac{1}{2}H_2(g,\,\mathcal{A}=1) \tag{14-11}$$

Question   What is the pH of the standard hydrogen electrode?

We will write all half-reactions as *reductions*. By convention, $E° = 0$ for S.H.E.

We *arbitrarily* assign a potential of zero to the standard hydrogen electrode. The voltage measured by the meter in Figure 14-6 can therefore be *assigned* to Reaction 14-10, which occurs in the right half-cell. The measured value $E° = +0.799$ V is the standard reduction potential for Reaction 14-10. The positive sign tells us that electrons flow from left to right through the meter.

The line notation for the cell in Figure 14-6 is

$$Pt(s) \mid H_2(g, \mathcal{A} = 1) \mid H^+(aq, \mathcal{A} = 1) \parallel Ag^+(aq, \mathcal{A} = 1) \mid Ag(s)$$

By convention, the left-hand electrode (Pt) is attached to the negative (reference) terminal of the potentiometer and the right-hand electrode is attached to the positive terminal. Because the standard hydrogen electrode is used frequently, we abbreviate it S.H.E.:

$$S.H.E. \parallel Ag^+(aq, \mathcal{A} = 1) \mid Ag(s)$$

Bear in mind that a standard reduction potential is really a potential *difference* between the standard potential of the reaction of interest and the potential of the S.H.E. half-reaction, which we have arbitrarily set to zero.

If we wanted to measure the standard potential of the half-reaction

$$Cd^{2+} + 2e^- \rightleftharpoons Cd(s) \tag{14-12}$$

we would construct the cell

$$S.H.E. \parallel Cd^{2+}(aq, \mathcal{A} = 1) \mid Cd(s)$$

with the cadmium half-cell connected to the positive terminal of the potentiometer. In this case, we would observe a *negative* cell voltage of -0.402 V. The negative sign means that electrons flow from Cd to Pt, a direction opposite to that of the cell in Figure 14-6.

Appendix H contains standard reduction potentials for many different half-reactions, arranged alphabetically by element. If the half-reactions were arranged according to descending value of $E°$ (as they are in Table 14-1), we would find the strongest oxidizing agents at the upper left and the strongest reducing agents at the lower right. If we connected the two half-cells represented by Reactions 14-10 and 14-12, Ag^+ would be reduced to $Ag(s)$ as $Cd(s)$ is oxidized to Cd^{2+}.

We will always attach the left-hand electrode to the negative terminal of the potentiometer and the right-hand electrode to the positive terminal. The voltage read on the meter is the difference:

voltage

= right-hand electrode potential

− left-hand electrode potential

Challenge Draw a picture of the cell S.H.E. $\parallel$ $Cd^{2+}(aq, \mathcal{A} = 1) \mid Cd(s)$ and show the direction of electron flow.

Question The potential for the reaction $K^+ + e^- \rightleftharpoons K(s)$ is -2.936 V. This means that K^+ is a very poor oxidizing agent. (It does not readily accept electrons.) Does this imply that K^+ is therefore a good reducing agent?

Answer: No! To be a good reducing agent, K^+ would have to give up electrons easily (forming K^{2+}), which it cannot do.

TABLE 14-1   Ordered redox potentials

Oxidizing agent	Reducing agent	$E°$ (V)
$F_2(g) + 2e^- \rightleftharpoons 2F^-$		2.890
$O_3(g) + 2H^+ + 2e^- \rightleftharpoons O_2(g) + H_2O$		2.075
$MnO_4^- + 8H^+ + 5e^- \rightleftharpoons Mn^{2+} + 4H_2O$		1.507
$Ag^+ + e^- \rightleftharpoons Ag(s)$		0.799
$Cu^{2+} + 2e^- \rightleftharpoons Cu(s)$		0.339
$2H^+ + 2e^- \rightleftharpoons H_2(g)$		0.000
$Cd^{2+} + 2e^- \rightleftharpoons Cd(s)$		−0.402
$K^+ + e^- \rightleftharpoons K(s)$		−2.936
$Li^+ + e^- \rightleftharpoons Li(s)$		−3.040

Oxidizing power increases →

Reducing power increases →

A reaction is spontaneous if ΔG is negative and E is positive. $\Delta G°$ and $E°$ refer to the free energy change and potential when the activities of reactants and products are unity.

$$\Delta G° = -nFE°.$$

Challenge Show that Le Châtelier's principle requires a negative sign in front of the reaction quotient term in the Nernst equation. *Hint:* The more favorable a reaction, the more positive is E.

Because we are dealing with a half-reaction, this Q applies to the half-reaction, not to a net cell reaction.

Appendix A tells how to convert ln to log.

14-4 The Nernst Equation

For reactions run under standard conditions (unit activities), the more positive the value of $E°$, the greater is the driving force. Le Châtelier's principle tells us that increasing reactant concentrations drives a reaction to the right and increasing the product concentrations drives a reaction to the left. The net driving force for a reaction is expressed by the **Nernst equation,** whose two terms include the driving force under standard conditions ($E°$) and a term that shows the dependence on concentrations.

Nernst Equation for a Half-Reaction

For the half-reaction

$$aA + ne^- \rightleftharpoons bB$$

the Nernst equation giving the half-cell potential, E, is

Nernst equation:
$$E = E° - \frac{RT}{nF} \ln \frac{\mathcal{A}_B^b}{\mathcal{A}_A^a} \qquad (14\text{-}13)$$

where

$E°$ = standard reduction potential ($\mathcal{A}_A = \mathcal{A}_B = 1$)

R = gas constant [8.314 510 J/(K · mol) = 8.314 510 (V · C)/(K · mol)]

T = temperature (K)

n = number of electrons in the half-reaction

F = Faraday constant (9.648 530 9 × 10^4 C/mol)

$\mathcal{A}_i$ = activity of species i

The logarithmic term in the Nernst equation is the **reaction quotient, Q.**

$$Q = \frac{\mathcal{A}_B^b}{\mathcal{A}_A^a} \qquad (14\text{-}14)$$

Q has the same form as the equilibrium constant, but the activities need not have their equilibrium values. Pure solids, pure liquids, and solvents are omitted from Q because their activities are unity (or close to unity); concentrations of solutes are expressed as moles per liter and concentrations of gases are expressed as pressures in atmospheres. When all activities are unity, $Q = 1$ and $\ln Q = 0$, thus giving $E = E°$.

Converting the natural logarithm in Equation 14-13 to the base 10 logarithm, and inserting $T = 298.15$ K (25.00°C) gives the most useful form of the Nernst equation:

Nernst equation at 25°C:
$$E = E° - \frac{0.059\,16 \text{ V}}{n} \log \frac{\mathcal{A}_B^b}{\mathcal{A}_A^a} \qquad (14\text{-}15)$$

The potential changes by $59.16/n$ mV for each factor-of-10 change in the value of Q.

EXAMPLE Writing the Nernst Equation for a Half-Reaction

Let's write the Nernst equation for the reduction of white phosphorus to phosphine gas:

$$\frac{1}{4} P_4(s, \text{white}) + 3H^+ + 3e^- \rightleftharpoons PH_3(g) \qquad E° = -0.046 \text{ V}$$

Solution We omit solids from the reaction quotient, and the concentration of a gas is expressed as the pressure of the gas. Therefore, the Nernst equation is

$$E = -0.046 - \frac{0.059\,16}{3} \log \frac{P_{PH_3}}{[H^+]^3}$$

. . . .

. . . .

EXAMPLE Multiplication of a Half-Reaction

If we multiply a half-reaction by any factor, the value of $E°$ does not change (Box 14-1). However, the factor n before the log term and the form of the reaction quotient, Q, do change. Lets's write the Nernst equation for the reaction in the previous example, multiplied by 2:

$$\frac{1}{2} P_4(s, \text{white}) + 6H^+ + 6e^- \rightleftharpoons 2PH_3(g) \qquad E° = -0.046 \text{ V}$$

Solution

$$E = -0.046 - \frac{0.059\,16}{6} \log \frac{P^2_{PH_3}}{[H^+]^6}$$

Even though this Nernst equation does not look like the one in the previous example, Box 14-1 shows that the numerical value of E is unchanged. The squared term in the reaction quotient cancels the doubled value of n in front of the log term.

. . . .

Nernst Equation for a Complete Reaction

Consider the experiment in Figure 14-5 in which the negative terminal of the potentiometer is connected to the Cd electrode and the positive terminal is connected to the Ag electrode. The voltage, E, is the difference between the potentials of the two electrodes:

Nernst equation for a complete cell: $\qquad E = E_+ - E_- \qquad$ (14-16)

where E_+ is the potential of the electrode attached to the positive input terminal of the potentiometer and E_- is the potential of the electrode attached to the negative terminal. The potential of each half-reaction (*written as a reduction*) is governed by a Nernst equation like Equation 14-13, and the voltage for the complete reaction is the difference between the two half-cell potentials.

Here is a procedure for writing a net cell reaction and finding its voltage:

Step 1. Write *reduction* half-reactions for both half-cells and find $E°$ for each in Appendix H. Multiply the half-reactions as necessary so that they each contain the same number of electrons. When you multiply a reaction by any number, you *do not* multiply $E°$.

Step 2.   Write a Nernst equation for the half-reaction in the right half-cell, which is attached to the positive terminal of the potentiometer. This is E_+.

Step 3.   Write a Nernst equation for the half-reaction in the left half-cell, which is attached to the negative terminal of the potentiometer. This is E_-.

Step 4.   Find the net cell voltage by subtraction: $E = E_+ - E_-$.

Step 5.   To write a balanced net cell reaction, subtract the left half-reaction from the right half-reaction. (This is equivalent to reversing the left-half reaction and adding.)

If the net cell voltage, $E (= E_+ - E_-)$, is positive, then the net cell reaction is spontaneous in the forward direction. If the net cell voltage is negative, then the reaction is spontaneous in the reverse direction.

- -

EXAMPLE   **Nernst Equation for a Complete Reaction**

Find the voltage of the cell in Figure 14-5 if the right half-cell contains 0.50 M $AgNO_3(aq)$ and the left half-cell contains 0.010 M $CdCl_2(aq)$. Write the net cell reaction and state whether it is spontaneous in the forward or reverse direction.

Solution

Step 1.   right electrode: $2Ag^+ + 2e^- \rightleftharpoons 2Ag(s)$ $E_+° = 0.799$ V

left electrode: $Cd^{2+} + 2e^- \rightleftharpoons Cd(s)$ $E_-° = -0.402$ V

Step 2.   Nernst equation for right electrode:

Pure solids, pure liquids, and solvents are omitted from Q.

$$E_+ = E_+° - \frac{0.059\,16}{2} \log \frac{1}{[Ag^+]^2} = 0.799 - \frac{0.059\,16}{2} \log \frac{1}{(0.50)^2} = 0.781 \text{ V}$$

Step 3.   Nernst equation for left electrode:

$$E_- = E_-° - \frac{0.059\,16}{2} \log \frac{1}{[Cd^{2+}]} = -0.402 - \frac{0.059\,16}{2} \log \frac{1}{0.010} = -0.461 \text{ V}$$

Step 4.   cell voltage: $E = E_+ - E_- = 0.781 - (-0.461) = +1.242$ V

Step 5.   net cell reaction:

Subtracting a reaction is the same as reversing the reaction and adding.

$$\begin{array}{r} 2Ag^+ + 2e^- \rightleftharpoons 2Ag(s) \\ - \quad Cd^{2+} + 2e^- \rightleftharpoons Cd(s) \\ \hline Cd(s) + 2Ag^+ \rightleftharpoons Cd^{2+} + 2Ag(s) \end{array}$$

Because the voltage is positive, the net reaction is spontaneous in the forward direction. $Cd(s)$ is oxidized and Ag^+ is reduced. Electrons flow from the left-hand electrode to the right-hand electrode.

- -

Box 14-1 $E°$ and the Cell Voltage Do Not Depend on How You Write the Cell Reaction

Multiplying a half-reaction by any number does not change the standard reduction potential, $E°$. The reason can be seen in Equation 14-3. The potential difference between two points is the work done *per coulomb of charge* carried through that potential difference ($E = \text{work}/q$). The work per coulomb is the same whether 0.1, 2.3, or 10^4 coulombs have been transferred. The total work is different in each case, but the work per coulomb is constant. Therefore, we do not double $E°$ if we multiply a half-reaction by 2.

Multiplying a half-reaction by any number does not change the half-cell potential, E. To see this, consider a silver half-cell reaction written with either one or two electrons:

$$Ag^+ + e^- \rightleftharpoons Ag(s) \qquad E = E° - 0.059\,16\log\frac{1}{[Ag^+]}$$

$$2Ag^+ + 2e^- \rightleftharpoons 2Ag(s) \qquad E = E° - \frac{0.059\,16}{2}\log\frac{1}{[Ag^+]^2}$$

These two expressions are equal because $\log a^b = b\log a$:

$$E° - \frac{0.059\,16}{2}\log\frac{1}{[Ag^+]^2} = E° - \frac{\cancel{2}\cdot 0.059\,16}{\cancel{2}}\log\frac{1}{[Ag^+]} = E° - 0.059\,16\log\frac{1}{[Ag^+]}$$

The exponent inside the log term is always canceled by the factor $1/n$ that precedes the log term. The cell potential cannot depend on how you write the reaction.

What if you had written the Nernst equation for the right half-cell with just one electron instead of two?

$$Ag^+ + e^- \rightleftharpoons Ag(s)$$

Would the net cell voltage be different from what we calculated? It better not be, because the chemistry is still the same. Box 14-1 shows that *neither $E°$ nor E depend on how we write the reaction.* Box 14-2 shows how to derive standard reduction potentials for half-reactions that are the sum of other half-reactions.

Different Descriptions of the Same Reaction

In Figure 14-3 the right half-reaction can be written

$$AgCl(s) + e^- \rightleftharpoons Ag(s) + Cl^- \qquad E_+° = 0.222 \text{ V} \qquad (14\text{-}17)$$

$$E_+ = E_+° - 0.059\,16\,\log[Cl^-]$$

$$= 0.222 - 0.059\,16\log(0.033\,4) = 0.309_3 \text{ V} \qquad (14\text{-}18)$$

The Cl^- concentration in the silver half-reaction was derived from 0.016 7 M $CdCl_2(aq)$.

Suppose that a different, less handsome, author had written this book and had chosen to describe the half-reaction differently:

$$Ag^+ + e^- \rightleftharpoons Ag(s) \qquad E_+° = 0.799 \text{ V} \qquad (14\text{-}19)$$

This description is just as valid as the previous one. In both cases, Ag(I) is being reduced to Ag(0).

Box 14-2 Latimer Diagrams

A **Latimer diagram** summarizes the standard reduction potential ($E°$) connecting various oxidation states of an element. For example, in acid solution the following standard reduction potentials are observed:

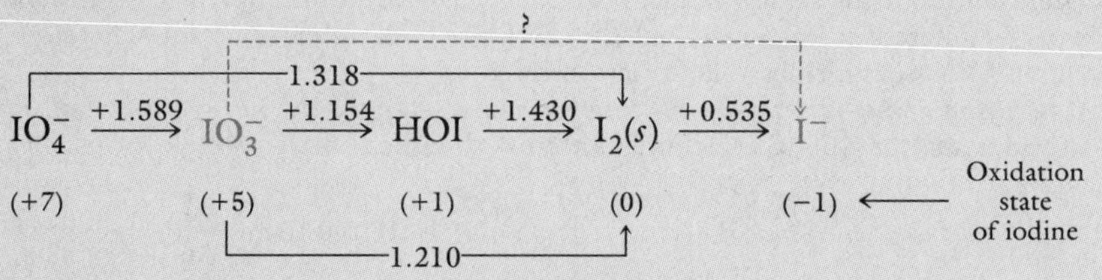

As an example of what each arrow means, let's write the balanced equation represented by the arrow connecting IO_3^- and HOI:

$$IO_3^- \xrightarrow{1.154} HOI$$

means

$$IO_3^- + 5H^+ + 4e^- \rightleftharpoons HOI + 2H_2O \qquad E° = +1.154 \text{ V}$$

It is possible to derive reduction potentials for arrows that are not shown in the diagram. Suppose we wish to determine $E°$ for the reaction shown by the dashed line in the Latimer diagram, which is

$$IO_3^- + 6H^+ + 6e^- \rightleftharpoons I^- + 3H_2O$$

We can use a thermodynamic cycle to find $E°$ for this reaction by expressing the desired reaction as a sum of reactions whose potentials are known.

The standard free energy change, $\Delta G°$, for a reaction is given by

$$\Delta G° = -nFE°$$

When two reactions are added to give a third reaction, the sum of the individual $\Delta G°$ values must equal the overall value of $\Delta G°$.

To apply free energy to the problem above, we write two reactions whose sum is the desired reaction:

$$IO_3^- + 6H^+ + 5e^- \xrightarrow{E_1° = 1.210} \tfrac{1}{2}I_2(s) + 3H_2O \qquad \Delta G_1° = -5F(1.210)$$
$$\tfrac{1}{2}I_2(s) + e^- \xrightarrow{E_2° = 0.535} I^- \qquad \Delta G_2° = -1F(0.535)$$
$$\overline{IO_3^- + 6H^+ + 6e^- \xrightarrow{E_3° = ?} I^- + 3H_2O \qquad \Delta G_3° = -6FE_3°}$$

But because $\Delta G_1° + \Delta G_2° = \Delta G_3°$, we may solve for $E_3°$:

$$\Delta G_3° = \Delta G_1° + \Delta G_2°$$
$$-6FE_3° = -5F(1.210) - 1F(0.535)$$
$$E_3° = \frac{5(1.210) + 1(0.535)}{6} = 1.098 \text{ V}$$

If the two descriptions are equally valid, then they should predict the same voltage. The Nernst equation for Reaction 14-19 is

$$E_+ = 0.799 - 0.059\,16 \log \frac{1}{[Ag^+]}$$

To find the concentration of Ag^+, we must use the solubility product for AgCl. Because the cell contains 0.033 4 M Cl^- and solid AgCl, we can say

$$[Ag^+] = \frac{K_{sp} \text{ (for AgCl)}}{[Cl^-]} = \frac{1.8 \times 10^{-10}}{0.033\,4} = 5.4 \times 10^{-9} \text{ M}$$

$K_{sp} = [Ag^+][Cl^-]$

Putting this value into the Nernst equation above gives

$$E_+ = 0.799 - 0.059\,16 \log \frac{1}{5.4 \times 10^{-9}} = 0.309_9 \text{ V}$$

which differs from the value calculated in Equation 14-18 only because of the accuracy of K_{sp}. Equations 14-17 and 14-19 give the same voltage because they describe the same cell.

The cell voltage cannot depend on how we write the reaction!

Advice for Finding Relevant Half-Reactions

When you are faced with a cell drawing or a line diagram, the first step is to write reduction reactions for each half-cell. To do this, *look for an element in the cell in two oxidation states.* For the cell

$$Pb(s) \mid PbF_2(s) \mid F^-(aq) \parallel Cu^{2+}(aq) \mid Cu(s)$$

we see Pb in two oxidation states, as $Pb(s)$ and $PbF_2(s)$, and Cu in two oxidation states, as Cu^{2+} and $Cu(s)$. Thus, the half-reactions are

right half-cell: $Cu^{2+} + 2e^- \rightleftharpoons Cu(s)$

left half-cell: $PbF_2(s) + 2e^- \rightleftharpoons Pb(s) + 2F^-$ (14-20)

You might have chosen to write the Pb half-reaction as

left half-cell: $Pb^{2+} + 2e^- \rightleftharpoons Pb(s)$ (14-21)

because you know that if $PbF_2(s)$ is present, there must be some Pb^{2+} in the solution. As we saw in the AgCl | Ag example, Reactions 14-20 and 14-21 are equally valid descriptions of the cell, and each should predict the same cell voltage. The decision to use Reaction 14-20 or Reaction 14-21 depends on whether the F^- or Pb^{2+} concentration is more easily known to you.

We described the left half-cell in terms of a redox reaction involving Pb because Pb is the element that appears in two oxidation states. We would not write a reaction such as $F_2(g) + 2e^- \rightleftharpoons 2F^-$ because $F_2(g)$ is not shown in the line diagram of the cell.

How to figure out the half-cell reactions.

Don't invent species not shown in the cell. Use what is shown in the line diagram to select the half-reactions.

The Nernst Equation Is Used in Measuring Standard Reduction Potentials

The standard reduction potential would be observed if the half-cell of interest (with unit activities) were connected to a standard hydrogen electrode, as it is in Figure 14-6. However, it is nearly impossible to construct such a cell because we have no way to adjust concentrations and ionic strength to give unit activities. In reality, activities less than unity are used in each half-cell, and the Nernst equation is used to extract the value of $E°$ from the cell voltage.[2] In the hydrogen electrode, standard buffers with known pH (Table 15-3) are used to obtain known values of $\mathcal{A}_{H^+}$.

Problem 18 gives an example of the use of the Nernst equation to find $E°$.

14-5 $E°$ and the Equilibrium Constant

A galvanic cell produces electricity because the cell reaction is not at equilibrium. The potentiometer allows negligible current to flow (Box 14-3), so the concentrations in each half-cell remain unchanged. If we replaced the potentiometer with a wire, much more current would flow, and concentrations would change until the cell reached equilibrium. At that point, nothing would be driving the reaction, and E would be zero.

At equilibrium, E (not $E°$) $= 0$.

Now let's relate E for a whole cell to the reaction quotient, Q, for the net cell reaction. For the two half-reactions

$$\text{right electrode:} \quad a\text{A} + ne^- \rightleftharpoons c\text{C} \qquad E_+°$$

$$\text{left electrode:} \quad d\text{D} + ne^- \rightleftharpoons b\text{B} \qquad E_-°$$

the Nernst equation looks like this:

$$E = E_+ - E_- = E_+° - \frac{0.05916}{n}\log\frac{\mathcal{A}_C^c}{\mathcal{A}_A^a} - \left(E_-° - \frac{0.05916}{n}\log\frac{\mathcal{A}_B^b}{\mathcal{A}_D^d}\right)$$

$\log a + \log b = \log ab$

$$E = \underbrace{(E_+° - E_-°)}_{E°} - \frac{0.05916}{n}\log\underbrace{\frac{\mathcal{A}_C^c\mathcal{A}_D^d}{\mathcal{A}_A^a\mathcal{A}_B^b}}_{Q} = E° - \frac{0.05916}{n}\log Q \qquad (14\text{-}22)$$

Equation 14-22 is true at any time. In the special case when the cell is at equilibrium, $E = 0$ and $Q = K$, the equilibrium constant. Therefore, Equation 14-22 is transformed to these most important forms at equilibrium:

To go from Equation 14-23 to 14-24:

$$\frac{0.05916}{n}\log K = E°$$

$$\log K = \frac{nE°}{0.05916}$$

$$10^{\log K} = 10^{nE°/0.05916}$$

$$K = 10^{nE°/0.05916}$$

Finding $E°$ from K: $\qquad E° = \dfrac{0.05916}{n}\log K \quad \text{(at 25°C)} \qquad (14\text{-}23)$

or

Finding K from $E°$: $\qquad K = 10^{nE°/0.05916} \quad \text{(at 25°C)} \qquad (14\text{-}24)$

Equation 14-24 allows us to deduce the equilibrium constant from $E°$. Alternatively, we can find $E°$ from K with Equation 14-23.

EXAMPLE Using $E°$ to Find the Equilibrium Constant

Find the equilibrium constant for the reaction

$$\text{Cu}(s) + 2\text{Fe}^{3+} \rightleftharpoons 2\text{Fe}^{2+} + \text{Cu}^{2+}$$

Solution The reaction is divided into two half-reactions found in Appendix H:

We associate $E_-°$ with the half-reaction that must be *reversed* to get the desired net reaction.

$$2\text{Fe}^{3+} + 2e^- \rightleftharpoons 2\text{Fe}^{2+} \qquad E_+° = 0.771 \text{ V}$$

$$-\quad \text{Cu}^{2+} + 2e^- \rightleftharpoons \text{Cu}(s) \qquad E_-° = 0.339 \text{ V}$$

$$\overline{\text{Cu}(s) + 2\text{Fe}^{3+} \rightleftharpoons 2\text{Fe}^{2+} + \text{Cu}^{2+}}$$

We find $E°$ for the net reaction

$$E° = E_+° - E_-° = 0.771 - 0.339 = 0.432 \text{ V}$$

Box 14-3 Concentrations in the Operating Cell

Doesn't operation of a cell change the concentrations in the cell? Yes, but cell voltage is measured under conditions of *negligible current flow*. For example, the resistance of a high-quality pH meter is $10^{13}\ \Omega$. If you use this meter to measure a potential of 1 V, the current is

$$I = \frac{E}{R} = \frac{1\ \text{V}}{10^{13}\ \Omega} = 10^{-13}\ \text{A}$$

If the cell in Figure 14-5 produces 50 mV, the current through the circuit is $0.050\ \text{V}/10^{13}\ \Omega$ $= 5 \times 10^{-15}$ A. This value corresponds to a flow of

$$\frac{5 \times 10^{-15}\ \text{C/s}}{9.649 \times 10^4\ \text{C/mol}} = 5 \times 10^{-20}\ \text{mol e}^-/\text{s}$$

The rate at which Cd^{2+} is produced is only 2.5×10^{-20} mol/s, which will have a negligible effect on the cadmium concentration in the cell. *The purpose of the potentiometer is to measure the voltage of the cell without affecting the concentrations in the cell.*

If the salt bridge were left in a real cell for very long, the concentrations and ionic strength would change because of diffusion between each compartment and the salt bridge. We assume that the cells are set up for a short enough time that mixing does not happen.

---

and compute the equilibrium constant with Equation 14-24:

$$K = 10^{(2)(0.432)/(0.059\,16)} = 4 \times 10^{14}$$

Note that a modest value of $E°$ produces a very large equilibrium constant. The value of K is correctly expressed with one significant figure, because $E°$ has three digits. Two are used for the exponent (14), and only one is left for the multiplier (4).

Significant figures for logs and exponents were discussed in Section 3-2.

· ·

Finding *K* for Net Reactions That Are Not Redox Reactions

Consider the following half-reactions whose difference gives the solubility equation for ferrous carbonate:

$$\begin{aligned} \text{FeCO}_3(s) + 2e^- &\rightleftharpoons \text{Fe}(s) + \text{CO}_3^{2-} & E°_+ &= -0.756\ \text{V} \\ \text{Fe}^{2+} + 2e^- &\rightleftharpoons \text{Fe}(s) & E°_- &= -0.44\ \text{V} \\ \hline \text{FeCO}_3(s) &\rightleftharpoons \text{Fe}^{2+} + \text{CO}_3^{2-} \quad (K = K_{sp}) & E° &= -0.756 - (-0.44) \\ & & &= -0.31_6\ \text{V} \end{aligned}$$

Ferrous
carbonate

$$K_{sp} = 10^{(2)(-0.31_6)/(0.059\,16)} = 10^{-11}$$

Potentiometric measurements provide one of the most useful means of measuring equilibrium constants that are too small or too large to measure by determining concentrations of reactants and products directly.

The general form of a problem involving the relation between $E°$ values for half-reactions and K for a net reaction is

$$\begin{aligned} \text{half-reaction:} \quad & E°_+ \\ \text{half-reaction:} \quad & E°_- \\ \hline \text{net reaction:} \quad & E° = E°_+ - E°_- \qquad K = 10^{nE°/0.059\,16} \end{aligned}$$

Any two of the three $E°$ values allow us to calculate the third value.

If you know $E°_-$ and $E°_+$, you can find $E°$ and K for the net cell reaction. Alternatively, if you know $E°$ and either $E°_-$ or $E°_+$, you can find the missing standard potential. If you know K, you can calculate $E°$ and use it to find either $E°_-$ or $E°_+$, provided you know one of them.

EXAMPLE Relating $E°$ and K

From the overall formation constant of $Ni(glycine)_2^{2+}$ plus the value of $E°$ for the $Ni^{2+} \mid Ni(s)$ couple,

$$Ni^{2+} + 2\,glycine \rightleftharpoons Ni(glycine)_2^{2+} \qquad K \equiv \beta_2 = 1.2 \times 10^{11}$$

$$Ni^{2+} + 2e^- \rightleftharpoons Ni(s) \qquad E° = -0.236\ V$$

deduce the value of $E°$ for the reaction

$$Ni(glycine)_2^{2+} + 2e^- \rightleftharpoons Ni(s) + 2\,glycine \qquad (14\text{-}25)$$

Solution We need to see the relation among the three reactions:

$$Ni^{2+} + 2e^- \rightleftharpoons Ni(s) \qquad\qquad E°_+ = -0.236\ V$$

$$\underline{-\quad Ni(glycine)_2^{2+} + 2e^- \rightleftharpoons Ni(s) + 2\,glycine \qquad E°_- = ?}$$

$$Ni^{2+} + 2\,glycine \rightleftharpoons Ni(glycine)_2^{2+} \qquad E° = ? \quad K = 1.2 \times 10^{11}$$

We know that $E°_+ - E°_-$ must equal $E°$, so we can deduce the value of $E°_-$ if we can find $E°$. But $E°$ can be determined from the equilibrium constant for the net reaction.

$$E° = \frac{0.059\,16}{n}\log K = \frac{0.059\,16}{2}\log(1.2 \times 10^{11}) = 0.328\ V$$

Hence, the standard reduction potential for half-reaction 14-25 is

$$E°_- = E°_+ - E° = -0.564\ V$$

14-6 Cells as Chemical Probes

It is essential to distinguish two classes of equilibria associated with galvanic cells:

1. Equilibrium *between* the two half-cells

2. Equilibrium *within* each half-cell

If a galvanic cell has a nonzero voltage, then the net cell reaction is not at equilibrium. We say that equilibrium *between* the two half-cells has not been established.

A chemical reaction that can occur *within one half-cell* will reach equilibrium and is assumed to remain at equilibrium. Such a reaction is not the net cell reaction.

However, *we do allow the half-cells to stand long enough to come to chemical equilibrium **within** each half-cell.* For example, in the right-hand half-cell in Figure 14-7, the reaction

$$AgCl(s) \rightleftharpoons Ag^+(aq) + Cl^-(aq)$$

comes to equilibrium with or without the presence of another half-cell. It is

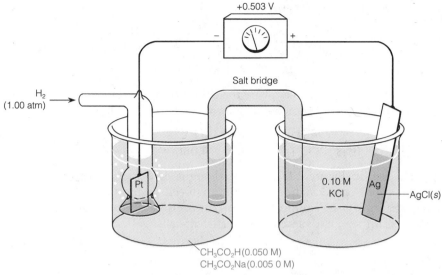

+0.503 V

Salt bridge

H_2
(1.00 atm)

Pt

0.10 M
KCl

Ag

AgCl(s)

CH_3CO_2H(0.050 M)
CH_3CO_2Na(0.005 0 M)

$Pt(s)|H_2(1.00\ atm)|CH_3CO_2H(0.050\ M),\ CH_3CO_2Na(0.005\ 0\ M)\ \|\ Cl^-(0.10\ M)|AgCl(s)|Ag(s)$

Aha! A buffer!

Figure 14-7 This galvanic cell can be used to measure the pH of the left half-cell.

not part of the net cell reaction. It is simply a chemical reaction whose equilibrium will be established when $AgCl(s)$ is in contact with an aqueous solution. In the left half-cell, the reaction

$$CH_3CO_2H \rightleftharpoons CH_3CO_2^- + H^+$$

has also come to equilibrium. Neither of these is a redox reaction involved in the net cell reaction.

The reaction for the right half-cell of Figure 14-7 is

$$AgCl(s) + e^- \rightleftharpoons Ag(s) + Cl^-(aq,\ 0.10\ M) \qquad E_+^\circ = 0.222\ V$$

But what is the reaction in the left half-cell? The only element we find in two oxidation states is hydrogen. We see that $H_2(g)$ bubbles into the cell, and we also realize that every aqueous solution contains H^+. Therefore, hydrogen is present in two oxidation states, and the half-reaction can be written as

$$2H^+(aq,\ ?\ M) + 2e^- \rightleftharpoons H_2(g,\ 1.00\ atm) \qquad E_-^\circ = 0$$

The net cell reaction is not at equilibrium, because the measured voltage is 0.503 V.

By doubling the silver half-reaction so that it contains $2e^-$, we can combine the log terms in the Nernst equation for the net cell reaction:

$$E = E_+ - E_- = \left(0.222 - \frac{0.059\ 16}{2}\log[Cl^-]^2\right) - \left(0 - \frac{0.059\ 16}{2}\log\frac{P_{H_2}}{[H^+]^2}\right)$$

$$E = 0.222 - \frac{0.059\ 16}{2}\log\underbrace{\frac{[H^+]^2[Cl^-]^2}{P_{H_2}}}_{\substack{\text{Reaction quotient for}\\\text{net cell reaction}}}$$

When we put in all of the quantities that are known, we discover that the only unknown is $[H^+]$. *The measured voltage therefore allows us to find the* concentration of H^+ in the left half-cell:

$$0.503 = 0.222 - \frac{0.059\,16}{2} \log \frac{[H^+]^2(0.10)^2}{(1.00)}$$

$$\Rightarrow [H^+] = 1.8 \times 10^{-4}\ M$$

This, in turn, allows us to evaluate the equilibrium constant for the acid-base reaction that has come to equilibrium in the left half-cell:

Question Why can we assume that the concentrations of acetic acid and acetate ion are equal to their initial (formal) concentrations?

$$K_a = \frac{[CH_3CO_2^-][H^+]}{[CH_3CO_2H]} = \frac{(0.005\,0)(1.8 \times 10^{-4})}{0.050} = 1.8 \times 10^{-5}$$

The cell in Figure 14-7 may be thought of as a *probe* to measure the unknown H^+ concentration in the left half-cell. Using this type of cell, we could determine the equilibrium constant for dissociation of any acid or the hydrolysis of any base placed in the left half-cell.

Survival Tips

The problems in this chapter include some brainbusters designed to bring together your knowledge of electrochemistry, chemical equilibrium, solubility, complex formation, and acid-base chemistry. They require you to find the equilibrium constant for a reaction that occurs in only one half-cell. The reaction of interest is not the net cell reaction and is not even a redox reaction. A good approach is outlined below:

The half-reactions that you write *must* involve species that appear in two oxidation states in the cell.

Step 1.   Write the two half-reactions and their standard potentials. If you choose a half-reaction for which you cannot find $E°$, then find another way to write the reaction.

Step 2.   Write a Nernst equation for the net reaction and put in all the known quantities. If all is well, there will be only one unknown in the equation.

Step 3.   Solve for the unknown concentration and use that concentration to solve the chemical equilibrium problem that was originally posed.

EXAMPLE   Analyzing a Very Complicated Cell

The cell in Figure 14-8 measures the formation constant (K_f) of $Hg(EDTA)^{2-}$. The solution in the right-hand compartment contains 0.500 mmol of Hg^{2+} and 2.00 mmol of EDTA in a volume of 0.100 L buffered to pH 6.00. If the voltage is $+0.331$ V, find the value of K_f for $Hg(EDTA)^{2-}$.

Solution

Step 1.   The left half-cell is a standard hydrogen electrode for which we can say $E_- = 0$. In the right half-cell, mercury is the element in two oxidation states. So let's write the half-reaction as

$$Hg^{2+} + 2e^- \rightleftharpoons 2Hg(l) \qquad E_+° = 0.852\ V$$

$$E_+ = 0.852 - \frac{0.059\,16}{2} \log \frac{1}{[Hg^{2+}]}$$

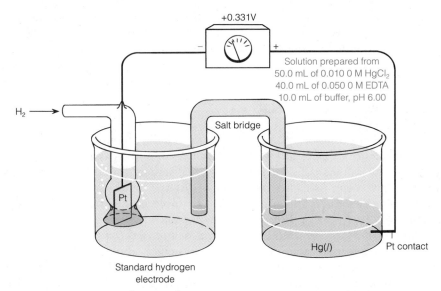

+0.331V

Solution prepared from
50.0 mL of 0.010 0 M $HgCl_2$
40.0 mL of 0.050 0 M EDTA
10.0 mL of buffer, pH 6.00

H_2

Salt bridge

Pt

Hg(*l*)   Pt contact

Standard hydrogen
electrode

S.H.E. $\|$ Hg(EDTA)$^{2-}$(*aq*, 0.005 00 M), EDTA(*aq*, 0.015 0 M) $|$ Hg(*l*)

Figure 14-8 A galvanic cell that can be used to measure the formation constant for $Hg(EDTA)^{2-}$.

In the right half-cell, the reaction between Hg^{2+} and EDTA is

$$Hg^{2+} + Y^{4-} \overset{K_f}{\rightleftharpoons} HgY^{2-}$$

Because we expect K_f to be large, we will assume that virtually all the Hg^{2+} has reacted to make HgY^{2-}. Therefore, the concentration of HgY^{2-} is 0.500 mmol/100 mL = 0.005 00 M. The remaining EDTA has a total concentration of (2.00 − 0.50) mmol/100 mL = 0.015 0 M. The right-hand compartment therefore contains 0.005 00 M HgY^{2-}, 0.015 0 M EDTA, and a small, unknown concentration of Hg^{2+}.

The formation constant for HgY^{2-} can be written

$$K_f = \frac{[HgY^{2-}]}{[Hg^{2+}][Y^{4-}]} = \frac{[HgY^{2-}]}{[Hg^{2+}]\alpha_{Y^{4-}}[EDTA]}$$

Recall that $[Y^{4-}] = \alpha_{Y^{4-}}[EDTA]$.

where [EDTA] is the formal concentration of EDTA not bound to metal. In this cell, [EDTA] = 0.015 0 M. The fraction of EDTA in the form Y^{4-} is $\alpha_{Y^{4-}}$, a parameter introduced in Section 13-2. Because we know that $[HgY^{2-}] = 0.005\,00$ M, all we need to find is $[Hg^{2+}]$ in order to evaluate K_f.

Step 2. The Nernst equation for the net cell reaction is

$$E = 0.331 = E_+ - E_- = \left(0.852 - \frac{0.059\,16}{2}\log\frac{1}{[Hg^{2+}]}\right) - (0)$$

in which the only unknown is $[Hg^{2+}]$.

Step 3. Now we solve the Nernst equation to find $[Hg^{2+}] = 2.4 \times 10^{-18}$ M, and this value of $[Hg^{2+}]$ allows us to evaluate the formation constant for HgY^{2-}:

$$K_f = \frac{[HgY^{2-}]}{[Hg^{2+}]\alpha_{Y^{4-}}[EDTA]} = \frac{(0.005\,00)}{(2.4 \times 10^{-18})(2.3 \times 10^{-5})(0.015\,0)}$$

$$= 6 \times 10^{21}$$

The value of $\alpha_{Y^{4-}}$ comes from Table 13-1.

The mixture of EDTA plus $Hg(EDTA)^{2-}$ in the cathode serves as a mercuric ion "buffer" that fixes the concentration of Hg^{2+}. This, in turn, determines the cell voltage.

14-7 Biochemists Use $E°'$

Perhaps the most important redox reactions in living organisms are involved in the respiratory process, in which molecules of food are oxidized by O_2 to yield energy or metabolic intermediates. The standard reduction potentials that we have been using so far apply to systems in which all activities of reactants and products are unity. If H^+ is involved in the reaction, $E°$ applies when pH = 0 ($\mathcal{A}_{H^+} = 1$). *Whenever H^+ appears in a redox reaction, or whenever reactants or products are acids or bases, reduction potentials are pH dependent.*

Because the pH inside a cell is in the neighborhood of 7, reduction potentials that apply at pH 0 are not particularly appropriate. For example, at pH 0, ascorbic acid (vitamin C) is a more powerful reducing agent than succinic acid is. However, at pH 7, this order is reversed. It is the reducing strength at pH 7, not at pH 0, that is relevant to a living cell.

The *standard potential* for a redox reaction is defined for a galvanic cell in which all activities are unity. The **formal potential** is the reduction potential that applies under a *specified* set of conditions (including pH, ionic strength, concentration of complexing agents, etc.). Biochemists call the formal potential at pH 7 $E°'$ (read "E zero prime"). Table 14-2 lists $E°'$ values for various biological redox couples.

> The formal potential at pH = 7 is called $E°'$.

Relation Between $E°$ and $E°'$

Consider the half-reaction

$$aA + ne^- \rightleftharpoons bB + mH^+ \qquad E°$$

in which A is an oxidized species and B is a reduced species. Both A and B might be acids or bases, as well. The Nernst equation for this half-reaction is

$$E = E° - \frac{0.059\,16}{n}\log\frac{[B]^b[H^+]^m}{[A]^a}$$

To find $E°'$, we must rearrange the Nernst equation to a form in which the log term contains only the *formal concentrations* of A and B raised to the powers a and b, respectively.

Recipe for $E°'$:
$$E = \underbrace{E° + \text{other terms}} - \frac{0.059\,16}{n}\log\frac{F_B^b}{F_A^a} \qquad (14\text{-}26)$$

All of this is called $E°'$ when pH = 7

The entire collection of terms over the brace, evaluated at pH = 7, is called $E°'$.

To convert [A] or [B] to F_A or F_B, we use fractional composition equations (Section 11-5), which relate the formal (i.e., total) concentration of *all* forms of an acid or a base to its concentration in a *particular* form:

> For a monoprotic acid:
> $$F = [HA] + [A^-]$$
> For a diprotic acid:
> $$F = [H_2A] + [HA^-] + [A^{2-}]$$

Monoprotic system:
$$[HA] = \alpha_{HA}F = \frac{[H^+]F}{[H^+] + K_a} \qquad (14\text{-}27)$$

TABLE 14-2   Reduction potentials of biological interest

Reaction	E° (V)	$E^{\circ\prime}$ (V)
$O_2 + 4H^+ + 4e^- \rightleftharpoons 2H_2O$	+1.229	+0.816
$Fe^{3+} + e^- \rightleftharpoons Fe^{2+}$	+0.771	+0.771
$I_2 + 2e^- \rightleftharpoons 2I^-$	+0.535	+0.535
Cytochrome a (Fe^{3+}) + $e^- \rightleftharpoons$ cytochrome a (Fe^{2+})	+0.290	+0.290
$O_2(g) + 2H^+ + 2e^- \rightleftharpoons H_2O_2$	+0.695	+0.281
Cytochrome c (Fe^{3+}) + $e^- \rightleftharpoons$ cytochrome c (Fe^{2+})	—	+0.254
2,6-Dichlorophenolindophenol + $2H^+$ + $2e^- \rightleftharpoons$ reduced 2,6-dichlorophenolindophenol	—	+0.22
Dehydroascorbate + $2H^+$ + $2e^- \rightleftharpoons$ ascorbate + H_2O	+0.390	+0.058
Fumarate + $2H^+$ + $2e^- \rightleftharpoons$ succinate	+0.433	+0.031
Methylene blue + $2H^+$ + $2e^- \rightleftharpoons$ reduced product	+0.532	+0.011
Glyoxylate + $2H^+$ + $2e^- \rightleftharpoons$ glycolate	—	−0.090
Oxaloacetate + $2H^+$ + $2e^- \rightleftharpoons$ malate	+0.330	−0.102
Pyruvate + $2H^+$ + $2e^- \rightleftharpoons$ lactate	+0.224	−0.190
Riboflavin + $2H^+$ + $2e^- \rightleftharpoons$ reduced riboflavin	—	−0.208
FAD + $2H^+$ + $2e^- \rightleftharpoons$ $FADH_2$	—	−0.219
(Glutathione-S)$_2$ + $2H^+$ + $2e^- \rightleftharpoons$ 2 glutathione-SH	—	−0.23
Safranine T + $2e^- \rightleftharpoons$ leucosafranine T	−0.235	−0.289
$(C_6H_5S)_2$ + $2H^+$ + $2e^- \rightleftharpoons 2C_6H_5SH$	—	−0.30
$NAD^+ + H^+ + 2e^- \rightleftharpoons$ NADH	−0.105	−0.320
$NADP^+ + H^+ + 2e^- \rightleftharpoons$ NADPH	—	−0.324
Cystine + $2H^+$ + $2e^- \rightleftharpoons$ 2 cysteine	—	−0.340
Acetoacetate + $2H^+$ + $2e^- \rightleftharpoons$ L-β-hydroxybutyrate	—	−0.346
Xanthine + $2H^+$ + $2e^- \rightleftharpoons$ hypoxanthine + H_2O	—	−0.371
$2H^+ + 2e^- \rightleftharpoons H_2$	0.000	−0.414
Gluconate + $2H^+$ + $2e^- \rightleftharpoons$ glucose + H_2O	—	−0.44
$SO_4^{2-} + 2e^- + 2H^+ \rightleftharpoons SO_3^{2-} + H_2O$	—	−0.454
$2SO_3^{2-} + 2e^- + 4H^+ \rightleftharpoons S_2O_4^{2-} + 2H_2O$	—	−0.527

$$[A^-] = \alpha_{A^-}F = \frac{K_a F}{[H^+] + K_a} \qquad (14\text{-}28)$$

Diprotic system: $\quad [H_2A] = \alpha_{H_2A}F = \dfrac{[H^+]^2 F}{[H^+]^2 + [H^+]K_1 + K_1K_2} \qquad (14\text{-}29)$

$$[HA^-] = \alpha_{HA^-}F = \frac{K_1[H^+]F}{[H^+]^2 + [H^+]K_1 + K_1K_2} \qquad (14\text{-}30)$$

$$[A^{2-}] = \alpha_{A^{2-}}F = \frac{K_1K_2 F}{[H^+]^2 + [H^+]K_1 + K_1K_2} \qquad (14\text{-}31)$$

where F is the formal concentration of HA or H_2A, K_a is the acid dissociation constant for HA, and K_1 and K_2 are the acid dissociation constants for H_2A.

EXAMPLE Finding the Formal Potential

Find $E^{\circ\prime}$ for the reaction[3]

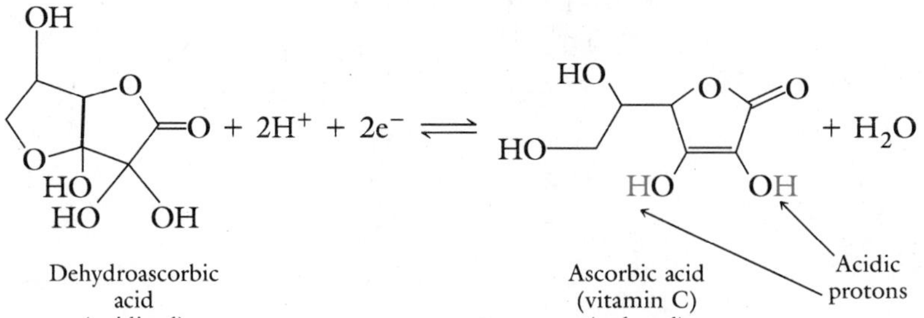

$$\text{[structure]} = O + 2H^+ + 2e^- \rightleftharpoons \text{[structure]} + H_2O \qquad E^\circ = 0.390 \text{ V} \qquad (14\text{-}32)$$

Dehydroascorbic
acid
(oxidized)

Ascorbic acid
(vitamin C)
(reduced)

Acidic
protons

$pK_1 = 4.10,\ pK_2 = 11.79$

Solution Abbreviating dehydroascorbic acid as D, and ascorbic acid as H_2A, we rewrite the reduction as

$$D + 2H^+ + 2e^- \rightleftharpoons H_2A + H_2O$$

for which the Nernst equation is

$$E = E^\circ - \frac{0.059\,16}{2}\log\frac{[H_2A]}{[D][H^+]^2} \qquad (14\text{-}33)$$

D is not an acid or a base, so its formal concentration equals its molar concentration: $F_D = [D]$. For the diprotic acid H_2A, we use Equation 14-29 to express $[H_2A]$ in terms of F_{H_2A}:

$$[H_2A] = \frac{[H^+]^2 F_{H_2A}}{[H^+]^2 + [H^+]K_1 + K_1K_2}$$

Putting these values into Equation 14-33 gives

$$E = E^\circ - \frac{0.059\,16}{2}\log\left(\frac{\dfrac{[H^+]^2 F_{H_2A}}{[H^+]^2 + [H^+]K_1 + K_1K_2}}{F_D[H^+]^2}\right)$$

which can be rearranged to the form

$$E = \underbrace{E^\circ - \frac{0.059\,16}{2}\log\frac{1}{[H^+]^2 + [H^+]K_1 + K_1K_2}}_{\substack{\text{Formal potential } (= E^{\circ\prime} \text{ if pH} = 7) \\ = +0.062 \text{ V}}} - \frac{0.059\,16}{2}\log\frac{F_{H_2A}}{F_D} \qquad (14\text{-}34)$$

Putting the values of E°, K_1, and K_2 into Equation 14-34 and setting $[H^+] = 10^{-7.00}$, we find $E^{\circ\prime} = +0.062$ V.

Curve (a) in Figure 14-9 shows how the calculated formal potential for Reaction 14-32 depends on pH. The potential decreases as the pH increases, until pH $\approx pK_2$. Above pK_2, A^{2-} is the dominant form of ascorbic acid, and no protons are involved in the net redox reaction. Therefore, the potential becomes independent of pH.

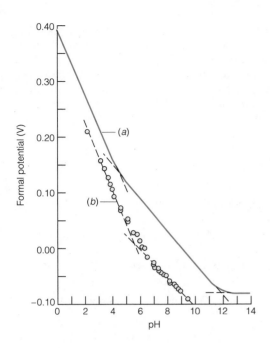

Figure 14-9 Formal reduction potential of ascorbic acid, showing its dependence on pH. (a) Graph of the function labeled formal potential in Equation 14-34. (b) Experimental polarographic half-wave reduction potential of ascorbic acid in a medium of ionic strength = 0.2 M. The half-wave potential, discussed in Chapter 18, is nearly the same as the formal potential. At high pH (>12), the half-wave potential does not level off to a slope of zero, as Equation 14-34 predicts. Instead, a hydrolysis reaction of ascorbic acid occurs and the chemistry is more complex than Reaction 14-32. [J. J. Ruiz, A. Aldaz, and M. Dominguez, *Can. J. Chem.* **1977,** *55,* 2799; *ibid.* **1978,** *56,* 1533.]

Terms to Understand

ampere	formal potential	oxidation	resistance
anode	galvanic cell	oxidizing agent	salt bridge
cathode	half-reaction	potentiometer	standard hydrogen electrode
coulomb	joule	power	standard reduction potential
current	Latimer diagram	reaction quotient	volt
$E°'$	Nernst equation	redox reaction	watt
electric potential	ohm	reducing agent	
electrode	Ohm's law	reductant	
Faraday constant	oxidant	reduction	

Summary

The work done when a charge of q coulombs passes through a potential difference of E volts is work = $E \cdot q$. The maximum work that can be done on the surroundings by a spontaneous chemical reaction is related to the free energy change for the reaction: work = $-\Delta G$. If the chemical change produces a potential difference, $E,$ the relation between free energy and the potential difference is $\Delta G = -nFE$. Ohm's law ($I = E/R$) describes the relation among current, voltage, and resistance in an electric circuit. It can be combined with the definitions of work and power (P = work per second) to give $P = E \cdot I = I^2 R$.

A galvanic cell uses a spontaneous redox reaction to produce electricity. The electrode at which oxidation occurs is the anode and the electrode at which reduction occurs is the cathode. The two half-cells are usually separated by a salt bridge that allows ions to migrate from one side to the other to maintain charge neutrality but prevents reactants in the two half-cells from mixing. The standard reduction potential of a half-reaction is measured by pairing that half-reaction with a standard hydrogen electrode. The term *standard* means that activities of reactants and products are unity. If several half-reactions are added to give another half-reaction, the standard potential of the net half-reaction can be found by equating the free energy of the net half-reaction to the sum of free energies of the component half-reactions.

The voltage for a complete reaction is the difference between the potentials of the two half-reactions: $E = E_+ - E_-$, where E_+ is the potential of the half-cell connected to the positive terminal of the potentiometer and E_- is the potential of the half-cell connected to the negative terminal. The potential of each half-reaction is given by the Nernst equation: $E = E° - (0.059\,16/n)\log Q$ (at 25°C), where Q is the reaction quotient. The reaction quotient has the same form as the equilibrium constant, but it is evaluated with concentrations existing at the time of interest.

Complex equilibria can be studied by making them part of an electrochemical cell. If we measure the voltage and know the concentrations (activities) of all but one of the reactants and products, the Nernst equation allows us to compute the concentration of the unknown species. The electrochemical cell serves as a probe for that species.

Biochemists use the formal potential of a half-reaction at pH 7 ($E^{\circ\prime}$) instead of the standard potential (E°), which applies at pH 0. $E^{\circ\prime}$ is found by writing the Nernst equation for the desired half-reaction and grouping together all terms except the logarithm containing the formal concentrations of reactant and product. The combination of terms, evaluated at pH 7, is $E^{\circ\prime}$.

Exercises

A. A mercury cell used to power heart pacemakers runs on the following reaction:

$$Zn(s) + HgO(s) \rightarrow ZnO(s) + Hg(l) \qquad E^{\circ} = 1.35 \text{ V}$$

If the power required to operate the pacemaker is 0.0100 W, how many kilograms of HgO (FW 216.59) will be consumed in 365 days? How many pounds of HgO is this? (1 pound = 453.6 g)

B. Calculate E° and K for each of the following reactions.

(a) $I_2(s) + 5Br_2(aq) + 6H_2O \rightleftharpoons$
$$2IO_3^- + 10Br^- + 12H^+$$

(b) $Cr^{2+} + Fe(s) \rightleftharpoons Fe^{2+} + Cr(s)$

(c) $Mg(s) + Cl_2(g) \rightleftharpoons Mg^{2+} + 2Cl^-$

(d) $5MnO_2(s) + 4H^+ \rightleftharpoons 3Mn^{2+} + 2MnO_4^- + 2H_2O$

(e) $Ag^+ + 2S_2O_3^{2-} \rightleftharpoons Ag(S_2O_3)_2^{3-}$

(f) $CuI(s) \rightleftharpoons Cu^+ + I^-$

C. Calculate the voltage of each of the following cells.

(a) $Fe(s) \mid FeBr_2(0.010 \text{ M}) \parallel NaBr(0.050 \text{ M}) \mid Br_2(l) \mid Pt(s)$

(b) $Cu(s) \mid Cu(NO_3)_2(0.020 \text{ M}) \parallel Fe(NO_3)_2(0.050 \text{ M}) \mid Fe(s)$

(c) $Hg(l) \mid Hg_2Cl_2(s) \mid KCl(0.060 \text{ M}) \parallel KCl(0.040 \text{ M}) \mid Cl_2(g, 0.50 \text{ atm}) \mid Pt(s)$

D. Consider the cell shown at the right. The reaction in the left half-cell can be written in *either* of two ways:

$$AgI(s) + e^- \rightleftharpoons Ag(s) + I^- \qquad (1)$$

or

$$Ag^+ + e^- \rightleftharpoons Ag(s) \qquad (2)$$

The right half-cell reaction is

$$H^+ + e^- \rightleftharpoons \tfrac{1}{2}H_2(g) \qquad (3)$$

(a) Using Reactions 2 and 3, calculate E° and write the Nernst equation for the cell.

(b) Use the value of K_{sp} for AgI to compute $[Ag^+]$ and find the cell voltage.

(c) Suppose, instead, that you wish to describe the cell with Reactions 1 and 3. We know that the cell voltage (E, not E°) must be the same, no matter which description we use. Write the Nernst equation for Reactions 1 and 3 and use it to find E° in Reaction 1. Compare your answer with the value in Appendix H.

E. Calculate the voltage of the cell

$$Cu(s) \mid Cu^{2+}(0.030 \text{ M}) \parallel K^+Ag(CN)_2^-(0.010 \text{ M}),$$
$$HCN(0.10 \text{ F}), \text{ buffer to pH } 8.21 \mid Ag(s)$$

You may wish to refer to the following reactions:

$$Ag(CN)_2^- + e^- \rightleftharpoons Ag(s) + 2CN^- \qquad E^{\circ} = -0.310 \text{ V}$$

$$HCN \rightleftharpoons H^+ + CN^- \qquad pK_a = 9.21$$

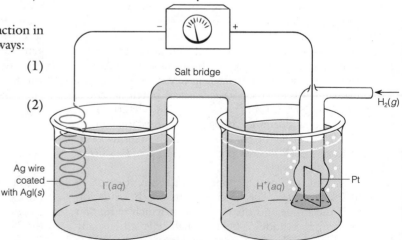

For Exercise D.

$Ag(s)|AgI(s)|NaI\ (0.10 \text{ M}) \parallel HCl\ (0.10 \text{ M})|H_2(g, 0.20 \text{ atm})|Pt(s)$

F. **(a)** Write a balanced equation for the reaction PuO_2^+ $\rightarrow Pu^{4+}$ and calculate $E°$ for the reaction.

$$PuO_2^{2+} \xrightarrow{+0.966} PuO_2^+ \xrightarrow{\ ?\ } Pu^{4+} \xrightarrow{+1.006} Pu^{3+}$$
$$\underset{1.021}{\underline{\hspace{8cm}}}$$

(b) Predict whether or not an equimolar mixture of PuO_2^{2+} and PuO_2^+ will oxidize H_2O to O_2 at a pH of 2.00. You may assume that $P_{O_2} = 0.20$ atm. Will O_2 be liberated at pH 7.00?

G. Calculate the voltage of the cell below, in which KHP is potassium hydrogen phthalate, the monopotassium salt of phthalic acid.

$$Hg(l) \mid Hg_2Cl_2(s) \mid KCl(0.10\ M) \parallel KHP(0.050\ M) \mid$$
$$H_2(g,\ 1.00\ atm) \mid Pt(s)$$

H. The cell below has a voltage of -0.321 V:

$$Hg(l) \mid Hg(NO_3)_2(0.0010\ M),\ KI(0.010\ M) \parallel S.H.E.$$

From this voltage, calculate the equilibrium constant for the reaction

$$Hg^{2+} + 4I^- \rightleftharpoons HgI_4^{2-}$$

Assume that the only forms of mercury in solution are Hg^{2+} and HgI_4^{2-}

I. The formation constant for $Cu(EDTA)^{2-}$ is 6.3×10^{18}, and the value of $E°$ for the reaction $Cu^{2+} + 2e^- \rightleftharpoons Cu(s)$ is $+0.339$ V. From this information, find $E°$ for the reaction

$$CuY^{2-} + 2e^- \rightleftharpoons Cu(s) + Y^{4-}$$

J. Using the reaction below, state which compound, $H_2(g)$ or glucose, is the more powerful reducing agent at pH = 0.00.

$$
\begin{array}{c}
\begin{array}{l}
CO_2H \\
\mid \\
HCOH \\
\mid \\
HOCH \\
\mid \\
HCOH \\
\mid \\
HCOH \\
\mid \\
CH_2OH \\
\\
\text{Gluconic acid} \\
pK_a = 3.56
\end{array}
\quad + 2H^+ + 2e^- \rightleftharpoons \quad
\begin{array}{l}
CHO \\
\mid \\
HCOH \\
\mid \\
HOCH \\
\mid \\
HCOH \\
\mid \\
HCOH \\
\mid \\
CH_2OH \\
\\
\text{Glucose} \\
\text{(no acidic protons)}
\end{array}
\quad + H_2O
\end{array}
$$

$$E°' = -0.45\ V$$

K. Living cells convert energy derived from sunlight or combustion of food into energy-rich ATP (adenosine triphosphate) molecules. For ATP synthesis, $\Delta G = +34.5$ kJ/mol. This energy is then made available to the cell when ATP is hydrolyzed to ADP (adenosine diphosphate). In animals, ATP is synthesized when protons pass through a complex enzyme in the mitochondrial membrane.[4] Two factors account for the movement of protons through this enzyme into the mitochondrion (see the figure). One factor is the gradient of concentration of H^+, higher outside the mitochondrion than inside. This gradient arises because protons are *pumped* out of the mitochondrion by enzymes involved in the oxidation of food molecules. A second factor is that the inside of the mitochondrion is negatively charged with respect to the outside. The synthesis of one ATP molecule requires two protons to pass through the phosphorylation enzyme.

(a) The difference in free energy when a molecule travels from a region of high activity to a region of low activity is given by

$$\Delta G = -RT\ln\frac{\mathscr{A}_{high}}{\mathscr{A}_{low}}$$

How big must the pH difference be (at 298 K) if the passage of two protons is to provide enough energy to synthesize one ATP molecule?

(b) pH differences this large have not been observed in mitochondria. How great an electric potential difference between inside and outside is necessary for the movement of two protons to provide energy to synthesize ATP? In answering this question, neglect any contribution from the pH difference.

(c) It is thought that the energy for ATP synthesis is provided by *both* the pH difference and the electric potential. If the pH difference is 1.00 pH unit, what is the magnitude of the potential difference?

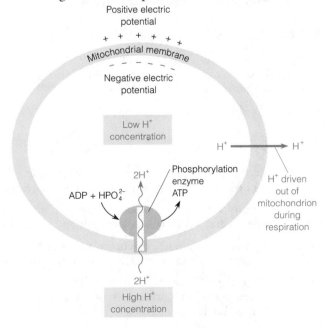

Problems

Basic Concepts

1. Explain the difference between electric charge (q, coulombs), electric current (I, amperes), and electric potential (E, volts).

2. (a) How many electrons are in one coulomb?

(b) How many coulombs are in one mole of charge?

3. The basal rate of consumption of O_2 by a 70-kg human is about 16 mol of O_2 per day. This O_2 oxidizes food and is reduced to H_2O, providing energy for the organism:

$$O_2 + 4H^+ + 4e^- \rightleftharpoons 2H_2O$$

(a) To what current (in amperes = C/s) does this respiration rate correspond? (Current is defined by the flow of electrons from food to O_2.)

(b) Compare your answer in (a) with the current drawn by a refrigerator using 5.00×10^2 W at 115 V. Remember that power (in watts) = work/s = $E \cdot I$.

(c) If the electrons flow from nicotinamide adenine dinucleotide (NADH) to O_2, they experience a potential drop of 1.1 V. What is the power output (in watts) of our human friend?

4. A 6.00-V battery is connected across a 2.00-kΩ resistor.

(a) How many electrons per second flow through the circuit?

(b) How many joules of heat are produced for each electron?

(c) If the circuit operates for 30.0 min, how many moles of electrons will have flowed through the resistor?

(d) What voltage would the battery need to deliver for the power to be 1.00×10^2 W?

5. Consider the redox reaction

$$I_2 + 2S_2O_3^{2-} \rightleftharpoons 2I^- + S_4O_6^{2-}$$
$$\text{Thiosulfate} \qquad\qquad \text{Tetrathionate}$$

(a) Identify the oxidizing agent on the left side of the reaction and write a balanced oxidation half-reaction.

(b) Identify the reducing agent on the left side of the reaction and write a balanced reduction half-reaction.

(c) How many coulombs of charge are passed from reductant to oxidant when 1.00 g of thiosulfate reacts?

(d) If the rate of reaction is 1.00 g of thiosulfate consumed per minute, what current (in amperes) flows from reductant to oxidant?

6. The space shuttle's expendable booster engines derive their power from solid reactants:

$$6NH_4ClO_4(s) + 10Al(s) \rightarrow$$
$$\text{FW 117.49}$$
$$3N_2(g) + 9H_2O(g) + 5Al_2O_3(s) + 6HCl(g)$$

(a) Find the oxidation numbers of the elements N, Cl, and Al in reactants and products. Which reactants act as reducing agents and which act as oxidants?

(b) The heat of reaction is $-9\,334$ kJ for every 10 mol of Al consumed. Express this as heat released per gram of total reactants.

Galvanic Cells

7. For each picture, a and b, at the top of the next page, write the line notation to describe the cell. Write an oxidation reaction for the left electrode and a reduction reaction for the right electrode.

8. (a) Draw a picture of the following cell, showing the location of each species:

$$Pt(s) \mid Fe^{3+}(aq), Fe^{2+}(aq) \parallel$$
$$Cr_2O_7^{2-}(aq), Cr^{3+}(aq), HA(aq) \mid Pt(s)$$

(b) Write an oxidation for the left half-cell and a reduction for the right half-cell.

(c) Write a balanced equation for the net cell reaction.

9. A lightweight rechargeable battery uses the following cell:

$$Zn(s) \mid ZnCl_2(aq) \parallel Cl^-(aq) \mid Cl_2(l) \mid C(s)$$

(a) Write an oxidation for the left half-cell and a reduction for the right half-cell.

(b) If the battery delivers a constant current of 1.00×10^3 A for 1.00 h, how many kilograms of Cl_2 will be consumed?

Standard Potentials

10. Which will be the strongest oxidizing agent under standard conditions (all activities = 1): HNO_2, Se, UO_2^{2+}, Cl_2, H_2SO_3, or MnO_2?

11. Use Le Châtelier's principle and half-reactions from Appendix H to find which of the following become stronger oxidizing agents as the pH is lowered. Which are unchanged, and which become weaker?

Cl_2	$Cr_2O_7^{2-}$	Fe^{3+}	MnO_4^-	IO_3^-
Chlorine	Dichromate	Ferric	Permanganate	Iodate

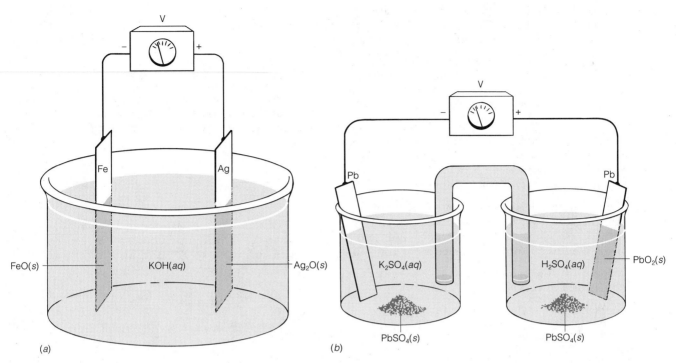

For Problem 7.

12. **(a)** In the presence of cyanide ion, the reduction potential of Fe(III) is decreased from 0.771 to 0.356 V.

$$Fe^{3+} + e^- \rightleftharpoons Fe^{2+} \qquad E° = 0.771 \text{ V}$$
Ferric   Ferrous

$$Fe(CN)_6^{3-} + e^- \rightleftharpoons Fe(CN)_6^{4-} \qquad E° = 0.356 \text{ V}$$
Ferricyanide   Ferrocyanide

Which ion, Fe^{3+} or Fe^{2+}, is stabilized more by complexing with CN^-?

(b) Using Appendix H, answer the same question when the ligand is phenanthroline instead of cyanide.

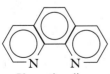

Phenanthronline

Nernst Equation

13. **(a)** Use the Nernst equation to write the spontaneous chemical reaction that occurs in the cell in Demonstration 14-1.

(b) If you use your fingers as a salt bridge in Demonstration 14-1, will your body take in Cu^{2+} or Zn^{2+}?

14. Consider the half-reaction

$$As(s) + 3H^+ + 3e^- \rightleftharpoons AsH_3(g) \qquad E° = -0.238 \text{ V}$$
                          Arsine

(a) Write the Nernst equation for the half-reaction.

(b) Find E (not $E°$) when pH = 3.00 and $P_{AsH_3} = 1.00$ torr. (Remember that 1 atm = 760 torr.)

15. **(a)** Write the line notation for the cell below.

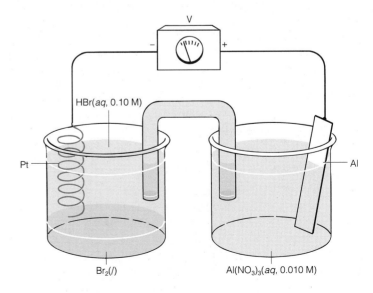

(b) Calculate the cell voltage (E, not $E°$), and state the direction in which electrons will flow through the potentiometer. Write the spontaneous net cell reaction.

(c) The left half-cell was loaded with 14.3 mL of $Br_2(l)$ (density = 3.12 g/mL). The aluminum electrode contains 12.0 g of Al. Which element, Br_2 or Al, is the limiting reagent in this cell? (That is, which reagent will be used up first?)

(d) If the cell is somehow operated under conditions in which it produces a constant voltage of 1.50 V, how much electrical work will have been done when 0.231 mL of $Br_2(l)$ has been consumed?

(e) If the potentiometer is replaced by a 1.20-kΩ resistor, and if the heat dissipated by the resistor is 1.00×10^{-4} J/s, at what rate (grams per second) is Al(s) dissolving? (In this question the voltage is not 1.50 V.)

16. Suppose that the concentrations of NaF and KCl were each 0.10 M in the cell

$$Pb(s) \mid PbF_2(s) \mid F^-(aq) \parallel Cl^-(aq) \mid AgCl(s) \mid Ag(s)$$

(a) Using the half-reactions $2AgCl(s) + 2e^- \rightleftharpoons 2Ag(s) + 2Cl^-$ and $PbF_2(s) + 2e^- \rightleftharpoons Pb(s) + 2F^-$, calculate the cell voltage.

(b) Now calculate the cell voltage by using the reactions $2Ag^+ + 2e^- \rightleftharpoons 2Ag(s)$ and $Pb^{2+} + 2e^- \rightleftharpoons Pb(s)$. For this part, you will need the solubility products for PbF_2 and AgCl.

17. Use the numerical values of R and F to derive Equation 14-15 from Equation 14-13. What would be the value of the numerical constant in front of the log term at 0°C? At 37°C?

18. The following cell was set up to measure the standard reduction potential of the $Ag^+ \mid Ag$ couple:

$$Pt(s) \mid HCl(0.010\,00\ M),\ H_2(g) \parallel$$

$$AgNO_3(0.010\,00\ M) \mid Ag(s)$$

The temperature was 25°C (the standard condition) and atmospheric pressure was 751.0 torr. Because the vapor pressure of water is 23.8 torr at 25°C, P_{H_2} in the cell was 751.0 − 23.8 = 727.2 torr. The Nernst equation for the cell, including activity coefficients, is derived as follows:

right electrode: $\quad Ag^+ + e^- \rightleftharpoons Ag(s) \qquad E_+^° = E_{Ag^+|Ag}^°$

left electrode: $\quad H^+ + e^- \rightleftharpoons \frac{1}{2} H_2(g) \qquad E_-^° = 0$ V

$$E_+ = E_{Ag^+|Ag}^° - 0.059\,16\log\frac{1}{[Ag^+]\gamma_{Ag^+}}$$

$$E_- = 0 - 0.059\,16\log\frac{P_{H_2}^{1/2}}{[H^+]\gamma_{H^+}}$$

$$E = E_+ - E_- = E_{Ag^+|Ag}^° - 0.059\,16\log\frac{[H^+]\gamma_{H^+}}{P_{H_2}^{1/2}[Ag^+]\gamma_{Ag^+}}$$

Given a measured cell voltage of +0.798 3 V, and using activity coefficients from Table 8-1, find $E_{Ag^+|Ag}^°$.

19. Write a balanced chemical equation (in acidic solution) for the reaction represented by the question mark on the lower arrow. Calculate $E°$ for the reaction.

$$\overset{\displaystyle 1.441}{\overbrace{BrO_3^- \xrightarrow{\ 1.491\ } HOBr \underset{\underset{\textstyle ?}{\overbrace{}}}{\xrightarrow{\ 1.584\ }} Br_2(aq) \xrightarrow{\ 1.098\ } Br^-}}$$

20. What must be the relationship between $E_1^°$ and $E_2^°$ if the species X^+ is to disproportionate spontaneously under standard conditions to X^{3+} and X(s)? Write a balanced equation for the disproportionation.

$$X^{3+} \overset{E_1^°}{\rightarrow} X^+ \overset{E_2^°}{\rightarrow} X(s)$$

21. *Without neglecting activities,* calculate the voltage of the cell

$$Ni(s) \mid NiSO_4(0.020\ M) \parallel CuCl_2(0.030\ M) \mid Cu(s)$$

Relation of E° and the Equilibrium Constant

22. The free energy change for the reaction $CO + \frac{1}{2} O_2 \rightleftharpoons CO_2$ is $\Delta G° = -257$ kJ per mole of CO at 298 K.

(a) Find $E°$ for the reaction.

(b) Find the equilibrium constant for the reaction.

23. Calculate $E°$, $\Delta G°$, and K for each of the following reactions.

(a) $4Co^{3+} + 2H_2O \rightleftharpoons 4Co^{2+} + O_2(g) + 4H^+$

(b) $Ag(S_2O_3)_2^{3-} + Fe(CN)_6^{4-} \rightleftharpoons Ag(s) + 2S_2O_3^{2-} + Fe(CN)_6^{3-}$

24. A solution contains 0.100 M Ce^{3+}, 1.00×10^{-4} M Ce^{4+}, 1.00×10^{-4} M Mn^{2+}, 0.100 M MnO_4^-, and 1.00 M $HClO_4$.

(a) Write a balanced net reaction that can occur between the species in this solution.

(b) Calculate $\Delta G°$ and K for the reaction.

(c) Calculate E for the conditions given above.

(d) Calculate ΔG for the conditions given above.

(e) At what pH would the concentrations of Ce^{4+}, Ce^{3+}, Mn^{2+}, and MnO_4^- listed above be in equilibrium at 298 K?

25. For the following cell, E (not $E°$) = −0.289 V. Write the net cell reaction and calculate its equilibrium constant. Do not use $E°$ values from Appendix H to answer this question.

$$\text{Pt}(s) \mid \text{VO}^{2+}(0.116 \text{ M}), \text{V}^{3+}(0.116 \text{ M}), \text{H}^+(1.57 \text{ M})$$
$$\parallel \text{Sn}^{2+}(0.0318 \text{ M}), \text{Sn}^{4+}(0.0318 \text{ M}) \mid \text{Pt}(s)$$

26. Calculate the standard potential for the half-reaction

$$\text{Pd(OH)}_2(s) + 2e^- \rightleftharpoons \text{Pd}(s) + 2\text{OH}^-$$

given that K_{sp} for Pd(OH)_2 is 3×10^{-28} and given that the standard potential for the reaction $\text{Pd}^{2+} + 2e^- \rightleftharpoons \text{Pd}(s)$ is 0.915 V.

27. From the standard potentials for reduction of $\text{Br}_2(aq)$ and $\text{Br}_2(l)$ in Appendix H, calculate the solubility of Br_2 in water at 25°C. Express your answer as g/L.

28. Given the following information

$\text{FeY}^- + e^- \rightleftharpoons \text{Fe}^{2+} + \text{Y}^{4-}$	$E° = -0.730 \text{ V}$	
FeY^{2-}:	$K_f = 2.1 \times 10^{14}$	
FeY^-:	$K_f = 1.3 \times 10^{25}$	

calculate the standard potential for the reaction

$$\text{FeY}^- + e^- \rightleftharpoons \text{FeY}^{2-}$$

where Y is EDTA.

29. Use the right-hand column of Appendix H to find the value of $E°$ for the reaction $\text{Al}^{3+} + 3e^- \rightleftharpoons \text{Al}(s)$ at 50°C. Refer to the footnote at the beginning of Appendix H.

30. Using the reaction

$$\text{HPO}_4^{2-} + 2\text{H}^+ + 2e^- \rightleftharpoons \text{HPO}_3^{2-} + \text{H}_2\text{O}$$
$$E° = -0.234 \text{ V}$$

and acid dissociation constants from Appendix G, calculate $E°$ for the reaction

$$\text{H}_2\text{PO}_4^- + \text{H}^+ + 2e^- \rightleftharpoons \text{HPO}_3^{2-} + \text{H}_2\text{O}$$

31. This problem is slightly tricky. Calculate $E°$, $\Delta G°$, and K for the reaction

which is the sum of *three* half-reactions listed in Appendix H. Use $\Delta G° (= -nFE°)$ for each of the half-reactions to find $\Delta G°$ for the net reaction. Note that if you reverse the direction of a reaction, you reverse the sign of $\Delta G°$.

Using Cells as Chemical Probes

32. Using the cell in Figure 14-8 as an example, explain what we mean when we say that there is equilibrium *within* each half-cell but not necessarily *between* the two half-cells.

33. In this problem we will use a cell similar to that in Figure 14-8 as a probe to find the concentration of Cl^- in the right compartment. The cell is

$$\text{Pt}(s) \mid \text{H}_2(g, 1.00 \text{ atm}) \mid \text{H}^+(aq, \text{pH} = 3.60) \parallel$$
$$\text{Cl}^-(aq, x \text{ M}) \mid \text{AgCl}(s) \mid \text{Ag}(s)$$

(a) Write reactions for each half-cell, a balanced net cell reaction, and the Nernst equation for the net cell reaction.

(b) Given a measured cell voltage of 0.485 V, find $[\text{Cl}^-]$ in the right compartment.

34. The quinhydrone electrode was introduced in 1921 as a means of measuring pH.[5]

$$\text{Pt}(s) \mid 1:1 \text{ mole ratio of quinone } (aq) \text{ and}$$
$$\text{hydroquinone}(aq), \text{ unknown pH} \parallel$$
$$\text{Cl}^-(aq, 0.50 \text{ M}) \mid \text{Hg}_2\text{Cl}_2(s) \mid \text{Hg}(l) \mid \text{Pt}(s)$$

The solution whose pH is to be measured is placed in the left half-cell, which also contains a 1:1 mole ratio of quinone and hydroquinone. The half-cell reaction is

(a) Write reactions for the right half-cell and for the whole cell. Write the Nernst equation for the whole cell.

(b) Ignoring activities and using the relation $\text{pH} = -\log[\text{H}^+]$, the Nernst equation can be changed into the form

$$E(\text{cell}) = A + B \cdot \text{pH}$$

where A and B are constants. Find the numerical values of A and B at 25°C.

(c) If the pH were 4.50, in which direction would electrons flow through the potentiometer.

35. The voltage for the cell below is 0.490 V. Find K_b for the organic base RNH_2.

$$\text{Pt}(s) \mid \text{H}_2(1.00 \text{ atm}) \mid \text{RNH}_2(aq, 0.10 \text{ M}),$$
$$\text{RNH}_3^+\text{Cl}^-(aq, 0.050 \text{ M}) \parallel \text{S.H.E.}$$

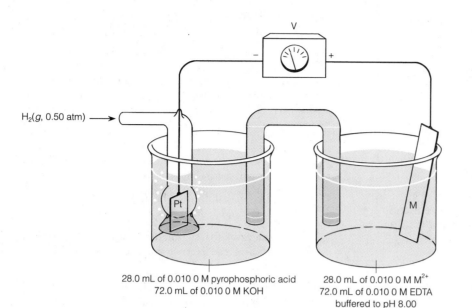

For Problem 36.

28.0 mL of 0.010 0 M pyrophosphoric acid
72.0 mL of 0.010 0 M KOH

28.0 mL of 0.010 0 M M^{2+}
72.0 mL of 0.010 0 M EDTA
buffered to pH 8.00

36. The voltage of the cell above is −0.246 V. The right half-cell contains the metal ion, M^{2+}, whose standard reduction potential is −0.266 V.

$$M^{2+} + 2e^- \rightleftharpoons M(s) \qquad E° = -0.266 \text{ V}$$

Calculate K_f for the metal-EDTA complex.

37. The following cell was constructed to find the difference in K_{sp} between two naturally occurring forms of $CaCO_3(s)$, called *calcite* and *aragonite*.[6]

$$Pb(s) \mid PbCO_3(s) \mid CaCO_3(s, \text{calcite}) \mid \text{aqueous}$$
$$\text{buffer(pH 7.00)} \parallel \text{aqueous buffer(pH 7.00)} \mid CaCO_3(s,$$
$$\text{aragonite}) \mid PbCO_3(s) \mid Pb(s)$$

Each compartment of the cell contains a mixture of solid $PbCO_3$ ($K_{sp} = 7.4 \times 10^{-14}$) and either calcite or aragonite, both of which have $K_{sp} \approx 5 \times 10^{-9}$. Each solution was buffered to pH 7.00 with an inert buffer, and the cell was completely isolated from atmospheric CO_2. The measured cell voltage was −1.8 mV. Find the ratio of solubility products for calcite and aragonite.

$$\frac{K_{sp} \text{ (for calcite)}}{K_{sp} \text{ (for aragonite)}} = ?$$

38. *Do not ignore activity coefficients in this problem.* If the voltage for the following cell is 0.489 V, find K_{sp} for $Cu(IO_3)_2$.

$$Ni(s) \mid NiSO_4(0.025 \text{ M}) \parallel$$
$$KIO_3(0.10 \text{ M}) \mid Cu(IO_3)_2(s) \mid Cu(s)$$

Biochemists Use $E°'$

39. Explain what $E°'$ is and why it is preferred over $E°$ in biochemistry.

40. In this problem we will find $E°'$ for the reaction $C_2H_2(g) + 2H^+ + 2e^- \rightleftharpoons C_2H_4(g)$.

(a) Write the Nernst equation for the half-reaction, using $E°$ from Appendix H.

(b) Rearrange the Nernst equation to the form

$$E = E° + \text{other terms} - \frac{0.059\,16}{2} \log \frac{P_{C_2H_4}}{P_{C_2H_2}}$$

(c) The quantity ($E°$ + other terms) is $E°'$. Evaluate $E°'$ for pH = 7.00.

41. Evaluate $E°'$ for the half-reaction $(CN)_2(g) + 2H^+ + 2e^- \rightleftharpoons 2HCN(aq)$.

42. Calculate $E°'$ for the reaction

$$H_2C_2O_4 + 2H^+ + 2e^- \rightleftharpoons 2HCO_2H \qquad E° = 0.204 \text{ V}$$

43. Suppose that HOx is a monoprotic acid with dissociation constant 1.4×10^{-5} and H_2Red^- is a diprotic acid with dissociation constants of 3.6×10^{-4} and 8.1×10^{-8}. Calculate $E°$ for the reaction

$$HOx + e^- \rightleftharpoons H_2Red^- \qquad E°' = 0.062 \text{ V}$$

44. Given the information below, find K_a for nitrous acid, HNO_2.

$$NO_3^- + 3H^+ + 2e^- \rightleftharpoons HNO_2 + H_2O$$

$$E° = 0.940 \text{ V} \quad E°' = 0.433 \text{ V}$$

45. The oxidized form (Ox) of a flavoprotein that functions as a one-electron reducing agent has a molar absorptivity (ϵ) of $1.12 \times 10^4 \text{ M}^{-1} \cdot \text{cm}^{-1}$ at 457 nm at pH = 7.00. For the reduced form (Red), $\epsilon = 3.82 \times 10^3$ at 457 nm at pH 7.00.

$$Ox + e^- \rightleftharpoons Red \qquad E^{\circ\prime} = -0.128 \text{ V}$$

The substrate (S) is the molecule reduced by the protein.

$$Red + S \rightleftharpoons Ox + S^-$$

Both S and S^- are colorless. A solution at pH 7.00 was prepared by mixing enough protein plus substrate (Red + S) so that the initial concentrations of Red and S are each 5.70×10^{-5} M. The absorbance at 457 mm was 0.500 in a 1.00-cm cell.

(a) Calculate the concentrations of Ox and Red from the absorbance data.

(b) Calculate the concentrations of S and S^-.

(c) Calculate the value of $E^{\circ\prime}$ for the reaction $S + e^- \rightleftharpoons S^-$.

- -

Notes and References

1. L. P. Silverman and B. B. Bunn, *J. Chem. Ed.* **1992,** *69,* 309.

2. A student experiment using the Nernst equation to measure E° has been described by A. Arévalo and G. Pastor, *J. Chem. Ed.* **1985,** *62,* 882.

3. The chemistry of ascorbic acid is discussed by D. T. Sawyer, G. Chiericato, Jr., and T. Tsuchiya, *J. Am. Chem. Soc.* **1982,** *104,* 6273.

4. See "How Cells Make ATP" by P. C. Hinkle and R. E. McCarty, *Scientific American* (March 1978), p. 104. An article describing how gradients of pH and electric potential across biological membranes are measured is that by D. F. Wilson and N. G. Forman, *Biochem.* **1982,** *21,* 1438.

5. For a history of the quinhydrone electrode and its inventor, E. C. S. Biilmann, see *J. Chem. Ed.* **1989,** *66,* 910.

6. The cell in this problem would not give an accurate result because of the *junction potential* at each liquid junction (Section 15-3). A clever way around this problem, using a cell without any liquid junctions, is described by P. A. Rock, *J. Chem. Ed.* **1975,** *52,* 787.

A Heparin Sensor

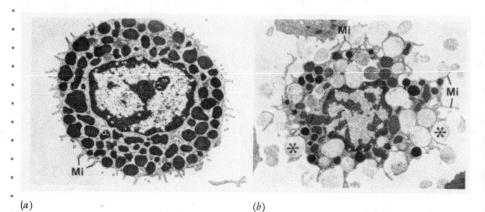

(a)   (b)

(*a*) Mast cell with numerous dark granules containing regulatory molecules such as heparin. (*b*) Granules expelled from the mast cell release heparin into a blood vessel. Two of the granules in the process of exiting the cell are denoted by asterisks. (Mi indicates details of a cell's surface called microvilli and microfolds.) [From R. G. Kessel and R. H. Kardon, *Tissues and Organs* (San Francisco: W. H. Freeman, 1978), p. 14; D. Lagunoff, *J. Invest. Dermatol.* **1972,** *58,* 296.]

Blood clotting must be precisely regulated in the human body to minimize *hemorrhage* (bleeding) in an injury and *thrombosis* (clotting) in uninjured vessels. The exquisitely complex clotting process is regulated by numerous substances, including *heparin,* a negatively charged molecule secreted by mast cells near the walls of blood vessels.

The concentration of heparin administered during surgery to prevent clotting must be monitored to prevent uncontrolled bleeding. Until recently, only clotting time could be measured; there was no direct, rapid determination of heparin at physiologic levels. Therefore, an ion-selective electrode described in this chapter was designed to respond to heparin. In this chapter we will study the principles of operation of different types of electrodes that respond to particular analytes.

Electrodes and Potentiometry

15

Analytical chemists design electrodes whose potentials vary in response to changes in the concentration of a specific analyte in solution or in the gas phase. These instruments range from straightforward galvanic cells of the type we studied in the last chapter, to ion-selective electrodes that are about the size of your pen, to ion-sensing field effect transistors that are just hundreds of micrometers in size and can be inserted into a blood vessel. The measurement of cell voltages to extract chemical information is called **potentiometry.**

In the simplest case, the species of interest participates in the chemistry of a galvanic cell. Imagine a solution containing an **electroactive species** whose activity (concentration) we wish to measure. An electroactive species is one that can donate or accept electrons from an electrode. We can turn the unknown into a half-cell by inserting an electrode (such as a Pt wire) into the solution to transfer electrons to or from the species of interest. Because this electrode responds directly to the analyte, it is called the **indicator electrode.** We then connect this half-cell to a second half-cell via a salt bridge. The second half-cell should have a known, fixed composition so that it has a known, constant potential. Because the second half-cell has a constant potential, it is called a **reference electrode.** The cell voltage is the difference between the variable potential that reflects changes in the analyte activity and the constant potential from the reference electrode.

Indicator electrode: responds to analyte activity

Reference electrode: maintains a fixed (reference) potential

15-1 Reference Electrodes

Suppose you have a solution containing Fe^{2+} and Fe^{3+}. If you are clever, you can make this solution part of a cell in such a way that the voltage will tell you the relative concentrations of these species—that is, the value of

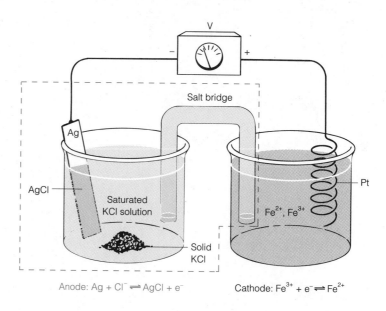

Figure 15-1 A galvanic cell that can be used to measure the quotient $[Fe^{2+}]/[Fe^{3+}]$ in the right half-cell. The Pt wire is the *indicator electrode,* and the entire left half-cell plus salt bridge (enclosed by the dashed line) can be considered to be a *reference electrode.*

Anode: $Ag + Cl^- \rightleftharpoons AgCl + e^-$ Cathode: $Fe^{3+} + e^- \rightleftharpoons Fe^{2+}$

$$Ag(s)\,|\,AgCl(s)\,|\,Cl^-(aq)\,\|\,Fe^{2+}(aq),\,Fe^{3+}(aq)\,|\,Pt(s)$$

Actually, the voltage tells us the quotient of activities, $\mathscr{A}_{Fe^{2+}}/\mathscr{A}_{Fe^{3+}}$. However, we will neglect activity coefficients and continue to write the Nernst equation in terms of concentrations instead of activities.

$[Fe^{2+}]/[Fe^{3+}]$. Figure 15-1 shows one way to do this. A Pt wire is inserted as an indicator electrode through which Fe^{3+} can receive electrons or Fe^{2+} can lose electrons. The left half-cell serves to complete the galvanic cell and has a known, constant potential.

The two half-reactions can be written as follows:

right electrode:	$Fe^{3+} + e^- \rightleftharpoons Fe^{2+}$	$E_+^\circ = 0.771$ V
left electrode:	$AgCl(s) + e^- \rightleftharpoons Ag(s) + Cl^-$	$E_-^\circ = 0.222$ V

The two electrode potentials are

E_+ is the potential of the electrode attached to the positive input of the potentiometer.
E_- is the potential of the electrode attached to the negative input of the potentiometer.

$$E_+ = 0.771 - 0.059\,16 \log \frac{[Fe^{2+}]}{[Fe^{3+}]}$$

$$E_- = 0.222 - 0.059\,16 \log\,[Cl^-]$$

and the cell voltage is just the difference between E_+ and E_-:

$$E = E_+ - E_-$$

$$= \left(0.771 - 0.059\,16 \log \frac{[Fe^{2+}]}{[Fe^{3+}]} \right) - \left(0.222 - 0.059\,16 \log[Cl^-] \right)$$

The concentration of Cl^- in the left half-cell is constant, fixed by the solubility of KCl, with which the solution is saturated. Therefore, the cell voltage changes only when the quotient $[Fe^{2+}]/[Fe^{3+}]$ changes.

The half-cell on the left in Figure 15-1 can be thought of as a *reference electrode.* We can picture the cell and salt bridge enclosed by the dashed line as a single unit dipped into the analyte solution, as shown in Figure 15-2. The Pt wire is the indicator electrode, whose potential responds to changes in the quotient $[Fe^{2+}]/[Fe^{3+}]$. The reference electrode completes the redox reaction and provides a *constant potential* to the left side of the potentiometer. Changes in the cell voltage can be unambiguously assigned to changes in the quotient $[Fe^{2+}]/[Fe^{3+}]$.

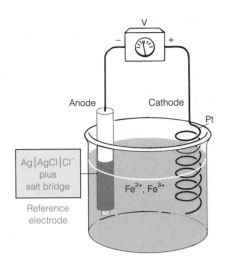

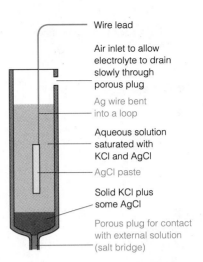

Figure 15-2 Another view of Figure 15-1. The contents of the colored box in Figure 15-1 are now considered to be a reference electrode dipped into the analyte solution.

Figure 15-3 Diagram of a silver-silver chloride reference electrode.

Silver-Silver Chloride Reference Electrode[1]

The reference electrode enclosed by the dashed line in Figure 15-1 is called a **silver-silver chloride electrode.** Figure 15-3 shows how the half-cell in Figure 15-1 is reconstructed as a thin, glass-enclosed electrode that can be dipped into the analyte solution, as shown in Figure 15-2.

The standard reduction potential for the AgCl | Ag couple is +0.222 V at 25°C. This would be the potential of a silver-silver chloride electrode if $\mathcal{A}_{Cl^-}$ were unity. But the activity of Cl^- in a saturated solution of KCl at 25°C is not unity, and the potential of the electrode in Figure 15-3 is found to be +0.197 V with respect to a standard hydrogen electrode at 25°C.

Ag | AgCl electrode: $\quad AgCl(s) + e^- \rightleftharpoons Ag(s) + Cl^- \quad E° = +0.222$ V

$$E(\text{saturated KCl}) = +0.197 \text{ V}$$

Figure 15-4 shows a versatile electrode that minimizes contact between the analyte solution and the KCl solution in the reference electrode. A problem with reference electrodes is that porous plugs become clogged, thereby causing sluggish, unstable electrical responses. Some designs incorporate a free-flow capillary in place of the porous plug. Other designs allow you to squirt fresh solution from the electrode through the electrode-analyte junction prior to a measurement.

Calomel Electrode

The **calomel electrode** in Figure 15-5 is based on the reaction

Calomel electrode: $\quad \frac{1}{2}Hg_2Cl_2(s) + e^- \rightleftharpoons Hg(l) + Cl^- \quad E° = +0.268$ V

Mercurous chloride (calomel)

$$E(\text{saturated KCl}) = +0.241 \text{ V}$$

The standard potential ($E°$) for this reaction is +0.268 V. If the cell is saturated with KCl at 25°C, the activity of Cl^- is such that the potential of the electrode is +0.241 V. The calomel electrode saturated with KCl is called the **saturated calomel electrode.** It is encountered so frequently that it is

The two most common reference electrodes are the silver-silver chloride electrode and the calomel electrode.

Figure 15-4 Double-junction silver-silver chloride reference electrode. The inner electrode is the same as the one in Figure 15-3. The outer compartment can contain KCl solution or any other solution that is more compatible with the analyte solution. For example, if you do not want Cl^- to contact the analyte, the outer electrode could be filled with KNO_3 solution. Because the inner and outer electrolyte solutions mix slowly, the outer electrode should be refilled occasionally with fresh KNO_3 solution. [Courtesy Fisher Scientific, Pittsburgh, PA.]

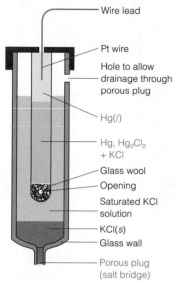

Figure 15-5 A saturated calomel electrode (S.C.E.).

abbreviated **S.C.E.** The advantage in using a saturated KCl solution is that the concentration of chloride does not change if some of the liquid evaporates.

Voltage Conversions Between Different Reference Scales

It is sometimes necessary to convert potentials from one reference scale to another. If an electrode has a potential of -0.461 V with respect to a calomel electrode, what is the potential with respect to a silver-silver chloride electrode? What would be the potential with respect to the standard hydrogen electrode?

To answer these questions, consider the following diagram, which shows the potentials of the calomel and silver-silver chloride electrodes on the standard hydrogen electrode potential scale:

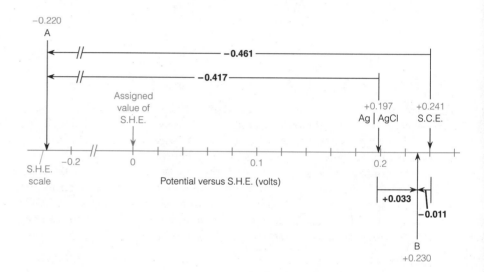

You can see that point A, which is -0.220 V on the S.H.E. scale, is -0.461 V from the S.C.E. potential and -0.417 V from the silver-silver chloride electrode potential. What about point B, whose potential is $+0.033$ V greater than the silver-silver chloride potential? Its position is -0.011 V from the S.C.E. potential and $+0.230$ V on the S.H.E. scale. By using the relations shown in this diagram, you can convert potentials from one scale to another.

15-2 Indicator Electrodes

The most common indicator electrode is made of platinum, which is relatively *inert*—it does not participate in many chemical reactions. Its purpose is simply to transmit electrons to or from a species in solution. Gold electrodes are even more inert than platinum. Various types of carbon also are used as indicator electrodes because the rates of many redox reactions on a carbon surface are fast. A metal electrode works best when its surface is large and clean. A brief dip in concentrated nitric acid, followed by rinsing with distilled water, is often an effective way to clean a metal electrode surface.

Figure 15-6 shows how a silver electrode can be used in conjunction with a calomel reference electrode to measure silver ion concentration (actually

Figure 15-6 Use of silver and calomel electrodes to measure the concentration of Ag^+ in a solution.

activity). The reaction at the silver indicator electrode is

$$Ag^+ + e^- \rightleftharpoons Ag(s) \qquad E_+^\circ = 0.799 \text{ V}$$

The reference cell reaction is

$$Hg_2Cl_2(s) + 2e^- \rightleftharpoons 2Hg(l) + 2Cl^- \qquad E_- = 0.241 \text{ V}$$

and the reference half-cell potential (E_-, not E_-°) is fixed at 0.241 V because the reference cell is saturated with KCl. The Nernst equation for the entire cell is therefore

$$E = E_+ - E_- = \underbrace{\left(0.799 - 0.059\,16 \log \frac{1}{[Ag^+]}\right)}_{\substack{\text{Potential of } Ag|Ag^+ \\ \text{indicator electrode}}} - \underset{\substack{\text{Potential of} \\ \text{S.C.E. reference electrode}}}{0.241}$$

$$E = 0.558 + 0.059\,16 \log[Ag^+] \qquad (15\text{-}1)$$

That is, the voltage of the cell in Figure 15-6 provides a direct measure of the concentration of Ag^+. Ideally, the voltage changes by 59.16 mV (at 25°C) for each factor-of-10 change in $[Ag^+]$.

In Figure 7-9 we used a silver indicator electrode and a *glass* reference electrode. The glass electrode responds to the pH of the solution, as will be discussed in Section 15-5. The cell in Figure 7-9 contains a buffer to maintain a constant pH. Therefore, the glass electrode remains at a constant potential; it is being used in an unconventional way as a reference electrode.

- -

EXAMPLE Potentiometric Precipitation Titration

A 100.0-mL solution containing 0.100 0 M NaCl was titrated with 0.100 0 M $AgNO_3$, and the voltage of the cell shown in Figure 15-6 was monitored. Calculate the voltage after the addition of 65.0, 100.0, and 103.0 mL of $AgNO_3$.

Solution The titration reaction is

$$Ag^+ + Cl^- \rightarrow AgCl(s)$$

for which V_e (the equivalence point) is 100.0 mL.

At 65.0 mL: 65.0% of the chloride has been precipitated and 35.0% remains in solution.

$$[Cl^-] = \underbrace{(0.350)}_{\substack{\text{Fraction} \\ \text{remaining}}} \underbrace{(0.100\,0 \text{ M})}_{\substack{\text{Original} \\ \text{concentration} \\ \text{of } Cl^-}} \underbrace{\left(\frac{100.0}{165.0}\right)}_{\substack{\text{Dilution} \\ \text{factor}}} = 0.021\,2 \text{ M}$$

where 100.0 is the Initial volume of Cl^- and 165.0 is the Total volume of solution.

To find the cell voltage in Equation 15-1, we need to know $[Ag^+]$:

$$[Ag^+] = \frac{K_{sp} \text{ (for AgCl)}}{[Cl^-]} = \frac{1.8 \times 10^{-10}}{0.021\,2} = 8.5 \times 10^{-9} \text{ M}$$

The cell voltage is therefore

$$E = 0.558 + 0.059\,16\log(8.5 \times 10^{-9}) = 0.081 \text{ V}$$

At 100.0 mL: This is the equivalence point, at which $[Ag^+] = [Cl^-]$.

$$[Ag^+][Cl^-] = [Ag^+]^2 = K_{sp}$$

$$[Ag^+] = \sqrt{K_{sp}} = 1.3_4 \times 10^{-5} \text{ M}$$

$$E = 0.558 + 0.059\,16\log(1.3_4 \times 10^{-5}) = 0.270 \text{ V}$$

At 103.0 mL: Now there is an excess of 3.0 mL of 0.100 0 M $AgNO_3$ in the solution. Therefore,

$$[Ag^+] = (0.100\,0 \text{ M}) \overbrace{\left(\frac{\overset{\text{Initial volume of } Ag^+}{3.0}}{203.0} \right)}^{} = 1.4_8 \times 10^{-3} \text{ M}$$

Initial concentration of Ag^+ — Dilution factor — Total volume of solution

and we can write

$$E = 0.588 + 0.059\,16 \log (1.4_8 \times 10^{-3}) = 0.391 \text{ V}$$

The cell responds to a change in Cl^- concentration because such a change necessarily changes the Ag^+ concentration: $[Ag^+][Cl^-] = K_{sp}$.

With a little imagination, we can say that a *silver electrode is also a halide electrode, if solid silver halide is present in the cell.* If the solution contains $AgCl(s)$, we can write

$$[Ag^+][Cl^-] = K_{sp} \Rightarrow [Ag^+] = \frac{K_{sp}}{[Cl^-]}$$

Putting this value of $[Ag^+]$ into Equation 15-1 gives

$$E = 0.558 + 0.059\,16 \log \frac{K_{sp}}{[Cl^-]}$$

That is, the voltage responds to changes in Cl^- concentration.

Some metals, such as Ag, Cu, Zn, Cd, and Hg, can be used as indicator electrodes for their aqueous ions. Most metals, however, are unsuitable for this purpose because the equilibrium $M^{n+} + ne^- \rightleftharpoons M$ is not readily established at the metal surface.

Demonstration 15-1 is a great example of indicator and reference electrodes. Box 15-1 describes how electrodes can be tailored for specific tasks.

15-3 What Is a Junction Potential?

$E_{observed} = E_{cell} + E_{junction}$

Because the junction potential is usually unknown, $E_{observed}$ is uncertain.

Any time two dissimilar electrolyte solutions are placed in contact, a voltage difference (called the **junction potential**) develops at their interface. This small, unknown voltage (usually a few millivolts) is found at each end of the salt bridge between two half-cells. *The junction potential puts a fundamental limitation on the accuracy of direct potentiometric measurements,* because usually we do not know the contribution of the junction to the measured voltage.

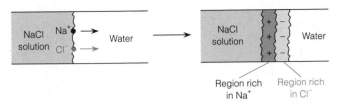

Figure 15-7 Development of the junction potential caused by unequal mobilities of Na^+ and Cl^-.

To see why the junction potential occurs, consider a solution containing NaCl in contact with distilled water (Figure 15-7). The Na^+ and Cl^- ions will begin to diffuse from the NaCl solution into the water phase. However, Cl^- ion has a greater **mobility** than Na^+. That is, Cl^- diffuses faster than Na^+. As a result, a region rich in Cl^-, with excess negative charge, develops at the front. Behind it is a positively charged region depleted of Cl^-. The result is an electric potential difference at the junction of the NaCl and H_2O phases. The junction potential opposes the movement of Cl^- and accelerates the movement of Na^+. The steady-state junction potential represents a balance between the unequal mobilities that create a charge imbalance and the tendency of the resulting charge imbalance to retard the movement of Cl^-.

Mobilities of several ions are shown in Table 15-1 and various liquid junction potentials are listed in Table 15-2. Because K^+ and Cl^- have similar mobilities, junction potentials at the two interfaces of a KCl salt bridge are slight. This is why saturated KCl is used in salt bridges.

. . . .

EXAMPLE Junction Potential

A 0.1-M NaCl solution was placed in contact with a 0.1 M $NaNO_3$ solution. Which side of the junction will be positive and which will be negative?

Solution   Because $[Na^+]$ is equal on both sides, there will be no net diffusion of Na^+ across the junction. However, Cl^- will diffuse into the $NaNO_3$, and NO_3^- will diffuse into the NaCl. Because the mobility of Cl^- is greater than that of NO_3^-, the NaCl region will be depleted of Cl^- faster than the $NaNO_3$ region will be depleted of NO_3^-. The $NaNO_3$ side will become negative, and the NaCl side will become positive.

. . . .

15-4   How Ion-Selective Electrodes Work

Ion-selective electrodes respond selectively to one species in a solution. All of the electrodes in the next two sections work on one principle. Differences in activity of a selected ion on either side of a membrane lead to an electric potential difference across the membrane. A difference of 59.16 mV (at 25°C) builds up across the glass pH electrode for every factor-of-10 difference

TABLE 15-1   Mobilities of ions in water at 25°C

Ion	Mobility $[m^2/(s \cdot V)]^a$
H^+	36.30×10^{-8}
Rb^+	7.92×10^{-8}
K^+	7.62×10^{-8}
NH_4^+	7.61×10^{-8}
La^{3+}	7.21×10^{-8}
Ba^{2+}	6.59×10^{-8}
Ag^+	6.42×10^{-8}
Ca^{2+}	6.12×10^{-8}
Cu^{2+}	5.56×10^{-8}
Na^+	5.19×10^{-8}
Li^+	4.01×10^{-8}
OH^-	20.50×10^{-8}
$Fe(CN)_6^{4-}$	11.45×10^{-8}
$Fe(CN)_6^{3-}$	10.47×10^{-8}
SO_4^{2-}	8.27×10^{-8}
Br^-	8.13×10^{-8}
I^-	7.96×10^{-8}
Cl^-	7.91×10^{-8}
NO_3^-	7.40×10^{-8}
ClO_4^-	7.05×10^{-8}
F^-	5.70×10^{-8}
HCO_3^-	4.61×10^{-8}
$CH_3CO_2^-$	4.24×10^{-8}

a. The mobility of an ion is the terminal velocity that the particle achieves in an electric field of 1 V/m. Mobility = velocity/field. The units of mobility are therefore $(m/s)/(V/m) = m^2/(s \cdot V)$.

. . . .

TABLE 15-2   Liquid junction potentials at 25°C

Junction	Potential (mV)
0.1 M NaCl $\mid$ 0.1 M KCl	−6.4
0.1 M NaCl $\mid$ 3.5 M KCl	−0.2
1 M NaCl $\mid$ 3.5 M KCl	−1.9
0.1 M HCl $\mid$ 0.1 M KCl	+27
0.1 M HCl $\mid$ 3.5 M KCl	+3.1

Note: A positive sign means that the right side of the junction becomes positive with respect to the left side.

Demonstration 15-1 Potentiometry with an Oscillating Reaction

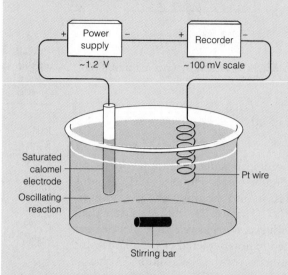

Apparatus used to monitor the quotient $[Ce^{3+}]/[Ce^{4+}]$ for an oscillating reaction. [The idea for this demonstration came from George Rossman, California Institute of Technology.]

The principles of potentiometry can be illustrated in a fascinating manner by using *oscillating reactions* in which the concentrations of various species oscillate between high and low values, instead of monotonically approaching their equilibrium values. In spite of the oscillatory behavior, the free energy of the system decreases throughout the entire reaction.[2]

One interesting example is the Belousov-Zhabotinskii reaction:

$$3CH_2(CO_2H)_2 + 2BrO_3^- + 2H^+ \rightarrow 2BrCH(CO_2H)_2 + 3CO_2 + 4H_2O$$

Malonic acid   Bromate   Bromomalonic acid

During this cerium-catalyzed oxidation of malonic acid by bromate, the quotient $[Ce^{3+}]/[Ce^{4+}]$ oscillates by a factor of 10 to 100.[3] When the Ce^{4+} concentration is high, the solution is yellow. When Ce^{3+} predominates, the solution is colorless. With redox indicators (Section 16-4), this reaction oscillates through a sequence of colors.[4]

Oscillation between yellow and colorless in a 300-mL beaker is set up with the following solutions:

160 mL of 1.5 M H_2SO_4

40 mL of 2 M malonic acid

30 mL of 0.5 M $NaBrO_3$ (or saturated $KBrO_3$)

4 mL of saturated ceric ammonium sulfate, $(Ce(SO_4)_2 \cdot 2(NH_4)_2SO_4 \cdot 2H_2O)$

in activity of H^+ across the electrode membrane. Because a factor-of-10 difference in activity of H^+ is one pH unit, a difference of, say, 4.00 pH units would lead to a potential difference of $4.00 \times 59.16 = 237$ mV.

To understand how an ion-selective electrode works, imagine a membrane separating two solutions containing $CaCl_2$ (Figure 15-8). The membrane contains a ligand that can bind and transport Ca^{2+}, but *not* Cl^-. Initially the potential difference across the membrane is zero, because both solutions are neutral (Figure 15-8a). However, there will be a tendency for the ions on the side of high activity (concentration) to diffuse to the side with low activity (concentration). Chloride has no way to cross the membrane, but

After an induction period of 5 to 10 min with magnetic stirring, oscillations can be initiated by adding 1 mL of ceric ammonium sulfate solution. The reaction is somewhat temperamental and may need more Ce^{4+} over a 5-min period to initiate oscillations.

A galvanic cell is built around the reaction as shown below. The value of $[Ce^{3+}]/[Ce^{4+}]$ is monitored by Pt and calomel electrodes. You should be able to write the cell reactions and a Nernst equation for this experiment.

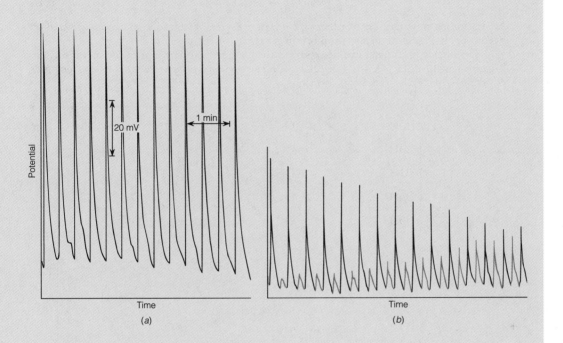

20 mV

1 min

Potential

Time

(a)

Time

(b)

In place of a potentiometer (a pH meter), we use a chart recorder to obtain a permanent record of the oscillations. Because the potential oscillates over a range of ~100 mV but is centered near ~1.2 V, the cell voltage is offset by ~1.2 V with any available power supply.

Trace a shows what is usually observed. The potential changes rapidly during the abrupt colorless-to-yellow change and gradually during the gentle yellow-to-colorless change. Trace b shows two different cycles superimposed in the same solution. This unusual event happened spontaneously in a reaction that had been oscillating normally for about 30 min.[5]

Ca^{2+} can bind to the ligand dissolved in the membrane and cross from one side to the other.

Thermodynamics tells us that the free energy difference between a species in solution at activity $\mathcal{A}_1$ in one place and activity $\mathcal{A}_2$ at another place is

Free energy difference due to concentration difference:

$$\Delta G = -RT\ln\frac{\mathcal{A}_1}{\mathcal{A}_2} \qquad (15\text{-}2)$$

As Ca^{2+} migrates from the region of high activity to the region of low activity, positive charge is built up on the low-activity side (Figure 15-8b). Eventually the charge buildup prevents further migration of Ca^{2+} to the

Mechanism of ion-selective electrode: Charge buildup balances tendency to move from high concentration to low concentration.

Box 15-1 Surface-Modified Electrodes[6]

Chemists modify electrode surfaces to accomplish specific tasks. For example, the reaction

$$HCO_3^- + 2e^- + 2H^+ \rightarrow HCO_2^- + H_2O \qquad \text{(A)}$$
$$\underset{\text{Bicarbonate}}{} \qquad\qquad\qquad \underset{\text{Formate}}{}$$

does not occur readily at most electrode surfaces. However, finely dispersed Pd is known to catalyze the hydrogenation reaction

$$H_2 + HCO_3^- \rightarrow HCO_2^- + H_2O \qquad \text{(B)}$$

An ingenious modification of otherwise unreactive metal electrodes allows Reaction B to occur near its thermodynamic equilibrium potential.

The surface of a metal can be oxidized to create hydroxyl groups at the metal surface. Treatment with $RSiX_3$ (X = Cl or OCH_3) in the presence of water gives a silicon polymer covalently attached to the metal surface:

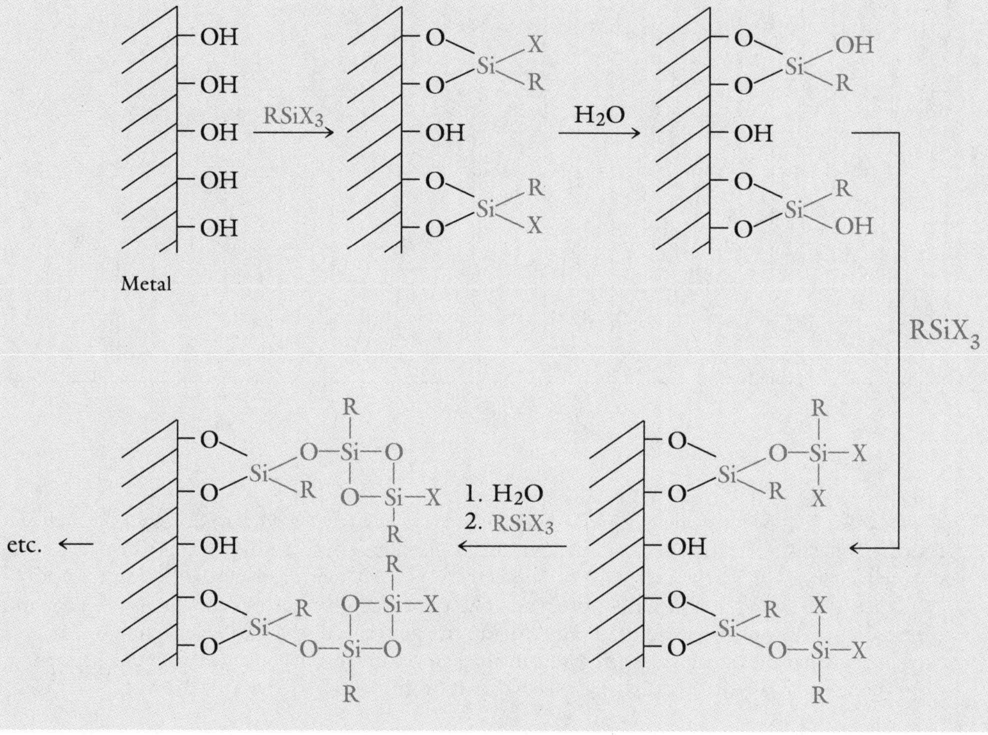

positively charged side. This is just what happened at a liquid junction, and a constant potential difference across the junction or membrane is reached.

The constant potential difference is such that the free energy decrease due to the activity difference is balanced by the free energy increase from repulsion of like charges:

$$\Delta G = -nFE$$

$$-RT \ln \frac{\mathcal{A}_1}{\mathcal{A}_2} = -nFE$$

$\ln x = (\ln 10)(\log x)$

Electric potential difference due to concentration difference:

$$E = \frac{RT}{nF} \ln \frac{\mathcal{A}_1}{\mathcal{A}_2} = \frac{0.059\,16}{n} \log \frac{\mathcal{A}_1}{\mathcal{A}_2} \text{ (volts at 25°C)} \qquad (15\text{-}3)$$

For bicarbonate reduction, the silicon compound contained a 4,4′-bipyridine moiety:

$$\left[(CH_3O)_3Si(CH_2)_3 \overset{+}{N} \underbrace{\bigcirc\!\!\!-\!\!\!\bigcirc} \overset{+}{N} (CH_2)_3Si(OCH_3)_3 \right]^{2+} \cdot 2Cl^-$$

4,4′-Bipyridine
group

(A *moiety* is a part of a molecule.) Tungsten and platinum electrodes were coated with a polymer formed from this compound, thereby giving a surface coverage of 10^{-7} mol of 4,4′-bipyridine groups per square centimeter of electrode surface. Then $PdCl_4^{2-}$ ions were exchanged for Cl^- in the polymer by soaking in K_2PdCl_4 solution. At electrode potentials of -0.30 to -0.75 V (versus S.C.E.), the 4,4′-bipyridine groups were reduced:

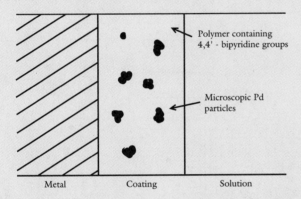

The radical cation thus formed was powerful enough to reduce $PdCl_4^{2-}$ to microscopic particles of metallic Pd.

Polymer containing
4,4′ - bipyridine groups

Microscopic Pd
particles

Metal   Coating   Solution

When a potential of -0.7 V (versus S.C.E.) was applied to the electrode immersed in $NaHCO_3$ solution, the 4,4′-bipyridine groups conducted electrons through the polymer into the Pd particles. There H_2O was reduced to H_2, which reacted with HCO_3^- at the Pd surface to give HCO_2^- (Reaction B). Up to 85% of the current went into reduction of HCO_3^-, and the reaction could be run long enough to reduce more than 1 000 HCO_3^- molecules per atom of Pd.

Because the charge of a calcium ion is $n = 2$, a potential difference of $59.16/2 = 29.58$ mV is expected for every factor-of-10 difference in activity of Ca^{2+} across the membrane. No charge buildup results from the Cl^- activity difference because Cl^- has no way to cross the membrane.

15-5 pH Measurement with a Glass Electrode

The most widely employed ion-selective electrode is the **glass electrode** for measuring pH.[7] A diagram of a typical **combination electrode,** incorporating both the glass and the reference electrodes in one body, is shown in Figure 15-9. A line diagram of this cell can be written as follows:

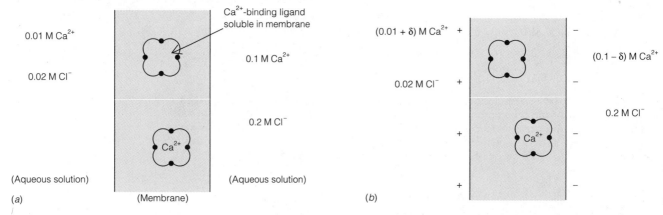

Figure 15-8 Mechanism of ion-selective electrode. (*a*) Initial conditions prior to Ca^{2+} migration across the membrane. (*b*) After δ moles of Ca^{2+} per liter have crossed the membrane, giving the left side a charge of $+2\delta$ mol/L and the right side a charge of -2δ mol/L.

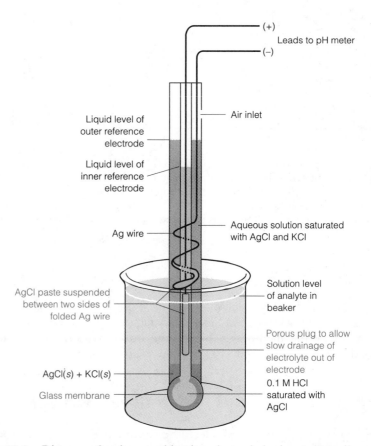

Figure 15-9 Diagram of a glass combination electrode having a silver-silver chloride reference electrode. The glass electrode is immersed in a solution of unknown pH so that the porous plug on the lower right is below the surface of the liquid. The two silver electrodes measure the voltage across the glass membrane.

$$\underbrace{Ag(s) \mid AgCl(s) \mid Cl^-(aq)}_{\substack{\text{Outer reference} \\ \text{electrode}}} \| \underbrace{H^+(aq, \text{ outside})}_{\substack{\text{H}^+ \text{ outside} \\ \text{glass electrode} \\ \text{(analyte solution)}}} \vdots \underbrace{H^+(aq, \text{ inside}),}_{\substack{\text{H}^+ \text{ inside} \\ \text{glass} \\ \text{electrode}}} \underbrace{Cl^-(aq) \mid AgCl(s) \mid Ag(s)}_{\substack{\text{Inner reference} \\ \text{electrode}}}$$

Glass
membrane

The pH-sensitive part of the electrode is the thin glass membrane in the shape of a bulb at the bottom of the electrodes in Figures 15-9 and 15-10.

Figure 15-11 shows the structure of the silicate lattice in glass. The irregular network contains negatively charged oxygen atoms that can bind to metal cations of suitable size. Monovalent cations, particularly Na^+, are able to move sluggishly through the silicate lattice. A schematic cross section of the glass membrane of the pH electrode is shown in Figure 15-12. The two exposed surfaces absorb water and become swollen. Most of the metal cations in these *hydrated gel* regions of the glass membrane diffuse out of the glass and into solution. Concomitantly, H^+ from solution can diffuse into the membrane. The reaction in which H^+ replaces metal cations in the glass is an **ion-exchange equilibrium.** Figure 15-13 depicts H^+ in equilibrium between the glass surface and the solution. The more H^+ in solution, the more H^+ will be bound to the glass surface.

Tracer studies with tritium (the radioactive isotope ^3H) show that H^+ does not cross the glass membrane of the pH electrode. However, the membrane is still in equilibrium with H^+ on each surface, because Na^+ ions do transport charge across the membrane—so one side still "senses" the other side. The mechanism of the glass electrode is therefore similar to that described for other ion-selective electrodes, except that the species that migrates across the membrane is not the same as the species selectively adsorbed at each surface. The H^+-sensitive membrane may be thought of as just the two surfaces—connected electrically by Na^+ transport. The resistance of the glass membrane is typically $10^8 \ \Omega$, so very little current actually flows across it.

The potential difference between the inner and outer silver-silver chloride electrodes in Figure 15-9 depends on the chloride concentration in each electrode compartment and on the potential difference across the glass membrane.[8] Because the chloride concentration is fixed in each electrode compartment and because the H^+ concentration is fixed on the inside of the glass membrane, the only variable factor is the pH of the analyte solution

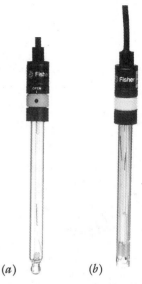

(a)   (b)

Figure 15-10 (a) Glass-body combination electrode with pH-sensitive glass bulb at the bottom. The circle near the bottom at the side of the electrode is the porous junction salt bridge to the reference electrode compartment. Two silver wires coated with AgCl are visible inside the electrode. (b) Polymer body surrounds glass electrode to protect the delicate bulb from breaking. [Courtesy Fisher Scientific, Pittsburgh, PA.]

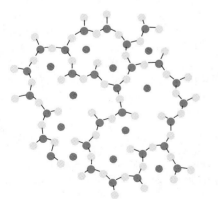

Figure 15-11 Schematic diagram of the structure of glass, which consists of an irregular network of SiO_4 tetrahedra connected through their oxygen atoms. ○ = O, ● = Si, ◉ = cation. Cations such as Li^+, Na^+, K^+, and Ca^{2+} are coordinated to the oxygen atoms. The silicate network is not planar. This diagram is a projection of each tetrahedron onto the plane of the page. [Adapted from G. A. Perley, *Anal. Chem.* **1949,** *21,* 394.]

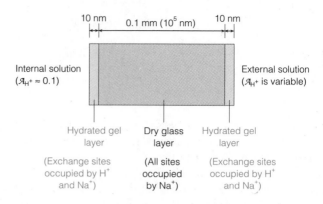

Figure 15-12 Schematic diagram showing a cross section of the glass membrane of a pH electrode.

outside the glass membrane. The thermodynamic argument in the preceding section states that *the voltage of the ideal pH electrode changes by 59.16 mV for every pH-unit change of the analyte.*

For real glass electrodes, the response can be described by the equation

Real electrodes come close to obeying the Nernst equation.

Response of glass electrode:
$$E = \text{constant} + \beta(0.059\,16)\log\frac{\mathcal{A}_{H^+}(\text{outside})}{\mathcal{A}_{H^+}(\text{inside})} \quad (\text{at } 25°C)$$
(15-4)

 A pH electrode **must** be calibrated before it can be used. It should be calibrated about every 2 h during sustained use.

The value of β, the *electromotive efficiency,* is close to 1.00 (typically >0.98). The constant term, called the *asymmetry potential,* arises because no two sides of a real object are identical, and a small voltage exists even if $\mathcal{A}_{H^+}$ is the same on both sides of the membrane. We correct for this asymmetry potential by calibrating the electrode in solutions of known pH.

Calibrating a Glass Electrode

The pH of the calibration standards should bracket the pH of the unknown.

A pH electrode should be calibrated with two (or more) standard buffers selected so that the pH of the unknown lies within the range of the standards. Standards in Table 15-3 are accurate to ±0.01 pH unit.[9,10]

Before calibrating the electrode, wash it with distilled water and gently *blot* it dry with a tissue. Do not *wipe* it because this might produce a static charge on the glass. Dip the electrode in a standard buffer whose pH is near 7 and allow the electrode to equilibrate for at least a minute. Solutions for calibration and measurement should be stirred during the measurement. Following the manufacturer's instructions, adjust the meter reading (usually with a knob labeled "Calibrate") to indicate the pH of the standard buffer. (If the meter has an adjustment of the "Isopotential point," it should be set

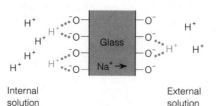

Figure 15-13 Ion-exchange equilibria on the inner and outer surfaces of the glass membrane. The pH of the internal solution is fixed. As the pH of the external solution (the sample) changes, the electric potential difference across the glass membrane changes.

to the pH of this first buffer.) The electrode is then washed, blotted dry, and immersed in a second standard whose pH is further from 7 than the pH of the first standard. If the electrode response were perfectly Nernstian, the voltage would change by 0.059 16 V per pH unit at 25°C. The actual change may be slightly less, so these two measurements establish the value of β in Equation 15-4. The pH of the second buffer is set on the meter with a knob that may be labeled "Slope" or "Temperature" on different instruments. If your meter does not have an isopotential point adjustment, it may be necessary to repeat the calibration with the two buffers to obtain correct readings for each. Finally, the electrode is dipped in the unknown and the voltage is translated directly into a pH reading by the meter.

A glass electrode is stored in aqueous solution to prevent dehydration of the glass. If the electrode has dried out, it should be reconditioned in water for several hours. If the electrode is to be used above pH 9, soak it in a high-pH buffer. (The field effect transistor pH electrode in Section 15-8 is stored dry. Prior to use, it should be scrubbed gently with a soft brush and soaked in pH 7 buffer for 10 min.)

If electrode response becomes sluggish or if the electrode cannot be calibrated properly, try soaking it in 6 M HCl, followed by a water soak. As a last resort, the electrode can be soaked in 20 wt% aqueous ammonium bifluoride, NH_4HF_2, for 1 min in a plastic beaker. This reagent dissolves a little of the glass and exposes fresh surface. Wash the electrode with water and try calibrating it again. *Caution: Ammonium bifluoride must not contact your skin, because it produces HF burns.*

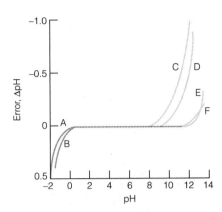

Figure 15-14 Acid and alkaline errors of some glass electrodes. A: Corning 015, H_2SO_4; B: Corning 015, HCl; C: Corning 015, 1 M Na^+; D: Beckman-GP, 1 M Na^+; E: L & N Black Dot, 1 M Na^+; F: Beckman Type E, 1 M Na^+. [R. G. Bates, *Determination of pH: Theory and Practice*, 2nd ed. (New York: Wiley, 1973).]

Errors in pH Measurement

To make intelligent use of a glass electrode, it is important to understand its limitations:

1. A pH measurement cannot be more accurate than our standards, which are typically accurate to ±0.01 pH unit.

2. A *junction potential* exists at the porous plug near the bottom of the electrode in Figure 15-9. If the ionic composition of the analyte solution is different from that of the standard buffer, the junction potential will change *even if the pH of the two solutions is the same*. This difference gives an uncertainty of at least ~0.01 pH unit. Box 15-2 describes measurements in which junction potential errors are significant.

> The apparent pH will change if the ionic composition of the analyte changes, even when the actual pH is constant.

3. When the concentration of H^+ is very low and the concentration of Na^+ is high, the electrode responds to Na^+ as well as to H^+. The electrode behaves as if Na^+ were H^+, and the apparent pH is lower than the true pH. This is called the **alkaline error,** or *sodium error* (Figure 15-14).

4. In strong acid, the measured pH is higher than the actual pH, for reasons that are not understood.

5. Time must be allowed for the electrode to equilibrate with each solution. In a well-buffered solution, this takes just seconds with adequate stirring. In a poorly buffered solution (such as near the equivalence point of a titration), it could take minutes.

6. A dry electrode requires several hours of soaking before it responds to H^+ correctly.

> Do not leave a glass electrode out of water (or in a nonaqueous solvent) any longer than necessary.

TABLE 15-3   pH values of National Institute of Standards and Technology buffers

Temperature (°C)	Saturated (25°C) potassium hydrogen tartrate (1)	0.05 m potassium dihydrogen citrate (2)	0.05 m potassium hydrogen phthalate (3)	0.08 m MOPSO 0.08 m NaMOPSO 0.08 m NaCl (4)
0	—	3.863	4.003	7.268
5	—	3.840	3.999	7.182
10	—	3.820	3.998	7.098
15	—	3.802	3.999	7.018
20	—	3.788	4.002	6.940
25	3.557	3.776	4.008	6.865
30	3.552	3.766	4.015	6.792
35	3.549	3.759	4.024	6.722
37	3.548	3.756	4.028	6.695
40	3.547	3.753	4.035	6.654
45	3.547	3.750	4.047	6.588
50	3.549	3.749	4.060	6.524
55	3.554	—	4.075	—
60	3.560	—	4.091	—
70	3.580	—	4.126	—
80	3.609	—	4.164	—
90	3.650	—	4.205	—
95	3.674	—	4.227	—

Note: The designation m stands for molality.

In the buffer solution preparations, it is essential to use high-purity materials and to employ freshly distilled or deionized water of specific conductivity not greater than 5 micromho/cm. Solutions having pH 6 or above should be stored in plastic containers, and preferably with an NaOH trap to prevent ingress of atmospheric carbon dioxide. They can normally be kept for 2–3 weeks, or slightly longer in a refrigerator. Buffer materials in this table are available as Standard Reference Materials from the National Institute of Standards and Technology (Standard Reference Materials Program, Room 204, Building 202, National Institute of Standards and Technology, Gaithersburg, MD 20899 (Phone: 301-975-6776).

1. Saturated (25°C) potassium hydrogen tartrate, $KHC_4H_4O_6$. An excess of the salt is shaken with water, and it can be stored in this way. Before use, it should be filtered or decanted at a temperature between 22° and 28°C.

2. 0.05 m potassium dihydrogen citrate, $KH_2C_6H_5O_7$. Dissolve 11.41 g of the salt in 1 L of solution at 25°C.

3. 0.05 m potassium hydrogen phthalate. Although this is not usually essential, the crystals may be dried at 110°C for 1 h, then cooled in a desiccator. At 25°C, 10.12 g $C_6H_4(CO_2H)(CO_2K)$ is dissolved in water, and the solution made up to 1 L.

4. 0.08 m MOPSO ((3-N-morpholino)-2-hydroxypropanesulfonic acid, Table 10-2), 0.08 m sodium salt of MOPSO, 0.08 m NaCl. Buffers 4 and 6 are recommended for 2-point standardization of electrodes for pH measurements of physiologic fluids. MOPSO is crystallized twice from 70 wt% ethanol and dried in vacuum at 50°C for 24 h. NaCl is dried at 110°C for 4 h. Na^+MOPSO^- may be prepared by neutralization of MOPSO with standard NaOH. The sodium salt is also available as a Standard Reference Material. Dissolve 18.021 g MOPSO, 19.780 g Na^+MOPSO^-, and 4.675 g NaCl in 1.000 kg H_2O.

0.025 m potassium dihydrogen phosphate 0.025 m disodium hydrogen phosphate (5)	0.08 m HEPES 0.08 m NaHEPES 0.08 m NaCl (6)	0.008 695 m potassium dihydrogen phosphate 0.030 43 m disodium hydrogen phosphate (7)	0.01 m borax (8)	0.025 m sodium bicarbonate 0.025 m sodium carbonate (9)
6.984	7.853	7.534	9.464	10.317
6.951	7.782	7.500	9.395	10.245
6.923	7.713	7.472	9.332	10.179
6.900	7.646	7.448	9.276	10.118
6.881	7.580	7.429	9.225	10.062
6.865	7.516	7.413	9.180	10.012
6.853	7.454	7.400	9.139	9.966
6.844	7.393	7.389	9.102	9.925
6.841	7.370	7.385	9.088	9.910
6.838	7.335	7.380	9.068	9.889
6.834	7.278	7.373	9.038	9.856
6.833	7.223	7.367	9.011	9.828
6.834	—	—	8.985	—
6.836	—	—	8.962	—
6.845	—	—	8.921	—
6.859	—	—	8.885	—
6.877	—	—	8.850	—
6.886	—	—	8.833	—

5. 0.025 m disodium hydrogen phosphate, 0.025 m potassium dihydrogen phosphate. The anhydrous salts are best; each should be dried for 2 h at 120°C and cooled in a desiccator, because they are slightly hygroscopic. Higher drying temperatures should be avoided to prevent formation of condensed phosphates. Dissolve 3.53 g Na_2HPO_4 and 3.39 g KH_2PO_4 in water to give 1 L of solution at 25°C.

6. 0.08 m HEPES (N-2-hydroxyethylpiperazine-N'-2-ethanesulfonic acid, Table 10-2), 0.08 m sodium salt of HEPES, 0.08 m NaCl. Buffers 4 and 6 are recommended for 2-point standardization of electrodes for pH measurements of physiologic fluids. HEPES is crystallized twice from 80 wt% ethanol and dried in vacuum at 50°C for 24 h. NaCl is dried at 110°C for 4 h. Na^+HEPES^- may be prepared by neutralization of HEPES with standard NaOH. The sodium salt is also available as a Standard Reference Material. Dissolve 19.065 g HEPES, 20.823 g Na^+HEPES^-, and 4.675 g NaCl in 1.000 kg H_2O.

7. 0.008 695 m potassium dihydrogen phosphate, 0.030 43 m disodium hydrogen phosphate. Prepare like Buffer 5; dissolve 1.179 g KH_2PO_4 and 4.30 g Na_2HPO_4 in water to give 1 L of solution at 25°C.

8. 0.01 m sodium tetraborate decahydrate. Dissolve 3.80 g $Na_2B_4O_7 \cdot 10H_2O$ in water to give 1 L of solution. This borax solution is particularly susceptible to pH change from carbon dioxide absorption, and it should be correspondingly protected.

9. 0.025 m sodium bicarbonate, 0.025 m sodium carbonate. Primary standard-grade Na_2CO_3 is dried at 250°C for 90 min and stored over $CaCl_2$ and Drierite. Reagent-grade $NaHCO_3$ is dried over molecular sieves and Drierite for 2 days at room temperature. Do not heat $NaHCO_3$, or it may decompose to Na_2CO_3. Dissolve 2.092 g of $NaHCO_3$ and 2.640 g of Na_2CO_3 in 1 L of solution at 25°C.

SOURCES: R. G. Bates, *J. Res. National Bureau of Standards,* **1962,** *66A,* 179; B. R. Staples and R. G. Bates, *J. Res. National Bureau of Standards,* **1969,** *73A,* 37. Data on HEPES and MOPSO are from Y. C. Wu, P. A. Berezansky, D. Feng, and W. F. Koch, *Anal. Chem.* **1993,** *65,* 1084 and D. Feng, W. F. Koch, and Y. C. Wu, *Anal. Chem.* **1989,** *61,* 1400. Instructions for preparing some of these solutions are from G. Mattock in C. N. Reilley, Ed., *Advances in Analytical Chemistry and Instrumentation* (New York: Wiley, 1963), Vol. 2, p. 45. See also R. G. Bates, *Determination of pH: Theory and Practice,* 2nd ed. (New York: Wiley, 1973), Chap. 4.

7. A pH meter should be calibrated at the same temperature at which the measurement will be made. You cannot calibrate your equipment at one temperature and then make an accurate measurement at a second temperature.

Challenge Show that the potential of the glass electrode described by Equation 15-4 changes by 1.3 mV when the analyte H^+ activity changes by 5.0%. Because 59 mV $\approx$ 1 pH unit, 1.3 mV = 0.02 pH unit.

Moral A small uncertainty in voltage (1.3 mV) or pH (0.02 units) corresponds to a large uncertainty (5%) in analyte concentration. Similar uncertainties arise in other potentiometric measurements.

Errors 1 and 2 limit the accuracy of pH measurement with the glass electrode to ± 0.02 pH unit, at best. Measurement of pH differences *between* solutions can be accurate to about ± 0.002 pH unit, but knowledge of the true pH will still be at least an order of magnitude more uncertain. An uncertainty of ± 0.02 pH unit corresponds to an uncertainty of $\pm 5\%$ in $\mathcal{A}_{H^+}$.

15-6 Types of Ion-Selective Electrodes

Most electrodes fall into one of the following classes:

1. *Glass membranes* for H^+ and certain monovalent cations

2. *Solid-state electrodes* based on inorganic salt crystals

3. *Liquid-based electrodes* using a hydrophobic polymer membrane saturated with a hydrophobic liquid ion exchanger

4. *Compound electrodes* with a species-selective electrode enclosed by a membrane that is able to separate that species from others or that generates the species in a chemical reaction

Selectivity Coefficient

No electrode responds exclusively to one kind of ion, but the glass electrode is among the most selective. A high-pH glass electrode responds to Na^+ only when $[H^+] \leq 10^{-12}$ M and $[Na^+] \geq 10^{-2}$ M (Figure 15-14). Glasses of different compositions are sensitive to different ions. By using different glass compositions, Li^+, Na^+, K^+, and Ag^+ ion-selective electrodes have been fabricated from glass membranes.

An electrode used for the measurement of ion X may also respond to ion Y. The selectivity of the electrode to different species with the same charge is given by the **selectivity coefficient,** defined as

Selectivity coefficient: $$k_{X,Y} = \frac{\text{response to Y}}{\text{response to X}} \tag{15-5}$$

Ideally, the selectivity coefficient should be very small ($k \ll 1$). One particular glass sodium ion-selective electrode has a selectivity coefficient $k_{Na^+,K^+} = 1/2800$ at pH 11, and $k_{Na^+,K^+} = 1/300$ at pH 7. The selectivity coefficient k_{Na^+,H^+} is 36, which means that the electrode is 36 times more sensitive to H^+ than to Na^+.

The behavior of most ion-selective electrodes can be described by the equation

Equation 15-6 describes the response of an electrode to its primary ion, X, and to all interfering ions, Y.

Response of ion-selective electrode: $$E = \text{constant} \pm \beta \frac{0.059\,16}{n_X} \log \left[\mathcal{A}_X + \sum_Y \left(k_{X,Y} \mathcal{A}_Y^{n_X/n_Y} \right) \right] \tag{15-6}$$

where $\mathcal{A}_X$ is the activity of the ion intended to be measured and $\mathcal{A}_Y$ is the activity of any interfering species (Y). The magnitude of the charge of each species is n_X or n_Y, and $k_{X,Y}$ is the selectivity coefficient. If the ion-selective electrode is connected to the positive terminal of the potentiometer, the sign before the log term is positive if X is a cation and negative if X is an anion. The value of β is near 1 for most electrodes.

EXAMPLE Using the Selectivity Coefficient

A Tl^{3+} ion-selective electrode has a selectivity coefficient $k_{Tl^{3+},K^+} = 3.4 \times 10^{-4}$. What will be the change in electrode potential when 10^{-1} M K^+ is added to 10^{-5} M Tl^{3+}?

Solution Using Equation 15-6 with $\beta = 1$, $n_X = +3$, and $n_Y = +1$, the potential without K^+ is

$$E = \text{constant} + \frac{0.059\,16}{3} \log[10^{-5}] = \text{constant} + 98.60 \text{ mV}$$

Addition of 10^{-1} M K^+ gives an electrode potential of

$$E = \text{constant} + \frac{0.059\,16}{3} \log[10^{-5} + (3.4 \times 10^{-4})(10^{-1})^{3/1}]$$

$$= \text{constant} + 98.31 \text{ mV}$$

The change is $98.31 - 98.60 = -0.29$ mV, barely enough to measure.

Solid-State Electrodes

A diagram of a **solid-state ion-selective electrode** based on an inorganic crystal is shown in Figure 15-15. One common electrode of this type is the fluoride electrode, employing a crystal of LaF_3 doped with Eu^{2+}. *Doping* means adding a small amount of Eu^{2+} in place of La^{3+}. The filling solution contains 0.1 M NaF and 0.1 M NaCl.

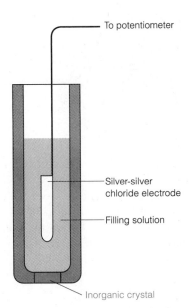

To potentiometer

Silver-silver chloride electrode

Filling solution

Inorganic crystal

Figure 15-15 Schematic diagram of an ion-selective electrode employing an inorganic salt crystal as the ion-selective membrane.

Box 15-2 Systematic Error in Rainwater pH Measurement: The Effect of Junction Potential

Combustion products from automobiles and factories include nitrogen oxides and sulfur dioxide, which react with water in the atmosphere to produce acids.

$$SO_2 + H_2O \rightarrow H_2SO_3 \xrightarrow{\text{oxidation}} H_2SO_4$$
$$\text{Sulfurous acid} \qquad \text{Sulfuric acid}$$

The map shows that *acid rain* in North America is most severe in the eastern half. This pollution is a serious threat to lakes and forests around the world. Monitoring the pH of rainwater is a critical component of programs to measure and reduce the production by mankind's activities of acid rain.

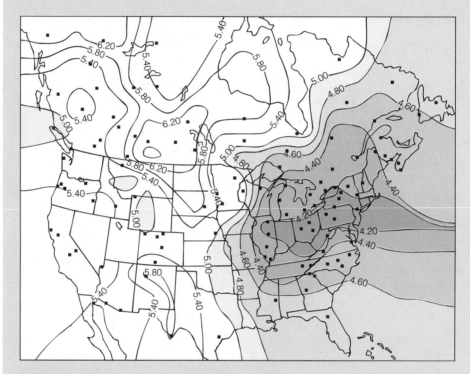

Precipitation-weighted annual average pH distribution over North America. [R. Semonin, *Study of Atmospheric Pollution Scavenging* (Washington, DC: U.S. Department of Energy, 1981, pp. 56–119.]

Fluoride ion from solution is selectively adsorbed on each surface of the crystal. Unlike H^+ in the glass pH electrode, F^- actually migrates through the LaF_3 crystal, as shown in Figure 15-16. By doping the LaF_3 with EuF_2, anion vacancies are created within the crystal. An adjacent fluoride ion can jump into the vacancy, thus leaving a new vacancy behind. In this manner F^- diffuses from one side to the other, thereby establishing the potential difference across the crystal required to make the electrode work.

To identify and correct systematic errors in the measurement of pH of rainwater, a careful study was conducted with 17 laboratories.[11] Eight samples were provided to each laboratory, along with explicit instructions for how to conduct the measurements. Each laboratory used two buffers to standardize pH meters. Sixteen laboratories successfully measured the pH of Unknown A (within ±0.02 pH units), which was 4.008 at 25°C. One lab whose measurement was 0.04 pH units low was later found to have a faulty commercial standardization buffer.

The figure below shows typical results for the pH of rainwater. The average of the 17 measurements is given by the horizontal line at pH 4.14 and the letters s, t, u, v, w, x, y, and z identify the type of pH electrode used for the measurements. Laboratories using electrode types s and w had relatively large systematic errors. The type s electrode was a combination electrode (Figure 15-9) containing a reference electrode liquid junction with an exceptionally large area. Electrode type w had a reference electrode filled with a gel.

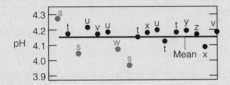

pH of rainwater from identical samples measured at 17 different labs using standard buffers for calibration. Letters designate different types of electrodes used for pH measurements.

It was hypothesized that variability in the liquid junction potential (Section 15-3) led to variability among the pH measurements. Buffers used for standardization typically have ionic strengths in the neighborhood of 0.05 M, whereas rainwater samples have ionic strengths two or more orders of magnitude lower. To test the hypothesis that junction potential was the cause of the systematic errors, a pure HCl solution with a concentration near 2×10^{-4} M was used as a pH standard in place of high ionic strength buffers. The data below were obtained, with good results from all but the first lab. The standard deviation of all 17 measurements was reduced from 0.077 pH units with the standard buffer to 0.029 pH units with the HCl standard. It was concluded that junction potential is the cause of most of the variability between labs and that a low ionic strength standard is appropriate for rainwater pH measurements.[12]

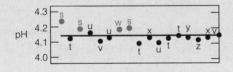

Rainwater pH measured after using low ionic strength HCl for calibration.

By analogy with the pH electrode, we can describe the response of the F^- electrode in the form

Response of F^- electrode: $E = \text{constant} - \beta(0.059\,16)\log \mathcal{A}_{F^-}(\text{outside})$ (15-7)

where β is close to 1.00. The F^- electrode gives a nearly Nernstian response over a F^- concentration range from about 10^{-6} M to 1 M. The electrode is more responsive to F^- than to other ions by a factor greater than 1 000. The

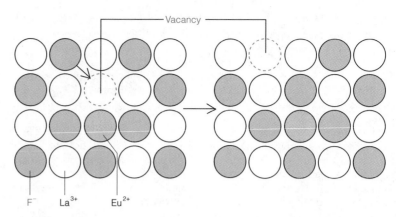

Figure 15-16   Migration of F^- through LaF_3 doped with EuF_2. Because Eu^{2+} has less charge than La^{3+}, an anion vacancy occurs for every Eu^{2+}. A neighboring F^- can jump into the vacancy, moving the vacancy to another site. Repetition of this process moves F^- through the lattice.

only interfering species is OH^-, for which the selectivity coefficient is $k_{F^-,OH^-} = 0.1$. At low pH, F^- is converted to HF ($pK_a = 3.17$), to which the electrode is insensitive. The F^- electrode is used to monitor continuously and control the fluoridation of municipal water supplies.

EXAMPLE   Response of an Ion-Selective Electrode

When a fluoride electrode was immersed in various standard solutions (maintained at a constant ionic strength of 0.1 M with $NaNO_3$), the following potentials (versus S.C.E.) were observed:

$[F^-]$ (M)	E (mV)
1.00×10^{-5}	100.00 } -58.5 mV
1.00×10^{-4}	41.5 } -58.5 mV
1.00×10^{-3}	-17.0

Because the ionic strength was constant, the response should depend upon the logarithm of the F^- *concentration*. What potential is expected if $[F^-] = 5.00 \times 10^{-5}$ M? What concentration of F^- will give a potential of 0.0 V?

Solution   We seek to fit the calibration data with Equation 15-7:

$$E = m \underbrace{\log [F^-]}_{x} + b$$
$$\underset{y}{}$$

Plotting E versus $\log [F^-]$ gives a straight line with a slope of -58.5 mV and a y-intercept of -192.5 mV. Setting $[F^-] = 5.00 \times 10^{-5}$ M gives

$$E = (-58.5)\log[5.00 \times 10^{-5}] - 192.5 = 59.1 \text{ mV}$$

If $E = 0.0$ mV, we can solve for the concentration of $[F^-]$:

$$0.0 = (-58.5)\log[F^-] - 192.5 \Rightarrow [F^-] = 5.1 \times 10^{-4} \text{ M}$$

TABLE 15-4   Properties of solid-state ion-selective electrodes

Ion	Concentration range (M)	Membrane material	pH range	Interfering species
F^-	10^{-6}–1	LaF_3	5–8	OH^-
Cl^-	10^{-4}–1	AgCl	2–11	CN^-, S^{2-}, I^-, $S_2O_3^{2-}$, Br^-
Br^-	10^{-5}–1	AgBr	2–12	CN^-, S^{2-}, I^-
I^-	10^{-6}–1	AgI	3–12	S^{2-}
SCN^-	10^{-5}–1	AgSCN	2–12	S^{2-}, I^-, CN^-, Br^-, $S_2O_3^{2-}$
CN^-	10^{-6}–10^{-2}	AgI	11–13	S^{2-}, I^-
S^{2-}	10^{-5}–1	Ag_2S	13–14	

Electrodes containing silver-based crystals such as Ag_2S should be stored in the dark and protected from light during use to prevent light-induced chemical degradation [R. DeMarco, R. W. Cattrall, J. Liesegang, G. L. Nyberg, and I. C. Hamilton, *Anal. Chem.* **1990**, *62*, 2339].

Another common inorganic crystal electrode uses Ag_2S for the membrane. This electrode responds to Ag^+ and to S^{2-}. By doping the electrode with CuS, CdS, or PbS, it is possible to prepare electrodes sensitive to Cu^{2+}, Cd^{2+}, or Pb^{2+}, respectively. Table 15-4 lists various ion-selective electrodes based on inorganic crystals.[13]

Figure 15-17 illustrates the mechanism by which a crystal responds selectively to certain ions. Hexagonal CdS crystals can be cleaved to expose

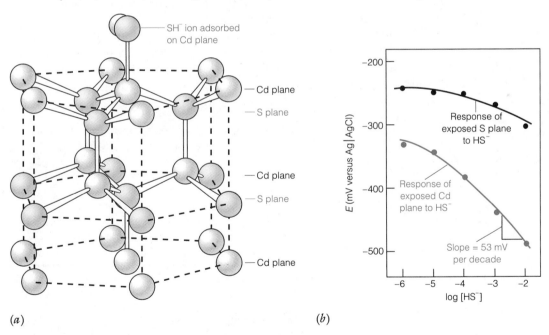

(a)   (b)

Figure 15-17   (a) Crystal structure of hexagonal CdS showing alternating planes of Cd and S along the vertical axis (the crystal *c* axis). A hydrogen sulfide ion is shown adsorbed to the uppermost Cd plane. (b) Potentiometric response of exposed crystal faces to HS^-. Because HS^- is selectively adsorbed on the Cd face, there is a near-Nernstian potentiometric response when the Cd face is exposed, but just a weak response when the S face is exposed. The partial response of the S face to SH^- in the upper curve is attributed to the fact that about 10% of the exposed atoms are actually Cd instead of S. [K. Uosaki, Y. Shigematsu, H. Kita, Y. Umezawa, and R. Souda, *Anal. Chem.* **1989**, *61*, 1980.]

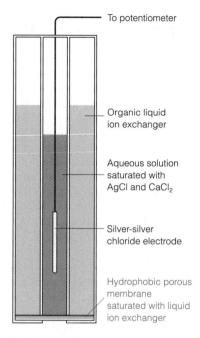

To potentiometer

Organic liquid ion exchanger

Aqueous solution saturated with AgCl and CaCl₂

Silver-silver chloride electrode

Hydrophobic porous membrane saturated with liquid ion exchanger

Figure 15-18 Schematic diagram of a calcium ion-selective electrode based on a liquid ion exchanger.

$$[(CH_3(CH_2)_8CH_2O)_2\overset{\overset{\textstyle O}{\|}}{P}O]_2Ca$$
Calcium didecylphosphate

$$(CH_3(CH_2)_6CH_2O)_2\overset{\overset{\textstyle O}{\|}}{P}C_6H_5$$
Dioctylphenylphosphonate

planes of cadmium atoms or sulfur atoms. The cadmium plane in Figure 15-17a selectively adsorbs SH^- ions, whereas an exposed sulfur plane does not interact strongly with HS^-. Figure 15-17b shows a near-ideal response of the exposed cadmium face to SH^-, but there is only weak response when the sulfur face is exposed. The opposite behavior is observed in response to Cd^{2+} ions.

Liquid-Based Ion-Selective Electrodes

The principle of the **liquid-based ion-selective electrode** was described in Figure 15-8.[14] Figure 15-18 shows the construction of a calcium ion-selective electrode, which is similar to the solid-state electrode in Figure 15-15. The difference is that the solid crystal is replaced by a membrane saturated with a hydrophobic liquid ion exchanger (a calcium chelator). A solution of the exchanger is housed in a reservoir surrounding the internal silver-silver chloride electrode. Calcium ion is selectively transported across the membrane to establish a voltage related to the difference in activity of Ca^{2+} between the sample and internal solution:

Response of Ca^{2+} electrode: $\quad E = \text{constant} + \beta\,\dfrac{0.059\,16}{2}\log\mathcal{A}_{Ca^{2+}}(\text{outside})$

$$(15\text{-}8)$$

where β is close to 1.00. Note that Equations 15-8 and 15-7 have different signs before the log term, because one involves an anion and the other a cation. Note also that the charge of the calcium ion requires a factor of 2 in the denominator before the logarithm.

The ion exchanger in the Ca^{2+} electrode is calcium didecylphosphate dissolved in dioctylphenylphosphonate. The didecylphosphate anion binds calcium ions at each surface of the membrane and transports Ca^{2+} across the membrane:

$$[(RO)_2PO_2^-]_2Ca \rightleftharpoons 2(RO)_2PO_2^- + Ca^{2+}$$

The most serious interference with the Ca^{2+} electrode comes from Zn^{2+}, Fe^{2+}, Pb^{2+}, and Cu^{2+}, but high concentrations of Sr^{2+}, Mg^{2+}, Ba^{2+}, and

TABLE 15-5 Properties of liquid-based ion-selective electrodes

Ion	Concentration range (M)	Carrier	Solvent for carrier	pH range	Interfering species
Ca^{2+}	10^{-5}–1	Calcium didecylphosphate	Dioctylphenyl-phosphonate	6–10	Zn^{2+}, Pb^{2+}, Fe^{2+}, Cu^{2+}
NO_3^-	10^{-5}–1	Tridodecylhexadecylammonium nitrate	Octyl-2-nitrophenyl ether	3–8	ClO_4^-, I^-, ClO_3^-, Br^-, HS^-, CN^-
ClO_4^-	10^{-5}–1	Tris(substituted 1,10-phenanthroline) iron(II) perchlorate	p-Nitrocymene	4–10	I^-, NO_3^-, Br^-
BF_4^-	10^{-5}–1	Tris(substituted 1,10-phenanthroline) nickel(II) tetrafluoroborate	p-Nitrocymene	2–12	NO_3^-
Cl^-	10^{-5}–1	Dimethyldioctadecylammonium chloride		3–10	ClO_4^-, I^-, NO_3^-, SO_4^{2-}, Br^-, OH^-, HCO_3^-, F^-, acetate

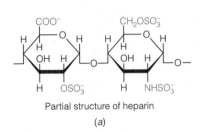

COO⁻ is shown — Partial structure of heparin

(a)

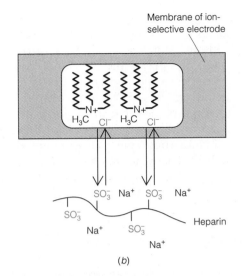

(b)

Figure 15-19 (*a*) Partial structure of heparin molecule, showing negatively charged substituents. Heparin belongs to a class of molecules called *proteoglycans* that contain about 95% polysaccharide (sugar) and 5% protein. (*b*) Schematic diagram showing ion exchange between the negatively charged heparin chain and Cl^- ions associated with tetraalkylammonium ions dissolved in the membrane of the ion-selective electrode. [S.-C. Ma, V. C. Yang, and M. E. Meyerhoff, *Anal. Chem.* **1992**, *64*, 694; S.-C. Ma, V. C. Yang, B. Fu, and M. E. Meyerhoff, *Anal. Chem.* **1993**, *65*, 2078.]

Na^+ also interfere. For one particular Ca^{2+} electrode, $k_{Ca^{2+},Fe^{2+}} = 0.8$ and $k_{Ca^{2+},Mg^{2+}} = 0.01$. Interference from H^+ is substantial below pH 4. Table 15-5 gives properties of other liquid-based ion-selective electrodes.

The heparin-sensitive electrode at the front of this chapter is similar to the Ca^{2+} electrode, but the liquid ion exchanger in the membrane is tridodecylmethylammonium chloride, $(C_{12}H_{25})_3NCH_3^+Cl^-$. Figure 15-19 shows exchange of chloride ions in the membrane with the negatively charged chain of the heparin molecule.

Compound Electrodes

Compound electrodes contain a conventional electrode surrounded by a membrane that isolates (or generates) the analyte to which the electrode responds. A CO_2 gas-sensing electrode is shown in Figure 15-20. It consists of an ordinary glass pH electrode surrounded by an electrolyte solution

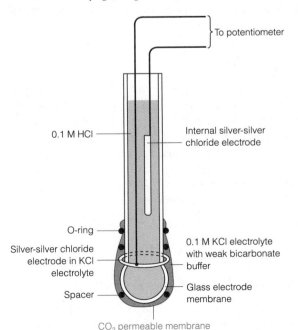

To potentiometer

0.1 M HCl

Internal silver-silver chloride electrode

O-ring

Silver-silver chloride electrode in KCl electrolyte

0.1 M KCl electrolyte with weak bicarbonate buffer

Spacer

Glass electrode membrane

CO_2 permeable membrane

Figure 15-20 Schematic diagram of a CO_2 gas-sensing electrode.

Acidic or basic gases are detected by a pH electrode surrounded by an electrolyte and enclosed in a gas-permeable membrane.

enclosed in a semipermeable membrane made of rubber, Teflon, or polyethylene. A silver-silver chloride reference electrode is immersed in the electrolyte solution. When CO_2 diffuses through the semipermeable membrane, it lowers the pH in the electrolyte compartment. The response of the glass electrode to the change in pH is measured. Other acidic or basic gases, including NH_3, SO_2, H_2S, NO_x (nitrogen oxides), and HN_3 (hydrazoic acid) can be detected in the same manner. These electrodes can be used to measure gases *in the gas phase* or dissolved in solution.

Numerous ingenious compound electrodes using *enzymes* have been built.[15] These devices contain a conventional electrode coated with an enzyme that catalyzes a reaction of the analyte. The product of the reaction is detected by the electrode.

15-7 Using Ion-Selective Electrodes

Advantages of ion-selective electrodes:
1. Wide range of linear response
2. Nondestructive
3. Noncontaminating
4. Short response time
5. Unaffected by color or turbidity

Ion-selective electrodes respond in a linear manner to the logarithm of the activity of analyte over a four to six order-of-magnitude range of activity. Electrodes do not consume unknown samples, and they introduce negligible contamination. Response time is seconds or minutes, so electrodes can be used to monitor flow streams in industrial applications. Color and turbidity do not hinder electrodes. Microelectrodes can be used in inaccessible locations such as the interior of living cells.

A 1-mV error in potential corresponds to a 4% error in monovalent ion activity. A 5-mV error corresponds to a 22% error. The relative error *doubles* for divalent ions and *triples* for trivalent ions.

Care is required to obtain reliable results. Precision is rarely better than 1% and is usually worse. Electrodes can be fouled by proteins or other organic solutes, leading to sluggish, drifting response. Certain ions interfere with or poison particular electrodes. Some electrodes are fragile and have limited shelf life.

Electrodes respond to the *activity* of *uncomplexed* ion. If the ionic strength is held constant, concentration is proportional to activity and the electrode measures concentration.

Electrodes respond to the *activity* of *uncomplexed* analyte ion. Therefore, ligands must be absent or masked. Because we usually wish to know concentrations, not activities, an inert salt is often used to bring all the standards and samples to a high and constant ionic strength. If the activity coefficients remain constant, the electrode potential gives concentrations directly.

Standard Addition with Ion-Selective Electrodes

Standard addition was introduced in Section 6-6.

When using ion-selective electrodes, it is important that the compositions of the standard solutions closely approximate the composition of the unknown. The medium in which the analyte exists is called the **matrix.** In cases where the matrix is complex or unknown, the **standard addition method** can be used. In this technique, the electrode is immersed in the unknown and the potential is recorded. Then a small volume of standard solution is added, so as not to perturb the ionic strength of the unknown. The change in potential tells how the electrode responds to analyte and how much analyte was in the original solution.

Standard addition is best if the addition increases analyte to 1.5 to 3 times its original concentration. Results are more accurate if the effects of several additions are averaged.

EXAMPLE **Standard Addition with an Ion-Selective Electrode**

A perchlorate ion-selective electrode immersed in 50.0 mL of unknown perchlorate solution gave a potential of 358.7 mV versus S.C.E. When 1.00 mL of 0.050 M $NaClO_4$ was added, the potential changed to 346.1 mV. Assuming that the electrode has a Nernstian response ($\beta = 1.00$), find the concentration of ClO_4^- in the unknown.

Solution   The unknown contains x moles of ClO_4^- in 0.0500 L. The standard addition adds $(0.00100\ L)(0.0500\ M) = 5.00 \times 10^{-5}$ mol of ClO_4^-. Therefore, the second solution contains $x + (5.00 \times 10^{-5})$ mol in 0.0510 L. We can write a Nernst equation for the first solution:

$$E_1 = \text{constant} - 0.05916 \log[ClO_4^-]_1$$

and another for the second solution:

$$E_2 = \text{constant} - 0.05916 \log[ClO_4^-]_2$$

We set $E_1 = 0.3587$ V, $E_2 = 0.3461$ V, $[ClO_4^-]_1 = x/0.0500$, and $[ClO_4^-]_2 = (x + 5.00 \times 10^{-5})/0.0510$. These assignments allow us to solve for x by subtracting one equation from the other:

$$0.3587 = \text{constant} - 0.05916 \log[x/0.0500]$$
$$-\ 0.3461 = \text{constant} - 0.05916 \log[(x + 5.00 \times 10^{-5})/0.0510]$$

$$0.0126 = -0.05916 \log \frac{x/0.0500}{(x + 5.00 \times 10^{-5})/0.0510}$$

$$\frac{x/0.0500}{(x + 5.00 \times 10^{-5})/0.0510} = 0.612 \Rightarrow x = 7.51 \times 10^{-5}\ \text{mol}$$

The original perchlorate concentration was therefore

$$\frac{7.51 \times 10^{-5}\ \text{mol}}{0.0500\ \text{L}} = 1.50\ \text{mM}$$

Metal Ion Buffers

It is pointless to dilute $CaCl_2$ to 10^{-6} M for standardizing an ion-selective electrode. At this low concentration, Ca^{2+} ion could be lost by adsorption on glass or reaction with impurities.

An alternative is to prepare a **metal ion buffer** from the metal and a suitable ligand. For example, consider the reaction of Ca^{2+} with nitrilotriacetic acid (NTA) at a pH high enough for NTA to be in its fully basic form (NTA^{3-}):

$$Ca^{2+} + NTA^{3-} \rightleftharpoons CaNTA^-$$

$$K_f = \frac{[CaNTA^-]}{[Ca^{2+}][NTA^{3-}]} = 10^{6.46} \quad \text{in 0.1 M } KNO_3 \quad (15\text{-}9)$$

If equal concentrations of NTA^{3-} and $CaNTA^-$ are present in a solution, the concentration of Ca^{2+} would be

$$[Ca^{2+}] = \frac{[\cancel{CaNTA^-}]}{K_f[\cancel{NTA^{3-}}]} = 10^{-6.46}\ \text{M}$$

Glass vessels are not used for very dilute standard solutions, because ions are adsorbed on the glass surface. Plastic bottles and beakers are much better than glass for dilute solutions. Adding strong acid (0.1–1 M) to any solution helps to minimize adsorption of cations on the walls of the container, because H^+ competes with the other cations for ion-exchange sites.

$$HN(CH_2CO_2H)_3^+$$
$$H_4NTA^+$$

$$pK_1 = 1.1 \qquad pK_3 = 2.940$$
$$pK_2 = 1.650 \quad pK_4 = 10.334$$

EXAMPLE   **Preparing a Metal Ion Buffer**

What concentration of NTA^{3-} should be added to 1.00×10^{-2} M $CaNTA^-$ in 0.1 M KNO_3 to give $[Ca^{2+}] = 1.00 \times 10^{-6}$ M?

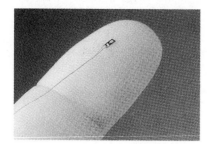

(a)

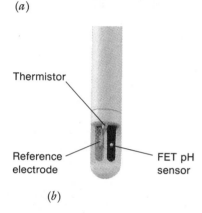

Thermistor

Reference electrode

FET pH sensor

(b)

Figure 15-21 (a) pH-sensitive field effect transistor. (b) Combination pH electrode based on field effect transistor. The thermistor senses temperature and is used for automatic temperature compensation. [Courtesy Sentron, Federal Way, WA.]

An activation energy is needed to get the charge carriers to move across the diode. For Si, ~0.6 V of forward bias is required before current will flow. For Ge, the requirement is ~0.2 V.

For moderate reverse bias voltages, no current flows. If the voltage is sufficiently negative, *breakdown* occurs and current flows in the reverse direction.

Solution Using Equation 15-9, we write

$$[NTA^{3-}] = \frac{[CaNTA^-]}{K_f[Ca^{2+}]} = \frac{1.00 \times 10^{-2}}{(10^{6.46})(1.00 \times 10^{-6})} = 3.47 \times 10^{-3} \text{ M}$$

These are practical concentrations of CaNTA$^-$ and NTA^{3-}.

· ·

15-8 Solid-State Chemical Sensors

Solid-state chemical sensors are fabricated by the same technology used for microelectronic chips.[16] The field effect transistor (FET) is the heart of commercially available sensors such as the pH electrode in Figure 15-21.

Semiconductors and Diodes

Semiconductors such as Si (Figure 15-22), Ge, and GaAs are materials whose electrical *resistivity*[17] lies between those of conductors and insulators. The four valence electrons of the pure materials are all involved in bonds between the atoms (Figure 15-23a). A phosphorus impurity, with five valence electrons, provides one extra **conduction electron** that is free to move through the crystal (Figure 15-23b). An aluminum impurity has one less bonding electron than is needed, thereby creating a vacancy called a **hole,** which behaves as a positive charge carrier. When a neighboring electron fills the hole, a new hole appears at an adjacent position (Figure 15-23c). A semiconductor with excess conduction electrons is called *n-type,* and one with excess holes is called *p-type.*

A **diode** is a *pn* junction (Figure 15-24). If *n*-Si is made negative with respect to *p*-Si, electrons flow from the external circuit into the *n*-Si. At the *pn* junction, electrons and holes combine. As electrons move from the *p*-Si into the circuit, a fresh supply of holes is created in the *p*-Si. The net result is that current flows when *n*-Si is made negative with respect to *p*-Si. The diode is said to be *forward biased.*

If the opposite polarity is applied, as depicted in Figure 15-24b, electrons are drawn out of the *n*-Si and holes are drawn out of the *p*-Si, leaving a thin *depletion region* devoid of charge carriers near the *pn* junction. The diode is *reverse biased* and does not conduct current in the reverse direction.

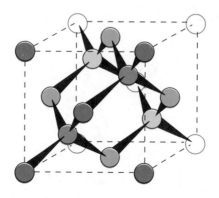

Figure 15-22 Diamondlike structure of silicon. Each atom is tetrahedrally bonded to four neighbors, with a Si—Si distance of 235 pm.

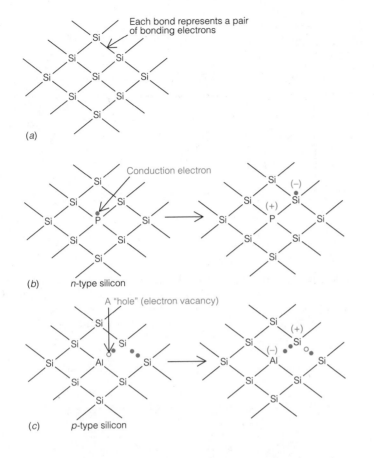

Figure 15-23 (*a*) The electrons of pure silicon are all involved in the sigma-bonding framework. (*b*) An impurity atom such as phosphorus adds one extra electron (●), which is relatively free to move through the crystal. (*c*) An aluminum impurity atom lacks one electron needed for the sigma-bonding framework. The hole (○) introduced by the Al atom can be occupied by an electron from a neighboring bond, effectively moving the hole to the neighboring bond.

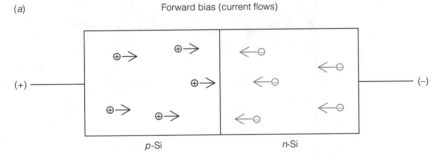

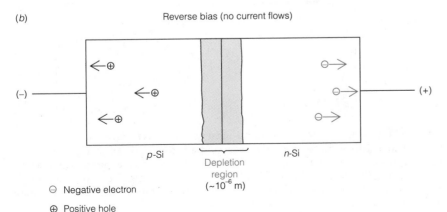

Figure 15-24 Behavior of a *pn* junction. Current can flow under forward bias conditions (*a*) but is prevented from flowing under reverse bias (*b*).

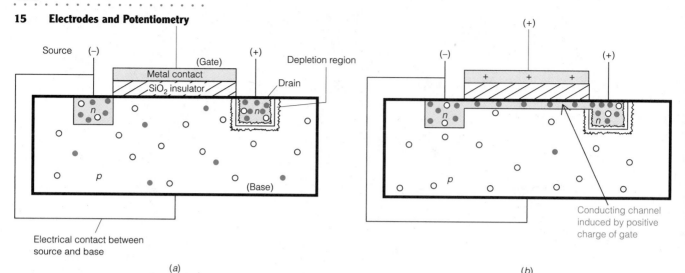

Figure 15-25   Operation of a field effect transistor. (*a*) Nearly random distribution of holes and electrons in the base in the absence of gate potential. (*b*) Positive gate potential attracts electrons that form a conductive channel beneath the gate. Current can flow through this channel between the source and drain.

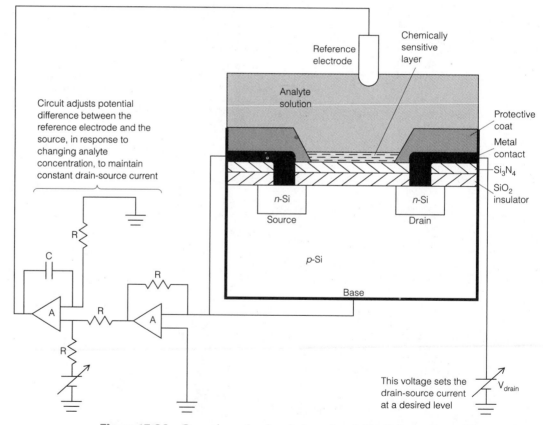

Figure 15-26   Operation of a chemical-sensing field effect transistor. The transistor is coated with an insulating SiO_2 layer and a second layer of Si_3N_4 (silicon nitride), which is impervious to ions and improves electrical stability. The circuit at the lower left adjusts the potential difference between the reference electrode and the source in response to changes in the analyte solution, such that a constant drain-source current is maintained.

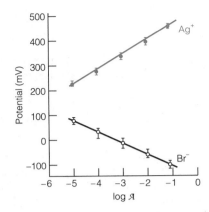

Figure 15-27 Response of a silver bromide-coated field effect transistor. Error bars are 95% confidence intervals for data obtained from 195 sensors prepared from different chips. [R. P. Buck and D. E. Hackleman, *Anal. Chem.* **1977,** *49,* 2315.]

Chemical-Sensing Field Effect Transistors

The *base* of the **field effect transistor** in Figure 15-25 is constructed of *p*-Si with two *n*-type regions called the *source* and the *drain*. An insulating surface layer of SiO_2 is overcoated by a conductive metal *gate* between the source and drain. The source and the base are held at the same electric potential. When a voltage is applied between the source and the drain (Figure 15-25a), little current flows because the drain-base interface is a *pn* junction in reverse bias.

When the gate is made positive, electrons from the base are attracted toward the gate and form a conductive channel between the source and the drain (Figure 15-25b). The current increases as the gate is made more positive. *The potential of the gate therefore regulates the current flow between source and drain.*

The more positive the gate, the more current can flow between source and drain.

The essential feature of the chemical-sensing field effect transistor in Figure 15-26 is the chemically sensitive layer over the gate. An example is a layer of AgBr. When exposed to silver nitrate solution, Ag^+ is adsorbed on the AgBr (Figure 7-12), thereby giving it a positive charge and increasing current between the source and the drain. *The voltage that must be applied by an external circuit to bring the current back to its initial value represents the response of the device to Ag^+.* Figure 15-27 shows that Ag^+ makes the gate more positive and Br^- makes the gate more negative. The response is close to 59 mV for a 10-fold concentration change. Advantages of the transistor are that it is smaller (Figure 15-21) and more rugged than ion-selective electrodes. The sensing surface is typically only 1 mm^2.

Terms to Understand

Summary

In potentiometric measurements, the indicator electrode responds to changes in the activity of analyte and the reference electrode is a self-contained half-cell producing a constant reference potential. The most common reference electrodes are calomel and silver-silver chloride. Common indicator electrodes include (1) the inert Pt electrode, (2) a silver electrode responsive to Ag^+, halides, and other ions that react with Ag^+, and (3) ion-selective electrodes. Small, unknown junction potentials at liquid-liquid interfaces limit the accuracy of most potentiometric measurements.

Ion-selective electrodes, including the glass pH electrode, respond selectively to one species in a solution. A gradient of activity produces a gradient of free energy, ΔG, equal to $-RT \ln \mathcal{A}_1/\mathcal{A}_2$. The electric potential difference corresponding to this free energy difference is $E =$ $-\Delta G/nF = (RT/nF) \ln \mathcal{A}_1/\mathcal{A}_2$. Most electrodes are sensitive to many species, and their net response can be described by the equation

$$E = \text{constant} \pm \beta \,(0.059\,16/n_X) \log$$
$$[\mathcal{A}_X + \sum(k_{X,Y}\mathcal{A}_Y{}^{n_X/n_Y})]$$

where $k_{X,Y}$ is the selectivity coefficient for each species. Common ion-selective electrodes can be classified as solid-state, liquid-based, and compound. Metal ion buffers are appropriate for establishing and maintaining low concentrations of ions. The field effect transistor is a solid-state device that uses a chemically sensitive coating to alter the electrical properties of a semiconductor in response to changes in the chemical environment.

Exercises

A.   The apparatus in Figure 7-9 was used to monitor the titration of 100.0 mL of solution containing 50.0 mL of 0.100 M $AgNO_3$ and 50.0 mL of 0.100 M $TlNO_3$. The titrant contained 0.200 M NaBr. Suppose that the glass electrode (used as a *reference electrode* in this experiment) gives a constant potential of +0.200 V. The glass electrode is attached to the *positive* terminal of the pH meter, and the silver wire to the negative terminal. Calculate the cell voltage at each of the following volumes of NaBr, and sketch the titration curve: 1.0, 15.0, 24.0, 24.9, 25.2, 35.0, 50.0, 60.0 mL.

B.   The apparatus shown on the next page can be used to follow the course of an EDTA titration and was used to generate the curves in Figure 13-8. The heart of the cell is a pool of liquid Hg in contact with the solution and with a Pt wire. A small amount of HgY^{2-} added to the analyte equilibrates with a very tiny amount of Hg^{2+}:

$$Hg^{2+} + Y^{4-} \rightleftharpoons HgY^{2-}$$

$$K_f = \frac{[HgY^{2-}]}{[Hg^{2+}][Y^{4-}]} = 5 \times 10^{21} \qquad (A)$$

The redox equilibrium $Hg^{2+} + 2e^- \rightleftharpoons Hg(l)$ is established rapidly at the surface of the Hg electrode, so the Nernst equation for the cell can be written in the form

$$E = E_+ - E_-$$
$$= \left(0.852 - \frac{0.059\,16}{2}\log\frac{1}{[Hg^{2+}]}\right) - E_- \qquad (B)$$

where E_- is the constant potential of the reference electrode. But from Equation A, we can write $[Hg^{2+}] = [HgY^{2-}]/K_f[Y^{4-}]$, and this can be substituted into Equation B to give

$$E = 0.852 - \frac{0.059\,16}{2}\log\frac{[Y^{4-}]K_f}{[HgY^{2-}]} - E_-$$
$$= 0.852 - E_- - \frac{0.059\,16}{2}\log\frac{K_f}{[HgY^{2-}]}$$
$$- \frac{0.059\,16}{2}\log[Y^{4-}] \qquad (C)$$

where K_f is the formation constant for HgY^{2-}. This apparatus thus responds to the changing EDTA concentration during an EDTA titration.

Suppose that 50.0 mL of 0.0100 M $MgSO_4$ is titrated with 0.0200 M EDTA at pH 10.0 by using the apparatus shown above with an S.C.E. reference electrode. Assume that the analyte contains 1.0×10^{-4} M $Hg(EDTA)^{2-}$ added at the beginning of the titration. Calculate the cell voltage at the following volumes of added EDTA, and draw a graph of millivolts versus milliliters: 0, 10.0, 20.0, 24.9, 25.0, and 26.0 mL

C.   A solid-state fluoride ion-selective electrode responds to F^-, but not to HF. It also responds to hydroxide ion at high concentration when $[OH^-] \approx [F^-]/10$. Suppose that such an electrode gave a potential of +100 mV (versus S.C.E.) in 10^{-5} M NaF and +41 mV in 10^{-4} M NaF. Sketch qualitatively how the potential would vary if the electrode were immersed in 10^{-5} M NaF and the pH ranged from 1 to 13.

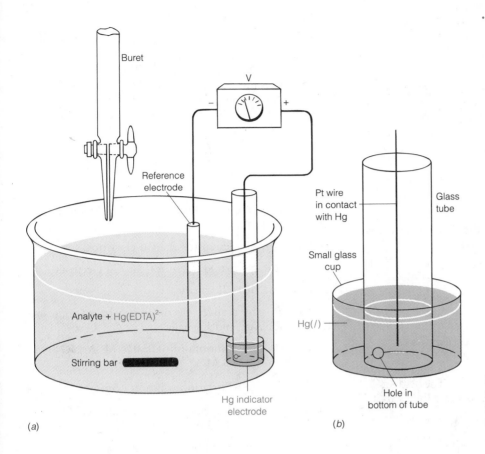

Diagram for Exercise B. (*a*) Apparatus for potentiometric EDTA titration. (*b*) Enlarged view of mercury electrode.

(*a*)

(*b*)

D. One commercial glass-membrane sodium ion-selective electrode has a selectivity coefficient $k_{Na^+, H^+} = 36$. When this electrode was immersed in 1.00 mM NaCl at pH 8.00, a potential of -38 mV (versus S.C.E.) was recorded.

(a) Neglecting activity coefficients and assuming $\beta = 1$ in Equation 15-6, calculate the potential if the electrode were immersed in 5.00 mM NaCl at pH 8.00.

(b) What would be the potential for 1.00 mM NaCl at pH 3.87?

After answering this question, you should realize that pH is a critical variable in the use of a sodium ion-selective electrode.

E. An ammonia gas-sensing electrode gave the following calibration points when all solutions contained 1 M NaOH.

NH_3 (M)	E (mV)	NH_3 (M)	E (mV)
1.00×10^{-5}	268.0	5.00×10^{-4}	368.0
5.00×10^{-5}	310.0	1.00×10^{-3}	386.4
1.00×10^{-4}	326.8	5.00×10^{-3}	427.6

A dry food sample weighing 312.4 mg was digested by the Kjeldahl procedure (Section 7-2) to convert all the nitrogen to NH_4^+. The digestion solution was diluted to 1.00 L, and 20.0 mL was transferred to a 100-mL volumetric flask. The 20.0-mL aliquot was treated with 10.0 mL of 10.0 M NaOH plus enough NaI to complex the Hg catalyst from the digestion, and diluted to 100.0 mL. When measured with the ammonia electrode, this solution gave a reading of 339.3 mV. Calculate the percentage of nitrogen in the food sample.

F. Cyanide ion was measured indirectly with a solid-state ion-selective electrode containing a silver sulfide membrane. Assuming that the electrode response is Nernstian and that the ionic strength of all solutions is constant, we can write the electrode response as follows:

$$E = \text{constant} + 0.059\,16 \log[Ag^+]$$

To an unknown cyanide solution was added $Ag(CN)_2^-$ such that $[Ag(CN)_2^-] = 1.0 \times 10^{-5}$ M in the final solution. This complex ion behaves as a silver ion buffer in the presence of CN^-:

$$Ag^+ + 2CN^- \rightleftharpoons Ag(CN)_2^-$$

$$K = \beta_2 = \frac{[Ag(CN)_2^-]}{[Ag^+][CN^-]^2} = 10^{19.85}$$

(a) Suppose that the unknown contained 8.0×10^{-6} M CN^- and the potential was found to be $+206.3$ mV.

Then a *standard addition* of CN^- was made to bring the CN^- concentration to 12.0×10^{-6} M. What will be the new potential?

(b) Now consider a real experiment in which 50.0 mL of unknown gave a potential of 134.8 mV before a standard addition of CN^- was made. After adding 1.00 mL of 2.50×10^{-4} M KCN, the potential dropped to 118.6 mV. What was the concentration of CN^- in the 50.0-mL sample? For this part, we do not know the value of the constant in the equation for E. That is, you cannot use the value of the constant derived from **(a)**.

Problems

Reference Electrodes

1. **(a)** Write the half-reactions for the silver-silver chloride and calomel reference electrodes.

(b) Predict the voltage for the cell below.

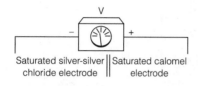

| Saturated silver-silver chloride electrode ‖ Saturated calomel electrode |

2. Convert the potentials listed below. The Ag|AgCl and calomel reference electrodes are saturated with KCl.

(a) 0.523 V versus S.H.E. = ? versus Ag|AgCl

(b) −0.111 V versus Ag|AgCl = ? versus S.H.E.

(c) −0.222 V versus S.C.E. = ? versus S.H.E.

(d) 0.023 V versus Ag|AgCl = ? versus S.C.E.

(e) −0.023 V versus calomel = ? versus Ag|AgCl

3. Suppose that the silver-silver chloride electrode in Figure 15-2 is replaced by a saturated calomel electrode. Calculate the cell voltage if $[Fe^{2+}]/[Fe^{3+}] = 2.5 \times 10^{-3}$.

4. For a silver-silver chloride electrode, the following potentials are observed:

$$E° = 0.222 \text{ V} \qquad E(\text{saturated KCl}) = 0.197 \text{ V}$$

Predict the value of E for a calomel electrode saturated with KCl, given that $E°$ for the calomel electrode is 0.268 V. (Your answer will not be exactly the value 0.241 V used in this text.)

5. Using the potential given below, calculate the *activity* of Cl^- in 1 M KCl.

$$E°(\text{calomel electrode}) = 0.268 \text{ V}$$

$$E(\text{calomel electrode, 1 M KCl}) = 0.280 \text{ V}$$

Indicator Electrodes

6. A cell was prepared by dipping a Cu wire and a saturated calomel electrode into 0.10 M $CuSO_4$ solution. The Cu wire was attached to the positive terminal of a potentiometer and the calomel electrode was attached to the negative terminal.

(a) Write a half-reaction for the Cu electrode.

(b) Write the Nernst equation for the Cu electrode.

(c) Calculate the cell voltage.

7. Explain why a silver electrode can be an indicator electrode for Ag^+ and for halides.

8. A 10.0 mL solution of 0.0500 M $AgNO_3$ was titrated with 0.025 0 M NaBr in the cell:

$$\text{S.C.E. ‖ titration solution | Ag}(s)$$

Find the cell voltage for the following volumes of titrant: 0.1, 10.0, 20.0, and 30.0 mL.

9. A solution prepared by mixing 25.0 mL of 0.200 M KI with 25.0 mL of 0.200 M NaCl was titrated with 0.100 M $AgNO_3$ in the following cell:

$$\text{S.C.E. ‖ titration solution | Ag}(s)$$

Call the solubility products of AgI and AgCl K_I and K_{Cl}, respectively. The answers to **(a)** and **(b)** should be expressions containing these constants.

(a) Calculate the concentration of $[Ag^+]$ in the solution when 25.0 mL of titrant has been added.

(b) Calculate the concentration of $[Ag^+]$ in the solution when 75.0 mL of titrant has been added.

(c) Write an expression showing how the cell voltage depends on $[Ag^+]$.

(d) The titration curve is shown on the next page. Calculate the numerical value of the quotient K_{Cl}/K_I.

10. A solution containing 50.0 mL of 0.100 M EDTA buffered to pH 10.00 was titrated with 50.0 mL of 0.020 0 M $Hg(ClO_4)_2$ in the cell below.

$$\text{S.C.E. ‖ titration solution | Hg}(l)$$

From the cell voltage of −0.036 V, calculate the formation constant of $Hg(EDTA)^{2-}$. (*Hint:* See Exercise B.)

11. Consider the cell S.C.E. ‖ cell solution | Pt(s) whose voltage is −0.126 V. The cell solution contains

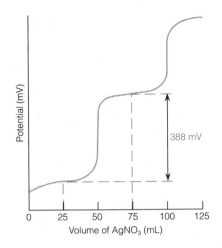

Titation curve for Problem 9.

2.00 mmol of $Fe(NH_4)_2(SO_4)_2$, 1.00 mmol of $FeCl_3$, 4.00 mmol of Na_2EDTA, and lots of buffer, pH 6.78, in a volume of 1.00 L.

(a) Write a reaction for the right half-cell.

(b) Find the value of $[Fe^{2+}]/[Fe^{3+}]$ in the cell solution. (This value is the ratio of *uncomplexed* ions.)

(c) Find the quotient of formation constants, $K_f(FeEDTA^-)/K_f(FeEDTA^{2-})$.

12. Here's a cell you'll really like:

$$\text{electrode} \parallel \text{cell solution} \parallel Cu(s)$$

where the electrode is a saturated silver-silver chloride electrode. The cell solution was made by mixing

25.0 mL of 4.00 mM KCN

25.0 mL of 4.00 mM $KCu(CN)_2$

25.0 mL of 0.400 M acid, HA, with $pK_a = 9.50$

25.0 mL of KOH solution

The measured voltage was -0.440 V. *Calculate the molarity of the KOH solution.* You may assume that essentially all the cuprous ion is present in the form $Cu(CN)_2^-$. For the right half-cell, the reaction is $Cu(CN)_2^- + e^- \rightleftharpoons Cu(s) + 2CN^-$ $(E° = -0.429$ V).

Junction Potential

13. What causes a junction potential? How does this limit the accuracy of potentiometric analyses? Identify a cell in the illustrations in Section 14-2 that has no junction potential.

14. Why is the 0.1 M HCl | 0.1 M KCl junction

potential of opposite sign and greater magnitude than the 0.1 M NaCl | 0.1 M KCl potential in Table 15-2?

15. Which side of the liquid junction 0.1 M KNO_3 | 0.1 M NaCl will be negative?

16. How many seconds will it take for (a) H^+ and (b) NO_3^- to migrate a distance of 12.0 cm in a field of 7.80×10^3 V/m?

17. Suppose that an ideal hypothetical cell such as that in Figure 14-6 were set up to measure $E°$ for the half-reaction $Ag^+ + e^- \rightleftharpoons Ag(s)$.

(a) Calculate the equilibrium constant for the net cell reaction.

(b) If there were a junction potential of $+2$ mV (increasing E from 0.799 to 0.801 V), by what percentage would the calculated equilibrium constant increase?

(c) Redo (a) and (b), using 0.100 V instead of 0.799 V for the value of $E°$ for the silver reaction.

18. Explain how the cell $Ag(s) \mid AgCl(s) \mid 0.1$ M HCl $\mid$ 0.1 M KCl $\mid AgCl(s) \mid Ag(s)$ can be used to measure the 0.1 M HCl $\mid$ 0.1 M KCl junction potential.

19. [icon] The junction potential, E_j, between solutions α and β can be estimated with the Henderson equation:

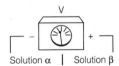

$$E_j \approx \frac{\sum_i \frac{|z_i|u_i}{z_i}[C_i(\beta) - C_i(\alpha)]}{\sum_i |z_i|u_i[C_i(\beta) - C_i(\alpha)]} \frac{RT}{F} \ln \frac{\sum_i |z_i|u_i C_i(\alpha)}{\sum_i |z_i|u_i C_i(\beta)}$$

where z_i is the charge of species i, u_i is the *mobility* of species i (Table 15-1), $C_i(\alpha)$ is the concentration of species i in phase α, and $C_i(\beta)$ is the concentration in phase β. (An approximation here is the neglect of activity coefficients.)

(a) Using your calculator, show that the junction potential of 0.1 M HCl | 0.1 M KCl is 26.9 mV at 25°C (where $(RT/F) \ln x = 0.059\,16 \log x$).

(b) Set up a spreadsheet to reproduce the result in (a). Then use your spreadsheet to compute and plot the junction potential for 0.1 M HCl | x M KCl, where x varies from 1 mM to 4 M.

(c) Use your spreadsheet to explore the behavior of the junction potential for y M HCl | x M KCl, where $y = 10^{-4}, 10^{-3}, 10^{-2},$ and 10^{-1} M and $x = 1$ mM or 4 M.

pH Measurement with a Glass Electrode

20. Describe how you would calibrate a pH electrode and measure the pH of blood (which is ~ 7.5) *at 37°C*. Use the standard buffers in Table 15-3.

21. List the sources of error associated with pH measurement using the glass electrode.

22. If electrode C in Figure 15-14 is placed in a solution of pH 11.0, what will the pH reading be?

23. Which National Institute of Standards and Technology buffer(s) would you use to calibrate an electrode for pH measurements in the range 3–4?

24. Why do glass pH electrodes tend to indicate a pH lower than the actual pH in strongly basic solution?

25. Suppose that the Ag | AgCl outer electrode in Figure 15-9 is filled with 0.1 M NaCl instead of saturated KCl. Suppose that the electrode is calibrated in a dilute buffer containing 0.1 M KCl at pH 6.54 at 25°C. The electrode is then dipped in a second buffer *at the same pH* and same temperature, but containing 3.5 M KCl. Use Table 15-2 to estimate how much the indicated pH will change.

26. (a) When the difference in pH across the membrane of a glass electrode at 25°C is 4.63 pH units, how much voltage is generated by the pH gradient?

(b) What would be the voltage for the same pH difference at 37°C?

Ion-Selective Electrodes

27. Explain the principle of operation of ion-selective electrodes. How does a compound electrode differ from a simple ion-selective electrode?

28. What does the selectivity coefficient tell us? Is it better to have a large or a small selectivity coefficient?

29. What feature gives a liquid-based ion-selective electrode specificity for a particular analyte?

30. Why is it preferable to use a metal ion buffer to achieve pM = 8, rather than just dissolving enough M to give a 10^{-8} M solution?

31. To determine the *concentration* of a dilute analyte with an ion-selective electrode, why is it desirable to use standards with a constant, high concentration of an inert salt?

32. A cyanide ion-selective electrode obeys the equation

$$E = \text{constant} - 0.059\,16\log[CN^-]$$

The electrode potential was -0.230 V when the electrode was immersed in 1.00×10^{-3} M NaCN.

(a) Evaluate the constant in the equation above.

(b) Using the result from **(a)**, find the concentration of CN^- if $E = -0.300$ V.

(c) Without using the constant from **(a)**, find the concentration of CN^- if $E = -0.300$V.

33. By how many volts will the potential of a Mg^{2+} ion-selective electrode change if the electrode is removed from 1.00×10^{-4} M $MgCl_2$ and placed in 1.00×10^{-3} M $MgCl_2$?

34. The selectivities of a lithium ion-selective electrode are indicated on the following diagram:

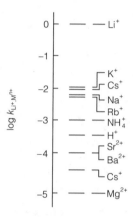

(a) Which alkali metal (Group I) ion causes the most interference?

(b) Do the alkali metal ions cause greater or less interference than the alkaline earth (Group II) ions?

35. A metal ion buffer was prepared from 0.030 0 M ML and 0.020 M L, where ML is a metal-ligand complex and L is free ligand.

$$M + L \rightleftharpoons ML \qquad K_f = 4.0 \times 10^8$$

Calculate the concentration of free metal ion, M, in this buffer.

36. The data below were obtained when a Ca^{2+} ion-selective electrode was immersed in a series of standard solutions whose ionic strength was constant at 2.0 M.

Ca^{2+} (M)	E (mV)
3.38×10^{-5}	-74.8
3.38×10^{-4}	-46.4
3.38×10^{-3}	-18.7
3.38×10^{-2}	$+10.0$
3.38×10^{-1}	$+37.7$

(a) Make a graph of the data and find the concentration of Ca^{2+} in a sample that gave a reading of -22.5 mV.

(b) Calculate the value of β in Equation 15-8.

37. (a) 🖳 Calculate the slope and the y-intercept (and their standard deviations) of the best straight line through the points in the previous problem, using your least-squares spreadsheet from Chapter 4.

(b) Calculate the concentration (and its associated uncertainty) of a sample that gave a reading of -22.5 (± 0.3) mV.

38. The selectivity coefficient, $k_{Li^+, Ca^{2+}}$, for a lithium ion-selective electrode is 5×10^{-5}. When this electrode is placed in 3.44×10^{-4} M Li^+ solution, the potential is -0.333 V versus S.C.E. What would the potential be if Ca^{2+} were added to give 0.100 M Ca^{2+}?

39. The following standard addition table assumes that an electrode has a Nernstian response to analyte:

Standard addition table: 10 mL of standard added to 100 mL of sample.

ΔE (mV)	Q	ΔE (mV)	Q	ΔE (mV)	Q
0	1.00	10	0.160	20	0.0716
1	0.696	11	0.145	21	0.0671
2	0.529	12	0.133	22	0.0629
3	0.423	13	0.121	23	0.0591
4	0.351	14	0.112	24	0.0556
5	0.297	15	0.1030	25	0.0523
6	0.257	16	0.0952	26	0.0494
7	0.225	17	0.0884	27	0.0466
8	0.199	18	0.0822	28	0.0440
9	0.178	19	0.0767	29	0.0416

To obtain sample concentration, multiply standard concentration by Q.

As an example, suppose chloride ion is measured with an ion-selective electrode and a reference electrode. The electrodes are placed in 100.0 mL of sample, and a reading of 228.0 mV is obtained. Then 10.0 mL of standard containing 100.0 ppm of Cl^- is added, and a new reading of 210.0 mV is observed. Because $|\Delta E| = 18.0$ mV, $Q = 0.0822$ in the table above. Therefore, the original concentration of Cl^- was $(0.0822)(100 \text{ ppm}) = 8.22$ ppm.

(a) What molarity of Cl^- is 8.22 ppm? Assume that the density of unknown is 1.00 g/mL.

(b) Use the original concentration of 8.22 ppm and the original potential of 228.0 mV to show that the second potential should be 210.0 mV if the electrode response obeys the equation

$$E = \text{constant} - 0.05916 \log [Cl^-]$$

(c) How would you change the table to use it for Ca^{2+} instead of Cl^-?

40. (a) 🖳 *Derive* the value of 0.696 for the second value of Q in the standard addition of the previous problem.

(b) Based on your derivation, create a spreadsheet that computes Q as a function of ΔE and reproduce the table in the previous problem.

41. A calcium ion-selective electrode was calibrated in metal ion buffers whose ionic strength was fixed at 0.50 M. Using the electrode readings below, write an equation for the response of the electrode to Ca^{2+} and Mg^{2+}.

$[Ca^{2+}]$ (M)	$[Mg^{2+}]$ (M)	mV
1.00×10^{-6}	0	-52.6
2.43×10^{-4}	0	$+16.1$
1.00×10^{-6}	3.68×10^{-3}	-38.0

42. An ion-selective electrode used to measure the cation M^{2+} obeys the equation

$$E - \text{constant} + \frac{0.0568 \text{ V}}{2} \log([M^{2+}] + 0.0013 \, [Na^+]^2)$$

When the electrode was immersed in 10.0 mL of unknown containing M^{2+} in 0.200 M $NaNO_3$, the reading was -163.3 mV (versus S.C.E.). When 1.00 mL of 1.07×10^{-3} M M^{2+} (in 0.200 M $NaNO_3$) was added to the unknown, the reading increased to -158.8 mV. Find the concentration of M^{2+} in the original unknown.

43. A lead ion buffer was prepared by mixing 0.100 mmol of $Pb(NO_3)_2$ with 2.00 mmol of $Na_2C_2O_4$ in a volume of 10.0 mL.

(a) Given the equilibrium below, find the concentration of free Pb^{2+} in this solution.

$$Pb^{2+} + 2C_2O_4^{2-} \rightleftharpoons Pb(C_2O_4)_2^{2-}$$

$$K = \beta_2 = 10^{6.54}$$

(b) How many millimoles of $Na_2C_2O_4$ should be used to give $[Pb^{2+}] = 1.00 \times 10^{-7}$ M?

44. Do not ignore activities in this problem. Citric acid is a triprotic acid (H_3A) whose anion (A^{3-}) forms stable complexes with many metal ions.

(a) A calcium ion-selective electrode gave a calibration curve similar to Figure B-2 in Appendix B. The slope of the curve was 29.58 mV. When the electrode was immersed in a solution having $\mathcal{A}_{Ca^{2+}} = 1.00 \times 10^{-3}$, the reading was $+2.06$ mV. When the electrode was immersed in the solution to be described in (b) of this problem, the reading was -25.90 mV. Calculate the activity of Ca^{2+} in the solution in (b).

(b) Ca^{2+} forms a 1:1 complex with citrate under the conditions of this problem.

$$Ca^{2+} + A^{3-} \overset{K_f}{\rightleftharpoons} CaA^-$$

A solution was prepared by mixing equal volumes of solutions 1 and 2 below.

solution 1: $[Ca^{2+}] = 1.00 \times 10^{-3}$ M
$$pH = 8.00, \quad \mu = 0.10 \text{ M}$$

solution 2: $[\text{citrate}]_{\text{total}} = 1.00 \times 10^{-3}$ M
$$pH = 8.00, \quad \mu = 0.10 \text{ M}$$

The activity of the calcium ion in the resulting solution was determined in (a) of this problem. Calculate the formation constant, K_f, for CaA^-. For the sake of this calculation, you may assume that the size of CaA^- is 500 pm. At pH 8.00 and $\mu = 0.10$ M, the fraction of free citrate in the form A^{3-} is 0.998.

Solid-State Chemical Sensors

45. What does analyte do to a chemical-sensing field effect transistor to produce a signal related to the activity of analyte?

Notes and References

1. Practical aspects of electrode fabrication are discussed in D. T. Sawyer and J. L. Roberts, Jr., *Experimental Electrochemistry for Chemists* (New York: Wiley, 1974); R. G. Bates, *Determination of pH* (New York: Wiley, 1984); and E. P. Serjeant, *Potentiometry and Potentiometric Titrations* (New York: Wiley, 1984). A versatile junction for reference electrodes can be prepared by attaching a molecular sieve to glass tubing with heat-shrinkable tubing as described by S. Yee and O.-K. Chang, *J. Chem. Ed.* **1988,** *65,* 129.

2. General discussions of oscillating reactions can be found in I. R. Epstein, K. Kustin, P. De Kepper, and M. Orbán, *Scientific American,* March 1983, p. 112; I. R. Epstein, *Chem. Eng. News,* 30 March 1987, p. 24; and H. Degn, *J. Chem. Ed.* **1972,** *49,* 302.

3. The mechanisms of oscillating reactions are discussed by R. J. Field and F. W. Schneider, *J. Chem. Ed.* **1989,** *66,* 195; R. M. Noyes, *J. Chem. Ed.* **1989,** *66,* 190; and P. Ruoff, M. Varga, and E. Körös, *Acc. Chem. Res.* **1988,** *21,* 326.

4. For some beautiful lecture demonstrations and experiments, see J. A. Pojman, R. Craven, D. C. Leard, *J. Chem. Ed.* **1994,** *71,* 84; D. Kolb, *J. Chem. Ed.* **1988,** *65,* 1004; R. J. Field, *J. Chem. Ed.* **1972,** *49,* 308; J. N. Demas and D. Diemente, *J. Chem. Ed.* **1973,** *50,* 357; and J. F. Lefelhocz, *J. Chem. Ed.* **1972,** *49,* 312. Computer monitoring of an oscillating reaction is described by P. Aroca, Jr., and R. Aroca, *J. Chem. Ed.* **1987,** *64,* 1017. Other systems of interest include a chemiluminescent oscillator [J. Amrehn, P. Resch, and F. W. Schneider, *J. Phys. Chem.* **1988,** *92,* 3318], the mercury beating heart [D. Avnir, *J. Chem. Ed.* **1989,** *66,* 211], a saltwater oscillator [K. Yoshikawa, S. Nakata, M. Yamanaka, and T. Waki, *J. Chem. Ed.* **1989,** *66,* 205], a gas flashback oscillator [L. J. Soltzberg, M. M. Boucher, D. M. Crane, and S. S. Pazar, *J. Chem. Ed.* **1987,** *64,* 1043], and a gas evolution oscillator [S. M. Kaushik, Z. Yuan, and R. M. Noyes, *J. Chem. Ed.* **1986,** *63,* 76; R. F. Melka, G. Olsen, L. Beavers, and J. A. Draeger, *J. Chem. Ed.* **1992,** *69,* 596].

5. This same experiment can be used to demonstrate a bromide ion-selective electrode, because the concentration of Br^- also oscillates in this solution. Simultaneous monitoring of $[Ce^{3+}]/[Ce^{4+}]$ and the Br^- concentration would be worthwhile. An even more striking oscillation, in which the concentration of I^- oscillates over four orders of magnitude, has been described [T. S. Briggs and W. C. Rauscher, *J. Chem. Ed.* **1973,** *50,* 496]. Monitoring the oscillating I^- concentration with a silver-silver iodide electrode is an informative classroom demonstration.

6. C. J. Stalder, S. Chao, and M. S. Wrighton, *J. Am. Chem. Soc.* **1984,** *106,* 3673. For reviews of surface-modified electrodes, see R. W. Murray, Ed., *Molecular Design of Electrode Surfaces* (New York: Wiley, 1992) and R. W. Murray, A. G. Ewing, and R. A. Durst, *Anal. Chem.* **1987,** *59,* 379A.

7. Although the glass electrode is by far the most widely used electrode for pH measurement, many other types of electrodes are also sensitive to pH and can withstand harsher environments than those tolerated by the glass electrode. An electrode of iridium coated with IrO_2 behaves as a pH sensor [S. Bordi, M. Carlà, and G. Papeschi, *Anal. Chem.* **1984,** *56,* 317; M. L. Hitchman and S. Ramanathan, *Anal. Chim. Acta* **1992,** *263,* 53]. An electrode based on ZrO_2 can be used to measure the pH of solutions up to 300°C [L. W. Niedrach, *Angew. Chem.* **1987,** *26,* 161]. Platinum and carbon surfaces are also pH sensitive, but they must be coated to make them selective for H^+ [G. Cheek, C. P. Wales, and R. J. Nowak, *Anal. Chem* **1983,** *55,* 380]. Ingenious surface-modified wire electrodes for pH measurement have also been described [I. Rubenstein, *Anal. Chem.* **1984,** *56,* 1135].

8. Cu wire attached to the inside surface of the glass electrode with silver paste is reported to work as well as the conventional aqueous Ag|AgCl inner reference electrode. This gives an all-solid-state pH electrode with no internal solution [K. L. Cheng and N. Ashraf, *Talanta* **1990,** *37,* 659].

9. pH standards are described by R. G. Bates, *Anal. Chem.* **1968** (No. 6), *40*, 28A and H. B. Kristensen, A. Salomon, and G. Kokholm, *Anal. Chem.* **1991**, *63*, 885A. When we calibrate a pH electrode with one standard buffer, we are making use of the *operational definition* of pH:

$$pH_{unknown} - pH_{standard} = \frac{E_{unknown} - E_{standard}}{(RT/F)\ln 10}$$

where $E_{unknown} - E_{standard}$ is the difference in voltage readings when the electrodes are immersed in unknown or standard, R is the gas constant, T is temperature, and F is the Faraday constant. This operational definition comes directly from Equation 15-3 and assumes an ideal Nernstian response of the electrode. A more accurate procedure is to calibrate with two standards, designated S1 and S2, whose pH values bracket that of the unknown. In this case, the operational definition for the pH of an unknown is

$$\frac{pH_{unknown} - pH_{S1}}{pH_{S2} - pH_{S1}} = \frac{E_{unknown} - E_{S1}}{E_{S2} - E_{S1}}$$

Using two buffers corrects for nonideal response of real electrodes.

10. Circuits for building inexpensive pH meters are described by D. L. Harris and D. C. Harris, *J. Chem. Ed.* **1992**, *69*, 563; M. R. Paris, D. J. Aymes, R. Poupon, and R. Gavasso, *J. Chem. Ed.* **1990**, *67*, 507; M. S. Caceci, *J. Chem. Ed.* **1984**, *61*, 935, and B. D. Warner, G. Boehme, and K. H. Pool, *J. Chem. Ed.* **1982**, *59*, 65.

11. W. F. Koch, G. Marinenko, and R. C. Paule, *J. Res. National Bureau of Standards* **1986**, *91*, 23.

12. A commercially available electrode is designed to minimize junction potentials induced by changing ionic strength. The *free diffusion junction* of this electrode is a Teflon capillary tube containing an electrolyte that is periodically renewed by injecting electrolyte with a syringe [A. Kopelove, S. Franklin, and G. M. Miller, *Am. Lab.* June 1989, p. 40].

13. Construction of inexpensive solid-state ion-selective electrodes for classroom experiments is described by A. Palanivel and P. Riyazuddin, *J. Chem. Ed.* **1984**, *61*, 920 and W. S. Selig, *J. Chem. Ed.* **1984**, *61*, 80. Many inorganic anions can be titrated with quaternary ammonium halides by using homemade indicator electrodes [W. S. Selig, *J. Chem. Ed.* **1987**, *64*, 141].

14. Construction of liquid-based ion-selective electrodes for student experiments is described by L. Ramaley, P. J. Wedge, and S. M. Crain, *J. Chem. Ed.* **1994**, *71*, 164.

15. Enzyme-based penicillin and urea electrodes for student experiments have been described by T. E. Mifflin, K. M. Andriano, and W. B. Robbins, *J. Chem. Ed.* **1984**, *61*, 638 and by T. L. Riechel, *J. Chem. Ed.* **1984**, *61*, 640. Compound electrodes with biological applications are described in a review by J. D. Czaban, *Anal. Chem.* **1985**, *57*, 345A. For a general review of biosensors, see G. A. Rechnitz, *Chem. Eng. News,* September 5, **1988**, p. 24.

16. H. Wohltjen, *Anal. Chem.* **1984**, *56*, 87A; J. Janata and R. J. Huber, *Solid State Chemical Sensors* (New York: Academic, 1985).

17. *Resistivity,* ρ, is a measure of how well a substance retards the flow of electric current when an electric field is applied: $J = E/\rho$, where J is the current density (current flowing through a unit cross section of the material, amperes per square meter) and E is the electric field (V/m). Units of resistivity are $V \cdot m/A$ or $\Omega \cdot m$, because $\Omega = V/A$, where $\Omega =$ ohm. Conductors have resistivities near 10^{-8} $\Omega \cdot m$; semiconductors have resistivities in the range 10^{-4}–10^{7} $\Omega \cdot m$; and insulators have resistivities in the range 10^{12}–10^{20} $\Omega \cdot m$. The reciprocal of resistivity is *conductivity*. Resistivity does not depend on the dimensions of the substance. Resistance, R, is related to resistivity by the equation $R = \rho l/A$, where l is the length and A is the cross-sectional area of the conducting substance.

Chemical Analysis of High-Temperature Superconductors

Permanent magnet levitates above superconducting disk cooled in a pool of liquid nitrogen. Redox titrations are crucial in measuring the chemical composition of a superconductor. [Photo courtesy D. Cornelius, Michelson Laboratory, with materials from T. Vanderah.]

Superconductors are materials that lose all electric resistance when cooled below a critical temperature. Prior to 1987, all known superconductors required cooling to temperatures near that of liquid helium (4 K), which is costly and impractical for all but a few applications. In 1987 a giant step was taken when "high-temperature" superconductors that retain their superconductivity above the boiling point of liquid nitrogen (77 K) were discovered. The most startling characteristic of a superconductor is magnetic levitation, which is shown here. When a magnetic field is applied to a superconductor, current flows in the outer skin of the material such that the applied magnetic field is exactly canceled by the induced magnetic field, and the net field inside the specimen is zero. The induced magnetic field repels the permanent magnet, which therefore floats above the superconductor. Expulsion of a magnetic field from a superconductor is called the *Meissner effect.*

A prototypical high-temperature superconductor is yttrium barium copper oxide, $YBa_2Cu_3O_7$, in which two-thirds of the copper is in the +2 oxidation state and one-third occurs in the unusual +3 state. Another example is $Bi_2Sr_2(Ca_{0.8}Y_{0.2})Cu_2O_{8.295}$, in which the average oxidation state of copper is +2.105 and the average oxidation state of bismuth is +3.090 (which is formally a mixture of Bi^{3+} and Bi^{5+}). The most reliable means to unravel these complex formulas is through "wet" oxidation-reduction titrations described in this chapter.

Redox Titrations

A redox titration is based on an oxidation-reduction reaction between analyte and titrant. In addition to the many common analytes in chemistry, biology, and environmental and materials science that can be measured by redox titrations, exotic oxidation states of elements in uncommon materials such as superconductors (see facing page) and laser materials are also most readily measured by redox titrations. For example, chromium added to laser crystals to increase their efficiency is commonly found in the oxidation states +3 and +6, and also takes on the unusual +4 oxidation number. A redox titration is a good way to unravel the nature of this complex mixture of chromium ions. In this chapter we deal with the theory of redox titrations and then discuss some common reagents. The theory is necessary to understand how potentiometric end-point detection and redox indicators work.

Even though many of the oxidizing agents listed in Table 16-1 can be used as titrants, few are primary standards.[1] Most reducing agents react with oxygen and require protection from air to be used as titrants. Box 16-1 provides a little background on four redox reagents.

16-1 The Shape of a Redox Titration Curve

Consider the titration of Fe^{2+} with standard Ce^{4+}, the course of which could be monitored potentiometrically as shown in Figure 16-1. The titration reaction is

$$\text{titration reaction:} \quad Ce^{4+} + Fe^{2+} \rightarrow Ce^{3+} + Fe^{3+} \qquad (16\text{-}1)$$

$$\underset{\substack{\text{Ceric} \\ \text{(titrant)}}}{} \quad \underset{\substack{\text{Ferrous} \\ \text{(analyte)}}}{} \quad \underset{\text{Cerous}}{} \quad \underset{\text{Ferric}}{}$$

The *titration reaction* is $Ce^{4+} + Fe^{2+} \rightarrow Ce^{3+} + Fe^{3+}$. It goes to completion after each addition of titrant. The equilibrium constant is $K = 10^{nE°/0.05916}$ at 25°C.

TABLE 16-1   **Oxidizing and reducing agents**

Oxidants		Reductants	
BiO_3^-	Bismuthate		
BrO_3^-	Bromate	AsO_3^{3-}	Arsenite
Br_2	Bromine		Ascorbic acid (vitamin C)
Ce^{4+}	Ceric	Cr^{2+}	Chromous
$CH_3-\bigcirc-SO_2NCl^-$	Chloramine T	$S_2O_4^{2-}$	Dithionite
Cl_2	Chlorine		Dithiothreitol
ClO_2	Chlorine dioxide		
$Cr_2O_7^{2-}$	Dichromate	$Fe(CN)_6^{4-}$	Ferrocyanide
H_2O_2	Hydrogen peroxide	Fe^{2+}	Ferrous
OCl^-	Hypochlorite	N_2H_4	Hydrazine
IO_3^-	Iodate	$HO-\bigcirc-OH$	Hydroquinone
I_2	Iodine	NH_2OH	Hydroxylamine
$Pb(acetate)_4$	Lead(IV) acetate	H_3PO_2	Hypophosphorous acid
HNO_3	Nitric acid	Hg_2^{2+}	Mercurous
O	Atomic oxygen		Retinol (vitamin A)
O_3	Ozone		
$HClO_4$	Perchloric acid		
IO_4^-	Periodate		
MnO_4^-	Permanganate	Sn^{2+}	Stannous
$S_2O_8^{2-}$	Peroxydisulfate	SO_3^{2-}	Sulfite
		SO_2	Sulfur dioxide
		$S_2O_3^{2-}$	Thiosulfate
			α-Tocopherol (vitamin E)

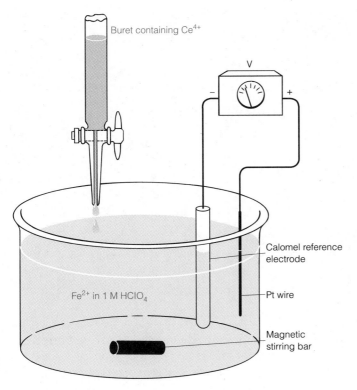

Figure 16-1 Apparatus for potentiometric titration of Fe^{2+} with Ce^{4+}.

for which $K \approx 10^{17}$ in 1 M $HClO_4$. Each mole of ceric ion oxidizes 1 mol of ferrous ion rapidly and quantitatively. The titration reaction thus creates a mixture of Ce^{4+}, Ce^{3+}, Fe^{2+}, and Fe^{3+} in the beaker in Figure 16-1.

To follow the course of the reaction, a pair of electrodes is inserted into the reaction mixture. At the *calomel reference electrode,* the reaction is

reference half-reaction: $2Hg(l) + 2Cl^- \rightleftharpoons Hg_2Cl_2(s) + 2e^-$

At the *Pt indicator electrode,* there are *two* reactions that come to equilibrium:

indicator half-reaction: $Fe^{3+} + e^- \rightleftharpoons Fe^{2+}$ $E° = 0.767$ V   (16-2)

indicator half-reaction: $Ce^{4+} + e^- \rightleftharpoons Ce^{3+}$ $E° = 1.70$ V   (16-3)

> These redox equilibria are established at the Pt electrode.

The potentials cited here are the formal potentials that apply in 1 M $HClO_4$.
The cell reaction can be described in either of two ways:

cell reaction: $2Fe^{3+} + 2Hg(l) + 2Cl^- \rightleftharpoons 2Fe^{2+} + Hg_2Cl_2(s)$ (16-4)

or

cell reaction: $2Ce^{4+} + 2Hg(l) + 2Cl^- \rightleftharpoons 2Ce^{3+} + Hg_2Cl_2(s)$ (16-5)

> The titration reaction goes to completion. The cell reactions proceed to a negligible extent. The potentiometer circuit *measures* the concentrations of species in solution but does not *change* them.

The cell reactions are not the same as the titration reaction (16-1). The potentiometer has no interest in how the Ce^{4+}, Ce^{3+}, Fe^{3+}, and Fe^{2+} happened to get into the beaker. What it does care about is that electrons want to flow from the anode to the cathode through the meter. If the

Box 16-1 Redox Tidbits

Cl_2 and ClO_2 are oxidants used in paper manufacturing as *bleaching agents* to make paper white. To reduce the quantity of toxic 2,3,7,8-tetrachlorodibenzo-*p*-dioxin by-product, Cl_2 and ClO_2 are being replaced by H_2O_2, O_2, and O_3. Dithiothreitol (and β-mercaptoethanol, $HSCH_2CH_2OH$) are used in biochemistry to reduce disulfide linkages in proteins.

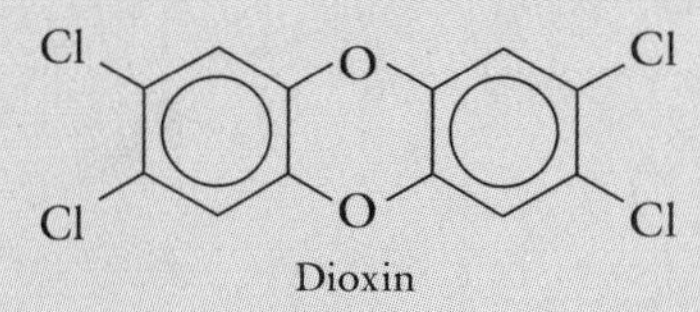

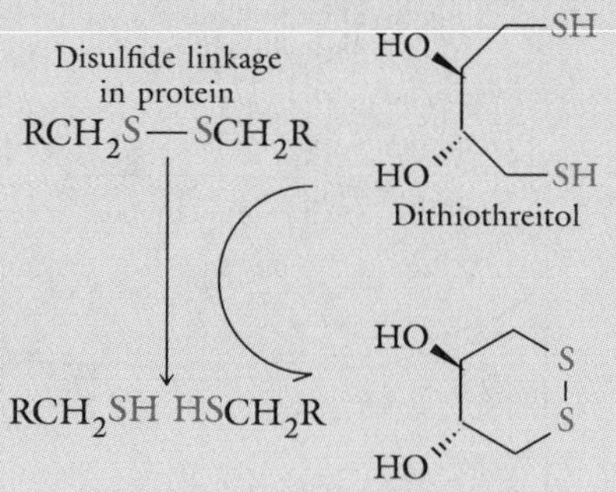

Dioxin

Disulfide linkage
in protein
RCH_2S — SCH_2R

Dithiothreitol

RCH_2SH $HSCH_2R$

solution has come to equilibrium, the potential driving both Reaction 16-4 and Reaction 16-5 must be the same. *We may describe the cell voltage with either Reaction 16-4 or Reaction 16-5, or both, however we please.*

To reiterate, the physical picture of the titration is this: Ce^{4+} is added from the buret to create a mixture of Ce^{4+}, Ce^{3+}, Fe^{3+}, and Fe^{2+} in the beaker. Because the equilibrium constant for Reaction 16-1 is large, the titration reaction goes to "completion" after each addition of Ce^{4+}. The potentiometer measures the voltage driving electrons from the reference electrode, through the meter, and out at the Pt electrode. That is, **the circuit measures the potential for reduction of Fe^{3+} or Ce^{4+} at the Pt surface by electrons from the $Hg|Hg_2Cl_2$ couple in the reference electrode.** The titration reaction, on the other hand, is an oxidation of Fe^{2+} and a reduction of Ce^{4+}. The titration reaction produces a certain mixture of Ce^{4+}, Ce^{3+}, Fe^{3+}, and Fe^{2+}. The circuit measures the potential for reduction of Ce^{4+} and Fe^{3+} by Hg. **The titration reaction goes to completion. The cell reaction is negligible. The cell is being used to measure activities, not to change them.**

We now set out to calculate how the cell voltage will change as Fe^{2+} is titrated with Ce^{4+}. The titration curve has three regions.

Region 1: Before the Equivalence Point

As each aliquot of Ce^{4+} is added, titration reaction 16-1 consumes the Ce^{4+} and creates an equal number of moles of Ce^{3+} and Fe^{3+}. Prior to the equivalence point, excess unreacted Fe^{2+} remains in the solution. Therefore, we can find the concentrations of Fe^{2+} and Fe^{3+} without difficulty. On the other hand, we cannot find the concentration of Ce^{4+} without solving a fancy little equilibrium problem. Because the amounts of Fe^{2+} and Fe^{3+} are both known, it is *convenient* to calculate the cell voltage by using Reaction 16-2 instead of Reaction 16-3.

$$E = E_+ - E_-$$

$$E = \left(0.767 - 0.059\,16 \log \frac{[Fe^{2+}]}{[Fe^{3+}]}\right) - 0.241$$

↑   ↑

<table>
<tr><td>Formal potential for
Fe^{3+} reduction in
1 M $HClO_4$</td><td>Potential of
saturated calomel
electrode</td></tr>
</table>

$$E = +0.526 - 0.059\,16 \log \frac{[Fe^{2+}]}{[Fe^{3+}]} \tag{16-6}$$

One special point is reached before the equivalence point. When the volume of titrant is one-half of the amount required to reach the equivalence point ($V = \frac{1}{2}V_e$), the concentrations of Fe^{3+} and Fe^{2+} are equal. In this case the log term in the equation above is zero, and $E_+ = E°$ for the $Fe^{3+}|Fe^{2+}$ couple. *The point at which $V = \frac{1}{2}V_e$ is analogous to the point at which $pH = pK_a$ when $V = \frac{1}{2}V_e$ in an acid-base titration.*

Region 2: At the Equivalence Point

Exactly enough Ce^{4+} has been added to react with all the Fe^{2+}. Virtually all cerium is in the form Ce^{3+}, and virtually all iron is in the form Fe^{3+}. Tiny amounts of Ce^{4+} and Fe^{2+} are present at equilibrium. From the stoichiometry of Reaction 16-1, we can say that

$$[Ce^{3+}] = [Fe^{3+}] \tag{16-7}$$

and

$$[Ce^{4+}] = [Fe^{2+}] \tag{16-8}$$

To understand why Equations 16-7 and 16-8 are true, imagine that *all* the cerium and the iron have been converted to Ce^{3+} and Fe^{3+}. Because we are at the equivalence point, $[Ce^{3+}] = [Fe^{3+}]$. Now let Reaction 16-1 come to equilibrium:

$$Fe^{3+} + Ce^{3+} \rightleftharpoons Fe^{2+} + Ce^{4+} \quad \text{(reverse of Reaction 16-1)}$$

If a little bit of Fe^{3+} goes back to Fe^{2+}, an equal number of moles of Ce^{4+} must be made. So $[Ce^{4+}] = [Fe^{2+}]$.

At any time, Reactions 16-2 and 16-3 are *both* in equilibrium at the cathode. At the equivalence point, it is *convenient* to use both reactions to describe the cell voltage. The Nernst equations for these reactions are

$$E_+ = 0.767 - 0.059\,16 \log \frac{[Fe^{2+}]}{[Fe^{3+}]} \tag{16-9}$$

$$E_+ = 1.70 - 0.059\,16 \log \frac{[Ce^{3+}]}{[Ce^{4+}]} \tag{16-10}$$

Here is where we stand: Each equation above is a statement of algebraic truth. But neither one alone allows us to find E_+ because we do not know exactly what tiny concentrations of Fe^{2+} and Ce^{4+} are present. It is possible

E_+ is the potential of the half-cell connected to the positive terminal of the potentiometer in Figure 16-1. E_- refers to the half-cell connected to the negative terminal.

For Reaction 16-2,
$$E_+ = E° (Fe^{3+} | Fe^{2+})$$
when $V = \frac{1}{2}V_e$.

At the equivalence point, it is convenient to use both Reactions 16-2 and 16-3 to calculate the cell voltage. This is strictly a matter of algebraic convenience.

to solve the four simultaneous equations 16-7 through 16-10, by first *adding* Equations 16-9 and 16-10. This is strictly an algebraic manipulation, the logic and beauty of which will become apparent soon.

Adding Equations 16-9 and 16-10 gives

$$2E_+ = 0.767 + 1.70 - 0.059\,16 \log \frac{[Fe^{2+}]}{[Fe^{3+}]} - 0.059\,16 \log \frac{[Ce^{3+}]}{[Ce^{4+}]}$$

$\log a + \log b = \log ab.$

$$2E_+ = 2.46_7 - 0.059\,16 \log \frac{[Fe^{2+}][Ce^{3+}]}{[Fe^{3+}][Ce^{4+}]}$$

But because $[Ce^{3+}] = [Fe^{3+}]$ and $[Ce^{4+}] = [Fe^{2+}]$ at the equivalence point, the ratio of concentrations in the log term is unity. Therefore, the logarithm is zero and

$$2E_+ = 2.46_7 \text{ V} \Rightarrow E_+ = 1.23 \text{ V}$$

E_+ is just the average of the standard potentials for the two half-reactions at the Pt electrode. The cell voltage is

$$E = E_+ - E(\text{calomel}) = 1.23 - 0.241 = 0.99 \text{ V} \qquad (16\text{-}11)$$

In this particular titration, the equivalence-point voltage is independent of the concentrations and volumes of the reactants.

Region 3: After the Equivalence Point

After the equivalence point, it is convenient to use Reaction 16-3 because we can easily calculate the concentrations of Ce^{3+} and Ce^{4+}. It is not convenient to use Reaction 16-2 because we do not know the concentration of Fe^{2+}, which has been "used up."

Now virtually all iron atoms are Fe^{3+}. The moles of Ce^{3+} equal the moles of Fe^{3+}, and there is a known excess of unreacted Ce^{4+}. Because we know both $[Ce^{3+}]$ and $[Ce^{4+}]$, it is *convenient* to use Reaction 16-3 to describe the chemistry at the Pt electrode:

$$E = E_+ - E(\text{calomel}) = \left(1.70 - 0.059\,16 \log \frac{[Ce^{3+}]}{[Ce^{4+}]} \right) - 0.241 \qquad (16\text{-}12)$$

At the special point when $V = 2V_e$, $[Ce^{3+}] = [Ce^{4+}]$ and $E_+ = E°(Ce^{4+}|Ce^{3+}) = 1.70$ V.

Before the equivalence point, the voltage is fairly steady near the value $E = E_+ - E(\text{calomel}) \approx E°(Fe^{3+}|Fe^{2+}) - 0.241$ V $= 0.53$ V. After the equivalence point, the voltage levels off near $E \approx E°(Ce^{4+}|Ce^{3+}) - 0.241$ V $= 1.46$ V. At the equivalence point, there is a rapid rise in voltage.

- -

EXAMPLE   Potentiometric Redox Titration

Suppose that we titrate 100.0 mL of 0.050 0 M Fe^{2+} with 0.100 M Ce^{4+}, using the cell in Figure 16-1. The equivalence point occurs when $V_{Ce^{4+}} = 50.0$ mL, because the Ce^{4+} is twice as concentrated as the Fe^{2+}. Calculate the cell voltage at 36.0, 50.0, and 63.0 mL.

Solution

At 36.0 mL: This is 36.0/50.0 of the way to the equivalence point. Therefore, 36.0/50.0 of the iron is in the form Fe^{3+}, and 14.0/50.0 is in the form Fe^{2+}. Putting the value $[Fe^{2+}]/[Fe^{3+}] = 14.0/36.0$ into Equation 16-6 gives a cell voltage of 0.550 V.

At 50.0 mL: This is the equivalence point. Equation 16-11 tells us that the cell voltage is 0.99 V, regardless of the concentrations of reagents for this particular titration.

At 63.0 mL: The first 50.0 mL of cerium has been converted to Ce^{3+}. Because 13.0 mL of excess Ce^{4+} has been added, the value of $[Ce^{3+}]/[Ce^{4+}]$ in Equation 16-12 is 50.0/13.0, and the cell voltage is 1.424 V.

The end point of the titration curve for Reaction 16-1 shown in Figure 16-2 is marked by a steep rise in the voltage. Because both reactants are one-electron redox reagents, the curve is symmetric near the equivalence point. This is not the case when the stoichiometry of reactants is not 1:1. The calculated value of E_+ at $\frac{1}{2}V_e$ is the formal potential of the $Fe^{3+}|Fe^{2+}$ couple, because the quotient $[Fe^{2+}]/[Fe^{3+}]$ is unity at this point. The calculated voltage at any point in this titration depends only on the *ratio* of reactants; their *concentrations* do not figure in any calculations in this example. We expect, therefore, that the curve in Figure 16-2 will be independent of dilution. We should observe the same curve if both reactants were diluted by a factor of 10.

The voltage at zero titrant volume cannot be calculated because we do not know how much Fe^{3+} is present. If $[Fe^{3+}] = 0$, the voltage calculated with Equation 16-6 would be $-\infty$. In fact, there must be some Fe^{3+} in each reagent, either as an impurity or from oxidation of Fe^{2+} by atmospheric oxygen. In any case, the voltage could never be lower than that needed to reduce the solvent $(H_2O + e^- \rightarrow \frac{1}{2}H_2 + OH^-)$.

16-2 🗔 General Approach to Redox Titration Curves

Now that we understand the chemistry on which redox titrations are based, it is time to introduce practical tools for calculating titration curves with spreadsheets. Our goal is a single equation that describes the entire titration curve with no approximations, except neglect of activity coefficients.

Titration with an Oxidizing Agent

Consider an oxidizing titrant such as Ce^{4+} added to a reducing analyte such as Fe^{2+}. Let T be the titrant and A be the analyte, and let the oxidation states change by one electron:

$$T + A \rightarrow T^- + A^+ \qquad (16\text{-}13)$$
$$\text{(titrant)} \quad \text{(analyte)}$$

The reduction half-reactions for the two reagents are

$$T + e^- \rightleftharpoons T^- \qquad E = E_T^\circ - 0.059\,16 \log \frac{[T^-]}{[T]}$$

$$A^+ + e^- \rightleftharpoons A \qquad E = E_A^\circ - 0.059\,16 \log \frac{[A]}{[A^+]}$$

Now we rearrange the two Nernst equations to find more useful relationships between the concentrations of reactants and products:

$$E = E_T^\circ - 0.059\,16 \log \frac{[T^-]}{[T]} \Rightarrow \frac{[T^-]}{[T]} = \underbrace{10^{(E_T^\circ - E)/0.059\,16}}_{\tau} \Rightarrow [T^-] = \tau[T]$$
$$(16\text{-}14)$$

The shape of the curve in Figure 16-2 is essentially independent of the concentrations of analyte and titrant. The curve is symmetric near V_e because the stoichiometry is 1:1.

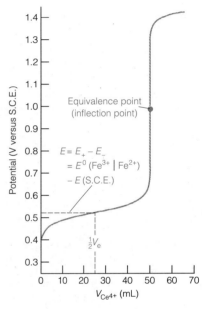

Figure 16-2 Theoretical curve for titration of 100.0 mL of 0.0500 M Fe^{2+} with 0.100 M Ce^{4+} in 1 M $HClO_4$.

For a temperature other than 25°C, the factor 0.059 16 V is really $(RT/F)\ln 10$, where R is the gas constant, T is temperature in kelvins, and F is the Faraday constant.

16 Redox Titrations

The algebra behind Equation 16-15:

$$E = E_A^\circ - 0.059\,16 \log \frac{[A]}{[A^+]}$$

$$\log \frac{[A]}{[A^+]} = \frac{E_A^\circ - E}{0.059\,16}$$

$$10^{\log [A]/[A^+]} = 10^{(E_A^\circ - E)/0.059\,16}$$

$$\frac{[A]}{[A^+]} = 10^{(E_A^\circ - E)/0.059\,16}$$

$$E = E_A^\circ - 0.059\,16 \log \frac{[A]}{[A^+]} \Rightarrow \frac{[A]}{[A^+]} = \underbrace{10^{(E_A^\circ - E)/0.059\,16}}_{\alpha} \Rightarrow [A] = \alpha[A^+] \tag{16-15}$$

where $\tau \equiv 10^{(E_T^\circ - E)/0.059\,16}$ and $\alpha \equiv 10^{(E_A^\circ - E)/0.059\,16}$. The letters tau and alpha were chosen as mnemonics for "titrant" and "analyte."

Next, we use Equations 16-14 and 16-15 and two mass balances to find expressions for the *products* of the titration reaction, $[T^-]$ and $[A^+]$.

mass balance for titrant: $[T] + [T^-] = T_{total}$

$$\frac{1}{\tau}[T^-] + [T^-] = T_{total}$$

$$[T^-]\left(\frac{1}{\tau} + 1\right) = T_{total} \qquad \Rightarrow [T^-] = \frac{\tau T_{total}}{1 + \tau}$$

mass balance for analyte: $[A] + [A^+] = A_{total} \qquad \Rightarrow \Rightarrow [A^+] = \frac{A_{total}}{1 + \alpha}$

Trying to contain our excitement, we carry on, for we have nearly derived a master equation for the titration. From the stoichiometry of Reaction 16-13, we know that $[T^-] = [A^+]$, *because they are created in a 1:1 mole ratio.* Equating $[T^-]$ and $[A^+]$, we find

$$[T^-] = [A^+]$$

$$\frac{\tau T_{total}}{1 + \tau} = \frac{A_{total}}{1 + \alpha} \tag{16-16}$$

But the fraction of the way (ϕ) to the equivalence point is just the quotient T_{total}/A_{total}. That is, when the total concentration of T ($= [T] + [T^-]$) equals the total concentration of A ($= [A] + [A^+]$), we are at the equivalence point.

fraction of titration: $\phi = \dfrac{T_{total}}{A_{total}}$ ($= 1$ at equivalence point)

If the stoichiometry required, say, 2 mol of T for 3 mol of A, the fraction of titration would be $\phi = 3T_{total}/2A_{total}$, because we demand that $\phi = 1$ at the equivalence point.

Rearranging Equation 16-16 to solve for the fraction of titration gives the master equation for the titration curve:

Titration with oxidizing titrant:

$$\phi = \frac{1 + \tau}{\tau(1 + \alpha)} \tag{16-17}$$

Equation 16-17 gives the fraction of titration as a function of the potential, *E*, which is buried in the numbers τ and α, defined in Equations 16-14 and 16-15. The titration curve is a graph of *E* versus ϕ. We will compute the curve in the reverse manner by inputting values of *E* and finding values of ϕ. This is analogous to what we did for acid-base titrations, in which we inserted a value of pH to find the value of ϕ in Section 12-9.

. . . .

EXAMPLE Spreadsheet Calculation of the Ce^{4+}/Fe^{2+} Titration Curve

Suppose that we titrate 100.0 mL of 0.0500 M Fe^{2+} with 0.100 M Ce^{4+}, using the cell in Figure 16-1. The equivalence volume is 50.0 mL, which means that $\phi = 1$ when $V_{Ce^{4+}} = 50.0$ mL. Use Equation 16-17 to compute the voltages in Figure 16-2 for titrant volumes of 50.0, 36.0, and 63.0 mL.

(a)   (b)   (c)

COLOR PLATE 1   HCl Fountain (Demonstration 5-2) (*a*) Basic indicator solution in beaker. (*b*) Indicator is drawn into flask and changes to acidic color. (*c*) Solution levels at end of experiment.

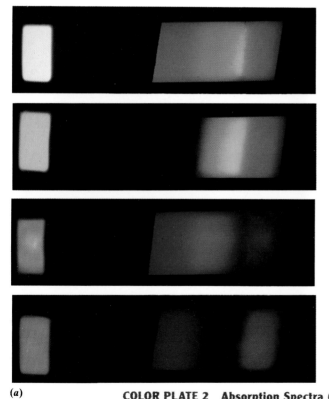

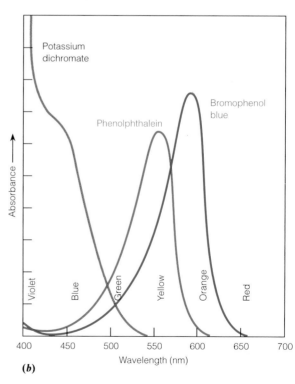

(*b*)

(*a*)

COLOR PLATE 2   Absorption Spectra (Demonstration 6-1) (*a*) Projected visible spectra of (from top to bottom) white light, potassium dichromate, bromophenol blue, and phenolphthalein. (*b*) Spectrophotometric visible spectra of colored compounds whose projected spectra are shown in (*a*).

COLOR PLATE 3 Grating Dispersion (Section 6-3) Visible spectrum produced by grating inside spectrophotometer.

(a) *(b)* *(c)*

COLOR PLATE 4 Fajans Titration of Cl⁻ with AgNO₃ Using Dichlorofluorescein (Demonstration 7-1) (*a*) Indicator before beginning titration. (*b*) AgCl precipitate before end point. (*c*) Indicator adsorbed on precipitate after end point.

(a)
(b)

COLOR PLATE 5   Effect of Ionic Strength on Ionic Dissociation (Demonstration 8-1) (*a*) Two beakers containing identical solutions with $FeSCN^{2+}$, Fe^{3+}, and SCN^-. (*b*) Change when KNO_3 is added to right-hand beaker.

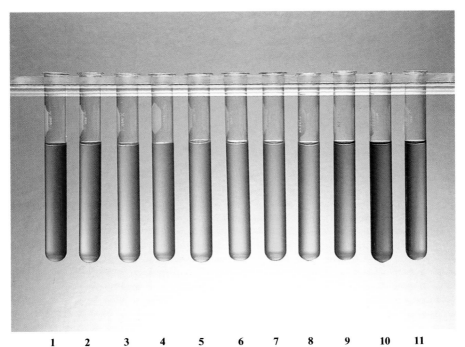

1   2   3   4   5   6   7   8   9   10   11

COLOR PLATE 6   Thymol Blue (Section 12-6) Acid-base indicator thymol blue between pH 1 (left) and 11 (right). The pK values are 1.7 and 8.9.

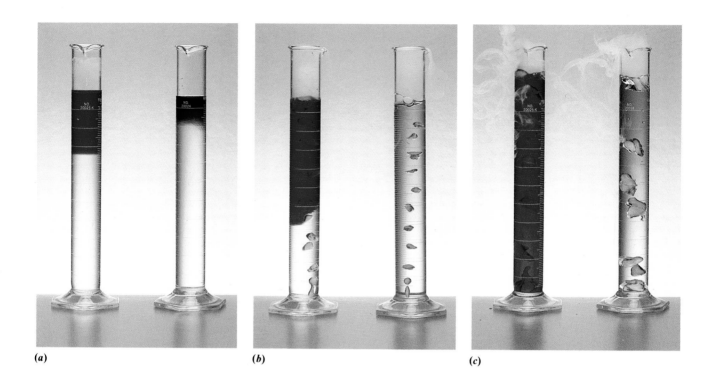

(a) (b) (c)

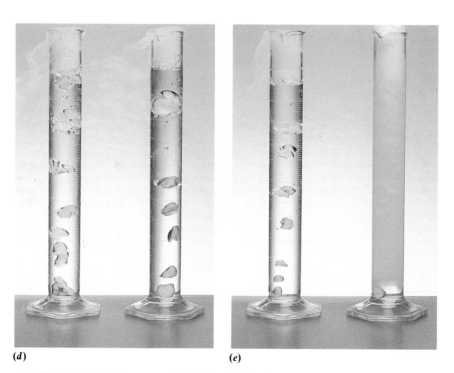

(d) (e)

COLOR PLATE 7 Indicators and Acidity of CO$_2$ (Demonstration 12-1) (*a*) Cylinders before adding Dry Ice. Ethanol indicator solutions of phenolphthalein (left) and bromothymol blue (right) have not yet mixed with entire cylinder. (*b*) Adding Dry Ice causes bubbling and mixing. (*c*) Further mixing. (*d*) Phenolphthalein changes to its colorless acidic form. Color of bromothymol blue is due to mixture of acidic and basic forms. (*e*) After addition of HCl and stirring of right-hand cylinder, bubbles of CO$_2$ can be seen leaving solution, and indicator changes completely to its acidic color.

COLOR PLATE 8 Titration of Cu(II) with EDTA Using Auxiliary Complexing Agent (Section 13-5) 0.02 M $CuSO_4$ before titration (left). Color of Cu(II)-ammonia complex after adding ammonia buffer, pH 10 (center). End-point color when all ammonia ligands have been displaced by EDTA (right).

(a)

(b)

COLOR PLATE 9 Titration of Mg^{2+} by EDTA Using Eriochrome Black T Indicator (Demonstration 13-1) (*a*) Before (left), near (center), and after (right) equivalence point. (*b*) Same titration with methyl red added as inert dye to alter colors.

COLOR PLATE 10 Titration of VO^{2+} with Potassium Permanganate (Section 16-6; Experiment 27-9) Blue VO^{2+} solution prior to titration (left). Mixture of blue VO^{2+} and yellow VO_2^+ observed during titration (center). Dark color of MnO_4^- at end point (right).

COLOR PLATE 11 Iodometric Titration (Section 16-9; Experiment 27-10) I_3^- solution (left). I_3^- solution before end point in titration with $S_2O_3^{2-}$ (left center). I_3^- solution immediately before end point with starch indicator present (right center). At the end point (right).

(a)

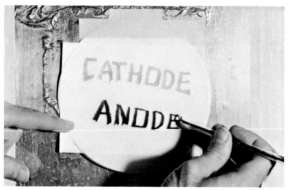

(b)

(c)

COLOR PLATE 12 Electrochemical Writing (Demonstration 17-1) (*a*) Stylus used as cathode. (*b*) Stylus used as anode. (*c*) Foil backing has a polarity opposite that of the stylus and produces reverse color on bottom sheet of paper.

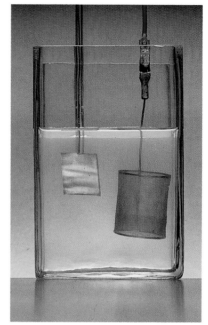

(a)

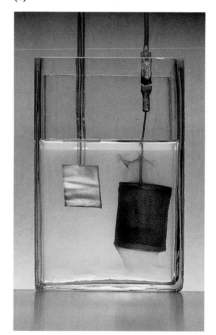

(b)

COLOR PLATE 13 Formation of Diffusion Layer During Electrolysis (Box 18-1) (*a*) Cu electrode (flat plate, left) and Pt electrode (mesh basket, right) immersed in solution containing KI and starch, with no electric current. (*b*) I_3^- starch complex forms at surface of Pt anode when current flows.

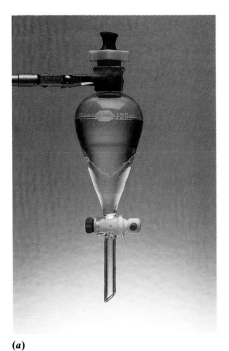 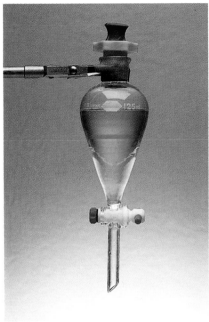 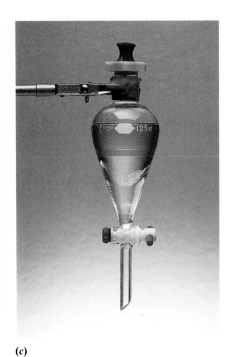

(a)   (b)   (c)

COLOR PLATE 14   Extraction of Uranyl Nitrate from Water into Ether (Section 22-1) (*a*) Separatory funnel with lower aqueous layer containing yellow 1 M $UO_2(NO_3)_2$ (plus 3 M HNO_3 and 4 M $Ca(NO_3)_2$) beneath colorless diethyl ether layer prior to mixing. (*b*) Yellow uranyl nitrate is distributed in both layers after shaking. (*c*) After eight extractions with ether, almost all yellow uranyl nitrate has been removed from the aqueous phase.

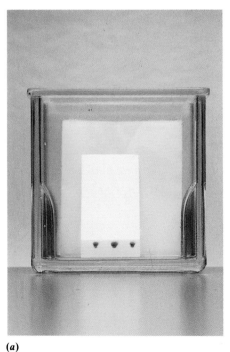 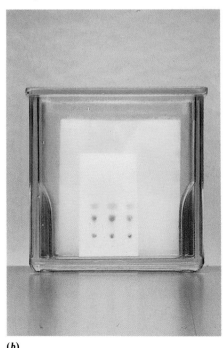 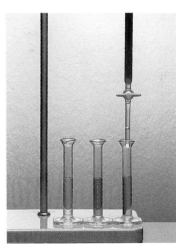

(a)   (b)

COLOR PLATE 16   Cation-Exchange Separation of Co(II) and Ni(II) (Section 24-1) A solution containing Co(II) and Ni(II) in 6 M HCl was applied to the top of the cation-exchange chromatography column. The blue Ni(II) was eluted first and collected in the graduated cylinders. Green Co(II) is retained more strongly and has not yet been eluted from the column.

COLOR PLATE 15   Thin-Layer Chromatography (Section 23-2) (*a*) Solvent ascends past mixture of dyes near bottom of plate. (*b*) Separation achieved after solvent has ascended most of the way up the plate.

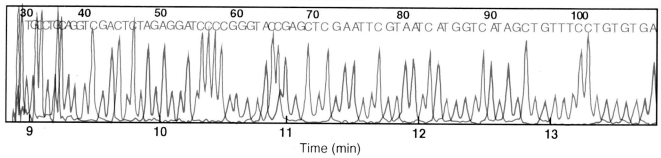

Base number

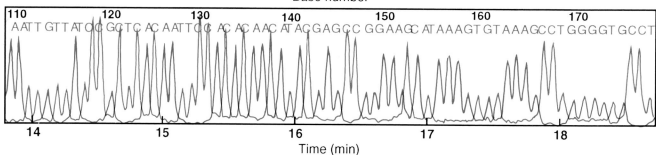

Base number

COLOR PLATE 17 DNA Sequencing by Capillary Gel Electrophoresis with Fluorescent Labels (Section 24-5) Tall red peaks correspond to chains terminating in cytosine and short red peaks correspond to thymine. Tall blue peaks arise from adenine and short blue peaks indicate guanine. Two different fluorescent labels and two fluorescence wavelengths were required to generate this information. [From M. C. Ruiz-Martinez, J. Berka, A. Belenkii, F. Foret, A. W. Miller, and B. L. Karger, *Anal. Chem.* **1993,** *65,* 2851.]

(a)

(b)

(c)

COLOR PLATE 18 Colloids and Dialysis (Demonstration 25-1) (*a*) Ordinary aqueous Fe(III) (right) and colloidal Fe(III) (left). (*b*) Dialysis bags containing colloidal Fe(III) (left) and a solution of Cu(II) (right) immediately after placement in flasks of water. (*c*) After 24 h of dialysis, the Cu(II) has diffused out and is dispersed uniformly between the bag and the flask, but the colloidal Fe(III) remains inside the bag.

COLOR PLATE 19 Fe(phenanthroline)$_3^{2+}$ Standards for Spectrophotometric Analysis (Experiment 27-16) Volumetric flasks containing Fe(phenanthroline)$_3^{2+}$ solutions with iron concentrations ranging from 1 mg/L (left) to 10 mg/L (right).

	A	B	C	D	E	F	G
1	E°(T) =	E (vs S.H.E.)	Tau	Alpha	Phi	E(vs S.C.E.)	Volume (mL)
2	1.7	0.600	3.92E+18	6.65E+02	0.00150	0.359	0.075
3	E°(A) =	0.700	8.00E+16	1.36E+01	0.06864	0.459	3.432
4	0.767	0.767	5.90E+15	1.00E+00	0.50000	0.526	25.000
5	Nernst =	0.800	1.63E+15	2.77E−01	0.78320	0.559	39.160
6	0.05916	1.000	6.80E+11	1.15E−04	0.99988	0.759	49.994
7	Ve =	1.200	2.83E+08	4.80E−08	1.00000	0.959	50.000
8	50	1.400	1.18E+05	2.00E−11	1.00001	1.159	50.000
9		1.600	4.90E+01	8.31E−15	1.02040	1.359	51.020
10		1.700	1.00E+00	1.70E−16	2.00000	1.459	100.000
11							
12		1.2335	7.68E+07	1.30E−08	1.00000	0.992	50.000
13		0.79127	2.29E+15	3.89E−01	0.72003	0.550	36.002
14		1.66539	3.85E+00	6.52E−16	1.26000	1.424	63.000
15							
16	C2 = 10^(($A$2–B2)/$A$6)				E2 = (1+C2)/(C2*(1+D2))		G2 = $A$8*E2
17	D2 = 10^(($A$4–B2)/$A$6)				F2 = B2–0.241		

Figure 16-3 Spreadsheet for the titration curve in Figure 16-2, based on Equation 16-17. Column B contains input and column E has the principal output. The fraction of titration in column E is converted to volume in column G, and potential (versus S.H.E.) in column B is converted to potential (versus S.C.E.) in column F. A titration curve is a graph of column F versus column G.

Solution The titration reaction is

$$Ce^{4+} + Fe^{2+} \rightarrow Ce^{3+} + Fe^{3+}$$

and the half-reactions, written as reductions, are

$$Ce^{4+} + e^- \rightleftharpoons Ce^{3+} \quad \Rightarrow \tau = 10^{(E_T^\circ - E)/0.059\,16} = 10^{(1.70 - E)/0.059\,16}$$

$$Fe^{3+} + e^- \rightleftharpoons Fe^{2+} \quad \Rightarrow \alpha = 10^{(E_A^\circ - E)/0.059\,16} = 10^{(0.767 - E)/0.059\,16}$$

The spreadsheet in Figure 16-3 is really simple. The constants in column A are the standard reduction potentials, the quantity 0.059 16 V in the Nernst equation, and the equivalence volume, 50 mL. The input to the spreadsheet is the value of E (versus S.H.E.) in column B. In columns C and D we compute τ and α, and in column E we calculate ϕ from Equation 16-17. Column F converts E (versus S.H.E.) to E (versus S.C.E.) by subtracting 0.241 V. Column G multiplies ϕ by 50 mL to convert the fraction of titration into volume of titrant.

To find the voltage for a particular volume of titrant, we vary E in column B until the desired volume appears in column G. Rows 12–14 in the spreadsheet compute voltages for 50.0, 36.0, and 63.0 mL.

• •

Because titration curves are steep near the equivalence point, *we recommend setting the precision of ϕ to 12 digits while varying E to search for the* *equivalence point potential.* The tables in this book were generated this way, and then the precision of φ was set back down to a reasonable number to display the results.

Equation 16-17 Applies to Many Oxidation Stoichiometries

Equation 16-17 applies to any titration in which T oxidizes A, providing neither reagent breaks into smaller fragments or associates into larger molecules. The stoichiometry of the reaction of T with A need not be 1:1, and there may be any number of electrons or other species (such as H^+ and H_2O) involved in the reaction. Thus, Equation 16-17 applies to the titration

$$MnO_4^- + 5Fe^{2+} + 8H^+ \rightarrow Mn^{2+} + 5Fe^{3+} + 4H_2O$$
$$\text{(titrant)} \quad \text{(analyte)}$$

but not to the titration

This reaction is treated in the Supplementary Problems in the Solutions Manual.

$$Cr_2O_7^{2-} + 6Fe^{2+} + 14H^+ \rightarrow 2Cr^{3+} + 6Fe^{3+} + 7H_2O$$
$$\text{(titrant)} \quad \text{(analyte)}$$

because $Cr_2O_7^{2-}$ breaks apart into two Cr^{3+} ions in the latter reaction.

When the Stoichiometry Is Not 1:1

Suppose that Cu^+ is titrated with Tl^{3+} in 1 M $HClO_4$:

$$Tl^{3+} + 2Cu^+ \rightarrow Tl^+ + 2Cu^{2+}$$
$$\text{(titrant)} \quad \text{(analyte)}$$

The values of τ and α are computed from the half-reactions:

$$Tl^{3+} + 2e^- \rightleftharpoons Tl^+ \qquad E = 1.26 - \frac{0.059\,16}{2} \log \frac{[Tl^+]}{[Tl^{3+}]}$$

Two electrons instead of one appear in τ.

$$\Rightarrow \tau \equiv \frac{[Tl^+]}{[Tl^{3+}]} = 10^{2(1.26 - E)/0.059\,16} \qquad (16\text{-}18)$$

$$Cu^{2+} + e^- \rightleftharpoons Cu^+ \qquad E = 0.339 - 0.059\,16 \log \frac{[Cu^+]}{[Cu^{2+}]}$$

$$\Rightarrow \alpha \equiv \frac{[Cu^+]}{[Cu^{2+}]} = 10^{(0.339 - E)/0.059\,16} \qquad (16\text{-}19)$$

Combining the mass balances with τ and α leads to the expressions

$$[Tl^+] = \frac{\tau Tl_{total}}{1 + \tau} \qquad [Cu^{2+}] = \frac{Cu_{total}}{1 + \alpha}$$

Now we know that $[Cu^{2+}] = 2[Tl^+]$, because these products are created in a 2:1 mole ratio. Inserting the expressions above for $[Tl^+]$ and $[Cu^{2+}]$ into this equality gives

$$\frac{Cu_{total}}{1 + \alpha} = \frac{2\tau Tl_{total}}{1 + \tau} \qquad (16\text{-}20)$$

Because the reaction requires one Tl^{3+} for two Cu^+ ions, the fraction of titration is

$$\phi \equiv \frac{2Tl_{total}}{Cu_{total}} \overset{\substack{\text{from} \\ \text{Eq. 16-20}}}{=} \frac{1 + \tau}{\tau(1 + \alpha)}$$

The factor of 2 appears in the definition of ϕ because the fraction of titration is defined as unity at the equivalence point. The final result is identical to Equation 16-17, which was derived for 1:1 stoichiometry.

Adding H$^+$ and Other Reactants

Now consider the titration of Tl$^+$ by iodate in HCl solution:

$$IO_3^- + 2Tl^+ + 2Cl^- + 6H^+ \rightarrow ICl_2^- + 2Tl^{3+} + 3H_2O$$

Iodate   Thallous   Thallic
(titrant) (analyte)

Thallium is the analyte in this example. Its half-reaction is the same as in the preceding example, with a formal potential of 0.77 V in 1 M HCl.

$$Tl^{3+} + 2e^- \rightleftharpoons Tl^+ \qquad E = 0.77 - \frac{0.059\,16}{2}\log\frac{[Tl^+]}{[Tl^{3+}]}$$

$$\Rightarrow \alpha \equiv \frac{[Tl^+]}{[Tl^{3+}]} = 10^{2(0.77 - E)/0.059\,16} \qquad (16\text{-}21)$$

The iodate half-reaction is more complicated:

$$IO_3^- + 2Cl^- + 6H^+ + 4e^- \rightleftharpoons ICl_2^- + 3H_2O \qquad E = 1.24 - \frac{0.059\,16}{4}\log\frac{[ICl_2^-]}{[IO_3^-][Cl^-]^2[H^+]^6}$$

We find τ by solving for the quotient $[ICl_2^-]/[IO_3^-]$, but now we must deal with extra terms involving $[Cl^-]$ and $[H^+]$:

$$\log\frac{[ICl_2^-]}{[IO_3^-][Cl^-]^2[H^+]^6} = \frac{4(1.24 - E)}{0.059\,16}$$

$$\log\frac{[ICl_2^-]}{[IO_3^-]} - 2\log[Cl^-] - 6\log[H^+] = \frac{4(1.24 - E)}{0.059\,16}$$

$$\log\frac{[ICl_2^-]}{[IO_3^-]} + 2\,pCl + 6\,pH = \frac{4(1.24 - E)}{0.059\,16} \qquad\qquad pH = -\log[H^+]$$
$$\qquad\qquad\qquad\qquad\qquad\qquad\qquad\qquad\qquad\qquad pCl = -\log[Cl^-]$$

$$\log\frac{[ICl_2^-]}{[IO_3^-]} = \frac{4(1.24 - E)}{0.059\,16} - 2\,pCl - 6\,pH$$

$$\tau \equiv \frac{[ICl_2^-]}{[IO_3^-]} = 10^{\{[4(1.24 - E)/0.059\,16] - 2\,pCl - 6\,pH\}} \qquad (16\text{-}22)$$

To complete the derivation, we equate $[Tl^{3+}]$ to $2[IO_3^-]$ and define ϕ as $2I_{total}/Tl_{total}$ to find the same expression we found before, Equation 16-17.

Equation 16-17 works for this case also—but the expression for τ is a little different.

- -

EXAMPLE Spreadsheet Calculation of the IO$_3^-$/Tl$^+$ Titration Curve

Suppose that we titrate 100.0 mL of 0.010 0 M Tl^{2+} with 0.010 0 M IO$_3^-$, using Pt and saturated calomel electrodes. Assume that all solutions contain 1.00 M HCl, which means that pH = pCl = $-\log(1.00)$ = 0.00 in Equation 16-22. Because 1 mol of IO$_3^-$ consumes 2 mol of Tl$^+$, the equivalence volume is 50.0 mL. Therefore, ϕ = 1 when $V_{IO_3^-}$ = 50.0 mL. Use Equation 16-17 to compute the titration curve.

Solution The formulas for α and τ are given in Equations 16-21 and 16-22. The work is set out in Figure 16-4, in which we have added the concentrations of H$^+$ and Cl$^-$ as constants in column A.

- -

	A	B	C	D	E	F	G
1	E°(T) =	E (vs S.H.E.)	Tau	Alpha	Phi	E (vs S.C.E.)	Volume (mL)
2	1.24	0.700	3.24E+36	2.33E+02	0.00428	0.459	0.214
3	E°(A) =	0.770	6.00E+31	1.00E+00	0.50000	0.529	25.000
4	0.77	0.800	5.62E+29	9.68E–02	0.91176	0.559	45.588
5	Nernst =	0.900	9.74E+22	4.03E–05	0.99996	0.659	49.998
6	0.05916	1.083	4.12E+10	2.62E–11	1.00000	0.842	50.000
7	Ve =	1.20	5.06E+02	2.91E–15	1.00197	0.959	50.099
8	50	1.24	1.00E+00	1.29E–16	2.00000	0.999	100.000
9	pCl =						
10	0		C2 = 10^(4*($A$2–B2)/$A$6–2*$A$10–6*$A$12)				
11	pH =		D2 = 10^(2*($A$4–B2)/$A$6)				
12	0		E2 = (1+C2)/(C2*(1+D2))				
13			F2 = B2–0.241				
14			G2 = $A$8*E2				

Figure 16-4 Spreadsheet for the titration curve in Figure 16-5, based on Equation 16-17. Input in column B is varied to obtain output in column G at any desired volume.

Figure 16-5 shows the curve calculated for the titration of Tl^+ by IO_3^-. Note that *the curve is not symmetric about the equivalence point* because the stoichiometry of reactants is not 1:1. Still, the curve is so steep near the equivalence point that negligible error is introduced if the center of the steepest portion is taken as the end point. Demonstration 16-1 illustrates another titration curve with an asymmetric end point.

The change in voltage near the equivalence point is smaller in Figure 16-5 than in Figure 16-2 because the standard reduction potentials of the

Figure 16-5 Theoretical curve for titration of 100.0 mL of 0.0100 M Tl^+ with 0.0100 M IO_3^- in 1.00 M HCl. The equivalence point at 0.842 V is not at the center of the steep part of the curve. When the stoichiometry of the reaction is not 1:1, the curve is not symmetric.

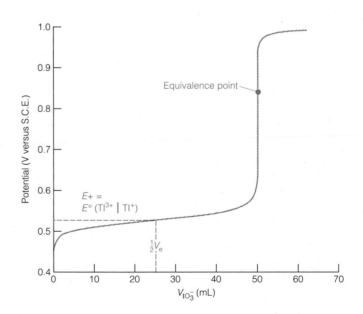

Potentiometric Titration of Fe^{2+} with MnO_4^-

The reaction of Fe^{2+} with $KMnO_4$ is an excellent illustration of the principles of potentiometric titrations. Dissolve 0.60 g of $Fe(NH_4)_2(SO_4)_2 \cdot 6H_2O$ (FW 392.13; 1.5 mmol) in 400 mL of 1 M H_2SO_4. Titrate the well-stirred solution with 0.02 M $KMnO_4$ (~15 mL will be required to reach the equivalence point), using Pt and calomel electrodes with a pH meter as a potentiometer. The reference socket of the pH meter is the negative input terminal. Before starting the titration, calibrate the meter by connecting the two input sockets directly to each other with a wire and setting the millivolt scale of the meter to zero.

Calculate some points on the theoretical titration curve before performing the experiment. The formal potential of the $Fe^{3+} \mid Fe^{2+}$ couple in 1 M H_2SO_4 is 0.68 V. Compare the theoretical and experimental results. Also note the coincidence of the potentiometric and visual end points.

Question Potassium permanganate is purple, and all the other species in this titration are colorless (or very faintly colored). What color change is expected at the equivalence point?

two half-reactions are closer to each other in Figure 16-5 than in Figure 16-2. Clearest results are achieved with the strongest oxidizing and reducing agents. The same rule applies to acid-base titrations, where we normally use a strong acid or strong base as titrant, to get the sharpest break at the equivalence point.

Titration with a Reducing Agent

Following the same reasoning that we have been using, we could show that if a reducing titrant is used, the general equation for the titration curve is

Titration with reducing titrant:
$$\phi = \frac{\alpha(1 + \tau)}{1 + \alpha} \qquad (16\text{-}23)$$

where τ applies to the titrant and α applies to the analyte.

16-3 Titration of a Mixture

The titration of two species will exhibit two breaks if the standard potentials of the redox couples are sufficiently different. Figure 16-6 shows the theoretical titration curve for an equimolar mixture of Tl^+ and Sn^{2+} titrated with IO_3^-. The two *titration reactions* are

first: $\quad IO_3^- + 2Sn^{2+} + 2Cl^- + 6H^+ \rightarrow ICl_2^- + 2Sn^{4+} + 3H_2O \quad (16\text{-}24)$

second: $\quad IO_3^- + 2Tl^+ + 2Cl^- + 6H^+ \rightarrow ICl_2^- + 2Tl^{3+} + 3H_2O \quad (16\text{-}25)$

and the relevant half-reactions are

$$IO_3^- + 2Cl^- + 6H^+ + 4e^- \rightleftharpoons ICl_2^- + 3H_2O \qquad E^\circ = 1.24 \text{ V}$$

$$\Rightarrow \tau \equiv \frac{[ICl_2^-]}{[IO_3^-]} = 10^{\{[4(1.24 - E)/0.059\,16] - 2\,pCl - 6\,pH\}}$$

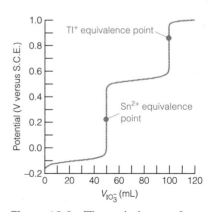

Figure 16-6 Theoretical curve for 100.0 mL containing 0.010 0 M Tl^+ plus 0.010 0 M Sn^{2+} titrated with 0.010 0 M IO_3^-, calculated with Equation 16-27. All solutions contain 1.00 M HCl.

$$Sn^{4+} + 2e^- \rightleftharpoons Sn^{2+} \qquad\qquad E^\circ = 0.139 \text{ V}$$

$$\Rightarrow \alpha_1 \equiv \frac{[Sn^{2+}]}{[Sn^{4+}]} = 10^{2(0.139\,-\,E)/0.059\,16}$$

$$Tl^{3+} + 2e^- \rightleftharpoons Tl^+ \qquad\qquad E^\circ = 0.77 \text{ V}$$

$$\Rightarrow \alpha_2 \equiv \frac{[Tl^+]}{[Tl^{3+}]} = 10^{2(0.77\,-\,E)/0.059\,16}$$

Question   Explain why the potential (versus S.H.E, not S.C.E.) in Figure 16-6 is 0.139 V at $\phi = 0.5$ and 0.77 V at $\phi = 1.5$.

Because the $Sn^{4+}|Sn^{2+}$ couple has a lower reduction potential than the $Tl^{3+}|Tl^+$ couple, Sn^{2+} will be oxidized before Tl^+. That is, the equilibrium constant for Reaction 16-24 is larger than for Reaction 16-25. This is another way of saying that Sn^{2+} is a stronger reducing agent than Tl^+.

To derive an equation for the titration of a mixture, the mass balance equates $[ICl_2^-]$ to the sum $\frac{1}{2}[Sn^{4+}] + \frac{1}{2}[Tl^{3+}]$, because 1 mol of ICl_2^- is generated for every 2 mol of Sn^{4+} and 1 mol of ICl_2^- is generated for every 2 mol of Tl^{3+}:

$$[ICl_2^-] = \tfrac{1}{2}[Sn^{4+}] + \tfrac{1}{2}[Tl^{3+}] \qquad (16\text{-}26)$$

$$\frac{\tau I_{total}}{1 + \tau} = \frac{1}{2}\left(\frac{Sn_{total}}{1 + \alpha_1}\right) + \frac{1}{2}\left(\frac{Tl_{total}}{1 + \alpha_2}\right)$$

Titration of a mixture: $\quad \phi \equiv 2\left(\dfrac{I_{total}}{Sn_{total}}\right) = \dfrac{1 + \tau}{\tau}\left(\dfrac{1}{1 + \alpha_1} + \dfrac{Tl_{total}/Sn_{total}}{1 + \alpha_2}\right) \qquad (16\text{-}27)$

Factor of 2 appears because 1 mol
of IO_3^- reacts with 2 mol of Sn^{2+} and
ϕ must be unity at the equivalence point

Equation 16-27 describes the titration curve in Figure 16-6. The factors of $\frac{1}{2}$ in Equation 16-26 arise from the stoichiometry of reaction of titrant with each analyte. In general, these two fractions will not be $\frac{1}{2}$ and will be different from each other. The factor Tl_{total}/Sn_{total} gives the ratio of moles of analytes in the original solution.

16-4   Redox Indicators

A chemical indicator may be used to detect the end point of a redox titration, just as an indicator may be used in an acid-base titration. A **redox indicator** changes color when it goes from its oxidized to its reduced state. One common indicator is ferroin, whose color change is from pale blue (almost colorless) to red.

Oxidized ferroin
(pale blue)
In(oxidized)

Reduced ferroin
(red)
In(reduced)

To predict the potential range over which the indicator color will change, we first write a Nernst equation for the indicator.

$$\text{In(oxidized)} + ne^- \rightleftharpoons \text{In(reduced)}$$

$$E = E° - \frac{0.059\,16}{n}\log\frac{[\text{In(reduced)}]}{[\text{In(oxidized)}]} \qquad (16\text{-}28)$$

As with acid-base indicators, the color of In(reduced) will be observed when

$$\frac{[\text{In(reduced)}]}{[\text{In(oxidized)}]} \gtrsim \frac{10}{1}$$

and the color of In(oxidized) will be observed when

$$\frac{[\text{In(reduced)}]}{[\text{In(oxidized)}]} \lesssim \frac{1}{10}$$

Putting these quotients into Equation 16-28 tells us that the color change will occur over the range

$$E = \left(E° \pm \frac{0.059\,16}{n}\right) \text{volts}$$

A redox indicator changes color over a range of $\pm(59/n)$ mV, centered at $E°$ for the indicator.

For ferroin, with $E° = 1.147$ V (Table 16-2), we expect the color change to occur in the approximate range 1.088 V to 1.206 V with respect to the standard hydrogen electrode. If a saturated calomel electrode is used as the reference instead, the indicator transition range will be

$$\begin{pmatrix} \text{indicator transition} \\ \text{range versus calomel} \\ \text{electrode (S.C.E.)} \end{pmatrix} = \begin{pmatrix} \text{transition range} \\ \text{versus standard hydrogen} \\ \text{electrode (S.H.E.)} \end{pmatrix} - E(\text{calomel}) \qquad (16\text{-}29)$$

See the diagram on page 382 for a better understanding of Equation 16-29.

$$= (1.088 \text{ to } 1.206) - (0.241)$$

$$= 0.847 \text{ to } 0.965 \text{ V (versus S.C.E.)}$$

Ferroin would therefore be a useful indicator for the titrations in Figures 16-2 and 16-5.

The indicator's transition range should overlap the steep part of the titration curve.

TABLE 16-2   Redox indicators

Indicator	Color Oxidized	Reduced	$E°$
Phenosafranine	Red	Colorless	0.28
Indigo tetrasulfonate	Blue	Colorless	0.36
Methylene blue	Blue	Colorless	0.53
Diphenylamine	Violet	Colorless	0.75
4'-Ethoxy-2,4-diaminoazobenzene	Yellow	Red	0.76
Diphenylamine sulfonic acid	Red-violet	Colorless	0.85
Diphenylbenzidine sulfonic acid	Violet	Colorless	0.87
Tris(2,2'-bipyridine)iron	Pale blue	Red	1.120
Tris(1,10-phenanthroline)iron (ferroin)	Pale blue	Red	1.147
Tris(5-nitro-1,10-phenanthroline)iron	Pale blue	Red-violet	1.25
Tris(2,2'-bipyridine)ruthenium	Pale blue	Yellow	1.29

Challenge Select suitable indicators from Table 16-2 for the two end points of the titration in Figure 16-6. What color change might you expect at each of the two end points? Remember that *both* indicators will be in the solution together.

The larger the difference in standard potential (or formal potential) between titrant and analyte, the sharper the break in the titration curve at the equivalence point. A redox titration is usually feasible if the difference between analyte and titrant is ≥ 0.2 V. However, the end point of such a titration is not very sharp and is best detected potentiometrically. If the difference in formal potentials is ≥ 0.4 V, then a redox indicator usually gives a satisfactory end point.

The Starch-Iodine Complex

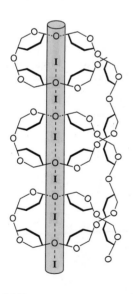

Numerous analytical procedures are based on titrations involving iodine. Starch is the indicator of choice for these procedures because it forms an intense blue complex with iodine. Starch is not a redox indicator because it responds specifically to the presence of I_2, not to a change in redox potential.

The active fraction of starch is amylose, a polymer of the sugar α-D-glucose, with the repeating unit shown in Figure 16-7. The polymer exists as a coiled helix into which small molecules can fit. In the presence of starch, iodine molecules form I_6 chains that occupy the center of the amylose helix (Figure 16-8).

$$I\cdot\cdot I\cdot\cdot I\cdot\cdot I\cdot\cdot I$$

Figure 16-7 Structure of the repeating unit of the sugar amylose.

Visible absorption by the I_6 chain bound within the helix gives rise to the characteristic starch-iodine color.

Starch is readily biodegraded, so it should either be freshly dissolved or the solution should contain a preservative, such as HgI_2 or thymol. A hydrolysis product of starch is glucose, which is a reducing agent. A partially hydrolyzed solution of starch could thus be a source of error in a redox titration.

16-5 Adjustment of Analyte Oxidation State

Sometimes it is necessary to adjust the oxidation state of the analyte before it can be titrated. For example, Mn^{2+} might be **preoxidized** to MnO_4^- and then titrated with standard Fe^{2+}. Preadjustment must be quantitative, and it must be possible to remove or destroy the excess preadjustment reagent so that it will not interfere in the subsequent titration.

Preoxidation

Several powerful oxidants can be easily removed after preoxidation. *Peroxydisulfate* ($S_2O_8^{2-}$, also called *persulfate*) is a strong oxidant that requires Ag^+ as a catalyst.

$$S_2O_8^{2-} + Ag^+ \rightarrow SO_4^{2-} + \underbrace{SO_4^- + Ag^{2+}}_{\text{Two powerful oxidants}}$$

Excess reagent is destroyed by boiling the solution after oxidation of analyte is complete.

$$2S_2O_8^{2-} + 2H_2O \xrightarrow{\text{boiling}} 4SO_4^{2-} + O_2 + 4H^+ \cdot$$

The $S_2O_8^{2-}/Ag^+$ mixture is able to oxidize Mn^{2+} to MnO_4^-, Ce^{3+} to Ce^{4+}, Cr^{3+} to $Cr_2O_7^{2-}$, and VO^{2+} to VO_2^+.

Figure 16-8 Schematic structure of the starch-iodine complex. The amylose sugar chain forms a helix around I_6 units. [Adapted from H. Davis, W. Skrzypek, and A. Khan, *J. Polymer Sci.,* **1994,** *A32,* 2267.]

Silver(II) oxide (AgO) dissolves in concentrated mineral acids to give Ag^{2+}, with oxidizing power similar to the $S_2O_8^{2-}/Ag^+$ combination. Excess Ag^{2+} can be removed by boiling:

$$4Ag^{2+} + 2H_2O \xrightarrow{\text{boiling}} 4Ag^+ + O_2 + 4H^+$$

Solid *sodium bismuthate* ($NaBiO_3$) has an oxidizing strength similar to that of Ag^{2+} and $S_2O_8^{2-}$. The excess solid oxidant is removed by filtration.

Hydrogen peroxide is a good oxidant in basic solution. It can transform Co^{2+} to Co^{3+}, Fe^{2+} to Fe^{3+}, and Mn^{2+} to MnO_2. In acidic solution it can *reduce* $Cr_2O_7^{2-}$ to Cr^{3+} and MnO_4^- to Mn^{2+}. Excess H_2O_2 spontaneously **disproportionates** in boiling water.

Do you remember what "disproportionation" means? Check the Glossary.

$$2H_2O_2 \xrightarrow{\text{boiling}} O_2 + 2H_2O$$

Challenge Write one half-reaction in which H_2O_2 behaves as an oxidant and one half-reaction in which it behaves as a reductant.

Prereduction

Stannous chloride ($SnCl_2$) can be used to prereduce Fe^{3+} to Fe^{2+} in hot HCl. Excess reductant is destroyed by adding excess $HgCl_2$:

$$Sn^{2+} + 2HgCl_2 \rightarrow Sn^{4+} + Hg_2Cl_2 + 2Cl^-$$

The Fe^{2+} is then titrated with an oxidant.

Chromous chloride is a powerful reductant sometimes used for prereduction. Any excess Cr^{2+} is oxidized by atmospheric oxygen. *Sulfur dioxide* and *hydrogen sulfide* are mild reducing agents that can be expelled by boiling an acidic solution after the reduction is complete.

An important prereduction technique uses a column packed with a solid reducing agent. Figure 16-9 shows the *Jones reductor*, which contains zinc coated with zinc **amalgam.** The amalgam is prepared by mixing granular zinc with 2 wt% aqueous $HgCl_2$ for 10 min, then washing with water. We can reduce a sample of Fe^{3+} to Fe^{2+} by passage through a Jones reductor, using 1 M H_2SO_4 as solvent. The column is washed well with water, and the combined washings are titrated with standard MnO_4^-, Ce^{4+}, or $Cr_2O_7^{2-}$. A blank determination is done with a solution passed through the reductor in the same manner as the unknown.

Most reduced analytes are easily reoxidized by atmospheric oxygen. To avoid oxidation, the reduced analyte can be collected in a solution containing excess Fe^{3+}. The ferric ion is immediately reduced to Fe^{2+}, which is stable in acid. The Fe^{2+} is then titrated with an oxidant. By this means, elements such as Cr, Ti, V, and Mo can be indirectly analyzed.

Zinc is such a powerful reducing agent that the Jones reductor is not very selective.

$$Zn^{2+} + 2e^- \rightleftharpoons Zn(s) \quad E° = -0.764 \text{ V}$$

More selective is the *Walden reductor*, filled with solid Ag and 1 M HCl. The reduction potential for the silver-silver chloride couple (0.222 V) is high enough that species such as Cr^{3+} and TiO^{2+} are not reduced and therefore

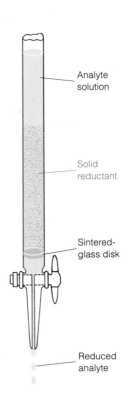

Analyte
solution

Solid
reductant

Sintered-
glass disk

Reduced
analyte

Figure 16-9 A column filled with a solid reagent used for prereduction of analyte is called a *reductor.* Often the analyte is drawn through by suction.

do not interfere in the analysis of a metal such as Fe^{3+}. Another selective reductor uses a filling of granular Cd metal. In determining levels of nitrogen oxides for air-pollution monitoring,[2] the gases are first converted to NO_3^-, which is not easy to analyze. Passing nitrate through a Cd-filled column reduces NO_3^- to NO_2^-, for which a convenient spectrophotometric analysis is available.

16-6 Oxidation with Potassium Permanganate

Potassium permanganate ($KMnO_4$) is an oxidizing agent with an intense violet color. In strongly acidic solutions (pH ≲ 1), it is reduced to colorless Mn^{2+}.

$$\underset{\text{Permanganate}}{MnO_4^-} + 8H^+ + 5e^- \rightleftharpoons \underset{\text{Manganous}}{Mn^{2+}} + 4H_2O \qquad E° = 1.507 \text{ V}$$

In neutral or alkaline solution, the product is the brown solid, MnO_2.

$$MnO_4^- + 4H^+ + 3e^- \rightleftharpoons \underset{\substack{\text{Manganese}\\\text{dioxide}}}{MnO_2(s)} + 2H_2O \qquad E° = 1.692 \text{ V}$$

In very strongly alkaline solution (2 M NaOH), green manganate ion is produced.

$$MnO_4^- + e^- \rightleftharpoons \underset{\text{Manganate}}{MnO_4^{2-}} \qquad E° = 0.56 \text{ V}$$

> $KMnO_4$ serves as its own indicator in acidic solution.

Representative permanganate titrations are listed in Table 16-3. For titrations in strongly acidic solution, $KMnO_4$ serves as its own indicator because the product, Mn^{2+}, is colorless (Color Plate 10). The end point is taken as the first persistent appearance of pale pink MnO_4^-. If the titrant is too dilute to be seen, an indicator such as ferroin can be used.

 Preparation and Standardization

> $KMnO_4$ is not a primary standard.

Potassium permanganate is not a primary standard, because traces of MnO_2 are invariably present. In addition, distilled water usually contains enough organic impurities to reduce some freshly dissolved MnO_4^- to MnO_2. To prepare a 0.02 M stock solution, $KMnO_4$ is dissolved in distilled water, boiled for an hour to hasten the reaction between MnO_4^- and organic impurities, and filtered through a clean, sintered-glass filter to remove precipitated MnO_2. Filter paper (organic matter!) should never be used. The reagent is stored in a dark glass bottle. Aqueous $KMnO_4$ is unstable by virtue of the reaction

$$4MnO_4^- + 2H_2O \rightarrow 4MnO_2(s) + 3O_2 + 4OH^-$$

which is slow in the absence of MnO_2, Mn^{2+}, heat, light, acids, and bases. Permanganate should be standardized often for the most accurate work. Dilute solutions are prepared fresh from 0.02 M stock solution (using water distilled from alkaline $KMnO_4$) and standardized.

Potassium permanganate can be standardized by titration of sodium oxalate ($Na_2C_2O_4$) or pure electrolytic iron wire. Dry (105°C, 2 h) sodium oxalate (available in a 99.9–99.95% pure form) is dissolved in 1 M H_2SO_4 and treated with 90–95% of the required $KMnO_4$ solution at room

TABLE 16-3   **Analytical applications of permanganate titrations**

Species analyzed	Oxidation reaction	Notes
Fe^{2+}	$Fe^{2+} \rightleftharpoons Fe^{3+} + e^-$	Fe^{3+} is reduced to Fe^{2+} with Sn^{2+} or a Jones reductor. Titration is carried out in 1 M H_2SO_4 or 1 M HCl containing Mn^{2+}, H_3PO_4, and H_2SO_4. Mn^{2+} inhibits oxidation of Cl^- by MnO_4^-. H_3PO_4 complexes Fe^{3+} to prevent formation of yellow Fe^{3+}-chloride complexes.
$H_2C_2O_4$	$H_2C_2O_4 \rightleftharpoons 2CO_2 + 2H^+ + 2e^-$	Add 95% of titrant at 25°C, then complete titration at 55°–60°C.
Br^-	$Br^- \rightleftharpoons \frac{1}{2}Br_2(g) + e^-$	Titrate in boiling 2 M H_2SO_4 to remove $Br_2(g)$.
H_2O_2	$H_2O_2 \rightleftharpoons O_2(g) + 2H^+ + 2e^-$	Titrate in 1 M H_2SO_4.
HNO_2	$HNO_2 + H_2O \rightleftharpoons NO_3^- + 3H^+ + 2e^-$	Add excess standard $KMnO_4$ and back-titrate after 15 min at 40°C with Fe^{2+}.
As^{3+}	$H_3AsO_3 + H_2O \rightleftharpoons H_3AsO_4 + 2H^+ + 2e^-$	Titrate in 1 M HCl with KI or ICl catalyst.
Sb^{3+}	$H_3SbO_3 + H_2O \rightleftharpoons H_3SbO_4 + 2H^+ + 2e^-$	Titrate in 2 M HCl.
Mo^{3+}	$Mo^{3+} + 2H_2O \rightleftharpoons MoO_2^{2+} + 4H^+ + 3e^-$	Reduce Mo in a Jones reductor, and run the Mo^{3+} into excess Fe^{3+} in 1 M H_2SO_4. Titrate the Fe^{2+} formed.
W^{3+}	$W^{3+} + 2H_2O \rightleftharpoons WO_2^{2+} + 4H^+ + 3e^-$	Reduce W with Pb(Hg) at 50°C and titrate in 1 M HCl.
U^{4+}	$U^{4+} + 2H_2O \rightleftharpoons UO_2^{2+} + 4H^+ + 2e^-$	Reduce U to U^{3+} with a Jones reductor. Expose to air to produce U^{4+}, which is titrated in 1 M H_2SO_4.
Ti^{3+}	$Ti^{3+} + H_2O \rightleftharpoons TiO^{2+} + 2H^+ + e^-$	Reduce Ti to Ti^{3+} with a Jones reductor, and run the Ti^{3+} into excess Fe^{3+} in 1 M H_2SO_4. Titrate the Fe^{2+} that is formed.
Mg^{2+}, Ca^{2+}, Sr^{2+}, Ba^{2+}, Zn^{2+}, Co^{2+}, La^{3+}, Th^{4+}, Pb^{2+}, Ce^{3+}, BiO^+, Ag^+	$H_2C_2O_4 \rightleftharpoons 2CO_2 + 2H^+ + 2e^-$	Precipitate the metal oxalate. Dissolve in acid and titrate the $H_2C_2O_4$.
K^+	$K_2NaCo(NO_2)_6 + 6H_2O \rightleftharpoons Co^{2+} + 6NO_3^- + 12H^+ + 2K^+ + Na^+ + 11e^-$	Precipitate potassium sodium cobaltinitrite. Dissolve in acid and titrate. Co^{3+} is *reduced* to Co^{2+}, and HNO_2 is oxidized to NO_3^-.
Na^+	$U^{4+} + 2H_2O \rightleftharpoons UO_2^{2+} + 4H^+ + 2e^-$	Precipitate $NaZn(UO_2)_3(acetate)_9$. Dissolve in acid, reduce the UO_2^{2+} (as above), and titrate.
$S_2O_8^{2-}$	$S_2O_8^{2-} + 2Fe^{2+} + 2H^+ \rightleftharpoons 2Fe^{3+} + 2HSO_4^-$	Peroxydisulfate is added to excess standard Fe^{2+} containing H_3PO_4. Unreacted Fe^{2+} is titrated with MnO_4^-.
PO_4^{3-}	$Mo^{3+} + 2H_2O \rightleftharpoons MoO_2^{2+} + 4H^+ + 3e^-$	$(NH_4)_3PO_4 \cdot 12MoO_3$ is precipitated and dissolved in H_2SO_4. The Mo(VI) is reduced (as above) and titrated.

temperature. The solution is then warmed to 55–60°C, and the titration is completed by slow addition of $KMnO_4$. A blank value is subtracted to account for the quantity of titrant (usually one drop) needed to impart a pink color to the solution.

$$2MnO_4^- + 5H_2C_2O_4 + 6H^+ \rightarrow 2Mn^{2+} + 10CO_2 + 8H_2O$$

If pure Fe wire is used as a standard, it is dissolved in warm 1.5 M H_2SO_4 under nitrogen. The product is Fe^{2+}, and the cooled solution can be used to standardize $KMnO_4$ (or other oxidants) with no special precautions. Addition of 5 mL of 86 wt% phosphoric acid per 100 mL of solution masks the yellow color of Fe^{3+} and makes the end point easier to see. Ferrous ammonium sulfate, $Fe(NH_4)_2(SO_4)_2 \cdot 6H_2O$, and ferrous ethylenediammonium sulfate, $Fe(H_3NCH_2CH_2NH_3)(SO_4)_2 \cdot 2H_2O$, are sufficiently pure to be used as standards for most purposes.

16-7 Oxidation with Ce^{4+}

The reduction of Ce^{4+} to Ce^{3+} proceeds cleanly in acidic solutions. The aquo ion, $Ce(H_2O)_n^{4+}$, probably does not exist in any of these solutions, because the cerium ion binds the acid counterion (ClO_4^-, SO_4^{2-}, NO_3^-, Cl^-) very strongly to give a variety of complexes. The variation of the $Ce^{4+}|Ce^{3+}$ formal potential with the medium is indicative of these interactions:

The variation of potential in each solvent implies that different species of cerium are present in each solvent.

$$Ce^{4+} + e^- \rightleftharpoons Ce^{3+} \qquad \text{Formal potential}$$

1.70 V in 1 F $HClO_4$
1.61 V in 1 F HNO_3
1.47 V in 1 F HCl
1.44 V in 1 F H_2SO_4

Ce^{4+} is yellow and Ce^{3+} is colorless, but the color change is not distinct enough for cerium to be its own indicator. Ferroin and other substituted phenanthroline redox indicators (Table 16-2) are well suited to titrations with Ce^{4+}.

Ce^{4+} can be used in place of $KMnO_4$ in most procedures. In addition, ceric ion finds applications in analysis of many organic compounds. In the oscillating reaction in Demonstration 15-1, Ce^{4+} oxidizes malonic acid to CO_2 and formic acid:

$$\underset{\text{Malonic acid}}{CH_2(CO_2H)_2} + 2H_2O + 6Ce^{4+} \rightarrow 2CO_2 + \underset{\text{Formic acid}}{HCO_2H} + 6Ce^{3+} + 6H^+$$

This reaction can be used for quantitative analysis of malonic acid by heating a sample in 4 M $HClO_4$ with excess standard Ce^{4+} and back-titrating the unreacted Ce^{4+} with Fe^{2+}. Analogous procedures are available for many alcohols, aldehydes, ketones, and carboxylic acids.

 Preparation and Standardization

$(NH_4)_2Ce(NO_3)_6$, primary standard-grade ammonium hexanitratocerate(IV), can be dissolved in 1 M H_2SO_4 and used directly. Although the oxidizing strength of Ce^{4+} is greater in $HClO_4$ or HNO_3, these solutions undergo slow photochemical decomposition with concomitant oxidation of water. Ce^{4+} in H_2SO_4 is stable indefinitely. Solutions in HCl are unstable because Cl^- is oxidized to Cl_2. Sulfuric acid solutions of Ce^{4+} can be used for titrations of unknowns in HCl because the reaction with analyte is fast, whereas reaction with Cl^- is slow. Less expensive salts, including $Ce(HSO_4)_4$, $(NH_4)_4Ce(SO_4)_4 \cdot 2H_2O$, and $CeO_2 \cdot xH_2O$ (also called $Ce(OH)_4$), are adequate for preparing titrants that are subsequently standardized with $Na_2C_2O_4$ or Fe as described for MnO_4^-.

16-8 Oxidation with Potassium Dichromate

In acidic solution, the orange dichromate ion is a powerful oxidant that is reduced to chromic ion, Cr^{3+}:

$$Cr_2O_7^{2-} + 14H^+ + 6e^- \rightleftharpoons 2Cr^{3+} + 7H_2O \qquad E° = 1.36 \text{ V}$$

In 1 M HCl, the formal potential is just 1.00 V, and in 2 M H_2SO_4, it is 1.11 V; so dichromate is a less powerful oxidizing agent than MnO_4^- or Ce^{4+}. In basic solution, $Cr_2O_7^{2-}$ is converted to the yellow chromate ion (CrO_4^{2-}), whose oxidizing power is nil:

$$CrO_4^{2-} + 4H_2O + 3e^- \rightleftharpoons Cr(OH)_3(s,\text{hydrated}) + 5OH^-$$

$$E° = -0.12 \text{ V}$$

Potassium dichromate, $K_2Cr_2O_7$, has several advantages: It is pure enough to be a primary standard, its solutions are stable, and it is cheap. Because $Cr_2O_7^{2-}$ is orange and complexes of Cr^{3+} range from green to violet, indicators with distinctive color changes, such as diphenylamine sulfonic acid or diphenylbenzidine sulfonic acid, must be used to find a dichromate end point. Alternatively, the reaction can be monitored with Pt and calomel electrodes.

Because potassium dichromate is not as strong an oxidant as $KMnO_4$ or Ce^{4+}, it is not used as widely. It is employed chiefly for the determination of Fe^{2+} and, indirectly, for many species that will oxidize Fe^{2+} to Fe^{3+}. For indirect analyses, the unknown is treated with a measured excess of Fe^{2+}, and the unreacted Fe^{2+} then titrated with $K_2Cr_2O_7$. Among the species that can be analyzed in this way are ClO_3^-, NO_3^-, MnO_4^-, and organic peroxides. Box 16-2 describes the use of dichromate in water pollution analysis.

Chromium waste is toxic and should not be poured down the drain. See Box 2-1.

Diphenylbenzidine sulfonate (reduced) (colorless)

Diphenylbenzidine sulfonate (oxidized) (violet)
+ $2H^+ + 2e^-$

16-9 Methods Involving Iodine

When a reducing analyte is titrated directly with iodine (to produce I^-), the method is called *iodimetry.* In *iodometry,* an oxidizing analyte is added to excess I^- to produce iodine, which is then titrated with standard thiosulfate solution.

Molecular iodine is only slightly soluble in water (1.3×10^{-3} M at 20°C), but its solubility is enhanced by complexation with iodide.

$$\underset{\text{Iodine}}{I_2(aq)} + \underset{\text{Iodide}}{I^-} \rightleftharpoons \underset{\text{Triiodide}}{I_3^-} \qquad K = 7 \times 10^2$$

A typical 0.05 M solution of I_3^- for titrations is prepared by dissolving 0.12 mol of KI plus 0.05 mol of I_2 in 1 L of water. When we speak of using iodine as a titrant, we almost always mean that we are using a solution of I_2 plus excess I^-.

Iodimetry: titration *with* iodine
Iodometry: titration *of* iodine produced by a chemical reaction

440

Box 16-2 Environmental Carbon Analysis and Oxygen Demand

Industrial waste streams are partially characterized and regulated on the basis of their carbon content or oxygen demand. *Total carbon* (*TC*) is defined by the amount of CO_2 evolved when a sample is completely oxidized at high temperature:

$$\text{total carbon analysis:} \quad \text{all carbon} \xrightarrow[\text{catalyst}]{O_2,\ 900°C} CO_2$$

Commercial instruments analyze 20-μL water samples in 3 min by this procedure, using infrared absorption to measure the CO_2.[3] Total carbon includes dissolved organic material (called *total organic carbon, TOC*) and dissolved CO_3^{2-} and HCO_3^- (called *inorganic carbon, IC*). By definition, TC = TOC + IC.

To distinguish TOC from IC, the pH of a fresh sample is lowered below 2 to convert CO_3^{2-} and HCO_3^- to CO_2, which is purged from the solution with N_2 prior to combustion analysis. Because IC has been removed, this procedure measures TOC only. The IC content is the difference between the two experiments. TOC is widely used to determine compliance with discharge laws.

Total oxygen demand (*TOD*) tells us how much O_2 is required for complete combustion of the pollutants in the waste stream. A volume of N_2 containing a known quantity of O_2 is mixed with the sample and complete combustion is carried out. The remaining O_2 is measured by a potentiometric sensor. This measurement is sensitive to the oxidation states of species in the waste stream. For example, urea consumes five times as much O_2 as formic acid does. Species such as NH_3 and H_2S also contribute to TOD.

Pollutants can be oxidized by refluxing with dichromate ($Cr_2O_7^{2-}$). *Chemical oxygen demand* (*COD*) is defined as the O_2 that is chemically equivalent to the $Cr_2O_7^{2-}$ consumed in this process. Because each $Cr_2O_7^{2-}$ consumes 6e$^-$ (to make 2Cr^{3+}), and because each O_2 consumes 4e$^-$ (to make H_2O), 1 mol of $Cr_2O_7^{2-}$ is chemically equivalent to 1.5 mol of O_2 for this computation. COD analysis is carried out by refluxing polluted water for 2 h with excess standard $Cr_2O_7^{2-}$ in H_2SO_4 solution containing Ag$^+$ catalyst. Unreacted $Cr_2O_7^{2-}$ is then measured by titration with standard Fe^{2+} or by spectrophotometry. Many permits for industrial operations are defined in terms of COD analysis of the waste streams.

Biochemical oxygen demand (*BOD*) measures the O_2 required for biochemical degradation of organic materials by microorganisms. BOD also measures species such as HS$^-$ and Fe^{2+} that may be in the water. Inhibitors are added to prevent oxidation of nitrogen species such as NH_3. The measurement requires 5 days of incubation at 20°C in the dark in a sealed container with no air space. The O_2 dissolved in the solution is measured before and after the incubation. The difference is BOD.[4]

Use of Starch Indicator

An alternative to using starch is to add a few milliliters of *p*-xylene to the vigorously stirred titration vessel. After each addition of reagent near the end point, stop stirring long enough to examine the color of the organic phase. I_2 is 400 times more soluble in *p*-xylene than in water, and its color is readily detected in the organic phase [S. C. Petrovic and G. M. Bodner, *J. Chem. Ed.* 1991, *68*, 509].

As described in Section 16-4, starch is used as an indicator for iodine. In a solution with no other colored species, it is possible to see the color of ~5 $\times$ 10^{-6} M I_3^-. With a starch indicator, the limit of detection is extended by about a factor of 10.

In iodimetry (titration *with* I_3^-), starch can be added at the beginning of the titration. The first drop of excess I_3^- after the equivalence point causes the solution to turn dark blue. In iodometry (titration *of* I_3^-), I_3^- is present throughout the reaction up to the equivalence point. *Starch should not be added to such a reaction until immediately before the equivalence point* (as detected visually, by fading of the I_3^-; Color Plate 11). Otherwise some iodine tends to remain bound to starch particles after the equivalence point is reached.

Starch-iodine complexation is temperature dependent. At 50°C, the color is only one-tenth as intense as at 25°C. If maximum sensitivity is required, cooling in ice water is recommended.[5] Organic solvents decrease the affinity of iodine for starch and markedly reduce the utility of the indicator.

Preparation and Standardization of I_3^- Solutions

Triiodide (I_3^-) is prepared by dissolving solid I_2 in excess KI. Sublimed I_2 is pure enough to be a primary standard, but it is seldom used as a standard because it vaporizes during weighing and some is lost. Instead, the approximate amount is rapidly weighed, and the solution of I_3^- is standardized with a pure sample of the intended analyte or with As_4O_6 or $Na_2S_2O_3$.

Acidic solutions of I_3^- are unstable because the excess I^- is slowly oxidized by air:

$$6I^- + O_2 + 4H^+ \rightarrow 2I_3^- + 2H_2O$$

In neutral solutions, oxidation is insignificant in the absence of heat, light, and metal ions.

Triiodide can be standardized by reaction with primary standard-grade arsenious oxide, As_4O_6 (Figure 16-10). When As_4O_6 is dissolved in acidic solution, arsenious acid is formed:

$$\underset{\text{Arsenious oxide}}{As_4O_6(s)} + 6H_2O \rightleftharpoons \underset{\text{Arsenious acid}}{4H_3AsO_3}$$

The latter reacts with I_3^- as follows:

$$H_3AsO_3 + I_3^- + H_2O \rightleftharpoons H_3AsO_4 + 3I^- + 2H^+$$

Challenge   Use standard potentials to show that the equilibrium constant for the reaction of H_3AsO_3 with I_3^- is 0.04.

Because the equilibrium constant is small, the concentration of H^+ must be kept low to ensure complete reaction. If $[H^+]$ is *too* small (pH $\gtrsim$ 11), triiodide disproportionates to hypoiodous acid, iodate, and iodide. Standardization is usuallly carried out at pH 7–8 in bicarbonate buffer.

An excellent way to make a standard solution of I_3^- is to add a weighed quantity of pure potassium iodate to a small excess of KI. Addition of excess strong acid (to give pH ≈ 1) produces I_3^- by quantitative reverse disproportionation:

$$\underset{\text{Iodate}}{IO_3^-} + 8I^- + 6H^+ \rightleftharpoons 3I_3^- + 3H_2O \qquad (16\text{-}30)$$

Freshly acidified iodate plus iodide can be used to standardize thiosulfate. The I_3^- reagent must be used immediately, or else it is oxidized by air. The only disadvantage of KIO_3 is its low molecular weight relative to the number of electrons it accepts. This property leads to a larger-than-desirable relative weighing error in preparing solutions.

There is a significant vapor pressure of toxic I_2 above solid I_2 and aqueous I_3^-. Vessels containing I_2 or I_3^- should be sealed or, better, kept in a fume hood. Waste solutions of I_3^- should not be dumped into a sink in the open lab.

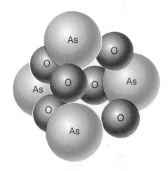

Figure 16-10   The As_4O_6 molecule consists of an As_4 tetrahedron with a bridging oxygen atom on each edge.

HOI: hypoiodous acid
IO_3^-: iodate

KIO_3 is a good primary standard for the generation of I_3^-.

TABLE 16-4   Titrations with standard triiodide (iodimetric titrations)

Species analyzed	Oxidation reaction	Notes
As^{3+}	$H_3AsO_3 + H_2O \rightleftharpoons H_3AsO_4 + 2H^+ + 2e^-$	Titrate directly in $NaHCO_3$ solution with I_3^-
Sb^{3+}	$SbO(O_2CCHOHCHOHCO_2)^- + H_2O \rightleftharpoons$ $SbO_2(O_2CCHOHCHOHCO_2)^- + 2H^+ + 2e^-$	Titrate directly in $NaHCO_3$ solution, using tartrate to mask As(III).
Sn^{2+}	$SnCl_4^{2-} + 2Cl^- \rightleftharpoons SnCl_6^{2-} + 2e^-$	Sn(IV) is reduced to Sn(II) with granular Pb or Ni in 1 M HCl and titrated in the absence of oxygen.
N_2H_4	$N_2H_4 \rightleftharpoons N_2 + 4H^+ + 4e^-$	Titrate in $NaHCO_3$ solution.
SO_2	$SO_2 + H_2O \rightleftharpoons H_2SO_3$ $H_2SO_3 + H_2O \rightleftharpoons SO_4^{2-} + 4H^+ + 2e^-$	Add SO_2 (or H_2SO_3 or HSO_3^- or SO_3^{2-}) to excess standard I_3^- in dilute acid and back-titrate unreacted I_3^- with standard thiosulfate.
H_2S	$H_2S \rightleftharpoons S(s) + 2H^+ + 2e^-$	Add H_2S to excess I_3^- in 1 M HCl and back-titrate with thiosulfate.
$Zn^{2+}, Cd^{2+}, Hg^{2+}, Pb^{2+}$, etc.	$M^{2+} + H_2S \rightarrow MS(s) + 2H^+$ $MS(s) \rightleftharpoons M^{2+} + S + 2e^-$	Precipitate and wash metal sulfide. Dissolve in 3 M HCl with excess standard I_3^- and back-titrate with thiosulfate.
Cysteine, glutathione, thioglycolic acid, mercaptoethanol	$2RSH \rightleftharpoons RSSR + 2H^+ + 2e^-$	Titrate the sulfhydryl compound at pH 4–5 with I_3^-.
HCN	$I_2 + HCN \rightleftharpoons ICN + I^- + H^+$	Titrate in carbonate-bicarbonate buffer, using $CHCl_3$ as an extraction indicator.
$H_2C{=}O$	$H_2CO + 3OH^- \rightleftharpoons HCO_2^- + 2H_2O + 2e^-$	Add excess I_3^- plus NaOH to the unknown. After 5 min, add HCl and back-titrate with thiosulfate.
Glucose (and other reducing sugars)	$\overset{\displaystyle O}{\overset{\displaystyle \|}{RCH}} + 3OH^- \rightleftharpoons RCO_2^- + 2H_2O + 2e^-$	Add excess I_3^- plus NaOH to the sample. After 5 min, add HCl and back-titrate with thiosulfate.
Ascorbic acid (vitamin C)	Ascorbate $+ H_2O \rightleftharpoons$ dehydroascorbate $+ 2H^+ + 2e^-$	Titrate directly with I_3^-.
H_3PO_3	$H_3PO_3 + H_2O \rightleftharpoons H_3PO_4 + 2H^+ + 2e^-$	Titrate in $NaHCO_3$ solution.

Use of Sodium Thiosulfate

Sodium thiosulfate is the almost universal titrant for triiodide. In neutral or acidic solution, triiodide oxidizes thiosulfate to tetrathionate:

Reaction of iodine with thiosulfate

$$I_3^- + 2\underset{\text{Thiosulfate}}{S_2O_3^{2-}} \rightleftharpoons 3I^- + \underset{\text{Tetrathionate}}{O{=}\overset{\displaystyle O}{\underset{\displaystyle O^-}{\overset{\displaystyle \|}{S}}}{-}S{-}S{-}\overset{\displaystyle O}{\underset{\displaystyle O^-}{\overset{\displaystyle \|}{S}}}{=}O} \qquad (16\text{-}31)$$

(In basic solution, I_3^- disproportionates to I^- and HOI. Because hypoiodite oxidizes thiosulfate to sulfate, the stoichiometry of Reaction 16-31 changes and the titration of I_3^- with thiosulfate is carried out below pH 9.) The

TABLE 16-5 Titration of I_3^- produced by analyte (iodometric titrations)

Species analyzed	Reaction	Notes
Cl_2	$Cl_2 + 3I^- \rightleftharpoons 2Cl^- + I_3^-$	Reaction in dilute acid.
HOCl	$HOCl + H^+ + 3I^- \rightleftharpoons Cl^- + I_3^- + H_2O$	Reaction in 0.5 M H_2SO_4.
Br_2	$Br_2 + 3I^- \rightleftharpoons 2Br^- + I_3^-$	Reaction in dilute acid.
BrO_3^-	$BrO_3^- + 6H^+ + 9I^- \rightleftharpoons Br^- + 3I_3^- + 3H_2O$	Reaction in 0.5 M H_2SO_4.
IO_3^-	$2IO_3^- + 16I^- + 12H^+ \rightleftharpoons 6I_3^- + 6H_2O$	Reaction in 0.5 M HCl.
IO_4^-	$2IO_4^- + 22I^- + 16H^+ \rightleftharpoons 8I_3^- + 8H_2O$	Reaction in 0.5 M HCl.
O_2	$O_2 + 4Mn(OH)_2 + 2H_2O \rightleftharpoons 4Mn(OH)_3$ $2Mn(OH)_3 + 6H^+ + 6I^- \rightleftharpoons 2Mn^{2+} + 2I_3^- + 6H_2O$	The sample is treated with Mn^{2+}, NaOH, and KI. After 1 min, it is acidified with H_2SO_4, and the I_3^- is titrated.
H_2O_2	$H_2O_2 + 3I^- + 2H^+ \rightleftharpoons I_3^- + 2H_2O$	Reaction in 1 M H_2SO_4 with NH_4MoO_3 catalyst.
O_3	$O_3 + 3I^- + 2H^+ \rightleftharpoons O_2 + I_3^- + H_2O$	O_3 is passed through neutral 2 wt% KI solution. Add H_2SO_4 and titrate.
NO_2^-	$2HNO_2 + 2H^+ + 3I^- \rightleftharpoons 2NO + I_3^- + 2H_2O$	The nitric oxide is removed (by bubbling CO_2 generated in situ) prior to titration of I_3^-.
As^{5+}	$H_3AsO_4 + 2H^+ + 3I^- \rightleftharpoons H_3AsO_3 + I_3^- + H_2O$	Reaction in 5 M HCl.
Sb^{5+}	$SbCl_6^- + 3I^- \rightleftharpoons SbCl_4^- + I_3^- + 2Cl^-$	Reaction in 5 M HCl
$S_2O_8^{2-}$	$S_2O_8^{2-} + 3I^- \rightleftharpoons 2SO_4^{2-} + I_3^-$	Reaction in neutral solution. Then acidify and titrate.
Cu^{2+}	$2Cu^{2+} + 5I^- \rightleftharpoons 2CuI(s) + I_3^-$	NH_4HF_2 is used as a buffer.
$Fe(CN)_6^{3-}$	$2Fe(CN)_6^{3-} + 3I^- \rightleftharpoons 2Fe(CN)_6^{4-} + I_3^-$	Reaction in 1 M HCl.
MnO_4^-	$2MnO_4^- + 16H^+ + 15I^- \rightleftharpoons 2Mn^{2+} + 5I_3^- + 8H_2O$	Reaction in 0.1 M HCl.
MnO_2	$MnO_2(s) + 4H^+ + 3I^- \rightleftharpoons Mn^{2+} + I_3^- + 2H_2O$	Reaction in 0.5 M H_3PO_4 or HCl.
$Cr_2O_7^{2-}$	$Cr_2O_7^{2-} + 14H^+ + 9I^- \rightleftharpoons 2Cr^{3+} + 3I_3^- + 7H_2O$	Reaction in 0.4 M HCl requires 5 min for completion and is particularly sensitive to air oxidation.
Ce^{4+}	$2Ce^{4+} + 3I^- \rightleftharpoons 2Ce^{3+} + I_3^-$	Reaction in 1 M H_2SO_4.

common form of thiosulfate, $Na_2S_2O_3 \cdot 5H_2O$, is not pure enough to be a primary standard. Instead, thiosulfate is usually standardized by reaction with a fresh solution of I_3^- prepared from KIO_3 plus KI or a solution of I_3^- standardized with As_4O_6.

A stable solution of $Na_2S_2O_3$ can be prepared by dissolving the reagent in high-quality, freshly boiled distilled water. Dissolved CO_2 promotes disproportionation of $S_2O_3^{2-}$:

$$S_2O_3^{2-} + H^+ \rightleftharpoons \underset{\text{Bisulfite}}{HSO_3^-} + \underset{\text{Sulfur}}{S(s)} \qquad (16\text{-}32)$$

and metal ions catalyze atmospheric oxidation of thiosulfate:

$$2Cu^{2+} + 2S_2O_3^{2-} \rightarrow 2Cu^+ + S_4O_6^{2-}$$

$$2Cu^+ + \tfrac{1}{2}O_2 + 2H^+ \rightarrow 2Cu^{2+} + H_2O$$

Thiosulfate solutions should be stored in the dark. Addition of 0.1 g of sodium carbonate per liter maintains the pH in an optimum range for stability of the solution. Three drops of chloroform should also be added to each bottle of thiosulfate solution to help prevent bacterial growth. An acidic solution of thiosulfate is unstable, but the reagent can be used to titrate I_3^- in

Box 16-3 Iodometric Analysis of High-Temperature Superconductors

An important application of superconductors (see beginning of this chapter) is in powerful electromagnets needed for medical magnetic resonance imaging. Ordinary conductors in such magnets require a huge amount of electric power. Because electricity moves through a superconductor with no resistance, the voltage can be removed from the electromagnetic coil once the current has started. Current continues to flow, and the power consumption is *zero* because the resistance is zero.

A breakthrough in superconductor technology came with the discovery[7] of yttrium barium copper oxide, $YBa_2Cu_3O_7$, whose crystal structure is shown here. When heated, the material readily loses oxygen atoms from the Cu—O chains, and any composition between $YBa_2Cu_3O_7$ and $YBa_2Cu_3O_6$ is observable.

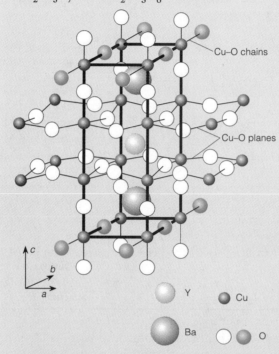

Cu–O chains

Cu–O planes

c
b
a

Y   Cu

Ba   O

Structure of $YBa_2Cu_3O_7$ reproduced from G. F. Holland and A. M. Stacy, *Acc. Chem. Res.* **1988**, *21*, 8. One-dimensional Cu—O chains (shown in color) run along the crystallographic *b* axis and two-dimensional Cu—O sheets lie in the *a–b* plane. Loss of colored oxygen atoms from the chains at elevated temperature results in $YBa_2Cu_3O_6$.

When high-temperature superconductors were discovered, the oxygen content in the formula $YBa_2Cu_3O_x$ was unknown. $YBa_2Cu_3O_7$ represents an unusual set of oxidation states, because the common states of yttrium and barium are Y^{3+} and Ba^{2+}, and the common states

acidic solution because the reaction with triiodide is faster than Reaction 16-32.

 Anhydrous sodium thiosulfate can be prepared from the pentahydrate and is suitable as a primary standard.[6] The standard is prepared by refluxing 21 g of $Na_2S_2O_3 \cdot 5H_2O$ with 100 mL of methanol for 20 min. The anhydrous salt is then filtered, washed with 20 mL of methanol, and dried at 70°C for 30 min.

Analytical Applications of Iodine

Reducing agent + $I_3^- \rightarrow 3I^-$

Reducing agents can be titrated directly with standard I_3^- in the presence of starch, until reaching the intense blue starch-iodine end point. An example is the iodimetric determination of vitamin C:

of copper are Cu^{2+} and Cu^+. If all the copper were Cu^{2+}, the formula of the superconductor would be $(Y^{3+})(Ba^{2+})_2(Cu^{2+})_3(O^{2-})_{6.5}$, with a cation charge of $+13$ and an anion charge of -13. If Cu^+ is present, the oxygen content would be less than 6.5 per formula unit. The composition $YBa_2Cu_3O_7$ requires Cu^{3+}, which is rather rare. Formally, $YBa_2Cu_3O_7$ can be thought of as $(Y^{3+})(Ba^{2+})_2(Cu^{2+})_2(Cu^{3+})(O^{2-})_7$ with a cation charge of $+14$ and an anion charge of -14.

Redox titrations proved to be the most reliable way to measure the oxidation state of copper and thereby deduce the oxygen content of $YBa_2Cu_3O_x$.[8] An iodometric method involves two experiments. In Experiment A, $YBa_2Cu_3O_x$ is dissolved in dilute acid, in which Cu^{3+} is converted to Cu^{2+}. For simplicity, we write the equations for the formula $YBa_2Cu_3O_7$, but you could balance these equations for $x \neq 7$.[9]

$$YBa_2Cu_3O_7 + 13H^+ \rightarrow Y^{3+} + 2Ba^{2+} + 3Cu^{2+} + \frac{13}{2} H_2O + \frac{1}{4} O_2 \qquad (1)$$

The total copper content is measured by treatment with iodide

$$3Cu^{2+} + \frac{15}{2} I^- \rightarrow 3CuI(s) + \frac{3}{2} I_3^- \qquad (2)$$

and titration of the liberated triiodide with standard thiosulfate by Reaction 16-31. Each mole of Cu in $YBa_2Cu_3O_7$ is equivalent to 1 mol of $S_2O_3^{2-}$ in Experiment A.

In Experiment B, $YBa_2Cu_3O_x$ is dissolved in dilute acid containing I^-. Each mole of Cu^{3+} produces 1 mol of I_3^-, and each mole of Cu^{2+} produces 0.5 mol of I_3^-.

$$Cu^{3+} + 4I^- \rightarrow CuI(s) + I_3^- \qquad (3)$$

$$Cu^{2+} + \frac{5}{2} I^- \rightarrow CuI(s) + \frac{1}{2} I_3^- \qquad (4)$$

The moles of thiosulfate required in Experiment A equal the total moles of Cu in the superconductor. The difference in thiosulfate required between Experiments B and A gives the Cu^{3+} content. From this difference, it is possible to calculate the value of x in the formula $YBa_2Cu_3O_x$. A procedure for this analysis is described in Experiment 27-11.[10]

Although we can balance cation and anion charges in the formula $YBa_2Cu_3O_7$ by including Cu^{3+} in the formula, there is no evidence for discrete Cu^{3+} ions in the crystal. There is also no evidence that some of the oxygen is in the form of peroxide, O_2^{2-}, which would also balance the cation and anion charges. The best description of the valence state in the solid crystal involves electrons and holes delocalized in the Cu—O planes and chains. Nonetheless, the formal designation of Cu^{3+} and the chemistry in Equations 1 through 4 accurately describe the redox chemistry of $YBa_2Cu_3O_7$. Problem 35 describes titrations that separately measure the oxidation numbers of Cu and Bi in superconductors such as $Bi_2Sr_2(Ca_{0.8}Y_{0.2})Cu_2O_{8.295}$.

Ascorbic acid
(vitamin C)

Dehydroascorbic
acid

Examples of iodimetric analyses are given in Table 16-4.

Oxidizing agents can be treated with excess I^- to produce I_3^- (Table 16-5, Box 16-3). The iodometric analysis is completed by titrating the liberated I_3^- with standard thiosulfate. Starch is not added until just before the end point.

Oxidizing agent $+ 3I^- \rightarrow I_3^-$

Terms to Understand

amalgam	preoxidation	redox indicator
disproportionation	prereduction	

Summary

Redox titrations are based on an oxidation-reduction reaction between analyte and titrant. Sometimes a quantitative chemical preoxidation (with reagents such as $S_2O_8^{2-}$, AgO, $NaBiO_3$, H_2O_2, or $HClO_4$) or prereduction (with reagents such as $SnCl_2$, $CrCl_2$, SO_2, H_2S, or a metallic reductor column) is necessary to adjust the oxidation state of the analyte prior to analysis. The end point of a redox titration is commonly detected with a redox indicator or by potentiometry. A useful indicator must have a transition range (= $E°$(indicator) ± $0.059\,16/n$ V) that overlaps the abrupt change in potential of the titration curve.

The greater the difference in reduction potential between the analyte and titrant, the sharper will be the end point. Plateaus before and after the equivalence point are centered near $E°$(analyte) and $E°$(titrant). Prior to the equivalence point, the half-reaction involving analyte is used to find the voltage because the concentrations of both the oxidized and the reduced forms of analyte are known. After the equivalence point, the half-reaction involving titrant is employed. At the equivalence point, both half-reactions are used simultaneously to find the voltage. Spreadsheets make these computations easier and more accurate.

Common oxidizing titrants include $KMnO_4$, Ce^{4+}, and $K_2Cr_2O_7$. A large number of procedures is based on oxidation with I_3^- or titration of I_3^- liberated in a chemical reaction.

Exercises

A. A 20.0-mL solution of 0.005 00 M Sn^{2+} in 1 M HCl was titrated with 0.020 0 M Ce^{4+} to give Sn^{4+} and Ce^{3+}. Calculate the potential (versus S.C.E.) at the following volumes of Ce^{4+}: 0.100, 1.00, 5.00, 9.50, 10.00, 10.10, and 12.00 mL. Sketch the titration curve.

B. Consider the titration of 50.0 mL of 1.00 mM I^- with 5.00 mM $Br_2(aq)$ to give $I_2(aq)$ and Br^-. Calculate the potential (versus S.C.E.) at the following volumes of Br_2: 0.100, 2.50, 4.99, 5.01, and 6.00 mL. Sketch the titration curve.

C. Vanadium(II) undergoes three stepwise oxidation reactions:

$$V^{2+} \rightarrow V^{3+} \rightarrow VO^{2+} \rightarrow VO_2^+$$

Calculate the potential (versus S.C.E.) at each of the following volumes when 10.0 mL of 0.0100 M V^{2+} is titrated with 0.0100 M Ce^{4+} in 1 M $HClO_4$: 5.0, 15.0, 25.0, and 35.0 mL. Sketch the titration curve.

D. Select an indicator from Table 16-2 that would be useful for finding **(a)** the second end point and **(b)** the third end point of the titration in Exercise **C**. What color change would you see at each point?

E. ▦ Prepare a spreadsheet to compute the titration curve for Demonstration 16-1, in which 400.0 mL of 3.75 mM Fe^{2+} is titrated with 20.0 mM MnO_4^{2-} at pH 0.00.

F. A 128.6-mg sample of a protein (MW 58 600) was treated with 2.000 mL of 0.048 7 M $NaIO_4$ to react with all the serine and threonine residues.

$$\begin{array}{c}
R \\
| \\
HOCH \\
\underset{+}{\overset{}{H_3NCHCO_2^-}}
\end{array}
+ IO_4^- \longrightarrow
\begin{array}{c}
R \\
| \\
O{=}CH \\
+ \\
O{=}CH \\
| \\
CO_2^-
\end{array}$$

Serine (R = H)
Threonine (R = CH_3)

$$+ NH_4^+ + IO_3^-$$

The solution was then treated with excess iodide to convert the unreacted periodate to iodine:

$$IO_4^- + 3I^- + H_2O \rightleftharpoons IO_3^- + I_3^- + 2OH^-$$

Titration of the iodine required 823 μL of 0.098 8 M thiosulfate.

(a) Calculate the number of serine plus threonine residues per molecule of protein. Answer to the nearest integer.

(b) How many milligrams of As_4O_6 (FW 395.68) would be required to react with the I_3^- liberated in this experiment?

Problems

Shape of a Redox Titration Curve

1. What is the difference between the *titration reaction* and the *cell reactions* in a potentiometric titration?

2. Consider the titration in Figure 16-2.

(a) Write a balanced titration reaction.

(b) Write two different half-reactions for the indicator electrode.

(c) Write two different Nernst equations for the net cell reaction.

(d) Calculate E at the following volumes of Ce^{4+}: 10.0, 25.0, 49.0, 50.0, 51.0, 60.0, and 100.0 mL. Compare your results to Figure 16-2.

3. Consider the titration of 100.0 mL of 0.010 0 M Ce^{4+} in 1 M $HClO_4$ by 0.040 0 M Cu^+ to give Ce^{3+} and Cu^{2+}, using Pt and saturated Ag|AgCl electrodes to find the end point.

(a) Write a balanced titration reaction.

(b) Write two different half-reactions for the indicator electrode.

(c) Write two different Nernst equations for the net cell reaction.

(d) Calculate E at the following volumes of Cu^+: 1.00, 12.5, 24.5, 25.0, 25.5, 30.0, and 50.0 mL. Sketch the titration curve.

4. Consider the titration of 25.0 mL of 0.010 0 M Sn^{2+} by 0.050 0 M Tl^{3+} in 1 M HCl, using Pt and saturated calomel electrodes to find the end point.

(a) Write a balanced titration reaction.

(b) Write two different half-reactions for the indicator electrode.

(c) Write two different Nernst equations for the net cell reaction.

(d) Calculate E at the following volumes of Tl^{3+}: 1.00, 2.50, 4.90, 5.00, 5.10, and 10.0 mL. Sketch the titration curve.

5. Ascorbic acid was added to Fe^{3+} in a solution buffered to pH 0.3, and the potential was monitored with Pt and Ag|AgCl electrodes.

dehydroascorbic acid $+ 2H^+ + 2e^-$
$\rightleftharpoons$ ascorbic acid $+ H_2O$ $E° = 0.390$ V

(a) Write a balanced equation for the titration reaction.

(b) Write two balanced equations to describe the cell reactions. (This question is asking for two complete reactions, not two half-reactions.)

General Approach to Redox Titration Curves

6. Use a spreadsheet to prepare a titration curve (potential referenced to saturated calomel electrode versus volume of titrant) for each case below. Compute the potential at $0.01V_e$, $0.5V_e$, $0.99V_e$, $0.999V_e$, V_e, $1.01V_e$, $1.1V_e$, and $2V_e$.

(a) Titration of 25.00 mL of 0.020 0 M Cr^{2+} with 0.010 00 M Fe^{3+} in 1 M $HClO_4$.

(b) Titration of 50.0 mL of 0.050 0 M Fe^{2+} ($E° = 0.68$ V) with 0.050 0 M MnO_4^- at pH 1.00 in H_2SO_4.

(c) Titration of 50 mL of 0.020 8 M Fe^{3+} with 0.017 3 M ascorbic acid at pH 1.00 in HCl.

(d) Titration of 50.0 mL of 0.050 M UO_2^{2+} in 1 M HCl with 0.100 M Sn^{2+} to give U^{4+} and Sn^{4+}.

7. Set up a spreadsheet to compute the curve for the titration of 100.0 mL of 0.010 0 M Tl^+ with 0.010 0 M IO_3^-. Investigate what happens if the constant concentration of HCl in both solutions is (a) 0.5, (b) 1.0, and (c) 2.0 M. Assume that the formal potential of the $Tl^{3+}|Tl^+$ couple remains at 0.77 V (which is a poor approximation).

8. Chromous ion (Cr^{2+}) was titrated with chlorate, ClO_3^-, at pH $= -0.30$ to give Cr^{3+} and Cl^-. The potential was measured with Pt and saturated Ag | AgCl electrodes.

(a) Write a balanced half-reaction for $ClO_3^- \rightarrow Cl^-$.

(b) Using just the information below, find $E°$ for the ClO_3^- half-reaction.

$$ClO_3^- + 6H^+ + 5e^- \rightleftharpoons \tfrac{1}{2}Cl_2(g) + 3H_2O$$
$$E° = 1.458 \text{ V}$$
$$Cl_2(g) + 2e^- \rightleftharpoons 2Cl^- \qquad E° = 1.360 \text{ V}$$

(c) Compute and graph the titration curve for titrant volume $= 0$ to volume $= 2V_e$.

Titration of a Mixture

9. Prepare a spreadsheet to reproduce the titration curve in Figure 16-6. Find the potential at 20.0, 25.0, 50.0, 60.0, 75.0, 100.0, and 110.0 mL.

10. Consider the titration of analytes A and B by 1.00 M oxidizing titrant T:

$$T(aq) + 3e^- \rightleftharpoons T^{3-}(aq) \qquad E°_T = 0.93 \text{ V}$$

$$\Rightarrow \tau = \frac{[T^{3-}]}{[T]} = 10^{3(0.93 - E)/0.059\,16}$$

$$A^{2+}(aq) + 2e^- \rightleftharpoons A(aq) \qquad E°_A = -0.13 \text{ V}$$

$$\Rightarrow \alpha = \frac{[A]}{[A^{2+}]} = 10^{2(-0.13 - E)/0.059\,16}$$

$$B^+(aq) + e^- \rightleftharpoons B(aq) \qquad E_B^\circ = 0.46 \text{ V}$$

$$\Rightarrow \beta = \frac{[B]}{[B^+]} = 10^{(0.46 - E)/0.059\,16}$$

The initial volume of 100.0 mL contains 0.300 M A and 0.0600 M B.

(a) Write the two titration reactions in the order in which they occur. Calculate the equilibrium constant for each one.

(b) Find the two equivalence volumes, designated V_{e1} and V_{e2}.

(c) Defining the fraction of titration (ϕ) to be unity at the first equivalence point, show that

$$\phi = \frac{3}{2} \frac{T_{total}}{A_{total}} = \left(\frac{1 + \tau}{\tau}\right)\left(\frac{1}{1 + \alpha} + \frac{1}{2}\frac{B_{total}/A_{total}}{1 + \beta}\right)$$

(d) Use a spreadsheet to prepare a graph of the titration curve (E versus S.H.E.).

(e) What are the potentials at the two equivalence points? Use 12 decimal places for V_e to obtain 3 digits for the equivalence-point potential.

11. A 25.0-mL solution containing a mixture of U^{4+} and Fe^{2+} in 1 M $HClO_4$ was titrated with 0.009 87 M $KMnO_4$ in 1 M $HClO_4$.

(a) Write balanced equations for the two titration reactions in the order in which they occur.

(b) Potentiometric end points were observed at 12.73 and 31.21 mL. Calculate the molarities of U^{4+} and Fe^{2+} in the unknown.

(c) Defining the fraction of titration, ϕ, such that $\phi = 1$ at the first equivalence point, show that

$$\phi = \frac{5}{2} \frac{Mn_{total}}{U_{total}} = \left(\frac{1 + \tau}{\tau}\right)\left(\frac{1}{1 + \alpha_1} + \frac{1}{2}\frac{Fe_{total}/U_{total}}{1 + \alpha_2}\right)$$

where α_1 refers to U^{4+} and α_2 refers to Fe^{2+}.

(d) Calculate the potential (versus S.H.E.) at $\frac{1}{2}V_{e1}$, V_{e1}, $V_{e1} + \frac{1}{2}V_{e2}$ (= 21.97 mL), and V_{e2} (= 31.21 mL), where V_{e1} and V_{e2} are the two equivalence volumes.

(e) Graph the titration curve up to 50 mL.

Redox Indicators

12. Select indicators from Table 16-2 that would be suitable for finding the end point in Figure 16-5. State what color changes would be observed.

13. Would indigo tetrasulfonate be a suitable redox indicator for the titration of $Fe(CN)_6^{4-}$ with Tl^{3+} in 1 M HCl?

14. Would tris(2,2'-bipyridine)iron be a useful indicator for the titration of Sn^{2+} with $Mn(EDTA)^-$?

15. Suppose that 100.0 mL of solution containing 0.100 M Fe^{2+} and 5.00×10^{-5} M tris(1,10-phenanthroline)Fe(II) (ferroin) is titrated with 0.0500 M Ce^{4+} in 1 M $HClO_4$. Calculate the potential (versus S.H.E.) at the following volumes of Ce^{4+} and prepare a graph of the entire titration curve: 1.0, 10.0, 100.0, 190.0, 199.0, 200.0, 200.05, 200.10, 200.15, 200.2, 201.0, and 210.0 mL. Is ferroin a suitable indicator for the titration?

16. An enzyme involved in the catalysis of redox reactions has an oxidized form and a reduced form differing by two electrons. The oxidized form of this enzyme was mixed at pH 7 with the oxidized form of a redox indicator whose two forms differ by one electron. The mixture was protected by an inert atmosphere and partially reduced with sodium dithionite. The equilibrium composition of the mixture was determined by spectroscopic analysis:

enzyme(oxidized)	4.2×10^{-5} M
indicator(oxidized)	3.9×10^{-5} M
enzyme(reduced)	1.8×10^{-5} M
indicator(reduced)	5.5×10^{-5} M

Given that $E^{\circ\prime} = -0.187$ V for the indicator, find $E^{\circ\prime}$ for the enzyme.

Adjustment of Analyte Oxidation State

17. Explain the terms *preoxidation* and *prereduction*. Why is it important to be able to destroy the reagents used for these purposes.

18. Write balanced reactions for the destruction of $S_2O_8^{2-}$, Ag^{2+}, and H_2O_2 by boiling.

19. What is a Jones reductor and what is it used for?

20. Why don't Cr^{3+} and TiO^{2+} interfere in the analysis of Fe^{3+} when a Walden reductor, instead of a Jones reductor, is used for prereduction?

Redox Reactions of $KMnO_4^-$, Ce^{4+}, and $K_2Cr_2O_7^{2-}$

21. Write balanced half-reactions in which MnO_4^- acts as an oxidant at (a) pH = 0, (b) pH = 10, and (c) pH = 15.

22. When 25.00 mL of unknown was passed through a Jones reductor, molybdate ion (MoO_4^{2-}) was converted to Mo^{3+}. The filtrate required 16.43 mL of 0.010 33 M $KMnO_4$ to reach the purple end point.

$$MnO_4^- + Mo^{3+} \rightarrow Mn^{2+} + MoO_2^{2+}$$

A blank required 0.04 mL. Find the molarity of molybdate in the unknown.

23. A 50.00-mL sample containing La^{3+} was treated with sodium oxalate to precipitate $La_2(C_2O_4)_3$, which was washed, dissolved in acid, and titrated with 18.04 mL of 0.006 363 M $KMnO_4$. Calculate the molarity of La^{3+} in the unknown.

24. An aqueous glycerol solution weighing 100.0 mg was treated with 50.0 mL of 0.083 7 M Ce^{4+} in 4 M $HClO_4$ at 60°C for 15 min to oxidize the glycerol to formic acid:

$$\underset{\substack{\text{Glycerol}\\\text{MW 92.095}}}{\overset{\displaystyle CH_2-CH-CH_2}{\underset{\displaystyle OH\quad OH\quad OH}{|\quad\;\;|\quad\;\;|}}} \qquad \underset{\text{Formic acid}}{HCO_2H}$$

The excess Ce^{4+} required 12.11 mL of 0.044 8 M Fe^{2+} to reach a ferroin end point. What is the weight percent of glycerol in the unknown?

25. Nitrite (NO_2^-) can be determined by oxidation with excess Ce^{4+}, followed by back-titration of the unreacted Ce^{4+}. A 4.030-g sample of solid containing only $NaNO_2$ (FW 68.995) and $NaNO_3$ was dissolved in 500.0 mL. A 25.00-mL sample of this solution was treated with 50.00 mL of 0.118 6 M Ce^{4+} in strong acid for 5 min, and the excess Ce^{4+} was back-titrated with 31.13 mL of 0.042 89 M ferrous ammonium sulfate.

$$2Ce^{4+} + NO_2^- + H_2O \rightarrow 2Ce^{3+} + NO_3^- + 2H^+$$
$$Ce^{4+} + Fe^{2+} \rightarrow Ce^{3+} + Fe^{3+}$$

Calculate the weight percent of $NaNO_2$ in the solid.

26. Calcium fluoroapatite ($Ca_{10}(PO_4)_6F_2$, FW 1 008.6) laser crystals were doped with chromium to improve their efficiency. It was suspected that the chromium could be in the +4 oxidation state.

1. To measure the total oxidizing power of chromium in the material, a crystal was dissolved in 2.9 M $HClO_4$ at 100°C, cooled to 20°C, and titrated with standard Fe^{2+} using Pt and Ag|AgCl electrodes to find the end point. Chromium oxidized above the +3 state should oxidize an equivalent amount of Fe^{2+} in this step. That is, Cr^{4+} would consume one Fe^{2+} and each atom of Cr^{6+} in $Cr_2O_7^{2-}$ would consume three Fe^{2+}:

$$Cr^{4+} + Fe^{2+} \rightarrow Cr^{3+} + Fe^{3+}$$
$$\tfrac{1}{2}Cr_2O_7^{2-} + 3Fe^{2+} \rightarrow Cr^{3+} + 3Fe^{3+}$$

2. In a second step, the total chromium content was measured by dissolving a crystal in 2.9 M $HClO_4$ at 100°C and cooling to 20°C. Excess $S_2O_8^{2-}$ and Ag^+ were then added to oxidize all chromium to $Cr_2O_7^{2-}$. Unreacted $S_2O_8^{2-}$ was destroyed by boiling, and the remaining solution was titrated with standard Fe^{2+}. In this step, each Cr in the original unknown reacts with three Fe^{2+}.

$$Cr^{x+} \xrightarrow{\;S_2O_8^{2-}\;} Cr_2O_7^{2-}$$
$$\tfrac{1}{2}Cr_2O_7^{2-} + 3Fe^{2+} \rightarrow Cr^{3+} + 3Fe^{3+}$$

In Step 1, 0.437 5 g of laser crystal required 0.498 mL of 2.786 mM Fe^{2+} (prepared by dissolving $Fe(NH_4)_2(SO_4)_2 \cdot 6H_2O$ in 2 M $HClO_4$). In Step 2, 0.156 6 g of crystal required 0.703 mL of the same Fe^{2+} solution. Find the average oxidation number of Cr in the crystal and find the total micrograms of Cr per gram of crystal.

Methods Involving Iodine

27. Why is iodine almost always used in a solution containing excess I^-?

28. List three ways to standardize triiodide solution.

29. In which technique, iodimetry or iodometry, is starch indicator not added until just before the end point? Why?

30. A solution of I_3^- was standardized by titrating freshly dissolved arsenious oxide (As_4O_6, FW 395.683). The titration of 25.00 mL of a solution prepared by dissolving 0.366 3 g of As_4O_6 in a volume of 100.0 mL required 31.77 mL of I_3^-.

(a) Calculate the molarity of the I_3^- solution.

(b) Does it matter whether starch indicator is added at the beginning or near the end point in this titration?

31. From the reduction potentials below

$$I_2(s) + 2e^- \rightleftharpoons 2I^- \qquad E° = 0.535 \text{ V}$$
$$I_2(aq) + 2e^- \rightleftharpoons 2I^- \qquad E° = 0.620 \text{ V}$$
$$I_3^- + 2e^- \rightleftharpoons 3I^- \qquad E° = 0.535 \text{ V}$$

(a) Calculate the equilibrium constant for the reaction $I_2(aq) + I^- \rightleftharpoons I_3^-$.

(b) Calculate the equilibrium constant for the reaction $I_2(s) + I^- \rightleftharpoons I_3^-$.

(c) Calculate the solubility (g/L) of $I_2(s)$ in water.

32. The Kjeldahl analysis in Section 7-2 is used to measure the nitrogen content of organic compounds, which are digested in boiling sulfuric acid to decompose to ammonia, which, in turn, is distilled into standard acid. The remaining acid is then back-titrated with base. Kjeldahl himself had difficulty discerning by lamplight in 1880 the methyl red indicator end point in the back titration. He could have refrained from working at night, but instead he chose to complete the analysis differently. After distilling the ammonia into standard sulfuric acid, he added a mixture of KIO_3 and KI to the acid. The liberated iodine was then titrated with thiosulfate, using starch for easy end-point detection—even by lamplight.[11] Explain how the thiosulfate titration is related to the nitrogen content of the unknown. Derive a relationship between moles of NH_3 liberated in the digestion and moles of thiosulfate required for titration of iodine.

33. Potassium bromate, $KBrO_3$, is a primary standard for the generation of Br_2 in acidic solution:

$$BrO_3^- + 5Br^- + 6H^+ \rightleftharpoons 3Br_2(aq) + 3H_2O$$

The Br_2 can be used to analyze many unsaturated organic compounds. Al^{3+} was analyzed as follows: An unknown was treated with 8-hydroxyquinoline (oxine) at pH 5 to precipitate aluminum oxinate, $Al(C_9H_6ON)_3$. The precipitate was washed, dissolved in warm HCl containing excess KBr, and treated with 25.00 mL of 0.020 00 M $KBrO_3$.

The excess Br_2 was reduced with KI, which was converted to I_3^-. The I_3^- required 8.83 mL of 0.051 13 M $Na_2S_2O_3$ to reach a starch end point. How many milligrams of Al were in the unknown?

34. *Iodometric analysis of high-temperature superconductor.* The procedure in Box 16-3 was carried out to find the effective copper oxidation state, and therefore the number of oxygen atoms, in the formula $YBa_2Cu_3O_{7-z}$, where z ranges from 0 to 0.5.

(a) In Experiment A of Box 16-3, 1.00 g of superconductor required 4.55 mmol of $S_2O_3^{2-}$. In Experiment B, 1.00 g of superconductor required 5.68 mmol of $S_2O_3^{2-}$. Calculate the value of z in the formula $YBa_2Cu_3O_{7-z}$ (FW 666.246 − 15.999 4 z).

(b) *Propagation of uncertainty.* In several replications of Experiment A, the thiosulfate required was 4.55 (±0.10) mmol of $S_2O_3^{2-}$ per gram of $YBa_2Cu_3O_{7-z}$.

In Experiment B, the thiosulfate required was 5.68 (±0.05) mmol of $S_2O_3^{2-}$ per gram. Calculate the uncertainty of x in the formula $YBa_2Cu_3O_x$.

35. *Warning! The Surgeon General has determined that this problem is hazardous to your health.* The oxidation numbers of Cu and Bi in high-temperature superconductors of the type $Bi_2Sr_2(Ca_{0.8}Y_{0.2})Cu_2O_x$ (which could contain Cu^{2+}, Cu^{3+}, Bi^{3+}, and Bi^{5+}) can be measured by the following procedure.[12] In Experiment A, the superconductor is dissolved in 1 M HCl containing excess 2 mM CuCl. Bi^{5+} (written as BiO_3^-) and Cu^{3+} consume Cu^+ to make Cu^{2+}:

$$BiO_3^- + 2Cu^+ + 4H^+ \rightarrow BiO^+ + 2Cu^{2+} + 2H_2O \quad (1)$$

$$Cu^{3+} + Cu^+ \rightarrow 2Cu^{2+} \quad (2)$$

The excess, unreacted Cu^+ is then titrated by a method called *coulometry,* which is described in Chapter 17. In Experiment B, the superconductor is dissolved in 1 M HCl containing excess 1 mM $FeCl_2 \cdot 4H_2O$. Bi^{5+} reacts with the Fe^{2+} but not with Cu^{3+}:[13]

$$BiO_3^- + 2Fe^{2+} + 4H^+ \rightarrow BiO^+ + 2Fe^{3+} + 2H_2O \quad (3)$$

$$Cu^{3+} + \tfrac{1}{2}H_2O \rightarrow Cu^{2+} + \tfrac{1}{4}O_2 + H^+ \quad (4)$$

The excess, unreacted Fe^{2+} is then titrated by coulometry. The total oxidation number of Cu + Bi is measured in Experiment A, and the oxidation number of Bi is determined in Experiment B. The difference gives the oxidation number of Cu.

(a) In Experiment A, a sample of $Bi_2Sr_2CaCu_2O_x$ (FW 760.37 + 15.999 4x) (containing no yttrium) weighing 102.3 mg was dissolved in 100.0 mL of 1 M HCl containing 2.000 mM CuCl. After reaction with the superconductor, coulometry detected 0.108 5 mmol of unreacted Cu^+ in the solution. In Experiment B, 94.6 mg of superconductor was dissolved in 100.0 mL of 1 M HCl containing 1.000 mM $FeCl_2 \cdot 4H_2O$. After reaction with the superconductor, coulometry detected 0.057 7 mmol of unreacted Fe^{2+}. Find the average oxidation numbers of Bi and Cu in the superconductor, and the oxygen stoichiometry coefficient, x.

(b) Find the uncertainties in the oxidation numbers and x if the quantities in Experiment A are 102.3 (±0.2) mg and 0.108 5 (±0.000 7) mmol and the quantities in Experiment B are 94.6 (±0.2) mg and 0.057 7 (±0.000 7) mmol. Assume negligible uncertainty in other quantities.

Notes and References

1. Sources of information on redox titrations include J. Bassett, R. C. Denney, G. H. Jeffery, and J. Mendham, *Vogel's Textbook of Inorganic Analysis,* 4th ed. (Essex, England: Longman, 1978); H. A. Laitinen and W. E.

Harris, *Chemical Analysis,* 2nd ed. (New York: McGraw-Hill, 1975); I. M. Kolthoff, R. Belcher, V. A. Stenger, and G. Matsuyama, *Volumetric Analysis,* Vol. 3 (New York: Wiley, 1957); A. Berka, J. Vulterin, and J. Zýka, *Newer Redox Titrants* (H. Weisz, trans., Oxford: Pergamon, 1965).

2. J. H. Margeson, J. C. Suggs, and M. R. Midgett, *Anal. Chem.* **1980,** *52,* 1955.

3. K. M. Queeney and F. B. Hoek, *Am. Lab.,* October **1989,** p. 26.

4. BOD and COD procedures are described in *Standard Methods for the Examination of Wastewater,* 18th ed. (Washington, DC: American Public Health Association, 1992), which is the standard reference for water analysis.

5. G. L. Hatch, *Anal. Chem.* **1984,** *56,* 2002.

6. A. A. Woolf, *Anal. Chem.* **1982,** *54,* 2134.

7. K. A. Müller and J. G. Bednorz, *Science* **1987,** *237,* 1133, and R. Pool, *Science* **1988,** *241,* 655.

8. D. C. Harris, M. E. Hills, and T. A. Hewston, *J. Chem. Ed.* **1987,** *64,* 847; D. C. Harris, "Oxidation State Chemical Analysis," in T. A. Vanderah, Ed., *Chemistry of Superconductor Materials* (Park Ridge, NJ: Noyes, 1992). Superconductor demonstration kits can be purchased from several vendors, including Sargent-Welch, 7400 N. Linder Ave., Skokie, IL 60077-1026.

9. Experiments with an ^{18}O-enriched superconductor show that the O_2 evolved in Reaction 1 is all derived from the solid, not from the solvent [M. W. Shafer, R. A. de Groot, M. M. Plechaty, G. J. Scilla, B. L. Olson, and E. I. Cooper, *Mater. Res. Bull.* **1989,** *24,* 687; P. Salvador, E. Fernandez-Sanchez, J. A. Garcia Dominguez, J. Amdor, C. Cascales, and I. Rasines, *Solid State Commun.* **1989,** *70,* 71].

10. A more sensitive and elegant iodometric procedure is described by E. H. Appelman, L. R. Morss, A. M. Kini, U. Geiser, A. Umezawa, G. W. Crabtree, and K. D. Carlson, *Inorg. Chem.* **1987,** *26,* 3237. This method can be modified by adding standard Br_2 to analyze superconductors with oxygen in the range 6.0–6.5, in which there is formally Cu^+ and Cu^{2+}. The use of electrodes instead of starch to find the end point in iodometric titrations of superconductors is recommended [P. Phinyocheep and I. M. Tang, *J. Chem. Ed.* **1994,** *71,* A115].

11. M. T. Garrett, Jr., and J. F. Stehlik, *Anal. Chem.* **1992,** *64,* 310A.

12. M. Karppinen, A. Fukuoka, J. Wang, S. Takano, M. Wakata, T. Ikemachi, and H. Yamauchi, *Physica* **1993,** *C208,* 130.

13. The oxygen liberated in Reaction 4 of this problem may be derived from the superconductor, not from solvent water. In any case, BiO_3^- reacts with Fe^{2+} and Cu^{3+} does not when the sample is dissolved in acid.

Marcus Theory of Electron Transfer

Life depends on the passage of electrons from molecules of food through a chain of redox proteins to molecules of O_2, storing metabolic potential energy along the way. In 1956 Rudolph Marcus predicted that for electron transfer from donor D^- to acceptor A, there exists an optimum free energy change, $\Delta G°$, for which the rate of electron transfer is maximum.[1]

$$D^- + A \rightarrow D + A^- \qquad (\Delta G° < 0 \text{ for spontaneous reaction})$$

The rate is slow when the magnitude of $\Delta G°$ is small. As $\Delta G°$ becomes more negative, the rate of electron transfer increases to a maximum. Then, when $\Delta G°$ becomes *too* negative, the rate *decreases,* even though the reaction becomes more thermodynamically favored. Experimental verification of this surprising prediction is shown below. Marcus received the Nobel Prize in 1992 for his work.

Rate of electron transfer from one donor to different acceptors held at a fixed distance by a rigid, nonconductive framework. Structures of eight different acceptors are shown. As $\Delta G°$ becomes more negative, the rate goes through a maximum and then decreases. [From J. R. Miller, L. T. Calcaterra, and G. L. Closs, *J. Am. Chem. Soc.* **1984,** *106,* 3047.]

We describe electron transfer in two steps in the diagram below. First, the electron is transferred from D^- to A without any change in bond lengths or geometry, and the surrounding solvent molecules do not move. In the second step, the system relaxes to its lowest energy state, with changes of bond lengths, geometry, and solvation. The energy of the second step is called the *reorganization energy,* λ. The Marcus theory predicts that the maximum rate of electron transfer occurs when the free energy change for the net reaction is equal to the reorganization energy ($\Delta G° = \lambda$).

Two steps in electron transfer. Marcus theory predicts that the maximum rate occurs when $\Delta G°$ for the first step is zero.

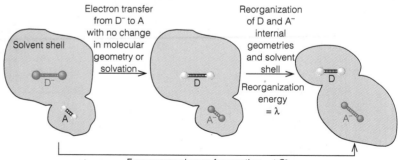

Electrogravimetric and Coulometric Analysis

17

In respiration and photosynthesis, electrons are transferred in a series of steps from one redox protein to another, during which energy is stored by generating molecules such as adenosine triphosphate (ATP). Respiration is *spontaneous,* driven by electron transfer from a reducing food molecule to an oxidizing molecule of O_2. Photosynthesis is *not spontaneous,* requiring energy from sunlight to convert CO_2 and H_2O into carbohydrates and O_2.

In potentiometry, we saw that spontaneous electrochemical reactions with negligible current flow were adapted for analytical purposes. Now we will see how nonspontaneous redox reactions, driven by an externally imposed current, are used in analytical chemistry.

In **electrogravimetric analysis,** the analyte is quantitatively deposited as a solid on the cathode or anode. The increase in mass of the electrode is a direct measure of the amount of analyte. **Coulometry** is an analytical technique based on measuring the number of electrons needed to complete a chemical reaction.

Electrolytic production of aluminum by the Hall-Heroult process consumes 4.5% of the electrical output of the United States! Al^{3+} in a molten solution of Al_2O_3 and cryolite (Na_3AlF_6) is reduced to aluminum metal at the cathode of a cell that typically draws 2.5×10^5 A. This process was invented by Charles Hall in 1886 when he was 22 years old, just after graduating from Oberlin College. [W. E. Haupin, *J. Chem. Ed.* **1983,** *60,* 279; N. C. Craig, *J. Chem. Ed.* **1986,** *63,* 557.]

Charles Martin Hall. [Photo courtesy Alcoa.]

17-1 Electrolysis: Putting Electrons to Work

In **electrolysis,** a reaction is driven in its nonspontaneous direction by an applied voltage. For example, the following reaction has a standard potential of -1.458 V, which means that it is spontaneous to the left, not the right.

$$Pb^{2+} + 2H_2O \rightleftharpoons PbO_2(s) + H_2(g) + 2H^+ \qquad E° = -1.458 \text{ V}$$

Suppose you wish to drive this reaction to the *right* in a solution containing 5.00 mM Pb^{2+}, 2.00 M HNO_3, and solid PbO_2 under 1.00 atm of $H_2(g)$.

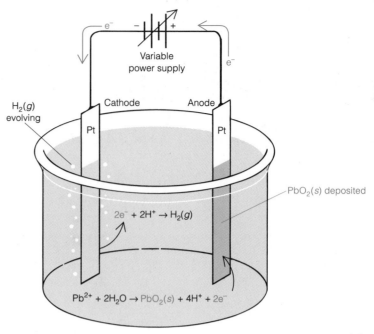

Figure 17-1 Electrolysis of Pb^{2+} in acidic solution. The symbol
represents a variable voltage source.

The equilibrium cell potential is

$$E_{eq} = E° - \frac{0.059\,16}{2}\log\frac{P_{H_2}[H^+]^2}{[Pb^{2+}]}$$

$$= -1.458 - \frac{0.059\,16}{2}\log\frac{(1.00)(2.00)^2}{(5.00 \times 10^{-3})} = -1.544\ \text{V}$$

$\Delta G = -nFE = -nF(-1.544\ \text{V}) =$
$+298\ \text{kJ/mol}$

A voltage whose magnitude is greater than 1.544 V must be *applied* to the
solution to force the reaction to proceed as written.

Figure 17-1 shows how this might be done. A pair of Pt electrodes is
immersed in the solution, and a voltage whose magnitude is greater than
1.544 V is imposed with an external power supply. At the cathode (where
reduction takes place), the reaction is

$$2H^+ + 2e^- \rightarrow H_2(g)$$

and at the anode (where oxidation occurs), the reaction is

$$Pb^{2+} + 2H_2O \rightarrow PbO_2(s) + 4H^+ + 2e^-$$

If a current I flows for a time t, the charge q passing any point in the
circuit is

amperes = coulombs/second
$F = 9.648\,530\,9 \times 10^4\ \text{C/mol}$

moles of electrons $= \dfrac{I \cdot t}{F}$

Relation of charge
to current and time:

$$q = I \cdot t \tag{17-1}$$

coulombs = amperes · seconds

The number of moles of electrons is

$$\text{moles of e}^- = \frac{\text{coulombs}}{\text{coulombs/mole}} = \frac{I \cdot t}{F} \tag{17-2}$$

If a species requires n electrons per molecule, the quantity reacting in time t is

Relation of moles to current and time:
$$\text{moles reacted} = \frac{I \cdot t}{nF} \qquad (17\text{-}3)$$

EXAMPLE Relating Current, Time, and Amount of Reaction

If a current of 0.17 A flows for 16 min through the cell in Figure 17-1, how many grams of PbO_2 will be deposited?

Solution We first calculate the moles of e^- flowing through the cell:

$$\text{moles of } e^- = \frac{I \cdot t}{F} = \frac{\left(0.17\frac{\cancel{C}}{\cancel{s}}\right)(16 \, \cancel{\text{min}})\left(60\frac{\cancel{s}}{\cancel{\text{min}}}\right)}{96\,485\left(\frac{\cancel{C}}{\text{mol}}\right)} = 1.6_9 \times 10^{-3} \, \text{mol}$$

For the half-reaction

$$Pb^{2+} + 2H_2O \rightleftharpoons PbO_2(s) + 4H^+ + 2e^-$$

2 mol of electrons are required for each mole of PbO_2 deposited. Therefore,

$$\text{moles of } PbO_2 = \tfrac{1}{2}(\text{moles of } e^-) = 8.4_5 \times 10^{-4} \, \text{mol}$$

The mass of PbO_2 is

$$\text{grams of } PbO_2 = (8.4_5 \times 10^{-4} \, \text{mol})(239.2 \, \text{g/mol}) = 0.20 \, \text{g}$$

Demonstration 17-1 illustrates electrolysis. Box 17-1 describes an approach to the conversion of solar energy into storable fuel by means of photoelectrolysis.

17-2 Why Voltage Changes When Current Flows

In previous chapters we considered the voltage of galvanic cells only under conditions of negligible current flow. For a cell to do useful work or for electrolysis to occur, significant current must flow. Whenever current flows, three factors decrease the magnitude of the output voltage of a galvanic cell and increase the magnitude of the applied voltage needed for electrolysis. These are the *ohmic potential, concentration polarization,* and *overpotential.*

Ohmic potential, concentration polarization, and overpotential are like *friction* in a mechanical system. They *decrease* the useful output of a galvanic cell and *increase* what must be applied to an electrolysis cell to make it work.

Ohmic Potential

Any cell has some electric resistance. The voltage needed to force current (ions) to flow through the cell is called the **ohmic potential** and is given by Ohm's law:

Ohmic potential:
$$E_{\text{ohmic}} = IR \qquad (17\text{-}4)$$

where I is the current and R is the resistance of the cell.

Demonstration 17-1 **Electrochemical Writing**[2]

Approximately 7% of the electric power output of the United States goes into electrolytic chemical production. The electrolysis apparatus below consists of a sheet of aluminum foil taped or cemented to a glass or wood surface. Any size will work, but an area about 15 cm on a side is convenient for a classroom demonstration. On the metal foil is taped (at one edge only) a sandwich consisting of filter paper, typing paper, and another sheet of filter paper. A stylus is prepared from a length of copper wire (18 gauge or thicker) looped at the end and passed through a length of glass tubing.

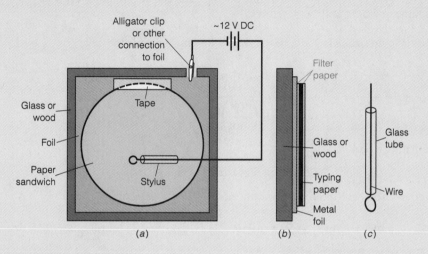

(*a*) Front view. (*b*) Side view. (*c*) Stylus.

A fresh solution is prepared from 1.6 g of KI, 20 mL of water, 5 mL of 1 wt% starch solution, and 5 mL of phenolphthalein indicator solution. (If the solution darkens after standing for several days, it can be decolorized by adding a few drops of dilute $Na_2S_2O_3$.) The three layers of paper are soaked with the KI-starch-phenolphthalein solution. Connect the stylus and foil to a 12-V DC power source, and write on the paper with the stylus.

When the stylus is the cathode, pink color appears from the reaction of OH^- with phenolphthalein:

$$\text{cathode:} \quad H_2O + e^- \rightarrow \frac{1}{2}H_2(g) + OH^-$$

When the polarity is reversed and the stylus is the anode, a black (very dark blue) color appears from the reaction of I_2 with starch:

$$\text{anode:} \quad I^- \rightarrow \frac{1}{2}I_2 + e^-$$

Pick up the top sheet of filter paper and the typing paper, and you will discover that the writing appears in the opposite color on the bottom sheet of filter paper. This sequence is shown in Color Plate 12.

The output voltage of a galvanic cell is decreased by $I \cdot R$.

In a galvanic cell at equilibrium, there is no ohmic potential because $I = 0$. If current is drawn from the cell, the output voltage decreases because part of the free energy released by the chemical reaction is needed to overcome resistance inside the cell. The voltage applied to an electrolysis cell must be great enough to provide the free energy for the chemical reaction and to overcome the cell resistance.

Let us denote the equilibrium voltage of a galvanic cell with zero current as E_{eq}, which is a positive number for a spontaneous reaction. If the reaction is reversed to make it an electrolysis, the voltage is $-E_{eq}$. In the absence of any other effects, *the voltage of a galvanic cell is decreased by IR, and the magnitude of the applied voltage in an electrolysis must be increased by IR for current to flow.*

The magnitude of the input voltage for electrolysis must be increased by $I \cdot R$.

Effect of ohmic potential $\begin{cases} \text{output of galvanic cell:} & E_{galvanic} = E_{eq} - IR \quad (17\text{-}5) \\ \\ \text{input to electrolysis cell:} & E_{electrolysis} = -E_{eq} - IR \quad (17\text{-}6) \end{cases}$

EXAMPLE Effect of Ohmic Potential

Consider the cell

$$Cd(s)\,|\,CdCl_2(aq, 0.167\text{ M})\,|\,AgCl(s)\,|\,Ag(s)$$

in which the spontaneous chemical reaction is

$$Cd(s) + 2AgCl(s) \rightarrow Cd^{2+} + 2Ag(s) + 2Cl^-$$

The Nernst equation tells us that the cell voltage is $E_{eq} = 0.764$ V. **(a)** If the cell has a resistance of 6.42 Ω and a current of 28.3 mA is drawn, what will be the cell voltage? **(b)** What voltage must be applied to operate the same cell in reverse as an electrolysis?

Solution **(a)** With a current of 28.3 mA, the voltage will *decrease* to

$$E_{galvanic} = E_{eq} - IR = 0.764 - (0.028\,3\text{ A})(6.42\ \Omega) = 0.582\text{ V}$$

volts = amperes · ohms

(b) The equilibrium voltage of the electrolysis cell is $-E_{eq} = -0.764$ V. The voltage needed to reverse the spontaneous reaction will be

$$E_{electrolysis} = -E_{eq} - IR = -0.764 - (0.028\,3\text{ A})(6.42\ \Omega) = -0.946\text{ V}$$

Notice that the *magnitude* of galvanic cell voltage is *decreased* by ohmic potential. The *magnitude* of the required voltage for electrolysis is *increased* by the ohmic potential. The voltage of a galvanic cell has a positive sign and the voltage for electrolysis has a negative sign.

Concentration Polarization

Concentration polarization refers to the situation in which the concentration of a species created or consumed at an electrode is not the same at the surface of the electrode as it is in bulk solution. *Concentration polarization decreases the magnitude of voltage available from a galvanic cell and increases the magnitude of the voltage required for electrolysis.*

Consider the cadmium anode in Figure 17-2, for which the reaction is

$$Cd(s) \rightarrow Cd^{2+} + 2e^- \quad (17\text{-}7)$$

If Cd^{2+} ions move rapidly away from the electrode by diffusion or convection, the concentration of Cd^{2+} is uniform in the entire solution. Let's

Box 17-1 Photoelectrolysis

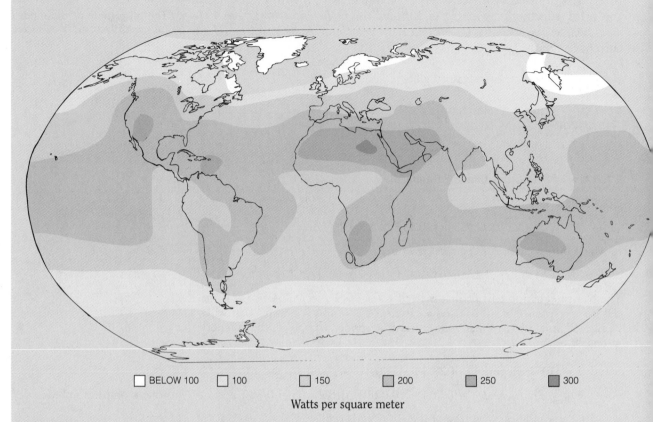

BELOW 100   100   150   200   250   300

Watts per square meter

Solar power incident on the earth's surface is very dilute. It must be concentrated and stored in chemical form for practical use. [From I. Dostrovsky, "Chemical Fuels from the Sun," *Scientific American,* December 1991.]

If solar energy could be used with 10% efficiency, just 3% of the sunlight falling on the earth's deserts would provide all the energy used in the world in 1980. A major research goal is to devise a means of converting solar energy into fuel. Nature accomplishes this in photosynthesis. Green leaves of plants use sunlight to reduce CO_2 to carbohydrates, which can be oxidized back to CO_2 by animals and plants to provide energy. The photocell described here produces H_2 and O_2 from sunlight.[3] Practical systems would store H_2 for use in an engine or fuel cell.

The cell contains thin semiconducting layers of CdSe and CoS. The valence electrons of a semiconductor lie in a band of energy levels called the *valence band.* Electrons at higher energy in the *conduction band* are free to roam through the bulk material. When the CdSe is irradiated with sunlight whose energy is greater than that of the band gap, valence electrons are promoted to the conduction band, leaving positively charged, mobile holes (electron vacancies, designated h^+) in the valence band.[4] Internal electric fields attract electrons to the CoS layer and holes to the CdSe|electrolyte interface.

The electrolyte solution contains 1 M KOH, 1 M Na_2S, and 1 M S, which generates Na_2S_2. At the CdSe|electrolyte interface, the holes oxidize S^{2-}:

$$S^{2-} + h^+ \rightarrow \frac{1}{2}S_2^{2-}$$

and at the CoS|electrolyte interface, electrons reduce S_2^{2-}:

$$\frac{1}{2}S_2^{2-} + e^- \rightarrow S^{2-}$$

If the switch between the Pt electrodes is closed and the stopcock at the bottom of the cell is closed, electricity flows across the load when the cell is irradiated. An array of six CdSe|CoS|electrolyte layers converts 1.9% of solar energy to electrical energy. If the switch and stopcock are open, electrons reduce water on the right side and holes oxidize water at the left electrode:

right-hand Pt electrode: $2H_2O + 2e^- \rightarrow H_2 + 2OH^-$

left-hand Pt electrode: $H_2O + 2h^+ \rightarrow \frac{1}{2}O_2 + 2H^+$

The efficiency of conversion of light to chemical energy is only 0.3%, but it can be improved with an anode material other than Pt, with a lower *overpotential* (Section 17-2) for O_2 production. Semiconductors with smaller band gaps also improve efficiency.

Major challenges in the conversion of sunlight to chemical fuels include increasing efficiency and preventing deterioration of the cell during operation. The sulfide electrolyte in this cell is chosen to prolong the life of the semiconductor materials.

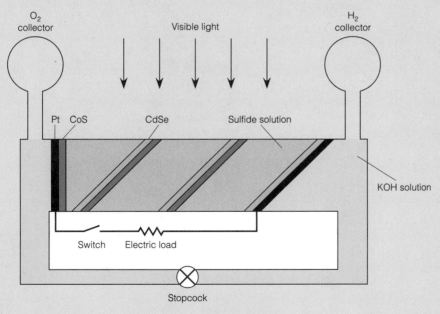

Schematic diagram of photolysis cell using sunlight to make H_2 and O_2, or to make electricity.

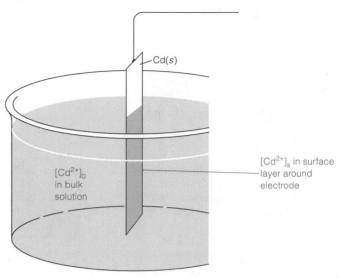

Figure 17-2   The potential of the $Cd^{2+}|Cd$ couple depends on the concentration of Cd^{2+} in the layer surrounding the electrode. The layer is greatly exaggerated in this drawing; it is only a few molecules thick.

The electrode potential depends on the concentration of species in the region immediately surrounding the electrode.

denote the concentration of Cd^{2+} in the bulk solution as $[Cd^{2+}]_0$ and the concentration near the electrode surface as $[Cd^{2+}]_s$. The anode potential depends on $[Cd^{2+}]_s$, not $[Cd^{2+}]_0$:

$$E(\text{anode}) = E°(\text{anode}) - \frac{0.059\,16}{2}\log\frac{1}{[Cd^{2+}]_s} \qquad (17\text{-}8)$$

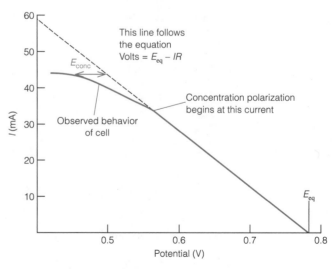

Figure 17-3   Behavior of a galvanic cell, illustrating concentration polarization that occurs when $[Cd^{2+}]_s > [Cd^{2+}]_0$. The resistance of the cell is $6.42\ \Omega$.

If $[Cd^{2+}]_s = [Cd^{2+}]_0$, the anode potential is consistent with the bulk Cd^{2+} concentration.

If current is flowing so fast that Cd^{2+} cannot escape from the vicinity of the electrode as fast as it is made, $[Cd^{2+}]_s$ will be greater than $[Cd^{2+}]_0$. This is *concentration polarization*. The anode potential in Equation 17-8 becomes more positive and the cell voltage [= E(cathode) − E(anode)] becomes more negative. The straight line in Figure 17-3 shows the behavior expected from ohmic potential. Deviation from the straight line at high currents is due to concentration polarization.

Concentration polarization decreases the magnitude of voltage available from a galvanic cell and increases the magnitude of the voltage required for electrolysis. Calling the additional voltage E_{conc}, we can write:

Effects of ohmic potential and concentration polarization:

output of galvanic cell: $\quad E_{galvanic} = E_{eq} - IR - E_{conc}$ (17-9)

input to electrolysis cell: $\quad E_{electrolysis} = -E_{eq} - IR - E_{conc}$ (17-10)

Ions move by diffusion, convection, and electrostatic forces. Raising the temperature increases the rate of diffusion and thereby decreases concentration polarization. Mechanical stirring transports species through the cell. Increasing ionic strength decreases electrostatic forces between ions and the electrode. These factors all affect the degree of polarization. Also, the greater the electrode surface area, the more current that can be passed without polarization.

Overpotential

Even when concentration polarization is absent and ohmic potential is taken into account, some electrolyses require a greater than expected applied voltage, and some galvanic cells produce less voltage than we anticipate. The difference between the expected voltage (after accounting for *IR* drop and concentration polarization) and the observed voltage is called the **overpotential** (E_{over}). *The faster you wish to drive an electrode reaction, the greater the overpotential that must be applied.*

Effects of ohmic potential, concentration polarization, and overpotential:

output of galvanic cell: $\quad E_{galvanic} = E_{eq} - IR - E_{conc} - E_{over}$ (17-11)

input to electrolysis cell: $\quad E_{electrolysis} = -E_{eq} - IR - E_{conc} - E_{over}$

(17-12)

Overpotential can be traced to the activation energy barrier for the electrode reaction.[5] The activation energy in Figure 17-4 is the barrier that must be overcome before reactants can be converted to products. The higher the temperature, the greater the number of molecules with sufficient energy to overcome the barrier, and the faster the reaction proceeds.

Now consider the reaction in which an electron from a metal electrode is transferred to H_3O^+ to initiate reduction to H_2:

$$H_3O^+ + e^- \rightarrow \tfrac{1}{2}H_2(g) + H_2O$$

When ions are not transported to or from an electrode as rapidly as they are consumed or created, *concentration polarization* exists and $[X]_s \neq [X]_0$.

Concentration polarization *decreases* the voltage output of a galvanic cell and *increases* the magnitude of the voltage input required for electrolysis.

To decrease concentration polarization:

1. Raise the temperature

2. Increase stirring

3. Increase electrode surface area

4. Change ionic strength to increase or decrease electrostatic interaction between the electrode and the reactive ion

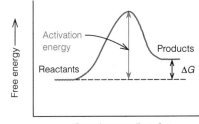

Figure 17-4 Schematic energy profile of a chemical reaction.

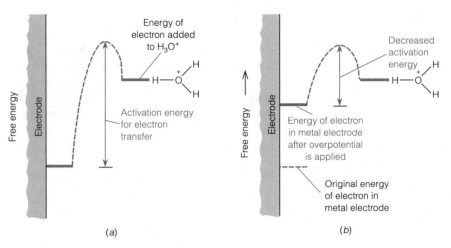

Figure 17-5 Schematic energy profile for electron transfer from a metal to H_3O^+ (a) with no applied potential; (b) after a potential is applied to the electrode. The overpotential increases the energy of the electrons in the electrode.

In Figure 17-5a there is a high barrier preventing electron transfer, and the rate is very slow. If an electric potential (the overpotential) is applied to the electrode, the energy of the electrons *in the metal electrode* is increased (Figure 17-5b). The applied potential decreases the barrier that must be overcome and increases the rate of electron transfer. *Overpotential is the voltage needed to sustain a particular rate of electron transfer.* The greater the rate, the higher must be the overpotential. Thus, *current density* (A/m^2) increases as overpotential increases (Table 17-1).

· ·

EXAMPLE Combined Effects of Ohmic Potential and Overpotential

Suppose we wish to electrolyze I^- to I_3^- in a 0.10 M KI solution containing 3.0×10^{-5} M I_3^- at pH 10.00 with P_{H_2} fixed at 1.00 atm.

$$3I^- + 2H_2O \rightarrow I_3^- + H_2(g) + 2OH^-$$

Suppose that the cell resistance is 2.0 Ω and the current is 63 mA. The cathode overpotential is 0.38 V and the anode overpotential is 0.02 V. What voltage is necessary to drive the reaction?

Solution The equilibrium cell voltage is found from the two half-reactions:

cathode: $2H_2O + 2e^- \rightarrow H_2(g) + 2OH^-$ $E° = -0.828$ V

anode (written as a reduction): $I_3^- + 2e^- \rightarrow 3I^-$ $E° = 0.535$ V

$$E(\text{cathode}) = -0.828 - \frac{0.059\,16}{2} \log P_{H_2} [OH^-]^2$$

$$= -0.828 - \frac{0.059\,16}{2} \log (1.00)(1.0 \times 10^{-4})^2 = -0.591 \text{ V}$$

TABLE 17-1 Overpotential (V) for gas evolution at various current densities at 25°C

Electrode	$10\ A/m^2$		$100\ A/m^2$		$1\,000\ A/m^2$		$10\,000\ A/m^2$	
	H_2	O_2	H_2	O_2	H_2	O_2	H_2	O_2
Platinized Pt	0.0154	0.398	0.0300	0.521	0.0405	0.638	0.0483	0.766
Smooth Pt	0.024	0.721	0.068	0.85	0.288	1.28	0.676	1.49
Cu	0.479	0.422	0.584	0.580	0.801	0.660	1.254	0.793
Ag	0.4751	0.580	0.7618	0.729	0.8749	0.984	1.0890	1.131
Au	0.241	0.673	0.390	0.963	0.588	1.244	0.798	1.63
Graphite	0.5995		0.7788		0.9774		1.2200	
Sn	0.8561		1.0767		1.2230		1.2306	
Pb	0.52		1.090		1.179		1.262	
Zn	0.716		0.746		1.064		1.229	
Cd	0.981		1.134		1.216		1.254	
Hg	0.9		1.0		1.1		1.1	
Fe	0.4036		0.5571		0.8184		1.2915	
Ni	0.563	0.353	0.747	0.519	1.048	0.726	1.241	0.853

SOURCE: *International Critical Tables* **1929**, *6*, 339. This reference also gives overpotentials for Cl_2, Br_2, and I_2.

$$E(\text{anode}) = 0.535 - \frac{0.05916}{2}\log\frac{[I^-]^3}{[I_3^-]}$$

$$= 0.535 - \frac{0.05916}{2}\log\frac{(0.10)^3}{3.0 \times 10^{-5}} = 0.490\ V$$

$$-E_{eq} = E(\text{cathode}) - E(\text{anode}) = -1.081\ V$$

This is the voltage that must be applied to drive the electrolysis in the absence of ohmic potential and overpotential. We call it $-E_{eq}$ (rather than E_{eq}) because we computed it for the electrolysis reaction, not the spontaneous galvanic reaction. The required electrolysis voltage is

$$E_{\text{electrolysis}} = -E_{eq} - I \cdot R - \text{overpotentials}$$

$$= -1.081 - (0.063\ A)(2.0\ \Omega) - 0.38 - 0.02 = -1.61\ V$$

To avoid confusion about signs, bear in mind that the effect of ohmic potential, concentration polarization, and overpotential is always to increase the *magnitude* of the applied voltage for electrolysis.

17-3 Electrogravimetric Analysis

In *electrogravimetric analysis* the analyte is electrolytically deposited as a solid on an electrode. The increase in the mass of the electrode tells us how much analyte was present.

EXAMPLE Electrogravimetric Deposition of Cobalt

A solution containing 0.40249 g of $CoCl_2 \cdot xH_2O$ was exhaustively electrolyzed to deposit 0.09937 g of metallic cobalt on a platinum cathode.

$$Co^{2+} + 2e^- \rightarrow Co(s)$$

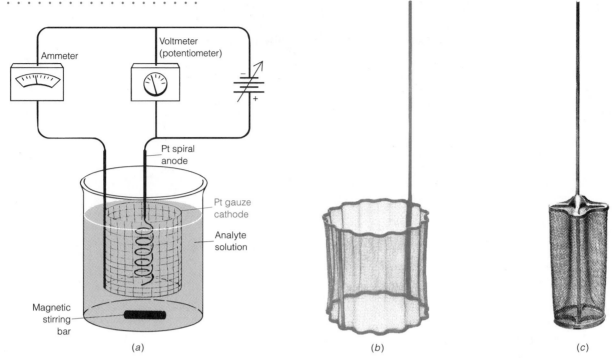

(a)　　　　　　　　　　(b)　　　　　　　　　　(c)

Figure 17-6 *(a)* Electrogravimetric analysis. Analyte is deposited on the large Pt gauze electrode. If analyte is to be oxidized, rather than re-duced, the polarity of the power supply is reversed so that deposition always occurs on the large electrode. *(b)* Outer Pt gauze electrode. *(c)* Optional inner Pt gauze electrode designed to be spun by a motor in place of magnetic stirring.

Calculate the number of moles of water per mole of cobalt in the reagent.

Solution If the reagent contains only $CoCl_2$ and H_2O, we can write

$$\text{grams of } CoCl_2 = \left(\frac{\text{grams of Co deposited}}{\text{atomic weight of Co}} \right) (\text{FW of } CoCl_2) = 0.218\,93 \text{ g}$$

$$\text{grams of } H_2O = 0.402\,49 - 0.218\,93 = 0.183\,56 \text{ g}$$

$$\frac{\text{moles of } H_2O}{\text{moles of Co}} = \frac{0.183\,56/\text{FW of } H_2O}{0.099\,37/\text{atomic weight of Co}} = 6.043$$

The reagent composition is close to $CoCl_2 \cdot 6H_2O$.

Apparatus for electrogravimetric analysis is shown in Figure 17-6. Analyte is typically deposited on a carefully cleaned, chemically inert Pt gauze cathode with a large surface area.

How do you know when an electrolysis is complete? One way is to observe the disappearance of color in a solution from which a colored species such as Co^{2+} or Cu^{2+} is removed. Another way is to expose most, but not all, of the surface of the cathode to the solution during electrolysis. To test whether or not the reaction is complete, raise the beaker or add water so that fresh surface of the cathode is exposed to the solution. After an additional period of electrolysis (15 min, say), see if the newly exposed electrode surface has a deposit. If it does, repeat the procedure. If not, the electrolysis is done. A third method is to remove a small sample of solution and perform a qualitative test for analyte.

Tests for completion of the deposi-tion:

1. Disappearance of color

2. Deposition on freshly exposed electrode surface

3. Qualitative test for analyte in solution

Electrogravimetric analysis would be simple if it merely involved a single analyte species in an otherwise inert solution. In practice, there may be other **electroactive species** that interfere by codeposition. Even the solvent (water) is electroactive, because it decomposes to $H_2 + \frac{1}{2}O_2$ at a sufficiently high voltage. Gas bubbles at the electrode interfere with deposition of solids. Because of these complications, control of electrode potential is important for successful analysis.

Electroactive species are those that can be oxidized or reduced at an electrode.

Current-Voltage Behavior During Electrolysis

Suppose that a solution containing 0.20 M Cu^{2+} and 1.0 M H^+ is electrolyzed to deposit $Cu(s)$ on a Pt cathode and to liberate O_2 at a smooth Pt anode.

cathode: $\qquad Cu^{2+} + 2e^- \rightleftharpoons Cu(s)$

anode: $\qquad\quad H_2O \rightleftharpoons \frac{1}{2}O_2(g) + 2H^+ + 2e^-$

net reaction: $\quad H_2O + Cu^{2+} \rightleftharpoons Cu(s) + \frac{1}{2}O_2(g) + 2H^+ \qquad$ (17-13)

$$E° = E°(\text{cathode}) - E°(\text{anode}) = 0.339 - 1.229 = -0.890$$

Electrolytic generation of arsine for the electronics industry. Arsine (AsH_3) is an extremely toxic gas that is essential to the electronics industry for manufacturing gallium arsenide and other semiconductors. To avoid shipping and storing this hazardous gas, AsH_3 is generated on site by electrolysis of elemental As in KOH solution:

$$As(s) + 3H_2O + 3e^- \rightarrow AsH_3(g) \\ + 3OH^-$$

Assuming that O_2 is liberated at a pressure of 0.20 atm, we naively calculate the voltage needed for the electrolysis as follows:

$$-E_{eq} = E° - \frac{0.059\,16}{n}\log\frac{P_{O_2}^{1/2}[H^+]^2}{[Cu^{2+}]}$$

$$= -0.890 - \frac{0.059\,16}{2}\log\frac{(0.20)^{1/2}(1.0)^2}{0.20} = -0.900\text{ V} \quad (17\text{-}14)$$

No reaction is expected if the applied voltage is more positive than -0.900 V.

The actual behavior of the electrolysis is shown in Figure 17-7. At low voltage, a small **residual current** is observed, even though none was expected. At -0.900 V, nothing special happens. Near -2 V, the reaction begins in earnest.

Why is it necessary to apply a voltage considerably more negative than -0.900 V to drive the reaction at an appreciable rate? The principal reason is that ~ 1 V of overpotential is required for O_2 formation at the smooth Pt electrode (Table 17-1).

The residual current observed prior to -2 V requires reduction at the cathode and an equal amount of oxidation at the anode. Reduction, for example, might involve dissolved O_2, dissolved Fe^{3+}, or surface oxide on the electrode.

Another reason why Equation 17-14 is naive is that when copper is deposited on a platinum electrode, the initial activity of $Cu(s)$ is infinitely small. Equation 17-14 assumes that copper is being deposited *on a copper surface,* for which the activity of copper is unity. If at some time early in the deposition, the activity of $Cu(s)$ on the Pt surface is 10^{-6}, then

It requires about 1 V of extra potential to overcome the barrier to O_2 formation at the anode.

$$-E_{eq} = E° - \frac{0.059\,16}{n}\log\frac{P_{O_2}^{1/2}[H^+]^2[Cu(s)]}{[Cu^{2+}]}$$

$$= -0.890 - \frac{0.059\,16}{2}\log\frac{(0.20)^{1/2}(1.0)^2(10^{-6})}{0.20} = -0.722\text{ V}$$

Note the appearance of $[Cu(s)]$ in this equation.

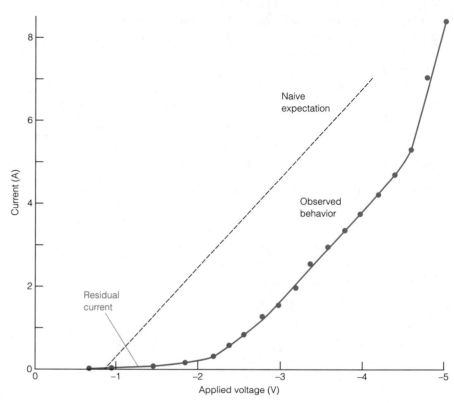

Figure 17-7   Observed current-voltage relationship for electrolysis of 0.2 M $CuSO_4$ in 1 M $HClO_4$ under N_2, using the apparatus shown in Figure 17-6.

When little Cu(*s*) has been deposited, the activity of Cu(*s*) is less than unity and it is not necessary to apply so much voltage for electrolysis.

Figures 17-8a and 17-8b show atomic force microscope images of less than a monolayer of Ag atoms deposited on a gold cathode at potentials less negative than that required for deposition of bulk Ag. When the activity of Ag(*s*) is less than unity, reaction occurs at less negative potentials than we expect for deposition onto bulk Ag(*s*).

The curve in Figure 17-7 can therefore be explained in the following way:

Residual current

1. Initially, a small residual current is present, owing both to oxidation and reduction of impurities and to a small amount of the intended electrolysis.

2. At a sufficiently negative voltage, the desired electrolysis is the main reaction. The voltage is shifted from $-E_{eq}$ by the overpotential for oxygen formation.

Ohmic potential

3. At more negative voltages, the current is linear with respect to voltage, according to Ohm's law. The slope tells us the resistance of the cell.

Electrolysis of solvent

4. At −4.6 V, reduction of water to H_2 begins, and the current shoots upward.

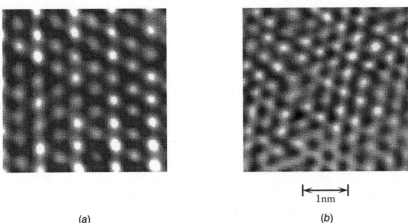

(a) |←——→| 1nm   (c)
                                          (b)

Figure 17-8 Atomic force microscope images of electrolytic deposition of Ag on a crystalline Au cathode surface from a solution containing 0.77 mM Ag_2SO_4 in 0.1 M H_2SO_4. Partially filled lattices in (a) and (b) are deposited at potentials less negative than that required for deposition of bulk Ag (c). [From C. Chen, S. M. Vesecky, and A. A. Gewirth, *J. Am. Chem. Soc.* **1992,** *114,* 451.]

. . . .

EXAMPLE   Current-Voltage Relation for Electrolysis

The resistance of the cell in Figure 17-7 is 0.44 Ω. Estimate the voltage needed to maintain a current of 2.0 A. Assume that the smooth Pt anode is supporting a current density of 1 000 A/m^2 and that there is no concentration polarization.

Solution   Our estimate includes E_{eq}, the ohmic potential, and the overpotential.

$$E_{electrolysis} = \underbrace{-E_{eq}}_{\substack{-0.900 \text{ V} \\ \text{from Equation 17-14}}} - IR - \underbrace{E_{conc}}_{0} - \underbrace{E_{over}}_{\substack{1.28V \\ \text{from Table 17-1}}}$$

$$= -0.900 - (2.0 \text{ A})(0.44 \ \Omega) - 1.28 \text{ V} = -3.06 \text{ V}$$

Question   If the cell in Figure 17-7 has a resistance of 0.44 Ω, what should be the slope (A/V) of the linear region between about -3.0 and -4.5 V? Measure this slope in the figure.

. . . .

Bad Things Can Happen in a Two-Electrode Cell

In the absence of concentration polarization, the voltage for an electrolysis includes contributions from ohmic potential and overpotential.

$$E_{electrolysis} = \underbrace{E(\text{cathode}) - E(\text{anode})}_{\substack{= -E_{eq} \text{ for} \\ \text{reversible cell with} \\ \text{negligible current flow}}} - IR - E_{over} \qquad (17\text{-}15)$$

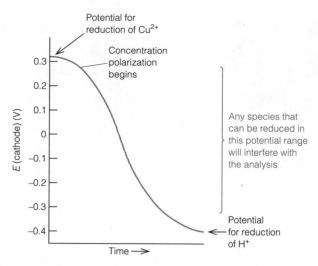

Figure 17-9 E(cathode) changes with time when electrolysis is conducted in a two-electrode cell with a constant voltage between the electrodes.

Question Why do ohmic potential and overpotential decrease as current decreases?

Suppose that we hold $E_{electrolysis}$ for 0.10 M Cu^{2+} in 1.0 M HNO_3 at -2.0 V. As Cu(s) is deposited, the concentration of Cu^{2+} decreases. Eventually, Cu^{2+} cannot be transported to the cathode rapidly enough to maintain the initial current. Because the current decreases, the magnitudes of the ohmic potential and overpotential in Equation 17-15 decrease. The value of E(anode) is fairly constant as a result of the high concentration of solvent being oxidized at the anode. (The anode oxidation reaction is $H_2O \rightarrow \frac{1}{2}O_2 + 2H^+ + 2e^-$.)

E(cathode) becomes more negative as analyte is consumed because of the analyte term in the Nernst equation and because of concentration polarization.

If $E_{electrolysis}$ and E(anode) are constant in Equation 17-15 and if IR and overpotential decrease in magnitude, *E(cathode) must become more negative* to maintain the algebraic equality. E(cathode) drops in Figure 17-9 to -0.4 V, at which H^+ is reduced to H_2:

$$H^+ + e^- \rightarrow \frac{1}{2} H_2(g)$$

As E(cathode) falls from $+0.3$ V to -0.4 V, other ions such as Co^{2+}, Sn^{2+}, and Ni^{2+} can be reduced. *In general, then, when $E_{electrolysis}$ is constant, the cathode potential drifts to negative values and any solute more easily reduced than H^+ will be electrolyzed.*

A *cathodic depolarizer* is reduced in preference to solvent. For *oxidation* reactions, *anodic depolarizers* include N_2H_4 (hydrazine) and NH_2OH (hydroxylamine).

To prevent the cathode potential from becoming so negative that unintended ions are reduced, a cathodic **depolarizer** such as NO_3^- is added to the solution. The cathodic depolarizer is more easily reduced than H^+:

$$NO_3^- + 10H^+ + 8e^- \rightarrow NH_4^+ + 3H_2O$$

Controlled-Potential Electrolysis with a Three-Electrode Cell

The three-electrode cell in Figure 17-10 maintains a *constant cathode potential* and thereby greatly increases the selectivity of electrolysis. The electrode at which the reaction of interest occurs is called the **working electrode.** A calomel electrode serves as the **reference electrode** against which the potential of the working electrode can be measured. The **auxiliary electrode** is the current-supporting partner of the working electrode. Significant current flows between the working and auxiliary electrodes. Negligible current flows at the reference electrode, so its potential is unaffected by ohmic potential, concentration polarization, and overpotential. It truly maintains a constant reference potential. In **controlled-potential electrolysis,** the voltage between the working and reference electrodes is maintained *constant* by an electronic device called a **potentiostat.**

To see why a constant cathode potential permits selective deposition of analyte, consider a solution containing 0.1 M Cu^{2+} and 0.1 M Sn^{2+}. From the standard potentials below, we expect Cu^{2+} to be reduced more easily than Sn^{2+}:

$$Cu^{2+} + 2e^- \rightleftharpoons Cu(s) \qquad E° = 0.339 \text{ V}$$

$$Sn^{2+} + 2e^- \rightleftharpoons Sn(s) \qquad E° = -0.141 \text{ V}$$

The cathode potential at which Cu^{2+} ought to be reduced is

$$E(\text{cathode}) = 0.339 - \frac{0.059\,16}{2}\log\frac{1}{0.1} = 0.31 \text{ V}$$

If 99.99% of the Cu^{2+} were deposited, the concentration of Cu^{2+} remaining in solution would be 10^{-5} M, and the cathode potential required to continue reduction would be

$$E(\text{cathode}) = 0.339 - \frac{0.059\,16}{2}\log\frac{1}{10^{-5}} = 0.19 \text{ V}$$

At a cathode potential of 0.19 V, then, rather complete deposition of copper is expected. Would Sn^{2+} be reduced at this potential? To deposit $Sn(s)$ from a solution containing 0.1 M Sn^{2+}, a cathode potential of -0.17 V is required:

$$E(\text{cathode, for reduction of } Sn^{2+}) = -0.141 - \frac{0.059\,16}{2}\log\frac{1}{[Sn^{2+}]}$$

$$= -0.141 - \frac{0.059\,16}{2}\log\frac{1}{0.1}$$

$$= -0.17 \text{ V} \tag{17-16}$$

We do not expect reduction of Sn^{2+} at a cathode potential more positive than -0.17 V.

If the *cathode* potential is kept near 0.19 V, more than 99.99% of the Cu^{2+} reacts without deposition of Sn^{2+}. On the other hand, if a two-electrode cell were run at constant voltage between the working and auxiliary electrodes, the cathode potential would behave as shown in Figure 17-9 and Sn^{2+} would be reduced. The price for achieving selective reduction with the three-electrode cell is that the reaction becomes slower as it nears completion.

Reduction reactions occur at working electrode potentials that are more *negative* than that required to start the reaction. *Oxidations* occur when the working electrode potential is more *positive* than that needed to start the reaction (Figure 17-11).

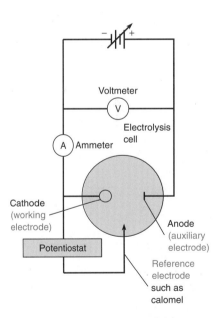

Figure 17-10 Circuit used for controlled-potential electrolysis with a three-electrode cell.

⎯O working electrode
⎯⏐ auxiliary electrode
⎯→ reference electrode

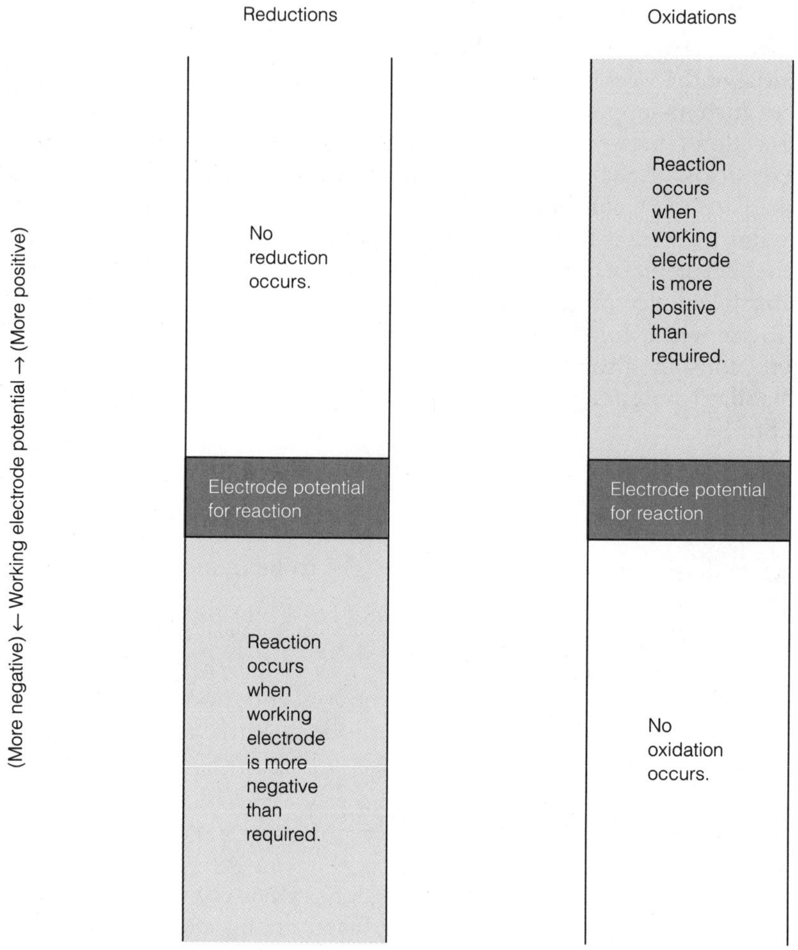

Figure 17-11 *Reduction* occurs at working electrode potentials more *negative* than that required to start the reaction. *Oxidation* occurs at working electrode potentials more *positive* than that required to start the reaction.

17-4 Coulometric Analysis

Coulometric methods are based on measurements of the number of electrons that participate in a chemical reaction.

Coulometry is based on counting the number of electrons used in a chemical reaction. For example, cyclohexene may be titrated with Br_2 generated by electrolytic oxidation of Br^-:

$$2Br^- \longrightarrow Br_2 + 2e^- \tag{17-17}$$

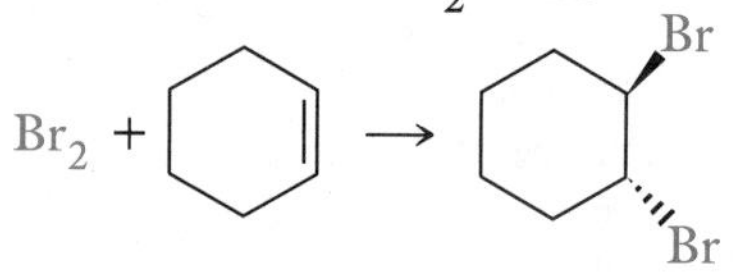

$$Br_2 + \text{Cyclohexene} \longrightarrow \textit{trans}\text{-1,2-Dibromocyclohexane} \tag{17-18}$$

Just enough Br_2 is generated to react with all of the cyclohexene. The moles of electrons liberated in Reaction 17-17 are equal to twice the moles of Br_2 and therefore twice the moles of cyclohexene.

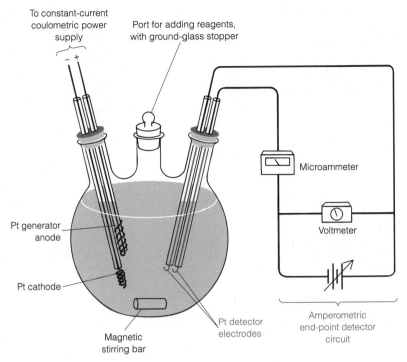

Figure 17-12 Apparatus for coulometric titration of cyclohexene with Br_2. The solution contains cyclohexene, 0.15 M KBr, and 3 mM mercuric acetate in a mixed solvent of acetic acid, methanol, and water. Mercuric acetate catalyzes the addition of Br_2 to the olefin. [Adapted from D. H. Evans, *J. Chem. Ed.* **1968,** *45,* 88.]

The reaction is conveniently carried out with the apparatus in Figure 17-12. Br_2 generated at the Pt anode at the left immediately reacts with cyclohexene. When all of the cyclohexene has been consumed, the concentration of Br_2 in the solution suddenly rises, signaling the end of the reaction.

The rise in Br_2 concentration is detected *amperometrically* by using the circuit shown on the right in Figure 17-12. A small voltage (~0.25 V) is applied between the two electrodes on the right. This voltage is not great enough to electrolyze any of the solutes, so only a small residual detector current of <1 μA flows through the sensitive ammeter (a device that measures electric current). When [Br_2] suddenly increases, detector current flows by virtue of the reactions:

> Amperometry is an analytical technique based on the measurement of electric current.

$$\text{detector anode:} \quad 2Br^- \rightarrow Br_2 + 2e^-$$

$$\text{detector cathode:} \quad Br_2 + 2e^- \rightarrow 2Br^-$$

The sudden increase of detector current is taken as the end point of the cyclohexene titration.

In practice, enough Br_2 is first generated in the absence of cyclohexene to give a detector current of 20.0 μA. When cyclohexene is added, the detector current decreases to a very small value because bromine is consumed. Bromine is then generated by the coulometric circuit, and the end point is taken when the detector again reaches 20.0 μA. Because the reaction is begun with Br_2 present, impurities that can react with Br_2 before analyte is added are eliminated.

The electrolysis current (not to be confused with the detector current) for the bromine-generating electrodes can be controlled by a hand-operated switch. As the detector current approaches 20.0 μA, you close the switch for shorter and shorter intervals. This is analogous to adding titrant dropwise from a buret near the end of a titration. The switch in the coulometer circuit serves as a "stopcock" for addition of Br_2 to the reaction.

EXAMPLE   Coulometry Calculation

A 2.000-mL volume of solution containing 0.611 3 mg of cyclohexene/mL is to be titrated in the apparatus shown in Figure 17-12. If the coulometer is operated at a constant current of 4.825 mA, how much time will be required for complete titration?

Solution   The quantity of cyclohexene is

$$\frac{(2.000 \text{ mL})(0.611 3 \text{ mg/mL})}{(82.146 \text{ mg/mmol})} = 0.014 88 \text{ mmol}$$

In Reactions 17-17 and 17-18, each mole of cyclohexene requires 1 mol of Br_2, which requires 2 mol of electrons. For 0.014 88 mmol of cyclohexene to react, 0.029 76 mmol of electrons must flow. From Equation 17-2, we can write

$$\text{moles of e}^- = \frac{I \cdot t}{F} \Rightarrow t = \frac{(\text{moles of e}^-)F}{I}$$

$$t = \frac{(0.029 76 \times 10^{-3} \text{ mol})(964 85 \text{ C/mol})}{(4.825 \times 10^{-3} \text{ C/s})} = 595.1 \text{ s}$$

It will require just under 10 min to complete the reaction if the current is constant.

Advantages of coulometry:

1. Accuracy

2. Sensitivity

3. Generation of unstable reagents in situ (in place)

This example illustrates the accuracy and sensitivity afforded by coulometric titrations. A commercial coulometer delivers current with an accuracy of ~0.1% (Figure 17-13). With extreme care, the value of the Faraday constant has been measured to within several parts per million by coulometry.[6] Unstable reagents such as Ag^{2+}, Cu^+, Mn^{3+}, and Ti^{3+} can be generated and used in a single vessel.

Types of Coulometry

Coulometry employs either a *constant current* or a *controlled potential*. Constant-current methods, like that in the preceding Br_2/cyclohexene example, are called **coulometric titrations.** If we know the current, to count the coulombs it is only necessary to measure the time needed for complete reaction:

$$q = I \cdot t \tag{17-1}$$

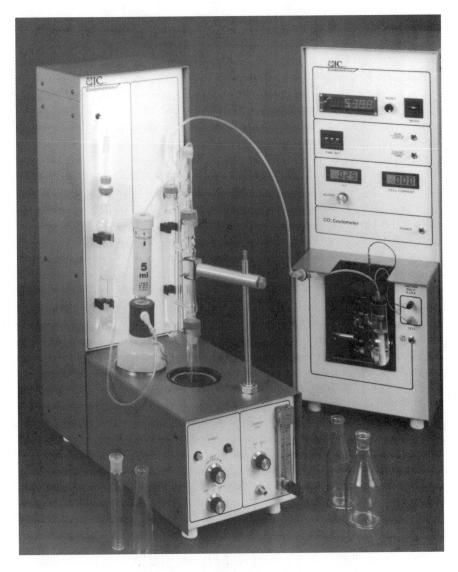

Figure 17-13 Commercial coulometer (right) for CO_2 detection. Unit at left is used to acidify inorganic carbonates and pass the released CO_2 into the coulometer. The coulometer can also be attached to a combustion train for analysis of carbon in organic compounds. A modification of the coulometer is used to measure sulfites or hydrogen sulfide [J. Greyson and S. Zeller, *Am. Lab.* July 1987, p. 44]. Sulfite measurement is important because sulfite preservatives in foods cause allergic reactions in 5–10% of people with asthma. [Photo courtesy UIC, Inc., Joliet, IL.]

Controlled-potential coulometry (with working, reference, and auxiliary electrodes) is more selective than constant-current coulometry. In controlled-potential coulometry, the initial current is high but decreases exponentially as analyte concentration decreases. Because current is not constant, charge is measured by integrating the current over the time of the reaction:

$$Q = \int_0^t I\,dt \qquad (17\text{-}19)$$

Question Why is the three-electrode cell more selective than the two-electrode cell?

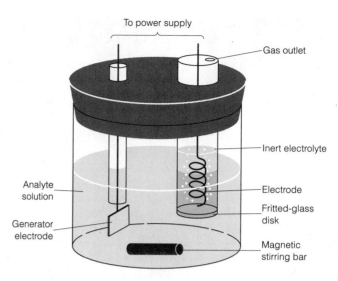

Figure 17-14   Cell showing how one electrode can be isolated from the analyte. Electrical contact is made through the porous fritted-glass disk.

In controlled-potential coulometry, the equivalence point is never reached because the current decays exponentially. However, you can approach the equivalence point by letting the current decay to an arbitrarily set value. For example, the current will ideally be 0.1% of its initial value when 99.9% of the analyte has been consumed. (This refers to current *above* the residual current.)

Separation of Anode and Cathode Reactions

In Figure 17-12 the reactive species (Br_2) is generated at the anode (the working anode). The cathode products (H_2 from solvent and Hg from the catalyst) do not interfere with the reaction of Br_2 and cyclohexene. Therefore, the cathode can be in the same compartment with the analyte. In some cases, however, the H_2 or Hg react with the analyte. Then it is desirable to separate the electrolysis product of the auxiliary electrode from the bulk solution. This can be done with the cell in Figure 17-14. An oxidizing agent, for instance, might be generated at the left-hand electrode operating as an anode. Gaseous H_2 produced at the cathode bubbles innocuously out of the cathode chamber without mixing with the bulk solution.

17-5 Mediators

A necessary condition for coulometric analysis is that the analytical reaction proceed with 100% electrochemical efficiency. If electrons were siphoned off into side reactions, the measurement of total coulombs would be meaningless. Although some reactions meet this requirement, most incorporate a **mediator** to improve the efficiency.

A *mediator* transports electrons quantitatively between analyte and the working electrode. The mediator undergoes no net reaction itself.

An example of the use of mediators is found in the technique used to monitor glucose levels in the blood of diabetic patients. An electrode used to detect glucose contains the *enzyme* glucose oxidase covalently attached to a

Box 17-2 Electrical Wiring of Redox Enzymes[8]

The enzyme glucose oxidase used in glucose-measuring electrodes has a net negative charge at physiologic pH. To transport electrons between the enzyme and a glassy carbon electrode, a polymer was designed that would have both covalent and electrostatic contact with the enzyme. Osmium complexes attached to the polymer cycle between the +2 and +3 oxidation states while transporting electrons between the enzyme and the electrode.

Schematic diagram showing redox polymer "wire" with covalent and electrostatic attachment to the negatively charged enzyme. Electrons can hop between the redox center in the enzyme and the electrode by moving from one osmium atom to another. In the actual polymer, about one side chain in five is an osmium redox center.

graphite surface.[7] This enzyme catalyzes oxidation of glucose inside living cells:

Glucose   Gluconolactone   Hydrogen
                                     peroxide

$$\text{(17-20)}$$

Hydrogen peroxide produced by the reaction is measured with an amperometric electrode system. The moles of hydrogen peroxide produced by the enzyme reaction equal the moles of glucose present in the blood sample.

The O_2 in Reaction 17-20 oxidizes the enzyme, which in turn oxidizes the glucose. Unfortunately, the reaction requires more O_2 than is generally present in blood. Therefore, the procedure makes use of the mediator 1,1'-dimethylferrocene to oxidize the enzyme.

Ferrocene contains flat five-membered aromatic carbon rings, similar to benzene. Each ring formally carries one negative charge, so the oxidation state of iron is $+2$. The iron atom sits in the middle of the face of each ring. Because of its shape, this type of molecule is called a *sandwich complex*.

1,1′-Dimethylferrocene → 1,1′-Dimethylferricinium cation

The enzyme cannot be oxidized at the electrode because the redox site is buried too deeply within the enzyme. Ferrocene is oxidized to ferricinium ion at the electrode. The ferricinium ion diffuses to and oxidizes the enzyme, which can then oxidize glucose. Ferricinium ion is reduced back to ferrocene, which can be oxidized again to repeat the process. Box 17-2 describes another way to provide electrical contact between an enzyme and an electrode.

Terms to Understand

ampere	coulometry	ohmic potential
amperometry	depolarizer	overpotential
auxiliary electrode	electroactive species	potentiostat
concentration polarization	electrogravimetric analysis	reference electrode
controlled-potential electrolysis	electrolysis	residual current
coulomb	mediator	working electrode
coulometric titration		

Summary

In electrolysis, a chemical reaction is forced to occur by the flow of electricity through a cell. The moles of electrons flowing through the cell are It/F, where I is current, t is time, and F is the Faraday constant. The voltage that must be applied to an electrolysis cell is greater than the equilibrium voltage predicted by the Nernst equation because of three factors:

1. Ohmic potential ($= IR$) is that voltage needed to overcome internal resistance of the cell.

2. Concentration polarization describes the condition in which the concentration of electroactive species near an electrode is not the same as the concentration in bulk solution. It always opposes the desired reaction and requires a voltage of greater magnitude to be applied.

3. Overpotential is the voltage required to overcome the activation energy of an electrode reaction. A greater overpotential is required to drive a reaction at a faster rate.

In electrogravimetric analysis, the analyte is electrolytically deposited on an electrode, whose increase in mass is then measured. With a constant voltage between the working and auxiliary electrodes of a two-electrode cell, electrolysis is not very selective. Greater selectivity results from controlled-potential electrolysis in a three-electrode cell, in which a constant voltage is imposed between working and reference electrodes.

In coulometry the quantity of electrons needed to carry out a chemical reaction is used to measure the quantity of analyte. Coulometric titrations are performed with a constant current. The time needed for complete reaction is a direct measure of the number of electrons consumed. The end point can be found by any conventional means, but amperometry is especially appropriate. Reactive mediators are often required to transfer electrons between the electrode and the analyte. Controlled-potential coulometry is inherently more selective than constant-current coulometry, but slower. Measuring the electrons consumed in the reaction is accomplished by electronic integration of the current-versus-time curve.

Exercises

A. Suppose that the following cell, producing a constant voltage of 1.02 V, is used to operate a light bulb with a resistance of 2.8 Ω:

$$Zn(s) \mid Zn^{2+}(aq) \parallel Cu^{2+}(aq) \mid Cu(s)$$

How many hours are required for 5.0 g of Zn to be consumed?

B. A dilute Na_2SO_4 solution is to be electrolyzed with a pair of smooth Pt electrodes at a current density of 100 A/m^2 and a current of 0.100 A. The products of the electrolysis are $H_2(g)$ and $O_2(g)$. Calculate the required voltage if the cell resistance is 2.00 Ω and there is no concentration polarization. Assume that H_2 and O_2 are both produced at 1.00 atm. What would your answer be if the Pt electrodes were replaced by Au electrodes?

C. **(a)** What percentage of 0.10 M Cu^{2+} could be reduced electrolytically before 0.010 M SbO^+ in the same solution begins to be reduced at pH 0.00? Consider the reaction

$$SbO^+ + 2H^+ + 3e^- \rightleftharpoons Sb(s) + H_2O$$
$$E° = 0.208 \text{ V}$$

(b) At what cathode potential (versus saturated Ag|AgCl) will Sb(s) deposition commence?

D. Calculate the cathode potential (versus S.C.E.) needed to reduce cobalt(II) to 1.0 μM in each solution below. In each case, Co(s) is the product of the reaction.

(a) A solution containing 0.10 M $HClO_4$.

(b) A solution containing 0.10 M $C_2O_4^{2-}$.

$$Co(C_2O_4)_2^{2-} + 2e^- \rightleftharpoons Co(s) + 2C_2O_4^{2-}$$
$$E° = -0.474 \text{ V}$$

This question is asking you to find the potential at which $[Co(C_2O_4)_2^{2-}]$ will be 1.0 μM.

(c) A solution containing 0.10 M EDTA at pH 7.00.

E. Ions that react with Ag^+ can be determined electrogravimetrically by deposition on a silver anode:

$$Ag(s) + X^- \rightarrow AgX(s) + e^-$$

(a) What will be the final mass of a silver anode used to electrolyze 75.00 mL of 0.023 80 M KSCN if the initial mass of the anode is 12.463 8 g?

(b) At what potential (versus S.C.E.) will AgBr(s) be deposited from 0.10 M Br^- on a silver anode?

(c) Is it theoretically possible to separate 99.99% of 0.10 M KI from 0.10 M KBr by controlled anode-potential electrolysis?

F. The galvanic cell pictured in the next column can be used to measure the concentration of O_2 in gases.

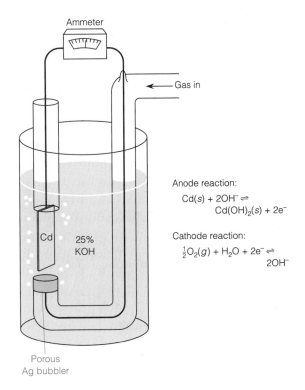

Anode reaction:
$$Cd(s) + 2OH^- \rightleftharpoons Cd(OH)_2(s) + 2e^-$$

Cathode reaction:
$$\tfrac{1}{2}O_2(g) + H_2O + 2e^- \rightleftharpoons 2OH^-$$

Oxygen is quantitatively reduced as it passes through the porous Ag bubbler, and Cd is oxidized to Cd^{2+} to complete the cell. Suppose that gas at 293 K and 1.00 atm is bubbled through the cell at a constant rate of 30.0 mL/min. How much current would be measured if the gas contains 1.00 ppt or 1.00 ppm (vol/vol) O_2?

G. The ligand R-(−)-1,2-propylenediaminetetraacetic acid (PDTA) is an optically active relative of EDTA.[9]

$$N(CH_2CO_2H)_2$$
$$\mid$$
$$C{\cdots}H$$
$$H_3C \qquad CH_2N(CH_2CO_2H)_2$$

R-(−)-1,2-Propylene-
diaminetetraacetic acid
(PDTA)

This chelate has a negative optical rotation at 365 nm and gives metal complexes with positive optical rotations. The reaction of PDTA with a metal can be studied by observing changes in the optical rotation of the solution. For the coulometric titration shown on the next page, the initial solution contains PDTA and excess Hg^{2+}. Between points A and B, excess Hg^{2+} is reduced to Hg(l), which has no effect on the optical rotation. Between points B and C, $Hg(PDTA)^{2-}$ is reduced to Hg(l) plus $PDTA^{4-}$, thereby decreasing the optical rotation. Excess $Hg(PDTA)^{2-}$ is still in solution at point C. When analyte solution containing Zn^{2+} is added, the Zn^{2+} displaces Hg^{2+} from PDTA, changing the optical rotation from C to D. Then coulometric reduction of the liberated Hg^{2+}

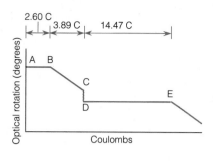

is continued beyond point D. At point E, the free Hg^{2+} is used up and $Hg(PDTA)^{2-}$ begins to be reduced, lowering the optical rotation once again. In a typical experiment, 2.000 mL of Zn^{2+} was added to the cell at point C. From the coulombs measured in each region, calculate the molarity of Zn^{2+} in the unknown.

Problems

Electrolysis and Why Voltage Changes
When Current Flows

1.   What is the difference between a galvanic cell and an electrolysis cell?

2.   Why is the ohmic potential subtracted in both Equations 17-5 and 17-6, instead of being added in one and subtracted in the other?

3.   Explain what concentration polarization is.

4.   Why does overpotential increase with current density?

5.   How many hours are required for 0.100 mol of electrons to flow through a circuit if the current is 1.00 A?

6.   The free energy change for the formation of $H_2(g)$ $+ \frac{1}{2}O_2(g)$ from $H_2O(l)$ is $\Delta G° = +237.19$ kJ. Calculate the standard voltage needed to decompose water into its elements by electrolysis.

7.   Consider the electrolysis reactions

cathode: $H_2O(l) + e^- \rightleftharpoons$

$$\frac{1}{2}H_2(g, 1.0 \text{ atm}) + OH^-(aq, 0.10 \text{ M})$$

anode: $Br^-(aq, 0.10 \text{ M}) \rightleftharpoons \frac{1}{2}Br_2(l) + e^-$

(a)   Calculate the equilibrium voltage needed to drive the net reaction.

(b)   Suppose the cell has a resistance of 2.0 Ω and a current of 100 mA is flowing. How much voltage is needed to overcome the cell resistance? This is the ohmic potential.

(c)   Suppose that the anode reaction has an overpotential (activation energy) of 0.20 V and that the cathode overpotential is 0.40 V. What voltage is necessary to overcome these effects combined with those of (a) and (b)?

(d)   Suppose that concentration polarization occurs. The concentration of OH^- at the cathode surface increases to 1.0 M and the concentration of Br^- at the anode surface decreases to 0.010 M. What voltage is necessary to overcome these effects combined with those of (b) and (c)?

8.   Consider the following cell, whose resistance is 3.50 Ω:

$Pt(s)|Fe^{2+}(0.10 \text{ M}), Fe^{3+}(0.10 \text{ M}), HClO_4 (1 \text{ M})||$
$Ce^{3+}(0.050 \text{ M}), Ce^{4+}(0.10 \text{ M}), HClO_4 (1 \text{ M})|Pt(s)$

Suppose that there is no concentration polarization or overpotential.

(a)   Calculate the voltage of the galvanic cell if it produces 30.0 mA.

(b)   Calculate the voltage that must be applied to run the reaction in reverse, as an electrolysis, at 30.0 mA.

9.   Suppose that the galvanic cell in the previous problem delivers 100 mA under the following conditions: $[Fe^{2+}]_s = 0.050$ M, $[Fe^{3+}]_s = 0.160$ M, $[Ce^{3+}]_s = 0.180$ M, $[Ce^{4+}]_s = 0.070$ M. Considering the ohmic potential and concentration polarization, calculate the cell voltage.

10.   The Weston cell, shown at the top of the next page, is a very stable voltage standard formerly used in potentiometers. (The potentiometer compares an unknown voltage to that of the standard. In contrast to the conditions of this problem, very little current may be drawn from the cell if it is to be an accurate voltage standard.)

(a)   How much work (J) can be done by the Weston cell if the voltage is 1.02 V and 1.00 mL of Hg (density = 13.53 g/mL) is deposited?

(b)   If the cell passes current through a 100-Ω resistor that dissipates heat at a rate of 0.209 J/min, how many grams of cadmium are oxidized each hour? (This part of the problem is not meant to be consistent with (a). The voltage is no longer 1.02 volts.)

11.   The chlor-alkali process,[10] in which seawater is electrolyzed to produce Cl_2 and NaOH, is the second

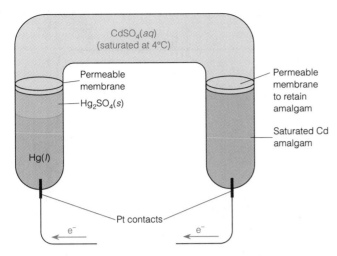

For Problem 10: A saturated Weston cell containing excess solid $CdSO_4$ is a more precise voltage standard than the unsaturated cell but is more sensitive to temperature and mechanical shock and cannot be easily incorporated into portable equipment.

most important commercial electrolysis, behind production of aluminum.

$$\text{anode:} \quad Cl^- \rightarrow \frac{1}{2}Cl_2 + e^-$$

$$\text{cathode:} \quad Na^+ + H_2O + e^- \rightarrow NaOH + \frac{1}{2}H_2$$

The semipermeable Nafion membrane used to separate the anode and cathode compartments is resistant to chemical attack. Its many anionic side chains permit conduction of Na^+, but not anions. The cathode compartment is loaded with pure water, and the anode compartment contains seawater from which Ca^{2+} and Mg^{2+} have been removed. Explain how the membrane allows NaOH free of NaCl to be formed.

$$\text{--}[(CF_2CF_2)_n\text{--}CFCF_2]_x\text{--}$$
$$|$$
$$O$$
$$|$$
$$CF_2 \qquad \text{Nafion}^{\textregistered}$$
$$|$$
$$CF\text{--}CF_3$$
$$|$$
$$O$$
$$|$$
$$CF_2CF_2SO_3^-Na^+$$

Electrogravimetric Analysis

12. Explain why *controlled-potential* electrolysis with a three-electrode cell is more selective (less prone to cause reactions of undesired species) than is a two-electrode cell with constant voltage between the two electrodes.

13. Which voltage, V_1 or V_2, in the diagram in the next column is constant in *controlled-potential* electrolysis?

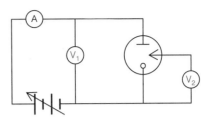

14. Would you use an anodic or a cathodic depolarizer to prevent the potential of the working electrode from becoming too negative during reduction of Cu^{2+} to $Cu(s)$?

15. Explain how Ag can be deposited in Figure 17-8 at a potential 420 mV more positive than E_{eq} for the reaction $Ag^+ + e^- \rightleftharpoons Ag(s)$.

16. A 0.326 8-g unknown containing lead lactate, $Pb(CH_3CHOHCO_2)_2$ (FW 385.3), plus inert material was electrolyzed to produce 0.111 1 g of PbO_2 (FW 239.2). Was the PbO_2 deposited at the anode or at the cathode? Find the weight percent of lead lactate in the unknown.

17. A solution of Sn^{2+} is to be electrolyzed to reduce the Sn^{2+} to $Sn(s)$. Calculate the cathode potential (versus S.H.E.) needed to reduce the Sn^{2+} concentration to 1.0×10^{-8} M if no concentration polarization occurs. What would be the potential versus S.C.E. instead of S.H.E? Would the potential be more positive or more negative if concentration polarization occurred?

18. A 1.00-L electrolysis cell initially containing 0.025 0 M Mn^{2+} and another metal ion, M^{3+}, is fitted with Mn and Pt electrodes. The reactions are

$$Mn(s) \rightarrow Mn^{2+} + 2e^-$$
$$M^{3+} + 3e^- \rightarrow M(s)$$

(a) Is the Mn electrode the anode or the cathode?

(b) A constant current of 2.60 A was passed through the cell for 18.0 min, causing 0.504 g of the metal M to plate out on the Pt electrode. What is the atomic weight of M?

(c) What will be the concentration of Mn^{2+} in the cell at the end of the experiment?

19. Calculate the initial voltage that should be applied to electrolyze 0.010 M $Zn(OH)_4^{2-}$ in 0.10 M NaOH, using Ni electrodes. Assume that the current is 0.20 A, the anode current density is 100 A/m², the cell resistance is 0.35 Ω, and O_2 is evolved at 0.20 atm. The reactions are

cathode: $Zn(OH)_4^{2-} + 2e^- \rightleftharpoons$
$\qquad\qquad Zn(s) + 4OH^- \qquad E° = -1.199$ V

anode: $H_2O \rightleftharpoons \frac{1}{2}O_2 + 2H^+ + 2e^-$

20. What equilibrium cathode potential (versus S.H.E.) is required to reduce 99.99% of Cd(II) from a solution containing 0.10 M Cd(II) in 1.0 M ammonia? Consider the following reactions and assume that nearly all Cd(II) is in the form $Cd(NH_3)_4^{2+}$.

$$Cd^{2+} + 4NH_3 \rightleftharpoons Cd(NH_3)_4^{2+} \qquad \beta_4 = 3.6 \times 10^6$$
$$Cd^{2+} + 2e^- \rightleftharpoons Cd(s) \qquad E° = -0.402 \text{ V}$$

21. Is it possible to remove 99% of a 1.0 μM CuY^{2-} impurity (by reduction to solid Cu) from a 10-mM CoY^{2-} solution at pH 4.0 without reducing any cobalt? Here, Y is EDTA and the total concentration of free EDTA is 10 mM.

Coulometric Analysis

22. Explain how the amperometric end-point detector in Figure 17-12 operates.

23. What does a mediator do?

24. The sensitivity of a coulometer is governed by the delivery of its minimum current for its minimum time. Suppose that 5 mA can be delivered for 0.1 s.

(a) To how many moles of electrons does this correspond?

(b) How many milliliters of a 0.01 M solution of a two-electron reducing agent are required to deliver the same number of electrons?

25. The experiment in Figure 17-12 required 5.32 mA for 964 s for complete reaction of a 5.00-mL aliquot of unknown cyclohexene solution.

(a) How many moles of electrons passed through the cell?

(b) How many moles of cyclohexene reacted?

(c) What was the molarity of cyclohexene in the unknown?

26. The electrolysis cell at the top of the next column was run at a constant current of 0.021 96 A. On one side, 49.22 mL of H_2 was produced (at 303 K and 0.983 atm); on the other side, Cu metal was oxidized to Cu^{2+}.

(a) How many moles of H_2 were produced?

(b) If 47.36 mL of EDTA were required to titrate the Cu^{2+} produced by the electrolysis, what was the molarity of the EDTA?

(c) For how many hours was the electrolysis run?

27. A mixture of trichloroacetate and dichloroacetate can be analyzed by selective reduction in a solution containing 2 M KCl, 2.5 M NH_3, and 1 M NH_4Cl. At a mercury cathode potential of -0.90 V (versus S.C.E.), only trichloroacetate is reduced:

$$Cl_3CCO_2^- + H_2O + 2e^- \rightarrow$$
$$Cl_2CHCO_2^- + OH^- + Cl^-$$

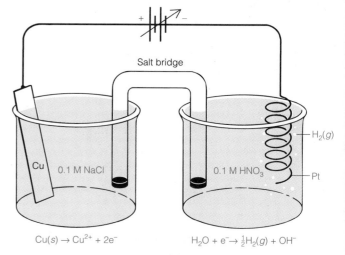

$$Cu(s) \rightarrow Cu^{2+} + 2e^- \qquad\qquad H_2O + e^- \rightarrow \tfrac{1}{2}H_2(g) + OH^-$$

For Problem 26.

At a potential of -1.65 V, dichloroacetate reacts:

$$Cl_2CHCO_2^- + H_2O + 2e^- \rightarrow$$
$$ClCH_2CO_2^- + OH^- + Cl^-$$

A hygroscopic mixture of trichloroacetic acid (MW 163.386) and dichloroacetic acid (MW 128.943) containing an unknown quantity of water weighed 0.721 g. Upon controlled-potential electrolysis, 224 C passed at -0.90 V, and 758 C was required to complete the electrolysis at -1.65 V. Calculate the weight percent of each acid in the mixture.

28. H_2S in aqueous solution can be analyzed by titration with coulometrically generated I_2:

$$H_2S + I_2 \rightarrow S(s) + 2H^+ + 2I^-$$

To 50.00 mL of sample was added 4 g of KI. Electrolysis required 812 s at a constant current of 52.6 mA. Calculate the concentration of H_2S (μg/mL) in the sample.

29. Ti^{3+} is to be generated in 0.10 M $HClO_4$ solution for coulometric reduction of azobenzene.

$$TiO^{2+} + 2H^+ + e^- \rightleftharpoons Ti^{3+} + H_2O$$
$$E° = 0.100 \text{ V}$$

$$4Ti^{3+} + C_6H_5N{=}NC_6H_5 + 4H_2O \rightarrow$$
$$\text{Azobenzene}$$
$$2C_6H_5NH_2 + 4TiO^{2+} + 4H^+$$
$$\text{Aniline}$$

At the counterelectrode, water is oxidized, and O_2 is liberated at a pressure of 0.20 atm. Both electrodes are made of smooth Pt, and each has a total surface area of 1.00 cm². The rate of reduction of the azobenzene is 25.9 nmol/s, and the resistance of the solution between the generator electrodes is 52.4 Ω.

(a) Calculate the current density (A/m^2) at the electrode surface. Use Table 17-1 to estimate the overpotential for O_2 liberation.

(b) Calculate the equilibrium cathode potential (versus S.H.E.), assuming that $[TiO^{2+}]_{surface} = [TiO^{2+}]_{bulk} = 0.050$ M and $[Ti^{3+}]_{surface} = 0.10$ M.

(c) Calculate the equilibrium anode potential (versus S.H.E.).

(d) What should the total applied voltage be?

30. In an extremely accurate measurement of the Faraday constant, a pure silver anode was oxidized to Ag^+ with a constant current of 0.203 639 0 ($\pm 0.000\,000\,4$) A for 18 000.075 (± 0.010) s to give a mass loss of 4.097 900 ($\pm 0.000\,003$) g from the anode. Given that the atomic weight of Ag is 107.868 2 ($\pm 0.000\,2$), find the value of the Faraday constant and its uncertainty.

Notes and References

1. G. McLendon and R. Hake, *Chem. Rev.* **1992**, *92*, 481; G. McLendon, *Acc. Chem. Res.* **1988**, *21*, 160.

2. E. C. Gilbert in H. N. Alyea and F. B. Dutton, Eds., *Tested Demonstrations in Chemistry* (Easton, PA: Journal of Chemical Education, 1965), p. 145.

3. E. S. Smotkin, S. Cervera-March, A. J. Bard, A. Campion, M. A. Fox, T. Mallouk, S. E. Webber, and J. M. White, *J. Phys. Chem.* **1987**, *91*, 6; *Chem. Eng. News* 26 June 1989, p. 37.

4. Articles dealing with photochemistry at semiconductor surfaces include B. Parkinson, *J. Chem. Ed.* **1983**, *60*, 338; M. S. Wrighton, *J. Chem. Ed.* **1983**, *60*, 335; H. O. Finklea, *J. Chem. Ed.* **1983**, *60*, 325; J. A. Turner, *J. Chem. Ed.* **1983**, *60*, 327; M. T. Spitler, *J. Chem. Ed.* **1983**, *60*, 330; and A. B. Ellis, *J. Chem. Ed.* **1983**, *60*, 332.

5. J. O'M. Bockris, *J. Chem. Ed.* **1971**, *48*, 352.

6. D. N. Craig, J. I. Hoffman, C. A. Law, and W. J. Hamer, *J. Res. Nat. Bur. Stand.* **1960**, *64A*, 381; H. Diehl, *Anal. Chem.* **1979**, *51*, 318A.

7. A. E. G. Cass, G. Davis, G. D. Francis, H. A. O. Hill, W. J. Aston, I. J. Higgins, E. V. Plotkin, L. D. L. Scott, and A. P. F. Turner, *Anal. Chem.* **1984**, *56*, 667. For a general review of electrochemical biosensors, see J. E. Frew and H. A. O. Hill, *Anal. Chem.* **1987**, *59*, 933A.

8. A. Heller, *Acc. Chem. Res.* **1990**, *23*, 128; A. Heller, *J. Phys. Chem.* **1992**, *96*, 3579; R. Maidan and A. Heller, *Anal. Chem.* **1992**, *64*, 2889.

9. R. A. Gibbs and R. J. Palma, Sr., *Anal. Chem.* **1976**, *48*, 1983.

10. S. Venkatesh and B. V. Tilak, *J. Chem. Ed.* **1983**, *60*, 276: S. C. Stinson, *Chem. Eng. News* 15 March 1982, p. 22; E. J. Taylor, R. Waterhouse, and A. Gelb, *Chemtech* **1987**, *17*, 316. Nafion is a trademark of Du Pont Co.

Oxygen Sensors for Automobile Pollution Control

The catalytic converter of an automobile reduces the quantity of carbon monoxide, hydrocarbons, and nitrogen oxides emitted by the car. When exhaust from the engine enters the converter, CO and unburned hydrocarbons are *oxidized* to CO_2 and nitrogen oxides (NO + NO_2, called NO_x) are *reduced* to N_2 by Pt and Rh catalysts. The balance between oxidation and reduction requires an optimized air-to-fuel ratio. Too much air prohibits the reduction of NO_x and too little air prevents complete combustion of CO and hydrocarbons.

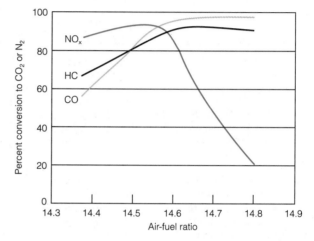

Efficiency of catalytic conversion of automobile exhaust to CO_2 and N_2 depends on precise balance of air and fuel. [From R. J. Farrauto, R. M. Heck, and B. K. Speronello, *Chemical Engineering News* 7 September 1992, p. 34.]

An electrochemical oxygen sensor developed by the National Aeronautics and Space Administration allows the correct air-to-fuel ratio to be maintained. Box 18-3 explains how this sensor works. In 1991 the worldwide market for automobile oxygen sensors was more than 150 million dollars.

Voltammetry

Voltammetry refers to techniques in which the relationship between voltage and current is observed during electrochemical processes. In this chapter we explore some powerful methods that are widely used in research and routine chemical analysis.

18-1 Why We Use the Dropping-Mercury Electrode in Polarography

In **polarography** the current flowing through an electrolysis cell is measured as a function of the potential of the working electrode. Usually, this current is proportional to the concentration of analyte. The working electrode in Figure 18-1 is a mercury droplet suspended from the bottom of a glass capillary tube. The current-carrying auxiliary electrode is a platinum wire, and the reference is a saturated calomel electrode through which negligible current flows.

The **dropping-mercury electrode** in Figure 18-1 consists of a capillary through which mercury drips from an adjustable reservoir. The height of mercury, measured from the outlet of the capillary tube, is typically 30 cm. With a capillary diameter of 0.05 mm, drops of mercury with a diameter of 0.5 mm are formed once every 3–6 s. The drop rate is controlled by raising or lowering the leveling bulb at the top right of Figure 18-1.

The dropping-mercury electrode is used because it yields reproducible current-potential data. Its reproducibility is attributed to the continuous exposure of fresh surface on the growing mercury drop. With any other electrode (such as Pt in various forms), the potential depends on its surface condition and therefore on its previous treatment.

In *polarography,* we measure the current as a function of the potential of the working electrode.

A *dropping-mercury electrode* is used because fresh Hg is continuously exposed to the analyte. This gives more reproducible behavior than does a static surface, whose characteristics change with use.

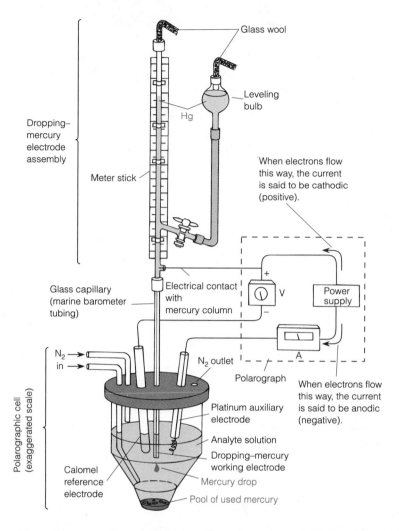

Figure 18-1 Apparatus for polarography. The current is considered to be positive when analyte is reduced at the Hg electrode. Polarography was invented by J. Heyrovský in 1922, for which he received the Nobel Prize in 1959.[1]

The vast majority of reactions studied with the mercury electrode are reductions. At a Pt surface, reduction of solvent competes with reduction of many analytes, especially in acidic solutions:

$$2H^+ + 2e^- \rightarrow H_2(g) \qquad E° = 0$$

The large overpotential for H$^+$ reduction at the Hg surface means that H$^+$ reduction does not occur and interfere with many reductions.

It is easier to reduce most metals to their amalgam than to the solid metal.

Question What do these potentials imply about the relative stabilities of K(*s*) and K(*in Hg*)?

Table 17-1 showed that there is a large *overpotential* for reduction of H$^+$ at the Hg surface. Therefore, reactions thermodynamically more difficult than reduction of H$^+$ can be carried out without competitive reduction of H$^+$. In neutral or basic solutions, even alkali metal cations can be reduced more easily than H$^+$, despite their lower standard potentials. This is partly because the potential for reduction of a metal into a mercury amalgam is more positive than its potential for reduction to the solid state:

$$K^+ + e^- \rightarrow K(s) \qquad E° = -2.936 \text{ V}$$
$$K^+ + e^- + Hg \rightarrow K(in\ Hg) \qquad E° = -1.975 \text{ V}$$

TABLE 18-1   Potential ranges (versus S.C.E.) of various solvents

Solvent	Supporting electrolyte[a]	Cathodic limit (V)[b]	Anodic limit (V)
Acetic acid	CH_3CO_2Na	-1.0	$+2.0$
Acetonitrile	$LiClO_4$	-3.0	$+2.5$
Ammonia	TBAI	$-2.8(Hg)$	$+0.3$
Dichloromethane	TBAP	-1.7	$+1.8$
Dimethylformamide	TBAP	-2.8	$+1.9$
Dimethylsulfoxide	$LiClO_4$	-3.4	$+1.3$
Hydrogen fluoride	NaF	-1.0	$+3.5$
Methanol	KOH	-1.0	$+0.6$
Methanol	KCN	-1.0	$+1.7$
Propylene carbonate	TEAP	-1.9	$+1.7$
Pyridine	TEAP	-2.2	$+3.3$
Sulfolane	TEAP	-2.9	$+3.0$
Tetrahydrofuran	$LiClO_4$	-3.2	$+1.6$
Trifluoroacetic acid	CF_3CO_2Na	-0.6	$+2.4$
Water	TBAP	$-2.7(Hg)$	$+1.3$

a. TBAI, tetrabutylammonium iodide; TBAP, tetrabutylammonium perchlorate; TEAP, tetraethylammonium perchlorate.
b. Electrodes are Pt unless Hg is indicated.
SOURCE: N. L. Weinberg, *J. Chem. Ed.* **1983,** *60,* 268.

Mercury is easily oxidized, so it is not very useful for oxidations. In a noncomplexing medium, Hg is oxidized near $+0.25$ V (versus S.C.E.). If the concentration of Cl^- is 1 M, Hg is oxidized near 0 V because the Hg(II) product is stabilized by Cl^-:

$$Hg(l) + 4Cl^- \rightleftharpoons HgCl_4^{2-} + 2e^-$$

With inert electrodes and appropriate solvents, a wide range of redox potentials is accessible (Table 18-1).

A capillary will function for years if it is not allowed to clog. Mercury flow should always be started prior to immersing the capillary in any solution. At the conclusion of an experiment, the electrode should be rinsed with distilled water with mercury still flowing. The clean electrode can be stored either dry or in a pool of liquid mercury.

Be nice to your electrode!

To clean up a mercury spill, consolidate the droplets as much as possible with a piece of cardboard. Then suck the mercury into a filter flask with an aspirator. A disposable Pasteur pipet attached to a hose makes a good vacuum cleaner. To remove residual mercury, sprinkle elemental zinc powder on the surface and dampen the powder with 5% aqueous H_2SO_4 to make a paste. The mercury dissolves in the zinc. After working the paste into contaminated areas with a sponge or brush, allow the paste to dry and then sweep it up. Discard the powder as contaminated mercury waste.[2]

18-2   The Polarogram

The graph of current versus potential in polarography is called a **polarogram.** Figure 18-2 shows the result from reduction of Cd^{2+} in HCl solution.

$$Cd^{2+} + 2e^- \rightleftharpoons Cd(in\ Hg)$$

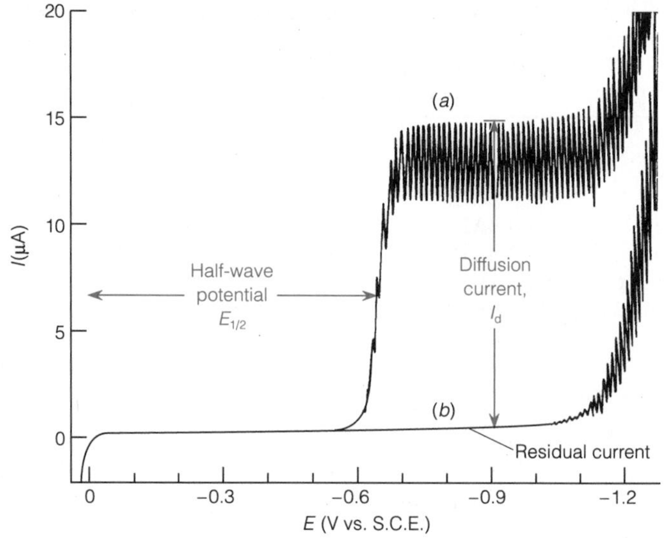

Figure 18-2 Polarograms for (a) 5×10^{-4} M Cd^{2+} in 1 M HCl and (b) 1 M HCl alone. Note that voltage becomes more negative to the *right* and the scale of current is μA. [From D. T. Sawyer and J. L. Roberts, Jr., *Experimental Electrochemistry for Chemists* (New York: Wiley, 1974).]

The proportionality between diffusion current and analyte concentration makes polarography useful for quantitative analysis.

At -0.3 V, only a small *residual current* flows. At -0.6 V, reduction of Cd^{2+} commences and the current increases. The reduced Cd dissolves in the Hg to form an amalgam. After a steep increase in current, *concentration polarization* sets in: Current becomes limited by the rate at which Cd^{2+} can diffuse from bulk solution to the surface of the electrode. The magnitude of this *diffusion current* (I_d) is proportional to Cd^{2+} concentration and is used for quantitative analysis. The upper trace in Figure 18-2 is called a **polarographic wave.**

Speaking of "reduction of H^+" is equivalent to speaking of "reduction of H_2O."

Around -1.2 V, reduction of H^+ begins and the curve rises steeply again. Near 0 V, oxidation of the Hg electrode produces a negative current. By convention, *a negative current means that the working electrode is the anode. A positive current means that the working electrode is the cathode.*

Oscillations in Figure 18-2 are due to the growth and fall of the Hg drops. As each drop begins to form, there is little Hg surface, and correspondingly little current. As the drop grows, its area increases, more solute can reach the surface in a given time, and more current flows. Finally, the drop falls off and the current decreases sharply.

Diffusion Current

Diffusion current is the limiting current when the rate of electrolysis is controlled by the rate of diffusion of species to the electrode.

Between -0.7 and -1.1 V in Figure 18-2, current is governed by the rate at which Cd^{2+} can reach the electrode. In unstirred solution, the rate of reduction is controlled by the rate of diffusion of analyte to the electrode. The limiting current is called the **diffusion current.**

The rate of diffusion of a solute from bulk solution to the surface of the electrode is proportional to the concentration difference between the two regions:

The symbol $\propto$ is read "is proportional to."

$$\text{current} \propto \text{rate of diffusion} \propto [C]_0 - [C]_s \qquad (18\text{-}1)$$

where $[C]_0$ is the concentration in bulk solution and $[C]_s$ is the concentration at the surface of the electrode (see Figure 17-2). At a sufficiently negative potential, the reduction is so fast that $[C]_s \ll [C]_0$ and Equation 18-1 reduces to

$$\text{limiting current} = \text{diffusion current} \propto [C]_0 \qquad (18\text{-}2)$$

The proportionality of the diffusion current to the bulk-solute concentration is the basis for quantitative applications of polarography.

The magnitude of the diffusion current, *measured at the top of each oscillation* in Figure 18-2, is given by the Ilkovič equation:

Diffusion current: $\qquad I_d = (7.08 \times 10^4)nCD^{1/2}m^{2/3}t^{1/6} \qquad (18\text{-}3)$

where I_d is the diffusion current, measured at the top of the oscillations in Figure 18-2, with the units μA; n is the number of electrons per molecule in the redox reaction; C is the concentration of electroactive species, with the units mmol/L; D is the diffusion coefficient[3] of electroactive species, with the units m^2/s; m is the rate of flow of Hg, in mg/s; and t is the drop interval, in s.

In quantitative polarography, it is important to control temperature within a few tenths of a degree, because diffusion coefficients increase by about 2% per degree. The quantity $m^{2/3}t^{1/6}$ in Equation 18-3 is called the *capillary constant* and is nearly proportional to the square root of the Hg column height (h), measured from the top of the Hg meniscus to the bottom of the Hg electrode in Figure 18-1. To demonstrate that the limiting current is indeed *diffusion*-controlled, you can measure the current at various heights of the Hg column and see whether the current is indeed proportional to $\sqrt{h}$.

> If the limiting current is not proportional to $\sqrt{h}$, effects other than diffusion are controlling the rate of reaction.

The transport of solute to the electrode depends on diffusion, mechanical transport (stirring and convection), and electrostatic attraction. In polarography, we try to minimize the latter two mechanisms so that limiting current is controlled solely by diffusion. To minimize mechanical transport, the solution is *not* stirred, and vibrations are reduced by setting the apparatus on a heavy base (such as the marble slab used for sensitive analytical balances). Electrostatic attraction (or repulsion) of analyte ions by the electrode is reduced by a high concentration of **supporting electrolyte,** such as 1 M HCl in Figure 18-2. Table 18-2 shows that supporting electrolyte

> Three ways that solute gets to the electrode: (**1**) diffusion (desired in polarography); (**2**) mechanical transport (not desired); (**3**) electrostatic attraction (not desired).

> Electrostatic attraction of analyte ions is minimized by using a high concentration of supporting electrolyte.

TABLE 18-2 Influence of supporting electrolyte on the limiting current for reduction of Pb^{2+}

Electrolyte concentration (M)[a]	Limiting current (μA)	
	KCl	KNO$_3$
0	17.6	17.6
0.000 1	16.3	16.2
0.000 2	15.9	15.0
0.000 5	13.3	13.4
0.001	11.8	12.0
0.005	9.8	9.8
0.1	8.35	8.45
1.0	8.00	8.45

a. 50 mL of 9.5×10^{-4} M $PbCl_2$ was analyzed at 25°C with 0.2 mL of 0.1 wt % sodium methyl red present as a maximum suppressor.

SOURCE: I. M. Kolthoff and J. J. Lingane, *Polarography,* Vol. I (New York: Wiley, 1952), p. 123.

concentration 50–100 times greater than the analyte concentration reduces electrostatic transport of analyte to a negligible level. (That is, 0.1 M supporting electrolyte reduces the limiting current to a value that is hardly reduced further by 1 M supporting electrolyte.)

Challenge   Explain why supporting electrolyte *increases* the magnitude of the limiting current for reduction of an *anion,* such as IO_3^-.

Residual Current

I_d = limiting current − residual current

To measure diffusion current (I_d in Figure 18-2) in a polarogram, subtract the *residual current* from the limiting current in the diffusion-controlled region. Residual current is measured with a blank solution containing the same supporting electrolyte as your sample. Reagents should come from the same stock solutions, because different batches have different impurities contributing to residual current.

An enlarged plot of residual current for 0.1 M HCl is shown in Figure 18-3. The dashed lines show that the residual current has an inflection point near −0.5 V in this case. At this point, called the *electrocapillary maximum,* the charge on the mercury drop is zero. At more positive potentials, the drop is positive with respect to the solution. At more negative potentials, the charge is negative. Because the mercury drop and the surrounding solution

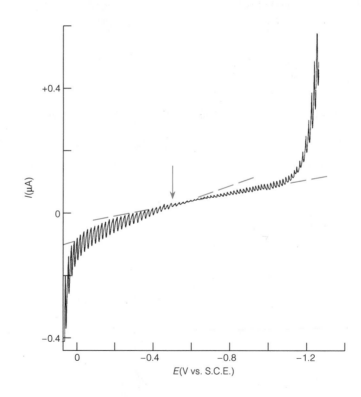

Figure 18-3   Residual current from 0.1 M HCl. The arrow marks the inflection point at the electrocapillary maximum. This occurs when the charge of the Hg drop is zero with respect to the solution. If there were no faradaic current, the net current would be zero at the electrocapillary maximum. [From L. Meites, *Polarographic Techniques,* 2nd ed. (New York: Wiley, 1965).]

can have different charges, the mercury-solution interface behaves as a capacitor. Box 18-1 describes the structure of the electrode-solution interface.

Residual current has two components: The **condenser current** (or *charging current*) is the current needed to charge or discharge the capacitor formed by the mercury-solution interface. As each drop of mercury falls, it carries its charge with it to the bottom of the cell. The new drop requires more current for charging. **Faradaic current** is the current that flows as a result of reduction or oxidation. The small faradaic component of residual current is due mainly to reactions of impurities in the supporting electrolyte.

Components of residual current:

1. condenser current (due to charging of Hg drops)
2. faradaic current (due to redox reactions)

The Shape of the Polarographic Wave

The two most common reactions at the dropping Hg electrode are reduction of an ion to an amalgam or reduction of a soluble ion to another soluble ion:

$$M^{n+} + ne^- + Hg \rightleftharpoons M(in\ Hg) \tag{18-4}$$

$$X^{a+} + ne^- \rightleftharpoons X^{(a-n)+} \tag{18-5}$$

If the reactions are reversible, the equation relating the current and potential in the polarographic wave is

Shape of polarographic wave:
$$E = E_{1/2} - \frac{0.059\,16}{n}\log\frac{I}{I_d - I} \tag{18-6}$$

$E_{1/2}$ is the **half-wave potential,** identified in Figure 18-2. It is the potential at which $I = \frac{1}{2}I_d$ (both corrected for residual current).

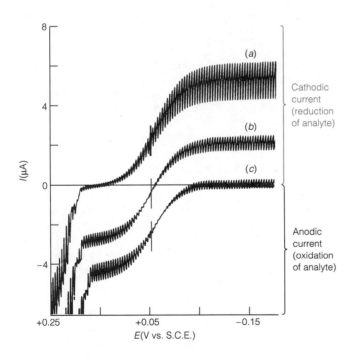

Figure 18-4 Polarograms of (*a*) 1.4 mM Fe^{3+}, (*b*) 0.7 mM Fe^{3+} plus 0.7 mM Fe^{2+}, and (*c*) 1.4 mM Fe^{2+}. Each solution contains saturated oxalic acid as supporting electrolyte and 0.000 2 wt% methyl red as maximum suppressor. Vertical lines mark the half-wave potentials. [From L. Meites, *Polarographic Techniques,* 2nd ed. (New York: Wiley, 1965).]

Box 18-1 The Electric Double Layer

When a power supply pumps electrons into or out of an electrode, the charged surface of the electrode attracts ions of opposite charge. The charged electrode and the oppositely charged ions next to it constitute the **electric double layer.** For a given solution composition, there is one *potential of zero charge* at which there is no excess charge on the electrode. This potential is −0.58 V (versus a calomel electrode containing 1 M KCl) for a mercury electrode immersed in 0.1 M KBr. It shifts to −0.72 V for the same electrode in 0.1 M KI.

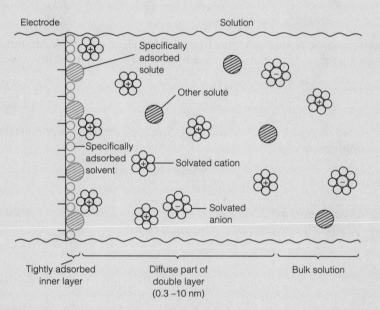

Electrode-solution interface. The tightly adsorbed inner layer (also called the *compact, Helmholtz,* or *Stern* layer) may include solvent and solute molecules. Cations in the inner layer do not completely balance the charge of the electrode. Therefore, excess cations are required in the *diffuse part of the double layer* for charge balance.

The first layer of molecules at the surface of the electrode is *specifically adsorbed* by van der Waals and electrostatic forces. The adsorbed solute could be neutral molecules, anions, or cations. Anions tend to be more *polarizable* than cations and therefore subject to stronger van der Waals forces. Iodide is more strongly adsorbed than bromide, because iodide is more polarizable. Therefore the potential of zero charge for KI is more negative than for KBr: A more negative potential is required to expel iodide from the electrode surface.

Cathodic current represents flow of electrons from the Hg electrode to the analyte (reduction). *Anodic current* represents flow of electrons from the analyte to the electrode (oxidation).

Reversible anodic and composite anodic-cathodic waves have the same shape, which is defined by Equation 18-6. Figure 18-4 shows polarograms of (a) ferric, (b) ferric plus ferrous, and (c) ferrous ions. In curve a, Fe^{3+} is reduced to Fe^{2+} at a half-wave potential of +0.05 V (versus S.C.E.). Only a residual current is observed at +0.15 V, because Fe^{3+} is not reduced at this potential. Curve c shows an anodic current at +0.15 V because Fe^{2+} is being oxidized. At −0.05 V, only residual current is observed in curve c because no Fe^{3+} is present to be reduced. Curve b shows an anodic current at +0.15 V

The next layer of molecules beyond the specifically adsorbed layer is rich in cations attracted by the negative electrode. The excess population of cations decreases with increasing distance from the electrode. This region, whose composition is different from that of bulk solution, is called the *diffuse part of the double layer* and is typically 0.3–10 nm thick. In the diffuse part of the double layer, there is a balance between attraction toward the electrode and randomization by thermal motion.

When a species is created or destroyed by an electrochemical reaction, its concentration near the electrode is different from its concentration in bulk solution (Color Plate 13). The region containing excess product or decreased reactant is called the *diffusion layer* (which should not be confused with the diffuse part of the double layer).

The following figure shows the concentration of tri(*p*-methoxyphenyl)amine cation in the diffusion layer at various times after stepping the potential of a platinum electrode from 0 to 0.8 V (versus S.C.E.). At 0 V, no cation is present. When the potential is suddenly changed to 0.8 V, the amine is oxidized to the cation at the electrode surface. As time goes on, more cation is generated, and it diffuses away from the electrode. The thickness of the diffusion layer is proportional to $\sqrt{Dt}$, where D is the diffusion coefficient of the cation and t is time.

$$\left(CH_3O-\bigcirc\!\!\!\!\!\bigcirc\!\!\!\!-\right)_3 N^{\bullet\,+} \qquad \text{Tri}(p\text{-methoxyphenyl)amine cation}$$

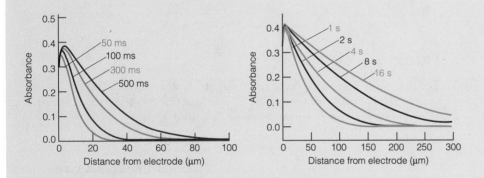

Diffusion profile of cation created by sudden oxidation of tri(*p*-methoxyphenyl)-amine at a platinum electrode. The ordinate scale (absorbance at 633 nm) is proportional to the concentration of the cation. Times are measured from the start of the oxidation. [From C.-C. Jan, R. L. McCreery, and F. T. Gamble, *Anal. Chem.* **1985,** *57,* 1763; *Anal. Chem.* **1986,** *58,* 2771.] Microelectrodes can also be used to probe the nature of the diffusion layer. See R. C. Engstrom, T. Meaney, R. Tople, and R. M. Wightman, *Anal. Chem.* **1987,** *59,* 2005.

due to oxidation of Fe^{2+} and a cathodic current at -0.05 V due to reduction of Fe^{3+}. All three curves have the same value of $E_{1/2}$ and the same shape.

The graph of Equation 18-6 in Figure 18-5 shows how the shape of a reversible polarographic wave depends on n, the number of electrons in the half-reaction. The larger the value of n, the steeper the polarographic wave. Equation 18-6 predicts that a graph of E versus $\log(I/I_d - I)$ should be linear, with a slope of $-0.059\,16/n$. Such a graph allows us to verify that the reaction is reversible and to find n.

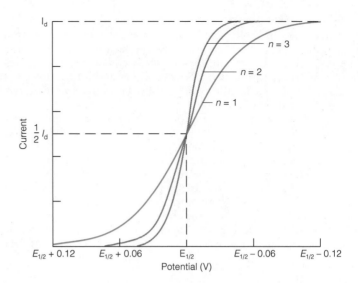

Figure 18-5   Graph of Equation 18-6 for $n = 1$, 2, and 3. The greater the value of n, the steeper is the polarographic wave.

Relation Between $E_{1/2}$ and $E°$

For reversible reactions, there is a straightforward thermodynamic interpretation of the half-wave potential, $E_{1/2}$. For Reaction 18-4, in which the product is an amalgam, the relation between $E_{1/2}$ and $E°$ is

$$E_{1/2} \approx E° + E_s + \frac{0.059\,16}{n}\log C\gamma \qquad (18\text{-}7)$$

The reaction $M^{n+} + ne^- \rightleftharpoons M(in\ Hg)$ is more favorable if

1. electrons flow spontaneously from M(s) to M(in Hg)
2. M is very soluble in liquid mercury

where $E°$ is the standard reduction potential for M^{n+} with respect to the reference electrode used in the polarographic cell. E_s is the standard potential for the cell $M(s)\,|\,M^{n+}\,|\,M(in\ Hg)_{saturated}$, in which the left electrode is solid metal and the right electrode is a saturated amalgam of the same metal. C is the concentration of M in the saturated amalgam, and γ is the activity coefficient of M in the saturated amalgam; therefore, $C\gamma$ is the activity of metal M in the saturated amalgam.

The term E_s in Equation 18-7 measures the tendency of electrons to flow between solid analyte and analyte amalgam. A positive value means that electrons flow spontaneously from pure reduced analyte to reduced analyte dissolved in mercury. The more positive the value of E_s, the easier it will be to reduce the species in the presence of mercury. The log term in Equation 18-7 gives the free energy change for dissolving reduced analyte in mercury. The more soluble the product is in mercury, the easier it will be to carry out the reduction. Table 18-3 gives data that demonstrate these effects.

Now consider Reaction 18-5, in which the oxidized and reduced species remain in solution. In this case,

For the reaction $X^{a+} + ne^- \rightleftharpoons X^{(a-n)+}$, $E_{1/2}$ is usually close to $E°$.

$$E_{1/2} = E° - \frac{0.059\,16}{n}\log\frac{\gamma_{red}D_{ox}^{1/2}}{\gamma_{ox}D_{red}^{1/2}} \qquad (18\text{-}8)$$

where $E°$ is the standard reduction potential for Reaction 18-5, with respect

TABLE 18-3   Test of Equation 18-7 for various metals at 25°C

Ion	$E°$ (V)^a	E_s (V)^a	C (M)	γ	$E_{1/2}$[Eq. 18-7]	$E_{1/2}$(observed)
Pb^{2+}	−0.372	0.006	0.96	0.72	−0.371	−0.388
Tl^+	−0.582	0.003	27.4	8.3	−0.440	−0.459
Cd^{2+}	−0.647	0.051	6.40	1.15	−0.570	−0.578
Zn^{2+}	−1.008	0	4.37	0.74	−0.993	−0.997
Na^+	−2.961	0.780	3.52	1.3	−2.14	−2.12
K^+	−3.170	1.001	1.69	5.6	−2.11	−2.14

a. With respect to S.C.E.

SOURCE: I. M. Kolthoff and J. L. Lingane, *Polarography,* Vol. I (New York: Wiley, 1952), p. 198.

to S.C.E. The activity coefficients of the reduced and oxidized species are γ_{red} and γ_{ox}, respectively. The corresponding diffusion coefficients are D_{red} and D_{ox}. The quotient in the log term is usually close to unity, so $E_{1/2}$ for soluble reactants and products is usually close to $E°$ for the redox couple.

Current Maxima

Figure 18-6 shows the polarograms of a mixture of Pb^{2+} and Zn^{2+} in 2 M NaOH. The wave near −0.8 V is due to reduction of Pb^{2+}, and the smaller wave near −1.5 V is due to reduction of Zn^{2+}. In the upper trace, the current "overshoots" the value of I_d and then settles back down. Current maxima are attributed to convection currents near the surface of the electrode. *Maximum suppressors* (such as gelatin, Triton X-100, or methyl red) eliminate this behavior, as shown in the lower trace, by altering the convective behavior of the solution. Too much suppressor distorts the polarogram.

Small quantities of surface-active agents are routinely used to suppress current maxima.

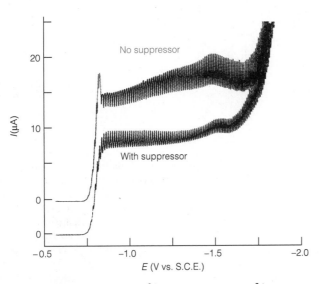

Figure 18-6 Polarograms of 3 mM Pb^{2+} and 0.25 mM Zn^{2+} in 2 M NaOH in the absence of a suppressor and in the presence of 0.002 wt% Triton X-100. [From L. Meites, *Polarographic Techniques,* 2nd ed. (New York: Wiley, 1965).]

Figure 18-7   Polarograms of 0.1 M KCl: (*a*) saturated with air; (*b*) after partial deaeration; (*c*) after further deaeration. [From L. Meites, *Polarographic Techniques*, 2nd ed. (New York: Wiley, 1965).]

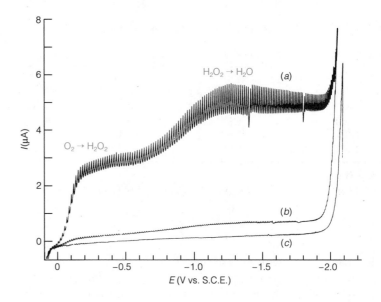

Why We Remove Oxygen

Figure 18-7 shows that O_2 gives rise to a pair of intense polarographic waves.

$$O_2 + 2H^+ + 2e^- \rightleftharpoons H_2O_2 \qquad E_{1/2} \approx -0.1 \text{ V (versus S.C.E.)}$$

$$H_2O_2 + 2H^+ + 2e^- \rightleftharpoons 2H_2O \qquad E_{1/2} \approx -0.9 \text{ V (versus S.C.E.)}$$

Therefore, N_2 gas is bubbled through solutions for a few minutes to remove O_2. The cell in Figure 18-1 has ports for purging the solution with N_2 and for maintaining N_2 over the solution during the experiment. Figure 18-8 shows how to remove traces of O_2 from N_2.

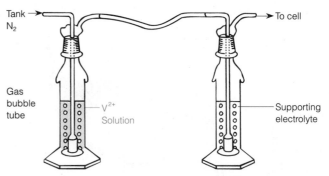

Figure 18-8   Purifying N_2 for polarography. The first chamber removes O_2 by reaction with V^{2+}, and the second chamber saturates the gas stream with water at the same vapor pressure as that in the polarography cell. The V^{2+} solution is prepared by boiling 2 g of NH_4VO_3 (ammonium metavanadate) with 25 mL of 12 M HCl and reducing to V^{2+} with zinc amalgam. Amalgam is prepared by covering granulated Zn with 2 wt% $HgCl_2$ solution and stirring for 10 min to reduce Hg^{2+} to Hg, which coats the Zn. The liquid is decanted, and the lustrous amalgam is washed three times with water by decantation. Amalgamation increases the overpotential for H^+ reduction at the Zn surface, so the Zn is not wasted by reaction with acid. The blue or green, oxidized vanadium solution turns violet upon reduction. When the violet color is exhausted through use, it can be regenerated by adding more zinc amalgam.

18-3 Polarography in Chemical Analysis

The principal use of polarography is in quantitative analysis. Because diffusion current is proportional to analyte concentration, the height of a polarographic wave tells how much analyte is present. A *calibration curve and curves for standard addition* and an *internal standard* are illustrated in Figure 18-9; their uses were described in Sections 6-5 and 6-6.

Polarography is also used for qualitative identification of an unknown. Tables of half-wave potentials[4] allow us to compare $E_{1/2}$ for an unknown with known values. Organic functional groups also give rise to polarographic waves that can be used for identification (Table 18-4). Synthetic organic reactions can be conducted electrochemically, instead of with oxidizing and reducing agents.[5]

To see how polarography is used in measuring equilibrium constants, consider a solution containing 1.00 mM Fe^{3+} and 1.00 mM Fe^{2+}. The potential for reduction of Fe^{3+} is given by

$$E = 0.771 - 0.059\,16\log\frac{[Fe^{2+}]}{[Fe^{3+}]} = 0.771\ \text{V}$$

Suppose that a ligand that binds only to Fe^{3+} is added to the solution. The potential needed for reduction of Fe^{3+} will change, because the quotient $[Fe^{2+}]/[Fe^{3+}]$ is no longer 1.00. Because the ligand decreases the concentration of Fe^{3+}, it will be harder to reduce Fe^{3+} after the ligand has been added.

The change in reduction potential can be predicted if we know the formation constant of the metal-ligand complex. Alternatively, measuring

In the older literature, an internal standard in polarography was called a "pilot ion."

For qualitative analysis, the half-wave potential of an unknown is measured in several different complexing media. Comparison with a table of half-wave potentials allows the unknown to be identified.

Any equilibrium that affects either $[Fe^{3+}]$ or $[Fe^{2+}]$ will alter the reduction potential of the solution.

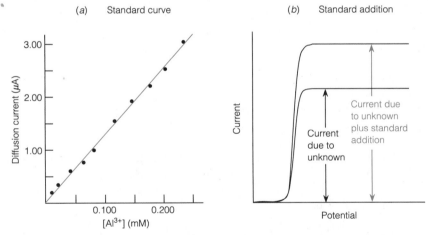

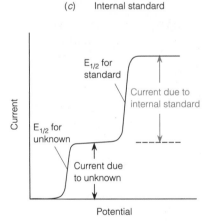

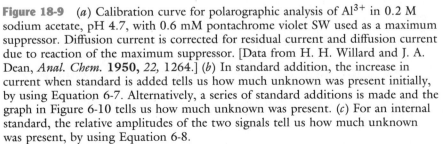

Figure 18-9 (*a*) Calibration curve for polarographic analysis of Al^{3+} in 0.2 M sodium acetate, pH 4.7, with 0.6 mM pontachrome violet SW used as a maximum suppressor. Diffusion current is corrected for residual current and diffusion current due to reaction of the maximum suppressor. [Data from H. H. Willard and J. A. Dean, *Anal. Chem.* **1950,** *22,* 1264.] (*b*) In standard addition, the increase in current when standard is added tells us how much unknown was present initially, by using Equation 6-7. Alternatively, a series of standard additions is made and the graph in Figure 6-10 tells us how much unknown was present. (*c*) For an internal standard, the relative amplitudes of the two signals tell us how much unknown was present, by using Equation 6-8.

TABLE 18-4   Polarographic behavior of various functional groups

Group	Reaction
C—C	RO_2C—⬡—$CN + H^+ + 2e^- \longrightarrow RO_2C$—⬡—$H + CN^-$
C=C	⬡—$CH{=}CH_2 + 2H^+ + 2e^- \longrightarrow$ ⬡—CH_2CH_3
C≡C	⬡—$C{\equiv}C{-}CHO + 2H^+ + 2e^- \longrightarrow$ ⬡—$CH{=}CH{-}CHO$
C—X	$RCH_2{-}Br + H^+ + 2e^- \longrightarrow RCH_3 + Br^-$
C=O	⬡—$\overset{\displaystyle O}{\overset{\|}{C}}R + 2H^+ + 2e^- \longrightarrow$ ⬡—$\overset{\displaystyle OH}{\underset{\displaystyle H}{\overset{\|}{\underset{\|}{C}}R}}$
C—N	$R\overset{\displaystyle O}{\overset{\|}{C}}CH_2{-}NR_2 + 2H^+ + 2e^- \longrightarrow R\overset{\displaystyle O}{\overset{\|}{C}}CH_3 + HNR_2$
C=N	⬡—$CH{=}CH{-}\overset{\displaystyle NH}{\underset{\displaystyle R}{C}} + 2H^+ + 2e^- \longrightarrow$ ⬡—$CH{=}CH{-}CHRNH_2$
C≡N	$R\overset{\displaystyle O}{\overset{\|}{C}}$—⬡—$C{\equiv}N + 4H^+ + 4e^- \longrightarrow R\overset{\displaystyle O}{\overset{\|}{C}}$—⬡—$CH_2NH_2$
N=N	⬡—$N{=}N$—⬡$ + 2H^+ + 2e^- \longrightarrow$ ⬡—$NH{-}NH$—⬡
N=O	$R{-}NO + 2H^+ + 2e^- \longrightarrow RNHOH$
NO₂	$RNO_2 + 4H^+ + 4e^- \longrightarrow RNHOH + H_2O$
O—O	$ROOR + 2H^+ + 2e^- \longrightarrow 2ROH$
S—S	$RSSR + 2H^+ + 2e^- \longrightarrow 2RSH$
S=O	$R_2S{=}O + 2H^+ + 2e^- \longrightarrow R_2S + H_2O$

the reduction potential tells us the formation constant. In polarography, the half-wave potential is sensitive to equilibria involving analyte.

For example, consider the reduction of a complex ion to yield an amalgam plus free ligand:

$$ML_p^{n-pb} + ne^- + Hg \rightleftharpoons M(in\ Hg) + pL^{-b} \qquad (18\text{-}9)$$

For such a reaction, it can be shown that $E_{1/2}$ is given by

Half-wave potential for complex ion:

A graph of $E_{1/2}$ versus $\log[L^{-b}]$ will give the values of p and the equilibrium constant.

$$E_{1/2} = E_{1/2}(\text{for free } M^{n+}) - \frac{0.059\,16}{n}\log\beta_p - \frac{0.059\,16p}{n}\log[L^{-b}] \quad (18\text{-}10)$$

where β_p is the equilibrium constant for the reaction $M^{n+} + pL^{-b} \rightleftharpoons ML_p^{n-pb}$. The value of $E_{1/2}(\text{for free } M^{n+})$ refers to $E_{1/2}$ in a noncomplexing medium. A graph of $E_{1/2}$ versus $\log[L^{-b}]$ should have a slope of $-0.059\,16p/n$ and a y-intercept of $E_{1/2} - (0.059\,16/n)\log\beta_p$.

Clever improvements have replaced the **direct current polarography** that we have been discussing. Our efforts were not wasted, however, because it was necessary to understand classical polarography before studying the improvements. In favorable cases, direct current polarography has a detection limit around 10^{-5} M and can resolve species with half-wave potentials differing by 0.2 V. **Differential pulse polarography** and **square wave polarography** provide detection limits near 10^{-7} M and resolutions of 0.05 V.

Differential Pulse Polarography

In direct current polarography, the voltage applied to the working electrode increases linearly with time, as shown in Figure 18-10a. The current is recorded continuously, and a polarogram such as that in Figure 18-2 results. Figure 18-10a is called a *linear voltage ramp*. In differential pulse polarography, small pulses are superimposed on the ramp, as shown in Figure 18-10b. The height of a pulse is called its *modulation amplitude*. Each pulse of magnitude 5–100 mV is applied during the last 60 ms of the life of each mercury drop. The drop is then mechanically dislodged. Current is measured once before the pulse and again for the last 17 ms of the pulse. The polarograph subtracts the first current from the second and plots this difference versus the applied potential (measured just before the voltage

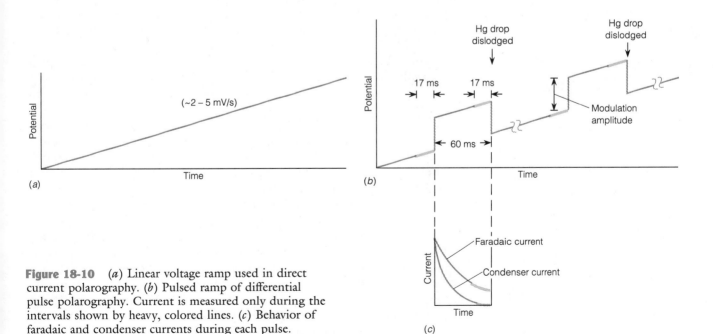

Figure 18-10 (*a*) Linear voltage ramp used in direct current polarography. (*b*) Pulsed ramp of differential pulse polarography. Current is measured only during the intervals shown by heavy, colored lines. (*c*) Behavior of faradaic and condenser currents during each pulse.

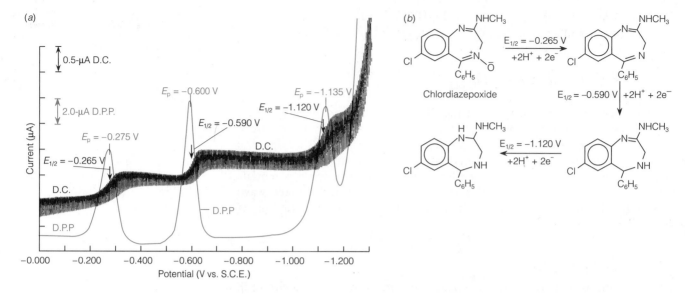

Figure 18-11 (*a*) Comparison of direct current (D.C.) and differential pulsed polarography (D.P.P.) of 1.2×10^{-4} M chlordiazepoxide (the drug Librium) in 3 mL of 0.05 M H_2SO_4. Modulation amplitude = 50 mV. Note the different current scales for the two curves. (*b*) The chemistry thought to be responsible for each wave. [From M. R. Hackman, M. A. Brooks, J. A. F. de Silva, and T. S. Ma, *Anal. Chem.* **1974,** *46,* 1075.]

Differential pulse polarography produces a curve that is nearly the derivative of an ordinary direct current polarogram.

pulse). The resulting differential pulse polarogram is nearly the *derivative* of a direct current polarogram, as shown in Figure 18-11a.

To see why a derivative is produced, consider the polarographic wave in Figure 18-12a. Periodically, the polarograph steps up the voltage and measures the resulting current increase. At potential V_1 in Figure 18-12, current increase ΔI_1 is rather small. At potential V_2, ΔI_2 is larger, and at the potential V_3, ΔI_3 is still larger. By plotting ΔI versus V, we produce a curve that is nearly the derivative of the direct current polarogram.

Increasing modulation amplitude increases the signal height but decreases resolution of neighboring peaks.

Figure 18-13 shows that as modulation amplitude is increased, signal increases, but resolution (separation of neighboring peaks) decreases. If the modulation amplitude is too great, the differential pulse polarogram no longer gives a faithful derivative shape.

The enhanced sensitivity of pulsed polarography relative to that of direct current polarography is due mainly to an increase in *faradaic current* and a decrease in *condenser current*. Consider the case in which the cathode potential is -0.200 V and the surface concentration of analyte is consistent with -0.200 V. Suddenly, a pulse is applied and the voltage changes to -0.250 V. If the voltage had been *slowly* changed from -0.200 to -0.250 V, the concentration of analyte near the electrode would have decreased. Instead, when the pulse is applied, the concentration of analyte is at the higher level consistent with -0.200 V. At the instant when each pulse is applied, there is a surge of electroactive species toward the working electrode, causing a surge in faradaic current. As analyte reacts, its concentration decreases, and faradaic current decays as shown in Figure 18-10c. At the instant the pulse is applied, the condenser current also increases to charge the drop. As the drop charges, the condenser current also decays, as shown in Figure 18-10c. *The condenser current decays faster than the faradaic current.* By

Faradaic current: redox reactions

Condenser current: charging of the Hg drop

Condenser current decays exponentially with time, whereas faradaic current decay is proportional to $1/\sqrt{time}$. The exponential decay is faster.

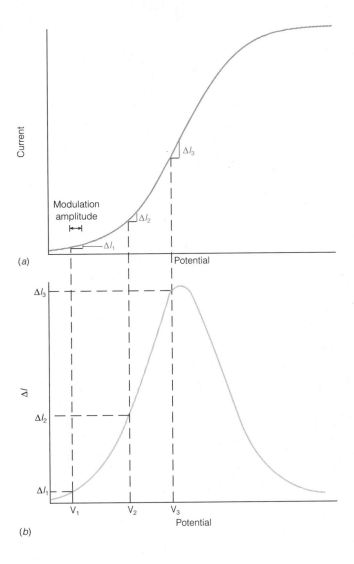

Figure 18-12 Illustration showing why the differential pulse polarogram (*b*) is very nearly the derivative of the direct current polarogram (*a*).

the time the total current is measured (~40 ms after applying the pulse), the condenser current has decayed to near zero, but the faradaic current is still substantial.

Relative to direct current polarography, pulsed measurement increases faradaic current and reduces condenser current, thereby enhancing sensitivity. Differential pulse polarography provides better resolution of adjacent waves because it is easier to distinguish partially overlapping derivative maxima than partially overlapping polarographic waves.

The dropping-mercury electrode of Figure 18-1 has been replaced by the electrode in Figure 18-14. This device uses an electrically controlled, mechanical mechanism to suspend a static drop of mercury at the base of the electrode. After recording current and potential, the drop is mechanically dislodged and a fresh drop is created. A new current is measured at the new potential of the new drop. Because the mercury drop does not change its size during the measurement, no oscillations are observed in the polarogram. By waiting a short time after the creation of each new drop, the condenser current decays to near zero (Figure 18-10c), and the signal-to-noise ratio is increased.

Differential pulse polarography provides greater sensitivity and better resolution than does direct current polarography.

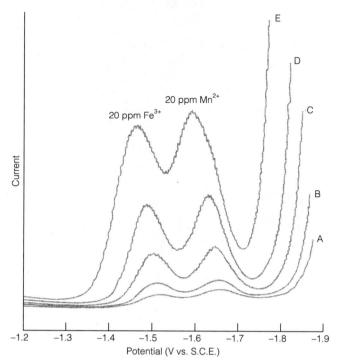

Figure 18-13 Effect of modulation amplitude on peak height and resolution in differential pulse polarography. Modulation amplitude: A = 5 mV; B = 10 mV; C = 25 mV; D = 50 mV; E = 100 mV. [Courtesy Princeton Applied Research Corp., Princeton, NJ, Application Note 151.]

Square Wave Polarography

The *square wave polarography* waveform in Figure 18-15 consists of a square wave (colored line) superimposed on a voltage staircase (dashed line). Current is measured in regions 1 and 2 in the figure and the difference ($I_1 - I_2$) is plotted versus staircase potential. The resulting polarogram is a

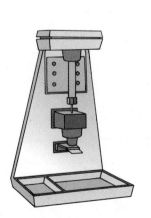

Figure 18-14 Mechanically controlled dropping-mercury electrode. [Courtesy EG&G Princeton Applied Research Corp., Princeton, NJ.]

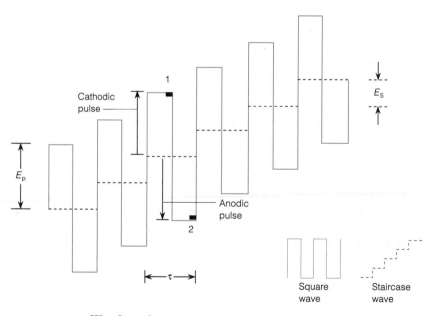

Figure 18-15 Waveform for square wave polarography. Typical parameters are pulse potential (E_p) = 25 mV, step height (E_s) = 10 mV, and pulse period (τ) = 5 ms. Regions of current measurement are labeled 1 and 2.

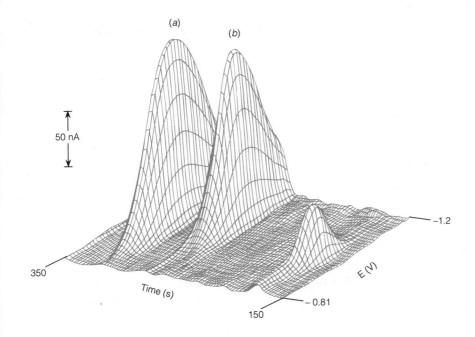

(a)

(b)

50 nA

350

Time (s)

150

−1.2

E (V)

− 0.81

Figure 18-16 Three-dimensional square wave polarograms used to monitor elution of (*a*) *N*-nitrosoproline and (*b*) *N*-nitrosodiethanolamine from a liquid chromatography column. Polarographic peaks occur when the electroactive compounds reach the detector at the end of the column. [From J. G. Osteryoung and R. A. Osteryoung, *Anal. Chem.* **1985,** *57,* 101A.]

N-Nitrosoproline

$(HOCH_2CH_2)_2N—N=O$

N-Nitrosodiethanolamine

symmetric peak similar to the differential pulse polarogram (Figure 18-12b).

Square wave polarography is more sensitive and faster than differential pulse polarography. The reverse (anodic) pulse in Figure 18-15 causes reoxidation of the product of each forward (cathodic) pulse. The polarographic signal is the difference between the two currents. It is larger than the signal in differential pulse polarography because the reverse current is not used in the differential pulse polarogram. The optimum pulse amplitude is about $50/n$ mV, where n is the number of electrons in the redox reaction. For reversible reactions, square wave polarography is about five times more sensitive than differential pulse polarography.

With a typical pulse period of 5 ms and step height of 10 mV, an entire square wave polarogram with a 1-V width is obtained with *one drop of Hg in 0.5 s.* The corresponding differential pulse experiment takes 100 times longer. Figure 18-16 shows how square wave polarography can be used for detection of compounds emerging from a liquid chromatograph. At each time interval, a 0.4-V scan is made and a complete polarogram is recorded. Compounds are distinguished by the times at which they emerge from the chromatography column and by their half-wave potentials.

18-5 Stripping Analysis

In **stripping analysis,** analyte from a dilute solution is concentrated into a single drop of Hg by electroreduction. The electroactive species is then *stripped* from the electrode by reversing the direction of the voltage sweep. The potential becomes more *positive, oxidizing* the species back into solution. Current measured during oxidation is related to the quantity of analyte that was initially deposited.[6] Figure 18-17 shows a differential pulse anodic stripping polarogram of trace elements in seawater at concentrations of 10^{-9} to 10^{-11} M.

Stripping is the most sensitive of all the polarographic techniques (Table 18-5), because analyte is concentrated from a dilute solution. The longer the

Stripping analysis:

1. concentrate analyte into a drop of Hg by reduction
2. reoxidize analyte by making the potential more positive
3. measure polarographic signal during oxidation

TABLE 18-5   **Detection limits for stripping analysis**

Analyte	Stripping mode	Detection limit
Ag^+	Anodic	2×10^{-12} M[a]
Testosterone	Anodic	2×10^{-10} M[b]
I^-	Cathodic	1×10^{-10} M[c]

a. S. Dong and Y. Wang, *Anal. Chim. Acta* **1988**, *212*, 341.
b. J. Wang, "Adsorptive Stripping Voltammetry," EG&G Princeton Applied Research Application Note A-7 (1985).
c. G. W. Luther III, C. Branson Swartz, and W. J. Ullman, *Anal. Chem.* **1988**, *60*, 1721. I^- is deposited onto the mercury drop by anodic oxidation:

$$Hg(l) + I^- \rightleftharpoons \frac{1}{2} Hg_2I_2(adsorbed\ on\ Hg) + e^-$$

period of concentration, the more sensitive is the analysis. Because only a fraction of analyte from the solution is deposited, deposition must be done for a reproducible time (such as 5 min) with reproducible stirring.

Potentiometric Stripping Analysis

Potentiometric stripping analysis:

1. concentrate analyte into Hg by reduction
2. turn off potentiostat and re-oxidize analyte with $Hg^{2+}(aq)$
3. record potential versus time during oxidation

Potentiometric stripping analysis is closely related to the stripping voltammetry just described, but the instrumentation for potentiometric stripping analysis is simpler. The procedure is so simple and inexpensive that it is being adapted for large-scale measurements of traces of lead in the blood of children, at one-tenth the cost of competing methods.[7]

In a typical procedure, a glassy carbon working electrode is coated with a film of liquid mercury by reduction of Hg^{2+} for a few minutes. The electrode is then washed with water and transferred to analyte containing ~1 mg of Hg^{2+} per liter. After purging the solution with N_2 and leaving it under a blanket of N_2, a controlled potential is applied for a fixed time to reduce some of the analyte into the mercury film. The potentiostat is then *disconnected* from the electrodes and replaced by a potentiometer. The voltage between the working and reference electrodes is then recorded as a function of time.

When the potentiostat is disconnected, analyte such as Pb^{2+} that was reduced to Pb and dissolved in the mercury film is oxidized by Hg^{2+} from the solution:

$$Pb(in\ Hg) + Hg^{2+} \rightarrow Pb^{2+} + Hg(l)$$

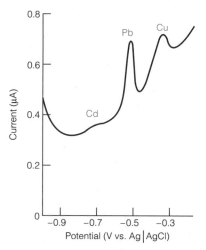

The amalgamated Pb is being titrated by Hg^{2+} from the solution and the potential of the working electrode remains close to $E_{1/2}$ for $Pb(in\ Hg)|Pb^{2+}$. When Pb is used up, the electrode potential suddenly becomes more positive. The time between disconnecting the potentiostat and the sudden potential change is proportional to the concentration of Pb^{2+} in the unknown.

Trace a of Figure 18-18 shows potentiometric stripping of Cd^{2+}, Pb^{2+}, Cu^{2+}, and Bi^{3+}. After 128 min of reduction at -0.95 V (versus S.C.E.), the potentiostat was disconnected and the voltage between the working and reference electrodes was measured as the reduced analytes reacted with Hg^{2+}. For 1.0 s, the potential remained near -0.72 V while the reaction

$$Cd(in\ Hg) + Hg^{2+} \rightarrow Cd^{2+} + Hg(l)$$

Figure 18-17   Anodic stripping voltammogram (differential pulse mode) of Sargasso seawater acidified to pH 2. Peaks for Cd and Cu correspond to 0.02 and 1.3 nmol/kg of seawater, respectively. Extreme precautions are required to avoid contamination when analyte concentrations are so low. [From S. R. Piotrowicz, M. Springer-Young, J. A. Puig, and M. J. Spencer, *Anal. Chem.* **1982**, *54*, 1367.]

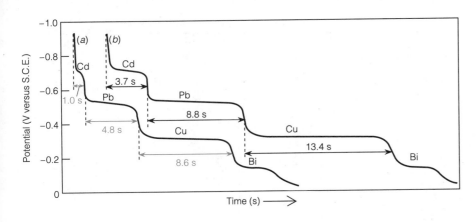

Figure 18-18 Potentiometric stripping curves from a sample containing Cd^{2+}, Pb^{2+}, Cu^{2+}, and Bi^{3+} (*a*) before and (*b*) after addition of 0.5 μg/L of Cd^{2+}, Pb^{2+}, and Cu^{2+}. There was no dilution caused by the standard addition. The sample contained 0.5 M NaCl, 0.05 M HCl, and 1 mg of Hg^{2+}/L. [From L. D. Jagner, *Anal. Chem.* **1978,** *50,* 1924.]

occurred. Then the voltage suddenly changed to −0.52 V for 4.8 s while Pb was oxidized. This was followed by 8.6 s at −0.31 V as Cu was oxidized to Cu^{2+} and ~4 s at −0.14 V as Bi was oxidized to Bi^{3+}.

EXAMPLE   **Standard Addition in Potentiometric Stripping Analysis**

In trace b of Figure 18-18, standard additions of 0.5 μg/L of Cd^{2+}, Pb^{2+}, and Cu^{2+} were made to the solution from trace a (with negligible dilution). Find the concentration of Pb^{2+} in the original unknown.

Solution   To find the unknown concentration, we set up a proportionality:

$$\frac{[Pb^{2+}]_{unknown}}{[Pb^{2+}]_{unknown} + [Pb^{2+}]_{standard}} = \frac{(\text{stripping time})_{unknown}}{(\text{stripping time})_{unknown + standard}}$$

$$\frac{[Pb^{2+}]_{unknown}}{[Pb^{2+}]_{unknown} + 0.5 \text{ μg/L}} = \frac{4.8 \text{ s}}{8.8 \text{ s}}$$

$$\Rightarrow [Pb^{2+}]_{unknown} = 0.6 \text{ μg/L} = 2.9 \times 10^{-9} \text{ M}$$

18-6 Cyclic Voltammetry

In **cyclic voltammetry,** the triangular waveform shown in Figure 18-19 is applied to the working electrode. After applying a linear voltage ramp between times t_0 and t_1 (typically a few seconds), the ramp is reversed to bring the potential back to its initial value at time t_2. The cycle may be repeated many more times.

Cyclic voltammetry uses a periodic, triangular waveform.

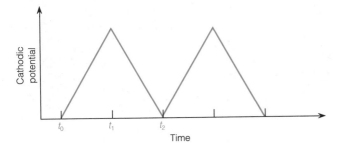

Figure 18-19   Waveform used in cyclic voltammetry. Corresponding times are indicated in Figure 18-20.

Figure 18-20 Cyclic voltammograms of (*a*) 1 mM O_2 in acetonitrile with 0.10 M $(C_2H_5)_4N^+ClO_4^-$ electrolyte and (*b*) 0.060 mM 2-nitropropane in acetonitrile with 0.10 M $(n\text{-}C_7H_{15})_4N^+ClO_4^-$ electrolyte. The reaction in curve (*a*) is

$$O_2 + e^- \rightleftharpoons O_2^-$$
$$\text{Superoxide}$$

Working electrode, Hg; reference electrode, Ag | 0.001 M $AgNO_3(aq)$ | 0.10 M $(C_2H_5)_4N^+ClO_4^-$ in acetonitrile; scan rate = 100 V/s. I_{pa} is the peak anodic current and I_{pc} is the peak cathodic current. E_{pa} and E_{pc} are the potentials at which these currents are observed. [From D. H. Evans, K. M. O'Connell, R. A. Petersen, and M. J. Kelly, *J. Chem. Ed.* **1983,** *60,* 290.]

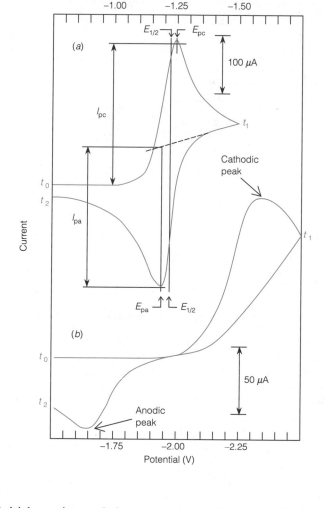

Current decreases after the cathodic peak because of *concentration polarization.*

The initial portions of the current-potential curves in Figure 18-20, beginning at t_0, look like ordinary polarograms, with a residual current followed by a *cathodic wave.* Instead of leveling off at the top of the wave, the current decreases as the potential is increased further. This happens because analyte becomes depleted near the electrode surface, and diffusion from bulk solution is too slow to replenish the concentration near the electrode. In the upper curve, at the time of the maximum voltage (t_1), the cathodic current has decayed to a fairly small value. After t_1, the potential is reversed, but a cathodic current continues to flow because the potential is still negative enough for reduction. When the potential becomes sufficiently less negative, reduced product near the electrode surface begins to be oxidized. This reaction gives rise to an *anodic wave.* Finally, as the reduced product is depleted, the anodic current decays back toward its initial value at t_2.

Figure 18-20a illustrates the behavior of a *reversible* reaction that is fast enough to maintain equilibrium concentrations of reactant and product *at the electrode surface.* The peak anodic and peak cathodic currents have equal magnitudes in a reversible process, and

$$E_{pa} - E_{pc} = \frac{2.22RT}{nF} = \frac{57.0}{n} \text{ (mV)} \quad \text{(at 25°C)} \qquad (18\text{-}11)$$

where E_{pa} and E_{pc} are the potentials at which the *peak anodic* and *peak cathodic* currents are observed and n is the number of electrons in the half-reaction. The half-wave potential, $E_{1/2}$, lies midway between the two peak potentials. For an irreversible reaction, the cathodic and anodic peaks are drawn out and more separated (Figure 18-20b). At the limit of irreversibility, where the oxidation is very slow, no anodic peak is seen.

For a reversible reaction, the peak current (i_p, amperes) for the forward sweep of the first cycle is proportional to the concentration of analyte and the square root of sweep rate:

$$i_p = (2.69 \times 10^8)n^{3/2}ACD^{1/2}v^{1/2} \quad \text{(at 25°C)} \qquad (18\text{-}12)$$

where n is the number of electrons in the half-reaction, A is the area of the electrode (m^2), C is the concentration (mol/L), D is the diffusion coefficient of the electroactive species (m^2/s), and v is scan rate (V/s). The faster the sweep rate, the greater the peak current, as long as the reaction remains reversible. If the electroactive species is confined close to the surface of the electrode and cannot diffuse freely into solution, the peak current is directly proportional to scan rate instead of to the square root of scan rate.

Cyclic voltammetry is used to characterize the redox behavior of compounds (such as C_{60} in Figure 18-21) and to elucidate the kinetics of electrode reactions.[8] Cleverly designed cells permit us to measure optical and magnetic resonance spectra of some intermediates, thereby helping to establish structures. Box 18-2 describes how the optical absorbance spectrum can be measured during a voltammetric experiment.

> For a reversible reaction, $E_{1/2}$ lies midway between the cathodic and anodic peaks.

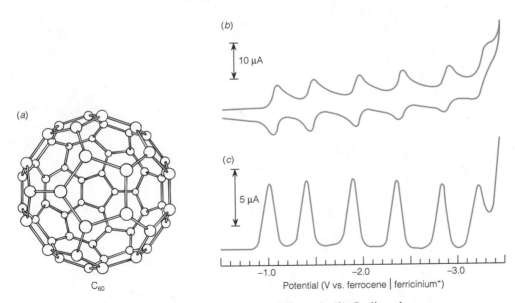

Figure 18-21 (*a*) Structure of C_{60} (buckminsterfullerene). (*b*) Cyclic voltammetry and (*c*) differential pulse polarography of 0.8 mM C_{60} showing six waves for reduction to C_{60}^-, $C_{60}^{2-}\cdots C_{60}^{6-}$. The acetonitrile/toluene solution was at −10°C with $(n-C_4H_9)_4N^+PF_6^-$ supporting electrolyte. The reference electrode contains the ferrocene|ferricinium redox couple (Section 17-5). [From Q. Xie, E. Pérez-Cordero, and L. Echegoyen, *J. Am. Chem. Soc.* **1992,** *114,* 3978.]

Box 18-2 An Optically Transparent Thin-Layer Electrode[9]

Clever cell design permits the simultaneous measurement of the optical and electrochemical properties of a solution. The key element is a thin gold screen, which is the working electrode. As shown here, the screen is sandwiched between two microscope slides, which form a thin-layer cell. The screen has an optical transmittance of 82%, which means that a visible spectrum of the solution can be recorded by placing the cell in a spectrophotometer. The working volume of the cell is 30–50 μL, and complete electrolysis of the solute requires 30–60 s.

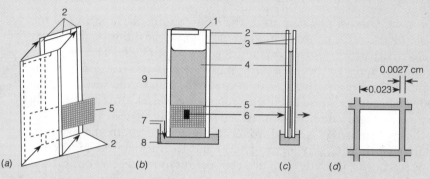

Diagram of an optically transparent thin-layer electrode. (*a*) Assembly of the cell. (*b*) Front view. (*c*) Side view. (*d*) Dimensions of gold grid. 1, Point where suction is applied to change solution; 2, teflon tape spacers; 3, microscope slides; 4, solution; 5, gold grid electrode; 6, optical path of spectrometer; 7, reference and auxiliary electrodes; 8, solution cup; 9, epoxy cement. A nickel grid electrode coated with pyrolytic graphite can be substituted for gold, thereby extending the range of working potentials 0.7 V more negative.[10]

The following figure shows a series of spectra recorded during the reduction of *o*-tolidine. From these spectra, the fraction of *o*-tolidine reduced at each potential can be measured.

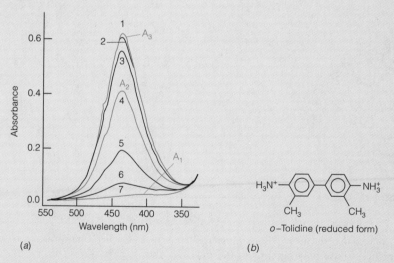

(*a*) Thin-layer spectra of 0.97 mM *o*-tolidine, 0.5 M acetic acid, and 1.0 M $HClO_4$ for different values of applied potential (versus S.C.E.): 1, 0.800; 2, 0.660; 3, 0.640; 4, 0.620; 5, 0.600; 6, 0.580; 7, 0.400 V. Cell thickness is 0.17 mm. (A_1, A_2, and A_3 refer to absorbances used in Problem 25.) (*b*) Structure of the reduced form of *o*-tolidine.

(a)

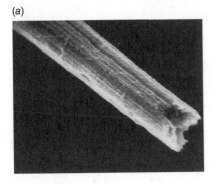

(b)

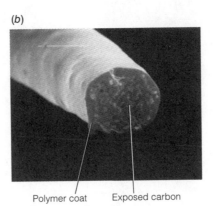

Polymer coat   Exposed carbon

Figure 18-22 (*a*) Scanning electron micrograph of a carbon fiber that has been etched to ~1 μm diameter by drawing through a flame. (*b*) Carbon fiber with thin, insulating coating formed by electrolytic copolymerization of phenol and 2-allylphenol. Once coated, the tip of the fiber is the only electrochemically active surface. [Photos courtesy Andrew Ewing, Pennsylvania State University. From T. G. Strein and A. G. Ewing, *Anal. Chem.* **1992,** *64,* 1368.]

Microelectrodes

Electrodes with diameters of ~1 μm (one-hundredth the diameter of a human hair) offer advantages such as being able to fit inside a living cell.[11] Because the surface area of the electrode is small, the number of molecules reaching the surface to be oxidized or reduced in a given time is low, and the current is tiny. With a small current, the ohmic drop ($= IR$) in a highly resistive medium is small. This allows microelectrodes to be used in poorly conducting nonaqueous solutions and even inside polymer membranes. The electrical capacitance of the double layer (Box 18-1) of a microelectrode is very small. Low capacitance enables the potential of the electrode to be varied at rates up to 500 kV/s, thus allowing short-lived species with lifetimes of less than 1 μs to be studied.

Figure 18-22 shows a carbon fiber electrode that is 1 μm in diameter and has an insulating polymer coating. Because carbon is readily fouled by adsorption of organic molecules inside living cells, a thin layer of Pt or Au is electrolytically plated onto the exposed carbon in Figure 18-22b. The metallized electrode is cleaned in situ (inside the cell during an experiment) by administration of an anodic voltage pulse that desorbs surface-bound species and produces an oxide layer on the metal, followed by a cathodic pulse that reduces the oxide.[12]

Figure 18-23 shows a carbon fiber microelectrode coated with a cation-exchange membrane called Nafion. This membrane, whose molecular structure was shown in problem 11 in Chapter 17, has fixed negative charges. Cations from solution rapidly diffuse through the membrane, but anions are excluded by the fixed negative charges. Figure 18-24 shows the cyclic voltammogram of the neurotransmitter dopamine, which is oxidized near

Advantages of microelectrodes:

1. They fit into small places.
2. They are useful in resistive, non-aqueous media (because of small ohmic losses).
3. Rapid voltage scans (possible because of small double-layer capacitance) allow short-lived species to be studied.

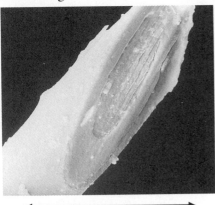

4 μm

Figure 18-23 Electron micrograph of the tip of a Nafion-coated carbon fiber electrode. The carbon inside the electrode has a diameter of 10 μm. Nafion permits cations to pass but excludes anions. [Photo courtesy R. M. Wightman. From R. M. Wightman, L. J. May, and A. C. Michael, *Anal. Chem.* **1988,** *60,* 769A.]

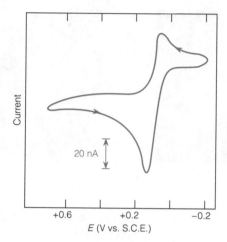

Figure 18-24 Cyclic voltammogram of dopamine using Nafion-coated carbon fiber electrode at pH 7.4. The scan begins and ends at -0.2 V. [From R. M. Wightman, L. J. May, and A. C. Michael, *Anal. Chem.* **1988**, *60*, 769A.] The oxidation half-reaction is

$+0.12$ V (versus S.C.E.). Dopamine concentrations inside the brain of a rat can be measured with a Nafion-coated carbon fiber electrode.[13] Ascorbate ordinarily interferes with the dopamine analysis. However, the Nafion membrane excludes ascorbate anion but admits dopamine cation. The response to dopamine is 1 000 times higher than the response to ascorbate at the same concentration.

18-7 Amperometric Titrations

Amperometry refers to the measurement of electric current. An **amperometric titration** employs a current measurement to detect the end point of a titration. In Figure 17-12 an amperometric circuit was used to detect excess Br_2 at the end of a coulometric titration of cyclohexene. The detection system was a pair of platinum electrodes with a voltage of 0.2 V applied between them. Only a small residual current was observed until the end point was reached; beyond the end point, the current increased markedly. Prior to the end point, only Br^- was present; beyond it, both Br_2 and Br^- were present. In the presence of both species, the following two reactions carry substantial current between the electrodes:

$$\text{cathode:} \quad Br_2 + 2e^- \rightarrow 2Br^-$$

$$\text{anode:} \quad 2Br^- \rightarrow Br_2 + 2e^-$$

A platinum electrode is said to be **polarizable** because its potential is easily changed when only a small current flows. In contrast, a calomel

The potential of a *polarizable electrode* is easily changed. The potential of a *nonpolarizable electrode* is hard to change.

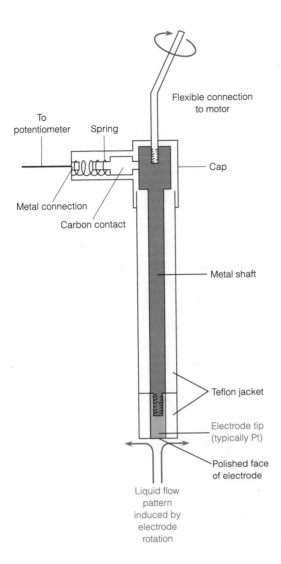

Figure 18-25 Rotating disk electrode.

electrode is said to be **nonpolarizable** because its potential remains constant unless a large current is flowing.

A common working electrode for an amperometric titration is the **rotating disk electrode** in Figure 18-25. The electrode spins at a constant rate of ~600 rpm, bringing analyte to the surface by convection as shown in the figure. The resulting current is much greater than that arising from diffusion alone. The polished bottom surface of the electrode is the only part in electrical contact with the solution. Although a typical electrode diameter is 5 mm, rotating microelectrodes with dimensions measured in micrometers are also used. A rotating Pt electrode is preferred over the dropping Hg electrode for easily reduced species—such as Ag^+, Br_2, and Fe^{3+}—and for anodic reactions. The Pt electrode has a less useful cathodic range than Hg because of the low overpotential for H^+ reduction at Pt.

Biamperometric Titrations

The procedure employing two polarizable electrodes for amperometric detection of an end point is called a **biamperometric titration.** Figure 18-26

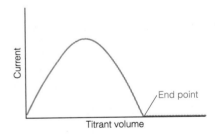

Figure 18-26 Schematic biamperometric titration curve for the addition of $S_2O_3^{2-}$ to I_2.

Biamperometric titration: Current is measured between two electrodes held at a constant potential difference.

Box 18-3 Oxygen Electrodes

The automobile oxygen sensor mentioned at the beginning of this chapter consists of a disk of Y_2O_3-doped ZrO_2 coated with porous Pt electrodes. Just as LaF_3 doped with EuF_2 contains mobile fluoride ions (Figure 15-16), ZrO_2 doped with Y_2O_3 has oxygen vacancies that allow oxide ions to diffuse through the solid at elevated temperature (~900°C). In the following diagram, one side of the sensor is exposed to unknown exhaust gas and the other side is exposed to a reference gas (such as the atmosphere). If the O_2 concentration is higher on the left side than on the right, O_2 diffuses through the porous Pt electrode at the left and is reduced to oxide ion at the surface of the Y_2O_3-doped ZrO_2. Oxide ions diffuse through the solid and are oxidized at the right-hand electrode. The redox chemistry gives a voltage difference between the two electrodes that is measured by a potentiometer.

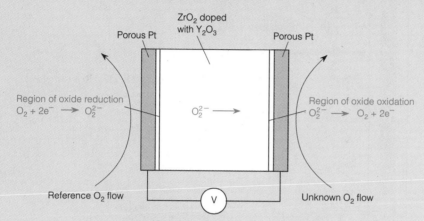

How the oxygen sensor works. The voltage difference between the two electrodes is governed by the Nernst equation: $\Delta V = (RT/2F)\ln\{P_{O_2}(\text{left})/P_{O_2}(\text{right})\}$, where R is the gas constant, T is the sensor temperature, and F is the Faraday constant.

Oxygen in solution is measured with a **Clark electrode**,[14] in which a Pt cathode is held at -0.6 V with respect to a Ag|AgCl anode. The cell is covered by a semipermeable Teflon membrane, across which oxygen can diffuse in a few seconds. The current between the two electrodes is proportional to the dissolved oxygen concentration:

Box 18-3 describes one of the most important applications of amperometry: the measurement of dissolved oxygen.

illustrates the shape of the curve obtained for the biamperometric titration of I_2 with $S_2O_3^{2-}$. The titration reaction is

$$2S_2O_3^{2-} + I_2 \rightarrow 2I^- + S_4O_6^{2-} \qquad (18\text{-}13)$$

Thiosulfate   Tetrathionate

The $I_2|I^-$ couple reacts reversibly at a Pt electrode, but the $S_2O_3^{2-}|S_4O_6^{2-}$ couple does not. That is, the reaction $I_2 + 2e^- \rightleftharpoons 2I^-$ occurs readily at the Pt surface, but the reaction $S_4O_6^{2-} + 2e^- \rightleftharpoons 2S_2O_3^{2-}$ does not.

Now consider what happens when $S_2O_3^{2-}$ is added to I_2. At first, no I^- is present and only residual current is observed. As the reaction proceeds, I^- is created and current is conducted between the two polarizable electrodes by means of the following reactions:

$$\text{cathode:} \quad I_2 + 2e^- \rightarrow 2I^- \qquad (18\text{-}14)$$

$$\text{anode:} \quad 2I^- \rightarrow I_2 + 2e^- \qquad (18\text{-}15)$$

The end point of a biamperometric titration is sometimes called a *dead stop* end point.

The current reaches a maximum near the middle of the titration, when both I_2 and I^- are present. The current then decreases as more I_2 is consumed by Reaction 18-13. Beyond the end point, no I_2 is present and only residual current flows.

cathode: $O_2 + 2H^+ + 2e^- \rightleftharpoons H_2O_2$

anode: $2Ag + 2Cl^- \rightleftharpoons 2AgCl + 2e^-$

The electrode is calibrated in solutions of known oxygen concentration.

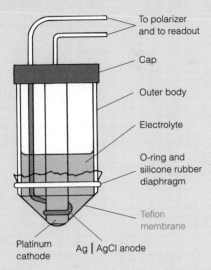

To polarizer
and to readout

Cap

Outer body

Electrolyte

O-ring and
silicone rubber
diaphragm

Teflon
membrane

Platinum
cathode

Ag | AgCl anode

Clark oxygen electrode. [From D. T. Sawyer and J. L. Roberts, Jr., *Experimental Electrochemistry for Chemists* (New York: Wiley, 1974).]

A Clark electrode can fit into the tip of a surgical catheter that is 1.5 mm in diameter and can be stored in a dry, sterile state.[15] When inserted through the umbilical artery of a newborn infant, water diffuses in and activates the electrode. By this means, blood oxygen concentration is monitored to detect respiratory distress. The sensor responds within 20–50 s to administration of O_2 for breathing or to mechanical ventilation of the lungs.

Challenge A schematic biamperometric titration curve for the addition of Ce^{4+} to Fe^{2+} is shown in Figure 18-27. The titration is

$$Ce^{4+} + Fe^{2+} \rightarrow Ce^{3+} + Fe^{3+}$$

and both couples ($Ce^{4+}|Ce^{3+}$ and $Fe^{3+}|Fe^{2+}$) react reversibly at the Pt electrodes. Explain the shape of the titration curve.

Karl Fischer Titration of H₂O

One of the most widely used procedures in all of analytical chemistry is the **Karl Fischer titration,**[16] which can measure residual water in purified solvents, foodstuffs, polymers, and numerous other substances. The main compartment of the titration cell in Figure 18-28 contains anode solution plus unknown. The smaller compartment at the left has an internal Pt electrode immersed in the cathode solution and an external Pt electrode immersed in the anode solution of the main compartment. The two compartments are separated by an ion-permeable membrane. At the right of the diagram is a pair of Pt electrodes used for end-point detection as described below. The septum at the right is for injecting sample by syringe without introducing unnecessary air containing traces of water.

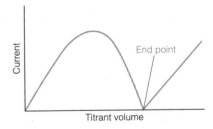

Current

End point

Titrant volume

Figure 18-27 Schematic biamperometric titration curve for the addition of Ce^{4+} to Fe^{2+}.

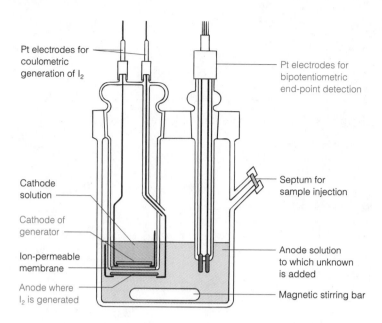

Figure 18-28 Apparatus for coulometric Karl-Fischer titration.

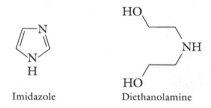

Imidazole   Diethanolamine

pH is maintained in the range 4 to 7. Above pH 8, nonstoichiometric side reactions occur. Below pH 3, the reaction is too slow.

Anode solution contains an alcohol, a base, SO_2, I^-, and other organic solvent. Methanol and diethylene glycol monomethyl ether ($CH_3OCH_2CH_2OCH_2CH_2OH$) are typical alcohols. Typical bases are imidazole and diethanolamine. Pyridine, which was used in Fischer's original recipe, has been replaced because of its toxicity and sluggish reaction rates. The organic solvent mixture may contain chloroform, methanol, carbon tetrachloride, formamide, or other solvents. The trend is to avoid chlorinated solvents because of their environmental hazards.

The anode at the lower left in Figure 18-28 generates I_2 by Reaction 18-15. In the presence of water, stoichiometric reactions occur between the alcohol (ROH), base (B), SO_2, and I_2:

$$B \cdot I_2 + B \cdot SO_2 + B + H_2O \rightarrow 2BH^+I^- + {}^+B\!-\!SO_3^- \quad (18\text{-}16)$$

$$^+B\!-\!SO_3^- + ROH \rightarrow BH^+ROSO_3^- \quad (18\text{-}17)$$

The net reaction is oxidation of SO_2 by I_2, with formation of an alkyl sulfate as product. One mole of I_2 is consumed for each mole of H_2O in the reaction medium.

In a typical procedure, the main compartment in Figure 18-28 is filled with anode solution and the coulometric generator is filled with cathode solution that may contain reagents designed to be reduced at the cathode. Coulometric current is run until all moisture in the main compartment is consumed, as indicated by the end-point detection system described below. An unknown sample is injected through the septum and the coulometer is run again until the injected moisture has been consumed. Two moles of electrons correspond to 1 mol of I_2, which corresponds to 1 mol of H_2O.

A **bipotentiometric** measurement is the most common way to detect the end point of a Karl Fischer titration. The detector circuit maintains a *constant current* (usually 5 or 10 μA) between the two detector electrodes at the right in Figure 18-28, while measuring the voltage needed to sustain the current. Prior to the equivalence point, the solution contains I^-, but little I_2

Most pH meters contain a pair of sockets at the back that are labeled "K-F" or "Karl Fischer." When the manufacturer's instructions are followed, a constant current (usually around 10 μA) is applied across these terminals. To perform a bipotentiometric titration, Pt electrodes are connected to the K-F sockets. The meter is set to the millivolt scale, which indicates the voltage needed to maintain the constant current between the electrodes immersed in a solution.

The figure here shows the results of a bipotentiometric titration of ascorbic acid with I_3^-. Ascorbic acid (146 mg) was dissolved in 200 mL of water in a 400-mL beaker. Two Pt electrodes were attached to the K-F outlets of the pH meter and spaced about 4 cm apart in the magnetically stirred solution. The solution was titrated with 0.04 M I_3^- (prepared by dissolving 2.4 g of KI plus 1.2 g of I_2 in 100 mL of water), and the voltage was recorded after each addition. Prior to the equivalence point, all the I_3^- is reduced to I^- by the excess ascorbic acid. Reaction 18-15 can occur, but Reaction 18-14 cannot. A voltage of about 300 mV is required to support a constant current of 10 μA. (The ascorbate | dehydroascorbate couple does not react at a Pt electrode and cannot carry current.) After the equivalence point, excess I_3^- is present, so Reactions 18-14 and 18-15 both occur, and the voltage drops precipitously.

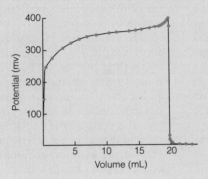

(which is consumed in Reaction 18-16 as fast as it is generated by the coulometer). To maintain a current of 10 μA, the cathode potential must be negative enough to reduce solvent:

$$CH_3OH + e^- \rightleftharpoons CH_3O^- + \tfrac{1}{2} H_2(g)$$

At the equivalence point, excess I_2 suddenly appears and current can be carried at very low voltage by Reactions 18-14 and 18-15, as shown in Demonstration 18-1. The abrupt voltage drop marks the end point.

End points in Karl Fischer titrations tend to drift because of slow chemical reactions and slow leakage of water into the cell from the air. Some instruments measure the rate at which I_2 must be generated to maintain the end point and compare this rate with that measured before sample was added. Other instruments allow you to set a "persistence of end point" time, typically 5 to 60 s, during which the detector voltage must be stable to define the end point.

Careful work with appropriate reagents and equipment gives results that systematically overestimate the H_2O content of the unknown by 0.5%, for H_2O in the range 10–20 mg. The precision of replicate analyses of 10–20 mg of H_2O is approximately 1 to 2%. Larger amounts of water give better precision and accuracy.

Terms to Understand

amperometric titration
amperometry
biamperometric titration
bipotentiometric titration
Clark electrode
condenser current
cyclic voltammetry
differential pulse polarography
diffusion current

direct current polarography
dropping-mercury electrode
electric double layer
faradaic current
half-wave potential
Karl Fischer titration
nonpolarizable electrode
polarizable electrode
polarogram

polarographic wave
polarography
potentiometric stripping analysis
rotating disk electrode
square wave polarography
stripping analysis
supporting electrolyte
voltammetry

Summary

In direct current polarography, we observe the current flowing through an electrochemical cell as a function of the potential applied to a dropping-mercury working electrode. This electrode gives reproducible polarograms because fresh surface is continuously exposed. The overpotential for H^+ reduction allows the observation of cathodic processes that are inherently less favorable than H^+ reduction. Most anodic processes must be observed with other electrodes because Hg is too readily oxidized. Polarography is useful in quantitative analysis because the diffusion current (= limiting current − residual current) is proportional to analyte concentration. Polarography is used in qualitative analysis, because $E_{1/2}$ is characteristic of a particular analyte in a particular medium. The shape of a polarographic wave can be used to find the number of electrons participating in a reversible electrode reaction. The dependence of $E_{1/2}$ on ligand concentration can be used to measure metal-ligand stability constants.

In differential pulse polarography, the voltage is periodically stepped up, and current is measured during the last part of the life of each mercury drop. This produces a derivative-shaped polarogram. Residual current is diminished because the condenser current for each mercury drop decreases more rapidly than the faradaic current. Signal can be increased by increasing the modulation amplitude, and the resolution of neighboring waves is better than in direct current polarography. Square wave polarography provides a further improvement in sensitivity because both reduction and oxidation reactions contribute to the observed signal. The increased speed of square wave polarography allows real-time measurements not possible with other electrochemical methods.

In anodic stripping voltammetry, the most sensitive form of polarography, analyte is concentrated into a single drop of mercury by reduction at a fixed voltage for a fixed time. The potential is then made more positive, and current is measured as analyte is spontaneously oxidized. Alternatively, in potentiometric stripping analysis, the potentiostat is disconnected after analyte reduction, and the time needed for analyte to be oxidized by Hg^{2+} from solution is monitored potentiometrically. In cyclic voltammetry, a triangular waveform is applied, and cathodic and anodic processes are observed in succession. Microelectrodes are useful because they fit into small places and their low current allows them to be used in resistive, nonaqueous media. Their low capacitance permits very rapid voltage scanning, which allows very short-lived species to be studied.

In amperometric titrations, the current flowing between a pair of electrodes is used to locate the end point. In a biamperometric titration, there is a fixed potential difference between two polarizable electrodes, and current is recorded. In a bipotentiometric titration, the voltage needed to maintain a constant current is measured. The current or the voltage changes abruptly at the equivalence point in these titrations, when one member of a redox couple is either created or destroyed. The Karl Fischer titration uses bipotentiometric end-point detection to measure water.

Exercises

A. The analysis of Ni^{2+} at the nanogram level is possible with differential pulse polarography.[17] Addition of dimethylglyoxime to an ammonium citrate buffer enhances the response by a factor of 15.

On the basis of a standard curve constructed from the data given in the following table, predict how much current is expected if 54.0 μL of solution containing 10.0 ppm Ni^{2+} is added to 5.00 mL of buffer.

Ni²⁺ (ppb)	Peak current (μA)
19.1	0.095
38.2	0.173
57.2	0.258
76.1	0.346
95.0	0.429
114	0.500
132	0.581
151	0.650
170	0.721

B. Cd^{2+} was used as an internal standard in the analysis of Pb^{2+} by differential pulse polarography. Cd^{2+} gives a reduction wave at -0.60 (±0.02) V and Pb^{2+} gives a reduction wave at -0.40 (±0.02) V. It was first verified that the ratio of peak heights is proportional to the ratio of concentrations over the whole range employed in the experiment. Results for known and unknown mixtures are given below.

Analyte	Concentration (M)	Current (μA)
Known		
Cd^{2+}	3.23 (±0.01) $\times 10^{-5}$	1.64 (±0.03)
Pb^{2+}	4.18 (±0.01) $\times 10^{-5}$	1.58 (±0.03)
Unknown + Internal Standard		
Cd^{2+}	?	2.00 (±0.03)
Pb^{2+}	?	3.00 (±0.03)

The unknown mixture was prepared by mixing 25.00 (±0.05) mL of unknown (containing only Pb^{2+}) plus 10.00 (±0.05) mL of 3.23 (±0.01) $\times 10^{-4}$ M Cd^{2+} and diluting to 50.00 (±0.05) mL.

(a) Disregarding uncertainties, find the concentration of Pb^{2+} in the undiluted unknown.

(b) Find the absolute uncertainty for the answer to **(a)**.

C. A large quantity of Fe^{3+} interferes with the polarographic analysis of Cu^{2+} because Fe^{3+} is reduced at less negative potentials than is Cu^{2+} in most supporting electrolytes. Interference is eliminated by addition of hydroxylamine (NH_2OH), which reduces Fe^{3+}, but not Cu^{2+}. The following figure shows the differential pulse polarogram of Cu^{2+} in the presence of 1 000 ppm of Fe^{3+} with saturated $NH_3OH^+Cl^-$ as supporting electrolyte. Each sample was made up to the *same final volume*. Averaging responses for the two standard additions, find $[Cu^{2+}]$ in the sample solution.

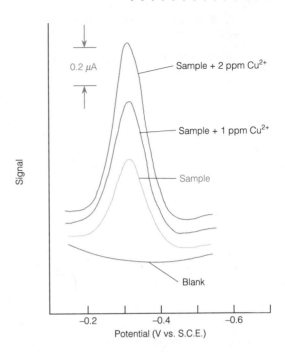

Differential pulse polarograms. Modulation amplitude, 25 mV; drop interval, 1 s; scan rate, 2 mV/s. Each scan is offset (displaced vertically) from the previous one. To measure peak height relative to the blank, subtract the vertical displacement at the left edge of the scan. [From EG&G Princeton Applied Research Corp., Princeton, NJ, Application Note 151.]

D. Polarographic data for the reaction

$$HPbO_2^- + H_2O + 2e^- + Hg \rightleftharpoons Pb(in\ Hg) + 3OH^-$$

are given below.[18]

$E_{1/2}$ (V vs. S.C.E.)	[OH⁻] (M, free OH⁻)
-0.603	0.011
-0.649	0.038
-0.666	0.060
-0.681	0.099
-0.708	0.201
-0.734	0.448
-0.747	0.702
-0.755	1.09

The species $HPbO_2^-$ can be treated as $Pb(OH)_3^-$ by virtue of the equilibrium

$$Pb(OH)_3^- \rightleftharpoons HPbO_2^- + H_2O$$

Use the data in the table to show that $p = 3$ in Equation 18-9, and find the value of β_3 in Equation 18-10 for $Pb(OH)_3^-$. The value of $E_{1/2}$(for free Pb^{2+}) in Equation

18-10 is -0.41 V. This is the half-wave potential for the reaction $Pb^{2+} + 2e^- \rightleftharpoons Pb(in\ Hg)$ in 1 M HNO_3.

E. A cyclic voltammogram of the Co^{3+} compound $Co(B_9C_2H_{11})_2^-$ is shown here.

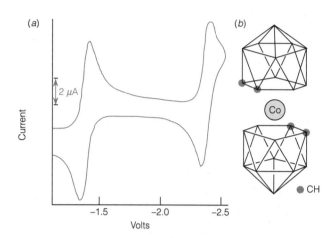

(a) Cyclic voltammogram and *(b)* structure of $Co(B_9C_2H_{11})_2^-$. [From W. E. Geiger, Jr., W. L. Bowden, and N. El Murr, *Inorg. Chem.* **1979**, *18*, 2358.]

$E_{1/2}$ (V vs. S.C.E.)	I_{pa}/I_{pc}	$E_{pa} - E_{pc}$ (mV)
-1.38	1.01	60
-2.38	1.00	60

Suggest a chemical reaction to account for each wave. Are the reactions reversible? How many electrons are involved in each step? Sketch the direct current and differential pulse polarograms expected for this compound.

F. Amperometric titration curves for the precipitation of Pb^{2+} with $Cr_2O_7^{2-}$ are shown below. The potential of the dropping-mercury electrode was -0.8 V versus S.C.E. for curve a and 0.0 V for curve b. Explain the shapes of the two curves.

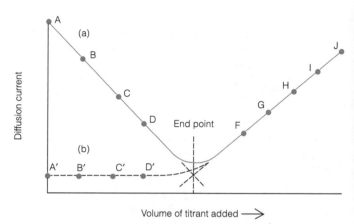

Amperometric titration curves. [From D. T. Sawyer and J. L. Roberts, Jr., *Experimental Electrochemistry for Chemists* (New York: Wiley, 1974).]

G. Sketch the shape of a bipotentiometric titration curve (*E* versus volume added) for the addition of Ce^{4+} to Fe^{2+}. Both the $Ce^{4+}|Ce^{3+}$ and $Fe^{3+}|Fe^{2+}$ couples react reversibly at Pt electrodes.

H. In a coulometric Karl Fischer water analysis, 25.00 mL of pure "dry" methanol required 4.23 C to generate enough I_2 to react with the residual H_2O in the methanol. A suspension of 0.8476 g of finely ground polymeric material in 25.00 mL of the same "dry" methanol required 63.16 C. Find the weight percent of H_2O in the polymer.

Problems

Polarography

1. Why does the reaction $Cu(I) \rightarrow Cu(0)$ have different half-wave potentials with Hg and Pt electrodes? Will the half-wave potential for Reaction 18-5 be dependent on the composition of the electrode?

2. In 1 M NH_3/1 M NH_4Cl solution, Cu^{2+} is reduced to Cu^+ near -0.3 V (versus S.C.E.), and Cu^+ is reduced to Cu (*in Hg*) near -0.6 V.

(a) Sketch a qualitative polarogram for a solution of Cu^+.

(b) Sketch a qualitative polarogram for a solution of Cu^{2+}.

(c) Suppose that Pt, instead of Hg, were used as the working electrode. Which, if any, reduction potential would you expect to change?

3. Solute is transported to an electrode by diffusion, convection, and electrostatic attraction. In polarography, we want the current to be limited by diffusion. Explain how convection and electrostatic attraction are minimized.

4. Draw graphs showing the voltage ramps used in direct current and differential pulse polarography. Label the axes. Label the modulation amplitude and show the two regions between which ΔI is measured on the differential pulse ramp.

5. **(a)** Explain the difference between condenser and faradaic currents.

(b) Why is differential pulse polarography more sensitive than direct current polarography?

6. The limiting current in Table 18-2 is given by Equation 18-3, except the constant 6.07×10^4 was used instead of 7.08×10^4. The reason for this substitution is that in the "old days" current was measured at the center of each oscillation, instead of at the top, using a damped ballistic galvanometer. For the data in Table 18-2, $m^{2/3}t^{1/6} = 2.28 \text{ mg}^{2/3}\text{s}^{1/2}$. Calculate the diffusion coefficient for Pb^{2+} in 1.0 M KNO_3.

7. For the polarogram in Figure 18-2, the diffusion current is 14 µA.

(a) If the solution contains 25 mL of 0.50 mM Cd^{2+}, calculate the fraction of Cd^{2+} reduced per minute of current passage.

(b) If the drop interval is 4 s, how many minutes were required to scan from −0.6 to −1.2 V?

(c) By what percentage has the concentration of Cd^{2+} changed during the scan from −0.6 to −1.2 V?

8. Use Equation 18-6 to derive an expression for $E_{3/4} - E_{1/4}$, where $E_{3/4}$ is the potential when $I = \frac{3}{4}I_d$ and $E_{1/4}$ applies when $I = \frac{1}{4}I_d$. The value of $E_{3/4} - E_{1/4}$ is used as a criterion for reversibility of a polarographic wave. If $E_{3/4} - E_{1/4}$ is close to the value you have calculated, the reaction is probably reversible.

9. A 2.0-µA signal was observed in polarography of a solution with the same composition as that in Figure 18-9a, prepared by mixing 10.0 mL of Al^{3+} unknown with buffer and diluting to 100.0 mL. Use Figure 18-9a to calculate the concentration of Al^{3+} in the unknown. Remember that the sample was diluted before making the measurement.

10. An unknown gave a polarographic signal of 10.0 µA. When 1.00 mL of 0.0500 M standard was added to 100.0 mL of unknown, the signal increased to 14.0 µA. Find the concentration of the original unknown.

11. Nitrite (NO_2^-) and nitrate (NO_3^-) in foods can be measured by the following procedure:[19]

> **1.** A 10-g food sample is blended with 82 mL of 0.07 M NaOH, transferred to a 200-mL volumetric flask, and heated on a steam bath for 1 h. Then 10 mL of 0.42 M $ZnSO_4$ is added; and after an additional 10 min of heating, the mixture is cooled and diluted to 200 mL.
> **2.** After the solids settle, an aliquot of supernatant solution is mixed with 200 mg of charcoal to remove some organic solutes and filtered.
> **3.** A 5.00-mL aliquot of the filtered solution is

placed in a test tube containing 5 mL of 9 M H_2SO_4 and 150 µL of 85 wt% NaBr, which reacts with nitrite, but not nitrate:

$$HNO_2 + Br^- \rightarrow NO(g) + Br_2$$

> **4.** The NO is transported by bubbling with purified N_2 that is subsequently passed through a trapping solution to convert NO into diphenylnitrosamine:

$$NO + (C_6H_5)_2NH \xrightarrow{H^+, \text{ catalyst}} (C_6H_5)_2N—N=O$$
Diphenylamine $\qquad\qquad$ Diphenylnitrosamine

> **5.** The acidic diphenylnitrosamine is analyzed by differential pulse polarography at −0.66 V (versus Ag|AgCl) to measure the NO liberated from the food sample:

$$(C_6H_5)_2N—N=OH^+ + 4H^+ + 4e^- \rightarrow$$
$$(C_6H_5)_2N—NH_3^+ + H_2O$$

> **6.** To measure nitrate, an additional 6 mL of 18 M H_2SO_4 is added to the sample tube in Step 3. The acid promotes reduction of HNO_3 to NO, which is then purged with N_2 and trapped and analyzed as in Steps 4 and 5.

$$HNO_3 + Br^- \xrightarrow{\text{strong acid}} NO(g) + Br_2$$

A 10.0-g bacon sample gave a peak height of 8.9 µA in Step 5, which increased to 23.2 µA in Step 6. (Step 6 measures the sum of signals from nitrite and nitrate, not just the signal from nitrite.) In a second experiment, the 5.00-mL sample in Step 3 was spiked with a standard addition of 5.00 µg of NO_2^- ion. The analysis was repeated to give a peak height of 14.6 µA in Step 5, which increased to 28.9 µA in Step 6.

(a) Find the nitrite content of the bacon, expressed as micrograms per gram of bacon.

(b) On the basis of the ratio of signals from nitrate and nitrite in the first experiment, find the micrograms of nitrate per gram of bacon.

12. The diffusion currents below were measured at −0.6 V for $CuSO_4$ in 2 M $NH_4Cl/2M$ NH_3. Use the method of least squares in Section 4-5 to estimate the molarity and uncertainty in molarity of an unknown solution giving $I_d = 15.6$ µA.

$[Cu^{2+}]$ (mM)	I_d (µA)	$[Cu^{2+}]$ (mM)	I_d (µA)
0.0393	0.256	0.990	6.37
0.0780	0.520	1.97	13.00
0.1585	1.058	3.83	25.0
0.489	3.06	8.43	55.8

13. The drug Librium gives a polarographic wave with $E_{1/2} = -0.265$ V (versus S.C.E.) in 0.05 M H_2SO_4. A 50.0-mL sample containing Librium gave a wave height of 0.37 µA. When 2.00 mL of 3.00 mM Librium in 0.05

M H_2SO_4 was added to the sample, the wave height increased to 0.80 μA. Find the molarity of Librium in the unknown.

14. The differential pulse polarogram of reagent-grade methanol is shown below, along with that of a standard made by adding an additional 0.001 00 wt% acetone, 0.001 00 wt% acetaldehyde, and 0.001 00 wt% formaldehyde to reagent-grade methanol. Estimate the weight percent of acetone in reagent-grade methanol.

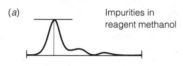

(a)   Impurities in reagent methanol

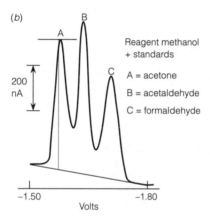

(b)

200 nA

Reagent methanol + standards

A = acetone
B = acetaldehyde
C = formaldehyde

−1.50   −1.80

Volts

Differential pulse polarograms of (a) reagent-grade methanol and (b) methanol containing added standards. [From D. B. Palladino, *Am. Lab.* August 1992, p. 56.] Solutions were prepared by diluting 25 mL of methanol (or methanol plus standards) up to 100 mL with water containing buffer and hydrazine sulfate. The latter reacts with carbonyl compounds to form hydrazones, which are the electroactive species in this analysis.

$(CH_3)_2C{=}O + H_2N{-}NH_2 \rightarrow$
Acetone   Hydrazine

$(CH_3)_2C{=}N{-}NH_2 \xrightarrow{2e^- + 2H^+}$
Acetone hydrazone

$(CH_3)_2CH{-}NH{-}NH_2$

15. Consider a copper wire immersed in ferric chloride solution. This *mixed* redox couple system can be described with the experimental voltammograms shown in the figure at the top of the next column. Curve a represents the response of a rotating copper disk anode immersed in a solution containing 3 M Cl^-. At potentials near 0 V, the copper is oxidized and a large current is observed. Curve b was generated by a rotating platinum

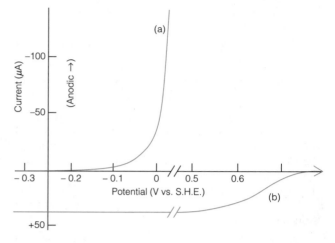

Voltammograms for Problem 15. [From G. P. Power and I. M. Ritchie, *J. Chem. Ed.* **1983**, *60*, 1022.]

disk electrode in a 3 M Cl^- solution containing Fe^{3+}. The experimentally observed current-voltage relation when the copper electrode is immersed in the Fe^{3+} solution is the same as the sum of curves a and b. Sketch this curve and estimate the *mixed potential* at which the current is zero. This is the potential at which the reaction $Fe^{3+} + Cu(s) \rightleftharpoons Fe^{2+} + Cu^+$ comes to equilibrium.

Stripping Analysis

16. Explain the difference between anodic stripping voltammetry and potentiometric stripping.

17. From the data for the standard addition shown in Fig 18-18, find the concentration of Cu^{2+} in the unknown.

18. In a standard addition experiment similar to the one in Figure 18-18, the following signals were observed for Pb^{2+}:

unknown: 4.5 s
unknown + 0.50 μg/L Pb^{2+}:   8.6 s
unknown + 1.00 μg/L Pb^{2+}: 12.9 s

Dilution of the unknown by the standards was negligible. Prepare a graph similar to Figure 6-10 to find the concentration of Pb^{2+} in the unknown.

Cyclic Voltammetry

19. The cyclic voltammogram of the antibiotic chloramphenicol (abbreviated RNO_2) is shown in the first figure on the next page. The scan was started at 0 V, and potential was swept toward negative voltage. The first cathodic wave, A, is from the reaction $RNO_2 + 4e^- + 4H^+ \rightarrow RNHOH + H_2O$. Explain what happens at peaks B and C by using the reaction $RNO + 2e^- + 2H^+ \rightleftharpoons RNHOH$. Why was peak C not seen in the initial scan?

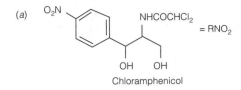

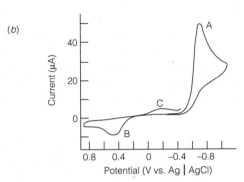

(*a*) Structure of chloramphenicol. (*b*) Cyclic voltammo-gram of 3.7×10^{-4} M chloramphenicol in 0.1 M acetate buffer, pH 4.62. The voltage of the carbon paste working electrode was scanned at a rate of 350 mV/s. [From P. T. Kissinger and W. R. Heineman, *J. Chem. Ed.* **1983**, *60*, 702.]

20. The cyclic voltammograms shown below are due to the irreversible reduction of *trans*-1,2-dibromocyclo-hexane. At room temperature, just one peak is seen (not shown). At low temperatures, two peaks are seen. At −60°C, the relative size of the peak near −3.1 V increases if the scan rate is increased. Explain these observations with the scheme here:

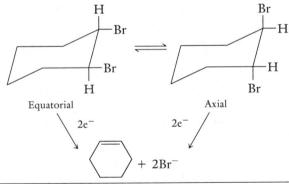

21. Tetraphenylporphyrin^{2-} binds a metal ion at the center of a planar ring containing four nitrogen ligands.

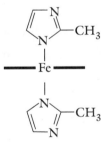

(*meso*-Tetraphenylporphyrinato)iron(II)
PFe

When the Fe^{3+} complex, PFe^+, is reduced at a glassy carbon electrode in dimethylformamide solvent contain-ing tetraethylammonium perchlorate electrolyte, reduc-tion waves are observed at -0.18 ($Fe^{3+} \rightarrow Fe^{2+}$) and -1.02 V ($Fe^{2+} \rightarrow Fe^+$) versus Ag|AgCl. When a 10-mM concentration of 2-methylimidazole is added to 1 mM PFe^+, the waves shift to -0.14 and -1.11 V.[20] 2-Methylimidazole can bind to one or both axial sites of the complex:

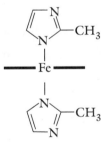

For Problem 20.

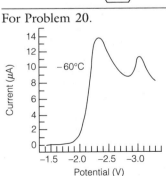

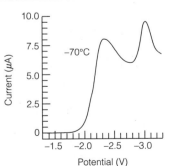

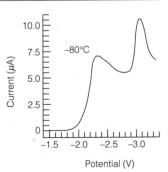

Voltammograms of 2.0 mM *trans*-1,2-dibromocyclohexane at a frozen mercury-drop electrode in butyroni-trile with 0.10 M (n-$C_4H_9)_4N^+ClO_4^-$ elec-trolyte. Scan rate, 1.00 V/s. [From D. H. Evans, K. M. O'Con-nell, R. A. Petersen, and M. J. Kelly, *J. Chem. Ed.* **1983**, *60*, 290.]

On the basis of these shifts, decide which oxidation state (Fe^{3+}, Fe^{2+}, or Fe^+) is most stabilized by the ligand. That is, which oxidation state has the greatest binding constant for 2-methylimidazole?

22. Peak current (i_p) and scan rate (v) are listed below for cyclic voltammetry (one-electron oxidation of Fe^{2+} to Fe^{3+}) of a water-soluble ferrocene derivative in 0.1 M NaCl.[21]

Scan rate (V/s)	Peak anodic current (μA)
0.019 2	2.18
0.048 9	3.46
0.075 1	4.17
0.125	5.66
0.175	6.54
0.251	7.55

If a graph of i_p versus $\sqrt{v}$ gives a straight line, then the reaction is diffusion controlled. Prepare such a graph and use it to find the diffusion coefficient of the reactant for this one-electron oxidation. The area of the working electrode is 0.020 1 cm^2 and the concentration of reactant is 1.00 mM.

23. What are the advantages of using a microelectrode for voltammetric measurements?

24. What is the purpose of the Nafion membrane in Figure 18-23?

25. A series of spectra for reduction of o-tolidine in an optically transparent, thin-layer electrode was shown in Box 18-2.

$$o\text{-tolidine(oxidized)} + ne^- \rightleftharpoons o\text{-tolidine(reduced)}$$

$$E_{applied} = E' - \frac{0.059\,16}{n}\log\frac{[\text{tolidine(reduced)}]}{[\text{tolidine(oxidized)}]}$$

E' is the formal reduction potential in the medium used for the experiment. The spectrum labeled 1 is that of the oxidized form of o-tolidine. The spectrum labeled 7 is that of the reduced form. Spectra 2 through 6 represent mixtures of both forms. The ratio of species can be calculated from Beer's law. For example, for spectrum 4,

$$\frac{[\text{tolidine(reduced)}]}{[\text{tolidine(oxidized)}]} = \frac{A_3 - A_2}{A_2 - A_1}$$

(a) Measure the quotient [tolidine(reduced)]/[tolidine (oxidized)] by using the absorption maxima for curves 2 through 6.

(b) Prepare a graph of $E_{applied}$ versus log([tolidine (reduced)]/[tolidine(oxidized)]).

(c) Use this graph to find E' (versus S.C.E.) and E' (versus S.H.E.).

(d) Use the slope of the graph to calculate the number of electrons in the half-reaction.

(e) Propose a structure for the oxidized form of o-tolidine.

Amperometric Titrations

26. Explain how the amperometric end-point detection in the experiment in Figure 17-12 works.

27. Why is a dropping-mercury electrode preferred for cathodic reactions in amperometric titrations, whereas a rotating platinum disk electrode is preferred for anodic reactions?

28. What is a Clark electrode and how does it work?

29. Write the chemical reactions that show that 1 mol of I_2 is required for 1 mol of H_2O in a Karl Fischer titration.

30. (a) Br_2 can be generated for quantitative analysis by addition of standard BrO_3^- to excess Br^- in acidic solution:

$$BrO_3^- + 5Br^- + 6H^+ \rightarrow 3H_2O + 3Br_2$$

Consider the biamperometric titration of H_3AsO_3 with Br_2:

$$Br_2 + H_3AsO_3 + H_2O \rightarrow$$
$$2Br^- + H_3AsO_4 + 2H^+$$

A solution containing H_3AsO_3 and Br^- is titrated with standard BrO_3^-. Given that the $H_3AsO_4|H_3AsO_3$ couple does not react at a Pt electrode, predict the shape of a graph of current versus volume of titrant.

(b) Sketch a graph of voltage versus volume of I_2 added for a bipotentiometric titration of H_3AsO_3 with I_2.

31. Hydrogen peroxide is found in concentrations of 10^{-8} to 10^{-4} M in water droplets in the troposphere— the lower 10 to 17 km of the atmosphere, where there is substantial vertical mixing (below the stratosphere, which is the next 30 to 40 km of the atmosphere). Oxidation of H_2O_2 with a rotating Pt electrode at $+0.4$ V (versus S.C.E.) gave the following calibration data:[22]

$[H_2O_2]$ (M)	Diffusion current (nA) ($\pm$ standard deviation)	
1.00×10^{-4}	6 995	$\pm$112
5.00×10^{-5}	3 510	$\pm$74
1.00×10^{-5}	698	$\pm$11
5.00×10^{-6}	345	$\pm$18
1.00×10^{-6}	64.4	$\pm$3.9
5.00×10^{-7}	32.4	$\pm$1.8
1.00×10^{-7}	6.88	$\pm$0.64
5.00×10^{-8}	3.17	$\pm$0.32
2.00×10^{-8}	1.03	$\pm$0.20

(a) Prepare a graph of log(current) versus log[H_2O_2] to show that the response is linear over four orders of magnitude.

(b) From a graph of current versus [H_2O_2] (not log(current) versus log[H_2O_2]), find the slope and intercept and their uncertainties by the method of least squares.

(c) Find [H_2O_2] (and its uncertainty) for an unknown whose diffusion current is 300 ± 15 nA. [Supplementary Problem S11 for Chapter 18 in the *Solutions Manual* shows how to use a *weighted least squares* algorithm, in which the uncertainties in the data are used to compute the equation of the straight line.]

Notes and References

1. For biographical information on Heyrovský, see L. R. Sherman, *Chem. Br.* **1990**, *26*, 1165 and J. Koryta, *Electrochim. Acta* **1991**, *36*, 221.

2. This procedure is better than sprinkling sulfur on the spill. Sulfur coats the mercury but does not react with the bulk of the droplet [D. N. Easton, *Am. Lab.* July 1988, p. 66].

3. The diffusion coefficient is defined from Fick's first law of diffusion: The rate (J) at which molecules diffuse across a plane of unit area is given by

$$J = -D\frac{dc}{dx}$$

where D is the diffusion coefficient and dc/dx is the concentration gradient in the direction of diffusion. The larger the diffusion coefficient, the more rapidly the molecules diffuse (see Section 22-4).

4. L. Meites, *Handbook of Analytical Chemistry* (New York: McGraw-Hill, 1963), pp. 5-53 to 5-103.

5. N. L. Weinberg, Ed., *Technique of Electroorganic Synthesis* (New York: Wiley, 1974); J. Chang, R. F. Large, and G. Popp in A. Weissberger and B. W. Rossiter, Eds., *Physical Methods of Chemistry*, Vol. I, Part IIB (New York: Wiley, 1971); Z. Nagy, *Electrochemical Synthesis of Inorganic Compounds: A Bibliography* (New York: Plenum Press, 1985); and J. H. Wagenknecht, *J. Chem. Ed.* **1983**, *60*, 271.

6. J. Wang, *Stripping Analysis: Principles, Instrumentation and Applications* (Deerfield Beach, FL: VCH Publishers, 1984). Stripping analysis is normally conducted with analytes whose reduced product dissolves in or adsorbs onto the working electrode. A clever modification allows an analyte such as $Ru(NH_3)_6^{3+}$, which cannot be deposited on the electrode, to reduce and deposit an analyte such as Ag^+, which is then measured by anodic stripping [T. Horiuchi, O. Niwa, M. Morita, and H. Tabei, *Anal. Chem.* **1992**, *64*, 3206].

7. D. Noble, *Anal. Chem.* **1993**, *65*, 265A.

8. G. A. Mabbott, *J. Chem. Ed.* **1983**, *60*, 697; P. T. Kissinger and W. R. Heineman, *J. Chem. Ed.* **1983**, *60*, 702; D. H. Evans, K. M. O'Connell, R. A. Petersen, and M. J. Kelly, *J. Chem. Ed.* **1983**, *60*, 290; H. H. Thorp, *J. Chem. Ed.* **1992**, *69*, 251; M. E. Gomez and A. E. Kaifer, *J. Chem. Ed.* **1992**, *69*, 502; M. D. Koppang and T. A. Holme, *J. Chem. Ed.* **1992**, *69*, 770.

9. T. P. DeAngelis and W. R. Heineman, *J. Chem. Ed.* **1976**, *53*, 594. A cell that uses an optical fiber and eliminates the gold screen is described by C. Zhang and S.-M. Park, *Anal. Chem.* **1988**, *60*, 1639.

10. M. Kummer and J. R. Kirchoff, *Anal. Chem.* **1993**, *65*, 3720.

11. R. M. Wightman, *Science* **1988**, *240*, 415.

12. T. G. Strein and A. G. Ewing, *Anal. Chem.* **1992**, *64*, 1368; T. G. Strein and A. G. Ewing, *Anal. Chem.* **1993**, *65*, 1203; T. K. Chen, Y. Y. Lau, D. K. Y. Wong, and A. G. Ewing, *Anal. Chem.* **1992**, *64*, 1264; B. D. Pendley and H. D. Abruña, *Anal. Chem.* **1990**, *62*, 782.

13. R. M. Wightman, L. J. May, and A. C. Michael, *Anal. Chem.* **1988**, *60*, 769A; A. J. Cunningham and J. B. Justice, Jr., *J. Chem. Ed.* **1987**, *64*, A34.

14. Construction of an inexpensive oxygen electrode has been described by J. E. Brunet, J. I. Gardiazabal, and R. Schrebler, *J. Chem. Ed.* **1983**, *60*, 677.

15. D. Parker, *J. Phys. E: Sci. Instrum.* **1987**, *20*, 1103.

16. S. K. MacLeod, *Anal. Chem.* **1991**, *63*, 557A; *Hydranal® Manual*, Hoechst-Celanese Corp. (Highway 43 North, Bucks, Alabama 36512). The *Hydranal® Manual* contains detailed instructions for numerous variations of the Karl Fischer procedure.

17. C. J. Flora and E. Neiboer, *Anal. Chem.* **1980**, *52*, 1013.

18. J. J. Lingane, *Chem. Rev.* **1941**, *29*, 1.

19. W. Holak and J. J. Specchio, *Anal. Chem.* **1992**, *64*, 1313.

20. D. K. Geiger, E. J. Pavlak, and L. T. Kass, *J. Chem. Ed.* **1991**, *68*, 337.

21. M. E. Gomez and A. E. Kaifer, *J. Chem. Ed.* **1992**, *69*, 502.

22. J. Lagrange and P. Lagrange, *Fresenius J. Anal. Chem.* **1991**, *339*, 452.

A Fiber-Optic Glucose Sensor

Biosensor contained in a hypodermic needle fits inside a patient's vein to measure blood sugar concentration. [From J. S. Schultz, *Scientific American*, August 1991.]

The optical sensor in this photograph was designed to measure glucose concentration and is thin enough to be inserted into a vein. Its principle of operation is shown in Figure 19-1.

The sensor wall is a *semipermeable membrane,* through which small molecules like glucose can diffuse but large molecules cannot (see Demonstration 25-1). The plant protein concanavalin A, bound on the inner wall of the membrane, binds sugars such as the tiny glucose molecule or the large, polymeric dextran molecule. Concanavalin is initially saturated with dextran to which is covalently attached a *fluorescent* chromophore. (A fluorescent substance emits light when it is irradiated with light.) Dextran is too large to diffuse through the semipermeable membrane. When inserted into a vein, glucose diffuses from the blood into the sensor, liberating some of the dextran from concanavalin. The greater the glucose concentration in the blood, the more glucose will bind to concanavalin and the more dextran will be released.

An optical fiber connected to a laser directs a thin beam of light along the center of the sensor. When all dextran is bound to the membrane, none of the fluorescent chromophore is in the path of the light, so no fluorescence occurs. Upon binding of glucose, some of the dextran diffuses into the light beam and begins to fluoresce. The same optical fiber that brought the laser light into the sensor carries the fluorescence out to a detector. The fluorescence intensity is proportional to the concentration of glucose in the patient's blood.

Applications of Spectrophotometry

In Chapter 6 we learned how Beer's law—the proportionality between absorbance and concentration—allows us to measure chemical concentrations. Later we saw how spectrophotometry can be used to monitor the course of a titration, how spectrophotometry with acid-base indicators can be used to measure pH or pK,[1] and how metal-ligand binding constants and concentrations in electrochemical cells can be measured. This chapter describes more applications of the absorption and emission of light in analytical chemistry.

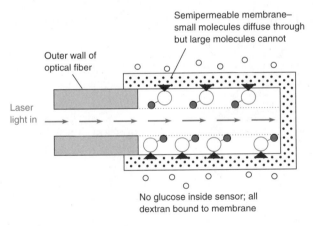
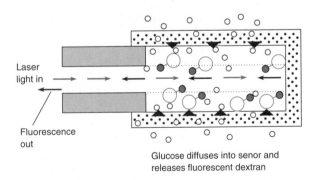

▲ Concanavalin A, a protein that binds sugars

◯● Large dextran molecule with covalently bound fluorescent label

○ Small glucose molecule

Figure 19-1 Operation of fiber-optic glucose sensor. Small glucose molecules diffusing through the semipermeable membrane displace fluorescently labeled dextran molecules from binding sites inside the membrane. The free dextran molecules diffuse into the path of the laser beam at the center of the sensor; their fluorescence is carried back to a detector by the optical fiber. Chapter 20 explains how an optical fiber works.

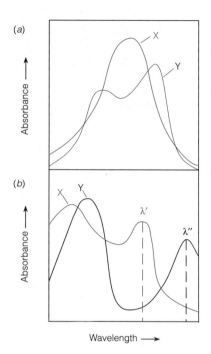

Figure 19-2 Two cases for analysis of a mixture. (*a*) There is significant overlap between the spectra of the two pure components in all regions. (*b*) Regions exist in which each component makes the major contribution.

19-1 Analysis of a Mixture[2]

When there is more than one absorbing species in a solution, the spectrophotometer sees the sum of the absorbances from all species. Our instrument cannot distinguish what fraction of the absorbance arises from each different molecule. However, if the different species have different absorptivities at different wavelengths and if we know what the spectra of the pure components look like, we can mathematically disassemble the spectrum of the mixture into those of its components. The key fact that permits us to analyze mixtures is that at each wavelength *the absorbance of a solution (containing species X, Y, Z · · ·) is the sum of the absorbances of each species:*

Absorbance of a mixture: $A = \epsilon_X b[X] + \epsilon_Y b[Y] + \epsilon_Z b[Z] + \cdots$ (19-1)

where ϵ is the molar absorptivity of each species at the wavelength in question and b is the cell pathlength (Figure 6-4).

For a mixture of compounds X and Y, two cases are distinguished in Figure 19-2. In Figure 19-2a the absorption bands of pure X and pure Y overlap significantly everywhere. This case is best treated by a graphical procedure that makes use of measurements at many wavelengths. In Figure 19-2b the bands of X and Y have relatively little overlap in some regions. We analyze this case by choosing wavelength λ' where X makes the major contribution and wavelength λ'' where Y makes the major contribution.

What to Do When the Individual Spectra Overlap

Let's apply Equation 19-1 to the analysis of a mixture containing two components whose spectra overlap a great deal. The problem is illustrated for a mixture of H_2O_2 complexes of Ti^{4+} and V^{5+} in H_2SO_4 solution in Figure 19-3. The absorbance of the mixture (A_m) at any chosen wavelength is

$$A_m = \epsilon_X b[X] + \epsilon_Y b[Y]$$ (19-2)

where X and Y refer to Ti^{4+} and V^{5+}, respectively. If a standard solution of species X with concentration $[X]_s$ is prepared, its absorbance will be

$$A_{X_s} = \epsilon_X b[X]_s$$ (19-3)

Similarly, a standard solution of species Y with concentration $[Y]_s$ will have absorbance

$$A_{Y_s} = \epsilon_Y b[Y]_s$$ (19-4)

Figure 19-3 Visible spectrum of 1.32 mM Ti^{4+}, 1.89 mM V^{5+}, and an unknown mixture containing both ions. All solutions contain 0.5 wt % H_2O_2 and ~0.01 M H_2SO_4. Absorbance values for points shown by dots are listed in Table 19-1. [From M. Blanco, H. Iturriaga, S. Maspoch, and P. Tarín, *J. Chem. Ed.* **1989,** *66,* 178.]

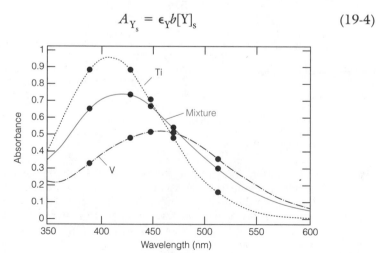

TABLE 19-1   Absorbance values and ratios for Figures 19-3 and 19-4

Wavelength (nm)	A_{X_s} titanium standard	A_{Y_s} vanadium standard	A_m mixture	A_m/A_{X_s}	A_{Y_s}/A_{X_s}
390	0.895	0.326	0.651	0.727_3	0.364_2
430	0.884	0.497	0.743	0.840_5	0.562_2
450	0.694	0.528	0.665	0.958_2	0.760_8
470	0.481	0.512	0.547	1.137_2	1.064_4
510	0.173	0.374	0.314	1.815_0	2.161_8

SOURCE: M. Blanco, H. Iturriaga, S. Maspoch, and P. Tarín, *J. Chem. Ed.* **1989**, *66*, 178.

Solving Equations 19-3 and 19-4 for ϵ_X and ϵ_Y, substituting these expressions into Equation 19-2, and doing a little algebra give

Analysis of a mixture when spectra overlap:

$$\frac{A_m}{A_{X_s}} = \underbrace{\frac{[Y]}{[Y]_s}\left(\frac{A_{Y_s}}{A_{X_s}}\right)}_{\text{Slope}} + \underbrace{\frac{[X]}{[X]_s}}_{\text{Intercept}} \qquad (19\text{-}5)$$

Analysis of a mixture:

plot A_m/A_{X_s} versus A_{Y_s}/A_{X_s}

slope = $[Y]/[Y]_s$

intercept = $[X]/[X]_s$

A graph of A_m/A_{X_s} versus A_{Y_s}/A_{X_s} (all of which we have measured) at various wavelengths has a slope of $[Y]/[Y]_s$ and an intercept of $[X]/[X]_s$. Because we know the concentrations $[X]_s$ and $[Y]_s$, we can find the concentrations $[X]$ and $[Y]$ in the unknown mixture.

Table 19-1 gives experimental absorbance values for the spectra in Figure 19-3. Figure 19-4 shows a graph of A_m/A_{X_s} versus A_{Y_s}/A_{X_s}. The least-squares straight line through the points is

$$\frac{A_m}{A_{X_s}} = (0.607)\frac{A_{Y_s}}{A_{X_s}} + 0.499$$

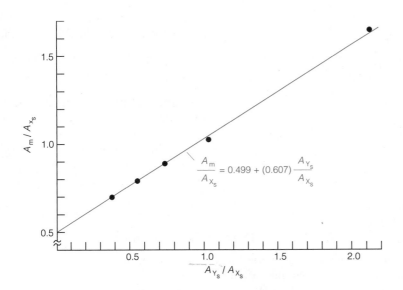

Figure 19-4   Graph of A_m/A_{X_s} versus A_{Y_s}/A_{X_s}, using data from Table 19-1.

from which we can say

$$\text{slope} = 0.607 = [Y]/[Y]_s$$

$$\Rightarrow [Y] = [V^{5+}] = (0.607)(1.89 \text{ mM}) = 1.15 \text{ mM}$$

$$\text{intercept} = 0.499 = [X]/[X]_s$$

$$\Rightarrow [X] = [Ti^{4+}] = (0.499)(1.32 \text{ mM}) = 0.659 \text{ mM}$$

In this method we used absorbances at many wavelengths to decompose the spectrum of the mixture into its two components. The best wavelength region to use is the one where the two individual spectra differ as much as possible in shape and where the errors in absorbance of the mixture are least.

What to Do When the Individual Spectra Are Well Resolved

Now we apply Equation 19-1 to a mixture of two components whose individual spectra do not overlap very much, as in Figure 19-2b. Suppose that species X has an absorbance maximum at wavelength λ' that is fairly well separated from the absorbance maximum of species Y at wavelength λ''. The absorbance at any wavelength is the sum of absorbances of each component at that wavelength. For the absorbance at wavelengths λ' and λ'', we can write

$$A' = \epsilon_X' b[X] + \epsilon_Y' b[Y] \qquad A'' = \epsilon_X'' b[X] + \epsilon_Y'' b[Y] \qquad (19\text{-}6)$$

where the ϵ values apply to each species at each wavelength. The absorptivities of X and Y at each wavelength must be measured in separate experiments.

We can solve the two Equations 19-6 for the two unknowns [X] and [Y]. The result is

The determinant $\begin{vmatrix} a & b \\ c & d \end{vmatrix}$ means $a \times d - b \times c$.

Analysis of a mixture when spectra are resolved:

$$[X] = \frac{\begin{vmatrix} A' & \epsilon_Y' b \\ A'' & \epsilon_Y'' b \end{vmatrix}}{\begin{vmatrix} \epsilon_X' b & \epsilon_Y' b \\ \epsilon_X'' b & \epsilon_Y'' b \end{vmatrix}} \qquad [Y] = \frac{\begin{vmatrix} \epsilon_X' b & A' \\ \epsilon_X'' b & A'' \end{vmatrix}}{\begin{vmatrix} \epsilon_X' b & \epsilon_Y' b \\ \epsilon_X'' b & \epsilon_Y'' b \end{vmatrix}} \qquad (19\text{-}7)$$

In Equation 19-7, each symbol $\begin{vmatrix} a & b \\ c & d \end{vmatrix}$ is called a *determinant*. It is a shorthand way of writing the product $a \times d$ minus the product $b \times c$. Thus the determinant $\begin{vmatrix} 1 & 2 \\ 3 & 4 \end{vmatrix}$ means $1 \times 4 - 2 \times 3 = -2$.

To analyze the mixture of two compounds in this case, it is necessary to measure the absorbances at two wavelengths and to know ϵ at each wavelength for each compound. Similarly, a mixture of n components can be analyzed by making n absorbance measurements at n wavelengths. Problem 6 gives a convenient template for solving simultaneous equations.

. . . .

E X A M P L E   Analysis of a Mixture, Using Equations 19-7

The molar absorptivities of compounds X and Y in the following table were measured with pure samples of each. A mixture of compounds X and Y in a 1.000-cm cell had an absorbance of 0.957 at 272 nm and 0.559 at 327 nm. Find the concentrations of X and Y in the mixture.

λ (nm)	ϵ (M^{-1} cm^{-1})	
	X	Y
272	16 440	3 870
327	3 990	6 420

Solution   Using Equations 19-7 and setting $b = 1.000$, we find

$$[X] = \frac{\begin{vmatrix} 0.957 & 3\,870 \\ 0.559 & 6\,420 \end{vmatrix}}{\begin{vmatrix} 16\,400 & 3\,870 \\ 3\,990 & 6\,420 \end{vmatrix}} = \frac{(0.957)(6\,420) - (3\,870)(0.559)}{(16\,400)(6\,420) - (3\,870)(3\,990)} = 4.43 \times 10^{-5} \text{ M}$$

$$[Y] = \frac{\begin{vmatrix} 16\,400 & 0.957 \\ 3\,990 & 0.559 \end{vmatrix}}{\begin{vmatrix} 16\,400 & 3\,870 \\ 3\,990 & 6\,420 \end{vmatrix}} = 5.95 \times 10^{-5} \text{ M}$$

· ·

Isosbestic Points

Often one absorbing species, X, is converted to another absorbing species, Y, during the course of a chemical reaction. This transformation leads to a very obvious and characteristic behavior, shown in Figure 19-5. If the spectra of

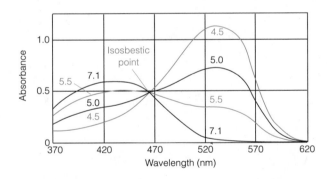

Figure 19-5   Absorption spectrum of 3.7×10^{-4} M methyl red as a function of pH between pH 4.5 and 7.1. [From E. J. King, *Acid-Base Equilibria* (Oxford: Pergamon Press, 1965).]

pure X and pure Y cross each other at any wavelength, then every spectrum recorded during this chemical reaction will cross at that same point, called an **isosbestic point**. *The observation of an isosbestic point during a chemical reaction is good evidence that only two principal species are present.*[3]

Consider methyl red, which is an indicator that changes between red (HIn) and yellow (In$^-$) near pH 5.1:

$$(CH_3)_2N - \bigcirc - N{=}\overset{+}{\underset{H}{N}} - \bigcirc {-}CO_2^- \quad \xrightarrow{\ pK_2 \approx 5.1\ } \quad (CH_3)_2N - \bigcirc - N{=}N - \bigcirc {-}CO_2^-$$

HIn
(red)

In$^-$
(yellow)

Because the spectra of HIn and In$^-$ (at the same concentration) happen to cross at 465 nm, all spectra in Figure 19-5 cross at this point. (If the spectra of HIn and In$^-$ crossed at several points, each would be an isosbestic point.)

To see why there is an isosbestic point, we write an equation for the absorbance of the solution at 465 nm:

$$A^{465} = \epsilon_{\text{HIn}}^{465} \, b[\text{HIn}] + \epsilon_{\text{In}^-}^{465} \, b[\text{In}^-] \qquad (19\text{-}8)$$

But because the spectra of pure HIn and pure In$^-$ (at the same concentration) cross at 465 nm, $\epsilon_{\text{HIn}}^{465}$ must be equal to $\epsilon_{\text{In}^-}^{465}$. Setting $\epsilon_{\text{HIn}}^{465} = \epsilon_{\text{In}^-}^{465} = \epsilon^{465}$, we can factor Equation 19-8:

$$A^{465} = \epsilon^{465} \, b([\text{HIn}] + [\text{In}^-]) \qquad (19\text{-}9)$$

> An isosbestic point occurs when $\epsilon_X = \epsilon_Y$ and $[X] + [Y]$ is constant.

In Figure 19-5 all solutions contain the same total concentration of methyl red (= [HIn] + [In$^-$]). Only the pH varies. Therefore, the sum of concentrations in Equation 19-9 is constant, and A^{465} is constant.

19-2 Measuring an Equilibrium Constant: The Scatchard Plot

To measure an equilibrium constant, we must measure the concentrations (actually activities) of the species involved in the equilibrium. Any physical property related to concentration or activity can be useful. We have seen how pH, other potentiometric measurements, and polarographic measurements can be used to find equilibrium constants. In this section, we will see how absorbance can be used to measure an equilibrium constant.[4]

> Because absorbance is proportional to *concentration* (not activity), concentrations must be converted to activities to get true equilibrium constants.

We confine our attention to the simplest equilibrium, in which the species P and X react to form PX.

$$P + X \rightleftharpoons PX \qquad (19\text{-}10)$$

Neglecting activity coefficients, we can write

$$K = \frac{[\text{PX}]}{[\text{P}][\text{X}]} \qquad (19\text{-}11)$$

Consider a series of solutions in which increments of X are added to a constant amount of P. Letting the total concentration of P (in the form P or PX) be called P_0, we can write

$$[\text{P}] = P_0 - [\text{PX}] \qquad (19\text{-}12)$$

> Clearing the cobwebs from your brain, you realize that Equation 19-12 is a mass balance.

Now the equilibrium expression, Equation 19-11, can be rearranged as follows:

$$\frac{[\text{PX}]}{[\text{X}]} = K[\text{P}] = K(P_0 - [\text{PX}]) \qquad (19\text{-}13)$$

> A *Scatchard plot* is a graph of [PX]/[X] versus [PX]. The slope is $-K$.

A graph of [PX]/[X] versus [PX] will have a slope of $-K$ and is called a **Scatchard plot.**[5] It is widely used to measure equilibrium constants, especially in biochemistry.

If we know [PX], we can find [X] with the mass balance

$$X_0 = [\text{total X}] = [\text{PX}] + [\text{X}]$$

To measure [PX], we might use spectrophotometric absorbance. Suppose that P and PX each have some absorbance at wavelength λ, but X has no absorbance at this wavelength. For simplicity, let all measurements be made in a cell of pathlength 1.000 cm. This condition will allow us to omit b (= 1.000 cm) when writing Beer's law.

The absorbance of the solution at some wavelength is the sum of absorbances of PX and P:

$$A = \epsilon_{PX}[PX] + \epsilon_P[P]$$

Substituting $[P] = P_0 - [PX]$, we can write

$$A = \epsilon_{PX}[PX] + \underbrace{\epsilon_P P_0}_{A_0} - \epsilon_P[PX] \qquad (19\text{-}14)$$

But $\epsilon_P P_0$ is A_0, the initial absorbance before any X is added. Rearranging Equation 19-14 gives

$$A = [PX](\epsilon_{PX} - \epsilon_P) + A_0 \Rightarrow [PX] = \frac{\Delta A}{\Delta \epsilon} \qquad (19\text{-}15)$$

where $\Delta\epsilon = \epsilon_{PX} - \epsilon_P$ and $\Delta A \ (= A - A_0)$ is the observed absorbance minus the initial absorbance for each point in the titration.

Substituting the expression for [PX] from Equation 19-15 into Equation 19-13 gives a useful result:

Scatchard equation:
$$\frac{\Delta A}{[X]} = K\,\Delta\epsilon P_0 - K\Delta A \qquad (19\text{-}16)$$

That is, a graph of $\Delta A/[X]$ versus ΔA should be a straight line with a slope of $-K$. In this way, absorbances measured while P is titrated with X can be used to find the equilibrium constant for the reaction of X with P.

Two cases commonly arise in the application of Equation 19-16. If the equilibrium constant is small, then large concentrations of X are needed to observe the formation of PX. Therefore, $X_0 \gg P_0$, and the concentration of unbound X in Equation 19-16 can be set equal to the total concentration, X_0. Alternatively, if K is not small, then [X] is not equal to X_0, and [X] must be measured. The best approach is to have an independent measurement of [X], either by measurement at another wavelength or by measurement of a different physical property.

In practice, the errors inherent in a Scatchard plot may be substantial and are sometimes overlooked. Defining the fraction of saturation of P as

$$\text{fraction of saturation} = S = \frac{[PX]}{P_0} \qquad (19\text{-}17)$$

it can be shown that the most accurate data are obtained for $0.2 \lesssim S \lesssim 0.8$.[6] Furthermore, data should be obtained throughout a range representing about 75% of the total saturation curve before it can be verified that the equilibrium (Equation 19-10) is obeyed. People have made mistakes by exploring too little of the binding curve and by not including the region $0.2 \lesssim S \lesssim 0.8$.

19-3 The Method of Continuous Variation

In the preceding section we considered the equilibrium

$$P + X \rightleftharpoons PX \qquad (19\text{-}18)$$

TABLE 19-2   Solutions for the method of continuous variation[a]

mL of 2.50 mM P	mL of 2.50 mM X	Mole ratio (X:P)	Mole fraction of X $\left(\dfrac{\text{mol X}}{\text{mol X + mol P}}\right)$
1.00	9.00	9.00:1	0.900
2.00	8.00	4.00:1	0.800
2.50	7.50	3.00:1	0.750
3.33	6.67	2.00:1	0.667
4.00	6.00	1.50:1	0.600
5.00	5.00	1.00:1	0.500
6.00	4.00	1:1.50	0.400
6.67	3.33	1:2.00	0.333
7.50	2.50	1:3.00	0.250
8.00	2.00	1:4.00	0.200
9.00	1.00	1:9.00	0.100

a. All solutions are diluted to a total volume of 25.0 mL with a buffer.

Suppose, however, that several complexes can form:

$$P + 2X \rightleftharpoons PX_2 \qquad (19\text{-}19)$$

$$P + 3X \rightleftharpoons PX_3 \qquad (19\text{-}20)$$

If one complex (say, PX_2) predominates, the **method of continuous variation** (also called *Job's method*) allows us to identify the stoichiometry of the predominant complex.

The classical procedure calls for mixing aliquots of equimolar solutions of P and X (perhaps followed by dilution to a constant volume) such that the total (formal) concentration of P + X remains constant. For example, stock solutions containing 2.50 mM P and 2.50 mM X could be mixed as shown in Table 19-2 to give various X:P ratios but a constant total concentration of 1.00 mM. The absorbance of each solution is then measured at a suitable wavelength, and a graph is made showing *corrected* absorbance (defined in Equation 19-21) versus mole fraction of X. *Maximum absorbance is reached at the composition corresponding to the stoichiometry of the predominant complex.*

The corrected absorbance is defined as the measured absorbance minus the absorbance that would be produced by free P and free X alone:

corrected absorbance = measured absorbance $- \epsilon_P b P_T - \epsilon_X b X_T$ (19-21)

where ϵ_P and ϵ_X are the molar absorptivities of pure P and pure X, b is the sample pathlength, and P_T and X_T are the total (formal) concentrations of P and X in the solution. For the first solution in Table 19-2, $P_T = (1.00/25.0)(2.50 \text{ mM}) = 0.100 \text{ mM}$ and $X_T = (9.00/25.0)(2.50 \text{ mM}) = 0.900 \text{ mM}$. If P and X do not absorb at the wavelength of interest, no absorbance correction is needed.

The maximum absorbance occurs at the mole fraction of X corresponding to the stoichiometry of the complex. If the predominant complex is PX_2, the maximum occurs at (mole fraction of X) = $2/(2 + 1) = 0.667$. If the predominant complex were P_3X, the maximum would occur at (mole fraction of X) = $1/(1 + 3) = 0.250$.

Precautions to take with this procedure include the following:

1.   Verify that the complex follows Beer's law.

For the reaction $P + nX \rightleftharpoons PX_n$, you could show that $[PX_n]$ reaches a maximum when the initial concentrations have the ratio $[X]_0 = n[P]_0$. To do this, write $K = [PX_n]/\{([P]_0 - [PX_n])([X]_0 - n[PX_n])\}$ and set the partial derivatives $\partial[PX_n]/\partial[P]_0$ and $\partial[PX_n]/\partial[X]_0$ equal to zero.

Method of continuous variation:

$$P + nX \rightleftharpoons PX_n$$

Maximum absorbance occurs when (mole fraction of X) = $n/(n + 1)$.

2. Use a constant ionic strength and pH, if applicable.

3. Take readings at more than one wavelength; the maximum should occur at the same mole fraction for each wavelength.

4. Do experiments at different total concentrations of P + X. If a second set of solutions were prepared in the proportions given in Table 19-2, but from stock concentrations of 5.00 mM, the maximum would still occur at the same mole fraction.

Although the method of continuous variation can be carried out with many separate solutions, like those in Table 19-2, it is more sensible to do one titration and plot the data in the format of Figure 19-6. Figure 19-6a shows results for the titration of EDTA with Cu^{2+}. In Figure 19-6b the abscissa has been transformed into mole fraction of Cu^{2+} (= [moles of Cu^{2+}]/[moles of Cu^{2+} + moles of EDTA]) instead of volume of Cu^{2+}. Each absorbance in the second curve is corrected for the absorbance of an equal amount of pure Cu^{2+} solution, which absorbs the wavelength of interest. EDTA is transparent at this wavelength.

The sharp maximum at a mole fraction of 0.5 indicates formation of a 1:1 complex. If the equilibrium constant is not large, the maximum is more curved than in Figure 19-6b. The curvature can be used to estimate the equilibrium constant.[7]

19-4 What Happens When a Molecule Absorbs Light?

In the next section we will study how *luminescence* (emission of photons) is used in analytical chemistry. First, we must discuss what happens to molecules when they absorb light. When a molecule absorbs a photon, the molecule is promoted to a more energetic *excited state*, as was shown in Figure 6-3. Conversely, when a molecule emits a photon, the energy of the molecule falls by an amount equal to the energy of the photon.

For a concrete example, consider formaldehyde in Figure 19-7a. In its ground state, the molecule is planar, with a double bond between carbon and oxygen. From the electron-dot description of formaldehyde, we expect two pairs of nonbonding electrons to be localized on the oxygen atom. The double bond consists of a sigma bond between carbon and oxygen and a pi bond made from the $2p_y$ (out-of-plane) atomic orbitals of carbon and oxygen.

Electronic States of Formaldehyde

Molecular orbitals describe the distribution of electrons in a molecule, just as *atomic orbitals* describe the distribution of electrons in an atom. In the molecular orbital diagram for formaldehyde in Figure 19-8, one of the nonbonding orbitals of oxygen is thoroughly mixed with the three sigma bonding orbitals. These four orbitals, labeled σ_1 through σ_4, are each occupied by a pair of electrons with opposite spin (spin quantum numbers = $+\frac{1}{2}$ and $-\frac{1}{2}$). At higher energy is an occupied pi bonding orbital (π), made of the p_y atomic orbitals of carbon and oxygen. The highest-energy occupied orbital is the nonbonding orbital (n), composed principally of the oxygen $2p_x$ atomic orbital. The lowest-energy unoccupied orbital is the pi antibonding orbital (π^*). An electron in this orbital produces repulsion, rather than attraction, between the carbon and oxygen atoms.

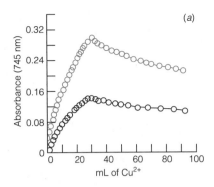

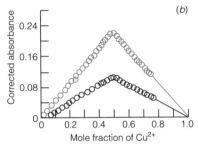

Figure 19-6 (*a*) Spectrophotometric titration of 30.0 mL of EDTA in acetate buffer with $CuSO_4$ in the same buffer. Upper curve: [EDTA] = [Cu^{2+}] = 5.00 mM. Lower curve: [EDTA] = [Cu^{2+}] = 2.50 mM. The absorbance has not been "corrected" in any way. (*b*) Transformation of data to mole fraction format. The absorbance of free $CuSO_4$ at the same formal concentration has been subtracted from each point in panel a. [From Z. D. Hill and P. MacCarthy, *J. Chem. Ed.* **1986,** *63,* 162.]

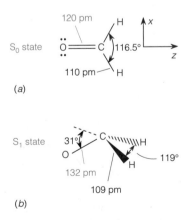

Figure 19-7 Geometry of formaldehyde. (*a*) Ground state. (*b*) Lowest excited singlet state.

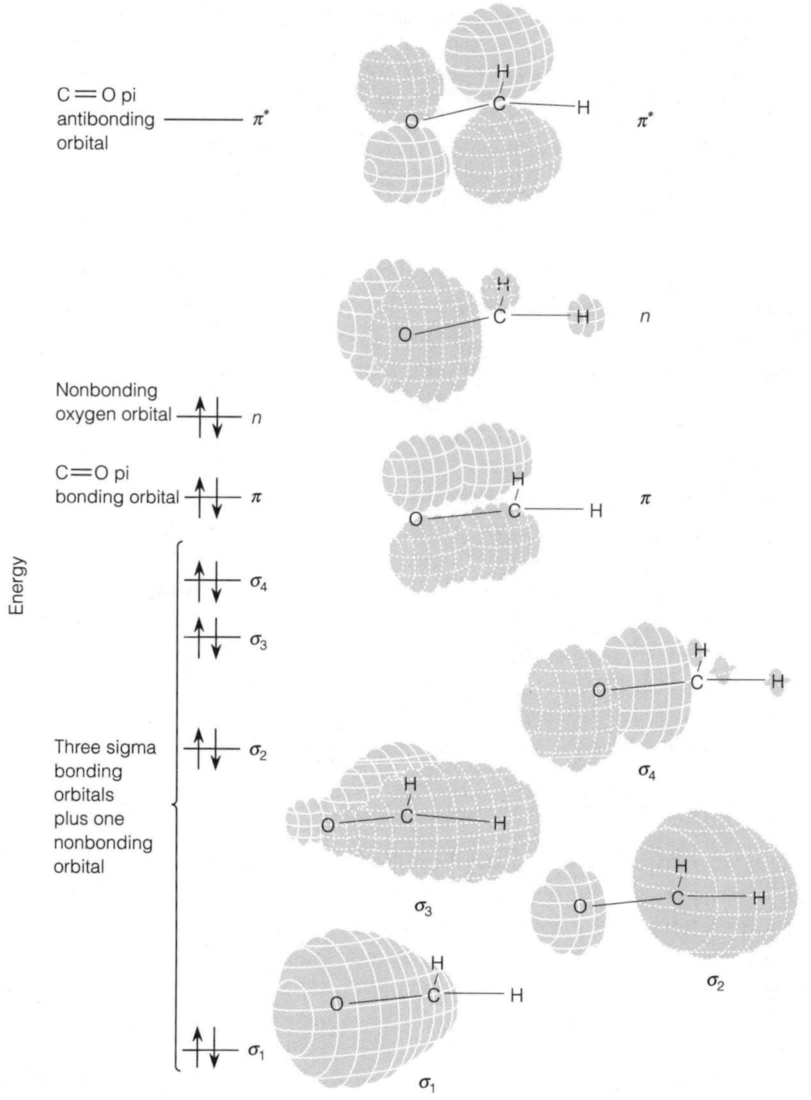

C=O pi
antibonding ——— π^*
orbital

Nonbonding
oxygen orbital ⊥↑↓⊥ n

C=O pi
bonding orbital ⊥↑↓⊥ π

Energy

⊥↑↓⊥ σ_4

⊥↑↓⊥ σ_3

Three sigma ⊥↑↓⊥ σ_2
bonding
orbitals
plus one
nonbonding
orbital

⊥↑↓⊥ σ_1

π^*

n

π

σ_4

σ_3

σ_2

σ_1

Figure 19-8   Molecular orbital dia-gram of formaldehyde, showing en-ergy levels and orbital shapes. The coordinate system was shown in the previous figure. [From W. L. Jor-gensen and L. Salem, *The Organic Chemist's Book of Orbitals* (New York: Academic Press, 1973).]

The terms *singlet* and *triplet* are used because a triplet state is split into three slightly different energy levels in the presence of a magnetic field, but a singlet state is not split.

The shorter the wavelength of light, the greater the energy.

In an **electronic transition,** an electron from one molecular orbital moves to another orbital, with a concomitant increase or decrease in the energy of the molecule. The lowest-energy electronic transition of formalde-hyde promotes a nonbonding (n) electron to the antibonding pi orbital (π^*).[8] There are actually two possible transitions, depending on the spin quantum numbers in the excited state (Figure 19-9). The state in which the spins are opposed is called a **singlet state.** If the spins are parallel, we call the excited state a **triplet state.**

The lowest-energy excited singlet and triplet states are called S_1 and T_1, respectively. In general, T_1 has lower energy than S_1. In formaldehyde, the transition $n \rightarrow \pi^*(T_1)$ requires the absorption of visible light with a wavelength of 397 nm. The $n \rightarrow \pi^*(S_1)$ transition occurs when ultraviolet radiation with a wavelength of 355 nm is absorbed.

With an electronic transition near 397 nm, you might expect solutions of formaldehyde to be green-yellow (Table 6-1) in appearance. In fact, formaldehyde is colorless, because the probability of undergoing any transition between singlet and triplet states (such as $n(S_0) \rightarrow \pi^*(T_1)$) is exceedingly small. The solution absorbs so little light at 397 nm that our eyes do not detect any absorbance at all. Singlet-to-singlet transitions such as $n(S_0) \rightarrow \pi^*(S_1)$ are much more probable, and the ultraviolet absorption is more intense.

Although formaldehyde is planar in its ground state, it has a pyramidal structure in both the S_1 (Figure 19-7) and T_1 excited states. Promotion of a nonbonding electron to an antibonding C—O orbital lengthens the C—O bond and changes the molecular geometry.

Vibrational and Rotational States of Formaldehyde

Absorption of visible or ultraviolet radiation promotes electrons to higher-energy orbitals in formaldehyde. Infrared and microwave radiation are not energetic enough to induce electronic transitions, but they can change the vibrational or rotational motion of the molecule.

Each atom of formaldehyde can move along three axes in space, so the entire molecule can move in $4 \times 3 = 12$ different ways. Three of these motions correspond to translation of the entire molecule in the x, y, and z directions. Another three motions correspond to rotation about the x, y, and z axes of the molecule. The remaining six motions represent vibrations of the molecule shown in Figure 19-10.

When formaldehyde absorbs an infrared photon with a wavenumber of $1\,251\ cm^{-1}$, the asymmetric bending vibration in Figure 19-10 is stimulated. Oscillations of the atoms are increased in amplitude, and the energy of the molecule increases.

. . . .

E X A M P L E   **Relation of Energy to Wavenumber**

By how many kilojoules per mole is the energy of formaldehyde increased when $1\,251\ cm^{-1}$ radiation is absorbed?

Solution   From Chapter 6, you should know that

$$\Delta E = h\nu = h\frac{c}{\lambda} = hc\tilde{\nu} \quad (\text{because } \tilde{\nu} = \frac{1}{\lambda})$$

where h is Planck's constant, ν is frequency, c is the speed of light, $\tilde{\nu}$ is wavenumber, and λ is wavelength. Plugging in the numbers gives

$$\Delta E = (6.626\,1 \times 10^{-34}\ J \cdot s)(2.997\,9 \times 10^8\ m/s)(1\,251\ cm^{-1})(100\ cm/m)$$

$$= 2.485 \times 10^{-20}\ J/molecule = 14.97\ kJ/mol$$

. . . .

Figure 19-10   The six kinds of vibrations of formaldehyde. The wavenumber of each vibration is given in units of cm^{-1}. These are the wavenumbers of infrared radiation needed to stimulate each kind of motion.

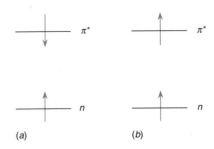

(a)   (b)

Figure 19-9   Diagram of two possible electronic states arising from an $n \rightarrow \pi^*$ transition. (a) Excited singlet state, S_1. (b) Excited triplet state, T_1.

A nonlinear molecule with n atoms has $3n - 6$ vibrational modes and three possible rotations. A linear molecule can rotate about only two axes; it therefore has $3n - 5$ vibrational modes and two rotations.

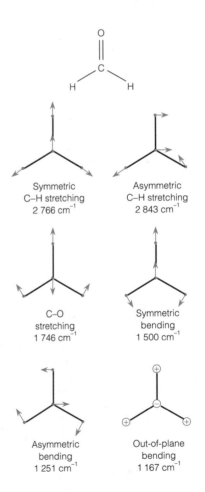

Symmetric C–H stretching $2\,766\ cm^{-1}$

Asymmetric C–H stretching $2\,843\ cm^{-1}$

C–O stretching $1\,746\ cm^{-1}$

Symmetric bending $1\,500\ cm^{-1}$

Asymmetric bending $1\,251\ cm^{-1}$

Out-of-plane bending $1\,167\ cm^{-1}$

Your microwave oven heats food by transferring rotational energy to water molecules in the food.

The spacings between rotational energy levels of a molecule are even smaller than vibrational energy spacings. A molecule in the rotational ground state could absorb microwave photons with energies of 0.029 07 or 0.087 16 kJ/mol (wavelengths of 4.115 or 1.372 mm) to be promoted to the two lowest excited states. Absorption of microwave radiation leads to rotational excitation of a molecule, in which the molecule rotates faster than it does in its ground state.

Combined Electronic, Vibrational, and Rotational Transitions

In general, when a molecule absorbs light having sufficient energy to cause an electronic transition, **vibrational** and **rotational transitions**—that is, changes in the vibrational and rotational states—can occur as well. Thus, for example, formaldehyde can absorb one photon with just the right energy to cause the following changes simultaneously: (1) a transition from the S_0 to the S_1 electronic state; (2) a change in vibrational energy from the ground vibrational state of S_0 to an excited vibrational state of S_1; and (3) a transition from one rotational state of S_0 to a different rotational state of S_1.

Vibrational transitions usually involve simultaneous rotational transitions.

Electronic transitions usually involve simultaneous vibrational and rotational transitions.

Electronic absorption bands are usually very broad (Color Plate 2) because many different vibrational and rotational levels are available at slightly different energies. A molecule can absorb photons with a wide range of energies and still be promoted from the ground electronic state to one particular excited electronic state.

What Happens to Absorbed Energy?

Suppose that absorption promotes the molecule from the ground electronic state, S_0, to a vibrationally and rotationally excited level of the excited electronic state S_1 (Figure 19-11). Usually, the first process following this absorption is *vibrational relaxation* to the ground vibrational level of S_1. This radiationless transition is labeled R_1 in Figure 19-11. The vibrational energy lost in this relaxation is transferred to other molecules (solvent, for example) through collisions. The net effect is to convert part of the energy of the absorbed photon into heat spread throughout the entire medium.

Figure 19-11 Diagram illustrating some of the physical processes that can occur after a molecule absorbs a photon. S_0 is the ground electronic state of the molecule. S_1 and T_1 are the lowest excited singlet and triplet states, respectively. Straight arrows represent processes involving photons, and wavy arrows represent radiationless transitions. A, absorption; F, fluorescence; P, phosphorescence; IC, internal conversion; ISC, intersystem crossing; R, vibrational relaxation.

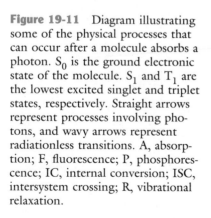

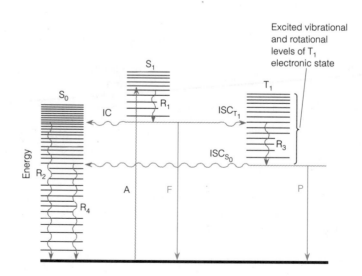

From the S_1 level, several events can happen. The molecule could enter a very highly excited vibrational state of S_0 having the same energy as S_1. This is called *internal conversion*. From this excited state, the molecule can relax back to the ground vibrational state and transfer its energy to neighboring molecules through collisions. This radiationless process is labeled R_2. If a molecule follows the path A-R_1-IC-R_2 in Figure 19-11, the entire energy of the photon will have been converted to heat.

Alternatively, the molecule could cross from S_1 into an excited vibrational level of T_1. Such an event is known as *intersystem crossing*. Following the radiationless vibrational relaxation R_3, the molecule finds itself at the lowest vibrational level of T_1. From here, the molecule might undergo a second intersystem crossing to S_0, followed by the radiationless relaxation R_4. All of the processes mentioned so far have the net effect of converting light to heat.

By contrast, from S_1 or T_1, a molecule could relax to S_0 by emitting a photon. The transition $S_1 \rightarrow S_0$ is called **fluorescence**, and the transition $T_1 \rightarrow S_0$ is called **phosphorescence**. The relative rates of internal conversion, intersystem crossing, fluorescence, and phosphorescence depend on the molecule, the solvent, and conditions such as temperature and pressure.

Fluorescence and phosphorescence are relatively rare. The *lifetime* of fluorescence is always very short (10^{-8} to 10^{-4} s). If a molecule is *not* to fluoresce, internal conversion or intersystem crossing must be even more rapid. The lifetime of phosphorescence is much longer (10^{-4} to 10^2 s). This means that phosphorescence is even rarer than fluorescence, because a molecule in the T_1 state has a good chance of undergoing intersystem crossing to S_0 before phosphorescence can occur.

In molecules containing transition metals, electronic states other than singlets and triplets are possible. Emission from a transition metal complex is usually called simply **luminescence**, which makes no distinction between fluorescence and phosphorescence. In the next section, luminescence refers to emission of light by any mechanism from any type of molecule.

Internal conversion is a radiationless transition between states with the same spin quantum numbers (e.g., $S_1 \rightarrow S_0$).

Intersystem crossing is a radiationless transition between states with different spin quantum numbers (e.g., $T_1 \rightarrow S_0$).

Fluorescence is a radiational transition between states with the same spin quantum numbers (e.g., $S_1 \rightarrow S_0$).
Phosphorescence is a radiational transition between states with different spin quantum numbers (e.g., $T_1 \rightarrow S_0$).

The *lifetime* of a state is the time needed for the population of that state to decay to $1/e$ of its initial value, where e is the base of natural logarithms.

19-5 Luminescence

Luminescence measurements are inherently more sensitive than absorption measurements. Imagine yourself in a stadium at night with the lights off, but each of the 50 000 raving fans is holding a lighted candle. If 500 people blow out their candles, you will hardly notice the difference. Now imagine that the stadium is completely dark, but 500 people suddenly light their candles. In this case, the effect is dramatic. The first example is analogous to changing the transmittance from 100% to 99%, which is equivalent to an absorbance of $-\log 0.99 = 0.0044$. It is hard to measure such a small absorbance because the background is so bright. The second example is analogous to observing fluorescence from 1% of the molecules in a sample. Against the black background, this fluorescence is easy to detect.

Relation Between Absorption and Emission Spectra

In general, fluorescence and phosphorescence are observed at a lower energy than that of the absorbed radiation (the *excitation* energy). That is, radiation emitted by molecules is of longer wavelength than that of the radiation they absorb. A typical example is shown in Figure 19-12. Let's try to understand why emission occurs at lower energy, why there is so much structure in

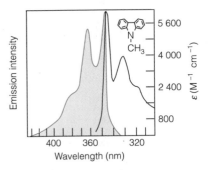

Figure 19-12 Absorption (black line) and emission (colored line) spectra of N-methlycarbazole in cyclohexane solution, illustrating the approximate mirror image relationship between absorption and emission. [From I. B. Berlman, *Handbook of Fluorescence Spectra of Aromatic Molecules* (New York: Academic Press, 1971).]

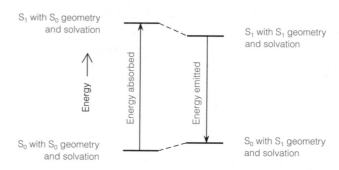

Figure 19-13   Diagram showing why the absorbed energy is greater than the emitted energy.

Figure 19-12, and why the emission spectrum is the approximate mirror image of the absorption spectrum.

A molecule absorbing radiation is initially in its electronic ground state, S_0. This molecule possesses a certain geometry and solvation. Suppose that the excited state is S_1. When radiation is first absorbed, the excited molecule still possesses its S_0 geometry and solvation (Figure 19-13). Very shortly after the excitation, the geometry and solvation revert to their most favorable values for the S_1 state. This must lower the energy of the excited molecule. When an S_1 molecule fluoresces, it returns to the S_0 state but retains the S_1 geometry and solvation. This unstable configuration must have a higher energy than that of an S_0 molecule with S_0 geometry and solvation. As shown in Figure 19-13, the net effect is that the emission energy is less than the excitation energy.

Figure 19-14 explains the structure in a spectrum and shows why the emission spectrum is roughly the mirror image of the absorption spectrum. In the absorption spectrum, wavelength λ_0 corresponds to a transition from the ground vibrational level of S_0 to the ground vibrational level of S_1. Absorption maxima at higher energy (shorter wavelength) correspond to the S_0 to S_1 transition accompanied by absorption of one or more quanta of vibrational energy. In polar solvents, the vibrational structure is often broadened beyond recognition, and only a broad envelope of absorption is observed. In Figure 19-12, the solvent is cyclohexane, which is nonpolar, and the vibrational structure is easily seen.

Following absorption, the vibrationally excited S_1 molecule relaxes back to the ground vibrational level of S_1 prior to emitting any radiation. As shown in Figure 19-14, emission from S_1 can go to any of the vibrational levels of S_0. The highest-energy transition comes at wavelength λ_0, with a series of peaks following at longer wavelength. The absorption and emission spectra will have an approximate mirror image relationship if the spacings between vibrational levels are roughly equal and if the transition probabilities are similar.[9]

Excitation and Emission Spectra

The general outline of an emission experiment is shown in Figure 19-15. An excitation wavelength is selected by one monochromator, and luminescence is observed through a second monochromator, usually positioned at 90° to the incident light. If we hold the excitation wavelength (λ_{ex}) fixed and scan

Electronic transitions are so fast, relative to nuclear motion, that each atom has nearly the same position and momentum before and after a transition. This is called the *Franck-Condon principle*.

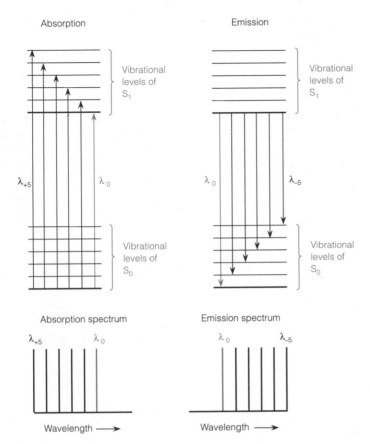

Figure 19-14 Energy-level diagram showing why structure is seen in the absorption and emission spectra, and why the spectra are roughly mirror images of each other. In absorption, wavelength λ_0 comes at lowest energy, and λ_{+5} is at highest energy. In emission, wavelength λ_0 comes at highest energy, and λ_{-5} is at lowest energy.

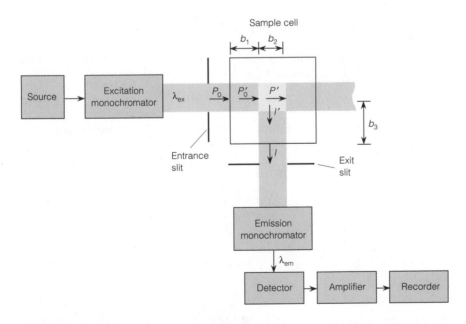

Figure 19-15 Block diagram of a fluorescence spectrophotometer, with the sample cell magnified to show various pathlengths.

Emission spectrum: constant λ_{ex} and variable λ_{em}

Excitation spectrum: variable λ_{ex} and constant λ_{em}

through the emitted radiation, an **emission spectrum** is produced. An emission spectrum is a graph of emission intensity versus emission wavelength.

If the emission wavelength (λ_{em}) is held constant and the excitation wavelength is varied, an **excitation spectrum** is produced. An excitation spectrum is a graph of emission intensity versus excitation wavelength. *An excitation spectrum looks very much like an absorption spectrum, because the greater the absorbance at the excitation wavelength, the more molecules are promoted to the excited state and the more emission will be observed.*

In either type of emission spectroscopy, we measure the intensity of emitted radiation, rather than the fraction of radiant power striking the detector. Because the response of any detector varies with wavelength, the recorded emission spectrum is not a true profile of emission intensity versus emission wavelength. For analytical measurements employing a single emission wavelength, this effect is inconsequential. If a true profile is required, it is necessary to calibrate the detector for the wavelength dependence of its response.

Emission Intensity

To derive a relation between the incident radiant power and the emission intensity, consider the sample cell in Figure 19-15. We expect the emission intensity to be proportional to the radiant power absorbed by the sample. That is, a certain proportion of the absorbed radiation will appear as emission under a given set of conditions (solvent, temperature, etc.). The exit slit in Figure 19-15 is set to observe emission from a region whose width is b_2.

Let the incident radiant power (W/m^2, also called *intensity*) striking the cell be called P_0. Some of this is absorbed by the sample over the pathlength b_1 in Figure 19-15, so the radiant power striking the central region of the cell is

Equation 19-22 follows from Beer's law:

$$\epsilon_{ex}b_1c = A = \log\left(\frac{P_0}{P_0'}\right)$$
$$\Rightarrow P_0' = P_0 10^{-\epsilon_{ex}b_1c}$$

$$\text{power striking central region} = P_0' = P_0 \cdot 10^{-\epsilon_{ex}b_1c} \qquad (19\text{-}22)$$

where ϵ_{ex} is the molar absorptivity at the wavelength λ_{ex}. The radiant power of the beam when it has traveled the additional distance b_2 is

$$P' = P_0' \cdot 10^{-\epsilon_{ex}b_2c} \qquad (19\text{-}23)$$

The emission intensity I is proportional to the radiant power absorbed in the central region of the cell:

$$\text{emission intensity} = I' = k'(P_0' - P') \qquad (19\text{-}24)$$

where k' is a constant of proportionality dependent on the emitting molecule and the conditions. Not all the radiation emitted from the center of the cell in the direction of the exit slit is observed. Some is absorbed by the solution between the center and the edge of the cell. The emission intensity I emerging from the cell is given by Beer's law:

$$I = I' \cdot 10^{-\epsilon_{em}b_3c} \qquad (19\text{-}25)$$

where ϵ_{em} is the molar absorptivity at the emission wavelength and b_3 is the distance from the center to the side of the cell (Figure 19-15).

Combining Equations 19-24 and 19-25 gives an expression for emission intensity:

$$I = k'(P_0' - P')10^{-\epsilon_{em}b_3 c} \qquad (19\text{-}26)$$

Substituting values of P_0' and P' from Equations 19-22 and 19-23, we obtain a relation between the incident radiant power and the emission intensity:

$$I = k'(P_0 \cdot 10^{-\epsilon_{ex}b_1 c} - P_0 \cdot 10^{-\epsilon_{ex}b_1 c} \cdot 10^{-\epsilon_{ex}b_2 c})\, 10^{-\epsilon_{em}b_3 c}$$

$$= k'P_0 \cdot \underbrace{10^{-\epsilon_{ex}b_1 c}}_{\substack{\text{Loss of intensity}\\\text{in region 1}}} \cdot \underbrace{(1 - 10^{-\epsilon_{ex}b_2 c})}_{\substack{\text{Emission is}\\\text{proportional to}\\\text{absorption of light}\\\text{in region 2}}} \cdot \underbrace{10^{-\epsilon_{em}b_3 c}}_{\substack{\text{Loss of intensity}\\\text{in region 3}}} \qquad (19\text{-}27)$$

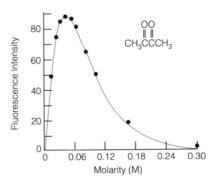

Figure 19-16 Concentration dependence of fluorescence intensity of biacetyl in CCl_4 with $\lambda_{ex} = 422$ nm and $\lambda_{em} = 464$ nm. [From G. Henderson, *J. Chem. Ed.* **1977,** *54,* 57.]

Equation 19-27 allows us to calculate the emission intensity as a function of solute concentration. At low concentration, the emission intensity increases with increasing concentration of analyte, because absorption is small and emission is proportional to emitter concentration. At high concentration, the emission intensity actually decreases, because the absorption increases more rapidly than the emission (Figure 19-16). We say the emission is *quenched* (decreased) by self-absorption. At high concentration, even the *shape* of the emission spectrum can change, because absorption and emission both depend on wavelength.

When the absorbance terms (the exponents) in Equation 19-27 are small, the equation can be greatly simplified. A modest derivation whose details we omit leads to the result:

Relation of emission intensity to concentration:
$$I = kP_0 c \qquad (19\text{-}28)$$

where $k = k'\epsilon_{ex}b_2 \ln 10$. That is, *when the absorbance is small, the emission intensity is directly proportional to the sample concentration (c) and to the incident radiant power (P_0).*

For most analytical applications, Equation 19-28 is obeyed and emission intensity is proportional to concentration. The linear relationship between I and P_0 means that doubling the incident radiant power will double the emission intensity (up to a point). In contrast, doubling P_0 has no effect on absorbance, which is a *ratio* of two intensities.

In analytical experiments, absorbance is low, and emission intensity is proportional to the incident radiant power and to the concentration of analyte: $I = kP_0 c$.

Luminescence in Analytical Chemistry

Some analytes are naturally fluorescent and can be analyzed directly. A typical procedure involves establishing a working curve of luminescence intensity versus analyte concentration. (Blank samples invariably scatter light and must be run in every analysis.) Among the more important naturally fluorescent compounds are riboflavin (vitamin B_2),[10] many drugs, polycyclic aromatic compounds (an important class of carcinogens), and proteins.

Most compounds are not naturally luminescent enough to be analyzed directly. However, coupling to a fluorescent moiety provides an easy route to sensitive analyses. *Fluorescein* is a strongly fluorescent compound that can be coupled to many molecules for analytical purposes. Box 19-1 describes how fluorescent labels are used for sequencing DNA. Fluorescent labeling of fingerprints is a powerful tool in forensic analysis.[11]

Riboflavin (vitamin B_2)

Box 19-1   DNA Sequencing with Fluorescent Labels

Genetic information is coded in the sequence of nucleotides in deoxyribonucleic acid (DNA). Biologists are beginning the arduous task of mapping human genes by learning the sequence of nucleotides in our entire set of chromosomes. This, in turn, will shed light at a molecular level on how living organisms operate. Fluorescent labels are central to automated DNA sequencing.[12]

DNA is made of two very long strands wrapped around each other to form a helix. The two strands are connected by hydrogen bonds as shown below.

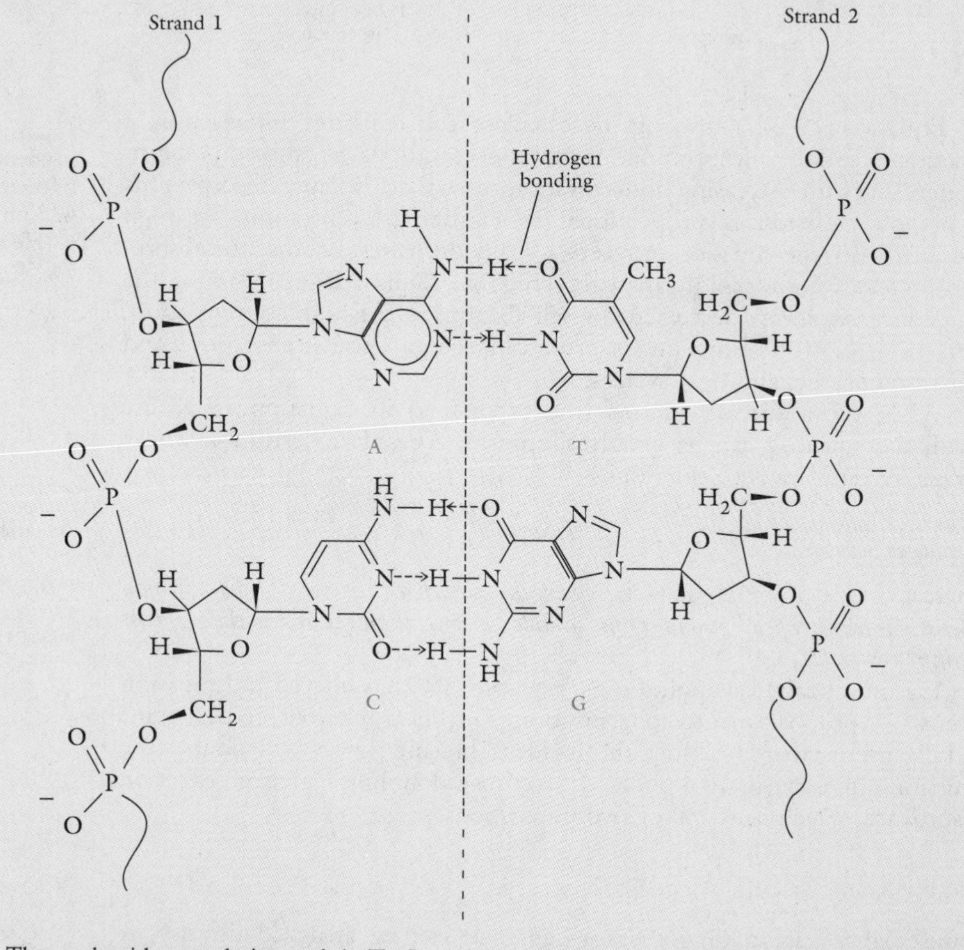

The nucleotides are designated A, T, C, and G, the letters standing for adenine, thymine, cytosine, and guanosine. A and T are always hydrogen bonded to each other, as are C and G. Sequencing begins by separating the two strands and selecting one.

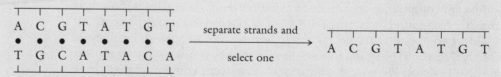

The single strand is then used as a *template* for constructing a new strand with a complementary sequence. For example, A on the original strand will give T on the complementary strand, and G on the original strand will give C on the complementary

Box 19-1 (continued)

strand. The enzyme DNA polymerase carries out the synthesis of the new strand, using the template strand and a supply of the four building blocks, designated dATP, dTTP, dCTP, and dGTP.

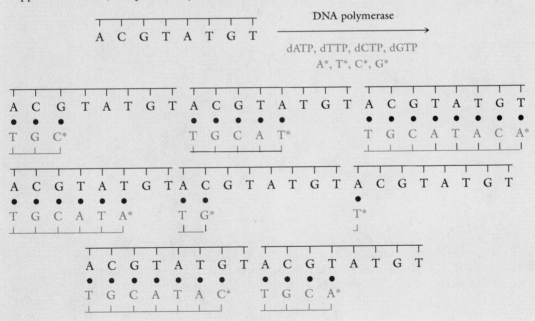

For DNA sequencing, *chain-terminating building blocks* are mixed with the supply of normal building blocks. The chain-terminating building blocks are similar to dATP, dTTP, dCTP, and dGTP, but they lack a hydroxyl group necessary to make the next bond in the chain.

When one of these chain-terminating building blocks, designated A*, T*, C*, and G*, is incorporated into the growing strand of DNA, no further growth can occur. Because this happens at random, complementary strands of DNA of every possible length are synthesized:

Box 19-1 (continued)

When the strands are separated again, pieces of DNA of every possible length (from one to eight nucleotides) are present, and each ends with one chain-terminating building block.

For sequencing DNA, the chain-terminating building blocks are labeled with fluorescent side groups. An example is G-505, whose side chain has a fluorescence maximum at 505 nm.

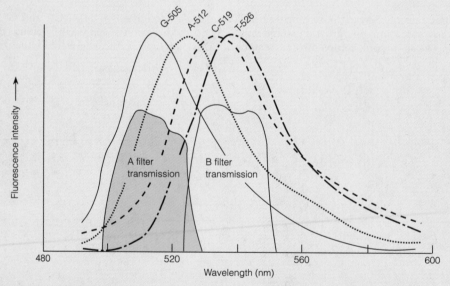

The side groups for the four terminators are selected so that they fluoresce at slightly different wavelengths.

Fluorescence spectra of chain-terminating building blocks used for DNA sequencing. [From J. M. Prober, G. L. Trainor, R. J. Dam, F. W. Hobbs, C. W. Robertson, R. J. Zagursky, A. J. Cocuzza, M. A. Jensen, and K. Baumeister, *Science* **1987,** *238,* 336.]

Box 19-1 (continued)

The mixture of terminated DNA strands now contains pieces of every possible length. Each one ending with G fluoresces at 505 nm. Each one terminated by A fluoresces at 512 nm. C- and T-terminated fragments fluoresce at 519 and 526 nm, respectively. By viewing the fluorescence through the two filters whose transmission is shown at the lower left, it is possible to determine which terminator is being observed. For example, a fragment ending in T* will have maximum fluorescence with the B filter and minimum transmission with the A filter. A fragment ending with A* will have about equal transmission through each filter.

The mixture of fragments is separated by *electrophoresis,* in which a strong electric field causes the negatively charged strands to migrate toward the positive pole. The longest strands are the slowest, giving a pattern that looks like this:

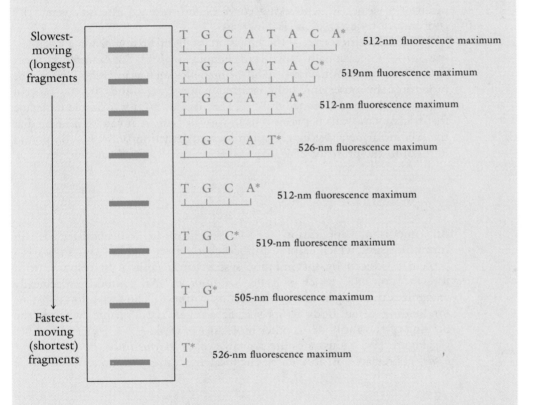

Slowest-moving (longest) fragments

T G C A T A C A*   512-nm fluorescence maximum

T G C A T A C*   519nm fluorescence maximum

T G C A T A*   512-nm fluorescence maximum

T G C A T*   526-nm fluorescence maximum

T G C A*   512-nm fluorescence maximum

T G C*   519-nm fluorescence maximum

T G*   505-nm fluorescence maximum

T*   526-nm fluorescence maximum

Fastest-moving (shortest) fragments

The bands in the electrophoretic experiment are invisible. However, when excited by a 488-nm laser, each fluoresces a color determined by its terminator. By scanning the gel from one end to the other with the laser, and observing the fluorescence through the two filters, it is possible to read off the exact sequence of nucleotides in the DNA fragment. By this means we hope to learn the sequence of the entire human genome, which consists of 46 chromosomes divided into 10^5 genes comprising $10^{9.5}$ nucleotides!

Fluorescein

Calcein

Chelating
aminodiacetate
group

Alizarin garnet R
(a fluorescent chelating agent)

Luminol
(used in chemiluminescent assay
of oxidizing agents)

Metal ions can be analyzed following their reaction with a fluorescent chelating agent. For example, a derivative of fluorescein called calcein (shown above) forms a fluorescent complex with calcium. Fluoride ion can be analyzed because of its ability to *quench* the fluorescence of the Al^{3+} complex of alizarin garnet R. A working curve of fluorescence intensity versus F^- concentration decreases as $[F^-]$ increases.

Chemiluminescence—the emission of light arising from a chemical reaction—is occasionally valuable in chemical analysis. For example, changes in Ca^{2+} concentration within cellular organelles such as mitochondria can be monitored by observing light emitted when Ca^{2+} binds to the protein aequorin, obtained from jellyfish.[13] Nitric oxide (NO) is a simple molecule that transmits signals between cells in various tissues. It can be measured at the picomolar level by its chemiluminescent reaction with the compound luminol in the presence of H_2O_2.[14]

19-6 Immunoassays

An important application of fluorescence in biochemistry is in **immunoassays**, which employ antibodies to detect analyte. An *antibody* is a protein produced by the immune system of an animal in response to a foreign molecule, which is called an *antigen*. An antibody specifically recognizes the antigen that stimulated its synthesis. The formation constant for binding of antibody to its specific antigen is very large, whereas the binding of the antibody to other molecules is weak.

Figure 19-17 illustrates the principle of an *enzyme-linked immunosorbent assay*, abbreviated ELISA in biochemical literature. Antibody 1, which is

1. Add sample containing analyte
2. Wash to remove unbound molecules

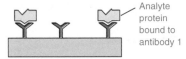

Analyte
protein
bound to
antibody 1

3. Add enzyme-labeled antibody 2
4. Wash to remove unbound antibody

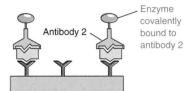

Antibody 2

Enzyme
covalently
bound to
antibody 2

Figure 19-17 Enzyme-linked immunosorbent assay. Antibody 1, which is specific for the analyte of interest, is bound to a polymer support and treated with unknown. After washing away excess, unbound molecules, the analyte remains bound to antibody 1. The bound analyte is then treated with antibody 2, which recognizes a different site on the analyte and to which an enzyme is covalently attached. After washing away unbound material, each molecule of analyte is coupled to an enzyme that will be used in Figure 19-18.

specific for the analyte of interest (the antigen), is bound to a polymeric support. In Steps 1 and 2, analyte is incubated with the polymer-bound antibody to form the antibody-antigen complex. The fraction of antibody sites that bind analyte is proportional to the concentration of analyte in the unknown. The surface is then washed to remove unbound substances. In Steps 3 and 4, the antibody-antigen complex is treated with antibody 2, which recognizes a different region of the analyte. (Prior to Step 3 an enzyme that will be used later was covalently attached to antibody 2.) Again, excess unbound substances are washed away.

The enzyme attached to antibody 2 is vital for quantitative analysis. Figure 19-18 shows two ways in which the enzyme may be used. At the top, the enzyme transforms a colorless reactant into a colored product. Because one enzyme molecule catalyzes the same reaction many times, many molecules of colored product are created for each molecule of antigen. The enzyme thereby *amplifies* the signal in the chemical analysis. The higher the concentration of analyte in the original unknown, the more enzyme is bound and the greater the extent of the enzyme-catalyzed reaction. At the bottom of Figure 19-18, the enzyme converts a nonfluorescent reactant into a fluorescent product. Colorimetric and fluorometric enzyme-linked immunosorbent assays are sensitive to less than a nanogram of analyte. Pregnancy tests are based on the immunoassay of a placental protein in urine.

Time-Resolved Fluorescence Immunoassays[15]

The sensitivity of fluorescence immunoassays can be enhanced by a factor of 100 (to detect 10^{-13} M analyte) with time-resolved measurements of luminescence from the lanthanide ion Eu^{3+}. Organic chromophores such as fluorescein are plagued by large background fluorescence from solvent, solutes, and particles. Figure 19-19 shows that fluorescence from common background sources typically decays to a negligible level within 100 μs after excitation by a short flash of laser light. Luminescence from Eu^{3+}, however,

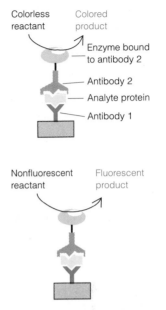

Figure 19-18 Enzyme bound to antibody 2 can catalyze reactions that produce colored or fluorescent products. Each molecule of analyte bound in the immunoassay leads to many molecules of colored or fluorescent product, which are easily measured.

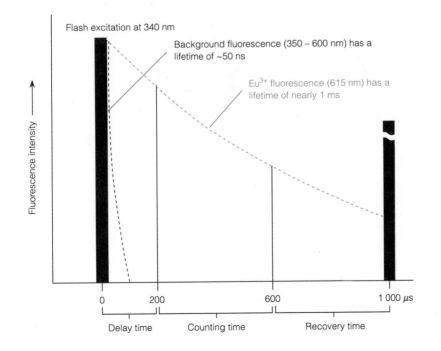

Flash excitation at 340 nm

Background fluorescence (350 – 600 nm) has a lifetime of ~50 ns

Eu^{3+} fluorescence (615 nm) has a lifetime of nearly 1 ms

Fluorescence intensity

0   200   600   1 000 μs

Delay time   Counting time   Recovery time

Figure 19-19 Emission intensity in a time-resolved fluorescence experiment. The sample is excited with a brief pulse of ultraviolet light at time = 0. Broadband background emission decays to near zero during the delay time of 200 μs. Long-lived Eu^{3+} luminescence is measured between 200 and 600 μs. The next pulse of ultraviolet light is flashed at 1 000 μs, and the cycle is repeated approximately 1 000 times per second.

Figure 19-20 (*a*) Antibody 2 in the immunosorbent assay of Figure 19-17 can be labeled with a Eu^{3+} ion that is not strongly luminescent when it is immobilized on the antibody. To complete the analysis, the pH of the solution is lowered to liberate a strongly luminescent Eu^{3+} ion. (*b*) In a more sensitive experiment, antibody 2 is labeled with several biotin molecules. Each biotin molecule then binds one molecule of the protein streptavidin, to which are bound several immobilized Eu^{3+} ions. Upon lowering the pH, strong luminescence is observed.

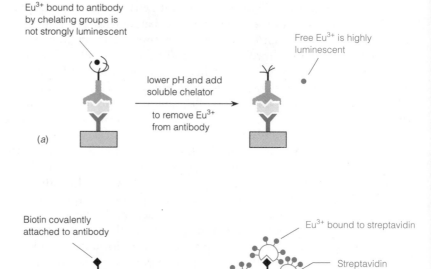

Eu^{3+} bound to antibody by chelating groups is not strongly luminescent

lower pH and add soluble chelator

to remove Eu^{3+} from antibody

Free Eu^{3+} is highly luminescent

(*a*)

Biotin covalently attached to antibody

add Eu^{3+}–labeled streptavidin

Eu^{3+} bound to streptavidin

Streptavidin protein binds to biotin

(*b*)

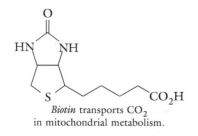

Biotin transports CO_2 in mitochondrial metabolism.

has a much longer lifetime, falling to $1/e$ (= 37%) of its initial intensity in approximately 730 μs. Luminescence from Eu^{3+} is further resolved from that of background because lanthanide emission is at longer wavelengths. Ultraviolet excitation at 340 nm typically gives short-lived, broad background fluorescence at all wavelengths between 350 and 600 nm, an emission that overlaps and masks the emission from many fluorescent molecules. In contrast, sharp Eu^{3+} luminescence is observed near 615 nm.

Figure 19-20a shows how Eu^{3+} can be incorporated into an immunoassay. A chelating group that binds lanthanide ions is attached to antibody 2 in Figure 19-17. While bound to the antibody, Eu^{3+} has weak luminescence. After completing all steps in Figure 19-17, the pH of the solution is lowered in the presence of a soluble chelator that extracts the metal ion into solution. Strong luminescence from the soluble metal ion is then easily detected by the time-resolved measurement in Figure 19-19.

A further improvement is shown in Figure 19-20b. In this case, antibody 2 is labeled by several small *biotin* molecules. After completing the steps of the immunoassay, a protein called *streptavidin*, extensively labeled with Eu^{3+}, is added. Streptavidin binds strongly and specifically to biotin, thereby attaching many lanthanide ions to the antibody complex. Upon lowering the pH in the presence of soluble chelator, the solution becomes strongly luminescent. Each molecule of analyte in the original unknown unleashes many Eu^{3+} ions in the last step of the analysis.

Terms to Understand

chemiluminescence	isosbestic point	rotational transition
electronic transition	luminescence	Scatchard plot
emission spectrum	method of continuous variation	singlet state
excitation spectrum	molecular orbital	triplet state
fluorescence	phosphorescence	vibrational transition
immunoassay		

Summary

The most common analytical application of spectropho-tometry relies on the proportionality between absorbance and concentration expressed in Beer's law. The absorbance spectrum of a mixture can be used to find the concentration of each component. If the spectra of individual components overlap a great deal, a graphical method of analysis is preferred. If the spectra overlap partially, then n measurements of absorbance at n wavelengths of maximum absorption are, in principle, sufficient to find the concentrations of the n absorbing components. A Scatchard plot is used to measure an equilibrium constant, and the method of continuous variation allows us to determine the stoichiometry of a complex.

When a molecule absorbs light, it is promoted to an excited state from which it may return to the ground state by radiationless processes or by fluorescence (singlet → singlet emission) or phosphorescence (triplet → singlet emission). Any form of luminescence is potentially useful for quantitative analysis, because emission intensity is proportional to sample concentration at low concentra-tion. An excitation spectrum (a graph of emission intensi-ty versus excitation wavelength) is similar to an absorp-tion spectrum (a graph of absorbance versus wavelength). An emission spectrum (a graph of emission intensity versus emission wavelength) is observed at lower energy than the absorption spectrum and tends to be the mirror image of the absorption spectrum.

A molecule that is not fluorescent can be analyzed by attaching a fluorescent group to it. Metal ions can be measured by their abilities to enhance or degrade fluores-cence of organic chelators. Light emitted from a chemical reaction—chemiluminescence—can also be used for quantitative analysis. Immunoassays use antibodies to detect the analyte of interest. In an enzyme-linked immu-nosorbent assay, signal is amplified by coupling analyte to an enzyme catalyzing many cycles of a reaction that produces a colored or fluorescent product. Time-resolved fluorescence measurements provide sensitivity by separat-ing the analyte fluorescence in time and wavelength from background fluorescence.

Exercises

A. Transferrin is the iron-transport protein found in blood. It has a molecular weight of 81 000 and carries two Fe^{3+} ions. Desferrioxamine B is a potent iron chelator used to treat patients with iron overload. It has a molecular weight of about 650 and can bind one Fe^{3+}. Desferrioxamine can take iron from many sites within the body and is excreted (with its iron) through the kidneys. The molar absorptivities of these compounds (saturated with iron) at two wavelengths are given below. Both compounds are colorless (no visible absorption) in the absence of iron.

λ (nm)	ϵ [M^{-1} cm^{-1}]	
	Transferrin	Desferrioxamine
428	3 540	2 730
470	4 170	2 290

(a) A solution of transferrin exhibits an absorbance of 0.463 at 470 nm in a 1.000-cm cell. Calculate the concentration of transferrin in milligrams per milliliter and the concentration of iron in micrograms per milliliter.

(b) A short time after adding desferrioxamine (which dilutes the sample), the absorbance at 470 nm was 0.424, and the absorbance at 428 nm was 0.401. Calculate the fraction of iron in transferrin and the fraction in desferrioxamine. Remember that transferrin binds two iron atoms and desferrioxamine binds only one.

B. Compound P, which absorbs light at 305 nm, was titrated with X, which does not absorb at this wavelength. The product, PX, also absorbs at 305 nm. The absorbance of each solution was measured in a 1.000-cm cell, and the concentration of free X was determined independently, with the results given below. Prepare a Scatchard plot and find the equilibrium constant for the reaction $X + P \rightleftharpoons PX$.

Exper-iment	P_0 (M)	X_0 (M)	A	[X] (M)
0	0.010 0	0	0.213	0
1	0.010 0	0.001 00	0.303	4.42×10^{-6}
2	0.010 0	0.002 00	0.394	9.10×10^{-6}
3	0.010 0	0.003 00	0.484	1.60×10^{-5}
4	0.010 0	0.004 00	0.574	2.47×10^{-5}
5	0.010 0	0.005 00	0.663	3.57×10^{-5}
6	0.010 0	0.006 00	0.752	5.52×10^{-5}
7	0.010 0	0.007 00	0.840	8.20×10^{-5}
8	0.010 0	0.008 00	0.926	1.42×10^{-4}
9	0.010 0	0.009 00	1.006	2.69×10^{-4}
10	0.010 0	0.010 0	1.066	5.87×10^{-4}
11	0.010 0	0.020 0	1.117	9.66×10^{-3}

C.   Complex formation by 3-aminopyridine and picric acid in chloroform solution gives a yellow product with an absorbance maximum at 400 nm. Neither starting material absorbs significantly at this wavelength. Stock solutions containing 1.00×10^{-4} M of each compound were mixed as follows and the absorbances of the mixtures were recorded:

Picric acid (mL)	3-Amino-pyridine (mL)	Absorbance at 400 nm
2.70	0.30	0.106
2.40	0.60	0.214
2.10	0.90	0.311
1.80	1.20	0.402
1.50	1.50	0.442
1.20	1.80	0.404
0.90	2.10	0.318
0.60	2.40	0.222
0.30	2.70	0.110

Data from E. Bruneau, D. Lavabre, G. Levy, and J. C. Micheau, *J. Chem. Ed.* **1992**, *69*, 833.

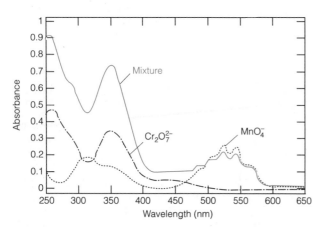

Picric acid   3-Aminopyridine

Prepare a graph of absorbance versus mole fraction of 3-aminopyridine and find the stoichiometry of the complex.

D.   Consider a fluorescence experiment in which the cell in Figure 19-15 is arranged so that b_1 and b_3 are negligible and, therefore, self-absorption can be neglected. To a first approximation, emission intensity is proportional to solute concentration. At what absorbance ($= \epsilon_{ex} b_2 c$) will the emission be 5% below the value expected if emission is proportional to concentration?

Problems

Analysis of a Mixture

1.   Consider compounds X and Y in the Example in Section 19-1 labeled "Analysis of a Mixture, Using Equations 19-7." A mixture of X and Y in a 0.100-cm cell had an absorbance of 0.233 at 272 nm and 0.200 at 327 nm. Find the concentrations of X and Y in the mixture.

2.   The figure below shows the spectra of 1.00×10^{-4} M MnO_4^-, 1.00×10^{-4} M $Cr_2O_7^{2-}$, and an unknown mixture of both. Absorbances at several wavelengths are given in the table. Use Equation 19-5 to find the concentration of each species in the mixture.

Wavelength (nm)	MnO_4^- standard	$Cr_2O_7^{2-}$ standard	Mixture
266	0.042	0.410	0.766
288	0.082	0.283	0.571
320	0.168	0.158	0.422
350	0.125	0.318	0.672
360	0.056	0.181	0.366

3.   When are isosbestic points observed and why?

4.   The metal ion indicator xylenol orange (Table 13-3) is yellow at pH 6 (λ_{max} = 439 nm). The spectral changes that occur as VO^{2+} is added to the indicator at pH 6 are shown in the figure at the top of the next page. The mole ratio VO^{2+}/xylenol orange at each point is

Trace	Ratio	Trace	Ratio
0	0	9	0.90
1	0.10	10	1.0
2	0.20	11	1.1
3	0.30	12	1.3
4	0.40	13	1.5
5	0.50	14	2.0
6	0.60	15	3.1
7	0.70	16	4.1
8	0.80		

Visible spectrum of MnO_4^-, $Cr_2O_7^{2-}$, and an unknown mixture containing both ions. [From M. Blanco, H. Iturriaga, S. Maspoch, and P. Tarín, *J. Chem. Ed.* **1989**, *66*, 178.]

Suggest a sequence of chemical reactions to explain the spectral changes, especially the isosbestic points at 457 and 528 nm.

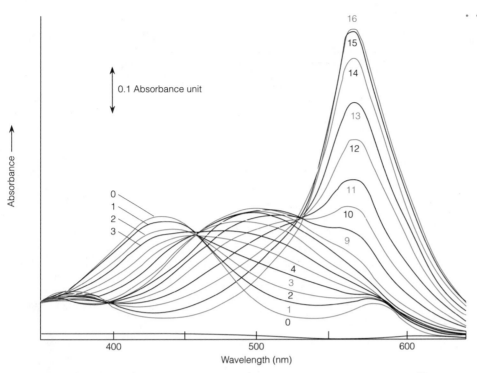

For Problem 4: Absorption spectra for the reaction of xylenol orange with VO^{2+} at pH 6.0. [From D. C. Harris and M. H. Gelb, *Biochim. Biophys. Acta* **1980**, *623*, 1.]

5. Infrared spectra are customarily recorded on a transmittance scale so that weak and strong bands can be displayed on the same scale. The region near $2\,000$ cm^{-1} in the infrared spectra of compounds A and B is shown at the right. Note that absorption corresponds to a downward peak on this scale. The spectra were recorded from a $0.010\,0$ M solution of each, in cells with $0.005\,00$-cm pathlengths. A mixture of A and B in a $0.005\,00$-cm cell gave a transmittance of 34.0% at $2\,022$ cm^{-1} and 38.3% at $1\,993$ cm^{-1}. Find the concentrations of A and B.

6. *Solving simultaneous linear equations.* You will find it useful for this and other courses to have a simple template into which you can plug the coefficients of simultaneous linear equations to be solved. For example, the equations

$$3x \quad -y \;=\; 5$$
$$-2x +10y \;=\; 6$$

$\underbrace{}_{\text{Coefficients}}\;\underbrace{}_{\text{Constants}}$

may be broken into a matrix of coefficients and a "vector" of constants:

$$\text{coefficient matrix} = \begin{bmatrix} 3 & -1 \\ -2 & 10 \end{bmatrix}$$

$$\text{constant vector} \;=\; \begin{bmatrix} 5 \\ 6 \end{bmatrix}$$

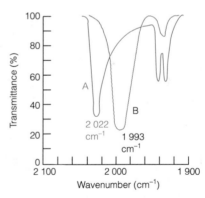

Wavenumber	Pure A	Pure B
$2\,022$ cm^{-1}	31.0% T	97.4% T
$1\,993$ cm^{-1}	79.7% T	20.0% T

The determinant of the coefficient matrix is

$$D = \begin{vmatrix} 3 & -1 \\ -2 & 10 \end{vmatrix} = 3{\cdot}10 - (-1){\cdot}(-2) = 28$$

and the solutions are given by Equations 19-7 in which the constant vector is used to replace either the first or the second column of the coefficient matrix in the numerator:

	A	**B**	**C**	**D**	**E**
1	Coefficient matrix (2x2)		Constant vector	Solution vector	
2	3	−1	5	2	
3	−2	10	6	1	
4					
5	Denominator = determinant of coefficient matrix =				
6	28				
7					
8	A6 = A2*B3–B2*A3				
9	D2 = (C2*B3–C3*B2)/A6				
10	D3 = (A2*C3–C2*A3)/A6				
11					
12					
13	Coefficient matrix (3x3)			Constant vector	Solution vector
14	7	2	−1	15	4
15	1	−1	15	112	−3
16	−9	0	2	−22	7
17					
18	Denominator = determinant of coefficient matrix				
19	−279				
20					
21	A19 = A14*B15*C16+A15*B16*C14+B14*C15*A16–C14*B15*A16				
22		−B14*A15*C16–C15*B16*A14			
23	E14 = (D14*B15*C16+D15*B16*C14+B14*C15*D16–C14*B15*D16				
24		−B14*D15*C16–C15*B16*D14)/A19			
25	E15 = (A14*D15*C16+A15*D16*C14+D14*C15*A16–C14*D15*A16				
26		−D14*A15*C16–C15*D16*A14)/A19			
27	E16 = (A14*B15*D16+A15*B16*D14+B14*D15*A16–D14*B15*A16				
28		−B14*A15*D16–D15*B16*A14)/A19			

Spreadsheet for solving systems of two or three simultaneous linear equations.

$$x = \begin{vmatrix} 5 & -1 \\ 6 & 10 \end{vmatrix} / D = [5 \cdot 10 - (-1) \cdot (6)]/28 = 2$$

$$y = \begin{vmatrix} 3 & 5 \\ -2 & 6 \end{vmatrix} / D = [3 \cdot 6 - (5) \cdot (-2)]/28 = 1$$

The procedure for using determinants to solve simultaneous equations is called *Cramer's rule*. For three simultaneous equations, Cramer's rule looks like this:

$$7x + 2y - z = 15$$
$$x - y + 15z = 112$$
$$-9x + 2z = -22$$

The 3×3 determinant of coefficients is evaluated in a pitchfork pattern:

Positive products

Negative products

positive terms:

$$(7) \cdot (-1) \cdot (2), \ (1) \cdot (0) \cdot (-1), \ (-9) \cdot (15) \cdot (2)$$

negative terms:

$$(-1) \cdot (-1) \cdot (-9), \ (2) \cdot (1) \cdot (2), \ (7) \cdot (0) \cdot (15)$$

$$D = \begin{vmatrix} 7 & 2 & -1 \\ 1 & -1 & 15 \\ -9 & 0 & 2 \end{vmatrix}$$

$$= (7) \cdot (-1) \cdot (2) + (1) \cdot (0) \cdot (-1) + (-9) \cdot (15) \cdot (2)$$
$$- (-1) \cdot (-1) \cdot (-9) - (2) \cdot (1) \cdot (2) - (7) \cdot (0) \cdot (15)$$
$$= -279$$

To solve three simultaneous equations, we replace the first, second, or third column in the coefficient matrix by the constant vector:

$$x = \frac{\begin{vmatrix} 15 & 2 & -1 \\ 112 & -1 & 15 \\ -22 & 0 & 2 \end{vmatrix}}{D} = \frac{-1\,116}{-279} = 4$$

$$y = \frac{\begin{vmatrix} 7 & 15 & -1 \\ 1 & 112 & 15 \\ -9 & -22 & 2 \end{vmatrix}}{D} = \frac{837}{-279} = -3$$

$$z = \frac{\begin{vmatrix} 7 & 2 & 15 \\ 1 & -1 & 112 \\ -9 & 0 & -22 \end{vmatrix}}{D} = \frac{-1\,953}{-279} = 7$$

(a) Rows 1 through 10 of the spreadsheet on the opposite page are for solving two simultaneous equations. The coefficient matrix is in columns A and B. The constant vector is in column C and the solution is in column D. The denominator is evaluated in cell A6. Set up a spreadsheet like this and use it to reproduce the Example following Equations 19-7. You can use this spreadsheet for any 2×2 problem.

(b) Rows 13 through 28 of the spreadsheet solve three simultaneous equations. The coefficient matrix is in columns A, B, and C. The constant vector is in column D and the solution is in column E. The denominator is evaluated in cell A19. Set up a spreadsheet like this to reproduce the 3×3 example above.

7. [spreadsheet icon] Spectrophotometric data for three compounds are given below. A solution containing X, Y, and Z in a 1.000-cm cuvet had absorbances of 0.846 at 246 nm, 0.400 at 298 nm, and 0.555 at 360 nm. Use your 3×3 spreadsheet from Problem 6 to find the concentrations of X, Y, and Z in the mixture.

	ϵ (M^{-1} cm^{-1})		
λ (nm)	X	Y	Z
246	12 200	3 210	290
298	4 140	6 550	990
360	3 000	2 780	8 080

Measuring an Equilibrium Constant

8. Compound P was titrated with X to form the complex PX. A series of solutions was prepared with the total concentration of P remaining fixed at 1.00×10^{-5} M. Both P and X have no visible absorbance, but PX has an absorption maximum at 437 nm. The following table shows how the absorbance at 437 nm in a 5.00-cm cell depends on the total concentration of added X ($X_T = [X] + [PX]$).

X_T (M)	A	X_T (M)	A
0	0.000	0.020 0	0.535
0.002 00	0.125	0.040 0	0.631
0.004 00	0.213	0.060 0	0.700
0.006 00	0.286	0.080 0	0.708
0.008 00	0.342	0.100	0.765
0.010 0	0.406		

(a) Make a Scatchard plot of $\Delta A/[X]$ versus ΔA. In this plot, [X] refers to the species X, not to X_T. However, since $X_T \gg [P]$, we can safely say that $[X] \approx X_T$ in this experiment.

(b) From the slope of the graph, find the equilibrium constant, K.

9. Iodine reacts with mesitylene to form a complex with an absorption maximum at 332 nm in CCl$_4$ solution:

$$K = \frac{[\text{complex}]}{[I_2][\text{mesitylene}]}$$

(a) Given that the product absorbs at 332 nm but neither reactant has significant absorbance at this wavelength, use the equilibrium constant, K, and Beer's law to show that

$$\frac{A}{[\text{mesitylene}][I_2]_{\text{tot}}} = K\epsilon - \frac{KA}{[I_2]_{\text{tot}}}$$

where A is the absorbance at 332 nm, ϵ is the molar absorptivity of the complex at 332 nm, [mesitylene] is the concentration of free mesitylene, and $[I_2]_{\text{tot}}$ is the total concentration of iodine in the solution ($= [I_2] + [\text{complex}]$). Assume that the cell pathlength is 1.000 cm.

(b) Here are spectrophotometric data for this reaction:

[Mesitylene]$_{tot}$ (M)	[I$_2$]$_{tot}$ (M)	Absorbance at 332 nm
1.690	7.817×10^{-5}	0.369
0.9218	2.558×10^{-4}	0.822
0.6338	3.224×10^{-4}	0.787
0.4829	3.573×10^{-4}	0.703
0.3900	3.788×10^{-4}	0.624
0.3271	3.934×10^{-4}	0.556

Data from P. J. Ogren and J. R. Norton, *J. Chem. Ed.* **1992**, *69*, A130.

Because [mesitylene]$_{tot}$ >> [I$_2$], we can say that [mesitylene] $\approx$ [mesitylene]$_{tot}$. Prepare a graph of $A/([mesitylene][I_2]_{tot})$ versus $A/[I_2]_{tot}$ and find the equilibrium constant and molar absorptivity.

Method of Continuous Variation

10. *Method of continuous variation.* Make a graph of absorbance versus mole fraction of thiocyanate for the data in the following table.

mL Fe^{3+} solution	mL SCN$^-$ solution	Absorbance
30.00	0	0.001
27.00	3.00	0.122
24.00	6.00	0.226
21.00	9.00	0.293
18.00	12.00	0.331
15.00	15.00	0.346
12.00	18.00	0.327
9.00	21.00	0.286
6.00	24.00	0.214
3.00	27.00	0.109
0	30.00	0.002

Note: Fe^{3+} solution: 1.00 mM Fe(NO$_3$)$_3$ + 10.0 mM HNO$_3$
SCN$^-$ solution: 1.00 mM KSCN + 15.0 mM HCl

Data from Z. D. Hill and P. MacCarthy, *J. Chem. Ed*, **1986**, *63*, 162.

(a) What is the stoichiometry of the predominant Fe(SCN)$_n^{3-n}$ species?

(b) Why is the peak not as sharp as those in Figure 19-6?

(c) Why does one solution contain 10.0 mM acid and the other 15.0 mM acid?

11. 📇 *Simulating a Job's plot.* Consider the reaction A + 2B ⇌ AB$_2$, for which $K = [AB_2]/[A][B]^2$. The following mixtures of A and B at a fixed total concentration of 10^{-4} M are prepared:

[A]$_{total}$ (M)	[B]$_{total}$ (M)	[A]$_{total}$ (M)	[B]$_{total}$ (M)
1.00×10^{-5}	9.00×10^{-5}	5.00×10^{-5}	5.00×10^{-5}
2.00×10^{-5}	8.00×10^{-5}	6.00×10^{-5}	4.00×10^{-5}
2.50×10^{-5}	7.50×10^{-5}	7.00×10^{-5}	3.00×10^{-5}
3.00×10^{-5}	7.00×10^{-5}	8.00×10^{-5}	2.00×10^{-5}
3.33×10^{-5}	6.67×10^{-5}	9.00×10^{-5}	1.00×10^{-5}
4.00×10^{-5}	6.00×10^{-5}		

(a) Prepare a spreadsheet to find the concentration of AB$_2$ for each mixture, for equilibrium constants of $K = 10^6$, 10^7, and 10^8. A good way to do this is with a circular definition. From the mass balance equations [A]$_{total}$ = [A] + [AB$_2$] and [B]$_{total}$ = [B] + 2[AB$_2$], we can write $K = [AB_2]/\{([A]_{total} - [AB_2])([B]_{total} - 2[AB_2])^2\}$. If you enter values of [A]$_{total}$ and [B]$_{total}$ in columns A and B of your spreadsheet, you can find the concentration of AB$_2$ in column C by writing the formula $K([A]_{total}-[AB_2])([B]_{total}-2[AB_2])^2$ for column C. That is, the function in cell C2 will be = $K*(A2-C2)*(B2-2*C2)\wedge 2$, where K is given the values 10^6, 10^7, or 10^8.

(b) Prepare a graph by the method of continuous variation in which you plot [AB$_2$] versus mole fraction of A for each equilibrium constant. Explain the shapes of the curves.

Luminescence

12. In formaldehyde, the transition $n \rightarrow \pi^*(T_1)$ occurs at 397 nm, while the $n \rightarrow \pi^*(S_1)$ transition comes at 355 nm. What is the difference in energy (kJ/mol) between the S$_1$ and T$_1$ states? This difference is due to the different electron spins in the two states.

13. What is the difference between fluorescence and phosphorescence?

14. Consider a molecule that can fluoresce from the S$_1$ state and phosphoresce from the T$_1$ state. Does fluorescence or phosphorescence come at the longer wavelength? Make a sketch showing absorption, fluorescence, and phosphorescence on a single spectrum.

15. What is the difference between a fluorescence excitation spectrum and a fluorescence emission spectrum? Which one resembles an absorption spectrum?

16. What does quenching mean in emission spectroscopy?

17. Explain how signal amplification is achieved in enzyme-linked immunosorbent assays.

18. What is the advantage of a time-resolved emission measurement with Eu^{3+} versus measurement of fluorescence from organic chromophores?

19. Adult humans require about 2 mg/day of riboflavin, a highly fluorescent, water-soluble vitamin. Riboflavin is stable to heat but decomposes in the presence of light, which is why milk should not be left in bright sunlight. From the fluorescence spectrum of riboflavin shown on the next page, sketch the absorption spectrum of riboflavin. Sketch the emission spectrum for an excitation wavelength of 400 nm.

20. This problem describes an "electronic dog" being developed to sniff trinitrotoluene (TNT) explosives at airports.[16]

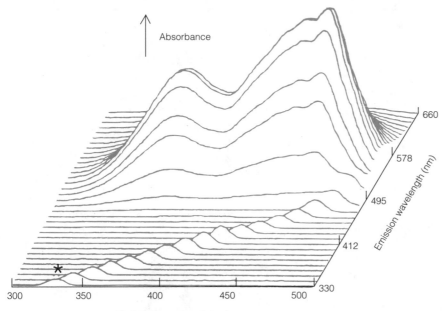

Absorbance

660
578
495
412
330

Emission wavelength (nm)

Excitation wavelength (nm)

300   350   400   450   500

★

For Problem 19: Three-dimensional fluorescence spectrum of riboflavin. For each trace, emission wavelength is held constant as excitation wavelength is varied. The small peak indicated by an asterisk in the lowest spectrum is an artifact that occurs when $\lambda_{excitation} = \lambda_{emission}$. [From R. E. Utecht, *J. Chem. Ed.* **1993**, *70*, 673.]

Trinitrotoluene (TNT)

NO_2

O_2N — CH_3

NO_2

Fluorescently labeled TNT

NO_2

O_2N — Flexible link

NO_2

Fluorescent group

1. A column was prepared that contained covalently bound antibodies to TNT. Fluorescently labeled TNT was passed through this column to saturate all of the antibodies with labeled TNT. The column was washed with excess solvent until no fluorescence was detected at the outlet.

2. Air containing TNT vapors was drawn through a sampling device that concentrated traces of the gaseous vapors into distilled water.

3. Aliquots of water from the sampler were injected into the antibody column. The fluorescence intensity of the liquid exiting the column was proportional to the concentration of TNT in the water from the sampler over the range 20 to 1 200 ng/mL.

Draw pictures showing the state of the column in Steps 1 and 3 and explain how this detector works.

21. The coenzyme $NADP^+$ can be assayed by a titration in which it is converted to the fluorescent product NADPH by the action of adenosine triphosphate (ATP) plus several enzymes. A hypothetical titration curve is

shown below. In this titration, the fluorescence intensity is plotted versus microliters of added ATP. The first eight points lie on one line, and the points from 90 to 140 μL lie on a second line. The uncertainties in slope and intercept of these lines are one standard deviation, as determined by the method of least squares. The end point of the titration lies at the intersection of the two lines. Using the equations for the two lines, *determine the volume of ATP (in microliters) at the equivalence point.* Also, use the standard deviations of the slopes and intercepts to *estimate the standard deviation of the volume of ATP at the equivalence point.* Express your answer (μL ± standard deviation) with an appropriate number of significant figures.

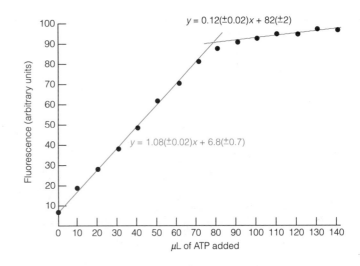

$y = 0.12(\pm0.02)x + 82(\pm2)$

$y = 1.08(\pm0.02)x + 6.8(\pm0.7)$

Fluorescence (arbitrary units)

100
90
80
70
60
50
40
30
20
10

0   10   20   30   40   50   60   70   80   90 100 110 120 130 140

μL of ATP added

22. How can the sample cell in Figure 19-15 be positioned to minimize the self-absorption expressed in the terms $10^{-\epsilon_{ex}b_1 c}$ and $10^{-\epsilon_{ex}b_3 c}$ in Equation 19-27?

23. Consider a fluorescence experiment in which $\epsilon_{ex} = 1530 \text{ M}^{-1} \text{ cm}^{-1}$, $\epsilon_{em} = 495 \text{ M}^{-1} \text{ cm}^{-1}$, $b_1 = 0.400 \text{ cm}$, $b_2 = 0.200 \text{ cm}$, and $b_3 = 0.500 \text{ cm}$ in Equation 19-27. Make a graph of relative fluorescence intensity versus concentration for the following concentrations of solute: 1.00×10^{-7}, 1.00×10^{-6}, 1.00×10^{-5}, 1.00×10^{-4}, 1.00×10^{-3}, and 1.00×10^{-2} M.

24. This problem describes how the fiber-optic pH sensor at the beginning of Chapter 10 works. The tip of the optical fiber contains the highly fluorescent chromophore, fluorescein, incorporated into a polymer. The absorption spectrum of the fluorescein chromophore has an isosbestic point at 461 nm. At shorter wavelengths, the absorbance decreases as pH increases. At longer wavelengths, the absorbance increases as pH increases. (See the figure below.)

(a) Fluorescence emission spectra of the polymer at the tip of the sensor are shown in the figures at the right. When the excitation wavelength is 442 nm, fluorescence intensity decreases with increasing pH. When the excitation wavelength is 488 nm, fluorescence intensity increases with increasing pH. Explain why.

(b) The graph on the next page shows the ratio of fluorescence intensities at two wavelengths, with excitation at 488 nm. (Ratios are used because they are fairly constant from experiment to experiment, whereas absolute intensity at one wavelength is difficult to reproduce.) Explain the shape of this curve and explain how the optical fiber can be used to measure pH.

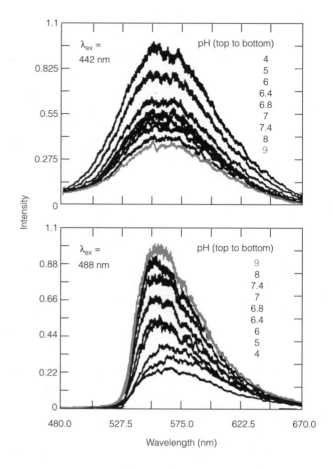

Fluorescence intensity from the polymer tip of the fiber-optic sensor shown at the beginning of Chapter 10. [From W. Tan, S.-Y. Shi, S. Smith, D. Birnbaum, and R. Kopelman, *Science* **1992,** *258,* 778.]

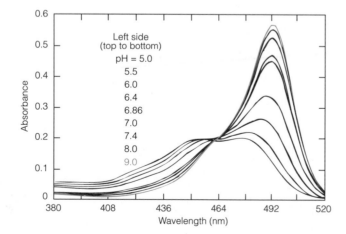

Absorption spectrum of *N*-fluoresceinylacrylamide monomer as a function of pH. [From W. Tan, S.-Y. Shi, S. Smith, D. Birnbaum, and R. Kopelman, *Science* **1992,** *258,* 778.]

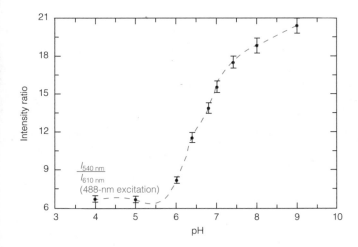

Calibration graph showing relation of fluorescence intensity ratio to pH. [From W. Tan, S.-Y. Shi, S. Smith, D. Birnbaum, and R. Kopelman, *Science* **1992**, *258*, 778.]

Notes and References

1. Precise spectrophotometric measurements of pH and pK with indicators have been described for laboratory measurements by H. Yamazaki, R. P. Sperline, and H. Freiser, *Anal. Chem.* **1992**, *64*, 2720 and for field measurements by J. A. Breland II and R. H. Byrne, *Anal. Chem.* **1992**, *64*, 2306.

2. A procedure (and student experiment) for simultaneous determination of more than two components by using measurements at more wavelengths than there are unknowns is described by G. Dado and J. Rosenthal, *J. Chem. Ed.* **1990**, *67*, 797.

3. Under certain conditions, it is possible for a solution with more than two principal species to exhibit an isosbestic point. See D. V. Stynes, *Inorg. Chem.* **1975**, *14*, 453.

4. R. W. Ramette, *J. Chem. Ed.* **1967**, *44*, 647.

5. G. Scatchard, *Ann. N.Y. Acad. Sci.* **1949**, *51*, 660.

6. D. A. Deranleau, *J. Am. Chem. Soc.* **1969**, *91*, 4044.

7. Obtaining equilibrium constants from continuous variation plots is described by E. Bruneau, D. Lavabre, G. Levy, and J. C. Micheau, *J. Chem. Ed.* **1992**, *69*, 833; W. Likussar and D. F. Boltz, *Anal. Chem.* **1971**, *43*, 1265; and V. M. S. Gil and N. C. Oliveira, *J. Chem. Ed.* **1990**, *67*, 473.

8. A pictorial description of the dynamics of the $n \rightarrow \pi^*$ and $\pi \rightarrow \pi^*$ transitions of formaldehyde is given by G. Henderson, *J. Chem. Ed.* **1990**, *67*, 392.

9. The pH dependence of emission intensity is a reflection of the acid-base properties of the excited state. An experiment to measure pK_a of the excited state of 2-naphthol is described by J. van Stam and J.-E. Löfroth, *J. Chem. Ed.* **1986**, *63*, 181.

10. For a lecture demonstration with riboflavin, see D. S. Chatellier and H. B. White III, *J. Chem. Ed.* **1988**, *65*, 814.

11. G. L. Trainor, *Anal. Chem.* **1990**, *62*, 418.

12. E. R. Menzel, *Anal. Chem.* **1989**, *61*, 557A.

13. R. Rizzuto, A. W. M. Simpson, M. Brini, and T. Pozzan, *Nature* **1992**, *358*, 325.

14. K. Kikuchi, T. Nagano, H. Hayakawa, Y. Hirata, and M. Hirobe, *Anal. Chem.* **1993**, *65*, 1794.

15. E. P. Diamandis and T. K. Christopoulos, *Anal. Chem.* **1990**, *62*, 1152A; T. Ståhlberg, E. Markela, H. Mikola, P. Mottram, and I. Hemmilä, *Am. Lab.* December 1993, p. 15.

16. J. P. Whelan, A. W. Kusterbeck, G. A. Wemhoff, R. Bredehorst, and F. S. Ligler, *Anal. Chem.* **1993**, *65*, 3561.

The Most Important Photoreceptor

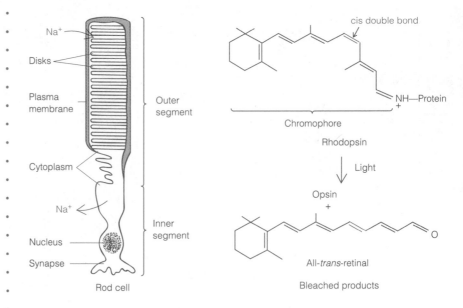

Rod cell diagram labels: Na^+, Disks, Plasma membrane, Cytoplasm, Na^+, Nucleus, Synapse, Outer segment, Inner segment

Rhodopsin chromophore diagram: cis double bond, NH—Protein, Chromophore, Rhodopsin, Light, Opsin +, All-*trans*-retinal, Bleached products

The retina at the back of each of your eyes contains photosensitive cells called *rods* and *cones* that are sensitive to levels of light varying over seven orders of magnitude. Light impinging on these cells is translated into nerve impulses that are transmitted via the optic nerve to the brain. Rod cells detect the dimmest light but cannot distinguish colors. Cone cells operate in bright light and give us color vision.

A stack of about 1 000 *disks* in each rod cell contains the light-sensing protein *rhodopsin,* in which the chromophore 11-*cis*-retinal (from vitamin A) is attached to the protein *opsin.* When light is absorbed by rhodopsin, a series of rapid transformations releases all-*trans*-retinal. At this stage, the pigment is *bleached* (loses all color) and cannot respond to more light until retinal isomerizes back to the 11-*cis* form and recombines with the protein.

In the dark, there is a continuous flow of 10^9 Na^+ ions per second out of the rod cell's inner segment, through the adjoining medium, and into the cell's outer segment. An energy-dependent process using adenosine triphosphate (ATP) and oxygen pumps Na^+ out of the cell. Another process involving a molecule called cyclic GMP keeps the gates of the outer segment open for ions to flow back into the cell. When light is absorbed and rhodopsin is bleached, a series of reactions leads to destruction of cyclic GMP and shutdown of the channels through which Na^+ flows into the cell. A single photon reduces the ion current by 3%—corresponding to a decreased current of 3×10^7 ions per second. This *amplification* is greater than that of a photomultiplier tube, which is one of the most sensitive man-made photodetectors! The ion current returns to its dark value as the protein and retinal recombine and cyclic GMP is restored to its initial concentration. A great deal of work remains to be done before the process of transduction of light into nerve impulses is fully understood.

Spectrophotometers

Just as your eye produces electrical signals when struck by photons, the detector in a spectrophotometer does the same. In Chapters 6 and 19 we learned how absorption and emission of light are used in chemical analysis. Now we examine the workings of the spectrophotometer and introduce Fourier transform spectroscopy. We begin with a discussion of the physics of the interaction of light with matter.

20-1 Interaction of Light with Matter

What happens to light when it strikes a sample? Consider the material in Figure 20-1 to be a uniform solid that absorbs some of the light striking it. The radiant power of the light incident on the first surface is called P_0. *Radiant power* (also called *intensity*) refers to the energy per second striking each unit area and has units of watts per square meter (W/m^2). At the first surface, some of the radiant power (R_1) is reflected back toward the source, and some (P_1) enters the sample. While this fraction of radiant power travels through the sample, some energy is absorbed, and radiant power P_2 ($< P_1$) strikes the second surface. At this point, some power (R_2) is reflected, and some (P) is transmitted out of the sample to emerge on the other side. The internal reflections continue indefinitely. Some of R_2 is absorbed as it travels back to the first surface. Here, some of this return ray is transmitted back out of the sample (toward the left in the figure) and some is reflected back into the sample. The power involved gets smaller with successive passes, becoming negligible after several cycles.

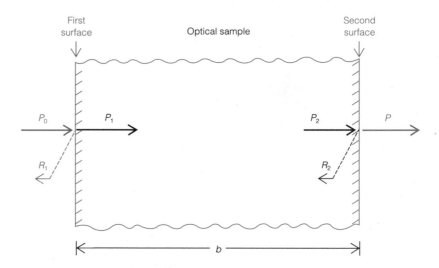

Figure 20-1 Light of radiant power P_0 strikes the first surface of sample whose thickness is b. Radiant power R_1 is reflected and radiant power P_1 enters the sample. Some of P_1 is absorbed as it travels through the sample and power P_2 arrives at the second surface. There, power P is transmitted and power R_2 is reflected.

Transmission, Absorption, Reflection, and Scatter

Light striking a sample can be

1. transmitted
2. absorbed
3. reflected
4. scattered

An example of a polycrystalline solid is a KBr pellet used for infrared spectroscopy, described in Section 20-4.

Four things can happen to light impinging on the sample in Figure 20-1:

1. Some light is *transmitted* through the sample. The **transmittance** measured in the lab is the fraction of incident light emerging from the other side:

   External transmittance: $\qquad T = \dfrac{P}{P_0} \qquad \leftarrow$ This is what we measure in the lab $\qquad$ (20-1)

2. Some light is *absorbed* by the material. Box 6-1 showed that there is an exponential relationship between power absorbed within the sample and the pathlength, b, of the sample:

   Internal transmittance: $\qquad \dfrac{P_2}{P_1} = e^{-\alpha b} \qquad$ (20-2)

   where α is a property of the sample called the **absorption coefficient** and customarily has units of cm^{-1}. The radiant powers P_1 and P_2, defined in Figure 20-1, represent light *inside* the sample. If $\alpha = 1\ cm^{-1}$, the radiant power inside the sample decays to $1/e$ times its initial value after traversing a pathlength of 1 cm. The greater the absorption coefficient, the more light is absorbed.

3. Some light is *reflected* at each surface and eventually leaves the sample in the direction of the light source.

4. Some light is *scattered* to the side. If the sample is a liquid, light can be scattered by suspended dust particles or by very large solute molecules such as proteins. If the sample is a solid made of microcrystalline particles compacted together, scattering occurs at boundaries between the particles

(called *grain boundaries*) or at microscopic voids between particles. If scattering is significant, then another term must be added to Equation 20-2, because radiant power is lost to both absorption and scattering:

$$\frac{P_2}{P_1} = e^{-(\alpha_a + \alpha_s)b} \tag{20-3}$$

where α_a is the true absorption coefficient and α_s describes the loss of transmitted light due to scattering.

The quantity P_2/P_1 in Equation 20-2 is the *internal transmittance* of the sample. It is the fraction of radiant power inside the sample that is transmitted through the material. The measured (*external*) transmittance is P/P_0. The **absorptance** is defined as the fraction of incident radiation that is absorbed by the sample. In the absence of scattering, absorptance is given by

Absorptance: $$a = \frac{P_1 - P_2}{P_0} \tag{20-4}$$

where $P_1 - P_2$ is the radiant power absorbed by the sample. In the absence of scattering, transmittance, absorptance, and **reflectance** (R, the fraction of P_0 that is reflected) must sum to unity:

$$T + a + R = 1 \tag{20-5}$$

Three confusing terms:

a = absorptance = fraction of light absorbed by sample
α = absorption coefficient (in Equation 20-2)
A = absorbance (= $-\log T$, Equation 6-5)

The Greater the Difference in Refractive Index at an Interface, the More Light Is Reflected

The speed of light in a medium of **refractive index** n is c/n, where c is the speed of light in vacuum. That is, for vacuum, $n = 1$. The refractive index of a material is a function of the wavelength of light (Figure 20-2) and the

Recall that $\lambda\nu = c$. When light travels from one medium to another, the frequency is constant, but λ changes if n changes.

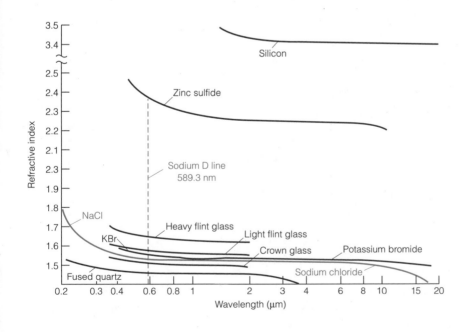

Figure 20-2 Dependence of refractive index on wavelength. Wavelength scale is logarithmic.

TABLE 20-1   Refractive index at 589.3 nm (sodium D line)

Substance	Refractive index
Vacuum	1
Air (0°C, 1 atm)	1.000 29
Water	1.33
Ethanol	1.36
Pentane	1.36
Magnesium fluoride	1.38
Fused quartz	1.46
Carbon tetrachloride	1.47
Benzene	1.50
Sodium chloride	1.54
Potassium bromide	1.56
Carbon disulfide	1.63
Bromine	1.66
Silver chloride	2.07
Zinc sulfide	2.36
Iodine	3.34

Refraction: bending of light when it passes between media with different refractive indexes

temperature of the material. The index is most often measured at 20°C at the wavelength of the sodium D line (λ = 589.3 nm). Representative values of refractive index are shown in Table 20-1.

When light passes from one medium to another, its path is bent, as shown in Figure 20-3. This phenomenon is called **refraction,** and the degree of bending is described by **Snell's law:**

Snell's law:
$$n_1 \sin\theta_1 = n_2 \sin\theta_2 \tag{20-6}$$

where n_1 and n_2 are the refractive indexes of the two media and θ_1 and θ_2 are angles defined in Figure 20-3.

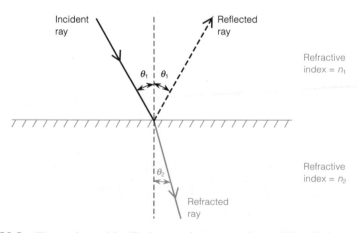

Figure 20-3   Illustration of Snell's law: $n_1 \sin\theta_1 = n_2 \sin\theta_2$. When light passes from air into any medium, the greater the refractive index of the medium, the smaller is θ_2.

EXAMPLE   Refraction of Light by Water

Visible light travels from air (medium 1) into water (medium 2) at a 45° angle (θ_1 in Figure 20-3).
   At what angle, θ_2, does the light ray pass through the water?

Solution   The refractive index for air is close to 1, and for water, ~1.33 (Table 20-1). Using Snell's law, we find

$$(1.00)(\sin 45°) = (1.33)(\sin\theta_2) \Rightarrow \theta_2 = 32°$$

What is θ_2 if the incident ray is perpendicular to the surface (i.e., $\theta_1 = 0°$)?

$$(1.00)(\sin 0°) = (1.33)(\sin\theta_2) \Rightarrow \theta_2 = 0°$$

A perpendicular ray is not refracted.

A perfectly smooth surface will reflect light at the same angle as the angle of incidence, as shown in Figure 20-3. The reflection of light at this angle is called **specular reflection.** Real surfaces are somewhat rough and reflect some light in all directions. Reflection at other than the angle of incidence is called **diffuse reflection.** The remainder of our discussion applies to specular reflection from a perfectly smooth surface.
   For a perpendicular light ray traveling from medium 1 to medium 2 ($\theta_1 = \theta_2 = 0°$), the ratio of radiant power reflected (P_r) to incident radiant power (P_0) is

Single-surface reflectance: $R = \dfrac{P_r}{P_0} = \left(\dfrac{n_1 - n_2}{n_1 + n_2}\right)^2$ (20-7)

EXAMPLE   Reflection of Light by Water

What fraction of visible light traveling from air to water at normal incidence ($\theta_1 = \theta_2 = 0°$) is reflected?

Solution   We use Equation 20-7 with $n_1 = 1.00$ and $n_2 = 1.33$:

$$R = \frac{P_r}{P_0} = \left(\frac{1.00 - 1.33}{1.00 + 1.33}\right)^2 = 0.020$$

The reflected radiant power is 2.0% of the incident power.

In Figure 20-1 we see that reflection occurs at each surface. The reflected ray bounces back and forth, undergoing partial transmission and partial reflection at each surface it strikes until its power is negligible. If the sample absorbs no light and the only processes that occur are reflection and transmission, then the theoretical transmittance of a ray striking perpendicular to the surface will be

Transmittance of a flat nonabsorbing plate: $T = \dfrac{P}{P_0} = \dfrac{1 - R}{1 + R}$ (20-8)

Reflectance of a perpendicular ray from a nonabsorbing, nonscattering plate with two surfaces.

where R is defined in Equation 20-7. The theoretical reflectance is $1 - T$, because we stipulated that all light is either reflected or transmitted. Therefore,

$$\text{total reflectance} = 1 - T = 1 - \frac{1 - R}{1 + R} = \frac{2R}{1 + R} \qquad (20\text{-}9)$$

· ·

EXAMPLE   Transmission of a Quartz Plate

What fraction of light striking a quartz plate in air at normal incidence (i.e., perpendicular to the surface) is transmitted?

Solution   The refractive index of quartz is 1.46. Putting $n_1 = 1.00$ and $n_2 = 1.46$ into Equation 20-7 gives $R = 0.035\,0$. This value of R in Equation 20-8 gives

$$T = \frac{1 - 0.035\,0}{1 + 0.035\,0} = 0.93$$

which tells us that quartz transmits 93% of the incident power and reflects 7%. This has a practical significance in spectrophotometry because a liquid sample must be contained in a cuvet that necessarily transmits less than 100% of the light that strikes the cuvet. If you want to measure the absorbance of a sample, you must have a reference that reflects the same fraction of light and therefore compensates for the reflectance of the cuvet.

· ·

From our discussion of reflection by transparent materials, you should realize that a significant fraction of the light in a spectrophotometer can be lost by reflection from the optical components. Fortunately, **antireflection coatings** can be applied to greatly decrease reflection. Ideally, the coating should have a refractive index given by $\sqrt{n_1 n_2}$, where n_1 is the refractive index of the external medium (usually air) and n_2 is the refractive index of the optical component (such as a lens). If the coating thickness is one-fourth of the wavelength of incident light (wavelength in the coating) and the refractive index is $\sqrt{n_1 n_2}$, reflection is theoretically reduced to zero. A homogeneous coating does not achieve zero reflection over a range of wavelengths because λ and n both vary. However, multilayer antireflection coatings can reduce reflection over a broad range of wavelengths.

Optical Fibers

A flexible **optical fiber** carries light from one place to another. It is especially useful for bringing an optical signal from inside a chemical reactor out to a spectrophotometer for process monitoring. The fiber in Figure 20-4a has a high-refractive-index, optically transmitting core (silica-based glass for visible light) enclosed in a lower-refractive-index cladding (a coating made of another glass or a polymer). The cladding is enclosed in a protective plastic jacket.

The principle of operation is shown in Figure 20-4b. Consider the ray striking the wall of the core at the angle of incidence θ_i. Part of the ray is reflected inside the core, and part might be transmitted into the cladding at

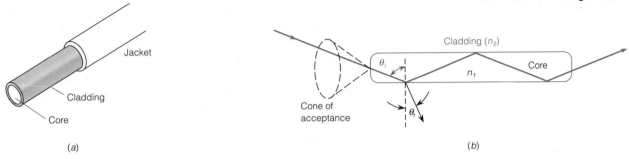

(a)   (b)

Figure 20-4 Optical fiber construction and principle of operation. Any light ray entering within the cone of acceptance will be totally reflected at the wall of the fiber.

the angle of refraction, θ_r. If the index of refraction of the core is n_1 and the index of the cladding is n_2, Snell's law (Equation 20-6) tells us that

$$n_1 \sin\theta_i = n_2 \sin\theta_r \Rightarrow \sin\theta_r = \frac{n_1}{n_2}\sin\theta_i \qquad (20\text{-}10)$$

If the product $(n_1/n_2)\sin\theta_i$ is greater than 1, no light is transmitted into the cladding because $\sin\theta_r$ cannot be greater than 1. In such a case, we say that θ_i exceeds the *critical angle* for total internal reflection. *If $n_1/n_2 > 1$, there is a range of angles θ_i in which essentially all light will be reflected at the walls of the core, and a negligible amount enters the cladding.* A ray entering one end of the fiber within the cone of acceptance will emerge from the other end of the fiber with little loss.

Figure 20-5 illustrates a sensor based on an optical fiber that is used to measure penicillin concentration inside a human vein.[1] The fiber carries light to a porous glass bead coated with a membrane containing *penicillinase*. This enzyme converts penicillin in the blood to penicilloic acid, which is fluorescent, and the fluorescence returning to the detector is proportional to the penicillin concentration in the blood.

A sensor based on an optical fiber is called an *optode* (or *optrode*), derived from the words "optical" and "electrode." Many optodes respond to optical absorbance changes at the tip of the fiber when a dye changes color in the presence of analytes such as sulfites in food or calcium ions in aqueous solution.[2]

Light traveling from a region of high refractive index (n_1) to a region of low refractive index (n_2) is totally reflected if the angle of incidence exceeds the *critical angle* given by $(n_1/n_2) \sin\theta_{critical} > 1$, or $\sin\theta_{critical} > n_2/n_1$.

Optical fibers are replacing electric wires for telephone communication. Fibers are immune to electrical noise, can transmit data at a higher rate, and can handle more signals simultaneously than can wires.

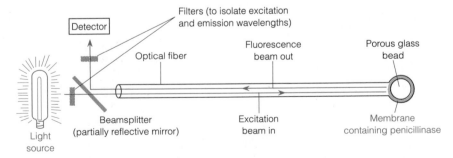

Figure 20-5 Optical fiber sensor used to measure penicillin concentration in the bloodstream.

Blackbody Radiation

When an object is heated, it emits radiation—it glows. Even at room temperature, all objects are glowing with infrared radiation. Imagine a hollow sphere whose inside surface is perfectly black. That is, the surface absorbs all radiation striking it. If the sphere is at constant temperature, it must emit as much radiation as it absorbs. If a small hole were made in the wall, we would observe that the escaping radiation has a continuous spectral distribution. The object is called a *blackbody,* and the radiation is called **blackbody radiation.** Emission from real objects such as the tungsten filament of a light bulb resembles that from an ideal blackbody.

The power per unit area radiating from the surface of an object is called the **exitance** (or *emittance*), *M,* and is given by

Exitance from blackbody: $$M = \sigma T^4 \qquad (20\text{-}11)$$

where

$$\sigma \equiv \text{Stefan-Boltzmann constant} = \frac{2\pi^5 k^4}{15 h^3 c^2} = 5.669\,8 \times 10^{-8}\ \frac{\text{W}}{\text{m}^2\text{K}^4} \qquad (20\text{-}12)$$

Here T is temperature, k is Boltzmann's constant, h is Planck's constant, and c is the speed of light. A blackbody whose temperature is $1\,000$ K radiates 5.67×10^4 watts per square meter of surface area. If the temperature is doubled, the exitance increases by a factor of $2^4 = 16$.

The emission spectrum changes with temperature (Figure 20-6). At higher temperatures, the maximum shifts to shorter wavelengths (higher

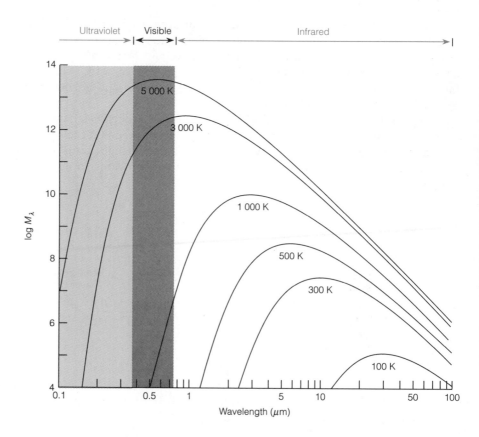

Figure 20-6 Spectral distribution of blackbody emission. This is a graph of Equation 20-13. The units of M_λ are W/m^3. Note that both axes are logarithmic.

energies). The wavelength dependence of the exitance is given by the *Planck distribution*:

Planck distribution for
blackbody radiation:
$$M_\lambda = \frac{2\pi hc^2}{\lambda^5}\left(\frac{1}{e^{hc/\lambda kT}-1}\right) \qquad (20\text{-}13)$$

where λ is wavelength, T is temperature in kelvins, and h, c, and k are the same as in Equation 20-12. The units of M_λ (W/m^3) can be thought of as watts per square meter of surface per meter of wavelength. The area under each curve between two wavelengths in Figure 20-6 is equal to the power per unit area (W/m^2) emitted between those two wavelengths.

At low temperature, the maximum emission in Figure 20-6 occurs at infrared wavelengths. At temperatures above several thousand degrees (such as the temperature of the sun's surface; Box 20-1), emission is greatest at visible wavelengths. For temperatures of 100 K and above, an excellent approximation for the wavelength of maximum emission (λ_{max}) in Figure 20-6 is given by the *Wien displacement law*:

Wavelength of
maximum emission:
$$\lambda_{max} \cdot T = 2.878 \times 10^{-3} \text{ m} \cdot \text{K} \qquad (20\text{-}14)$$

where T is temperature in kelvins.

The **emissivity** of an object is defined as the radiant power emitted by the object divided by the radiant power emitted by a blackbody at the same temperature:

$$\text{emissivity} = \frac{\text{radiant power emitted by object}}{\text{radiant power emitted by blackbody}} \qquad (20\text{-}15)$$

Emissivity is a fraction between zero and one.

20-2 The Spectrophotometer

Figure 6-4 described the operation of a *single-beam* spectrophotometer. Light from an appropriate *source* is passed through a *monochromator* that selects one narrow band of wavelengths to pass through the sample. The *detector* measures the radiant power after it has passed through the sample. The instrument computes transmittance by dividing the radiant power that has passed through the sample by the radiant power that has passed through a blank sample in a well-matched cell. Ideally, the blank has the same reflection and scatter as the sample, and the only difference is due to absorption. Single-beam instruments normally measure transmittance at just one wavelength at a time.

In the *double-beam* spectrophotometer in Figure 6-5, a rotating *beam chopper* alternately directs the light through the sample or reference cells. The monochromator is mechanically scanned to produce a continuous measurement of transmittance versus wavelength. By comparing the sample and reference energies many times each second, the double-beam instrument compensates for drift in the intensity of the source and response of the detector with time and wavelength. The spectrum produced by the instrument is a graph of transmission or absorbance versus wavelength.

Figure 20-7 shows the optical schematic of a double-beam infrared spectrophotometer. The light source is a resistively heated rod (that glows like a blackbody) near the right side of the diagram. Two mirrors at the far right direct polychromatic radiation through the sample and reference

A *two-color pyrometer* (infrared thermometer) indicates the temperature of an object by measuring the ratio of infrared energy emitted by the object at two wavelengths. Using Equation 20-13, this ratio gives the temperature of the object.

Challenge Explain how an infrared camera can "see" at night. Why is nighttime aerial infrared photography used to search for a lost person in the wilderness?

Monochromatic: one wavelength

Polychromatic: many wavelengths

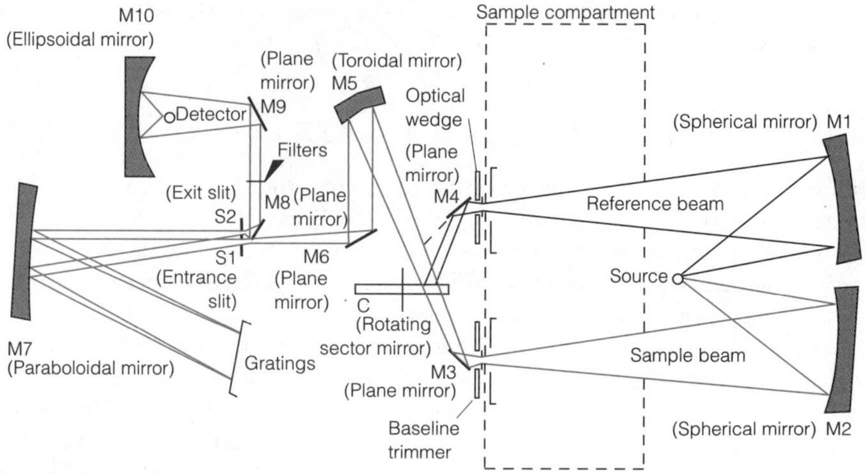

Figure 20-7 Optical schematic of the Perkin-Elmer 1320 infrared spectrophotometer. [Courtesy Perkin-Elmer Corp., Norwalk, CT.]

compartments. The rotating mirror at the center of the diagram alternately admits beams from the sample or reference into the monochromator at the left. The key component of the monochromator is a *grating* that reflects just one of the many wavelengths from the entrance slit back to the exit slit. The narrower the exit slit, the narrower the band of wavelengths sent to the detector.

All spectrophotometers must have some type of readout device. In the least expensive instruments, this might be a meter or digital display. More expensive instruments display data with a strip chart or computer. Now that we have had a brief overview of how spectrophotometers work, we will describe the various components in some detail.

20-3 Lamps and Lasers: Sources of Light

A *tungsten lamp* is an excellent source of visible and near-infrared radiation. A typical tungsten filament operates at a temperature near 3 000 K and produces useful radiation in the range 320 to 2 500 nm (Figure 20-8). This range covers the entire visible region and parts of the infrared and ultraviolet

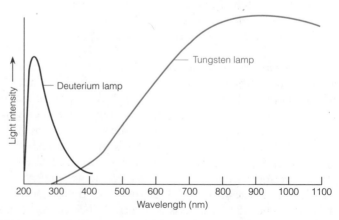

Figure 20-8 Intensity of a tungsten filament at 3 200 K and a deuterium arc lamp.

regions as well. Ultraviolet spectroscopy normally employs a *deuterium arc lamp* in which an electric discharge (a spark) causes D_2 to dissociate and emit ultraviolet radiation in the approximate range 200 to 400 nm (Figure 20-8). In a typical ultraviolet-visible spectrophotometer, a change is made between deuterium and tungsten lamps when passing through 360 nm, so that the source giving the best signal-to-noise ratio is always employed. For both visible and ultraviolet regions, electric discharge (spark) lamps filled with mercury vapor or xenon gas are also widely used.

Mid-infrared radiation (4 000 to 200 cm^{-1}) is commonly obtained from a silicon carbide rod called a *globar*, heated to near 1 500 K by passage of an electric current through the rod. The hot globar emits radiation with approximately the same spectrum as a blackbody at 1 000 K (Figure 20-6).

Lasers provide monochromatic radiation for many applications. How monochromatic is a laser? A laser with a wavelength of 3 μm might have a **bandwidth** (range of wavelengths) of 3×10^{-14} to 3×10^{-8} μm. The bandwidth is measured at the frequencies at which the power falls to half of its maximum value. The brightness of a low-power laser at its frequency of operation is 10^{13} times greater than that of the sun at its brightest (yellow) wavelength. (Of course, the sun emits all wavelengths, whereas the laser emits only one. The total brightness of the sun is much greater than that of the laser.) The angular divergence of the laser beam from its direction of travel is typically less than 0.05°. Laser light is typically *plane polarized*, with the electric field oscillating in one plane perpendicular to the direction of travel (Figure 6-1). Another characteristic of laser light is *coherence*, which means that all light waves emerging from the laser oscillate in phase with one another.

A necessary condition for lasing is *population inversion*, in which a higher energy state has a greater population (n) than a lower energy state in the lasing medium. In Figure 20-9a, this occurs when the population of state E_2 exceeds that of E_1. Molecules in ground state E_0 of the lasing medium are *pumped* to excited state E_3 by broadband radiation from a powerful lamp or by an electric discharge. Molecules in state E_3 rapidly relax to E_2, which has a

Ultraviolet radiation is harmful to the naked eye. Do not view an ultraviolet source without protection.

Properties of laser light:

monochromatic:	one wavelength
extremely bright:	high power at one wavelength
collimated:	parallel rays
polarized:	electric field of waves oscillating in one plane
coherent:	all waves in phase

Disadvantages of a laser:

expensive
high maintenance
limited wavelengths

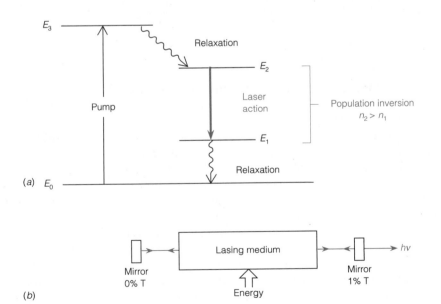

(a)

(b)

Figure 20-9 (*a*) Energy-level diagram illustrating the principle of operation of a laser. (*b*) Basic components of a laser. The population inversion is created in the lasing medium. Pump energy might be derived from intense lamps or an electric discharge.

Box 20-1 Blackbody Radiation and the Greenhouse Effect

The outer region of the sun behaves very much like a blackbody with a temperature near 5 800 K, emitting mainly *visible* light. Of the solar flux of 1 368 W/m² reaching the upper atmosphere, 23% is absorbed by the atmosphere and 25% is reflected back into space. The earth absorbs 48% of the solar flux and reflects 4%. Radiation reaching the earth should be just enough to keep the surface temperature at 254 K, which would not support life as we know it. Why does the average temperature of the earth's surface stay at a comfortable 287 K?

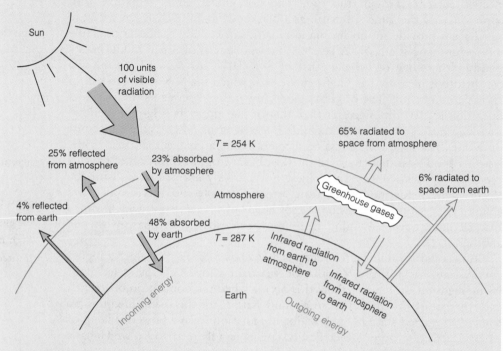

Balance between energy reaching the earth from the sun and energy reradiated to space. Exchange of infrared radiation between the earth and its atmosphere keeps the earth's surface 33 K warmer than the upper atmosphere.

relatively long lifetime. After a molecule in E_2 decays to E_1, it rapidly relaxes to the ground state, E_0 (thereby keeping the population of E_2 greater than the population of E_1).

A photon with an energy that exactly spans two states can be absorbed to raise a molecule to an excited state. Alternatively, that same photon can stimulate the excited molecule to emit a photon and return to the lower state. This is called *stimulated emission*. When a photon emitted by a molecule falling from E_2 to E_1 strikes another molecule in E_2, a second photon can be emitted with the same phase and polarization as the incident photon. If there is a population inversion ($n_2 > n_1$), one photon stimulates the emission of many photons as it travels through the laser.

Figure 20-9b shows the essential components of a laser. Pump energy directed through the side of the lasing medium creates the population inversion. On one end of the laser cavity is a mirror that reflects all light. On

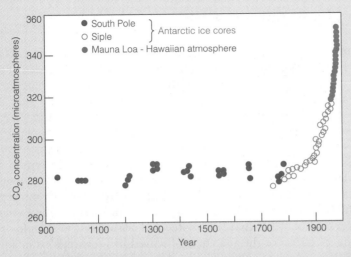

Thousand-year record documenting rise in atmospheric CO_2 concentration is derived from measurements of CO_2 trapped in ice cores in Antarctica and from direct atmospheric measurements at Mauna Loa in Hawaii. [From J. L. Sarmiento, *Chem. Eng. News* 31 May 1993, p. 30.]

The blackbody radiation curves in Figure 20-6 tell us that the earth radiates mainly *infrared* light, rather than visible light. Although the atmosphere is transparent to incoming visible light, it strongly absorbs outgoing infrared radiation. The main absorbers, called *greenhouse gases,* are H_2O, CO_2, O_3, CH_4, chlorofluorocarbons, and N_2O. Radiation emitted from the earth is absorbed by the atmosphere and part of it is reradiated back to the earth. The atmosphere thereby behaves like an insulating blanket, maintaining the earth's surface temperature 33 K warmer than the temperature of the upper atmosphere.

Man's activities since the dawn of the Industrial Revolution have increased atmospheric carbon dioxide levels markedly through the burning of fossil fuel. Our ability to predict how the earth will respond to the rise in greenhouse gas concentration is primitive. Will there be disastrous climatic changes? Will there be compensating responses that lead to little temperature change? We cannot accurately answer these questions, but prudence suggests that we should avoid making such large relative changes in our atmosphere.

the other end is a partially transparent mirror that reflects most light. Photons with energy $E_2 - E_1$ that bounce back and forth between the mirrors stimulate an avalanche of new photons. The small fraction of light passing through the partially transparent mirror at the right is the useful output of the laser.

A helium-neon laser is a common source of red light with a wavelength of 632.8 nm and an output power of 0.1–25 mW. An electric discharge pumps helium atoms to state E_3 in Figure 20-9. The excited helium transfers energy by colliding with a neon atom, raising the neon to state E_2. The high concentration of helium and intense electric pumping create a population inversion among neon atoms.

Another convenient source of laser light is the *laser diode,* in which population inversion of charge carriers in a semiconductor is achieved by a very high electric potential across a *p-n* junction in gallium arsenide.[3]

Currently available diodes operate at wavelengths in the range 0.68 to 1.55 μm.

20-4 Liquid Sample Cells

Box 20-2 describes a continuous analytical method that makes use of flow cells.

Approximate low-energy cutoff for common infrared windows:

sapphire (Al_2O_3)	1 500 cm^{-1}
NaCl	650 cm^{-1}
KBr	350 cm^{-1}
AgCl	350 cm^{-1}
CsBr	250 cm^{-1}
CsI	200 cm^{-1}

Cells come in all sizes and shapes appropriate for various experiments (Figure 20-10). The most common cuvets for visible and ultraviolet spectra are made of quartz, have a 1.000-cm pathlength, and are sold in matched sets (for sample and reference). Glass cells are suitable for visible but not for ultraviolet spectroscopy (because they absorb ultraviolet radiation). Quartz is transparent through the normally accessible visible and ultraviolet regions.

Cells for infrared measurements on liquids are commonly constructed of NaCl, KBr, or AgCl. For measurements in the 400–50 cm^{-1} region, polyethylene is a suitable transparent window material. Solid samples are commonly ground to a fine powder, which can be added to mineral oil (a viscous hydrocarbon also called Nujol) to give a dispersion called a *mull*. The mull is pressed between two KBr plates and placed in the sample compartment. The spectrum of the analyte is obscured in a few regions in which the mineral oil absorbs infrared radiation. Alternatively, the fine powder can be combined with excess KBr (KBr/solid ≈ 100 g/g), and the resulting mixture pressed into a translucent pellet at a pressure of ~60 MPa (600 atm). (The best KBr pellets are transparent if the KBr and sample are both dry.) Solids and powders also can be examined by *diffuse reflectance,* in which reflected infrared radiation, instead of transmitted infrared radiation, is observed. Wavelengths absorbed by the sample will not be reflected as well as

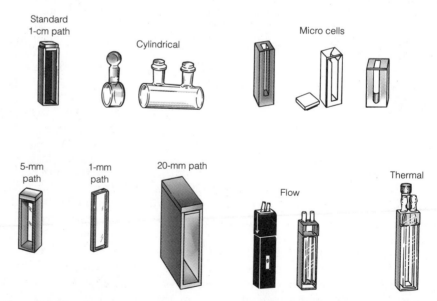

Figure 20-10 Common cuvets for visible and ultraviolet spectroscopy. Flow cells allow for continuous flow of solution through the cell. They are especially useful for measuring the absorbance of a solution emerging from a chromatography column. The thermal cell uses liquid from a constant-temperature bath to flow through the cell jacket and maintain the contents of the cell at a desired temperature. [Courtesy A. H. Thomas Co., Philadelphia, PA.]

other wavelengths. This technique is sensitive only to the surface of the sample.

Quantitative spectrophotometry is dependent on the availability of a suitable reference sample. The reference might be solvent or a reagent blank (containing all reagents except analyte) in a cell that is identical to the sample cell. Reflection, scattering, and absorption by the cell, solvent, and other reagents should be virtually the same in both samples. The differences in transmittance between the sample and the reference, then, will be due to the analyte. Only under these conditions can the absorbance of the sample be equated to $\log P_0/P$.

20-5 Monochromators

Light of one wavelength (or color) is said to be *monochromatic*. A *grating* is the most common device used to disperse light into its component wavelengths. *Prisms* were used in older instruments, and *filters* are widely used to select or reject relatively broad bands of radiation.

Gratings[4]

A typical design of a grating monochromator is shown in Figure 20-11. *Polychromatic* radiation from the entrance slit is *collimated* into a beam of parallel rays by a concave mirror. These rays fall on a reflection **grating,** whereupon different wavelengths are *diffracted* at different angles. The light strikes a second concave mirror, which focuses each wavelength at a different point on the focal plane. The orientation of the reflection grating directs only one narrow band of wavelengths to the exit slit of the monochromator. By rotation of the grating, different wavelengths are allowed to pass through the exit slit.

The principle of the **diffraction** grating is shown in Figure 20-12. The grating is ruled with a series of closely spaced parallel grooves with a repeat

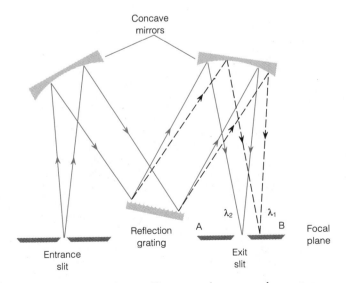

Figure 20-11 Diagram of a Czerny-Turner grating monochromator.

Box 20-2 Flow Injection Analysis

In **flow injection analysis,** a sample is injected into a liquid carrier stream to which various reagents can be added. After a suitable time, the reacted sample reaches a detector, which is usually a spectrophotometric cell. In the first schematic diagram here, a solvent carrier stream (usually an aqueous solution) is pumped continuously through the sample injector, in which 40- to 200-μL volumes of sample can be added to the stream. A reagent stream is then combined with the carrier stream, and the resulting solution passes through a coil to give the reagent and sample time to react. By measuring the absorbance of the stream in a cell like that shown in Figure 23-15, the concentration of analyte in the sample is determined. Typical flow rates are 0.5–2.5 mL/min and the diameter of the inert Teflon tubing from which the system is constructed is about 0.5 mm. Coils are 10–200 cm in length to allow suitable reaction times. Replicate analyses can be completed in 20–60 s.

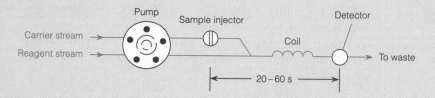

The next schematic gives an idea of variations possible with flow injection analysis. In this example, reagents 1 and 2 are mixed and added to the sample in the carrier before reagent 3 is added. Commercial equipment allows different flow paths to be readily assembled. Streams can be passed through reactive columns, ion exchangers, dialysis tubes, gas diffusers, and solvent extractors. Detectors might measure absorbance, fluorescence, or luminescence, or they could utilize potentiometry or amperometry.

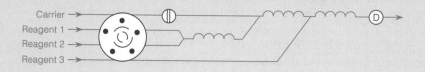

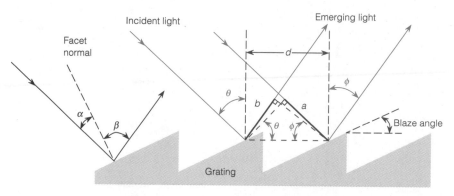

Figure 20-12 Diagram illustrating the principle of a reflection grating.

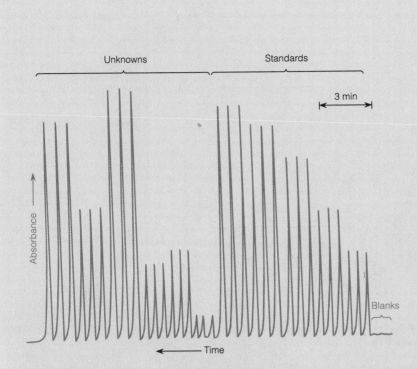

Continuous flow analysis of ethanol in beverages. The time axis goes from right to left. [From P. J. Worsfold, *Chem. Br.* **1988,** *24,* 1215.]

A key feature of flow injection is rapid, repetitive analysis. The chart above is an example in which standards and samples were run in triplicate to measure the ethanol content of beverages. The method makes use of an enzymatic reaction that produces a colored product. Flow injection is widely used in medical and pharmaceutical analysis, water analysis, and industrial process control.

distance *d*. The grating is coated with aluminum to make it reflective. On top of the aluminum is a thin protective layer of silica (SiO_2) to prevent the metal surface from tarnishing (oxidizing), a reaction that would reduce its reflectivity. When light is reflected from the grating, each groove behaves as a source of radiation. When adjacent light rays are in phase, they reinforce each other. When they are not in phase, they partially or completely cancel each other (Figure 20-13).

Consider the two rays shown in Figure 20-12. Fully constructive interference will occur only if the difference in length of the two paths is exactly equal to the wavelength of light. The difference in path is equal to the distance *a* − *b* in Figure 20-12. Constructive interference occurs if

$$n\lambda = a - b \qquad (20\text{-}16)$$

where $n = \pm 1, \pm 2, \pm 3, \pm 4, \cdots$. The interference maximum for which $n =$

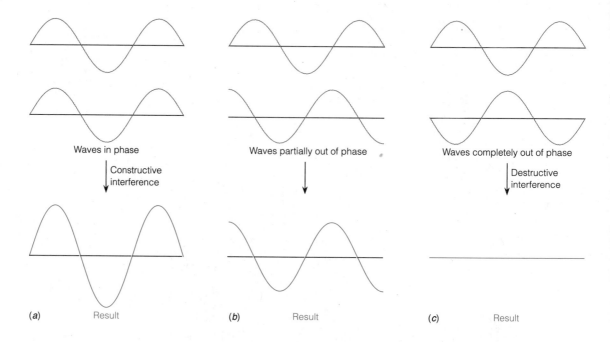

Figure 20-13 Interference of adjacent waves that are (*a*) 0°, (*b*) 90°, and (*c*) 180° out of phase.

±1 is called *first-order diffraction*. When $n = 2$, we have *second-order diffraction,* and so on.

From the geometry in Figure 20-12, we see that $a = d \sin \theta$ and $b = d \sin \phi$. Therefore, the condition for constructive interference is

Grating equation:
$$n\lambda = d(\sin \theta - \sin \phi) \qquad (20\text{-}17)$$

For each incident angle θ, there is a series of reflection angles ϕ, at which a given wavelength will produce maximum constructive interference. Some spectrophotometers contain two monochromators in series (a *double monochromator*) to reduce unwanted radiation by several orders of magnitude.

Resolution, Dispersion, and Efficiency of a Grating

Resolution measures the ability to separate two closely spaced peaks. The greater the resolution, the smaller is the difference ($\Delta \lambda$) between two wavelengths that can be distinguished from each other. The precise definition (which is beyond the scope of our discussion) means that the valley between the two peaks is about three-fourths of the height of the peaks when they are just barely resolved. The resolution of a grating is given by

Resolution of grating:
$$\frac{\lambda}{\Delta \lambda} = nN \qquad (20\text{-}18)$$

For example, a resolution of 10^4 means that two peaks with a wavelength near 1 μm are just resolved if they differ in wavelength by 10^{-4} μm.

where λ is wavelength, n is the diffraction order in Equation 20-17, and N is the number of grooves of the grating that are illuminated. The larger the number of grooves in a grating, the more closely can different wavelengths be resolved. Equation 20-18 tells us that if we desire a first-order resolution of 10^4, there must be 10^4 grooves in the grating. If the grating has a ruled length of 10 cm, we require 10^3 grooves/cm.

Dispersion measures the ability to separate wavelengths differing by $\Delta\lambda$ through the difference in angle, $\Delta\phi$ (radians). For the grating in Figure 20-12, the dispersion is

Dispersion of grating:
$$\frac{\Delta\phi}{\Delta\lambda} = \frac{n}{d\cos\phi} \qquad (20\text{-}19)$$

Dispersion increases with decreasing groove spacing, which increases resolution. Equation 20-19 tells us that a grating with 10^3 grooves/cm provides a resolution of 0.102 radians (5.8°) per micrometer of wavelength if $n = 1$ and $\phi = 10°$. Wavelengths differing by 1 μm would be separated by an angle of 5.8°.

Decreasing the monochromator exit slit width decreases the bandwidth of radiation and decreases the energy reaching the detector. Thus, *resolution of closely spaced absorption bands is achieved at the expense of decreased signal-to-noise ratio.* For quantitative analysis, a monochromator bandwidth that is $\leq 1/5$ of the width of the absorption band is reasonable. (Figure 20-22 shows that an even narrower bandwidth is required to avoid distorting the spectrum.)

The relative *efficiency* of a grating (typically 45–80%) is defined as

$$\text{relative efficiency} = \frac{E_\lambda^n(\text{grating})}{E_\lambda(\text{mirror})} \qquad (20\text{-}20)$$

where $E_\lambda^n(\text{grating})$ is the radiant energy at a particular wavelength diffracted in the order of interest (n), and E_λ (mirror) is the radiant energy at the same wavelength that would be reflected by a mirror with the same coating as the grating. Efficiency is partially controlled by the *blaze angle* at which the grooves are cut in Figure 20-12. To direct a certain wavelength into the diffraction order of interest, the blaze angle is chosen to give *specular reflection* ($\alpha = \beta$ in Figure 20-12). Because each grating is optimized for a limited range of wavelengths, a spectrophotometer may use several different gratings to scan through its entire spectral range.

Filters

It is frequently necessary to filter (remove) wide bands of radiation from a signal. For example, the grating monochromator in Figure 20-11 directs first-order diffraction of a small wavelength band to the exit slit. (By "first-order," we mean diffraction for which $n = 1$ in Equation 20-17.) Let λ_1 be the wavelength whose first-order diffraction reaches the exit slit. Inspection of Equation 20-17 shows that if $n = 2$, the wavelength $\frac{1}{2}\lambda_1$ also reaches the same exit slit (because the wavelength $\frac{1}{2}\lambda_1$ also gives constructive interference at the same angle as λ_1 does). For $n = 3$, the wavelength $\frac{1}{3}\lambda_1$ also reaches the slit. One solution for selecting just λ_1 is to place a **band-pass filter** in the beam, so that only some range of wavelengths around λ_1 can pass through the filter. The very different wavelengths $\frac{1}{2}\lambda_1$ and $\frac{1}{3}\lambda_1$ will be blocked by the filter. To cover a wide range of wavelengths, it may be necessary to use several filters and to change them as the desired wavelength region changes.

The simplest kind of filter is just colored glass, in which the coloring species absorbs a broad portion of the spectrum and transmits other portions. Table 6-1 showed the relation between absorbed and transmitted colors. For finer control of filtering behavior, specially designed *interference filters* are constructed to pass radiation in the region of interest and reflect

Resolution: ability to distinguish two closely spaced peaks

Dispersion: ability to produce angular separation of adjacent wavelengths

Trade-off between resolution and signal: The narrower the exit slit, the greater the resolution and the noisier the signal.

Filters permit certain bands of wavelength to pass through.

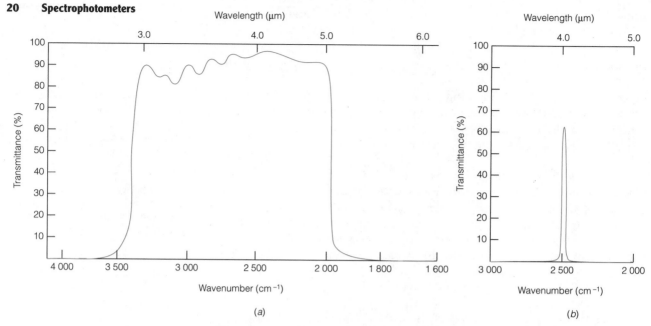

Figure 20-14 Transmission spectra of infrared interference filters. (*a*) Wide band-pass filter has ~90% transmission in the 3- to 5-μm-wavelength range but <0.01% transmittance outside this range. (*b*) Narrow band-pass filter has a width of 0.1 μm centered around 4 μm. [Courtesy Barr Associates, Inc., Westford, MA.]

other wavelengths (Figure 20-14). These devices derive their performance from constructive or destructive interference of light waves within the filter.

20-6 Detectors

Detector response is a function of wavelength of incident light.

A detector produces an electric signal when it is struck by photons. For example, a **phototube** emits electrons from a photosensitive, negatively charged surface when struck by visible light or ultraviolet radiation. The electrons flow through a vacuum to a positively charged collector whose current is proportional to the radiation intensity.

Figure 20-15 shows that detector response depends on the wavelength of the incident photons. For example, for a given energy flux (W/m^2) of 420-

Figure 20-15 Response of several different detectors. The greater the sensitivity, the greater the output (current or voltage) of the detector for a given incident power (watts per unit area) of photons. Each curve is normalized to a maximum value of 1. [Courtesy Barr Associates, Inc., Westford, MA.]

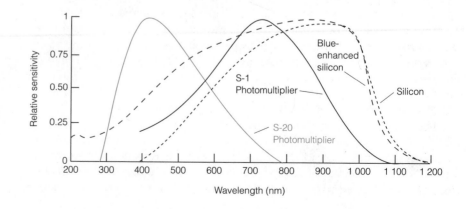

nm light, the S-20 photomultiplier produces a current about four times greater than the current produced for the same energy flux of 300-nm radiation. The response below 280 nm and above 800 nm is essentially zero. In a single-beam spectrophotometer, the 100% transmittance control must be readjusted each time the wavelength is changed. This calibration adjusts the spectrophotometer to the maximum detector output that can be obtained at each wavelength. Subsequent readings are scaled to the 100% reading.

Photomultiplier Tube

A **photomultiplier tube** (Figure 20-16) is a very sensitive device in which electrons emitted from the photosensitive surface strike a second surface, called a *dynode,* which is kept positive with respect to the photosensitive emitter. Electrons are accelerated and strike the dynode with more than their original kinetic energy. These energetic electrons cause more electrons to be emitted from the dynode than those striking the dynode. These new electrons are accelerated toward a second dynode, which is more positive than the first dynode. Upon striking the second dynode, even more electrons are knocked off and accelerated toward a third dynode. This process is repeated several times so that more than 10^6 electrons are finally collected for each photon striking the first surface. Extremely low light intensities are translated into measurable electric signals.

Photodiode Array[5]

Conventional spectrophotometers scan through a spectrum one wavelength at a time. Newer instruments record the entire spectrum at once in a fraction of a second. One application of rapid scanning is chromatography, in which

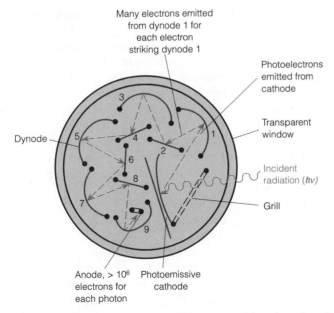

Figure 20-16 Diagram of a photomultiplier tube with nine dynodes. Amplification of the signal occurs at each dynode, which is approximately 90 volts more positive than the previous dynode.

Figure 20-17 (*a*) Schematic cross-sectional view of photodiode array. (*b*) Photograph of array with 1 024 elements, each 25 μm wide and 2.5 mm high. The central black rectangle is the photosensitive area. The entire chip is 5 cm in length. [Courtesy Oriel Corp., Stratford, CT.]

For a refresher on semiconductors, see Section 15-8.

A typical photodiode array has a response curve similar to that of the blue-enhanced silicon in Figure 20-15.

the full spectrum of a compound is recorded in seconds as it emerges from the chromatography column.

At the heart of rapid spectroscopy is the **photodiode array** shown in Figure 20-17. Rows of *p*-type silicon on a substrate (the underlying body) of *n*-type silicon create a series of *p-n* junction diodes. A reverse bias is applied to each diode, drawing electrons and holes away from the junction. There is a depletion region at each junction, in which there are few electrons and holes. The junction functions as a capacitor, with charge stored on either side of the depletion region. At the beginning of the measurement cycle, each diode is fully charged.

When radiation strikes the semiconductor, free electrons and holes are created and migrate to regions of opposite charge, partially discharging the capacitor. The more radiation that strikes each diode, the less charge remains at the end of the measurement. The longer the array is irradiated between readings, the more each capacitor is discharged. The state of each capacitor is determined at the end of the cycle by measuring the current needed to recharge the capacitor.

The key feature of a spectrophotometer utilizing a photodiode array is that *white light* (with all wavelengths) is passed through the sample (Figure 20-18). The light then enters a **polychromator,** which disperses the light into its component wavelengths and directs the light at the diode array. *A different wavelength band strikes each diode,* and the resolution depends on how closely spaced the diodes are and how much dispersion is produced by

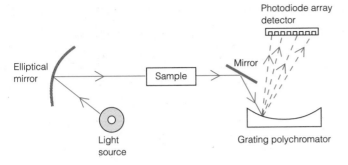

Figure 20-18 Schematic design of photodiode array spectrophotometer.

the polychromator. The spectrum in Figure 21-23 (in the next chapter) was produced with a photodiode array detector.

Charge Coupled Device

A **charge coupled device** is an extremely sensitive detector that stores photo-generated charge in a two-dimensional array. The device in Figure 20-19a is constructed of p-doped Si on an n-doped substrate. This diode structure is capped with an insulating layer of SiO_2, on top of which is placed a pattern of conducting Si electrodes. When light is absorbed in the p-doped region, an electron is introduced into the conduction band and a hole is left in the valence band. The electron is attracted to the region beneath the positive electrode, where it is stored. The hole migrates to the n-doped substrate, where it combines with an electron. Each electrode can store 10^5 electrons before electrons spill out into adjacent elements.

The charge coupled device is a two-dimensional array, as shown in Figure 20-19b. After the desired observation time, electrons stored in each *pixel*

Very sensitive television cameras use charge coupled devices to record the video image.

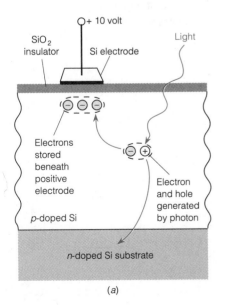

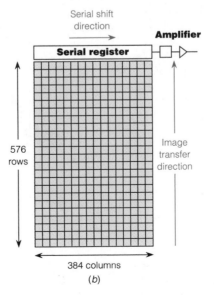

Figure 20-19 Schematic representation of a charge coupled device. (*a*) Cross-sectional view indicating charge generation and storage in each pixel. (*b*) Top view showing two-dimensional nature of one particular array. An actual array is about the size of a postage stamp.

TABLE 20-2 Minimum detectable signal (photons/s/detector element) of ultraviolet/visible detectors

Signal acquisition time (s)	Photodiode array		Photomultiplier tube		Charge coupled device	
	Ultraviolet	Visible	Ultraviolet	Visible	Ultraviolet	Visible
1	6 000	3 300	30	122	31	17
10	671	363	6.3	26	3.1	1.7
100	112	62	1.8	7.3	0.3	0.2

SOURCE: R. B. Bilhorn, J. V. Sweedler, P. M. Epperson, and M. B. Denton, *Appl. Spec.* **1987,** *41,* 1114.

Electrons from several adjacent pixels can be combined to create a single, larger picture element. This process, called *binning,* increases the sensitivity of the charge coupled device at the expense of resolution.

(picture element) of the top row are moved into the serial register at the top, and then moved, one pixel at a time, to the top right position, where the stored charge is read out. Then the next row is moved up and read out, and the sequence is repeated until the entire array has been read. The transfer of stored charges is carried out by an array of electrodes considerably more complex than we have indicated in Figure 20-19a. The transfer of charge from one pixel to the next is extremely efficient, with a loss of approximately five of every million electrons.

The minimum detectable signal for visible light in Table 20-2 is 17 photons per second. The sensitivity of the charge coupled device is derived from its high *quantum efficiency* (electrons generated per incident photon), low background electrical noise (thermally generated free electrons), and low noise associated with readout. Figure 20-20 compares spectra recorded under the same conditions by a photomultiplier tube and by a charge coupled device. Although a photomultiplier is very sensitive, the charge coupled device is even better.

Figure 20-21 shows an example of how a charge coupled device can be used to record a rapidly changing event. The grating polychromator illuminates one masked row of the charge coupled device. Each pixel of that row records the light intensity at a different wavelength. After a preselected time, the second row is unmasked and the first row is masked. The second row then records the next spectrum. Each row of the device records a spectrum at a different time. When the process is finished, the rows are read sequentially to reconstruct the event.

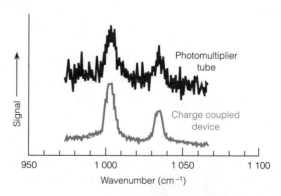

Figure 20-20 Comparison of spectra recorded in 5 min by a photomultiplier tube and a charge coupled device. [From P. M. Epperson, J. V. Sweedler, R. B. Bilhorn, G. R. Sims, and M. B. Denton, *Anal. Chem.* **1988,** *60,* 327A.]

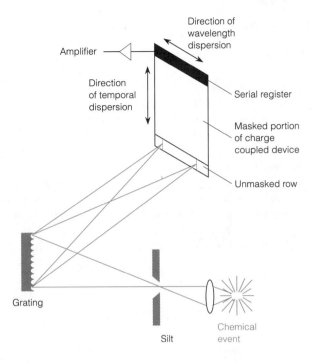

Amplifier

Direction of
wavelength
dispersion

Serial register

Direction
of temporal
dispersion

Masked portion
of charge
coupled device

Unmasked row

Grating

Silt

Chemical
event

Figure 20-21 Use of a charge coupled device for time-resolved spectroscopy. The mask is moved periodically, so that each row of the detector records a spectrum at a different time. [From P. M. Epperson, J. V. Sweedler, R. B. Bilhorn, G. R. Sims, and M. B. Denton, *Anal. Chem.* **1988,** *60,* 327A.]

Infrared Detectors

Several kinds of devices are used for infrared detection. A **thermocouple** is a junction between two different electrical conductors. Electrons have a lower free energy in one conductor than in the other, so they flow from one to the other until the resulting small voltage difference prevents further flow. The junction potential is temperature dependent because electrons can flow back to the high-energy conductor at higher temperature. When the thermocouple is blackened so that it absorbs radiation, its temperature (and hence the voltage) becomes sensitive to radiation. A typical sensitivity is 6 V per watt of radiation absorbed.

A **ferroelectric material,** such as deuterated triglycine sulfate, has a permanent electric polarization because of alignment of the molecules in the crystal. One face of the crystal is positively charged and the opposite face is negative. The polarization is temperature dependent, and the variation with temperature is called the *pyroelectric effect.* When the crystal absorbs infrared radiation, its temperature and polarization change. This change is the signal in a pyroelectric detector.

A **Golay cell** is an infrared detector that uses thermal expansion of a gas in a blackened chamber to deform a flexible mirror. A beam of light is reflected from the mirror onto a photocell. The change in radiant power reaching the photocell as the mirror is deflected is the signal in this device.

A **bolometer,** also called a *thermistor,* is a device whose electrical resistance changes with temperature. For example, the resistance of flakes of oxides of Ni, Co, or Mn decreases by about 4% per degree Celsius. In a bolometer, the temperature of a flake increases when infrared radiation is absorbed by the flake, and the change in resistance is the detector signal. These oxides are semiconductors whose electrical conductivity increases when heat excites electrons from the valence band to the conduction band. A

In a *ferroelectric material,* the dipole moments of the molecules remain aligned in the absence of an external field. This gives the material a permanent electric polarization.

high-temperature superconductor (Box 16-3) poised at its critical temperature is an extremely sensitive bolometer. A slight increase in temperature destroys the superconductivity, and a temperature decrease reduces the resistance to near zero.

A **photoconductive detector** is a semiconductor whose electrical conductivity increases when infrared radiation directly excites electrons from the valence band to the conduction band. **Photovoltaic detectors** contain *p-n* semiconductor junctions, across which an electric field exists. Absorption of infrared radiation creates more electrons and holes, which are attracted to opposite sides of the junction and which change the voltage across the junction. This voltage change is the detector signal. Mercury cadmium telluride ($Hg_{1-x}Cd_xTe$, $0 < x < 1$) is a detector material whose sensitivity to different wavelengths of radiation is affected by the stoichiometry coefficient, *x*. Cooling photoconductive and photovoltaic devices to liquid nitrogen temperature reduces thermal electric noise by more than an order of magnitude.

20-7 Errors in Spectrophotometry

The manufacturer of a spectrophotometer tells us how much instrumental error is to be expected. Additional errors associated with sample preparation and cell positioning will increase the uncertainty in any measurement.

Choosing the Wavelength and Bandwidth

If spectrum 1 in Figure 20-22 is to be used for single-wavelength analytical measurements, it is sensible to choose a spectral peak to obtain the greatest absorbance for a given concentration of analyte. Beer's law (Equation 6-6), which states that absorbance is proportional to concentration, applies to monochromatic radiation. At the absorbance peak, the variation of absorbance with wavelength ($dA/d\lambda$) is a minimum. Therefore, the effect of using light that is not perfectly monochromatic will be minor, because the molar absorptivity (ϵ in Beer's law) is fairly constant over a small range of wavelengths.

The breadth of a spectral peak determines how great a monochromator bandwidth can be tolerated. We want to keep the bandwidth as wide as the spectrum permits, because the narrower the bandwidth, the less light will reach the detector. Less light gives a smaller signal-to-noise ratio and reduces the precision of the measurement. Figure 20-22 shows how widening the monochromator bandwidth distorts the peak shapes.

Instrumental Errors

Spectrophotometric measurements are most reliable for absorbances in the range 0.4 to 0.9. At lower absorbance, the radiant power coming through the sample is similar to the power coming through the reference, and there is a large relative uncertainty in the difference between these two numbers. At high absorbance, little light reaches the detector, thereby decreasing the signal-to-noise ratio.

Figure 20-23 shows errors measured for a research-quality spectrophotometer. Instrumental electrical noise was only modestly dependent on

For quantitative analysis:

1. Absorbance should not be too large or too small.
2. Absorbance should be measured at a peak or shoulder where $dA/d\lambda$ is small.
3. Monochromator bandwidth should be as large as possible, but small compared with the band being measured.

The most accurate spectrophotometric measurements generally occur when the absorbance lies between 0.4 and 0.9.

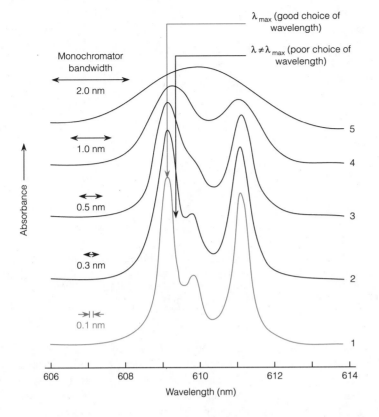

Figure 20-22 Choosing wavelength and monochromator bandwidth. Increasing the monochromator bandwidth broadens the bands and decreases the apparent absorbance of Pr^{3+} in a crystal of yttrium aluminum garnet (a laser material). In spectrum 1, the bandwidth is one-fifth of the absorption bandwidth (measured at half the peak height), which is a reasonable setting for negligible distortion. [Courtesy M. D. Seltzer, Michelson Laboratory.]

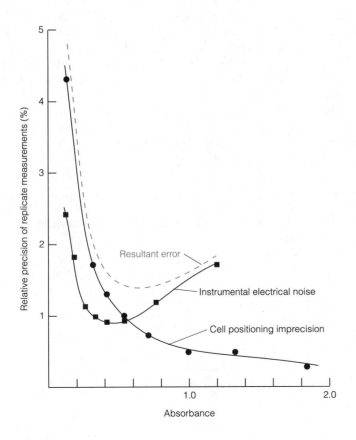

Figure 20-23 Errors in spectrophotometric measurements due to instrumental electrical noise and cell positioning imprecision. [Data from L. D. Rothman, S. R. Crouch, and J. D. Ingle, Jr., *Anal. Chem.* **1975,** *47,* 1226.]

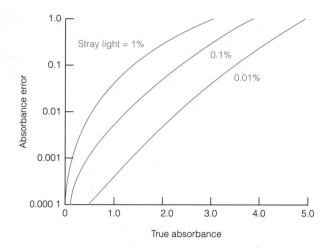

Figure 20-24 Absorbance error introduced by different levels of stray light. The stray light is expressed as a percentage of the radiant power incident on the sample. This illustration is from M. R. Sharp, *Anal. Chem.* **1984,** *53,* 339A, which provides an excellent discussion of stray light.

For greatest precision, samples should be free of dust, and cuvets must be free of fingerprints or other contamination.

sample absorbance. The largest source of imprecision for $A < 0.6$ was irreproducible positioning of the cuvet in the sample holder, despite care in placement and cleaning of the cuvet. The resulting error curve reaches a minimum near $A = 0.6$. At higher absorbances, instrumental noise is most important. Problem 19 in Chapter 4 showed that *the detection limit for an analytical procedure is determined by the reproducibility of the measurement.*

In every instrument, inadvertent **stray light** (with wavelengths outside the bandwidth expected from the monochromator) reaches the detector. Stray light coming through the monochromator from the light source arises from diffraction into unwanted orders and angles and unintended scattering from optical components and walls. Stray light can also come from outside the instrument if the sample compartment is not closed tightly or from openings into the sample chamber for tubing or wires. Figure 20-24 shows that the error introduced by stray light in quantitative analysis is most serious when the absorbance is high. The absorbance standard in Table 20-3 can be used to check the accuracy of your instrument.

TABLE 20-3 **Calibration standard for ultraviolet absorbance**

Wavelength (nm)	Absorbance of $K_2Cr_2O_7$ (60.06 mg/L) in 5.0 mM H_2SO_4 in 1-cm cell
235	0.748 ± 0.010
257	0.865 ± 0.010
313	0.292 ± 0.010
350	0.640 ± 0.010

SOURCE: S. Ebel, *Fresenius J. Anal. Chem.* **1992,** *342,* 769.

EXAMPLE Stray Light

If the true absorbance of a sample is 2.00 and there is 1.0% stray light, find the apparent absorbance.

Solution A true absorbance of 2.00 gives a true transmittance of $T = 10^{-A} = 10^{-2.00} = 0.010 = 1.0\%$. If, in addition, there is 1.0% stray light, then the sample appears to have a transmittance of 2.0% instead of 1.0%. The apparent absorbance is $-\log T = -\log(0.020) = 1.69$, instead of 2.00.

. . . .

20-8 Fourier Transform Infrared Spectroscopy[6]

We have seen that a photodiode array or charge coupled device can be used to measure an entire spectrum at once. The spectrum is spread into its component wavelengths, and each small band of wavelengths is directed onto one detector element. For the infrared region, the most important and widely used method for observing the entire spectrum at once is *Fourier transform spectroscopy*.

Fourier Analysis

Fourier analysis is a procedure in which a curve is decomposed into a sum of sine and cosine terms, called a *Fourier series*. This procedure forms the basis for Fourier transform infrared spectroscopy. To analyze the curve in Figure 20-25, which spans the interval $x_1 = 0$ to $x_2 = 10$, the Fourier series has the form

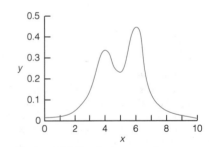

Figure 20-25 A curve to be decomposed into a sum of sine and cosine terms by Fourier analysis.

Fourier series:
$$y = a_0 \sin(0\omega x) + b_0 \cos(0\omega x) + a_1 \sin(1\omega x) + b_1 \cos(1\omega x)$$
$$+ a_2 \sin(2\omega x) + b_2 \cos(2\omega x) + \cdots$$
$$= \sum_{n=0}^{\infty} [a_n \sin(n\omega x) + b_n \cos(n\omega x)] \qquad (20\text{-}21)$$

where

$$\omega = \frac{2\pi}{x_2 - x_1} = \frac{2\pi}{10 - 0} = \frac{\pi}{5} \qquad (20\text{-}22)$$

Equation 20-21 says that the value of y for any value of x can be expressed by an infinite sum of sine and cosine waves. Successive terms correspond to sine or cosine waves with increasing frequency.

Figure 20-26 shows how sequences of three, five, or nine sine and cosine waves give better and better approximations to the curve in Figure 20-25. The coefficients a_n and b_n required to construct the curves in Figure 20-26 are given in Table 20-4. The larger the coefficient, the more that term contributes to the sum. The table tells us that the functions $\sin(2\omega x)$ and $\sin(3\omega x)$ contribute much more than $\sin(5\omega x)$.

Challenge Enter the data from Table 20-4 into a spreadsheet and confirm that the sine and cosine waves add together as shown in Figure 20-26.

TABLE 20-4   Fourier coefficients for Figure 20-26

n	a_n	b_n
0	0	0.136912
1	−0.006906	−0.160994
2	0.015185	0.037705
3	−0.014397	0.024718
4	0.007860	−0.043718
5	0.000089	0.034864
6	−0.004813	−0.018858
7	0.006059	0.004580
8	−0.004399	0.003019

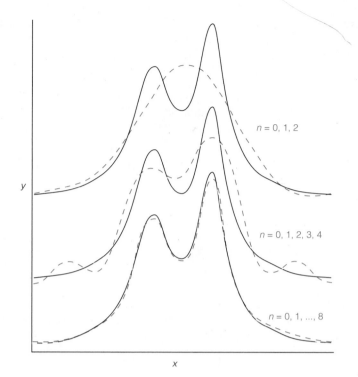

Figure 20-26   Fourier series reconstruction of the curve in Figure 20-25. Solid line is the original curve and dashed lines are made from a series of $n = 0$ to $n = 2$, 4, or 8 in Equation 20-21. Coefficients a_n and b_n are given in Table 20-4.

Albert Michelson developed the interferometer around 1880 and conducted the Michelson-Morley experiment in 1887, in which it was found that the speed of light is independent of the motion of the source and the observer. This crucial experiment led Einstein to the theory of relativity. Michelson also used the interferometer to create the predecessor of today's length standard based on the wavelength of light. He received the Nobel Prize in 1907 "for precision optical instruments and the spectroscopic and metrological investigations carried out with their aid."

Interferometry

The heart of a Fourier transform infrared spectrophotometer is the **interferometer** in Figure 20-27. Radiation from the source at the left strikes a *beamsplitter,* which transmits some light and reflects some light (Figure 20-27). For the sake of this discussion, let us consider a narrow beam of monochromatic radiation. For simplicity, suppose that the beamsplitter reflects half of the light and transmits half. When light strikes the beamsplitter at point O, some is reflected to a stationary mirror at a distance OS and some is transmitted to a movable mirror at a distance OM. The rays reflected by the mirrors travel back to the beamsplitter, where half of each ray is transmitted and half is reflected. One recombined ray travels in the direction of the detector, and another heads back to the source.

In general, the paths OM and OS are not equal, so the two waves reaching the detector are not in phase. As shown in Figure 20-13, if the two waves are in phase, they interfere constructively to give a wave with twice the amplitude. If the waves are one-half wavelength (180°) out of phase, they interfere destructively and cancel. For any intermediate-phase difference, there is partial cancellation.

The difference in pathlength followed by the two waves in the interferometer in Figure 20-27 is 2(OM − OS). This difference is called the *retardation,* δ. Constructive interference occurs whenever δ is an integral

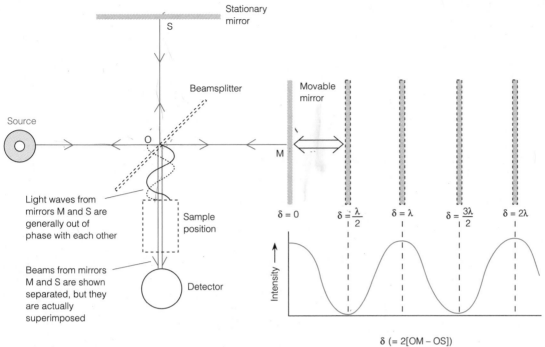

Figure 20-27 Schematic diagram of Michelson interferometer. Detector response as a function of retardation (= 2[OM − OS]) is shown for the case of monochromatic incident radiation of wavelength λ.

multiple of the wavelength (λ) of the light. A minimum appears when δ is a half-integral multiple of λ. If one mirror moves away from the beamsplitter at a constant speed, light reaching the detector goes through a steady sequence of maxima and minima as the interference alternates between constructive and destructive phases.

A graph of output light intensity versus retardation, δ, is called an **interferogram.** If the light from the source is monochromatic, the interferogram is a simple cosine wave:

$$I(\delta) = B(\tilde{\nu})\cos\left(\frac{2\pi\delta}{\lambda}\right) = B(\tilde{\nu})\cos(2\pi\tilde{\nu}\delta) \qquad (20\text{-}23)$$

where $I(\delta)$ is the intensity of light reaching the detector and $\tilde{\nu}$ is the wavenumber ($= 1/\lambda$) of the light. Clearly, I is a function of the retardation, δ. $B(\tilde{\nu})$ is a constant that accounts for the intensity of the light source, efficiency of the beamsplitter (which never gives exactly 50% reflection and 50% transmission), and response of the detector. All these factors depend on $\tilde{\nu}$. In the case of monochromatic light, there is only one value of $\tilde{\nu}$.

Figure 20-28a shows the interferogram produced by monochromatic radiation of wavenumber $\tilde{\nu}_0 = 2$ cm^{-1}. The wavelength (repeat distance) of the interferogram can be seen in the figure to be λ = 0.5 cm, which is equal to $1/\tilde{\nu}_0 = 1/(2$ cm$^{-1})$. Figure 20-28b shows the interferogram that results from a source with two monochromatic waves ($\tilde{\nu}_0 = 2$ and $\tilde{\nu}_0 = 8$ cm^{-1})

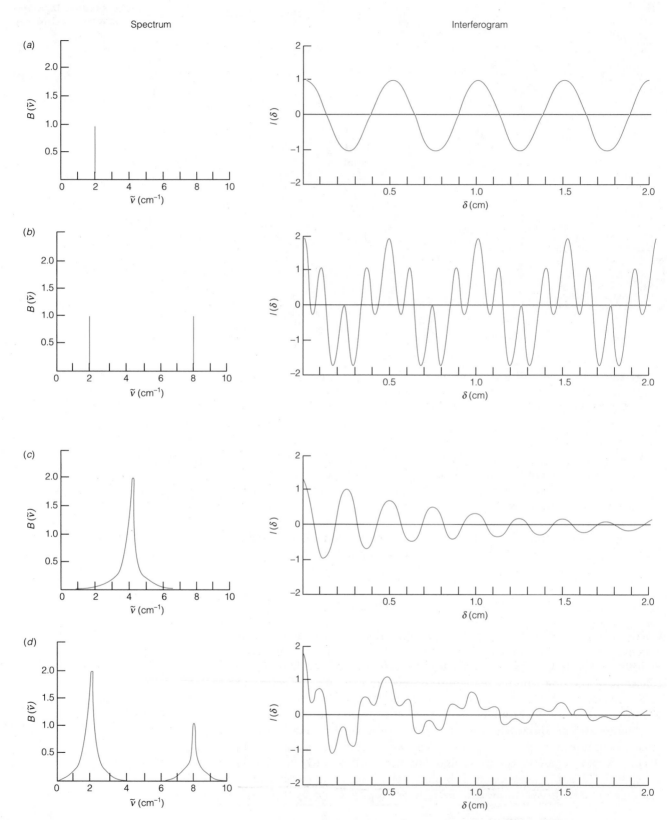

Spectrum   Interferogram

Figure 20-28 Interferograms produced by different spectra.

with relative intensities 1:1. The interferogram contains a short wave oscillation ($\lambda = \frac{1}{8}$ cm) superimposed on a long wave oscillation ($\lambda = \frac{1}{2}$ cm). The interferogram is a sum of two terms in this case:

$$I(\delta) = B_1 \cos(2\pi\tilde{\nu}_1\delta) + B_2\cos(2\pi\tilde{\nu}_2\delta) \qquad (20\text{-}24)$$

where $B_1 = 1$, $\tilde{\nu}_1 = 2$ cm^{-1}, $B_2 = 1$, and $\tilde{\nu}_2 = 8$ cm^{-1}.

Fourier analysis is a way to decompose a curve into its component wavelengths. Fourier analysis of the interferogram in Figure 20-28a gives the (trivial) result that the interferogram is made from a single wavelength function, with $\lambda = \frac{1}{2}$ cm. Fourier analysis of the interferogram in Figure 20-28b gives the slightly more interesting result that the interferogram is composed of two wavelengths ($\lambda = \frac{1}{2}$ and $\lambda = \frac{1}{8}$ cm) with relative contributions 1:1. We say that the spectrum is the *Fourier transform* of the interferogram.

The interferogram in Figure 20-28c is a less trivial case in which the spectrum consists of a packet of wavelengths centered around $\tilde{\nu}_0 = 4$ cm^{-1}. The interferogram is the sum of contributions from all source wavelengths. The Fourier transform of the interferogram in Figure 20-28c is indeed the spectrum in Figure 20-28c. That is, decomposition of the interferogram into its component wavelengths gives the packet of wavelengths centered around $\tilde{\nu}_0 = 4$ cm^{-1}. *Fourier analysis of the interferogram gives back the intensities of its component wavelengths.* The interferogram in Figure 20-28d is obtained from the two wavelength packets in the spectrum at the left. The Fourier transform of this interferogram gives back the spectrum to its left.

> Fourier analysis of the interferogram gives back the spectrum from which the interferogram is made. *The spectrum is the Fourier transform of the interferogram.*

Fourier Transform Spectroscopy

In a Fourier transform spectrophotometer, the sample is usually placed between the output of the interferometer and the detector, as shown in Figure 20-27. Because the sample absorbs certain wavelengths of light, *the interferogram contains the spectrum of the source minus the spectrum of the sample.* An interferogram of a reference sample containing the cell and solvent is first recorded and transformed into a spectrum. Then the interferogram of a sample in the same kind of solvent and cell is recorded and transformed into a spectrum. The quotient of the second spectrum divided by the first is the infrared spectrum of the sample (Figure 20-29). Expressing the quotient of the two spectra is the same as computing P/P_0 to find transmittance. P_0 is the radiant power received at the detector through the reference and P is the radiant power received after passage through the sample.

> The interferogram loses intensity at those wavelengths absorbed by the sample.

The interferogram is not recorded continuously, but at discrete intervals. The greater the number of data points, the more time and memory is consumed by the Fourier transform computation. The resolution of the spectrum (ability to discern closely spaced peaks) is approximately equal to $1/\Delta$ cm^{-1}, where Δ is the maximum retardation. If the mirror travel is ± 2 cm, the retardation is ± 4 cm and the resolution is 0.25 cm^{-1}.

> Resolution $\approx 1/\Delta$ cm^{-1}
> Δ = maximum retardation

The mathematics of the Fourier transform dictate that the wavelength range of the spectrum is determined by how the interferogram is sampled. The closer the spacing between data points, the greater the wavelength range of the spectrum. To cover a range of $\Delta\tilde{\nu}$ wavenumbers requires sampling the interferogram at retardation intervals of $\delta = 1/(2\Delta\tilde{\nu})$. If $\Delta\tilde{\nu}$ is $4\,000$ cm^{-1}, sampling must occur at intervals of $\delta = 1/(2 \cdot 4\,000 \text{ cm}^{-1}) = 1.25 \times 10^{-4}$ cm = 1.25 μm. This sampling interval corresponds to a mirror motion of

> For a spectral range of $\Delta\tilde{\nu}$ cm^{-1}, points must be taken at retardation intervals of $1/(2\,\Delta\tilde{\nu})$.

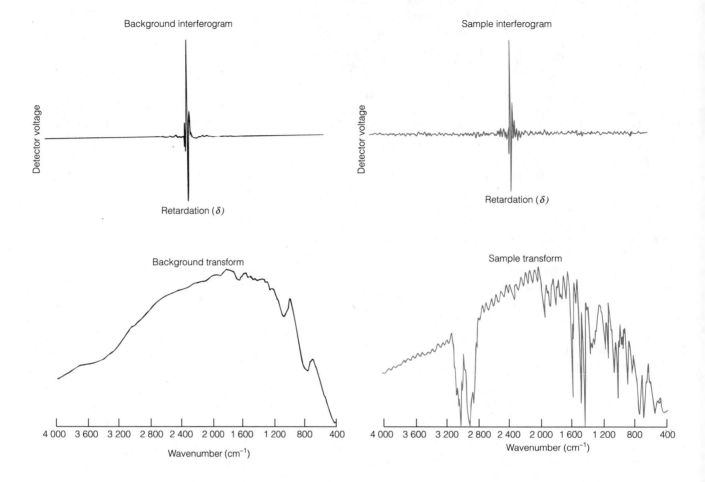

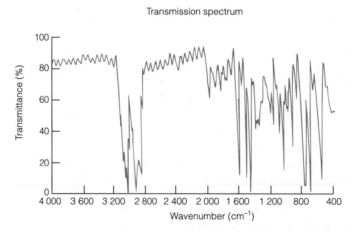

Figure 20-29 Fourier transform infrared spectrum of polystyrene film. The Fourier transform of the background interferogram gives a spectrum determined by the source intensity, beamsplitter efficiency, and detector response. The transform of the sample interferogram is a measure of all the instrumental factors, plus absorption by the sample. The transmission spectrum is obtained by dividing the sample transform by the background transform. Each interferogram is an average of 32 scans and contains 4 096 data points, giving a resolution of 4 cm^{-1}. The mirror velocity was 0.693 cm/s. [Courtesy M. P. Nadler, Michelson Laboratory.]

0.625 μm. For every centimeter of mirror travel, 1.6×10^4 data points must be collected. If the mirror moves at a rate of 0.2 cm per second, the data collection rate would be 3 200 points per second.

The source, beamsplitter, and detector each place limitations on the wavelength range as well. Clearly, the instrument cannot respond to a wavelength that is absorbed by the beamsplitter or to which the detector does not respond. The beamsplitter for the mid-infrared region (~4 000 to 400 cm^{-1}) is typically a layer of germanium evaporated onto a KBr plate. For longer wavelengths ($\tilde{v} < 400$ cm^{-1}), a film of the organic polymer Mylar is a suitable beamsplitter.

To control the sampling interval for the interferogram, a monochromatic visible laser beam is passed through the interferometer along with the polychromatic infrared light. The laser beam gives destructive interference whenever the retardation is a half-integral multiple of the laser wavelength. These zeros in the laser signal, observed with a detector of visible light, are used to control sampling of the infrared interferogram. For example, an infrared data point might be taken at every second zero point of the visible-light interferogram. The precision with which the laser frequency is known gives an accuracy of 0.01 cm^{-1} in the infrared spectrum, which is a two-orders-of-magnitude improvement over that available from dispersive (grating or prism) instruments.

Advantages of Fourier Transform Spectroscopy

Compared with dispersive instruments, the Fourier transform spectrometer offers improved signal-to-noise ratio at a given resolution, much better frequency accuracy, speed, and built-in data-handling capabilities. The signal-to-noise improvement comes mainly because the Fourier transform spectrometer uses energy from the entire spectrum, instead of analyzing a sequence of small wavebands available from a monochromator. Precise reproduction of wavenumber position from one spectrum to the next allows Fourier transform instruments to average signals from multiple scans to further improve the signal-to-noise ratio. Wavenumber precision and low noise levels allow spectra with slight differences to be subtracted from each other to expose those differences.

20-9 Dealing with Noise

An advantage of Fourier transform spectroscopy is that the entire interferogram is recorded in a few seconds and stored in a computer. The signal-to-noise ratio can be improved by collecting tens or hundreds of interferograms and averaging them.

Signal Averaging

Signal averaging is used in many kinds of experiments to improve the quality of the data, as illustrated in Figure 20-30. The lowest trace is a simulated spectrum containing a great deal of noise. One simple way to estimate the noise level is to measure the maximum amplitude of the noise in a region free of signal. Such a measurement in the lowest trace of Figure 20-30 gives a signal-to-noise ratio of $14/9 = 1.6$.

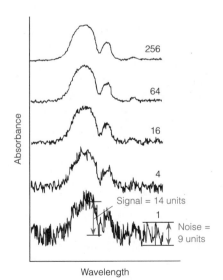

Figure 20-30 Effect of signal averaging on a simulated noisy spectrum. Labels refer to number of scans averaged. [From R. Q. Thompson, *J. Chem. Ed.* **1985**, *62*, 866.]

To improve the signal-to-noise ratio by a factor of n requires averaging n^2 spectra.

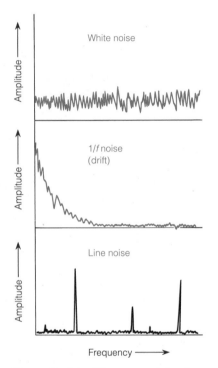

Figure 20-31   Three types of noise in electrical instruments. White noise is always present. A beam chopping frequency can be selected to reduce $1/f$ and line noise to insignificant values.

Consider what happens if you record the spectrum twice and add the results. The signal is the same in both spectra and adds to give twice the value of each spectrum. If n spectra are added, the signal will be n times as large as in the first spectrum. Noise is random, so it may be positive or negative at any point in any spectrum. It turns out that if n spectra are added, the noise increases in proportion to $\sqrt{n}$. Because the signal increases in proportion to n, the signal-to-noise ratio increases in proportion to $n/\sqrt{n} = \sqrt{n}$.

By averaging n spectra, the signal-to-noise ratio is improved by $\sqrt{n}$. To improve the ratio by a factor of 2 requires averaging four spectra. To improve the ratio by a factor of 10 requires averaging 100 spectra. Spectroscopists record as many as 10^4 to 10^5 scans to observe weak signals. It is rarely possible to do better than this (or even this well) because of instrumental instabilities that cause steady drift in addition to random noise. Other powerful methods for improving signal-to-noise ratio are available.[7]

Question   By what factor should the signal-to-noise ratio be improved when 16 spectra are averaged? Does Figure 20-30 agree?

Beam Chopping

The spectrophotometers in Figures 6-5, 6-6, and 20-7 each have a rotating mirror called a *beam chopper* that alternately sends light through the sample and reference cells. (In Figure 20-7, the chopper is mirror C at the center of the diagram.) Chopping allows both cells to be sampled almost continuously. However, it also serves a second purpose related to noise reduction.

Figure 20-31 shows three common types of noise in electrical instruments. The upper trace is ordinary, random *white noise* arising from such causes as the random motion of electrons in a circuit. The second trace shows *1/f noise,* also called *drift,* which is greatest at zero frequency and decreases in proportion to $1/$frequency. An example of low-frequency noise in laboratory instruments is flickering or drifting of a light source in a spectrophotometer or a flame in atomic spectroscopy. The origin of $1/f$ noise in electronic circuits is not well understood. The bottom trace contains *line noise* (also called *interference*) at discrete frequencies such as the 60-Hz transmission line frequency or the 0.2-Hz vibrational frequency when elephants walk through the basement of your building.

Beam chopping in a spectrophotometer moves the analytical signal from zero frequency to the frequency of the chopper. The chopping frequency can be selected so that $1/f$ and line noise are minimal. High-frequency detector circuits are required to take advantage of beam chopping.

Terms to Understand

absorptance
absorption coefficient
antireflection coating
band-pass filter
bandwidth
beam chopping
blackbody radiation
bolometer
charge coupled device
diffraction

diffuse reflection
dispersion
emissivity
exitance
ferroelectric material
flow injection analysis
Fourier analysis
Golay cell
grating
interferogram

interferometer
laser
optical fiber
photoconductive detector
photodiode array
photomultiplier tube
phototube
photovoltaic detector
polychromator
reflectance

refraction
refractive index
resolution
signal averaging
Snell's law
specular reflection
stray light
thermocouple
transmittance

Summary

Light striking a sample can be reflected, absorbed, scattered, or transmitted. Light whose angle of reflection is equal to the angle of incidence is called specular reflection. Rough surfaces give diffuse reflection in all directions, as well. Reflection can be decreased by an antireflection coating. Once light enters the sample, the radiant power decreases exponentially: $P_2/P_1 = e^{-\alpha b}$, where P_2 is the radiant power reaching a depth b, P_1 is the radiant power that penetrated the first surface, and α is the absorption coefficient. This exponential relationship is the basis for Beer's law, which states that absorbance is proportional to the concentration of chromophore and to pathlength. Absorptance is defined as the fraction of incident radiant power that is absorbed by the sample. If the sample does not scatter light, the sum of absorptance, transmittance, and reflectance is unity. When light passes from a region of refractive index n_1 to a region of refractive index n_2, the angle of refraction (θ_2) is related to the angle of incidence (θ_1) by Snell's law: $n_1 \sin\theta_1 = n_2 \sin\theta_2$. The greater the difference in refractive index between two media, the more light is reflected at their interface. Optical fibers transmit light by a series of total internal reflections.

A blackbody is an object that absorbs all light striking it. Emission of radiant energy from the surface of the blackbody is proportional to the fourth power of temperature and follows the Planck distribution. The emission maximum shifts to shorter wavelengths as the temperature increases, as given by the Wien displacement law. Emissivity is the ratio of radiant power emitted by an object to the radiant power emitted by a blackbody at the same temperature.

Components of spectrophotometers include the source, sample cell, monochromator, and detector. Tungsten and deuterium lamps provide visible and ultraviolet radiation; a silicon carbide globar is a good infrared source. Lasers provide high-intensity, coherent, monochromatic radiation by stimulated emission from a medium in which an excited state has been pumped to a higher population than that of a lower state. Sample cells must be transparent to the radiation of interest. A reference sample compensates for reflection and scattering by the cell and solvent. In flow injection analysis, sample injected into a carrier stream is mixed with reagent and passes into a flow-through detector. A grating monochromator disperses light into its component wavelengths. The finer

a grating is ruled, the higher the resolution and the greater the dispersion of wavelengths over angles. Narrow slits improve resolution but increase noise, because less light reaches the detector. Filters pass entire bands of wavelength and reject other bands.

A photomultiplier tube is a sensitive detector of visible and ultraviolet radiation; photons cause electrons to be ejected from a metallic cathode within the device. The signal is amplified at each successive dynode on which the photoelectrons impinge. Photodiode arrays and charge coupled devices are solid-state detectors in which photons create electrons and holes in semiconductor materials. Coupled to a polychromator, these devices can record all wavelengths of a spectrum simultaneously, with resolution limited by the number and spacing of detector elements. Infrared detectors include thermocouples, ferroelectric materials, Golay cells, thermistors, and photoconductive and photovoltaic devices. The absorbance maximum is usually chosen for a spectrophotometric analysis to maximize sensitivity and minimize the effects of imperfectly monochromatic light. Reproducible cell positioning and stray light are important factors in quantitative analysis.

Fourier analysis is a mathematical way to decompose a signal into its component wavelengths. An interferometer contains a beamsplitter, a stationary mirror, and a movable mirror. Reflection of light from the two mirrors creates an interferogram. Fourier analysis of the interferogram tells us what frequencies went into the interferogram. In a Fourier transform spectrophotometer, the interferogram of the source is first measured without a sample present. Then the sample is placed in the beam and a second interferogram is recorded. The transforms of the interferograms tell what frequencies of light reach the detector with and without the sample present. The quotient of the two transforms is the transmission spectrum. The resolution of a Fourier transform spectrum is approximately $1/\Delta$, where Δ is the maximum retardation. To cover a wavenumber range $\Delta\bar{\nu}$ requires sampling the interferogram at intervals of $\delta = 1/(2\Delta\bar{\nu})$. Signal-to-noise ratio can be improved by averaging many spectra. The theoretical signal-to-noise equals $\sqrt{n}$, where n is the number of scans that are averaged. Beam chopping in a dual-beam spectrophotometer reduces $1/f$ and line noise.

Exercises

A. A plate that is 4.00 mm thick and does not absorb or scatter light transmits 82.0% of monochromatic light striking the plate at normal incidence ($\theta_1 = 0°$ in Figure 20-3) in air. Calculate the refractive index of the plate. You will be dealing with a quadratic equation that has two roots. Choose the root that is greater than 1.

B. The transmittance of a solid that both absorbs and reflects light is given by

$$T = \frac{(1 - R)^2 \, e^{-\alpha b}}{1 - R^2 \, e^{-2\alpha b}}$$

where the single-surface reflectance, R, was given in Equation 20-7. The pathlength is b and the absorption coefficient is α.

(a) Estimate the transmittance of a plate of ZnS 1.20 cm thick in the air at a wavelength of 12.0 μm, at which $\alpha = 0.47 \text{ cm}^{-1}$ and the refractive index is 2.17.

(b) What would be the transmittance of a sample 1.20 mm thick?

(c) If the material scatters light as well as absorbing it, then the coefficient α should be modified to the form $\alpha = \alpha_a + \alpha_s$, where α_a is the contribution of absorption and α_s is the contribution of scattering. Find the transmittance of a 1.20-cm-thick plate of ZnS if $\alpha_a = 0.47 \text{ cm}^{-1}$ and $\alpha_s = 0.20 \text{ cm}^{-1}$.

C. *Derivation of Wien displacement law.*

(a) The term $e^{hc/\lambda kT}$ in the Planck distribution for blackbody radiation (Equation 20-13) is much larger than 1 at the wavelength of maximum emission at most temperatures. Under this condition, the denominator is approximately $e^{hc/\lambda kT}$ and the equation can be written

$$M_\lambda \approx 2\pi h c^2 \lambda^{-5} e^{-hc/\lambda kT}$$

Find the wavelengths of maximum emission, λ_{max}, by setting the derivative $dM_\lambda/d\lambda$ equal to zero and solving for $\lambda_{max} \cdot T$. When you evaluate the constants in this expression, your answer should reproduce Equation 20-14.

(b) Calculate λ_{max} (in μm) at 100, 500, and 5 000 K and compare your results to Figure 20-6.

D. (a) If a diffraction grating has a resolution of 10^4, is it possible to distinguish two spectral lines with wavelengths of 10.00 and 10.01 μm?

(b) With a resolution of 10^4, how close in wavenumbers (cm^{-1}) is the closest line to $1\,000 \text{ cm}^{-1}$ that can barely be resolved?

(c) Calculate the resolution of a grating 5.0 cm long ruled at 250 lines/mm for first-order $(n = 1)$ diffraction and tenth-order $(n = 10)$ diffraction.

(d) Find the angular dispersion ($\Delta\phi$, in radians and degrees) between light rays with wavenumbers of $1\,000$ and $1\,001 \text{ cm}^{-1}$ for second-order diffraction $(n = 2)$ from a grating with 250 lines/mm and $\phi = 30°$.

E. The true absorbance of a sample is 1.000, but the monochromator passes 1.0% stray light. Add the light coming through the sample to the stray light to find the apparent transmittance of the sample. Convert this back to absorbance and find the relative error in the calculated concentration of the sample.

F. Refer to the Fourier transform infrared spectrum in Figure 20-29.

(a) The interferogram was sampled at retardation intervals of $1.266\,0 \times 10^{-4}$ cm. What is the theoretical wavenumber range (0 to ?) of the spectrum?

(b) A total of $4\,096$ data points was collected from $\delta = -\Delta$ to $\delta = +\Delta$. Compute the value of Δ, the maximum retardation.

(c) Calculate the approximate resolution of the spectrum.

(d) How many microseconds elapse between each datum?

(e) How many seconds were required to record each interferogram once?

(f) What kind of beamsplitter is typically used for the region $400–4\,000 \text{ cm}^{-1}$ covered in Figure 20-29? Explain why the region below 400 cm^{-1} was not observed. Does the background transform agree with your explanation?

G. The table below shows signal-to-noise ratios recorded in a nuclear magnetic resonance experiment. Construct graphs of (a) signal-to-noise ratio versus n and (b) signal-to-noise ratio versus $\sqrt{n}$, where n is the number of scans. Draw error bars corresponding to the standard deviation at each point. Is the signal-to-noise ratio proportional to $\sqrt{n}$?

Signal-to-noise ratio at the aromatic protons of 1% ethylbenzene in CCl_4

Number of experiments	Number of accumulations (n)	Signal-to-noise ratio	Standard deviation
8	1	18.9	1.9
6	4	36.4	3.7
6	9	47.3	4.9
8	16	66.7	7.0
6	25	84.6	8.6
6	36	107.2	10.7
6	49	130.3	13.3
4	64	143.4	15.1
4	81	146.2	15.0
4	100	159.4	17.1

Data from M. Henner, P. Levoir, and B. Ancian, *J. Chem. Ed.* **1979**, *56*, 685.

Problems

Absorption, Reflection, Refraction, and Emission of Light

1. Distinguish the terms absorbance, absorption coefficient, absorptance, and molar absorptivity.

2. What is the difference between specular and diffuse reflection?

3. A sample that does not scatter light has an absorptance of 6% and a reflectance of 16%. Calculate the transmittance.

4. Calculate the internal transmittance (P_2/P_1 in Figure 20-1) of a 3.00-mm-thick plate with absorption coefficient $\alpha = 0.100 \text{ cm}^{-1}$.

5. Light passes from benzene (medium 1) to water (medium 2) in Figure 20-3 at (a) $\theta_1 = 30°$ or (b) $\theta_1 = 0°$. Find the angle θ_2 in each case.

6. Calculate the power per unit area (the exitance, W/m^2) radiating from a blackbody at 77 K (liquid nitrogen temperature) and at 298 K (room temperature).

7. Diamond windows (refractive index = 2.4) are sufficiently inert to be used in process monitoring where they are in contact with molten plastics, alkalis, and salts. Diamond windows of the Pioneer Venus probe launched in 1978 had to survive in the Venusian atmosphere. Find the external transmittance of a diamond window. If the window has negligible absorption, does the thickness of the window affect the external transmittance?

8. Explain how an optical fiber works. Why does the fiber still work when it is bent?

9. (a) Find the critical value of θ_i in Figure 20-4 beyond which there is total internal reflection in a ZrF_4-based infrared optical fiber whose core refractive index is 1.52 and whose cladding refractive index is 1.50.

(b) The loss of radiant power (due to absorption and scatter) in an optical fiber is expressed in decibels per meter (dB/m), defined as

$$\frac{dB}{m} = \frac{-10\log\left(\dfrac{\text{power out}}{\text{power in}}\right)}{\text{length of fiber (m)}}$$

Calculate the quotient power out/power in for a 20.0-m-long fiber with a loss of 0.0100 dB/m.

10. At a wavelength of 24 μm, the absorption coefficient (α) for KBr is 0.25 cm^{-1} and the refractive index (n) is 1.47.

(a) What wavenumber ($\tilde{\nu}$) corresponds to $\lambda = 24$ μm?

(b) What fraction of radiant flux is reflected when 24-μm infrared light strikes a KBr surface from air at normal incidence? Consider reflection just from this one surface.

(c) The fraction of radiant power that is transmitted after the light enters the sample is $e^{-\alpha b}$, where b is the pathlength. Calculate this internal transmittance for a KBr plate 6.0 mm thick.

(d) Use the equation in Exercise **B** to calculate the transmittance of the 6.0-mm KBr plate.

11. A 1.00-cm-thick plate absorbs no light but scatters 5.0% of the light passing through it. Find the value of the scattering coefficient, α_s, in Equation 20-3.

12. When light from the sodium D line passes through the interface between benzene and sodium chloride at normal incidence, what fraction of the radiant power is reflected?

13. The variation of refractive index (n) with wavelength for fused quartz is given by

$$n^2 - 1 = \frac{(0.696\,166\,3)\lambda^2}{\lambda^2 - (0.068\,404\,3)^2} + $$
$$\frac{(0.407\,942\,6)\lambda^2}{\lambda^2 - (0.116\,241\,4)^2} + \frac{(0.897\,479\,4)\lambda^2}{\lambda^2 - (9.896\,161)^2}$$

where λ is expressed in μm.

(a) Make a graph of n versus λ with points at the following wavelengths: 0.2, 0.4, 0.6, 0.8, 1, 2, 3, 4, 5, and 6 μm.

(b) The ability of a prism to spread apart (disperse) neighboring wavelengths increases as the slope $dn/d\lambda$ increases. Is the dispersion of fused quartz greater for blue light or red light?

14. Find the minimum angle θ_i for total reflection in the optical fiber in Figure 20-4b if the index of refraction of the cladding is 1.400 and the index of refraction of the core is (a) 1.600 or (b) 1.800.

15. The prism below is used to reflect light at a 90° angle. No surface of this prism is silvered. Use Snell's law

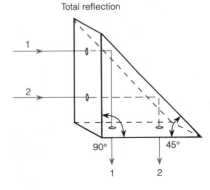

Total reflection

to explain why total reflection occurs. What is the minimum refractive index of the prism for total reflection?

16. Which material(s) in Table 20-1 might be useful for making an antireflection coating for silver chloride?

17. The power radiating from one square meter of a blackbody surface in the wavelength range λ_1 to λ_2 is obtained by integrating the Planck distribution function in Equation 20-13:

$$\text{power emitted} = \int_{\lambda_1}^{\lambda_2} M_\lambda \, d\lambda$$

For a narrow wavelength range, $\Delta\lambda$, the value of M_λ is nearly constant and the power emitted is simply the product $M_\lambda \Delta\lambda$.

(a) Evaluate M_λ at $\lambda = 2.00$ μm and at $\lambda = 10.00$ μm at $T = 1\,000$ K.

(b) Calculate the power emitted per square meter at $1\,000$ K in the interval $\lambda = 1.99$ to $\lambda = 2.01$ μm by evaluating the product $M_\lambda \Delta\lambda$, where $\Delta\lambda = 0.02$ μm.

(c) Repeat **(b)** for the interval 9.99 to 10.01 μm.

(d) The quantity $[M_\lambda (\lambda = 2 \text{ μm})]/[M_\lambda (\lambda = 10 \text{ μm})]$ is the relative exitance at the two wavelengths. Compare the relative exitance at these two wavelengths at $1\,000$ K to the relative exitance at 100 K. What does your answer mean?

The Spectrophotometer

18. Explain how a laser generates light. List important properties of laser light.

19. Would you use a tungsten or a deuterium lamp as a source of 300-nm radiation?

20. What variables increase the resolution of a grating?

21. What is the role of a filter in a grating monochromator?

22. What are the advantages and disadvantages of decreasing monochromator slit width?

23. Deuterated triglycine sulfate (abbreviated DTGS) is a common ferroelectric infrared detector material. Explain how it works.

24. Refer to Problem 19 in Chapter 4 to explain why the detection limit for a spectrophotometric assay depends on the reproducibility of the measurement.

25. Consider a reflection grating operating with an incident angle of 40° in Figure 20-12.

(a) How many lines per centimeter should be etched in the grating if the first-order diffraction angle for 600 nm (visible) light is to be 30°?

(b) Answer (*a*) for $1\,000$ cm^{-1} (infrared) light.

26. Show that a grating with 10^3 grooves/cm provides a dispersion of 5.8° per μm of wavelength if $n = 1$ and $\phi = 10°$ in Equation 20-19.

27. **(a)** What resolution is required for a diffraction grating to resolve wavelengths of 512.23 and 512.26 nm?

(b) With a resolution of 10^4, how close in wavenumbers (cm^{-1}) is the closest line to 512.23 nm that can barely be resolved?

(c) Calculate the fourth-order resolution of a grating that is 8.00 cm long and is ruled at 185 lines/mm.

(d) Find the angular dispersion ($\Delta\phi$) between light rays with wavelengths of 512.23 and 512.26 nm for first-order diffraction ($n = 1$) and thirtieth-order diffraction from a grating with 250 lines/mm and $\phi = 3.0°$.

28. The true absorbance of a sample is 1.500, but 0.50% stray light reaches the detector. Find the apparent transmittance and apparent absorbance of the sample.

29. The pathlength of a cell for infrared spectroscopy can be measured by counting *interference fringes* (ripples in the transmission spectrum). The spectrum below shows 30 interference maxima between $1\,906$ and 698 cm^{-1} obtained by placing an empty KBr cell in a

For Problem 29.

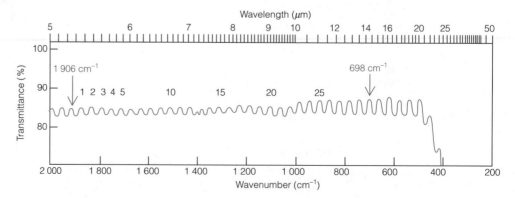

spectrophotometer. The fringes arise because light reflected from the cell compartment interferes constructively or destructively with the unreflected beam (as shown below).

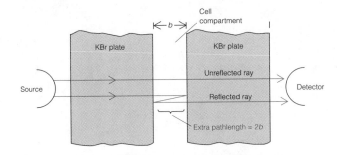

If the reflected beam travels an extra distance λ, it will interfere constructively with the unreflected beam. If the reflection pathlength is $\lambda/2$, destructive interference occurs. Peaks therefore arise when $m\lambda = 2b$ and troughs occur when $m\lambda/2 = 2b$, where m is an integer. If the medium between the KBr plates has refractive index n, the wavelength in the medium is λ/n, so the equations become $m\lambda/n = 2b$ and $m\lambda/2n = 2b$. It can be shown that the cell pathlength is given by

$$b = \frac{N}{2n} \cdot \frac{\lambda_1 \lambda_2}{\lambda_2 - \lambda_1} = \frac{N}{2n} \cdot \frac{1}{\tilde{\nu}_2 - \tilde{\nu}_1}$$

where N maxima occur between wavelengths λ_1 and λ_2.

Calculate the pathlength of the cell that gave the interference fringes shown earlier.

Fourier Transform Spectroscopy

30. The interferometer mirror of a Fourier transform infrared spectrophotometer travels ± 1 cm.

(a) How many centimeters is the maximum retardation, Δ?

(b) State what is meant by resolution.

(c) What is the approximate resolution (cm^{-1}) of the instrument?

(d) At what retardation interval, δ, must the interferogram be sampled (converted to digital form) to cover a spectral range of 0 to $2\,000\ cm^{-1}$?

31. Explain why the transmission spectrum in Figure 20-29 is calculated from the quotient (sample transform)/(background transform) instead of the difference (sample transform) − (background transform).

Signal Averaging

32. A spectrum has a signal-to-noise ratio of $8/1$. How many spectra must be averaged to increase the signal-to-noise ratio to $20/1$?

33. A measurement with a signal-to-noise ratio of $100/1$ can be thought of as a signal (S) with 1% uncertainty (e). That is, the measurement is $S \pm e = 100 \pm 1$.

(a) Use the rules for propagation of uncertainty to show that if you add two such signals the result is: total signal $= 200 \pm \sqrt{2}$, giving a signal-to-noise ratio of $200/\sqrt{2} = 141/1$.

(b) Show that if you add four such measurements, the signal-to-noise ratio increases to $200/1$.

(c) Show that averaging n measurements increases the signal-to-noise ratio by a factor of $\sqrt{n}$ compared with the value for one measurement.

Notes and References

1. K. J. Skogerboe, *Anal. Chem.* **1988**, *60*, 1271A; R. E. Dessy, *Anal. Chem.* **1989**, *61*, 1079A.

2. W. R. Seitz, *Anal. Chem.* **1984**, *56*, 16A; M. A. Arnold, *Anal. Chem.* **1992**, *64*, 1015A; M. Kuratli, M. Badertscher, B. Rusterholz, and W. Simon, *Anal. Chem.* **1993**, *65*, 3473; M. Kuratli and E. Pretsch, *Anal. Chem.* **1994**, *66*, 85.

3. M. G. D. Baumann, J. C. Wright, A. B. Ellis, T. Kuech, and G. C. Lisensky, *J. Chem. Ed.* **1992**, *69*, 89; T. Imasaka and N. Ishibashi, *Anal. Chem.* **1990**, *62*, 363A.

4. W. E. L. Grossman, *J. Chem. Ed.* **1993**, *70*, 741.

5. D. G. Jones, *Anal. Chem.* **1985**, *57*, 1057A, 1207A.

6. P. R. Griffiths, *Chemical Infrared Fourier Transform Spectroscopy* (New York: Wiley, 1975) and W. D. Perkins, *J. Chem. Ed.* **1986**, *63*, A5; **1987**, *64*, A269, A296.

7. There are many excellent digital and electronic techniques for improving the quality of a spectrum without recording numerous scans. See T. C. O'Haver, *J. Chem. Ed.* **1991**, *68*, A147; M. G. Prais, *J. Chem. Ed.* **1992**, *69*, 488 (provides spreadsheet exercises); R. Q. Thompson, *J. Chem. Ed.* **1985**, *62*, 866; M. P. Eastman, G. Kostal, and T. Mayhew, *J. Chem. Ed.* **1986**, *63*, 453; and B. H. Vassos and L. López, *J. Chem. Ed.* **1985**, *62*, 543. A signal averaging experiment including a circuit to generate noise is described by D. C. Tardy, *J. Chem. Ed.* **1986**, *63*, 648.

Atomic Spectroscopy: Old and New

Hollow–cathode lamp

Flame

Grating Monochromator

Photomultiplier tube

Lamp power supply

Sample is sucked into the flame through a tube

Detector power supply and readout

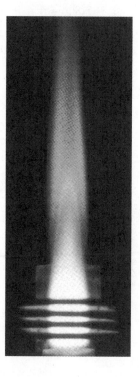

The upper photo shows one of the earliest atomic absorption spectrophotometers assembled in the laboratory of Alan Walsh in Australia in the late 1950s.[1] The photo at the left shows an inductively coupled plasma torch, which is widely used for atomic emission measurements today. Atomic spectroscopy owes its beginnings to the work of Kirchoff and Bunsen in the 1850s, who observed that atoms in flames absorb or emit radiation that is characteristic of each element.

Atomic Spectroscopy

21

$\mathbf{A}$t sufficiently high temperature, most compounds break apart into atoms in the gas phase. Unlike optical spectra from condensed phases, whose bandwidths are ~100 nm, the spectra of gaseous atoms consist of sharp lines with widths of ~0.001 nm. The sharp absorptions in Figure 21-1 arise from transitions between different electronic states of Fe, Ni, and Cr atoms. Because the lines are so sharp, there is usually little overlap between the spectra of different elements in the same sample.

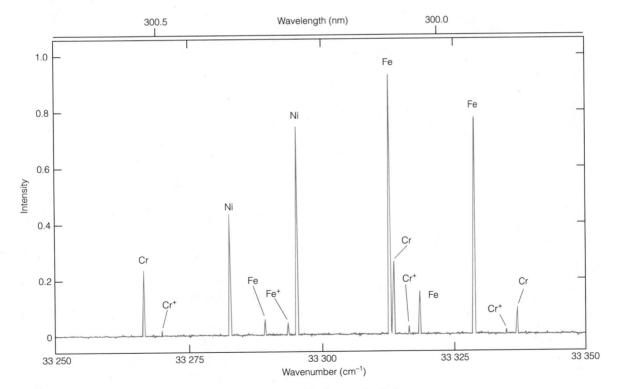

Figure 21-1 A tiny portion of the spectrum of a steel hollow-cathode lamp, showing sharp lines characteristic of gaseous Fe, Ni, and Cr atoms and weak lines from Cr^+ and Fe^+ ions. The resolution is 0.001 nm, which is comparable to the true linewidths. [From A. P. Thorne, *Anal. Chem.* **1991,** *63,* 57A.]

The unit ppm (parts per million) refers to micrograms of solute per gram of solution. Because the density of dilute aqueous solutions is close to 1 g/mL, ppm usually refers to µg/mL. A concentration of 1.00 ppm of Fe corresponds to 1.00×10^{-6} g Fe/mL = 1.79×10^{-5} M.

In *atomic spectroscopy*, samples are vaporized at 2 000–6 000 K and atomic concentrations are determined by measuring absorption or emission at characteristic wavelengths. Because of its high sensitivity and the ease with which many samples can be examined, atomic spectroscopy is a principal tool of analytical chemistry, especially in industrial settings. Measuring analyte concentrations at the parts-per-million level is routine, and parts-per-trillion levels are sometimes amenable to analysis. In analyzing the major constituents of an unknown, the sample is usually diluted to reduce concentrations to the parts-per-million level. Atomic spectroscopy is not as accurate as some wet chemical methods, because its precision is rarely better than 1–2%. The equipment is expensive, but widely available. Because atomic lines are so sharp and have such little overlap, some instruments can measure over 60 elements in a sample simultaneously.

21-1 Absorption, Emission, and Fluorescence

In conventional molecular spectroscopy, the absorbance of a sample placed in the beam of light is measured. Alternatively, the sample is irradiated, and its luminescence is measured in a direction perpendicular to the incident beam. Both of these experiments can also be done with an atomic vapor (Figure 21-2). At the high temperature of the vapor, many atoms are already in thermally populated, excited electronic states. They can spontaneously emit photons and return to a lower state (from which they will be excited

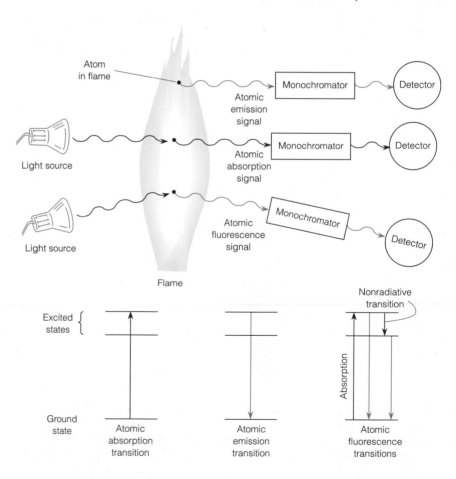

Figure 21-2 Schematic representation of absorption, emission, and fluorescence by atoms in a flame. In atomic absorption, the atoms absorb part of the light from the source and the unabsorbed light reaches the detector. Atomic emission comes from atoms that are in an excited state because of the high thermal energy of the flame. Atomic fluorescence arises from atoms that are excited by an external lamp or laser. The atom can emit the same wavelength that was absorbed, or it can fall to other states and emit other wavelengths.

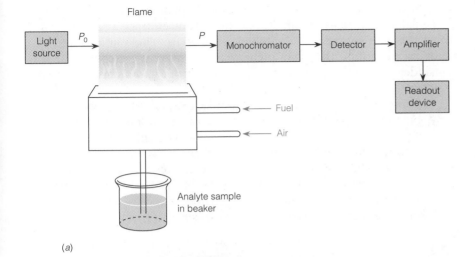

(a)

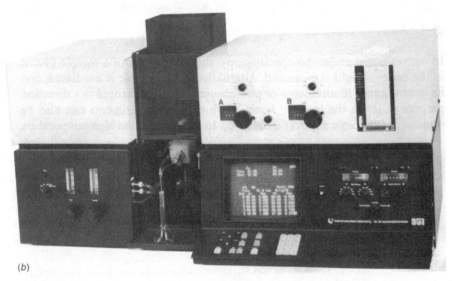

(b)

Figure 21-3 (*a*) Outline of an atomic absorption spectrophotometer. Compare this block diagram with the apparatus in the photograph at the front of this chapter. (*b*) A research-quality instrument for atomic absorption and emission. The sample in the flask is aspirated into the burner, which is behind the metal cage. Valves on the left control the flow of fuel and oxidizer. Dials on the right select wavelengths, monochromator bandwidth, and observation modes. Results are displayed on the video tube and a printer. [Courtesy Instrumentation Laboratory, Wilmington, MA.]

again by collisions). Therefore, atomic spectroscopy falls into three classes commonly designated as *absorption, fluorescence,* and *emission.* Routinely available instruments perform absorption and emission experiments with equal ease. **Atomic fluorescence spectroscopy,** a technique potentially a thousand times more sensitive (and even able to count atoms!), is not yet in widespread use.

Apparatus for **atomic absorption spectroscopy** is shown in Figure 21-3. A liquid sample is aspirated (sucked) into a flame whose temperature is 2 000–3 000 K. The sample is **atomized** (broken into atoms) in the flame,

Atomic spectroscopy:
1. absorption
2. emission (luminescence from a thermally populated excited state)
3. fluorescence (luminescence following absorption of radiation)

Atomic emission requires equipment similar to that for atomic absorption, but the lamp is not used.

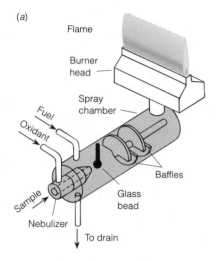

(a)

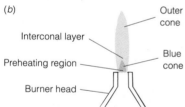

(b)

Figure 21-4 (a) Schematic diagram of a premix burner. (b) End view of flame. The slot in the burner head is about 0.5 mm wide.

which replaces the cuvet in conventional spectrophotometry. The pathlength of the flame is typically 10 cm. To measure the absorbance of Fe atoms in a flame, the light source uses a cathode made of Fe. This source produces a vapor of Fe atoms that emit light with characteristic frequencies. The remainder of the apparatus is not very different from an ordinary spectrophotometer.

Atomic emission spectroscopy is similar to atomic absorption spectroscopy, but no light source is needed. Some of the atoms in the flame are promoted to excited electronic states by collisions with other atoms. The excited atoms emit their characteristic radiation as they return to their ground state. The emission intensity at a characteristic wavelength of an element is proportional to the concentration of the element in the sample. For both absorption and emission, standard curves are usually used to establish the relation between signal and concentration.

21-2 Atomization: Flames, Furnaces, and Plasmas

The essential feature of atomic spectroscopy is the atomization of the sample in a flame, an electrically heated oven, or a radio-frequency plasma. Sensitivity and interferences observed in atomic spectroscopy depend on the details of atomization.

Flames

Most flame spectrometers use a **premix burner,** such as that in Figure 21-4, in which the sample, oxidant, and fuel are mixed before introduction into the flame. Sample solution (which need not be aqueous) is drawn into the *nebulizer* by the rapid flow of oxidant (usually air) past the tip of the sample capillary. The liquid breaks into a fine mist as it leaves the tip of the nebulizer. The spray is directed at high speed against a glass bead, upon which the droplets are broken into even smaller particles. The formation of small droplets is termed **nebulization.** Then the mist, oxidant, and fuel flow past a series of baffles that promotes further mixing and blocks large droplets of liquid. Liquid that collects at the bottom of the spray chamber flows out to a drain. Only a very fine mist containing about 5% of the initial sample reaches the flame. Figure 21-5 shows the distribution of droplet sizes from three different kinds of nebulizers. In general, the narrowest size distribution and smallest sizes are most desirable.

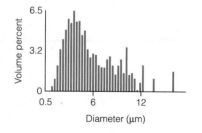

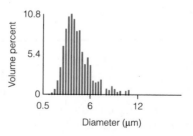

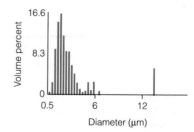

Figure 21-5 Distribution of droplet sizes produced by different types of nebulizers. [From R. H. Clifford, I. Ishii, A. Montaser, and G. A. Meyer, *Anal. Chem.* **1990,** *62,* 390.]

TABLE 21-1   Maximum flame temperatures

Fuel	Oxidant	Temperature (K)
Acetylene	Air	2 400–2 700
Acetylene	Nitrous oxide	2 900–3 100
Acetylene	Oxygen	3 300–3 400
Hydrogen	Air	2 300–2 400
Hydrogen	Oxygen	2 800–3 000
Cyanogen	Oxygen	4 800

The most common fuel-oxidizer combination is acetylene and air, which produces a flame temperature of ~2 400–2 700 K (Table 21-1). When a hotter flame is required, the acetylene-nitrous oxide combination is usually used. In the flame profile in Figure 21-4, gas that enters the preheating region from the burner head is heated by downward conduction and radiation from the primary reaction zone (the blue cone). Combustion is completed in the outer cone, where surrounding air is drawn into the flame. Flames emit light that must be subtracted from the total signal to obtain the analyte signal.

Droplets entering the flame lose their water through evaporation; then the remaining sample vaporizes and decomposes into atoms. Many elements form oxides and hydroxides as they rise through the outer cone. Molecules do not have the same spectra as atoms, so the atomic signal is lowered. If the flame is relatively rich in fuel (a "rich" flame), excess carbon species tend to reduce the metal oxides and hydroxides and thereby increase sensitivity. The opposite of a rich flame is a "lean" flame, which has excess oxidant and is hotter. Lean or rich flames are recommended in the analysis of different elements.

The position in the flame at which maximum atomic absorption or emission is observed depends on the element being measured as well as the flow rates of sample, fuel, and oxidizer. A profile of emission from Ca atoms in a cyanogen ($N\equiv C—C\equiv N$)-oxygen flame is shown in Figure 21-6. The

Organic solvents with surface tensions lower than that of water are excellent for atomic spectroscopy because they form smaller droplets, thus leading to more efficient atomization.

Hotter flames are needed for *refractory* elements (those with high vaporization temperature) or to decompose species such as metal oxides formed during passage through the flame.

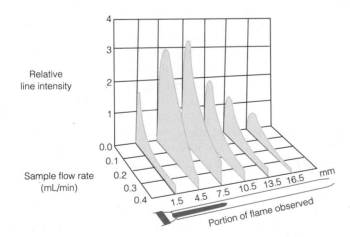

Figure 21-6   Profile of Ca emission line in a cyanogen-oxygen flame. [From K. Fuwa, R. E. Thiers, B. L. Vallee, and M. R. Baker, *Anal. Chem.* **1959**, *31*, 2039.]

decreasing intensity at higher sample flow rate is attributed to cooling of the flame by water from the sample. Different elements have different absorption and emission profiles. The sample, fuel, and oxidant flow rates and the level at which the flame is observed can be optimized for each element.

Furnaces

Furnaces offer increased sensitivity and require less sample than a flame.

The electrically heated furnace offers greater sensitivity than that afforded by flames and requires a smaller volume of sample. Figure 21-7 shows a **graphite furnace** mounted in the beam of a spectrometer. From 1 to 100 μL of sample is injected into the oven through the hole at the center. The light beam travels through windows at each end of the tube. The maximum recommended temperature for a graphite furnace is 2 550°C for not more than 7 s. A surrounding atmosphere of Ar helps prevent graphite oxidation.

A graphite furnace confines the atomized sample in the optical path for a *residence time* of several seconds, resulting in high sensitivity. In flame spectroscopy, the sample is diluted by the time it has been nebulized, and its residence time in the optical path is only a fraction of a second as it rises through the flame. Flames also require a much larger volume of sample, because sample is constantly flowing into the flame. Whereas 1–2 mL is the minimum necessary for flame analysis, as little as 1 μL is adequate for a furnace. In one extreme example where only nanoliter volumes of kidney tubular fluid were available, a method was devised to reproducibly deliver 0.1-nL volumes to a graphite furnace for analysis of Na and K.[2] Precision with a furnace is rarely better than 5–10% with manual sample introduction, but automated sample injection improves reproducibility.

The operator must determine reasonable time and temperature for each stage of the analysis. Once a program is established, it can be applied to a large number of similar samples.

Compared with flames, furnaces require more operator skill to determine the proper conditions for each type of sample. The furnace must be heated in three or more steps to properly atomize the sample. To analyze iron in the iron-storage protein ferritin, 10 μL of sample containing ~0.1 ppm Fe is injected into the cold oven. The furnace is programmed to *dry* the sample at 125°C for 20 s to remove solvent. Drying is followed by 60 s of *charring* (also called *pyrolysis*) at 1 400°C to destroy organic matter, which would otherwise create a great deal of smoke and interfere with the Fe determination. Finally, *atomization* is carried out at 2 100°C for 10 s, during which the absorbance reaches a maximum and then decreases as Fe evaporates from the oven. The time-integrated absorbance is taken as the analytical signal. Following atomization, the furnace is heated to 2 400°–2 500°C for 3 s to clean out any remaining residue. When developing a method for a new kind of sample, it is important to record the signal as a function of time, because signals are also observed from smoke produced during charring and from the glow of the red-hot oven in the latter part of atomization. A skilled operator must interpret which signal is due to sample and which to other effects.

Improved performance is obtained with a *L'vov platform* in the furnace (Figure 21-8a). Because the temperature of the sample on the platform lags behind the rising temperature of the walls of the furnace, the analyte does not vaporize until the walls reach constant temperature (Figure 21-8b). At constant furnace temperature, the area of the signal-versus-time curve is a reliable measure of the total analyte vaporized from the sample. A heating rate of 2 000 K/s minimizes the time that the sample spends in an unatomized state and increases the concentration of free atoms during the short time that the vapor is in the furnace.

Figure 21-7 An electrically heated graphite-rod furnace used for flameless atomic spectroscopy. Light travels the length of the furnace (~38 mm in this case), and sample is injected through the hole at the top. [Courtesy Instrumentation Laboratory, Wilmington, MA.]

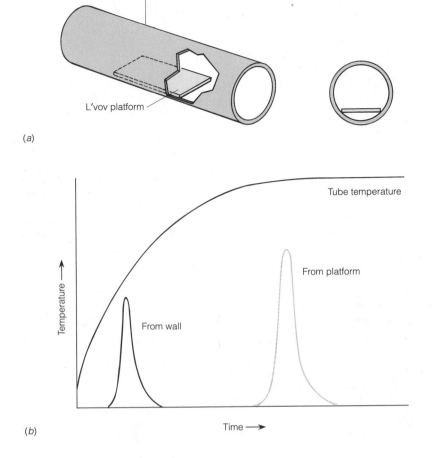

(a)

(b)

Figure 21-8 (*a*) L'vov platform in graphite furnace. (*b*) Heating profile comparing analyte evaporation from wall and from platform. [From W. Slavin, *Anal. Chem.* **1982,** *54,* 685A.]

With *longitudinal* (end-to-end) heating (as in Figure 21-7), the center of the furnace is hotter than the ends. Atoms from the central region condense at the ends, where they can vaporize during the next sample run. Interference from previous runs is called a *memory effect*. Memory effects are reduced in the *transversely* (side-to-side) heated furnace shown in Figure 21-9, because the temperature is nearly constant for the whole length of the graphite tube. Ordinary graphite is also coated with a layer of pyrolytic graphite to further reduce memory effects. This dense layer of carbon, formed by thermal decomposition of an organic vapor, seals the relatively porous graphite so it cannot absorb foreign atoms and molecules.

The temperature needed to char the sample **matrix** (the medium containing the analyte) may cause analyte evaporation. Appropriate **matrix modifiers** can retard evaporation of analyte until the matrix has charred away. For example, a modifier consisting of $Mg(NO_3)_2$ and $Pd(NO_3)_2$ added to the sample solution allows the charring temperature for a sample containing tin to be raised from 800° to 1 400°C without evaporation of the analyte.

A study of the effect of $Mg(NO_3)_2$ in Al determination showed that at high temperature $Mg(NO_3)_2$ is converted to $MgO(s)$, which steadily evaporates and maintains some vapor pressure of $MgO(g)$.[3] Aluminum is con-

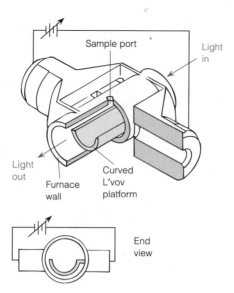

Sample port

Light in

Light out

Furnace wall

Curved L'vov platform

End view

Figure 21-9 Transversely heated graphite furnace maintains nearly constant temperature over its whole length, thereby reducing memory effect from previous runs. The curved platform is uniformly heated by radiation from the outer wall, not by conduction. The platform is attached to the wall by one small connection that is hidden from view. [Courtesy Perkin-Elmer Corp., Norwalk, CT.]

verted to Al_2O_3 during heating. At a sufficiently high temperature, Al_2O_3 decomposes to Al and O, and the Al evaporates. However, evaporation of Al is retarded as long as MgO is present, by virtue of the reaction

$$3MgO(g) + 2Al(s) \rightleftharpoons 3Mg(g) + Al_2O_3(s) \qquad (21\text{-}1)$$

When all of the MgO has evaporated, Reaction 21-1 no longer occurs and Al_2O_3 finally decomposes and evaporates. The $Mg(NO_3)_2$ matrix modifier prevents Al from evaporating until a high temperature is reached.

Inductively Coupled Plasmas

The **inductively coupled plasma** at the opening of this chapter reaches a much higher temperature than ordinary combustion flames. Its high temperature and stability eliminate much of the interference encountered with conventional flames. Because of these desirable features, the plasma is replacing flames for emission spectroscopy. The plasma's principal disadvantage is its expense to purchase and operate.

The cross-sectional view of an inductively coupled plasma burner in Figure 21-10 shows two turns of a radio-frequency induction coil wrapped around the upper opening of the quartz apparatus. High-purity argon gas is fed through the plasma gas inlet. After a spark from a Tesla coil ionizes the Ar gas, free electrons are immediately accelerated by the powerful radio-frequency field that oscillates about the load coil at a frequency of 27 MHz. Accelerated electrons transfer energy to the entire gas by collisions with atoms. Once the process is begun, the electrons absorb enough energy from the electric field to maintain a temperature of 6 000 to 10 000 K in the plasma (Figure 21-11). It is so hot (especially near the coils) that the quartz burner must be protected by argon coolant gas flowing around the outer edge of the apparatus.

The concentration of analyte needed for adequate signal can be reduced by a factor of 10–20 with an *ultrasonic nebulizer,* in which sample solution is directed onto a piezoelectric crystal (Box 2-2) oscillating at 1 MHz. This creates a very fine **aerosol** (a suspension of solid or liquid particles in a gas)

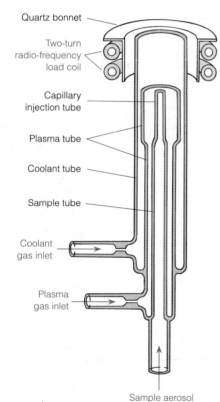

Quartz bonnet

Two-turn radio-frequency load coil

Capillary injection tube

Plasma tube

Coolant tube

Sample tube

Coolant gas inlet

Plasma gas inlet

Sample aerosol inlet

Figure 21-10 Diagram of an inductively coupled plasma burner head. [From R. N. Savage and G. M. Hieftje, *Anal. Chem.* **1979,** *51,* 408.]

that is carried by a stream of argon through a heated tube, where solvent is vaporized. The stream then passes through a refrigerated zone in which solvent condenses and is removed. Analyte reaches the plasma flame as a fine dry cloud. Flame energy is not needed to evaporate solvent, so more energy is available for atomization. A larger fraction of the original sample reaches the flame than with a conventional nebulizer.

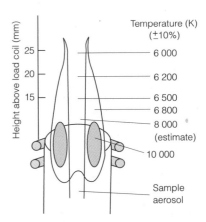

Figure 21-11 Temperature profile of a typical inductively coupled plasma used in analytic spectroscopy. [From V. A. Fassel, *Anal. Chem.* **1979**, *51*, 1290A.]

21-3 How Temperature Affects Atomic Spectroscopy

Temperature determines the degree to which a sample breaks down to atoms and the extent to which a given atom is found in its ground, excited, or ionized state. These effects influence the strength of the signal we observe.

The Boltzmann Distribution

Consider a molecule with two available energy levels (Figure 21-12) separated by energy ΔE (which is a positive number). Call the lower level E_0 and the upper level E^*. In general, an atom (or molecule) may have more than one state available at a given energy level. In Figure 21-12, we show three states available at E^* and two available at E_0. The number of states available at each energy level is called the *degeneracy* of the level. We will call the degeneracies g_0 and g^*.

The **Boltzmann distribution** describes the relative populations of different states at thermal equilibrium. If equilibrium exists (which is not true in the blue cone of a flame but is probably true above the blue cone), the relative population (N^*/N_0) of any two states is

The Boltzmann distribution applies to a system at thermal equilibrium.

Boltzmann distribution:
$$\frac{N^*}{N_0} = \frac{g^*}{g_0}e^{-\Delta E/kT} \tag{21-2}$$

where T is temperature (K) and k is Boltzmann's constant ($1.380\,658 \times 10^{-23}$ J/K).

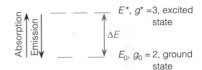

Figure 21-12 Two energy levels with different degeneracies. Ground-state atoms can absorb light to be promoted to the excited state. Excited-state atoms can emit light to return to the ground state.

The Effect of Temperature on the Excited-State Population

The lowest excited state of a sodium atom lies 3.371×10^{-19} J/atom above the ground state. The degeneracy of the excited state is 2, whereas that of the ground state is 1. Let's calculate the fraction of sodium atoms in the excited state in an acetylene-air flame at $2\,600$ K. Using Equation 21-2, we find

$$\frac{N^*}{N_0} = \left(\frac{2}{1}\right)e^{-(3.371 \times 10^{-19}\text{ J})/[(1.381 \times 10^{-23}\text{ J/K})(2\,600\text{ K})]} = 1.67 \times 10^{-4} \tag{21-3}$$

That is, fewer than 0.02% of the atoms are in the excited state.

How would the fraction of atoms in the excited state change if the temperature were $2\,610$ K instead?

$$\frac{N^*}{N_0} = \left(\frac{2}{1}\right)e^{-3.371 \times 10^{-19}/(1.381 \times 10^{-23} \cdot 2610)} = 1.74 \times 10^{-4} \tag{21-4}$$

A 10 K temperature rise changes the excited-state population by 4% in this example.

The fraction of atoms in the excited state is still less than 0.02%, but that fraction has increased by $100(1.74 - 1.67)/1.67 = 4\%$.

TABLE 21-2   Effect of energy separation and temperature on population of excited states

Wavelength separation of states (nm)	Energy separation of states (J/atom)	Excited-state fraction $(N^*/N_0)^a$	
		2 500 K	6 000 K
250	7.95×10^{-19}	1.0×10^{-10}	6.8×10^{-5}
500	3.97×10^{-19}	1.0×10^{-5}	8.3×10^{-3}
750	2.65×10^{-19}	4.6×10^{-4}	4.1×10^{-2}

a. Based on the equation $N^*/N_0 = (g^*/g_0)e^{-\Delta E/KT}$ in which $g^* = g_0 = 1$.

The Effect of Temperature on Absorption and Emission

We see that more than 99.98% of the sodium atoms are in their ground state at 2 600 K. *Varying the temperature by 10 K hardly affects the ground-state population and would not noticeably affect the signal in an atomic absorption experiment.*

How would emission intensity be affected by a 10-K rise in temperature? In Figure 21-12 we see that absorption arises from ground-state atoms, but emission arises from excited-state atoms. Emission intensity is proportional to the population of the excited state. *Because the excited-state population changes by 4% when the temperature rises 10 K, the emission intensity rises by 4%.* It is critical in atomic *emission* spectroscopy that the flame be very stable, or the emission intensity will vary significantly. In atomic *absorption* spectroscopy, flame temperature variation is important, but not as critical.

The inductively coupled plasma is almost always used for emission, not absorption, because it is so hot that there is a substantial population of excited-state atoms and ions. Table 21-2 compares excited-state populations for a flame at 2 500 K and a plasma at 6 000 K. Although the fraction of excited atoms is small, each atom emits many photons per second because it is rapidly promoted back to the excited state by collisions.

Halogen atoms (F, Cl, Br, I) are not observed by atomic absorption because their lowest excited states lie at energies so high that far-ultraviolet radiation is required for excitation. (Nitrogen and oxygen absorb far-ultraviolet energy and would have to be excluded from the optical path.) The excited states lie at such high energy that they are not sufficiently populated to provide adequate intensity for optical emission, even in an inductively coupled plasma. However, in a helium microwave plasma (instead of an argon radio-frequency plasma), nonequilibrium energy transfer occurs and halogen atoms are excited. Halogen-sensitive detectors for gas chromatography are based on atomic emission from a helium plasma.

The magnitude of atomic absorption is not as sensitive to temperature as the intensity of atomic emission, which is exponentially sensitive to temperature.

21-4 Instrumentation

Fundamental requirements for an atomic absorption experiment were shown in Figure 21-3. The principal differences between atomic and ordinary molecular spectroscopy lie in the light source, the sample container (the flame or furnace), and the need to subtract background emission from the observed signal.

The Linewidth Problem

Beer's law applies to monochromatic radiation. In practical terms, the linewidth of the radiation being measured should be substantially narrower than the linewidth of the absorbing sample. Otherwise, the measured absorbance will not be proportional to the sample concentration.

Atomic absorption lines are very sharp, with an inherent width of only $\sim10^{-4}$ nm. Two mechanisms serve to broaden the lines in atomic spectroscopy. One is the **Doppler effect.** An atom moving toward the radiation source samples the oscillating electromagnetic wave more frequently than one moving away from the source (Figure 21-13). That is, an atom moving toward the source "sees" higher-frequency light than that encountered by one moving away. In the laboratory frame of reference, the atom moving toward the source absorbs lower-frequency light than that absorbed by the one moving away. The linewidth, $\Delta\nu$, due to the Doppler effect, is given approximately by

Doppler linewidth: $$\Delta\nu \approx \nu(7 \times 10^{-7}) \sqrt{\frac{T}{M}} \qquad (21\text{-}5)$$

where ν is the frequency (in Hz) of the peak, T is temperature (K), and M is the mass of the atom in atomic mass units; $\Delta\nu$ is the width of the absorption line measured at half the height of the peak.

Linewidth is also affected by **pressure broadening** from collisions between atoms. Atomic energy levels are perturbed during a collision, so the colliding atom does not absorb the same frequency of radiation as an isolated atom. Pressure broadening, $\Delta\nu_p$, is roughly numerically equal to the collision frequency and is proportional to pressure. The Doppler effect and pressure broadening are similar in magnitude and yield linewidths of 10^{-3} to 10^{-2} nm in atomic spectroscopy.

Hollow-Cathode Lamps

Monochromators cannot isolate lines narrower than 10^{-3} to 10^{-2} nm. To produce narrow lines of the correct frequency, we use a **hollow-cathode lamp** containing a vapor of the same element as that being analyzed.

A hollow-cathode lamp, such as that shown in Figure 21-14, is filled with Ne or Ar at a pressure of $\sim130-700$ Pa (1–5 torr). When a high voltage is applied between the anode and cathode, the filler gas is ionized and positive

The linewidth of the source must be narrower than the linewidth of the atomic vapor for Beer's law to be obeyed. "Linewidth" and "bandwidth" are used interchangeably, but "lines" are narrower than "bands."

Doppler and pressure effects broaden the atomic lines by 1–2 orders of magnitude relative to their inherent linewidths.

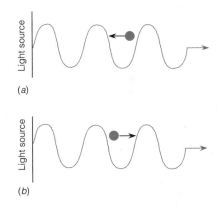

(a)

(b)

Figure 21-13 The Doppler effect. A molecule moving (*a*) toward the radiation source "feels" the electromagnetic field oscillate more often than one moving (*b*) away from the source.

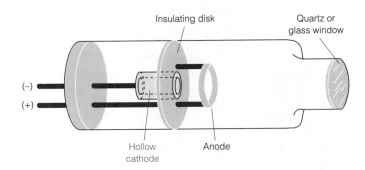

Figure 21-14 A hollow-cathode lamp.

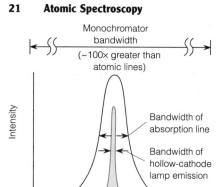

Figure 21-15 Relative linewidths of hollow-cathode emission, atomic absorption, and a monochromator. Linewidths are measured at half the signal height. The linewidth from the hollow cathode is relatively narrow because the gas temperature in the lamp is lower than a flame temperature (so there is less Doppler broadening) and the pressure in the lamp is lower than the pressure in a flame (so there is less pressure broadening).

ions are accelerated toward the cathode. They strike the cathode with enough energy to "sputter" metal atoms from the cathode into the gas phase. The free atoms are excited by collisions with high-energy electrons and then emit photons to return to the ground state. This atomic radiation (shown in Figure 21-1) has the same frequency as that absorbed by analyte atoms in the flame or furnace. The linewidth is sufficiently narrow, with respect to that of the high-temperature analyte, to be nearly "monochromatic" (Figure 21-15). A different lamp is usually required for each element, although some lamps are made with more than one element in the cathode.

Spectrophotometers

There are two unusual features of an atomic absorption spectrophotometer: The sample "container" is a flame or furnace and the lamp emits a few sharp lines at precise frequencies absorbed by the analyte. The remainder of the spectrophotometer is not very different from one used for ordinary absorption spectroscopy. The purpose of the monochromator is to select one line from the hollow-cathode lamp and to reject as much emission from the flame or furnace as possible.

Figure 21-16 illustrates a significant advance in detection technology applied to plasma atomic emission from a torch that would be located at the

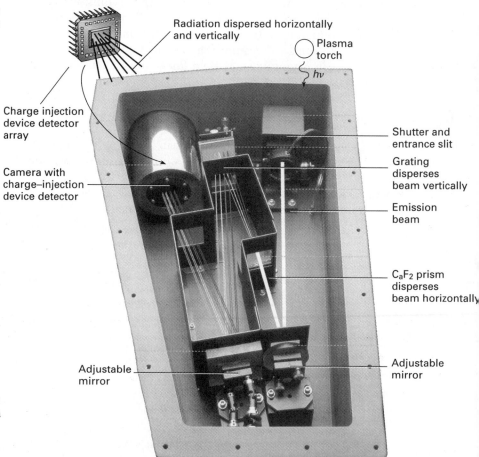

Figure 21-16 Monochromator of plasma atomic emission spectrometer disperses light horizontally (prism) and vertically (grating) to create a two-dimensional pattern. The detector is a 388×244 charge injection device array, in which each pixel receives a different portion of the spectrum. [Courtesy Thermo Jarrell Ash Corp., Franklin, MA.]

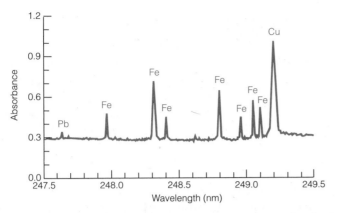

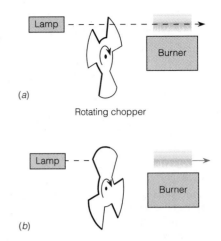

(a)

Rotating chopper

(b)

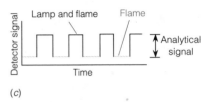

(c)

Figure 21-17 Graphite furnace absorption spectrum of bronze dissolved in HNO_3. [Reproduced from B. T. Jones, B. W. Smith, and J. D. Winefordner, *Anal. Chem.* **1989,** *61,* 1670.]

Figure 21-18 Operation of a beam chopper for subtracting the signal due to flame background emission. (*a*) Lamp and flame emission reach detector. (*b*) Only flame emission reaches detector. (*c*) Resulting square-wave signal.

upper right of the diagram. The emission beam passes through a CaF_2 prism that disperses the beam in the horizontal plane. A diffraction grating then disperses each band in the vertical direction, creating a two-dimensional pattern. The detector is a 388×244 charge injection device array (related to the charge coupled device in Figure 20-19), in which each of the 94 672 pixels receives a different portion of the spectrum. A computer displays the entire spectrum, and multiple lines from multiple analytes are rapidly identified.

Background Correction

Unlike ordinary spectrophotometers, those used in atomic spectroscopy must incorporate **background correction** to distinguish analyte signal from absorption, emission, and optical scattering of the sample matrix, the flame, plasma, or white-hot graphite furnace. For example, Figure 21-17 shows the absorption spectrum of Fe, Cu, and Pb in a graphite furnace. The sharp atomic signals with a maximum absorbance near 1.0 are superimposed on a broad background with an absorbance of 0.3. If we did not measure the background absorbance, significant errors would result. Background correction is most critical for graphite furnaces that tend to be filled with smoke from the charring step. Optical scatter from smoke must somehow be distinguished from optical absorption by analyte.

For atomic absorption, *beam chopping* or electrical *modulation* of the hollow-cathode lamp are used to distinguish the signal of the flame from the desired atomic line at the same wavelength. Figure 21-18 shows light from the lamp being periodically blocked by a rotating chopper. Signal reaching the detector while the beam is blocked must be from flame emission. Signal reaching the detector when the beam is not blocked is from the lamp and the flame. The difference between these two signals is the desired analytical signal.

Beam chopping corrects for flame emission, but not for scattering. Most spectrometers provide an additional means to correct for scattering and broad background absorption. Deuterium lamps, Smith-Hieftje, and Zeeman correction systems are most common.

Background signal arises from absorption, emission, or scatter by everything in the sample besides analyte (the *matrix*), as well as absorption, emission, or scatter by the flame, the plasma, or the furnace.

Modulation means applying an oscillating voltage to the lamp to pulse it on and off.

Background correction methods:
beam chopping
D_2 lamp
Smith-Hieftje
Zeeman

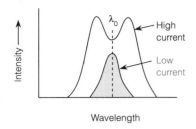

Figure 21-19 Distortion of hollow-cathode lamp emission at elevated current. Analyte atomic absorption is centered at wavelength λ_0.

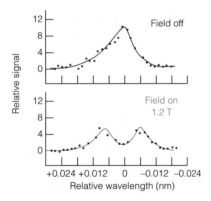

Figure 21-20 Zeeman effect on Co fluorescence in a graphite furnace with excitation at 301 nm and detection at 341 nm. [Reproduced from J. P. Dougherty, F. R. Preli, Jr., J. T. McCaffrey, M. D. Seltzer, and R. G. Michel, *Anal. Chem.* **1987,** *59,* 1112.]

Detection limits vary from instrument to instrument. Adding an ultrasonic nebulizer improves the detection limit for some elements by more than a factor of 10.

For *deuterium lamp background correction,* broad emission from a D_2 lamp (Figure 20-8) is passed through the flame in alternation with that from the hollow cathode. The monochromator bandwidth is so wide that a negligible fraction of D_2 radiation is absorbed by the analyte atomic absorption line. Light from the hollow-cathode lamp is absorbed by analyte and absorbed and scattered by background. Light from the D_2 lamp is absorbed and scattered only by background. The difference between absorbance measured with the hollow-cathode and absorbance measured with the D_2 lamp is the absorbance due to analyte.

A superior and less expensive procedure is called *Smith-Hieftje background correction* (pronounced HEEF-yeh). When a hollow-cathode lamp is run at high electric current, the output is broadened and a dip develops at the central wavelength due to absorption by free ground-state atoms located between the window of the lamp and the hot atoms near the cathode (Figure 21-19). For background correction, the lamp is first run at low current to measure absorbance due to analyte plus background. Then the lamp is pulsed with a high current. During the pulse, analyte absorbance is reduced—but not to zero, because the two-humped output in Figure 21-19 still has significant intensity at the analytical wavelength. Most of the absorbance during the pulse is due to the broad background, because most of the lamp intensity no longer coincides with the sharp atomic absorption line of analyte. The difference between the absorbances measured before and during the pulse is due to analyte. Smith-Hieftje background correction is more reliable than D_2 lamp correction, but the price is a 10–75% loss of sensitivity because analyte absorbs some fraction of the light during the high-current pulse.

An excellent, but expensive, background correction technique for many elements relies on the *Zeeman effect* (pronounced ZAY-mon). This term refers to the shifting of energy levels of atoms and molecules in a magnetic field. When the field is applied parallel to the light path through a flame or furnace, the absorption (or emission) line of analyte atoms is split into three components. Two are shifted to slightly lower and slightly higher wavelengths (Figure 21-20), and one component is unshifted. The unshifted component does not have the correct electromagnetic polarization to absorb light traveling parallel to the magnetic field and is therefore "invisible."

To use the Zeeman effect for background correction, a strong magnetic field is pulsed on and off. Sample and background are observed when the field is off, and background alone is observed when the field is on. The difference is the corrected signal.

The advantage of Zeeman and Smith-Hieftje background corrections is that they operate at the analytical wavelength. In contrast, D_2 background correction is made over a broad band. A structured or sloping background is averaged by this process, potentially misrepresenting the true background signal at the analytical wavelength.

Detection Limits

The **detection limit** is the concentration of an element that gives a signal equal to twice the peak-to-peak noise level of the baseline (Figure 21-21).[4] The baseline noise level should be measured while a blank sample is being aspirated into the flame.

Figure 21-22 compares detection limits for flame, furnace, and inductively coupled plasma analyses on instruments from one manufacturer. The

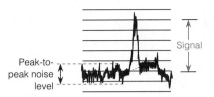

Figure 21-21 Measurement of peak-to-peak noise level and signal level. The signal is measured from its base at the midpoint of the noise component along the slightly slanted baseline. This sample exhibits a signal-to-noise ratio of 2.4.

detection limit for furnaces is typically two orders of magnitude lower than that observed with a flame. The reason for this is that the sample is confined in a small volume for a relatively long time in the furnace, compared with its fleeting moment in a flame. Detection limits for the inductively coupled plasma are intermediate between the flame and furnace. Although most elements can be analyzed by either absorption or emission, alkali metals have especially strong emission and are commonly measured in clinical analysis with relatively simple equipment (Box 21-1).

Commercial standard solutions of elements for flame atomic absorption are not necessarily suitable for more sensitive plasma and furnace analyses. The latter methods require purer grades of water and acids for standard solutions and, especially, for dilutions in your laboratory. For the most sensitive analyses, solutions are prepared in a dust-free environment (a clean room with a filtered air supply) to reduce background contamination that *will be* detected by your instruments.

Detection limits (ng/mL)
— Inductively coupled plasma emission
— Flame atomic absorption
— Graphite furnace atomic absorption

Key element box example (Fe): 0.7 / 5 / 0.02

Li	Be											B	C	N	O	F	Ne
0.7	0.07											1	10				
2	1											500	—				
0.1	0.02											15	—				
Na	**Mg**											**Al**	**Si**	**P**	**S**	**Cl**	**Ar**
3	0.08											2	5	7	3		
0.2	0.3											30	100	40,000	—		
0.005	0.004											0.01	0.1	30	—		
K	**Ca**	**Sc**	**Ti**	**V**	**Cr**	**Mn**	**Fe**	**Co**	**Ni**	**Cu**	**Zn**	**Ga**	**Ge**	**As**	**Se**	**Br**	**Kr**
20	0.07	0.3	0.4	0.7	2	0.2	0.7	1	3	0.9	0.6	10	20	7	10		
3	0.5	40	70	50	3	2	5	4	90	1	0.5	60	200	200	250		
0.1	0.01	—	0.5	0.2	0.01	0.01	0.02	0.02	0.1	0.02	0.001	0.5	—	0.2	0.5		
Rb	**Sr**	**Y**	**Zr**	**Nb**	**Mo**	**Tc**	**Ru**	**Rh**	**Pd**	**Ag**	**Cd**	**In**	**Sn**	**Sb**	**Te**	**I**	**Xe**
1	0.2	0.6	2	5	3		10	20	4	0.8	0.5	20	9	9	4		
7	2	200	1000	2000	20		60	4	10	0.4	0.5	40	30	40	30		
0.05	0.1	—	—	—	0.02		1	—	0.3	0.005	0.003	1	0.2	0.15	0.1		
Cs	**Ba**	**La**	**Hf**	**Ta**	**W**	**Re**	**Os**	**Ir**	**Pt**	**Au**	**Hg**	**Tl**	**Pb**	**Bi**	**Po**	**At**	**Rn**
40,000	0.6	1	4	10	8	3	0.2	7	2	7	7	10	10	7			
4	10	2000	2000	2000	1000	600	100	400	100	10	150	20	10	40			
0.2	0.04	—	—	—	—	—	—	—	0.2	0.1	2	0.1	0.05	0.1			

Ce	**Pr**	**Nd**	**Pm**	**Sm**	**Eu**	**Gd**	**Tb**	**Dy**	**Ho**	**Er**	**Tm**	**Yb**	**Lu**
2	9	10		10	0.9	5	6	2	2	0.7	2	0.3	0.3
—	6000	1000		1000	20	2000	500	30	40	30	900	4	300
—	—	—		—	0.5	—	0.1	1	—	2	—	—	—
Th	**Pa**	**U**	**Np**	**Pu**	**Am**	**Cm**	**Bk**	**Cf**	**Es**	**Fm**	**Md**	**No**	**Lr**
7		60											
—		40,000											
—		—											

— Requires N_2O/C_2H_2 flame and is therefore better analyzed by inductively coupled plasma
— Best analyzed by emission

Figure 21-22 Flame, furnace, and inductively coupled plasma detection limits (ng/mL) with instruments from GBC Scientific Equipment, Australia. [From R. J. Gill, *Am. Lab.* November 1993, p. 24F.] Quantitative analysis usually requires concentrations 10–100 times greater than the detection limit.

Box 21-1 The Flame Photometer in Clinical Chemistry

Before the advent of atomic spectroscopy, laborious gravimetric procedures were required to measure Na^+ and K^+ in serum and urine. Today these measurements are routine in clinical analysis.

    The typical flame photometer shown here operates on air and natural gas. Optical filters isolate the strong emission lines of Na, K, Li, Ca, and Ba. Full-scale sensitivity for Na and K corresponds to 3 µg/mL, or 50 µg/mL for Ca. The digital readout is calibrated by aspirating standard samples through the narrow tube immersed in the beaker. Lithium is often used as an internal standard for sodium and potassium.

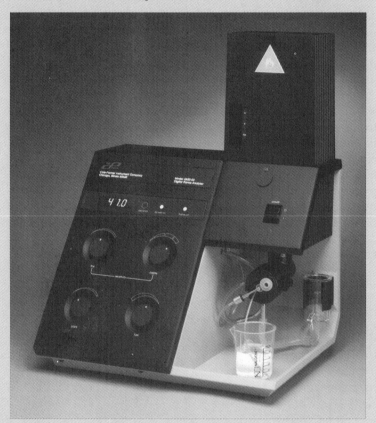

Digital flame photometer. [Courtesy Cole-Parmer Instrument Co., Chicago, IL.]

21-5 Interference

Types of interference:

    spectral: unwanted signals overlapping analyte signal

    chemical: chemical reactions decreasing the concentration of analyte atoms

    ionization: ionization of analyte atoms decreases the concentration of neutral atoms

Interference is any effect that changes the signal when analyte concentration remains unchanged. In atomic spectroscopy, interference is widespread and easy to overlook. If you are clever enough to discern that interference is occurring, it may be corrected by counteracting the source of interference or by preparing standards that exhibit the same interference.

Types of Interference

Spectral interference refers to the overlap of analyte signal with signals due to other elements or molecules in the sample or with signals due to the flame or furnace. Interference from the flame can be subtracted by using D_2, Smith-Hieftje, or Zeeman background correction. The best means of dealing with overlap between lines of different elements in the sample is to choose another wavelength for analysis.

Elements that form very stable diatomic oxides are said to be *refractory* because they are incompletely atomized at the temperature of the flame or furnace. The spectrum of a molecule is much broader and more complex than that of an atom, because vibrational and rotational transitions are combined with electronic transitions (as described in Section 19-4). The broad spectrum leads to spectral interference at many wavelengths. Figure 21-23 shows an example of a plasma containing Y and Ba atoms, as well as YO molecules. Note how broad the molecular emission is relative to the atomic emission.

Chemical interference is caused by any component of the sample that decreases the extent of atomization of analyte. For example, SO_4^{2-} and PO_4^{3-} hinder the atomization of Ca^{2+}, perhaps by forming nonvolatile salts. **Releasing agents** are chemicals that can be added to a sample to decrease chemical interference. EDTA and 8-hydroxyquinoline protect Ca^{2+} from the interfering effects of SO_4^{2-} and PO_4^{3-}. La^{3+} can also be used as a releasing agent, apparently because it preferentially reacts with PO_4^{3-} and frees the Ca^{2+}. A fuel-rich flame is recommended to reduce certain oxidized analyte species that would otherwise hinder atomization. Higher flame temperatures eliminate many kinds of chemical interference.

Ionization interference can be a problem in the analysis of alkali metals at relatively low flame temperature and in the analyses of other elements at higher temperature. For any element, we can write a gas-phase ionization reaction:

$$M(g) \rightarrow M^+(g) + e^-(g) \qquad K = \frac{[M^+][e^-]}{[M]} \qquad (21-6)$$

Because the alkali metals have the lowest ionization potentials, they are most extensively ionized in a flame. At 2 450 K and a pressure of 0.1 Pa, sodium is 5% ionized. With its lower ionization potential, potassium is 33% ionized under the same conditions. Because ionized atoms have energy levels different from those of neutral atoms, the desired signal is decreased.

An **ionization suppressor** is an element added to a sample to decrease the extent of ionization of analyte. For example, in the analysis of potassium, it is recommended that solutions contain 1 000 ppm of CsCl, because cesium is more easily ionized than potassium. By producing a high concentration of electrons in the flame, ionization of Cs suppresses ionization of K. Ionization suppression is desirable in a low-temperature flame in which we want to observe neutral atoms.

Virtues of the Inductively Coupled Plasma

An inductively coupled argon plasma eliminates many common interferences. The plasma is twice as hot as a conventional flame, and the residence time of analyte in the plasma is about twice as long. Therefore, atomization is more complete and the signal is correspondingly enhanced. Formation of analyte oxides and hydroxides is negligible. The plasma is remarkably free of background radiation 15–35 mm above the load coil where sample emission is observed.

A problem in flame emission spectroscopy is that the concentration of electronically excited atoms in the cooler, outer part of the flame is lower

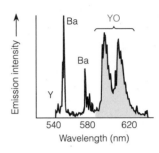

Figure 21-23 Emission from a plasma produced by laser irradiation of the high-temperature superconductor $YBa_2Cu_3O_7$. The solid is vaporized by the laser, and excited atoms and molecules in the gas phase emit light at their characteristic wavelengths. [Reproduced from W. A. Weimer, *Appl. Phys. Lett.* **1988,** *52,* 2171.]

This is an application of Le Châtelier's principle to Reaction 21-6.

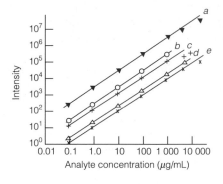

Figure 21-24   Analytical calibration curves for emission from (a) Ba^+, (b) Cu, (c) Na, (d) Fe, and (e) Ba in an inductively coupled plasma. Note that both axes are logarithmic. [From R. N. Savage and G. M. Hieftje, *Anal. Chem.* **1979**, *51*, 408.]

than in the warmer, central part of the flame. Emission from the central region is absorbed in the outer region. **Self-absorption** increases with increasing concentration of analyte and gives nonlinear calibration curves. In a plasma, the temperature is more uniform, and self-absorption is not nearly so important. Figure 21-24 shows plasma emission calibration curves that are linear over nearly five orders of magnitude. In flames and furnaces, the linear range covers about two orders of magnitude.

Terms to Understand

aerosol	detection limit	matrix modifier
atomic absorption spectroscopy	Doppler effect	nebulization
atomic emission spectroscopy	graphite furnace	premix burner
atomic fluorescence spectroscopy	hollow-cathode lamp	pressure broadening
atomization	inductively coupled plasma	releasing agent
background correction	ionization interference	self-absorption
Boltzmann distribution	ionization suppressor	spectral interference
chemical interference	matrix	

Summary

In atomic spectroscopy, absorption, emission, or fluorescence from gaseous atoms is measured. Liquids may be atomized by a flame, furnace, or plasma. In a premix burner, the sample is mixed with fuel and oxidant before it flows into the flame, whose temperature is usually in the range $2\,300–3\,400$ K. The fuel and oxidant chosen determine the temperature of the flame and affect the extent of spectral, chemical, or ionization interference that will be encountered. Temperature instability affects atomization in atomic absorption and has an even larger effect on atomic emission, because the excited-state population is exponentially sensitive to temperature. An electrically heated furnace requires less sample than a flame and has a lower detection limit. In an inductively coupled plasma, a radio-frequency induction coil is used to heat Ar^+ ions in an argon gas stream to $6\,000–10\,000$ K. At this high temperature, emission is observed from electronically excited atoms and ions. There is little chemical interference in an inductively coupled plasma, the temperature is very stable, and little self-absorption is observed.

Instrumentation for atomic spectroscopy is similar to that for molecular spectroscopy, but the sample must be atomized, the radiation source must be highly monochromatic, and the background signal must be subtracted. Lamps with a hollow cathode made of the analyte element provide spectral lines sharper than those of the atomic vapor (whose lines are broadened by collisions and by the Doppler effect). Correction for background emission from the flame is possible by electrically pulsing the lamp on and off or mechanically chopping the beam. Light scattering and spectral background can be subtracted by measuring absorption of a deuterium lamp or a pulsed hollow-cathode lamp (Smith-Hieftje correction) or by Zeeman background correction, in which the atomic energy levels are alternately shifted in and out of resonance with the lamp frequency by a magnetic field. Chemical interference may be reduced by addition of releasing agents, which prevent the analyte from reacting with interfering species. Ionization interference in flames may be suppressed by adding easily ionized elements such as Cs.

Exercises

A. Li was determined by atomic emission using the method of standard additions. From the data in the following table, use a graph similar to Figure 6-10 to find the concentration of Li in pure unknown. The Li standard contained 1.62 µg Li/mL.

Unknown (mL)	Standard (mL)	Final volume (mL)	Emission intensity (arbitrary units)
10.00	0.00	100.0	309
10.00	5.00	100.0	452
10.00	10.00	100.0	600
10.00	15.00	100.0	765
10.00	20.00	100.0	906

B. Mn was used as an internal standard for measuring Fe by atomic absorption. A standard mixture containing 2.00 µg Mn/mL and 2.50 µg Fe/mL gave a quotient (Fe signal/Mn signal) = 1.05. A mixture with a volume of 6.00 mL was prepared by mixing 5.00 mL of unknown Fe solution with 1.00 mL containing 13.5 µg Mn/mL. The absorbance of this mixture at the Mn wavelength was 0.128, and the absorbance at the Fe wavelength was 0.185. Find the molarity of the unknown Fe solution.

C. The atomic absorption signal shown below was obtained with 0.048 5 µg Fe/mL in a graphite furnace. Estimate the detection limit for Fe.

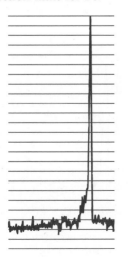

D. The laser atomic fluorescence excitation and emission spectra of sodium in an air-acetylene flame are shown below. In the *excitation* spectrum, the laser (bandwidth = 0.03 nm) was scanned through various wavelengths, while the detector monochromator (bandwidth = 1.6 nm) was held fixed near 589 nm. In the *emission* spectrum, the laser was fixed at 589.0 nm, and the detector monochromator wavelength was varied. Explain why the emission spectrum gives one broad band, whereas the excitation spectrum gives two sharp lines. How can the excitation linewidths be much narrower than the detector monochromator bandwidth?

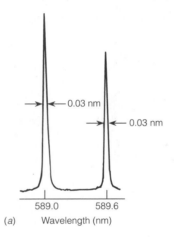

(a)   Wavelength (nm)

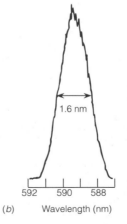

(b)   Wavelength (nm)

Fluorescence excitation and emission spectra of the two sodium D lines in an air-acetylene flame. (a) In the excitation spectrum, the laser was scanned. (b) In the emission spectrum, the monochromator was scanned. The monochromator slit width was the same for both spectra. [From S. J. Weeks, H. Haraguchi, and J. D. Winefordner, *Anal. Chem.* **1978**, *50*, 360.]

Problems

Techniques of Atomic Spectroscopy

1. In which technique, atomic absorption or atomic emission, is flame temperature stability more critical? Why?

2. State the advantages and disadvantages of furnaces compared with flames in atomic absorption spectroscopy.

3. State the advantages and disadvantages of the inductively coupled plasma compared with conventional flames in atomic spectroscopy.

4. Explain what is meant by the Doppler effect. Rationalize why Doppler broadening increases with increasing temperature and decreasing mass in Equation 21-5.

5. Explain how the following background correction techniques work:

(a) beam chopping   **(c)** Smith-Hieftje

(b) deuterium lamp   **(d)** Zeeman

6. Explain what is meant by spectral, chemical, and ionization interference.

7. Calculate the emission wavelength (nm) of excited atoms that lie 3.371×10^{-19} J per molecule above the ground state.

8. Derive the entries for 500 nm in Table 21-2. What would be the value of N^*/N_0 at $6\,000$ K if $g^* = 3$ and $g_0 = 1$?

9. Calculate the Doppler linewidth in Hz $(= s^{-1})$ for the 589-nm line of Na and for the 254-nm line of Hg, both at $2\,000$ K.

10. The first excited state of Ca is reached by absorption of 422.7-nm light.

(a) What is the energy difference (kJ/mol) between the ground and excited states?

(b) The relative degeneracies are $g^*/g_0 = 3$ for Ca. What is the ratio N^*/N_0 at $2\,500$ K?

(c) By what percentage will the fraction in **(b)** be changed by a 15-K rise in temperature?

(d) What will be the ratio N^*/N_0 at $6\,000$ K?

11. An *electron volt* (eV) is the energy change of an electron moved through a potential difference of 1 volt:

$$eV = (1.602 \times 10^{-19} \text{ C})(1 \text{ V})$$

$$= 1.602 \times 10^{-19} \text{ J per electron}$$

$$= 96.49 \text{ kJ per mole of electrons}$$

Use the Boltzmann distribution to fill in the following table and explain why Br is not readily observed in atomic absorption or atomic emission.

Element:	Na	Cu	Br
Excited state energy (eV):	2.10	3.78	8.04
Wavelength (nm):			
Degeneracy ratio (g^*/g_0):	3	3	2/3
N^*/N_0 at $2\,600$ K in flame:			
N^*/N_0 at $6\,000$ K in plasma:			

12. *Sensitivity* in atomic absorption is defined as the concentration of analyte that absorbs 1% of the light from the lamp and therefore gives 99% transmittance (corresponding to an absorbance of $-\log(0.99) = 0.004\,36$). A sample containing 1.00 μg Fe/mL gave an absorbance of 0.055. Estimate the sensitivity for Fe.

13. The text explained how a matrix modifier could prevent premature evaporation of Al in a furnace by maintaining the aluminum in the form Al_2O_3. Another type of matrix modifier prevents loss of signal from the atom X that readily forms the molecular carbide XC in a graphite furnace (which is a source of carbon). For example, adding yttrium to a sample containing barium increases the Ba signal by 30%. The bond dissociation energy of YC is greater than that of BaC. Explain what is happening to increase the Ba signal.

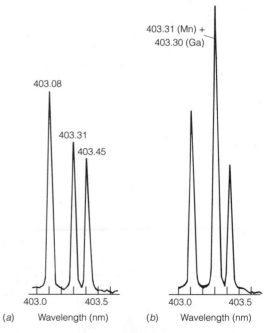

For Problem 14: Fluorescence excitation of solution containing (a) 1 μg Mn/mL and (b) 1 μg Mn/mL plus 5 μg Ga/mL. [From S. J. Weeks, H. Haraguchi, and J. D. Winefordner, *Anal. Chem.* **1978**, *50*, 360.]

14. Fluorescence excitation spectra of Mn and Mn + Ga solutions are shown on page 618. The Ga line is within 0.2 nm of all three Mn lines. Explain why there is no spectral interference by Ga in the laser atomic fluorescence excitation analysis of Mn, even though the detector monochromator bandwidth is 1.0 nm. You may wish to refer to Exercise D.

Quantitative Analysis by Atomic Spectroscopy

15. Why is an internal standard most appropriate for quantitative analysis when unavoidable sample losses are expected during sample preparation?

16. *Standard addition.* An unknown containing element X was mixed with aliquots of a standard solution of element X for atomic absorption spectroscopy. The standard solution contained $1\,000.0$ μg of X per milliliter.

Volume of unknown (mL)	Volume of standard (mL)	Total volume (mL)	Absorbance
10.00	0	100.0	0.163
10.00	1.00	100.0	0.240
10.00	2.00	100.0	0.319
10.00	3.00	100.0	0.402
10.00	4.00	100.0	0.478

(a) Calculate the concentration (μg X/mL) of added standard in each solution.

(b) Prepare a graph similar to Figure 6-10 to determine the concentration of X in the unknown.

17. *Internal standard.* A solution was prepared by mixing 10.00 mL of unknown (X) with 5.00 mL of standard (S) containing 8.24 μg S/mL, and diluting to 50.0 mL. The measured signal quotient was (signal due to X/signal due to S) = 1.69.

(a) In a separate experiment it was found that for equal concentrations of X and S, the signal due to X was 0.93 times as intense as the signal due to S. Find the concentration of X in the unknown.

(b) Answer the same question if in a separate experiment it was found that for the concentration of X equal to 3.42 times the concentration of S, the signal due to X was 0.93 times as intense as the signal due to S.

18. A series of potassium standards gave the following emission intensities at 404.3 nm. Find the concentration of potassium in the unknown.

Sample (μg K/mL)	Relative emission
Blank	0
5.00	124
10.00	243
20.0	486
30.0	712
Unknown	417

19. Free cyanide in aqueous solution can be determined indirectly by atomic absorption based on its ability to dissolve silver as it passes through a porous silver membrane filter at pH 12:[5]

$$4Ag(s) + 8CN^- + 2H_2O + O_2 \rightarrow 4Ag(CN)_2^- + 4OH^-$$

A series of silver standards gave a linear calibration curve in flame atomic absorption with a slope of 807 meter units per ppm Ag in the standard. (The meter units are arbitrary numbers related to absorbance, and ppm refers to μg Ag/mL.) An unknown cyanide solution passed through the silver membrane gave a meter reading of 198 units. Find the molarity of CN^- in the unknown.

Notes and References

1. The development of atomic absorption instruments is described by A. Walsh, *Anal. Chem.* **1991**, *63*, 933A. Evolution of the graphite furnace is recounted by B. V. L'vov, *Anal. Chem.* **1991**, *63*, 924A.

2. L. A. Nash, L. N. Peterson, S. P. Nadler, and D. Z. Levine, *Anal. Chem.* **1988**, *60*, 2413.

3. D. L. Styris and D. A. Redfield, *Anal. Chem.* **1987**, *59*, 2891.

4. Detection limit is really the lowest concentration that can be "reliably" detected. Problem 19 in Chapter 4 defined detection limit as a signal that has a 99% probability of being greater than zero. For various statistical definitions of detection limit, see G. L. Long and J. D. Winefordner, *Anal. Chem.* **1983**, *55*, 712A; W. R. Porter, *Anal. Chem.* **1983**, *55*, 1290A; and J. E. Knoll, *J. Chromatographic Sci.* **1985**, *23*, 422.

5. J. J. Rosentreter and R. K. Skogerboe, *Anal. Chem.* **1991**, *63*, 682.

Analytical Separations and Chemical Problem Solving

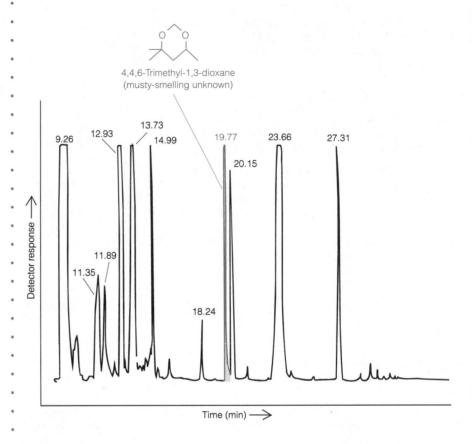

4,4,6-Trimethyl-1,3-dioxane
(musty-smelling unknown)

Different volatile compounds released from plastic food wrapping emerge from the gas chromatography column at different times. [From R. J. McGorrin, T. R. Pofahl, and W. R. Croasmun, *Anal. Chem.* **1987,** *59,* 1109A.]

When plastic packaging with a musty odor arrived at a Kraft food plant, the Basic Flavor Chemistry Group was asked to find the source of the odor. The team of chemists evaporated volatile compounds from the plastic at 100°C and used gas chromatography to separate the mixture into its components. The musty odor was associated only with the compound emerging at 19.77 min.

Infrared spectroscopy and mass spectrometry provided clues to the structure of the offending compound. An authentic sample of the suspected compound was then synthesized and found to have the same chromatographic and spectroscopic properties—and odor—as the unknown. The musty smell was derived from a chemical reaction of a coating used to help ink adhere to the plastic. Once the source of the odor was discovered, the plastic manufacturer could take steps to eliminate its future occurrence.

Introduction to Analytical Separations

In the vast majority of real analytical problems, we must identify and quantitate one or more components from a complex mixture. Isolating the desired unknown is the challenging first step. This chapter discusses the fundamentals of analytical separations, and the next two chapters describe specific methods.

22-1 Solvent Extraction

Extraction refers to the transfer of a solute from one phase to another. The most common case is the extraction of an aqueous solution with an organic solvent. Diethyl ether, toluene, and hexane are common solvents that are *immiscible* with and less dense than water. They form a separate phase that floats on top of the aqueous phase, as shown in Color Plate 14. Chloroform, dichloromethane, and carbon tetrachloride are common solvents that are denser than water.[†] In the two-phase mixture, one phase is predominantly water and the other phase is predominantly organic.

To be **miscible** means that the two liquids form a single phase when they are mixed in any ratio. *Immiscible* liquids remain in separate phases.

Suppose that solute S is partitioned between phases 1 and 2, as depicted in Figure 22-1. The **partition coefficient,** K, is the equilibrium constant for the reaction

$$S \text{ (in phase 1)} \rightleftharpoons S \text{ (in phase 2)}$$

Partition coefficient:
$$K = \frac{\mathcal{A}_{S_2}}{\mathcal{A}_{S_1}} \approx \frac{[S]_2}{[S]_1} \qquad (22\text{-}1)$$

where $\mathcal{A}_{S_1}$ refers to the activity of solute in phase 1. Lacking knowledge of the

. . . .
[†]Whenever a choice exists between $CHCl_3$ and CCl_4, the less toxic $CHCl_3$ should be chosen. Hexane and toluene are greatly preferred over benzene, which is a carcinogen.

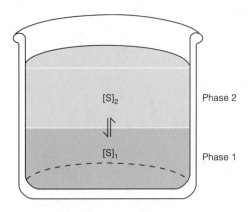

Figure 22-1 Partitioning of a solute between two liquid phases.

activity coefficients, we will write the partition coefficient in terms of concentrations.

Suppose that solute S in V_1 mL of solvent 1 is extracted with V_2 mL of solvent 2. Let m be the number of moles of S in the system and let q be the fraction of S remaining in phase 1 at equilibrium. The molarity in phase 1 is therefore qm/V_1. The fraction of total solute transferred to phase 2 is $(1 - q)$, and the molarity in phase 2 is $(1 - q)m/V_2$. Therefore,

$$K = \frac{[S]_2}{[S]_1} = \frac{(1 - q)m/V_2}{qm/V_1}$$

from which we can solve for q:

The larger the partition coefficient, the less solute remains in phase 1.

$$\text{fraction remaining in phase 1 after 1 extraction} = q = \frac{V_1}{V_1 + KV_2} \tag{22-2}$$

Equation 22-2 says that the fraction of solute remaining in phase 1 depends on the partition coefficient and the two volumes. If the phases are separated and fresh solvent 2 is mixed with phase 1, the fraction of solute remaining in phase 1 at equilibrium will be

Example of extraction: If $q = \frac{1}{4}$, then $\frac{1}{4}$ of the solute remains in phase 1 after one extraction. A second extraction reduces the concentration to $\frac{1}{4}$ of the value after the first extraction $= \left(\frac{1}{4}\right)\left(\frac{1}{4}\right) = \frac{1}{16}$ of the initial concentration.

$$\text{fraction remaining in phase 1 after 2 extractions} = q \cdot q = \left(\frac{V_1}{V_1 + KV_2}\right)^2$$

After n extractions with volume V_2, the fraction remaining in phase 1 is

$$\text{fraction remaining in phase 1 after } n \text{ extractions} = q^n = \left(\frac{V_1}{V_1 + KV_2}\right)^n \tag{22-3}$$

· ·

EXAMPLE Extraction Efficiency

Solute A has a partition coefficient of 3 between toluene and water (with 3 times as much in the toluene phase.) Suppose that 100 mL of a 0.01 M aqueous solution of A is extracted with toluene. What fraction of A remains in the aqueous phase **(a)** if one extraction with 500 mL is performed and **(b)** if five extractions with 100 mL are performed?

Solution **(a)** Taking water as phase 1 and toluene as phase 2, Equation 22-2 says that after a 500-mL extraction, the fraction remaining in the aqueous phase is

$$q = \frac{100}{100 + (3)(500)} = 0.062 \approx 6\%$$

(b) With five 100-mL extractions, the fraction remaining is given by Equation 22-3:

$$\text{fraction remaining} = \left(\frac{100}{100 + (3)(100)}\right)^5 = 0.000\,98 \approx 0.1\%$$

It is more efficient to do several small extractions than one big extraction.

Many small extractions are much more effective than a few large extractions.

pH Effects

The distribution of a solute such as a basic amine between two phases is pH dependent. Suppose that the neutral form, B, has partition coefficient, K, between phases 1 and 2. Suppose that the conjugate acid BH^+ is soluble *only* in the aqueous phase (1), and denote its acid dissociation constant as K_a. The **distribution coefficient**, D, is defined as

Distribution coefficient: $\qquad D = \dfrac{\text{total concentration in phase 2}}{\text{total concentration in phase 1}} \qquad$ (22-4)

which becomes

$$D = \frac{[B]_2}{[B]_1 + [BH^+]_1} \qquad (22\text{-}5)$$

Substituting $K = [B]_2/[B]_1$ and $K_a = [H^+][B]/[BH^+]$ into Equation 22-5 leads to

Distribution of base between two phases: $\qquad D = \dfrac{K \cdot K_a}{K_a + [H^+]} = K \cdot \alpha_B \qquad$ (22-6)

where α_B is the fraction of weak base in the neutral form, B, in the aqueous phase. *The distribution coefficient D is used in place of the partition coefficient, K, in Equation 22-2 when dealing with a species that has more than one chemical form, such as B and BH^+.*

To extract a base into water, you should use a pH low enough to convert it to BH^+ (Figure 22-2). By the same reasoning, to extract the acid HA into water, you should use a high enough pH to convert the acid to A^-.

Challenge Suppose that the acid HA (with dissociation constant K_a) is partitioned between aqueous phase 1 and organic phase 2. Calling the partition coefficient K for HA and assuming that A^- is not soluble in the organic phase, show that the distribution coefficient is given by

Distribution of acid between two phases: $\qquad D = \dfrac{K \cdot [H^+]}{[H^+] + K_a} = K \cdot \alpha_{HA} \qquad$ (22-7)

where α_{HA} is the fraction of weak acid in the form HA in the aqueous phase.

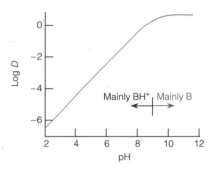

Figure 22-2 Effect of pH on the distribution coefficient for the extraction of a base into an organic solvent. In this example, $K = 3.0$ and pK_a for BH^+ is 9.00.

$$\alpha_B = \frac{[B]_{aq}}{[B]_{aq} + [BH^+]_{aq}}$$

α_B is the same as α_{A^-} in Equation 11-18.

EXAMPLE   Effect of pH on Extraction

Suppose that the partition coefficient for an amine, B, is $K = 3.0$ and the acid dissociation constant of BH^+ is $K_a = 1.0 \times 10^{-9}$. If 50 mL of 0.010 M aqueous amine is extracted with 100 mL of solvent, what will be the formal concentration remaining in the aqueous phase **(a)** at pH 10.00 and **(b)** at pH 8.00?

Solution   **(a)** At pH 10.00, $D = KK_a/(K_a + [H^+]) = (3.0)(1.0 \times 10^{-9})/(1.0 \times 10^{-9} + 1.0 \times 10^{-10}) = 2.73$. Equation 22-2 says that the fraction remaining in the aqueous phase is

$$q = \frac{50}{50 + (2.73)(100)} = 0.15 \Rightarrow 15\% \text{ left in water}$$

The concentration of amine in the aqueous phase is 15% of 0.010 M = 0.001 5 M. *In the preceding equation, we have used the distribution coefficient, D, in place of the partition coefficient, K, in Equation 22-2.*
   (b) At pH 8.00, $D = (3.0)(1.0 \times 10^{-9})/(1.0 \times 10^{-9} + 1.0 \times 10^{-8}) = 0.273$. Therefore,

$$q = \frac{50}{50 + (0.273)(100)} = 0.65 \Rightarrow 65\% \text{ left in water}$$

The concentration in the aqueous phase is 0.006 5 M. At pH 10, the base is predominantly in the form B and is extracted into the organic solvent. At pH 8, it is in the form BH^+ and remains in the water.

Extraction with a Metal Chelator

Most complexes that can be extracted into organic solvents must be neutral. Charged complexes, such as $Fe(EDTA)^-$ or $Fe(1,10\text{-phenanthroline})_3^{2+}$, are not very soluble in organic solvents. One scheme for separating metal ions from one another is to selectively complex one ion with an organic ligand and extract it into an organic solvent. Three ligands commonly employed for this purpose are shown below.

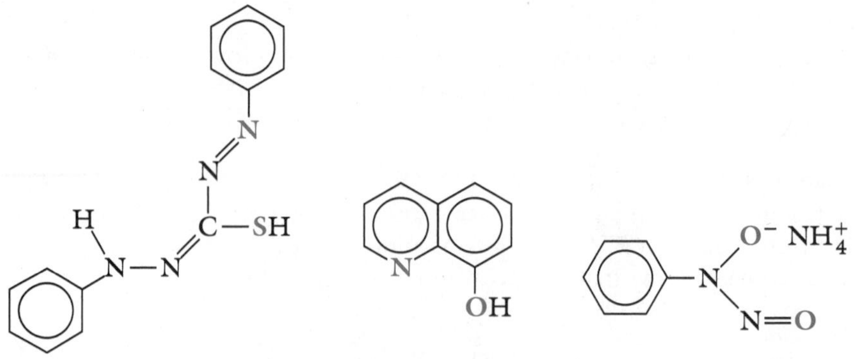

Dithizone
(diphenylthiocarbazone)

8-Hydroxyquinoline
(oxine)

Cupferron

Each ligand can be represented as a weak acid, HL, which loses one proton when it binds to a metal ion through the atoms shown in **bold** type.

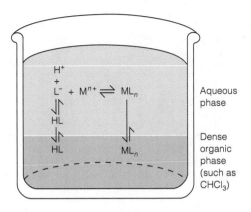

Figure 22-3 Extraction of a metal ion with a chelator. The predominant form of metal in the aqueous phase is M^{n+}, and the predominant form in the organic phase is ML_n.

$$HL(aq) \rightleftharpoons H^+(aq) + L^-(aq) \qquad K_a = \frac{[H^+]_{aq}[L^-]_{aq}}{[HL]_{aq}} \qquad (22\text{-}8)$$

$$nL^-(aq) + M^{n+}(aq) \rightleftharpoons ML_n(aq) \qquad \beta = \frac{[ML_n]_{aq}}{[M^{n+}]_{aq}[L^-]_{aq}^n} \qquad (22\text{-}9)$$

Each of these ligands can react with many different metal ions, but some selectivity is achieved by controlling the pH.

Let's derive an equation for the distribution coefficient of a metal between two phases when essentially all of the metal in the aqueous phase (aq) is in the form M^{n+} and all of the metal in the organic phase (org) is in the form ML_n (Figure 22-3). We define the partition coefficients for ligand and complex as follows:

M^{n+} is in the aqueous phase and ML_n is in the organic phase.

$$HL(aq) \rightleftharpoons HL(org) \qquad K_L = \frac{[HL]_{org}}{[HL]_{aq}} \qquad (22\text{-}10)$$

$$ML_n(aq) \rightleftharpoons ML_n(org) \qquad K_M = \frac{[ML_n]_{org}}{[ML_n]_{aq}} \qquad (22\text{-}11)$$

The distribution coefficient we seek is

$$D = \frac{[\text{total metal}]_{org}}{[\text{total metal}]_{aq}} \approx \frac{[ML_n]_{org}}{[M^{n+}]_{aq}} \qquad (22\text{-}12)$$

From Equations 22-11 and 22-9, we can write

$$[ML_n]_{org} = K_M[ML_n]_{aq} = K_M\beta[M^{n+}]_{aq}[L^-]_{aq}^n$$

Using $[L^-]_{aq}$ from Equation 22-8 gives

$$[ML_n]_{org} = \frac{K_M\beta[M^{n+}]_{aq}K_a^n[HL]_{aq}^n}{[H^+]_{aq}^n}$$

Putting this value of $[ML_n]_{org}$ into Equation 22-12 gives

$$D = \frac{K_M\beta K_a^n[HL]_{aq}^n}{[H^+]_{aq}^n}$$

Because most HL is in the organic phase, we substitute $[HL]_{aq} = [HL]_{org}/K_L$ to produce the most useful expression for the distribution coefficient:

Distribution of metal-chelate complex between phases: $\quad D \approx \dfrac{K_M\beta K_a^n}{K_L^n}\dfrac{[HL]_{org}^n}{[H^+]_{aq}^n} \qquad (22\text{-}13)$

Figure 22-4 Extraction of metal ions by dithizone into CCl_4. At pH 5, Cu^{2+} is completely extracted into CCl_4, whereas Pb^{2+} and Zn^{2+} remain in the aqueous phase. [Adapted from G. H. Morrison and H. Freiser in C. L. Wilson and D. Wilson, Eds., *Comprehensive Analytical Chemistry*, Vol. IA (New York: Elsevier, 1959).]

You can select a pH to bring the metal into either phase.

We see that the distribution coefficient for metal ion extraction depends on pH and ligand concentration. It is often possible to select a pH where D is large for one metal and small for another. For example, Figure 22-4 shows that Cu^{2+} could be separated from Pb^{2+} and Zn^{2+} by extraction with dithizone at pH 5. Demonstration 22-1 illustrates the pH dependence of an extraction with dithizone. Box 22-1 describes *crown ethers* that are used to extract polar reagents into nonpolar solvents for chemical reactions.

Box 22-1 Crown Ethers

Crown ethers are a class of synthetic compounds that envelop metal ions (especially alkali metal cations) in a pocket of oxygen ligands. Crown ethers are used as *phase transfer catalysts* because they can extract water-soluble ionic reagents into nonpolar solvents, where reaction with hydrophobic compounds can occur. The following structure of the potassium complex of dibenzo-30-crown-10 shows that the K^+ ion is engulfed by 10 oxygen atoms with K—O distances averaging 288 pm. Only the hydrophobic outside of the complex is exposed to solvent.

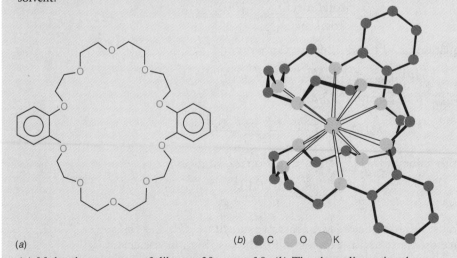

(*a*) Molecular structure of dibenzo-30-crown-10. (*b*) The three-dimensional structure of its K^+ complex. [Adapted from M. A. Bush and M. R. Truter, *J. Chem. Soc. Chem. Commun.* **1970**, 1439.]

Demonstration 22-1 Extraction with Dithizone

Dithizone (diphenylthiocarbazone) is a green compound, soluble in nonpolar organic solvents and insoluble in water below pH 7. It forms red, hydrophobic complexes with most di- and trivalent metal ions.

Dithizone
(green)

(colorless)

Metal complex
(red)

Dithizone is widely used for analytical extractions, for colorimetric determinations of metal ions, and for removing traces of metals from aqueous buffers.

In the latter application, an aqueous buffer is extracted repeatedly with a green solution of dithizone in $CHCl_3$. As long as the organic phase turns red, metal ions are being extracted from the buffer. When the extracts are green, the last traces of metal ions have been removed. In Figure 22-4 we see that only certain metal ions can be extracted at a given pH.[1]

Equilibrium between the green ligand and red complex is demonstrated by using three large test tubes sealed with tightly fitting rubber stoppers. In each tube is placed some hexane plus a few milliliters of dithizone solution (prepared by dissolving 1 mg of dithizone in 100 mL of $CHCl_3$). To tube A is added distilled water; to tube B is added tap water; and to tube C is added 2 mM $Pb(NO_3)_2$. After shaking and settling, tubes B and C contain a red upper phase, whereas A remains green.

The proton equilibrium in the reaction above is shown by adding a few drops of 1 M HCl to tube C. After shaking, the dithizone turns green again. Competition with a stronger ligand is shown by adding a few drops of 0.05 M EDTA solution to tube B. Again, shaking causes a reversion to the green color.

22-2 What Is Chromatography?

Chromatography operates on the same principle as extraction, but one phase is held in place while the other moves past it. Figure 22-5 shows an experiment in which solution containing solutes A and B is placed on top of a column packed with solid particles and filled with solvent. When the outlet is opened, solutes A and B flow down into the column. Fresh solvent is then applied to the top of the column and the mixture is washed down the column by continuous solvent flow. If solute A is more strongly adsorbed than solute B on the solid particles, then solute A spends a smaller fraction of the time free in solution. Solute A moves down the column more slowly than solute B and emerges at the bottom after solute B. We have just separated a mixture into its components by *chromatography*.

The **mobile phase** (the solvent moving through the column) in chromatography is either a liquid or a gas. The **stationary phase** (the one that stays in place inside the column) is most commonly a viscous liquid coated on the inside of a capillary tube or on the surface of solid particles packed into the

In 1903 M. Tswett first applied adsorption chromatography to the separation of plant pigments, using a hydrocarbon solvent and $CaCO_3$ stationary phase. The separation of colored bands led to the name *chromatography*, from the Greek word *khrōma*, meaning color.

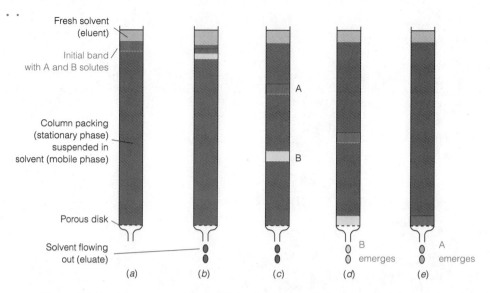

Fresh solvent
(eluent)

Initial band
with A and B solutes

Column packing
(stationary phase)
suspended in
solvent (mobile phase)

Porous disk

Solvent flowing
out (eluate)

(a)   (b)   (c)   (d)   (e)

Figure 22-5 Schematic representation of a chromatographic separation. Solute A, with a greater affinity than solute B for the stationary phase, remains on the column longer.

column. Alternatively, as in Figure 22-5, the solid particles themselves may be the stationary phase. In any case, the partitioning of solutes between the mobile and stationary phases gives rise to separation.

Fluid entering the column is called **eluent.** Fluid emerging from the end of the column is called **eluate:**

$$\text{eluent} \atop \text{in} \longrightarrow \boxed{\text{COLUMN}} \longrightarrow \text{eluate} \atop \text{out}$$

The process of passing liquid or gas through a chromatography column is called **elution.**

Columns are either **packed** or **open tubular.** A packed column is filled with particles containing stationary phase, as in Figure 22-5. An open tubular column is a narrow capillary with stationary phase coated on the inside walls. The center of the column is hollow.

Types of Chromatography

Chromatography is divided into categories based on the mechanism of interaction of the solute with the stationary phase, as shown in Figure 22-6.

Adsorption chromatography. The oldest form of chromatography, this makes use of a solid stationary phase and a liquid or gaseous mobile phase. Solute is adsorbed on the surface of the solid particles. Equilibration between the stationary phase and the mobile phase accounts for separation of different solutes.

Partition chromatography. In this technique, a liquid stationary phase forms a thin film on the surface of a solid support. Solute equilibrates between the stationary liquid and the mobile phase.

Ion-exchange chromatography. Anions such as $-SO_3^-$ or cations such as $-N(CH_3)_3^+$ are covalently attached to the stationary solid phase, usually a

Eluent—in.
Eluate—out.

For their pioneering work on liquid-liquid partition chromatography in 1941, A. J. P. Martin and R. L. M. Synge received the Nobel Prize in 1952.

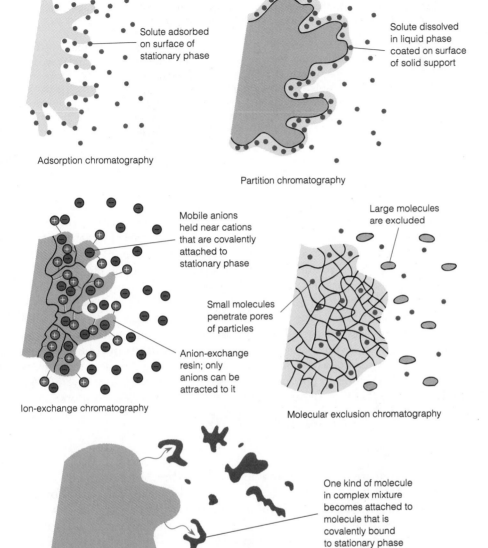

Adsorption chromatography

Solute adsorbed on surface of stationary phase

Partition chromatography

Solute dissolved in liquid phase coated on surface of solid support

Ion-exchange chromatography

Mobile anions held near cations that are covalently attached to stationary phase

Anion-exchange resin; only anions can be attracted to it

Molecular exclusion chromatography

Large molecules are excluded

Small molecules penetrate pores of particles

Affinity chromatography

One kind of molecule in complex mixture becomes attached to molecule that is covalently bound to stationary phase

All other molecules simply wash through

Figure 22-6 Major types of chromatography.

resin, in this type of chromatography. Solute ions of the opposite charge are attracted to the stationary phase by electrostatic force. The mobile phase is a liquid.

Molecular exclusion chromatography. Also called **gel filtration** or **gel permeation** chromatography, this technique separates molecules by size, with the larger solutes passing through most quickly. Unlike other forms of chromatography, there is no attractive interaction between the "stationary phase" and the solute in the ideal case of molecular exclusion. Rather, the liquid or gaseous mobile phase passes through a porous gel. The pores are small

B. A. Adams and E. L. Holmes developed the first synthetic ion-exchange resins in 1935. *Resins* are relatively hard, amorphous organic solids. *Gels* are relatively soft.

Large molecules pass through the column *faster* than small molecules.

enough to exclude large solute molecules but not small ones. The large molecules stream past without entering the pores. The small molecules take longer to pass through the column because they enter the gel and therefore must flow through a larger volume before leaving the column.

Affinity chromatography. This most selective kind of chromatography employs specific interactions between one kind of solute molecule and a second molecule that is covalently attached (immobilized) to the stationary phase. For example, the immobilized molecule might be an antibody to a particular protein. When a mixture containing a thousand proteins is passed through the column, only the one protein that reacts with the antibody is bound to the column. After washing all the other solutes from the column, the desired protein is dislodged by changing the pH or ionic strength.

22-3 A Plumber's View of Chromatography

The speed of the mobile phase passing through a chromatography column is expressed either as a volume flow rate or as a linear flow rate. Consider a liquid chromatography experiment in which the column has an inner diameter of 0.60 cm (radius $\equiv r = 0.30$ cm) and the mobile phase occupies 20% of the column volume. Each centimeter of column length has a volume of ($\pi r^2 \times$ length) $= \pi(0.30$ cm$)^2$ (1 cm) $= 0.283$ mL, of which 20% (= 0.0565 mL) is mobile phase (solvent). The **volume flow rate,** such as 0.3 mL/min, tells how many milliliters of solvent per minute travel through the column. The **linear flow rate** tells how many centimeters of column length are traveled in 1 min by the solvent. Because 1 cm of column length contains 0.0565 mL of mobile phase, 0.3 mL would occupy (0.3 mL)/(0.0565 mL/cm) = 5.3 cm of column length. The linear flow rate corresponding to 0.3 mL/min is 5.3 cm/min.

> Volume flow rate = volume of solvent per unit time traveling through column
>
> Linear flow rate = distance per unit time traveled by solvent

The Chromatogram

Solutes eluted from a chromatography column are observed with various detectors described in the next chapter. A **chromatogram** is a graph showing the detector response as a function of elution time. Figure 22-7 shows what might be observed when a mixture of octane, nonane, and an unknown are separated by gas chromatography, which is described in the next chapter. The **retention time,** t_r, for each component is the time needed after injection of the mixture onto the column until that component reaches the detector.

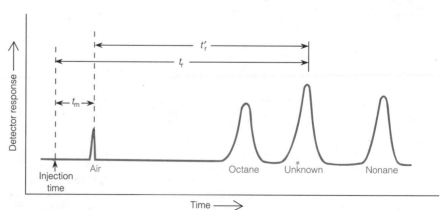

Figure 22-7 Schematic gas chromatogram showing measurement of retention times.

Unretained mobile phase travels through the column in the minimum possible time, designated t_m. The **adjusted retention time** for a solute is the additional time required for solute to travel the length of the column, beyond the time required by unretained solvent:

Adjusted retention time:
$$t_r' = t_r - t_m \qquad (22\text{-}14)$$

If a thermal conductivity detector (Figure 23-10) is used in gas chromatography, t_m is usually taken as the time needed for air to travel through the column (Figure 22-7). This can be determined by injecting a few microliters of air along with the 0.1- to 2-μL sample. Even columns that retain components of air usually display a discontinuity in the baseline of the chromatogram at time t_m because of the pressure burst associated with sample injection.

For any two components 1 and 2, the **relative retention, α,** is

Relative retention:
$$\alpha = \frac{t_{r2}'}{t_{r1}'} \qquad (22\text{-}15)$$

where $t_{r2}' > t_{r1}'$, so $\alpha > 1$. The greater the relative retention, the greater the separation between two components. Relative retention is fairly independent of flow rate and can therefore be used to help identify peaks when the flow rate changes.

For each peak in the chromatogram, the **capacity factor, k',** is defined as

Capacity factor:
$$k' = \frac{t_r - t_m}{t_m} \qquad (22\text{-}16)$$

Capacity factor is also called *retention factor, capacity ratio,* or *partition ratio.*

The longer a component is retained by the column, the greater is the capacity factor. To monitor the performance of a particular column, it is good practice to periodically measure the capacity factor of a standard, the number of plates (Equation 22-28), and peak asymmetry (Figure 22-12). Changes reflect degradation of the column.

. . . .

EXAMPLE Retention Parameters

A mixture of benzene, toluene, and air was injected into a gas chromatograph. Air gave a sharp spike in 42 s, whereas benzene required 251 s and toluene was eluted in 333 s. Find the adjusted retention time and capacity factor for each solute. Also, find the relative retention of the two solutes.

Solution The adjusted retention times are

benzene: $t_r' = t_r - t_m = 251 - 42 = 209$ s   toluene: $t_r' = 333 - 42 = 291$ s

The capacity factors are

benzene: $k' = \dfrac{t_r - t_m}{t_m} = \dfrac{251 - 42}{42} = 5.0$ toluene: $k' = \dfrac{333 - 42}{42} = 6.9$

The relative retention is always expressed as a number greater than unity:

$$\alpha = \frac{t_r'(\text{toluene})}{t_r'(\text{benzene})} = \frac{333 - 42}{251 - 42} = 1.39$$

Relation Between Retention Time and the Partition Coefficient

The definition of capacity factor in Equation 22-16 is equivalent to saying

$$k' = \frac{\text{time solute spends in stationary phase}}{\text{time solute spends in mobile phase}} \qquad (22\text{-}17)$$

Let's see why this is true. If the solute spends all of its time in the mobile phase and none in the stationary phase, it would be eluted in time t_m, by definition. Putting $t_r = t_m$ into Equation 22-16 gives $k' = 0$, because solute spends no time in the stationary phase. Suppose that solute spends equal time in the stationary and mobile phases. The retention time would then be $t_r = 2t_m$ and $k' = (2t_m - t_m)/t_m = 1$. If solute spends three times as much time in the stationary phase as in the mobile phase, $t_r = 4t_m$ and $k' = (4t_m - t_m)/t_m = 3$.

If solute spends three times as much time in the stationary phase as in the mobile phase, there will be three times as many moles of solute in the stationary phase as in the mobile phase at any time. The quotient in Equation 22-17 is equivalent to

$$\frac{\text{time solute spends in stationary phase}}{\text{time solute spends in mobile phase}} = \frac{\text{moles of solute in stationary phase}}{\text{moles of solute in mobile phase}}$$

$$k' = \frac{C_s V_s}{C_m V_m} \qquad (22\text{-}18)$$

where C_s is the concentration of solute in the stationary phase, V_s is the volume of the stationary phase, C_m is the concentration of solute in the mobile phase, and V_m is the volume of the mobile phase.

The quotient C_s/C_m is the ratio of concentrations of solute in the stationary and mobile phases. If the column is run slowly enough to be near equilibrium, the quotient C_s/C_m is the *partition coefficient, K,* introduced in connection with solvent extraction. Therefore, we cast Equation 22-18 in the form

Partition coefficient = $K = \dfrac{C_s}{C_m}$

Relation of retention time to partition coefficient: $\quad k' = K\dfrac{V_s}{V_m} \overset{\text{Eq. 22-16}}{=} \dfrac{t_r - t_m}{t_m} = \dfrac{t_r'}{t_m} \qquad (22\text{-}19)$

which relates the retention time to the partition coefficient and the volumes of stationary and mobile phases. Because $t_r' \propto k' \propto K$, relative retention can also be expressed as

Relative retention: $\quad \alpha = \dfrac{t_{r2}'}{t_{r1}'} = \dfrac{k_2'}{k_1'} = \dfrac{K_2}{K_1} \qquad (22\text{-}20)$

That is, the relative retention of two solutes is proportional to the ratio of their partition coefficients. This is the physical basis of chromatography.

Physical basis of chromatography: The greater the ratio of partition coefficients between mobile and stationary phases, the greater the separation between two components of a mixture.

EXAMPLE Retention Time and the Partition Coefficient

In the preceding example, air gave a sharp spike in 42 s, whereas benzene required 251 s. The open tubular chromatography column has an inner diameter of 250 µm and is coated on the inside with a layer of stationary phase 1.0 µm thick. Estimate the partition coefficient ($K = C_s/C_m$) for benzene between the stationary and mobile phases and state what fraction of the time benzene spends in the mobile phase.

Solution   We need to calculate the relative volumes of the stationary and mobile phases. The column is an open tube with a thin coating of stationary phase on the inside wall.

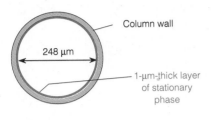

Column wall

248 μm

1-μm-thick layer
of stationary
phase

radius of hollow cavity: $r_1 = 124$ μm
radius to middle of stationary phase: $r_2 = 124.5$ μm

cross-sectional area of column $= \pi r_1^2$
$$= \pi(124)^2 = 4.83 \times 10^4 \ \mu m^2$$

cross-sectional area of coating
$$\approx 2\pi r_2 \times \text{thickness}$$
$$= 2\pi(124.5) \times (1.0)$$
$$= 7.8 \times 10^2 \ \mu m^2$$

The relative volumes of the phases are proportional to the relative cross-sectional areas of the phases, because each phase extends for the full length of the tube. Therefore, $V_s/V_m = (7.8 \times 10^2 \ \mu m^2)/(4.83 \times 10^4 \ \mu m^2) = 0.0161$. In the previous example, we found that the capacity factor for benzene is

$$k' = \frac{t_r - t_m}{t_m} = \frac{251 - 42}{42} = 5.0$$

Substituting this value into Equation 22-19 gives the partition coefficient:

$$k' = K\frac{V_s}{V_m} \Rightarrow 5.0 = K(0.0161) \Rightarrow K = 310$$

To find the fraction of time spent in the mobile phase, we use the Equations 22-16 and 22-17:

$$k' = \frac{\text{time in stationary phase}}{\text{time in mobile phase}} = \frac{t_r - t_m}{t_m} = \frac{t_s}{t_m} \Rightarrow t_s = k't_m$$

where t_s is the time in the stationary phase. The fraction of time in the mobile phase is

$$\text{fraction of time in mobile phase} = \frac{t_m}{t_s + t_m} = \frac{t_m}{k't_m + t_m} =$$

$$\frac{1}{k' + 1} = \frac{1}{5.0 + 1} = 0.17$$

Relation Between Retention Volume and the Partition Coefficient

Retention volume, V_r, is the volume of mobile phase required to elute a particular solute from the column:

Retention volume:
$$V_r = t_r \cdot u_v$$

where u_v is the volume flow rate (volume per unit time) of the mobile phase. The retention volume of a particular solute is constant over a wide range of flow rates.

Because volume is proportional to time, any relation expressed as a ratio of times can be written as the corresponding ratio of volumes. For example,

$$k' = \frac{t_r - t_m}{t_m} = \frac{V_r - V_m}{V_m}$$

Now we can derive one more equation that will be useful in Chapter 24. First solve Equation 22-19 for t_r:

$$t_r = t_m \left(K \frac{V_s}{V_m} + 1 \right)$$

Multiplying both sides by u_v gives

$$V_r = t_r \cdot u_v = \underbrace{t_m \cdot u_v}_{V_m} \left(K \frac{V_s}{V_m} + 1 \right)$$

But $t_m \cdot u_v$ is just V_m, the volume of mobile phase in the column. Replacing $t_m \cdot u_v$ by V_m gives an equation that we will use for molecular exclusion chromatography:

*Relation of retention volume
to partition coefficient:* $$V_r = KV_s + V_m \qquad (22\text{-}21)$$

Scaling Up

If you have developed a chromatographic procedure to separate 2 mg of a mixture on a column with a diameter of 1.0 cm, what size column should you use to separate 20 mg of the mixture? The most straightforward way to scale up is to maintain the same column length and to increase the cross-sectional area to maintain a constant ratio of unknown to column volume. Because cross-sectional area is πr^2, where r is the column radius, the desired diameter is given by

Scaling rules:

$$\frac{load_2}{load_1} = \left(\frac{radius_2}{radius_1} \right)^2$$

Maintain constant linear flow rate:

$$\frac{volume\ flow_2}{volume\ flow_1} = \frac{load_2}{load_1}$$

Scaling equation: $$\frac{\text{large load}}{\text{small load}} = \left(\frac{\text{large column radius}}{\text{small column radius}} \right)^2 \qquad (22\text{-}22)$$

$$\frac{20\ mg}{2\ mg} = \left(\frac{\text{large column radius}}{0.50\ cm} \right)^2$$

large column radius = 1.58 cm

A column with a diameter near 3 cm would be appropriate.

To reproduce the conditions of the smaller column in the larger column, the *linear flow rate* (not the volume flow rate) should be kept constant. Because the area (and hence volume) of the large column is 10 times greater than that of the small column in this example, the volume flow rate should be 10 times greater to maintain a constant linear flow rate. If the small column had a volume flow rate of 0.3 mL/min, the large column should be run at 3 mL/min.

22-4 Efficiency of Separation

Two factors contribute to how well compounds are separated by chromatography. One is the difference in elution times between peaks: The further apart, the better their separation. The other factor is how broad the peaks are: The wider the peaks, the poorer their separation. Having considered the relationship between partition coefficient and elution time in the previous section, we now consider factors affecting peak width in chromatography.[2]

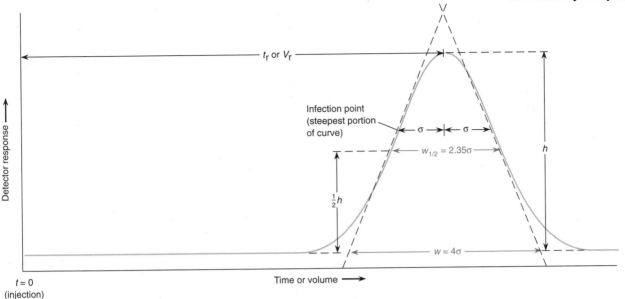

Resolution

Solute moving through a chromatography column tends to spread into a Gaussian shape with standard deviation σ (Figure 22-8). Common measures of breadth are (1) the width $w_{1/2}$ measured at a height equal to half of the peak height and (2) the width w at the baseline between tangents drawn to the steepest parts of the peak. From the analytical expression for a Gaussian peak (Equation 4-3), it is possible to show that $w_{1/2} = 2.35\sigma$ and $w = 4\sigma$.

In chromatography, the **resolution** of two peaks from each other is defined as

Resolution:

$$\text{resolution} = \frac{\Delta t_r}{w_{av}} = \frac{\Delta V_r}{w_{av}} \qquad (22\text{-}23)$$

where Δt_r or ΔV_r is the separation between peaks (in units of time or volume) and w_{av} is the average width of the two peaks in corresponding units. (Peak width is measured at the base, as shown in Figure 22-8.) Figure 22-9 shows that when two peaks have a resolution of 0.50, the overlap of area is 16%. When resolution = 1.00, the overlap is 2.3%, and the overlap is reduced to 0.1% at a resolution of 1.50.

Figure 22-8 Idealized Gaussian chromatogram showing how w and $w_{1/2}$ are measured. The value w is obtained by extrapolating the tangents to the inflection points down to the baseline.

. . . .

EXAMPLE Measuring Resolution

A solute with a retention time of 407 s has a width at the base of 13 s. A neighboring peak is eluted at 424 s with a width of 16 s. Find the resolution for these two components.

Solution

$$\text{resolution} = \frac{\Delta t_r}{w_{av}} = \frac{424 - 407}{\frac{1}{2}(13 + 16)} = 1.1_7$$

. . . .

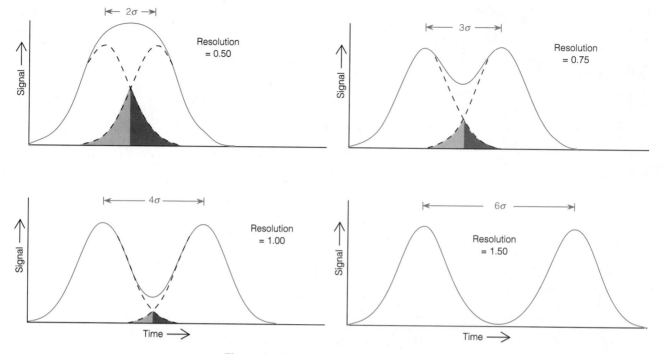

Figure 22-9 Resolution of Gaussian peaks of equal area and amplitude. Dashed lines show individual peaks, and solid lines are the sum of two peaks. Overlapping area is shaded.

Diffusion

A band of solute broadens as it moves through a chromatography column. Ideally, an infinitely narrow band applied to the inlet of the column emerges with a Gaussian shape at the outlet (Figure 22-10). In less ideal circumstances, the band becomes asymmetric.

One main cause of band spreading is *diffusion*. The **diffusion coefficient** measures the rate at which a substance moves from a region of high concentration to a region of lower concentration. Figure 22-11 shows spontaneous diffusion of solute across a plane with a concentration gradient dc/dx. The number of moles crossing each square meter per second, called the *flux* (J), is proportional to the concentration gradient:

Equation 22-24 is called *Fick's first law of diffusion*. If concentration is expressed as mol/m³, the units of D are m²/s.

Definition of diffusion coefficient:

$$\text{flux} \left(\frac{\text{mol}}{\text{m}^2 \cdot \text{s}} \right) \equiv J = -D \frac{dc}{dx} \tag{22-24}$$

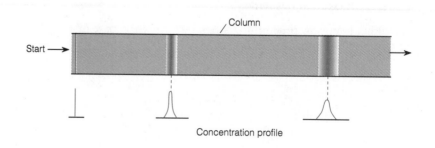

Figure 22-10 Broadening of an initially sharp band of solute as it moves through a chromatography column.

The constant of proportionality (D) is the diffusion coefficient, and the negative sign is necessary because the net flux is from the region of high concentration to the region of low concentration. Box 22-2 draws an analogy between molecular diffusion and a random walk, and Table 22-1 lists some representative diffusion coefficients. Diffusion in liquids is 10^4 times slower than diffusion in gases. Macromolecules such as ribonuclease and albumin diffuse 10 to 100 times slower than small molecules.

If solute begins its journey through a column in an infinitely sharp layer with m moles per unit cross-sectional area of the column and spreads by diffusive mechanisms as it travels, then the Gaussian profile of the band is described by

Broadening of chromatography band by diffusion:

$$c = \frac{m}{\sqrt{4\pi Dt}} e^{-x^2/(4Dt)} \qquad (22\text{-}25)$$

where c is concentration (mol/m^3), t is time, and x is the distance along the column from the current center of the band. (The band center is always $x = 0$ in this equation.) Comparison of Equation 22-25 to Equation 4-3 for a Gaussian curve shows that the standard deviation of the band is

Standard deviation of band:

$$\sigma = \sqrt{2Dt} \qquad (22\text{-}26)$$

Bandwidth $\propto \sqrt{t}$. To double the width of a band requires four times as much elution time.

Plate Height: A Measure of Column Efficiency

Band spreading in chromatography is described by terms evolved from separations that were carried out in discrete steps, such as a series of extractions. Equation 22-26 tells us that the standard deviation for diffusive band spreading is $\sqrt{2Dt}$. If solute has traveled a distance x at the linear flow

TABLE 22-1   Representative diffusion coefficients at 298 K

Solute	Solvent	Diffusion coefficient (m^2/s)
H_2O	H_2O	2.3×10^{-9}
Sucrose	H_2O	0.52×10^{-9}
Glycine	H_2O	1.1×10^{-9}
CH_3OH	H_2O	1.6×10^{-9}
Ribonuclease (MW 13 700)	H_2O (293 K)	0.12×10^{-9}
Serum albumin (MW 65 000)	H_2O (293 K)	0.059×10^{-9}
I_2	Hexane	4.0×10^{-9}
CCl_4	Heptane	3.2×10^{-9}
N_2	CCl_4	3.4×10^{-9}
$CS_2(g)$	Air (293 K)	1.0×10^{-5}
$O_2(g)$	Air (273 K)	1.8×10^{-5}
H^+	H_2O	9.3×10^{-9}
OH^-	H_2O	5.3×10^{-9}
Li^+	H_2O	1.0×10^{-9}
Na^+	H_2O	1.3×10^{-9}
K^+	H_2O	2.0×10^{-9}
Cl^-	H_2O	2.0×10^{-9}
I^-	H_2O	2.0×10^{-9}

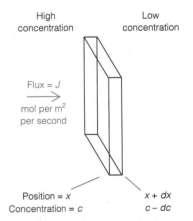

Figure 22-11   The flux of molecules diffusing across a plane of unit area is proportional to the concentration gradient and to the diffusion coefficient: $J = -D(dc/dx)$.

Box 22-2 Molecular Diffusion and the Random Walk

Consider a particle moving along the x axis in discrete steps of length, l, that may be left or right, chosen at random. This classic problem is described as a *random walk*. If the initial position is $x = 0$, then after the particle has taken n steps it is not likely to be at the origin. We will write the position in our lab notebook and call it x_1, which is just as likely to be positive as negative. Now we put the particle back at $x = 0$ and repeat the experiment with n more steps and find it at position x_2, which we dutifully record. If we repeat the experiment many times, it should not surprise us that the positions x_i have a Gaussian distribution about the origin.

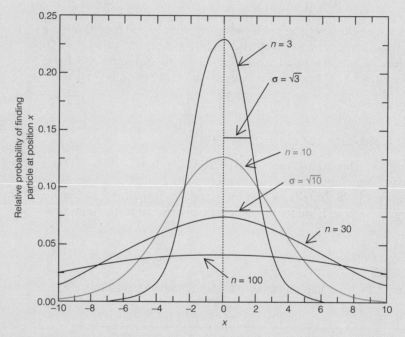

Gaussian distributions obtained from random walks of 3, 10, 30, and 100 steps of length $l = 1$.

The equation for the Gaussian curve turns out to be

$$y = \frac{1}{l\sqrt{n}\sqrt{2\pi}} e^{-x^2/(2l^2n)}$$

Comparing this with Equation 4-3 for a general Gaussian curve, we see that the standard deviation for the random walk is

$$\text{standard deviation} = l\sqrt{n}$$

The standard deviation for n steps is proportional to $\sqrt{n}$.

Diffusion can be idealized as a random walk by molecules. The step length is the mean free path traveled by a molecule before colliding with a neighbor. The number of steps taken is proportional to time. Because the standard deviation for a random walk is proportional to $\sqrt{n}$, the standard deviation for diffusional spreading of molecules is proportional to $\sqrt{\text{time}}$.

rate u_x (m/s), then the time it has been on the column is $t = x/u_x$. Therefore

$$\sigma^2 = 2Dt = 2D\,\frac{x}{u_x} = \left(\frac{2D}{u_x}\right)x = Hx$$

↑
Plate height $\equiv H$

Plate height:
$$H = \sigma^2/x \qquad\qquad (22\text{-}27)$$

u_v = volume flow rate (volume/ time)

u_x = linear flow rate (distance/time)

Plate height is the constant of proportionality between the variance (σ^2) of the band and the distance it has traveled (x). The name comes from separation methods that employed discrete stages called plates. Plate height is also called the *height equivalent to a theoretical plate*.

Plate height is simply a quantity relating the width of a band to the distance traveled through the column. *The smaller the plate height, the narrower the bandwidth*. The ability of a column to separate components of a mixture is improved by decreasing plate height. We say that an efficient column has more theoretical plates than an inefficient column. Different solutes passing through the same column have different plate heights because they have different diffusion coefficients. Plate heights are in the neighborhood of ~0.1 to 1 mm in gas chromatography. Plate height is ~10 μm in high-performance liquid chromatography and <1 μm in capillary electrophoresis.

The plate height is the length σ^2/x, where σ is the standard deviation of the Gaussian band in Figure 22-8 and x is the distance traveled through the column. For solute emerging from a column of length L, the number of theoretical plates (N) in the entire column is the length L divided by the plate height:

$$N = \frac{L}{H} = \frac{Lx}{\sigma^2} = \frac{L^2}{\sigma^2} = \frac{16L^2}{w^2}$$

because $x = L$ and $\sigma = w/4$. In this expression, w has units of length and the number of plates is dimensionless. If we express L and w (or σ) in units of time instead of length, N is still dimensionless. We obtain the most useful expression for N by writing

Number of plates on column:
$$N = \frac{16t_r^2}{w^2} = \left(\frac{t_r^2}{\sigma^2}\right) \qquad\qquad (22\text{-}28\text{a})$$

where t_r is the retention time of the peak and w is the width at the base in Figure 22-8 *in units of time*. If we use the width at half-height instead of the width at the base, we get

Number of plates on column:
$$N = \frac{5.55t_r^2}{w_{1/2}^2} \qquad\qquad (22\text{-}28\text{b})$$

Recall that variance is the square of the standard deviation.

Small plate height ⇒ narrow peaks ⇒ better separations

Choose a peak with a capacity factor greater than 5 when you measure plate height for a column.

Challenge If N is constant, show that the width of a chromatographic peak increases with increasing retention time. That is, successive peaks on a chromatogram should be increasingly broad.

. . . .

EXAMPLE Measuring Plates

A solute with a retention time of 407 s has a width at the base of 13 s on a column 12.2 m long. Find the number of plates and plate height.

Solution $N = \dfrac{16 \cdot 407^2}{13^2} = 1.57 \times 10^4$

$H = \dfrac{L}{N} = \dfrac{12.2\text{ m}}{1.57 \times 10^4} = 0.78\text{ mm}$

. . . .

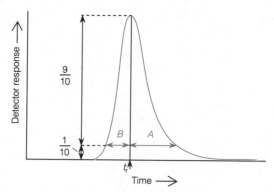

Figure 22-12 Asymmetric peak showing parameters used to estimate the number of theoretical plates.

To estimate the number of theoretical plates for the asymmetric peak in Figure 22-12, draw a horizontal line across the band at one-tenth of the maximum height. The quantities A and B can then be measured, and the number of plates is[3]

All quantities must be measured in the same units, such as minutes or centimeters of chart paper.

$$N \approx \frac{41.7 \, (t_r/w_{0.1})^2}{(A/B + 1.25)} \qquad (22\text{-}29)$$

where $w_{0.1}$ is the width at one-tenth-height ($= A + B$).

Factors Affecting Resolution

The greater the resolution (Equation 22-23), the better the separation between two peaks. The relation between the number of plates on a column and the resolution is[4]

$$\text{resolution} = \frac{\sqrt{N}}{4} \left(\frac{\alpha - 1}{\alpha}\right) \left(\frac{k_2'}{1 + k_{av}'}\right) \qquad (22\text{-}30)$$

where N is the number of theoretical plates in the column, α is the relative retention of the two peaks (Equation 22-15), k_2' is the capacity factor for the more retained component (Equation 22-16), and k_{av}' is the average capacity factor for both components. If the number of plates for the two peaks is not the same—say, N_1 and N_2—we replace N by $\sqrt{N_1 N_2}$ and replace k_2' by k_{av}'.

Resolution $\propto \sqrt{N} \propto \sqrt{L}$

The most important feature of Equation 22-30 is that resolution is proportional to $\sqrt{N}$. Separation of components increases in proportion to the number of plates (N), but band spreading is proportional to $\sqrt{N}$ (Box 22-2). The net result is that resolution is proportional to $\sqrt{N}$. Because the number of theoretical plates in a column is proportional to its length, *doubling the column length increases resolution by $\sqrt{2}$.*

Equation 22-30 also tells us that resolution increases as α increases and as the capacity factor k_2' increases. Increasing the capacity factor is equivalent to increasing the fraction of time spent by the solute in the stationary phase. There is a practical limit to increasing the capacity factor, because retention times become too long and peaks become too broad. Important equations from chromatography are summarized in Table 22-2.

TABLE 22-2 Summary of chromatography equations

Quantity	Equation	Parameters
Partition coefficient	$K = C_s/C_m$	C_s = concentration of solute in stationary phase C_m = concentration of solute in mobile phase
Adjusted retention time	$t'_r = t_r - t_m$	t_r = retention time of solute of interest t_m = retention time of unretained solute
Retention volume	$V_r = t_r \cdot u_v = KV_s + V_m$	u_v = volume flow rate = volume/unit time
Capacity factor	$k' = t'_r/t_m = KV_s/V_m$ $k' = \dfrac{t_s}{t_m}$	V_s = volume of stationary phase V_m = volume of mobile phase t_s = time solute spends in stationary phase t_m = time solute spends in mobile phase
Relative retention	$\alpha = \dfrac{t'_{r2}}{t'_{r1}} = \dfrac{k'_2}{k'_1} = \dfrac{K_2}{K_1}$	Subscripts 1 and 2 refer to two solutes
Number of plates	$N = \dfrac{16t_r^2}{w^2} = \dfrac{5.55t_r^2}{w_{1/2}^2}$	w = width at base $w_{1/2}$ = width at half-height
Plate height	$H = \dfrac{\sigma^2}{x} = \dfrac{L}{N}$	σ = standard deviation of band x = distance traveled by center of band L = length of column N = number of plates on column
Resolution	$\text{Resolution} = \dfrac{\Delta t_r}{w_{av}} = \dfrac{\Delta V_r}{w_{av}}$	Δt_r = difference in retention times ΔV_r = difference in retention volumes w_{av} = average width measured at baseline in same units as numerator (time or volume)
	$\text{Resolution} = \dfrac{\sqrt{N}}{4}\left(\dfrac{\alpha - 1}{\alpha}\right)\left(\dfrac{k'_2}{1 + k'_{av}}\right)$	N = number of plates α = relative retention k'_2 = capacity factor for second peak k'_{av} = average capacity factor

EXAMPLE Plates Needed for Desired Resolution

Two solutes have a relative retention of $\alpha = 1.08$ and capacity factors $k'_1 = 5.0$ and $k'_2 = 5.4$. The number of theoretical plates is nearly the same for both compounds. How many plates are required to give a resolution of 1.5? Of 3.0? If the plate height is 0.20 mm, how long must the column be for a resolution of 1.5?

Solution We use Equation 22-30:

$$\text{resolution} = \frac{\sqrt{N}}{4}\left(\frac{1.08 - 1}{1.08}\right)\left(\frac{5.4}{1 + 5.2}\right) = 1.5 \Rightarrow N = 8.6_5 \times 10^3 \text{ plates}$$

To double the resolution to 3.0 would require four times as many plates = $3.4_6 \times 10^4$ plates. For a resolution of 1.5, the length of column required is $(0.20 \text{ mm/plate})(8.6_5 \times 10^3 \text{ plates}) = 1.73 \text{ m}$.

22-5 Why Bands Spread

A band of solute invariably spreads apart as it travels through a chromatography column (Figure 22-10) and emerges at the detector with a standard deviation σ. Each mechanism contributing to broadening produces a standard deviation σ_i. The observed variance (σ_{obs}^2) of a band is the sum of the variances from all contributing mechanisms:

Variance is additive, but standard deviation is not.

Variance is additive:

$$\sigma_{obs}^2 = \sigma_1^2 + \sigma_2^2 + \sigma_3^2 + \cdots = \sum \sigma_i^2 \qquad (22\text{-}31)$$

Broadening Outside the Column

Some factors outside the column contribute to broadening. For example, solute cannot be applied to the column in an infinitesimally thin zone, so the band has a finite width even before it begins spreading on the column. If the band is applied as a sharp-edged zone of width Δt (measured in units of time), the contribution to the variance of the final bandwidth is

Variance due to injection or detection width:

$$\sigma_{injection}^2 = \frac{(\Delta t)^2}{12} \qquad (22\text{-}32)$$

A similar relationship holds for broadening in a detector that requires a time Δt for the sample to pass through. Sometimes on-column detection is possible, which eliminates band spreading caused by a separate detector cavity.

. . . .

EXAMPLE   Band Spreading Before and After the Column

A band from a column eluted at a rate of 1.35 mL/min has a width at half-height of 16.3 s. The sample was applied as a sharp plug with a volume of 0.30 mL, and the detector volume is 0.20 mL. Find the variances introduced by injection and detection. What would be the width at half-height if broadening only occurred on the column?

Solution   Figure 22-8 tells us that the width at half-height is $w_{1/2} = 2.35\sigma$. Therefore, the observed total variance is

$$\sigma_{obs}^2 = \left(\frac{w_{1/2}}{2.35}\right)^2 = \left(\frac{16.3}{2.35}\right)^2 = 48.11 \text{ s}^2$$

The contribution from injection is $\sigma_{injection}^2 = (\Delta t)_{injection}^2/12$. The time of injection is $\Delta t_{injection} = (0.30 \text{ mL})/(1.35 \text{ mL/min}) = 0.222 \text{ min} = 13.3 \text{ s}$. Therefore,

$$\sigma_{injection}^2 = \frac{\Delta t_{injection}^2}{12} = \frac{13.3^2}{12} = 14.81 \text{ s}^2$$

Similarly, the time spent in the detector is $\Delta t_{detector} = (0.20 \text{ mL})/(1.35 \text{ mL/min}) = 8.89 \text{ s}$ and $\sigma_{detector}^2 = (\Delta t)_{detector}^2/12 = 6.58 \text{ s}^2$. The observed variance is

$$\sigma_{obs}^2 = \sigma_{column}^2 + \sigma_{injection}^2 + \sigma_{detector}^2$$

$$48.11 = \sigma_{column}^2 + 14.81 + 6.58 \Rightarrow \sigma_{column} = 5.17 \text{ s}$$

The width due to column broadening alone is $w_{1/2} = 2.35\sigma_{column} = 12.1 \text{ s}$, which is about 3/4 of the observed width.

. . . .

Spreading would be much worse at an outlet where each new drop mixed totally with many previous drops. This happens in the large dead space beneath some rudimentary benchtop chromatography columns. To minimize band spreading, dead spaces and tubing lengths should be minimized. Sample should be applied uniformly in a narrow zone and allowed to enter the column before mixing with eluent.

Plate Height Equation[2,5]

Plate height (H) is proportional to the variance of a chromatographic band: The smaller the plate height, the narrower the band. The **van Deemter equation** summarizes on-column effects (not including the injector and detector) that contribute to plate height:

van Deemter equation for plate height:

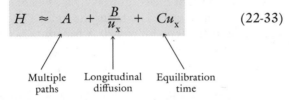

$$H \approx A + \frac{B}{u_x} + Cu_x \qquad (22\text{-}33)$$

Multiple paths   Longitudinal diffusion   Equilibration time

where u_x is the linear flow rate and A, B, and C are constants. There are band broadening mechanisms that are proportional to flow rate, inversely proportional to flow rate, and independent of flow rate (Figure 22-13).

In packed columns, all three terms contribute to band broadening. For open tubular columns, the multiple path term (A) is zero, so bandwidth decreases and resolution increases. In capillary electrophoresis (Chapter 24), both A and C go to zero, thereby reducing plate height to submicron values and providing extraordinary separation powers.

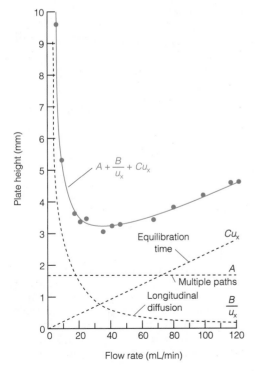

Figure 22-13 Application of van Deemter equation to gas chromatography. Experimental points are data from H. W. Moody, *J. Chem. Ed.* **1982,** *59,* 290. Van Deemter parameters are $A = 1.65$ mm, $B = 25.8$ mm · mL/min, and $C = 0.0236$ mm · min/mL.

Longitudinal Diffusion

The term B/u_x in Equation 22-33 arises from **longitudinal diffusion,** which means that solute spreads out along the length of the column—mainly by diffusion in the mobile phase (Figure 22-14). Equation 22-26 told us that the variance resulting from diffusion is

Longitudinal diffusion: $\qquad \sigma^2 = 2D_m t = \dfrac{2D_m L}{u_x}$

$$H_D = \frac{\sigma^2}{L} = \frac{2D_m}{u_x} \equiv \frac{B}{u_x} \qquad (22\text{-}34)$$

where D_m is the diffusion coefficient of solute in the mobile phase, t is time, and H_D is the plate height due to longitudinal diffusion. The time needed to travel the length of the column is L/u_x, where L is the column length and u_x is the linear flow rate. The faster the linear flow, the less time is spent in the column and the less diffusional broadening occurs.

Because longitudinal diffusion in a gas is much faster than diffusion in a liquid, the optimum flow rate in gas chromatography is higher than in liquid chromatography.

Finite Equilibration Time Between Phases

The term Cu_x in Equation 22-33 comes from the finite time required for solute to equilibrate between the mobile and stationary phases. Although some solute is stuck in the stationary phase, the remainder in the mobile phase moves forward, thereby resulting in spreading of the overall zone of solute (Figure 22-15).

The plate height from finite equilibration time, also called the *mass transfer term,* is

Finite equilibration time: $\qquad H_s = \dfrac{qk'}{(k'+1)^2}\ \dfrac{d^2 u_x}{D_s} \equiv Cu_x \qquad (22\text{-}35)$

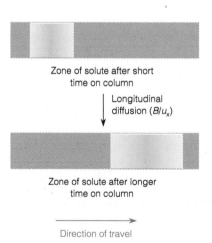

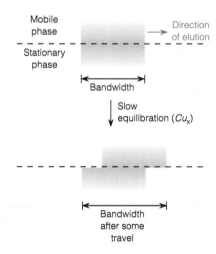

Figure 22-14 Longitudinal diffusion gives rise to the term B/u_x in the van Deemter equation. Solute continuously diffuses away from the concentrated center of its zone. The greater the flow rate, the less time is spent on the column and the less longitudinal diffusion occurs.

Figure 22-15 The finite time required for solute to equilibrate between the mobile and stationary phases gives rise to the term Cu_x in the van Deemter equation. The slower the linear flow, the more complete is equilibration and the less zone broadening occurs.

where D_s is the diffusion coefficient of solute in the stationary phase, k' is the capacity factor, d is the thickness of stationary phase, and q is a fudge factor of order unity (i.e., approximately 1) accounting for the shape of the stationary phase.

The mass transfer plate height is reduced by decreasing the thickness of stationary phase and by increasing temperature, both of which increase the diffusion coefficient of solute in the stationary phase. In the example depicted in Figure 22-16, raising the temperature allowed the linear flow rate to be increased by a factor of 4, while maintaining constant resolution. This is possible because of the increased rate of mass transfer between phases at elevated temperature.

More Fudge

The term A (formerly called the *eddy diffusion* term) in the van Deemter equation (22-33) arises from multiple effects for which the theory is murky. Figure 22-17 is a pictorial explanation of one effect. Because some flow paths are longer than others, solute molecules entering the column at the same time on the left are eluted at different times on the right. For simplicity, we approximate many different effects by the constant A in Equation 22-33.

Implications for Analytical and Preparative Separations

Chromatography is performed for *analytical* or *preparative* purposes. An analytical separation is designed to obtain clean separation of the components of a mixture. Therefore, we use a long, thin column with a thin stationary phase for minimal plate height. Because the capacity of the stationary phase is small, we are limited to small quantities of unknown. A preparative separation is intended to purify useful quantities of one or more components of a mixture. Therefore, we use a fat column with a thicker layer of stationary phase for increased capacity. The thick stationary phase broadens bands and decreases resolution. We usually compromise resolution to separate large quantities of material.

Advantages of Open Tubular Columns

Open tubular columns provide higher resolution than packed columns, shorter analysis times, and increased sensitivity to small quantities of analyte. Open tubular columns have small sample capacity, so they are not useful for preparative separations.

Particles in a packed column resist flow of the mobile phase, so the linear flow rate cannot be very fast. For the same length of column and applied pressure, the linear flow rate in an open tubular column is much higher than that of a packed column. Therefore, the open tubular column can be made 100 times longer than the packed column, to give a similar pressure drop and linear flow rate. If all else is the same, the longer column provides 100 times more theoretical plates, yielding greatly increased resolution.

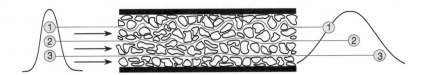

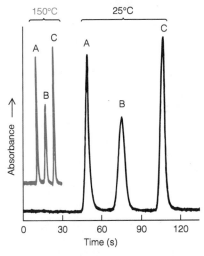

Figure 22-16 Molecular exclusion chromatography showing the decrease in analysis time at constant resolution when temperature is raised from 25° to 150°C. A, polystyrene (MW 111 000); B, polystyrene (MW 9 000); C, toluene (MW 92). The 1-mm-diameter × 25-cm-long column packed with 5-μm-diameter silica with 6-nm pores was eluted with dichloromethane at a rate of 0.10 mL/min at 25° (28 MPa pressure) or 0.40 mL/min at 150° (35 MPa pressure). [From C. N. Renn and R. E. Synovec, *Anal. Chem.* **1992,** *64,* 479.]

Compared with packed columns, open tubular columns can provide

1. higher resolution

2. shorter analysis time

3. increased sensitivity

4. lower sample capacity

Figure 22-17 Band spreading from multiple flow paths. The smaller the stationary phase particles, the less serious is this problem. This process is absent in an open tubular column.

TABLE 22-3   Comparison of packed and wall-coated open tubular column performance[a]

Property	Packed	Open tubular
Column length, L	2.4 m	100 m
Linear gas velocity	8 cm/s	16 cm/s
Plate height for methyl oleate	0.73 mm	0.34 mm
Capacity factor, k', for methyl oleate	58.6	2.7
Theoretical plates, N	3 290	294 000
Resolution of methyl stearate and methyl oleate	1.5	10.6
Retention time of methyl oleate	29.8 min	38.5 min

a. Methyl stearate ($CH_3(CH_2)_{16}CO_2CH_3$) and methyl oleate (cis-$CH_3(CH_2)_7CH{=}CH$-$(CH_2)_7CO_2CH_3$) were separated on columns with poly(diethylene glycol succinate) stationary phase at 180°C.

SOURCE: L. S. Ettre, *Introduction to Open Tubular Columns* (Norwalk, CT: Perkin-Elmer Corp., 1979), p. 26.

For a given pressure, flow rate is proportional to the cross-sectional area of the column and inversely proportional to column length:

$$\text{flow} \propto \frac{\text{area}}{\text{length}}$$

Open tubes give

1. increased linear flow rate and/or a longer column

2. decreased plate height, which means higher resolution

C_s = concentration of solute in stationary phase

C_m = concentration of solute in mobile phase

Overloading produces a gradual rise and an abrupt fall of the chromatographic peak.

Plate height is reduced in an open tubular column because multiple flow paths do not contribute to band spreading (Figure 22-17). In the van Deemter curve for a packed column in Figure 22-13, the A term accounts for half of the plate height at the most efficient flow rate (minimum H) near 30 mL/min. If the A term were deleted, the number of plates on the column would be doubled. To obtain high performance from an open tubular column, the radius of the column must be small and the stationary phase must be as thin as possible to ensure rapid exchange of solute between the mobile and stationary phases.

Table 22-3 compares the performance of packed and open tubular gas chromatography columns with the same stationary phase. For similar analysis times, the open tubular column gives 7 times better resolution (10.6 versus 1.5) than the packed column. Alternatively, speed could be traded for resolution. If the open tubular column were reduced to 5 m in length, the same solutes would be separated with a resolution of 1.5, but the time would be reduced from 38.5 to 0.83 min.

A Touch of Reality

A Gaussian band shape results when the partition coefficient, K ($= C_s/C_m$), is independent of the concentration of solute on the column. In real columns, K changes as the concentration of solute increases, and band shapes are skewed. A graph of C_s versus C_m (at a given temperature) is called an *isotherm*. Three common isotherms and their resulting band shapes are shown in Figure 22-18. The center isotherm is the ideal one leading to a symmetric peak.

The upper isotherm in Figure 22-18 arises from an *overloaded* column in which so much solute has been applied that the stationary phase resembles a liquid phase made of the solute. Because solute is most soluble in itself ("like dissolves like"), the value C_s/C_m increases with increased solute loading. The front of an overloaded peak has gradually increasing concentration. As the concentration increases, the band becomes overloaded and the partition coefficient ($= C_s/C_m$) increases. The solute is so soluble in its zone on the stationary phase that little solute trails behind the peak. The band emerges gradually but ends abruptly.

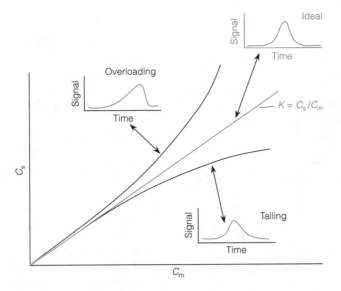

Figure 22-18 Common isotherms and their resulting chromatographic band shapes.

The lower isotherm in Figure 22-18 arises when small quantities of solute are retained more strongly than large quantities. It leads to an abrupt rise of concentration at the beginning of the band and a long "tail" of gradually decreasing concentration after the peak.

Sites that bind solute strongly cause tailing. In gas chromatography, a common solid support is *diatomite*—hydrated silica skeletons of algae. Surface hydroxyl groups of diatomite form hydrogen bonds with polar solutes, thereby leading to serious tailing. **Silanization** reduces tailing by blocking the hydroxyl groups with nonpolar trimethylsilyl groups:

A long tail occurs when some sites retain solute more strongly than other sites.

$$\begin{array}{c}
\text{OH} \quad \text{OH} \\
| \quad\quad | \\
-\text{Si}-\text{O}-\text{Si}- + (CH_3)_3\text{SiNHSi}(CH_3)_3 \rightarrow \\
| \quad\quad | \\
\end{array}
\begin{array}{c}
(CH_3)_3\text{Si} \quad\quad \text{Si}(CH_3)_3 \\
| \quad\quad\quad | \\
\text{O} \quad\quad \text{O} \\
| \quad\quad\quad | \\
-\text{Si}-\text{O}-\text{S}- + NH_3 \\
| \quad\quad\quad |
\end{array}$$

Solid phase with exposed —OH groups

Hexamethyldisilazine

Protected surface

(22-36)

Glass and silica columns used for gas and liquid chromatography can also be silanized to minimize interaction of the solute with active sites on the walls.

....

Terms to Understand

adjusted retention time
adsorption chromatography
affinity chromatography
capacity factor
chromatogram
diffusion coefficient
distribution coefficient
eluate
eluent
elution
extraction

gel filtration chromatography
gel permeation chromatography
ion-exchange chromatography
linear flow rate
longitudinal diffusion
miscible
mobile phase
molecular exclusion chromatography
open tubular column
packed column
partition chromatography

partition coefficient
plate height
relative retention
resolution
retention time
retention volume
silanization
stationary phase
van Deemter equation
volume flow rate

Summary

A solute can be extracted from one phase into another in which it is more soluble. The ratio of solute concentrations in each phase at equilibrium is called the partition coefficient. If more than one form of the solute exists, we use a distribution coefficient instead of a partition coefficient. We derived equations relating the fraction of solute extracted to the partition coefficient, volumes, and pH. A series of small extractions is more effective than a few large extractions. A metal chelator, soluble only in organic solvents, may extract metal ions from aqueous solutions, with selectivity achieved by adjusting pH.

In adsorption and partition chromatography, a continuous equilibration of solute between the mobile and stationary phases occurs. Eluent goes into a column and eluate comes out. Columns may be packed with stationary phase or may be open tubular, with stationary phase coated on the inner wall. In ion-exchange chromatography, the solute is attracted to the stationary phase by coulombic forces. In molecular exclusion chromatography, the fraction of stationary-phase volume available to solute decreases as the size of the solute molecules increases. Affinity chromatography relies on specific, noncovalent interactions between the stationary phase and one solute in a complex mixture.

The relative retention of two components is the quotient of their adjusted retention times. The capacity factor for a single component is the adjusted retention time divided by the elution time for solvent. Capacity factor gives the ratio of time spent by solute in the stationary phase to time spent in the mobile phase. When scaling up from a small load to a large load, the cross-sectional area of the column should be increased in proportion to the loading. Column length and linear flow rate are held constant.

Plate height ($H = \sigma^2/x$) is a measure of how well bands can be separated on a column. The number of plates for a Gaussian peak is $N = 16t_r^2/w^2$. Resolution of neighboring peaks is the difference in retention time divided by the average width (measured at the baseline, $w = 4\sigma$). Resolution is proportional to $\sqrt{N}$ and also increases with relative retention and capacity factor. Doubling the length of a column increases resolution by $\sqrt{2}$.

The standard deviation of a diffusing band of solute is $\sigma = \sqrt{2Dt}$, where D is the diffusion coefficient and t is time. The van Deemter equation describes band broadening on the chromatographic column: $H \approx A + B/u_x + Cu_x$, where H is plate height, u_x is linear flow rate, and A, B, and C are constants. The first term represents irregular flow paths, the second longitudinal diffusion, and the third the finite rate of transfer of solute between mobile and stationary phases. The optimum flow rate, which minimizes plate height, is faster for gas chromatography than for liquid chromatography. The number of plates and the optimal flow rate increase as the stationary-phase particle size decreases. Open tubular columns provide higher resolution and shorter analysis times than packed columns. Bands spread not only on the column but also during injection and detection. Overloading and tailing can be corrected by using smaller samples and by masking strong adsorption sites on the stationary phase.

Exercises

A. Consider a chromatography experiment in which two components with capacity factors $k_1' = 4.00$ and $k_2' = 5.00$ are injected into a column with $N = 1.00 \times 10^3$ theoretical plates. The retention time for the less-retained component is $t_{r1} = 10.0$ min.

(a) Calculate t_m and t_{r2}. Find $w_{1/2}$ (width at half-height) and w (width at the base) for each peak.

(b) Using graph paper, sketch the chromatogram analogous to Figure 22-7, supposing that the two peaks have the same amplitude (height). Draw the half-widths accurately.

(c) Calculate the resolution of the two peaks and compare this value with those drawn in Figure 22-9.

B. A solute with a partition coefficient of 4.0 is extracted from 10 mL of phase 1 into phase 2.

(a) What volume of phase 2 is needed to extract 99% of the solute in one extraction?

(b) What is the total volume of solvent 2 needed to remove 99% of the solute in three equal extractions instead?

C. (a) Find the capacity factors for octane and nonane in Figure 22-7.

(b) Find the ratio

$$\frac{\text{time octane spends in stationary phase}}{\text{total time octane spends on column}}$$

(c) Find the relative retention for octane and nonane.

(d) If the volume of the stationary phase equals half the volume of the mobile phase, find the partition coefficient for octane.

D. A gas chromatogram of a mixture of toluene and ethyl acetate is shown at the top of the next page.

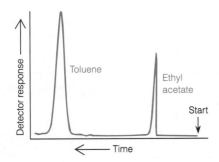

(a) Use the width of each peak (measured at the base) to calculate the number of theoretical plates in the column. Estimate all lengths to the nearest 0.1 mm.

(b) Using the width of the toluene peak at its base, calculate the width expected at half-height. Compare the measured and calculated values. When the thickness of the pen trace is significant relative to the length being measured, it is important to take the pen width into account. It is best to measure from the edge of one trace to the corresponding edge of the other trace, as shown below.

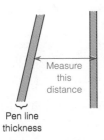

E. The three chromatograms shown at the right were obtained with 2.5, 1.0, and 0.4 μL of ethyl acetate injected on the same column under the same conditions. Explain why the asymmetry of the peak decreases as the sample size decreases.

F. The relative retention for two compounds in gas chromatography is 1.068 on a column with a plate height of 0.520 mm. The capacity factor for compound 1 is 5.16.

(a) What length of column will separate the compounds with a resolution of 1.00?

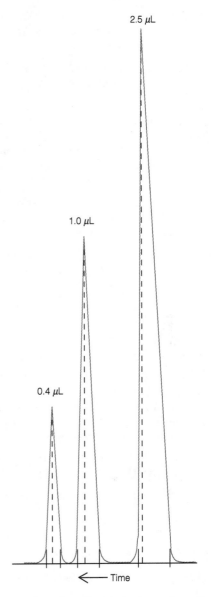

(b) The retention time for air (t_m) on the column in (a) is 2.00 min. If the number of plates is the same for both compounds, find the retention times and widths of each peak.

(c) If the ratio of stationary phase to mobile phase is 0.30, find the partition coefficient for component 1.

Problems

Solvent Extraction

1. If you are extracting a substance from water into ether, is it more effective to do one extraction with 300 mL of ether or three extractions with 100 mL?

2. If you wish to extract aqueous acetic acid into hexane, is it more effective to adjust the aqueous phase to pH 3 or pH 8?

3. Why is it difficult to extract the EDTA complex of aluminum into organic solution and easy to extract the 8-hydroxyquinoline complex?

4. Why is the extraction of a metal ion into an organic solvent with 8-hydroxyquinoline more complete at higher pH?

5. The distribution coefficient for extraction of a metal complex from aqueous to organic solvents is $D = $ [total metal]$_{org}$/[total metal]$_{aq}$. Give physical reasons why β and K_a appear in the numerator of Equation 22-13 but K_L and [H$^+$]$_{aq}$ appear in the denominator.

6. Give a physical interpretation of Equations 22-6 and 22-7 in terms of the fractional composition equations for a monoprotic acid discussed in Section 11-5.

7. Solute S has a partition coefficient of 4.0 between water (phase 1) and chloroform (phase 2) in Equation 22-1.

(a) Calculate the concentration of S in chloroform if [S(aq)] is 0.020 M.

(b) If the volume of water is 80.0 mL and the volume of chloroform is 10.0 mL, find the quotient (mol S in chloroform)/(mol S in water).

8. The solute in the previous problem is initially dissolved in 80.0 mL of water. It is then extracted six times with 10.0 mL portions of chloroform. Find the fraction of solute remaining in the aqueous phase.

9. The weak base B ($K_b = 1.0 \times 10^{-5}$) equilibrates between water (phase 1) and benzene (phase 2).

(a) Define the distribution coefficient, D, for this system.

(b) Explain the difference between D and K, the partition coefficient.

(c) Calculate D at pH 8.00 if $K = 50.0$.

(d) Will D be greater or less at pH 10 than at pH 8? Explain why.

10. Consider the extraction of M^{n+} from aqueous solution into organic solution by reaction with protonated ligand, HL:

$$M^{n+}(aq) + nHL(org) \rightleftharpoons ML_n(org) + nH^+(aq)$$

$$K_{extraction} = \frac{[ML_n]_{org}[H^+]^n_{aq}}{[M^{n+}]_{aq}[HL]^n_{org}}$$

Rewrite Equation 22-13 in terms of $K_{extraction}$ and express $K_{extraction}$ in terms of the constants in Equation 22-13. Give a physical reason why each constant increases or decreases $K_{extraction}$.

11. Butanoic acid has a partition coefficient of 3.0 (favoring benzene) when distributed between water and benzene. Find the formal concentration of butanoic acid

in each phase when 100 mL of 0.10 M aqueous butanoic acid is extracted with 25 mL of benzene **(a)** at pH 4.00 and **(b)** at pH 10.00.

12. For a given value of [HL]$_{org}$ in Equation 22-13, over what pH range (how many pH units) will D change from 0.01 to 100 if $n = 2$?

13. For the extraction of Cu^{2+} by dithizone in CCl$_4$, $K_L = 1.1 \times 10^4$, $K_M = 7 \times 10^4$, $K_a = 3 \times 10^{-5}$, $\beta = 5 \times 10^{22}$, and $n = 2$.

(a) Calculate the distribution coefficient for extraction of 0.1 μM Cu^{2+} into CCl$_4$ by 0.1 mM dithizone at pH 1.0 and pH 4.0.

(b) If 100 mL of 0.1 μM aqueous Cu^{2+} is extracted once with 10 mL of 0.1 mM dithizone at pH 1.0, what fraction of Cu^{2+} remains in the aqueous phase?

14. ▦ Consider the extraction of 100.0 mL of M^{2+}(aq) by 2.0 mL of 1×10^{-5} M dithizone in CHCl$_3$, for which $K_L = 1.1 \times 10^4$, $K_M = 7 \times 10^4$, $K_a = 3 \times 10^{-5}$, $\beta = 5 \times 10^{18}$, and $n = 2$.

(a) Derive an expression for the fraction of metal ion extracted into the organic phase, in terms of the distribution coefficient and volumes of the two phases.

(b) Prepare a graph of the percentage of metal ion extracted over the pH range 0 to 5.

A Plumber's View of Chromatography

15. Match the terms in the first list with the characteristics in the second list.

1. adsorption chromatography

2. partition chromatography

3. ion-exchange chromatography

4. molecular exclusion chromatography

5. affinity chromatography

A. Ions in mobile phase are attracted to counterions covalently attached to stationary phase.

B. Solute in mobile phase is attracted to specific groups covalently attached to stationary phase.

C. Solute equilibrates between mobile phase and surface of stationary phase.

D. Solute equilibrates between mobile phase and film of liquid attached to stationary phase.

E. Different-sized solutes penetrate voids in stationary phase to different extents. Largest solutes are eluted first.

16. The partition coefficient for a solute in chromatography is $K = C_s/C_m$, where C_s is the concentration in the

stationary phase and C_m is the concentration in the mobile phase. Explain why the larger the partition coefficient, the longer it takes a solute to be eluted.

17. **(a)** Write the meaning of the capacity factor, k', in terms of time spent by solute in each phase.

(b) Write an expression in terms of k' for the fraction of time spent by a solute molecule in the mobile phase.

(c) The *retention ratio* in chromatography is defined as

$$R = \frac{\text{time for solvent to pass through column}}{\text{time for solute to pass through column}} = \frac{t_m}{t_r}$$

Show that R is related to the capacity factor by the equation $R = 1/(k' + 1)$.

18. **(a)** A chromatography column with a length of 10.3 cm and inner diameter of 4.61 mm is packed with a stationary phase that occupies 61.0% of the volume. If the volume flow rate is 1.13 mL/min, find the linear flow rate in cm/min.

(b) How long does it take for solvent (which is the same as unretained solute) to pass through the column?

(c) Find the retention time for a solute with a capacity factor of 10.0.

19. An open tubular column is 30.1 m long and has an inner diameter of 0.530 mm. It is coated on the inside wall with a layer of stationary phase 3.1 μm thick. Unretained solute passes through in 2.16 min, whereas a particular solute has a retention time of 17.32 min.

(a) Find the linear and volume flow rates.

(b) Find the capacity factor for the solute and the fraction of time spent in the stationary phase.

(c) Find the partition coefficient, C_s/C_m, for this solute.

20. A chromatographic procedure separates 4.0 mg of unknown mixture on a column with a length of 40 cm and a diameter of 0.85 cm.

(a) What size column would you use to separate 100 mg of the same mixture?

(b) If the flow is 0.22 mL/min on the small column, what volume flow rate should be used on the large column?

(c) If mobile phase occupies 35% of the column volume, calculate the linear flow rate for the small column and the large column.

21. Solvent passes through a column in 3.0 min but solute requires 9.0 min.

(a) Calculate the capacity factor, k'.

(b) What fraction of time is the solute in the mobile phase as it passes through the column?

(c) The volume of stationary phase is one-tenth of the volume of the mobile phase in the column ($V_s = 0.10\ V_m$). Find the partition coefficient, K, for this system.

22. Solvent occupies 15% of the volume of a chromatography column whose inner diameter is 3.0 mm. If the volume flow rate is 0.2 mL/min, find the linear flow rate.

23. Consider a chromatography column in which $V_s = V_m/5$. Find the capacity factor if $K = 3$ and if $K = 30$.

24. The retention volume of a solute is 76.2 mL for a column with $V_m = 16.6$ mL and $V_s = 12.7$ mL. Calculate the partition coefficient and capacity factor for this solute.

25. An open tubular column has a diameter of 207 μm and the thickness of the stationary phase on the inner wall is 0.50 μm. Unretained solute passes through in 63 s and a particular solute emerges in 433 s. Find the partition coefficient for this solute and find the fraction of time spent in the stationary phase.

Efficiency and Band Spreading

26. Why does plate height depend on linear flow rate, not volume flow rate?

27. Which column is more efficient: plate height = **(a)** 0.1 mm or **(b)** 1 mm?

28. Why is longitudinal diffusion a more serious problem in gas chromatography than in liquid chromatography?

29. Why is the optimal flow rate greater for a chromatographic column if the stationary-phase particle size is smaller?

30. What is the optimal flow rate in Figure 22-13 for best separation of solutes?

31. Explain how silanization reduces tailing of chromatographic peaks.

32. Describe how nonlinear partition isotherms lead to non-Gaussian band shapes. Draw the band shape produced by an overloaded column and a column with tailing.

33. A separation of 2.5 mg of an unknown mixture has been optimized on a column of length L and diameter d. Explain why you might not achieve the same resolution for 5.0 mg on a column of length $2L$ and diameter d.

34. An infinitely sharp zone of solute is placed at the center of a column at time $t = 0$. After diffusion for time t_1, the standard deviation of the Gaussian band is 1.0 mm. After 20 min more, at time t_2, the standard deviation is 2.0 mm. What will be the width after another 20 min, at time t_3?

35. A chromatogram with ideal Gaussian bands has $t_r = 9.0$ min and $w_{1/2} = 2.0$ min.

(a) How many theoretical plates are present?

(b) Find the plate height if the column is 10 cm long.

36. The asymmetric chromatogram in Figure 22-12 has a retention time equal to 15 cm of chart paper, and the values of A and B are 3.0 and 1.0 cm, respectively. Find the number of theoretical plates.

37. Two chromatographic peaks with widths of 6 min are eluted at 24 and 29 min. Which diagram in Figure 22-9 will most closely resemble the chromatogram?

38. A band having a width of 4.0 mL and a retention volume of 49 mL was eluted from a chromatography column. What width is expected for a band with a retention volume of 127 mL? Assume that the only band spreading occurs on the column itself.

39. A band from a column eluted at a rate of 0.66 mL/min has a width at half-height of 39.6 s. The sample was applied as a sharp plug with a volume of 0.40 mL and the detector volume is 0.25 mL. Find the variances introduced by injection and detection. What would be the width at half-height if the only broadening occurred on the column?

40. Two compounds with partition coefficients of 0.15 and 0.18 are to be separated on a column with $V_m/V_s = 3.0$. Calculate the number of theoretical plates needed to produce a resolution of 1.5.

41. Write the relation between α, k'_1, and k'_2 for two chromatographic peaks. Calculate the number of theoretical plates needed to achieve a resolution of 1.0 if

(a) $\alpha = 1.05$ and $k'_2 = 5.00$

(b) $\alpha = 1.10$ and $k'_2 = 5.00$

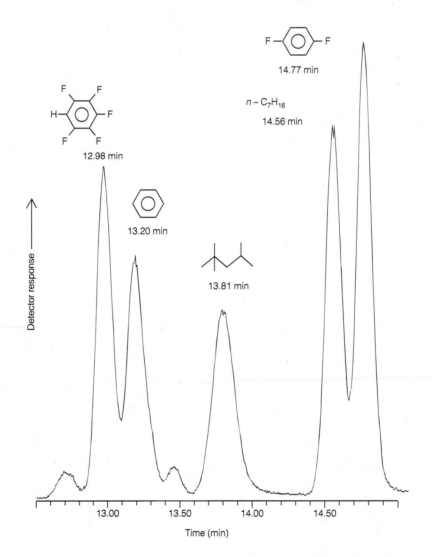

Chromatogram for Problem 42

(c) $\alpha = 1.05$ and $k'_2 = 10.00$

(d) How can you increase N, α, and k'_2 in a chromatography experiment? In this problem, which has a larger effect on resolution, α or k'_2?

42. Consider the peaks for pentafluorobenzene and benzene in the chromatogram shown on the preceding page. The elution time for unretained solute is 1.06 min. The open tubular column is 30.0 m in length and 0.530 mm in diameter, with a layer of stationary phase 3.0 μm thick on the inner wall.

(a) Find the adjusted retention times and capacity factors for both compounds.

(b) Find the relative retention.

(c) Measuring $w_{1/2}$ on the chromatogram, find the number of plates (N_1 and N_2) and the plate height for these two compounds.

(d) Measuring the width (w) at the baseline on the chromatogram, find the number of plates for these two compounds.

(e) Use your answer to (d) to find the resolution between the two peaks.

(f) Using the number of plates ($N = \sqrt{N_1 N_2}$, with values from (d)), the relative retention, and the

capacity factors, calculate what the resolution should be and compare your answer with the measured resolution in (e).

43. A layer with negligible thickness containing 10.0 nmol of methanol ($D = 1.6 \times 10^{-9}$ m^2/s) was placed in a tube of water 5.00 cm in diameter and allowed to spread by diffusion. Using Equation 22-25, prepare a graph showing the Gaussian concentration profile of the methanol zone after 1.00, 10.0, and 100 min. Prepare a second graph showing the same experiment with ribonuclease ($D = 0.12 \times 10^{-9}$ m^2/s).

44. An open tubular gas chromatographic column is coated with a layer of stationary phase that is 2.0 μm thick. The diffusion coefficient for a compound with a capacity factor $k' = 10$ is $D_m = 1.0 \times 10^{-5}$ m^2/s in the gas phase and $D_s = 1.0 \times 10^{-9}$ m^2/s in the stationary phase. Considering only longitudinal diffusion and finite equilibration time as sources of broadening, prepare a graph showing the plate height from each of these mechanisms and the total plate height as a function of linear flow rate (from 2 cm/s to 1 m/s). Assume that the factor q in Equation 22-35 is 1.

45. Consider two Gaussian chromatographic peaks with relative areas 4:1. Construct a set of graphs to show the overlapping peaks if the resolution is 0.5, 1, or 2.

Notes and References

1. An experiment using a pH-dependent separation of radioactive $^{212}Bi^{3+}$ from $^{212}Pb^{2+}$ with dithizone at pH ≈ 3 has been described by D. M. Downey, D. D. Farnsworth, and P. G. Lee, *J. Chem. Ed.* **1984,** *61,* 259.

2. An excellent discussion of band broadening in chromatography is found in J. C. Giddings, *Unified Separation Science* (New York: Wiley, 1991).

3. J. P. Foley and J. G. Dorsey, *Anal. Chem.* **1983,** *55,* 730; B. A. Bidlingmeyer and F. V. Warren, Jr., *Anal. Chem.* **1984,** *56,* 1583A.

4. J. P. Foley, *Analyst* **1991,** *116,* 1275.

5. S. J. Hawkes, *J. Chem. Ed.* **1983,** *60,* 393.

Separation of Optical Isomers by Chromatography

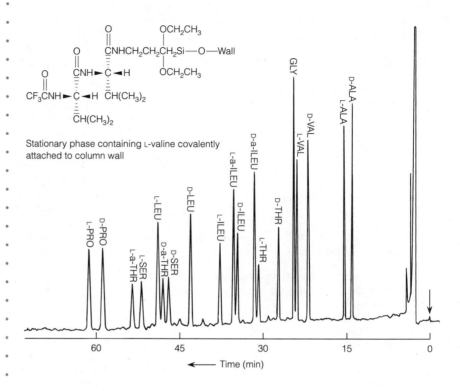

Stationary phase containing L-valine covalently attached to column wall

← Time (min)

Separation of amino acid enantiomers at 110°C on a 0.30-mm (inner diameter) × 39-m open tubular column with 10^5 plates. [From W. A. Koenig and G. J. Nicholson, *Anal. Chem.* **1975**, *47*, 951.] Amino acids are not volatile enough to be chromatographed directly, so the more volatile *N*-trifluoroacetyl isopropyl esters are prepared:

$$\underset{\text{Amino acid}}{\overset{\displaystyle R}{\underset{\displaystyle H}{H_3\overset{+}{N}CCO_2^-}}} \qquad \underset{\text{Volatile derivative}}{\overset{\displaystyle O \quad R}{\underset{\displaystyle H}{CF_3CNHCCO_2CH(CH_3)_2}}}$$

Classical methods for separating mirror image isomers (*enantiomers*), such as D- and L-amino acids, required the preparation of chemical derivatives, followed by tedious crystallizations that did not always work. Gas or liquid chromatography with an optically active stationary phase rapidly separates enantiomers.

Gas and Liquid Chromatography

In this chapter we discuss fundamental details of gas and liquid chromatography, which are among our most powerful methods of separation. For example, Figure 23-1 shows the chromatographic separation of compounds from the gas phase in a can of beer. Drink up!

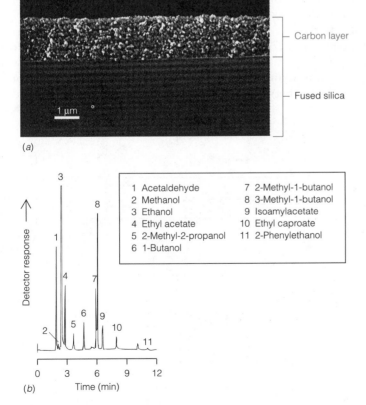

- Carbon layer

- Fused silica

1 μm

(a)

1	Acetaldehyde	7	2-Methyl-1-butanol
2	Methanol	8	3-Methyl-1-butanol
3	Ethanol	9	Isoamylacetate
4	Ethyl acetate	10	Ethyl caproate
5	2-Methyl-2-propanol	11	2-Phenylethanol
6	1-Butanol		

Detector response →

Time (min)

(b)

Figure 23-1 (*a*) Porous carbon stationary phase (2 μm thick) on the inside wall of a fused silica open tubular capillary gas chromatography column. (*b*) Chromatogram of compounds in the headspace of a beer can, obtained with a 0.25-mm-diameter × 30-m-long porous carbon column operated at 30°C for 2 min and then ramped up to 160°C at 20°/min. [Courtesy Alltech Associates, State College, PA.]

23-1 Gas Chromatography

In **gas chromatography,** a gaseous solute (or vapor from a volatile liquid) is carried by a gaseous mobile phase. In *gas-liquid partition chromatography,* the stationary phase is a nonvolatile liquid coated on the inside of the column or on a fine solid support, like that shown at the upper right in Figure 22-6. In *gas-solid adsorption chromatography* (Figure 23-1), solid particles that adsorb solute serve as the stationary phase.[1]

A schematic diagram of a gas chromatograph is shown in Figure 23-2. A volatile liquid is injected through a rubber **septum** (a thin disk) into a heated port, which vaporizes the sample. The sample is swept through the column by He, N_2, or H_2 *carrier gas,* and the separated solutes flow through a detector, whose response is displayed on a recorder or computer. The column must be hot enough to produce sufficient vapor pressure for each solute to be eluted in a reasonable time. The detector is maintained at a higher temperature than the column, so that all solutes are gaseous.

The choice of carrier gas depends on the detector and the desired separation efficiency and speed.

Packed Columns

Packed columns contain a fine solid support coated with a nonvolatile liquid stationary phase, or the solid itself may be the stationary phase. Despite their inferior resolution, packed columns are useful for preparative separations, when a great deal of stationary phase is required, or to separate low-boiling gases that are poorly retained. Columns made of stainless steel, nickel (which is less catalytically active than steel), or glass are typically 3–6 mm in diameter and 1–5 m in length. The solid support is often diatomite—the silica skeletons of algae—that is silanized (Reaction 22-36) to reduce hydrogen bonding to polar solutes. For solutes that bind more tenaciously, Teflon is a useful support, but it is limited to <200°C.

Raising the percentage of stationary phase leads to

1. greater capacity for solute
2. longer retention time
3. increased plate height

Teflon is a chemically inert polymer with the structure

$$-CF_2-CF_2-CF_2-CF_2-.$$

Uniform particle size decreases the multiple path term in the van Deemter equation (22-33), thereby reducing the plate height and increasing resolution. Small particle size decreases the time required for solute equilibration, thereby improving column efficiency. However, the smaller the particle size, the less space between particles and the more pressure required to force

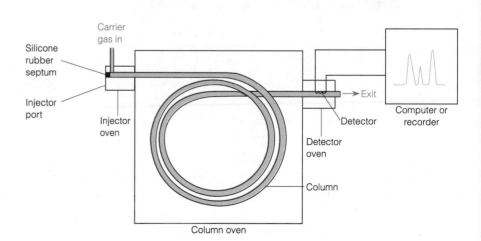

Figure 23-2 Schematic diagram of a gas chromatograph.

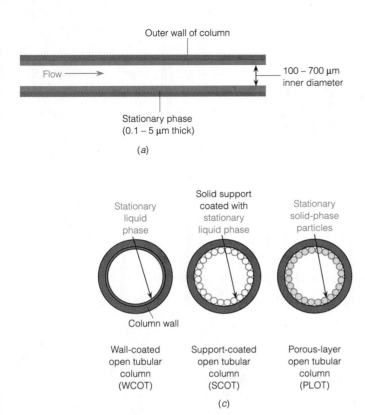

Outer wall of column

Flow →

100 – 700 µm inner diameter

Stationary phase (0.1 – 5 µm thick)

(a)

(b)

Stationary liquid phase

Solid support coated with stationary liquid phase

Stationary solid-phase particles

Column wall

Wall-coated open tubular column (WCOT)

Support-coated open tubular column (SCOT)

Porous-layer open tubular column (PLOT)

(c)

Figure 23-3 (*a*) Typical dimensions of open tubular column for gas chromatography. (*b*) Aluminum-clad fused silica chromatography column. The diameter of the cage is 18 cm and typical column lengths are 15–100 m. (*c*) Cross-sectional view of wall-coated, support-coated, and porous-layer columns.

mobile phase through the column. Particle size is expressed in micrometers or as a *mesh size,* which refers to the size of screens through which the particles are passed or retained (Table 26-2). A 100/200 mesh particle passes through a 100 mesh screen, but not through a 200 mesh screen. The mesh gives the number of openings per linear inch of screen.

Open Tubular Columns

The vast majority of analyses use long, narrow **open tubular columns** (Figure 23-3). As discussed in Section 22-5, the open tubular design offers higher resolution, shorter analysis time, and greater sensitivity than packed columns, but lower capacity for sample (Figure 23-4).

The *wall-coated* column in Figure 23-3c features a thin, uniform film of stationary liquid phase on the inner wall of the column. The *support-coated* design has solid particles attached to the inner wall of the column, coated with stationary liquid phase. In the *porous-layer* design, the solid particles *are* the active stationary phase, as in Figure 23-1a. Because of their increased surface area, support-coated columns can handle larger samples than can wall-coated columns. The performance of support-coated columns is intermediate between those of wall-coated columns and packed columns.

Open tubular columns are usually made of fused silica (SiO_2) coated with polyimide (a plastic capable of withstanding 350°C) or aluminum for support and protection from atmospheric moisture. Some columns are made

Wall-coated open tubular column (WCOT): liquid stationary phase on inside wall of column

Support-coated open tubular column (SCOT): liquid stationary phase coated on solid support attached to inside wall of column

Porous-layer open tubular column (PLOT): solid stationary phase on inside wall of column

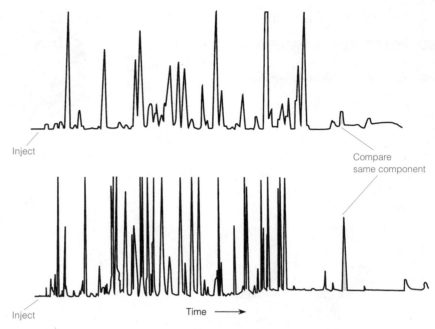

Inject

Compare same component

Inject   Time ⟶

Figure 23-4 Gas chromatographic separation of a perfume oil on a 2-mm-diameter × 1.5-m-long packed column (upper trace) and a 0.25-mm-diameter × 30-m-long open tubular column (lower trace), both using Carbowax 20 M (Table 23-1) as stationary phase. The taller peaks in both chromatograms are clipped off so that the smaller peaks can be seen. One component eluted near the end of each trace is shaded to emphasize how much sharper the trace from the open tubular column is. [From R. R. Freeman, Ed., *High Resolution Gas Chromatography* (Palo Alto, CA: Hewlett Packard Co., 1981).]

Box 23-1 describes a *chiral* (optically active) bonded phase used to separate optical isomers.

of stainless steel and others have a thin layer of fused silica on the inside of the steel. As the column ages, stationary phase is baked off, surface silanol groups (Si—O—H) are exposed, and *tailing* (Figure 22-18) increases. Exposure to oxygen at high temperatures also leads to degradation and tailing. To reduce the tendency of stationary phase to bleed from the column at elevated temperature, it may be *bonded* (covalently attached) to the silica surface or covalently *crosslinked* to itself.

Four common liquid stationary phases are listed in Table 23-1. The choice of liquid phase for a given problem is based on the rule "like dissolves like." Nonpolar columns are best for the nonpolar solutes in Table 23-2. Columns of intermediate polarity are best for intermediate polarity solutes, and strongly polar columns are best for strongly polar solutes. Alumina (Al_2O_3) is a solid stationary phase that separates hydrocarbons in gas-solid partition chromatography.

Molecular sieves (Figure 23-5) are inorganic materials or organic polymers with large cavities into which small molecules enter and are partially retained. Molecules such as H_2, O_2, N_2, CO_2, and CH_4 can be separated from one another. Gases are dried by passage through traps containing molecular sieves because water is strongly retained by sieves. Inorganic sieves can be regenerated (freed of water) by heating to 300°C in vacuum or under flowing N_2.

TABLE 23-1 Common stationary phases in capillary gas chromatography

Structure	Polarity	Temperature range (°C)
(Diphenyl)$_x$(dimethyl)$_{1-x}$ polysiloxane — $x = 0$	Nonpolar	$-60°$–$360°$
$x = 0.05$	Nonpolar	$-60°$–$360°$
$x = 0.35$	Intermediate polarity	$0°$–$300°$
$x = 0.65$	Intermediate polarity	$50°$–$370°$
(Cyanopropylphenyl)$_{0.14}$ (dimethyl)$_{0.86}$ polysiloxane	Intermediate polarity	$-20°$–$280°$
Carbowax (poly(ethylene glycol))	Strongly polar	$40°$–$250°$
(Biscyanopropyl)$_{0.9}$ (cyanopropylphenyl)$_{0.1}$ polysiloxane	Strongly polar	$0°$–$275°$

The Retention Index

Figure 23-6 illustrates how the relative retention times of polar and nonpolar solutes change as the polarity of the stationary phase changes. In Figure 23-6a, 10 compounds are eluted nearly in order of increasing boiling point from a nonpolar stationary phase. The principal determinant of retention on this column is the volatility of the solutes. In Figure 23-6b, the strongly polar stationary phase strongly retains the polar solutes. The three alcohols are the

Box 23-1 Cyclodextrins

A versatile class of optically active stationary phases in gas and liquid chromatography and capillary electrophoresis is based on *cyclodextrins,* which are naturally occurring cyclic sugars:

β-Cyclodextrin has a 0.78-nm-diameter opening into a chiral, hydrophobic cavity. (The opening on the opposite face is smaller.) Cyclodextrins can be bonded to the stationary phase for separation of chiral molecules. The hydroxyls may be capped with groups such as pentyl ($-C_5H_{11}$) or trifluoroacetyl ($[-C(=O)CF_3]$ to decrease the polarity of the faces.[2]

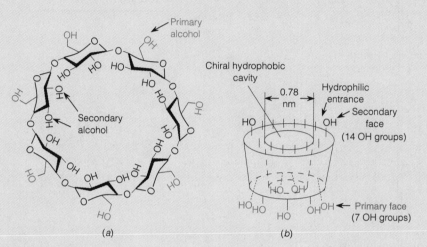

(*a*) (*b*)

(*a*) Structure of β-cyclodextrin, a cyclic sugar made of seven glucose molecules. (α-Cyclodextrin contains six monomers and γ-cyclodextrin contains eight.) (*b*) The primary hydroxyl groups lie on one face and the secondary hydroxyl groups lie on the other face.

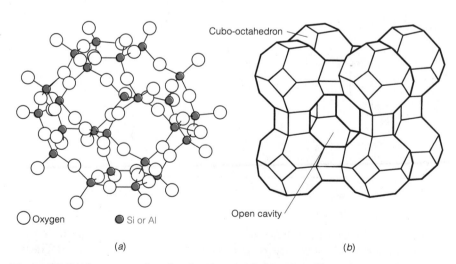

Oxygen   Si or Al

(*a*) (*b*)

Figure 23-5 Structure of molecular sieves. (*a*) Aluminosilicate framework of one cubo-octahedron of a mineral class called *zeolites.* (*b*) Interconnection of eight cubo-octahedra to produce a cavity into which small molecules can enter. The synthetic zeolite $Na_{12}(Al_{12}Si_{12}O_{48}) \cdot 27H_2O$ has this structure.

TABLE 23-2   Polarity of solutes

Nonpolar	Weak intermediate polarity
Saturated hydrocarbons	Ethers
Olefinic hydrocarbons	Ketones
Aromatic hydrocarbons	Aldehydes
Halocarbons	Esters
Mercaptans	Tertiary amines
Sulfides	Nitro compounds (without α-H atoms)
CS_2	Nitriles (without α-atoms)

Strong intermediate polarity	Strongly polar
Alcohols	Polyhydroxyalcohols
Carboxylic acids	Amino alcohols
Phenols	Hydroxy acids
Primary and secondary amines	Polyprotic acids
Oximes	Polyphenols
Nitro compounds (with α-H atoms)	
Nitriles (with α-H atoms)	

SOURCE: Adapted from H. M. McNair and E. J. Bonelli, *Basic Gas Chromatography* (Palo Alto, CA: Varian Instrument Division, 1968).

last to be eluted, following the three ketones, which follow four alkanes. Hydrogen bonding to the stationary phase is probably the strongest force leading to retention. Dipole interactions of the ketones are the second strongest force.

The Kovats **retention index** (I) for a linear alkane equals 100 times the number of carbon atoms. For octane, $I = 800$; and for nonane, $I = 900$. A

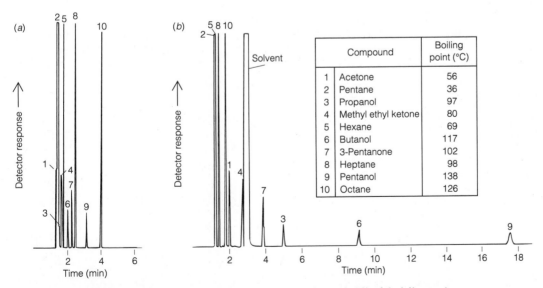

	Compound	Boiling point (°C)
1	Acetone	56
2	Pentane	36
3	Propanol	97
4	Methyl ethyl ketone	80
5	Hexane	69
6	Butanol	117
7	3-Pentanone	102
8	Heptane	98
9	Pentanol	138
10	Octane	126

Figure 23-6 Separation of 10 compounds on (*a*) nonpolar poly(dimethylsiloxane) and (*b*) strongly polar poly(ethylene glycol) 1-μm-thick stationary phases in 0.32-mm-diameter × 30-m-long open tubular columns at 70°C. [Courtesy Restek Co., Bellefonte, PA.]

Retention index relates the retention time of a solute to the retention times of linear alkanes.

compound eluted between octane and nonane (Figure 22-7) has a retention index that lies between 800 and 900 and is computed by the formula

$$\text{Retention index: } I = 100\left[n + (N - n)\frac{\log t'_r(\text{unknown}) - \log t'_r(n)}{\log t'_r(N) - \log t'_r(n)}\right] \quad (23\text{-}1)$$

where n is the number of carbon atoms in the *smaller* alkane; N is the number of carbon atoms in the *larger* alkane; $t'_r(n)$ is the adjusted retention time of the *smaller* alkane; and $t'_r(N)$ is the adjusted retention time of the *larger* alkane.

EXAMPLE   Retention Index

If the retention times in Figure 22-7 are $t_r(\text{air}) = 0.5$ min, $t_r(\text{octane}) = 14.3$ min, $t_r(\text{unknown}) = 15.7$ min, and $t_r(\text{nonane}) = 18.5$ min, find the retention index for the unknown.

Solution   The index is computed with Equation 23-1:

$$I = 100\left[8 + (9 - 8)\frac{\log 15.2 - \log 13.8}{\log 18.0 - \log 13.8}\right] = 836$$

The retention index of 657 for benzene on poly(dimethylsiloxane) in Table 23-3 tells us that benzene is eluted between hexane and heptane from

TABLE 23-3   Retention indexes for several compounds on common stationary phases

Phase	Retention index[a]				
	Benzene b.p. 80°C	Butanol b.p. 117°C	2-Pentanone b.p. 102°C	1-Nitropropane b.p. 132°C	Pyridine b.p. 116°C
Poly(dimethylsiloxane)	657	648	670	708	737
(Diphenyl)$_{0.05}$(dimethyl)$_{0.95}$- polysiloxane	672	664	691	745	761
(Diphenyl)$_{0.35}$(dimethyl)$_{0.65}$- polysiloxane	754	717	777	871	879
(Cyanopropylphenyl)$_{0.14}$- (dimethyl)$_{0.86}$polysiloxane	726	773	784	880	852
(Diphenyl)$_{0.65}$(dimethyl)$_{0.35}$- polysiloxane	797	779	824	941	943
Poly(ethylene glycol)	956	1142	987	1217	1185
(Biscyanopropyl)$_{0.9}$- (cyanopropylphenyl)$_{0.1}$- polysiloxane	1061	1232	1174	1409	1331

a. For reference, boiling points for various alkanes are hexane, 69°C; heptane, 98°C; octane, 126°C; nonane, 151°C; decane, 174°C; undecane, 196°C.

SOURCE: Restek *Chromatography Products Catalog,* 1993–94, Bellefonte, PA.

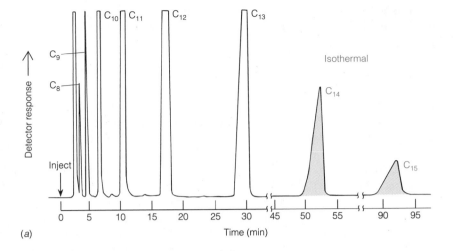

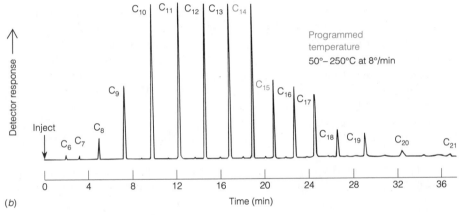

Figure 23-7 Comparison of (*a*) isothermal and (*b*) programmed temperature chromatography. Each sample contains linear alkanes run on a 1.6-mm-diameter × 6-m-long column containing 3% Apiezon L (liquid phase) on 100/120 mesh VarAport 30 solid support with He flow rate of 10 mL/min. Detector sensitivity is 16 times greater in (*a*) than (*b*). [From H. M. McNair and E. J. Bonelli, *Basic Gas Chromatography* (Palo Alto, CA: Varian Instrument Division, 1968).]

this nonpolar stationary phase. 1-Nitropropane is eluted just after heptane on the same column. As we go down the table, the stationary phases become more polar. For (biscyanopropyl)$_{0.9}$(cyanopropylphenyl)$_{0.1}$polysiloxane at the bottom of the table, benzene is eluted after decane and nitropropane is eluted after n-$C_{14}H_{30}$.

The retention index of an unknown measured on several different columns is useful for identifying the unknown by comparison with tabulated retention indexes. For a homologous series of compounds (those with similar structures, but differing by the number of CH_2 groups in a chain, such as $CH_3CO(CH_2)_nCO_2CH_2CH_3$), log t_r' is usually a linear function of the number of carbon atoms.

Temperature Programming

Increasing the temperature of a column increases solute vapor pressure and thereby decreases retention time. When separating compounds with a wide range of boiling points or polarities, it helps to raise the column temperature *during* the separation. An example is shown in Figure 23-7. At a constant temperature of 150°C, the more volatile compounds emerge close together, and the less volatile compounds may not even be eluted from the column. If the temperature is increased from 50° to 250°C at a rate of 8°/min, all of the

Raising column temperature

1. decreases retention time
2. sharpens peaks

compounds are eluted and the separation of peaks is fairly uniform. Care should be exercised to prevent thermal decomposition of analytes during chromatography.

Carrier Gas

Column and detector performance depend on the identity of the carrier gas. Figure 23-8 shows that H_2, He, and N_2 give optimal plate height (0.3 mm) at significantly different flow rates. Although N_2 requires relatively slow flow for best results, the performance of H_2 is hardly degraded at high flow rates.

H_2 and He give optimal resolution at higher flow rates than N_2.

H_2 and He give better resolution (smaller plate height) than N_2 at high flow rate because solutes diffuse more rapidly through H_2 and He than through N_2. The more rapidly a solute diffuses between phases, the smaller is the mass transfer (Cu_x) term in the van Deemter equation (22-33). Contrary to the discussion below Equation 22-35, mass transfer for the column in Figure 23-8 is dominated by slow diffusion through the *mobile phase,* rather than through the *stationary phase.*

For column efficiency and speed of analysis, H_2 is a good carrier gas. Because H_2 forms explosive mixtures with air, it must be vented safely. A possible disadvantage of H_2 is that it may hydrogenate C=C bonds of analytes at elevated temperature, especially in metal columns. He is frequently used instead of H_2 because of the safety hazard.

Gas added to the stream after the column is called *makeup gas.*

Different detectors work best with different gases. In capillary chromatography, the flow through the column may be too low for the best detector performance. Therefore, the optimum gas for separation is used in the column, and the best gas for detection is added between the column and the detector.

Impurities in the carrier gas degrade the stationary phase. High-quality gases should be used, and even these are passed through purifiers to remove oxygen, water, and traces of organic compounds.

Sample Injection

Liquids are introduced into a column through the rubber septum of a port leading to a glass tube inside a hot metal block (Figure 23-9a). (Some manufacturers are switching to the septumless port shown in Figure 23-9b, thereby eliminating septum replacement and bleeding of volatile constituents from the septum.) Carrier gas sweeps the vaporized sample out of the port and into the chromatography column. Analytical columns typically require 0.1–2 μL of liquid sample, whereas preparative columns can handle 20–1 000 μL. Gases are introduced by a gas-tight syringe or a gas-sampling valve. Analytical work usually requires 0.5–10 mL of gas, whereas preparative columns can handle up to 1 L of gas.

A complete injection contains too much material for a capillary column. In *split injection,* only 0.1–10% of the injected sample reaches the column. In Figure 23-9a, sample is injected through the septum into the evaporation zone, with needle valve 1 closed. Carrier gas from the flow controller sweeps the sample through silanized glass wool, where complete vaporization and good mixing occur. At the split point, some of the sample enters the chromatography column and the remainder passes through needle valve 2 to a waste vent. The pressure regulator leading to needle valve 2 controls the fraction of sample discarded. If some components are not completely

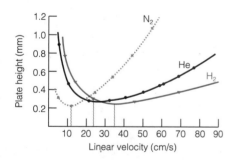

Figure 23-8 Van Deemter curves for gas chromatography of *n*-$C_{17}H_{36}$ at 175°C, using N_2, He, or H_2 in a 0.25-mm-diameter × 25-m-long wall-coated column with OV-101 stationary phase. [From R. R. Freeman, Ed., *High Resolution Gas Chromatography* (Palo Alto, CA: Hewlett Packard Co., 1981).]

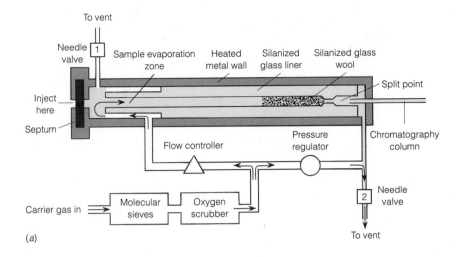

(a)

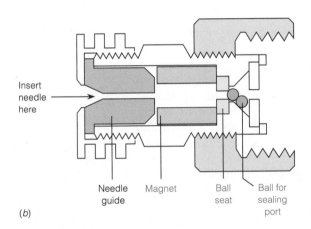

(b)

Figure 23-9 (*a*) Port for split or splitless injection into an open tubular column. The glass liner and glass wool are slowly contaminated by nonvolatile and decomposed samples and must be replaced periodically. (*b*) Septumless port sealed by a metal ball held against a precisely machined ring by a magnet. [Courtesy Alltech Associates, State College, PA.]

vaporized, fractionation of a sample during injection leads to errors in quantitative analysis. If the injector temperature is too high, decomposition can occur.

For quantitative analysis and for trace analysis, *splitless injection* is appropriate. A dilute solution in a low-boiling solvent is injected into the port in Figure 23-9a, with needle valves 1 and 2 closed. The initial column temperature is 20°–25°C lower than the boiling point of the solvent, which therefore condenses at the beginning of the column. As solutes catch up with the condensed plug of solvent, they are trapped in a small band at the beginning of the column. This *solvent trapping* leads to sharp chromatographic peaks. Some sample vapor remains near the septum, which would lead to tailing. Therefore, after 20–60 s, needle valve 1 is opened and vapors near the septum are purged from the port. In splitless injection, ~80% of the sample is applied to the column, and little fractionation occurs during injection.

Injection into open tubular columns:

split: routine means of introducing small sample volume into open tubular column

splitless: best for trace levels of high-boiling solutes in low-boiling solvents; better than split injection for quantitative analysis

on-column: best for thermally unstable solutes and high-boiling solvents

TABLE 23-4   Thermal conductivity at 273 K and 1 atm

Gas	Thermal conductivity $J/(K \cdot m \cdot s)$	Molecular weight
H_2	0.170	2
He	0.141	4
NH_3	0.0215	17
N_2	0.0243	28
C_2H_4	0.0170	28
O_2	0.0246	32
Ar	0.0162	40
CO_2	0.0144	44
C_3H_8	0.0151	44
Cl_2	0.0076	71

The energy per unit area per unit time flowing from a hot region to a cold region is given by

$$\text{energy flux } (J/m^2 \cdot s) = -\kappa(dT/dx)$$

where κ is the thermal conductivity [units = $J/(K \cdot m \cdot s)$] and dT/dx is the temperature gradient (K/m). Thermal conductivity is to energy flux as the diffusion coefficient is to mass flux.

An alternative means of condensing solutes in a narrow band at the beginning of the column is called *cold trapping*. The initial column temperature is 150° lower than the boiling points of the solutes of interest. Solvent and low-boiling components are eluted rapidly, but high-boiling solutes remain in a narrow band. We then rapidly warm the column to initiate chromatography of the high-boiling solutes.

On-column injection is used for samples that decompose above their boiling point. Solution is inserted directly into the column, without going through a hot injector. The initial column temperature is low enough to condense solutes in a narrow zone. Warming the column initiates chromatography. Samples are subjected to the lowest possible temperature in this procedure, and little loss of any solute occurs.

Thermal Conductivity Detector

Thermal conductivity measures the ability of a substance to transport heat from a hot region to a cold region (Table 23-4). In the **thermal conductivity detector** in Figure 23-10, gas emerging from the chromatography column flows over a hot tungsten-rhenium filament. When solute emerges from the column, the thermal conductivity of the gas stream decreases, the filament gets hotter, its electrical resistance increases, and the voltage drop through the filament changes.

The detector responds to *changes* in thermal conductivity, so the conductivities of solute and carrier gas should be as different as possible. Because H_2 and He have the highest thermal conductivity, these two are the carriers of choice for thermal conductivity detection.

It is common practice to split the carrier gas into two streams, sending part through the analytical column and then to the working filament. The other part is passed through a matched reference column leading to a

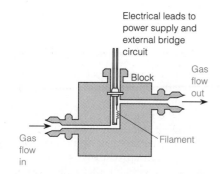

Electrical leads to power supply and external bridge circuit

Block

Gas flow out

Gas flow in

Filament

Figure 23-10 Thermal conductivity detector. [Courtesy Varian Associates, Palo Alto, CA.]

reference filament. The resistance of the working filament is measured with respect to that of the reference filament. The reference column minimizes flow differences when temperature is changed. Sensitivity increases with the square of the filament current. However, the maximum recommended current should not be exceeded, to avoid burning out the filament. To prevent overheating and oxidation, the filament should never be left on when carrier gas is not flowing.

The sensitivity of a thermal conductivity detector (but *not* the flame ionization detector described next) is inversely proportional to flow rate: It is more sensitive at a lower flow rate. Sensitivity also increases with increasing temperature differences between the filament and the surrounding block in Figure 23-10. The block should therefore be maintained at the lowest temperature that allows all solutes to remain gaseous.

Flame Ionization Detector

In the **flame ionization detector** in Figure 23-11, eluate is burned in a mixture of H_2 and air. Carbon atoms (except carbonyl and carboxyl carbons) produce CH radicals, which go on to produce CHO^+ ions in the flame.

$$CH + O \rightarrow CHO^+ + e^- \tag{23-2}$$

Only about 1 in 10^5 carbon atoms produces an ion, but ion production is

Thermal conductivity detector:

1. 10^4 linear response range
2. H_2 and He give lowest detection limit
3. sensitivity increases with
   a. increasing filament current
   b. decreasing flow rate
   c. lower detector block temperature

Eluent: mobile phase entering the column
Eluate: what comes out of the column

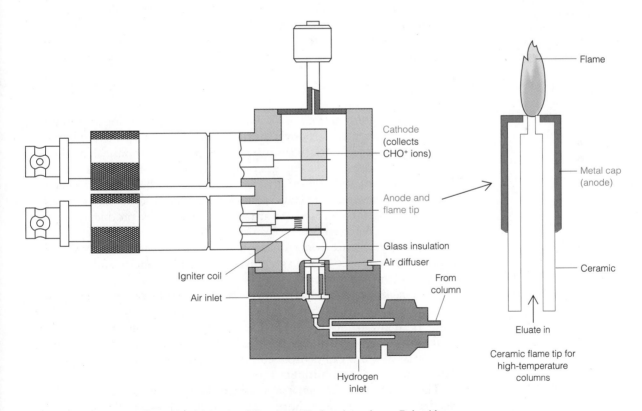

Figure 23-11 Flame ionization detector. [Courtesy Varian Associates, Palo Alto, CA.] In columns operating up to 470°C, the metal flame tip adsorbs solutes as they emerge from the column and leads to tailing. This problem was solved by a ceramic-lined flame tip in which solutes never contact the hot metal.[3]

TABLE 23-5   Detection limits and linear ranges of gas chromatography detectors

Detector	Approximate detection limit	Linear range
Thermal conductivity	400 pg/mL (propane)	$>10^5$
Flame ionization	2 pg/s	$>10^7$
Electron capture	As low as 5 fg/s	10^4
Flame photometric	<1 pg/s (phosphorus)	$>10^4$
	<10 pg/s (sulfur)	$>10^3$
Alkali flame	100 fg/s	10^5
Sulfur chemiluminescence	100 fg/s (sulfur)	10^5
Fourier transform infrared	200 pg to 40 ng	10^4
Mass spectrometric	25 fg to 100 pg	10^5

SOURCE: Most data are from D. G. Westmorland and G. R. Rhodes, *Pure. Appl. Chem.* **1989,** *61,* 1147.

strictly proportional to the number of susceptible carbon atoms entering the flame. The flame ionization detector is relatively insensitive to O_2, CO_2, H_2O, and NH_3.

The CHO^+ produced in the flame is collected at the cathode above the flame. Response to organic compounds is directly proportional to solute mass over seven orders of magnitude. In the absence of organic solutes, the current is almost zero. The detection limit is 100 times smaller than that of the thermal conductivity detector (Table 23-5) and is reduced by 50% when N_2 carrier gas is used instead of He. For open tubular columns, N_2 makeup gas is added to the H_2 or He eluate before it enters the detector.

Flame ionization detector:

1. N_2 gives lowest detection limit
2. signal proportional to number of susceptible carbon atoms
3. 100-fold lower detection limit than thermal conductivity
4. 10^7 linear response range

Other Detectors

The *electron capture detector* is particularly sensitive to halogen-containing molecules, conjugated carbonyls, nitriles, nitro compounds, and organometallic compounds but relatively insensitive to hydrocarbons, alcohols, and ketones. The carrier or makeup gas must be either N_2 or 5% methane in Ar. Moisture decreases the sensitivity. Gas entering the detector is ionized by high-energy electrons ("beta rays") emitted from a foil containing radioactive ^{63}Ni. Electrons thus formed are attracted to an anode, producing a small, steady current. When analyte molecules with a high electron affinity enter the detector, they capture some of the electrons. The detector responds by varying the frequency of voltage pulses between the anode and cathode to maintain a constant current.

A *flame photometric detector* measures optical emission from phosphorus or sulfur compounds, such as pesticides. When eluate passes through a H_2-air flame, as in the flame ionization detector, excited sulfur and phosphorus species emit characteristic light. Phosphorus emission at 536 nm or sulfur emission at 394 nm can be isolated by a narrow-band interference filter and detected with a photomultiplier tube.

The *alkali flame detector* is a modified flame ionization detector that is selectively sensitive to phosphorus and nitrogen. It is especially important for drug analysis. Ions produced by these elements when they contact a Rb_2SO_4-containing glass bead at the burner tip create the current that is measured. N_2 carrier gas, of course, cannot be used for nitrogen-containing samples.

Other gas chromatography detectors:

electron capture: halogens, conjugated
  $C=O$, $-C\equiv N$, $-NO_2$
flame photometer: P and S
alkali flame: P and N
sulfur chemiluminescence: S
atomic emission: most elements

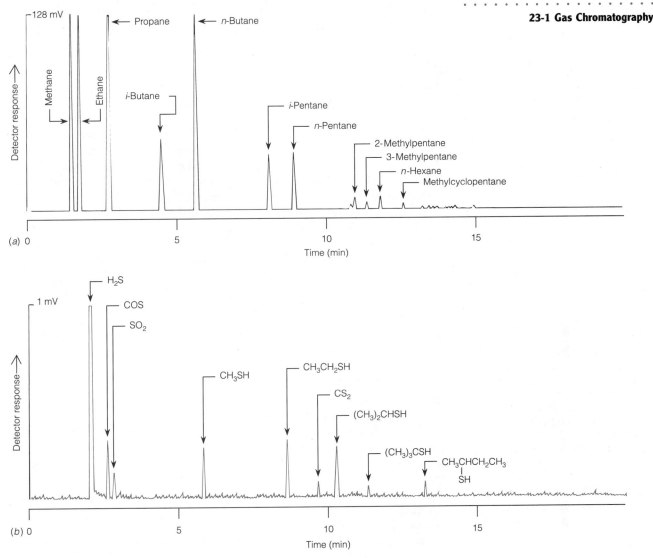

Figure 23-12 Gas chromatograms showing sulfur compounds in natural gas: (a) flame ionization detector response and (b) sulfur chemiluminescence detector response. The organosulfur compounds are too dilute to be seen in flame ionization, and the sulfur chemiluminescence is insensitive to hydrocarbons. [From N. G. Johansen and J. W. Birks, *Am. Lab.* February 1991, 112. See also R. L. Shearer, *Am. Lab.* April 1994, 34P.]

A *sulfur chemiluminescence detector* takes the exhaust from a flame ionization detector, in which sulfur has been oxidized to SO, and mixes it with ozone to form an excited state of SO_2 that emits blue light and ultraviolet radiation. Emission intensity is proportional to the mass of sulfur eluted, regardless of the source, and the sensitivity to sulfur is 10^7 times greater than the sensitivity to hydrocarbons (Figure 23-12).

An *atomic emission detector* directs eluate through a helium plasma in a microwave cavity. Every element of the periodic table produces characteristic atomic emission that can be detected by a photodiode array polychromator

Reactions giving sulfur chemiluminescence:

$$\text{sulfur compound} \xrightarrow{\text{H}_2\text{-O}_2 \text{ flame}} \text{SO} + \text{products}$$

$$SO + O_3 \rightarrow SO_2^* + O_2$$
$$(SO_2^* = \text{excited state})$$

$$SO_2^* \rightarrow SO_2 + h\nu$$

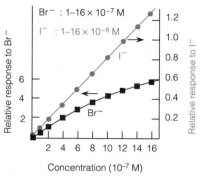

Figure 23-13 Calibration curve for electron capture detector response to iodide and bromide in natural waters. Iodate that may be present is first reduced to I^- with thiosulfate. Then Br^- and I^- are oxidized to Br_2 and I_2 with chromic acid in the presence of acetone, to give $BrCH_2COCH_3$ and ICH_2COCH_3. Products are extracted into benzene and analyzed by gas chromatography, using *p*-dichlorobenzene as *internal standard*. The graph compares detector response for each compound with that of the internal standard. [From L. Maros, M. Káldy, and S. Igaz, *Anal. Chem.* **1989**, *61*, 733.]

Linear detector response means that peak area is proportional to analyte concentration. For very narrow peaks, peak height is often substituted for peak area.

(Section 20-6). You can set the detector to observe any element you choose in each analyte as it emerges from the column.

Qualitative Analysis

Each chromatographic peak can be directed into a mass spectrometer or Fourier transform infrared spectrophotometer (Section 20-8) to record a spectrum as the substance is eluted from the column. The spectrum can be identified by comparison with a library of spectra stored in a computer. Packed columns allow us to use larger samples, from which microliter quantities of eluate can be condensed in a Dry Ice–cooled, U-shaped glass tube inserted into the exit port. The sample can then be subjected to any desired analysis, such as nuclear magnetic resonance. Thermal conductivity detection is usually used, because flames burn the sample.

The simplest method of identifying a chromatographic peak is comparison of its retention time with that of an authentic sample of the suspected compound. The most reliable way to do this is by **co-chromatography,** in which authentic sample is added to the unknown. If the added compound is identical to one component of the unknown, then the relative area of that one peak will increase. Identification is only tentative when carried out with one column, but it is nearly conclusive when carried out with several different kinds of columns.

Quantitative Analysis

The area of a chromatographic peak is related to the quantity of that component in the mixture. We normally choose conditions under which the response is *linear,* which means that *the area of a peak is proportional to the quantity of that component.* Figure 23-13 illustrates linear response to I^- and nonlinear response to Br^- by an electron capture detector. In the discussion that follows, it is assumed that response is linear.

Peak areas are measured automatically if the chromatograph is computer controlled.[4] If you must make a measurement by hand, a triangle can be drawn with two sides tangent to the inflection points on each side of the peak. When the third side is drawn along the baseline, the area of the triangle ($= \frac{1}{2}$ (base × height)) is 96% of the area of a Gaussian peak.

The data in Figure 23-13 were obtained with *p*-dichlorobenzene as an **internal standard.** A constant concentration of *p*-dichlorobenzene (designated S for standard) was added to each sample and the area of the peak for each analyte (designated X) was compared with that of the standard. Calculations are based on the proportionality between area and concentration:

$$\frac{\text{concentration ratio (X/S) in unknown}}{\text{concentration ratio in standard mixture}} = \frac{\text{area ratio (X/S) in unknown}}{\text{area ratio in standard mixture}}$$

(23-3)

EXAMPLE Using an Internal Standard

A mixture containing 52.4 nM iodoacetone (X) and 38.9 nM *p*-dichlorobenzene (S) gave the relative detector response (peak area of X)/(peak area of S) = 0.644. A solution containing an unknown quantity of X plus 742 nM S gave a relative detector response (peak area of X)/(peak area of S) = 1.093. Find the concentration of X in the unknown.

Solution

$$\frac{[X]/[S] \text{ in unknown}}{[X]/[S] \text{ in standard}} = \frac{\text{area ratio in unknown}}{\text{area ratio in standard}}$$

$$\frac{[X]/(742 \text{ nM})}{(52.4 \text{ nM}/38.9 \text{ nM})} = \frac{1.093}{0.644} \Rightarrow [X] = 1.70 \times 10^3 \text{ nM} = 1.70 \text{ } \mu\text{M}$$

. . . .

23-2 Classical Liquid Chromatography

Liquid chromatography is vital, because most compounds are not sufficiently volatile for gas chromatography. Modern chromatography evolved from the experiment in Figure 22-5, in which analyte and eluent are applied to the top of an open, gravity-fed column containing stationary phase. This technique is widely used for preparative chemistry and biochemistry, in which milligrams to grams of material are isolated. Open columns are used in analytical chemistry for sample cleanup and preconcentration (Section 26-3). For analytical purposes, closed high-pressure columns are best (Section 23-3).

Band broadening during chromatography is a natural consequence of diffusion and the finite rate of mass transfer between the mobile and stationary phases. Broadening from imperfect application of analyte to the column and mixing of zones during flow to the detector causes unnecessary loss of resolution. The column in Figure 23-14 minimizes broadening because the stationary phase is supported by a porous nylon net beneath which is very little dead space. An adjustable *flow adaptor* pressed tightly to the top of the solid phase prevents sample and solvent from mixing above the column. Inlet and outlet tubing have a 1-mm diameter to minimize mixing of liquids within the tubing. Each of these features decreases band spreading and increases resolution. Eluted solutes are usually detected by ultraviolet absorption as they pass through a flow cell (Figure 23-15).

The Stationary Phase

A common stationary phase is *silica gel* ($SiO_2 \cdot xH_2O$, also called silicic acid), obtained by acid precipitation of silicate solutions. The active adsorption sites are surface Si—O—H (silanol) groups, which can adsorb water from the air and are thereby slowly deactivated. Silica is activated by heating to 200°C to drive off water. Higher temperatures cause irreversible dehydration and loss of surface area. Silica is weakly acidic and interacts most strongly with basic solutes.[5] The surface can be chemically modified to alter its interaction with solutes.

Alumina ($Al_2O_3 \cdot xH_2O$) is the other most common adsorbent. Overnight heating to 400°C in air produces highly adsorbent *activity grade I* alumina. Equilibration with various quantities of water (for 1 day with occasional shaking) reduces the activity. Alumina is typically sold in acidic (pH 4), neutral (pH 7), or basic (pH 10) forms. Neutral alumina is used for most nonaqueous separations. It has basic sites that *strongly* adsorb acidic solutes. It also has Lewis acid sites that adsorb unsaturated compounds. Activated alumina may react with esters, anhydrides, aldehydes, and ketones and may catalyze elimination or isomerization reactions. Basic alumina has cation-exchange properties in aqueous solution, and acidic alumina is an anion exchanger.

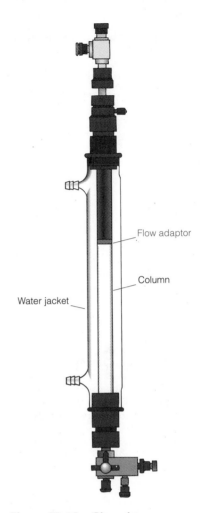

Flow adaptor

Column

Water jacket

Figure 23-14 Glass chromatography column with a flow adaptor at the top. The column is enclosed by a jacket through which water from a constant-temperature bath may be circulated to keep delicate biological samples refrigerated. [Courtesy Glenco Scientific, Houston, TX.]

Alumina grades:

I	anhydrous
II	3% H_2O
III	6% H_2O
IV	10% H_2O
V	15% H_2O

Cellulose is used as an adsorbent for compounds that are too polar to be eluted from silica gel or alumina. Other adsorbents include activated charcoal, Florisil (a coprecipitate of silica and magnesium oxide), and magnesia ($MgO \cdot xH_2O$). The latter is particularly useful for olefins and aromatic compounds.

Solvents

In adsorption chromatography, the solvent competes with the solute for adsorption sites on the stationary phase. The relative abilities of different solvents to elute a given solute from the column are nearly independent of the nature of the solute. Elution can be described as a displacement of solute from the adsorbent by solvent.

An *eluotropic series* ranks solvents by their relative abilities to displace solutes from a given adsorbent. The **eluent strength** ($\epsilon°$) is a measure of the solvent adsorption energy, with the value for pentane defined as zero. The more polar the solvent, the greater is its eluent strength. Table 23-6 applies to alumina, but a similar ranking is observed for silica gel. In general, the greater the eluent strength, the more rapidly will solutes be eluted from the column.

> Gradient elution in liquid chromatography is analogous to temperature programming in gas chromatography. Increased eluent strength is required to elute more strongly retained solutes.

A *gradient* of eluent strength is used for many separations. First, weakly retained solutes are eluted with a solvent of low eluent strength. Then a second solvent is mixed with the first, either in discrete steps or continuously, to increase eluent strength and elute more strongly adsorbed solutes. A small amount of polar solvent will markedly increase the eluent strength of a nonpolar solvent.

Choosing Conditions

We often use **thin-layer chromatography** to select conditions for a chromatographic separation. In this method, a tiny spot of sample is applied near the bottom of a sheet of glass or plastic coated with a thin layer of stationary phase. When the plate is placed in a shallow pool of solvent in a closed chamber, solvent ascends into the stationary phase by capillary action and performs a chromatographic separation (Color Plate 15). When solvent nears the top, the plate is withdrawn and dried. Spots can be made visible by placing the plate in a warm chamber containing a crystal of I_2. The I_2 vapor is reversibly adsorbed on most substances and creates a dark spot wherever a compound is located.

Alternatively, many thin-layer chromatography plates contain a fluorescent material whose emission is *quenched* (reduced) by most solutes. After solvent evaporates, view the plate under an ultraviolet lamp in the dark. Solute spots appear dark, whereas the rest of the plate is bright. With thin-layer chromatography, you can rapidly test different solvents and stationary phases to find a suitable combination for chromatography.[6]

A small column with a total volume of $1-2$ mL is an excellent way to screen different combinations of stationary phase and solvent, if a suitable method is available to detect components as they are eluted. You can run several such columns in a short time to select conditions for a larger-scale separation.

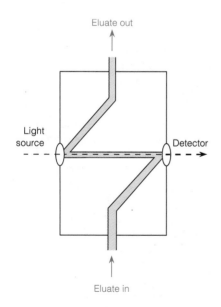

Figure 23-15 Light path in a micro flow cell of a spectrophotometric detector. Cells that have a 0.5-cm pathlength and contain only 10 μL of liquid are available.

TABLE 23-6 **Eluotropic series**

Solvent	$\epsilon°$ (for alumina)	Solvent	$\epsilon°$ (for alumina)
Fluoroalkanes	−0.25	Dichloromethane	0.42
n-Pentane	0.00	Tetrahydrofuran	0.45
i-Octane	0.01	1,2-Dichloroethane	0.49
n-Heptane	0.01	2-Butanone	0.51
n-Decane	0.04	Acetone	0.56
Cyclohexane	0.04	Dioxane	0.56
Cyclopentane	0.05	Ethyl acetate	0.58
Carbon disulfide	0.15	Methyl acetate	0.60
Carbon tetrachloride	0.18	1-Pentanol	0.61
1-Chloropentane	0.26	Dimethyl sulfoxide	0.62
i-Propyl ether	0.28	Aniline	0.62
i-Propyl chloride	0.29	Nitromethane	0.64
Toluene	0.29	Acetonitrile	0.65
1-Chloropropane	0.30	Pyridine	0.71
Chlorobenzene	0.30	2-Propanol	0.82
Benzene	0.32	Ethanol	0.88
Bromoethane	0.37	Methanol	0.95
Diethyl ether	0.38	1,2-Ethanediol	1.11
Chloroform	0.40	Acetic acid	Large

SOURCE: S. G. Perry, R. Amos, and P. I. Brewer, *Practical Liquid Chromatography* (New York: Plenum Press, 1972); and L. R. Snyder, *Principles of Adsorption Chromatography* (New York: Marcel Dekker, 1968).

23-3 High-Performance Liquid Chromatography

High-performance liquid chromatography (HPLC) uses high pressure to force eluent through a closed column packed with micron-size particles that provide exquisite separations of picograms to micrograms of analyte. The HPLC equipment in Figure 23-16 uses columns that are 5–30 cm in length,

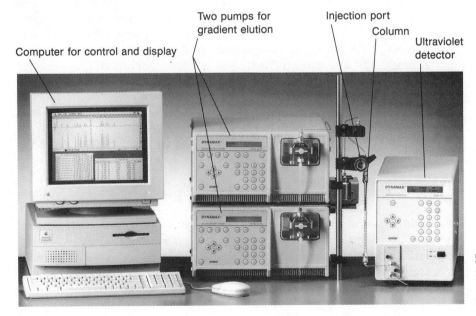

Computer for control and display

Two pumps for gradient elution

Injection port

Column

Ultraviolet detector

Figure 23-16 Equipment for high-performance liquid chromatography (HPLC). [Courtesy Rainin Instrument Co., Emeryville, CA.]

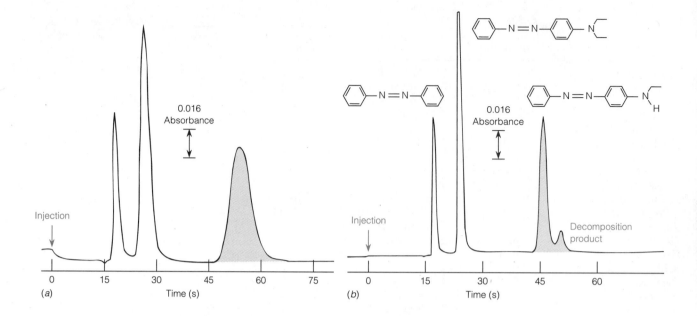

(a) Time (s)   (b) Time (s)

Injection — 0.016 Absorbance

Figure 23-17 Chromatograms of the same sample run on columns packed with (*a*) 10-μm and (*b*) 5-μm particle diameter silica. [From R. E. Majors, *J. Chromatogr. Sci.* **1973**, *11*, 88.]

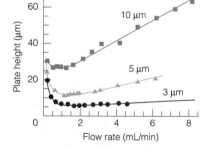

Figure 23-18 Plate height as a function of flow rate for stationary-phase particle sizes of 10, 5, and 3 μm. [Courtesy Perkin-Elmer Corp., Norwalk, CT.]

with an inner diameter of 1–5 mm. Common stationary phases provide 50 000 to 100 000 plates per meter. Essential components include a solvent delivery system, a sample injection valve, a detector, and a recorder or computer to display results.

If solute can diffuse rapidly between the mobile and stationary phases, then plate height is decreased and resolution increases. In capillary gas chromatography, we promote rapid mass transfer between the two phases by reducing the diameter of the column so that molecules can diffuse quickly from the center to the wall and by decreasing the thickness of the stationary phase on the wall. In liquid chromatography, we increase the rate of mass transfer by reducing the dimensions of the stationary-phase particles, thereby reducing the distance through which solute must diffuse in both phases. Smaller particles also make migration paths more uniform, which decreases the *A* term of the van Deemter equation (22-33).[7]

Figure 23-17 illustrates the increased resolution afforded by decreasing the stationary-phase particle size from 10 to 5 μm. Notice how much sharper the peaks become and how a decomposition product is resolved from the slow-moving component. Figure 23-18 shows that smaller particle size reduces plate height, even at higher flow rates. The penalty for using very fine particles is resistance to solvent flow. Pressures of ~7–40 MPa (70–400 atm) are routinely required to attain flow rates of ~0.5–5 mL/min.

The Column

The HPLC column in Figure 23-19 uses porous titanium frits to provide uniform flow at the junction between narrow tubing and the relatively wide column. HPLC columns are expensive and easily degraded by irreversible adsorption of impurities from samples and solvents. Therefore, the entrance to the main column is protected by a **guard column** that is 1 cm long and contains the same stationary phase as the main column. The guard column collects irreversibly adsorbed solutes and is periodically replaced. Figure 23-20 shows a column with a piston that you can tighten as needed to remove the void space that forms as the particles pack together. The void is a dead space that is extremely detrimental to resolution, because incoming solvent mixes with sample in the void.

A chromatography column may be heated to lower the viscosity of the solvent. Reducing solvent viscosity reduces the required pressure or permits faster flow. Increased temperature decreases retention volumes and improves resolution by speeding the diffusion of solutes. However, increased temperature may degrade the stationary phase and decrease column lifetime.

Stationary phases for liquid-solid adsorption, liquid-liquid partition, ion-exchange, molecular exclusion, and affinity chromatography are available for HPLC. A common support is **microporous particles** of silica with diameters of 3–10 μm. Microporous particles are permeable to solvent and have a surface area up to 500 m^2 per gram of silica. Compared with the irregularly shaped particles in Figure 23-21a, more expensive spherical particles give 10–20% smaller plate height, produce more symmetric peaks, allow about 10% less volume of the mobile phase on the column (called the *void volume*), pack in a more stable manner, and require 10–30% less pressure for a given flow rate. Polystyrene (Figure 23-21b–d, subunit structure in Figure 24-1) is a less rigid support than silica and has favorable pore sizes for molecular exclusion separations. Polymer stationary phases have a smaller operating pressure range and swell to different extents in different solvents.

Adsorption chromatography occurs directly on the surface of silica particles. More commonly, liquid-liquid partition chromatography is con-

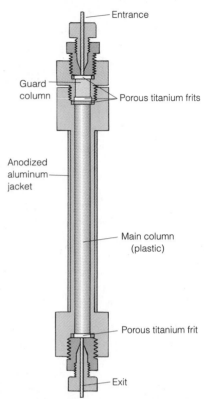

Figure 23-19 HPLC column with replaceable guard column to collect irreversibly adsorbed impurities. Titanium frits distribute the liquid evenly over the diameter of the column. [Courtesy Upchurch Scientific, Oak Harbor, WA.]

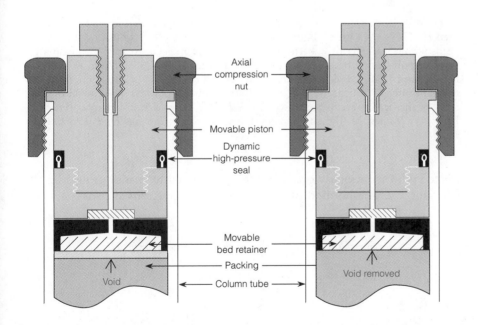

Figure 23-20 Adjustable end fitting can be tightened to remove the void that forms when the stationary phase packs together during routine use. [Courtesy Rainin Instrument Co., Emeryville, CA.]

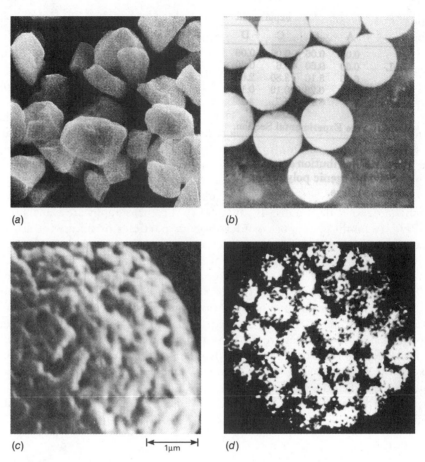

Figure 23-21 (*a*) Microporous silica particles with a mean diameter of 10 μm; used for HPLC. [Courtesy Phenomenex, Torrance, CA.] (*b*) Porous polystyrene beads with a diameter of 7.4 μm; used for molecular exclusion HPLC. (*c*) Enlargement of polystyrene bead, showing porous surface. [From Q. C. Wang, K. Hosoya, F. Svec, and J. M. J. Fréchet, *Anal. Chem.* **1992,** *64,* 1232.] (*d*) Magnetic resonance image of benzene inside a 500-μm-diameter polystyrene bead shows that most of the bead is accessible to solvent. White areas are rich in benzene. [From E. Bayer, W. Müller, M. Ilg, and K. Albert, *Angew. Chem. Int. Ed. Engl.* **1989,** *28,* 1029.]

ducted with a stationary phase covalently attached to silanol groups on the silica surface. The **bonded phase** is prepared by reactions such as

Residual silanol groups on the silica surface are capped with trimethylsilyl groups by reaction with ClSi(CH$_3$)$_3$ to eliminate polar adsorption sites that promote peak tailing.

Common polar phases		Common nonpolar phases	
$R = (CH_2)_3NH_2$	Amino	$R = (CH_2)_{17}CH_3$	Octadecyl
$R = (CH_2)_3C{\equiv}N$	Cyano	$R = (CH_2)_7CH_3$	Octyl
$R = (CH_2)_2OCH_2CH(OH)CH_2OH$	Diol	$R = (CH_2)_3C_6H_5$	Phenyl

There are ~4 μmol of R groups per square meter of support surface area, with very little bleeding of the stationary phase from the column during chromatography. The siloxane (Si—O—Si) bond is only stable over the pH range 2–8. For chromatography of basic compounds at pH 8–12, we can use polystyrene beads with R groups covalently attached by C—C bonds to benzene rings.[8]

Normal-phase chromatography refers to the use of a polar stationary phase and a less polar solvent. Eluent strength is increased by adding a more polar solvent. **Reverse-phase chromatography** is the more common scheme in which the stationary phase is nonpolar or weakly polar and the solvent is more polar. Eluent strength is increased by addition of *less polar* solvent. Reverse-phase chromatography provides excellent separations and eliminates peak tailing associated with adsorption of polar compounds by polar packings. Reverse-phase chromatography is also less sensitive to polar impurities (such as water) in the eluent.

Selecting the Separation Mode

There are usually many ways to separate the components of a given mixture. Figure 23-22 shows a decision tree for choosing a starting point. If the molecular weight of the analyte is below 2 000, we use the upper part of the figure; if the molecular weight is greater than 2 000, we use the lower part. In either part, the first question is whether the solutes dissolve in water or in organic solvents. Suppose we have a mixture of small molecules (molecular weight <2 000) soluble in dichloromethane. Consulting the list of solvents in Table 23-6, we see that nonpolar and weakly polar solvents are in the left half of the table, and moderately polar to polar solvents are in the right half. The eluent strength of dichloromethane (0.42) is closer to that of $CHCl_3$ (0.40) than it is to those of alcohols, acetonitrile, or ethyl acetate (>0.58). Therefore, Figure 23-22 suggests that we try adsorption chromatography, for which microporous silica is by far the most common HPLC stationary phase.

If solutes only dissolve in hydrocarbons, the decision tree suggests that we try bonded reverse-phase chromatography. Our choices include bonded phases containing octadecyl (n-$C_{18}H_{37}$), octyl, butyl, ethyl, methyl, phenyl, and cyano groups. The most common reverse-phase supports contain octadecyl groups.

If the molecular weights of solutes are >2 000 and if they are soluble in organic solvents and their molecular diameter is >30 nm, Figure 23-22 tells us to try molecular exclusion chromatography. Stationary phases for this type of separation are described in the next chapter. If the molecular weight of solutes is >2 000, and they are soluble in water, but not ionic, and have diameters <30 nm, the decision tree says to use bonded reverse-phase chromatography (such as C_{18}-silica), or *hydrophobic interaction chromatography*.

Hydrophobic interaction chromatography is based on the interaction of a hydrophobic stationary phase with a hydrophobic solute. (**Hydrophobic**

The octadecyl (C_{18}) stationary phase is, by far, the most common in HPLC.

Box 23-2 describes a stationary phase that resembles a cell membrane.

There are no hard and fast rules in Figure 23-22. Methods in either part of the diagram may work perfectly well for molecules whose size belongs to the other part.

Glass is an example of a hydrophilic substance that is not soluble in water, but whose surface attracts (is *wetted* by) water. Teflon is an insoluble material whose hydrophobic surface is not wetted by water.

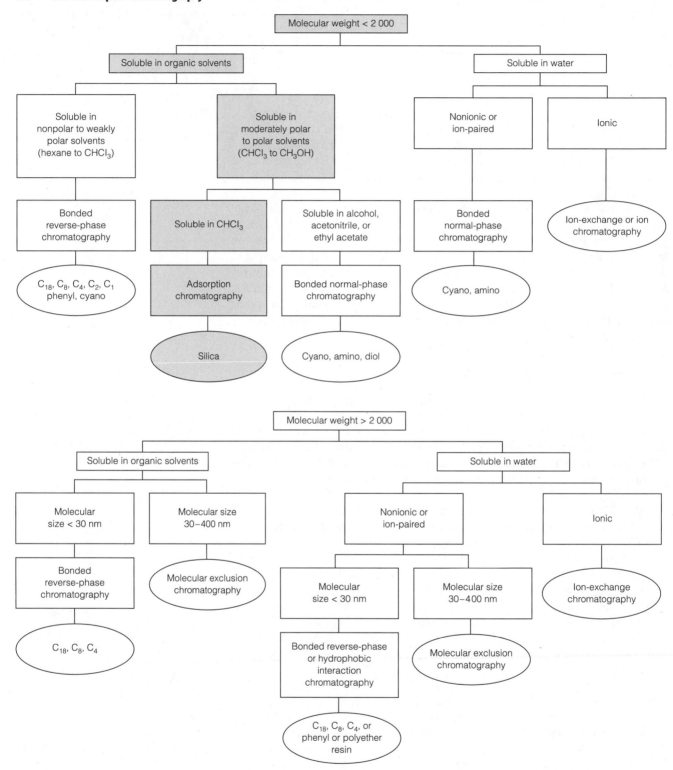

Figure 23-22   Guide to HPLC mode selection.

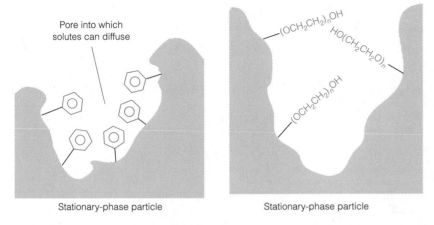

Figure 23-23 Two stationary phases for hydrophobic interaction chromatography.

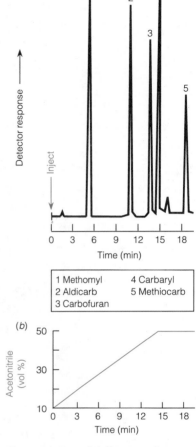

1 Methomyl	4 Carbaryl
2 Aldicarb	5 Methiocarb
3 Carbofuran	

Figure 23-24 (*a*) Reverse-phase gradient separation of carbamate insecticides using a flow rate of 1.6 mL/min in a 4.6-mm-diameter × 15-cm-long column containing octadecyl groups on spherical 5-μm-diameter microporous particles. (*b*) Solvent composition was changed linearly from 10 vol % CH_3CN in water to 50% CH_3CN in water during the first 15 min. [Courtesy Anspec Co., Ann Arbor, MI.]

substances repel water and their surfaces are not wetted by water. **Hydrophilic substances** are soluble in water or attract water to their surface.) Two hydrophobic stationary phases are shown in Figure 23-23. The bulk particle is polystyrene (Figure 24-1) with pores approximately 100 nm in diameter, through which most molecules can diffuse. The surface is coated with phenyl groups or with poly(ethylene glycol), both of which attract hydrophobic solutes and provide a means of chromatographic separation.

Solvents

To avoid degrading expensive columns and to minimize background ultraviolet absorbance, solvents for HPLC must be very pure. Intake tubing in the solvent reservoir is protected by a fine filter that rejects dust particles larger than 2–5 μm. Sample and solvent are passed through a short, expendable *guard column* (Figure 23-19) containing the same stationary phase as the analytical column. Even with a guard column, routine washing with solvents is recommended to prolong the life of HPLC columns.[9] Solvents are degassed by evacuation, boiling, or purging with He, because gas bubbles interfere with pumps, columns, and detectors.

Isocratic elution is performed with a single solvent (or homogeneous solvent mixture). If one solvent does not discriminate adequately between several components of a mixture or if the solvent does not provide sufficiently rapid elution of all components, then **gradient elution** can be used. In this case, increasing amounts of solvent B are added to solvent A to create a continuous gradient. Figure 23-24 shows an example in which increasing the fraction of acetonitrile in a water-acetonitrile mixture elutes solutes that were initially strongly retained by a C_{18}-silica column.

Pumps and Injection Valves

The quality of a pump is measured by how steady and reproducible a flow it can produce. A fluctuating flow rate creates detector noise that degrades weak signals.

Elution can be

isocratic: one solvent
gradient: continuous change of solvent composition to increase eluent strength
stepwise: discontinuous change to increase eluent strength

Box 23-2 **A Stationary Phase That Mimics a Cell Membrane**

Animal cells are enclosed by a flexible membrane that limits contact between inside and outside, and in which are embedded numerous proteins, such as transport and receptor molecules. A typical membrane component is a *phospholipid,* which has a highly polar headgroup and a long hydrocarbon tail. Such molecules assemble into *bilayers,* in which the tails form a central nonpolar layer and the headgroups face the aqueous environment.

Membrane proteins consist of nonpolar regions that lodge in the central hydrocarbon portion of the membrane and polar regions projecting out of the membrane. Biochemists use *detergents* (also called *surfactants*) to extract and dissolve membrane proteins, because detergents have water-soluble ionic headgroups attached to long hydrocarbon tails.

The bonded chromatographic stationary phase shown on the next page resembles half of a lipid bilayer.[10] The membrane protein cytochrome P-450 can be purified from a crude extract by a single pass through a column containing this stationary phase. After proteins that are soluble in dilute detergent solution are eluted, more concentrated detergent is used to elute cytochrome P-450. The bonded stationary phase is also used to predict how well drugs will be transported across biological membranes.

Figure 23-25 shows a pump with two sapphire pistons that produce a programmable, constant flow rate up to 10 mL/min at pressures up to 40 MPa (400 atm). Solvent at the left passes through an electronic inlet valve synchronized with the large piston and designed to minimize the formation of solvent vapor bubbles during the intake stroke. The spring-loaded outlet valve maintains a constant outlet pressure, and the damper further reduces pressure surges. Pressure surges from the first piston are decreased in the damper that "breathes" against a constant outside pressure. Pressure pulses are typically <1% of the operating pressure. As the large piston draws in liquid, the small piston propels liquid into the chromatograph. During the return stroke of the small piston, the large piston delivers solvent into the expanding chamber of the small piston. Part of the solvent fills the chamber, while the remainder flows into the chromatograph. Delivery rate is controlled by the stroke volumes.

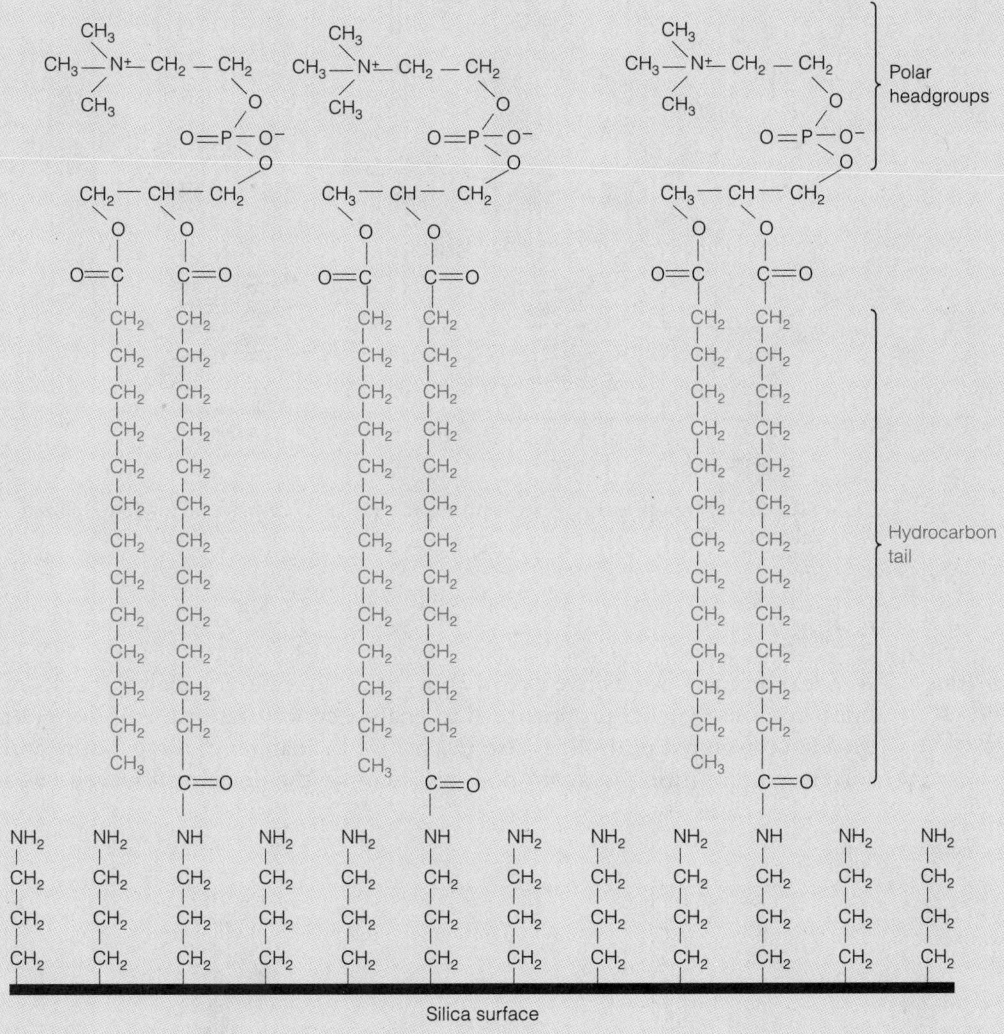

Polar headgroups

Hydrocarbon tail

Silica surface

Gradients made from up to four solvents are constructed by proportioning the liquids through a four-way valve at low pressure and then pumping the mixture at high pressure into the column. The gradient-making process is electronically controlled and programmable in 0.1 vol % increments.

The *injection valve* in Figure 23-26 has interchangeable steel sample loops, each of which holds a fixed volume. Loops of different sizes hold volumes that range from 2 to 1 000 μL. In the load position, a syringe is used to wash and load the loop with fresh sample at atmospheric pressure. High-pressure flow from the pump to the column passes through the segment of valve at the lower left. When the valve is rotated 60° counterclockwise, the content of the sample loop is injected into the column at high pressure.

Samples should be passed through a 0.5- to 2-μm filter prior to injection, to avoid contaminating the column with particles, plugging the tubing, and damaging the pump.

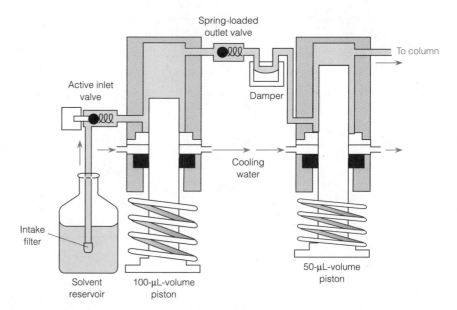

Figure 23-25 High-pressure piston pump for HPLC. [Courtesy Hewlett-Packard Co., Palo Alto, CA.]

Detectors

Linear range: analyte concentration range over which detector response is proportional to concentration

Dynamic range: range over which detector responds in any manner (not necessarily linearly) to changes in analyte concentration

Detection limit: concentration of analyte that gives a signal-to-noise ratio of 2 (Figure 21-21)

An ideal detector is sensitive to low concentrations of every analyte, provides linear response (signal proportional to analyte concentration), and does not broaden the eluted peaks. It is also insensitive to changes in temperature and solvent composition. To avoid peak broadening, the detector flow cell has a

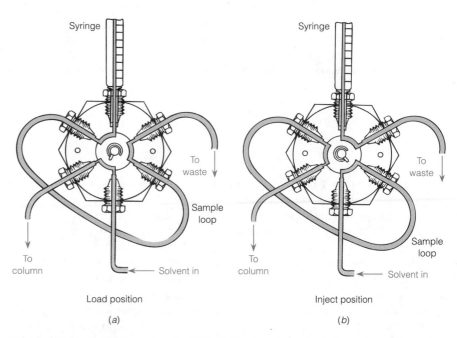

Figure 23-26 Injection valve for HPLC. Replaceable sample loop comes in various fixed-volume sizes.

volume that is <20% of the volume of the chromatographic band. Cell volumes of 1–20 μL are common. Because gas bubbles create noise, back pressure may be applied to the detector to prevent bubble formation during depressurization of eluate.

The **ultraviolet detector** is most common, because many solutes absorb ultraviolet light. The flow cell in Figure 23-15 has a volume of 10 μL for a pathlength of 1 cm. Simple systems employ the intense 254-nm emission of a mercury lamp. More versatile instruments use a deuterium or xenon lamp and monochromator, so you can choose the optimum wavelength for your analytes. High-quality detectors provide full-scale absorbance ranges from 0.000 5 to 3 absorbance units. On the most sensitive scale, an absorbance of 0.000 5 would give a 100% signal, with a noise level near 1% of full scale. The linear range extends over five orders of magnitude of solute concentration (which is another way of saying that Beer's law is obeyed over this range). Ultraviolet detectors are good for gradient elution with nonabsorbing solvents. The system in Figure 23-27 uses a photodiode array to record the entire spectrum of each solute as it passes through the detector.

The **refractive index detector** responds to almost every solute, but its detection limit is about 1 000 times poorer than that of the ultraviolet

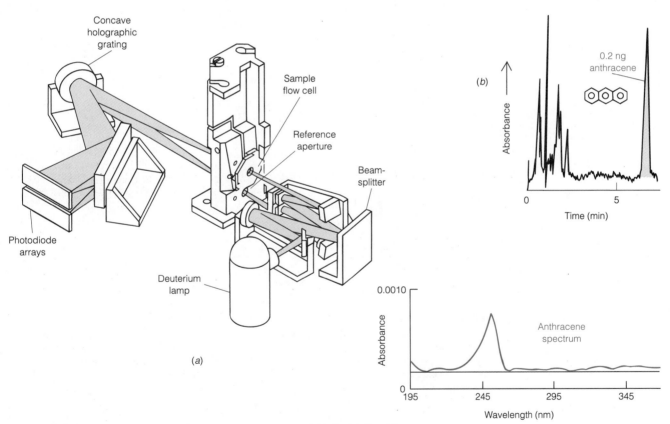

Figure 23-27 Photodiode array ultraviolet detector for HPLC. (*a*) Dual-beam optical system uses grating polychromator, one diode array for the sample spectrum, and another diode array for the reference spectrum. (*b*) Reverse-phase chromatography (using C_{18}-silica) of sample containing 0.2 ng of anthracene, with detection at 250 nm. Full-scale absorbance is 0.001. (*c*) Spectrum of anthracene recorded as it emerged from the column. [Courtesy Perkin-Elmer Corp., Norwalk, CT.] Photodiode arrays are described in Section 20-6.

Box 23-3 Supercritical Fluid Chromatography

This technique fills a gap between gas and liquid chromatography, because the solvent properties are between those of gas and liquid. Consider the phase diagram of carbon dioxide presented here. At a temperature of $-78°C$, solid CO_2 (Dry Ice) is in equilibrium with gaseous CO_2 at 1 atm. The solid *sublimes* without turning to liquid. At any temperature above the *triple point* at $-56.6°C$, there is a pressure at which liquid and vapor coexist as separate phases. For example, at $0°C$, liquid is in equilibrium with gas at 34.4 atm. Moving up the liquid-gas boundary, two phases always exist until the *critical point* is reached at $31.3°C$ and 72.9 atm. *Above this temperature, only one phase exists, no matter what the pressure*. We call this phase a **supercritical fluid.** Its density and viscosity are between those of the gas and liquid, as is its ability to act as a solvent.

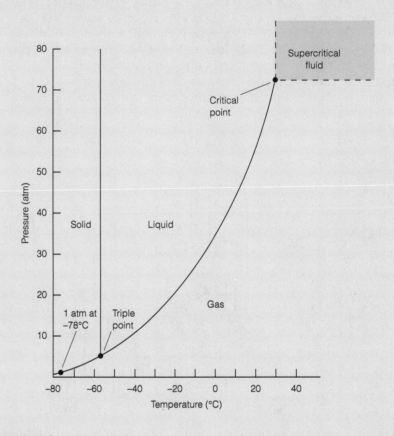

Critical constants†

Compound	Critical temperature (C°)	Critical pressure (atm)	Critical density (g/mL)
Carbon dioxide	31.3	72.9	0.448
Ammonia	132.3	111.3	0.24
Water	374.4	226.8	0.344
Methanol	240.5	78.9	0.272
Diethyl ether	193.6	36.3	0.267

. . . .

†Note that supercritical N_2O presents an explosion hazard in the presence of organic matter [D. E. Raynie, *Anal. Chem.* **1993,** *65,* 3127].

Supercritical fluid chromatography provides increased speed and resolution, relative to liquid chromatography, because of the increased diffusion coefficients of solutes in supercritical fluids. (However, speed and resolution are lower than those of gas chromatography.) Unlike gases, supercritical fluids can dissolve nonvolatile solutes. When the pressure on the supercritical solution is released, the solvent turns to gas, leaving the solute in the gas phase for easy detection. Carbon dioxide has been the supercritical fluid of choice for chromatography because it is compatible with the versatile flame ionization detector of gas chromatography, it has a low critical temperature, and it is nontoxic. Unfortunately, it is not a particularly good solvent for highly polar or high molecular weight solutes.

Equipment for supercritical fluid chromatography is similar to that for HPLC, with open tubular columns similar to those of gas chromatography. A bonded stationary phase is required, so that it does not dissolve in the solvent. Most detectors compatible with HPLC or gas chromatography are useful, but flame ionization and ultraviolet absorption are most common. Eluent strength is increased in HPLC by gradient elution and in gas chromatography by raising the temperature. In supercritical fluid chromatography, eluent strength is increased by making the solvent *denser* (by increasing the pressure). The chromatogram below illustrates density gradient elution.

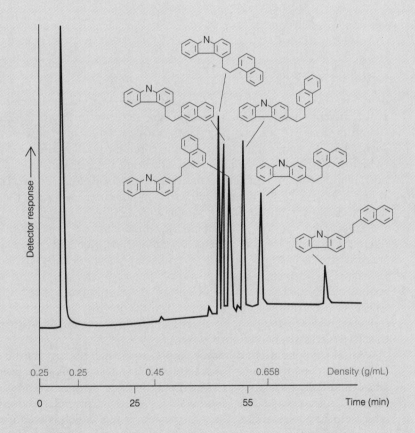

Capillary supercritical chromatogram of aromatic compounds with CO_2 using density gradient elution at 140°C. [From R. D. Smith, B. W. Wright, and C. R. Yonker, *Anal. Chem.* **1988,** *60,* 1323A.]

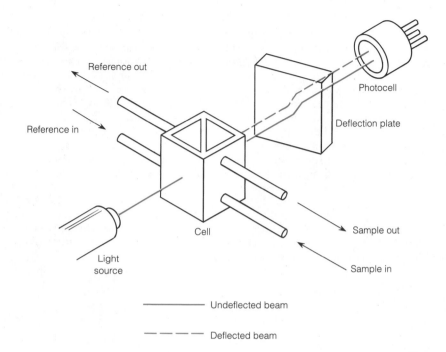

Reference out

Reference in

Photocell

Deflection plate

Cell

Sample out

Sample in

Light source

———— Undeflected beam

– – – – Deflected beam

Figure 23-28   Deflection-type refractive index detector.

The relation between angle of refraction and refractive index is described in Section 20-1.

detector, and it is not useful for gradient elution because the baseline changes as the solvent changes. The deflection-type detector in Figure 23-28 has two triangular 5- to 10-μL compartments through which pure solvent or eluate passes. Collimated (parallel) visible light filtered to remove infrared radiation (which would heat the sample) passes through the cell with pure solvent in both compartments and is directed to the photocell by the deflection plate. When solute with a different refractive index enters the cell, the beam is deflected and the photocell output changes.

In gradient elution, it is impossible to match exactly the sample and the reference, so refractive index detectors are useless. Refractive index detectors are sensitive to changes in pressure and temperature (~0.01°C). Because of their low sensitivity, refractive index detectors are not useful for trace analysis. They also have a small linear range, spanning only a factor of 500 in solute concentration. The primary appeal of this detector is its nearly universal response to all solutes, such as carbohydrates and aliphatic polymers that have little ultraviolet absorption.

An **electrochemical detector** responds to analytes that can be oxidized or reduced, such as phenols, aromatic amines, peroxides, mercaptans, ketones, aldehydes, conjugated nitriles, aromatic halogen compounds, and aromatic nitro compounds. Figure 23-29 shows a three-electrode system in which eluate passes over a glassy carbon electrode, which is commonly used for oxidizable solutes. The potential is maintained at a selected value with respect to the silver-silver chloride reference electrode, and current is measured between the glassy carbon working electrode and the stainless steel auxiliary electrode. For reducible solutes, a drop of mercury is a good working electrode. Current is proportional to solute concentration over six orders of magnitude. Aqueous or other polar solvents containing dissolved electrolytes are required, and they must be rigorously free of oxygen. Metal

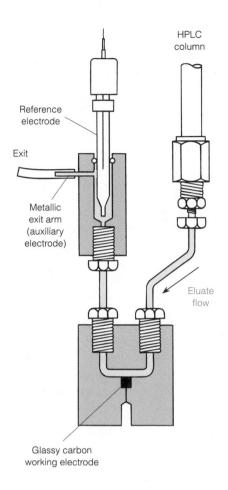

Figure 23-29 Cross-sectional view of an electrochemical detector. A thin layer of eluate passes over the top surface of the working electrode. The auxiliary electrode is the hollow metal sidearm at the exit port on the left. [Courtesy Bioanalytical Systems, West Lafayette, IN.]

ions extracted from tubing may be masked by adding EDTA to the solvent. The detector is sensitive to flow rate and temperature changes.

Pulsed electrochemical measurements at Au or Pt working electrodes expand the classes of detectable compounds to include alcohols, carbohydrates, and sulfur compounds. The electrode is held at +0.8 V (versus S.C.E.) for 120 ms to oxidatively desorb organic compounds from the electrode surface and to oxidize the metal surface. Then the electrode is brought to −0.6 V for 200 ms to reduce the oxide to pristine metal. The electrode is held at a constant working potential (typically in the range +0.4 to −0.4 V), at which analyte is oxidized or reduced. After waiting 400 ms for the condenser current (Figure 18-10) to decay to zero, current is integrated for the next 200 ms to measure analyte. Ethylene glycol ($HOCH_2CH_2OH$) in Figure 23-30 gives a signal-to-noise ratio of 3 at a concentration of 10 ppb.

Fluorescence detectors are especially sensitive but respond only to the few analytes that fluoresce. To increase the utility of fluorescence and electrochemical detectors, fluorescent or electroactive groups can be chemically attached to the desired analytes.[11] Such **derivatization** can be performed on the mixture prior to chromatography, or by addition of reagents to the eluate between the column and the detector (called *post-column derivatization*). Other detectors for ionic species are based on *ion-selective electrodes* and *conductivity measurement. Inductively coupled plasma atomic emission, mass*

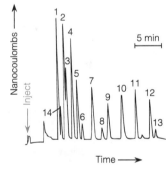

Figure 23-30 Pulsed electrochemical detection of alcohols separated on Dionex AS-1 anion-exchange column with 0.05 M $HClO_4$. Peaks: 1, glycerol; 2, ethylene glycol; 3, propylene glycol; 4, methanol; 5, ethanol; 6, 2-propanol; 7, 1-propanol; 8, 2-butanol; 9, 2-methyl-1-propanol; 10, 1-butanol; 11, 3-methyl-1-butanol; 12, 1-pentanol; 13, cyclohexanol; 14, diethylene glycol. [From D. C. Johnson and W. R. LaCourse, *Anal. Chem.* **1990,** *62,* 589A.]

TABLE 23-7   Comparison of commercial HPLC detectors

Detector	Approximate limit of detection[a] (ng)	Useful with gradient?
Ultraviolet	0.1–1	Yes
Refractive index	100–1 000	No
Electrochemical	0.01–1	No
Fluorescence	0.001–0.01	Yes
Conductivity	0.5–1	No
Mass spectrometry	0.1–1	Yes
Fourier transform infrared	1 000	Yes

a. Detection limits from E. W. Yeung and R. E. Synovec, *Anal. Chem.* **1986,** *58,* 1237A.

spectrometry, and *Fourier transform infrared spectroscopy* have a variety of HPLC detector applications. The *flame ionization detector* of gas chromatography (Section 23-1) can be adapted to HPLC by depositing eluate on a moving wire that goes into the flame. Solvent evaporates as the wire moves, allowing only nonvolatile solutes to reach the detector. *Low-angle laser light scattering* detects high molecular weight polymers in molecular exclusion chromatography and characterizes molecular size distribution. Table 23-7 compares various common detectors.

23-4 Practical Notes for Classical Liquid Chromatography

There is an art to pouring uniform open columns, applying samples evenly, and obtaining symmetric elution bands. This section describes some practical aspects of chromatography.

Preparing the Stationary Phase

A *slurry* is a suspension of a solid in a liquid.

Add gel *to* solvent. *Never* use magnetic stirring.

The stationary phase must be equilibrated with solvent before pouring the column. For alumina and silica gel, make a slurry of the solid in the appropriate solvent. For ion-exchange and molecular exclusion gels, follow the manufacturer's instructions concerning time, temperature, and solution composition to hydrate the gel. When mixing dry gel with solvent, sprinkle the gel on top of the liquid and allow it to settle while gently stirring with a glass rod. *Never* use magnetic stirring, which breaks and destroys gel particles.

Hydrated gel should be suspended in 2–5 volumes of solvent and allowed to stand until all but the smaller particles (called *fines*) have settled. Remove the fines by decanting or suction and repeat the process several times to remove more fines. If not discarded, the fines reduce the column flow rate. Many gels are sold in a pre-equilibrated form from which the fines have already been removed.

Pouring the Column

Next, suspend the stationary phase in enough solvent to allow the solid to occupy one-half to two-thirds of the total volume after it settles. Place a few centimeters of liquid in the chromatography column and pour the suspension gently onto a glass rod leading to the wall of the column (Figure 23-31). If most of the column is to be filled with stationary phase, additional volume must be available to hold the slurry. Either a wide-stem funnel held to the top of the column by a rubber stopper or a reservoir that screws onto the top of the column is adequate. Avoid the creation of a discontinuity that arises from pouring part of a column, allowing it to settle, and then pouring the rest of the column.

When the entire slurry has been added to the column, allow it to stand until a few centimeters have settled. Then begin a slow flow of solvent to pack the remainder. Never exceed the hydrostatic pressure limit (discussed below) set by the manufacturer. Do not allow liquid to drain below the top of the gel, because this creates air spaces and irregular flow patterns. Solvent should be directed gently down the wall of the column or on top of a few centimeters of liquid above the packing. In *no* case should the solvent be allowed to dig a channel into the gel.

Applying the Sample

To apply sample, drain the solvent to the top of the gel and halt the flow. Apply the sample gently down the wall of the column (evenly around the whole column) by pipet. Resume flow, drain the sample into the gel, and halt the flow again. Then add *small* quantities of solvent down the walls by pipet to finish washing the sample into the gel. Finally, build up a layer of solvent above the gel by pipet.

Running the Column

Maintain flow by siphoning solvent from a reservoir into the column. The proper flow rate depends on the column diameter and the degree of separation between solutes. Maximum resolution demands a slow flow rate. The manufacturer's bulletin for the particular stationary phase should be consulted to select a flow rate.

Flow rate is governed by hydrostatic pressure, expressed as the distance between the level of liquid in the reservoir and the level of the column outlet (Figure 23-32a). If a *Mariotte flask* is used as a reservoir, the pressure is measured between the bottom of the air inlet and the level of the column outlet (Figure 23-32b). Ordinarily, the hydrostatic pressure decreases as liquid is drained from the reservoir. A Mariotte flask maintains constant pressure because the pressure at the bottom of the air inlet tube must be equal to atmospheric pressure.

Tubing is a potential source of contamination because plasticizers used in tubing are leached out by some solvents. When using organic solvents, avoid common tubing such as Tygon® and use fluoroplastic or glass tubing instead.

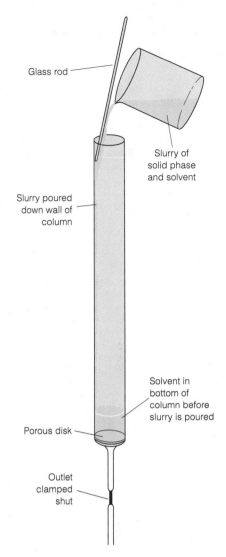

Glass rod

Slurry of solid phase and solvent

Slurry poured down wall of column

Solvent in bottom of column before slurry is poured

Porous disk

Outlet clamped shut

Figure 23-31 A slurry of the stationary phase in starting solvent should be poured gently down the wall of the column.

Pressure is measured from the inlet level to the outlet level.

A Mariotte flask maintains constant pressure and therefore constant flow rate.

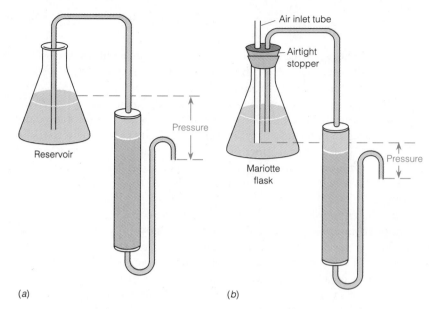

(a)   (b)

Figure 23-32   Measurement of the hydrostatic pressure with (*a*) an ordinary solvent reservoir and (*b*) a Mariotte reservoir.

Gradients

In ion-exchange chromatography (Chapter 24), a gradient of ionic strength may be needed to elute strongly retained solutes. In the gradient maker in Figure 23-33, a solution of low ionic strength is placed in the right reservoir and an equal volume at high ionic strength is placed in the left reservoir. The first drop of eluent leaving the right side has low ionic strength. As liquid from the left side flows to the right to keep the levels equal, the ionic strength

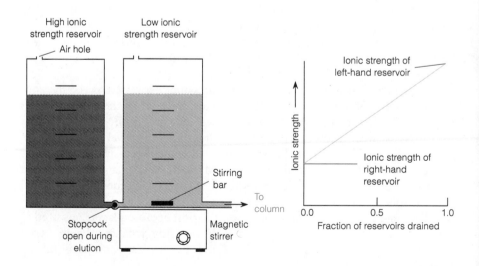

Figure 23-33   A device for gradient elution.

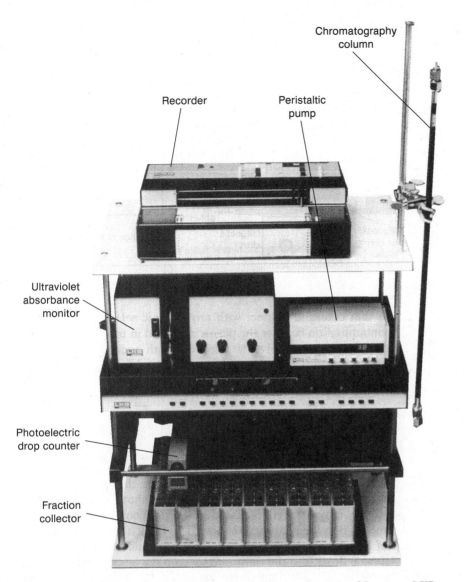

Chromatography column

Recorder

Peristaltic pump

Ultraviolet absorbance monitor

Photoelectric drop counter

Fraction collector

Figure 23-34 Automated apparatus for liquid chromatography. [Courtesy LKB-Produkter AB, Pleasant Hill, CA.]

in the right compartment increases. The last drop of eluent has the ionic strength of the left-hand reservoir.

Accessories

Figure 23-34 shows a complete setup for liquid chromatography. Eluent is fed to the column at the desired flow rate by a peristaltic pump, so regulation of hydrostatic pressure is not necessary. Eluate passes through a spectrophotometric flow cell on its way to a fraction collector, which automatically changes test tubes when a preset volume has been collected. Absorbance is displayed on a recorder, which also marks when the fraction collector changes tubes. All chromatographic devices require loving attention, but the automatic features shown in Figure 23-34 allow you to go home and sleep fitfully as the column runs through the night.

Terms to Understand

bonded phase	high-performance liquid chromatography	packed column
co-chromatography	hydrophilic substance	refractive index detector
derivatization	hydrophobic substance	retention index
electrochemical detector	internal standard	reverse-phase chromatography
eluent strength	isocratic elution	septum
flame ionization detector	microporous particles	supercritical fluid
gas chromatography	molecular sieve	thermal conductivity detector
gradient elution	normal-phase chromatography	thin-layer chromatography
guard column	open tubular column	ultraviolet detector

Summary

In gas chromatography, a volatile liquid or gaseous solute is carried by a gaseous mobile phase over a stationary liquid coated on the inside of an open tubular column or on a solid support. Long, narrow open tubular columns have low capacity but give excellent separation. Packed columns provide high capacity but poor resolution. Fused silica open tubular columns may be wall-coated, support-coated, or porous-layer. Each liquid stationary phase most strongly retains solutes in its own polarity class. Solid stationary phases include porous carbon, alumina, and molecular sieves. Temperature programming reduces elution times of highly retained components. Without compromising separation efficiency, the linear flow rate may be increased when H_2 or He, instead of N_2, is used as carrier gas. Split injection is routine, whereas splitless injection is better for quantitative analysis. On-column injection is best for thermally unstable solutes. Thermal conductivity detection has universal response, but flame ionization is more sensitive. Electron capture, flame photometry, alkali flame detectors, sulfur chemiluminescence, mass spectrometry, Fourier transform infrared spectroscopy, and atomic emission are also used for detection. A compound can be identified by its retention time on different columns and quantified by the area of its elution peak. Response factors relate the area to the concentration of different compounds.

Classical liquid chromatography utilizes large, packed columns with gravity-fed (or very low pressure, pumped) mobile phase. Common stationary phases for adsorption chromatography are silica gel and alumina. The eluent strength describes the ability of a given solvent to elute solutes from the column. Thin-layer chromatography and small-scale columns are used to select conditions for preparative separations by large columns.

In high-performance liquid chromatography (HPLC), solvent is pumped at high pressure through a small column containing particles of stationary phase 3- to 10-μm in diameter. The smaller the particle size, the more efficient the column, but the greater the resistance to flow. Spherical and irregular microporous particles with an adsorptive surface or a bonded liquid phase are most common. In normal-phase chromatography, the stationary phase is polar and a less polar solvent is used. The more common reverse-phase chromatography employs a nonpolar stationary phase and polar solvent. The choice of separation procedure is based on the size, polarity, and ionic nature of the solute. Isocratic or gradient elution may be used. Injection valves allow rapid, precise sample introduction, preferably through a guard column. Ultraviolet detection is most common; refractive index is more universal but less sensitive. Electrochemical and fluorescence detectors are extremely sensitive, but selective. In supercritical fluid chromatography, nonvolatile solutes are separated by a process whose efficiency, speed, and detectors more closely resemble those of gas chromatography than of liquid chromatography.

Exercises

A. When 1.06 mmol of 1-pentanol and 1.53 mmol of 1-hexanol were separated by gas chromatography, they gave relative peak areas of 922 and 1570 units, respectively. When 0.57 mmol of pentanol was added to an unknown containing hexanol, the relative chromatographic peak areas were 843:816 (pentanol:hexanol). How much hexanol did the unknown contain?

B. A compound was known to be a member of the family $(CH_3)_2CH(CH_2)_nCH_2OSi(CH_3)_3$.

(a) From the gas chromatographic retention times below, prepare a graph of log t_r' versus n and estimate the value of n in the chemical formula.

$n = 7$:	4.0 min	air:	1.1 min
$n = 8$:	6.5 min	unknown:	42.5 min
$n = 14$:	86.9 min		

(b) Calculate the capacity factor for the unknown.

C. Make a graph showing the qualitative shape of the ionic-strength gradient that would be produced by each device shown here.

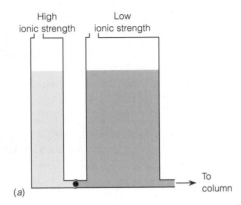

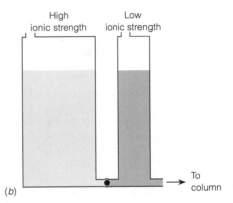

Problems

Gas Chromatography

1. What is the advantage of temperature programming in gas chromatography?

2. **(a)** What are the relative advantages and disadvantages of packed and open tubular columns in gas chromatography?

(b) Explain the difference between wall-coated, support-coated, and porous-layer open tubular columns.

(c) What is the advantage of a bonded stationary phase in gas chromatography?

3. **(a)** Why do open tubular columns provide greater resolution than packed columns in gas chromatography?

(b) Why do H_2 and He allow more rapid linear flow rates in gas chromatography than does N_2, without loss of column efficiency (Figure 23-8)?

4. **(a)** When would you use split, splitless, or on-column injection in gas chromatography?

(b) Explain how solvent trapping and cold trapping work in splitless injection.

5. To which kinds of analytes do the following gas chromatography detectors respond?

(a) thermal conductivity **(d)** flame photometric

(b) flame ionization **(e)** alkali flame

(c) electron capture **(f)** sulfur chemiluminescence

6. What does the designation 200/400 mesh mean on a bottle of chromatography stationary phase? What is the size range of such particles? (See Table 26-2.) Which particles are smaller, 100/200 or 200/400 mesh?

7. Using Table 23-3, predict the order of elution of the following compounds from columns containing **(a)** poly-(dimethylsiloxane), **(b)** (diphenyl)$_{0.35}$(dimethyl)$_{0.65}$poly-siloxane, and **(c)** poly(ethylene glycol): hexane, heptane, octane, benzene, butanol, 2-pentanone.

8. Using Table 23-3, predict the order of elution of the following compounds from columns containing **(a)** poly-(dimethylsiloxane), **(b)** (diphenyl)$_{0.35}$(dimethyl)$_{0.65}$poly-siloxane, and **(c)** poly(ethylene glycol):

    1. 1-pentanol (n-$C_5H_{11}OH$, b.p. 138°C)

    2. 2-hexanone ($CH_3C(=O)C_4H_9$, b.p. 128°C)

  3. heptane (n-C_7H_{16}, b.p. 98°C)
  4. octane (n-C_8H_{18}, b.p. 126°C)
  5. nonane (n-C_9H_{20}, b.p. 151°C)
  6. decane (n-$C_{10}H_{22}$, b.p. 174°C)

9. This problem reviews some concepts from Chapter 22. An unretained solute passes through a chromatography column in 3.7 min and analyte requires 8.4 min.

(a) Find the adjusted retention time and capacity factor for the analyte.

(b) The volume of the mobile phase is 1.4 times the volume of the stationary phase. Find the partition coefficient for the analyte.

10. The unadjusted retention times in Figure 22-7 are 1.0 min for air, 12.0 min for octane, 13.0 min for the unknown, and 15.0 min for nonane. Find the Kovats retention index for the unknown.

11. Retention time depends on temperature (T) according to the equation $\log t'_r = (a/T) + b$, where a and b are constants for a specific compound on a specific column. A compound is eluted from a gas chromatography column at an adjusted retention time $t'_r = 15.0$ min when the column temperature is 373 K. At 363 K, $t'_r = 20.0$ min. Find the parameters a and b and predict t'_r at 353 K.

12. Suppose that a solution containing 6.3×10^{-8} M I^- and 2.0×10^{-7} M p-dichlorobenzene gave peak areas of 395 and 787, respectively, in the experiment in Figure 23-13. A 3.00-mL solution of unknown containing I^- was treated with 0.100 mL of 1.6×10^{-5} M p-dichlorobenzene and the mixture was diluted to 10.00 mL. Gas chromatography gave peak areas of 633 and 520 for I^- and p-dichlorobenzene, respectively. Find the concentration of I^- in the 3.00 mL of original unknown.

13. An unknown compound was co-chromatographed with heptane and decane. The adjusted retention times were heptane, 12.6 min; decane, 22.9 min; unknown, 20.0 min. Realizing that the scale of the retention index is logarithmic and that the indexes for heptane and decane are 700 and 1 000, respectively, find the retention index for the unknown.

14. *Theoretical performance in gas chromatography.* As the inside radius of a gas chromatography column is decreased, the maximum possible column efficiency increases and the sample capacity decreases. For a very thin stationary phase (that equilibrates rapidly with analyte), the minimum theoretical plate height is given by

$$\frac{H_{min}}{r} = \sqrt{\frac{1 + 6k' + 11k'^2}{3(1 + k')^2}}$$

where r is the inside radius of the column and k' is the capacity factor.

(a) Find the limit of the square root term as $k' \to 0$ (unretained solute) and as $k' \to \infty$ (infinitely retained solute).

(b) If the column radius is 0.10 mm, find H_{min} for the two cases in **(a)**.

(c) What is the maximum number of theoretical plates in a column 50 m long with a 0.10-mm radius if $k' = 5$?

(d) The relation between the capacity factor k' and partition coefficient K (Equation 22-19) can also be written $k' = 2dK/r$, where d is the thickness of the stationary phase in a wall-coated column and r is the inside diameter of the column. Find k' if $K = 1\,000$, $d = 0.2\ \mu m$, and $r = 0.10$ mm.

15. *Separation number.* In chromatography the separation number for peaks 1 and 2 is defined as

$$\text{separation number} = TZ = \frac{t_{r2} - t_{r1}}{w_2 + w_1} - 1$$

where t_{ri} is the retention time of peak i and w_i is the width of peak i at half-height. (The symbol TZ comes from the German word *trennzahl*, meaning "separation number.") If all peak widths are equal, TZ tells us how many peaks can be placed between peaks 1 and 2, with a resolution (Figure 22-9) of 1.17 between all peaks. In Figure 22-7, suppose that $t_r(\text{octane}) = 14.3$ min, $t_r(\text{nonane}) = 18.5$ min, and $w = 0.5$ min for both peaks. Calculate the separation number. Does it look as though this many peaks can be placed between octane and nonane in Figure 22-7?

Liquid Chromatography

16. **(a)** Why does the eluent strength increase as solvent becomes less polar in reverse-phase chromatography, whereas the eluent strength increases as solvent becomes more polar in normal-phase chromatography?

(b) What kind of gradient is used in supercritical fluid chromatography?

17. Name two small-scale methods that can be used to select a suitable solvent and stationary phase for large-scale chromatographic separations.

18. Why are the relative eluent strengths of solvents in adsorption chromatography fairly independent of solute?

19. **(a)** Why is high pressure needed in HPLC?

(b) What is a bonded phase in liquid chromatography?

(c) Why does the efficiency (decreased plate height) of liquid chromatography increase as the stationary-phase particle size is reduced?

20. Using Figure 23-22, suggest which type of liquid chromatography you could use to separate compounds in each of the following categories:

(a) MW <2 000, soluble in octane

(b) MW <2 000, soluble in dioxane

(c) MW <2 000, ionic

(d) MW >2 000, soluble in water, nonionic, size 50 nm

(e) MW >2 000, soluble in water, ionic

(f) MW >2 000, soluble in tetrahydrofuran, size 50 nm

21. What is the purpose of the Mariotte flask in Section 23-4?

22. (a) Nonpolar aromatic compounds were separated by HPLC on a bonded phase containing octadecyl groups $[-(CH_2)_{17}CH_3]$ covalently attached to silica particles. The eluent was 65 vol % methanol in water. How would the retention times be affected if 90% methanol were used instead?

(b) Octanoic acid and 1-aminooctane were passed through the same column described in **(a)**, using an eluent of 20 vol % methanol in water adjusted to pH 3.0 with HCl. State which compound is expected to be eluted first and why.

$$CH_3CH_2CH_2CH_2CH_2CH_2CH_2CO_2H$$
Octanoic acid

$$CH_3CH_2CH_2CH_2CH_2CH_2CH_2CH_2NH_2$$
1-Aminooctane

23. Suppose the HPLC column produces ideal Gaussian profiles for which

$$\text{area of peak} = (1.19)hw_{1/2}$$

where h is peak height and $w_{1/2}$ is the width at half the maximum peak height. The detector measures absorbance at 254 nm. A sample containing equal moles of compounds A and B was injected into the column. Suppose that compound A ($\epsilon_{254} = 2.26 \times 10^4$ M^{-1} cm^{-1}) has $h = 128$ mm and $w_{1/2} = 10.1$ mm. Compound B ($\epsilon_{254} = 1.68 \times 10^4$ M^{-1} cm^{-1}) has $w_{1/2} = 7.6$ mm. What is the height of peak B in millimeters?

24. A known mixture of compounds C and D gave the following HPLC results:

Compound	Concentration (mg/mL) in mixture	Peak area (cm^2)
C	1.03	10.86
D	1.16	4.37

A solution was prepared by mixing 12.49 mg of D plus 10.00 mL of unknown containing just C, and diluting to 25.00 mL. Peak areas of 5.97 and 6.38 cm^2 were observed for C and D, respectively. Find the concentration of C (mg/mL) in the unknown.

25. The graph below shows the dependence of plate height on flow rate for the separation of three compounds by liquid chromatography.

(a) Explain why the plate height is different for each compound.

(b) The Blue Dextran data are flat, but, for tryptophan and acetone, plate height rises with increasing flow rate. Rationalize these behaviors in terms of the van Deemter equation.

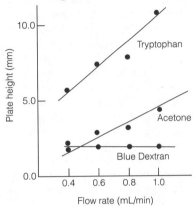

Graph of plate height versus flow rate for separation of Blue Dextran (a high molecular weight polymer), acetone, and tryptophan on a 4-mm-diameter × 50-cm-diameter column of Sephadex G-25. [From J. M. Sosa, *Anal. Chem.* **1980,** *52,* 910.]

26. *Eluent strength and selectivity in reverse-phase HPLC.* Eluent strength (also called *solvent strength*) is a measure of the ability of a solvent to elute solutes from a chromatography column. The greater the eluent strength, the more easily solutes are eluted and the lower the capacity factors are. *Selectivity* is the ability of a solvent to distinguish between different solutes and separate them. Different mixtures of solvents A or B with water have the same approximate eluent strength when eluent strength = $\phi_A \cdot \epsilon_A = \phi_B \cdot \epsilon_B$, where ϕ_A and ϕ_B are the volume fractions of A or B in the mixture, and ϵ comes from the table below.

Eluent strength factors for reverse-phase chromatography[12] Solvent	ϵ
Methanol	3.0
Acetonitrile	3.1
Acetone	3.4
Dioxane	3.5
Ethanol	3.6
2-Propanol	4.2
Tetrahydrofuran	4.4

Thus, a 40 vol% mixture of 2-propanol in water has the same solvent strength as a 50% mixture of acetone in water because $\phi_A \cdot \epsilon_A = (0.40)(4.2) = \phi_B \cdot \epsilon_B = (0.50)(3.4)$.

(a) Which mixture has a greater eluent strength for reverse-phase chromatography, 30% tetrahydrofuran in water or 40% methanol in water?

(b) Which solvent is expected to produce lower capacity factors for most solutes in reverse-phase chromatography, 30% tetrahydrofuran in water or 40% methanol in water?

(c) For similar solutes, it is desirable to keep the eluent strength approximately constant, so that one component does not require much longer than another to be eluted. However, it is necessary to have some difference in retention, or else the peaks are not separated. If two solutes have a reasonable capacity factor (say, between 1 and 5), but are not resolved from each other, you can try a different solvent mixture that ought to give similar capacity factors but might yield better resolution. For example, two components not separated by acetonitrile in water might be separated by tetrahydrofuran in water. What volume percent of tetrahydrofuran in water has the same eluent strength as 20% acetonitrile in water?

27. A bonded stationary phase for the separation of optical isomers has the structure

To resolve the enantiomers of amines, alcohols, or thiols, the compounds are first attached to a nitroaromatic group that increases their interaction with the bonded phase and makes them observable with a spectrophotometric detector.

When the mixture is eluted with 20 vol % 2-propanol in hexane, the R enantiomer is eluted before the S enantiomer, with the following chromatographic parameters:

$$\text{resolution} = \frac{\Delta t_r}{\overline{w}} = 7.7$$

$$\text{capacity factor for R isomer} = 1.35$$

$$\text{relative retention} = 4.53$$

where $\overline{w}$ is the average width of the two peaks at their base. A schematic chromatogram is shown here. Find t_1, t_2, and $\overline{w}$, with units of minutes.

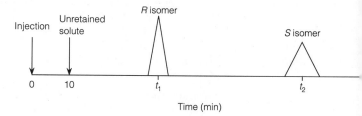

28. Microporous silica particles with a density of 2.2 g/mL and a diameter of 10 μm have a measured surface area of 300 m²/g. Calculate the surface area if the spherical silica were solid particles. What does this tell you about the shape or porosity of the particles?

Notes and References

1. Classroom demonstrations of both types of chromatography have been described by A. Wollrab, *J. Chem. Ed.* **1982**, *59*, 1042 and C. E. Bricker, M. A. Taylor, and K. E. Kolb, *J. Chem. Ed.* **1981**, *58*, 41.

2. A. Berthod, W. Li, and D. W. Armstrong, *Anal. Chem.* **1992**, *64*, 873.

3. J. R. Berg and T. Hawkins, *Am. Lab.* February 1992, 44.

4. Do not place blind faith in an integrator. For a discussion of integration errors, see A. N. Papas and M. F. Delaney, *Anal. Chem.* **1987**, *59*, 55A.

5. To avoid contamination from previous use, researchers rarely reuse silica gel. However, regeneration and reuse is economical for student work. To a boiling mixture of 500 g of used silica gel in 500 mL of water is added dropwise 50 mL of 5 wt% aqueous sodium hypochlorite (commercial bleach solution). The pH should be maintained below 7 with HCl during the addition. After 1 h of further boiling, cool, decant, and wash the gel with three 500-mL volumes of water, three 500-mL volumes of 1 M HCl, and sufficient water to bring the pH to 5.5–6. Air dry and then activate by heating to 100°–200°C for 1 h. [A. E. Guarconi and V. F. Ferriera, *J. Chem. Ed.* **1988**, *65*, 891.]

6. High-resolution thin-layer chromatography is possible with proper equipment and forced flow of solvent; see C. F. Poole and S. K. Poole, *Anal. Chem.* **1989**, *61*, 1257A.

7. Reducing the HPLC column diameter to values as small as 20 μm also increases efficiency, but such columns are not in common use. See R. T. Kennedy and J. W. Jorgenson, *Anal. Chem.* **1989**, *61*, 1128.

8. G. T. Marshall, *Am. Lab.* September 1991, 36D.

9. HPLC columns should be periodically cleaned to prevent impurities from migrating irreversibly into the inner pores of the supports. Silica gel and cyano- and diol-bonded phases are washed (in order) with heptane, chloroform, ethyl acetate, acetone, ethanol, and water. Then the order is reversed, using dried solvents, to reactivate the column. Use 10 empty-column volumes of each solvent. Amino-bonded phases are washed in the same manner as silica gel, but a 0.5 M ammonia wash is used after water. C_{18} and other nonpolar phases are washed with water, acetonitrile, and chloroform, and then the order is reversed. If this is insufficient, wash with 0.5 M sulfuric acid, and then water. [F. Rabel and K. Palmer, *Am. Lab.* August 1992, 65.]

10. *Chem. Eng. News,* 12 December 1988, p. 23.

11. H. Lingeman and W. J. M. Underberg, Eds., *Detection-Oriented Derivatization Techniques in Liquid Chromatography* (New York: Marcel Dekker, 1990).

12. From L. R. Snyder and J. J. Kirkland, *Introduction to Modern Liquid Chromatography,* 2nd ed. (New York: Wiley, 1979), p. 264.

The Shape of the Future: Analysis on a Chip

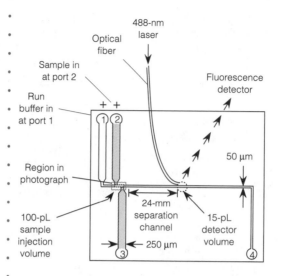

488-nm laser

Optical fiber

Sample in at port 2

Run buffer in at port 1

Fluorescence detector

Region in photograph

50 μm

100-pL sample injection volume

24-mm separation channel

15-pL detector volume

250 μm

Design of capillary electrophoresis chip. Sample introduced by electrokinetic injection (described in this chapter) between ports 2 and 3 gives a 100-pL injection plug that would reside in the section of channel shown in the micrograph. A potential difference of 11 250 V applied between ports 1 and 4 separates the components. [From C. S. Effenhauser, A. Manz, and H. M. Widmer, *Anal. Chem.* **1993,** *65,* 2637. Photo courtesy C. S. Effenhauser, Ciba-Geigy.]

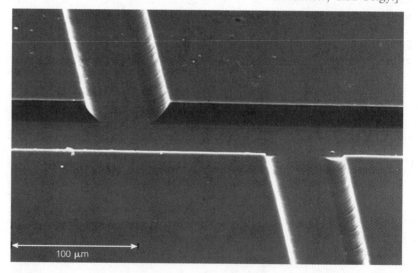

100 μm

With lithographic techniques borrowed from the microelectronics industry, capillary electrophoresis can be adapted for miniature instruments that can be used in the field. A 24-mm length of the 12-μm-deep × 50-μm-wide channel etched into the glass chip in the diagram above provides 45 000 to 75 000 theoretical plates for the separation of fluorescence-labeled amino acids in 14 s. (The *electropherogram* is shown in Problem 21.) Ingenious ideas such as electrophoresis on a chip and chemical-sensing field effect transistors (Section 15-8), devised by cross-disciplinary teams of scientists, provide revolutionary advances in chemical instrumentation.

Challenge   After you have read the section on capillary electrophoresis, come back to this page and explain how this chip works.

Chromatographic Methods and Capillary Electrophoresis

This chapter continues our detailed study of chromatographic methods. We examine ion-exchange chromatography, molecular exclusion chromatography, and affinity chromatography, followed by the newest kid on the block—capillary electrophoresis.

24-1 Ion-Exchange Chromatography

In **ion-exchange chromatography,** retention is based on the attraction between solute ions and charged sites bound to the stationary phase (Figure 22-6 and Color Plate 16). In **anion exchangers,** positively charged groups on the stationary phase attract solute <u>anions</u>. **Cation exchangers** contain covalently bound, negatively charged sites that attract solute <u>cations</u>.

Anion exchangers contain bound *positive* groups.
Cation exchangers contain bound *negative* groups.

Ion Exchangers

Resins are amorphous (noncrystalline) particles of organic material. *Polystyrene resins* for ion exchange are made by the copolymerization of styrene and divinylbenzene (Figure 24-1). Divinylbenzene content is varied from 1 to 16% to increase the extent of **crosslinking** of the insoluble hydrocarbon polymer. The benzene rings can be modified to produce a cation-exchange resin, containing sulfonate groups ($-SO_3^-$), or an anion-exchange resin, containing ammonium groups ($-NR_3^+$). If methacrylic acid is used in place of styrene, a polymer with carboxyl groups results.

Ion exchangers are classified as being strongly or weakly acidic or basic, as indicated in Table 24-1. Sulfonate groups ($-SO_3^-$) of strongly acidic

Methacrylic acid

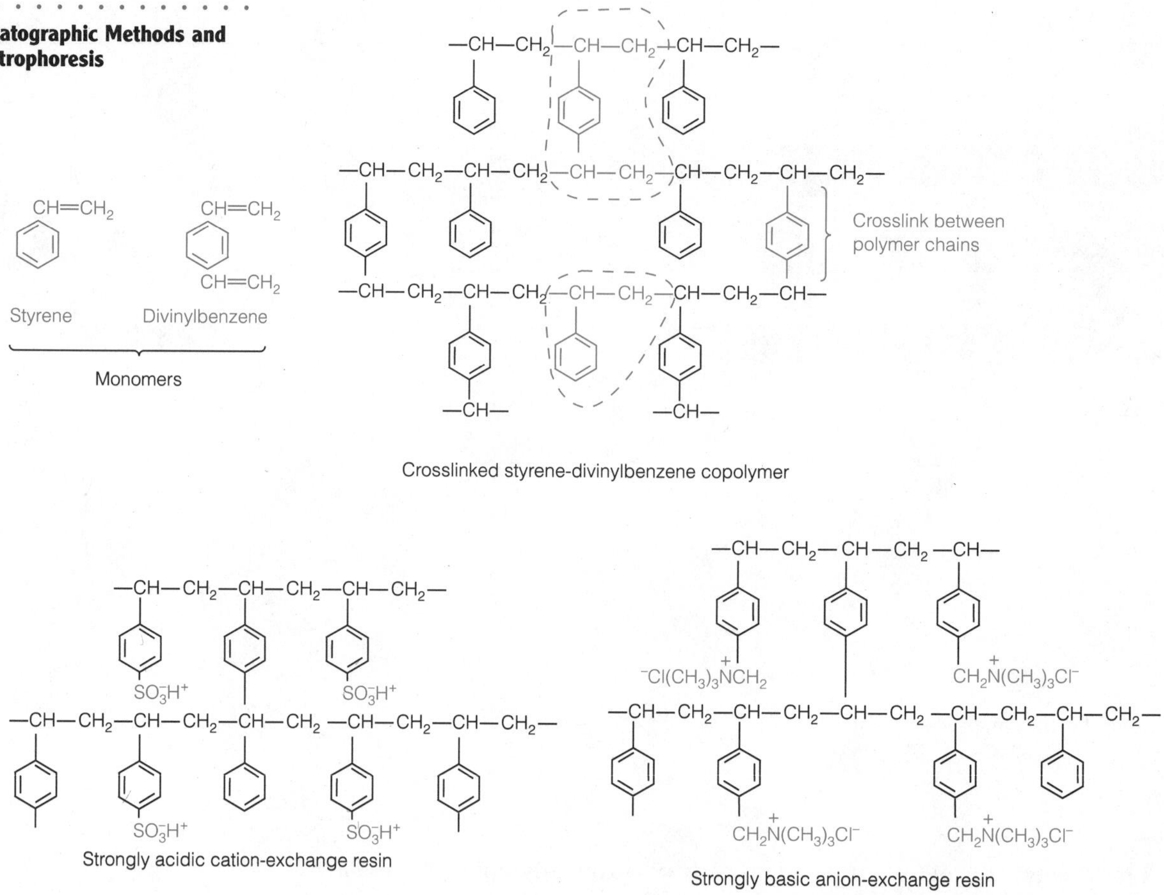

Styrene   Divinylbenzene

Monomers

Crosslink between
polymer chains

Crosslinked styrene-divinylbenzene copolymer

Strongly acidic cation-exchange resin

Strongly basic anion-exchange resin

Figure 24-1   Structures of styrene-divinylbenzene crosslinked ion-exchange resins.

Strongly acidic cation exchangers:
  RSO_3^-
Weakly acidic cation exchangers:
  RCO_2^-
"Strongly basic" anion exchangers:
  $RNR_3'^+$
Weakly basic anion exchangers:
  $RNR_2'H^+$

resins remain ionized even in strongly acidic solutions. Carboxyl groups ($-CO_2^-$) of the weakly acidic resins are protonated near pH 4 and lose their cation-exchange capacity. "Strongly basic" quaternary ammonium groups ($-CH_2NR_3^+$) (which are not really basic at all) remain cationic at all values of pH. Weakly basic tertiary ammonium ($-CH_2NHR_2^+$) anion exchangers are deprotonated in moderately basic solution and lose their ability to bind anions.

The extent of crosslinking is indicated by the notation "-XN" after the name of the resin. For example, Dowex 1-X4 contains 4% divinylbenzene, and Bio-Rad AG 50W-X12 contains 12% divinylbenzene. The resin becomes more rigid and less porous as crosslinking increases. Lightly crosslinked resins permit rapid equilibration of solute between the inside and outside of the particle. On the other hand, resins with little crosslinking swell in water. This hydration decreases both the density of ion-exchange sites and the selectivity of the resin for different ions. More heavily crosslinked resins exhibit less swelling and higher exchange capacity and selectivity, but have longer equilibration times. The charge density of polystyrene ion exchangers is so great that highly charged macromolecules such as proteins may be irreversibly bound.

TABLE 24-1 Ion-exchange resins

Resin type	Chemical constitution	Usual form as purchased	Common trade names		Selectivity	Thermal stability
			Rohm & Haas	Dow Chemical		
Strongly acidic cation exchanger	Sulfonic acid groups attached to styrene and divinylbenzene copolymer	$\phi-SO_3^-H^+$	Amberlite IR-120	Dowex 50W	$Ag^+ > Rb^+ > Cs^+ > K^+ > NH_4^+ > Na^+ > H^+ > Li^+$ $Zn^{2+} > Cu^{2+} > Ni^{2+} > Co^{2+}$	Good up to 150°C
Weakly acidic cation exchanger	Carboxylic acid groups attached to acrylic and divinylbenzene copolymer	$R-COO^-Na^+$	Amberlite IRC-50	—	$H^+ >> Ag^+ > K^+ > Na^+ > Li^+$ $H^+ >> Fe^{2+} > Ba^{2+} Sr^{2+} > Ca^{2+} > Mg^{2+}$	Good up to 100°C
Strongly basic anion exchanger	Quaternary ammonium groups attached to styrene and divinylbenzene copolymer	$\phi-CH_2N(CH_3)_3^+Cl^-$	Amberlite IRA-400	Dowex 1	$I^- > phenolate^- > HSO_4^- > ClO_3^- > NO_3^- > Br^- > CN^- > HSO_3^- > NO_2^- > Cl^- > HCO_3^- > IO_3^- > HCOO^- > acetate^- > OH^- > F^-$	OH^- form fair up to 50°C Cl^- and other forms good up to 150°C
Weakly basic anion exchanger	Polyalkylamine groups attached to styrene and divinylbenzene copolymer	$\phi-NH(R)_2^+Cl^-$	Amberlite IR-45	Dowex 3	$\phi SO_3H > citric > CrO_3 > H_2SO_4 > tartaric > oxalic > H_3PO_4 > H_3AsO_4 > HNO_3 > HI > HBr > HCl > HF > HCO_2H > CH_3CO_2H > H_2CO_3$	Extensive information not available; tentatively limited to 65°C

SOURCE: Adapted from J. X. Khym, *Analytical Ion-Exchange Procedures in Chemistry and Biology* (Englewood Cliffs, NJ: Prentice Hall, 1974).

Cellulose and dextran ion exchangers, which are polymers of the sugar glucose, possess larger pore sizes and lower charge densities. They are well suited to ion exchange of macromolecules, such as proteins. Dextran, crosslinked by glycerin, is sold under the name Sephadex (Figure 24-2). Other macroporous ion exchangers are based on the polysaccharide agarose and on polyacrylamide. Charged functional groups attached to occasional hydroxyl groups of the polysaccharides are listed in Table 24-2. DEAE-Sephadex, for example, refers to an anion-exchange Sephadex containing diethylaminoethyl groups.

Because they are much softer than polystyrene *resins,* dextran and its relatives are called **gels.**

See Figure 24-8 for the structure of polyacrylamide.

Ion-Exchange Selectivity

Consider the competition of Na^+ and Li^+ for sites on the cation-exchange resin, R^-:

Selectivity coefficient: $R^-Na^+ + Li^+ \rightleftharpoons R^-Li^+ + Na^+$ $\qquad K = \dfrac{[R^-Li^+][Na^+]}{[R^-Na^+][Li^+]}$ (24-1)

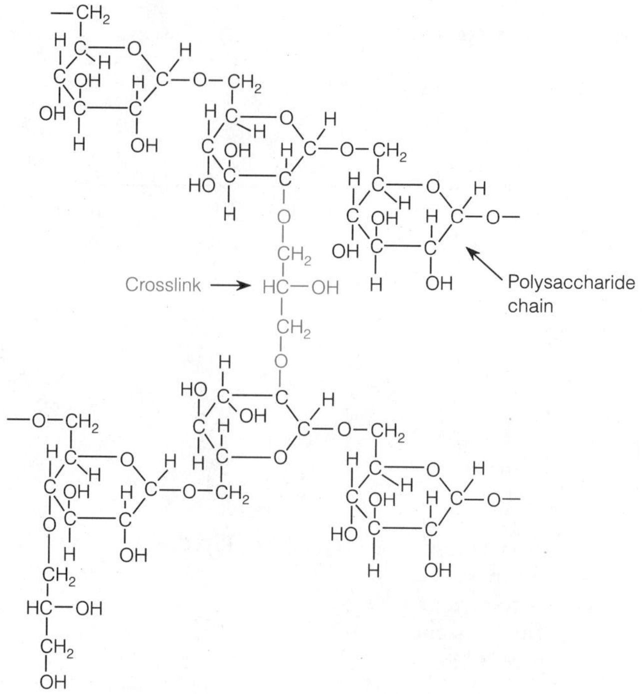

Figure 24-2 Structure of Sephadex, a crosslinked dextran sold by Pharmacia Fine Chemicals, Piscataway, NJ.

TABLE 24-2 **Common active groups of ion-exchange gels**

Type	Abbreviation	Name	Structure
Cation exchangers			
Strong acid	SP	Sulfopropyl	$-OCH_2CH_2CH_2SO_3H$
	SE	Sulfoethyl	$-OCH_2CH_2SO_3H$
Intermediate acid	P	Phosphate	$-OPO_3H_2$
Weak acid	CM	Carboxymethyl	$-OCH_2CO_2H$
Anion exchangers			
Strong base	TEAE	Triethylaminoethyl	$-OCH_2CH_2\overset{+}{N}(CH_2CH_3)_3$
	QAE	Diethyl(2-hydroxypropyl) quaternary amino	$-OCH_2CH_2\overset{+}{N}(CH_2CH_3)_2$ $\mid$ $CH_2CHOHCH_3$
Intermediate base	DEAE	Diethylaminoethyl	$-OCH_2CH_2N(CH_2CH_3)_2$
	ECTEOLA	Triethanolamine coupled to cellulose through glyceryl chains	
	BD	Benzoylated DEAE groups	
Weak base	PAB	*p*-Aminobenzyl	$-O-CH_2-\langle\bigcirc\rangle-NH_2$

TABLE 24-3   Relative selectivity coefficients of ion-exchange resins

Sulfonic acid cation-exchange resin				Quaternary ammonium anion-exchange resin	
Cation	Divinylbenzene content (%)			Anion	Relative selectivity
	4	8	10		
Li^+	1.00	1.00	1.00	F^-	0.09
H^+	1.30	1.26	1.45	OH^-	0.09
Na^+	1.49	1.88	2.23	Cl^-	1.0
NH_4^+	1.75	2.22	3.07	Br^-	2.8
K^+	2.09	2.63	4.15	NO_3^-	3.8
Rb^+	2.22	2.89	4.19	I^-	8.7
Cs^+	2.37	2.91	4.15	ClO_4^-	10.0
Ag^+	4.00	7.36	19.4		
Tl^+	5.20	9.66	22.2		

SOURCE: *Amberlite Ion Exchange Resins—Laboratory Guide* (Rohm & Haas Co., 1979).

The equilibrium constant is called the **selectivity coefficient,** because it describes the relative selectivity of the resin for Li^+ and Na^+. Selectivities of polystyrene resins in Table 24-3 tend to increase with the extent of crosslinking, because the pore size of the resin shrinks as crosslinking increases. Ions such as Li^+, with a large hydrated radius (Figure 8-2, Table 8-1), do not have as much access to the resin as smaller ions, such as Cs^+.

In general, ion exchangers favor the binding of ions of higher charge, decreased hydrated radius, and increased *polarizability*. A fairly general order of selectivity for cations is the following:

$$Pu^{4+} \gg La^{3+} > Ce^{3+} > Pr^{3+} > Eu^{3+} > Y^{3+} > Sc^{3+} > Al^{3+} \gg$$
$$Ba^{2+} > Pb^{2+} > Sr^{2+} > Ca^{2+} > Ni^{2+} > Cd^{2+} > Cu^{2+} >$$
$$Co^{2+} > Zn^{2+} > Mg^{2+} > UO_2^{2+} \gg Tl^+ > Ag^+ > Rb^+ > K^+ >$$
$$NH_4^+ > Na^+ > H^+ > Li^+$$

Polarizability refers to the ability of an ion's electron cloud to be deformed by nearby charges. Deformation of the electron cloud induces a dipole in the ion. The attraction between the induced dipole and the nearby charge increases the binding of the ion to the resin.

Reaction 24-1 can be driven in either direction, even though Na^+ is bound more tightly than Li^+. Washing a column containing Na^+ with a substantial excess of Li^+ will replace Na^+ with Li^+. Washing a column in the Li^+ form with Na^+ will convert it to the Na^+ form.

Remember that Na^+ has a smaller hydrated radius than Li^+.

Ion exchangers loaded with one kind of ion bind small amounts of a different ion nearly quantitatively. Na^+-loaded resin will bind small amounts of Li^+ nearly quantitatively, even though the selectivity is greater for Na^+. The same column binds large quantities of Ni^{2+} or Fe^{3+}, because the resin has greater selectivity for these ions than for Na^+. Even though Fe^{3+} is bound more tightly than H^+, Fe^{3+} can be quantitatively removed from the resin by washing with excess of acid.

The Donnan Equilibrium

When an ion exchanger is placed in an electrolyte solution, *the concentration of electrolyte is higher outside the resin than inside it.* The equilibrium between ions in solution and ions inside the resin is called the **Donnan equilibrium.**

A phase containing bound charges tends to exclude electrolyte.

Consider a quaternary ammonium anion-exchange resin (R^+) in its Cl^- form immersed in a solution of KCl. Let the concentration of an ion inside the membrane be $[X]_i$ and the concentration outside the membrane be $[X]_o$. It can be shown from thermodynamics that the ion product inside the resin is approximately equal to the product outside the resin:

$$[K^+]_i[Cl^-]_i = [K^+]_o[Cl^-]_o \qquad (24\text{-}2)$$

From considerations of charge balance, we know that

$$[K^+]_o = [Cl^-]_o \qquad (24\text{-}3)$$

We are ignoring H^+ and OH^-, which are assumed to be negligible.

Inside the resin, there are three charged species, and the charge balance is

$$[R^+]_i + [K^+]_i = [Cl^-]_i \qquad (24\text{-}4)$$

where $[R^+]$ is the concentration of quaternary ammonium ions attached to the resin. Substituting Equations 24-3 and 24-4 into Equation 24-2 gives

$$[K^+]_i([K^+]_i + [R^+]_i) = [K^+]_o^2 \qquad (24\text{-}5)$$

which says that $[K^+]_o$ must be greater than $[K^+]_i$.

EXAMPLE Exclusion of Cations by an Anion Exchanger

Suppose that the concentration of cationic sites in the resin is 6.0 M. When the Cl^- form of this resin is immersed in 0.050 M KCl, what will be the ratio $[K^+]_o/[K^+]_i$?

Solution Let us assume that $[K^+]_o$ remains 0.050 M. Equation 24-5 gives

$$[K^+]_i \, ([K^+]_i + 6.0) = (0.050)^2 \Rightarrow [K^+]_i = 0.000\,42 \text{ M}$$

The concentration of K^+ inside the resin is less than 1% of that outside the resin.

The high concentration of positive charges within the resin repels cations from the resin.

Ions with the *same* charge as the resin are excluded. (The quaternary ammonium resin excludes K^+.) The counterion, Cl^- in the above example, is *not excluded* from the resin. There is no electrostatic barrier to penetration of an anion into the resin. Anion exchange takes place freely in the quaternary ammonium resin even though cations are repelled from the resin.

The volume V_m is available to an electrolyte. The volume $V_m + V_s$ is available to a nonelectrolyte.

The Donnan equilibrium is the basis of *ion-exclusion chromatography*. Because dilute electrolytes are excluded from the resin, they pass through a column when the volume of mobile phase (V_m) has been eluted. Nonelectrolytes, such as sugar, freely penetrate the resin. They are not eluted until a volume $V_m + V_s$ (where V_s is the volume of liquid inside the resin) has passed. When a solution of NaCl and sugar is applied to an ion-exchange column, NaCl emerges from the column *before* the sugar.

Conducting Ion-Exchange Chromatography

Three classes of ion exchangers:

1. resins
2. gels
3. inorganic exchangers

Ion-exchange *resins* are used for applications involving small molecules (MW $\leq$ 500), which can penetrate the small pores of the resin. A mesh size of 100/200 is suitable for most work. Higher mesh numbers (smaller particle

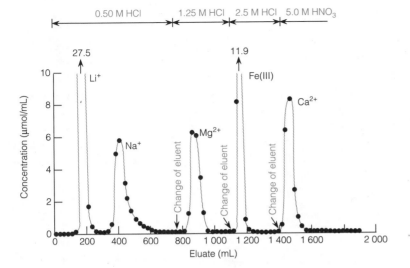

Figure 24-3 Elution of a mixture containing 0.5 mmol Na^+, Mg^{2+}, Ca^{2+}, and Fe^{3+} plus 1 mmol Li^+ from a 2.1 × 16-cm column of Bio-Rad AG MP-50 resin (200/400 mesh) at a rate of 120 mL/h. [From F. W. E. Strelow, *Anal. Chem.* **1984,** *56,* 1053.]

size) lead to finer separations but slower column operation. For preparative separations, the sample may occupy 10–20% of the column volume. Ion-exchange *gels* are used for large molecules (such as proteins and nucleic acids), which cannot penetrate the pores of resins. Separations involving harsh chemical conditions (high temperature, high radiation levels, strongly basic solution, powerful oxidizing agents) employ *inorganic ion exchangers,* such as hydrous oxides of Zr, Ti, Sn, and W.

Gradient elution with increasing ionic strength or changing pH is extremely valuable in ion-exchange chromatography. Consider a column to which anion A^- is bound more tightly than anion B^- is. We separate A^- from B^- by elution with C^-, which is less tightly bound than either A^- or B^-. As the concentration of C^- is increased, B^- is eventually displaced and moves down the column. At a still higher concentration of C^-, the anion A^- is also eluted.

An ionic strength gradient is analogous to a solvent or temperature gradient.

Figure 24-3 shows a chromatogram in which discontinuous HCl concentrations were used for a cation-exchange separation. Low concentrations of H^+ eluted Li^+ and Na^+, but high concentrations were required for Ca^{2+}. At sufficiently high Cl^- concentration, Fe(III) is converted to $FeCl_4^-$, which is not retained by a cation exchanger.

Applications of Ion Exchange

Ion exchange is used to purify water for laboratory and industrial use. **Deionized water** is prepared by passing water through an anion-exchange resin in its OH^- form and a cation-exchange resin in its H^+ form. Suppose, for example, that $Cu(NO_3)_2$ is present in the solution. The cation-exchange resin binds Cu^{2+} and replaces it with $2H^+$. The anion-exchange resin binds NO_3^- and replaces it with OH^-. The eluate is pure water:

Water softeners use ion exchange to remove Ca^{2+} and Mg^{2+} from "hard" water.

$$Cu^{2+} \xrightarrow{H^+ \text{ ion exchange}} 2H^+$$
$$2NO_3^- \xrightarrow{OH^- \text{ ion exchange}} 2OH^- \Big\} \longrightarrow \text{ pure } H_2O$$

Ion exchange can be used to convert one salt to another. For example, we can prepare tetrapropylammonium hydroxide from a tetrapropylammonium salt of some other anion:

$$(CH_3CH_2CH_2)_4N^+I^- \xrightarrow[\text{OH}^- \text{ form}]{\text{anion exchanger}} (CH_3CH_2CH_2)_4N^+OH^-$$

Tetrapropylammonium
iodide   Tetrapropylammonium
                                          hydroxide

Ion exchange is used for **preconcentration** of trace components of a solution to obtain enough for analysis. For example, we can pass a large volume of fresh lake water through a cation-exchange resin in the H^+ form to concentrate metal ions onto the resin. Chelex 100, a styrene-divinylbenzene resin containing iminodiacetic acid groups, is noteworthy for its ability to bind transition metal ions.

$$\text{resin} - N \underset{CH_2CO_2H}{\overset{CH_2CO_2H}{\diagdown}}$$ Chelex 100

Iminodiacetic
acid

The metals are eluted with a small volume of 2 M HNO_3, which protonates the iminodiacetate groups.

24-2 Ion Chromatography

Ion chromatography, a high-performance version of ion-exchange chromatography, has become the method of choice for anion analysis.[1] It is used in the semiconductor industry to monitor anions and cations at 0.1 ppb levels in deionized water.

Suppressed-Ion Anion and Cation Chromatography

In **suppressed-ion anion chromatography** (Figure 24-4a), a mixture of anions is separated by ion exchange and detected by electrical conductivity. The key feature of suppressed-ion chromatography is removal of unwanted electrolyte prior to conductivity measurement.

> The separator column separates the analytes, and the suppressor replaces the ionic eluent with a nonionic species.

For the sake of illustration, consider a sample containing KNO_3 and $CaSO_4$ injected into the *separator column*—an anion-exchange column in the carbonate form—followed by elution with Na_2CO_3. NO_3^- and SO_4^{2-} equilibrate with the resin and are slowly displaced by the CO_3^{2-} eluent. K^+ and Ca^{2+} cations are not retained and simply wash through. After a period of time, $NaNO_3$ and Na_2SO_4 are eluted from the separator column, as shown in the upper graph of Figure 24-4a. These species cannot be easily detected, however, because the solvent contains a high concentration of Na_2CO_3, whose high conductivity obscures that of the analyte species.

> The Donnan equilibrium tells us that ions with the same charge as those covalently attached to the membrane are excluded from the membrane.

To remedy this problem, the solution is next passed through a *membrane ion suppressor,* in which cations are replaced by H^+. The suppressor is a narrow-diameter cylinder or thin, flat channel enclosed by a thin, semipermeable cation-exchange membrane made of sulfonated polyethylene. The sulfonate groups ($-SO_3^-$) in the membrane repel anions but allow cations to pass freely. The outside of the membrane is bathed in H_2SO_4 solution. When $NaNO_3$ and Na_2CO_3 from the separator column pass through the suppres-

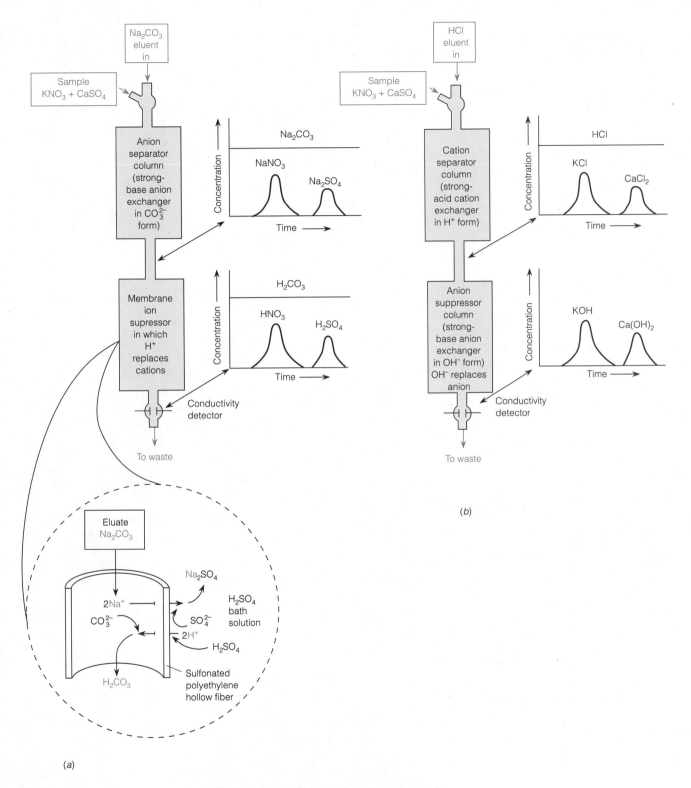

Figure 24-4 Schematic illustrations of (*a*) suppressed-ion anion chromatography and (*b*) suppressed-ion cation chromatography. [Adapted from H. Small, *Anal. Chem.* **1983,** *55,* 235A.]

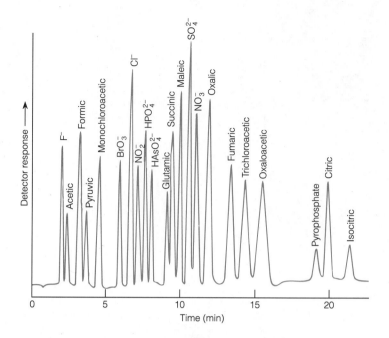

Figure 24-5 Chromatogram illustrating resolving power of anion chromatography. The HPLC apparatus used in this experiment is capable of drawing solvents from four reservoirs, giving gradients of solvent composition, ionic strength, and pH. Sodium carbonate background is suppressed by the hollow-fiber ion-exchange membrane. [Courtesy Dionex Corp., Sunnyvale, CA.]

sor, Na^+ is replaced by H^+, making a solution of HNO_3 and H_2CO_3 ($\rightleftharpoons$ $CO_2 + H_2O$). Na_2SO_4 formed in the outside bath is washed away. The process in analogous to kidney dialysis (Demonstration 25-1), in which large protein molecules are retained while small waste products diffuse into the surrounding medium.

In the absence of analyte, only H_2CO_3, which has very low conductivity, emerges from the suppressor. When analyte is present, HNO_3 or H_2SO_4 with high conductivity is produced and detected. Figure 24-5 shows the resolving power of anion chromatography.

Suppressed-ion cation chromatography is conducted in a similar manner, but the membrane suppressor is replaced by an anion-exchange suppressor column loaded with OH^-. Figure 24-4b illustrates the separation of KNO_3 and $CaSO_4$. With HCl as eluent, KCl and $CaCl_2$ emerge from the cation-exchange separator column, and KOH and $Ca(OH)_2$ emerge from the suppressor column. The HCl eluate is converted to H_2O in the suppressor. Self-regenerating suppressor units generate acid or base in the external compartment electrolytically.

Single-Column Ion Chromatography

If the ion-exchange capacity of the separator column is sufficiently low and if dilute eluent is used, ion suppression is unnecessary. For *single-column anion chromatography,* we use a resin with an exchange capacity near 5 μequiv/g, with 10^{-4} M Na^+ or K^+ salts of benzoic, *p*-hydroxybenzoic, or phthalic acid as eluent. These eluents give a low background conductivity, and analyte anions are detected by a small *change* in conductivity as they emerge from the column. By judicious choice of pH, an average eluent charge between 0 and -2 can be obtained, which allows control of eluent strength. Even dilute carboxylic acids (which are slightly ionized) are suitable eluents for some separations. *Single-column cation chromatography* is conducted with dilute

Benzene-1,4-diammonium cation is a stronger eluent that can be used instead of H^+ for suppressed-ion cation chromatography. After passing through the suppressor column, a neutral product is formed:

$$H_3\overset{+}{N}\!-\!\!\bigcirc\!\!-\!\overset{+}{N}H_3 \quad \overset{OH^-}{\longrightarrow}$$

Benzene-1,4-diammonium ion

$$H_2N\!-\!\!\bigcirc\!\!-\!NH_2$$

HNO$_3$ eluent for monovalent ions, and ethylenediammonium salts ($^+$H$_2$NCH$_2$CH$_2$NH$_2^+$) for divalent ions.

Benzoate — $-CO_2^-$

p-Hydroxybenzoate — HO—⟨⟩—CO_2^-

Phthalate — with CO_2^- and CO_2^-

Detectors

Conductivity detectors respond to all ions. In suppressed-ion chromatography, it is easy to measure analyte because eluent conductivity is reduced to near zero by the suppression step. Suppression also allows us to use eluent concentration gradients.

In single-column anion chromatography, the conductivity of the analyte anion is higher than that of the eluent, so conductivity increases when analyte emerges from the column. Detection limits are normally in the mid-ppb to low-ppm range but can be lowered by a factor of 10 by using carboxylic acid eluents instead of carboxylate salts.

Using benzoate or phthalate eluents provides for sensitive (<1 ppm) **indirect detection** of anions. In Figure 24-6 the eluate has a strong, constant ultraviolet absorption. In each emerging peak, nonabsorbing analyte anion replaces an equivalent amount of the absorbing eluent anion. Absorbance therefore *decreases* when analyte appears. For cation chromatography, CuSO$_4$ is a suitable ultraviolet-absorbing eluent.

Ion-Pair Chromatography

Ion-pair chromatography (also called *ion-interaction chromatography*) uses a reverse-phase HPLC column instead of an ion-exchange column. To separate a mixture of sodium salts of different anions, we add a hydrophobic cation such as Fe(phenanthroline)$_3^{2+}$. The *ion pairs* [Fe(phenanthroline)$_3^{2+}$][anion] are retained by the hydrophobic stationary phase. (Ion pairs do not exist in the aqueous solution, but pairing occurs in the nonpolar stationary phase.) Because the stationary phase retains ion pairs, the reverse-phase column effectively functions as an ion-exchange column.

Fe(phenanthroline)$_3^{2+}$ has a strong visible absorption, so it is used for indirect spectrophotometric detection of anions. Each anion eluted from the column is accompanied by an equivalent quantity of the colored cation.

24-3 Molecular Exclusion Chromatography

In **molecular exclusion chromatography** (also called **gel filtration** or *gel permeation chromatography*), molecules are separated according to their size. Small molecules penetrate into the small pores in the stationary phase, but large molecules do not (Figure 22-6). Because small molecules must pass through an effectively larger volume in the column, large molecules are eluted first (Figure 24-7). This technique is widely used in biochemistry to purify macromolecules.

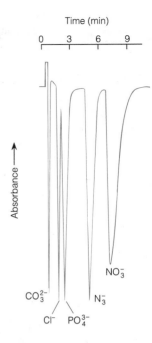

Figure 24-6 Indirect spectrophotometric detection of transparent ions. Column was eluted with 1 mM sodium phthalate plus 1 mM borate buffer, pH 10. [Reproduced from H. Small, *Anal. Chem.* **1982,** *54,* 462.] The principle of indirect detection is illustrated in Figure 24-23.

Mobile-phase additives for ion-pair chromatography

For separating anions:

Fe(phenanthroline)$_3^{2+}$

tetraalkylammonium (R$_4$N$^+$) with long-chain R group

For separating cations: alkylsulfonate (RSO$_3^-$) with long-chain R group

Large molecules pass through the column faster than small molecules do.

Salts of low molecular weight (or any small molecule) can be removed from solutions of large molecules by gel filtration. This technique, called *desalting,* is useful for changing the buffer composition of a macromolecule solution.

The Elution Equation

In molecular exclusion chromatography, the volume of mobile phase (V_m) is usually called the **void volume,** V_0. In Chapter 22 we derived the relationship

$$V_r = V_m + KV_s \qquad (22\text{-}21)$$

where V_r is the retention volume, V_s is the volume of stationary phase, and K is the partition coefficient (= concentration of solute in stationary phase / concentration in mobile phase). Equation 22-21 can be rearranged to the form

> The void volume is the same as the volume of mobile phase: $V_0 = V_m$.

$$K = \frac{V_r - V_m}{V_s} = \frac{V_r - V_0}{V_s}$$

The volume of solvent inside the gel particles is V_s. If the gel matrix occupied no volume, then V_s would be $V_t - V_0$, where V_t is the total volume of the column. The gel matrix does occupy some space, so $V_t - V_0$ is greater than V_s. However, V_s is proportional to $V_t - V_0$, because the liquid inside the particles occupies a constant fraction of the particles' volume.

The quantity K_{av} (read "K average") is defined as

$$K_{av} = \frac{V_r - V_0}{V_t - V_0} \qquad (24\text{-}6)$$

For a large molecule that does not penetrate the gel, $V_r = V_0$, and $K_{av} = 0$. For a small molecule that freely penetrates the gel, $V_r \approx V_t$, and $K_{av} \approx 1$. Molecules of intermediate size penetrate some gel pores, but not others, so K_{av} is between 0 and 1. Ideally, gel penetration is the only mechanism by which molecules are retained in this type of chromatography. In fact, there is always some adsorption, so K_{av} can be greater than 1.

Void volume is measured by passing a large, inert molecule through the column. Its elution volume is defined as V_0. Blue Dextran 2000, a blue dye of molecular weight 2×10^6, is most commonly used for this purpose. The total volume, V_t, is calculated from the column dimensions ($V_t = \pi r^2 \times$ length, where r = radius).

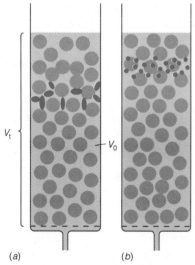

(a) (b)

Figure 24-7 (*a*) Large molecules cannot penetrate the pores of the stationary phase. They are eluted by a volume of solvent equal to the volume of mobile phase. (*b*) Small molecules that can be found inside or outside the gel require a larger volume for elution. V_t is the total column volume occupied by gel plus solvent. V_0 (= V_m) is the volume of the mobile phase. $V_t - V_0$ is the volume occupied by the gel plus its internal liquid phase.

Types of Gels

Gels for open column, preparative-scale molecular exclusion include Sephadex, whose structure was given in Figure 24-2, and Bio-Gel P, which is a polyacrylamide crosslinked by N,N'-methylenebisacrylamide (Figure 24-8). The smallest pore sizes in highly crosslinked gels exclude molecules with a molecular weight ≥ 700, whereas the largest pore sizes exclude molecules with molecular weights $\geq 10^8$ (Table 24-4). The finer the particle size of the gel, the greater the resolution and the slower the flow rate of the column.

For HPLC, polystyrene spheres (Figures 23-21 and 24-1) have pore sizes ranging from 5 nm up to hundreds of nanometers. Particles with 5-µm diameter yield up to 80 000 plates per meter of column length. Microporous

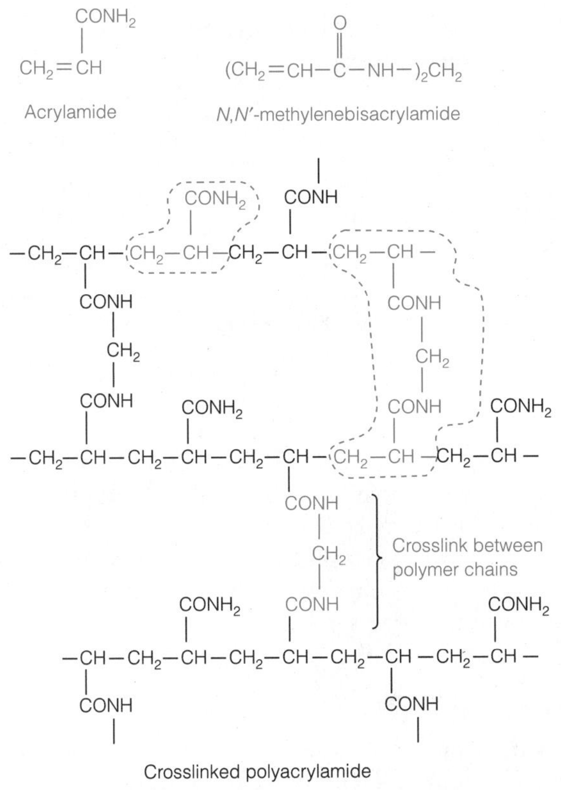

Figure 24-8 Structure of polyacrylamide.

TABLE 24-4   **Gel-filtration media**

Name[a]	Fractionation range (MW) for globular proteins	Name[a]	Fractionation range (MW) for globular proteins
Sephadex G-10	to 700	Bio-Gel P-2	100–1 800
Sephadex G-15	to 1 500	Bio-Gel P-4	800–4 000
Sephadex G-25	1 000–5 000	Bio-Gel P-6	1 000–6 000
Sephadex G-50	1 500–30 000	Bio-Gel P-10	1 500–20 000
Sephadex G-75	3 000–80 000	Bio-Gel P-30	2 500–40 000
Sephadex G-100	4 000–150 000	Bio-Gel P-60	3 000–60 000
Sephadex G-150	5 000–300 000	Bio-Gel P-100	5 000–100 000
Sephadex G-200	5 000–600 000	Bio-Gel P-150	15 000–150 000
		Bio-Gel P-200	30 000–200 000
		Bio-Gel P-300	60 000–400 000
Sephacryl S-200	5 000–250 000		
Sephacryl S-300	10 000–1 500 000	Bio-Gel A-0.5 m	<10 000–500 000
		Bio-Gel A-1.5 m	<10 000–1 500 000
		Bio-Gel A-5 m	10 000–5 000 000
Sepharose 2B	70 000–40 000 000	Bio-Gel A-15 m	40 000–15 000 000
Sepharose 4B	60 000–20 000 000	Bio-Gel A-50 m	100 000–50 000 000
Sepharose 6B	10 000–4 000 000	Bio-Gel A-150 m	1 000 000–150 000 000

a. Sephadex and Sephacryl are manufactured by Pharmacia Fine Chemical Co., Piscataway, NJ. Bio-Gel is sold by Bio-Rad Laboratories, Richmond, CA.

SOURCE: The information in this table was taken from the manufacturers' bulletins, which provide a great deal of useful information about these products.

1 Glutamate dehydrogenase (290 000)
2 Lactate dehydrogenase (140 000)
3 Enolase kinase (67 000)
4 Adenylate kinase (32 000)
5 Cytochrome *c* (12 400)

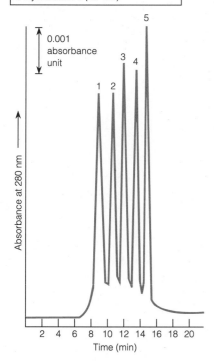

Figure 24-9 Separation of proteins by molecular exclusion chromatography with TSK 3000SW column. [Courtesy Varian Associates, Palo Alto, CA.]

TABLE 24-5 TSK SW silica for HPLC molecular exclusion chromatography

Designation	Pore size (nm)	Molecular weight range for globular proteins
G2000SW	13	500–60 000
G3000SW	24	1 000–300 000
G4000SW	45	5 000–1 000 000
G5000SW	100	>1 500 000

SOURCE: Perkin-Elmer liquid chromatography catalog.

silica with controlled pore size provides 10 000–16 000 plates per meter (Table 24-5). The silica is coated with a hydrophilic phase that minimizes solute adsorption. A hydroxylated polyether resin with a well-defined pore size can be used over the pH range 2–12, whereas silica phases generally cannot be used above pH 8. Particles with different pore sizes can be mixed to give a wider molecular size separation range.

Molecular Weight Determination

Gel filtration is used mainly to separate molecules of significantly different molecular weights (Figure 24-9). For each stationary phase, there is a range over which there is a logarithmic relation between molecular weight and elution volume (Figure 24-10). We can estimate the molecular weight of an unknown by comparing its elution volume with those of standards. We must exercise caution in interpreting our results, however, because molecules with the same molecular weight but different shapes exhibit different elution characteristics. For proteins, it is important to use an ionic strength high enough (>0.05 M) to eliminate electrostatic adsorption of solute by occasional charged sites on the gel.

Figure 24-10 Molecular weight calibration graph for polystyrene on Beckman μSpherogel™ molecular exclusion column (7.7-mm diameter × 30-cm length). The various gels used had pore sizes ranging from 5 nm to 100 μm. [Courtesy Anspec Co., Ann Arbor, MI.]

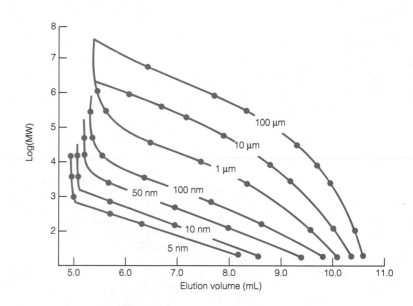

24-4 Affinity Chromatography

Affinity chromatography is used to isolate a single compound from a complex mixture. The technique is based on specific binding of that one compound to the stationary phase (Figure 22-6). When sample is passed through the column, only one solute is bound. After everything else has washed through, the one adhering solute is eluted by changing conditions to weaken its binding.

The technique is especially applicable to biochemistry, because of specific interactions between enzymes and substrates, antibodies and antigens, or receptors and hormones. Affinity chromatography has even been used to separate one type of cell from others.

As an example, nucleotides with coplanar *cis*-diol groups $\overset{\overset{\displaystyle OH}{|}}{-}C\overset{\overset{\displaystyle OH}{|}}{-}C-$ can be isolated by passage through Affi-Gel 601, which contains phenyl-boronic acid:

$$\text{Bio-Gel P-6}-\underbrace{\overset{O}{\overset{||}{C}}NHCH_2CH_2NH\overset{O}{\overset{||}{C}}CH_2CH_2\overset{O}{\overset{||}{C}}NH}_{\text{Spacer arm}}-\bigcirc\!\!-\underset{\underset{\text{Phenylboronic acid}}{HO\diagdown\quad\diagup OH}}{\overset{|}{B}}$$

Boronic acid covalently binds nucleotides, but not deoxynucleotides or cyclic nucleotides, which do not have *cis*-diols. The nucleotide can be eluted from the column by citrate-containing buffer.

The spacer arm in Affi-Gel extends the boronic acid away from the bulk of the gel particle. Spacer arms permit binding to macromolecules that are sterically hindered from approaching the gel particle too closely (Figure 24-11).

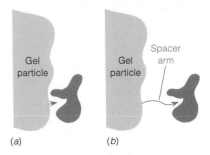

Figure 24-11 Spacer arm allows solute-stationary phase interactions that might otherwise be sterically forbidden.

24-5 Capillary Electrophoresis[2]

Electrophoresis is the migration of ions in solution under the influence of an electric field. In the **capillary electrophoresis** experiment shown in Figure 24-12, we use an electric field of ~30 kV to separate the components of a solution inside a fused silica (SiO_2) capillary tube that is 50 cm long and has an inner diameter of 25–75 µm. Different solutes have different *mobilities* and therefore migrate through the capillary at different speeds. Clever modifications of this experiment (described later) allow neutral molecules, as well as ions, to be separated. Box 24-1 shows how capillary electrophoresis is used to analyze the contents of a single cell.[3]

Capillary electrophoresis provides unprecedented resolution. An open tubular column reduces plate height and improves resolution (relative to that of a packed column) by eliminating the multiple path term (A) in the van Deemter equation (22-33). Capillary electrophoresis reduces plate height further by knocking out the mass transfer term (Cu_x) that comes from the finite time needed for solute to equilibrate between the mobile and stationary

Cations are attracted to the negative terminal (the cathode).

Anions are attracted to the positive terminal (the anode).

Electric potential difference = 30 kV

$$Electric\ field = \frac{30\ kV}{0.50\ m} = 60\frac{kV}{m}$$

Figure 24-12 Apparatus for capillary electrophoresis. One way to inject sample is to place the capillary in a sample vial and apply pressure to the vial or suction at the outlet of the capillary. The use of an electric field for sample injection is described in the text.

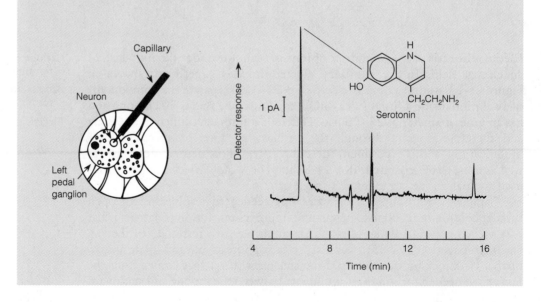

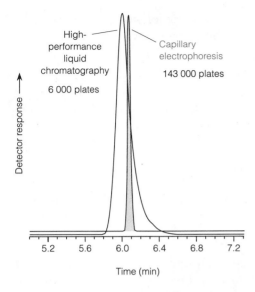

Figure 24-13 Comparison of peak widths for benzyl alcohol ($C_6H_5CH_2OH$) in capillary electrophoresis and HPLC. [From S. Fazio, R. Vivilecchia, L. Lesueur, and J. Sheridan, *Am. Biotech. Lab.* January 1990, p. 10.] Separations with ~3 million plates have been demonstrated [R. D. Smith, J. O. Olivares, N. T. Nguyen, and H. R. Udseth, *Anal. Chem.* **1988,** *60,* 436].

phases. In capillary electrophoresis *there is no stationary phase*. The only fundamental source of broadening under ideal conditions is longitudinal diffusion (B/u_x).

$$H = \cancel{\bigcirc{A}} + \frac{B}{u_x} + \cancel{\bigcirc{Cu_x}} \qquad (24\text{-}7)$$

Multiple path term eliminated by open tubular column   Mass transfer term eliminated because there is no stationary phase

(Other sources of broadening in real systems are mentioned later.) We routinely observe 50 000 to 500 000 theoretical plates in capillary electrophoresis (Figure 24-13), which is an order-of-magnitude better performance than chromatography.

Electrophoresis

When an ion with charge q (coulombs) is placed in an electric field E (V/m), the force on the ion is qE (newtons). In solution, the retarding frictional force is fu_{ep}, where u_{ep} is the velocity of the ion and f is the *friction coefficient*. The subscript ep refers to electrophoresis. The ion almost instantly reaches a steady speed when the accelerating force equals the frictional force:

$$
\begin{array}{c}
+ \\ + \\ +
\end{array}
\Bigg|
\quad \xleftarrow{fu_{ep}} \; \oplus \; \xrightarrow{qE} \quad
\Bigg|
\begin{array}{c}
- \\ - \\ -
\end{array}
\qquad
\underset{\substack{\text{Accelerating} \\ \text{force}}}{qE} = \underset{\substack{\text{Frictional} \\ \text{force}}}{fu_{ep}}
$$

$$\text{Ion}$$

Electrophoretic mobility:
$$u_{ep} = \frac{q}{f}E \equiv \mu_{ep}E \qquad (24\text{-}8)$$

Electrophoretic mobility

The *electrophoretic* **mobility** is the constant of proportionality between the speed of the ion and the electric field. Mobility is proportional to the charge

We encountered mobility earlier in connection with junction potentials (Table 15-1).

of the ion and inversely proportional to the friction coefficient. For molecules of similar size, the magnitude of the mobility increases with charge:

$$\mu_{ep} = -2.54 \times 10^{-8} \frac{m^2}{V \cdot s} \qquad \mu_{ep} = -4.69 \times 10^{-8} \frac{m^2}{V \cdot s} \qquad \mu_{ep} = -5.95 \times 10^{-8} \frac{m^2}{V \cdot s}$$

(solvent is H_2O at 25°C)

Viscosity measures resistance to flow in a fluid. The units are kg m^{-1} s^{-1}. Maple syrup is very viscous relative to water, which is more viscous than alcohol.

For a spherical particle of radius r moving through a fluid of viscosity η, the friction coefficient (f) is

Stokes equation: $$f = 6\pi\eta r \qquad (24\text{-}9)$$

Because mobility is q/f, the greater the radius, the lower the mobility. Most molecules are not spherical, but Equation 24-9 can be used to define an effective *hydrodynamic radius* of a molecule based on its mobility.

Electroosmosis

The inside wall of a fused silica capillary is covered with silanol (Si—OH) groups with a negative charge (Si—O$^-$) above pH $\approx$ 2. Figure 24-14a shows the *electric double layer* (Box 18-1) at the wall of the capillary. A tightly adsorbed, immobile layer of cations adjacent to the negative surface partially neutralizes the negative charge. The remaining negative charge is neutralized by excess mobile solvated cations in the *diffuse part of the double layer* in solution near the wall. The thickness of the diffuse part of the double layer ranges from ~10 nm when the ionic strength is 1 mM to ~0.3 nm when the ionic strength is 1 M.

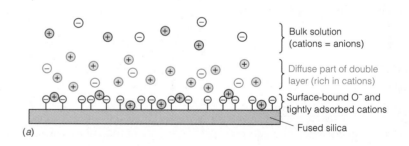

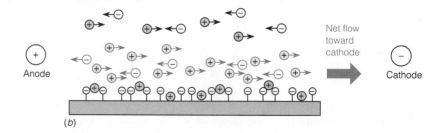

Figure 24-14 (*a*) Electric double layer created by negatively charged silica surface and nearby cations. (*b*) Predominance of cations in diffuse part of the double layer produces net electroosmotic flow toward the cathode when an external field is applied.

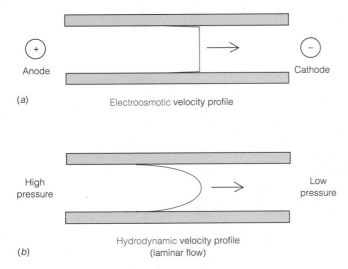

Figure 24-15   (*a*) Electroosmosis gives uniform flow over more than 99.9% of the cross section of the capillary. The speed decreases immediately adjacent to the capillary wall. (*b*) Parabolic velocity profile of hydrodynamic flow (also called *laminar flow*), with the highest velocity at the center of the tube and zero velocity at the walls.

In an electric field, cations are attracted to the cathode and anions are attracted to the anode (Figure 24-14b). Excess cations in the diffuse part of the double layer impart net momentum toward the cathode. This pumping action, called **electroosmosis** (also called *electroendosmosis*), is driven by solvated cations within ~10 nm of the walls and creates uniform pluglike *electroosmotic flow* of the entire solution toward the cathode (Figure 24-15a). This is in sharp contrast to ordinary *hydrodynamic flow,* which is driven by a pressure difference between the ends of the capillary. In hydrodynamic flow, the velocity profile through a cross section of the fluid is parabolic: It is fastest at the center and slows to zero at the walls (Figure 24-15b).

Ions in the diffuse part of the double layer adjacent to the capillary wall are the "pump" that drives electroosmotic flow.

The constant of proportionality between electroosmotic velocity (u_{eo}) and applied field is called the *electroosmotic mobility* (μ_{eo}).

Electroosmotic mobility:
$$u_{eo} = \mu_{eo}E \qquad (24\text{-}10)$$

$\uparrow$
Electroosmotic mobility
(units = $m^2/[V \cdot s]$)

Electroosmotic mobility is proportional to the surface charge density on the silica and inversely proportional to the square root of ionic strength. Electroosmosis decreases at low pH (because $Si\text{—}O^- \rightarrow Si\text{—}OH$ decreases surface charge density) and high ionic strength. At pH 9 in 20 mM borate buffer, electroosmotic flow is typically around 2 mm/s. At pH 3, flow is reduced by an order of magnitude.

Uniform electroosmotic flow contributes to the high resolution of capillary electrophoresis. Any effect that decreases uniformity creates band broadening and decreases resolution. Electric current in the capillary (flow of ions) generates heat (called *Joule heating*) at a rate of I^2R joules per second (Section 14-1), where I is current (A) and R is the resistance of the solution (ohms). Table 24-6 shows that under typical conditions, the centerline of the capillary channel is 0.02 to 0.3 K hotter than the edge of the channel. The viscosity of the solution is lower in the warmer region, which disturbs the flat electroosmotic profile of the fluid. This is not a serious problem in a 50-μm-

The capillary must be thin enough to dissipate heat rapidly. Temperature gradients disturb the flow and reduce resolution.

TABLE 24-6   Heat generation in capillary electrophoresis[a]

Buffer	Current density (A/cm^2)	Temperature difference (K) (capillary center to capillary wall)
100 mM sodium phosphate, pH 7.0	4.0	0.30
50 mM sodium citrate, pH 2.5	0.90	0.066
20 mM 3-(cyclohexylamino)propane sulfonate (CAPS), pH 11.0	0.31	0.024

a. Fused silica capillary with 50-μm diameter and electric field of 2.5×10^4 V/m.
SOURCE: Data from P. D. Grossman and J. C. Colburn, *Capillary Electrophoresis: Theory and Practice* (San Diego: Academic Press, 1992), Chapter 1.

diameter capillary tube, but the temperature gradient would be prohibitive in millimeter-diameter tubing. Fluid cooling in some instruments reduces the conductivity of solution inside the capillary and helps prevent runaway Joule heating.

Apparent Mobility

The apparent (or observed) mobility (μ_{ap}) of an ion is the sum of the electrophoretic mobility of the ion plus the electroosmotic mobility of the solution.

Apparent mobility:
$$\mu_{app} = \mu_{ep} + \mu_{eo} \qquad (24\text{-}11)$$

For an analyte *cation* moving in the same direction as the electroosmotic flow, μ_{ep} and μ_{eo} have the same sign, so μ_{app} is greater than μ_{ep}. Electrophoresis transports *anions* in the direction opposite from electroosmosis (Figure 24-14b), so for anions the two terms in Equation 24-11 have opposite signs. At neutral or high pH, brisk electroosmosis transports anions to the *cathode* because electroosmosis is usually faster than electrophoresis. At low pH, electroosmosis is weak and anions may never reach the detector. (If you want to separate anions at low pH, the polarity of the instrument can be reversed to make the sample side negative and the detector side positive.)

The apparent mobility (μ_{app}) of a particular species is the net speed (u_{net}) of the species divided by the electric field (E):

Apparent mobility:
$$\mu_{app} = \frac{u_{net}}{E} = \frac{L_d / t}{V / L_t} \qquad (24\text{-}12)$$

where L_d is the length of column from injection to the detector, L_t is the total length of the column from end to end, V is the voltage applied between the two ends, and t is the time required for solute to migrate from the injection end to the detector. Electroosmotic flow is measured by adding an ultraviolet-absorbing neutral solute to the sample and measuring its migration time ($t_{neutral}$) to the detector. Electroosmotic mobility is the speed of the neutral species ($u_{neutral}$) divided by the electric field:

Electroosmotic mobility:

$$\mu_{eo} = \frac{u_{neutral}}{E} = \frac{L_d / t_{neutral}}{V / L_t} \qquad (24\text{-}13)$$

The electrophoretic mobility of an analyte is the difference $\mu_{app} - \mu_{eo}$.

For maximum precision, mobilities are measured relative to an internal standard. Absolute variation from run to run should not affect relative mobilities, unless there are time-dependent (nonequilibrium) interactions of the solute with the wall.

Theoretical Plates

Consider a capillary of length L. In Section 22-4, we defined the number of theoretical plates as $N = L^2/\sigma^2$, where σ is the standard deviation of the band. If the only mechanism of zone broadening is longitudinal diffusion, the standard deviation was given by Equation 22-26: $\sigma = \sqrt{2Dt}$, where D is the diffusion coefficient and t is the migration time ($= L/u_{net} = L/[\mu_{app}E]$). Combining these expressions with the definition of electric field ($E = V/L$, where V is the applied voltage) gives an expression for the number of plates in capillary electrophoresis:

Number of plates:

$$N = \frac{\mu_{app}V}{2D} \qquad (24\text{-}14)$$

Number of plates: $N = \dfrac{L^2}{\sigma^2}$

L = distance traveled
σ = standard deviation of Gaussian band

How many theoretical plates might we hope to attain? Using a typical value of $\mu_{app} = 2 \times 10^{-8} \text{ m}^2/(\text{V·s})$ (derived for a 10-min migration time in a 55-cm-long capillary with 25 kV), and using diffusion coefficients from Table 22-1, we find

for K^+:

$$N = \frac{[2 \times 10^{-8} \text{ m}^2/(\text{V·s})][25\,000 \text{ V}]}{2(2 \times 10^{-9} \text{ m}^2/\text{s})} = 125\,000 \text{ plates}$$

for serum albumin:

$$N = \frac{[2 \times 10^{-8} \text{ m}^2/(\text{V·s})][25\,000 \text{ V}]}{2(0.059 \times 10^{-9} \text{ m}^2/\text{s})}$$

$$= 4.2 \times 10^6 \text{ plates}$$

For the small, rapidly diffusing K^+ ion, we expect 125 000 plates. For the slowly diffusing protein serum albumin (MW 65 000), we expect 4 million plates! High plate count means that bands are very narrow and resolution between adjacent bands is excellent.

In reality, additional sources of zone broadening include the finite width of the solute band applied to the capillary (Equation 22-32), a parabolic flow profile from heat produced inside the capillary tube, adsorption of solute on the capillary wall (which therefore serves as a stationary phase), the finite length of the detection zone, and mobility mismatch of solute and buffer ions that leads to less than ideal electrophoretic behavior. If these other factors are properly controlled, $\sim 10^5$ plates are routinely achieved.

Equation 24-14 says that the higher the voltage, the greater the number of plates. This equation also tells us that plate count is independent of capillary length at constant voltage. However, a longer capillary allows us to use a higher voltage, which increases the plate count. Voltage is ultimately limited by capillary heating, which produces a parabolic temperature profile that gives band broadening. The optimum voltage is found by making an

Background electrolyte (the solution in the capillary and the electrode reservoirs) controls pH and electrolyte composition in the capillary. Background electrolyte is also called *run buffer*.

Ohm's law plot of current versus voltage with *background electrolyte* in the capillary. In the absence of overheating, this should be a straight line. The maximum allowable voltage is the value at which the curve deviates from linearity (by, say, 5%). Buffer concentration and composition, thermostat temperature, and active cooling all play roles in how much voltage can be tolerated. Up to a point, higher voltage gives better resolution and faster separations.

Controlling the Environment Inside the Capillary

 A fused silica capillary is prepared for use the first time by washing for 15 min each with 1 M NaOH and 0.1 M NaOH, followed by the background electrolyte (typically containing 20 mM buffer) that will be used for the analytical separation. For subsequent use, a 2-min wash with 0.1 M NaOH is followed by equilibration for at least 5 min with background electrolyte. Commercial instruments force fluid through the capillary either by applying pressure to the source (Figure 24-12) or by reducing pressure at the outlet. When changing buffers, allow at least 5 min of flow for equilibration. Periodically, buffer in both reservoirs should be replaced because ions become depleted and electrolysis raises the pH at the cathode and lowers the pH at the anode. For work in the pH range 4–6, at which equilibration of the column wall with buffer is critical and very slow, the capillary needs frequent regeneration with 0.1 M NaOH if migration times become erratic.

The capillary may be left filled with buffer in the instrument. If you need to remove the capillary from the instrument, wash the capillary with distilled water for 5 min and then draw air through for 5 min. If the capillary is not in a preassembled cartridge, the optical window should be covered with protective tubing. When storing a capillary, it should be filled with distilled water.

Different separations require more or less electroosmotic flow. For example, small anions with high mobility and highly negatively charged proteins require a brisk electroosmotic flow or they will not migrate toward the cathode. At pH 2, there is little charge on the silanol groups and little electroosmotic flow. At pH 11, the wall is highly charged and electroosmotic flow is strong. Proteins, with many positively charged substituents, may bind tightly to negatively charged silica. To control this, 30–60 mM diaminopropane (which gives $^+H_3NCH_2CH_2CH_2NH_3^+$) may be added to the background electrolyte to neutralize the charge on the wall. The wall can be modified by covalent attachment of silanes with neutral, hydrophilic substituents to reduce the wall charge to near zero (Figure 24-16). However, many coatings are unstable under alkaline conditions.

We can add a cationic surfactant, such as cetyltrimethylammonium bromide, to the background electrolyte to reverse the direction of electroosmotic flow. This molecule has a positive charge at one end and a long hydrocarbon tail at the other. The surfactant coats the negatively charged silica with the tails pointing away from the surface (Figure 24-17). A second layer of surfactant orients itself in the opposite direction so that the tails form a nonpolar hydrocarbon layer. This *bilayer* adheres tightly to the wall of the capillary and effectively reverses the charge of the wall from positive to negative. Buffer anions coating the bilayer create electroosmotic flow from the cathode to the anode when a voltage is applied. Best results are obtained when the capillary environment is freshly regenerated for each run.

Example of covalent wall coating:

$$\begin{array}{c} -O \\ | \\ -O \end{array} \overset{\textstyle OCH_3}{\underset{\textstyle O}{\overset{|}{\underset{|}{Si}}}} O\text{—}O\text{—}\overset{\text{Poly-}}{\text{acrylamide}}$$

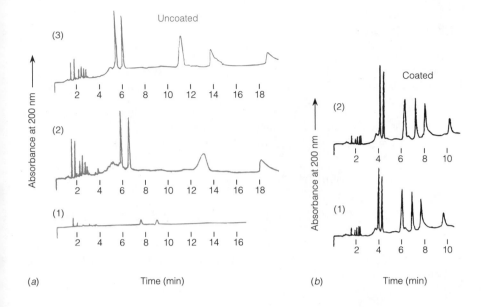

(a)   Time (min)   (b)   Time (min)

Figure 24-16 *Wall effects:* Consecutive injections of proteins in (*a*) uncoated and (*b*) covalently coated capillaries at pH 8.5. The coating is a hydrophilic polymer. Most of the protein from the first injection into the uncoated column stuck to the walls and never reached the detector. Adsorption gives irreproducible migration times and peak areas, as well as asymmetric peak shapes. Migration is from the cathode to the anode, which means that electrophoresis of the negatively charged proteins is faster than electroosmosis toward the cathode. [Courtesy Bio-Rad Laboratories, Richmond, CA.]

Sample Injection and Composition

Injection may be *hydrodynamic* (using a pressure difference between the two ends of the capillary) or *electrokinetic* (using an electric field to drive sample into the capillary). For hydrodynamic injection (Figure 24-12), the capillary is dipped into a sample solution and the injected volume is

Hydrodynamic injection: $\text{volume} = \dfrac{\Delta P \pi d^4 t}{128 \eta L_t}$ (24-15)

where ΔP is the pressure difference between the ends of the capillary, d is the inner diameter of the capillary, t is the injection time, η is the sample viscosity, and L_t is the total length of the capillary.

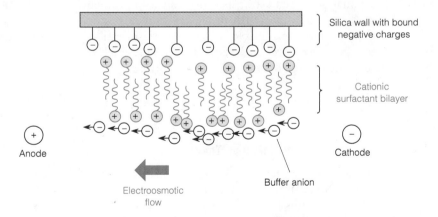

Figure 24-17 Charge reversal created by a cationic surfactant bilayer coated on the capillary wall. The diffuse part of the double layer contains predominantly anions, and electroosmotic flow is in the direction opposite that shown in Figure 24-14. The surfactant is the cetyltrimethylammonium ion, $n\text{-}C_{16}H_{33}N(CH_3)_3^+$, represented as $\sim\!\!\oplus$ in the illustration.

EXAMPLE Hydrodynamic Injection Time

How much time is required to inject a sample equal to 1.0% of the length of a 50-cm capillary if the diameter is 50 μm and the pressure difference is 2.0×10^4 Pa (0.20 atm)? Assume that the viscosity is 0.001 0 kg/(m·s), which is close to the viscosity of water.

Solution The injection plug will be 0.50 cm long and occupy a volume of $\pi r^2 \times$ length $= \pi(25 \times 10^{-6}$ m$)^2(0.5 \times 10^{-2}$ m$) = 9.8 \times 10^{-12}$ m^3. The required time is

$$t = \frac{128\eta L_t(\text{volume})}{\Delta P \pi d^4}$$

$$= \frac{128[0.001\,0\ \text{kg}/(\text{m·s})](0.50\ \text{m})(9.8 \times 10^{-12}\ \text{m}^3)}{(2.0 \times 10^4\ \text{Pa})\pi(50 \times 10^{-6}\ \text{m})^4} = 5.0\ \text{s}$$

The units work out when we realize that Pa = force/area = $(\text{kg·m/s}^2)/\text{m}^2 =$ kg/(m·s^2).

For electrokinetic injection, we dip the capillary in sample solution and apply a voltage between the ends of the column. The moles of each ion taken into the capillary in t seconds are

Electrokinetic injection: $\quad$ moles injected $= \mu_{app} \underbrace{\left(E\,\frac{\kappa_b}{\kappa_s}\right)}_{\text{Effective electric field}} t\pi r^2 C$ $\quad$ (24-16)

where μ_{app} is the apparent mobility of analyte ($= \mu_{ep} + \mu_{eo}$), E is the applied electric field (V/m), r is the capillary radius, C is the sample concentration (mol/m^3), and κ_b/κ_s is the ratio of conductivities of the buffer and sample. One problem with electrokinetic injection is that each analyte has a different mobility. For qualitative analysis, this is not usually a problem. For quantitative analysis, the sample injected into the column does not have the same composition as the original sample. Electrokinetic injection is most useful for capillary gel electrophoresis (described later), in which the liquid in the capillary is too viscous for hydrodynamic injection.

EXAMPLE Electrokinetic Injection Time

How much time is required to inject a sample equal to 1.0% of the length of a 50-cm capillary if the diameter is 50 μm and the injection electric field is 10 kV/m? Assume that the sample has 1/10 of the conductivity of background electrolyte and $\mu_{app} = 2.0 \times 10^{-8}$ m^2/(V·s).

Solution Equation 24-16 is simpler than it looks. The factor κ_b/κ_s is 10 in this case. The length of sample plug injected onto the column is just (sample speed) × (time) = $\mu_{app}Et$. The desired injection plug will be 0.50 cm long.

The required time is

$$t = \frac{\text{plug length}}{\text{speed}} = \frac{\text{plug length}}{\mu_{app}\left(E\frac{\kappa_b}{\kappa_s}\right)}$$

$$= \frac{0.0050 \text{ m}}{[2.0 \times 10^{-8} \text{ m}^2/(\text{V·s})](100\,000 \text{ V/m})} = 2.5 \text{ s}$$

Equation 24-16 multiplies the plug length times its cross-sectional area to find its volume and then multiplies by concentration to find the moles in that volume.

. . . .

Optimal buffer concentration in the injection solution is 1/10 of the background electrolyte concentration, and the sample concentration is 1/500 of the background electrolyte concentration. Because the injection solution has a lower ionic strength, its conductivity is lower and its resistance is greater than that of the background electrolyte. The electric field across the sample plug inside the capillary is much higher than the electric field in the background electrolyte. Figure 24-18 shows ions in the sample plug migrating very fast, because the electric field is very high. When ions reach the zone boundary, they slow down because the field is lower outside of the sample plug. This process, called **stacking,** continues until most of the analyte

Electroneutrality is maintained throughout the sample plug in Figure 24-18 by background electrolyte ions that are not shown.

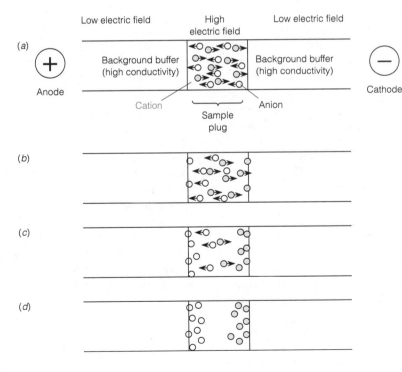

Figure 24-18 Stacking of anions and cations at opposite ends of the low-conductivity sample plug (zone) occurs because the electric field in the sample plug is much higher than the electric field in the background electrolyte. Time increases from (*a*) to (*d*). Electroneutrality is maintained by migration of background electrolyte ions that are not shown.

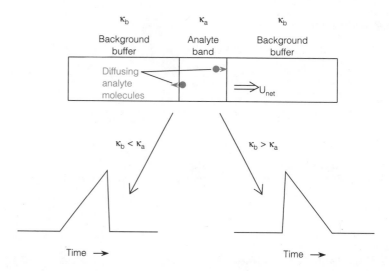

Figure 24-19   Irregular peak shapes arise when the conductivity of the analyte band (κ_a) is not the same as the background conductivity (κ_b).

cations are concentrated in a narrow zone at one end of the sample plug and most of the analyte anions are at the other end. By this means, we can concentrate a broad injection into a very narrow band. The injection zone can be as large as 10–20% of the capillary length, thus allowing dilute solutions to be analyzed.[5]

If the conductivity of an analyte band is significantly different from the conductivity of the background electrolyte, peak distortion occurs. Figure 24-19 shows a band containing one analyte (as opposed to the sample plug in Figure 24-18, which contains all analytes in the entire injection). Electric field strength is inversely proportional to conductivity: The higher the conductivity, the lower the electric field. Suppose that the background conductivity is greater than the analyte conductivity ($\kappa_b > \kappa_a$). This means that the electric field is lower outside the analyte band than inside the band. The band migrates to the right in Figure 24-19. An analyte molecule that diffuses past the front on the right suddenly encounters a lower electric field and its electrophoretic speed decreases. Soon, the analyte zone catches up with the molecule and it is back in the zone. A molecule that diffuses out of the zone on the left encounters a lower electric field and its electrophoretic speed also decreases. The analyte zone is moving faster than the wayward molecule, and pulls away. This condition leads to a sharp front and a broad tail, as shown in the lower right electropherogram in Figure 24-19. When $\kappa_b < \kappa_a$, we observe the opposite electropherogram.

To minimize band distortion, sample concentration must be much less than the background electrolyte concentration. Otherwise it is necessary to choose a buffer co-ion that has the same electrophoretic mobility as the analyte ion. (The *co-ion* is the buffer ion with the same charge as analyte. The *counterion* has the opposite charge.)

Detectors

Because water is so transparent, *ultraviolet detectors* can operate at wavelengths as short as 185 nm, where most solutes have strong absorption.

(a)

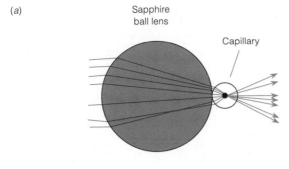

Sapphire
ball lens

Capillary

(b)

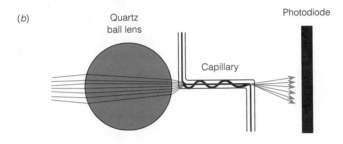

Quartz
ball lens

Photodiode

Capillary

(c)

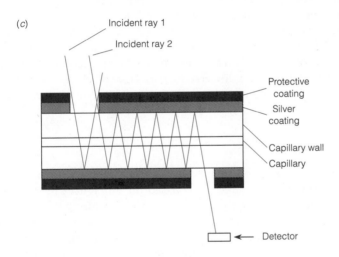

Incident ray 1

Incident ray 2

Protective
coating

Silver
coating

Capillary wall

Capillary

Detector

Figure 24-20 Optical designs for ultraviolet absorbance detection in capillary electrophoresis. (*a*) Sapphire ball lens concentrates illumination in the capillary. (*b*) Bent capillary increases pathlength. (*c*) Multiple reflections from silver-coated capillary increase pathlength. In very high resolution separations, increased pathlengths give peak broadening if the analyte zone is narrower than the detection zone. [From M. Albin, P. D. Grossman, and S. E. Moring, *Anal. Chem.* **1993,** *65,* 489A.]

Figure 24-20 shows strategies to increase the optical pathlength for higher sensitivity. *Fluorescence detection* is sensitive to naturally fluorescent analytes or to fluorescent derivatives, as shown in Problem 21. *Amperometric detection* is sensitive to analytes that can be oxidized or reduced at an electrode. A carbon fiber microelectrode (Figure 18-22) can be placed in the end of the capillary,[6] or an ordinary electrode can be positioned at the outlet (Figure 24-21). Liquid emerging from capillary electrophoresis can be directed into a *mass spectrometer* that not only responds to analyte but also provides information on molecular structure.[7] *Conductivity detection* with ion-exchange suppression of the background electrolyte (as in Figure 24-4) gives 1–10 ppb sensitivity for small analyte ions (Figure 24-22).

Figure 24-23 shows the principle of *indirect detection,*[8] which applies to fluorescence, absorbance, amperometry, conductivity, and other forms of

With a microelectrode in a 9-μm-inner-diameter capillary, the detection limit for the neurotransmitter serotonin is 700 zmol (7×10^{-19} mol).

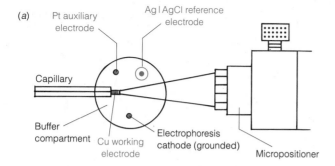

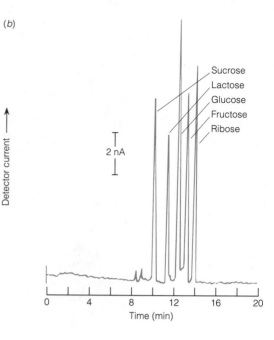

Figure 24-21 (*a*) Amperometric detection with macroscopic working electrode at the outlet of the capillary. (*b*) Electropherogram of sugars separated in 0.1 M NaOH, in which OH groups are partially ionized, so the molecules are anions. [From J. Ye and R. P. Baldwin, *Anal. Chem.* **1993,** *65,* 3525.]

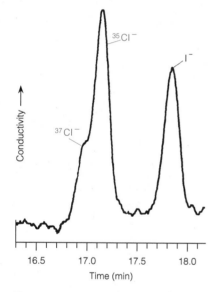

Figure 24-22 Partial separation of isotopes of 1 µM chloride by capillary electrophoresis with conductivity detection. Prior to detection, eluate is passed through a cation-exchange membrane to convert the conductive 2 mM sodium borate background electrolyte into poorly conductive boric acid. [From N. Avdalovic, C. A. Pohl, R. D. Rocklin, and J. R. Stillian, *Anal. Chem.* **1993,** *65,* 1470.]

detection. A component with a steady background signal is added to the background electrolyte. In the analyte band, analyte molecules displace the chromophoric component, so the detector signal decreases when analyte passes by. Figure 24-24 shows indirect detection of anions in the presence of the ultraviolet-absorbing anion, chromate. Electroneutrality dictates that analyte bands must have a lower concentration of chromate anion than is found in the background electrolyte. Benzoate and phthalate are other anions useful for this purpose.

Modes of Separation

The type of electrophoresis we have been discussing so far is called **capillary zone electrophoresis.** Separation is based on differences in electrophoretic mobility. If the capillary wall is negative, electroosmotic flow is toward the cathode (Figure 24-14) and the order of elution is cations < neutrals < anions. If the capillary wall charge is reversed by coating it with a cationic surfactant (Figure 24-17) and the instrument polarity is reversed, the order of elution is anions < neutrals < cations. Neither scheme separates neutral molecules.

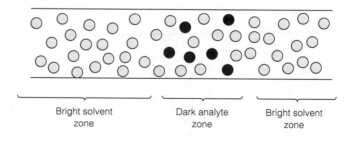

Figure 24-23 Principle of indirect detection. When analyte emerges from the capillary, the strong background signal decreases.

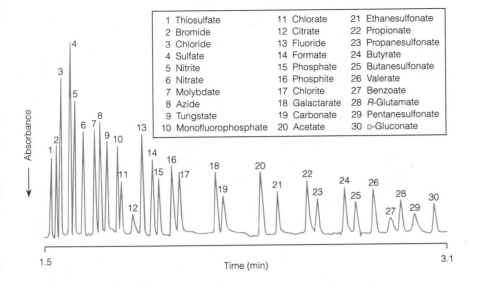

1 Thiosulfate	11 Chlorate	21 Ethanesulfonate
2 Bromide	12 Citrate	22 Propionate
3 Chloride	13 Fluoride	23 Propanesulfonate
4 Sulfate	14 Formate	24 Butyrate
5 Nitrite	15 Phosphate	25 Butanesulfonate
6 Nitrate	16 Phosphite	26 Valerate
7 Molybdate	17 Chlorite	27 Benzoate
8 Azide	18 Galactarate	28 R-Glutamate
9 Tungstate	19 Carbonate	29 Pentanesulfonate
10 Monofluorophosphate	20 Acetate	30 D-Gluconate

Figure 24-24 Anion separation with indirect ultraviolet absorbance detection (254 nm) of CrO_4^{2-} in the background electrolyte. Thirty anions were separated in 3 min on a 50-μm-diameter × 60-cm-long capillary at 30 kV. [From W. R. Jones, P. Jandik, and R. Pfeifer, *Am. Lab.* May 1991, p. 40.]

Sodium dodecyl sulfate (n-$C_{12}H_{25}OSO_3^-Na^+$)

Micelles were described in Section 12-8 and Figure 12-11.

Micellar electrokinetic capillary chromatography separates neutral molecules and ions. We illustrate a case in which the anionic surfactant sodium dodecyl sulfate is present above its *critical micelle concentration,* so that negatively charged micelles are formed.[9] In Figure 24-25, electroosmotic flow is to the right. Electrophoretic migration of the negatively charged micelles is to the left, but net motion is to the right because the strong electroosmotic flow dominates.

In the absence of the micelles, all neutral molecules would reach the detector in time t_0. Micelles injected with the sample reach the detector in time t_{mc}, which is longer than t_0 because the micelles migrate upstream. If a neutral molecule equilibrates between free solution and the inside of the micelles, its migration time is increased, because it migrates at the slower rate of the micelle part of the time. In this case, the neutral molecule reaches the detector at a time between t_0 and t_{mc}. *The more time the neutral molecule spends inside the micelle, the longer is its migration time.* This is the basis for micellar electrokinetic capillary chromatography. Migration times of cations and anions are also affected by micelles, because ions partition between the solution and the micelles and interact electrostatically with the micelles.

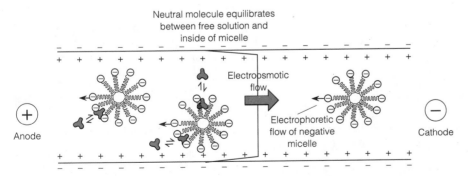

Figure 24-25 Negatively charged sodium dodecyl sulfate micelles migrate upstream against the electroosmotic flow. Neutral molecules are in dynamic equilibrium between free solution and the inside of the micelle. The more time spent in the micelle, the more the neutral molecule lags behind the electroosmotic flow.

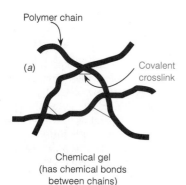

Polymer chain

(a)

Covalent crosslink

Chemical gel
(has chemical bonds
between chains)

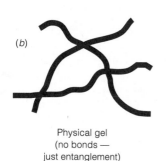

(b)

Physical gel
(no bonds —
just entanglement)

Figure 24-26 (*a*) A chemical gel contains covalent crosslinks between different polymer chains. (*b*) A physical gel is not crosslinked but derives its properties from physical entanglement of the polymers.

Micellar electrokinetic capillary chromatography is truly a form of chromatography because the micelle behaves as a pseudostationary phase. Separation of neutral molecules is based on partitioning between the solution and the "stationary" phase. The mass transfer term Cu_x is no longer zero in the van Deemter equation 24-7, but mass transfer into the micelles is so fast that band broadening is negligible.

Your imagination can run wild with the extraordinary range of variables in micellar electrokinetic capillary chromatography. We can add anionic, cationic, zwitterionic, and neutral surfactants to change the partition coefficients of analytes. (Cationic surfactants also change the charge on the wall and the direction of electroosmotic flow.) We can add solvents such as acetonitrile and *N*-methylformamide to increase the solubility of organic analytes and to change the partition coefficient between the solution and the micelles. We can add cyclodextrins (Box 23-1), with an optically active cavity into which small molecules fit, to separate optical isomers (enantiomers) that spend different fractions of the time associated with the cyclodextrins.

Capillary gel electrophoresis is a variant of gel electrophoresis, which has been a primary tool in biochemistry for three decades. We illustrate one example involving sequencing of DNA, which was introduced in Box 19-1. Polymer gels used to separate macromolecules according to size have customarily been chemical gels, in which chains are crosslinked by chemical bonds (Figure 24-26a). Chemical gels are hard to flush from the capillary when a problem develops, so physical gels (Figure 24-26b) in which the linear polymers are simply entangled are now common. Physical gels can be flushed and reloaded to generate a fresh capillary for each separation.

Macromolecules are separated in a gel by *sieving,* in which smaller molecules migrate faster than large molecules through the entangled polymer network. Color Plate 17 shows part of a DNA sequence analysis in which a mixture of fluorescence-labeled fragments with up to 400 nucleotides was separated in a capillary containing 6% polyacrylamide (Figure 24-8, with no crosslinks). DNA with 30 nucleotides had a migration time of 9 min, and DNA with 400 nucleotides required 34 min. The terminal nucleotide of each fragment was labeled with one of two fluorescent labels that was detected by fluorescence at two wavelengths. The combination of two labels and two wavelengths allows a unique assignment of each of the four possible nucleotides.

Terms to Understand

affinity chromatography
anion exchanger
capillary electrophoresis
capillary gel electrophoresis
capillary zone electrophoresis
cation exchanger
crosslinking
deionized water
Donnan equilibrium

electroosmosis
electrophoresis
gel
gel filtration
gradient elution
indirect detection
ion chromatography
ion-exchange chromatography
ion-pair chromatography

micellar electrokinetic capillary electrophoresis
mobility
molecular exclusion chromatography
preconcentration
resin
selectivity coefficient
stacking
suppressed-ion chromatography
void volume

Summary

Ion-exchange resins and gels contain covalently bound charged groups that attract solute counterions (and that exclude ions having the same charge as the resin). Polystyrene resins are useful for small ions. Greater crosslinking of the resin increases the capacity, selectivity, and time needed for equilibration. Ion-exchange gels based on cellulose and dextran have large pore sizes and low charge densities, suitable for the separation of macromolecules. Certain inorganic solids have ion-exchange properties and are useful at extremes of temperature or radiation. Ion exchangers operate by the principle of mass action, with a gradient of increasing ionic strength most commonly used to effect a separation.

In suppressed-ion chromatography, a separator column separates ions of interest, and a suppressor (membrane or column) converts eluent to a nonionic form so that analytes can be detected by their conductivity. Alternatively, single-column ion chromatography uses an ion-exchange column and low-concentration eluent. If the eluent absorbs light, indirect spectrophotometric detection is convenient and sensitive. Ion-pair chromatography utilizes a hydrophobic ionic additive in the eluent to make a reverse-phase column function as an ion-exchange column.

Molecular exclusion chromatography is based on the relative inability of large molecules to enter pores in the stationary phase. Small molecules enter these spaces and therefore exhibit longer elution times than large molecules. Exclusion chromatography is used for separations based on size and for molecular weight determinations of macromolecules. In affinity chromatography, we design a stationary phase to interact with one particular solute in a complex mixture. After all other components have been eluted, the desired species is liberated by a change in conditions.

In capillary zone electrophoresis, ions are separated by differences in mobility in a strong electric field applied between the ends of a silica capillary tube. The greater the charge and the smaller the hydrodynamic radius, the greater the electrophoretic mobility. Normally, the capillary wall is negative, and solution is transported from anode to cathode by electroosmosis of cations in the electric double layer. Solute cations arrive first, followed by neutral species, followed by solute anions (if electroosmosis is stronger than electrophoresis). Apparent mobility is the sum of electrophoretic mobility and electroosmotic mobility (which is the same for all species). Zone dispersion (broadening) arises mainly from longitudinal diffusion and the finite length of the injected sample. Stacking of solute ions in the capillary occurs when the sample has a low conductivity. Electroosmotic flow is reduced at low pH, because surface $Si-O^-$ groups are protonated. $Si-O^-$ groups can be masked by polyamine cations, and the wall charge can be reversed by a cationic surfactant that forms a bilayer along the wall. Covalent coatings reduce electroosmosis and wall adsorption. Hydrodynamic sample injection uses pressure or siphoning; electrokinetic injection uses an electric field. Short-wavelength ultraviolet absorbance is commonly used for detection. Micellar electrophoretic capillary chromatography uses micelles as a pseudostationary phase to separate neutral molecules and ions. Capillary gel electrophoresis uses physical or chemical gels to separate macromolecules by sieving. In contrast to molecular exclusion chromatography, small molecules move fastest in gel electrophoresis.

Exercises

A. Vanadyl sulfate ($VOSO_4$), as supplied commercially, is contaminated with H_2SO_4 and H_2O. A solution was prepared by dissolving 0.2447 g of impure $VOSO_4$ in 50.0 mL of water. Spectrophotometric analysis indicated that the concentration of the blue VO^{2+} ion was 0.0243 M. A 5.00-mL sample was passed through a cation-exchange column loaded with H^+. When VO^{2+} from the 5.00-mL sample became bound to the column, the H^+ released required 13.03 mL of 0.02274 M NaOH for titration. Find the weight percent of each component ($VOSO_4$, H_2SO_4, and H_2O) in the vanadyl sulfate.

B. Blue Dextran 2000 was eluted in gel filtration in a volume of 36.4 mL from a 2.0 × 40-cm (diameter × length) column of Sephadex G-50 (fractionation range MW 1500 to MW 30000).

(a) At what retention volume would hemoglobin (MW 64000) be expected?

(b) At what volume is $^{22}NaCl$ expected?

(c) What would be the retention volume of a molecule with $K_{av} = 0.65$?

C. The capillary electrophoresis experiment in Figure 24-22 was conducted near pH 9, where the electroosmotic flow is stronger than the electrophoretic flow.

(a) Draw a picture of the capillary, showing the placement of the anode, cathode, injector, and detector. Show the direction of electroosmotic flow and the direction of electrophoretic flow of a cation and an anion. Show the direction of net flow.

(b) Using Table 15-1, explain why Cl^- has a shorter migration time than I^-. Predict whether Br^- will have a shorter migration time than Cl^- or a greater migration time than I^-.

(c) Give a physical explanation of why the mobility of I^- is greater than that of Cl^-.

(d) Explain why $^{37}Cl^-$ has a shorter migration time than $^{35}Cl^-$.

Problems

Ion-Exchange and Ion Chromatography

1. State the purpose of the separator column and suppressor in suppressed-ion chromatography. For cation chromatography, why does the suppressor contain anion-exchange resin?

2. State the effects of increasing crosslinking on an ion-exchange column.

3. What is deionized water? What kind of impurities are not removed by deionization?

4. The exchange capacity of an ion-exchange resin can be defined as the number of moles of charged sites per gram of dry resin. Describe how you would measure the exchange capacity of an anion-exchange resin by using standard NaOH, standard HCl, or any other reagent.

5. Consider a protein with a net negative charge tightly adsorbed on an anion-exchange gel at pH 8.

(a) How will a gradient of eluent pH (from pH 8 to some lower pH) be useful for eluting the protein? Assume that the ionic strength of the eluent is kept constant.

(b) How would a gradient of ionic strength (at constant pH) be useful for eluting the protein?

6. How should Figure 23-33 be modified to produce a gradient of decreasing ionic strength?

7. Propose a scheme for separating trimethylamine, dimethylamine, methylamine, and ammonia from each other by ion-exchange chromatography.

8. Suppose that an ion-exchange resin (R^-Na^+) is immersed in a solution of NaCl. Let the concentration of R^- in the resin be 3.0 M.

(a) What will be the ratio $[Cl^-]_o/[Cl^-]_i$ if $[Cl^-]_o$ is 0.10 M?

(b) What will be the ratio $[Cl^-]_o/[Cl^-]_i$ if $[Cl^-]_o$ is 1.0 M?

(c) Will the fraction of electrolyte inside the resin increase or decrease as the outside concentration of electrolyte increases?

9. Compounds with $-\overset{\overset{\displaystyle OH}{|}}{C}-\overset{\overset{\displaystyle OH}{|}}{C}-$ or $-\overset{\overset{\displaystyle OH}{|}}{C}-\overset{\overset{\displaystyle NH_2}{|}}{C}-$ linkages can be analyzed by cleavage with periodate. One mole of 1,2-ethanediol consumes 1 mol of periodate:

$$\begin{array}{c} CH_2OH \\ | \\ CH_2OH \end{array} + IO_4^- \longrightarrow 2CH_2{=}O + H_2O + IO_3^-$$

1,2-Ethanediol   Periodate   Formaldehyde   Iodate

To analyze 1,2-ethanediol, oxidation with excess IO_4^- is followed by passage of the whole reaction solution through an anion-exchange resin that binds both IO_4^- and IO_3^-. The IO_3^- is then selectively and quantitatively removed from the resin by elution with NH_4Cl. The absorbance of the eluate is measured at 232 nm ($\epsilon = 900 \text{ M}^{-1} \text{ cm}^{-1}$) to find the quantity of IO_3^- produced by the reaction. In one experiment, 0.2139 g of aqueous 1,2-ethanediol was dissolved in 10.00 mL. Then 1.000 mL of the solution was treated with 3 mL of 0.15 M KIO_4 and subjected to ion-exchange separation of IO_3^- from unreacted IO_4^-. The eluate (diluted to 250.0 mL) gave $A_{232} = 0.521$ in a 1.000-cm cell, and a blank gave $A_{232} = 0.049$. Find the weight percent of 1,2-ethanediol in the original sample.

10. In ion-exclusion chromatography, ions are separated from nonelectrolytes by an ion-exchange column. Nonelectrolytes penetrate the stationary phase, whereas (half of) the ions are repelled by the fixed charges. Because electrolytes have access to less of the column volume, they are eluted before the nonelectrolytes. The chromatogram below shows the separation of trichloroacetic acid (TCA, $pK_a = 0.66$), dichloroacetic acid (DCA,

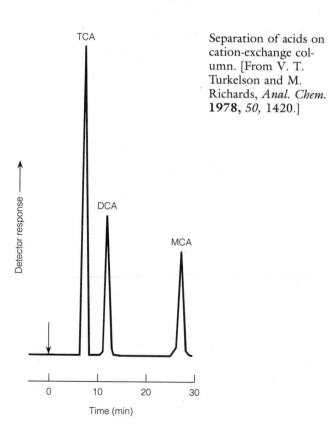

Separation of acids on cation-exchange column. [From V. T. Turkelson and M. Richards, *Anal. Chem.* **1978**, *50*, 1420.]

pK$_a$ 1.30), and monochloroacetic acid (MCA, pK$_a$ 2.86) by passage through a cation-exchange resin eluted with 0.01 M HCl. Explain why the three acids are separated and the order of elution.

11. Norepinephrine (NE) in human urine can be assayed by ion-pair chromatography by using an octadecylsilane stationary phase and sodium octyl sulfate as the mobile-phase additive. Electrochemical detection (oxidation at 0.65 V versus Ag|AgCl) is used, with 2,3-dihydroxybenzylamine (DHBA) as internal standard,

Norepinephrine
cation
(NE)

2,3-Dihydroxybenzylamine
cation
(DHBA)

$$CH_3CH_2CH_2CH_2CH_2CH_2CH_2CH_2OSO_3^-Na^+$$
Sodium octyl sulfate

(a) Explain the physical mechanism by which an ion-pair separation works.

(b) A urine sample containing an unknown amount of NE and a fixed, added concentration of DHBA gave a detector peak height ratio NE/DHBA = 0.298. Then small standard additions of NE were made, with the following results:

Added concentration of NE (ng/mL)	Peak height ratio NE/DHBA
12	0.414
24	0.554
36	0.664
48	0.792

Using the graphical treatment shown in Figure 6-10, find the original concentration of NE in the specimen.

12. Decomposition of dithionite ($S_2O_4^{2-}$) was studied by chromatography on an anion-exchange column eluted with 20 mM trisodium 1,3,6-naphthalenetrisulfonate in 90 vol % H_2O / 10 vol % CH_3CN with ultraviolet detection at 280 nm. A solution of sodium dithionite stored for 34 days in the absence of air gave five peaks identified as SO_3^{2-}, SO_4^{2-}, $S_2O_3^{2-}$, $S_2O_4^{2-}$, and $S_2O_5^{2-}$. All of the peaks had a *negative* absorbance. Explain why.

Molecular Exclusion and Affinity Chromatography

13. (a) How can molecular exclusion chromatography be used to measure the molecular weight of a protein?

(b) Which pore size in Figure 24-10 is most suitable for chromatography of molecules with molecular weight near 100 000?

14. A gel-filtration column has a radius (r) of 0.80 cm and a length (l) of 20.0 cm.

(a) Calculate the volume (V_t) of the column, which is equal to $\pi r^2 l$.

(b) The void volume (V_0) was found to be 18.1 mL and a solute was eluted at 27.4 mL. Find K_{av} for the solute.

15. The compounds ferritin (MW 450 000), transferrin (MW 80 000), and ferric citrate were separated by molecular exclusion chromatography on Bio-Gel P-300. The column had a length of 37 cm and a 1.5-cm diameter. Eluate fractions of 0.65 mL were collected. The maximum of each peak came at the following fractions: ferritin, 22; transferrin, 32; and ferric citrate, 84. (That is, the ferritin peak came at an elution volume of $22 \times 0.65 = 14.3$ mL.) Assuming that ferritin is eluted at the void volume, find K_{av} for transferrin and for ferric citrate.

16. (a) The void volume in Figure 24-10 is the volume at which the curves rise vertically at the left. What is the smallest molecular weight of molecules excluded from the 10-nm pore size column?

(b) What is the molecular weight of molecules eluted at 6.5 mL from the 10-nm column?

17. A polystyrene resin molecular exclusion HPLC column has a diameter of 7.8 mm and a length of 30 cm. The solid portion of the gel particles occupies 20% of the volume, the pores occupy 40%, and the volume between particles occupies 40%.

(a) At what volume would totally excluded molecules be expected to emerge?

(b) At what volume would the smallest molecules be expected?

(c) A mixture of poly(ethylene glycol)s of various molecular weights is eluted between 23 and 27 mL. What does this imply about the retention mechanism for these solutes on the column?

18. The substances below were chromatographed on a gel-filtration column. Estimate the molecular weight of the unknown.

Compound	V_r (mL)	Molecular weight
Blue Dextran 2000	17.7	2×10^6
Aldolase	35.6	158 000
Catalase	32.3	210 000
Ferritin	28.6	440 000
Thyroglobulin	25.1	669 000
Unknown	30.3	?

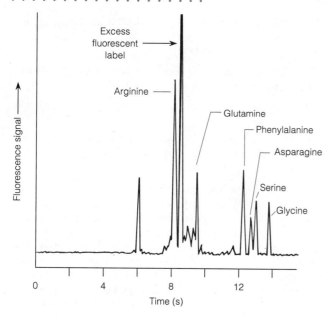

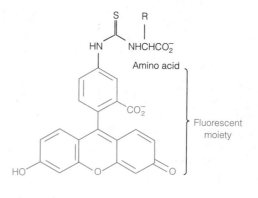

Electropherogram and chemical structure for Problem 21.

Capillary Electrophoresis

19. What is electroosmosis?

20. Explain how sample is loaded in the "analysis on a chip" at the beginning of this chapter.

21. The electropherogram above, showing the separation of amino acids labeled with a fluorescent group, was obtained with the glass chip described at the beginning of this chapter. Explain why arginine has the shortest migration time.

22. What is the principal source of zone broadening in ideal capillary electrophoresis?

23. **(a)** The capillary shown at the beginning of this chapter has a rectangular 12×50-μm cross section. If the sample volume is 100 pL, how long is the sample zone (in mm)?

(b) If sample travels 24 mm in 8 s to reach the detector, what is the standard deviation in bandwidth contributed by the finite length of the injection zone. (*Hint:* See Equation 22-32.)

(c) If the diffusion coefficient of a solute is 1.0×10^{-8} m^2/s, what is the diffusional contribution (standard deviation) to band broadening?

(d) What is the expected total bandwidth at the baseline ($w = 4\sigma$), based on your answers to the previous two questions?

24. State three different methods to reduce electroosmotic flow.

25. Explain how neutral molecules can be separated by micellar electrokinetic capillary chromatography. Why is this a form of chromatography?

26. **(a)** What pressure difference is required to inject a sample equal to 1.0% of the length of a 60.0-cm capillary in 4.0 s if the diameter is 50 μm? Assume that the viscosity of the solution is $0.001\ 0$ kg/(m·s).

(b) The pressure exerted by a column of water of height h is $h\rho g$, where ρ is the density of water and g is the acceleration of gravity (9.8 m/s^2). To what height would you need to raise the sample vial to create the necessary pressure to load the sample in 4.0 s? Is it possible to raise the inlet of this column to this height? How could you obtain the desired pressure?

27. **(a)** How many moles of analyte are present in a 10.0 μM solution that occupies 1.0% of the length of a 25-μm $\times$ 60.0-cm capillary?

(b) What voltage is required to inject this many moles into a capillary in 4.0 s if the sample has $1/10$ of the conductivity of background electrolyte and μ_{app} = 3.0×10^{-8} $m^2/(V \cdot s)$ and the sample concentration is 10.0 μM?

28. Measure the number of plates for the electrophoretic peak in Figure 24-13. Use the formula for asymmetric peaks to find the number of plates for the chromatographic peak.

29. **(a)** Predict whether fumarate or maleate will have greater electrophoretic mobility and state your reason.

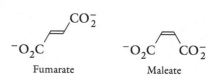

(b) At pH 8.5, both anions have a charge of -2. The electroosmotic flow from the positive terminal to the negative terminal is greater than the electrophoretic flow, so these two anions have a net migration from the positive to the negative end of the capillary in electrophoresis. Based on your answer to **(a)**, predict the order of elution of these two species.

(c) At pH 4.0, both anions have a charge close to -1 and the electroosmotic flow is weak. The anions migrate from the negative end of the capillary to the positive end. Predict the order of elution.

30. **(a)** A particular solution in a particular capillary has an electroosmotic mobility of 1.3×10^{-8} $m^2/(V \cdot s)$ at pH 2 and 8.1×10^{-8} $m^2/(V \cdot s)$ at pH 12. How long will it take a neutral solute to travel 52 cm from the injector to the detector if 27 kV is applied across the 62-cm-long capillary tube at pH 2? At pH 12?

(b) An analyte anion has an electrophoretic mobility of -1.6×10^{-8} $m^2/(V \cdot s)$. How long will it take to reach the detector at pH 2 and at pH 12?

31. The observed behavior of benzyl alcohol $(C_6H_5CH_2OH)$ in capillary electrophoresis is given below.[10] Explain what happens as voltage is increased.

Electric field (V/m)	Number of plates
6 400	38 000
12 700	78 000
19 000	96 000
25 500	124 000
31 700	124 000
38 000	96 000

32. Measure the width of the $^{35}Cl^-$ peak at half-height in Figure 24-22 and calculate the number of theoretical plates. The capillary was 60.0 cm in length. Find the plate height.

33. The migration time for $^{35}Cl^-$ in Figure 24-22 is 17.12 min and the migration time for I^- is 17.78 min. Using mobilities in Table 15-1, predict the migration time of Br^-. (The observed value is 19.6 min.)

34. Resolution of neighboring peaks in capillary zone electrophoresis is described by the equation

$$\text{resolution} = \frac{1}{4} \frac{\Delta \mu}{\mu_{av}} \sqrt{N}$$

where $\Delta \mu$ is the difference in apparent mobilities (which is the same as the difference in electrophoretic mobilities), μ_{av} is the average apparent mobility, and N is the number of theoretical plates ($= \sqrt{N_1 N_2}$ if the two plate counts are not equal). Suppose that the electroosmotic mobility of a solution is $+1.61 \times 10^{-7}$ $m^2/(V \cdot s)$. How many plates are required to separate sulfate from bromide with a resolution of 2.0? Refer to Table 15-1 for mobilities.

35. The water-soluble vitamins niacinamide (a neutral compound), riboflavin (a neutral compound), niacin (an anion), and thiamine (a cation) were separated by micellar electrokinetic capillary chromatography in 15 mM borate buffer (pH 8.0) with 50 mM sodium dodecyl sulfate. The migration times were niacinamide (8.1 min), riboflavin (13.0 min), niacin (14.3 min), and thiamine (21.9 min). What would the order have been in the absence of sodium dodecyl sulfate? Which compound is most soluble in the micelles?

36. To obtain the best separation of two weak acids in capillary electrophoresis, it makes sense to use the pH at which their charge difference is greatest. Prepare a spreadsheet to examine the charges of malonic and phthalic acid as a function of pH. At what pH is the difference greatest?

Notes and References

1. P. K. Dasgupta, *Anal. Chem.* **1992,** *64,* 775A.

2. R. Weinberger, *Practical Capillary Electrophoresis* (Boston: Academic Press, 1993); P. D. Grossman and J. C. Colburn, *Capillary Electrophoresis: Theory and Practice* (San Diego: Academic Press, 1992).

3. For examples of protein analysis of single cells, see B. L. Hogan and E. S. Yeung, *Anal. Chem.* **1992,** *64,* 2841; T. L. Lee and E. S. Yeung, *Anal. Chem.* **1992,** *64,* 3045.

4. From T. M. Olefirowicz and A. G. Ewing, *Anal. Chem.* **1990,** *62,* 1872.

5. The theory of stacking and concentration of extremely large volume samples by stacking is described by R.-L. Chien and D. S. Burgi, *Anal. Chem.* **1992,** *64,* 489A.

6. A. G. Ewing, R. A. Wallingford, and T. M. Olefirowicz, *Anal. Chem.* **1989,** *61,* 293A.

7. R. D. Smith, J. H. Wahl, D. R. Goodlett, and S. A. Hofstadler, *Anal. Chem.* **1993,** *65,* 574A.

8. E. S. Yeung and W. G. Kuhr, *Anal. Chem.* **1991,** *63,* 275A.

9. For a classroom demonstration of micelles, see C. J. Marzzacco, *J. Chem. Ed.* **1992,** *69,* 1024.

10. Data from S. Fazio, R. Vivilecchia, L. Lesueur, and J. Sheridan, *Am. Biotech. Lab.* January 1990, 10.

When All Else Fails

The objective of the space shuttle *Endeavor* in May 1992 was to retrieve a satellite from the wrong orbit, attach a rocket engine, and launch the satellite into the correct orbit. Expensive hardware designed for the capture was unsuccessful after two exhausting days of trying. The men and women of the crew decided to make one last try—just using their hands! Three astronauts succeeded in doing by hand what a multi-million-dollar instrument failed to do.

Gravimetric analysis is like the astronauts' hands. It is likely to be the last choice in the analyst's arsenal of methods, because it is so tedious. But when it is applicable, gravimetry is one of the most accurate methods of analysis. Commercial standards used to calibrate our most elaborate instruments are frequently derived from gravimetric or titrimetric procedures.

Gravimetric and Combustion Analysis

In **gravimetric analysis,** the mass of a product is used to calculate the quantity of the original analyte (the species being analyzed). Exceedingly careful gravimetric analysis by T. W. Richards and his colleagues early in this century determined the atomic weights of Ag, Cl, and N to six-figure accuracy.[1] This Nobel Prize–winning research formed the basis for accurate determination of the atomic weights of many elements. In **combustion analysis,** a sample is burned in excess oxygen and the products are analyzed. Combustion is typically used to measure C, H, N, S, and halogens in organic compounds.

Gravimetric procedures were the mainstay of chemical analysis of ores and industrial materials in the eighteenth and nineteenth centuries, long before the chemical basis for the procedures was understood.

25-1 An Example of Gravimetric Analysis

A familiar example of gravimetric analysis is the determination of Cl^- by precipitation with Ag^+:

$$Ag^+ + Cl^- \rightarrow AgCl(s) \qquad (25\text{-}1)$$

The weight of AgCl produced tells us how much Cl^- was originally present.

EXAMPLE A Simple Gravimetric Calculation

A 10.00-mL solution containing Cl^- was treated with excess $AgNO_3$ to precipitate 0.436 8 g of AgCl. What was the molarity of Cl^- in the unknown?

Solution The formula weight of AgCl is 143.321. A precipitate weighing 0.463 8 g contains

$$\frac{0.436\,8 \text{ g AgCl}}{143.321 \text{ g AgCl/mol AgCl}} = 3.048 \times 10^{-3} \text{ mol AgCl}$$

Because 1 mol of AgCl contains 1 mol of Cl^-, there must have been 3.048×10^{-3} mol of Cl^- in the unknown.

$$[Cl^-] = \frac{3.048 \times 10^{-3} \text{ mol}}{0.010\,00 \text{ L}} = 0.304\,8 \text{ M}$$

Representative analytical precipitations are listed in Table 25-1. A few common organic **precipitants** (agents that cause precipitation) are listed in Table 25-2. Conditions must be controlled to selectively precipitate one species. Potentially interfering substances may need to be removed prior to analysis.

TABLE 25-1 Representative gravimetric analyses

Species analyzed	Precipitated form	Form weighed	Some interfering species
K^+	$KB(C_6H_5)_4$	$KB(C_6H_5)_4$	NH_4^+, Ag^+, Hg^{2+}, Tl^+, Rb^+, Cs^+
Mg^{2+}	$Mg(NH_4)PO_4 \cdot 6H_2O$	$Mg_2P_2O_7$	Many metals except Na^+ and K^+
Ca^{2+}	$CaC_2O_4 \cdot H_2O$	$CaCO_3$ or CaO	Many metals except Mg^{2+}, Na^+, K^+
Ba^{2+}	$BaSO_4$	$BaSO_4$	Na^+, K^+, Li^+, Ca^{2+}, Al^{3+}, Cr^{3+}, Fe^{3+}, Sr^{2+}, Pb^{2+}, NO_3^-
Ti^{4+}	$TiO(5,7\text{-dibromo-}$ $8\text{-hydroxyquinoline})_2$	Same	Fe^{3+}, Zr^{4+}, Cu^{2+}, $C_2O_4^{2-}$, citrate, HF
VO_4^{3-}	Hg_3VO_4	V_2O_5	Cl^-, Br^-, I^-, SO_4^{2-}, CrO_4^{2-}, AsO_4^{3-}, PO_4^{3-}
Cr^{3+}	$PbCrO_4$	$PbCrO_4$	Ag^+, NH_4^+
Mn^{2+}	$Mn(NH_4)PO_4 \cdot H_2O$	$Mn_2P_2O_7$	Many metals
Fe^{3+}	$Fe(HCO_2)_3$	Fe_2O_3	Many metals
Co^{2+}	$Co(1\text{-nitroso-2-naphtholate})_3$	$CoSO_4$ (by reaction with H_2SO_4)	Fe^{3+}, Pd^{2+}, Zr^{4+}
Ni^{2+}	$Ni(dimethylglyoximate)_2$	Same	Pd^{2+}, Pt^{2+}, Bi^{3+}, Au^{3+}
Cu^{2+}	$CuSCN$	$CuSCN$	NH_4^+, Pb^{2+}, Hg^{2+}, Ag^+
Zn^{2+}	$Zn(NH_4)PO_4 \cdot H_2O$	$Zn_2P_2O_7$	Many metals
Ce^{4+}	$Ce(IO_3)_4$	CeO_2	Th^{4+}, Ti^{4+}, Zr^{4+}
Al^{3+}	$Al(8\text{-hydroxyquinolate})_3$	Same	Many metals
Sn^{4+}	$Sn(cupferron)_4$	SnO_2	Cu^{2+}, Pb^{2+}, As(III)
Pb^{2+}	$PbSO_4$	$PbSO_4$	Ca^{2+}, Sr^{2+}, Ba^{2+}, Hg^{2+}, Ag^+, HCl, HNO_3
NH_4^+	$NH_4B(C_6H_5)_4$	$NH_4B(C_6H_5)_4$	K^+, Rb^+, Cs^+
Cl^-	$AgCl$	$AgCl$	Br^-, I^-, SCN^-, S^{2-}, $S_2O_3^{2-}$, CN^-
Br^-	$AgBr$	$AgBr$	Cl^-, I^-, SCN^-, S^{2-}, $S_2O_3^{2-}$, CN^-
I^-	AgI	AgI	Cl^-, Br^-, SCN^-, S^{2-}, $S_2O_3^{2-}$, CN^-
SCN^-	$CuSCN$	$CuSCN$	NH_4^+, Pb^{2+}, Hg^{2+}, Ag^+
CN^-	$AgCN$	$AgCN$	Cl^-, Br^-, I^-, SCN^-, S^{2-}, $S_2O_3^{2-}$
F^-	$(C_6H_5)_3SnF$	$(C_6H_5)_3SnF$	Many metals (except alkali metals), SiO_4^{4-}, CO_3^{2-}
ClO_4^-	$KClO_4$	$KClO_4$	
SO_4^{2-}	$BaSO_4$	$BaSO_4$	Na^+, K^+, Li^+, Ca^{2+}, Al^{3+}, Cr^{3+}, Fe^{3+}, Sr^{2+}, Pb^{2+}, NO_3^-
PO_4^{3-}	$Mg(NH_4)PO_4 \cdot 6H_2O$	$Mg_2P_2O_7$	Many metals except Na^+, K^+
NO_3^-	Nitron nitrate	Nitron nitrate	ClO_4^-, I^-, SCN^-, CrO_4^{2-}, ClO_3^-, NO_2^-, Br^-, $C_2O_4^{2-}$
CO_3^{2-}	CO_2 (by acidification)	CO_2	(The liberated CO_2 is trapped with Ascarite and weighed.)

25-2 Precipitation

The ideal product of a gravimetric analysis should be insoluble, easily filterable, and very pure, and should possess a known composition. Although few substances meet all these requirements, appropriate techniques can help optimize the properties of gravimetric precipitates. For example, the solubility of a precipitate may be decreased by cooling the solution, because the solubility of most compounds decreases as the temperature is lowered.

Particles of precipitate need to be large enough to be easily separated by filtration; they should not be so small that they clog or pass through the filter. Large crystals also have less surface area to which foreign species may become attached. At the other extreme is a *colloidal suspension* of particles that have diameters in the range 1–100 nm and pass through most filters (Demonstration 25-1). Precipitation conditions have much to do with the resulting particle size.

Crystal Growth

Crystallization occurs in two phases: nucleation and particle growth. During **nucleation,** molecules in solution come together randomly and form small aggregates. *Particle growth* involves the addition of more molecules to the nucleus to form a crystal. When a solution contains more solute than should be present at equilibrium, the solution is said to be **supersaturated.** *Relative supersaturation* is $(Q - S)/S$, where Q is the actual concentration of solute and S is the concentration at equilibrium.

Nucleation proceeds faster than particle growth in a highly supersaturated solution. The result is a suspension of tiny particles or, worse, a colloid. In a less supersaturated solution, nucleation is slower, and the nuclei have a chance to grow into larger, more tractable particles.

Techniques that promote particle growth include

Q	Relative supersaturation
S	0
$2S$	1
$3S$	2

Supersaturation decreases particle size.

1. Raising the temperature to increase S and thereby decrease relative supersaturation

2. Adding precipitant slowly with vigorous mixing, to avoid a local, highly supersaturated condition where the stream of precipitant first enters the analyte

3. Keeping the volume of solution large so that the concentrations of analyte and precipitant are low

Homogeneous Precipitation

In **homogeneous precipitation,** the precipitant is generated slowly by a chemical reaction. For example, urea decomposes in boiling water to produce OH^-:

$$\underset{\text{Urea}}{H_2N-\overset{\overset{\textstyle O}{\|}}{C}-NH_2} + 3H_2O \xrightarrow{\text{heat}} CO_2 + 2NH_4^+ + 2OH^- \qquad (25\text{-}2)$$

TABLE 25-2 Common organic precipitating agents

Name	Structure	Some ions precipitated
Dimethylglyoxime	(structure)	Ni^{2+}, Pd^{2+}, Pt^{2+}
Cupferron	(structure)	Fe^{3+}, VO_2^+, Ti^{4+}, Zr^{4+}, Ce^{4+}, Ga^{3+}, Sn^{4+}
8-Hydroxyquinoline (oxine)	(structure)	Mg^{2+}, Zn^{2+}, Cu^{2+}, Cd^{2+}, Pb^{2+}, Al^{3+}, Fe^{3+}, Bi^{3+}, Ga^{3+}, Th^{4+}, Zr^{4+}, UO_2^{2+}, TiO^{2+}
Salicylaldoxime	(structure)	Cu^{2+}, Pb^{2+}, Bi^{3+}, Zn^{2+}, Ni^{2+}, Pd^{2+}
1-Nitroso-2-naphthol	(structure)	Co^{2+}, Fe^{3+}, Pd^{2+}, Zr^{4+}
Nitron	(structure)	NO_3^-, ClO_4^-, BF_4^-, WO_4^{2-}
Sodium tetraphenylborate	$Na^+B(C_6H_5)_4^-$	K^+, Rb^+, Cs^+, NH_4^+, Ag^+, organic ammonium ions
Tetraphenylarsonium chloride	$(C_6H_5)_4As^+Cl^-$	$Cr_2O_7^{2-}$, MnO_4^-, ReO_4^-, MoO_4^{2-}, WO_4^{2-}, ClO_4^-, I_3^-

By this means, the pH of a solution can be raised very gradually, and the slow OH^- formation enhances the particle size of ferric formate precipitate:

$$\text{Formic acid: } HC(=O)OH + OH^- \longrightarrow HCO_2^- + H_2O \tag{25-3}$$

$$3HCO_2^- + Fe^{3+} \longrightarrow Fe(HCO_2)_3 \cdot nH_2O(s) \downarrow \tag{25-4}$$

Ferric formate

Table 25-3 lists reagents commonly employed in homogeneous precipitation.

Demonstration 25-1   Colloids, Dialysis, and Perfusion

Colloids are particles with diameters in the range 1–100 nm. They are larger than molecules but too small to precipitate. Colloids remain in solution indefinitely, suspended by the Brownian motion (random movement) of solvent molecules.

Heat a beaker containing 200 mL of distilled water to 70°–90°C and leave an identical beaker of water at room temperature. Add 1 mL of 1 M $FeCl_3$ to each beaker and stir. The warm solution turns brown-red in a few seconds, whereas the cold solution remains yellow (Color Plate 18). The yellow color is characteristic of low molecular weight Fe^{3+} compounds. The red color results from colloidal aggregates of Fe^{3+} ions held together by hydroxide, oxide, and some chloride ions. These particles have a molecular weight of ~10^5 and a diameter of ~10 nm and contain ~10^3 atoms of Fe.[2]

To demonstrate the size of colloidal particles, we can perform a **dialysis** experiment in which two solutions are separated by a *semipermeable membrane*. A semipermeable membrane is one with holes through which some molecules, but not others, can diffuse. Dialysis tubing is made of cellulose and has pore diameters of 1–5 nm.[3] Small molecules diffuse through these pores, but large molecules (such as proteins or colloids) cannot.

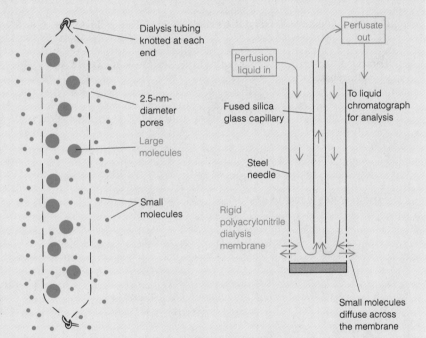

(Left) Large molecules remain trapped inside a dialysis bag, whereas small molecules diffuse through the membrane in both directions.

(Right) *Microdialysis probe* (also called a *perfusion tube*) can be inserted into an organ of an animal to collect molecules small enough to pass through dialysis membrane.

Dialysis tubing knotted at each end

2.5-nm-diameter pores

Large molecules

Small molecules

Perfusion liquid in

Perfusate out

Fused silica glass capillary

To liquid chromatograph for analysis

Steel needle

Rigid polyacrylonitrile dialysis membrane

Small molecules diffuse across the membrane

Pour some of the brown-red colloidal Fe solution into a dialysis tube knotted at one end; then tie off the other end. Drop the tube into a flask of distilled water to show that the color remains entirely within the bag even after several days (Color Plate 18). For comparison, an identical bag containing a dark blue solution of 1 M $CuSO_4 \cdot 5H_2O$ is left in another flask of water. Cu^{2+} diffuses out of the bag, and the solution in the flask will be a uniform light blue color in 24 h. Alternatively, the yellow food coloring, tartrazine, is recommended in place of Cu^{2+}. If dialysis is conducted in hot water, it is completed during one class period.[4]

Dialysis is used to treat patients suffering from kidney failure. Blood is run over a dialysis membrane having a very large surface area. The small molecules of metabolic waste products in the blood diffuse across the membrane and are diluted into a large volume of liquid going out as waste. Protein molecules, which are a necessary part of the blood plasma, are too large to cross the membrane and are retained in the blood.

TABLE 25-3 Common reagents used for homogeneous precipitation

Precipitant	Reagent	Reaction	Some elements precipitated
OH^-	Urea	$(H_2N)_2CO + 3H_2O \rightarrow CO_2 + 2NH_4^+ + 2OH^-$	Al, Ga, Th, Bi, Fe, Sn
OH^-	Potassium cyanate	$HOCN + 2H_2O \rightarrow NH_4^+ + CO_2 + OH^-$ Hydrogen cyanate	Cr, Fe
S^{2-}	Thioacetamide[a]	$\underset{\text{S}}{\overset{\text{S}}{\underset{\parallel}{CH_3CNH_2}}} + H_2O \rightarrow \overset{O}{\overset{\parallel}{CH_3CNH_2}} + H_2S$	Sb, Mo, Cu, Cd
SO_4^{2-}	Sulfamic acid	$H_3\overset{+}{N}SO_3^- + H_2O \rightarrow NH_4^+ + SO_4^{2-} + H^+$	Ba, Ca, Sr, Pb
$C_2O_4^{2-}$	Dimethyl oxalate	$\overset{OO}{\overset{\parallel\parallel}{CH_3OCCOCH_3}} + 2H_2O \rightarrow 2CH_3OH + C_2O_4^{2-} + 2H^+$	Ca, Mg, Zn
PO_4^{3-}	Trimethyl phosphate	$(CH_3O)_3P{=}O + 3H_2O \rightarrow 3CH_3OH + PO_4^{3-} + 3H^+$	Zr, Hf
CrO_4^{2-}	Chromic ion plus bromate	$2Cr^{3+} + BrO_3^- + 5H_2O \rightarrow 2CrO_4^{2-} + Br^- + 10H^+$	Pb
8-Hydroxyquinoline	8-Acetoxyquinoline		Al, U, Mg, Zn

[a]. Hydrogen sulfide is volatile and toxic; it should be handled only in a well-vented hood. Thioacetamide is a carcinogen that should be handled with gloves. If thioacetamide contacts your skin, wash yourself thoroughly immediately. Leftover reagent is destroyed by heating at 50°C with 5 mol of NaOCl per mole of thioacetamide, and then washing down the drain. [H. Elo, *J. Chem. Ed.* **1987,** *64,* A144.]

Precipitation in the Presence of Electrolyte

An *electrolyte* is a compound that dissociates into ions when it dissolves.

Ionic compounds are usually precipitated in the presence of added electrolyte. To understand why, we must discuss how tiny colloidal crystallites *coagulate* (come together) into larger crystals. We will illustrate the case of AgCl, which is commonly formed in the presence of 0.1 M HNO_3.

Figure 25-1 shows a colloidal particle of AgCl growing in a solution containing excess Ag^+, H^+, and NO_3^-. The surface of the particle has an

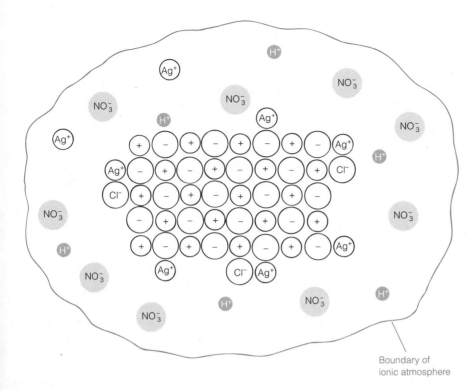

Figure 25-1 Schematic diagram showing a colloidal particle of AgCl growing in a solution containing excess Ag^+, H^+, and NO_3^-. The particle has a net positive charge because of adsorbed Ag^+ ions. The region of solution surrounding the particle is called the *ionic atmosphere*. It has a net negative charge, because the particle attracts anions and repels cations.

Boundary of
ionic atmosphere

excess positive charge due to the **adsorption** of extra silver ions on exposed chloride ions. (To be adsorbed means to be attached to the surface. In contrast, **absorption** involves penetration beyond the surface, to the inside.) The positively charged surface attracts anions and repels cations from the *ionic atmosphere* (Figure 8-1) surrounding the particle. The positively charged particle and the negatively charged ionic atmosphere together are called the **electric double layer.**

Colloidal particles must collide with each other to coalesce. However, the negatively charged ionic atmospheres of the particles repel one another. The particles, therefore, must have enough kinetic energy to overcome electrostatic repulsion before they can coalesce.

Heating the solution promotes coalescence by increasing the particles' kinetic energy. Increasing electrolyte concentration (HNO_3 for AgCl) decreases the volume of the ionic atmosphere and allows particles to come closer together before electrostatic repulsion becomes significant. This is why most gravimetric precipitations are done in the presence of an electrolyte.

Although it is common to find the excess common ion adsorbed on the crystal surface, it is also possible to find other ions selectively adsorbed. In the presence of citrate and sulfate, there is more citrate than sulfate adsorbed on a particle of $BaSO_4$.

Digestion

Following precipitation, most procedures call for a period of standing in the presence of the mother liquor, usually with heating. This treatment, called **digestion,** promotes slow recrystallization of the precipitate. Particle size increases and impurities tend to be expelled from the crystal.

Purity

Adsorbed impurities are bound to the surface of a crystal. *Absorbed* impurities (within the crystal) are classified as *inclusions* or *occlusions*. Inclusions are impurity ions that randomly occupy sites in the crystal lattice normally occupied by ions that belong in the crystal. Inclusions are more likely when the impurity ion has a size and charge similar to those of one of the ions that belongs to the product. Occlusions are pockets of impurity that are literally trapped inside the growing crystal.

Adsorbed, occluded, and included impurities are said to be **coprecipitated.** That is, the impurity is precipitated along with the desired product, even though the solubility of the impurity has not been exceeded. Coprecipitation tends to be worst in colloidal precipitates (which have a large surface area), such as $BaSO_4$, $Al(OH)_3$, and $Fe(OH)_3$. Many procedures call for washing away the mother liquor, redissolving the precipitate, and *reprecipitating* the product. During the second precipitation, the concentration of impurities in the solution is lower than during the first precipitation, and the degree of coprecipitation therefore tends to be lower. Occasionally, a trace component is intentionally isolated by coprecipitation with a major component of the solution. The precipitate used to collect the trace component is said to be a *gathering agent,* and the process is called **gathering.**

Reprecipitation improves the purity of some precipitates.

For example, Se(IV) is *gathered* by coprecipitation with $Fe(OH)_3$. This allows Se(IV) at a concentration of 25 ng/L to be analyzed with a precision of 6%. [K. W. M. Siu and S. S. Berman, *Anal. Chem.* **1984,** *56,* 1806.]

Some impurities can be treated with a **masking agent** to prevent them from reacting with the precipitant. In the gravimetric analysis of Be^{2+}, Mg^{2+}, Ca^{2+}, or Ba^{2+} with the reagent N-p-chlorophenylcinnamohydroxamic acid, impurities such as Ag^+, Mn^{2+}, Zn^{2+}, Cd^{2+}, Hg^{2+}, Fe^{2+}, and Ga^{3+} are kept in solution by excess KCN. The ions Pb^{2+}, Pd^{2+}, Sb^{3+}, Sn^{2+}, Bi^{3+}, Zr^{4+}, Ti^{4+}, V^{5+}, and Mo^{6+} are masked with a mixture of citrate and oxalate.

N-p-Chlorophenylcinnamohydroxamic acid

Precipitate

Cyanide *masks* Mn^{2+} from the precipitating agent.

$$Mn^{2+} + 6CN^- \longrightarrow Mn(CN)_6^{4-}$$
$$\text{(from KCN)} \qquad \text{(stays in solution)}$$

Even when a precipitate forms in a pure state, impurities might collect on the product while it is standing in the mother liquor. This is called *postprecipitation* and usually involves a supersaturated impurity that does not readily crystallize. An example is the crystallization of MgC_2O_4 on CaC_2O_4.

Washing a precipitate on a filter helps remove droplets of liquid containing excess solute. Some precipitates can be washed with water, but many require electrolyte to maintain coherence. For these precipitates, the ionic atmosphere is required to neutralize the surface charge of the tiny particles. If the electrolyte is washed away with water, the charged solid particles repel each other and the product breaks up. This breaking up is called **peptization** and can result in loss of product through the filter. Silver chloride will peptize if washed with water, so it is washed with dilute HNO_3 instead. Electrolyte used for washing must be volatile, so that it will be lost during drying. Volatile electrolytes include HNO_3, HCl, NH_4NO_3, NH_4Cl, and $(NH_4)_2CO_3$.

Product Composition

The final product must have a known, stable composition. A **hygroscopic substance** is one that picks up water from the air and is therefore difficult to weight accurately. Many precipitates contain a variable quantity of water and must be dried under conditions that give a known (possibly zero) stoichiometry of H_2O.

Ignition (strong heating) is used to change the chemical form of some precipitates. For example, $Fe(HCO_2)_3 \cdot nH_2O$ is ignited at 850°C for 1 h to give Fe_2O_3, and $Mg(NH_4)PO_4 \cdot 6H_2O$ is ignited at 1100°C to give $Mg_2P_2O_7$.

In **thermogravimetric analysis,** a sample is heated, and its mass is measured as a function of temperature. Figure 25-2 shows how the composition of calcium salicylate changes in four stages:

Ammonium chloride, for example, decomposes as follows when it is heated:

$$NH_4Cl(s) \rightarrow NH_3(g) + HCl(g)$$

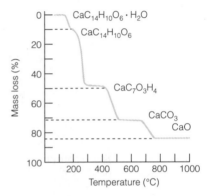

Figure 25-2 Thermogravimetric curve for calcium salicylate. [From G. Liptay, Ed., *Atlas of Thermo-analytical Curves* (London: Heyden and Son, 1976).]

$$(25\text{-}5)$$

The composition of the product depends on the temperature and duration of heating.

25-3 Examples of Gravimetric Calculations

We now examine some examples that illustrate how to relate the mass of a gravimetric precipitate to the quantity of the original analyte. The general approach is to relate the moles of product to the moles of reactant.

EXAMPLE Relating Mass of Product to Mass of Reactant

The piperazine content of an impure commercial material can be determined by precipitating and weighing the diacetate:[5]

$$HN \underset{\text{Piperazine}}{\underset{\text{FW 86.137}}{\text{:}\quad\text{:NH}}} + \underset{\substack{\text{Acetic acid} \\ \text{FW 60.053}}}{2CH_3CO_2H} \rightarrow \underset{\substack{\text{Piperazine diacetate} \\ \text{FW 206.243}}}{H_2\overset{+}{N}\quad\overset{+}{N}H_2\,(CH_3CO_2^-)_2} \qquad (25\text{-}6)$$

If you were performing this analysis, it would be important to determine that impurities in the piperazine are not precipitated.

In one experiment, 0.3126 g of the sample was dissolved in 25 mL of acetone, and 1 mL of acetic acid was added. After 5 min, the precipitate was filtered, washed with acetone, dried at 110°C, and found to weigh 0.7121 g. What is the weight percent of piperazine in the commercial material?

Solution For each mole of piperazine in the impure material, 1 mol of product is formed.

$$\text{moles of product} = \frac{0.7121 \text{ g}}{206.243 \text{ g/mol}} = 3.453 \times 10^{-3} \text{ mol}$$

This many moles of piperazine corresponds to

$$\text{grams of piperazine} = (3.453 \times 10^{-3} \text{ mol})\left(86.137 \frac{\text{g}}{\text{mol}}\right) = 0.2974 \text{ g}$$

which gives

$$\text{percentage of piperazine in analyte} = \frac{0.2974 \text{ g}}{0.3126 \text{ g}} \times 100 = 95.14\%$$

An alternative (but equivalent) way to work this problem is to realize that 206.243 g (1 mol) of product will be formed for every 86.137 g (1 mol) of piperazine analyzed. Because 0.7121 g of product was formed, the amount of reactant is given by

$$\frac{x \text{ g piperazine}}{0.7121 \text{ g product}} = \frac{86.137 \text{ g piperazine}}{206.243 \text{ g product}}$$

$$\Rightarrow x = \left(\frac{86.137 \text{ g piperazine}}{206.243 \text{ g product}}\right) 0.7121 \text{ g product} = 0.2974 \text{ g piperazine}$$

The gravimetric factor relates mass of product to mass of analyte.

The quantity 86.137/206.243 is the *gravimetric factor* relating the mass of starting material to the mass of product.

For a reaction in which the stoichiometric relationship between analyte and product is not 1:1, we must use the correct stoichiometry in formulating the gravimetric factor. For example, an unknown containing Mg^{2+} (atomic weight = 24.3050) can be analyzed gravimetrically to produce $Mg_2P_2O_7$ (FW 222.553). The gravimetric factor would be

$$\frac{\text{grams of Mg in analyte}}{\text{grams of } Mg_2P_2O_7 \text{ formed}} = \frac{2 \times (24.3050)}{222.553}$$

because it takes 2 mol of Mg^{2+} to make 1 mol of $Mg_2P_2O_7$.

EXAMPLE Calculating How Much Precipitant to Use

(a) To measure the nickel content in steel, the alloy is dissolved in 12 M HCl and neutralized in the presence of citrate ion, which maintains iron in solution. The slightly basic solution is warmed, and dimethylglyoxime (DMG) is added to precipitate the red DMG-nickel complex quantitatively. The product is filtered, washed with cold water, and dried at 110°C.

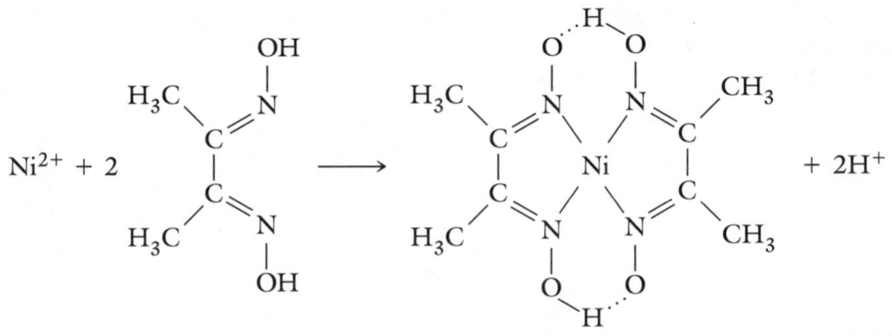

Dimethylglyoxime   Bis(dimethylglyoximate)nickel(II)   (25-7)
FW 58.69   FW 116.12   FW 288.91

If the nickel content is known to be near 3 wt% and you wish to analyze 1.0 g of the steel, what volume of 1.0 wt% alcoholic DMG solution should be used to give a 50% excess of DMG for the analysis? Assume that the density of the alcohol solution is 0.79 g/mL.

Solution Because the Ni content is around 3%, 1.0 g of steel will contain about 0.03 g of Ni, which corresponds to

$$\frac{0.03 \text{ g Ni}}{58.69 \text{ g Ni/mol Ni}} = 5.11 \times 10^{-4} \text{ mol Ni}$$

This amount of metal requires

$$2(5.11 \times 10^{-4} \text{ mol Ni})(116.12 \text{ g DMG/mol Ni}) = 0.119 \text{ g DMG}$$

because 1 mol of Ni^{2+} requires 2 mol of DMG. A 50% excess of DMG would be $(1.5)(0.119 \text{ g}) = 0.178$ g. This much DMG is contained in

$$\left(\frac{0.178 \text{ g DMG}}{0.010 \text{ g DMG/g solution}}\right) = 17.8 \text{ g solution}$$

which occupies a volume of

$$\frac{17.8 \text{ g solution}}{0.79 \text{ g solution/mL}} = 23 \text{ mL}$$

(b) If 1.1634 g of steel gave 0.1795 g of precipitate, what is the percentage of Ni in the steel?

Solution For each mole of Ni in the steel, 1 mol of precipitate will be formed. Therefore, 0.1795 g of precipitate corresponds to

$$\frac{0.1795 \text{ g Ni(DMG)}_2}{288.91 \text{ g Ni(DMG)}_2/\text{mol Ni(DMG)}_2} = 6.213 \times 10^{-4} \text{ mol Ni(DMG)}_2$$

The Ni in the alloy must therefore be

$$(6.213 \times 10^{-4} \text{ mol Ni})\left(58.69 \frac{\text{g}}{\text{mol Ni}}\right) = 0.03646 \text{ g}$$

The weight percent of Ni in steel is

$$\frac{0.03646 \text{ g Ni}}{1.1634 \text{ g steel}} \times 100 = 3.134\%$$

A slightly simpler way to approach this problem comes from realizing that 58.69 g of Ni (1 mol) would give 288.91 g (1 mol) of product. Calling the mass of Ni in the sample x, we can write

$$\frac{\text{grams of Ni analyzed}}{\text{grams of product formed}} = \frac{x}{0.1795} = \frac{58.69}{288.91} \Rightarrow \text{Ni} = 0.03646 \text{ g}$$

· ·

EXAMPLE   A Problem with Two Components

A mixture of the 8-hydroxyquinoline complexes of aluminum and magnesium weighed 1.0843 g. When ignited in a furnace open to the air, the mixture decomposed, leaving a residue of Al_2O_3 and MgO weighing 0.1344 g. Find the weight percent of $Al(C_9H_6NO)_3$ in the original mixture.

FW 459.441   FW 312.611   FW 101.961   FW 40.304

Solution   We will abbreviate the 8-hydroxyquinoline anion as Q. Letting the mass of AlQ_3 be x and the mass of MgQ_2 be y, we can write

$$\underset{\substack{\text{Mass of} \\ AlQ_3}}{x} + \underset{\substack{\text{Mass of} \\ MgQ_2}}{y} = 1.0843 \text{ g}$$

The moles of Al will be $x/459.441$, and the moles of Mg will be $y/312.611$. The moles of Al_2O_3 will be one-half of the total moles of Al, because it takes 2 mol of Al to make 1 mol of Al_2O_3.

$$\text{moles of } Al_2O_3 = \left(\frac{1}{2}\right)\frac{x}{459.441}$$

The moles of MgO will equal the moles of Mg = $y/312.611$. Now we can write

$$\overbrace{\underbrace{\left(\frac{1}{2}\right)\frac{x}{459.441}}_{\text{mol } Al_2O_3} \underbrace{(101.961)}_{\frac{\text{g } Al_2O_3}{\text{mol } Al_2O_3}}}^{\text{mass of } Al_2O_3} + \overbrace{\underbrace{\frac{y}{312.611}}_{\text{mol MgO}} \underbrace{(40.304)}_{\frac{\text{g MgO}}{\text{mol MgO}}}}^{\text{mass of MgO}} = 0.1344 \text{ g}$$

Substituting $y = 1.0843 - x$ into the equation above gives

$$\left(\frac{1}{2}\right)\left(\frac{x}{459.441}\right)(101.961) + \left(\frac{1.0843 - x}{312.611}\right)(40.304) = 0.1344 \text{ g}$$

from which we find $x = 0.3003$ g, which is 27.7% of the original mixture.

· ·

25-4 Combustion Analysis

A historically important form of gravimetric analysis was *combustion analysis,* used to determine the carbon and hydrogen content of organic compounds burned in excess O_2 (Figure 25-3). Instead of weighing combustion products, modern instruments use thermal conductivity, infrared absorption, or coulometry (with electrochemically generated reagents) to measure the products.

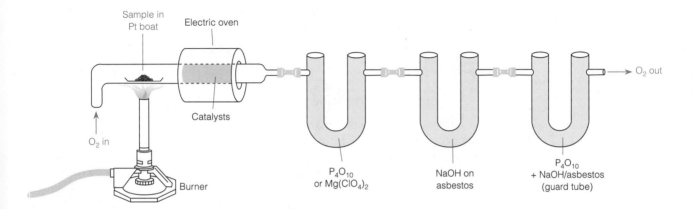

Figure 25-3   Gravimetric combustion analysis for carbon and hydrogen.

Gravimetric Combustion Analysis

In gravimetric combustion analysis, partially combusted product is passed through catalysts such as Pt gauze, CuO, PbO_2, or MnO_2 at elevated temperature to complete the oxidation to CO_2 and H_2O. The combustion products are flushed through a chamber containing P_4O_{10} ("phosphorus pentoxide"), which absorbs water, and then through a chamber of Ascarite® (NaOH on asbestos), which absorbs CO_2. The increase in mass of each chamber tells how much hydrogen and carbon, respectively, were initially present. A guard tube downstream of the two chambers prevents atmospheric H_2O or CO_2 from entering the chambers.

· ·

EXAMPLE   Combustion Analysis Calculations

A compound weighing 5.714 mg produced 14.414 mg of CO_2 and 2.529 mg of H_2O upon combustion. Find the weight percent of C and H in the sample.

Solution   One mole of CO_2 contains 1 mol of carbon. Therefore,

moles of C in sample = moles of CO_2 produced

$$= \frac{14.414 \times 10^{-3} \text{ g } CO_2}{44.010 \text{ g/mol } CO_2} = 3.275 \times 10^{-4} \text{ mol}$$

mass of C in sample = $(3.275 \times 10^{-4} \text{ mol C})(12.011 \text{ g/mol C}) = 3.934$ mg

$$\text{weight percent of C} = \frac{3.934 \text{ mg C}}{5.714 \text{ mg sample}} \times 100 = 68.84\%$$

One mole of H_2O contains 2 mol of H. Therefore,

moles of H in sample = 2(moles of H_2O produced)

$$= 2\left(\frac{2.529 \times 10^{-3} \text{ g } H_2O}{18.0152 \text{ g/mol } H_2O}\right) = 2.808 \times 10^{-4} \text{ mol}$$

mass of H in sample = $(2.808 \times 10^{-4} \text{ mol H})(1.0079 \text{ g/mol H}) = 2.830 \times 10$

$$\text{weight percent of H} = \frac{0.2830 \text{ mg H}}{5.714 \text{ mg sample}} \times 100 = 4.952\%$$

Combustion Analysis Today[6]

Figure 25-4 shows an instrument that measures C, H, N, and S in a single operation. First, a ~2-mg sample is accurately weighed and sealed in a tin or silver capsule. The analyzer is swept with He gas that has been treated to remove traces of O_2, H_2O, and CO_2. At the start of a run, a measured excess volume of O_2 is added to the He stream. Then the sample capsule is dropped into a preheated ceramic crucible where the capsule melts and the sample is rapidly oxidized.

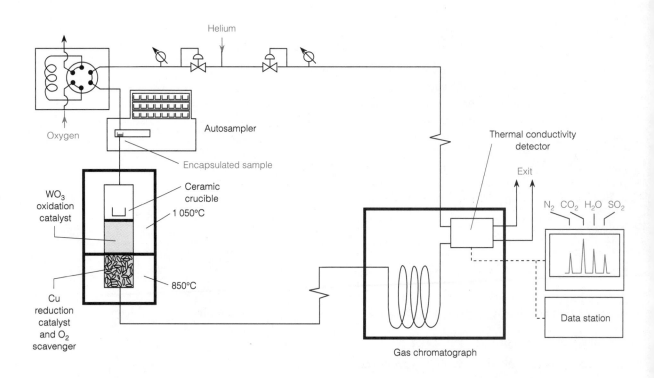

Figure 25-4 Schematic diagram of C, H, N, S elemental analyzer that uses gas chromatographic separation and thermal conductivity detection. [Adapted from E. Pella, *Am. Lab.* August 1990, p. 28.]

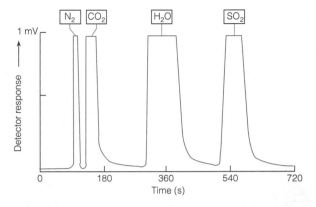

Figure 25-5 Gas chromatographic trace from elemental analyzer, showing substantially complete separation of combustion products. The area of each peak (when they are not off scale) is proportional to the mass of each product. [From E. Pella, *Am. Lab.* August 1990, p. 28.]

$$C, H, N, S \xrightarrow{1\,050°C/O_2} CO_2(g) + H_2O(g) + N_2(g) + \underbrace{SO_2(g) + SO_3(g)}_{95\% \; SO_2}$$

The products pass through hot WO_3 catalyst to complete the combustion of carbon to CO_2. In the next zone, metallic Cu at 850°C converts SO_3 to SO_2 and removes excess O_2:

$$Cu + SO_3 \xrightarrow{850°C} SO_2 + CuO(s)$$

$$Cu + \tfrac{1}{2}O_2 \xrightarrow{850°C} CuO(s)$$

The mixture of CO_2, H_2O, N_2, and SO_2 is separated by gas chromatography, and each component is measured with a thermal conductivity detector, described in Section 23-1 (Figure 25-5). Figure 25-6 shows a different C,H,N,S analyzer that uses infrared absorbance to measure CO_2, H_2O, and SO_2 and thermal conductivity for N_2.

A key attribute of elemental analyzers is *dynamic flash combustion,* which creates a short burst of gaseous products, instead of slowly bleeding products out over several minutes. This feature is important because chromatographic analysis requires that the whole sample be injected at once. Otherwise, the injection zone is so broad that the products cannot be separated.

In dynamic flash combustion, the tin-encapsulated sample is dropped into the preheated furnace shortly after the flow of a 50 vol % O_2/50 vol % He mixture is started (Figure 25-7). The Sn capsule melts at 235°C and is instantly oxidized to SnO_2, thereby liberating 594 kJ/mol and heating the sample to 1 700°–1 800°C. By dropping the sample in before very much O_2 is present (Figure 25-7), pyrolytic decomposition (cracking) of the sample occurs prior to oxidation, which minimizes the formation of nitrogen oxides. (Flammable liquid samples are admitted prior to any O_2 to prevent explosions.)

Analyzers that measure C, H, and N (but not S) use catalysts that are better optimized for this process. The oxidation catalyst is Cr_2O_3. The gas then passes through hot Co_3O_4 coated with Ag to absorb halogens and sulfur. A hot Cu reduction column then scavenges excess O_2.

Elemental analyzers use an *oxidation catalyst* to complete the oxidation of the sample and a *reduction catalyst* to carry out any required reduction and to remove excess O_2.

The Sn capsule is oxidized to SnO_2, which

1. liberates heat to vaporize and crack sample
2. uses available oxygen immediately
3. ensures that sample oxidation occurs in gas phase
4. acts as an oxidation catalyst

Figure 25-6 Combustion analyzer that uses infrared absorbance to measure CO_2, H_2O, and SO_2 and thermal conductivity to measure N_2. Three separate infrared cells in series are equipped with filters that isolate wavelengths absorbed by one of the products. Absorbance is integrated over time as the combustion product mixture is swept through each cell. [Courtesy Leco Corp., St. Joseph, MI.]

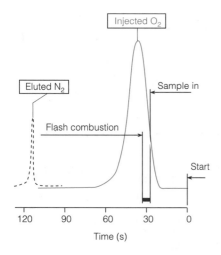

Figure 25-7 Sequence of events in dynamic flash combustion. [From E. Pella, *Am. Lab.* February 1990, p. 116.]

Oxygen analysis requires a different strategy. The sample is thermally decomposed (a process called **pyrolysis**) in the absence of added oxygen. The gaseous products are passed through nickelized carbon at 1 075°C to convert oxygen from the compound into CO (not CO_2). Other products include N_2, H_2, CH_4, and hydrogen halides. Acidic products are absorbed by NaOH/asbestos, and the remaining gases are separated and measured by gas chromatography with a thermal conductivity detector.

For halogen analysis, combustion is carried out to give CO_2, H_2O, N_2, and HX (X = halogen). The HX is trapped in aqueous solution and titrated with Ag^+ ions generated by oxidation of Ag in a *coulometer* (Section 17-4). This electrochemical instrument counts the number of electrons produced (one electron for each Ag^+) during complete reaction with the HX.

Table 25-4 shows the averages of five consecutive analyses of the pure compound acetanilide on two different commercial instruments. Chemists normally consider a result within ±0.3 of the theoretical percentage of an element to be good evidence that the compound has the expected formula.

TABLE 25-4   C, H, and N in acetanilide:
$$\text{C}_6\text{H}_5-\overset{\overset{\displaystyle O}{\displaystyle \|}}{\text{NHCCH}_3} \quad (\text{C}_8\text{H}_9\text{NO})$$

Element	Theoretical value (wt%)	Instrument 1	Instrument 2
C	71.09	71.17 ± 0.41	71.22 ± 1.1
H	6.71	6.76 ± 0.12	6.84 ± 0.10
N	10.36	10.34 ± 0.08	10.33 ± 0.13

Data from E. M. Hodge, H. P. Patterson, M. C. Williams, and E. S. Gladney, *Am. Lab.* June 1991, p. 34. Uncertainties are standard deviations from five replicate determinations.

For nitrogen in acetanilide, ±0.3 corresponds to a relative error of $0.3/10.36 = 3\%$, which is not hard to achieve. For carbon, ±0.3 corresponds to a relative error of $0.3/71.09 = 0.4\%$, which is not so easy. Notice that the standard deviation for carbon in Instrument 1 is $0.41/71.17 = 0.6\%$, and for Instrument 2 it is $1.1/71.22 = 1.5\%$.

Terms to Understand

absorption	electric double layer	nucleation
adsorption	gathering	peptization
colloid	gravimetric analysis	precipitant
combustion analysis	homogeneous precipitation	pyrolysis
coprecipitation	hygroscopic substance	supersaturated solution
dialysis	ignition	thermogravimetric analysis
digestion	masking agent	

Summary

Gravimetric analysis is based on the formation of a product whose mass can be related to the mass of analyte. Most commonly, analyte ion is precipitated by a suitable counterion. Measures taken to reduce supersaturation and promote the formation of large, easily filtered particles (as opposed to colloids) include (1) raising the temperature during precipitation, (2) slow addition and vigorous mixing of reagents, (3) maintaining a large sample volume, and (4) use of homogeneous precipitation. Precipitates are usually digested in hot mother liquor to promote particle growth and recrystallization. All precipitates are then filtered and washed; some must be washed with a volatile electrolyte to prevent peptization. The product is heated to dryness or ignited to achieve a reproducible, stable composition. Gravimetric calculations relate moles of product to moles of analyte.

In combustion analysis, an organic compound in a tin capsule is rapidly heated with excess oxygen to give predominantly CO_2, H_2O, N_2, SO_2, and HX (hydrogen halides). A hot oxidation catalyst completes the process, and hot Cu scavenges excess oxygen. For sulfur analysis, hot copper also converts SO_3 to SO_2. Products may be separated by gas chromatography and measured by their thermal conductivity. Some instruments use infrared absorption or coulometric reactions (counting electrons with an electronic circuit) to measure the products. Oxygen analysis is done by pyrolysis in the absence of oxygen, a process that ultimately converts oxygen from the compound into CO.

Exercises

A. An organic compound with a molecular weight of 417 was analyzed for ethoxyl (CH_3CH_2O-) groups by the reactions

$$ROCH_2CH_3 + HI \rightarrow ROH + CH_3CH_2I$$

$$CH_3CH_2I + Ag^+ + OH^- \rightarrow AgI(s) + CH_3CH_2OH$$

A 25.42-mg sample of compound produced 29.03 mg of AgI. How many ethoxyl groups are there in each molecule?

B. A 0.649-g sample containing only K_2SO_4 (FW 174.27) and $(NH_4)_2SO_4$ (FW 132.14) was dissolved in water and treated with $Ba(NO_3)_2$ to precipitate all of the SO_4^{2-} as $BaSO_4$ (FW 233.40). Find the weight percent of K_2SO_4 in the sample if 0.977 g of precipitate was formed.

C. Consider a mixture of the two solids $BaCl_2 \cdot 2H_2O$ (FW 244.26) and KCl (FW 74.551), in an unknown ratio. (The notation $BaCl_2 \cdot 2H_2O$ means that a crystal is formed with two water molecules for each $BaCl_2$.) When the unknown is heated to 160°C for 1 h, the water of crystallization is driven off:

$$BaCl_2 \cdot 2H_2O(s) \xrightarrow{160°C} BaCl_2(s) + 2H_2O(g)$$

A sample originally weighing 1.7839 g weighed 1.5623 g after heating. Calculate the weight percent of Ba, K, and Cl in the original sample.

D. A mixture containing only aluminum tetrafluoroborate, $Al(BF_4)_3$ (FW 287.39), and magnesium nitrate, $Mg(NO_3)_2$ (FW 148.31), weighed 0.2828 g. It was dissolved in 1 wt% aqueous HF and treated with nitron solution to precipitate a mixture of nitron tetrafluoroborate and nitron nitrate weighing 1.322 g. Find the weight percent of Mg in the original solid mixture.

Nitron
$C_{20}H_{16}N_4$
FW 312.37

Nitron tetrafluoroborate
$C_{20}H_{17}N_4BF_4$
FW 400.18

Nitron nitrate
$C_{20}H_{17}N_5O_3$
FW 375.39

Problems

Gravimetric Analysis

1. **(a)** What is the difference between absorption and adsorption?

(b) How is an inclusion different from an occlusion?

2. State four desirable properties of a gravimetric precipitate.

3. Why is high relative supersaturation undesirable in a gravimetric precipitation?

4. What measures can be taken to decrease the relative supersaturation during a precipitation?

5. Why are many ionic precipitates washed with electrolyte solution instead of pure water?

6. Why is it less desirable to wash AgCl precipitate with aqueous $NaNO_3$ than with HNO_3 solution?

7. Why would a reprecipitation be employed in a gravimetric analysis?

8. Explain what is done in thermogravimetric analysis.

9. A 50.00-mL solution containing NaBr was treated with excess $AgNO_3$ to precipitate 0.2146 g of AgBr (MW 187.772). What was the molarity of NaBr in the solution?

10. To find the Ce^{4+} content of a solid, 4.37 g was dissolved and treated with excess iodate to precipitate $Ce(IO_3)_4$. The precipitate was collected, washed well, dried, and ignited to produce 0.104 g of CeO_2 (MW 172.114). What was the weight percent of Ce in the original solid?

11. A 0.05002-g sample of impure piperazine contains 71.29 wt% piperazine (MW 86.137). How many grams of product (MW 206.243) will be formed when this sample is analyzed by Reaction 25-6?

12. A 1.000-g sample of unknown gave 2.500 g of bis(dimethylglyoximate)nickel(II) (FW 288.91) when analyzed by Reaction 25-7. Find the weight percent of Ni in the unknown.

13. Referring to Figure 25-2, name the product when calcium salicylate monohydrate is heated to 550°C or 1 000°C. Using the formula weights of these products, calculate what mass is expected to remain when 0.635 6 g of calcium salicylate monohydrate is heated to 550°C or 1 000°C.

14. A method for measurement of soluble organic carbon in seawater involves oxidation of the organic materials to CO_2 with $K_2S_2O_8$, followed by gravimetric determination of the CO_2 trapped by a column of NaOH-coated asbestos. A water sample weighing 6.234 g produced 2.378 mg of CO_2 (FW 44.010). Calculate the ppm carbon in the seawater.

15. How many milliliters of 2.15% alcoholic dimethyl-glyoxime should be used to provide a 50.0% excess for Reaction 25-7 with 0.998 4 g of steel containing 2.07 wt% Ni? Assume that the density of the dimethyl-glyoxime solution is 0.790 g/mL.

16. Twenty dietary iron tablets with a total mass of 22.131 g were ground and mixed thoroughly. Then 2.998 g of the powder were dissolved in HNO_3 and heated to convert all the iron to Fe^{3+}. Addition of NH_3 caused quantitative precipitation of $Fe_2O_3 \cdot xH_2O$, which was ignited to give 0.264 g of Fe_2O_3 (FW 159.69). What is the average mass of $FeSO_4 \cdot 7H_2O$ (FW 278.01) in each tablet?

17. A 1.475-g sample containing NH_4Cl (FW 53.492), K_2CO_3 (FW 138.21), and inert ingredients was dissolved to give 0.100 L of solution. A 25.0-mL aliquot was acidified and treated with excess sodium tetraphenyl-borate, $Na^+B(C_6H_5)_4^-$, to precipitate K^+ and NH_4^+ ions completely:

$$(C_6H_5)_4B^- + K^+ \rightarrow (C_6H_5)_4BK(s)$$
$$\text{FW 358.33}$$

$$(C_6H_5)_4B^- + NH_4^+ \rightarrow (C_6H_5)_4BNH_4(s)$$
$$\text{FW 337.27}$$

The resulting precipitate amounted to 0.617 g. A fresh 50.0-mL aliquot of the original solution was made alkaline and heated to drive off all the NH_3:

$$NH_4^+ + OH^- \rightarrow NH_3(g) + H_2O$$

It was then acidified and treated with sodium tetraphenyl-borate to give 0.554 g of precipitate. Find the weight percent of NH_4Cl and K_2CO_3 in the original solid.

18. A mixture containing only Al_2O_3 (FW 101.96) and Fe_2O_3 (FW 159.69) weighs 2.019 g. When heated under a stream of H_2, the Al_2O_3 is unchanged, but the Fe_2O_3 is converted to metallic Fe plus $H_2O(g)$. If the residue weighs 1.774 g, what is the weight percent of Fe_2O_3 in the original mixture?

19. A solid mixture weighing 0.548 5 g contained only ferrous ammonium sulfate and ferrous chloride. The

sample was dissolved in 1 M H_2SO_4, oxidized to Fe^{3+} with H_2O_2, and precipitated with cupferron. The ferric cupferron complex was ignited to produce 0.167 8 g of ferric oxide, Fe_2O_3 (FW 159.69). Calculate the weight percent of Cl in the original sample.

$$FeSO_4 \cdot (NH_4)_2SO_4 \cdot 6H_2O \qquad FeCl_2 \cdot 6H_2O$$
Ferrous ammonium sulfate   Ferrous chloride
FW 392.13   FW 234.84

Cupferron
FW 155.16

20. *Propagation of error.* A mixture containing only silver nitrate and mercurous nitrate was dissolved in water and treated with excess sodium hexacyanocobaltate(III) ($Na_3[Co(CN)_6]$) to precipitate both hexacyanocobaltate(III) salts:

$AgNO_3$	FW 169.873
$Ag_3[Co(CN)_6]$	FW 538.643
$Hg_2(NO_3)_2$	FW 525.19
$(Hg_2)_3[Co(CN)_6]_2$	FW 1 633.62

(a) The unknown weighed 0.432 1 g and the product weighed 0.451 5 g. Find the weight percent of silver nitrate in the unknown. *Caution:* In this type of arithmetic, you should keep all the digits in your calculator or else serious rounding errors may occur. Do not round off until the end of the calculation.

(b) Even a skilled analyst is not likely to have less than a 0.3% error in isolating the precipitate. Suppose that there is negligible error in all quantities, except the mass of product. Suppose that the mass of product has an uncertainty of 0.30%. Calculate the relative uncertainty in the mass of silver nitrate in the unknown.

21. The thermogravimetric trace on page 754 shows mass loss by $Y_2(OH)_5Cl \cdot xH_2O$ upon heating. In the first step, waters of hydration are lost to give ~8.1% mass loss. After a second decomposition step, 19.2% of the original mass is lost. Finally, the composition stabilizes at Y_2O_3 above 800°C.

(a) Find the value of x in the formula $Y_2(OH)_5Cl \cdot xH_2O$. Because the 8.1% mass loss is not accurately defined in the experiment, use the 31.8% total mass loss for your calculation.

(b) Suggest a formula for the material remaining at the 19.2% plateau. Be sure that the charges of all ions in your formula sum to zero. The cation is Y^{3+}.

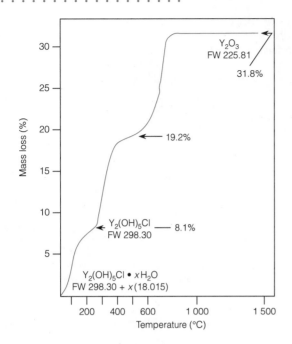

Thermogravimetric analysis of $Y_2(OH)_5Cl \cdot xH_2O$. [From T. Hours, P. Bergez, J. Charpin, A. Larbot, C. Guizard, and L. Cot, *Ceramic Bull.* **1992**, *71*, 200.]

22. When the *high-temperature superconductor* yttrium barium copper oxide (see the beginning of Chapter 16 and Box 16-3) is heated under flowing H_2, the solid remaining at $1\,000°C$ is a mixture of Y_2O_3, BaO, and Cu. The starting material has the formula $YBa_2Cu_3O_{7-x}$, in which the oxygen stoichiometry varies between 7 and 6.5 ($x = 0$ to 0.5).

$$YBa_2Cu_3O_{7-x}(s) + (3.5 - x)H_2(g) \xrightarrow{1\,000°C}$$
$$\text{FW } 666.19 - 16.00x$$

$$\tfrac{1}{2}Y_2O_3(s) + 2BaO(s) + 3Cu(s) + (3.5 - x)H_2O(g)$$
$$\underbrace{\phantom{\tfrac{1}{2}Y_2O_3(s) + 2BaO(s) + 3Cu(s)}}_{\displaystyle YBa_2Cu_3O_{3.5}}$$

(a) *Thermogravimetric analysis.* When 34.397 mg of superconductor was subjected to this analysis, 31.661 mg of solid remained after heating to $1\,000°C$. Find the value of x in the formula $YBa_2Cu_3O_{7-x}$ of the original material.

(b) *Propagation of error.* Suppose that the uncertainty in each mass in **(a)** is ±0.002 mg. Find the uncertainty in the value of x.

23. *The man in the vat problem.*[7] Long ago a workman at a dye factory fell into a vat containing a hot concentrated mixture of sulfuric and nitric acids. He dissolved completely! Because nobody witnessed the accident, it was necessary to prove that he fell in so that the man's wife

could collect his insurance money. The man weighed 70 kg, and a human body contains about 6.3 parts per thousand phosphorus. The acid in the vat was analyzed for phosphorus to see if it contained a dissolved human.

(a) The vat contained 8.00×10^3 L of liquid, and 100.0 mL was analyzed. If the man did fall into the vat, what is the expected quantity of phosphorus in 100.0 mL?

(b) The 100.0 mL was treated with a molybdate reagent that caused ammonium phosphomolybdate, $(NH_4)_3[P(Mo_{12}O_{40})] \cdot 12H_2O$, to precipitate. This substance was dried at 110°C to remove waters of hydration and heated to 400°C until it reached a constant composition corresponding to the formula $P_2O_5 \cdot 24MoO_3$, which weighed 0.371 8 g. When a fresh mixture of the same acids (not from the vat) was treated in the same manner, 0.033 1 g of $P_2O_5 \cdot 24MoO_3$ (FW 3 596.46) was produced. This *blank determination* gives the amount of phosphorus in the starting reagents. The $P_2O_5 \cdot 24MoO_3$ that could have come from the dissolved man is therefore $0.371\,8 - 0.033\,1 = 0.338\,7$ g. How much phosphorus was present in the 100.0-mL sample? Is this quantity consistent with a dissolved man?

Combustion Analysis

24. What is the difference between combustion and pyrolysis?

25. What is the purpose of the WO_3 and Cu in Figure 25-4?

26. Why is tin used to encapsulate a sample for combustion analysis?

27. Why is sample dropped into the preheated furnace before the oxygen concentration reaches its peak in Figure 25-7?

28. Write a balanced equation for the combustion of benzoic acid, $C_6H_5CO_2H$, to give CO_2 and H_2O. How many milligrams of CO_2 and of H_2O will be produced by the combustion of 4.635 mg of benzoic acid?

29. Write a balanced equation for the combustion of $C_8H_7NO_2SBrCl$ in a C,H,N,S elemental analyzer.

30. Combustion analysis of a compound known to contain just C, H, N, and O demonstrated that it contains 46.21 wt% C, 9.02 wt% H, 13.74 wt% N, and, by difference, $100 - 46.21 - 9.02 - 13.74 = 31.04$ wt% O. This means that 100 g of unknown would contain 46.21 g of C, 9.02 g of H, etc. Find the atomic ratio C:H:N:O and express this as the lowest reasonable integer ratio.

31. A mixture weighing 7.290 mg contained only cyclohexane, C_6H_{12} (FW 84.161), and oxirane, C_2H_4O

(FW 44.053). When the mixture was analyzed by combustion analysis, 21.999 mg of CO_2 (FW 44.010) was produced. Find the weight percent of oxirane in the mixture.

32. Use the uncertainties from Instrument 1 in Table 25-4 to estimate the uncertainties in the stoichiometry coefficients in the formula $C_8H_{h \pm x}N_{n \pm y}$.

33. One way to determine sulfur is by combustion analysis, which produces a mixture of SO_2 and SO_3 that can be passed through H_2O_2 to convert both oxides to H_2SO_4, which is titrated with standard base. When 6.123 mg of a substance was burned, the H_2SO_4 required 3.01 mL of 0.015 76 M NaOH for titration. What is the weight percent of sulfur in the sample?

Notes and References

1. T. W. Richards, *Chem. Rev.* **1925**, *1*, 1. For some history of classical gravimetric and titrimetric analysis, see C. M. Beck II, *Anal. Chem.* **1991**, *63*, 993A; C. M. Beck II, *Anal. Chem.* **1994**, *66*, 225A; I. M. Kolthoff, *Anal. Chem.* **1994**, *66*, 241A; D. T. Burns, *Fresenius J. Anal. Chem.* **1990**, *337*, 205; L. Niiniströ, *Fresenius J. Anal. Chem.* **1990**, *337*, 213; R. Stanley, *Lab. Prac.* **1987**, *36*, 27.

2. R. N. Silva, *Rev. Pure Appl. Chem.* **1972**, *22*, 115; K. M. Towe and W. F. Bradley, *J. Colloid Interface Sci.* **1967**, *24*, 384; T. G. Spiro, S. E. Allerton, J. Renner, A. Terzis, R. Bils, and P. Saltman, *J. Am. Chem. Soc.* **1966**, *88*, 2721.

3. Tubing such as catalog number 3787, sold by A. H. Thomas Co., P. O. Box 99, Swedesboro, NJ 08085-0099, is adequate for this demonstration.

4. M. Suzuki, *J. Chem. Ed.* **1993**, *70*, 821.

5. G. W. Latimer, Jr., *J. Chem. Ed.* **1966**, *43*, 148; G. R. Bond, *Anal. Chem.* **1962**, *32*, 1332.

6. E. Pella, *Am. Lab.* February 1990, p. 116; August 1990, p. 28.

7. R. W. Ramette, *J. Chem. Ed.* **1988**, *65*, 800.

Extraction Membranes

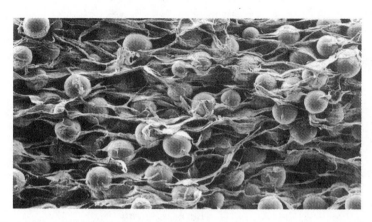

Is it alive? Ion-exchange beads enmeshed in a polytetrafluoroethylene membrane. [Photograph courtesy Bio-Rad Laboratories, Hercules, CA. Diagram below courtesy Alltech Associates, Deerfield, IL.]

Desired analytes or unwanted impurities in liquid samples can be trapped in solid-phase extraction membranes containing just about any stationary phase used in high-performance liquid chromatography. The micrograph shows 25-μm-diameter ion-exchange beads entangled in a loose membrane of polytetrafluoroethylene (Teflon). The membrane is a disk that fits into a plastic holder (see below) attached to a syringe. To conduct an extraction, solution from the syringe is simply forced through the disk.

A sandwich made of a cation-exchange membrane loaded with H^+ and an anion-exchange membrane loaded with OH^- deionizes a solution, because all entering cations are exchanged for outgoing H^+ and all incoming anions are exchanged for outgoing OH^-. A cation-exchange membrane loaded with Ag^+ selectively removes halides (X^-) from a sample, forming $AgX(s)$ on the membrane. A cation-exchange membrane loaded with Ba^{2+} selectively removes sulfate by forming $BaSO_4(s)$.

Sample Preparation

26

In Chapter 1 we described the steps in a chemical analysis. Because real objects are rarely homogeneous, samples must be carefully selected and prepared for analysis. From the *lot* (the total material, such as a dinosaur bone or a warehouse full of corn), we gather a representative *bulk sample* (Figure 26-1). Then a homogeneous *laboratory sample* designed to have the same composition as the bulk sample is formed. Small *aliquots* (test samples) of the laboratory sample are used for replicate measurements of the analyte. This chapter discusses methods for converting dinosaur bones, rocks, and cupcakes into forms that are suitable for analysis.

Besides being careful how we choose a sample, we must be careful about storing the sample. The composition may change with time after collection because of internal chemical changes, reaction with the air, or interaction of the sample with its container. Glass surfaces are notorious for ion-exchange reactions that can alter the concentrations of trace ionic species in a solution. Therefore, plastic (especially Teflon) collection bottles are frequently employed. Even plastic containers can be a source of contamination, especially if they are not washed before use. Table 26-1 shows that the manganese content of blood serum samples increased by a factor of 7 when stored in unwashed polyethylene containers prior to analysis. Steel needles are a common source of metal contamination in biochemical analysis.

"Unless the complete history of any sample is known with certainty, the analyst is well advised not to spend his [or her] time in analyzing it."[1] Your laboratory notebook ought to describe how a sample was collected and stored, as well as stating how it was analyzed.

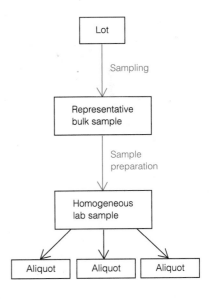

Figure 26-1 *Sampling* selects a representative bulk sample from the lot. *Sample preparation* converts the bulk sample into a homogeneous laboratory sample. Sample preparation may also refer to further steps to eliminate interference or concentrate the analyte.

variance = (standard deviation)^2

$$\frac{\text{total}}{\text{variance}} = \frac{\text{analytical}}{\text{variance}} + \frac{\text{sampling}}{\text{variance}}$$

The *Solutions Manual* contains a statistics supplement that shows you how to dissect the total variance into its two components.

26-1 Statistics of Sampling[2]

For random errors, the overall *variance*, s_o^2, is the sum of the variance of the analytical procedure, s_a^2, and the variance of the sampling operation, s_s^2:

Additivity of variance:
$$s_o^2 = s_a^2 + s_s^2 \tag{26-1}$$

If either s_a or s_s is sufficiently smaller than the other, there is little point trying to reduce the smaller one. For example, if s_s is 10% and s_a is 5%, the overall standard deviation is 11% ($\sqrt{0.10^2 + 0.05^2} = 0.11$). A more expensive and time-consuming analytical procedure that reduces s_a to 1% only improves s_o from 11 to 10% ($\sqrt{0.10^2 + 0.01^2} = 0.10$).

Origin of Sampling Variance

To understand the nature of the uncertainty in selecting a sample for analysis, consider a random mixture of two kinds of solid particles. The theory of probability allows us to state the likelihood that a randomly drawn sample has the same composition as the bulk sample. It may surprise you to learn how large a sample is required for accurate sampling.

Suppose that the mixture contains n_A particles of type A and n_B particles of type B. The probabilities of drawing A or B from the mixture are

$$p = \text{probability of drawing A} = \frac{n_A}{n_A + n_B} \tag{26-2}$$

$$q = \text{probability of drawing B} = \frac{n_B}{n_A + n_B} = 1 - p \tag{26-3}$$

If n particles are drawn at random, the expected number of particles of type A is np and the standard deviation of many drawings is known from the binomial distribution to be

Standard deviation in sampling operation:
$$\sigma_n = \sqrt{npq} \tag{26-4}$$

TABLE 26-1   **Manganese concentration of serum stored in washed and unwashed polyethylene containers**

Container[a]	Mn (ng/mL)
Unwashed	0.85
Unwashed	0.55
Unwashed	0.20
Unwashed	0.67
Average	0.57 ± 0.27
Washed	0.096
Washed	0.018
Washed	0.12
Washed	0.10
Average	0.084 ± 0.045

a. Washed containers were rinsed with water distilled twice from quartz vessels, which introduce less contamination into water than does glass.

SOURCE: J. Versieck, *Trends in Anal. Chem.* **1983**, *2*, 110.

EXAMPLE   Statistics of Drawing Particles

A mixture contains 1% KCl particles and 99% KNO_3 particles. If 10^4 particles are taken, what is the expected number of KCl particles, and what will be the standard deviation if the experiment is repeated many times?

Solution   The expected number is just

expected number of KCl particles = np = $(10^4)(0.01)$ = 100 particles

and the standard deviation will be

standard deviation = $\sqrt{npq}$ = $\sqrt{(10^4)(0.01)(0.99)}$ = 9.9

The standard deviation $\sqrt{npq}$ applies to both kinds of particles. The standard deviation is 9.9% of the expected number of KCl particles, but only 0.1% of the expected number of KNO_3 particles (nq = 9900). If you want to know how much nitrate is in the mixture, this sample is probably sufficient. For chloride, 9.9% uncertainty may not be acceptable.

TABLE 26-2 Standard test sieves

Sieve number	Screen opening (mm)	Sieve number	Screen opening (mm)
5	4.00	45	0.355
6	3.35	50	0.300
7	2.80	60	0.250
8	2.36	70	0.212
10	2.00	80	0.180
12	1.70	100	0.150
14	1.40	120	0.125
16	1.18	140	0.106
18	1.00	170	0.090
20	0.850	200	0.075
25	0.710	230	0.063
30	0.600	270	0.053
35	0.500	325	0.045
40	0.425	400	0.038

How much sample corresponds to 10^4 particles? Suppose that the particles are 1-mm-diameter spheres. The volume of a 1-mm-diameter sphere is $\frac{4}{3}\pi(0.5 \text{ mm})^3 = 0.524 \ \mu\text{L}$. The density of KCl is 1.984 g/mL and that of KNO_3 is 2.109 g/mL, so the average density of the mixture is $(0.01)(1.984) + (0.99)(2.109) = 2.108$ g/mL. The mass of mixture containing 10^4 particles is $(10^4)(0.524 \times 10^{-3} \text{ mL})(2.108 \text{ g/mL}) = 11.0$ g. *If you take 11.0-g test portions from a larger laboratory sample, the expected sampling standard deviation for chloride is 9.9%!* The sampling standard deviation for nitrate will be only 0.1%.

How can you prepare a mixture of 1-mm-diameter particles? You might make such a mixture by grinding larger particles and passing them through a 16 mesh sieve, whose screen openings are squares with sides of length 1.18 mm (Table 26-2). Particles that pass through the screen are then passed through a 20 mesh sieve, whose openings are 0.85 mm, and material that does not pass is retained for your sample. This gives particles whose diameters are in the range 0.85–1.18 mm. We refer to the size range as *16/20 mesh*.

Suppose that much finer particles of 80/120 mesh size (average diameter = 152 μm, average volume = 1.84 nL) were used instead. Now the mass containing 10^4 particles is reduced from 11.0 g to 0.0388 g. We could analyze a larger sample to reduce the sampling uncertainty for chloride.

Wow! I'm surprised it's so big.

EXAMPLE Reducing Sample Uncertainty with a Larger Test Portion

How many grams of 80/120 mesh sample are required to reduce the chloride sampling uncertainty to 1%?

Solution We are looking for a standard deviation of 1% of the number of KCl particles:

$$\sigma_n = \sqrt{npq} = (0.01)np$$

Using $p = 0.01$ and $q = 0.99$, we find $n = 9.9 \times 10^5$ particles. With a particle volume of 1.84 nL and an average density of 2.108 g/mL, the mass required for 1% chloride sampling uncertainty is

$$\text{mass} = (9.9 \times 10^5 \text{ particles}) \left(1.84 \times 10^{-6} \frac{\text{mL}}{\text{particle}}\right) \left(2.108 \frac{\text{g}}{\text{mL}}\right) = 3.84 \text{ g}$$

Even with an average particle diameter of 152 μm, we must analyze 3.84 g to reduce the sampling uncertainty to 1%. There is no point using an expensive analytical method with a precision of 0.1%, because the overall uncertainty will still be 1% on the basis of sampling uncertainty.

> There is no advantage to reducing the analytical uncertainty if the sampling uncertainty is high, and vice versa.

You should now see that sampling uncertainty arises from the random nature of drawing particles from a mixture. If the mixture is a liquid and the particles are molecules, there are around 10^{22} particles/mL. It will not require much volume of homogeneous liquid solution to reduce the sampling error to a negligible value. Solids, however, must be ground to very fine dimensions, and large quantities must be used to ensure a small sampling variance. Grinding invariably contaminates the sample with material from the grinding apparatus.

Choosing a Sample Size

A well-mixed powder containing KCl and KNO_3 is an example of a heterogeneous material in which the variation from place to place is random. *How much of a random mixture should be analyzed to reduce the sampling variance for one analysis to a desired level?*

To answer this question, consider Figure 26-2, which shows results for sampling the radioisotope ^{24}Na in human liver. The tissue has been "homogenized" in a blender but is not truly homogeneous because it is a suspension of small particles in water. The average number of radioactive counts per second per gram of sample is about 237. When the mass of sample for each analysis was about 0.09 g, the standard deviation (shown by the error bar at the left in the diagram) was ±31 counts per second per gram of homogenate, which is ±13.1% of the mean value (237). When the sample size was increased to about 1.3 g, the standard deviation was decreased to ±13 counts per second per gram, or ±5.5% of the mean. For a sample size

Figure 26-2 Sampling diagram of experimental results for ^{24}Na in liver homogenate. Dots are experimental points, and error bars extend ±1 standard deviation about the mean. Note that abscissa is logarithmic. [From B. Kratochvil and J. K. Taylor, *Anal. Chem.* **1981**, *53*, 925A; National Bureau of Standards Internal Report 80-2164, 1980, p. 66.]

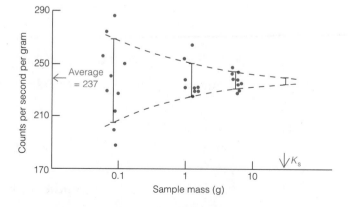

near 5.8 g, the standard deviation was reduced to ± 5.7 counts per second per gram, or $\pm 2.4\%$ of the mean.

Equation 26-4 told us that when n particles are drawn from a mixture of two kinds of particles (such as liver tissue particles and droplets of water), the sampling standard deviation will be $\sigma_n = \sqrt{npq}$, where p and q are the fraction of each kind of particle present. The relative standard deviation is $\sigma_n/n = \sqrt{npq}\,/\,n = \sqrt{pq/n}$. The relative variance, $(\sigma_n/n)^2$, is therefore

$$\text{relative variance} \equiv R^2 = \left(\frac{\sigma_n}{n}\right)^2 = \frac{pq}{n} \Rightarrow nR^2 = pq \qquad (26\text{-}5)$$

Noting that the mass of sample drawn, m, is proportional to the number of particles drawn, we can rewrite Equation 26-5 in the form

Sampling constant: $\qquad\qquad mR^2 = K_s \qquad\qquad\qquad (26\text{-}6)$

in which R is the relative standard deviation (expressed in percent), and K_s is called the *sampling constant*. K_s is the mass of sample required to reduce the relative sampling standard deviation to 1%.

Let's see if Equation 26-6 describes Figure 26-2. Table 26-3 shows that mR^2 is approximately constant for large samples, but agreement is poor for the smallest sample (0.09 g). Attributing the poor agreement at low mass to random sampling variation, we assign $K_s \approx 36$ g in Equation 26-6. This is the average from the 1.3- and 5.8-g samples in Table 26-3.

TABLE 26-3   Calculation of sampling constant for Figure 26-2

Sample mass, m (g)	Relative standard deviation (%)	mR^2 (g)
0.09	13.1	15.4
1.3	5.5	39.3
5.8	2.4	33.4

EXAMPLE   Mass of Sample Required to Produce a Given Sampling Variance

What mass in Figure 26-2 will give a sampling standard deviation of $\pm 7\%$?"

Solution   With the sampling constant $K_s \approx 36$ g, the answer is

$$m = \frac{36\ \text{g}}{7^2} = 0.73\ \text{g}$$

A 0.7-g sample should give ~7% sampling standard deviation. This is strictly a sampling standard deviation. If the analytical procedure also has some uncertainty, the net variance will depend on both factors, as described by Equation 26-1.

Choosing the Number of Replicate Analyses

We just saw that a single 0.7-g sample is expected to give a sampling standard deviation of $\pm 7\%$. *How many 0.7-g samples must be analyzed to give 95% confidence that the mean is known to within $\pm 4\%$?* The meaning of 95% confidence is that there is only a 5% chance that the true mean lies more than $\pm 4\%$ away from the measured mean. The question we just asked refers only to sampling uncertainty and assumes that analytical uncertainty is much smaller than sampling uncertainty.

Rearranging Student's t equation 4-6 allows us to answer the question:

Required number of replicate analyses: $\qquad \underbrace{\mu - \bar{x}}_{e} = \frac{ts_s}{\sqrt{n}} \Rightarrow n = \frac{t^2 s_s^2}{e^2} \qquad (26\text{-}7)$

The sampling contribution to the overall uncertainty can be reduced by analyzing more samples.

in which μ is the true population mean, $\bar{x}$ is the measured mean, n is the number of samples needed, s_s^2 is the variance of the sampling operation, and e is the sought-for uncertainty. The quantities s_s and e must both be expressed as absolute uncertainties or both as relative uncertainties. Student's t is taken from Table 4-2 for 95% confidence at $n - 1$ degrees of freedom. Because n is not yet known, the value of t for $n = \infty$ can be used to estimate n. After a value of n is calculated, the process is repeated a few times until a constant value of n is found.

EXAMPLE   Sampling a Random Bulk Material

How many 0.7-g samples must be analyzed to give 95% confidence that the mean is known to within ±4%?

Solution   A 0.7-g sample gives $s_s = 7\%$, and we are seeking $e = 4\%$. We will express both uncertainties in relative form. Taking $t = 1.960$ (from Table 4-2 for 95% confidence and ∞ degrees of freedom) as a starting value, we find

$$n \approx \frac{(1.960)^2(0.07)^2}{(0.04)^2} = 11.8 \approx 12$$

For $n = 12$, there are 11 degrees of freedom, so a second trial value of Student's t (interpolated from Table 4-2) is 2.209. A second cycle of calculation gives

$$n \approx \frac{(2.209)^2(0.07)^2}{(0.04)^2} = 14.9 \approx 15$$

For $n = 15$, there are 14 degrees of freedom and $t = 2.150$, which gives

$$n \approx \frac{(2.150)^2(0.07)^2}{(0.04)^2} = 14.2 \approx 14$$

For $n = 14$, there are 13 degrees of freedom and $t = 2.170$, which gives

$$n \approx \frac{(2.170)^2(0.07)^2}{(0.04)^2} = 14.4 \approx 14$$

The calculations reach a constant value near $n \approx 14$, so we need about 14 samples of 0.7-g size to determine the mean value to within 4% with 95% confidence.

For the preceding calculations, we needed some prior knowledge of the standard deviation. Some preliminary study of the sample must be made before the remainder of the analysis can be planned. If there are many similar samples to be analyzed, a thorough analysis of one sample might allow you to plan less thorough—but adequate—analyses of the remainder.

26-2   Preparing Samples for Analysis [3]

Once a bulk sample is selected, a laboratory sample must be prepared. A coarse solid must be ground and mixed so that the laboratory sample has the same composition as the bulk sample. Solids are typically dried at 110°C at atmospheric pressure to remove adsorbed water prior to analysis. Delicate samples may simply be stored in an environment that brings them to a constant, reproducible moisture level.

The laboratory sample is usually dissolved for analysis. It is important to dissolve the entire sample, or else we cannot be sure that some of the intended analyte did not dissolve. If the sample does not dissolve under mild conditions, *acid digestion* or *fusion* may be used. Organic material may be destroyed by *combustion* (also called *dry ashing*) or *wet ashing* (oxidation with liquid reagents) to place inorganic elements in suitable form for analysis.

Grinding

Solids can be ground in a **mortar and pestle** like those shown in Figure 26-3. The steel mortar (also called a percussion mortar or "diamond" mortar) is a hardened steel tool into which the sleeve and pestle fit snugly. Hard, brittle materials such as ores and minerals can be crushed by striking the pestle lightly with a hammer. The *agate* mortar (or similar ones made of porcelain, mullite, or alumina) is designed for grinding small particles into a fine powder. Less expensive mortars tend to be more porous and more easily scratched, which leads to contamination of the sample with mortar material or portions of previously ground samples. If the ground material comes off easily, a ceramic mortar can be cleaned by wiping with a wet tissue and washing with distilled water. More difficult sample residues can be cleaned up by grinding with 4 M HCl in the mortar or by grinding an abrasive cleaner (such as Ajax®), followed by washing with HCl and water. A *boron carbide* mortar and pestle is five times harder than agate and less prone to contaminate the sample.

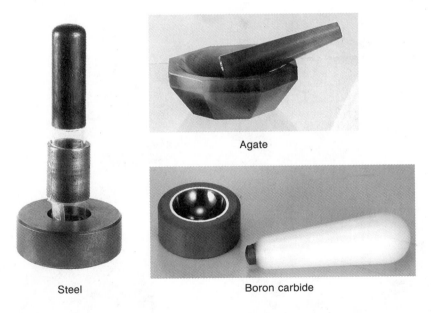

Agate

Steel

Boron carbide

Figure 26-3 Steel, agate, and boron carbide mortars and pestles. The mortar is the base and the pestle is the grinding tool. In the case of boron carbide, the mortar is a hemispheric shell enclosed in a plastic or aluminum body. The pestle has a boron carbide button at the tip of a plastic handle. [Courtesy Thomas Scientific, Swedesboro, NJ, and Spex Industries, Edison, NJ.]

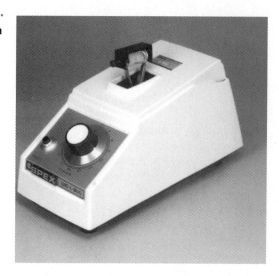

Figure 26-4 Wig-L-Bug® sample shaker and polystyrene vial and ball for pulverizing soft materials. [Courtesy Spex Industries, Edison, NJ.]

A **ball mill** is a grinding device in which steel or ceramic balls are rotated inside a drum to crush the sample to a fine powder. Figure 26-4 shows a Wig-L-Bug®, which pulverizes a sample by rapidly shaking it in a vial containing a ball that moves back and forth. For soft materials, plastic vials and balls are appropriate. For harder materials, steel, agate, and tungsten carbide are used. A Shatterbox® laboratory mill spins a puck and ring inside a grinding container at 825 revolutions per minute to pulverize up to 100 g of material (Figure 26-5). Tungsten carbide and zirconia containers are used for very hard samples.

Shatterbox® grinding action

Figure 26-5 Shatterbox® laboratory mill spins a puck and ring inside a container at high speed to grind up to 100 mL of sample to a fine powder. [Courtesy Spex Industries, Edison, NJ.]

TABLE 26-4 Acids for sample dissolution

Acid	Typical composition (wt% and density)	Notes
HCl	37% 1.19 g/mL	Nonoxidizing acid useful for many metals, oxides, sulfides, carbonates, and phosphates. Constant boiling composition at 109°C is 20% HCl. As, Sb, Ge, and Pb form volatile chlorides that may be lost from an open vessel.
HBr	48–65%	Similar to HCl in solvent properties. Constant boiling composition at 124°C is 48% HBr.
H_2SO_4	95–98% 1.84 g/mL	Good solvent at its boiling point of 338°C. Attacks metals and dehydrates and oxidizes organic compounds.
H_3PO_4	85% 1.70 g/mL	Hot acid dissolves refractory oxides insoluble in other acids. Becomes anhydrous above 150°C. Dehydrates to pyrophosphoric acid (H_2PO_3—O—PO_3H_2) above 200°C and dehydrates further to metaphosphoric acid ($[HPO_3]_n$) above 300°C.
HF	50% 1.16 g/mL	Used primarily to dissolve silicates, making volatile SiF_4. This product and excess HF are removed by adding H_2SO_4 or $HClO_4$ and heating. Constant boiling composition at 112°C is 38% HF. Used in Teflon, silver, or platinum containers. Extremely harmful upon contact or inhalation. Fluorides of As, B, Ge, Se, Ta, Nb, Ti, and Te are volatile. LaF_3, CaF_2, and YF_3 precipitate. F^- is removed by adding H_3BO_3 and taking to dryness with H_2SO_4 present.
$HClO_4$	60–72% 1.54–1.67 g/mL	Cold and dilute acid are not oxidizing, but hot, concentrated acid is an extremely powerful, explosive oxidant, especially useful for organic matter that has already been partially oxidized by hot HNO_3. Constant boiling composition at 203°C is 72%. **Before using $HClO_4$, evaporate the sample to near dryness several times with hot HNO_3 to destroy as much organic material as possible.**

Dissolving Inorganic Materials with Acids

Table 26-4 lists common acids for dissolving inorganic materials. The nonoxidizing acids HCl, HBr, HF, H_3PO_4, dilute H_2SO_4, and dilute $HClO_4$ dissolve metals by the redox reaction

$$M + nH^+ \rightarrow M^{n+} + \frac{n}{2}H_2 \qquad (26\text{-}8)$$

Metals with negative reduction potentials should dissolve, although some, such as Al, form a protective oxide coat that inhibits dissolution. Volatile species formed by protonation of such anions as carbonate ($CO_3^{2-} \rightarrow H_2CO_3 \rightarrow CO_2$), sulfide ($S^{2-} \rightarrow H_2S$), phosphide ($P^{3-} \rightarrow PH_3$), fluoride ($F^- \rightarrow$

HF is extremely harmful to touch or breathe. Flood the affected area with water, coat the skin with calcium gluconate (or another calcium salt), and seek medical help.

HF), and borate ($BO_3^{3-} \rightarrow H_3BO_3$) will be lost from hot acids in open vessels. Volatile metal halides such as $SnCl_4$ and $HgCl_2$ and some molecular oxides such as OsO_4 and RuO_4 can also be lost. Hot hydrofluoric acid is especially useful for dissolving silicates. Glass or platinum vessels can be used for HCl, HBr, H_2SO_4, H_3PO_4, and $HClO_4$. HF should be used in Teflon, polyethylene, silver, or platinum vessels. The highest quality acids must be used to minimize contamination by the concentrated reagent.

Substances that do not dissolve in nonoxidizing acids may dissolve in the oxidizing acids HNO_3; hot, concentrated H_2SO_4; or hot, concentrated $HClO_4$. Nitric acid attacks most metals, but not Au and Pt, which dissolve in the 3:1 (vol/vol) mixture of $HCl:HNO_3$ called **aqua regia.** Strong oxidants such as Cl_2 or $HClO_4$ in HCl dissolve difficult materials such as Ir at elevated temperature. A mixture of HNO_3 and HF attacks the refractory carbides, nitrides, and borides of Ti, Zr, Ta, and W. Hot, concentrated $HClO_4$ (described later for organic substances) is a **DANGEROUS,** powerful oxidant whose oxidizing power is increased by adding concentrated H_2SO_4 and catalysts such as V_2O_5 or CrO_3.

Digestion is conveniently carried out in a Teflon-lined **bomb** (a sealed vessel) heated in a microwave oven.[4] The vessel in Figure 26-6 has a volume of 23 mL and digests up to 1 g of inorganic material (or 0.1 g of organic material, which releases a great deal of $CO_2(g)$) in up to 15 mL of concentrated acid. A microwave oven heats the contents to 200°C in a minute. To prevent explosions, the lid releases gas from the vessel if the internal pressure exceeds 8 MPa (80 atm). The bomb cannot be made of metal, which absorbs microwaves. An advantage of a bomb is that it is cooled prior to opening, thus preventing loss of volatile products.

Dissolving Inorganic Materials by Fusion

Substances that will not dissolve in acid can usually be dissolved by a hot, molten inorganic **flux** (Table 26-5). The finely powdered unknown is mixed with 2–20 times its mass of solid flux, and **fusion** (melting) is carried out in

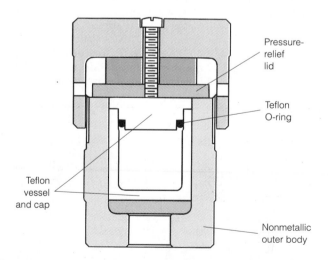

Figure 26-6   Microwave digestion bomb lined with Teflon. The outer container retains strength to 150°C, but rarely reaches 50°C. Deformation of the cap releases pressure exceeding 80 atm from inside the vessel, thereby preventing explosions. [Courtesy Parr Instrument Co., Moline, IL.]

Pressure-relief lid

Teflon O-ring

Teflon vessel and cap

Nonmetallic outer body

TABLE 26-5   Fluxes for sample dissolution

Flux	Crucible	Uses
Na_2CO_3	Pt	For dissolving silicates (clays, rocks, minerals, glasses), refractory oxides, insoluble phosphates, and sulfates.
$Li_2B_4O_7$ or $LiBO_2$ or $Na_2B_4O_7$	Pt, graphite Au-Pt alloy, Au-Rh-Pt alloy	Individual or mixed borates are used to dissolve aluminosilicates, carbonates, and samples with high concentrations of basic oxides. $B_4O_7^{2-}$ is called tetraborate and BO_2^- is metaborate.
NaOH or KOH	Au, Ag	Dissolves silicates and SiC. Frothing occurs when H_2O is eliminated from flux, so it is best to prefuse flux and then add sample. Analytical capabilities are limited by impurities in NaOH and KOH.
Na_2O_2	Zr, Ni	Strong base and powerful oxidant good for silicates not dissolved by Na_2CO_3. Useful for iron and chromium alloys. Because it slowly attacks crucibles, a good procedure is to coat the inside of a Ni crucible with molten Na_2CO_3, cool, and add Na_2O_2. The peroxide melts at lower temperature than the carbonate, which shields crucible from the melt.
$K_2S_2O_7$	Porcelain, SiO_2, Au, Pt	Potassium pyrosulfate ($K_2S_2O_7$) is prepared by heating $KHSO_4$ until all water is lost and foaming ceases. Alternatively, potassium persulfate ($K_2S_2O_8$) decomposes to $K_2S_2O_7$ upon heating. Good for refractory oxides, not silicates.
B_2O_3	Pt	Useful for oxides and silicates. Principal advantage is that flux can be completely removed as volatile methyl borate ($[CH_3O]_3B$) by several treatments with HCl in methanol.
$Li_2B_4O_7$ + Li_2SO_4 (2:1 wt/wt)	Pt	Example of a powerful mixture for dissolving refractory silicates and oxides in 10–20 min at 1 000°C. One gram of flux dissolves 0.1 g of sample. The solidified melt dissolves readily in 20 mL of hot 1.2 M HCl.

a platinum-gold alloy crucible at 300°–1200°C in a furnace or over a burner. The automated apparatus in Figure 26-7 carries out three fusions at once over propane burners with mechanical agitation of the crucibles. When the samples are homogeneous, the molten flux is poured into beakers containing 10 wt% aqueous HNO_3 to dissolve the product.

Most fusions use lithium tetraborate ($Li_2B_4O_7$, m.p. 930°C), lithium metaborate ($LiBO_2$, m.p. 845°C), or a mixture of the two. A nonwetting agent such as KI can be added to prevent the flux from sticking to the crucible. For example, 0.2 g of cement might be fused with 2 g of $Li_2B_4O_7$ and 30 mg of KI.

Figure 26-7 Automated apparatus that fuses three samples at once over propane burners. It also provides mechanical agitation of the Pt/Au crucibles. Crucibles are shown in the tipped position used to pour out contents after fusion is complete. [Courtesy Spex Industries, Edison, NJ.]

Fusion is a last resort, because the flux can introduce impurities.

A disadvantage of a flux is that impurities are introduced by the large mass of solid reagent. If part of the unknown can be dissolved with acid prior to fusion, it should be dissolved. Then the insoluble component is dissolved with flux and the two portions are combined for analysis.

Basic fluxes in Table 26-5 ($LiBO_2$, Na_2CO_3, NaOH, KOH, and Na_2O_2) are best used to dissolve acidic oxides of Si and P. Acidic fluxes ($Li_2B_4O_7$, $Na_2B_4O_7$, $K_2S_2O_7$, and B_2O_3) are most suitable for basic oxides (including cements and ores) of the alkali metals, alkaline earths, lanthanides, and Al. KHF_2 is useful for lanthanide oxides. Sulfides and some oxides, some iron and platinum alloys, and some silicates require an oxidizing flux for dissolution. For this purpose, pure Na_2O_2 may be suitable, or oxidants such as KNO_3, $KClO_3$, or Na_2O_2 can be added to Na_2CO_3. Boric oxide can be converted to $B(OCH_3)_3$ after fusion and completely evaporated. The solidified flux is treated with 100 mL of methanol saturated with HCl gas and heated gently. The procedure is repeated several times, if necessary, to remove all of the boron.

Decomposition of Organic Substances

Box 26-1 describes a simple benchtop combustion analysis of wood.

Digestion of organic material is classified as **dry ashing** if the procedure does not involve liquid, or **wet ashing** if liquid is used. Occasionally, fusion with Na_2O_2 (called Parr oxidation) or alkali metals may be carried out in a sealed bomb. Section 25-4 discussed **combustion analysis,** in which C, H, N, S, and halogens are measured.

Convenient *wet ashing* procedures involve microwave digestion with acid in a Teflon bomb (Figure 26-6). For example, 0.25 g of animal tissue can be digested for metals analysis by placing it in a 60-mL Teflon vessel containing 1.5 mL of high-purity 70 wt% HNO_3 plus 1.5 mL of high-purity 96 wt% H_2SO_4 and heating in a 700-W kitchen microwave oven for 1 min.[5] An important wet ashing process is *Kjeldahl* digestion with H_2SO_4 for nitrogen analysis (Section 7-2). Digestion with fuming HNO_3 (which contains excess dissolved NO_2) in a sealed, heavy-walled glass tube at 200°–300°C is called the *Carius* method.

Figure 26-8 shows microwave wet ashing apparatus. Sulfuric acid or a mixture of H_2SO_4 and HNO_3 (~15 mL of acid per gram of unknown) are added to an organic substance in a glass digestion tube fitted with the reflux cap. In the first step, the sample is *carbonized* for 10 to 20 min at gentle reflux until all particles have dissolved and the solution has a uniform black appearance. Power is turned off and the sample is allowed to cool for 1–2 min. *Oxidation* is carried out next by adding H_2O_2 or HNO_3 through the reflux cap until the color disappears or the solution is just barely tinted. If the solution is not homogeneous, the power is turned up and the sample is heated to bring all solids into solution. Repeated cycles of oxidation and solubilization may be required. Once conditions for a particular type of material are worked out, the procedure is automated, with power levels and reagent delivery (via the peristaltic pump) programmed into the controller.

$HClO_4$ with organic material is an extreme **explosion hazard.** Always oxidize first with HNO_3. Always use a blast shield for $HClO_4$.

Wet ashing with refluxing HNO_3-$HClO_4$ (Figure 26-9) is a widely applicable, but hazardous, procedure.[6] **Perchloric acid has caused numerous explosions.** Use a good blast shield in a metal-lined fume hood designed for $HClO_4$. First, heat the sample slowly to boiling with HNO_3 but *without*

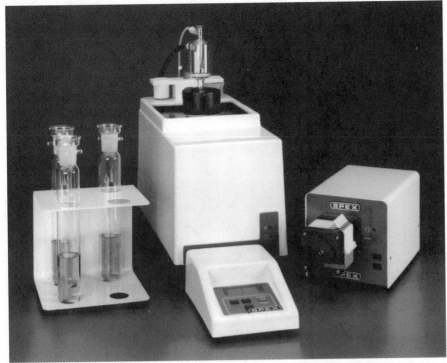

Digestion   Microwave heater   Peristaltic
tubes   and controller   pump

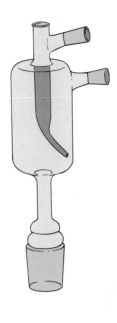

Reflux cap

Figure 26-8 Microwave apparatus for digesting organic materials by wet ashing. [Courtesy Spex Industries, Edison, NJ.]

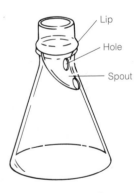

Figure 26-9 Reflux cap for wet ashing in an Erlenmeyer flask. The hole allows vapor to escape, and the spout is curved to contact the side of the flask. [From D. D. Siemer and H. G. Brinkley, *Anal. Chem.* **1981,** *53,* 750.]

$HClO_4$. Boil to near dryness to complete the oxidation of easily oxidized material that might explode in the presence of $HClO_4$. Add fresh HNO_3 and repeat the evaporation several times. After cooling to room temperature, $HClO_4$ is added and heating is begun again. If possible, HNO_3 should be present during the $HClO_4$ treatment. A large excess of HNO_3 should be present when oxidizing inorganic materials. A reviewer of this book once wrote, "I have seen someone substitute perchloric acid for sulfuric acid in a Jones reductor experiment with spectacular results—no explosion but the tube melted!" Bottles of $HClO_4$ should not be stored on wooden shelves, because acid spilled on wood can form explosive cellulose perchlorate esters. Perchloric acid also should not be stored near organic reagents or reducing agents.

The combination of Fe^{2+} and H_2O_2, called *Fenton's reagent,* can oxidize organic material in dilute aqueous solutions. For example, the organic components of urine were destroyed in 30 min at 50°C to release traces of mercury for analysis.[7] A 50-mL sample was adjusted to pH 3–4 with 0.5 M H_2SO_4. Then 50 μL of saturated aqueous ferrous ammonium sulfate, $Fe(NH_4)_2(SO_4)_2$, was added, followed by 100 μL of 30 wt% H_2O_2.

Fenton's reagent was formerly thought to generate OH· radical. It is now believed that $Fe^{II}OOH$ is the active species.[8]

Box 26-1 Chemical Analysis of a Tree

A benchtop combustion procedure was employed to measure Cl, P, and S in pine tree rings in southern Italy. Wood was taken from the tree with a core drill and rings were sliced apart and dried at 75°C for 3 days. Samples weighing 10–100 mg were wrapped in ashless filter paper, leaving an unfolded paper tail as shown in the diagram. The wrapped sample was then placed in the Pt gauze basket of the disassembled *Schöniger flask*.

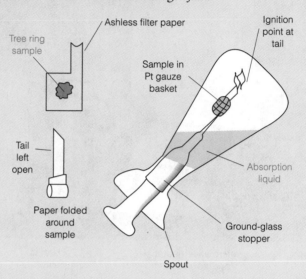

Schöniger combustion procedure. The Pt gauze basket helps catalyze complete combustion of the sample, which is carried out behind a shield. The ashless filter paper had to be washed several times with distilled water to remove traces of Cl^- and SO_4^{2-} prior to analysis. To analyze a liquid sample, a polycarbonate capsule with a 0.2-mL capacity is used instead of filter paper.

A mixture of 20 mL of water plus 0.5 mL of 30 wt% H_2O_2 was placed in the 250-mL flask, which was then flushed with O_2 gas. The paper tail of the wrapped sample was ignited and the stopper-sample assembly was immediately inserted into the flask. The flask was

26-3 Preconcentration, Cleanup, and Derivatization

Trace analysis may require **preconcentration** of the analyte to bring it to a greater concentration prior to analysis. Metal ions in natural waters can be preconcentrated with cation-exchange resin. For example,[9] a 500-mL volume of seawater adjusted to pH 6.5 with ammonium acetate and ammonia was passed through 2 g of Chelex-100 in the Mg^{2+} form to trap all the trace metal ions. Washing with 2 M HNO_3 eluted the metals in a total volume of 10 mL, thereby giving a concentration increase of $500/10 = 50$. Metals in the HNO_3 solution were then analyzed by graphite furnace atomic absorption, with a typical detection limit for Pb being 15 pg/mL. The detection limit for Pb in the seawater is therefore 50 times lower, or 0.3 pg/mL. Figure 26-10

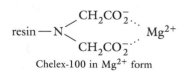

Chelex-100 in Mg^{2+} form

inverted as shown in the diagram to prevent gaseous products from leaking out. Combustion was rapid, but 30 min were allowed for absorption of the gaseous products by the solution. The elements Cl, P, and S in the wood were converted to Cl^-, PO_4^{3-}, and SO_4^{2-} in the H_2O_2 solution.

The solution was then analyzed by ion chromatography (Section 24-2), with results shown in the following chart. The average Cl uptake by the tree in the period 1971–1986 was 10 times greater than the average uptake in 1956–1970. Uptake of P and S increased by a factor of 2 in the recent 15-year period. The chemical content of the rings presumably mirrors chemical changes in the environment.

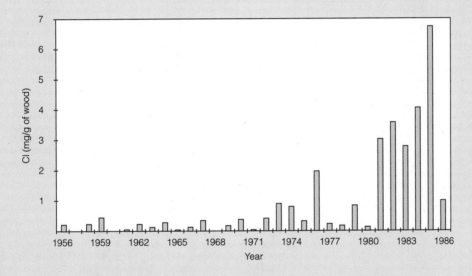

Chlorine content over a 30-year period of rings from a tree in southern Italy. [M. Ferretti, R. Udisti, and E. Barbolani, *Fresenius J. Anal. Chem.* **1992**, *343*, 607.]

shows the effect of pH on the recovery of metals from seawater. At low pH, H^+ competes with the metal ions for ion-exchange sites and prevents complete recovery.

Ion-exchange resins can capture basic or acidic gases. Carbonate liberated as CO_2 from $(ZrO)_2CO_3(OH)_2 \cdot xH_2O$ used in nuclear fuel reprocessing can be measured by placing a known amount of powdered solid in the test tube in Figure 26-11 and adding 3 M HNO_3. When the solution is purged with N_2, CO_2 is captured quantitatively by moist anion-exchange resin in the sidearm:

$$CO_2 + H_2O \rightarrow H_2CO_3$$
$$2resin^+OH^- + H_2CO_3 \rightarrow (resin^+)_2CO_3^{2-} + H_2O$$

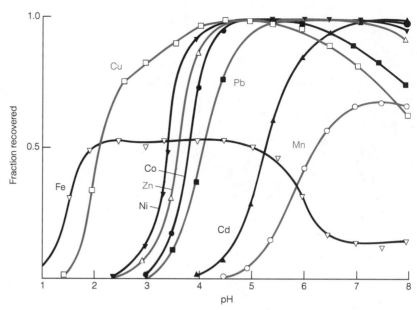

Figure 26-10   The pH dependence of the recovery of trace metals from seawater by Chelex-100. The graph shows the pH of the seawater when it was passed through the column. [From S.-C. Pai, *Anal. Chim. Acta* **1988,** *211,* 271.]

Carbonate is eluted from the resin with 1 M $NaNO_3$ and measured by titration with acid. Table 26-6 gives other applications of this technique.

The **purge and trap method** is used to preconcentrate volatile substances from a solution for gas chromatographic analysis. Helium gas bubbled

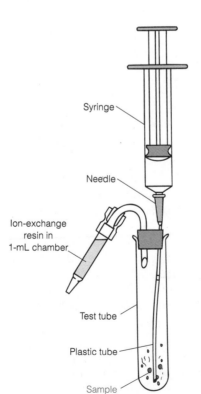

Figure 26-11   Apparatus for trapping basic or acidic gases by ion exchange. [From D. D. Siemer, *Anal. Chem.* **1987,** *59,* 2439.]

TABLE 26-6   Use of ion-exchange resin for trapping gases

Gas	Species trapped	Eluent	Analytical method
CO_2	CO_3^{2-}	1 M $NaNO_3$	Titrate with acid
H_2S	S^{2-}	0.5 M Na_2CO_3 + H_2O_2	S^{2-} is oxidized to SO_4^{2-} by H_2O_2. The sulfate is measured by ion chromatography.
SO_2	SO_3^{2-}	0.5 M Na_2CO_3 + H_2O_2	SO_3^{2-} is oxidized to SO_4^{2-} by H_2O_2. The sulfate is measured by ion chromatography.
HCN	CN^-	1 M Na_2SO_4	Titration of CN^- with hypobromite: $$CN^- + OBr^- \rightarrow CNO^- + Br^-$$
NH_3	NH_4^+	1 M $NaNO_3$	Colorimetric assay with Nessler's reagent: $$2K_2HgI_4 + 2NH_3 \rightarrow NH_2Hg_2I_3 + 4KI + NH_4I$$

Nessler's reagent   Absorbs strongly at 400–425 nm

SOURCE: D. D. Siemer, *Anal. Chem.* **1987,** *59,* 2439.

through the unknown solution transports volatile species into a tube containing a series of increasingly stronger adsorbents, such as silicones, silica gel, and activated carbon. Each gaseous analyte is trapped in the first zone that is adsorbent enough to retain it. The trap can be stored for many days in a freezer without losing analytes. To analyze the volatile species, the trap is heated rapidly and backflushed with He into a gas chromatograph.

Coprecipitation with a large excess of highly adsorbent, high-surface-area solid provides significant preconcentration. For example, many metal ions in seawater coprecipitate with $Ga(OH)_3$. To accomplish this, 200 μL of HCl solution containing 50 μg of Ga^{3+} is added to 10.00 mL of the water sample. When the pH is brought to 9.1 with NaOH, a jellylike precipitate forms, on which is adsorbed a large fraction of the cations from the solution (Table 26-7). After centrifugation to pack the precipitate, the water is removed and the gel washed with water. Then the gel is dissolved in 50 μL of 1 M HNO_3 and aspirated into an inductively coupled plasma for atomic emission analysis. The preconcentration factor is 10 mL/50 μL = 200, and the detection limits for the elements in Table 26-7 in the original seawater range from 15 pg/mL for Zn to 3 ng/mL for Mo.

The substance used to coprecipitate traces of analyte is called a *gathering agent.*

TABLE 26-7   Recovery of elements by coprecipitation with $Ga(OH)_3$

Element	Recovery (%)	Element	Recovery (%)	Element	Recovery (%)
Al	101.1 ± 2.1	Ti	98.8 ± 0.6	V	33.5 ± 4.2
Cr	100.7 ± 7.9	Mn	78.3 ± 8.9	Fe	95.9 ± 6.3
Co	92.4 ± 4.6	Ni	93.5 ± 3.9	Cu	94.0 ± 4.7
Zn	98.3 ± 5.6	Y	91.0 ± 5.4	Mo	1.3 ± 0.4
Cd	42.4 ± 7.2	Pb	98.8 ± 11.2		

Standard deviation is derived from five replicate analyses of artificial samples.
SOURCE: T. Akagi and H. Haraguchi, *Anal. Chem.* **1990,** *62,* 81.

Solid-phase extraction preconcentrates trace organic analytes from a large volume of water by passage through a short, disposable reverse-phase chromatography column containing octadecyl ($-(CH_2)_{17}CH_3$) groups. Nonpolar molecules retained by the column are then eluted in a small volume of nonpolar solvent, such as hexane or dichloromethane. Figure 26-12 shows the recovery of steroids from urine by this method. Every common stationary phase for liquid chromatography is available for solid-phase extraction.[10]

Similar approaches are useful for **sample cleanup,** in which undesirable components of the unknown are removed prior to analysis. If the impurities are more polar than the analyte, the unknown is dissolved in a polar solvent (such as aqueous methanol) and passed through a short, disposable, reverse-phase column. Polar impurities pass through the column, whereas the analyte and nonpolar impurities are retained. Elution with a nonpolar solvent gives analyte free of polar impurities. If the undesired impurities are less polar than the analyte, sample dissolved in nonpolar solvent is passed through a plain silica column. Impurities pass through and analyte is retained. Finally, analyte is eluted with a more polar solvent.

Derivatization is a procedure in which analyte is chemically modified to make it easier to detect or separate. For example, a fluorescent group can be attached to a nonfluorescent molecule so that it can be easily detected. In biochemistry, an enzyme that is easily detected by the reaction it catalyzes can be covalently bound to another large molecule that is otherwise difficult to observe. The large molecule is then located wherever the enzyme is seen.

Nanogram quantities of nonvolatile carboxylic acids can be converted to more volatile pentafluorobenzyl esters that are suitable for gas chromatography and readily detected with an electron capture detector (Section 23-1).

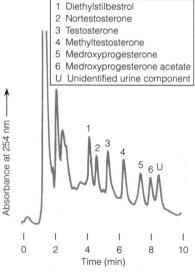

> Preliminary *cleanup* removes a large fraction of material that might interfere in the analysis.

Figure 26-12 High-performance liquid chromatography of anabolic steroids preconcentrated from urine by solid-phase extraction with 1 mL of C_{18} silica. The extractor was first washed with 2 mL of methanol followed by 2 mL of H_2O. Then 10 mL of urine containing 10 ppb of several steroids was passed through. To remove unwanted components, the tube was washed with 4 mL of 25 mM borate buffer at pH 8, followed by 4 mL of 40% methanol in water, followed by 4 mL of 20% acetone in water. Finally, steroids were eluted with two 500-µL aliquots of 73% methanol in water. [Courtesy Supelco, Inc., Bellefonte, PA.]

$$CH_3(CH_2)_{10}CO_2H + F\text{—}\underset{\substack{\text{Pentafluorobenzyl}\\\text{bromide}}}{\overset{\substack{F\quad F}}{\bigcirc}}\text{—}CH_2Br \xrightarrow{-HBr}$$

Nonvolatile
carboxylic acid

$$F\text{—}\underset{\substack{F\quad F}}{\overset{\substack{F\quad F}}{\bigcirc}}\text{—}CH_2O_2C(CH_2)_{10}CH_3$$

Volatile derivative with pentafluorophenyl group easily observed by electron capture

A *Keele microreactor* is useful for this microscale derivatization. Into the bottom of the reactor in Figure 26-13a is placed 10 µL of 0.1 M aqueous tetrabutylammonium hydroxide, 10 µL of 0.1 M NaOH, 100 ng of carboxylic acid in 10 µL of hexane, and 2 µL of pentafluorobenzyl bromide. After the mixture has been blended by being taken up in a syringe several times, the vial is sealed with its Teflon-lined screw cap and shaken for 20 min. Figure 26-13b shows how the sample is manipulated to extract the product

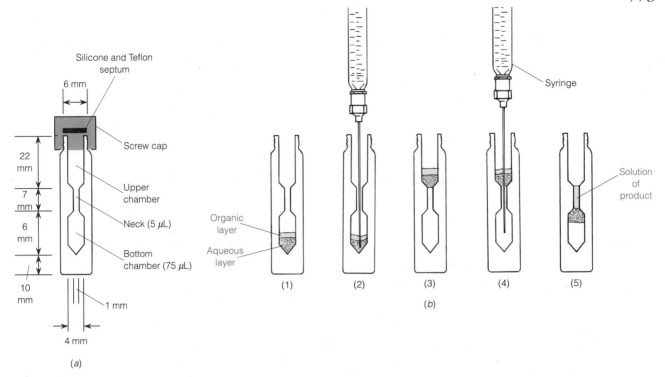

Figure 26-13 (*a*) Dimensions of a Keele microreactor. (*b*) Procedure for derivatization and extraction of product. [From A. B. Attygalle and E. D. Morgan, *Anal. Chem.* **1986,** *58,* 3054.]

for chromatography. First, 20 μL of water and 10 μL of hexane are added to give the two phases in diagram 1. The mixture is then withdrawn by syringe (2) and placed above the thin neck of the vial (3). Withdrawal of air from the lower chamber with the syringe (4) causes the hexane layer containing organic product to move to the thin neck (5), from which it can be removed by syringe.

26-4 Analytical Standards

Analytical methods are validated by demonstrating that accurate results are obtained with known samples. The more complex the unknown, the harder it is to prepare suitable knowns in which the *matrix* (the bulk of the sample) and potential interfering species are similar to those of the unknown. The method of *standard addition* (Chapter 6), in which standards are added to the unknown, helps ensure that matrix effects and interference are under control. *Standard Reference Materials* from the National Institute of Standards and Technology (Box 3-1) are an important means by which procedures can be tested on samples with known composition. The validity of analytical methods can also be checked by showing that different methods produce similar results with the same unknown.

Table 26-8 recommends primary standards for many elements. As an *elemental assay standard,* the material must contain a known amount of the

Validation of analytical method:

1. Analyze standard reference materials
2. Verify results by different methods
3. Use standard additions

TABLE 26-8 Calibration standards

Element	Source[a]	Purity	Comments[b]
Li	SRM 924 (Li_2CO_3)	$100.05 \pm 0.02\%$	E; dry at 200°C for 4 h.
	Li_2CO_3	five–six 9s	M; purity calculated from impurities. Stoichiometry unknown.
Na	SRM 919 or 2201 (NaCl)	99.9%	E; dry for 24 h over $Mg(ClO_4)_2$.
	Na_2CO_3	three 9s	M; purity based on metallic impurities.
K	SRM 918 (KCl)	99.9%	E; dry for 24 h over $Mg(ClO_4)_2$.
	SRM 999 (KCl)	$52.435 \pm 0.004\%$ K	E; ignite at 500°C for 4 h.
	K_2CO_3	five–six 9s	M; purity based on metallic impurities.
Rb	SRM 984 (RbCl)	$99.90 \pm 0.02\%$	E; hygroscopic. Dry for 24 h over $Mg(ClO_4)_2$.
	Rb_2CO_3		M
Cs	Cs_2CO_3		M
Be	metal	three 9s	E, M; purity based on metallic impurities.
Mg	SRM 929	$100.1 \pm 0.4\%$	E; magnesium gluconate clinical standard.
		$5.403 \pm 0.022\%$ Mg	Dry for 24 h over $Mg(ClO_4)_2$.
	metal	five 9s	E; purity based on metallic impurities.
Ca	SRM 915 ($CaCO_3$)	three 9s	E; use without drying.
	$CaCO_3$	five 9s	E, M; dry at 200°C for 4 h in CO_2. User must determine stoichiometry.
Sr	SRM 987 ($SrCO_3$)	99.8%	E; ignite to establish stoichiometry. Dry at 110° for 1 h.
	$SrCO_3$	five 9s	M; up to 1% off stoichiometry. Ignite to establish stoichiometry. Dry at 200° for 4 h.
Ba	$BaCO_3$	four–five 9s	M; dry at 200°C for 4 h.
B	SRM 951 (H_3BO_3)	100.00 ± 0.01	E; expose to room humidity ($\sim$35%) for 30 min before use.
Al	metal	five 9s	E, M; SRM 1257 Al metal available.
Ga	metal	five 9s	E, M; SRM 994 Ga metal available.
In	metal	five 9s	E, M

Transition metals: Use pure metals (usually $\geq$four 9s) for elemental and matrix standards. Assays are based on impurities and do not include dissolved gases.

Lanthanides: Use pure metals (usually $\geq$four 9s) for elemental standards and oxides as matrix standards. Oxides may be difficult to dry and stoichiometry is not certain.

a. SRM is the National Institute of Standards and Technology designation for a standard reference material.
b. E means elemental assay standard; M means matrix matching standard.
SOURCE: J. R. Moody, R. R. Greenberg, K. W. Pratt, and T. C. Rains, *Anal. Chem.* **1988**, *60*, 1203A.

Elemental standard: provides known amount of analyte

Matrix matching standard: must be free of desired analyte

desired element. As a *matrix matching standard,* the most important (and demanding) characteristic is that the material must contain extremely low concentrations of undesired impurities, such as the analyte. If you want to prepare 10 ppm Fe in 10% aqueous NaCl, the NaCl cannot contain significant Fe impurity, or the Fe impurity would have a higher concentration than the added Fe.

Manufacturers frequently indicate elemental purity by some number of 9s. This deceptive nomenclature is based on the measurement of certain impurities. For example, 99.999% (five 9s) pure Al is certified to contain $\leq$0.001% *metallic* impurities, on the basis of the analysis of other metals present. However, C, H, N, and O are not measured. The Al might contain 0.1% Al_2O_3 and still be "five 9s pure." For the most accurate work, the dissolved gas content in solid elements may also be a source of error.

TABLE 26-8 (continued)

Element	Source[a]	Purity	Comments[b]
Tl	metal	five 9s	E, M; SRM 997 Tl metal available.
C			No recommendation.
Si	metal	six 9s	E, M; SRM 990 SiO_2 available.
Ge	metal	five 9s	E, M
Sn	metal	six 9s	E, M; SRM 741 Sn metal available.
Pb	metal	five 9s	E, M; several SRMs available.
N	NH_4Cl	six 9s	E; can be prepared from $HCl + NH_3$.
	N_2	>three 9s	E
	HNO_3	six 9s	M; contaminated with NO_x. Purity based on impurities.
P	SRM 194 ($NH_4H_2PO_4$)	three 9s	E
	P_2O_5	five 9s	E, M; difficult to keep dry.
	H_3PO_4	four 9s	E; must titrate 2 hydrogens to be certain of stoichiometry.
As	metal	five 9s	E, M
	SRM 83d (As_2O_3)	99.9926 ± 0.0030%	Redox standard. As assay not ensured.
Sb	metal	four 9s	E, M
Bi	metal	five 9s	E, M
O	H_2O	eight 9s	E, M; contains dissolved gases.
	O_2	>four 9s	E
S	element	six 9s	E, M; difficult to dry. Other sources are H_2SO_4, Na_2SO_4, and K_2SO_4. Stoichiometry must be proved (e.g., no SO_3^{2-} present).
Se	metal	five 9s	E, M; SRM 726 Se metal available.
Te	metal	five 9s	E, M
F	NaF	four 9s	E, M; no good directions for drying.
Cl	NaCl	four 9s	E, M; dry for 24 h over $Mg(ClO_4)_2$. Several SRMs (NaCl and KCl) available.
Br	KBr	four 9s	E, M; need to dry and demonstrate stoichiometry.
	Br_2	four 9s	E
I	sublimed I_2	six 9s	E
	KI	three 9s	E, M
	KIO_3	three 9s	Stoichiometry not ensured.

Carbonates, oxides, and other compounds may not possess the stoichiometry on the label. For example, TbO_2 will have a high Tb content if some Tb_4O_7 is present. Ignition in an O_2 atmosphere may be helpful, but the final stoichiometry is never guaranteed. Carbonates may contain traces of bicarbonate, oxide, and hydroxide. Firing in a CO_2 atmosphere may improve the stoichiometry. Sulfates may contain some HSO_4^-. Some form of chemical analysis may be required to ensure that you know what you are really working with.

Most metals dissolve in 6 M HCl or HNO_3, or a mixture of the two, possibly with heating. Frothing accompanies dissolution of metals or carbonates in acid, so vessels should be loosely covered by a watchglass or Teflon lid to prevent loss of material. Concentrated HNO_3 (16 M) may *passivate* some metals, forming an insoluble oxide coat that prevents dissolution. If you have a choice between a bulk element and a powder, the bulk form is preferred because it has a smaller surface area on which oxides can form and impurities can be adsorbed. After cutting a pure metal to be used as a standard, it should be etched ("pickled") in a dilute solution of the acid in which it will be dissolved to remove surface oxides and contamination from the cutter. The material is then washed well with water after etching and dried in a vacuum desiccator.

Use buoyancy corrections (Equation 2-1) for accurate weighing.

Dilute solutions are best prepared in Teflon or plastic vessels, because glass is an ion exchanger that can replace analyte species. Because volumetric dilutions are rarely more accurate than 0.1%, gravimetric dilutions are required for greater accuracy. Evaporation of standard solutions is a source of error that is avoided if the mass of the reagent bottle is recorded after each use. If the mass changes between uses, the contents are evaporating.

The path to accurate analysis is fraught with obstacles that must be appreciated by those who use analytical results and that challenge the skill and imagination of the dedicated analyst.

Terms to Understand

aqua regia	derivatization	preconcentration
ball mill	dry ashing	purge and trap method
bomb	flux	sample cleanup
combustion analysis	fusion	solid-phase extraction
coprecipitation	mortar and pestle	wet ashing

Summary

The variance of an analysis is the sum of the sampling variance and the analytical variance. Sampling variance can be understood in terms of the statistics of selecting particles from a heterogeneous mixture. If the probabilities of selecting two kinds of particles from a two-particle mixture are p and q, the standard deviation in selecting n particles is $\sqrt{npq}$. You should be able to use this relationship to estimate how large a sample is required to reduce the sampling variance to a desired level. Student's t can be used to estimate how many repetitions of the analysis are required to reach a certain level of confidence in the final result.

Relatively insoluble inorganic materials can be dissolved in strong acids with heating. Glass vessels are often useful, but Teflon, platinum, or silver are required for HF, which dissolves silicates. If a nonoxidizing acid is insufficient, aqua regia or other oxidizing acids may do the job. A Teflon-lined bomb heated in a microwave oven is a particularly convenient means of dissolving difficult

samples. If acid digestion fails, fusion in a molten salt will usually work, but the large quantity of flux adds trace impurities. Organic materials are decomposed by wet ashing with hot concentrated acids or dry ashing with heat.

Dilute analyte can be preconcentrated by ion exchange, chromatographic adsorption, purge and trap, or coprecipitation. Sample cleanup to remove potential interference can be accomplished by a short solid-phase extraction column. Derivatization transforms the analyte into a more easily detected or separated form. Analytical methods can be validated with standard reference materials, by using different methods to produce the same results, and by standard addition. Elemental standards provide a known amount of analyte, whereas matrix matching standards must be completely free of the intended analyte. Reagent purity is affected by factors such as oxide coatings and nonstoichiometry. Care and thought are required in every phase of chemical analysis.

Exercises

A. A box contains 120 000 red marbles and 880 000 yellow marbles.

(a) If you draw a random sample of 1 000 marbles from the box, what are the expected numbers of red and yellow marbles?

(b) Now put those marbles back in the box and repeat the experiment again. What will be the absolute and relative standard deviations for the numbers in **(a)** after many drawings of 1 000 marbles?

(c) What will be the absolute and relative standard deviations after many drawings of 4 000 marbles?

(d) Fill in the blanks: If you quadruple the size of the sample, you decrease the sampling standard deviation by a factor of _____ . If you increase the sample size by a factor of n, you decrease the sampling standard deviation by a factor of _____ .

(e) What sample size is required to reduce the sampling standard deviation of red marbles to ±2%?

B. **(a)** What mass of sample in Figure 26-2 is expected to give a sampling standard deviation of ±10%?

(b) With the mass from **(a)**, how many samples should be taken to assure 95% confidence that the mean is known to within ±20 counts per second per gram?

C. Because they were used in gasoline to boost octane rating, alkyl lead compounds found their way into the environment as toxic pollutants. One way to measure lead compounds in natural waters is by anodic stripping voltammetry, in which the lead is first reduced to the element at a mercury electrode and dissolves in the mercury. Reoxidation occurs when the electrode potential is made sufficiently positive, with current proportional to the concentration of dissolved Pb.

reduction at −1.2 V: $(CH_3CH_2)_3PbCl \rightarrow Pb$ (*in Hg*)

oxidation at −0.5 V: Pb (*in Hg*) $\rightarrow Pb^{2+}$ (*aq*)

(see Figure 18-17)

Inorganic (ionic) forms of lead are present in much higher concentrations than alkyl lead in natural samples, so inorganic lead must be removed before analysis of

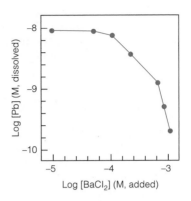

Coprecipitation of Pb^{2+} with $BaSO_4$ from distilled water at pH 2. [Data from N. Mikac and M. Branica, *Anal. Chim. Acta* **1988**, *212*, 349.]

alkyl lead. One way to do this is by *coprecipitation* with $BaSO_4$, as shown above.

In this experiment, $BaCl_2$ added to Pb^{2+} in distilled water at the concentration on the abscissa was precipitated with a 20% excess of Na_2SO_4 and the residual concentration of lead was measured by stripping voltammetry. What control experiments should be performed to show that coprecipitation can be used to remove a large amount of inorganic lead from a small quantity of $(CH_3CH_2)_3PbCl$ in seawater, without removing the alkyl lead compound?

D. A soil sample contains some acid-soluble inorganic matter, some organic material, and some minerals that do not dissolve in any combination of hot acids that you try. Suggest a procedure for dissolving the entire sample.

E. What is meant when we say that a standard may not have the expected stoichiometry? Give an example.

Problems

Statistics of Sampling

1. Review Section 1-1 for this question.

(a) What is the difference between homogeneous and heterogeneous samples? Give an example of each.

(b) What is the difference between a random heterogeneous material and a segregated heterogeneous material?

2. Explain what is meant by the statement "Unless the complete history of any sample is known with certainty, the analyst is well advised not to spend his or her time in analyzing it."

3. **(a)** In the analysis of a barrel of powder, the standard deviation of the sampling operation is ±4% and the standard deviation of the analytical procedure is ±3%. What is the overall standard deviation?

(b) To what value must the sampling standard deviation be reduced so that the overall standard deviation is ±4%?

4. After reviewing Section 1-1, describe how you would prepare a composite sample of a segregated material.

5. What mass of sample in Figure 26-2 is expected to give a sampling standard deviation of ±6%?

6. Explain how to prepare a powder with an average particle diameter near 100 μm by using the sieves in Table 26-2. How would such a particle mesh size be designated?

7. Following Equation 26-4, we have an example of a mixture of 1-mm-diameter particles of KCl and KNO_3 in a number ratio 1:99. A sample containing 10^4 particles weighs 11.0 g. What is the expected number and relative standard deviation of KCl particles in a sample weighing 11.0×10^2 g?

8. When you flip a coin, the probability of its landing on each side is $p = q = \frac{1}{2}$ in Equations 26-2 and 26-3. If you flip it n times, the expected number of heads equals the expected number of tails = $np = nq = \frac{1}{2} n$. The expected standard deviation for n flips is $\sigma_n = \sqrt{npq}$. From Table 4-1, we expect that 68.3% of the results will lie within $\pm 1\sigma_n$ and 95.5% of the results will lie within $\pm 2\sigma_n$.

(a) Find the expected standard deviation for the number of heads in 1 000 coin flips.

(b) By interpolation in Table 4-1, find the value of z that includes 90% of the area of the Gaussian curve. We expect that 90% of the results will lie within this number of standard deviations from the mean.

(c) If you repeat the 1 000 coin flips many times, what is the expected range for the number of heads that includes 90% of the results? (For example, your answer might be, "The range 490 to 510 will be observed 90% of the time.")

9. In analyzing a lot with random sample variation, there is a sampling standard deviation of $\pm 5\%$. Assuming negligible error in the analytical procedure, how many samples must be analyzed to give 95% confidence that the error in the mean is within $\pm 4\%$ of the true value? Answer the same question for a confidence level of 90%.

10. In an experiment analogous to that in Figure 26-2, the sampling constant is found to be $K_s = 20$ g.

(a) What mass of sample is required for a $\pm 2\%$ sampling standard deviation?

(b) How many samples of the size in (a) are required to produce 90% confidence that the mean is known to within 1.5%?

11. Consider a random mixture containing 4.00 g of Na_2CO_3 (density 2.532 g/mL) and 96.00 g of K_2CO_3 (density 2.428 g/mL) with a uniform spherical particle radius of 0.075 mm.

(a) Calculate the mass of a single particle of Na_2CO_3 and the number of particles of Na_2CO_3 in the mixture. Do the same for K_2CO_3.

(b) What is the expected number of particles in 0.100 g of the mixture?

(c) Calculate the relative sampling standard deviation in the number of particles of each type in a 0.100-g sample of the mixture.

Sample Preparation

12. Based on standard reduction potentials, specify which of the following metals you would expect to dissolve in HCl by the reaction $M + nH^+ \rightarrow M^{n+} + \frac{n}{2} H_2$: Zn, Fe, Co, Al, Hg, Cu, Pt, Au. (When the potential predicts that the element will not dissolve, it probably will not. If it is expected to dissolve, it may dissolve if some other process does not interfere. Predictions based on standard reduction potentials at 25°C are only tentative, because the potentials and activities in hot, concentrated solutions vary widely from those in the table of standard potentials.)

13. The following wet ashing procedure was used to measure arsenic in organic soil samples by atomic absorption spectroscopy: A 0.1- to 0.5-g sample was heated in a 150-mL Teflon bomb (Figure 26-6) in a microwave oven for 2.5 min with 3.5 mL of 70% HNO_3. After cooling, a mixture containing 3.5 mL of 70% HNO_3, 1.5 mL of 70% $HClO_4$, and 1.0 mL of H_2SO_4 was added and the sample was reheated for three 2.5-min intervals with 2-min unheated periods in between. The final solution was diluted with 0.2 M HCl for analysis. Explain why $HClO_4$ was not introduced until the second heating.

14. Barbital can be extracted from urine by passage through a reverse-phase octadecyl extraction tube. The barbital is then eluted with 1:1 acetone:chloroform. Explain how this procedure works.

Barbital

15. Explain the difference between an elemental assay standard and a matrix matching standard.

16. Why is a metal "pickled" before being used as a standard?

17. Referring to Table 26-6, explain how an anion-exchange resin can be used for absorption and analysis of SO_2 released by combustion.

18. The figure on the next page shows elemental concentrations in seawater as a function of depth near hydrothermal vents. Analysis was carried out by the $Ga(OH)_3$ coprecipitation method described in Section 26-3.

(a) What is the atomic ratio (Ga added):(Ni in seawater) for the sample with the highest concentration of Ni?

(b) The results given by colored lines were obtained with seawater samples that were not filtered prior to coprecipitation. The black-line results refer to filtered

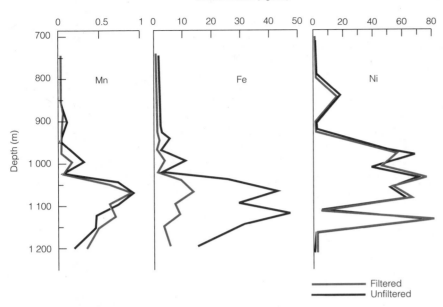

Depth profile of elements in seawater near hydrothermal vents. [From T. Akagi and H. Haraguchi, *Anal. Chem.* **1990,** *62,* 81.]

samples. The results for Mn and Ni do not vary between the two procedures, but the results for Fe do vary. Explain what this means.

19. Barium titanate, a ceramic used in electronics, was analyzed by the following procedure: Into a Pt crucible was placed 1.2 g of Na_2CO_3 and 0.8 g of $Na_2B_4O_7$ plus 0.3146 g of unknown. After fusion at 1 000°C in a furnace for 30 min, the cooled solid was extracted with 50 mL of 6 M HCl, transferred to a 100-mL volumetric flask, and diluted to the mark. A 25.00-mL aliquot was treated with 5 mL of 15% tartaric acid (which complexes Ti^{4+} and keeps it in aqueous solution) and 25 mL of ammonia buffer, pH 9.5. The solution was treated with organic reagents that complex Ba^{2+} and the Ba complex extracted into CCl_4. After acidification (to release the Ba^{2+} from its organic complex), the Ba^{2+} was back-extracted into 0.1 M HCl. The final aqueous sample was treated with ammonia buffer and methylthymol blue (a metal ion indicator) and titrated with 32.49 mL of 0.011 44 M EDTA. Find the weight percent of Ba in the ceramic.

20. How many milliliters of $O_2(g)$ at 298 K and 1.00 atm are required for complete combustion of 0.100 g of paper (cellulose, empirical formula CH_2O) in the Schöniger flask in Box 26-1? Remember that the paper provides some oxygen.

Notes and References

1. R. E. Thiers, *Methods of Biochemical Analysis* (D. Glick, Ed.), Vol. 5 (New York: Interscience, 1957), p. 274.

2. B. Kratochvil and J. K. Taylor, *Anal. Chem.* **1981,** *53,* 924A; H. A. Laitinen and W. E. Harris, *Chemical Analysis,* 2nd ed. (New York: McGraw-Hill, 1975), Chap. 27.

3. D. C. Bogen, *Treatise on Analytical Chemistry,* 2nd ed. (P. J. Elving, E. Grushka, and I. M. Kolthoff, Eds.), Part I, Vol. 5 (New York: Wiley, 1982), Chap. 1; E. C. Dunlop and C. R. Ginnard, *ibid.,* Chap. 2.

4. Procedures for inorganic materials are described by B. D. Zehr, *Am. Lab.* December 1992, p. 24.

5. P. Aysola, P. Anderson, and C. H. Langford, *Anal. Chem.* **1987,** *59,* 1582.

6. A. A. Schilt, *Perchloric Acid and Perchlorates* (Columbus, OH: G. F. Smith Chemical Co., 1979).

7. L. Ping and P. K. Dasgupta, *Anal. Chem.* **1989,** *61,* 1230.

8. D. T. Sawyer, C. Kang, A. Llobet, and C. Redman, *J. Am. Chem. Soc.* **1993,** *115,* 5817.

9. S.-C. Pai, P.-Y. Whung, and R. L. Lai, *Anal. Chim. Acta* **1988,** *211,* 257, 271.

10. N. Simpson, *Am. Lab.* August 1992, p. 37; M. Zief and R. Kiser, *Am. Lab.* January 1990, p. 70.

CHAPTER

27

Experiments

The experiments described in this chapter are intended to illustrate the major analytical techniques described in the text.[1] These procedures are organized roughly in the same order as the topics in the text.

Although these procedures are safe when carried out with reasonable care, *all chemical experiments are potentially hazardous*. Any solution that fumes (such as concentrated HCl), and all nonaqueous solvents, should be handled in a fume hood. Pipetting should never be done by mouth. Spills on your body should immediately be flooded with water and washed with soap and water; your instructor should be notified for possible further action. Spills on the benchtop should be cleaned immediately. Toxic chemicals should not be flushed down the drain. Your instructor should establish a safe procedure for disposing of each chemical that you use (see Box 2-1).

27-1 Calibration of Volumetric Glassware

An important trait of good analysts is the ability to extract the best possible data from their equipment. For this purpose, it is desirable to calibrate your own volumetric glassware (burets, pipets, flasks, etc.) to measure the exact volumes delivered or contained. This experiment also promotes improved technique in handling volumetric glassware. Before beginning this experiment, you should study Section 2-9.

Calibration of 50-mL Buret

In this procedure, we will construct a graph (such as Figure 3-3) needed to convert the measured volume delivered by a buret to the true volume delivered at 20°C.

PROCEDURE

1. Fill the buret with distilled water and force any air bubbles out the tip. See that the buret drains without leaving drops on the walls. If drops are left, clean the buret with soap and water or soak it with peroxydisulfate-sulfuric acid cleaning solution (see footnote on page 28). Adjust the meniscus to be at or slightly below 0.00 mL, and touch the buret tip to a beaker to remove the suspended drop of water. Allow the buret to stand for 5 min while you weigh a 125-mL flask fitted with a rubber stopper. (Hold the flask with a tissue or paper towel, not with your hands, to avoid changing its mass with fingerprint residue.) If the level of the liquid in the buret has changed, tighten the stopcock and repeat the procedure.

2. Drain approximately 10 mL of water (at a rate of <20 mL/min) into the weighed flask, and cap it tightly to prevent evaporation. Allow about 30 s for the film of liquid on the walls to descend before you read the buret. Estimate all readings to the nearest 0.01 mL. Weigh the flask again to determine the mass of water delivered.

3. Now drain the buret from 10 to 20 mL, and measure the mass of water delivered. Repeat the procedure for 30, 40, and 50 mL. Then do the entire procedure (10, 20, 30, 40, 50 mL) a second time.

4. Use Table 2-6 to convert the mass of water to the volume delivered. Repeat any set of duplicate buret corrections that do not agree to within 0.04 mL. Prepare a calibration graph such as Figure 3-3, showing the correction factor at each 10-mL interval.

EXAMPLE   Buret Calibration

When draining the buret at 24°C, you observe the following values:

Final reading	10.01	10.08 mL
Initial reading	0.03	0.04
Difference	9.98	10.04 mL
Mass	9.984	10.056 g
Actual volume delivered	10.02	10.09 mL
Correction	+0.04	+0.05 mL
Average correction		+0.045 mL

To calculate the actual volume delivered when 9.984 g of water is delivered at 24°C, look at the column of Table 2-6 headed "Corrected to 20°C." Across from 24°C, you find that 1.000 0 g of water occupies 1.003 8 mL. Therefore, 9.984 g occupies (9.984 g)(1.003 8 mL/g) = 10.02 mL. The average correction for both sets of data is +0.045 mL.

To obtain the correction for a volume greater than 10 mL, add successive masses of water collected in the flask. Suppose that the following masses were measured:

Volume interval (mL)	Mass delivered (g)
0.03–10.01	9.984
10.01–19.90	9.835
19.90–30.06	10.071
Sum 30.03 mL	29.890 g

The total mass of water delivered corresponds to $(29.890 \text{ g})(1.003\,8 \text{ mL/g})$ = 30.00 mL. Because the indicated volume is 30.03 mL, the buret correction at 30 mL is −0.03 mL.

What does this mean? Suppose that Figure 3-3 applies to your buret. If you begin a titration at 0.04 mL and end at 29.00 mL, you would deliver 28.96 mL if the buret were perfect. In fact, Figure 3-3 tells you that the buret delivers 0.03 mL less than the indicated amount, so only 28.93 mL was actually delivered. To use the calibration curve, either begin all titrations near 0.00 mL or correct both the initial and the final readings. Use the calibration curve whenever you use your buret.

Other Calibrations

Pipets can be calibrated by weighing the water delivered from them. A volumetric flask can be calibrated by weighing it empty and then weighing it filled to the mark with distilled water. Perform each procedure at least twice. Compare your results with the tolerances in Tables 2-2, 2-3, and 2-4.

27-2 Gravimetric Determination of Calcium as $CaC_2O_4 \cdot H_2O$ [2]

Calcium ion can be analyzed by precipitation with oxalate in basic solution to form $CaC_2O_4 \cdot H_2O$. The precipitate is soluble in acidic solution because the oxalate anion is a weak base. Large, easily filtered, relatively pure crystals of product will be obtained if the precipitation is carried out slowly. This can be done by dissolving Ca^{2+} and $C_2O_4^{2-}$ in acidic solution and gradually raising the pH by thermal decomposition of urea (Reaction 25-2).

REAGENTS

Ammonium oxalate solution: Make 1 L of solution containing 40 g of $(NH_4)_2C_2O_4$ plus 25 mL of 12 M HCl. Each student will need 80 mL of this solution.

Unknowns: Prepare 1 L of solution containing 15–18 g of $CaCO_3$ plus 38 mL of 12 M HCl. Each student will need 100 mL of this solution. Alternatively, solid unknowns are available from Thorn Smith.[3]

PROCEDURE

1. Dry three medium-porosity, sintered-glass funnels for 1–2 h at 105°C; cool them in a desiccator for 30 min and weigh them. Repeat the procedure with 30-min heating periods until successive weighings agree to within 0.3 mg. Use a paper towel or tongs, not your fingers, to handle the funnels. An alternative method of drying the crucible and precipitate is with a microwave oven.[4] A 900-W kitchen microwave oven dries the crucible to constant mass in two heating periods of 4 min and 2 min (with 15 min allowed for cooldown after each cycle). You will need to experiment with your oven to find appropriate heating times.

2. Use a few small portions of unknown to rinse a 25-mL transfer pipet, and discard the washings. *Use a rubber bulb, not your mouth, to provide suction.* Transfer exactly 25 mL of unknown to each of three 250- to 400-mL beakers, and dilute each with ~75 mL of 0.1 M HCl. Add 5 drops of methyl red indicator solution (Table 12-4) to each beaker. This indicator is red below pH 4.8 and yellow above pH 6.0.

3. Add ~25 mL of ammonium oxalate solution to each beaker while stirring with a glass rod. Remove the rod and rinse it into the beaker. Add ~15 g of solid urea to each sample, cover it with a watchglass, and boil gently for ~30 min until the indicator turns yellow.

4. Filter each hot solution through a weighed funnel using suction (Figure 2-16). Add ~3 mL of ice-cold water to the beaker, and use a rubber policeman to help transfer the remaining solid to the funnel. Repeat this procedure with small portions of ice-cold water until all of the precipitate has been transferred. Finally, use two 10-mL portions of ice-cold water to rinse each beaker, and pour the washings over the precipitate.

5. Dry the precipitate, first with aspirator suction for 1 min, then in an oven at 105°C for 1–2 h. Bring each filter to constant weight. The product is somewhat hygroscopic, so only one filter at a time should be removed from the desiccator, and weighings should be done rapidly. Alternatively, the precipitate can be dried in a microwave oven once for 4 min, followed by several 2-min periods, with cooling for 15 min before weighing. The water of crystallization is not lost.

6. Calculate the molarity of Ca^{2+} in the unknown solution or the weight percent of Ca in the unknown solid. Report the standard deviation and relative standard deviation ($s/\bar{x}$ = standard deviation/average).

27-3 Gravimetric Determination of Iron as Fe₂O₃ [5]

A sample containing iron can be analyzed by precipitation of the hydrous oxide from basic solution, followed by ignition to produce Fe_2O_3:

$$Fe^{3+} + (2 + x)H_2O \xrightarrow{\text{base}} FeOOH \cdot xH_2O(s) + 3H^+$$

$$FeOOH \cdot xH_2O \xrightarrow{900°C} Fe_2O_3(s)$$

The hydrous oxide is gelatinous and may occlude impurities. If this is suspected, the initial precipitate is dissolved in acid and reprecipitated. Because the concentration of impurities is lower during the second precipitation, occlusion is diminished. Solid unknowns can be prepared from reagent ferrous ammonium sulfate or purchased from Thorn Smith.[3]

PROCEDURE

1. Bring three porcelain crucibles and caps to constant weight by heating to redness for 15 min over a burner (Figure 27-1). Cool for 30 min in a desiccator and weigh each crucible. Repeat this procedure until successive weighings agree within 0.3 mg. Be sure that all oxidizable substances on the entire surface of each crucible have burned off.

2. Accurately weigh three samples of unknown containing enough Fe to produce ~0.3 g of Fe_2O_3. Dissolve each sample in 10 mL of 3 M HCl (with heating, if necessary). If there are insoluble impurities, filter through qualitative filter paper and wash the filter very well with distilled water.

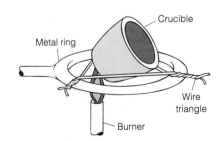

Figure 27-1 Positioning a crucible above a burner.

Add 5 mL of 6 M HNO_3 to the filtrate, and boil for a few minutes to ensure that all iron is oxidized to Fe(III).

3. Dilute the sample to 200 mL with distilled water and add 3 M ammonia[†] with constant stirring until the solution is basic (as determined with litmus paper or pH indicator paper). Digest the precipitate by boiling for 5 min and allow the precipitate to settle.

4. Decant the supernatant liquid through coarse, ashless filter paper (Whatman 41 or Schleicher and Schuell Black Ribbon, as in Figures 2-17 and 2-18). Do not pour liquid higher than 1 cm from the top of the funnel. Proceed to Step 5 if a reprecipitation is desired. Wash the precipitate repeatedly with hot 1% NH_4NO_3 until little or no Cl^- is detected in the filtered supernate. (Test for Cl^- by acidifying a few milliliters of filtrate with 1 mL of dilute HNO_3 and adding a few drops of 0.1 M $AgNO_3$.) Finally, transfer the solid to the filter with the aid of a rubber policeman and more hot liquid. Proceed to Step 6 if a reprecipitation is not used.

5. Wash the gelatinous mass twice with 30 mL of boiling 1% aqueous NH_4NO_3, decanting the supernate through the filter. Then put the filter paper back into the beaker with the precipitate, add 5 mL of 12 M HCl to dissolve the iron, and tear the filter paper into small pieces with a glass rod. Add ammonia with stirring and reprecipitate the iron. Decant through a funnel fitted with a fresh sheet of ashless filter paper. Wash the solid repeatedly with hot 1% NH_4NO_3 until little or no Cl^- is detected in the filtered supernate. Then transfer all the solid to the filter with the aid of a rubber policeman and more hot liquid.

6. Allow the filter to drain overnight—if possible, protected from dust. Carefully lift the paper out of the funnel, fold it (see Figure 27-2), and transfer it to a porcelain crucible that has been brought to constant weight.

7. Dry the crucible cautiously with a small flame, as shown in Figure 27-1. The flame should be directed at the top of the container, and the lid should be off. Avoid spattering. After it is dry, *char* the filter paper by increasing the flame temperature. The crucible should have free access to air to avoid reduction of iron by carbon. (The lid should be kept handy to smother the crucible if the paper inflames.) Any carbon left on the crucible or lid should be removed by directing the burner flame at it. Use tongs to manipulate the crucible. Finally, *ignite* the product for 15 min with the full heat of the burner.

8. Cool the crucible briefly in air and then in a desiccator for 30 min. Weigh the crucible and the lid, reignite, and bring to constant weight (within 0.3 mg) with repeated heatings.

9. Calculate the weight percent of iron in each sample, the average, the standard deviation, and the relative standard deviation ($s/\bar{x}$).

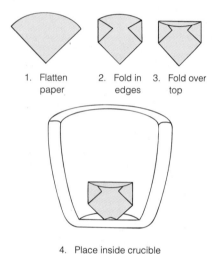

1. Flatten paper   2. Fold in edges   3. Fold over top

4. Place inside crucible with point pushed against bottom

Figure 27-2 Folding filter paper and placing it inside a crucible for ignition. Continue folding the paper so that the entire package fits at the bottom of the crucible. Be careful not to puncture the paper.

27-4 Preparing Standard Acid and Base

Section 12-7 provides background information related to the procedures described below. Unknown samples of potassium hydrogen phthalate or

---

[†]Basic reagents should not be stored in glass bottles because they will slowly dissolve the glass. If ammonia from a glass bottle is used, it may contain silica particles and should be freshly filtered.

sodium carbonate (available from Thorn Smith[3]) can be analyzed by the procedures described in this section.

Standard NaOH

P R O C E D U R E

1. A 50 wt % aqueous NaOH solution must be prepared in advance and allowed to settle. Sodium carbonate is insoluble in this solution and precipitates. The solution is stored in a tightly sealed polyethylene bottle and handled gently to avoid stirring the precipitate when supernate is taken. The density is close to 1.50 g of solution per milliliter.

2. Primary standard-grade potassium hydrogen phthalate should be dried for 1 h at 110°C and stored in a desiccator.

FW 204.223

3. Boil 1 L of water for 5 min to expel CO_2. Pour the water into a polyethylene bottle, which should be tightly capped whenever possible. Calculate the volume of aqueous 50 wt% NaOH needed to produce 1 L of ~0.1 M NaOH. Use a graduated cylinder to transfer this much concentrated NaOH to the bottle of water. Mix well and allow the solution to cool to room temperature (preferably overnight).

4. Weigh four samples of solid potassium hydrogen phthalate and dissolve each in ~25 mL of distilled water in a 125-mL flask. Each sample should contain enough solid to react with ~25 mL of 0.1 M NaOH. Add 3 drops of phenolphthalein indicator (Table 12-4) to each, and titrate one of them rapidly to find the approximate end point. The buret should have a loosely fitted cap to minimize entry of CO_2.

5. Calculate the volume of NaOH required for each of the other three samples and titrate them carefully. During each titration, you should periodically tilt and rotate the flask to wash all liquid from the walls into the bulk solution. When very near the end, you should deliver less than 1 drop of titrant at a time. To do this, carefully suspend a fraction of a drop from the buret tip, touch it to the inside wall of the flask, wash it into the bulk solution by careful tilting, and swirl the solution. The end point is the first appearance of faint pink color that persists for 15 s. (The color will slowly fade as CO_2 from the air dissolves in the solution.)

6. Calculate the average molarity, the standard deviation, and the relative standard deviation $(s/\bar{x})$. If you have used some care, the relative standard deviation should be <0.2%.

Standard HCl

P R O C E D U R E

1. Use the information in the table on the inside cover of this text to calculate the volume of concentrated (~37 wt %) HCl that should be added to 1 L of distilled water to produce ~0.1 M HCl, and prepare this solution.

2. Dry primary standard-grade sodium carbonate for 1 h at 110°C and cool it in a desiccator.

3. Weigh four samples containing enough Na_2CO_3 to react with ~25 mL of 0.1 M HCl and place each in a 125-mL flask. As you are ready to titrate each one, dissolve it in ~25 mL of distilled water.

$$2HCl + Na_2CO_3 \rightarrow CO_2 + 2NaCl$$
$$\text{FW } 105.989$$

Add 3 drops of bromocresol green indicator (Table 12-4) to each and titrate one rapidly (to a green color) to find the approximate end point.

4. Carefully titrate each sample until it just turns from blue to green. Then boil the solution to expel CO_2. The solution should return to a blue color.

5. Carefully add HCl from the buret until the solution turns green again. As desired, a blank titration can be performed, by using 3 drops of indicator in 50 mL of 0.05 M NaCl. Subtract the volume of HCl needed for the blank titration from that required to titrate Na_2CO_3.

6. Calculate the mean HCl molarity, standard deviation, and relative standard deviation.

27-5 Analysis of a Mixture of Carbonate and Bicarbonate[6]

This procedure involves two titrations. First, total alkalinity ($= [HCO_3^-] + 2[CO_3^{2-}]$) is measured by titrating the mixture with standard HCl to a bromocresol green end point:

$$HCO_3^- + H^+ \rightarrow H_2CO_3$$
$$CO_3^{2-} + 2H^+ \rightarrow H_2CO_3$$

A separate aliquot of unknown is treated with excess standard NaOH to convert HCO_3^- to CO_3^{2-}:

$$HCO_3^- + OH^- \rightarrow CO_3^{2-} + H_2O$$

Then all the carbonate is precipitated with $BaCl_2$:

$$Ba^{2+} + CO_3^{2-} \rightarrow BaCO_3(s)$$

The excess NaOH is immediately titrated with standard HCl to determine how much HCO_3^- was present. From the total alkalinity and bicarbonate concentration, we can calculate the original carbonate concentration. Solid unknowns may be prepared from either reagent-grade sodium or potassium carbonate and bicarbonate.

PROCEDURE

1. Solid unknown should be stored in a desiccator to keep it dry, but should not be heated. Even mild heating at 50°–100°C converts $NaHCO_3$ to Na_2CO_3. Accurately weigh 2.0–2.5 g of unknown into a 250-mL volumetric flask. This is conveniently done by weighing the sample in a capped weighing bottle, delivering some to a funnel in the volumetric flask, and

reweighing the bottle. Continue this process until the desired mass of reagent has been transferred to the funnel. Rinse the funnel repeatedly with small portions of freshly boiled and cooled water to dissolve the sample. Remove the funnel, dilute to the mark, and mix well.

2. Pipet a 25.00-mL aliquot of the unknown solution into a 250-mL flask and titrate with standard 0.1 M HCl, using bromocresol green indicator as described in Section 27-4. Repeat this procedure with two more 25.00-mL samples of unknown.

3. Pipet a 25.00-mL aliquot of unknown and 50.00 mL of standard 0.1 M NaOH into a 250-mL flask. Swirl and add 10 mL of 10 wt % $BaCl_2$, using a graduated cylinder. Swirl again to precipitate $BaCO_3$, add 2 drops of phenolphthalein indicator (Table 12-4), and immediately titrate with standard 0.1 M HCl. Repeat this procedure with two more 25.00-mL samples of unknown.

4. From the results of Step 2, calculate the total alkalinity and its standard deviation. From the results of Step 3, calculate the bicarbonate concentration and its standard deviation. Using the standard deviations as estimates of uncertainty, calculate the concentration (and uncertainty) of carbonate in the sample. Express the composition of the solid unknown as weight percent (± uncertainty) for each component. For example, your final result might be written 63.4 (±0.5) wt % K_2CO_3 and 36.6 (±0.2) wt % $NaHCO_3$.

27-6 Kjeldahl Nitrogen Analysis

The Kjeldahl nitrogen analysis described in Section 7-2 is widely applicable to pure organic compounds and complex substances such as milk, cereal, and flour. Modifications of the procedure described below are necessary for nitro compounds, nitrites, azo compounds, cyanides, and derivatives of hydrazine.[7] Kjeldahl digestion is also recommended for destruction of highly toxic organic waste.[†]

After digestion converts organic nitrogen into NH_4^+ (Reaction 7-2), the solution is made basic and the liberated NH_3 is distilled into a receiver containing a known amount of HCl (Reactions 7-3 to 7-5). Unreacted HCl is titrated with NaOH to determine how much HCl was consumed by NH_3. Because the solution to be titrated contains both HCl and NH_4^+, we must choose an indicator that permits titration of HCl without beginning to titrate NH_4^+. Bromocresol green (transition range 3.8–5.4) fulfills this purpose. (See Problem 40 in Chapter 12.)

Unknowns for this experiment can be prepared from pure acetanilide, N-2-hydroxyethylpiperazine-N'-2-ethanesulfonic acid (HEPES buffer, Table 10-2), tris(hydroxymethyl)aminomethane (tris buffer, Table 10-2), or the p-toluenesulfonic acid salts of ammonia, glycine, or nicotinic acid.

. . . .

[†]To destroy highly toxic organic compounds, place 10 g of K_2SO_4 and 100 mL of 98 wt % H_2SO_4 in a 500-mL flask fitted with a reflux condenser. Add up to 5 g of waste and a few glass beads and reflux until the solution is clear. If clarification does not occur in 1 h, cool the flask, slowly add 10 mL of 30 wt % H_2O_2 with stirring, add a fresh glass bead, and reflux again. Organic nitro, azo, and peroxide compounds should be reduced prior to this destruction procedure. The H_2SO_4 mixture can be reused until it solidifies when cooled.

Figure 27-3   Kjeldahl digestion flask. The long neck prevents loss of sample by spattering.

PROCEDURE

1. Dry your unknown at 105°C for 45 min and accurately weigh an amount that will produce 2–3 mmol of NH_3. Place the sample in a *dry* 500-mL Kjeldahl flask (Figure 27-3), so that as little as possible sticks to the walls. Add 10 g of K_2SO_4 and three selenium-coated boiling chips.[†] Pour in 25 mL of 98 wt % H_2SO_4, washing down any solid from the walls. (*Caution:* Concentrated H_2SO_4 eats people. If you get any on your skin, flood it immediately with water, followed by soap and water.)

2. In a fume hood, clamp the flask at a 30° angle away from you. Heat gently with a burner until foaming ceases and the solution becomes homogeneous. Gentle boiling should then be continued for an additional 30 min.

3. Cool the flask for 30 min *in the air,* and then in an ice bath for 15 min. Slowly, and with constant stirring, add 50 mL of ice-cold water. Dissolve any solids that have crystallized. Transfer the liquid to a 500-mL distillation flask (Figure 27-4), wash the Kjeldahl flask five times with 10-mL portions of water, and pour the washings into the distillation flask.

Distillation

PROCEDURE

1. Apparatus such as that in Figure 27-4 should be set up and the connections made airtight. Pipet 50.00 mL of standard 0.1 M HCl into the receiving beaker and clamp the funnel in place below the liquid level.

2. Add 5–10 drops of phenolphthalein indicator (Table 12-4) to the 3-neck flask in Figure 27-4 and secure the stoppers. Pour 60 mL of 50 wt % NaOH into the adding funnel and drip this into the distillation flask over a period of 1 min until the indicator turns pink. (*Caution:* 50 wt% NaOH eats people. Flood any spills on your skin with water.) Do not let the last 1 mL through the stopcock, so that gas cannot escape from the flask. Close the stopcock and heat the flask gently until two-thirds of the liquid has distilled.

3. Remove the funnel from the receiving beaker *before* removing the burner from the flask (to avoid sucking distillate back into the condenser). Rinse the funnel well with distilled water and catch the rinses in the beaker. Add 6 drops of bromocresol green indicator solution (Table 12-4) to the beaker and carefully titrate to the blue end point with standard 0.1 M NaOH. You are looking for the first appearance of light blue color. (Several practice titrations with HCl and NaOH will familiarize you with the end point.)

4. Calculate the weight percent of nitrogen in the unknown.

27-7   Analysis of an Acid-Base Titration Curve: The Gran Plot[8]

In this experiment, you will titrate a sample of pure potassium hydrogen phthalate (Table 12-5) with standard NaOH. The Gran plot (which you

---

[†]Hengar selenium-coated granules are convenient. Alternatively, the catalyst can be 0.1 g of Se, 0.2 g of $CuSeO_3$, or a crystal of $CuSO_4$. If a mercury salt must be used, distillation should be carried out in a hood, because elemental mercury may be released.

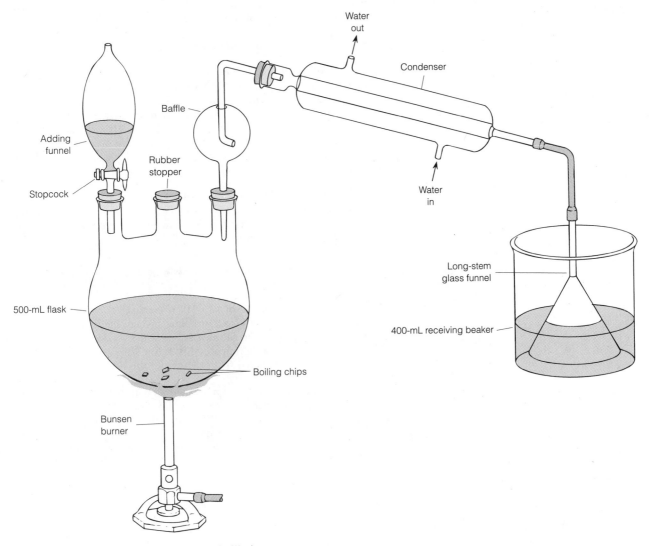

Figure 27-4 Apparatus for Kjeldahl distillation.

should read about in Section 12-5) will be used to find the equivalence point and K_a. Activity coefficients are used in the calculations of this experiment.

PROCEDURE (AN EASY MATTER)

1. Dry about 1.5 g of potassium hydrogen phthalate at 105°C for 1 h and cool it in a desiccator for 20 min. Accurately weigh out ∼1.5 g and dissolve it in water in a 250-mL volumetric flask. Dilute to the mark and mix well.

2. Following the instructions for your particular pH meter, calibrate a meter and glass electrode, using buffers with pH values near 7 and 4. Rinse the electrodes well with distilled water and blot them dry with a tissue before immersing in a solution.

3. Pipet 100.0 mL of phthalate solution into a 250-mL beaker containing a magnetic stirring bar. Position the electrode(s) in the liquid so that the

stirring bar will not strike the electrode. If a combination electrode is used, the small hole near the bottom on the side must be immersed in the solution. This hole is the reference electrode salt bridge. Allow the electrodes to equilibrate for 1 min (with stirring) and record the pH.

4. Add 1 drop of phenolphthalein indicator (Table 12-4) and titrate the solution with standard ~0.1 M NaOH. Until you are within 4 mL of the theoretical equivalence point, add base in ~1.5-mL aliquots, recording the volume and pH 30 s after each addition. Thereafter, use 0.4-mL aliquots until you are within 1 mL of the equivalence point. After that, add base 1 drop at a time until you have passed the pink end point by a few tenths of a milliliter. (Record the volume at which the pink color is observed.) Then add five more 1-mL aliquots.

5. Construct a graph of pH versus V_b (volume of added base). Locate the equivalence volume (V_e) as the point of maximum slope or zero second derivative, as described in Section 12-5. Compare this with the theoretical and phenolphthalein end points.

CALCULATIONS (THE WORK BEGINS!)

1. Construct a Gran plot (a graph of $V_b \cdot 10^{-pH}$ versus V_b) by using the data collected between $0.9V_e$ and V_e. Draw a line through the linear portion of this curve and extrapolate it to the abscissa to find V_e. Use this value of V_e in the calculations below. Compare this value with those found with phenolphthalein and estimated from the graph of pH versus V_b.

2. Compute the slope of the Gran plot and use Equation 12-5 to find K_a for potassium hydrogen phthalate as follows: The slope of the Gran plot is $-K_a\gamma_{HP^-}/\gamma_{P^{2-}}$. In this equation, P^{2-} is the phthalate anion and HP^- is monohydrogen phthalate. Because the ionic strength changes slightly as the titration proceeds, so also do the activity coefficients. Calculate the ionic strength at $0.95V_e$ and use this "average" ionic strength to find the activity coefficients.

. . . .

EXAMPLE Calculating Activity Coefficients

Find γ_{HP^-} and $\gamma_{P^{2-}}$ at $0.95V_e$ in the titration of 100.0 mL of 0.020 0 M potassium hydrogen phthalate with 0.100 M NaOH.

Solution The equivalence point is 20.0 mL, so $0.95V_e = 19.0$ mL. The concentrations of H^+ and OH^- are negligible compared with those of K^+, Na^+, HP^-, and P^{2-}, whose concentrations are

$$[K^+] = \left(\frac{100}{119}\right)(0.020\,0) = 0.016\,8 \text{ M}$$

$$[Na^+] = \left(\frac{19}{119}\right)(0.100) = 0.016\,0 \text{ M}$$

$$[HP^-] = (0.050)\left(\frac{100}{119}\right)(0.020\,0) = 0.000\,84 \text{ M}$$

$$[P^{2-}] = (0.95)\left(\frac{100}{119}\right)(0.020\,0) = 0.016\,0 \text{ M}$$

The ionic strength is

$$\mu = \tfrac{1}{2} \sum c_i z_i^2$$

$$= \tfrac{1}{2}[(0.0168) \cdot 1^2 + (0.0160) \cdot 1^2 + (0.00084) \cdot 1^2 + (0.0160) \cdot 2^2]$$

$$= 0.0488 \text{ M}$$

To estimate $\gamma_{P^{2-}}$ and γ_{HP^-} at $\mu = 0.0488$ M, interpolate in Table 8-1. In this table, we find that the hydrated radius of P^{2-}—phthalate, $C_6H_4(CO_2^-)_2$—is 600 pm. The size of HP^- is not listed, but we will suppose that it is also 600 pm. An ion with charge ± 2 and a size of 600 pm has $\gamma = 0.485$ at $\mu = 0.05$ and $\gamma = 0.675$ at $\mu = 0.01$. Interpolating between these values, we estimate $\gamma_{P^{2-}} = 0.49$ when $\mu = 0.0488$ M. Similarly, we estimate $\gamma_{HP^-} = 0.84$ at this same ionic strength.

. . . .

3. From the measured slope of the Gran plot and the values of γ_{HP^-} and $\gamma_{P^{2-}}$, calculate pK_a. Then choose an experimental point near $\frac{1}{3}V_e$ and one near $\frac{2}{3}V_e$. Use Equation 10-18 to find pK_a with each point. (You will have to calculate $[P^{2-}]$, $[HP^-]$, $\gamma_{P^{2-}}$, and γ_{HP^-} at each point.) Compare the average value of pK_a from your experiment with pK_2 for phthalic acid listed in Appendix G.

27-8 EDTA Titration of Ca²⁺ and Mg²⁺ in Natural Waters

The most common multivalent metal ions in natural waters are Ca^{2+} and Mg^{2+}. In this experiment, we will find the total concentration of metal ions that can react with EDTA, and we will assume that this equals the concentration of Ca^{2+} and Mg^{2+}. In a second experiment, Ca^{2+} is analyzed separately by precipitating $Mg(OH)_2$ with strong base.

REAGENTS

Buffer (pH 10): Add 142 mL of 28 wt % aqueous NH_3 to 17.5 g of NH_4Cl and dilute to 250 mL with water.

Eriochrome black T indicator: Dissolve 0.2 g of the solid indicator in 15 mL of triethanolamine plus 5 mL of absolute ethanol.

PROCEDURE

1. Dry $Na_2H_2EDTA \cdot 2H_2O$ (FW 372.25) at 80°C for 1 h and cool in the desiccator. Accurately weigh out ~0.6 g and dissolve it with heating in 400 mL of water in a 500-mL volumetric flask. Cool to room temperature, dilute to the mark, and mix well.

2. Pipet a sample of unknown into a 250-mL flask. A 1.000-mL sample of seawater or a 50.00-mL sample of tap water is usually reasonable. If you use 1.000 mL of seawater, add 50 mL of distilled water. To each sample, add 3 mL of pH 10 buffer and 6 drops of Eriochrome black T indicator. Titrate

with EDTA from a 50-mL buret and note when the color changes from wine red to blue. You may need to practice finding the end point several times by adding a little tap water and titrating with more EDTA. Save a solution at the end point to use as a color comparison for other titrations.

3. Repeat the titration with three samples to find an accurate value of the total $Ca^{2+} + Mg^{2+}$ concentration. Perform a blank titration with 50 mL of distilled water and subtract the value of the blank from each result.

4. For the determination of Ca^{2+}, pipet four samples of unknown into clean flasks (adding 50 mL of distilled water if you use 1.000 mL of seawater). Add 30 drops of 50 wt% NaOH to each solution and swirl for 2 min to precipitate $Mg(OH)_2$ (which may not be visible). Add ~0.1 g of solid hydroxynaphthol blue to each flask. (This indicator is used because it remains blue at higher pH than does Eriochrome black T.) Titrate one sample rapidly to find the end point; practice finding it several times, if necessary.

5. Titrate the other three samples carefully. After reaching the blue end point, allow each sample to sit for 5 min with occasional swirling so that any $Ca(OH)_2$ precipitate may redissolve. Then titrate back to the blue end point. (Repeat this procedure if the blue color turns to red upon standing.) Perform a blank titration with 50 mL of distilled water.

6. Calculate the total concentration of Ca^{2+} and Mg^{2+}, as well as the individual concentrations of each ion. Calculate the relative standard deviation of replicate titrations.

27-9 Synthesis and Analysis of Ammonium Decavanadate

The balance of species in a solution of V^{5+} is a delicate function of both pH and concentration.

$$VO_4^{3-} \rightleftharpoons HVO_4^{2-} \rightleftharpoons V_2O_7^{4-} \rightleftharpoons H_2V_2O_7^{2-}$$
In strong base

$$VO_2^+ \rightleftharpoons H_2V_{10}O_{28}^{4-} \rightleftharpoons \begin{bmatrix} V_4O_{12}^{4-} \\ V_3O_9^{3-} \end{bmatrix}$$
In strong acid

$\downarrow$ NH_4^+ alcohol

$$(NH_4)_6V_{10}O_{28} \cdot 6H_2O$$
FW 1173.7

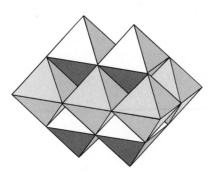

Figure 27-5 Structure of $V_{10}O_{28}^{6-}$ anion, consisting of VO_6 octahedra sharing edges with one another.

The decavanadate ion ($V_{10}O_{28}^{6-}$), which we will isolate in this experiment as the ammonium salt, consists of 10 VO_6 octahedra sharing edges with one another (Figure 27-5).

After preparing this salt, we will determine the vanadium content by a redox titration and NH_4^+ by the Kjeldahl method.[9] In the redox titration,

V^{5+} will first be reduced to V^{4+} with sulfurous acid and then titrated with standard permanganate (Color Plate 10).

$$V_{10}O_{28}^{6-} + H_2SO_3 \rightarrow VO^{2+} + H_2SO_4$$

$$VO^{2+} + MnO_4^- \rightarrow VO_2^+ + Mn^{2+}$$

$$\text{(blue)} \quad \text{(purple)} \quad \text{(yellow)} \quad \text{(colorless)}$$

Synthesis

PROCEDURE

1. Heat 3.0 g of ammonium metavanadate (NH_4VO_3) in 100 mL of water with constant stirring (but not boiling) until most or all of the solid has dissolved. Filter the solution and add 4 mL of 50 vol % aqueous acetic acid with stirring.

2. Add 150 mL of 95% ethanol with stirring and then cool the solution in a refrigerator or ice bath.

3. After maintaining a temperature of 0°–10°C for 15 min, filter the orange product with suction and wash with two 15-mL portions of ice-cold 95% ethanol.

4. Dry the product in the air (protected from dust) for 2 days. Typical yield is 2.0–2.5 g.

Analysis of Vanadium with $KMnO_4$

Preparation and standardization of $KMnO_4$ [10]

See Section 16–6.

PROCEDURE

1. Prepare a 0.02 M permanganate solution by dissolving 1.6 g of $KMnO_4$ in 500 mL of distilled water. Boil gently for 1 h, cover, and allow the solution to cool overnight. Filter through a clean, fine sintered-glass funnel, discarding the first 20 mL of filtrate. Store the solution in a clean glass amber bottle. Do not let the solution touch the cap.

2. Dry sodium oxalate ($Na_2C_2O_4$) at 105°C for 1 h, cool in a desiccator, and weigh three ~0.25-g samples into 500-mL flasks or 400-mL beakers. To each, add 250 mL of 0.9 M H_2SO_4 that has been recently boiled and cooled to room temperature. Stir with a thermometer to dissolve the sample, and add 90–95% of the theoretical amount of $KMnO_4$ solution needed for the titration. (This can be calculated from the mass of $KMnO_4$ used to prepare the permanganate solution. The chemical reaction is given by Equation 7-1.)

3. Leave the solution at room temperature until it is colorless. Then heat it to 55°–60°C and complete the titration by adding $KMnO_4$ until the first pale pink color persists. Proceed slowly near the end, allowing 30 s for each drop to lose its color. As a blank, titrate 250 mL of 0.9 M H_2SO_4 to the same pale pink color.

Vanadium analysis

PROCEDURE

1. Accurately weigh two 0.3-g samples of ammonium decavanadate into 250-mL flasks and dissolve each in 40 mL of 1.5 M H_2SO_4 (with warming, if necessary).

2. In a fume hood, add 50 mL of water and 1 g of $NaHSO_3$ to each and dissolve with swirling. After 5 min, boil the solution gently for 15 min to remove SO_2.

3. Titrate the warm solution with standard 0.02 M $KMnO_4$ from a 50-mL buret. The end point is taken when the yellow color of VO_2^+ takes on a dark shade (from excess MnO_4^-) that persists for 15 s.

Analysis of Ammonium Ion

Ammonium ion is analyzed by a modification of the Kjeldahl procedure. Transfer 0.6 g of accurately weighed ammonium decavanadate to the distillation flask in Figure 27-4 and add 200 mL of water. Then proceed as described under "Distillation" in Section 27-6.

27-10 Iodimetric Titration of Vitamin C[11]

Ascorbic acid (vitamin C) is a mild reducing agent that reacts rapidly with triiodide (Section 16-9). In this experiment, we will generate a known excess of I_3^- by the reaction of iodate with iodide (Reaction 16-30), allow the reaction with ascorbic acid to proceed, and then back-titrate the excess I_3^- with thiosulfate (Reaction 16-31 and Color Plate 11).

Preparation and Standardization of Thiosulfate Solution

PROCEDURE

1. Starch indicator is prepared by making a paste of 5 g of soluble starch and 5 mg of HgI_2 in 50 mL of water. Pour the paste into 500 mL of boiling water and boil until it is clear.

2. Prepare 0.07 M $Na_2S_2O_3$[†] by dissolving ~8.7 g of $Na_2S_2O_3 \cdot 5H_2O$ in 500 mL of freshly boiled water containing 0.05 g of Na_2CO_3. Store this solution in a tightly capped amber bottle. Prepare ~0.01 M KIO_3 by accurately weighing ~1 g of solid reagent and dissolving it in a 500-mL volumetric flask.

3. Standardize the thiosulfate solution as follows: Pipet 50.00 mL of KIO_3 solution into a flask. Add 2 g of solid KI and 10 mL of 0.5 M H_2SO_4. Immediately titrate with thiosulfate until the solution has lost almost all its color (pale yellow). Then add 2 mL of starch indicator and complete the titration. Repeat the titration with two additional 50.00-mL volumes of KIO_3 solution.

. . . .

[†]An alternative to standardizing $Na_2S_2O_3$ solution is to use anhydrous primary standard $Na_2S_2O_3$, as described on page **444.**

Analysis of Vitamin C

Commercial vitamin C containing 100 mg per tablet can be used. Perform the following analysis three times, and find the mean value (and relative standard deviation) for the number of milligrams of vitamin C per tablet.

PROCEDURE

1. Dissolve two tablets in 60 mL of 0.3 M H_2SO_4, using a glass rod to help break the solid. (Some solid binding material will not dissolve.)

2. Add 2 g of solid KI and 50.00 mL of standard KIO_3. Then titrate with standard thiosulfate as above. Add 2 mL of starch indicator just before the end point.

27-11 Preparation and Iodometric Analysis of High-Temperature Superconductor[12]

In this experiment we will determine the oxygen content of a high-temperature superconductor, yttrium barium copper oxide ($YBa_2Cu_3O_x$). This material is an example of a *nonstoichiometric solid,* in which the value of x is variable, but near 7. Some of the copper in the formula $YBa_2Cu_3O_7$ is in the unusual high-oxidation state, Cu^{3+}. This experiment describes the synthesis of $YBa_2Cu_3O_x$ (which can also be purchased) and then gives two alternative procedures to measure the Cu^{3+} content. The *iodometric method* is based on the reactions in Box 16-3. The more elegant *citrate-complexed copper titration*[13] eliminates many of the experimental errors associated with the simpler iodometric method and gives more accurate and precise results.

Preparation of $YBa_2Cu_3O_x$

PROCEDURE

1. Place in a mortar 0.750 g of Y_2O_3, 2.622 g of $BaCO_3$, and 1.581 g of CuO (atomic ratio Y:Ba:Cu = 1:2:3). Grind the mixture well with a pestle for 20 min and transfer the powder to a porcelain crucible or boat. Heat in the air in a furnace at 920°–930°C for 12 h or longer. Turn off the furnace and allow the sample to cool slowly *in the furnace.* This slow cooling step is critical for achieving an oxygen content in the range $x = 6.5$–7 in the formula $YBa_2Cu_3O_x$. The crucible may be removed when the temperature is below 100°C.

2. The black solid mass is gently dislodged from the crucible and ground to a fine powder in a mortar and pestle. It can now be used for this experiment, but better quality material is produced if the powder is heated again to 920°–930°C and cooled slowly as in Step 1. If the powder from Step 1 is green instead of black, raise the temperature of the furnace by 20°C and repeat Step 1. The final product must be black, or it is not the correct compound.

PROCEDURE

1. *Sodium Thiosulfate and Starch Indicator.* Prepare 0.03 M $Na_2S_2O_3$ as described in Experiment 27-10, using 3.7 g of $Na_2S_2O_3 \cdot 5H_2O$ instead of 8.7 g. The starch indicator solution is the same one used in Experiment 27-10.

2. *Standard Cu^{2+}.* Weigh accurately 0.5–0.6 g of reagent Cu wire into a 100-mL volumetric flask. In a fume hood, add 6 mL of distilled water and 3 mL of 70 wt % nitric acid, and boil gently on a hot plate until the solid has dissolved. Add 10 mL of distilled water and boil gently. Add 1.0 g of urea or 0.5 g of sulfamic acid and boil for 1 min to destroy HNO_2 and oxides of nitrogen that would interfere with the iodometric titration. Cool to room temperature and dilute to the mark with 1.0 M HCl.

3. *Standardization of $Na_2S_2O_3$ with Cu^{2+}.* The titration should be carried out as rapidly as possible under a brisk flow of N_2, because I^- is oxidized to I_2 in acid solution by atmospheric oxygen. Use a 180-mL tall-form beaker (or a 150-mL standard beaker) with a loosely fitting two-hole cork at the top. One hole serves as the inert gas inlet and the other is for the buret. Pipet 10.00 mL of standard Cu^{2+} into the beaker and flush with N_2. Remove the cork just long enough to pour in 10 mL of distilled water containing 1.0–1.5 g of KI (freshly dissolved) and begin magnetic stirring. In addition to the dark color of iodine in the solution, suspended solid CuI will be present. Titrate with $Na_2S_2O_3$ solution from a 50-mL buret, adding 2 drops of starch solution just before the last trace of I_2 color disappears. If starch is added too soon, there can be irreversible attachment of I_2 to the starch and the end point is harder to detect. You may want to practice this titration several times to learn to distinguish the colors of I_2 and I_2/starch from the color of suspended CuI(*s*). (Alternatively, Pt and calomel electrodes can be used instead of starch to eliminate subjective judgment of color in finding the end point.[14]) Repeat this standardization two more times and use the average $Na_2S_2O_3$ molarity from the three determinations.

4. *Superconductor Experiment A.* Dissolve an accurately weighed 150- to 200-mg sample of powdered $YBa_2Cu_3O_x$ in 10 mL of 1.0 M $HClO_4$ in a titration beaker in a fume hood.[†] Boil gently for 10 min, so that Reaction 1 in Box 16-3 goes to completion. Cool to room temperature, cap with the two-hole-stopper–buret assembly, and begin N_2 flow. Dissolve 1.0–1.5 g of KI in 10 mL of distilled water and immediately add the solution to the beaker. Titrate rapidly with magnetic stirring as described in Step 3. Repeat this procedure two more times.

5. *Superconductor Experiment B.* Place an accurately weighed 150- to 200-mg sample of powdered $YBa_2Cu_3O_x$ in the titration beaker and begin N_2 flow. Dissolve 1.0–1.5 g of KI in 10 mL of 1.0 M $HClO_4$ and immediately add the solution to the titration beaker. Stir magnetically for 1 min, so that the Reactions 3 and 4 of Box 16-3 occur. Add 10 mL of water and rapidly complete the titration. Repeat this procedure two more times.

CALCULATIONS

1. Suppose that mass, m_A, is analyzed in Experiment A and the volume, V_A, of standard thiosulfate is required for titration. Let the corresponding

. . . .

[†]Perchloric acid is recommended because it is inert to reaction with superconductor, which might oxidize HCl to Cl_2. We have used HCl instead of $HClO_4$ with no significant interference in the analysis. Solutions of $HClO_4$ should not be boiled to dryness because of their explosion hazard.

quantities in Experiment B be m_B and V_B. Let the average oxidation state of Cu in the superconductor be $2 + p$. Show that p is given by

$$p = \frac{(V_B/m_B) - (V_A/m_A)}{V_A/m_A} \tag{27-1}$$

and x in the formula $YBa_2Cu_3O_x$ is related to p as follows:

$$x = \frac{7}{2} + \frac{3}{2}(2 + p) \tag{27-2}$$

For example, if the superconductor contains one Cu^{3+} and two Cu^{2+}, the average oxidation state of copper is $\frac{7}{3}$ and the value of p is $\frac{1}{3}$. Setting $p = \frac{1}{3}$ in Equation 27-2 gives $x = 7$. Equation 27-1 does not depend on the metal stoichiometry being exactly Y:Ba:Cu = 1:2:3, but Equation 27-2 does require this exact stoichiometry.

2. Use the average results of Experiments A and B to calculate the values of p and x in Equations 27-1 and 27-2.

3. Suppose that the uncertainty in mass of superconductor analyzed is 1 in the last decimal place. Calculate the standard deviations for Steps 3, 4, and 5 of the iodometric analysis. Using these standard deviations as uncertainties in volume, calculate the uncertainties in the values of p and x in Equations 27-1 and 27-2.

Citrate-Complexed Copper Titration

This procedure directly measures Cu^{3+}. The sample is first dissolved in a closed container with 4.4 M HBr, in which Cu^{3+} oxidizes Br^- to Br_3^-:

$$Cu^{3+} + \frac{11}{2}Br^- \rightarrow CuBr_4^{2-} + \frac{1}{2}Br_3^- \tag{27-3}$$

The solution is transferred to a vessel containing excess I^-, excess citrate, and enough NH_3 to neutralize most of the acid. Cu^{2+} is complexed by citrate and is not further reduced to $CuI(s)$. (This eliminates the problem in the iodometric titration of performing a titration in the presence of colored solid.) The Br_3^- from Reaction 27-3 oxidizes I^- to I_3^-:

$$Br_3^- + 3I^- \rightarrow 3Br^- + I_3^- \tag{27-4}$$

and the I_3^- is titrated with thiosulfate.

Our experience with the iodometric procedure is that the precision in oxygen content of $YBa_2Cu_3O_x$ is ± 0.04 in the value of x. The uncertainty is reduced to ± 0.01 by the citrate-complexed copper procedure.

PROCEDURE

1. Prepare and standardize sodium thiosulfate solution as described in Steps 1–3 of the iodometric analysis in the previous section.

2. Place an accurately weighed 20–50 mg sample of superconductor in a 4-mL screw cap vial with a Teflon cap liner and add 2.00 mL of ice-cold 4.4 M HBr by pipet. (The HBr is prepared by diluting 50 mL of 48 wt % HBr to 100 mL.) Cap tightly and gently agitate the vial for 15 min as it warms to room temperature. (We use a mechanical stirring motor to rotate the sample slowly for 15 min.)

3. Cool the solution back to 0°C and carefully transfer it to the titration beaker (used in the thiosulfate standardization) containing an ice-cold, freshly prepared solution made from 0.7 g of KI, 20 mL of water, 5 mL of 1.0 M trisodium citrate, and approximately 0.5 mL of 28 wt % NH_3. The exact amount of NH_3 should be enough to neutralize all but 1 mmol of acid present in the sample. When calculating the acid content, remember that each mole of $YBa_2Cu_3O_x$ consumes $2x$ moles of HBr. (You can estimate that x is close to 7.) Wash the vial with three 1-ml aliquots of 2 M HBr to complete the quantitative transfer to the beaker.

4. Add 0.1 mL of 1 wt % starch solution and titrate with 0.1 M standard $Na_2S_2O_3$ under a brisk flow of N_2 using a 250-μL Hamilton syringe to deliver titrant. The end point is marked by a change from dark blue (I_2-starch) to the light blue-green of the Cu^{2+}-citrate complex.

5. Run a blank reaction with $CuSO_4$ in place of superconductor. The moles of Cu in the blank should be the same as the moles of Cu in the superconductor. In a typical experiment, 30 mg of $YBa_2Cu_3O_{6.88}$ required approximately 350 μL of $Na_2S_2O_3$ and the blank required 10 μL of $Na_2S_2O_3$. If time permits, run two more blanks. Subtract the average blank from the titrant in Step 4.

6. Repeat the analysis with two more samples of superconductor.

CALCULATIONS

1. From the thiosulfate required to titrate I_3^- released in Reaction 27-4, find the average moles of Cu^{3+} per gram of superconductor (1 mol $S_2O_3^{2-}$ = 1 mol Cu^{3+}) and the standard deviation for your three samples.

2. Defining R as (mol Cu^{3+})/(g superconductor), show that z in the formula $YBa_2Cu_3O_{7-z}$ is

$$z = \frac{1 - 666.20\,R}{2 - 15.9994\,R} \tag{27-5}$$

where 666.20 is the formula mass of $YBa_2Cu_3O_7$ and 15.9994 is the atomic mass of oxygen.

3. Using your average value of R and using its standard deviation as an estimate of uncertainty, calculate the average value of z and its uncertainty. Find the average value of x and its uncertainty in the formula $YBa_2Cu_3O_x$. If you are really daring, you should use Equation C-1 in Appendix C for propagation of uncertainty in Equation 27-5.

27-12 Potentiometric Halide Titration with Ag^+

Mixtures of halides may be titrated with $AgNO_3$ solution as described in Section 7-5. In this experiment, we will use the apparatus in Figure 7-9 to monitor the activity of Ag^+ as the titration proceeds. The theory of the potentiometric measurement is described in Section 15-2.

Each student is given a vial containing 0.22–0.44 g of KCl plus 0.50–1.00 g of KI (both weighed accurately). The object is to determine the quantity of each salt in the mixture. A 0.4 M bisulfate buffer (pH 2) should be available in the lab. This is prepared by titrating 1 M H_2SO_4 with 1 M NaOH to a pH near 2.0.

PROCEDURE

1. Pour your unknown carefully into a 50- or 100-mL beaker. Dissolve the solid in ~20 mL of water and pour it into a 100-mL volumetric flask. Rinse the sample vial and beaker many times with small portions of H_2O and transfer the washings to the flask. Dilute to the mark and mix well.

2. Dry 1.2 g of $AgNO_3$ at 105°C for 1 h and cool in a dessicator for 30 min with minimal exposure to light. Some discoloration is normal (and tolerable in this experiment) but should be minimized. Accurately weigh 1.2 g and dissolve it in a 100-mL volumetric flask.

3. The apparatus in Figure 7-9 should be set up. The silver electrode is simply a 3-cm length of silver wire connected to copper wire. (Fancier electrodes can be prepared by housing the connection in a glass tube sealed with epoxy at the lower end. Only the silver should protrude from the epoxy.) The copper wire is fitted with a jack that goes to the reference socket of a pH meter. The reference electrode for this titration is a glass pH electrode connected to its usual socket on the meter. If a combination pH electrode is employed, the reference jack of the combination electrode is not used. The silver electrode should be taped to the inside of the 100-mL beaker so that the Ag/Cu junction remains dry for the entire titration. The stirring bar should not hit either electrode.

4. Pipet 25.00 mL of unknown into the beaker, add 3 mL of bisulfate buffer, and begin magnetic stirring. Record the initial level of $AgNO_3$ in a 50-mL buret and add ~1 mL of titrant to the beaker. Turn the pH meter to the millivolt scale and record the volume and voltage. It is convenient (but is not essential) to set the initial reading to +800 mV by adjusting the meter.

5. Titrate the solution with ~1-mL aliquots until 50 mL of titrant has been added or until you can see two clear potentiometric end points. You need not allow more than 15–30 s for each point. Record the volume and voltage at each point. Make a graph of millivolts versus milliliters to find the approximate positions (±1 mL) of the two end points.

6. Turn the pH meter to standby, remove the beaker, rinse the electrodes well with water, and blot them dry with a tissue.[†] Clean the beaker and set up the titration apparatus again. (The beaker need not be dry.)

7. Now perform an accurate titration, using 1-drop aliquots near the end points (and 1-mL aliquots elsewhere). You need not allow more than 30 s per point for equilibration.

8. Prepare a graph of millivolts versus milliliters and locate the end points as in Figure 7-8. The I^- end point is taken as the intersection of the two dashed lines in the inset of Figure 7-8. The Cl^- end point is the inflection point at the second break. Calculate milligrams of KI and milligrams of KCl in your solid unknown.

[†]Silver halide adhering to the glass electrode can be removed by soaking in concentrated sodium thiosulfate solution. This thorough cleaning is not necessary between Steps 6 and 7 in this experiment. The silver halides in the titration beaker can be saved and converted back to pure $AgNO_3$ by using the procedure of E. Thall, *J. Chem. Ed.* **1981,** *58,* 561.

27-13 Electrogravimetric Analysis of Copper

Most copper-containing compounds can be electrolyzed in acidic solution, with quantitative deposition of Cu at the cathode. Sections 17-1 to 17-3 discuss the theory of this technique.

Students may analyze a preparation of their own (such as copper acetylsalicylate[15]) or be given unknowns prepared from $CuSO_4 \cdot 5H_2O$ or metallic Cu. In the latter case, dissolve the metal in 8 M HNO_3, boil to remove HNO_2, neutralize with ammonia, and *barely* acidify the solution with dilute H_2SO_4 (using litmus paper to test for acidity). Samples must be free of chloride and nitrous acid.[16] Copper oxide unknowns (soluble in acid) are available from Thorn Smith.[3]

The apparatus shown in Figure 17-6 uses any 6–12 V direct-current power supply. A tall-form 150-mL beaker is the reaction vessel.

PROCEDURE

1. Handle the Pt gauze cathode with a tissue, touching only the thick stem, not the wire gauze. Immerse the electrode in hot 8 M HNO_3 to remove previous deposits, rinse with water and alcohol, dry at 110°C for 5 min, cool for 5 min, and weigh accurately. If the electrode contains any grease, it can be heated to red heat over a burner after the treatment above.[†]

2. The sample should contain 0.2–0.3 g of Cu in 100 mL. Add 3 mL of 98 wt % H_2SO_4 and 2 mL of freshly boiled 8 M HNO_3. Position the cathode so that the top 5 mm are above the liquid level after magnetic stirring is begun. Adjust the current to 2 A, which should require 3–4 V. When the blue color of Cu(II) has disappeared, add some distilled water so that new Pt surface is exposed to the solution. If no further deposition of Cu occurs on the fresh surface in 15 min at a current of 0.5 A, the electrolysis is complete. If deposition is observed, continue electrolysis and test the reaction for completeness again.

3. *Without* turning off the power, lower the beaker while washing the electrode with a squirt bottle. Then the current can be turned off. (If current is disconnected before removing the cathode from the liquid and rinsing off the acid, some Cu could redissolve.) Wash the cathode gently with water and alcohol, dry at 110°C for 3 min, cool in a desiccator for 5 min, and weigh.

27-14 Polarographic Measurement of an Equilibrium Constant[17]

In this experiment, we will find the overall formation constant and stoichiometry for the reaction of oxalate with Pb^{2+}:

[†]Some metals, such as Zn, Ga, and Bi, form alloys with Pt and should not be deposited directly on the Pt surface. The electrode should be coated first with Cu, dried, and then used. Alternatively, Ag may be used in place of Pt for depositing these metals. Platinum anodes are attacked by Cl_2 formed by electrolysis of Cl^- solutions. To prevent this, 1–3 g of a hydrazinium salt (per 100 mL of solution) may be used as an *anodic depolarizer,* because hydrazine is more readily oxidized than Cl^-: $N_2H_4 \rightarrow N_2 + 4H^+ + 4e^-$.

$$Pb^{2+} + pC_2O_4^{2-} \rightleftharpoons Pb(C_2O_4)_p^{2-2p}$$

$$\beta_p = \frac{[Pb(C_2O_4)_p^{2-2p}]}{[Pb^{2+}][C_2O_4^{2-}]^p}$$

where p is a stoichiometry coefficient. We will do this by measuring the polarographic half-wave potential for solutions containing Pb^{2+} and various amounts of oxalate. According to Equation 18-10, the change of half-wave potential, $\Delta E_{1/2}$ $[= E_{1/2}(\text{observed}) - E_{1/2}(\text{for } Pb^{2+} \text{ without oxalate})]$ should obey the equation

$$\Delta E_{1/2} = -\frac{RT}{nF} \ln \beta_p - \frac{pRT}{nF} \ln [C_2O_4^{2-}] \qquad (27\text{-}6)$$

Here we have converted $0.059\,16 \log x$ in Equation 18-10 to $(RT/F) \ln x$ in Equation 27-6. R is the gas constant, F is the Faraday constant, and T is temperature in kelvins. You should measure the lab temperature at the time of the experiment or use a thermostatically controlled cell.

PROCEDURE

1. Pipet 1.00 mL of 0.020 M $Pb(NO_3)_2$ into each of five 50-mL volumetric flasks labeled A–E and add 1 drop of 1 wt % Triton X-100 to each. Then add the following solutions and dilute to the mark with water. The KNO_3 may be delivered carefully with a graduated cylinder. The oxalate should be pipetted.

   A: Add nothing else. Dilute to the mark with 1.20 M KNO_3.
   B: Add 5.00 mL of 1.00 M $K_2C_2O_4$ and 37.5 mL of 1.20 M KNO_3.
   C: Add 10.00 mL of 1.00 M $K_2C_2O_4$ and 25.0 mL of 1.20 M KNO_3.
   D: Add 15.00 mL of 1.00 M $K_2C_2O_4$ and 12.5 mL of 1.20 M KNO_3.
   E: Add 20.00 mL of 1.00 M $K_2C_2O_4$.

2. Transfer each solution to a polarographic cell, deoxygenate for 10 min, and record the polarogram from -0.20 to -0.95 V (versus S.C.E.). Measure the residual current, using the same settings and a solution containing just 1.20 M KNO_3 (plus 1 drop of 1 wt % Triton X-100). Record each polarogram on a scale sufficiently expanded to allow accurate measurements. (Use a sweep rate of 0.05 V/min and a chart speed of 2.5 cm/min, with a mercury drop interval of 4 s.)

3. For each polarogram, make a graph of E versus $\log [I/(I_d - I)]$, using 6–8 points for each graph. According to Equation 18-6, $E = E_{1/2}$ when $\log [I/(I_d - I)] = 0$. Use this condition to locate $E_{1/2}$ on each graph. When you measure currents on the polarograms, be sure to subtract the residual current at each potential. Current on the polarographic wave is measured at the top of each oscillation, which corresponds to the maximum size of each mercury drop. Residual current is measured as follows: The electrocapillary maximum is the potential at which the residual current shows no oscillations. If the potential is more *positive* than the electrocapillary maximum, use the top of the oscillation, which corresponds to the maximum mercury drop size. If the potential is more *negative* than the electrocapillary maximum, use the bottom of each oscillation, which also corresponds to the maximum mercury drop size.

4. Make a graph of $\Delta E_{1/2}$ versus $\ln [C_2O_4^{2-}]$. Use Equation 27-6 to find p, the stoichiometry coefficient, from the slope of the graph. Then use the intercept to find the value of β_p. A worthwhile exercise is to use the method of least squares (Section 4-5) to find the standard deviations of the slope and intercept. From the standard deviations, find the uncertainties in p and β_p and express each with the correct number of significant figures.

27-15 Coulometric Titration of Cyclohexene with Bromine[18]

This experiment is described in Section 17-4, and the apparatus is shown in Figure 17-12. A conventional coulometric power supply can be employed, but we use the homemade circuits in Figure 27-6. A stopwatch is manually started as the generator switch is closed. Alternatively, a double-pole, double-throw switch can be used to simultaneously start the generator circuit and an electric clock.

P R O C E D U R E

1. The electrolyte is a 60:26:14 (vol/vol/vol) mixture of acetic acid, methanol, and water. The solution contains 0.15 M KBr and 0.1 g of mercuric acetate per 100 mL. (The latter catalyzes the reaction between Br_2 and cyclohexene.) The electrodes should be covered with electrolyte. Begin vigorous magnetic stirring (without spattering) and adjust the voltage of the detector circuit to 0.25 V.

2. Generate Br_2 with the generator circuit until the detector current is 20.0 μA. (The generator current is 5–10 mA.)

3. Pipet 2–5 mL of unknown (containing 1–5 mg of cyclohexene in methanol) into the flask and set the clock or coulometer to zero. The detector current should drop to near zero because the cyclohexene consumes the Br_2.

4. Turn the generator circuit on and simultaneously begin timing. While the reaction is in progress, measure the voltage across the precision resistor (100.0 ± 0.1 Ω) to find the exact current flowing through the cell ($I = E/R$). Continue the electrolysis until the detector current rises to 20.0 μA. Then stop the coulometer and record the time.

5. Repeat the procedure two more times and find the average molarity (and relative standard deviation) of cyclohexene.

6. When you are finished, be sure all switches are off. The generator electrodes should be soaked in 8 M HNO_3 to dissolve Hg that is deposited during the electrolysis.

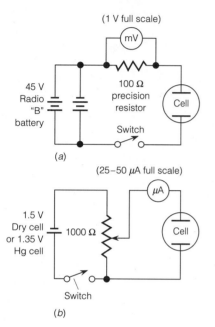

Figure 27-6 Circuits for coulometric titrations. (a) Generator circuit. (b) Detector circuit.

27-16 Spectrophotometric Determination of Iron in Vitamin Tablets[19]

In this procedure, iron from a vitamin supplement tablet is dissolved in acid, reduced to Fe^{2+} with hydroquinone, and complexed with *o*-phenanthroline to form an intensely colored complex (Color Plate 19).

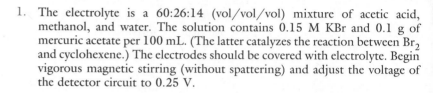

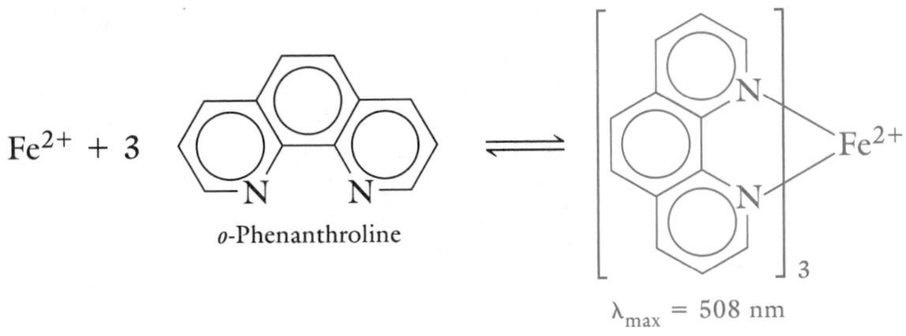

$\lambda_{max} = 508$ nm

o-Phenanthroline

REAGENTS

Hydroquinone: Freshly prepared solution containing 10 g/L in water. Store in an amber bottle.

Trisodium citrate: 25 g/L in water.

o-Phenanthroline: Dissolve 2.5 g in 100 mL of ethanol and add 900 mL of water. Store in an amber bottle.

Standard Fe (0.04 mg Fe/mL): Prepare by dissolving 0.281 g of reagent-grade $Fe(NH_4)_2(SO_4)_2 \cdot 6H_2O$ in water in a 1-L volumetric flask containing 1 mL of 98 wt % H_2SO_4.

PROCEDURE

1. Place one tablet of the iron-containing vitamin in a 125-mL flask or 100-mL beaker and boil gently (*in a fume hood*) with 25 mL of 6 M HCl for 15 min. Filter the solution directly into a 100-mL volumetric flask. Wash the beaker and filter several times with small portions of water to complete a quantitative transfer. Allow the solution to cool, dilute to the mark, and mix well. Dilute 5.00 mL of this solution to 100.0 mL in a fresh volumetric flask. If the label indicates that the tablet contains <15 mg of Fe, use 10.00 mL instead of 5.00 mL.

2. Pipet 10.00 mL of standard Fe solution into a beaker and measure the pH (with pH paper or a glass electrode). Add sodium citrate solution 1 drop at a time until a pH of ~3.5 is reached. Count the drops needed. (It will require about 30 drops.)

3. Pipet a fresh 10.00-mL aliquot of Fe standard into a 100-mL volumetric flask and add the same number of drops of citrate solution as required in Step 2. Add 2.00 mL of hydroquinone solution and 3.00 mL of *o*-phenanthroline solution, dilute to the mark with water, and mix well.

4. Prepare three more solutions from 5.00, 2.00, and 1.00 mL of Fe standard and prepare a blank containing no Fe. Use sodium citrate solution in proportion to the volume of Fe solution. (If 10 mL of Fe requires 30 drops of citrate solution, 5 mL of Fe requires 15 drops of citrate solution.)

5. Find out how many drops of citrate solution are needed to bring 10.00 mL of the iron supplement tablet solution to pH 3.5. This will require about 3.5 or 7 mL of citrate, depending on whether 5 or 10 mL of unknown was diluted in the second part of Step 1.

6. Transfer 10.00 mL of the solution containing the dissolved tablet to a 100-mL volumetric flask. Add the required amount of citrate solution, 2.00 mL

of hydroquinone solution, and 3.0 mL of *o*-phenanthroline solution; dilute to the mark and mix well.

7. Allow the solutions to stand for at least 10 min. Then measure the absorbance of each solution at 508 nm. (The color is stable, so all solutions may be prepared and all the absorbances measured at once.) Use distilled water in the reference cuvette and subtract the absorbance of the blank from the absorbance of the Fe standards.

8. Make a graph of absorbance versus micrograms of Fe in the standards. If desired, find the slope and intercept (and standard deviations) by the method of least squares, as described in Section 4-5. Calculate the molarity of Fe(*o*-phenanthroline)$_3^{2+}$ in each solution and find the average molar absorptivity (ϵ in Beer's law) from the four absorbances. (Remember that all the iron has been converted to the phenanthroline complex.) If a Spectronic 20 is used for absorbance measurements, assume that the pathlength is 1.1 cm for the sake of this calculation.

9. Using the calibration curve (or its least-squares parameters), find the number of milligrams of Fe in the tablet.

27-17 Spectrophotometric Measurement of an Equilibrium Constant[20]

In this experiment, we will use the Scatchard plot described in Section 19-2 to find the equilibrium constant for the formation of a complex between iodine and pyridine in cyclohexane:

Both I_2 and $I_2 \cdot$ pyridine absorb visible radiation, but pyridine is colorless. Analysis of the spectral changes associated with variation of pyridine concentration (with a constant total concentration of iodine) will allow us to evaluate K for the reaction. The experiment is best performed with a recording spectrophotometer, but single-wavelength measurements can be used.

PROCEDURE

All operations described below should be carried out *in a fume hood,* including pouring solutions into and out of the spectrophotometer cell. Only a *capped* cuvette containing the solution whose spectrum is to be measured should be taken from the hood. Do not spill solvent on your hands or breathe the vapors. Used solutions should be discarded in a waste container *in the hood,* not down the drain.

1. The following stock solutions should be available in the lab:
   (a) 0.050–0.055 M pyridine in cyclohexane (40 mL for each student, concentration known accurately).
   (b) 0.0120–0.0125 M I_2 in cyclohexane (10 mL for each student, concentration known accurately).

2. Pipet the following volumes of stock solutions into six 25-mL volumetric flasks A–F, dilute to the mark with cyclohexane, and mix well.

Flask	Pyridine stock solution (mL)	I_2 stock solution (mL)
A	0	1.00
B	1.00	1.00
C	2.00	1.00
D	4.00	1.00
E	5.00	1.00
F	10.00	1.00

3. Using glass or quartz cells, record a baseline between 350 and 600 nm with solvent in both the sample and the reference cells. Subtract the absorbance of the baseline from all future absorbances. If possible, record all spectra, including the baseline, on one sheet of chart paper. (If a fixed-wavelength instrument is used, first find the positions of the two absorbance maxima in solution E. Then make all measurements at these two wavelengths.)

4. Record the spectrum of each solution A–F or measure the absorbance at each maximum if a fixed-wavelength instrument is used.

DATA ANALYSIS

1. Measure the absorbance at the wavelengths of the two maxima in each spectrum. Be sure to subtract the absorbance of the blank from each.

2. The analysis of this problem follows that of Reaction 19-10, in which P is iodine and X is pyridine. As a first approximation, assume that the concentration of free pyridine equals the total concentration of pyridine in the solution (because [pyridine] $>>$ [I_2]). Prepare a graph of ΔA/[free pyridine] versus ΔA (a Scatchard plot), using the absorbance at the $I_2 \cdot$ pyridine maximum.

3. From the slope of the graph, find the equilibrium constant using Equation 19-16. From the intercept, find $\Delta\epsilon$ ($= \epsilon_{PX} - \epsilon_X$).

4. Now refine the values of K and $\Delta\epsilon$. Use $\Delta\epsilon$ to find ϵ_{PX}. Then use the absorbance at the wavelength of the $I_2 \cdot$ pyridine maximum to find the concentration of bound and free pyridine in each solution. Make a new graph of ΔA/[free pyridine] versus ΔA using the new values of [free pyridine]. Find a new value of K and $\Delta\epsilon$. If you feel it is justified, perform another cycle of refinement.

5. Using the values of free pyridine concentration from your last refinement and the values of absorbance at the I_2 maximum, prepare another Scatchard plot and see if you get the same value of K.

6. Explain why an isosbestic point is observed in this experiment.

27-18 Properties of an Ion-Exchange Resin[21]

In this experiment, we explore the properties of a cation-exchange resin, which is an organic polymer containing many sulfonic acid groups ($-SO_3H$). When a cation, such as Cu^{2+}, flows into the resin, the cation is tightly bound by sulfonate groups, and one H^+ is released for each positive charge bound to the resin (Figure 27-7). The bound cation can be displaced from the resin by a large excess of H^+ or by an excess of any other cation for which the resin has some affinity.

Figure 27-7 Stoichiometry of ion exchange.

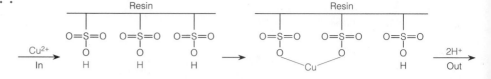

First, known quantities of NaCl, $Fe(NO_3)_3$, and NaOH will be passed through the resin in the H^+ form. The H^+ released by each cation will be measured by titration with NaOH.

In the second part of the experiment, we will analyze a sample of impure vanadyl sulfate ($VOSO_4 \cdot 2H_2O$). As supplied commercially, this salt contains $VOSO_4$, H_2SO_4, and H_2O. A solution will be prepared from a known mass of reagent. The VO^{2+} content can be assayed spectrophotometrically, and the total cation (VO^{2+} and H^+) content can be assayed by ion exchange. Together, these measurements enable us to establish the quantities of $VOSO_4$, H_2SO_4, and H_2O in the sample.

REAGENTS

0.3 M NaCl: A bottle containing 5–10 mL per student, with an accurately known concentration.

0.1 M Fe(NO$_3$)$_3$ · 6 H$_2$O: A bottle containing 5–10 mL per student, with an accurately known concentration.

VOSO$_4$: The commonly available grade (usually designated "purified") is used for this experiment. Students can make their own solutions and measure the absorbance at 750 nm, or a bottle of stock solution (25 mL per student) can be supplied. The stock should contain 8 g/L (accurately weighed) and be labeled with the absorbance.

0.02 M NaOH: Each student should prepare an accurate $\frac{1}{5}$ dilution of standard 0.1 M NaOH.

PROCEDURE

1. Prepare a chromatography column from a 0.7-cm diameter, 15-cm length of glass tubing, fitted at the bottom with a cork having a small hole that serves as the outlet. Place a small ball of glass wool above the cork to retain the resin. Use a small glass rod to plug the outlet and shut off the column. (Alternatively, an inexpensive column such as 0.7 × 15 cm Econo-Column from Bio-Rad Laboratories[22] works well in this experiment.) Fill the column with water, close it off, and test for leaks. Then drain the water until 2 cm remains and close the column again.

2. Make a slurry of 1.1 g of Bio-Rad Dowex 50W-X2 (100/200 mesh) cation-exchange resin in 5 mL of water and pour it into the column. If the resin cannot be poured all at once, allow some to settle, remove the supernatant liquid with a pipet, and pour in the rest of the resin. If the column is stored between laboratory periods, it should be upright, capped, and contain water above the level of the resin.[†]

- -

[†]When the experiment is finished, the resin can be collected, washed with 1 M HCl and water, and reused.

3. The general procedure for analysis of a sample is as follows:
   (a) Generate the H^+-saturated resin by passing ~10 mL of 1 M HCl through the column. Apply the liquid sample to the glass wall so as not to disturb the resin.
   (b) Wash the column with ~15 mL of water. Use the first few milliliters to wash the glass walls and allow the water to soak into the resin before continuing the washing.[†]
   (c) Place a clean 125-mL flask under the outlet and pipet the sample onto the column.
   (d) After the reagent has soaked in, wash it through with 10 mL of H_2O, collecting all eluate.
   (e) Add 3 drops of phenolphthalein indicator (Table 12-4) to the flask and titrate with standard 0.02 M NaOH.

4. Analyze 2.000-mL aliquots of 0.3 M NaCl and 0.1 M $Fe(NO_3)_3$, following the procedures in Step 3. Calculate the theoretical volume of NaOH needed for each titration. If you do not come within 2% of this volume, repeat the analysis.

5. Pass 10.0 mL of your 0.02 M NaOH through the column as in Step 3, and analyze the eluate. Explain what you observe.

6. Analyze 10.00 mL of $VOSO_4$ solution as described in Step 3.

7. Using the molar absorptivity of vanadyl ion ($\epsilon = 18.0$ $M^{-1} \cdot cm^{-1}$ at 750 nm) and the results of Step 6, express the composition of the vanadyl sulfate in the form $(VOSO_4)_{1.00}(H_2SO_4)_x(H_2O)_y$.

(27-19) Quantitative Analysis by Gas Chromatography or HPLC[23]

We now describe an experiment illustrating the use of internal standards for quantitative analysis. The directions are purposely general, because the specific mixture and analytical conditions depend on the equipment and columns employed. The procedure can be modified for gas or liquid chromatography.

You will receive an unknown solution containing two or three compounds from a specified list of possibilities.[‡] Pure samples of each possible component of the unknown should also be available.

PROCEDURE

1. Identify the components of your mixture. Run a chromatogram of the pure unknown. Then mix pure samples of suspected components with samples of the unknown and run a new chromatogram of each. By observing which peaks grow or whether new peaks appear, you should be able to identify each species in your unknown.

. . . .

[†]Unlike most other chromatography resins, the one used in this experiment retains water when allowed to run "dry." Ordinarily, you must not let liquid fall below the top of the solid phase in a chromatography column.

[‡]For example, we have done this experiment with Carle student gas chromatographs having columns packed with Carbowax or dinonyl phthalate. We used dichloromethane, chloroform, ethyl acetate, 1-propanol, and toluene as unknowns. We have also done this experiment by HPLC, using an octadecyl-bonded phase eluted with 65 vol % methanol/water. The unknowns included toluene, biphenyl, naphthalene, 2-naphthol, and 9-fluorenone.

2. Perform a quantitative analysis of one component (designated A) of the unknown, which is well separated from all other components. Find a compound (designated S) that is not a component of the unknown and is separated from all other peaks in the chromatogram. Prepare a mixture of unknown plus S such that the peak areas (or heights, if the peaks are sharp) of A and S are within a factor of 2 of each other. This is done by trial and error. The mass (or volume) of S and the mass of unknown (not A, but total unknown) in the mixture must be accurately known.

3. Prepare known mixtures of pure A and pure S to establish a calibration curve. If a solvent is used, the mixtures should have the same concentration of S required in Step 2, plus variable quantities of unknown. If pure liquids are used without a solvent, the concentrations of both S and A will necessarily vary. Prepare a calibration curve in which you plot the signal ratio (peak area of A/peak area of S) versus the known concentration ratio (concentration of A/concentration of S). The graph should span the range including the signal ratio measured for the unknown mixture plus standard.

4. From your graph, calculate the concentration of species A in the unknown.

Notes and References

1. An excellent source of further experiments involving instrumental methods of analysis is D. T. Sawyer, W. R. Heineman, and J. M. Beebe, *Chemistry Experiments for Instrumental Methods* (New York: Wiley, 1984). For a computerized list of experiments published in the *Journal of Chemical Education,* see *J. Chem. Ed.* **1994,** *71,* 450.

2. C. H. Hendrickson and P. R. Robinson, *J. Chem. Ed.* **1979,** *56,* 341. For an experiment in thermogravimetric analysis, see J. O. Hill and R. J. Magee, *J. Chem. Ed.* **1988,** *65,* 1024.

3. Thorn Smith, Inc., 7755 Narrow Gauge Road, Beulah, MI 49617.

4. R. Q. Thompson and M. Ghadiali, *J. Chem. Ed.* **1993,** *70,* 170. This article describes microwave heating of gravimetric precipitates of AgCl, $BaSO_4$, and $CaC_2O_4 \cdot H_2O$.

5. Adapted from D. A. Skoog and D. M. West, *Fundamentals of Analytical Chemistry,* 3d ed. (New York: Holt, Rinehart and Winston, 1976).

6. For a related experiment, see V. T. Lieu and G. E. Kalbus, "Potentiometric Titration of Acidic and Basic Compounds in Household Cleaners," *J. Chem. Ed.* **1988,** *65,* 184; L. H. Kalbus, R. H. Petrucci, J. E. Forman, and G. E. Kalbus, "Titration of Chromate-Dichromate Mixtures," *J. Chem. Ed.* **1991,** *68,* 677; D. T. Harvey, "Statistical Evaluation of Acid/Base Indicators," *J. Chem. Ed.* **1991,** *68,* 329.

7. H. C. Moore, *Ind. Eng. Chem.* **1920,** *12,* 669; *J. Assoc. Offic. Agr. Chemists* **1924–1925,** *8,* 411; I. K. Phelps and H. W. Daudt, *J. Assoc. Offic. Agr. Chemists* **1920,** *3,* 306; **1921,** *4,* 72.

8. For a related experiment, see C. A. Castillo S. and A. Jaramillo A., "An Alternative Procedure for Titration Curves of a Mixture of Acids of Different Strengths," *J. Chem. Ed.* **1989,** *66,* 341; A. Vesala, "Estimation of Carbonate Content of Base by Gran's Method," *J. Chem. Ed.* **1992,** *69,* 577; J. R. Amend, K. A. Tucker, and R. P. Furstenau, "Computer Interfacing: A New Look at Acid-Base Titrations," *J. Chem. Ed.* **1991,** *68,* 857.

9. G. G. Long, R. L. Stanfield, and F. C. Hentz, Jr., *J. Chem. Ed.* **1979,** *56,* 195. For related experiments involving synthesis followed by several analyses, see G. H. Searle, G. S. Bull, and D. A. House, "Reinecke's Salt Revisited," *J. Chem. Ed.* **1989,** *66,* 605; and E. P. Dudek, "A Project Lab for an Advanced General Chemistry Course Featuring the Amino Acid Glycine," *J. Chem. Ed.* **1987,** *64,* 899.

10. R. M. Fowler and H. A. Bright, *J. Res. National Bureau of Standards* **1935,** *15,* 493.

11. D. N. Bailey, *J. Chem. Ed.* **1974,** *51,* 488. For related experiments involving redox titrations, see S. Kaufman and H. DeVoe, "Iron Analysis by Redox Titration," *J. Chem. Ed.* **1988,** *65,* 183; and W. B. Guenther, "Supertitrations: High Precision Methods," *J. Chem. Ed.* **1988,** *65,* 1097. The latter is an extremely precise mass titration that stresses careful laboratory technique.

12. D. C. Harris, M. E. Hills, and T. A. Hewston, *J. Chem. Ed.* **1987,** *64,* 847. An alternative synthesis utilizing homogeneously coprecipitated reactants is described

by P. I. Djurovich and R. J. Watts, *J. Chem. Ed.* **1993**, *70*, 497.

13. E. H. Appelman, L. R. Morss, A. M. Kini, U. Geiser, A. Umezawa, G. W. Crabtree, and K. D. Carlson, *Inorg. Chem.* **1987**, *26*, 3237.

14. P. Phinyocheep and I. M. Tang, *J. Chem. Ed.* **1994**, *71*, A115.

15. E. Dudek, *J. Chem. Ed.* **1977**, *54*, 329.

16. J. F. Owen, C. S. Patterson, and G. S. Rice, *Anal. Chem.* **1983**, *55*, 990, describe simple procedures for the removal of chloride from Cu, Ni, and Co samples prior to electrogravimetric analysis.

17. W. C. Hoyle and T. M. Thorpe, *J. Chem. Ed.* **1978**, *55*, A229. For related electrochemistry experiments, see N. Radić and J. Komijenović, "Potentiometric Determination of an Overall Formation Constant Using an Ion-Selective Membrane Electrode," *J. Chem. Ed.* **1993**, *70*, 509; J. L. Town, F. MacLaren, and H. D. Dewald, "Rotating Disk Voltammetry Experiment," *J. Chem. Ed.* **1991**, *68*, 352; P. Lanza, "Multiple Analysis by Differential Pulse Polarography," *J. Chem. Ed.* **1990**, *67*, 704; D. Martin and F. Menduciti, "Polarographic Determination of Composition and Thermodynamic Stability Constant of a Complex Metal Ion," *J. Chem. Ed.* **1988**, *65*, 916; J. G. Ibáñez, I. González, and M. A. Cárdenas, "The Effect of Complex Formation upon the Redox Potentials of Metallic Ions," *J. Chem. Ed.* **1988**, *65*, 173; T. J. Farrell, R. J. Laub, and E. P. Wadsworth, Jr., "Anodic Polarography of Cyanide in Foodstuffs," *J. Chem. Ed.* **1987**, *64*, 635; E. Briullas, J. A. Garrido, R. M. Rodríguez, and J. Doménech, "A Cyclic Voltammetry Experiment Using a Mercury Electrode," *J. Chem. Ed.* **1987**, *64*, 189; L. Piszczek, A. Ignatowicz, and K. Kielbasa, "Application of Cyclic Voltammetry for Stoichiometry Determination of Ni(II), Co(II), and Cd(II) Complex Compounds with Polyaminopolycarboxylic Acids," *J. Chem. Ed.* **1988**, *65*, 171; R. S. Pomeroy, M. B. Denton, and N. R. Armstrong, "Voltammetry at the Thin-Film Mercury Electrode," *J. Chem. Ed.* **1989**, *66*, 877; W. H. Chan, M. S. Wong, and C. W. Yip, "Ion-Selective Electrode in Organic Analysis: A Salicylate Electrode," *J. Chem. Ed.* **1986**, *63*, 915.

18. D. H. Evans, *J. Chem. Ed.* **1968**, *45*, 88. A similar experiment in which vitamin C is analyzed has been described by D. G. Marsh, D. L. Jacobs, and H. Veening, *J. Chem. Ed.* **1973**, *50*, 626. Home-built circuits for constant-current or controlled-potential coulometry have been described by E. Grimsrud and J. Amend, *J. Chem. Ed.* **1979**, *56*, 131.

19. R. C. Atkins, *J. Chem. Ed.* **1975**, *52*, 550. For other experiments involving spectrophotometry, see M. A. Grompone, "Determination of Iron in a Bar of Soap," *J. Chem. Ed.* **1987**, *64*, 1057; K. W. Street, "Method Development for Analysis of Aspirin Tablets," *J. Chem. Ed.* **1988**, *65*, 915; T. Matsuo, A. Muromatsu, K. Katayama, and M. Mori, "Construction of a Photoelectric Colorimeter and Application to Students' Experiments," *J. Chem. Ed.* **1989**, *66*, 329, 848; M. A. Williamson, "Determination of Copper by Graphite Furnace Atomic Absorption Spectrophotometry," *J. Chem. Ed.* **1989**, *66*, 261; A. Ríos, M. Dolores, L. de Castro, and M. Valcárcel, "Determination of Reaction Stoichiometries by Flow Injection Analysis," *J. Chem. Ed.* **1986**, *63*, 552; C. L. Stults, A. P. Wade, and S. R. Crouch, "Investigation of Temperature Effects on Dispersion in a Flow Injection Analyzer," *J. Chem. Ed.* **1988**, *65*, 645.

20. For related experiments, see S. A. Tucker, V. L. Amszi, and W. E. Acree, Jr., "Primary and Secondary Inner Filtering: Effect of $K_2Cr_2O_7$ on Fluorescence Emission Intensities of Quinine Sulfate," *J. Chem. Ed.* **1992**, *69*, A8; S. W. Bigger, K. P. Ghiggino, G. A. Meilak, and B. Verity, "Illustration of the Principles of Fluorimetry," *J. Chem. Ed.* **1992**, *69*, 675; C. Cappas, N. Hoffman, J. Jones, and S. Young, "Determination of Concentrations of Species Whose Absorption Bands Overlap Extensively," *J. Chem. Ed.* **1991**, *68*, 300; G. Dado and J. Rosenthal, "Simultaneous Determination of Cobalt, Copper, and Nickel by Multivariate Linear Regression," *J. Chem. Ed.* **1990**, *67*, 797; J. Lieberman, Jr., and K. J. Yun, "A Semimicro Spectrophotometric Determination of the K_{sp} of Silver Acetate at Various Temperatures," *J. Chem. Ed.* **1988**, *65*, 729; J. J. Cruywagen and J. B. B. Heyns, "Spectrophotometric Determination of the Thermodynamic Parameters for the First Two Protonation Reactions of Molybdate," *J. Chem. Ed.* **1989**, *66*, 861; and H. A. Rowe and M. Brown, "Practical Enzyme Kinetics," *J. Chem. Ed.* **1988**, *65*, 548.

21. Part of this experiment is taken from M. W. Olson and J. M. Crawford, *J. Chem. Ed.* **1975**, *52*, 546. For a related experiment, see S. F. Russo and A. Radcliffe, "Separations Utilizing Gel Filtration and Ion-Exchange Chromatography," *J. Chem. Ed.* **1991**, *68*, 168.

22. Bio-Rad Laboratories, 2000 Alfred Nobel Drive, Hercules, CA 94547.

23. For related experiments, see V. T. Remcho, H. M. McNair, and H. T. Rasmussen, "HPLC Method Development with the Photodiode Array Detector," *J. Chem. Ed.* **1992**, *69*, A117; C. H. Clapp, J. S. Swan, and J. L. Poechmann, "Identification of Amino Acids in Unknown Dipeptides" *J. Chem. Ed.* **1992**, *69*, A122. R. J. Kominar, "The Preparation and Testing of a Fused-Silica Gas Chromatography Capillary Column," *J. Chem. Ed.* **1991**, *68*, A249; D. E. Goodney, "Analysis of Vitamin C by High-Pressure Liquid Chromatography," *J. Chem. Ed.* **1987**, *64*, 187; R. C. Graham and J. K. Robertson, "Analysis of Trihalomethanes in Soft Drinks," *J. Chem. Ed.* **1988**, *65*, 735; G. W. Rice, "Determination of Impurities in Whiskey Using Internal Standard Techniques," *J. Chem. Ed.* **1987**, *64*, 1055; and S. L. Tackett, "Determination of Methanol in Gasoline by Gas Chromatography," *J. Chem. Ed.* **1987**, *64*, 1059.

Glossary

absolute uncertainty An expression of the margin of uncertainty associated with a measurement. Absolute error could also refer to the difference between a measured value and the "true" value.

absorbance, A, or optical density, OD Defined as $A = \log (P_0/P)$, where P_0 is the radiant power of light striking the sample on one side and P is the radiant power emerging from the other side.

absorptance, a Fraction of incident radiant power absorbed by a sample.

absorption Occurs when a substance is taken up *inside* another. See also **adsorption.**

absorption coefficient Light absorbed by a sample is attenuated at the rate $P_2/P_1 = e^{-\alpha b}$, where P_1 is the initial radiant power, P_2 is the power after traversing a pathlength b, and α is called the absorption coefficient.

absorption spectrum A graph of absorbance or transmittance of light versus wavelength, frequency, or wavenumber.

accuracy A measure of how close a measured value is to the "true" value.

acid A substance that increases the concentration of H^+ when added to water.

acid-base titration One in which the reaction between analyte and titrant is an acid-base reaction.

acid dissociation constant, K_a The equilibrium constant for the reaction of an acid, HA, with H_2O:

$$HA + H_2O \rightleftharpoons A^- + H_3O^+ \qquad K_a = \frac{\mathcal{A}_{A^-} \, \mathcal{A}_{H_3O^+}}{\mathcal{A}_{HA}}$$

acid error Occurs in strongly acidic solutions, where glass electrodes tend to indicate a value of pH that is too high.

acid wash Procedure in which glassware is soaked in 3–6 M HCl for >1 h (followed by rinsing well with distilled water and soaking in distilled water) to remove traces of cations adsorbed on the surface of the glass and to replace them with H^+.

acidic solution One in which the activity of H^+ is greater than the activity of OH^-.

acidity In natural waters, the quantity of carbonic acid and other dissolved acids that react with strong base when the pH of the sample is raised to 8.3. Expressed as mmol OH^- needed to raise the pH of 1 L to pH 8.3.

activation energy, E_a Energy needed for a process to overcome a barrier that otherwise prevents the process from occurring.

activity, $\mathcal{A}$ The value that replaces concentration in a thermodynamically correct equilibrium expression. The activity of X is given by $\mathcal{A}_X = [X]\gamma_X$, where γ_X is the activity coefficient and [X] is the concentration.

activity coefficient, γ The number by which the concentration must be multiplied to give activity.

adduct The product formed when a Lewis base combines with a Lewis acid.

adjusted retention time, t'_r In chromatography, this is given by $t'_r = t_r - t_m$, where t_r is the retention time of a solute and t_m is the time needed for mobile phase to travel the length of the column.

adsorption Occurs when a substance becomes attached to the *surface* of another substance. See also **absorption.**

adsorption chromatography A technique in which the solute equilibrates between the mobile phase and adsorption sites on the stationary phase.

adsorption indicator Used for precipitation titrations, it becomes attached to a precipitate and changes color when the surface charge of the precipitate changes sign at the equivalence point.

aerosol A suspension of very fine liquid or solid particles in air or gas. Examples include fog and smoke.

affinity chromatography A technique in which a particular solute is retained by a column by virtue of a specific interaction with a molecule covalently bound to the stationary phase.

aliquot Portion.

alkali flame detector Modified flame ionization detector that responds to N and P, which produce ions when they contact a Rb_2SO_4-containing glass bead in the flame.

alkalimetric titration With reference to EDTA titrations, this involves titration of the protons liberated from EDTA upon binding to a metal.

alkaline error Occurs when a glass pH electrode is placed in a strongly basic solution containing very little H^+ and a high concentration of Na^+. The electrode begins to respond to Na^+ as if it were H^+, so that the pH reading is lower than the actual pH.

alkalinity In natural waters, the quantity of base (mainly HCO_3^-, CO_3^{2-}, and OH^-) that reacts with strong acid when the pH of the sample is lowered to 4.5. Expressed as mmol H^+ needed to lower the pH of 1 L to pH 4.5.

allosteric interaction An effect at one part of a molecule caused by a chemical reaction or conformational change at another part of the molecule.

amalgam A solution of anything in mercury.

amine A compound with the general formula RNH_2, R_2NH, or R_3N, where R is any group of atoms.

amino acid One of 20 building blocks of proteins, having the general structure

$$\overset{\displaystyle R}{\underset{\displaystyle +H_3NCHCO_2^-}{\mid}}$$

where R is a different substituent for each acid.

ammonium ion *The* ammonium ion is NH_4^+. *An* ammonium ion is any ion of the type RNH_3^+, $R_2NH_2^+$, R_3NH^+, or R_4N^+, where R is an organic substituent.

ampere, A One ampere is the current that will produce a force of exactly 2×10^{-7} N/m when that current flows through two "infinitely" long, parallel conductors of negligible cross section, with a spacing of 1 m, in a vacuum.

amperometric titration One in which the end point is determined by monitoring the current passing between two electrodes immersed in the sample solution and maintained at a constant potential difference.

amperometry The measurement of electric current for analytical purposes.

amphiprotic molecule One that can act as both a proton donor and a proton acceptor. The intermediate species of polyprotic acids are amphiprotic.

analyte The substance being analyzed.

analytical concentration See **formal concentration.**

anhydrous Adjective describing a substance from which all water has been removed.

anion exchanger An ion exchanger with positively charged groups covalently attached to the support. It can reversibly bind anions.

anode The electrode at which oxidation occurs.

anodic depolarizer A molecule that is easily oxidized, thereby preventing the anode potential of an electrochemical cell from becoming too positive.

anodic wave In polarography, a flow of current due to oxidation of analyte.

anolyte The solution present in the anode chamber of an electrochemical cell.

antibody A protein manufactured by an organism to sequester and mark for destruction foreign molecules.

antigen A molecule that is foreign to an organism and that causes antibodies to be made.

antilogarithm The antilogarithm of a is b if $10^a = b$.

antireflection coating A coating placed on an optical component to diminish reflection. Ideally, the index of refraction of the coating should be $\sqrt{n_1 n_2}$, where n_1 is the refractive index of the surrounding medium and n_2 is the refractive index of the optical component. The thickness of the coating should be one-fourth of the wavelength of light inside the coating. Antireflection coatings are also made by layers that produce a gradation of refractive index.

apodization Mathematical treatment of an interferogram prior to computing the spectrum by calculating the Fourier transform. Apodization reduces the wiggly feet that result from the finite extent of the interferogram.

apparent mobility The constant of proportionality (μ_{app}) between the net speed (u_{net}) of an ion in solution and the applied electric field (E): $u_{net} = \mu_{app}E$. Apparent mobility is the sum of the electrophoretic and electroosmotic mobilities.

aprotic solvent One that cannot donate protons (hydrogen ions) in an acid-base reaction.

aqua regia 3:1 (vol/vol) mixture of concentrated (37 wt%) HCl and concentrated (70 wt%) HNO_3.

aquo ion The species $M(H_2O)_n^{m+}$, containing just the cation M and its tightly bound water ligands.

argentometric titration One using Ag^+ ion.

Arrhenius equation Empirical relation between the rate constant, k, for a chemical reaction and temperature, T, in kelvins: $k = Be^{-\Delta G^\ddagger/RT}$, where R is the gas constant, $\Delta G^\ddagger$ has the dimensions of energy and is taken as the free energy of activation for the chemical reaction, and B is a constant called the preexponential factor. This equation is more commonly written in the form $k = Ae^{-E_a/RT}$, where E_a is called the activation energy.

ashless filter paper   Specially treated paper that leaves a negligible residue after ignition. It is used for gravimetric analysis.

asymmetry potential   When the activity of analyte is the same on the inside and outside of an ion-selective electrode, there should be no voltage across the membrane. In fact, the two surfaces are never identical, and some voltage (called the asymmetry potential) is usually observed. The asymmetry potential changes with time and leads to electrode drift.

atmosphere, atm   One atm is defined as a pressure of 101 325 Pa. It is also equal to the pressure exerted by a column of Hg 760 mm in height at the earth's surface.

atomic absorption spectroscopy   A technique in which the absorption of light by free gaseous atoms in a flame or furnace is used to measure the concentration of atoms.

atomic emission spectroscopy   A technique in which the emission of light by thermally excited atoms in a flame or furnace is used to measure the concentration of atoms.

atomic fluorescence spectroscopy   A technique in which electronic transitions of atoms in a flame, furnace, or plasma are excited by light, and the fluorescence is observed at a right angle to the incident beam.

atomic weight   The number of grams of an element containing Avogadro's number of atoms.

atomization   The process in which a compound is decomposed into its atoms at high temperature.

autoprotolysis   The reaction of a neutral solvent, in which two molecules of the same species transfer a proton from one to the other, thereby producing ions; e.g., $CH_3OH + CH_3OH \rightleftharpoons CH_3OH_2^+ + CH_3O^-$.

auxiliary complexing agent   A species, such as ammonia, that is added to a solution to stabilize another species and keep that other species in solution. It binds loosely enough to be displaced by a titrant.

auxiliary electrode   The current-carrying partner of the working electrode in an electrolysis.

average   The sum of several values divided by the number of values.

azeotrope   The distillate produced by two liquids. It is of constant composition, containing both substances.

back titration   One in which an excess of standard reagent is added to react with analyte. Then the excess reagent is trtrated with a second reagent or with a standard solution of analyte.

background buffer   In capillary electrophoresis, the buffer in which separation is carried out.

background correction   In atomic spectroscopy, a means of distinguishing signal due to analyte from signal due to absorption, emission, or scattering by the flame, furnace, plasma, or sample matrix.

ball mill   Ceramic drum in which solid sample is ground to fine powder by tumbling with hard ceramic balls.

band gap   Energy separating the valence band and conduction band in a semiconductor.

band-pass filter   A filter that allows a band of wavelengths to pass through, while absorbing or reflecting other wavelengths.

bandwidth   Usually, the range of wavelengths or frequencies of an absorption or emission band at a height equal to half of the peak height. It also refers to the width of radiation emerging from the exit slit of a monochromator.

base   A substance that decreases the concentration of H^+ when added to water.

base "dissociation" constant, K_b   The equilibrium constant for the reaction of a base, B, with H_2O:

$$B + H_2O \rightleftharpoons BH^+ + OH^- \qquad K_b = \frac{\mathcal{A}_{BH^+}\mathcal{A}_{OH^-}}{\mathcal{A}_B}$$

base hydrolysis constant   Same as base "dissociation" constant, K_b.

basic solution   One in which the activity of OH^- is greater than the activity of H^+.

beam chopper   A rotating mirror that directs light alternately through the sample and reference cells of a double-beam spectrophotometer. In atomic absorption, periodic blocking of the beam allows a distinction to be made between light from the source and light from the flame.

beam chopping   A technique using a rotating *beam chopper* to modulate the signal in a spectrophotometer at a frequency at which noise is reduced.

beamsplitter   A partially reflective, partially transparent plate that reflects some light and transmits some light.

Beer's law   Relates the absorbance (A) of a sample to its concentration (c), pathlength (b), and molar absorptivity (ϵ): $A = \epsilon bc$.

biamperometric titration   An amperometric titration conducted with two polarizable electrodes held at a constant potential difference.

bilayer   Two-dimensional membrane structure that is formed by a surfactant and in which polar or ionic head groups are pointing outward and nonpolar tails are pointing inward.

biochemical oxygen demand, BOD   In a water sample, the quantity of dissolved oxygen consumed by microorganisms during a 5-day incubation is a sealed vessel at 20°C. Oxygen consumption is limited by organic nutrients, so BOD is a measure of pollutant concentration.

bipotentiometric titration   A potentiometric titration in which a constant current is passed between two polarizable electrodes immersed in the sample solution. An abrupt change in potential characterizes the end point.

blackbody An ideal surface that absorbs all photons striking it.

blackbody radiation Radiation emitted by a blackbody. The energy and spectral distribution of the emission depend only on the temperature of the blackbody.

blank titration One in which a solution containing all of the reagents except analyte is titrated. The volume of titrant needed in the blank titration should be subtracted from the volume needed to titrate unknown.

blocking Occurs when metal ion binds tightly to a metal ion indicator. A blocked indicator is unsuitable for a titration because no color change is observed at the end point.

bolometer An infrared detector whose electrical resistance changes when it is heated by infrared radiation.

Boltzmann distribution The relative population of two states at thermal equilibrium:

$$\frac{N_2}{N_1} = \frac{g_2}{g_1} e^{-(E_2 - E_1)/kT}$$

where N_i is the population of the state, g_i is the degeneracy of the state, E_i is the energy of the state, k is Boltzmann's constant, and T is temperature in kelvins; degeneracy refers to the number of states with the same energy.

bomb Sealed vessel for conducting high-temperature, high-pressure reactions.

bonded phase In HPLC, a stationary liquid phase covalently attached to the solid support.

boxcar truncation In Fourier transform spectroscopy, using the entire interferogram up to a predefined limit and then rejecting everything outside that limit.

Brewster window A flat optical window tilted at an angle such that light whose electric vector is polarized parallel to the plane of the window is 100% transmitted. Light polarized perpendicular to the window is partially reflected. It is used on the ends of a laser to produce light whose electric field oscillates perpendicular to the long axis of the laser.

Brønsted-Lowry acid A proton (hydrogen ion) donor.

Brønsted-Lowry base A proton (hydrogen ion) acceptor.

buffer A mixture of an acid and its conjugate base. A buffered solution is one that resists changes in pH when acids or bases are added.

buffer capacity, β, or **buffer intensity** A measure of the ability of a buffer to resist changes in pH. The larger the buffer capacity, the greater the resistance to pH change. The definition of buffer capacity is $\beta = dC_b/d\text{pH} = -dC_a/d\text{pH}$, where C_a and C_b are the number of moles of strong acid or base per liter needed to produce a unit change in pH.

bulk sample Material taken from lot being analyzed—usually chosen to be representative of the entire lot. Also called *gross sample*.

bulk solution Chemists' jargon referring to the main body of a solution. In electrochemistry, we distinguish properties of the bulk solution from properties that may be different in the immediate vicinity of an electrode.

buoyancy Occurs when an object is weighed in air and the observed mass is less than the true mass because the object has displaced an equal volume of air from the balance pan.

buret A calibrated glass tube with a stopcock at the bottom. Used to deliver known volumes of liquid.

calibration curve A graph showing the value of some property versus concentration of analyte. When the corresponding property of an unknown is measured, its concentration can be determined from the graph.

calomel electrode A common reference electrode based on the half-reaction $Hg_2Cl_2(s) + 2e^- \rightleftharpoons 2Hg(l) + 2Cl^-$.

candela, cd Luminous intensity, in a given direction, of a source that emits monochromatic radiation of frequency 540 THz and that has a radiant intensity of $\frac{1}{683}$ W/sr in that direction.

capacity factor, k′ In chromatography, the adjusted retention time for a peak divided by the time for the mobile phase to travel through the column. Capacity factor is also equal to the ratio of the time spent by the solute in the stationary phase to the time spent in the mobile phase. Also called *retention factor, capacity ratio,* and *partition ratio.*

capacity ratio See **capacity factor.**

capillary constant The quantity $m^{2/3}t^{1/6}$ characteristic of each dropping Hg electrode. The rate of flow is m (mg/s) and t is the drop interval (s). The capillary constant is proportional to the square root of the Hg height.

capillary electrophoresis Separation of a mixture into its components by using a strong electric field imposed between the two ends of a narrow capillary tube filled with electrolyte solution.

capillary gel electrophoresis A form of capillary electrophoresis in which the tube is filled with a polymer gel that serves as a sieve for macromolecules. The largest molecules migrate slowest through the gel.

capillary zone electrophoresis A form of capillary electrophoresis in which ionic solutes are separated because of differences in their electrophoretic mobility.

carboxylate anion The conjugate base (RCO_2^-) of a carboxylic acid.

carboxylic acid A molecule with the general structure RCO_2H, where R is any group of atoms.

carcinogen A cancer-causing agent.

carrier gas The mobile phase gas in gas chromatography.

catalytic wave One that results when the product of a polarographic reaction is rapidly regenerated by reaction with another species and the polarographic wave height increases.

cathode The electrode at which reduction occurs.

cathodic depolarizer A molecule that is easily reduced, thereby preventing the cathode potential of an electrochemical cell from becoming very low.

catholyte The solution present in the cathode chamber of an electrochemical cell.

cation exchanger An ion exchanger with negatively charged groups covalently attached to the support. It can reversibly bind cations.

character The part of a logarithm to the left of the decimal point.

charge balance A statement that the sum of all positive charge in solution equals the magnitude of the sum of all negative charge in solution.

charge coupled device An extremely sensitive detector in which light creates electrons and holes in a semiconductor material. The electrons are attracted to regions near positive electrodes, where the electrons are "stored" until they are ready to be counted. The number of electrons in each pixel (picture element) is proportional to the number of photons striking the pixel.

charge effect With respect to the strength of acids and bases, the repulsion between H^+ and a positive charge in the same molecule. It could also refer to the attraction between H^+ and a negative charge. A positive charge within a molecule increases the acidity, and a negative charge decreases the acidity.

charging current Electric current arising from charging or discharging of the electric double layer at the electrode-solution interface.

charring In a gravimetric analysis, the precipitate (and filter paper) are first dried gently. Then the filter paper is *charred* at intermediate temperature to destroy the paper without letting it inflame. Finally, the precipitate is ignited at high temperature to convert it to its analytical form.

chelate effect The observation that a single multidentate ligand forms metal complexes that are more stable than those formed by several individual ligands with the same ligand atoms.

chelating ligand A ligand that binds to a metal through more than one atom.

chemical coulometer A device that measures the yield of an electrolysis reaction in order to determine how much electricity has flowed through a circuit.

chemical interference In atomic spectroscopy, any chemical reaction that decreases the efficiency of atomization.

chemical oxygen demand, COD In a natural water or industrial effluent sample, the quantity of O_2 equivalent to the quantity of $K_2Cr_2O_7$ consumed by refluxing the sample with a standard dichromate-sulfuric acid solution containing Ag^+ catalyst. Because 1 mol of $K_2Cr_2O_7$ consumes $6e^-$ ($Cr^{6+} \rightarrow Cr^{3+}$), it is equivalent to 1.5 mol of O_2 ($O \rightarrow O^{2-}$).

chemiluminescence Emission of light by an excited-state product of a chemical reaction.

chromatogram A graph showing the concentration of solutes emerging from a chromatography column as a function of elution time or volume.

chromatograph A machine used to perform chromatography.

chromatography A technique in which molecules in a mobile phase are separated because of their different affinities for a stationary phase. The greater the affinity for the stationary phase, the longer the molecule is retained.

chromophore The part of a molecule responsible for absorption of light of a particular frequency.

chronoamperometry A technique in which the potential of a working electrode in an unstirred solution is varied rapidly while the current between the working and auxiliary electrodes is measured. Suppose that the analyte is reducible and that the potential of the working electrode is made more negative. Initially, no reduction occurs. At a certain potential, the analyte begins to be reduced and the current increases. As the potential becomes more negative, the current increases further until the concentration of analyte at the surface of the electrode is sufficiently depleted. Then the current decreases, even though the potential becomes more negative. The maximum current is proportional to the concentration of analyte in bulk solution.

chronopotentiometry A technique in which a constant current is forced to flow between two electrodes. The voltage remains fairly steady until the concentration of an electroactive species becomes depleted. Then the voltage changes rapidly as a new redox reaction assumes the burden of current flow. The elapsed time when the voltage suddenly changes is proportional to the concentration of the initial electroactive species in bulk solution.

Clark electrode One that measures the activity of dissolved oxygen by amperometry.

coagulation With respect to gravimetric analysis, small crystallites coming together to form larger crystals.

co-chromatography Simultaneous chromatography of known compounds with an unknown. If a known and an unknown have the same retention time on several columns, they are probably identical.

coherence The property of laser light that all waves are in phase with one another.

co-ion An ion with the same charge as the ion of interest.

cold trapping Splitless gas chromatography injection technique in which solute is condensed far below its boiling point in a narrow band at the start of the column.

collimated light Light in which all rays travel in parallel paths.

collimation The process of making light rays travel parallel to each other.

colloid A dissolved particle with a diameter in the approximate range 1–100 nm. It is too large to be considered one molecule but too small to simply precipitate.

combination electrode Consists of a glass electrode with a concentric reference electrode built on the same body.

combustion analysis A technique in which a sample is heated in an atmosphere of O_2 to oxidize it to CO_2 and H_2O, which are collected and weighed or measured by gas chromatography. Modifications permit the simultaneous analysis of N, S, and halogens.

common ion effect Occurs when a salt is dissolved in a solution already containing one of the ions of the salt. The salt is less soluble than it would be in a solution without that ion. An application of Le Châtelier's principle.

complex ion Historical name for any ion containing two or more ions or molecules that are each stable by themselves; e.g., $CuCl_2^-$ contains $Cu^+ + 2Cl^-$.

complexometric titration One in which the reaction between analyte and titrant involves complex formation.

composite sample A representative sample prepared from a heterogeneous material. If the material consists of distinct regions, the composite is made of portions of each region, with relative amounts proportional to the size of each region.

compound electrode An ion-selective electrode consisting of a conventional electrode surrounded by a barrier that is selectively permeable to the analyte of interest. Alternatively, the barrier region might convert external analyte into a different species, to which the inner electrode is sensitive.

concentration An expression of the quantity per unit volume or unit mass of a substance. Common measures of concentration are molarity (mol/L) and molality (mol/kg of solvent).

concentration cell A galvanic cell in which both half-reactions are the same, but the concentrations in each half-cell are not identical. The cell reaction increases the concentration of species in one half-cell and decreases the concentration in the other.

concentration polarization Occurs when an electrode reaction occurs so rapidly that the concentration of solute near the surface of the electrode is not the same as the concentration in bulk solution.

condenser current Electric current arising from charging or discharging of the electric double layer at the electrode-solution interface.

conditional formation constant The equilibrium con-

stant for formation of a complex under a particular stated set of conditions, such as pH, ionic strength, and concentration of auxiliary complexing species. Also called *effective formation constant.*

conduction band Energy levels containing conduction electrons in a semiconductor.

conduction electron An electron relatively free to move about within a solid and carry electric current. In a semiconductor, the energies of the conduction electrons are above those of the valence electrons that are localized in chemical bonds. The energy separating the valence and conduction bands is called the band gap.

conductivity, σ Proportionality constant between electric current density, J (A/m^2), and electric field, E (V/m): $J = \sigma E$. Units are $\Omega^{-1} \cdot m^{-1}$. Conductivity is the reciprocal of resistivity.

confidence interval The range of values within which there is a specified probability that the true value will occur.

conjugate acid-base pair An acid and a base that differ only through the gain or loss of a single proton.

constant-current electrolysis Electrolysis in which a constant current flows between the working and auxiliary electrodes. Because an ever-increasing voltage is required to keep the current flowing, this is the least selective form of electrolysis.

constant mass In gravimetric analysis, the product is heated and cooled to room temperature in a desiccator until successive weighings are "constant." There is no standard definition of "constant mass"; but for ordinary work, it is usually taken as about ±0.3 mg. Constancy is usually limited by the irreproducible regain of moisture picked up by the sample during cooling in the desiccator and during weighing.

constant-voltage electrolysis Electrolysis in which a constant voltage is maintained between the working and auxiliary electrodes. This is less selective than controlled-potential electrolysis because the potential of the working electrode becomes steadily more extreme as ohmic potential and overpotential change.

control chart A graph in which periodic observations of a process are recorded to see if the process is within specified control limits.

controlled-potential electrolysis A technique for selective reduction (or oxidation), in which the voltage between the working and reference electrodes is held constant.

convection Process in which solute is carried from one place to another by bulk motion of the solution.

cooperativity Interaction between two parts of a molecule such that an event at one part affects the behavior of the other. Consider a molecule, M, with two sites to which two identical species, S, can bind to form MS and MS_2. When binding at one site makes binding at the second site more favorable than in the absence of the first binding, there is *positive cooperativity.* When the first binding makes the second binding less favorable, there is *negative cooperativity*. In *noncooperative* binding, neither site affects the other.

coprecipitation   Occurs when a substance whose solubility is not exceeded precipitates along with one whose solubility is exceeded.

coulomb, C   The amount of charge per second that flows past any point in a circuit when the current is one ampere. There are 96 485.309 coulombs in a mole of electrons.

coulometric titration   One conducted with a constant current for a measured time.

coulometry   A technique in which the quantity of analyte is determined by measuring the number of coulombs needed for complete electrolysis.

countercurrent distribution   A technique in which a series of solvent extractions is used to separate solutes from one another.

counterion   An ion with a charge opposite to that of the ion of interest.

critical temperature   Temperature above which a fluid cannot be condensed to two phases (liquid and gas), no matter how great a pressure is applied.

crosslinking   The covalent linkage between different strands of a polymer.

crystallization   Process in which a substance comes out of solution slowly to form a solid with a regular arrangement of atoms.

cumulative formation constant, β_n   The equilibrium constant for a reaction of the type $M + nX \rightleftharpoons MX_n$. Synonym for *overall formation constant*.

current, I   The amount of charge flowing through a circuit per unit time.

current density   Electric current per unit area (A/m^2)

current-overpotential equation   Relation between current and overpotential in a redox process occurring at an electrode.

cuvet   A cell used to hold samples for spectrophotometric measurements.

cyclic voltammetry   A polarographic technique in which a triangular waveform is applied within a period of a few seconds. Both cathodic and anodic currents are observed for reversible reactions.

dead stop end point   The end point of a biamperometric titration.

Debye-Hückel equation   Gives the activity coefficient (γ) as a function of ionic strength (μ). The extended Debye-Hückel equation, applicable to ionic strengths up to about 0.1 M, is $\log\gamma = [-0.51z^2\sqrt{\mu}]/[1 + (\alpha\sqrt{\mu}/305)]$, where z is the ionic charge and α is the effective hydrated radius in picometers.

decant   To pour liquid off a solid or, perhaps, a denser liquid. The denser phase is left behind.

decomposition potential   In an electrolysis, that voltage at which rapid reaction first begins.

degrees of freedom   In statistics, this is the number of independent observations upon which a result is based.

deionized water   Water that has been passed through a cation exchanger (in the H^+ form) and an anion exchanger (in the OH^- form) to remove ions from the solution.

deliquescent substance   Like a hygroscopic substance, one that spontaneously picks up water from the air. It can eventually absorb so much water that the substance completely dissolves.

demasking   The removal of a masking agent from the species protected by the masking agent.

density   The mass per unit volume of a substance.

depolarizer   A molecule that is oxidized or reduced at a modest potential. It is added to an electrolytic cell to prevent the cathode or anode potential from becoming too extreme.

derivatization   Chemical alteration to attach a group to a molecule so that it can be detected conveniently. Alternatively, treatment can alter volatility or solubility.

desalting   The removal of salts (or any small molecules) from a solution of macromolecules. Gel filtration is conveniently used for desalting.

desiccant   A drying agent.

desiccator   A sealed chamber in which samples can be dried in the presence of a desiccant and/or by vacuum pumping.

detection limit   That concentration of an element which gives a signal equal to twice the peak-to-peak noise level of the baseline.

determinant   The value of the two-dimensional determinant $\begin{vmatrix} a & b \\ c & d \end{vmatrix}$ is the difference $ad - bc$.

determinate error   See **systematic error.**

deuterium arc lamp   Source of broadband ultraviolet radiation. An electric discharge (a spark) in deuterium gas causes D_2 molecules to dissociate and emit many wavelengths of radiation.

dialysis   A technique in which solutions are placed on either side of a semipermeable membrane that allows small molecules, but not large molecules, to cross. The small molecules in the two solutions diffuse across and equilibrate with each other. The large molecules are retained on their original side.

dielectric constant, ϵ   The electrostatic force, F, between two charged particles is given by $F = kq_1q_2/\epsilon r^2$, where k is a constant, q_1 and q_2 are the charges, r is the separation between particles, and ϵ is the dielectric constant of the medium. The higher the dielectric constant, the less force is exerted by one charged particle on another.

differential pulse polarography   A technique in which

current is measured before and at the end of pulses of potential superimposed on the ordinary waveform. It is more sensitive than ordinary polarography, and the signal closely approximates the derivative of a polarographic wave.

diffraction Occurs when electromagnetic radiation passes through slits with a spacing comparable to the wavelength. Interference of waves from adjacent slits produces a spectrum of radiation, with each wavelength at a different angle.

diffuse part of the double layer Region of solution near a charged surface in which there are excess counterions attracted to the charge. The thickness of this layer is 0.3–10 nm.

diffuse reflection Reflection of light in all directions by a rough surface.

diffusion Random motion of molecules in a liquid or gas (or, very slowly, in a solid).

diffusion coefficient, D Defined by Fick's first law of diffusion: $J = -D(dc/dx)$, where J is the rate at which molecules diffuse across a plane of unit area and dc/dx is the concentration gradient in the direction of diffusion.

diffusion current In polarography, the current observed when the rate of reaction is limited by the rate of diffusion of analyte to the electrode. See also **limiting current.**

diffusion layer Region near an electrode containing excess product or decreased reactant involved in the electrode reaction. The thickness of this layer can be hundreds of micrometers.

digestion The process in which a precipitate is left (usually warm) in the presence of mother liquor to promote particle recrystallization and growth. Purer, more easily filterable crystals result. Also used to describe any chemical treatment in which a substance is decomposed to transform the analyte into a form suitable for analysis.

dilution factor The factor (initial volume of reagent)/(total volume of solution) used to multiply the initial concentration of reagent to find the diluted concentration.

dimer A molecule made from two identical units.

diode A semiconductor device consisting of a p-n junction through which current can pass in only one direction. Current flows when the n-type material is made negative and the p-type material is made positive. A voltage sufficient to overcome the activation energy for carrier movement must be supplied before any current flows. For silicon diodes, this is approximately 0.6 V. If a sufficiently large voltage, called the breakdown voltage, is applied in the reverse direction, current will flow in the wrong direction through the diode.

diprotic acid One that can donate two protons.

direct current polarography The classical form of

polarography, in which a linear voltage ramp is applied to the working electrode.

direct titration One in which the analyte is treated with titrant, and the volume of titrant required for complete reaction is measured.

dispersion A measure of the ability of a monochromator to separate wavelengths differing by $\Delta\lambda$ through the angle $\Delta\phi$. The greater the dispersion, the greater the angle separating two closely spaced wavelengths. For a prism, dispersion refers to the rate of change of refractive index with wavelength, $dn/d\lambda$.

displacement titration An EDTA titration procedure in which analyte is treated with excess $MgEDTA^{2-}$ to displace Mg^{2+}: $M^{n+} + MgEDTA^{2-} \rightleftharpoons MEDTA^{n-4} + Mg^{2+}$. The liberated Mg^{2+} is then titrated with EDTA. This procedure is useful if there is not a suitable indicator for direct titration of M^{n+}.

disproportionation A reaction in which an element in one oxidation state gives products containing that element in both higher and lower oxidation states; e.g., $Cu^+ \rightleftharpoons Cu^{2+} + Cu(s)$.

distribution coefficient, D Describes the distribution of a solute partitioned between two phases. The distribution coefficient is defined as the total concentration of all forms of solute in phase 2 divided by the total concentration in phase 1.

Donnan equilibrium The phenomenon that ions of the same charge as those fixed on an exchange resin are repelled from the resin. Thus, anions do not readily penetrate a cation-exchange resin, and cations are repelled from an anion-exchange resin.

dopant When small amounts of substance B are added to substance A, we call B a dopant and say that A is doped with B. Doping is done to alter the properties of A.

Doppler effect The phenomenon that a molecule moving toward a source of radiation experiences a higher frequency than one moving away from the source.

double layer Heterogeneous region consisting of a charged surface and oppositely charged region of solution adjacent to the surface.

dropping-mercury electrode One that delivers fresh drops of Hg to a polarographic cell.

dry ashing Oxidation of organic matter with O_2 at high temperature to leave behind inorganic components for analysis.

dynamic range The range of analyte concentration over which a change in concentration gives a change in detector response.

$E°$ The standard reduction potential.

$E°'$ The effective standard reduction potential at pH 7 (or at some other specified conditions).

EDTA (ethylenediaminetetraacetic acid) The compound $(HO_2CCH_2)_2NCH_2CH_2N(CH_2CO_2H)_2$—

the most widely used reagent for complexometric titrations. It forms 1:1 complexes with virtually all cations with a charge of 2 or more.

effective formation constant   The equilibrium constant for formation of a complex under a particular stated set of conditions, such as pH, ionic strength, and concentration of auxiliary complexing species. Also called *conditional formation constant.*

effervescence   Rapid release of gas with bubbling and hissing.

efflorescence   The property in which the outer surface or entire mass of a substance turns to powder from loss of water of crystallization.

effluent   See **eluate.**

einstein   A mole of photons.

electric discharge emission spectroscopy   A technique in which atomization and excitation are stimulated by an electric arc, a spark, or microwave radiation.

electric double layer   Region comprising the charged surface of an electrode or a particle, plus the oppositely charged region of solution adjacent to the surface.

electric potential   The electric potential (in volts) at a point is the energy (in joules) needed to bring one coulomb of positive charge from infinity to that point. The potential difference between two points is the energy needed to transport one coulomb of positive charge from the negative point to the positive point.

electroactive species   Any species that can be oxidized or reduced at an electrode.

electrocapillary maximum   The potential at which the net charge on a mercury drop from a dropping-mercury electrode is zero (and the surface tension of the drop is maximal).

electrochemical detector   Liquid chromatography detector that measures current when an electroactive solute emerges from the column and passes over a working electrode held at a fixed potential with respect to a reference electrode. Also called *amperometric detector.*

electrochemistry   Use of electrical measurements on a chemical system for analytical purposes. Also refers to use of electricity to drive a chemical reaction or use of a chemical reaction to produce electricity.

electrode   A device at which or through which electrons flow into or out of chemical species involved in a redox reaction.

electroendosmosis   A synonym for **electroosmosis.**

electrogravimetric analysis   A technique in which the mass of an electrolytic deposit is used to quantify the analyte.

electrokinetic injection   In capillary electrophoresis, the use of an electric field to inject sample into the capillary. Because different species have different mobilities, the injected sample does not have the same composition as the original sample.

electrolysis   The process in which the passage of electric current causes a chemical reaction to occur.

electrolyte   A substance that produces ions when dissolved.

electromagnetic spectrum   The spectrum of all electromagnetic radiation (visible light, radio waves, X-rays, etc.).

electron capture detector   Gas chromatography detector in which analytes with high electron affinity capture electrons from radioactively ionized N_2 and decrease the ion current.

electronic balance   A balance that uses an electromagnetic servomotor to balance the load on the pan. The mass of the load is proportional to the current needed to balance it.

electronic transition   One in which an electron is promoted from one energy level to another.

electroosmosis   Bulk flow of fluid in a capillary tube induced by an electric field. Mobile ions in the diffuse part of the double layer at the wall of the capillary serve as the "pump."

electroosmotic flow   Uniform, pluglike flow of fluid in a capillary tube under the influence of an electric field. The greater the charge on the wall of the capillary, the greater the number of counterions in the double layer and the stronger is the electroosmotic flow.

electroosmotic mobility, μ_{eo}   The constant of proportionality between the electroosmotic speed (u_{eo}) of a fluid in a capillary and the applied electric field (E): $u_{eo} = \mu_{eo}E$.

electropherogram   A graph of detector response versus time for electrophoresis.

electrophoresis   Migration of ions in solution in an electric field. Cations move toward the cathode and anions move toward the anode.

electrophoretic mobility, μ_{ep}   The constant of proportionality between the electrophoretic speed (u_{ep}) of an ion in solution and the applied electric field (E): $u_{ep} = \mu_{ep}E$.

eluate or **effluent**   What comes out of a chromatography column.

eluent   The solvent applied to the beginning of a chromatography column.

eluent strength, $\epsilon°$ A measure of the absorption energy of a solvent on the stationary phase in chromatography. The greater the eluent strength, the more rapidly will the solvent elute solutes from the column. Also called *solvent strength.*

eluotropic series   Ranks solvents according to their ability to displace solutes from the stationary phase in adsorption chromatography.

elution   The process of passing a liquid or a gas through a chromatography column.

emission spectrum   A graph of luminescence intensity versus luminescence wavelength (or frequency or wavenumber), using a fixed excitation wavelength.

emissivity A quotient given by the radiant emission from a real object divided by the radiant emission of a blackbody at the same temperature.

enantiomers Mirror image isomers.

end point The point in a titration at which there is a sudden change in a physical property, such as indicator color, pH, conductivity, or absorbance. Used as a measure of the equivalence point.

endergonic reaction One for which ΔG is positive; it is not spontaneous.

endothermic reaction One for which ΔH is positive; heat must be supplied to reactants for them to react.

enthalpy change, ΔH The heat absorbed or released when a reaction occurs at constant pressure.

enthalpy of hydration The heat liberated when a gaseous species is transferred to water.

entropy A measure of the "disorder" of a substance.

enzyme A protein that catalyzes a chemical reaction.

equilibrium The state in which the forward and reverse rates of all reactions are equal, so that the concentrations of all species remain constant.

equilibrium constant, K For the reaction $a\text{A} + b\text{B} \rightleftharpoons c\text{C} + d\text{D}$, $K = \mathcal{A}_\text{C}^c \mathcal{A}_\text{D}^d / \mathcal{A}_\text{A}^a \mathcal{A}_\text{B}^b$, where $\mathcal{A}_i$ is the activity of the ith species.

equimolar mixture of compounds One that contains an equal number of moles of each compound.

equivalence point The point in a titration at which the quantity of titrant is exactly sufficient for stoichiometric reaction with the analyte.

equivalent For a redox reaction, the amount of reagent that can donate or accept one mole of electrons. For an acid-base reaction, the amount of reagent that can donate or accept one mole of protons.

equivalent weight The mass of substance containing one equivalent.

excitation spectrum A graph of luminescence (measured at a fixed wavelength) versus excitation frequency or wavelength. It closely corresponds to an absorption spectrum because the luminescence is generally proportional to the absorbance.

excited state Any state of an atom or a molecule having more than the minimum possible energy.

exergonic reaction One for which ΔG is negative; it is spontaneous.

exitance, M Power per unit area radiating from the surface of an object.

exothermic reaction One for which ΔH is negative; heat is liberated when products are formed.

extended Debye-Hückel equation See **Debye-Hückel equation.**

extensive property A property of a system or chemical reaction, such as entropy, that depends on the amount of matter in the system; e.g., ΔG, which is twice as large when two moles of product are formed as it is when one mole is formed. See also **intensive property.**

extinction coefficient See **molar absorptivity.**

extraction The process in which a solute is allowed to equilibrate between two phases, usually for the purpose of separating solutes from one another.

extrapolation The estimation of a value that lies beyond the range of measured data.

F test For two variances, s_1^2 and s_2^2 (with s_1 chosen to be the larger of the two), the statistic F is defined as $F = \dfrac{s_1^2}{s_2^2}$. To decide whether s_1 is significantly greater than s_2, we compare F with the critical values in a table based on a certain confidence level. If the calculated value of F is greater than the value in the table, the difference is significant.

Fajans titration A precipitation titration in which the end point is signaled by adsorption of a colored indicator on the precipitate.

farad Unit of electrical capacitance. One farad of capacitance will store one coulomb of charge in a potential difference of one volt.

faradaic current That component of current in an electrochemical cell due to oxidation and reduction reactions.

Faraday constant $9.648\,530\,9 \times 10^4$ C/mol of charge.

Faraday's laws These two laws state that the extent of an electrochemical reaction is directly proportional to the quantity of electricity that has passed through the cell. The mass of substance that reacts is proportional to its formula weight and inversely proportional to the number of electrons required in its half-reaction.

ferroelectric material A solid with a permanent electric polarization (dipole) in the absence of an external electric field. The polarization results from alignment of molecules within the solid.

field effect transistor A semiconductor device in which the electric field between the gate and base governs the flow of current between the source and drain.

filtrate The liquid that passes through a filter.

fines The smallest particles of stationary phase used for chromatography. It is desirable to remove the fines before packing a column because they plug the column and retard solvent flow.

Fischer titration See **Karl Fischer titration.**

flame ionization detector A gas chromatography detector in which solute is burned in a H_2/air flame to produce CHO^+ ions. The current carried through the flame by these ions is proportional to the concentration of susceptible species in the eluate.

flame photometer A device that uses flame atomic emission and a filter photometer to quantify Li, Na, K, and Ca in liquid samples. It is widely used in clinical laboratories.

flame photometric detector Gas chromatography detector that measures emission from S and P in H_2-O_2 flame.

flow adaptor An adjustable plungerlike device that may be used on either side of a chromatographic bed to support the bed and to minimize the dead space through which liquid can flow outside of the column bed.

flow injection analysis Analytical technique in which sample is injected into a flowing stream. Other reagents can also be injected into the stream, and some type of measurement, such as absorption of light, is made downstream. Because the sample spreads out as it travels, different concentrations of sample are available in different parts of the band when it reaches the detector.

fluorescence The process in which a molecule emits a photon 10^{-8} to 10^{-4} s after absorbing a photon. It results from a transition between states of the same spin multiplicity (e.g., singlet → singlet).

flux In sample preparation, flux is the agent used as the medium for a fusion. In transport phenomena, flux is the quantity of whatever you like crossing each unit area in one unit of time. For example, the flux of diffusing molecules could be mol/(m$^2 \cdot$ s). Heat flux would be J/(m$^2 \cdot$ s).

formal concentration or **analytical concentration** The molarity of a substance if it did not change its chemical form upon being dissolved. It represents the total number of moles of substance dissolved in a liter of solution, regardless of any reactions that take place when the solute is dissolved.

formality, F Same as formal concentration.

formal potential The potential of a half-reaction (relative to a standard hydrogen electrode) when the formal concentrations of reactants and products are unity. Any other conditions (such as pH, ionic strength, and concentrations of ligands) must also be specified.

formation constant or **stability constant** The equilibrium constant for the reaction of a metal with its ligands to form a metal-ligand complex.

formula weight, FW The mass containing one mole of the indicated chemical formula of a substance. For example, the formula weight of $CuSO_4 \cdot 5H_2O$ is the sum of the masses of copper, sulfate, and five water molecules.

Fourier analysis Process of decomposing a function into an infinite series of sine and cosine terms. Because each term represents a certain frequency or wavelength, Fourier analysis decomposes a function into its component frequencies or wavelengths.

Fourier series Infinite sum of sine and cosine terms that add to give a particular function in a particular interval.

fraction of association, α For the reaction of a base (B) with H_2O, the fraction of base in the form BH^+.

fraction of dissociation, α For the dissociation of an acid (HA), the fraction of acid in the form A^-.

frequency The number of oscillations of a wave per second.

friction coefficient A molecule migrating through a solution is retarded by a force that is proportional to its speed. The constant of proportionality is the friction coefficient.

fugacity The activity of a gas. The activity coefficient for a gas is called the **fugacity coefficient.**

fusion The process in which an otherwise insoluble substance is dissolved in a molten salt such as Na_2CO_3, Na_2O_2, or KOH. Once the substance has dissolved, the melt is cooled, dissolved in aqueous solution, and analyzed.

galvanic cell One that produces electricity by means of a spontaneous chemical reaction.

gas chromatography A form of chromatography in which the mobile phase is a gas.

gathering A process in which a trace constituent of a solution is intentionally coprecipitated with a major constituent.

Gaussian distribution Theoretical bell-shaped distribution of measurements when all error is random. The center of the curve is the mean (μ) and the width is characterized by the standard deviation (σ). A *normalized* Gaussian distribution, also called the *normal error curve,* has an area of unity and is given by

$$y = \frac{1}{\sigma\sqrt{2\pi}} e^{-(x-\mu)^2/2\sigma^2}$$

gel Chromatographic stationary-phase particles, such as Sephadex or polyacrylamide, which are soft and pliable.

gel filtration or **gel permeation chromatography** See **molecular exclusion chromatography.**

geometric mean For a series of n measurements with the values x_i, $[\Pi_i(x_i)]^{1/n}$.

Gibbs free energy, G The change in Gibbs free energy (ΔG) for any process at constant temperature is related to the change in enthalpy (ΔH) and entropy (ΔS) by the equation $\Delta G = \Delta H - T \Delta S$, where T is temperature in kelvins. A process is spontaneous (thermodynamically favorable) if ΔG is negative.

glass electrode One that has a thin glass membrane across which a pH-dependent voltage develops. The voltage (and hence pH) is measured by a pair of reference electrodes on either side of the membrane.

glassy carbon electrode An inert carbon electrode, impermeable to gas, and especially well suited as an anode. The isotropic (same in all directions) structure is thought to consist of tangled ribbons of graphitelike sheets, with some crosslinking.

globar An infrared radiation source made of a ceramic such as silicon carbide heated by passage of electricity.

Golay cell Infrared detector that uses expansion of a gas in a blackened chamber to deform a flexible mirror. Deflection of a beam of light by the mirror changes the power impinging on a phototube.

Gooch crucible A short, cup-shaped container with holes at the bottom, used for filtration and ignition of precipitates. For ignition, the crucible is made of porcelain or platinum and lined with a mat of purified asbestos to retain the precipitate. For precipitates that do not need ignition, the crucible is made of glass and has a porous glass disk instead of holes at the bottom.

gradient elution Chromatography in which the composition of the mobile phase is progressively changed to increase the eluent strength of the solvent.

graduated cylinder or **graduate** A tube with volume calibrations along its length.

gram-atom The amount of an element containing Avogadro's number of atoms; it is the same as a mole of the element.

Gran plot A graph such as the plot of $V_b \cdot 10^{-pH}$ versus volume used to find the end point of a titration.

graphite furnace A hollow graphite rod that can be heated electrically to about $2\,500$ K to decompose and atomize a sample for atomic spectroscopy.

grating Either a reflective or a transmitting surface etched with closely spaced lines; used to disperse light into its component wavelengths.

gravimetric analysis Any analytical method that relies on measuring the mass of a substance (such as a precipitate) to complete the analysis.

gross sample Same as **bulk sample.**

ground state The state of an atom or a molecule with the minimum possible energy.

guard column In HPLC, a short column packed with the same material as the main column, placed between the injector and main column. The guard column removes impurities that might irreversibly bind to and degrade the main column. Also called *precolumn.*

half-height Half of the maximum amplitude of a signal.

half-reaction Any redox reaction can be conceptually broken into two half-reactions, one involving only oxidation and one involving only reduction.

half-wave potential The potential at the midpoint of the rise in the current of a polarographic wave.

half-width The width of a signal at its half-height.

Hall process Electrolytic production of aluminum metal from a molten solution of Al_2O_3 and cryolite (Na_3AlF_6).

Hammett acidity function The acidity of a solvent that protonates the weak base, B, is called the Hammett acidity function (H_0) and is given by

$$H_0 = pK_a \text{ (for BH}^+) + \log \frac{[B]}{[BH^+]}$$

For dilute aqueous solutions, H_0 approaches pH.

hanging-drop electrode One with a stationary drop of Hg that is used for stripping analysis.

hardness Total concentration of alkaline earth ions in natural water. Expressed as mg $CaCO_3$ equivalent to the same number of moles of alkaline earth cations per liter of water.

Henderson-Hasselbalch equation A logarithmic rearranged form of the acid dissociation equilibrium equation:

$$pH = pK_a + \log \frac{[A^-]}{[HA]}$$

Henry's law The partial pressure of a gas in equilibrium with gas dissolved in a solution is proportional to the concentration of dissolved gas: $P = k[\text{dissolved gas}]$. The constant k is called the *Henry's law constant.* It is a function of the gas, the liquid, and the temperature.

hertz, Hz The unit of frequency, s^{-1}.

heterogeneous Not uniform throughout.

HETP (height equivalent to a theoretical plate) The length of a chromatography column divided by the number of theoretical plates in the column.

hexadentate ligand One that binds to a metal atom through six ligand atoms.

high-performance liquid chromatography (HPLC) A chromatographic technique using very small stationary-phase particles and high pressure to force solvent through the column.

hole Absence of an electron in a semiconductor. When a neighboring electron moves into the hole, a new hole is created where the electron came from. By this means, a hole can move through a solid just as an electron can move through a solid.

hollow-cathode lamp One that emits sharp atomic lines characteristic of the element from which the cathode is made.

homogeneous Having the same composition everywhere.

homogeneous precipitation A technique in which a precipitating agent is generated slowly by a reaction in homogeneous solution, effecting a slow crystallization instead of a rapid precipitation of product.

hydrated radius, α The effective size of an ion or a molecule plus its associated water molecules in solution.

hydrodynamic flow Motion of fluid through a tube, driven by a pressure difference. Hydrodynamic flow is usually laminar, in which there is a parabolic profile of velocity vectors, with the highest velocity at the center of the stream and zero velocity at the walls.

hydrodynamic injection In capillary electrophoresis,

the use of a pressure difference between the two ends of the capillary to inject sample into the capillary. This can be done by applying pressure on one end, applying suction on one end, or by siphoning.

hydrodynamic radius   The effective radius of a molecule migrating through a fluid. It is defined by the Stokes equation, in which the friction coefficient is $6\pi\eta r$, where η is the viscosity of the fluid and r is the hydrodynamic radius of the molecule.

hydrolysis   "Reaction with water." The reaction $B + H_2O \rightleftharpoons BH^+ + OH^-$ is often called hydrolysis of a base.

hydronium ion, H_3O^+   What we really mean when we write H^+ (*aq*).

hydrophilic substance   One that is soluble in water or attracts water to its surface.

hydrophobic interaction chromatography   Chromatographic separation based on the interaction of a hydrophobic solute with a hydrophobic stationary phase.

hydrophobic substance   One that is insoluble in water or repels water from its surface.

hygroscopic substance   One that readily picks up water from the atmosphere.

ignition   The heating to high temperature of some gravimetric precipitates to convert them to a known, constant composition that can be weighed.

immunoassay   An analytical measurement using antibodies.

inclusion   An impurity that occupies lattice sites in a crystal.

indeterminate error   See **random error.**

indicator   A compound having a physical property (usually color) that changes abruptly near the equivalence point of a chemical reaction.

indicator electrode   One that develops a potential whose magnitude depends on the activity of one or more species in contact with the electrode.

indicator error   The difference between the indicator end point of a titration and the true equivalence point.

indirect detection   Chromatographic detection based on the *absence* of signal from a background species. For example, in ion chromatography a light-absorbing ionic species can be added to the eluent. Nonabsorbing analyte replaces an equivalent amount of light-absorbing eluent when analyte emerges from the column, thereby decreasing the absorbance of eluate.

indirect titration   One that is used when the analyte cannot be directly titrated. For example, analyte A may be precipitated with excess reagent R. The product is filtered, and the excess R washed away. Then AR is dissolved in a new solution, and R can be titrated.

inductive effect   The attraction of electrons by an electronegative element through the sigma-bonding framework of a molecule.

inductively coupled plasma   A high-temperature plasma that derives its energy from an oscillating radio-frequency field. It is used to atomize a sample for atomic emission spectroscopy.

inflection point   One at which the derivative of the slope is zero: $d^2y/dx^2 = 0$. That is, the slope reaches a maximum or minimum value.

inner Helmholtz plane   Imaginary plane going through the centers of ions or molecules specifically adsorbed on an electrode.

inorganic carbon   In a natural water or industrial effluent sample, the quantity of dissolved carbonate and bicarbonate.

intensity   See **radiant power.**

intensive property   A property of a system or chemical reaction that does not depend on the amount of matter in the system; e.g., temperature and electric potential. See also **extensive property.**

intercept   For a straight line whose equation is $y = mx + b$, the value of b is the intercept. It is the value of y when $x = 0$.

interference   The effect when the presence of one substance changes the signal in the analysis of another substance.

interference filter   A filter that transmits a particular band of wavelengths and reflects others. Transmitted light interferes constructively within the filter, whereas light that is reflected interferes destructively.

interferogram   A graph of light intensity versus retardation (or time) for the radiation emerging from an interferometer.

interferometer   A device with a beamsplitter, fixed mirror, and moving mirror that breaks input light into two beams that interfere with each other. The degree of interference depends on the difference in pathlength of the two beams.

internal conversion   A radiationless, isoenergetic, electronic transition between states of the same electron-spin multiplicity.

internal standard   A known quantity of a compound added to a solution containing an unknown quantity of analyte. The concentration of analyte is then measured relative to that of the internal standard.

interpolation   The estimation of the value of a quantity that lies between two known values.

intersystem crossing   A radiationless, isoenergetic, electronic transition between states of different electron-spin multiplicity.

iodimetry   The use of triiodide (or iodine) as a titrant.

iodometry   A technique in which an oxidant is treated with I^- to produce I_3^-, which is then titrated (usually with thiosulfate).

ion chromatography   HPLC ion-exchange separation of ions. See also **suppressed-ion chromatography** and **single-column ion chromatography.**

ion-exchange chromatography   A technique in which

solute ions are retained by oppositely charged sites in the stationary phase.

ion-exchange equilibrium An equilibrium involving replacement of a cation by another cation or replacement of an anion by another anion. Usually the ions in these reactions are bound by electrostatic forces.

ion-exchange membrane Membrane containing covalently bound charged groups. Oppositely charged ions in solution penetrate the membrane freely, but similarly charged ions tend to be excluded from the membrane by the bound charges.

ion-exclusion chromatography A technique in which electrolytes are separated from nonelectrolytes by means of an ion-exchange resin.

ionic atmosphere The region of solution around an ion or a charged particle. It contains an excess of oppositely charged ions.

ionic radius The effective size of an ion in a crystal.

ionic strength, μ Given by $\mu = \frac{1}{2} \sum_i c_i z_i^2$, where c_i is the concentration of the ith ion in solution and z_i is the charge on that ion. The sum extends over all ions in solution, including the ions whose activity coefficients are being calculated.

ionization interference In atomic spectroscopy, a lowering of signal intensity as a result of ionization of analyte atoms.

ionization suppressor An element used in atomic spectroscopy to decrease the extent of ionization of the analyte.

ionophore A molecule with a hydrophobic outside and a polar inside that can engulf an ion and carry the ion through a hydrophobic phase (such as a cell membrane).

ion pair A closely associated anion and cation, held together by electrostatic attraction. In solvents less polar than water, ions are usually found as ion pairs.

ion-pair chromatography Separation of ions on reverse-phase HPLC column by adding to the eluent a hydrophobic counterion that pairs with analyte ion and is attracted to stationary phase.

ion-selective electrode One whose potential is selectively dependent on the concentration of one particular ion in solution.

isocratic elution Chromatography using a single solvent for the mobile phase.

isoelectric focusing A technique in which a sample containing polyprotic molecules is subjected to a strong electric field in a medium with a pH gradient. Each species migrates until it reaches the region of its isoelectric pH. In that region, the molecule has no net charge, ceases to migrate, and remains focused in a narrow band.

isoelectric pH or **isoelectric point** That pH at which the average charge of a polyprotic species is zero.

isoionic pH or **isoionic point** The pH of a pure solution of a neutral, polyprotic molecule. The only ions present are H^+, OH^-, and those derived from the polyprotic species.

isosbestic point A wavelength at which the absorbance spectra of two species cross each other. The appearance of isosbestic points in a solution in which a chemical reaction is occurring is evidence that there are only two components present, with a constant total concentration.

Job's method See **method of continuous variation.**

Jones reductor A column packed with zinc amalgam. An oxidized analyte is passed through to reduce the analyte, which is then titrated with an oxidizing agent.

joule, J SI unit of energy. One joule is expended when a force of 1 N acts over a distance of 1 m. This energy is equivalent to that required to raise 102 g (about $\frac{1}{4}$ pound) by 1 m.

Joule heating Heat produced in an electric curcuit by the flow of electricity. Power $(J/s) = I^2 R$, where I is the current (A) and R is the resistance (ohms).

junction potential An electric potential that exists at the junction between two different electrolyte solutions or substances. It arises in solutions as a result of unequal rates of diffusion of different ions.

Karl Fischer titration A sensitive technique for determining water, based on the reaction of H_2O with amine I_2, SO_2, and an alcohol.

kelvin, K The absolute unit of temperature defined such that the temperature of water at its triple point (where water, ice, and water vapor are at equilibrium) is 273.16 K and the absolute zero of temperature is 0 K.

Kieselguhr The German term for diatomaceous earth, which is used as a solid support in gas chromatography.

kilogram, kg The mass of a particular Pt-Ir cylinder kept at the International Bureau of Weights and Measures, Sèvres, France.

kinetic current A polarographic wave that is affected by the rate of a chemical reaction involving the analyte and some species in the solution.

kinetic polarization Occurs whenever an overpotential is associated with an electrode process.

Kjeldahl nitrogen analysis Procedure for the analysis of nitrogen in organic compounds. The compound is digested with boiling H_2SO_4 to convert nitrogen to NH_4^+, which is treated with base and distilled as NH_3 into a standard acid solution. The number of moles of acid consumed equals the number of moles of NH_3 liberated from the compound.

Kovats index In chromatography, a retention index comparing the adjusted retention time of an unknown with those of linear alkanes eluted before and after the unknown.

laboratory sample   Portion of bulk sample taken to the lab for analysis. Must have the same composition as the bulk sample.

laminar flow   Motion with a parabolic velocity profile of fluid through a tube. Motion is fastest at the center, and the flow is zero at the walls.

laser   Source of intense, coherent monochromatic radiation. Light is produced by stimulated emission of radiation from a medium in which an excited state has been pumped to a high population. Coherence means that all light exiting the laser has the same phase.

Latimer diagram   One that shows the reduction potentials connecting a series of species containing an element in different oxidation states.

law of mass action   States that for the chemical reaction $aA + bB \rightleftharpoons cC + dD$, the condition at equilibrium is $K = \mathcal{A}_C^c \mathcal{A}_D^d / \mathcal{A}_A^a \mathcal{A}_B^b$, where $\mathcal{A}_i$ is the activity of the ith species. The law is usually used in approximate form, in which the activities are replaced by concentrations.

Le Châtelier's principle   States that if a system at equilibrium is disturbed, the direction in which it proceeds back to equilibrium is such that the disturbance is partially offset.

least squares   Process of fitting a mathematical function to a set of measured points by minimizing the sum of the squares of the distances from the points to the curve.

leveling effect   The strongest acid that can exist in solution is the protonated form of the solvent. Any acid stronger than this will donate its proton to the solvent and be leveled to the acid strength of the protonated solvent. Similarly, the strongest base that can exist in a solvent is the deprotonated form of the solvent.

Lewis acid   One that can form a chemical bond by sharing a pair of electrons donated by another species.

Lewis base   One that can form a chemical bond by sharing a pair of its electrons with another species.

ligand   An atom or a group attached to a central atom in a molecule. The term is often used to mean any group attached to anything else of interest.

limiting current   In a polarographic experiment, the current that is reached at the plateau of a polarographic wave. See also **diffusion current.**

linear flow rate   In chromatography, the distance per unit time traveled by the mobile phase.

linear interpolation   A form of interpolation in which it is assumed that the variation in some quantity is linear. For example, to find the value of b when $a = 32.4$ in the table below

a:	32	32.4	33
b:	12.85	x	17.96

you can set up the proportion

$$\frac{32.4 - 32}{33 - 32} = \frac{x - 12.85}{17.96 - 12.85}$$

which gives $x = 14.89$.

linear range   The concentration range over which the change in detector response is proportional to the change in analyte concentration.

linear response   The case in which the analytical signal is directly proportional to the concentration of analyte.

linear voltage ramp   The linearly increasing potential that is applied to the working electrode in polarography.

lipid bilayer   Double layer formed by molecules containing hydrophilic headgroup and hydrophobic tail. The tails of the two layers associate with each other, while the headgroups face the aqueous solvent.

liquid-based ion-selective electrode   One that has a hydrophobic membrane separating an inner reference electrode from the analyte solution. The membrane is saturated with a liquid ion exchanger dissolved in a nonpolar solvent. The ion-exchange equilibrium of analyte between the liquid ion exchanger and the aqueous solution gives rise to the electrode potential.

liter, L   Defined in 1964 as exactly $1\,000\ cm^3$.

Littrow prism   A prism with a reflecting back surface.

logarithm   The base 10 logarithm of n is a if $10^a = n$ (which means $\log n = a$). The natural logarithm of n is a if $e^a = n$ (which means $\ln n = a$). The number e (= $2.718\,28\ldots$) is called the base of the natural logarithm.

longitudinal diffusion   Diffusion of solute molecules parallel to the direction of travel through a chromatography column.

Lorentzian   Refers to an analytical function commonly used to describe the shape of a spectroscopic band: amplitude = $A_{max}\Gamma^2 / [\Gamma^2 + (\nu - \nu_0)^2]$, where ν is the frequency (or wavenumber), ν_0 is the frequency (or wavenumber) of the center of the band, Γ is half the width at half-height, and A_{max} is the maximum amplitude.

lot   Entire material that is to be analyzed. Examples are a bottle of reagent, a lake, or a truckload of gravel.

luminescence   Any emission of light by a molecule.

L'vov platform   Platform on which sample is placed in a graphite-rod furnace for atomic spectroscopy to prevent sample vaporization before the walls reach constant temperature.

mantissa   The part of a logarithm to the right of the decimal point.

Mariotte flask   A reservoir that maintains a constant hydrostatic pressure for liquid chromatography.

masking   The process of adding a chemical substance (a masking agent) to a sample to prevent one or more components from interfering in a chemical analysis.

masking agent   A reagent that selectively reacts with one (or more) component(s) of a solution to prevent the component(s) from interfering in a chemical analysis.

mass balance A statement that the sum of the moles of any element in all of its forms in a solution must equal the moles of that element delivered to the solution.

mass spectrograph An apparatus in which a sample is bombarded with electrons to produce charged molecular fragments that are then separated according to their mass in a magnetic field.

mass titration One in which the mass of titrant (instead of the volume) is measured.

matrix The medium containing analyte. For many analyses, it is important that standards be prepared in the same matrix as the unknown.

matrix effect A change in analytical signal caused by anything in the sample other than analyte.

matrix modifier Substance added to sample for atomic spectroscopy to retard analyte evaporation until the matrix is fully charred.

maximum suppressor A surface-active agent (such as the detergent Triton X-100) used to eliminate current maxima in polarography.

mean The average of a set of all results.

mean activity coefficient For the salt (cation)$_m$ (anion)$_n$, the mean activity coefficient, $\gamma_\pm$, is related to the individual ion activity coefficients (γ_+ and γ_-) by the equation $\gamma_\pm = (\gamma_+^m \gamma_-^n)^{1/(m+n)}$.

mechanical balance A balance having a beam that pivots on a fulcrum. Standard masses are used to measure the mass of an unknown.

median For a set of data, that value above and below which there is an equal number of data.

mediator With reference to electrolysis, a molecule added to a solution to carry electrons between the electrode and a dissolved species. Used when the target species cannot react directly at the electrode or when the target concentration is so low that other reagents react instead.

meniscus The curved surface of a liquid.

mesh size The number of spacings per linear inch in a standard screen used to sort particles.

metal ion buffer Consists of a metal-ligand complex plus excess free ligand. The two serve to fix the concentration of free metal ion through the reaction $M + nL \rightleftharpoons ML_n$.

metal ion indicator A compound whose color changes when it binds to a metal ion.

meter, m Defined as the distance light travels in a vacuum during $\frac{1}{299\,792\,458}$ of a second.

method of continuous variation or **Job's method** Procedure for finding the stoichiometry of a complex by preparing a series of solutions with different metal-to-ligand ratios. The ratio at which the extreme response (such as spectrophotometric absorbance) occurs corresponds to the stoichiometry of the complex.

method of least squares Process of fitting a mathematical function to a set of measured points by minimizing the sum of the squares of the distances from the points to the curve.

micellar electrokinetic capillary chromatography A form of capillary electrophoresis in which a micelle-forming surfactant is present. Migration times of solutes depend on the fraction of time spent in the micelles.

micelle An aggregate of molecules with ionic head groups and long, nonpolar tails. The inside of the micelle resembles hydrocarbon solvent, whereas the outside interacts strongly with aqueous solution.

microelectrode A very tiny electrode, with a diameter of perhaps $10\ \mu m$. Microelectrodes fit into very small places, such as living cells. Their small current gives rise to little ohmic loss, so they can be used in resistive, nonaqueous media. Small double-layer capacitance allows their voltage to be changed rapidly, permitting short-lived species to be studied.

microequilibrium constant An equilibrium constant that describes the reaction of a chemically distinct site in a molecule. For example, a base may be protonated at two distinct sites, each of which has a different equilibrium constant.

microporous particles Chromatographic stationary phase consisting of porous particles 3–$10\ \mu m$ in diameter, with high efficiency and high capacity for solute.

migration Electrostatically induced motion of ions in a solution under the influence of an electric field.

miscible liquids Two liquids that form a single phase when mixed in any ratio.

mobile phase In chromatography, the phase that travels through the column.

mobility The terminal velocity that an ion reaches in a field of $1\ V/m$. Velocity = mobility × field.

modulation amplitude In pulsed polarography, the magnitude of the voltage pulse.

Mohr titration Argentometric titration in which the end point is signaled by the formation of red $Ag_2CrO_4(s)$.

molality The number of moles of solute per kilogram of solvent.

molar absorptivity, ϵ, or **extinction coefficient** The constant of proportionality in Beer's law: $A = \epsilon bc$, where A is absorbance, b is pathlength, and c is the molarity of the absorbing species.

molarity, M The number of moles of solute per liter of solution.

mole, mol The amount of substance that contains as many molecules as there are atoms in 12 g of ^{12}C. There are approximately $6.022\,136\,7 \times 10^{23}$ molecules per mole.

molecular exclusion chromatography or **gel filtration** or **gel permeation chromatography** A technique in which the stationary phase has a porous structure into which small molecules can enter but large molecules cannot. Molecules are separated by size, with larger molecules moving faster than smaller ones.

molecular orbital Describes the distribution of an electron within a molecule.

molecular sieve A solid particle with pores the size of small molecules. Zeolites (sodium aluminosilicates) are a common type.

molecular weight, MW The number of grams of a substance that contains Avogadro's number of molecules.

mole fraction The number of moles of a substance in a mixture divided by the total number of moles of all components present.

monochromatic light Light of a single wavelength (color).

monochromator A device (usually a prism, grating, or filter) for selecting a single wavelength of light.

monodentate ligand One that binds to a metal ion through only one atom.

mortar and pestle A mortar is a hard ceramic or steel vessel in which a solid sample is ground with a hard pestle.

mother liquor The solution from which a substance has crystallized.

mull A fine dispersion of a solid in an oil.

multidentate ligand One that binds to a metal ion through more than one atom.

natural logarithm The natural logarithm (ln) of a is b if $e^b = a$. See also **logarithm**

nebulization The process of breaking a liquid into a mist of fine droplets.

nebulizer In atomic spectroscopy, this device breaks the liquid sample into a mist of fine droplets.

needle valve A valve with a tapered plunger that fits into a small orifice to constrict flow.

nephelometry A technique in which the intensity of light scattered by a suspension is measured to determine the concentration of suspended particles.

Nernst equation Relates the voltage of a cell (E) to the activities of reactants and products:

$$E = E° - \frac{RT}{nF} \ln Q$$

where R is the gas constant, T is temperature in kelvins, F is the Faraday constant, Q is the reaction quotient, and n is the number of electrons transferred in the balanced reaction. E° is the cell voltage when all activities are unity.

neutralization The process in which a stoichiometric equivalent of acid (or base) is added to a base (or acid).

neutron-activation analysis A technique in which radiation is observed from a sample bombarded by slow neutrons. The radiation gives both qualitative and quantitative information about the sample composition.

newton, N SI unit of force. One newton will accelerate a mass of 1 kg by 1 m/s^2.

nonelectrolyte A substance that does not dissociate into ions when dissolved.

nonpolarizable electrode One whose potential remains nearly constant, even when current flows; e.g., a saturated calomel electrode.

normal error curve A Gaussian distribution whose area is unity.

normal hydrogen electrode (N.H.E.) See **standard hydrogen electrode (S.H.E)**

normality n times the molarity of a redox reagent, where n is the number of electrons donated or accepted by that species in a particular chemical reaction. For acids and bases, it is also n times the molarity, but n is the number of protons donated or accepted by the species.

normal-phase chromatography A chromatographic separation utilizing a polar stationary phase and a less polar mobile phase.

nucleation The process whereby molecules in solution come together randomly to form small aggregates.

occlusion An impurity that becomes trapped (sometimes with solvent) in a pocket within a growing crystal.

ohm, Ω SI unit of electrical resistance. A current of 1 A flows across a potential difference of 1 V if the resistance of the circuit is 1 Ω.

ohmic potential Voltage required to overcome the electric resistance of an electrochemical cell.

Ohm's law States that the current (I) in a circuit is proportional to voltage (E) and inversely proportional to resistance (R): $I = E/R$.

Ohm's law plot In capillary electrophoresis, a graph of current versus applied voltage. The graph deviates from a straight line when Joule heating becomes significant.

on-column injection Used in gas chromatography to place a thermally unstable sample directly on the column without excessive heating in an injection port. Solute is condensed at the start of the column by low temperature, which is raised to initiate chromatography.

open tubular column In chromatography, a capillary column whose walls are coated with stationary phase.

optical density, OD See **absorbance.**

optical fiber Fiber that carries light by total internal reflection because the transparent core has a higher refractive index than the surrounding cladding.

optode or **optrode** A sensor based on an optical fiber.

osmolarity An expression of concentration that gives the total number of particles (ions and molecules) per liter of solution. For nonelectrolytes such as glucose, the osmolarity equals the molarity. For the strong electrolyte $CaCl_2$, the osmolarity is 3 times the molarity, because each mole of $CaCl_2$ provides 3 mol of ions ($Ca^{2+} + 2Cl^-$).

outer Helmholtz plane Imaginary plane passing through centers of hydrated ions just outside the layer of specifically adsorbed molecules on the surface of an electrode.

overall formation constant, β_n The equilibrium constant for a reaction of the type $M + nX \rightleftharpoons MX_n$.

overpotential The potential above that expected from the equilibrium potential, concentration polarization, and ohmic potential needed to cause an electrolytic reaction to occur at a given rate. It is zero for a reversible reaction.

oxidant See **oxidizing agent.**

oxidation A loss of electrons or a raising of the oxidation state.

oxidation state (number) A bookkeeping device used to tell how many electrons have been gained or lost by a neutral atom when it forms a compound.

oxidizing agent or **oxidant** A substance that takes electrons in a chemical reaction.

packed column A chromatography column filled with stationary-phase particles.

parallax The apparent displacement of an object when the observer changes position. Occurs when the scale of an instrument is viewed from a position that is not perpendicular to the scale, so that the apparent reading is not the true reading.

particle growth The process in which molecules become attached to a crystal to form a larger crystal.

partition chromatography A technique in which separation is achieved by equilibrium of solute between two phases.

partition coefficient, K The equilibrium constant for the reaction in which a solute is partitioned between two phases: solute (in phase 1) $\rightleftharpoons$ solute (in phase 2).

partition ratio See **capacity factor.**

pascal, Pa A unit of pressure equal to 1 N/m. There are 101 325 Pa in 1 atm.

pellicular particles A type of stationary phase used in liquid chromatography. Contains a thin layer of liquid coated on a spherical bead. It has high efficiency (low plate height), but low capacity.

peptization Occurs when washing some ionic precipitates with distilled water causes the ions that neutralize the charges of individual particles and thereby help to hold the particles together to be washed away. The particles then disintegrate and pass through the filter with the wash liquid.

permanent hardness Component of water hardness not due to dissolved alkaline earth bicarbonates. This hardness remains in the water after boiling.

p function The negative logarithm (base 10) of a quantity: $pX = -\log X$.

pH Defined as $pH = -\log \mathcal{A}_{H^+}$, where $\mathcal{A}_{H^+}$ is the activity of H^+. In most approximate applications, the pH is taken as $-\log [H^+]$.

phase correction In Fourier transform spectroscopy, correcting for the slight asymmetry of the interferogram prior to computing the Fourier transform.

phase-transfer catalysis A technique in which a compound such as a crown ether is used to extract a reactant from one phase into another in which a chemical reaction can occur.

pH meter A very sensitive potentiometer used in conjunction with a glass electrode to measure pH.

phospholipid A molecule with a phosphate-containing polar headgroup and long hydrocarbon (lipid) tail.

phosphorescence Emission of light during a transition between states of different spin multiplicity (e.g., triplet $\rightarrow$ singlet). Phosphorescence is slower than fluorescence, with emission occurring $\sim 10^{-4}$ to 10^2 s after absorption of a photon.

photoconductive detector A detector whose conductivity changes when light is absorbed by the detector material.

photodiode array An array of semiconductor diodes used to detect light. The array is normally used to detect light that has been spread into its component wavelengths. One small band of wavelengths falls on each detector.

photomultiplier tube One in which the cathode emits electrons when struck by light. The electrons then strike a series of dynodes (plates that are positive with respect to the cathode), and more electrons are released each time a dynode is struck. As a result, more than 10^6 electrons may reach the anode for every photon striking the cathode.

photon A "particle" of light with energy $h\nu$, where h is Planck's constant and ν is the frequency of the light.

phototube A vacuum tube with a photoemissive cathode. The electric current flowing between the cathode and the anode is proportional to the intensity of light striking the cathode.

photovoltaic detector A photodetector with a junction across which the voltage changes when light is absorbed by the detector material.

pH-stat A device that maintains a constant pH in a solution by continually injecting (or electrochemically generating) acid or base to counteract pH changes.

piezoelectric effect Development of electric charge on the surface of certain crystals when subjected to pressure. Conversely, application of an electric field can cause the crystal to deform.

pilot ion In polarography, an internal standard.

pipet   A glass tube calibrated to deliver a fixed or variable volume of liquid.

pK   The negative logarithm (base 10) of an equilibrium constant: $pK = -\log K$.

Planck distribution   Equation giving the spectral distribution of blackbody radiation:

$$M_\lambda = \frac{2\pi hc^2}{\lambda^5}\left(\frac{1}{e^{hc/\lambda kT}-1}\right)$$

where h is Planck's constant, c is the speed of light, λ is the wavelength of light, k is Boltzmann's constant, and T is temperature in kelvins. M_λ is the power (watts) per square meter of surface per meter of wavelength radiating from the surface. The integral $\int_{\lambda_1}^{\lambda_2} M_\lambda\, d\lambda$ gives the power emitted per unit area in the wavelength interval from λ_1 to λ_2.

Planck's constant   Fundamental constant of nature equal to the energy of light divided by its frequency: $h = E/\nu = 6.6260755 \times 10^{-34}$ J·s.

plate height, H   The length of a chromatography column divided by the number of theoretical plates in the column.

polarizability   The proportionality constant relating the induced dipole to the strength of the electric field. When a molecule is placed in an electric field, a dipole is induced in the molecule by attraction of the electrons toward the positive pole and attraction of the nuclei toward the negative pole.

polarizable electrode   One whose potential can change readily when a small current flows. Examples are Pt or Ag wires used as indicator electrodes.

polarogram   A graph showing the relation between current and potential during a polarographic experiment.

polarograph   An instrument used to obtain and record a polarogram.

polarographic wave   The S-shaped increase in current during a redox reaction in polarography.

polarography   A technique in which the current flowing into an electrolysis cell is measured as a function of the applied potential.

polychromator   A device that spreads light into its component wavelengths and directs each small band of wavelengths to a different region.

polyprotic acids and bases   Compounds that can donate or accept more than one proton.

population inversion   A necessary condition for laser operation in which the population of an excited energy level is greater than that of a lower energy level.

porous-layer column   Gas chromatography column containing an adsorptive solid phase coated on the inside surface of its wall.

postprecipitation   The adsorption of otherwise soluble impurities on the surface of a precipitate after the precipitation is over.

potential   See **electric potential.**

potentiometer   A device that measures electric potential by balancing it with a known potential of the opposite sign. A potentiometer measures the same quantity as that measured by a voltmeter, but the potentiometer is designed to draw much less current from the circuit being measured.

potentiometric stripping analysis   Technique in which analyte is electrochemically concentrated onto the working electrode at a controlled potential. The potentiostat is then disconnected and the voltage of the working electrode is measured as a function of time, during which a species in solution oxidizes or reduces the concentrated analyte back into solution. The time required to reach the potentiometric end point is proportional to the quantity of analyte.

potentiometry   An analytical method in which an electric potential difference (a voltage) of a cell is measured.

potentiostat   An electronic device that maintains a constant voltage between a pair of electrodes.

power   The amount of energy per unit time (J/s) being expended.

ppb (parts per billion)   An expression of concentration that refers to nanograms (10^{-9} g) of solute per gram of solution.

ppm (parts per million)   An expression of concentration that refers to micrograms (10^{-6} g) of solute per gram of solution.

ppt (parts per thousand)   An expression of concentration that refers to milligrams (10^{-3} g) of solute per gram of solution.

precipitant   A substance that precipitates a species from solution.

precipitation   Occurs when a substance leaves solution rapidly (to form either microcrystalline or amorphous solid).

precipitation titration   One in which the analyte forms a precipitate with the titrant.

precision   A measure of the reproducibility of a measurement.

precolumn   See **guard column.**

preconcentration   The process of concentrating trace components of a mixture prior to their analysis.

premix burner   In atomic spectroscopy, one in which the sample is nebulized and simultaneously mixed with fuel and oxidant before being fed into the flame.

preoxidation   In some redox titrations, adjustment of the analyte oxidation state to a higher value so that it can be titrated with a reducing agent.

prereduction   The process of reducing an analyte to a lower oxidation state prior to performing a titration with an oxidizing agent.

pressure   Force per unit area, commonly measured in pascals (N/m) or atmospheres.

pressure broadening   In spectroscopy, line broadening due to collisions between molecules.

primary standard A reagent that is pure enough and stable enough to be used directly after weighing. The entire mass is considered to be pure reagent.

prism A transparent, triangular solid. Each wavelength of light passing through the prism is bent at a different angle. Therefore, light is dispersed into its component wavelengths by the prism.

protic solvent One with an acidic hydrogen atom.

proton The ion H^+.

proton acceptor A Brønsted-Lowry base: a molecule that combines with H^+.

proton donor A Brønsted-Lowry acid: a molecule that can give up H^+ to another molecule.

purge and trap method Technique in which volatile materials are purged from a solution by flowing gas and trapped in a column containing a series of increasingly stronger adsorbants. To analyze the trapped substances, the column is attached to a gas chromatograph, heated rapidly, and backflushed with inert gas.

pyroelectric effect Variation of electric polarization with temperature of a ferroelectric material.

pyrolysis Thermal decomposition of a substance.

Q **test** Used to decide whether to discard a datum that appears discrepant.

qualitative analysis The process of determining the identity of the constituents of a substance.

quantitative analysis The process of measuring how much of a constituent is present in substance.

quaternary ammonium ion A cation containing four substituents attached to a nitrogen atom; e.g., $(CH_3CH_2)_4N^+$, the tetraethylammonium ion.

radian, rad Unit of plane angle. There are 2π radians in a complete circle.

radiant power, *P,* or **intensity** The energy per unit time per unit area (W/m^2) carried by a beam of light.

random error or **indeterminate error** A type of error, which can be either positive or negative and cannot be eliminated, based on the ultimate limitations on a physical measurement.

random sample Bulk sample constructed by taking portions of the entire lot at random.

range or **spread** The difference between the highest and the lowest value in a set of data.

reactant The species that is consumed in a chemical reaction. It appears on the left side of a chemical equation.

reaction quotient, *Q* Has the same form as the equilibrium constant for a reaction. However, the reaction quotient is evaluated for a particular set of existing activities (concentrations), which are generally not the equilibrium values. At equilibrium, $Q = K$.

reagent blank A solution prepared from all of the reagents, but no analyte. The blank measures the response of the analytical method to impurities in the reagents or any other effects caused by any component other than the analyte.

redox couple A pair of reagents related by electron transfer; e.g., $Fe^{3+} \mid Fe^{2+}$ or $MnO_4^- \mid Mn^{2+}$.

redox indicator A compound used to find the end point of a redox titration because its different oxidation states have different colors. The standard potential of the indicator must be such that its color changes near the equivalence point of the titration.

redox reaction A chemical reaction involving transfer of electrons from one element to another.

redox titration One in which the reaction between analyte and titrant is an oxidation-reduction reaction.

reduced plate height In chromatography, the quotient plate height/d, where the numerator is the height equivalent to a theoretical plate and the denominator is the diameter of stationary-phase particles.

reducing agent or **reductant** A substance that donates electrons in a chemical reaction.

reduction A gain of electrons or a lowering of the oxidation state.

reference electrode One that maintains a constant potential against which the potential of another half-cell may be measured.

reflectance, *R* Fraction of incident radiant power reflected by an object.

refraction Bending of light when it passes between media with different refractive indexes.

refractive index, *n* The speed of light in any medium is c/n, where c is the speed of light in vacuum and n is the refractive index of the medium. The refractive index also measures the angle at which a light ray will be bent when it passes from one medium into another. Snell's law states that $n_1 \sin \theta_1 = n_2 \sin \theta_2$, where n_i is the refractive index for each medium and θ_i is the angle of the ray with respect to a normal between the two media.

refractive index detector Liquid chromatography detector that measures the change in the refractive index of a solution as solutes emerge from column.

relative retention In chromatography, this is the ratio of adjusted retention times for two components. If component 1 has an adjusted retention time of t'_{r1} and component 2 has an adjusted retention time of t'_{r2} ($> t'_{r1}$), the relative retention is $\alpha = t'_{r2}/t'_{r1}$.

relative supersaturation Defined as $(Q - S)/S$, where S is the concentration of solute in a saturated solution and Q is the concentration in a particular supersaturated solution.

relative uncertainty The uncertainty of a quantity divided by the value of the quantity. It is usually expressed as a percentage of the measured quantity.

releasing agent In atomic spectroscopy, a substance that prevents chemical interference.

reprecipitation Sometimes a gravimetric precipitate can be freed of impurities only by redissolving it and reprecipitating it. The impurities are present at lower concentration during the second precipitation and are less likely to coprecipitate.

residual current The small current that is observed prior to the decomposition potential in an electrolysis.

resin An ion exchanger, such as polystyrene, which exists as small, hard particles.

resistance, R A measure of the retarding force opposing the flow of electric current.

resistivity, ρ A measure of the ability of a material to retard the flow of electric current. $J = E/\rho$, where J is the current density (A/m^2) and E is electric field (V/m). Units of resistivity are $V \cdot m/A = \Omega \cdot m$. The resistance ($\Omega$) of a conductor with a given length and cross-sectional area is given by $R = \rho \cdot \text{length}/\text{area}$.

resolution How close two bands in a spectrum or a chromatogram can be to each other and still be seen as two peaks. In chromatography, it is defined as the difference in retention times of adjacent peaks divided by their width.

resonance effect Contribution to a physical property by delocalization of electrons through the pi orbitals of a molecule.

response factor An empirical factor measuring the relative response of a detector to a given compound. It is used in gas chromatography to determine the amount of unknown relative to an internal standard. For a known mixture of analyte X and standard S, the response factor (F) may be defined through the equation $[X]/[S] = F(\text{signal from X})/(\text{signal from S})$. Once you have measured F with a standard mixture, you can use it to find $[X]$ in an unknown if you know $[S]$ and the quotient (signal from X)/(signal from S).

retardation, δ The difference in pathlength between light striking the stationary and moving mirrors of an interferometer.

retention factor See **capacity factor**.

retention index, I In gas chromatography, the Kovats retention index is a logarithmic scale that relates the retention time of a compound to those of linear alkanes.

retention ratio In chromatography, the time required for solvent to pass through the column divided by the time required for solute to pass through the column.

retention time The time, measured from injection, needed for a solute to be eluted from a chromatography column.

retention volume The volume of solvent needed to elute a solute from a chromatography column.

reverse-phase chromatography A technique in which the stationary phase is less polar than the mobile phase.

root-mean-square (rms) noise The quantity

$$\text{rms noise} = \left(\frac{1}{|\lambda_2 - \lambda_1|} \int_{\lambda_1}^{\lambda_2} [N(\lambda) - \overline{N}]^2 \, d\lambda \right)^{1/2}$$

where $N(\lambda)$ is the noise at wavelength λ and $\overline{N}$ is the average noise in the interval λ_1 to λ_2.

rotating disk electrode A motor-driven electrode with a smooth flat face in contact with the solution. Rapid convection created by rotation brings fresh analyte to the surface of the electrode. A Pt electrode is especially suitable for studying anodic processes, in which a mercury electrode would be too easily oxidized.

rotational transition Occurs when a molecule changes its rotation energy.

rubber policeman A glass rod with a flattened piece of rubber on the tip. The rubber is used to scrape solid particles from glass surfaces in gravimetric analysis.

salt An ionic solid.

salt bridge A conducting ionic medium in contact with two electrolyte solutions. It allows an ionic current to flow without allowing immediate diffusion of one electrolyte solution into the other.

sample cleanup Removal of portions of the sample that do not contain analyte and may interfere with analysis.

sampling variance The square of the standard deviation associated with the heterogeneity of the sample itself, not the analytical procedure. For inhomogeneous materials, different samples will have different compositions. It is necessary to take larger portions or more portions to reduce the uncertainty of composition due to variation from one region of the sample to another. The total variance of an analysis is the sum of the variances due to sampling and due to the analytical procedure.

saturated solution One that contains the maximum amount of a compound that can dissolve at equilibrium.

Scatchard plot A graph used to find the equilibrium constant for a reaction such as $X + P \rightleftharpoons PX$. It is a graph of $[PX]/[X]$ versus $[PX]$, or any functions proportional to these quantities. The magnitude of the slope of the graph is the equilibrium constant.

S.C.E. (saturated calomel electrode) A calomel electrode saturated with KCl. The electrode half-reaction is $Hg_2Cl_2(s) + 2e^- \rightleftharpoons 2Hg(l) + 2Cl^-$.

second, s The duration of 9 192 631 770 periods of the radiation corresponding to the transition between two hyperfine levels of the ground state of ^{133}Cs.

segregated material A material containing distinct regions of different composition.

selectivity coefficient, K With respect to an ion-selective electrode, a measure of the relative response of the electrode to two different ions. In ion-exchange chromatography, this is the equilibrium constant for displacement of one ion by another from the resin.

self-absorption In flame emission atomic spectroscopy, there is a lower concentration of excited-state atoms in the cool, outer part of the flame than in the hot, inner flame. The cool atoms can absorb emission from the hot ones and thereby decrease the observed signal.

semiconductor A material whose conductivity ($10^{-7} - 10^4 \ \Omega^{-1} \cdot m^{-1}$) is intermediate between that of good conductors ($10^8 \ \Omega^{-1} \cdot m^{-1}$) and that of insulators ($10^{-20} - 10^{-12} \ \Omega^{-1} \cdot m^{-1}$).

sensitivity The response of an instrument or method to a given amount of analyte. In spectrophotometric analysis, sensitivity is the concentration of analyte necessary to produce 99% T (or an absorbance of 0.004 4). For a balance, sensitivity is the deflection of the pointer divided by the difference in mass on the two pans. The greater the sensitivity, the greater the deflection.

separator column Ion-exchange column used to separate analyte species in ion chromatography.

septum A disk, usually made of silicone rubber, covering the injection port of a gas chromatograph. The sample is injected by syringe through the septum.

SI units The units of an international system of measurement based on the meter, kilogram, second, ampere, kelvin, candela, mole, radian, and steradian.

sieving In electrophoresis, the separation of macromolecules by migration through a polymer gel. The smallest molecules move fastest and the largest move slowest.

signal averaging Improvement of a signal by averaging successive scans. The signal increases in proportion to the number of scans accumulated. The noise increases in proportion to the square root of the number of scans. Therefore, the signal-to-noise ratio improves in proportion to the square root of the number of scans collected.

significant figure The number of significant figures in a quantity is the minimum number of figures needed to express the quantity in scientific notation. In experimental data, the first uncertain figure is the last significant figure.

silanization The treatment of a chromatographic solid support or glass column with silicon compounds that bind to the most reactive Si—OH groups. It reduces irreversible adsorption and tailing of polar solutes.

silver-silver chloride electrode A common reference electrode containing a silver wire coated with AgCl paste and dipped in a solution saturated with AgCl and (usually) KCl. The half-reaction is $AgCl(s) + e^- \rightleftharpoons Ag(s) + Cl^-$.

single-column ion chromatography Separation of ions on a low-capacity ion-exchange column using low ionic strength eluent.

single-electrode potential The voltage measured when the electrode of interest is connected to the positive terminal of a potentiometer and a standard hydrogen electrode is connected to the negative terminal.

singlet state One in which all electron spins are paired.

slope For a straight line whose equation is $y = mx + b$, the value of m is the slope. It is the ratio $\Delta y / \Delta x$ for any segment of the line.

slurry A suspension of a solid in a solvent.

Smith-Hieftje background correction In atomic absorption spectroscopy, a method of distinguishing analyte signal from background signal, based on applying a periodic pulse of high current to the hollow-cathode lamp to distort the lamp signal. The signal detected during the current pulse is subtracted from the signal detected without the pulse to obtain the corrected response.

smoothing Use of a mathematical procedure or electrical filtering to improve the quality of a signal.

Snell's law Relates angle of refraction (θ_2) to angle of incidence (θ_1) for light passing from a medium with refractive index n_1 to a medium of refractive index n_2: $n_1 \sin \theta_1 = n_2 \sin \theta_2$. Angles are measured with respect to the normal to the surface between the two media.

solid-phase extraction Preconcentration procedure in which a solution is passed through a short column of chromatographic stationary phase, such as C_{18} on silica. Trace solutes adsorbed on the column can be eluted with a small volume of solvent with a high eluent strength.

solid-state ion-selective electrode A type of ion-selective electrode that has a solid membrane made of an inorganic salt crystal. Ion-exchange equilibria between the solution and the surface of the crystal account for the electrode potential.

solubility product, K_{sp} The equilibrium constant for the dissolution of a solid salt to give its ions in solution. For the reaction $M_m N_n(s) \rightleftharpoons mM^{n+} + nN^{m-}$, $K_{sp} = \mathcal{A}_{M^{n+}}^m \mathcal{A}_{N^{m-}}^n$ where $\mathcal{A}$ is the activity of each species.

solute A minor component of a solution.

solvation The interaction of solvent molecules with solute. In general, solvent molecules will orient themselves around a solute to minimize the energy of the solution through dipole and van der Waals forces.

solvent The major constituent of a solution.

solvent extraction A method in which a chemical species is transferred from one liquid phase to another. It is used to separate components of a mixture.

solvent trapping Splitless gas chromatography injection technique in which solvent is condensed near its boiling point at the start of the column. Solutes dissolve in a narrow band in the condensed solvent.

specific adsorption Process in which molecules are held tightly to a surface by van der Waals forces.

specific gravity A dimensionless quantity equal to the mass of a substance divided by the mass of an equal volume of water at 4°C.

spectral interference In atomic spectroscopy, any physical process that affects the light intensity at the analytical wavelength. Created by substances that absorb, scatter, or emit light of the analytical wavelength.

spectrophotometer   A device used to measure absorption of light. It includes a source of light, a wavelength selector (monochromator), and an electrical means of detecting light.

spectrophotometric analysis   Any method in which light absorption, emission, or scattering is used to measure chemical concentrations.

spectrophotometric titration   One in which absorption of light is used to monitor the progress of the chemical reaction.

spectrophotometry   In a broad sense, any method using light to measure chemical concentrations.

specular reflection   Reflection of light at an angle equal to the angle of incidence.

split injection   Used in capillary gas chromatography to inject a small fraction of sample onto the column, while the rest of the sample is discarded.

splitless injection   Used in capillary gas chromatography for trace analysis and quantitative analysis. The entire sample in a low-boiling solvent is directed to the column, where the sample is concentrated by solvent trapping (condensing the solvent below its boiling point) or cold trapping (condensing solutes far below their boiling range). The column is then warmed to initiate separation.

spontaneous process   One that is energetically favorable. It will eventually occur, but thermodynamics makes no prediction as to how long it will take.

spread   See **range**.

square wave polarography   Form of polarography with a waveform consisting of a square wave superimposed on a staircase wave. The technique is faster and more sensitive than differential pulse polarography.

stability constant   See **formation constant**.

stacking   In electrophoresis, the concentration of a dilute electrolyte into a narrow band by an electric field. Stacking occurs because the electric field in the dilute electrolyte is stronger than the field in more concentrated surrounding electrolyte.

standard addition method   A technique in which an analytical signal due to an unknown is first measured. Then a known quantity of analyte is added, and the increase in signal is recorded. Assuming linear response, it is possible to calculate what quantity of analyte must have been present in the unknown.

standard curve   A graph showing the response of an analytical technique to known quantities of analyte.

standard deviation   Measures how closely data are clustered about the mean value. For a finite set of data, the standard deviation, s, is computed from the formula

$$s = \sqrt{\frac{\sum(x_i - \bar{x})^2}{n - 1}}$$

where n is the number of results, x_i is an individual result, and $\bar{x}$ is the mean result. For a large number of

measurements, s approaches σ, the true standard deviation of the population.

standard hydrogen electrode (S.H.E.) or **normal hydrogen electrode (N.H.E.)**   One that contains $H_2(g)$ bubbling over a catalytic Pt surface in contact with aqueous H^+. The activities of H_2 and H^+ are both unity in the hypothetical standard electrode. The cell reaction is $H^+ + e^- \rightleftharpoons \frac{1}{2}H_2(g)$.

standardization   The process whereby the concentration of a reagent is determined by reaction with a known quantity of a second reagent.

standard reduction potential, $E°$ The voltage that would be measured when a hypothetical cell containing the desired half-reaction (with all species present at unit activity) is connected to a standard hydrogen electrode anode.

standard reference materials   Certified samples sold by the U.S. National Institute of Standards and Technology containing known concentrations or quantities of particular analytes. Used to standardize testing procedures in different laboratories.

standard solution   A solution whose composition is known by virtue of the way it was made from a reagent of known purity or by virtue of its reaction with a known quantity of a standard reagent.

standard state   When writing equilibrium constants, the standard state of a solute is 1 M and the standard state of a gas is 1 atm. Pure solids and liquids are considered to be in their standard states.

stationary phase   In chromatography, the phase that does not move through the column.

stepwise formation constant, K_n   The equilibrium constant for a reaction of the type $ML_{n-1} + L \rightleftharpoons ML_n$.

steradian, sr   Unit of solid angle. There are 4π steradians in a complete sphere.

stimulated emission   Emission of a photon induced by the passage of another photon of the same wavelength.

Stokes equation   The friction coefficient for a molecule migrating through solution is $6\pi\eta r$, where η is the viscosity of the fluid and r is the hydrodynamic radius (equivalent spherical radius) of the molecule.

stray light   In spectrophotometry, light reaching the detector that is not part of the narrow waveband expected from the monochromator.

stripping analysis   A very sensitive polarographic technique in which analyte is concentrated from a dilute solution by reduction into a single drop of Hg. It is then analyzed polarographically during an anodic redissolution process.

strong acids and bases   Those that are completely dissociated (to H^+ or OH^-) in water.

strong electrolyte   One that dissociates completely into its ions when dissolved.

Student's t   A statistical tool used to express confidence intervals and to compare results from different experiments.

superconductor A material that loses all electric resistance when cooled below a critical temperature.

supercritical fluid A fluid above its critical temperature.

supercritical fluid chromatography Chromatography using supercritical fluid as the mobile phase. Capable of highly efficient separations of nonvolatile solutes and able to use detectors suitable for gas or liquid.

supernatant liquid Liquid remaining above the solid after a precipitation. Also called *supernate*.

supersaturated solution One that contains more dissolved solute than would be present at equilibrium.

support-coated column Open tubular gas chromatographic column in which the stationary phase is coated on solid support particles attached to the inside wall of the column.

supporting electrolyte An unreactive salt added in high concentration to most solutions for voltammetric measurements (such as polarography). The supporting electrolyte carries most of the ion-migration current and therefore decreases the coulombic migration of electroactive species to a negligible level. The electrolyte also decreases the resistance of the solution.

suppressed-ion chromatography Separation of ions using an ion-exchange column followed by suppressor (membrane or column) to remove ionic eluent.

suppressor column Ion-exchange column used in ion chromatography to transform ionic eluent into a nonionic form.

surface-modified electrode An electrode whose surface has been changed by a chemical reaction. For example, electroactive materials that react specifically with certain solutes can be attached to the electrode.

surfactant A molecule with an ionic or polar headgroup and a long, nonpolar tail. Surfactants may aggregate in aqueous solution to form micelles. Surfactants derive their name from the fact that they accumulate at boundaries between polar and nonpolar phases and modify the surface tension, which is the free energy of formation of the surface. Soaps are surfactants.

syringe A device having a calibrated barrel into which liquid is sucked by a plunger. The liquid is expelled through a needle by pushing on the plunger.

systematic error or **determinate error** A type of error due to procedural or instrumental factors that cause a measurement to be systematically too large or too small. The error can, in principle, be discovered and corrected.

systematic treatment of equilibrium A method that uses the charge balance, mass balance(s), and equilibria to completely specify the system's composition.

tailing An asymmetric chromatographic elution band in which the later part of the band is drawn out. It often results from adsorption of a solute to a few active adsorption sites on the stationary phase.

tare The mass of an empty vessel used to receive a substance to be weighed. Many balances can be tared. That is, with the empty receiver in place, the balance can be set to read zero grams.

temporary hardness Component of water hardness due to dissolved alkaline earth bicarbonates. It is temporary because boiling causes precipitation of the carbonates.

test portion Part of the laboratory sample used for one analysis. Also called *aliquot*.

theoretical plate An imaginary construct in chromatography denoting a segment of a column in which one equilibration of solute between stationary and mobile phases occurs. The number of theoretical plates on a column with Gaussian bandshapes is defined as $N = t_r^2/\sigma^2$, where t_r is the retention time of a peak and σ is the standard deviation of the band.

thermal conductivity, κ Rate at which a substance transports heat (energy per unit time per unit area) through a temperature gradient (degrees per unit distance). Energy flow $[J/(s \cdot m^2)] = -\kappa(dT/dx)$, where κ is the thermal conductivity $[W/(m \cdot K)]$ and dT/dx is the temperature gradient (K/m).

thermal conductivity detector A device that detects bands eluted from a gas chromatography column by measuring changes in the thermal conductivity of the gas stream.

thermistor A device whose electrical resistance changes markedly with changes in temperature.

thermocouple An electrical junction across which a temperature-dependent voltage exists. Thermocouples are calibrated for measurement of temperature and usually consist of two dissimilar metals in contact with each other.

thermogravimetric analysis A technique in which the mass of a substance is measured as the substance is heated. Changes in mass reflect decomposition of the substance, often to well-defined products.

thermometric titration One in which the temperature is measured to determine the end point. Most titration reactions are exothermic, so the temperature rises during the reaction and suddenly stops rising when the equivalence point is reached.

thin-layer chromatography A technique in which the stationary phase is coated on a flat glass or plastic plate. Solute is spotted near the bottom of the plate. The bottom edge of the plate is placed in contact with solvent, which is allowed to creep up the plate by capillary action.

titer A measure of concentration, usually defined as how many milligrams of reagent B will react with 1 mL of reagent A. Consider a $AgNO_3$ solution with a titer of 1.28 mg of NaCl per milliliter of $AgNO_3$. The reaction is $Ag^+ + Cl^- \rightarrow AgCl(s)$. Because 1.28 mg of NaCl = 2.19×10^{-5} mol, the concentration of Ag^+ is

2.19 × 10^{-5} mol/mL = 0.0219 M. The same solution of $AgNO_3$ has a titer of 0.993 mg of KH_2PO_4, because 3 mol of Ag^+ react with 1 mol of PO_4^{3-} (to precipitate Ag_3PO_4), and 0.993 mg of KH_2PO_4 equals $\frac{1}{3}$(2.19 × 10^{-5} mol).

titrant   The substance added to the analyte in a titration.

titration   A procedure in which one substance (titrant) is carefully added to another (analyte) until complete reaction has occurred. The quantity of titrant required for complete reaction tells how much analyte is present.

titration error   The difference between the observed end point and the true equivalence point in a titration.

tolerance   Manufacturer's acceptable uncertainty in the accuracy of a device such as a buret or volumetric flask. A 100-mL flask with a tolerance of ± 0.08 mL may contain 99.92 to 100.08 mL and be within tolerance.

total carbon   In a natural water or industrial effluent sample, the quantity of CO_2 produced when the sample is completely oxidized by oxygen at 900°C in the presence of a catalyst.

total organic carbon   In a natural water or industrial effluent sample, the quantity of CO_2 produced when the sample is first acidified and purged to remove carbonate and bicarbonate and then completely oxidized by oxygen at 900°C in the presence of a catalyst.

total oxygen demand   In a natural water or industrial effluent sample, the quantity of O_2 required for complete oxidation of species in the water at 900°C in the presence of a catalyst.

transmittance, T   Defined as $T = P/P_0$, where P_0 is the radiant power of light striking the sample on one side and P is the radiant power of light emerging from the other side of the sample.

triangular apodization   In Fourier transform spectroscopy, multiplying the interferogram by a symmetric function that decreases linearly from 1 at the center to 0 at predefined limits on either side of 0.

triple point   The one temperature and pressure at which the solid, liquid, and gaseous forms of a substance are in equilibrium with one another.

triplet state   An electronic state in which there are two unpaired electrons.

truncation   The process of cutting off abruptly.

t test   Used to decide whether the results of two experiments are within experimental uncertainty of each other. The uncertainty must be specified to within a certain probability.

tungsten lamp   An ordinary light bulb in which electricity passing through a tungsten filament heats the wire and causes it to emit visible light.

turbidimetry   A technique in which the decrease in radiant power of light traveling through a turbid solution is measured.

turbidity   The light-scattering property associated with suspended particles in a liquid. A turbid solution appears cloudy.

turbidity coefficient   The transmittance of a turbid solution is given by $P/P_0 = e^{-\tau b}$, where P is the transmitted radiant power, P_0 is the incident radiant power, b is the pathlength, and τ is the turbidity coefficient.

ultraviolet detector   Liquid chromatography detector that measures ultraviolet absorbance of solutes emerging from the column.

valence band   Energy levels containing valence electrons in a semiconductor. The electrons in these levels are localized in chemical bonds.

van Deemter equation   Describes the dependence of chromatographic plate height (H) on the linear flow rate of (u_x) elution: $H = A + B/u_x + Cu_x$.

variance, σ^2   The square of the standard deviation.

vibrational transition   Occurs when a molecule changes its vibrational energy.

viscosity   Resistance to flow in a fluid.

void volume, V_0   The volume of the mobile phase, V_m.

volatile   Easily vaporized.

volatilization   The selective removal of a component from a mixture by transforming the component into a volatile (low-boiling) species and removing it by heating, pumping, or bubbling a gas through the mixture.

Volhard titration   That of Ag^+ with SCN^-, in which the formation of the red complex $Fe(SCN)^{2+}$ marks the end point.

volt, V   Unit of electric potential or electric potential difference between two points. If the potential difference between two points is one volt, it requires one joule of energy to move one coulomb of charge between the two points.

voltammetry   An analytical method in which the relationship between current and voltage is observed during an electrochemical reaction.

volume flow rate   In chromatography, the volume of mobile phase per unit time eluted from the column.

volume percent   Defined as (volume of solute/volume of solution) × 100.

volumetric analysis   A technique in which the volume of material needed to react with the analyte is measured.

volumetric flask   One having a tall, thin neck with a calibration mark. When the liquid level is at the calibration mark, the flask contains its specified volume of liquid.

von Weimarn ratio   The quotient $(Q - S)/S$, where Q is the concentration of a solute and S is the concentration at equilibrium. A large value of this ratio means that the solution is highly supersaturated.

Walden reductor A column packed with silver and eluted with HCl. An oxidized analyte is reduced upon passage through the column. The reduced product is titrated with an oxidizing agent.

wall-coated column Hollow chromatographic column in which the stationary phase is coated on the inside surface of its wall.

watt, W The SI unit of power, equal to an energy flow of one joule per second. When an electric current of one ampere flows through a potential difference of one volt, the power is one watt.

wavelength, λ The distance between consecutive crests of a wave.

wavenumber, $\tilde{v}$ The reciprocal of the wavelength, λ.

weak acids and bases Those whose dissociation constants are not large.

weak electrolyte One that only partially dissociates into ions when it dissolves.

weighing paper Used as a base on which to place a solid reagent on a balance. The paper has a very smooth surface, from which solids fall easily for transfer to a vessel.

weight percent Defined as (mass of solute/mass of solution) $\times$ 100.

weight/volume percent Defined as (mass of solute/volume of solution) $\times$ 100.

Weston cell An extremely stable voltage source, based on the reaction $Cd(s) + HgSO_4(aq) \rightleftharpoons CdSO_4(aq) + Hg(l)$. It is often used to standardize a potentiometer.

wet ashing The destruction of organic matter in a sample by a liquid reagent (such as boiling aqueous $HClO_4$) prior to analysis of an inorganic component.

white light Light of all different wavelengths.

Wien displacement law Approximate formula for the wavelength (λ_{max}) of maximum blackbody emission: $\lambda_{max} \cdot T \approx hc/5k = 2.878 \times 10^{-3}$ m $\cdot$ K, where T is temperature in kelvins, h is Planck's constant, c is the speed of light, and k is Boltzmann's constant. Valid for $T > 100$ K.

working electrode The one at which the reaction of analytical interest in coulometry or polarography occurs.

Zeeman background correction Technique used in atomic spectroscopy in which analyte signals are shifted outside the detector monochromator range by applying a strong magnetic field to the sample. Signal that remains is the background.

Zeeman effect Shifting of atomic energy levels in a magnetic field.

zwitterion A molecule with a positive charge localized in one position and a negative charge localized at another position.

Logarithms and Exponents

If a is the base 10 logarithm of n ($a = \log n$), then $n = 10^a$. On a calculator, you find the logarithm of a number by pressing the "log" button. If you know $a = \log n$ and you wish to find n, use the "antilog" button or raise 10 to the power a:

$$a = \log n$$

$$10^a = 10^{\log n} = n \; (\Rightarrow n = \text{antilog} \, a)$$

Natural logarithms (ln) are based on the number e (= 2.718 281 ...) instead of 10:

$$b = \ln n$$

$$e^b = e^{\ln n} = n$$

On a calculator, you find the ln of n with the "ln" button. To find n when you know $b = \ln n$, use the e^x key.

Here are some useful properties to know:

$$\log (a \cdot b) = \log a + \log b \qquad \log (a^b) = b \log a \qquad a^b \cdot a^c = a^{(b + c)}$$

$$\log \left(\frac{a}{b} \right) = \log a - \log b \qquad \log 10^a = a \qquad \frac{a^b}{a^c} = a^{(b - c)}$$

Problems

Test yourself by simplifying each expression as much as possible:

(a) $e^{\ln a}$

(b) $10^{\log a}$

(c) $\log 10^a$

(d) $10^{-\log a}$

(e) $e^{-\ln a^3}$

(f) $e^{\ln a^{-3}}$

(g) $\log (10^{1/a^3})$

(h) $\log (10^{-a^2})$

(i) $\log (10^{a^2 - b})$

(j) $\log (2a^3 10^{b^2})$

(k) $e^{(a + \ln b)}$

(l) $10^{[(\log 3) - (4 \log 2)]}$

Solving a logarithmic equation: In working with the Nernst and Henderson-Hasselbalch equations, we will need to solve equations such as

$$a = b - c \log \frac{d}{gx}$$

for the variable, x. First isolate the log term:

$$\log \frac{d}{gx} = \frac{(b - a)}{c}$$

Then raise 10 to the value of each side of the equation:

$$10^{\log (d/gx)} = 10^{(b - a)/c}$$

But $10^{\log (d/gx)}$ is just d/gx, so

$$\frac{d}{gx} = 10^{(b - a)/c} \Rightarrow x = \frac{d}{g10^{(b - a)/c}}$$

Converting between ln x and log x: The relation between these is derived by writing $x = 10^{\log x}$ and taking ln of both sides:

$$\ln x = \ln (10^{\log x}) = (\log x)(\ln 10)$$

because $\ln a^b = b \ln a$.

Answers

(a) a

(b) a

(c) a

(d) $1/a$

(e) $1/a^3$

(f) $1/a^3$

(g) $1/a^3$

(h) $-a^2$

(i) $a^2 - b$

(j) $b^2 + \log (2a^3)$

(k) be^a

(l) $3/16$

Graphs of Straight Lines

The general form of the equation of a straight line is

$$y = mx + b$$

where m = slope = $\dfrac{\Delta y}{\Delta x} = \dfrac{y_2 - y_1}{x_2 - x_1}$

b = intercept on y axis

The meanings of slope and intercept are illustrated in Figure B-1.

If you know two points $[(x_1, y_1)$ and $(x_2, y_2)]$ that lie on the line, you can generate the equation of the line by noting that the slope is the same for every pair of points on the line. Calling some general point on the line (x, y), we can write

$$\frac{y - y_1}{x - x_1} = \frac{y_2 - y_1}{x_2 - x_1} = m \qquad \text{(B-1)}$$

which can be rearranged to the form

$$y - y_1 = \left(\frac{y_2 - y_1}{x_2 - x_1}\right)(x - x_1)$$

$$y = \underbrace{\left(\frac{y_2 - y_1}{x_2 - x_1}\right)}_{m} x + \underbrace{y_1 - \left(\frac{y_2 - y_1}{x_2 - x_1}\right)x_1}_{b}$$

When you have a series of experimental points that should lie on a line, the best line is generally obtained by the method of least squares, described in Chapter 4. This method gives the slope and the intercept directly. If, instead, you wish to draw the "best" line by eye, you can derive the equation of the line by selecting two points *that lie on the line* and applying Equation B-1.

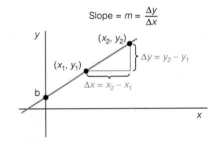

Figure B-1
Parameters of a straight line.

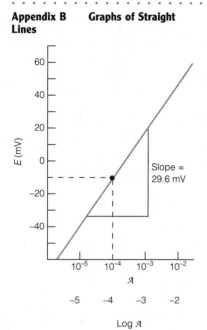

Figure B-2   A linear graph in which one axis is a logarithmic function.

Sometimes you are presented with a linear plot in which x and/or y are nonlinear functions. An example is shown in Figure B-2, in which the potential of an electrode is expressed as a function of the activity of analyte. Given that the slope is 29.6 mV and the line passes through the point ($\mathcal{A} = 10^{-4}$, $E = -10.2$), find the equation of the line. To do this, first note that the y axis is linear but the x axis is *logarithmic*. That is, the function E versus $\mathcal{A}$ is *not* linear, but E versus log $\mathcal{A}$ is linear. The form of the straight line should therefore be

$$E = (29.6) \underbrace{\log \mathcal{A}}_{} + b$$
$$\quad\quad \uparrow \quad\quad \uparrow \quad\quad \uparrow$$
$$\quad\quad y \quad\quad m \quad\quad x$$

To find b, we can use the coordinates of the one known point in Equation B-1:

$$\frac{y - y_1}{x - x_1} = \frac{E - E_1}{\log \mathcal{A} - \log \mathcal{A}_1} = \frac{E - (-10.2)}{\log \mathcal{A} - \log (10^{-4})} = m = 29.6$$

or

$$E + 10.2 = 29.6 \log \mathcal{A} + (29.6)(4)$$

$$E \text{ (mV)} = 29.6 \text{ (mV)} \log \mathcal{A} + 108.2 \text{ (mV)}$$

Propagation of Uncertainty

The rules given for propagation of uncertainty in Table 3-1 are special cases of a general formula. Suppose you wish to calculate the function, F, of several experimental quantities, x, y, z, ... If the errors $(e_x, e_y, e_z, \ldots)$ in measuring x, y, z, ... are small, random, and independent of one another, then the uncertainty (e_F) in the function F is approximately:

$$e_F = \sqrt{\left(\frac{\partial F}{\partial x}\right)^2 e_x^2 + \left(\frac{\partial F}{\partial y}\right)^2 e_y^2 + \left(\frac{\partial F}{\partial z}\right)^2 e_z^2 + \ \cdots} \qquad \text{(C-1)}$$

The quantities in parentheses are partial derivatives, which are calculated in the same manner as ordinary derivatives, except that all but one variable are treated as constants. For example, if $F = 3xy^2$, $\partial F/\partial x = 3y^2$ and $\partial F/\partial y = (3x)(2y) = 6xy$.

As an example of using Equation C-1, let's find the uncertainty in the function

$$F = x^y = (2.00 \pm 0.02)^{3.00 \pm 0.09}$$

The partial derivatives are

$$\frac{\partial F}{\partial x} = yx^{y-1} \qquad \frac{\partial F}{\partial y} = x^y \ln x$$

Putting these quantities into Equation C-1 gives

$$e_F = \sqrt{(yx^{y-1})^2 e_x^2 + (x^y \ln x)^2 e_y^2}$$

$$= \sqrt{y^2 x^{2y-2} e_x^2 + x^{2y} (\ln x)^2 e_y^2}$$

$$= \sqrt{y^2 x^{2y} \left(\frac{e_x}{x}\right)^2 + x^{2y} (\ln x)^2 e_y^2}$$

Multiplying and dividing the second term by y^2 allows us to rearrange to a more pleasant form:

$$e_F = \sqrt{y^2 x^{2y} \left(\frac{e_x}{x}\right)^2 + y^2 x^{2y} (\ln x)^2 \left(\frac{e_y}{y}\right)^2}$$

Removing $\sqrt{y^2 x^{2y}} = yF$ from both terms gives

$$e_F = yF \sqrt{\left(\frac{e_x}{x}\right)^2 + (\ln x)^2 \left(\frac{e_y}{y}\right)^2} \tag{C-2}$$

Now for the number crunching. Disregarding uncertainties for a moment, we know that $F = 2.00^{3.00} = 8.00 \pm$? The uncertainty is given by Equation C-2:

$$e_F = 3.00 \cdot 8.00 \sqrt{\left(\frac{0.02}{2.00}\right)^2 + (\ln 2.00)^2 \left(\frac{0.09}{3.00}\right)^2} = 0.55$$

Reasonable answers are therefore $F = 8.0_0 \pm 0.5_5$ or 8.0 ± 0.6.

Exercises

C-1. Verify the following calculations.

(a) $2.36^{4.39 \pm 0.08} = 43._4 \pm 3._0$

(b) $(2.36 \pm 0.06)^{4.39 \pm 0.08} = 43._4 \pm 5._7$

C-2. For $F = \sin(2\pi xy)$, show that $e_F = 2\pi xy \cos(2\pi xy) \sqrt{\left(\frac{e_x}{x}\right)^2 + \left(\frac{e_y}{y}\right)^2}$.

Oxidation Numbers and Balancing Redox Equations

The *oxidation number,* or *oxidation state,* is a bookkeeping device used to keep track of the number of electrons formally associated with a particular element. The oxidation number is meant to tell how many electrons have been lost or gained by a neutral atom when it forms a compound. Because oxidation numbers have no real physical meaning, they are somewhat arbitrary, and not all chemists will assign the same oxidation number to a given element in an unusual compound. However, there are some ground rules that provide a useful start.

1. The oxidation number of an element by itself—e.g., $Cu(s)$ or $Cl_2(g)$—is zero.

2. The oxidation number of H is almost always $+1$, except in metal hydrides—e.g., NaH—in which H is -1.

3. The oxidation number of oxygen is almost always -2. The only common exceptions are peroxides, in which two oxygen atoms are connected and each has an oxidation number of -1. Two examples are hydrogen peroxide $(H—O—O—H)$ and its anion $(H—O—O^-)$. The oxidation number of oxygen in gaseous O_2 is, of course, zero.

4. The alkali metals (Li, Na, K, Rb, Cs, Fr) almost always have an oxidation number $+1$. The alkaline earth metals (Be, Mg, Ca, Sr, Ba, Ra) are almost always in the $+2$ oxidation state.

5. The halogens (F, Cl, Br, I) are usually in the -1 oxidation state. Exceptions occur when two different halogens are bound to each other or when a halogen is bound to more than one atom. When different halogens are bound to each other, we assign the oxidation number -1 to the more electronegative halogen.

The sum of the oxidation numbers of each atom in a molecule must equal the charge of the molecule. In H_2O, for example, we have

$$2 \text{ hydrogen} = 2(+1) = +2$$
$$\text{oxygen} \qquad\qquad = -2$$
$$\text{net charge} \qquad\qquad\quad 0$$

In SO_4^{2-}, sulfur must have an oxidation number of $+6$ so that the sum of the oxidation numbers will be -2:

$$\text{oxygen} = 4(-2) = -8$$
$$\text{sulfur} \qquad\quad = +6$$
$$\text{net charge} \qquad\qquad -2$$

In benzene (C_6H_6), the oxidation number of each carbon must be -1 if hydrogen is assigned the number $+1$. In cyclohexane (C_6H_{12}), the oxidation number of each carbon must be -2, for the same reason. The carbons in benzene are in a higher oxidation state than those in cyclohexane.

The oxidation number of iodine in ICl_2^- is $+1$. This is unusual, because halogens are usually -1. However, because chlorine is more electronegative than iodine, we assign Cl as -1, thereby forcing I to be $+1$.

The oxidation number of As in As_2S_3 is $+3$, and the value for S is -2. This is arbitrary, but reasonable. Because S is more electronegative than As, we make S negative and As positive; and because S is in the same family as oxygen, which is usually -2, we assign S as -2, thus leaving As as $+3$.

The oxidation number of S in $S_4O_6^{2-}$ (tetrathionate) is $+2.5$. The *fractional oxidation state* comes about because six O atoms contribute -12. Because the charge is -2, the four S atoms must contribute $+10$. The average oxidation number of S must be $+\frac{10}{4} = 2.5$.

The oxidation number of Fe in $K_3Fe(CN)_6$ is $+3$. To make this assignment, we first recognize cyanide (CN^-) as a common ion that carries a charge of -1. Six cyanide ions give -6, and three potassium ions (K^+) give $+3$. Therefore, Fe should have an oxidation number of $+3$ for the whole formula to be neutral. In this approach, it is not necessary to assign individual oxidation numbers to carbon and nitrogen, as long as we recognize that the charge of CN is -1.

Problems

Answers are given at the end of this appendix.

D-1.   Write the oxidation state of the boldface atom in each of the following species.

(a) **Ag**Br

(b) **S**$_2$O$_3^{2-}$

(c) **Se**F$_6$

(d) H**S**$_2$O$_3^-$

(e) H**O**$_2$

(f) **N**O

(g) **Cr**$^{3+}$

(h) **Mn**O$_2$

(i) **Pb**(OH)$_3^-$

(j) **Fe**(OH)$_3$

(k) **Cl**O$^-$

(l) K$_4$**Fe**(CN)$_6$

(m) **Cl**O$_2$

(n) **Cl**O$_2^-$

(o) **Mn**(CN)$_6^{4-}$

(p) **N**$_2$

(q) **N**H$_4^+$

(r) **N**$_2$H$_5^+$

(s) HA**s**O$_3^{2-}$

(t) **Co**$_2$(CO)$_8$ (CO group is neutral)

(u) (**C**H$_3$)$_4$Li$_4$

(v) **P**$_4$O$_{10}$

(w) **C**$_2$H$_6$O (ethanol, CH$_3$CH$_2$OH)

(x) **V**O(SO$_4$)

(y) **Fe**$_3$O$_4$

(z) **C**$_3$H$_3^+$

Structure: $H-C \overset{\oplus}{\underset{|}{\big\langle}} \begin{matrix} C-H \\ \\ C-H \end{matrix}$

D-2. Identify the oxidizing agent and the reducing agent on the left side of each of the following reactions.

(a) $Cr_2O_7^{2-} + 3Sn^{2+} + 14H^+ \rightarrow$
$$2Cr^{3+} + 3Sn^{4+} + 7H_2O$$

(b) $4I^- + O_2 + 4H^+ \rightarrow 2I_2 + 2H_2O$

(c) $5CH_3\overset{\overset{\displaystyle O}{\|}}{C}H + 2MnO_4^- + 6H^+ \rightarrow$
$$5CH_3\overset{\overset{\displaystyle O}{\|}}{C}OH + 2Mn^{2+} + 3H_2O$$

(d) $HOCH_2CHOHCH_2OH + 2IO_4^- \rightarrow$
 Glycerol
$$2H_2C{=}O + HCO_2H + 2IO_3^- + H_2O$$
   Formaldehyde   Formic
                      acid

(e) $C_8H_8 + 2Na \rightarrow C_8H_8^{2-} + 2Na^+$
C_8H_8 is cyclooctatetraene with the structure

(f) $I_2 + OH^- \rightarrow HOI + I^-$
              Hypoiodous
              acid

Balancing Redox Reactions

To balance a reaction involving oxidation and reduction, we must first identify which element is oxidized and which is reduced. We then break the net reaction into two imaginary *half-reactions,* one of which involves only oxidation and the other only reduction. Although free electrons never appear in a balanced net reaction, they do appear in balanced half-reactions. If we are dealing with aqueous solutions, we proceed to balance each half-reaction, using H_2O and either H^+ or OH^-, as necessary. *A reaction is balanced when the number of atoms of each element is the same on both sides, and the net charge is the same on both sides.*[†]

Acidic Solutions

Here are the steps we will follow:

1. Assign oxidation numbers to the elements that are oxidized or reduced.

2. Break the reaction into two half-reactions, one involving oxidation and the other reduction.

3. For each half-reaction, balance the number of atoms that are oxidized or reduced.

. . . .

[†]A completely different method for balancing complex redox equations by inspection has been described by D. Kolb, *J. Chem. Ed.* **1981,** *58,* 642.

4. Balance the electrons to account for the change in oxidation number by adding electrons to one side of each half-reaction.

5. Balance oxygen atoms by adding H_2O to one side of each half-reaction.

6. Balance the H atoms by adding H^+ to one side of each half-reaction.

7. Multiply each half-reaction by the number of electrons in the other half-reaction so that the number of electrons on each side of the total reaction will cancel. Then add the two half-reactions and simplify to the smallest integral coefficients.

EXAMPLE   Balancing a Redox Equation

Balance the following equation using H^+, but not OH^-:

$$\underset{+2}{Fe^{2+}} + \underset{+7}{MnO_4^-} \rightleftharpoons \underset{+3}{Fe^{3+}} + \underset{+2}{Mn^{2+}}$$

Permanganate

Solution

1. *Assign oxidation numbers.* These are assigned for Fe and Mn in each species in the above reaction.

2. *Break the reaction into two half-reactions.*

   oxidation half-reaction: $\underset{+2}{Fe^{2+}} \rightleftharpoons \underset{+3}{Fe^{3+}}$

   reduction half-reaction: $\underset{+7}{MnO_4^-} \rightleftharpoons \underset{+2}{Mn^{2+}}$

3. *Balance the atoms that are oxidized or reduced.* Because there is only one Fe or Mn in each species on each side of the equation, the atoms of Fe and Mn are already balanced.

4. *Balance electrons.* Electrons are added to account for the change in each oxidation state.

$$Fe^{2+} \rightleftharpoons Fe^{3+} + e^-$$

$$MnO_4^- + 5e^- \rightleftharpoons Mn^{2+}$$

   In the second case, we need $5e^-$ on the left side to take Mn from $+7$ to $+2$.

5. *Balance oxygen atoms.* There are no oxygen atoms in the Fe half-reactions. There are four oxygen atoms on the left side of the Mn reaction, so we add four molecules of H_2O to the right side:

$$MnO_4^- + 5e^- \rightleftharpoons Mn^{2+} + 4H_2O$$

6. *Balance hydrogen atoms.* The Fe equation is already balanced. The Mn equation needs $8H^+$ on the left.

$$MnO_4^- + 5e^- + 8H^+ \rightarrow Mn^{2+} + 4H_2O$$

   At this point, each half-reaction must be completely balanced (the same number of atoms and charge on each side), *or you have made a mistake.*

7. *Multiply and add the reactions.* We multiply the Fe equation by 5 and the Mn equation by 1 and add:

$$5Fe^{2+} \rightleftharpoons 5Fe^{3+} + 5e^-$$

$$MnO_4^- + 5e^- + 8H^+ \rightleftharpoons Mn^{2+} + 4H_2O$$

$$\overline{5Fe^{2+} + MnO_4^- + 8H^+ \rightleftharpoons 5Fe^{3+} + Mn^{2+} + 4H_2O}$$

The total charge on each side is $+17$, and we find the same number of atoms of each element on each side. The equation is balanced.

· ·

EXAMPLE A Reverse Disproportionation

Now try the next reaction, which represents the reverse of a *disproportionation.* (In a disproportionation, an element in one oxidation state reacts to give the same element in higher and lower oxidation states.)

$$\underset{\substack{0 \\ \text{Iodine}}}{I_2} + \underset{\substack{+5 \\ \text{Iodate}}}{IO_3^-} + \underset{-1}{Cl^-} \rightleftharpoons \underset{+1\ -1}{ICl_2^-}$$

Solution

1. The oxidation numbers are assigned above. Note that chlorine has an oxidation number of -1 on both sides of the equation. Only iodine is involved in electron transfer.

2. oxidation half-reaction: $\underset{\substack{0 \quad\ +1}}{I_2 \rightleftharpoons ICl_2^-}$

   reduction half-reaction: $\underset{\substack{+5 \qquad +1}}{IO_3^- \rightleftharpoons ICl_2^-}$

3. We need to balance I atoms in the first reaction and add Cl^- to each reaction to balance Cl.

$$I_2 + 4Cl^- \rightleftharpoons 2ICl_2^-$$

$$IO_3^- + 2Cl^- \rightleftharpoons ICl_2^-$$

4. Now add electrons to each.

$$I_2 + 4Cl^- \rightleftharpoons 2ICl_2^- + 2e^-$$

$$IO_3^- + 2Cl^- + 4e^- \rightleftharpoons ICl_2^-$$

The first reaction needs $2e^-$ because there are two I atoms, each of which changes from 0 to $+1$.

5. The second reaction needs $3H_2O$ on the right side to balance oxygen atoms.

$$IO_3^- + 2Cl^- + 4e^- \rightleftharpoons ICl_2^- + 3H_2O$$

6. The first reaction is balanced, but the second needs $6H^+$ on the left.

$$IO_3^- + 2Cl^- + 4e^- + 6H^+ \rightleftharpoons ICl_2^- + 3H_2O$$

As a check, the charge on each side of this half-reaction is -1, and all atoms are balanced.

7. Multiply and add.

$$2(I_2 + 4Cl^- \qquad\qquad \rightleftharpoons 2ICl^- + 2e^-)$$

$$IO_3^- + 2Cl^- + 4e^- + 6H^+ \rightleftharpoons ICl_2^- + 3H_2O$$

$$2I_2 + IO_3^- + 10Cl^- + 6H^+ \rightleftharpoons 5ICl_2^- + 3H_2O \qquad (D\text{-}1)$$

We multiplied the first reaction by 2 so that there would be the same number of electrons in each half-reaction. You could have multiplied the first reaction by 4 and the second by 2, but then all coefficients would simply be doubled. We customarily write the smallest coefficients.

Basic Solutions

The method many people prefer for basic solutions is to balance the equation first with H^+. The answer can then be converted to one in which OH^- is used instead. This is done by adding to each side of the equation a number of hydroxide ions equal to the number of H^+ ions appearing in the equation. For example, to balance Equation D-1 with OH^- instead of H^+, proceed as follows:

$$2I_2 + IO_3^- + 10Cl^- + 6H^+ \rightleftharpoons 5ICl_2^- + 3H_2O$$
$$+6OH^- \qquad\qquad\qquad + 6OH^-$$

$$2I_2 + IO_3^- + 10Cl^- + \underbrace{6H^+ + 6OH^-} \rightleftharpoons 5ICl_2^- + 3H_2O + 6OH^-$$
$$6H_2O$$
$$\Downarrow$$
$$3H_2O$$

Realizing that $6H^+ + 6OH^- = 6H_2O$, and canceling $3H_2O$ on each side, gives the final result:

$$2I_2 + IO_3^- + 10Cl^- + 3H_2O \rightleftharpoons 5ICl_2^- + 6OH^-$$

Problems

D-3. Balance the following reactions by using H^+, but not OH^-.

(a) $Fe^{3+} + Hg_2^{2+} \rightleftharpoons Fe^{2+} + Hg^{2+}$

(b) $Ag + NO_3^- \rightleftharpoons Ag^+ + NO$

(c) $VO^{2+} + Sn^{2+} \rightleftharpoons V^{3+} + Sn^{4+}$

(d) $SeO_4^{2-} + Hg + Cl^- \rightleftharpoons SeO_3^{2-} + Hg_2Cl_2$

(e) $CuS + NO_3^- \rightleftharpoons Cu^{2+} + SO_4^{2-} + NO$

(f) $S_2O_3^{2-} + I_2 \rightleftharpoons I^- + S_4O_6^{2-}$

(g) $ClO_3^- + As_2S_3 \rightleftharpoons Cl^- + H_2AsO_4^- + SO_4^{2-}$

(h) $Cr_2O_7^{2-} + CH_3\overset{\displaystyle O}{\overset{\|}{C}}H \rightleftharpoons CH_3\overset{\displaystyle O}{\overset{\|}{C}}OH + Cr^{3+}$

(i) $MnO_4^{2-} \rightleftharpoons MnO_2 + MnO_4^-$

(j) $Hg_2SO_4 + Ca^{2+} + S_8 \rightleftharpoons Hg_2^{2+} + CaS_2O_3$

(k) $ClO_3^- \rightleftharpoons Cl_2 + O_2$

D-4. Balance the following equations by using OH^-, but not H^+.

(a) $PbO_2 + Cl^- \rightleftharpoons ClO^- + Pb(OH)_3^-$

(b) $HNO_2 + SbO^+ \rightleftharpoons NO + Sb_2O_5$

(c) $Ag_2S + CN^- + O_2 \rightleftharpoons S + Ag(CN)_2^- + OH^-$

(d) $HO_2^- + Cr(OH)_3 \rightleftharpoons CrO_4^{2-} + OH^-$

(e) $ClO_2 + OH^- \rightleftharpoons ClO_2^- + ClO_3^-$

(f) $WO_3^- + O_2 \rightleftharpoons HW_6O_{21}^{5-} + OH^-$

(g) $Mn_2O_3 + CN^- \rightleftharpoons Mn(CN)_6^{4-} + (CN)_2$

(h) $Cu^{2+} + H_2 \rightleftharpoons Cu + H_2O$

(i) $BH_4^- + H_2O \rightleftharpoons H_3BO_3 + H_2$

(j) $Mn_2O_3 + Hg + CN^- \rightleftharpoons Mn(CN)_6^{4-} + Hg(CN)_2$

(k) $MnO_4^- + \overset{\overset{\textstyle O}{\textstyle \|}}{H}CCH_2CH_2OH \rightleftharpoons CH_2(CO_2^-)_2 + MnO_2$

(l) $K_3V_5O_{14} + HOCH_2CHOHCH_2OH \rightleftharpoons VO(OH)_2 + HCO_2^- + K^+$

Answers

D-1.

(a) $+1$	**(j)** $+3$	**(s)** $+3$
(b) $+2$	**(k)** $+1$	**(t)** 0
(c) $+6$	**(l)** $+2$	**(u)** -4
(d) $+2$	**(m)** $+4$	**(v)** $+5$
(e) $-\dfrac{1}{2}$	**(n)** $+3$	**(w)** -2
(f) $+2$	**(o)** $+2$	**(x)** $+4$
(g) $+3$	**(p)** 0	**(y)** $+8/3$
(h) $+4$	**(q)** -3	**(z)** $-2/3$
(i) $+2$	**(r)** -2	

D-2.

	Oxidizing agent	Reducing agent
(a)	$Cr_2O_7^{2-}$	Sn^{2+}
(b)	O_2	I^-
(c)	MnO_4^-	CH_3CHO
(d)	IO_4^-	Glycerol
(e)	C_8H_8	Na
(f)	I_2	I_2

Reaction **(f)** is called a *disproportionation*, because an element in one oxidation state is transformed into two different oxidation states—one higher and one lower than the original oxidation state.

D-3. **(a)** $2Fe^{3+} + Hg_2^{2+} \rightleftharpoons 2Fe^{2+} + 2Hg^{2+}$

(b) $3Ag + NO_3^- + 4H^+ \rightleftharpoons 3Ag^+ + NO + 2H_2O$

(c) $4H^+ + 2VO^{2+} + Sn^{2+} \rightleftharpoons 2V^{3+} + Sn^{4+} + 2H_2O$

(d) $2Hg + 2Cl^- + SeO_4^{2-} + 2H^+ \rightleftharpoons Hg_2Cl_2 + SeO_3^{2-} + H_2O$

(e) $3CuS + 8NO_3^- + 8H^+ \rightleftharpoons 3Cu^{2+} + 3SO_4^{2-} + 8NO + 4H_2O$

(f) $2S_2O_3^{2-} + I_2 \rightleftharpoons S_4O_6^{2-} + 2I^-$

(g) $14ClO_3^- + 3As_2S_3 + 18H_2O \rightleftharpoons 14Cl^- + 6H_2AsO_4^- + 9SO_4^{2-} + 24H^+$

(h) $Cr_2O_7^{2-} + 3CH_3CHO + 8H^+ \rightleftharpoons 2Cr^{3+} + 3CH_3CO_2H + 4H_2O$

(i) $4H^+ + 3MnO_4^{2-} \rightleftharpoons MnO_2 + 2MnO_4^- + 2H_2O$

(j) $2Hg_2SO_4 + 3Ca^{2+} + \dfrac{1}{2}S_8 + H_2O \rightleftharpoons 2Hg_2^{2+} + 3CaS_2O_3 + 2H^+$

(k) $2H^+ + 2ClO_3^- \rightleftharpoons Cl_2 + \dfrac{5}{2}O_2 + H_2O$

The balanced half-reaction for As_2S_3 in **(g)** is

$As_2S_3 + 20H_2O \rightleftharpoons$
$\underset{+3\ -2}{}$
$\qquad 2\underset{+5}{H_2AsO_4^-} + 3\underset{+6}{SO_4^{2-}} + 28e^- + 36H^+$

Because As_2S_3 is a single compound, we must consider the $As_2S_3 \rightarrow H_2AsO_4^-$ and $As_2S_3 \rightarrow SO_4^{2-}$ reactions together. The net change in oxidation number for the *two* As atoms is $2(5-3) = +4$. The net change in oxidation number for the *three* S atoms is $3[6-(-2)] = +24$. Therefore, $24 + 4 = 28e^-$ are involved in the half-reaction.

D-4. **(a)** $H_2O + OH^- + PbO_2 + Cl^- \rightleftharpoons Pb(OH)_3^- + ClO^-$

(b) $4HNO_2 + 2SbO^+ + 2OH^- \rightleftharpoons 4NO + Sb_2O_5 + 3H_2O$

(c) $Ag_2S + 4CN^- + \dfrac{1}{2}O_2 + H_2O \rightleftharpoons S + 2Ag(CN)_2^- + 2OH^-$

(d) $2HO_2^- + Cr(OH)_3^- \rightleftharpoons CrO_4^{2-} + OH^- + 2H_2O$

(e) $2ClO_2 + 2OH^- \rightleftharpoons ClO_2^- + ClO_3^- + H_2O$

(f) $12WO_3^- + 3O_2 + 2H_2O \rightleftharpoons 2HW_6O_{21}^{5-} + 2OH^-$

(g) $Mn_2O_3 + 14CN^- + 3H_2O \rightleftharpoons 2Mn(CN)_6^{4-} + (CN)_2 + 6OH^-$

(h) $Cu^{2+} + H_2 + 2OH^- \rightleftharpoons Cu + 2H_2O$

(i) $BH_4^- + 4H_2O \rightleftharpoons H_3BO_3 + 4H_2 + OH^-$

(j) $3H_2O + Mn_2O_3 + Hg + 14CN^- \rightleftharpoons 2Mn(CN)_6^{4-} + Hg(CN)_2 + 6OH^-$

(k) $2MnO_4^- + \overset{\overset{\textstyle O}{\textstyle \|}}{H}CCH_2CH_2OH \rightleftharpoons 2MnO_2 + 2H_2O + CH(CO_2^-)_2$

For **(k)**, the organic half-reaction is $8OH^- + C_3H_6O_2 \rightleftharpoons C_3H_2O_4^{2-} + 6e^- + 6H_2O$.

(l) $32H_2O + 8K_3V_5O_{14} + 5HOCH_2CHOHCH_2OH \rightleftharpoons 40VO(OH)_2 + 15HCO_2^- + 9OH^- + 24K^+$

For **(l)**, the two half-reactions are $K_3V_5O_{14} + 9H_2O + 5e^- \rightleftharpoons 5VO(OH)_2 + 8OH^- + 3K^+$ and $C_3H_8O_3 + 11OH^- \rightleftharpoons 3HCO_2^- + 8e^- + 8H_2O$.

Normality

The *normality,* N, of a redox reagent is *n* times the molarity, where *n* is the number of electrons donated or accepted by that species in a chemical reaction.

$$N = nM \tag{E-1}$$

For example, in the half-reaction

$$MnO_4^- + 8H^+ + 5e^- \rightleftharpoons Mn^{2+} + 4H_2O \tag{E-2}$$

the normality of permanganate ion is five times its molarity, because each MnO_4^- accepts $5e^-$. If the molarity of permanganate is 0.1 M, the normality for the reaction

$$MnO_4^- + 5Fe^{2+} + 8H^+ \rightleftharpoons Mn^{2+} + 5Fe^{3+} + 4H_2O \tag{E-3}$$

is $5 \times 0.1 = 0.5$ N (read "0.5 normal"). In this reaction, each Fe^{2+} ion donates one electron. The normality of ferrous ion *equals* the molarity of ferrous ion, even though it takes five ferrous ions to balance the reaction.

In the half-reaction

$$MnO_4^- + 4H^+ + 3e^- \rightleftharpoons MnO_2 + 2H_2O \tag{E-4}$$

each MnO_4^- ion accepts only *three* electrons. The normality of permanganate for this reaction is equal to three times the molarity of permanganate. A 0.06 N permanganate solution for this reaction contains 0.02 M MnO_4^-.

The normality of a solution is a statement of the moles of "reacting units" per liter. One mole of reacting units is called one *equivalent.* Therefore, the units of normality are equivalents per liter (equiv/L). For redox reagents, *one equivalent is the amount of substance that can donate or accept one electron.* It is possible to speak of equivalents only with respect to a particular half-reaction. For example, in Reaction E-2, there are five equivalents per mole of MnO_4^-; but in Reaction E-4, there are only three equivalents per mole of MnO_4^-. The mass of substance containing one equivalent is called the *equivalent weight.* The formula weight of $KMnO_4$ is

158.0339. The equivalent weight of $KMnO_4$ for Reaction E-2 is $158.0339/5 = 31.6068$ g/equiv. The equivalent weight of $KMnO_4$ for Reaction E-4 is $158.0339/3 = 52.6780$ g/equiv.

EXAMPLE Finding Normality

Find the normality of a solution containing 6.34 g of ascorbic acid in 250.0 mL if the relevant half-reaction is

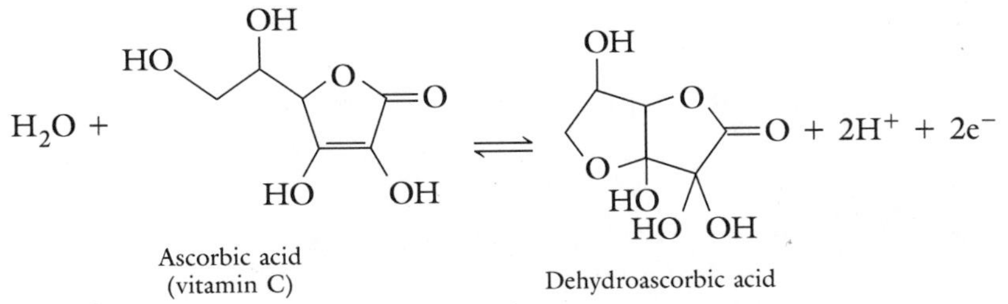

Ascorbic acid
(vitamin C)

Dehydroascorbic acid

Solution The formula weight of ascorbic acid ($C_6H_8O_6$) is 176.126. In 6.34 g, there are $(6.34 \text{ g})/(176.126 \text{ g/mol}) = 3.60 \times 10^{-2}$ mol. Because each mole contains 2 equivalents in this example, $6.34 \text{ g} = (2 \text{ equiv/mol})(3.60 \times 10^{-2} \text{ mol}) = 7.20 \times 10^{-2}$ equivalent. The normality is $(7.20 \times 10^{-2} \text{ equiv})/(0.2500 \text{ L}) = 0.288$ N.

EXAMPLE Using Normality

How many grams of potassium oxalate should be dissolved in 500.0 mL to make a 0.100 N solution for titration of MnO_4^-?

$$5H_2C_2O_4 + 2MnO_4^- + 6H^+ \rightleftharpoons 2Mn^{2+} + 10CO_2 + 8H_2O \quad \text{(E-5)}$$

Solution It is first necessary to write the oxalic acid half-reaction:

$$H_2C_2O_4 \rightleftharpoons 2CO_2 + 2H^+ + 2e^-$$

It is apparent that there are two equivalents per mole of oxalic acid. Hence, a 0.100 N solution will be 0.0500 M:

$$\frac{0.100 \text{ equiv/L}}{2 \text{ equiv/mol}} = 0.0500 \text{ mol/L} = 0.0500 \text{ M}$$

Therefore, we must dissolve $(0.0500 \text{ mol/L})(0.5000 \text{ L}) = 0.0250$ mol in 500.0 mL. Because the formula weight of $K_2C_2O_4$ is 166.216, we should make use of $(0.0250 \text{ mol}) \times (166.216 \text{ g/mol}) = 4.15$ g of potassium oxalate.

The utility of normality in volumetric analysis lies in the equation

$$N_1V_1 = N_2V_2 \quad \text{(E-6)}$$

where N_1 is the normality of reagent 1, V_1 is the volume of reagent 1, N_2 is the normality of reagent 2, and V_2 is the volume of reagent 2. V_1 and V_2 may be expressed in any units, as long as the same units are used for both.

EXAMPLE   A solution containing 25.0 mL of oxalic acid required 13.78 mL of 0.041 62 N $KMnO_4$ for titration, according to Reaction E-5. Find the normality and molarity of the oxalic acid.

Solution   Setting up Equation E-6, we write

$$N_1(25.0 \text{ mL}) = (0.041 62 \text{ N})(13.78 \text{ mL})$$

$$N_1 = 0.022 94 \text{ equiv/L}$$

Because there are two equivalents per mole of oxalic acid in Reaction E-5,

$$M = \frac{N}{n} = \frac{0.022 94}{2} = 0.011 47 \text{ M}$$

Normality is sometimes used in acid-base or ion-exchange chemistry. With respect to acids and bases, the equivalent weight of a reagent is the amount that can donate or accept one mole of H^+. With respect to ion exchange, the equivalent weight is the mass of reagent containing one mole of charge.

Solubility Products†

Formula	pK_{sp}	K_{sp}	Temperature (°C)	Ionic strength (M)
Azides: $L = N_3^-$				
CuL	8.31	4.9×10^{-9}	25	0
AgL	8.56	2.8×10^{-9}	25	0
Hg_2L_2	9.15	7.1×10^{-10}	25	0
TlL	3.66	2.2×10^{-4}	25	0
$PdL_2(\alpha)$	8.57	2.7×10^{-9}	25	0
Bromates: $L = BrO_3^-$				
$BaL \cdot H_2O$	5.11	7.8×10^{-6}	25	0.5
AgL	4.26	5.5×10^{-5}	25	0
TlL	3.78	1.7×10^{-4}	25	0
PbL_2	5.10	7.9×10^{-6}	25	0
Bromides: $L = Br^-$				
CuL	8.3	5×10^{-9}	25	0
AgL	12.30	5.0×10^{-13}	25	0
Hg_2L_2	22.25	5.6×10^{-23}	25	0
TlL	5.44	3.6×10^{-6}	25	0
HgL_2	18.9	1.3×10^{-19}	25	0.5
PbL_2	5.68	2.1×10^{-6}	25	0
Carbonates: $L = CO_3^{2-}$				
MgL	7.46	3.5×10^{-8}	25	0
CaL(calcite)	8.35	4.5×10^{-9}	25	0
CaL(aragonite)	8.22	6.0×10^{-9}	25	0
SrL	9.03	9.3×10^{-10}	25	0
BaL	8.30	5.0×10^{-9}	25	0
Y_2L_3	30.6	2.5×10^{-31}	25	0
La_2L_3	33.4	4.0×10^{-34}	25	0
MnL	9.30	5.0×10^{-10}	25	0
FeL	10.68	2.1×10^{-11}	25	0
CoL	9.98	1.0×10^{-10}	25	0
NiL	6.87	1.3×10^{-7}	25	0
CuL	9.63	2.3×10^{-10}	25	0
Ag_2L	11.09	8.1×10^{-12}	25	0
Hg_2L	16.05	8.9×10^{-17}	25	0
ZnL	10.00	1.0×10^{-10}	25	0
CdL	13.74	1.8×10^{-14}	25	0
PbL	13.13	7.4×10^{-14}	25	0

†The designations α, β, or γ after some formulas refer to particular crystalline forms (which are customarily identified by Greek letters). Data for salts except oxalates are taken mainly from A. E. Martell and R. M. Smith, *Critical Stability Constants,* Vol. 4 (New York: Plenum Press, 1976). Data for oxalates are from L. G. Sillén and A. E. Martell, *Stability Constants of Metal-Ion Complexes,* Supplement No. 1 (London: The Chemical Society, Special Publication No. 25, 1971). Another source: R. M. H. Verbeeck et al., *Inorg. Chem.* **1984,** *23,* 1922.

(Continued)

Formula	pK_{sp}	K_{sp}	Temperature (°C)	Ionic strength (M)
Chlorides: L = Cl$^-$				
CuL	6.73	1.9×10^{-7}	25	0
AgL	9.74	1.8×10^{-10}	25	0
Hg$_2$L$_2$	17.91	1.2×10^{-18}	25	0
TlL	3.74	1.8×10^{-4}	25	0
PbL$_2$	4.78	1.7×10^{-5}	25	0
Chromates: L = CrO$_4^{2-}$				
BaL	9.67	2.1×10^{-10}	25	0
CuL	5.44	3.6×10^{-6}	25	0
Ag$_2$L	11.92	1.2×10^{-12}	25	0
Hg$_2$L	8.70	2.0×10^{-9}	25	0
Tl$_2$L	12.01	9.8×10^{-13}	25	0
Cobalticyanides: L = Co(CN)$_6^{3-}$				
Ag$_3$L	25.41	3.9×10^{-26}	25	0
(Hg$_2$)$_3$L$_2$	36.72	1.9×10^{-37}	25	0
Cyanides: L = CN$^-$				
AgL	15.66	2.2×10^{-16}	25	0
Hg$_2$L$_2$	39.3	5×10^{-40}	25	0
ZnL$_2$	15.5	3×10^{-16}	25	3
Ferrocyanides: L = Fe(CN)$_6^{4-}$				
Ag$_4$L	44.07	8.5×10^{-45}	25	0
Zn$_2$L	15.68	2.1×10^{-16}	25	0
Cd$_2$L	17.38	4.2×10^{-18}	25	0
Pb$_2$L	18.02	9.5×10^{-19}	25	0
Fluorides: L = F$^-$				
LiL	2.77	1.7×10^{-3}	25	0
MgL$_2$	8.18	6.6×10^{-9}	25	0
CaL$_2$	10.41	3.9×10^{-11}	25	0
SrL$_2$	8.54	2.9×10^{-9}	25	0
BaL$_2$	5.76	1.7×10^{-6}	25	0
ThL$_4$	28.3	5×10^{-29}	25	0
PbL$_2$	7.44	3.6×10^{-8}	25	0
Hydroxides: L = OH$^-$				
MgL$_2$	11.15	7.1×10^{-12}	25	0
CaL$_2$	5.19	6.5×10^{-6}	25	0
BaL$_2 \cdot$ 8H$_2$O	3.6	3×10^{-4}	25	0
YL$_3$	23.2	6×10^{-24}	25	0
LaL$_3$	20.7	2×10^{-21}	25	0
CeL$_3$	21.2	6×10^{-22}	25	0
UO$_2$ ($\rightleftharpoons$ U^{4+} + 4OH$^-$)	56.2	6×10^{-57}	25	0
UO$_2$L$_2$($\rightleftharpoons$ UO$_2^{2+}$ + 2OH$^-$)	22.4	4×10^{-23}	25	0
MnL$_2$	12.8	1.6×10^{-13}	25	0
FeL$_2$	15.1	7.9×10^{-16}	25	0
CoL$_2$	14.9	1.3×10^{-15}	25	0
NiL$_2$	15.2	6×10^{-16}	25	0
CuL$_2$	19.32	4.8×10^{-20}	25	0
VL$_3$	34.4	4.0×10^{-35}	25	0
CrL$_3$	29.8	1.6×10^{-30}	25	0.1
FeL$_3$	38.8	1.6×10^{-39}	25	0
CoL$_3$	44.5	3×10^{-45}	19	0
VOL$_2$($\rightleftharpoons$ VO^{2+} + 2OH$^-$)	23.5	3×10^{-24}	25	0
PdL$_2$	28.5	3×10^{-29}	25	0
ZnL$_2$(amorphous)	15.52	3.0×10^{-16}	25	0

(Continued)

Formula	pK_{sp}	K_{sp}	Temperature (°C)	Ionic strength (M)
$CdL_2(\beta)$	14.35	4.5×10^{-15}	25	0
HgO (red) ($\rightleftharpoons Hg^{2+} + 2OH^-$)	25.44	3.6×10^{-26}	25	0
$Cu_2O (\rightleftharpoons 2Cu^+ + 2OH^-)$	29.4	4×10^{-30}	25	0
$Ag_2O (\rightleftharpoons 2Ag^+ + 2OH^-)$	15.42	3.8×10^{-16}	25	0
AuL_3	5.5	3×10^{-6}	25	0
$AlL_3(\alpha)$	33.5	3×10^{-34}	25	0
GaL_3(amorphous)	37	10^{-37}	25	0
InL_3	36.9	1.3×10^{-37}	25	0
SnO ($\rightleftharpoons Sn^{2+} + 2OH^-$)	26.2	6×10^{-27}	25	0
PbO (yellow) ($\rightleftharpoons Pb^{2+} + 2OH^-$)	15.1	8×10^{-16}	25	0
PbO (red) ($\rightleftharpoons Pb^{2+} + 2OH^-$)	15.3	5×10^{-16}	25	0
Iodates: L = IO_3^-				
CaL_2	6.15	7.1×10^{-7}	25	0
SrL_2	6.48	3.3×10^{-7}	25	0
BaL_2	8.81	1.5×10^{-9}	25	0
YL_3	10.15	7.1×10^{-11}	25	0
LaL_3	10.99	1.0×10^{-11}	25	0
CeL_3	10.86	1.4×10^{-11}	25	0
ThL_4	14.62	2.4×10^{-15}	25	0.5
$UO_2L_2 (\rightleftharpoons UO_2^{2+} + 2IO_3^-)$	7.01	9.8×10^{-8}	25	0.2
CrL_3	5.3	5×10^{-6}	25	0.5
AgL	7.51	3.1×10^{-8}	25	0
Hg_2L_2	17.89	1.3×10^{-18}	25	0
TlL	5.51	3.1×10^{-6}	25	0
ZnL_2	5.41	3.9×10^{-6}	25	0
CdL_2	7.64	2.3×10^{-8}	25	0
PbL_2	12.61	2.5×10^{-13}	25	0
Iodides: L = I^-				
CuL	12.0	1×10^{-12}	25	0
AgL	16.08	8.3×10^{-17}	25	0
$CH_3HgL (\rightleftharpoons CH_3Hg^+ + I^-)$	11.46	3.5×10^{-12}	20	1
$CH_3CH_2HgL (\rightleftharpoons CH_3CH_2Hg^+ + I^-)$	4.11	7.8×10^{-5}	25	1
TlL	7.23	5.9×10^{-8}	25	0
Hg_2L_2	27.95	1.1×10^{-28}	25	0.5
SnL_2	5.08	8.3×10^{-6}	25	4
PbL_2	8.10	7.9×10^{-9}	25	0
Oxalates: L = $C_2O_4^{2-}$				
CaL	7.9	1.3×10^{-8}	20	0.1
SrL	6.4	4×10^{-7}	20	0.1
BaL	6.0	1×10^{-6}	20	0.1
La_2L_3	25.0	1×10^{-25}	20	0.1
ThL_2	21.38	4.2×10^{-22}	25	1
$UO_2L (\rightleftharpoons UO_2^{2+} + C_2O_4^{2-})$	8.66	2.2×10^{-9}	20	0.1
Phosphates: L = PO_4^{3-}				
$MgHL \cdot 3H_2O (\rightleftharpoons Mg^{2+} + HL^{2-})$	5.78	1.7×10^{-6}	25	0
$CaHL \cdot 2H_2O (\rightleftharpoons Ca^{2+} + HL^{2-})$	6.58	2.6×10^{-7}	25	0
$SrHL (\rightleftharpoons Sr^{2+} + HL^{2-})$	6.92	1.2×10^{-7}	20	0
$BaHL (\rightleftharpoons Ba^{2+} + HL^{2-})$	7.40	4.0×10^{-8}	20	0
LaL	22.43	3.7×10^{-23}	25	0.5
$Fe_3L_2 \cdot 8H_2O$	36.0	1×10^{-36}	25	0
$FeL \cdot 2H_2O$	26.4	4×10^{-27}	25	0
$(VO)_3L_2 (\rightleftharpoons 3VO^{2+} + 2L^{3-})$	25.1	8×10^{-26}	25	0
Ag_3L	17.55	2.8×10^{-18}	25	0
$Hg_2HL (\rightleftharpoons Hg_2^{2+} + HL^{2-})$	12.40	4.0×10^{-13}	25	0

(Continued)

Formula	pK_{sp}	K_{sp}	Temperature (°C)	Ionic strength (M)
$Zn_3L_2 \cdot 4H_2O$	35.3	5×10^{-36}	25	0
Pb_3L_2	43.53	3.0×10^{-44}	38	0
GaL	21.0	1×10^{-21}	25	1
InL	21.63	2.3×10^{-22}	25	1
Sulfates: L = SO_4^{2-}				
CaL	4.62	2.4×10^{-5}	25	0
SrL	6.50	3.2×10^{-7}	25	0
BaL	9.96	1.1×10^{-10}	25	0
RaL	10.37	4.3×10^{-11}	20	0
Ag_2L	4.83	1.5×10^{-5}	25	0
Hg_2L	6.13	7.4×10^{-7}	25	0
PbL	6.20	6.3×10^{-7}	25	0
Sulfides: L = S^{2-}				
MnL (pink)	10.5	3×10^{-11}	25	0
MnL (green)	13.5	3×10^{-14}	25	0
FeL	18.1	8×10^{-19}	25	0
$CoL(\alpha)$	21.3	5×10^{-22}	25	0
$CoL(\beta)$	25.6	3×10^{-26}	25	0
$NiL(\alpha)$	19.4	4×10^{-20}	25	0
$NiL(\beta)$	24.9	1.3×10^{-25}	25	0
$NiL(\gamma)$	26.6	3×10^{-27}	25	0
CuL	36.1	8×10^{-37}	25	0
Cu_2L	48.5	3×10^{-49}	25	0
Ag_2L	50.1	8×10^{-51}	25	0
Tl_2L	21.2	6×10^{-22}	25	0
$ZnL(\alpha)$	24.7	2×10^{-25}	25	0
$ZnL(\beta)$	22.5	3×10^{-23}	25	0
CdL	27.0	1×10^{-27}	25	0
HgL (black)	52.7	2×10^{-53}	25	0
HgL (red)	53.3	5×10^{-54}	25	0
SnL	25.9	1.3×10^{-26}	25	0
PbL	27.5	3×10^{-28}	25	0
In_2L_3	69.4	4×10^{-70}	25	0
Thiocyanates: L = SCN^-				
CuL	13.40	4.0×10^{-14}	25	5
AgL	11.97	1.1×10^{-12}	25	0
Hg_2L_2	19.52	3.0×10^{-20}	25	0
TlL	3.79	1.6×10^{-4}	25	0
HgL_2	19.56	2.8×10^{-20}	25	1

Acid Dissociation Constants

·A·P·P·E·N·D·I·X·

G

Name	Structure[†]	pK_a[‡]	K_a
Acetic acid (ethanoic acid)	CH_3CO_2H	4.757	1.75×10^{-5}
Alanine	$\overset{\displaystyle NH_3^+}{\underset{\displaystyle CO_2H}{\mid}}\!\!CHCH_3$	2.348 (CO_2H) 9.867 (NH_3)	4.49×10^{-3} 1.36×10^{-10}
Aminobenzene (aniline)	$\text{C}_6\text{H}_5\text{—}NH_3^+$	4.601	2.51×10^{-5}
4-Aminobenzenesulfonic acid (sulfanilic acid)	$^-O_3S\text{—}\text{C}_6\text{H}_4\text{—}NH_3^+$	3.232	5.86×10^{-4}
2-Aminobenzoic acid (anthranilic acid)	(benzene ring with NH_3^+ and CO_2H)	2.08 (CO_2H) 4.96 (NH_3)	8.3×10^{-3} 1.10×10^{-5}
2-Aminoethanethiol (2-mercaptoethylamine)	$HSCH_2CH_2NH_3^+$	8.21 (SH) ($\mu = 0.1$) 10.71 (NH_3) ($\mu = 0.1$)	6.2×10^{-9} 1.95×10^{-11}
2-Aminoethanol (ethanolamine)	$HOCH_2CH_2NH_3^+$	9.498	3.18×10^{-10}
2-Aminophenol	(benzene ring with OH and NH_3^+)	4.78 (NH_3) (20°) 9.97 (OH) (20°)	1.66×10^{-5} 1.05×10^{-10}
Ammonia	NH_4^+	9.244	5.70×10^{-10}
Arginine	$\overset{\displaystyle NH_3^+}{\underset{\displaystyle CO_2H}{\mid}}\!\!CHCH_2CH_2CH_2NHC\overset{\displaystyle NH_2^+}{\underset{\displaystyle NH_2}{<}}$	1.823 (CO_2H) 8.991 (NH_3) (12.48) (NH_2)	1.50×10^{-2} 1.02×10^{-9} 3.3×10^{-13}

[†] Each acid is written in its protonated form. The acidic protons are indicated in bold type.
[‡] pK_a values refer to 25°C and zero ionic strength unless otherwise indicated. Values in parentheses are considered to be less reliable. Data are from A. E. Martell and R. M. Smith, *Critical Stability Constants* (New York: Plenum Press, 1974).

(Continued)

Name	Structure	pK_a	K_a
Arsenic acid (hydrogen arsenate)	HO—As(=O)—OH, OH	2.24 6.96 11.50	5.8×10^{-3} 1.10×10^{-7} 3.2×10^{-12}
Arsenious acid (hydrogen arsenite)	$As(OH)_3$	9.29	5.1×10^{-10}
Asparagine	NH_3^+ , CHCH$_2$CNH$_2$ (=O), CO$_2$H	2.14 (CO$_2$H) ($\mu = 0.1$) 8.72 (NH$_3$) ($\mu = 0.1$)	7.2×10^{-3} 1.9×10^{-9}
Aspartic acid	NH_3^+ , CHCH$_2$CO$_2$H , CO$_2$H	1.990 (α-CO$_2$H) 3.900 (β-CO$_2$H) 10.002 (NH$_3$)	1.02×10^{-2} 1.26×10^{-4} 9.95×10^{-11}
Aziridine (dimethyleneimine)	$\triangle NH_2^+$	8.04	9.1×10^{-9}
Benzene-1,2,3-tricarboxylic acid (hemimellitic acid)	CO$_2$H, CO$_2$H, CO$_2$H (benzene ring)	2.88 4.75 7.13	1.32×10^{-3} 1.78×10^{-5} 7.4×10^{-8}
Benzoic acid	C$_6$H$_5$—CO$_2$H	4.202	6.28×10^{-5}
Benzylamine	C$_6$H$_5$—CH$_2$NH$_3^+$	9.35	4.5×10^{-10}
2,2'-Bipyridine	(bipyridine structure) N, N, H$^+$	4.35	4.5×10^{-5}
Boric acid (hydrogen borate)	$B(OH)_3$	9.236 (12.74) (20°) (13.80) (20°)	5.81×10^{-10} 1.82×10^{-13} 1.58×10^{-14}
Bromoacetic acid	$BrCH_2CO_2H$	2.902	1.25×10^{-3}
Butane-2,3-dione dioxime (dimethylglyoxime)	HON, NOH, CH$_3$, CH$_3$	10.66 12.0	2.2×10^{-11} 1×10^{-12}
Butanoic acid	$CH_3CH_2CH_2CO_2H$	4.819	1.52×10^{-5}
cis-Butenedioic acid (maleic acid)	CO$_2$H, CO$_2$H (cis structure)	1.910 6.332	1.23×10^{-2} 4.66×10^{-7}

(Continued)

Name	Structure	pK_a	K_a
trans-Butenedioic acid (fumaric acid)	CO$_2$H / HO$_2$C (structure)	3.053 4.494	8.85×10^{-4} 3.21×10^{-5}
Butylamine	$CH_3CH_2CH_2CH_2NH_3^+$	10.640	2.29×10^{-11}
Carbonic acid[†] (hydrogen carbonate)	HO—C(=O)—OH	6.352 10.329	4.45×10^{-7} 4.69×10^{-11}
Chloroacetic acid	$ClCH_2CO_2H$	2.865	1.36×10^{-3}
3-Chloropropanoic acid	$ClCH_2CH_2CO_2H$	4.11	7.8×10^{-5}
Chlorous acid (hydrogen chlorite)	HOCl=O	1.95	1.12×10^{-2}
Chromic acid (hydrogen chromate)	HO—Cr(=O)(=O)—OH	−0.2 (20°) 6.51	1.6 3.1×10^{-7}
Citric acid (2-hydroxypropane-1,2,3-tricarboxylic acid)	$HO_2CCH_2CCH_2CO_2H$ with CO$_2$H and OH	3.128 4.761 6.396	7.44×10^{-4} 1.73×10^{-5} 4.02×10^{-7}
Cyanoacetic acid	$NCCH_2CO_2H$	2.472	3.37×10^{-3}
Cyclohexylamine	(cyclohexyl)—NH$_3^+$	10.64	2.3×10^{-11}
Cysteine	NH$_3^+$ / CHCH$_2$SH / CO$_2$H	(1.71) (CO$_2$H) 8.36 (SH) 10.77 (NH$_3$)	1.95×10^{-2} 4.4×10^{-9} 1.70×10^{-11}
Dichloroacetic acid	Cl_2CHCO_2H	1.30	5.0×10^{-2}
Diethylamine	$(CH_3CH_2)_2NH_2^+$	10.933	1.17×10^{-11}
1,2-Dihydroxybenzene (catechol)	benzene with OH, OH	9.40 12.8	4.0×10^{-10} 1.6×10^{-13}
1,3-Dihydroxybenzene (resorcinol)	benzene with OH, OH	9.30 11.06	5.0×10^{-10} 8.7×10^{-12}

[†] The concentration of "carbonic acid" is considered to be the sum $[H_2CO_3] + [CO_2(aq)]$. See Box 5-4.

(Continued)

Name	Structure	pK_a	K_a
D-2,3-Dihydroxybutanedioc acid (D-tartaric acid)	$\overset{\displaystyle OH}{\underset{\displaystyle OH}{\overset{\displaystyle \vert}{\underset{\displaystyle \vert}{HO_2CCHCHCO_2H}}}}$	3.036 4.366	9.20×10^{-4} 4.31×10^{-5}
2,3-Dimercaptopropanol	$\underset{\displaystyle SH}{\overset{\displaystyle \vert}{HOCH_2CHCH_2SH}}$	8.58 ($\mu = 0.1$) 10.68 ($\mu = 0.1$)	2.6×10^{-9} 2.1×10^{-11}
Dimethylamine	$(CH_3)_2NH_2^+$	10.774	1.68×10^{-11}
2,4-Dinitrophenol	$O_2N-\underset{\displaystyle}{\bigcirc}\overset{\displaystyle NO_2}{}-OH$	4.11	7.8×10^{-5}
Ethane-1,2-dithiol	$HSCH_2CH_2SH$	8.85 (30°, $\mu = 0.1$) 10.43 (30°, $\mu = 0.1$)	1.4×10^{-9} 3.7×10^{-11}
Ethylamine	$CH_3CH_2NH_3^+$	10.636	2.31×10^{-11}
Ethylenediamine (1,2-diaminoethane)	$H_3\overset{+}{N}CH_2CH_2\overset{+}{N}H_3$	6.848 9.928	1.42×10^{-7} 1.18×10^{-10}
Ethylenedinitrilotetraacetic acid (EDTA)	$(HO_2CCH_2)_2\overset{+}{N}HCH_2CH_2\overset{+}{N}H(CH_2CO_2H)_2$	0.0 (CO_2H) ($\mu = 1.0$) 1.5 (CO_2H) ($\mu = 0.1$) 2.0 (CO_2H) ($\mu = 0.1$) 2.66 (CO_2H) ($\mu = 0.1$) 6.16 (NH) ($\mu = 0.1$) 10.24 (NH) ($\mu = 0.1$)	1.0 0.032 0.010 0.002 2 6.9×10^{-7} 5.8×10^{-11}
Formic acid (methanoic acid)	HCO_2H	3.745	1.80×10^{-4}
Glutamic acid	$\overset{\displaystyle NH_3^+}{\underset{\displaystyle CO_2H}{\overset{\displaystyle \vert}{\underset{\displaystyle \vert}{CHCH_2CH_2CO_2H}}}}$	2.23 (α-CO_2H) 4.42 (γ-CO_2H) 9.95 (NH_3)	5.9×10^{-3} 3.8×10^{-5} 1.12×10^{-10}
Glutamine	$\overset{\displaystyle NH_3^+}{\underset{\displaystyle CO_2H}{\overset{\displaystyle \vert}{\underset{\displaystyle \vert}{CHCH_2CH_2\overset{\overset{\displaystyle O}{\displaystyle \Vert}}{C}NH_2}}}}$	2.17 (CO_2H) ($\mu = 0.1$) 9.01 (NH_3) ($\mu = 0.1$)	6.8×10^{-3} 9.8×10^{-10}
Glycine (aminoacetic acid)	$\overset{\displaystyle NH_3^+}{\underset{\displaystyle CO_2H}{\overset{\displaystyle \vert}{\underset{\displaystyle \vert}{CH_2}}}}$	2.350 (CO_2H) 9.778 (NH_3)	4.47×10^{-3} 1.67×10^{-10}
Guanidine	$H_2N-\overset{\overset{\displaystyle {}^+NH_2}{\displaystyle \Vert}}{C}-NH_2$	13.54 (27°, $\mu = 1.0$)	2.9×10^{-14}
1,6-Hexanedioic acid (adipic acid)	$HO_2CCH_2CH_2CH_2CH_2CO_2H$	4.42 5.42	3.8×10^{-5} 3.8×10^{-6}
Hexane-2,4-dione	$CH_3\overset{\overset{\displaystyle O}{\displaystyle \Vert}}{C}CH_2\overset{\overset{\displaystyle O}{\displaystyle \Vert}}{C}CH_2CH_3$	9.38	4.2×10^{-10}

(Continued)

Name	Structure	pK_a	K_a
Histidine	$\overset{NH_3^+}{\underset{\underset{CO_2H}{\mid}}{\underset{\mid}{CHCH_2}}}$ (imidazole ring)	1.7 (CO_2H) ($\mu = 0.1$) 6.02 (NH) ($\mu = 0.1$) 9.08 (NH_3) ($\mu = 0.1$)	2×10^{-2} 9.5×10^{-7} 8.3×10^{-10}
Hydrazoic acid (hydrogen azide)	$HN\overset{+}{=}N\overset{-}{=}N$	4.65	2.2×10^{-5}
Hydrogen cyanate	$HOC\equiv N$	3.48	3.3×10^{-4}
Hydrogen cyanide	$HC\equiv N$	9.21	6.2×10^{-10}
Hydrogen fluoride	HF	3.17	6.8×10^{-4}
Hydrogen peroxide	$HOOH$	11.65	2.2×10^{-12}
Hydrogen sulfide	H_2S	7.02 13.9†	9.5×10^{-8} 1.3×10^{-14}†
Hydrogen thiocyanate	$HSC\equiv N$	0.9	0.13
Hydroxyacetic acid (glycolic acid)	$HOCH_2CO_2H$	3.831	1.48×10^{-4}
Hydroxybenzene (phenol)	(benzene ring)—OH	9.98	1.05×10^{-10}
2-Hydroxybenzoic acid (salicylic acid)	(benzene ring with CO_2H and OH)	2.97 (CO_2H) 13.74 (OH)	1.07×10^{-3} 1.82×10^{-14}
L-Hydroxybutanedioic acid (malic acid)	$\overset{OH}{\underset{\mid}{HO_2CCH_2CHCO_2H}}$	3.459 5.097	3.48×10^{-4} 8.00×10^{-6}
Hydroxylamine	$HON\overset{+}{H}_3$	5.96	1.10×10^{-6}
8-Hydroxyquinoline (oxine)	(quinoline ring with N^+H and HO)	4.91 (NH) 9.81 (OH)	1.23×10^{-5} 1.55×10^{-10}
Hypobromous acid (hydrogen hypobromite)	$HOBr$	8.63	2.3×10^{-9}
Hypochlorous acid (hydrogen hypochlorite)	$HOCl$	7.53	3.0×10^{-8}
Hypoiodous acid (hydrogen hypoiodite)	HOI	10.64	2.3×10^{-11}
Hypophosphorous acid (hydrogen hypophosphite)	$\overset{O}{\underset{H_2POH}{\parallel}}$	1.23	5.9×10^{-2}

† A better value of pK_2 for H_2S is probably in the range 17–19 [R. J. Myers, *J. Chem. Ed.*, **1986**, *63*, 687; S. Licht, F. Forouzan, and K. Longo, *Anal. Chem.* **1990**, *62*, 1356]. The value $pK_2 = 13.9$ was used for problems in this text.

(Continued)

Name	Structure	pK_a	K_a
Imidazole (1,3-diazole)	(ring structure with $-NH^+$, N, H)	6.993	1.02×10^{-7}
Iminodiacetic acid	$H_2\overset{+}{N}(CH_2CO_2H)_2$	1.82 (CO_2H) ($\mu = 0.1$) 2.84 (CO_2H) 9.79 (NH_2)	1.51×10^{-2} 1.45×10^{-3} 1.62×10^{-10}
Iodic acid (hydrogen iodate)	$HOI{=}O$ (with O double bond)	0.77	0.17
Iodoacetic acid	ICH_2CO_2H	3.175	6.68×10^{-4}
Isoleucine	$\overset{NH_3^+}{\underset{CO_2H}{\overset{\mid}{\underset{\mid}{CHCH(CH_3)CH_2CH_3}}}}$	2.319 (CO_2H) 9.754 (NH_3)	4.80×10^{-3} 1.76×10^{-10}
Leucine	$\overset{NH_3^+}{\underset{CO_2H}{\overset{\mid}{\underset{\mid}{CHCH_2CH(CH_3)_2}}}}$	2.329 (CO_2) 9.747 (NH_3)	4.69×10^{-3} 1.79×10^{-10}
Lysine	$\overset{NH_3^+}{\underset{CO_2H}{\overset{\mid}{\underset{\mid}{CHCH_2CH_2CH_2CH_2NH_3^+}}}}$	2.04 (CO_2H) ($\mu = 0.1$) 9.08 (α-NH_3) ($\mu = 0.1$) 10.69 (ϵ-NH_3) ($\mu = 0.1$)	9.1×10^{-3} 8.3×10^{-10} 2.0×10^{-11}
Malonic acid (propanedioic acid)	$HO_2CCH_2CO_2H$	2.847 5.696	1.42×10^{-3} 2.01×10^{-6}
Mercaptoacetic acid (thioglycolic acid)	$HSCH_2CO_2H$	(3.60) (CO_2H) 10.55 (SH)	2.5×10^{-4} 2.82×10^{-11}
2-Mercaptoethanol	$HSCH_2CH_2OH$	9.72	1.91×10^{-10}
Methionine	$\overset{NH_3^+}{\underset{CO_2H}{\overset{\mid}{\underset{\mid}{CHCH_2CH_2SCH_3}}}}$	2.20 (CO_2H) ($\mu = 0.1$) 9.05 (NH_3) ($\mu = 0.1$)	6.3×10^{-3} 8.9×10^{-10}
2-Methoxyaniline (o-anisidine)	(benzene ring with OCH_3 and $\overset{+}{N}H_3$)	4.527	2.97×10^{-5}
4-Methoxyaniline (p-anisidine)	CH_3O-(benzene ring)$-\overset{+}{N}H_3$	5.357	4.40×10^{-6}
Methylamine	$CH_3\overset{+}{N}H_3$	10.64	2.3×10^{-11}
2-Methylaniline (o-toluidine)	(benzene ring with CH_3 and $\overset{+}{N}H_3$)	4.447	3.57×10^{-5}

(Continued)

Name	Structure	pK_a	K_a
4-Methylaniline (*p*-toluidine)		5.084	8.24×10^{-6}
2-Methylphenol (*o*-cresol)		10.09	8.1×10^{-11}
4-Methylphenol (*p*-cresol)		10.26	5.5×10^{-11}
Morpholine (perhydro-1,4-oxazine)		8.492	3.22×10^{-9}
1-Naphthoic acid		3.70	2.0×10^{-4}
2-Naphthoic acid		4.16	6.9×10^{-5}
1-Naphthol		9.34	4.6×10^{-10}
2-Naphthol		9.51	3.1×10^{-10}
Nitrilotriacetic acid	$H\overset{+}{N}(CH_2CO_2H)_3$	1.1 (CO_2H) (20°, $\mu = 1.0$) 1.650 (CO_2H) (20°) 2.940 (CO_2H) (20°) 10.334 (NH) (20°)	8×10^{-2} 2.24×10^{-2} 1.15×10^{-3} 4.63×10^{-11}
2-Nitrobenzoic acid		2.179	6.62×10^{-3}
3-Nitrobenzoic acid		3.449	3.56×10^{-4}

(Continued)

Name	Structure	pK_a	K_a
4-Nitrobenzoic acid	O_2N—⟨⟩—CO_2H	3.442	3.61×10^{-4}
Nitroethane	$CH_3CH_2NO_2$	8.57	2.7×10^{-9}
2-Nitrophenol	(structure: benzene with NO_2 and OH)	7.21	6.2×10^{-8}
3-Nitrophenol	(structure: benzene with NO_2 and OH)	8.39	4.1×10^{-9}
4-Nitrophenol	O_2N—⟨⟩—OH	7.15	7.1×10^{-8}
N-Nitrosophenylhydroxylamine (cupferron)	(structure: phenyl–N(NO)–OH)	4.16 ($\mu = 0.1$)	6.9×10^{-5}
Nitrous acid	$HON{=}O$	3.15	7.1×10^{-4}
Oxalic acid (ethanedioic acid)	HO_2CCO_2H	1.252 4.266	5.60×10^{-2} 5.42×10^{-5}
Oxoacetic acid (glyoxylic acid)	$HCCO_2H$ (with $=O$)	3.46	3.5×10^{-4}
Oxobutanedioic acid (oxaloacetic acid)	$HO_2CCH_2CCO_2H$ (with $=O$)	2.56 4.37	2.8×10^{-3} 4.3×10^{-5}
2-Oxopentanedioic (α-ketoglutaric acid)	$HO_2CCH_2CH_2CCO_2H$ (with $=O$)	1.85 ($\mu = 0.5$) 4.44 ($\mu = 0.5$)	1.41×10^{-2} 3.6×10^{-5}
2-Oxopropanoic acid (pyruvic acid)	CH_3CCO_2H (with $=O$)	2.55	2.8×10^{-3}
1,5-Pentanedioic acid (glutaric acid)	$HO_2CCH_2CH_2CH_2CO_2H$	4.34 5.43	4.6×10^{-5} 3.7×10^{-6}
Pentanoic acid (valeric acid)	$CH_3CH_2CH_2CH_2CO_2H$	4.843	1.44×10^{-5}
1,10-Phenanthroline	(structure: phenanthroline with NH^+ and N)	4.86	1.38×10^{-5}
Phenylacetic acid	⟨⟩—CH_2CO_2H	4.310	4.90×10^{-5}

(Continued)

Name	Structure	pK_a	K_a
Phenylalanine	$\overset{\overset{\displaystyle NH_3^+}{\vert}}{\underset{\underset{\displaystyle CO_2H}{\vert}}{CHCH_2}}$—⬡	2.20 (CO_2H) 9.31 (NH_3)	6.3×10^{-3} 4.9×10^{-10}
Phosphoric acid[†] (hydrogen phosphate)	$HO\overset{\overset{\displaystyle O}{\Vert}}{\underset{\underset{\displaystyle OH}{\vert}}{P}}OH$	2.148 7.199 12.15	7.11×10^{-3} 6.32×10^{-8} 7.1×10^{-13}
Phosphorous acid (hydrogen phosphite)	$HP\overset{\overset{\displaystyle O}{\Vert}}{\underset{\underset{\displaystyle OH}{\vert}}{}}OH$	1.5 6.79	3×10^{-2} 1.62×10^{-7}
Phthalic acid (benzene-1,2-dicarboxylic acid)	⬡$\begin{smallmatrix} CO_2H \\ CO_2H \end{smallmatrix}$	2.950 5.408	1.12×10^{-3} 3.90×10^{-6}
Piperazine (perhydro-1,4-diazine)	H_2^+N⬡NH_2^+	5.333 9.731	4.65×10^{-6} 1.86×10^{-10}
Piperidine	⬡NH_2^+	11.123	7.53×10^{-12}
Proline	⬠CO_2H $\overset{N}{\underset{+H_2}{}}$	1.952 (CO_2H) 10.640 (NH_2)	1.12×10^{-2} 2.29×10^{-11}
Propanoic acid	$CH_3CH_2CO_2H$	4.874	1.34×10^{-5}
Propenoic acid (acrylic acid)	$H_2C=CHCO_2H$	4.258	5.52×10^{-5}
Propylamine	$CH_3CH_2CH_2NH_3^+$	10.566	2.72×10^{-11}
Pyridine (azine)	⬡NH^+	5.229	5.90×10^{-6}
Pyridine-2-carboxylic acid (picolinic acid)	⬡NH^+ CO_2H	1.01 5.39	9.8×10^{-2} 4.1×10^{-6}
Pyridine-3-carboxylic acid (nicotinic acid)	HO_2C⬡NH^+	2.05 4.81	8.9×10^{-3} 1.55×10^{-5}

[†] pK_3 from A. G. Miller and J. W. Macklin, *Anal. Chem.* **1983**, *55*, 684.

(Continued)

Name	Structure	pK_a	K_a
Pyridoxal-5-phosphate		1.4 (POH) ($\mu = 0.1$) 3.44 (OH) ($\mu = 0.1$) 6.01 (POH) ($\mu = 0.1$) 8.45 (NH) ($\mu = 0.1$)	0.04 3.6×10^{-4} 9.8×10^{-7} 3.5×10^{-9}
Pyrophosphoric acid (hydrogen diphosphate)	$(HO)_2POP(OH)_2$	0.8 2.2 6.70 9.40	0.16 6×10^{-3} 2.0×10^{-7} 4.0×10^{-10}
Pyrrolidine		11.305	4.95×10^{-12}
Serine	$\begin{array}{c} NH_3^+ \\ | \\ CHCH_2OH \\ | \\ CO_2H \end{array}$	2.187 (CO_2H) 9.209 (NH_3)	6.50×10^{-3} 6.18×10^{-10}
Succinic acid (butanedioic acid)	$HO_2CCH_2CH_2CO_2H$	4.207 5.636	6.21×10^{-5} 2.31×10^{-6}
Sulfuric acid (hydrogen sulfate)		1.99 (pK_2)	1.02×10^{-2}
Sulfurous acid (hydrogen sulfite)		1.91 7.18	1.23×10^{-2} 6.6×10^{-8}
Thiosulfuric acid (hydrogen thiosulfate)		0.6 1.6	0.3 0.03
Threonine	$\begin{array}{c} NH_3^+ \\ | \\ CHCHOHCH_3 \\ | \\ CO_2H \end{array}$	2.088 (CO_2H) 9.100 (NH_3)	8.17×10^{-3} 7.94×10^{-10}
Trichloroacetic acid	Cl_3CCO_2H	0.66 ($\mu = 0.1$)	0.22
Triethanolamine	$(HOCH_2CH_2)_3NH^+$	7.762	1.73×10^{-8}
Triethylamine	$(CH_3CH_2)_3NH^+$	10.715	1.93×10^{-11}
1,2,3-Trihydroxybenzene (pyrogallol)		8.94 11.08 (14)	1.15×10^{-9} 8.3×10^{-12} 10^{-14}

(Continued)

Name	Structure	pK_a	K_a
Trimethylamine	$(CH_3)_3NH^+$	9.800	1.58×10^{-10}
Tris(hydroxymethyl)aminomethane (tris or tham)	$(HOCH_2)_3CNH_3^+$	8.075	8.41×10^{-9}
Tryptophan		2.35 (CO_2H) $(\mu = 0.1)$ 9.33 (NH_3) $(\mu = 0.1)$	4.5×10^{-3} 4.7×10^{-10}
Tyrosine		2.17 (CO_2H) $(\mu = 0.1)$ 9.19 (NH_3) 10.47 (OH)	6.8×10^{-3} 6.5×10^{-10} 3.4×10^{-11}
Valine		2.286 (CO_2H) 9.718 (NH_3)	5.18×10^{-3} 1.91×10^{-10}
Water[†]	H_2O	13.996	1.01×10^{-14}

[†] The constant given for water is K_w.

Standard Reduction Potentials[†]

Reaction	$E°$ (volts)	$dE°/dT$ (mV/K)
Aluminum		
$Al^{3+} + 3e^- \rightleftharpoons Al(s)$	-1.677	0.533
$AlCl^{2+} + 3e^- \rightleftharpoons Al(s) + Cl^-$	-1.802	
$AlF_6^{3-} + 3e^- \rightleftharpoons Al(s) + 6F^-$	-2.069	
$Al(OH)_4^- + 3e^- \rightleftharpoons Al(s) + 4OH^-$	-2.328	-1.13
Antimony		
$SbO^+ + 2H^+ + 3e^- \rightleftharpoons Sb(s) + H_2O$	0.208	
$Sb_2O_3(s) + 6H^+ + 6e^- \rightleftharpoons 2Sb(s) + 3H_2O$	0.147	-0.369
$Sb(s) + 3H^+ + 3e^- \rightleftharpoons SbH_3(g)$	-0.510	-0.030
Arsenic		
$H_3AsO_4 + 2H^+ + 2e^- \rightleftharpoons H_3AsO_3 + H_2O$	0.575	-0.257
$H_3AsO_3 + 3H^+ + 3e^- \rightleftharpoons As(s) + 3H_2O$	0.2475	-0.505
$As(s) + 3H^+ + 3e^- \rightleftharpoons AsH_3(g)$	-0.238	-0.029
Barium		
$Ba^{2+} + 2e^- + Hg \rightleftharpoons Ba(in\ Hg)$	-1.717	
$Ba^{2+} + 2e^- \rightleftharpoons Ba(s)$	-2.906	-0.401
Beryllium		
$Be^{2+} + 2e^- \rightleftharpoons Be(s)$	-1.968	0.60
Bismuth		
$Bi^{3+} + 3e^- \rightleftharpoons Bi(s)$	0.308	0.18
$BiCl_4^- + 3e^- \rightleftharpoons Bi(s) + 4Cl^-$	0.16	
$BiOCl(s) + 2H^+ + 3e^- \rightleftharpoons Bi(s) + H_2O + Cl^-$	0.160	
Boron		
$2B(s) + 6H^+ + 6e^- \rightleftharpoons B_2H_6(g)$	-0.150	-0.296
$B_4O_7^{2-} + 14H^+ + 12e^- \rightleftharpoons 4B(s) + 7H_2O$	-0.792	
$B(OH)_3 + 3H^+ + 3e^- \rightleftharpoons B(s) + 3H_2O$	-0.889	-0.492

[†]All species are aqueous unless otherwise indicated. The reference state for amalgams is an infinitely dilute solution of the element in Hg. The temperature coefficient, $dE°/dT$, allows us to calculate the standard potential, $E°(T)$, at temperature T: $E°(T) = E° + (dE°/dT)\ \Delta T$, where ΔT is $T - 298.15$ K. Note the units mV/K for $dE°/dT$. Once you know the $E°$ for a net cell reaction at temperature T, you can find the equilibrium constant, K, for the reaction from the formula $K = 10^{nFE°/RT \ln 10}$, where n is the number of electrons in each half-reaction, F is the Faraday constant, and R is the gas constant.

SOURCE: The most authoritative source is S. G. Bratsch, *J. Phys. Chem. Ref. Data* **1989**, *18*, 1. Additional data come from L. G. Sillen and A. E. Martell, *Stability Constants of Metal-Ion Complexes* (London: The Chemical Society, Special Publications No. 17 and 25, 1964 and 1971); G. Milazzo and S. Caroli, *Tables of Standard Electrode Potentials* (New York: Wiley, 1978); T. Mussini, P. Longhi, and S. Rondinini, *Pure Appl. Chem.* **1985**, *57*, 169. Another good source is A. J. Bard, R. Parsons, and J. Jordan, *Standard Potentials in Aqueous Solution* (New York: Marcel Dekker, 1985). Reduction potentials for 1 200 free radical reactions are given by P. Wardman, *J. Phys. Chem. Ref. Data* **1989**, *18*, 1637.

Reaction	$E°$ (volts)	$dE°/dT$ (mV/K)
Bromine		
$BrO_4^- + 2H^+ + 2e^- \rightleftharpoons BrO_3^- + H_2O$	1.745	−0.511
$HOBr + H^+ + e^- \rightleftharpoons \frac{1}{2}Br_2(l) + H_2O$	1.584	−0.75
$BrO_3^- + 6H^+ + 5e^- \rightleftharpoons \frac{1}{2}Br_2(l) + 3H_2O$	1.513	−0.419
$Br_2(aq) + 2e^- \rightleftharpoons 2Br^-$	1.098	−0.499
$Br_2(l) + 2e^- \rightleftharpoons 2Br^-$	1.078	−0.611
$Br_3^- + 2e^- \rightleftharpoons 3Br^-$	1.062	−0.512
$BrO^- + H_2O + 2e^- \rightleftharpoons Br^- + 2OH^-$	0.766	−0.94
$BrO_3^- + 3H_2O + 6e^- \rightleftharpoons Br^- + 6OH^-$	0.613	−1.287
Cadmium		
$Cd^{2+} + 2e^- + Hg \rightleftharpoons Cd(in\ Hg)$	−0.380	
$Cd^{2+} + 2e^- \rightleftharpoons Cd(s)$	−0.402	−0.029
$Cd(C_2O_4)(s) + 2e^- \rightleftharpoons Cd(s) + C_2O_4^{2-}$	−0.522	
$Cd(C_2O_4)_2^{2-} + 2e^- \rightleftharpoons Cd(s) + 2C_2O_4^{2-}$	−0.572	
$Cd(NH_3)_4^{2+} + 2e^- \rightleftharpoons Cd(s) + 4NH_3$	−0.613	
$CdS(s) + 2e^- \rightleftharpoons Cd(s) + S^{2-}$	−1.175	
Calcium		
$Ca(s) + 2H^+ + 2e^- \rightleftharpoons CaH_2(s)$	0.776	
$Ca^{2+} + 2e^- + Hg \rightleftharpoons Ca(in\ Hg)$	−2.003	
$Ca^{2+} + 2e^- \rightleftharpoons Ca(s)$	−2.868	−0.186
$Ca(acetate)^+ + 2e^- \rightleftharpoons Ca(s) + acetate^-$	−2.891	
$CaSO_4(s) + 2e^- \rightleftharpoons Ca(s) + SO_4^{2-}$	−2.936	
$Ca(malonate)(s) + 2e^- \rightleftharpoons Ca(s) + malonate^{2-}$	−3.608	
Carbon		
$C_2H_2(g) + 2H^+ + 2e^- \rightleftharpoons C_2H_4(g)$	0.731	
$O{=}\bigcirc{=}O + 2H^+ + 2e^- \rightleftharpoons HO{-}\bigcirc{-}OH$	0.700	
$CH_3OH + 2H^+ + 2e^- \rightleftharpoons CH_4(g) + H_2O$	0.583	−0.039
dehydroascorbic acid + $2H^+ + 2e^- \rightleftharpoons$ ascorbic acid + H_2O	0.390	
$(CN)_2(g) + 2H^+ + 2e^- \rightleftharpoons 2HCN(aq)$	0.373	
$H_2CO + 2H^+ + 2e^- \rightleftharpoons CH_3OH$	0.237	−0.51
$C(s) + 4H^+ + 4e^- \rightleftharpoons CH_4(g)$	0.131 5	−0.209 2
$HCO_2H + 2H^+ + 2e^- \rightleftharpoons H_2CO + H_2O$	−0.029	−0.63
$CO_2(g) + 2H^+ + 2e^- \rightleftharpoons CO(g) + H_2O$	−0.103 8	−0.397 7
$CO_2(g) + 2H^+ + 2e^- \rightleftharpoons HCO_2H$	−0.114	−0.94
$2CO_2(g) + 2H^+ + 2e^- \rightleftharpoons H_2C_2O_4$	−0.432	−1.76
Cerium		
	1.72	1.54
	1.70 1 F $HClO_4$	
$Ce^{4+} + e^- \rightleftharpoons Ce^{3+}$	1.44 1 F H_2SO_4	
	1.61 1 F HNO_3	
	1.47 1 F HCl	
$Ce^{3+} + 3e^- \rightleftharpoons Ce(s)$	−2.336	0.280
Cesium		
$Cs^+ + e^- + Hg \rightleftharpoons Cs(in\ Hg)$	−1.950	
$Cs^+ + e^- \rightleftharpoons Cs(s)$	−3.026	−1.172
Chlorine		
$HClO_2 + 2H^+ + 2e^- \rightleftharpoons HOCl + H_2O$	1.674	0.55
$HClO + H^+ + e^- \rightleftharpoons \frac{1}{2}Cl_2(g) + H_2O$	1.630	−0.27
$ClO_3^- + 6H^+ + 5e^- \rightleftharpoons \frac{1}{2}Cl_2(g) + 3H_2O$	1.458	−0.347
$Cl_2(aq) + 2e^- \rightleftharpoons 2Cl^-$	1.396	−0.72
$Cl_2(g) + 2e^- \rightleftharpoons 2Cl^-$	1.360 4	−1.248
$ClO_4^- + 2H^+ + 2e^- \rightleftharpoons ClO_3^- + H_2O$	1.226	−0.416
$ClO_3^- + 3H^+ + 2e^- \rightleftharpoons HClO_2 + H_2O$	1.157	−0.180
$ClO_3^- + 2H^+ + e^- \rightleftharpoons ClO_2 + H_2O$	1.130	0.074
$ClO_2 + e^- \rightleftharpoons ClO_2^-$	1.068	−1.335

(Continued)

Reaction	$E°$ (volts)	$dE°/dT$ (mV/K)
Chromium		
$Cr_2O_7^{2-} + 14H^+ + 6e^- \rightleftharpoons 2Cr^{3+} + 7H_2O$	1.36	-1.32
$CrO_4^{2-} + 4H_2O + 3e^- \rightleftharpoons Cr(OH)_3 \text{ (s, hydrated)} + 5OH^-$	-0.12	-1.62
$Cr^{3+} + e^- \rightleftharpoons Cr^{2+}$	-0.42	1.4
$Cr^{3+} + 3e^- \rightleftharpoons Cr(s)$	-0.74	0.44
$Cr^{2+} + 2e^- \rightleftharpoons Cr(s)$	-0.89	-0.04
Cobalt		
$Co^{3+} + e^- \rightleftharpoons Co^{2+}$	$\begin{cases} 1.92 \\ 1.817 \quad 8\text{ F }H_2SO_4 \\ 1.850 \quad 4\text{ F }HNO_3 \end{cases}$	1.23
$Co(NH_3)_5(H_2O)^{3+} + e^- \rightleftharpoons Co(NH_3)_5(H_2O)^{2+}$	0.37 1 F NH_4NO_3	
$Co(NH_3)_6^{3+} + e^- \rightleftharpoons Co(NH_3)_6^{2+}$	0.1	
$CoOH^+ + H^+ + 2e^- \rightleftharpoons Co(s) + H_2O$	0.003	-0.04
$Co^{2+} + 2e^- \rightleftharpoons Co(s)$	-0.282	0.065
$Co(OH)_2(s) + 2e^- \rightleftharpoons Co(s) + 2OH^-$	-0.746	-1.02
Copper		
$Cu^+ + e^- \rightleftharpoons Cu(s)$	0.518	-0.754
$Cu^{2+} + 2e^- \rightleftharpoons Cu(s)$	0.339	0.011
$Cu^{2+} + e^- \rightleftharpoons Cu^+$	0.161	0.776
$CuCl(s) + e^- \rightleftharpoons Cu(s) + Cl^-$	0.137	
$Cu(IO_3)_2(s) + 2e^- \rightleftharpoons Cu(s) + 2IO_3^-$	-0.079	
$Cu(\text{ethylenediamine})_2^+ + e^- \rightleftharpoons Cu(s) + 2\text{ ethylenediamine}$	-0.119	
$CuI(s) + e^- \rightleftharpoons Cu(s) + I^-$	-0.185	
$Cu(EDTA)^{2-} + 2e^- \rightleftharpoons Cu(s) + EDTA^{4-}$	-0.216	
$Cu(OH)_2(s) + 2e^- \rightleftharpoons Cu(s) + 2OH^-$	-0.222	
$Cu(CN)_2^- + e^- \rightleftharpoons Cu(s) + 2CN^-$	-0.429	
$CuCN(s) + e^- \rightleftharpoons Cu(s) + CN^-$	-0.639	
Dysprosium		
$Dy^{3+} + 3e^- \rightleftharpoons Dy(s)$	-2.295	0.373
Erbium		
$Er^{3+} + 3e^- \rightleftharpoons Er(s)$	-2.331	0.388
Europium		
$Eu^{3+} + e^- \rightleftharpoons Eu^{2+}$	-0.35	1.53
$Eu^{3+} + 3e^- \rightleftharpoons Eu(s)$	-1.991	0.338
$Eu^{2+} + 2e^- \rightleftharpoons Eu(s)$	-2.812	-0.26
Fluorine		
$F_2(g) + 2e^- \rightleftharpoons 2F^-$	2.890	-1.870
$F_2O(g) + 2H^+ + 4e^- \rightleftharpoons 2F^- + H_2O$	2.168	-1.208
Gadolinium		
$Gd^{3+} + 3e^- \rightleftharpoons Gd(s)$	-2.279	0.315
Gallium		
$Ga^{3+} + 3e^- \rightleftharpoons Ga(s)$	-0.549	0.61
$GaOOH(s) + 3H_2O + 3e^- \rightleftharpoons Ga(s) + 3OH^-$	-1.320	-1.08
Germanium		
$Ge^{2+} + 2e^- \rightleftharpoons Ge(s)$	0.1	
$H_4GeO_4 + 4H^+ + 4e^- \rightleftharpoons Ge(s) + 4H_2O$	-0.039	-0.429
Gold		
$Au^+ + e^- \rightleftharpoons Au(s)$	1.69	-1.1
$Au^{3+} + 2e^- \rightleftharpoons Au^+$	1.41	
$AuCl_2^- + e^- \rightleftharpoons Au(s) + 2Cl^-$	1.154	
$AuCl_4^- + 2e^- \rightleftharpoons AuCl_2^- + 2Cl^-$	0.926	

(Continued)

Reaction	$E°$ (volts)		$dE°/dT$ (mV/K)
Hafnium			
$Hf^{4+} + 4e^- \rightleftharpoons Hf(s)$	−1.55		0.68
$HfO_2(s) + 4H^+ + 4e^- \rightleftharpoons Hf(s) + 2H_2O$	−1.591		−0.355
Holmium			
$Ho^{3+} + 3e^- \rightleftharpoons Ho(s)$	−2.33		0.371
Hydrogen			
$2H^+ + 2e^- \rightleftharpoons H_2(g)$	0.0000		0
$H_2O + e^- = \frac{1}{2}H_2(g) + OH^-$	−0.8280		−0.8360
Indium			
$In^{3+} + 3e^- + Hg \rightleftharpoons In(in\ Hg)$	−0.313		
$In^{3+} + 3e^- \rightleftharpoons In(s)$	−0.338		0.42
$In^{3+} + 2e^- \rightleftharpoons In^+$	−0.444		
$In(OH)_3(s) + 3e^- \rightleftharpoons In(s) + 3OH^-$	−0.99		−0.95
Iodine			
$IO_4^- + 2H^+ + 2e^- \rightleftharpoons IO_3^- + H_2O$	1.589		−0.85
$H_5IO_6 + 2H^+ + 2e^- \rightleftharpoons HIO_3 + 3H_2O$	1.567		−0.12
$HOI + H^+ + e^- \rightleftharpoons \frac{1}{2}I_2(s) + H_2O$	1.430		−0.339
$ICl_3(s) + 3e^- \rightleftharpoons \frac{1}{2}I_2(s) + 3Cl^-$	1.28		
$ICl(s) + e^- \rightleftharpoons \frac{1}{2}I_2(s) + Cl^-$	1.22		
$IO_3^- + 6H^+ + 5e^- \rightleftharpoons \frac{1}{2}I_2(s) + 3H_2O$	1.210		−0.367
$IO_3^- + 5H^+ + 4e^- \rightleftharpoons HOI + 2H_2O$	1.154		−0.374
$I_2(aq) + 2e^- \rightleftharpoons 2I^-$	0.620		−0.234
$I_2(s) + 2e^- \rightleftharpoons 2I^-$	0.535		−0.125
$I_3^- + 2e^- \rightleftharpoons 3I^-$	0.535		−0.186
$IO_3^- + 3H_2O + 6e^- \rightleftharpoons I^- + 6OH^-$	0.269		−1.163
Iridium			
$IrCl_6^{2-} + e^- \rightleftharpoons IrCl_6^{3-}$	1.026	1 F HCl	
$IrBr_6^{2-} + e^- \rightleftharpoons IrBr_6^{3-}$	0.947	2 F NaBr	
$IrCl_6^{2-} + 4e^- \rightleftharpoons Ir(s) + 6Cl^-$	0.835		
$IrO_2(s) + 4H^+ + 4e^- \rightleftharpoons Ir(s) + 2H_2O$	0.73		−0.36
$IrI_6^{2-} + e^- \rightleftharpoons IrI_6^{3-}$	0.485	1 F KI	
Iron			
$Fe(phenanthroline)_3^{3+} + e^- \rightleftharpoons Fe(phenanthroline)_3^{2+}$	1.147		
$Fe(bipyridyl)_2^{3+} + e^- \rightleftharpoons Fe(bipyridyl)_3^{2+}$	1.120		
$FeOH^{2+} + H^+ + e^- \rightleftharpoons Fe^{2+} + H_2O$	0.900		0.096
$FeO_4^{2-} + 3H_2O + 3e^- \rightleftharpoons FeOOH(s) + 5OH^-$	0.80		−1.59
$Fe^{3+} + e^- \rightleftharpoons Fe^{2+}$	$\begin{cases} 0.771 \\ 0.732 \\ 0.767 \\ 0.746 \end{cases}$	$\begin{array}{l} \\ 1\ F\ HCl \\ 1\ F\ HClO_4 \\ 1\ F\ HNO_3 \end{array}$	1.175
$FeOOH(s) + 3H^+ + e^- \rightleftharpoons Fe^{2+} + 2H_2O$	0.74		−1.05
$ferricinium^+ + e^- \rightleftharpoons ferrocene$	0.400		
$Fe(CN)_6^{3-} + e^- \rightleftharpoons Fe(CN)_6^{4-}$	0.356		
$Fe(glutamate)^{3+} + e^- \rightleftharpoons Fe(glutamate)^{2+}$	0.240		
$FeOH^+ + H^+ + 2e^- \rightleftharpoons Fe(s) + H_2O$	−0.16		0.07
$Fe^{2+} + 2e^- \rightleftharpoons Fe(s)$	−0.44		0.07
$FeCO_3(s) + 2e^- \rightleftharpoons Fe(s) + CO_3^{2-}$	−0.756		−1.293
Lanthanum			
$La^{3+} + 3e^- \rightleftharpoons La(s)$	−2.379		0.242
$La(succinate)^+ + 3e^- \rightleftharpoons La(s) + succinate^{2-}$	−2.601		

(Continued)

Reaction	$E°$ (volts)	$dE°/dT$ (mV/K)
Lead		
$Pb^{4+} + 2e^- \rightleftharpoons Pb^{2+}$	1.69 1 F HNO_3	
$PbO_2(s) + 4H^+ + SO_4^{2-} + 2e^- \rightleftharpoons PbSO_4(s) + 2H_2O$	1.685	
$PbO_2(s) + 4H^+ + 2e^- \rightleftharpoons Pb^{2+} + 2H_2O$	1.458	-0.253
$3PbO_2(s) + 2H_2O + 4e^- \rightleftharpoons Pb_3O_4(s) + 4OH^-$	0.269	-1.136
$Pb_3O_4(s) + H_2O + 2e^- \rightleftharpoons 3PbO(s, red) + 2OH^-$	0.224	-1.211
$Pb_3O_4(s) + H_2O + 2e^- \rightleftharpoons 3PbO(s, yellow) + 2OH^-$	0.207	-1.177
$Pb^{2+} + 2e^- \rightleftharpoons Pb(s)$	-0.126	-0.395
$PbF_2(s) + 2e^- \rightleftharpoons Pb(s) + 2F^-$	-0.350	
$PbSO_4(s) + 2e^- \rightleftharpoons Pb(s) + SO_4^{2-}$	-0.355	
Lithium		
$Li^+ + e^- + Hg \rightleftharpoons Li(in\ Hg)$	-2.195	
$Li^+ + e^- \rightleftharpoons Li(s)$	-3.040	-0.514
Lutetium		
$Lu^{3+} + 3e^- \rightleftharpoons Lu(s)$	-2.28	0.412
Magnesium		
$Mg^{2+} + 2e^- + Hg \rightleftharpoons Mg(in\ Hg)$	-1.980	
$Mg(OH)^+ + H^+ + 2e^- \rightleftharpoons Mg(s) + H_2O$	-2.022	0.25
$Mg^{2+} + 2e^- \rightleftharpoons Mg(s)$	-2.360	0.199
$Mg(C_2O_4)(s) + 2e^- \rightleftharpoons Mg(s) + C_2O_4^{2-}$	-2.493	
$Mg(OH)_2(s) + 2e^- \rightleftharpoons Mg(s) + 2OH^-$	-2.690	-0.946
Manganese		
$MnO_4^- + 4H^+ + 3e^- \rightleftharpoons MnO_2(s) + 2H_2O$	1.692	-0.671
$Mn^{3+} + e^- \rightleftharpoons Mn^{2+}$	1.56	1.8
$MnO_4^- + 8H^+ + 5e^- \rightleftharpoons Mn^{2+} + 4H_2O$	1.507	-0.646
$Mn_2O_3(s) + 6H^+ + 2e^- \rightleftharpoons 2Mn^{2+} + 3H_2O$	1.485	-0.926
$MnO_2(s) + 4H^+ + 2e^- \rightleftharpoons Mn^{2+} + 2H_2O$	1.230	-0.609
$Mn(EDTA)^- + e^- \rightleftharpoons Mn(EDTA)^{2-}$	0.825	-1.10
$MnO_4^- + e^- \rightleftharpoons MnO_4^{2-}$	0.56	-2.05
$3Mn_2O_3(s) + H_2O + 2e^- \rightleftharpoons 2Mn_3O_4(s) + 2OH^-$	0.002	-1.256
$Mn_3O_4(s) + 4H_2O + 2e^- \rightleftharpoons 3Mn(OH)_2(s) + 2OH^-$	-0.352	-1.61
$Mn^{2+} + 2e^- \rightleftharpoons Mn(s)$	-1.182	-1.129
$Mn(OH)_2(s) + 2e^- \rightleftharpoons Mn(s) + 2OH^-$	-1.565	-1.10
Mercury		
$2Hg^{2+} + 2e^- \rightleftharpoons Hg_2^{2+}$	0.908	0.095
$Hg^{2+} + 2e^- \rightleftharpoons Hg(l)$	0.852	-0.116
$Hg_2^{2+} + 2e^- \rightleftharpoons 2Hg(l)$	0.796	-0.327
$Hg_2SO_4(s) + 2e^- \rightleftharpoons 2Hg(l) + SO_4^{2-}$	0.614	
$Hg_2Cl_2(s) + 2e^- \rightleftharpoons 2Hg(l) + 2Cl^-$	$\begin{cases} 0.268 \\ 0.241 \text{ (saturated calomel electrode)} \end{cases}$	
$Hg(OH)_3^- + 2e^- \rightleftharpoons Hg(l) + 3OH^-$	0.231	
$Hg(OH)_2 + 2e^- \rightleftharpoons Hg(l) + 2OH^-$	0.206	-1.24
$Hg_2Br_2(s) + 2e^- \rightleftharpoons 2Hg(l) + 2Br^-$	0.140	
$HgO(s, yellow) + H_2O + 2e^- \rightleftharpoons Hg(l) + 2OH^-$	0.0983	-1.125
$HgO(s, red) + H_2O + 2e^- \rightleftharpoons Hg(l) + 2OH^-$	0.0977	-1.1206
Molybdenum		
$MoO_4^{2-} + 2H_2O + 2e^- \rightleftharpoons MoO_2(s) + 4OH^-$	-0.818	-1.69
$MoO_4^{2-} + 4H_2O + 6e^- \rightleftharpoons Mo(s) + 8OH^-$	-0.926	-1.36
$MoO_2(s) + 2H_2O + 4e^- \rightleftharpoons Mo(s) + 4OH^-$	-0.980	-1.196
Neodymium		
$Nd^{3+} + 3e^- \rightleftharpoons Nd(s)$	-2.323	0.282

(Continued)

Reaction	$E°$ (volts)	$dE°/dT$ (mV/K)
Neptunium		
$NpO_3^+ + 2H^+ + e^- \rightleftharpoons NpO_2^{2+} + H_2O$	2.04	
$NpO_2^{2+} + e^- \rightleftharpoons NpO_2^+$	1.236	0.058
$NpO_2^+ + 4H^+ + e^- \rightleftharpoons Np^{4+} + 2H_2O$	0.567	−3.30
$Np^{4+} + e^- \rightleftharpoons Np^{3+}$	0.157	1.53
$Np^{3+} + 3e^- \rightleftharpoons Np(s)$	−1.768	0.18
Nickel		
$NiOOH(s) + 3H^+ + e^- \rightleftharpoons Ni^{2+} + 2H_2O$	2.05	−1.17
$Ni^{2+} + 2e^- \rightleftharpoons Ni(s)$	−0.236	0.146
$Ni(CN)_4^{2-} + e^- \rightleftharpoons Ni(CN)_3^{2-} + CN^-$	−0.401	
$Ni(OH)_2(s) + 2e^- \rightleftharpoons Ni(s) + 2OH^-$	−0.714	−1.02
Niobium		
$\frac{1}{2}Nb_2O_5(s) + H^+ + e^- \rightleftharpoons NbO_2(s) + \frac{1}{2}H_2O$	−0.248	−0.460
$\frac{1}{2}Nb_2O_5(s) + 5H^+ + 5e^- \rightleftharpoons Nb(s) + \frac{5}{2}H_2O$	−0.601	−0.381
$NbO_2(s) + 2H^+ + 2e^- \rightleftharpoons NbO(s) + H_2O$	−0.646	−0.347
$NbO_2(s) + 4H^+ + 4e^- \rightleftharpoons Nb(s) + 2H_2O$	−0.690	−0.361
Nitrogen		
$HN_3 + 3H^+ + 2e^- \rightleftharpoons N_2(g) + NH_4^+$	2.079	0.147
$N_2O(g) + 2H^+ + 2e^- \rightleftharpoons N_2(g) + H_2O$	1.769	−0.461
$2NO(g) + 2H^+ + 2e^- \rightleftharpoons N_2O(g) + H_2O$	1.587	−1.359
$NO^+ + e^- \rightleftharpoons NO(g)$	1.46	
$2NH_3OH^+ + H^+ + 2e^- \rightleftharpoons N_2H_5^+ + 2H_2O$	1.40	−0.60
$NH_3OH^+ + 2H^+ + 2e^- \rightleftharpoons NH_4^+ + H_2O$	1.33	−0.44
$N_2H_5^+ + 3H^+ + 2e^- \rightleftharpoons 2NH_4^+$	1.250	−0.28
$HNO_2 + H^+ + e^- \rightleftharpoons NO(g) + H_2O$	0.984	0.649
$NO_3^- + 4H^+ + 3e^- \rightleftharpoons NO(g) + 2H_2O$	0.955	0.028
$NO_3^- + 3H^+ + 2e^- \rightleftharpoons HNO_2 + H_2O$	0.940	−0.282
$NO_3^- + 2H^+ + e^- \rightleftharpoons \frac{1}{2}N_2O_4(g) + H_2O$	0.798	0.107
$N_2(g) + 8H^+ + 6e^- \rightleftharpoons 2NH_4^+$	0.274	−0.616
$N_2(g) + 5H^+ + 4e^- \rightleftharpoons N_2H_5^+$	−0.214	−0.78
$N_2(g) + 2H_2O + 4H^+ + 2e^- \rightleftharpoons 2NH_3OH^+$	−1.83	−0.96
$\frac{3}{2}N_2(g) + H^+ + e^- \rightleftharpoons HN_3$	−3.334	−2.141
Osmium		
$OsO_4(s) + 8H^+ + 8e^- \rightleftharpoons Os(s) + 4H_2O$	0.834	−0.458
$OsCl_6^{2-} + e^- \rightleftharpoons OsCl_6^{3-}$	0.85 1 F HCl	
Oxygen		
$OH + H^+ + e^- \rightleftharpoons H_2O$	2.56	−1.0
$O(g) + 2H^+ + 2e^- \rightleftharpoons H_2O$	2.430 1	−1.148 4
$O_3(g) + 2H^+ + 2e^- \rightleftharpoons O_2(g) + H_2O$	2.075	−0.489
$H_2O_2 + 2H^+ + 2e^- \rightleftharpoons 2H_2O$	1.763	−0.698
$HO_2 + H^+ + e^- \rightleftharpoons H_2O_2$	1.44	−0.7
$\frac{1}{2}O_2(g) + 2H^+ + 2e^- \rightleftharpoons H_2O$	1.229 1	−0.845 6
$O_2(g) + 2H^+ + 2e^- \rightleftharpoons H_2O_2$	0.695	−0.993
$O_2(g) + H^+ + e^- \rightleftharpoons HO_2$	−0.05	−1.3
Palladium		
$Pd^{2+} + 2e^- \rightleftharpoons Pd(s)$	0.915	0.12
$PdO(s) + 2H^+ + 2e^- \rightleftharpoons Pd(s) + H_2O$	0.79	−0.33
$PdCl_6^{4-} + 2e^- \rightleftharpoons Pd(s) + 6Cl^-$	0.615	
$PdO_2(s) + H_2O + 2e^- \rightleftharpoons PdO(s) + 2OH^-$	0.64	−1.2

(Continued)

Reaction	$E°$ (volts)	$dE°/dT$ (mV/K)
Phosphorus		
$\frac{1}{4}P_4(s, \text{white}) + 3H^+ + 3e^- \rightleftharpoons PH_3(g)$	−0.046	−0.093
$\frac{1}{4}P_4(s, \text{red}) + 3H^+ + 3e^- \rightleftharpoons PH_3(g)$	−0.088	−0.030
$H_3PO_4 + 2H^+ + 2e^- \rightleftharpoons H_3PO_3 + H_2O$	−0.30	−0.36
$H_3PO_4 + 5H^+ + 5e^- \rightleftharpoons \frac{1}{4}P_4(s, \text{white}) + 4H_2O$	−0.402	−0.340
$H_3PO_3 + 2H^+ + 2e^- \rightleftharpoons H_3PO_2 + H_2O$	−0.48	−0.37
$H_3PO_2 + H^+ + e^- \rightleftharpoons \frac{1}{4}P_4(s) + 2H_2O$	−0.51	
Platinum		
$Pt^{2+} + 2e^- \rightleftharpoons Pt(s)$	1.18	−0.05
$PtO_2(s) + 4H^+ + 4e^- \rightleftharpoons Pt(s) + 2H_2O$	0.92	−0.36
$PtCl_4^{2-} + 2e^- \rightleftharpoons Pt(s) + 4Cl^-$	0.755	
$PtCl_6^{2-} + 2e^- \rightleftharpoons PtCl_4^{2-} + 2Cl^-$	0.68	
Plutonium		
$PuO_2^+ + e^- \rightleftharpoons PuO_2(s)$	1.585	0.39
$PuO_2^{2+} + 4H^+ + 2e^- \rightleftharpoons Pu^{4+} + 2H_2O$	1.000	−1.615
$Pu^{4+} + e^- \rightleftharpoons Pu^{3+}$	1.006	1.441
$PuO_2^{2+} + e^- \rightleftharpoons PuO_2^+$	0.966	0.03
$PuO_2(s) + 4H^+ + 4e^- \rightleftharpoons Pu(s) + 2H_2O$	−1.369	−0.38
$Pu^{3+} + 3e^- \rightleftharpoons Pu(s)$	−1.978	0.23
Potassium		
$K^+ + e^- + Hg \rightleftharpoons K(\text{in } Hg)$	−1.975	
$K^+ + e^- \rightleftharpoons K(s)$	−2.936	−1.074
Praseodymium		
$Pr^{4+} + e^- \rightleftharpoons Pr^{3+}$	3.2	1.4
$Pr^{3+} + 3e^- \rightleftharpoons Pr(s)$	−2.353	0.291
Promethium		
$Pm^3 + 3e^- \rightleftharpoons Pm(s)$	−2.30	0.29
Radium		
$Ra^{2+} + 2e^- \rightleftharpoons Ra(s)$	−2.80	−0.44
Rhenium		
$ReO_4^- + 2H^+ + e^- \rightleftharpoons ReO_3(s) + H_2O$	0.72	−1.17
$ReO_4^- + 4H^+ + 3e^- \rightleftharpoons ReO_2(s) + 2H_2O$	0.510	−0.70
Rhodium		
$Rh^{6+} + 3e^- \rightleftharpoons Rh^{3+}$	1.48 1 F HClO$_4$	
$Rh^{4+} + e^- \rightleftharpoons Rh^{3+}$	1.44 3 F H$_2$SO$_4$	
$RhCl_6^{2-} + e^- \rightleftharpoons RhCl_6^{3-}$	1.2	
$Rh^{3+} + 3e^- \rightleftharpoons Rh(s)$	0.76	0.4
$2Rh^{3+} + 2e^- \rightleftharpoons Rh_2^{4+}$	0.7	
$RhCl_6^{3-} + 3e^- \rightleftharpoons Rh(s) + 6Cl^-$	0.44	
Rubidium		
$Rb^+ + e^- + Hg \rightleftharpoons Rb(\text{in } Hg)$	−1.970	
$Rb^+ + e^- \rightleftharpoons Rb(s)$	−2.943	−1.140
Ruthenium		
$RuO_4^- + 6H^+ + 3e^- \rightleftharpoons Ru(OH)_2^{2+} + 2H_2O$	1.53	
$Ru(\text{dipyridyl})_3^{3+} + e^- \rightleftharpoons Ru(\text{dipyridyl})_3^{2+}$	1.29	
$RuO_4(s) + 8H^+ + 8e^- \rightleftharpoons Ru(s) + 4H_2O$	1.032	−0.467
$Ru^{2+} + 2e^- \rightleftharpoons Ru(s)$	0.8	
$Ru^{3+} + 3e^- \rightleftharpoons Ru(s)$	0.60	
$Ru^{3+} + e^- \rightleftharpoons Ru^{2+}$	0.24	
$Ru(NH_3)_6^{3+} + e^- \rightleftharpoons Ru(NH_3)_6^{2+}$	0.214	

(Continued)

Reaction	$E°$ (volts)	$dE°/dT$ (mV/K)
Samarium		
$Sm^{3+} + 3e^- \rightleftharpoons Sm(s)$	-2.304	0.279
$Sm^{2+} + 2e^- \rightleftharpoons Sm(s)$	-2.68	-0.28
Scandium		
$Sc^{3+} + 3e^- \rightleftharpoons Sc(s)$	-2.09	0.41
Selenium		
$SeO_4^{2-} + 4H^+ + 2e^- \rightleftharpoons H_2SeO_3 + H_2O$	1.150	0.483
$H_2SeO_3 + 4H^+ + 4e^- \rightleftharpoons Se(s) + 3H_2O$	0.739	-0.562
$Se(s) + 2H^+ + 2e^- \rightleftharpoons H_2Se(g)$	-0.082	0.238
$Se(s) + 2e^- \rightleftharpoons Se^{2-}$	-0.67	-1.2
Silicon		
$Si(s) + 4H^+ + 4e^- \rightleftharpoons SiH_4(g)$	-0.147	-0.196
$SiO_2(s, \text{quartz}) + 4H^+ + 4e^- \rightleftharpoons Si(s) + 2H_2O$	-0.990	-0.374
$SiF_6^{2-} + 4e^- \rightleftharpoons Si(s) + 6F^-$	-1.24	
Silver		
$Ag^{2+} + e^- \rightleftharpoons Ag^+$	$\begin{cases} 2.000 & \text{4 F HClO}_4 \\ 1.989 & \\ 1.929 & \text{4 F HNO}_3 \end{cases}$	0.99
$Ag^{3+} + 2e^- \rightleftharpoons Ag^+$	1.9	
$AgO(s) + H^+ + e^- \rightleftharpoons \frac{1}{2}Ag_2O(s) + \frac{1}{2}H_2O$	1.40	
$Ag^+ + e^- \rightleftharpoons Ag(s)$	$0.799\,3$	-0.989
$Ag_2C_2O_4(s) + 2e^- \rightleftharpoons 2Ag(s) + C_2O_4^{2-}$	0.465	
$AgN_3(s) + e^- \rightleftharpoons Ag(s) + N_3^-$	0.293	
$AgCl(s) + e^- \rightleftharpoons Ag(s) + Cl^-$	$\begin{cases} 0.222 & \\ 0.197 & \text{saturated KCl} \end{cases}$	
$AgBr(s) + e^- \rightleftharpoons Ag(s) + Br^-$	0.071	
$Ag(S_2O_3)_2^{3-} + e^- \rightleftharpoons Ag(s) + 2S_2O_3^{2-}$	0.017	
$AgI(s) + e^- \rightleftharpoons Ag(s) + I^-$	-0.152	
$Ag_2S(s) + H^+ + 2e^- \rightleftharpoons 2Ag(s) + SH^-$	-0.272	
Sodium		
$Na^+ + e^- + Hg \rightleftharpoons Na(in\ Hg)$	-1.959	
$Na^+ + \frac{1}{2}H_2(g) + e^- \rightleftharpoons NaH(s)$	-2.367	-1.550
$Na^+ + e^- \rightleftharpoons Na(s)$	$-2.714\,3$	-0.757
Strontium		
$Sr^{2+} + 2e^- \rightleftharpoons Sr(s)$	-2.889	-0.237
Sulfur		
$S_2O_8^{2-} + 2e^- \rightleftharpoons 2SO_4^{2-}$	2.01	
$S_2O_6^{2-} + 4H^+ + 2e^- \rightleftharpoons 2H_2SO_3$	0.57	
$4SO_2 + 4H^+ + 6e^- \rightleftharpoons S_4O_6^{2-} + 2H_2O$	0.539	-1.11
$SO_2 + 4H^+ + 4e^- \rightleftharpoons S(s) + 2H_2O$	0.450	-0.652
$2H_2SO_3 + 2H^+ + 4e^- \rightleftharpoons S_2O_3^{2-} + 3H_2O$	0.40	
$S(s) + 2H^+ + 2e^- \rightleftharpoons H_2S(g)$	0.174	0.224
$S(s) + 2H^+ + 2e^- \rightleftharpoons H_2S(aq)$	0.144	-0.21
$S_4O_6^{2-} + 2H^+ + 2e^- \rightleftharpoons 2HS_2O_3^-$	0.10	-0.23
$5S(s) + 2e^- \rightleftharpoons S_5^{2-}$	-0.340	
$2S(s) + 2e^- \rightleftharpoons S_2^{2-}$	-0.50	-1.16
$2SO_3^{2-} + 3H_2O + 4e^- \rightleftharpoons S_2O_3^{2-} + 6OH^-$	-0.566	-1.06
$SO_3^{2-} + 3H_2O + 4e^- \rightleftharpoons S(s) + 6OH^-$	-0.659	-1.23
$SO_4^{2-} + 4H_2O + 6e^- \rightleftharpoons S(s) + 8OH^-$	-0.751	-1.288
$SO_4^{2-} + H_2O + 2e^- \rightleftharpoons SO_3^{2-} + 2OH^-$	-0.936	-1.41
$2SO_3^{2-} + 2H_2O + 2e^- \rightleftharpoons S_2O_4^{2-} + 4OH^-$	-1.130	-0.85
$2SO_4^{2-} + 2H_2O + 2e^- \rightleftharpoons S_2O_6^{2-} + 4OH^-$	-1.71	-1.00

(Continued)

Reaction	$E°$ (volts)		$dE°/dT$ (mV/K)

Tantalum

$Ta_2O_5(s) + 10H^+ + 10e^- \rightleftharpoons 2Ta(s) + 5H_2O$ — -0.752 — -0.377

Technetium

$TcO_4^- + 2H_2O + 3e^- \rightleftharpoons TcO_2(s) + 4OH^-$ — -0.366 — -1.82

$TcO_4^- + 4H_2O + 7e^- \rightleftharpoons Tc(s) + 8OH^-$ — -0.474 — -1.46

Tellurium

$TeO_3^{2-} + 3H_2O + 4e^- \rightleftharpoons Te(s) + 6OH^-$ — -0.47 — -1.39

$2Te(s) + 2e^- \rightleftharpoons Te_2^{2-}$ — -0.84

$Te(s) + 2e^- \rightleftharpoons Te^{2-}$ — -0.90 — -1.0

Terbium

$Tb^{4+} + e^- \rightleftharpoons Tb^{3+}$ — 3.1 — 1.5

$Tb^{3+} + 3e^- \rightleftharpoons Tb(s)$ — -2.28 — 0.350

Thallium

$Tl^{3+} + 2e^- \rightleftharpoons Tl^+$

$\begin{cases} 1.280 \\ 0.77 & 1\ F\ HCl \\ 1.22 & 1\ F\ H_2SO_4 \\ 1.23 & 1\ F\ HNO_3 \\ 1.26 & 1\ F\ HClO_4 \end{cases}$ — 0.97

$Tl^+ + e^- + Hg \rightleftharpoons Tl(in\ Hg)$ — -0.294

$Tl^+ + e^- \rightleftharpoons Tl(s)$ — -0.336 — -1.312

$TlCl(s) + e^- \rightleftharpoons Tl(s) + Cl^-$ — -0.557

Thorium

$Th^{4+} + 4e^- \rightleftharpoons Th(s)$ — -1.826 — 0.557

Thullium

$Tm^{3+} + 3e^- \rightleftharpoons Tm(s)$ — -2.319 — 0.394

Tin

$Sn(OH)_3^+ + 3H^+ + 2e^- \rightleftharpoons Sn^{2+} + 6H_2O$ — 0.142

$Sn^{4+} + 2e^- \rightleftharpoons Sn^{2+}$ — 0.139 $1\ F\ HCl$

$SnO_2(s) + 4H^+ + 2e^- \rightleftharpoons Sn^{2+} + 2H_2O$ — -0.094 — -0.31

$Sn^{2+} + 2e^- \rightleftharpoons Sn(s)$ — -0.141 — -0.32

$SnF_6^{2-} + 4e^- \rightleftharpoons Sn(s) + 6F^-$ — -0.25

$Sn(OH)_6^{2-} + 2e^- \rightleftharpoons Sn(OH)_3^- + 3OH^-$ — -0.93

$Sn(s) + 4H_2O + 4e^- \rightleftharpoons SnH_4(g) + 4OH^-$ — -1.316 — -1.057

$SnO_2(s) + H_2O + 2e^- \rightleftharpoons SnO(s) + 2OH^-$ — -0.961 — -1.129

Titanium

$TiO^{2+} + 2H^+ + e^- \rightleftharpoons Ti^{3+} + H_2O$ — 0.1 — -0.6

$Ti^{3+} + e^- \rightleftharpoons Ti^{2+}$ — -0.9 — 1.5

$TiO_2(s) + 4H^+ + 4e^- \rightleftharpoons Ti(s) + 2H_2O$ — -1.076 — 0.365

$TiF_6^{2-} + 4e^- \rightleftharpoons Ti(s) + 6F^-$ — -1.191

$Ti^{2+} + 2e^- \rightleftharpoons Ti(s)$ — -1.60 — -0.16

Tungsten

$W(CN)_8^{3-} + e^- \rightleftharpoons W(CN)_8^{4-}$ — 0.457

$W^{6+} + e^- \rightleftharpoons W^{5+}$ — 0.26 $12\ F\ HCl$

$WO_3(s) + 6H^+ + 6e^- \rightleftharpoons W(s) + 3H_2O$ — -0.091 — -0.389

$W^{5+} + e^- \rightleftharpoons W^{4+}$ — -0.3 $12\ F\ HCl$

$WO_2(s) + 2H_2O + 4e^- \rightleftharpoons W(s) + 4OH^-$ — -0.982 — -1.197

$WO_4^{2-} + 4H_2O + 6e^- \rightleftharpoons W(s) + 8OH^-$ — -1.060 — -1.36

(Continued)

Reaction	$E°$ (volts)	$dE°/dT$ (mV/K)
Uranium		
$UO_2^+ + 4H^+ + e^- \rightleftharpoons U^{4+} + 2H_2O$	0.39	-3.4
$UO_2^{2+} + 4H^+ + 2e^- \rightleftharpoons U^{4+} + 2H_2O$	0.273	-1.582
$UO_2^{2+} + e^- \rightleftharpoons UO_2^+$	0.16	0.2
$U^{4+} + e^- \rightleftharpoons U^{3+}$	-0.577	1.61
$U^{3+} + 3e^- \rightleftharpoons U(s)$	-1.642	0.16
Vanadium		
$VO_2^+ + 2H^+ + e^- \rightleftharpoons VO^{2+} + H_2O$	1.001	-0.901
$VO^{2+} + 2H^+ + e^- \rightleftharpoons V^{3+} + H_2O$	0.337	-1.6
$V^{3+} + e^- \rightleftharpoons V^{2+}$	-0.255	1.5
$V^{2+} + 2e^- \rightleftharpoons V(s)$	-1.125	-0.11
Xenon		
$H_4XeO_6 + 2H^+ + 2e^- \rightleftharpoons XeO_3 + 3H_2O$	2.38	0.0
$XeF_2 + 2H^+ + 2e^- \rightleftharpoons Xe(g) + 2HF$	2.2	
$XeO_3 + 6H^+ + 6e^- \rightleftharpoons Xe(g) + 3H_2O$	2.1	-0.34
Ytterbium		
$Yb^{3+} + 3e^- \rightleftharpoons Yb(s)$	-2.19	0.363
$Yb^{2+} + 2e^- \rightleftharpoons Yb(s)$	-2.76	-0.16
Yttrium		
$Y^{3+} + 3e^- \rightleftharpoons Y(s)$	-2.38	0.034
Zinc		
$ZnOH^+ + H^+ + 2e^- \rightleftharpoons Zn(s) + 2H_2O$	-0.497	0.03
$Zn^{2+} + 2e^- \rightleftharpoons Zn(s)$	-0.762	0.119
$Zn^{2+} + 2e^- + Hg \rightleftharpoons Zn(in\ Hg)$	-0.801	
$Zn(NH_3)_4^{2+} + 2e^- \rightleftharpoons Zn(s) + 4NH_3$	-1.04	
$ZnCO_3(s) + 2e^- \rightleftharpoons Zn(s) + CO_3^{2-}$	-1.06	
$Zn(OH)_3^- + 2e^- \rightleftharpoons Zn(s) + 3OH^-$	-1.183	
$Zn(OH)_4^{2-} + 2e^- \rightleftharpoons Zn(s) + 4OH^-$	-1.199	
$Zn(OH)_2(s) + 2e^- \rightleftharpoons Zn(s) + 2OH^-$	-1.249	-0.999
$ZnO(s) + H_2O + 2e^- \rightleftharpoons Zn(s) + 2OH^-$	-1.260	-1.160
$ZnS(s) + 2e^- \rightleftharpoons Zn(s) + S^{2-}$	-1.405	
Zirconium		
$Zr^{4+} + 4e^- \rightleftharpoons Zr(s)$	-1.45	0.67
$ZrO_2(s) + 4H^+ + 4e^- \rightleftharpoons Zr(s) + 2H_2O$	-1.473	-0.344

Stepwise Formation Constants[†]

Reacting ions	$\log K_1$	$\log K_2$	$\log K_3$	$\log K_4$	Temperature (°C)	Ionic strength (M)
Acetate, $CH_3CO_2^-$						
Ag^+	0.73	−0.09			25	0
Ca^{2+}	1.24				25	0
Cd^{2+}	1.93	1.22			25	0
Cu^{2+}	2.23	1.40			25	0
Fe^{2+}	1.82				25	0.5
Fe^{3+}	3.38	3.7	2.6		20	0.1
Mg^{2+}	1.25				25	0
Mn^{2+}	1.40				25	0
Na^+	−0.18				25	0
Ni^{2+}	1.43				25	0
Zn^{2+}	1.28	0.81			20	0.1
Ammonia, NH_3						
Ag^+	3.31	3.92			25	0
Cd^{2+}	2.51	1.96	1.30	0.79	30	0
Co^{2+}	1.99	1.51	0.93	0.64	30	0
	($\log K_5 = 0.06$, $\log K_6 = -0.74$)					
Cu^{2+}	3.99	3.34	2.73	1.97	30	0
Hg^{2+}	8.8	8.7	1.00	0.78	22	2
Ni^{2+}	2.67	2.12	1.61	1.07	30	0
	($\log K_5 = 0.63$, $\log K_6 = -0.09$)					
Zn^{2+}	2.18	2.25	2.31	1.96	30	0
Cyanide, CN^-						
Ag^+	($\log \beta_2 = 20$)		0.95		20	0
Cd^{2+}	5.18	4.42	4.32	3.19	25	?
Cu^+	($\log \beta_2 = 24$)		4.6	1.7	25	0
Ni^{2+}	($\log \beta_4 = 30$)				25	0
Tl^{3+}	13.21	13.29	8.67	7.44	25	4[‡]
Zn^{2+}	($\log \beta_2 = 11.07$)		4.98	3.57	25	0

[†]Stepwise formation constants, K_i, and overall (or cumulative) formation constants, β_i, are defined in Box 5-2. For example, K_2 for the reaction of Ag^+ with acetate refers to the reaction $Ag(CH_3CO_2) + CH_3CO_2^- \rightleftharpoons Ag(CH_3CO_2)_2^-$. The overall formation constant is related to the stepwise formation constants by $\beta_n = K_1 K_2 \ldots K_n$. Data from L. G. Sillen and A. E. Martell, *Stability Constants of Metal-Ion Complexes* (London: The Chemical Society, Special Publications No. 17 and 25, 1964 and 1971).
[‡]J. Blixt, B. Györi, and J. Glaser, *J. Am. Chem. Soc.* **1989**, *111,* 7784.

Reacting ions	$\log K_1$	$\log K_2$	$\log K_3$	$\log K_4$	Temperature (°C)	Ionic strength (M)
Ethylenediamine (1,2-diaminoethane), $H_2NCH_2CH_2NH_2$						
Ag^+	4.70	3.00	2.0		20	0.1
Cd^{2+}	5.69	4.67	2.44		25	0.5
Cu^{2+}	10.66	9.33			20	0
Hg^{2+}	14.3	9.0	−0.1		25	0.1
Ni^{2+}	7.52	6.32	4.49		20	0
Zn^{2+}	5.77	5.06	3.28		20	0
Nitrilotriacetate, $N(CH_2CO_2^-)_3$						
Ag^+	5.16				20	0.1
Al^{3+}	9.5				20	0.1
Ba^{2+}	4.83				20	0.1
Ca^{2+}	6.46				20	0.1
Cd^{2+}	10.0	4.6			20	0.1
Co^{2+}	10.0	3.9			20	0.1
Cu^{2+}	11.5	3.3			20	0.1
Fe^{3+}	15.91	8.70			20	0.1
Ga^{3+}	13.6	8.2			20	0.1
In^{3+}	16.9				20	0.1
Mg^{2+}	5.46				20	0.1
Mn^{2+}	7.4				20	0.1
Ni^{2+}	11.54				20	0.1
Pb^{2+}	11.47				20	0.1
Tl^+	4.75				20	0.1
Zn^{2+}	10.44				20	0.1
Oxalate, $^-O_2CCO_2^-$						
Al^{3+}	($\log \beta_3 = 15.60$)				20	0.1
Ba^{2+}	2.31				18	0
Ca^{2+}	1.66	1.03			25	1
Cd^{2+}	3.71				20	0.1
Co^{2+}	4.69	2.46			25	0
Cu^{2+}	6.23	4.04			25	0
Fe^{3+}	7.54	7.05	5.41		?	0.5
Ni^{2+}	5.16	1.3			25	0
Zn^{2+}	4.85	2.7			25	0

Reacting ions	$\log K_1$	$\log K_2$	$\log K_3$	$\log K_4$	Temperature (°C)	Ionic strength (M)
1,10-Phenanthroline,						
Ag^+	5.02	7.05			25	0.1
Ca^{2+}	0.7				20	0.1
Cd^{2+}	5.17	4.83	4.25		25	0.1
Co^{2+}	7.02	6.70	6.38		25	0.1
Cu^{2+}	8.82	6.57	5.02		25	0.1
Fe^{2+}	5.86	5.25	10.03		25	0.1
Fe^{3+}	($\log \beta_3 = 14.10$)				25	0.1
Hg^{2+}	($\log \beta_2 = 19.65$)		3.7		20	0.1
Mn^{2+}	4.50	4.15	4.05		25	0.1
Ni^{2+}	8.0	8.0	7.9		25	0.1
Zn^{2+}	6.30	5.65	5.10		25	0.1

Solutions to Exercises

Chapter 1

A. (a)

$$\frac{(25.00 \text{ mL})(0.791\,4 \text{ g/mL})/(32.042 \text{ g/mol})}{0.500\,0 \text{ L}}$$

$$= 1.235 \text{ M}$$

(b) 500.0 mL of solution weighs $(1.454 \text{ g/mL}) \times (500.0 \text{ mL}) = 727.0$ g and contains 25.00 mL ($= 19.78$ g) of methanol. The mass of chloroform in 500 mL must be $727.0 - 19.78 = 707.2$ g. The molality of methanol is

$$\text{molality} = \frac{\text{mol methanol}}{\text{kg chloroform}}$$

$$= \frac{(19.78 \text{ g})/(32.042 \text{ g/mol})}{0.707\,2 \text{ kg}}$$

$$= 0.872\,9 \text{ m}$$

B. (a) $\left(\dfrac{48.0 \text{ g HBr}}{100 \text{ g solution}}\right)\left(1.50 \dfrac{\text{g solution}}{\text{mL solution}}\right)$

$$= \frac{0.720 \text{ g HBr}}{\text{mL solution}}$$

$$= \frac{720 \text{ g HBr}}{\text{L solution}}$$

$$= 8.90 \text{ F}$$

(b) $\dfrac{36.0 \text{ g HBr}}{0.480 \text{ g HBr/g solution}} = 75.0 \text{ g solution}$

(c) $233 \text{ mmol} = 0.233 \text{ mol}$

$$\frac{0.233 \text{ mol}}{8.90 \text{ mol/L}} = 26.2 \text{ mL}$$

(d) $V = 250 \text{ mL}\left(\dfrac{0.160 \text{ M}}{8.90 \text{ M}}\right) = 4.49 \text{ mL}$

C. Fraction of mass that is nitrate is

$$\frac{2 \times \text{FW (NO}_3)}{\text{FW (Ca(NO}_3)_2)} = \frac{2 \times 62.005}{164.088}$$

$$= 0.755\,7$$

If $\text{Ca(NO}_3)_2 = 12.6 \text{ ppm}$, $\text{NO}_3^- = (0.755\,7)(12.6) = 9.52 \text{ ppm}$.

Chapter 2

A. (a) At 15°C, water density $= 0.999\,102\,6 \text{ g/mL}$

$$m = \frac{(5.397\,4 \text{ g})\left(1 - \dfrac{0.001\,2 \text{ g/mL}}{8.0 \text{ g/mL}}\right)}{\left(1 - \dfrac{0.001\,2 \text{ g/mL}}{0.999\,102\,6 \text{ g/mL}}\right)} = 5.403\,1 \text{ g}$$

(b) At 25°C, water density $= 0.997\,047\,9 \text{ g/mL}$ and $m = 5.403\,1$ g.

B. Use Equation 2-1 with $m' = 0.2961$ g, $d_a = 0.0012$ g/mL, $d_w = 8.0$ g/mL, and $d = 5.24$ g/mL $\Rightarrow m = 0.2961$ g.

C. $\dfrac{c'}{d'} = \dfrac{c}{d}$

Let the primes stand for 16°C:

$$\Rightarrow \frac{c' \text{ at } 16°C}{0.9989460 \text{ g/mL}} = \frac{0.05138 \text{ M}}{0.9972995 \text{ g/mL}}$$

$$\Rightarrow c' \text{ at } 16° = 0.05146 \text{ M}$$

D. Column 3 of Table 2-6 tells us that water occupies 1.0033 mL/g at 22°C. Therefore $(15.569 \text{ g}) \times (1.0033 \text{ mL/g}) = 15.620$ mL.

Chapter 3

A. (a) 21.0_9 ($\pm 0.1_6$) or 21.1 (± 0.2); relative uncertainty = 0.8%

(b) 27.4_3 ($\pm 0.8_6$); relative uncertainty = $3._2$%

(c) $(14._9 \pm 1._3) \times 10^4$ or $(15 \pm 1) \times 10^4$; relative uncertainty = 9%

(d) $\%e_y = \frac{1}{2}\%e_x = \frac{1}{2}\left(\dfrac{0.08}{3.24} \times 100\right) = 1.235\%$

$(3.24 \pm 0.08)^{1/2} = 1.80 \pm 1.235\% = 1.80 \pm 0.02_2$ ($\pm 1._2\%$)

(e) $\%e_y = 4\%e_x = 4\left(\dfrac{0.08}{3.24} \times 100\right) = 9.877\%$

$(3.24 \pm 0.08)^4 = 110.20 \pm 9.877\% = 1.10$ (± 0.11) $\times 10^2$ ($\pm 9._9\%$)

(f) $\%e_y = 0.43429\dfrac{e_x}{x} = 0.43429\left(\dfrac{0.08}{3.24}\right) = 0.0107$

$\log(3.24 \pm 0.08) = 0.5105 \pm 0.0107 = 0.51 \pm 0.01$ ($\pm 2._1\%$)

(g) $\dfrac{e_y}{y} = 2.3026\, e_x = 2.3026\,(0.08) = 0.184$

$10^{3.24 \pm 0.08} = 1.74 \times 10^3 \pm 18.4\% = 1.7_4$ ($\pm 0.3_2$) $\times 10^3$ ($\pm 18\%$)

B. (a) 2.000 L of 0.169 M NaOH (FW = 39.9971) requires 0.338 mol = 13.52 g NaOH

$$\frac{13.52 \text{ g NaOH}}{0.534 \text{ g NaOH/g solution}} = 25.32 \text{ g solution}$$

$$\frac{25.32 \text{ g solution}}{1.52 \text{ g solution/mL solution}} = 16.6_6 \text{ mL}$$

(b) molarity =

$$[16.66\,(\pm 0.10)\text{ mL}]\left[1.52\,(\pm 0.01)\frac{\text{g solution}}{\text{mL}}\right]$$

$$\frac{\times \left[0.534\,(\pm 0.004)\dfrac{\text{g NaOH}}{\text{g solution}}\right]}{\left(39.9971\dfrac{\text{g NaOH}}{\text{mol}}\right)(2.000 \text{ L})}$$

Because the relative errors in molecular weight and final volume are negligible (≈ 0), we can write

$$\begin{array}{l}\text{relative} \\ \text{error in} \\ \text{molarity}\end{array} = \sqrt{\left(\frac{0.10}{16.66}\right)^2 + \left(\frac{0.01}{1.52}\right)^2 + \left(\frac{0.004}{0.534}\right)^2}$$

$$= 1.16\%$$

molarity = 0.169 (± 0.002)

C. 0.0500 ($\pm 2\%$) mol =

$$[4.18\,(\pm x)\text{ mL}]\left[1.18\,(\pm 0.01)\frac{\text{g solution}}{\text{mL}}\right]$$

$$\frac{\times \left[0.370\,(\pm 0.005)\dfrac{\text{g HCl}}{\text{g solution}}\right]}{36.461\dfrac{\text{g HCl}}{\text{mol}}}$$

Error analysis:

$$(0.02)^2 = \left(\frac{x}{4.18}\right)^2 + \left(\frac{0.01}{1.18}\right)^2 + \left(\frac{0.005}{0.370}\right)^2$$

$$x = 0.05 \text{ mL}$$

Chapter 4

A. mean = $\frac{1}{5}(116.0 + 97.9 + 114.2 + 106.8 + 108.3)$

$= 108.6_4$

standard deviation

$$= \sqrt{\frac{(116.0 - 108.6_4)^2 + \cdots + (108.3 - 108.6_4)^2}{5 - 1}}$$

$= 7.1_4$

90% confidence interval

$$= 108.6_4 \pm \frac{(2.132)(7.1_4)}{\sqrt{5}} = 108.6_4 \pm 6.8_1$$

$$Q = \frac{106.8 - 97.9}{116.0 - 97.9} = 0.49 < [Q(\text{Table 4-6}) = 0.64]$$

Therefore, 97.9 should be retained.

B. **(a)** $z = \dfrac{(45\,800 - 62\,700)}{10\,400} = -1.625$

In Table 4-1, we see that the area listed for $|z| = 1.6$ is 0.445 2, and the area for $|z| = 1.7$ is 0.455 4. By linear interpolation, the area for 1.625 is

$$0.445\,2 + \left(\dfrac{1.625 - 1.6}{1.7 - 1.6}\right)(0.455\,4 - 0.445\,2)$$
$$= 0.447\,8$$

The area *beyond* $|z| = 1.625$ must be $0.500\,0 - 0.447\,8 = 0.052\,2$. The fraction of brakes expected to wear out in less than 45 800 miles is 0.052 2 (or 5.22%).

(b) 60 000 miles lies at $z = -0.259\,6$. 70 000 miles lies at $z = +0.701\,9$. By linear interpolation, the area from $\bar{x}$ to $z = -0.259\,6$ is 0.102 3. The area from $\bar{x}$ to $z = 0.701\,9$ is 0.258 6. Total area $= 0.102\,3 + 0.258\,6 = 0.360\,9$. Thus, approximately 36% of the brakes are expected to wear out between 60 000 and 70 000 miles.

C. confidence 95%: $\mu = \bar{x} \pm \dfrac{ts}{\sqrt{n}}$

$$= 116._4 \pm \dfrac{(2.776)(3._{58})}{\sqrt{5}}$$

$$= 112._0 \text{ to } 120._8$$

You can be 95% sure that the result is different.

D.

$$t = \dfrac{0.027\,5_6 - 0.026\,9_0}{0.000\,4_5}\sqrt{\dfrac{5\cdot5}{5+5}} = 2._{32}$$

Because $t_{\text{calculated}} = 2._{32} > t_{\text{table}}\,(95\%) = 2.306$, the difference is significant.

E. **(a)**

x_i	y_i	x_iy_i	x_i^2	d_i	d_i^2
0.00	0.466	0	0	−0.004 6	2.12×10^{-5}
9.36	0.676	6.327	87.61	+0.001 6	2.58×10^{-6}
18.72	0.883	16.530	350.44	+0.004 8	2.31×10^{-5}
28.08	1.086	30.495	788.49	+0.004 0	1.61×10^{-5}
37.44	1.280	47.923	1 401.75	−0.005 8	3.34×10^{-5}
Sum: 93.60	4.391	101.275	2 628.29		9.64×10^{-5}

$$D = \begin{vmatrix} \sum(x_i^2) & \sum x_i \\ \sum x_i & n \end{vmatrix}$$

$$= (2\,628.29)(5) - (93.60)(93.60) = 4\,380.5$$

$$m = \begin{vmatrix} \sum x_iy_i & \sum x_i \\ \sum y_i & n \end{vmatrix} \div D$$

$$= \dfrac{(101.275)(5) - (93.60)(4.391)}{D}$$

$$= 95.377 \div 4\,380.5 = 0.021\,773$$

$$b = \begin{vmatrix} \sum(x_i^2) & \sum x_iy_i \\ \sum x_i & \sum y_i \end{vmatrix} \div D$$

$$= \dfrac{(2\,628.29)(4.391) - (101.275)(93.60)}{D}$$

$$= 2\,061.48 \div 4\,380.5 = 0.470\,60$$

$$\sigma_y^2 \approx s_y^2 = \dfrac{\sum(d_i^2)}{n-2} = \dfrac{9.64 \times 10^{-5}}{3}$$

$$= 3.21 \times 10^{-5}; \quad \sigma_y = 0.005\,7$$

$$\sigma_m = \sqrt{\dfrac{\sigma_y^2 n}{D}} = \sqrt{\dfrac{(3.21 \times 10^{-5})5}{4\,380.5}} = 0.000\,191$$

$$\sigma_b = \sqrt{\dfrac{\sigma_y^2 \sum(x_i^2)}{D}} = \sqrt{\dfrac{(3.21 \times 10^{-5})(2\,628.29)}{4\,380.5}}$$

$$= 0.004\,39$$

Equation of the best line:

$$y = [0.021\,8\,(\pm 0.000\,2)]x + [0.471\,(\pm 0.004)]$$

(c) $x = \dfrac{y(\pm\sigma_y) - b(\pm\sigma_b)}{m(\pm\sigma_m)}$

$$= \dfrac{0.973\,(\pm 0.005_7) - 0.471\,(\pm 0.004_4)}{0.021\,8\,(\pm 0.000\,1_9)}$$

$$= 23.0 \pm 0.4 \text{ μg}$$

Chapter 5

A. **(a)** $\mathrm{A\!\!\!/g^+ + C\!\!\!/l^- \rightleftharpoons AgCl(aq)}$
$\mathrm{AgCl(s) \rightleftharpoons A\!\!\!/g^+ + C\!\!\!/l^-}$
$$\overline{\mathrm{AgCl(s) \rightleftharpoons AgCl(aq)}}$$

$K_1 = 2.0 \times 10^3$
$K_2 = 1.8 \times 10^{-10}$
$$\overline{K_3 = K_1K_2 = 3.6 \times 10^{-7}}$$

(b) The answer to **(a)** tells us $[\mathrm{AgCl}(aq)] = 3.6 \times 10^{-7}$ M.

(c) $\mathrm{AgCl_2^- \rightleftharpoons A\!\!\!/gCl(aq) + Cl^-}$
$\mathrm{A\!\!\!/g^+ + C\!\!\!/l^- \rightleftharpoons AgCl(s)}$
$\mathrm{A\!\!\!/gCl(aq) \rightleftharpoons A\!\!\!/g^+ + C\!\!\!/l^-}$
$$\overline{\mathrm{AgCl_2^- \rightleftharpoons AgCl(s) + Cl^-}}$$

$$K_1 = 1/(9.3 \times 10^1)$$
$$K_2 = 1/(1.8 \times 10^{-10})$$
$$\underline{K_3 = 1/(2.0 \times 10^3)}$$
$$K_4 = K_1 K_2 K_3 = 3.0 \times 10^4$$

B. (a)

	BrO_3^-	Cr^{3+}	Br^-	$Cr_2O_7^{2-}$	H^+
Initial concentration:	0.0100	0.0100	0	0	1.00
Final concentration:	$0.0100 - x$	$0.0100 - 2x$	x	x	$1.00 + 8x$

$$\frac{(x)(x)(1.00 + 8x)^8}{(0.0100 - x)(0.0100 - 2x)^2} = 1 \times 10^{11}$$

(b) $[Br^-]$ and $[Cr_2O_7^{2-}]$ will both be 0.005 00 M because Cr^{3+} is the *limiting reagent*. Reaction 5-7 requires two moles of Cr^{3+} per mole of BrO_3^-. The Cr^{3+} will be used up first, making one mole of Br^- and one mole of $Cr_2O_7^{2-}$ per two moles of Cr^{3+} consumed. To solve the equation above, we set $x = 0.005\,00$ M in all terms except $[Cr^{3+}]$. The concentration of $[Cr^{3+}]$ will be a small, unknown quantity.

$$\frac{(0.005\,00)(0.005\,00)[1.00 + 8(0.005\,00)]^8}{(0.0100 - 0.005\,00)[Cr^{3+}]^2}$$

$$= 1 \times 10^{11}$$

$$[Cr^{3+}] = 2.6 \times 10^{-7} \text{ M}$$

$$[BrO_3^-] = 0.0100 - 0.005\,00 = 0.005\,00 \text{ M}$$

C. (a) $[La^{3+}][IO_3^-]^3 = x(3x)^3 = 1.0 \times 10^{-11}$

$$\Rightarrow x = 7.8 \times 10^{-4} \text{ M}$$

$$= 0.13 \text{ g La(IO}_3)_3/250 \text{ mL}$$

(b) $[La^{3+}][IO_3^-]^3 = x(3x + 0.050)^3$

$$\approx x(0.050)^3 = 1.0 \times 10^{-11}$$

$$\Rightarrow x = 8 \times 10^{-8}$$

$$= 1.3 \times 10^{-5} \text{ g}$$

D. (a) $Ca(IO_3)_2$ (because it has a larger K_{sp})

(b) The two salts do not have the same stoichiometry. Therefore, K_{sp} values cannot be compared. For $TlIO_3$, $x^2 = K_{sp} \Rightarrow x = 1.8 \times 10^{-3}$. For $Sr(IO_3)_2$, $x(2x)^2 = K_{sp} \Rightarrow x = 4.4 \times 10^{-3}$. $Sr(IO_3)_2$ is more soluble.

E. $[Fe^{3-}][OH^-]^3 = (10^{-10})[OH^-]^3 = 1.6 \times 10^{-39}$

$$\Rightarrow [OH^-] = 2.5 \times 10^{-10}$$

$$[Fe^{2+}][OH^-]^2 = (10^{-10})[OH^-]^2 = 7.9 \times 10^{-16}$$

$$\Rightarrow [OH^-] = 2.8 \times 10^{-3}$$

F. First we need to find which salt will precipitate at the lowest $[C_2O_4^{2-}]$ concentration:

$$[Ca^{2+}][C_2O_4^{2-}] = x^2 = 1.3 \times 10^{-8}$$

$$\Rightarrow [C_2O_4^{2-}] = x$$

$$= 1.1 \times 10^{-4} \text{ M}$$

$$[Ce^{3+}]^2[C_2O_4^{2-}]^3 = (2x)^2(3x)^3 = 3 \times 10^{-29}$$

$$\Rightarrow [C_2O_4^{2-}] = 3x$$

$$= 2.3 \times 10^{-7} \text{ M}$$

$Ce_2(C_2O_4)_3$ is less soluble than CaC_2O_4. The concentration of $C_2O_4^{2-}$ needed to reduce Ce^{3+} to 1% of 0.010 M is

$$[C_2O_4^{2-}] = \left[\frac{K_{sp}}{(0.000\,10)^2}\right]^{1/3} = 1.4 \times 10^{-7}$$

This concentration of $C_2O_4^{2-}$ will not precipitate Ca^{2+} because

$$Q = [Ca^{2+}][C_2O_4^{2-}] = (0.010)(1.4 \times 10^{-7})$$

$$= 1.4 \times 10^{-9} < K_{sp}$$

The separation is feasible.

G. Assuming that all of the Ni is in the form $Ni(en)_3^{2+}$, $[Ni(en)_3^{2+}] = 1.00 \times 10^{-5}$ M. This uses up just 3×10^{-5} mol of en, which leaves the en concentration at 0.100 M. Adding the three equations gives

$$Ni^{2+} + 3en \rightleftharpoons Ni(en)_3^{2+}$$

$$K = K_1 K_2 K_3 = 2.1_4 \times 10^{18}$$

$$[Ni^{2+}] = \frac{[Ni(en)_3^{2+}]}{K[en]^3}$$

$$= \frac{(1.00 \times 10^{-5})}{(2.1_4 \times 10^{18})(0.100)^3} = 4.7 \times 10^{-21} \text{ M}$$

Now we verify that $[Ni(en)^{2+}]$ and $[Ni(en)_2^{2+}] \ll 10^{-5}$ M:

$$[Ni(en)^{2+}] = K_1[Ni^{2+}][en] = 1.6 \times 10^{-14} \text{ M}$$

$$[Ni(en)_2^{2+}] = K_2[Ni(en)^{2+}][en] = 3.3 \times 10^{-9} \text{ M}$$

H. (a) Neutral—neither Na^+ nor Br^- has any acidic or basic properties.

(b) Basic—$CH_3CO_2^-$ is the conjugate base of acetic acid, and Na^+ is neither acidic nor basic.

(c) Acidic—NH_4^+ is the conjugate acid of NH_3, and Cl^- is neither acidic nor basic.

(d) Basic—PO_4^{3-} is a base, and K^+ is neither acidic nor basic.

(e) Neutral—Neither ion is acidic or basic.

(f) Basic—The quaternary ammonium ion is neither acidic nor basic, and the $C_6H_5CO_2^-$ anion is the conjugate base of benzoic acid.

I. $K_{b1} = K_w/K_{a2} = 4.4 \times 10^{-9}$
$K_{b2} = K_w/K_{a1} = 1.6 \times 10^{-10}$

J. $K = K_{b2} = K_w/K_{a2} = 1.0 \times 10^{-8}$

K. **(a)** $[H^+][OH^-] = x^2 = K_w \Rightarrow x = \sqrt{K_w} \Rightarrow pH = -\log\sqrt{K_w} = 7.472$ at 0°C, 7.083 at 20°C, and 6.767 at 40°C.

(b) Because $[D^+] = [OD^-]$ in pure D_2O, $K = 1.35 \times 10^{-15} = [D^+][OD^-] = [D^+]^2 \Rightarrow [D^+] = 3.67 \times 10^{-8}$ M $\Rightarrow$ pD = 7.43.

Chapter 6

A. **(a)** $A = -\log P/P_0 = -\log T = -\log (0.45) = 0.347$

(b) The absorbance will double to 0.694, giving $T = 10^{-A} = 10^{-0.694} = 0.202 \Rightarrow \%T = 20.2\%$.

B. **(a)** $\epsilon = \dfrac{A}{cb} = \dfrac{0.624 - 0.029}{(3.96 \times 10^{-4}\text{ M})(1.000\text{ cm})}$
$= 1.50 \times 10^3 \text{ M}^{-1}\text{ cm}^{-1}$

(b) $c = \dfrac{A}{\epsilon b} = \dfrac{0.375 - 0.029}{(1.50 \times 10^3 \text{ M}^{-1}\text{ cm}^{-1})(1.000\text{ cm})}$
$= 2.31 \times 10^{-4}$ M

(c) $c = \underbrace{\left(\dfrac{25.00\text{ mL}}{2.00\text{ mL}}\right)}_{\substack{\text{Dilution}\\\text{factor}}} \dfrac{0.733 - 0.029}{(1.50 \times 10^3 \text{ M}^{-1}\text{ cm}^{-1})(1.000\text{ cm})}$
$= 5.87 \times 10^{-3}$ M

C. **(a)** 1.00×10^{-2} g of NH_4Cl in 1.00 L $= 1.869 \times 10^{-4}$ M. In the colored solution, the concentration is $(\frac{10}{50})(1.869 \times 10^{-4}$ M$) = 3.739 \times 10^{-5}$ M. $\epsilon = A/bc = (0.308 - 0.140)/(1.00)(3.739 \times 10^{-5}) = 4.49_3 \times 10^3 \text{ M}^{-1}\text{ cm}^{-1}$.

(b) $\dfrac{\text{absorbance of unknown}}{\text{absorbance of reference}}$
$= \dfrac{0.592 - 0.140}{0.308 - 0.140}$
$= \dfrac{\text{concentration of unknown}}{\text{concentration of reference}}$
$\Rightarrow$ concentration of NH_3 in unknown
$= \left(\dfrac{0.452}{0.168}\right)(1.869 \times 10^{-4})$
$= 5.028 \times 10^{-4}$ M

100.00 mL of unknown
$= 5.028 \times 10^{-5}$ mol of N
$= 7.043 \times 10^{-4}$ g of N
$\Rightarrow$ weight % of N
$= (7.043 \times 10^{-4}\text{ g})/(4.37 \times 10^{-3}\text{ g})$
$= 16.1\%$

D. **(a)** Milligrams of Cu in flask C $= (1.00)(\frac{10}{250})(\frac{15}{30}) = 0.020\,0$ mg. This entire quantity is in the isoamyl alcohol (20.00 mL), so the concentration is $(2.00 \times 10^{-5}$ g$)/(0.020\,0$ L$)(63.546$ g/mol$) = 1.57 \times 10^{-5}$ M.

(b) observed absorbance
$=$ absorbance due to Cu in rock
$+$ blank absorbance
$= \epsilon bc + 0.056$
$= (7.90 \times 10^3)(1.00)(1.574 \times 10^{-5}) + 0.056$
$= 0.180$

Note that the observed absorbance is equal to the absorbance from Cu in the rock *plus* the blank absorbance. In the lab we measure the observed absorbance and subtract the blank absorbance from it to find the absorbance due to copper.

(c) $\dfrac{\text{Cu in unknown}}{\text{Cu in known}} = \dfrac{A \text{ of unknown}}{A \text{ of known}}$
$\dfrac{x\text{ mg}}{1.00\text{ mg}} = \dfrac{0.874 - 0.056}{0.180 - 0.056} \Rightarrow x = 6.60$ mg Cu

E. **(c)**

Vol% acetone	Corrected absorbance
10	0.215
20	0.411
30	0.578
40	0.745
50	0.913

(Your answers will be somewhat different from Dan's, depending on where you draw the baselines and how you measure the peaks.)

(e) $y = 0.0173_0 \,(\pm0.000\,3_3)\, x + 0.05_3 \,(\pm0.01_1)$
$(s_y = 0.0104)$

(f) vol % $= \dfrac{0.611\,(\pm0.0104) - 0.053}{0.017\,30} = 32.3\%$

(g) vol % $= \dfrac{0.611\,(\pm0.0104) - 0.053\,(\pm0.011)}{0.017\,30\,(\pm0.000\,33)}$
$= 32._3\,(\pm1._1)\%$

F. (a) $[Ni^{2+}]_f = [Ni^{2+}]_i \dfrac{V_i}{V_f} = [Ni^{2+}]_i \left(\dfrac{25.0}{25.5}\right)$

$$= 0.980_4[Ni^{2+}]_i$$

(b) $[S]_f = (0.028\,7\text{ M}) \left(\dfrac{0.500}{25.5}\right) = 0.000\,562_7$ M

(c) $\dfrac{[Ni^{2+}]_i}{0.000\,562\,7 + 0.980\,4[Ni^{2+}]_i} = \dfrac{2.36\ \mu A}{3.79\ \mu A}$

$$\Rightarrow [Ni^{2+}]_i = 9.00 \times 10^{-4}\text{ M}$$

G. [S] in mixture $= \left(4.13\ \dfrac{\mu g}{mL}\right)\left(\dfrac{2.00}{10.0}\right) = 0.826\ \dfrac{\mu g}{mL}$

$\dfrac{\text{concentration ratio (X/S) in unknown}}{\text{concentration ratio in standard mixture}}$

$\quad = \dfrac{\text{signal ratio (X/S) in unknown}}{\text{signal ratio in standard mixture}}$

$\dfrac{[X]/(0.826\ \mu g/mL)}{(1/1)} = \dfrac{0.808}{1.31}$

$$\Rightarrow [X] = 0.509_5\ \mu g/mL$$

But X had been diluted by a factor of $(5.00/10.0)$ when it was mixed with S. Therefore, initial concentration of X was $\left(\frac{10.0}{5.00}\right)$ $(0.509\,5\ \mu g/mL) = 1.02\ \mu g/mL$.

Chapter 7

A. (a) formula weight of ascorbic acid = 176.126

$\quad$ 0.197 0 g of ascorbic acid = 1.118 5 mmol

$\quad$ molarity of $I_3^- = 1.118\,5$ mmol/29.41 mL

$$= 0.038\,03\text{ M}$$

(b) 31.63 mL of $I_3^- = 1.203$ mmol of I_3^-

$$= 1.203\text{ mmol of ascorbic acid}$$

$$= 0.211\,9\text{ g} = 49.94\%\text{ of the tablet}$$

B. (a) $\dfrac{0.824\text{ g acid}}{204.233\text{ g/mol}} = 4.03$ mmol. This many mmol of NaOH is contained in 0.038 314 kg of NaOH solution

$\Rightarrow$ concentration $= \dfrac{4.03 \times 10^{-3}\text{ mol NaOH}}{0.038\,314\text{ kg solution}}$

$$= 0.105_3\text{ mol/kg solution}$$

(b) mol NaOH $= (0.057\,911\text{ kg})(0.105\,3\text{ mol/kg}) = 6.10$ mmol. Because 2 mol NaOH reacts with 1 mol H_2SO_4,

$$[H_2SO_4] = \dfrac{3.05\text{ mmol}}{10.00\text{ mL}} = 0.305\text{ M}$$

C. 34.02 mL of 0.087 71 M NaOH = 2.983 9 mmol of OH^-. Let x be the mass of malonic acid and y be the mass of anilinium chloride. Then $x + y = 0.2376$ g and

$\quad$ (moles of anilinium chloride)
$\quad$ $\quad$ + 2(moles of malonic acid) $= 0.002\,983\,9$

$$\dfrac{y}{129.59} + 2\left(\dfrac{x}{104.06}\right) = 0.002\,983\,9$$

Substituting $y = 0.2376 - x$ gives $x = 0.099\,97$ g = 42.07% malonic acid. Anilinium chloride = 57.93%.

D. The absorbance must be corrected by multiplying each observed absorbance by (total volume/initial volume). For example, at 36.0 μL, A(corrected) = $(0.399)[(2\,025 + 36)/2\,025] = 0.406$. A graph of corrected absorbance versus volume of Pb^{2+} (μL) is similar to Figure 7-5, with the end point at 46.7 μL. The moles of Pb^{2+} in this volume are $(46.7 \times 10^{-6}$ L$)(7.515 \times 10^{-4}$ M$) = 3.510 \times 10^{-8}$ mol. The concentration of semi-xylenol orange is $(3.510 \times 10^{-8}$ mol$)/(2.025 \times 10^{-3}$ L$) = 1.73 \times 10^{-5}$ M.

E. The reaction is $SCN^- + Cu^+ \rightarrow CuSCN(s)$. The equivalence point occurs when moles of Cu^+ = moles of $SCN^- \Rightarrow V_e = 100.0$ mL. Before the equivalence point, there is excess SCN^- remaining in the solution. We calculate the molarity of SCN^- and then find $[Cu^+]$ from the relation $[Cu^+] = K_{sp}/[SCN^-]$. For example, when 0.10 mL of Cu^+ has been added,

$$[SCN^-] = \left(\dfrac{100.0 - 0.10}{100.0}\right)(0.080\,0)\left(\dfrac{50.0}{50.1}\right)$$

$$= 7.98 \times 10^{-2}\text{ M}$$

$$[Cu^+] = 4.8 \times 10^{-15}/7.98 \times 10^{-2}$$

$$= 6.0 \times 10^{-14}$$

$$pCu^+ = 13.22$$

At the equivalence point, $[Cu^+][SCN^-] = x^2 = K_{sp}$ $\Rightarrow x = [Cu^+] = 6.9 \times 10^{-8} \Rightarrow pCu^+ = 7.16$.

$\quad$ Past the equivalence point, there is excess Cu^+. For example, when $V = 101.0$ mL,

$$[Cu^+] = (0.040\,0)\left(\dfrac{101.0 - 100.0}{151.0}\right) = 2.6 \times 10^{-4}\text{ M}$$

$$pCu^+ = 3.58$$

mL	pCu	mL	pCu	mL	pCu
0.10	13.22	75.0	12.22	100.0	7.16
10.0	13.10	95.0	11.46	100.1	4.57
25.0	12.92	99.0	10.75	101.0	3.58
50.0	12.62	99.9	9.75	110.0	2.60

F. V_e = 23.66 mL for AgBr. At 2.00, 10.00, 22.00, and 23.00 mL, AgBr is partially precipitated and excess Br^- remains.

At 2.00 mL:

$$[Ag^+] = \frac{K_{sp}(\text{for AgBr})}{[Br^-]}$$

$$= \frac{5.0 \times 10^{-13}}{\underbrace{\left(\dfrac{23.66 - 2.00}{23.66}\right)}_{\substack{\text{Fraction} \\ \text{remaining}}} \underbrace{(0.050\,00)}_{\substack{\text{Original} \\ \text{molarity} \\ \text{of } Br^-}} \underbrace{\left(\dfrac{40.00}{42.00}\right)}_{\substack{\text{Dilution} \\ \text{factor}}}}$$

$$= 1.15 \times 10^{-11} \text{ M} \Rightarrow pAg^+ = 10.94$$

By similar reasoning, we find

at 10.00 mL: pAg^+ = 10.66

at 22.00 mL: pAg^+ = 9.66

at 23.00 mL: pAg^+ = 9.25

At 24.00, 30.00, and 40.00 mL, AgCl is precipitating and excess Cl^- remains in solution.

At 24.00 mL:

$$[Ag^+] = \frac{K_{sp}(\text{for AgCl})}{[Cl^-]}$$

$$= \frac{1.8 \times 10^{-10}}{\left(\dfrac{47.32 - 24.00}{23.66}\right)(0.050\,00)\left(\dfrac{40.00}{64.00}\right)}$$

$$= 5.8 \times 10^{-9} \text{ M} \Rightarrow pAg^+ = 8.23$$

By similar reasoning, we find

at 30.00 mL: pAg^+ = 8.07

at 40.00 mL: pAg^+ = 7.63

At the second equivalence point (47.32 mL), $[Ag^+] = [Cl^-]$, and we can write

$$[Ag^+][Cl^-] = x^2 = K_{sp}(\text{for AgCl})$$

$$\Rightarrow [Ag^+] = 1.34 \times 10^{-5} \text{ M}$$

$$pAg^+ = 4.87$$

At 50.00 mL, there is an excess of (50.00 − 47.32) = 2.68 mL of Ag^+.

$$[Ag^+] = \left(\frac{2.68}{90.00}\right)(0.084\,54 \text{ M}) = 2.5 \times 10^{-3} \text{ M}$$

$$pAg^+ = 2.60$$

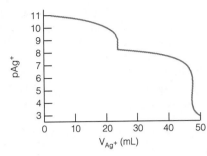

G. (a) 12.6 mL of Ag^+ is required to precipitate I^-. (27.7 − 12.6) = 15.1 mL is required to precipitate SCN^-.

$$[SCN^-] =$$

$$\frac{\text{moles of } Ag^+ \text{ needed to react with } SCN^-}{\text{original volume of } SCN^-}$$

$$= \frac{[27.7\,(\pm0.3) - 12.6\,(\pm0.4)][0.068\,3\,(\pm0.000\,1)]}{50.00\,(\pm0.05)}$$

$$= \frac{[15.1\,(\pm0.5)][0.068\,3\,(\pm0.000\,1)]}{50.00\,(\pm0.05)}$$

$$= \frac{[15.1\,(\pm3.31\%)][0.068\,3\,(\pm0.146\%)]}{50.00\,(\pm0.100\%)}$$

$$= 0.020\,6\,(\pm0.000\,7) \text{ M}$$

(b) $[SCN^-]\,(\pm4.0\%) =$

$$\frac{[27.7\,(\pm0.3) - 12.6\,(\pm?)][0.068\,3\,(\pm0.000\,1)]}{50.00\,(\pm0.05)}$$

Let the error in 15.1 mL be $y\%$:

$$(4.0\%)^2 = (y\%)^2 + (0.146\%)^2 + (0.100\%)^2$$

$$\Rightarrow y = 4.00\% = 0.603 \text{ mL}$$

$$27.7\,(\pm0.3) - 12.6\,(\pm?) = 15.1\,(\pm0.603)$$

$$\Rightarrow 0.3^2 + ?^2 = 0.603^2 \Rightarrow ? = 0.5 \text{ mL}$$

Chapter 8

A. (a) $\mu = \frac{1}{2}([K^+] \cdot 1^2 + [Br^-] \cdot (-1)^2) = 0.02$ M

(b) $\mu = \frac{1}{2}([Cs^+] \cdot 1^2 + [CrO_4^{2-}] \cdot (-2)^2)$

$$= \frac{1}{2}([0.04] \cdot 1 + [0.02] \cdot 4) = 0.06 \text{ M}$$

(c) $\mu = \frac{1}{2}([Mg^{2+}] \cdot 2^2 + [Cl^-] \cdot (-1)^2 + [Al^{3+}] \cdot 3^2)$

$$= \frac{1}{2}([0.02] \cdot 4 +$$
$$\left[\underset{\substack{\uparrow \\ \text{From } MgCl_2}}{0.04} + \underset{\substack{\uparrow \\ \text{From } AlCl_3}}{0.09}\right] \cdot 1 + [0.03] \cdot 9)$$

$$= 0.24 \text{ M}$$

B. For 0.005 0 M $(CH_3CH_2CH_2)_4N^+Br^-$ plus 0.005 0 M $(CH_3)_4N^+Cl^-$, $\mu = 0.010$ M. The size of the ion $(CH_3CH_2CH_2)_4N^+$ is 800 pm. At $\mu = 0.01$ M, $\gamma = 0.912$ for an ion of charge ± 1 with $\alpha = 800$ pm. $\mathcal{A} = (0.005\,0)(0.912) = 0.0046$.

C. (a) $\mu = 0.060$ M from KNO_3 (assuming that AgSCN has negligible solubility)

$$[Ag^+]\gamma_{Ag^+}[SCN^-]\gamma_{SCN^-} = K_{sp}$$

$$[x](0.79)[x](0.80) = 1.1 \times 10^{-12}$$

$$\Rightarrow x = [Ag^+] = 1.3 \times 10^{-6} \text{ M}$$

(b) $\mu = 0.060$ M from KSCN

$$[Ag^+]\gamma_{Ag^+}[SCN^-]\gamma_{SCN^-} = K_{sp}$$

$$[x](0.79)\underbrace{[x + 0.060]}_{\approx 0.060}(0.80) = 1.1 \times 10^{-12}$$

$$\Rightarrow x = [Ag^+] = 2.9 \times 10^{-11} \text{ M}$$

D. Assuming that $Mn(OH)_2$ gives a negligible concentration of ions, $\mu = 0.075$ from $CaCl_2$.

$$[Mn^{2+}]\gamma_{Mn^{2+}}[OH^-]^2\gamma_{OH^-}^2 = K_{sp}$$

$$[x](0.445)[2x]^2(0.785)^2 = 1.6 \times 10^{-13}$$

$$\Rightarrow 2x = [OH^-] = 1.1 \times 10^{-4} \text{ M}$$

E. For 0.02 M $MgCl_2$ (which dissociates into 0.020 M Mg^{2+} and 0.040 M Cl^-), $\mu = 0.06$ M. At this ionic strength, interpolation in Table 8-1 gives $\gamma_{Mg^{2+}} = 0.506$ and $\gamma_{Cl^-} = 0.795$.

$$\gamma_\pm = (\gamma_{Mg^{2+}}\gamma_{Cl^-}^2)^{1/(1+2)} = 0.68_4$$

F. $[H^+]\gamma_{H^+}[OH^-]\gamma_{OH^-} = (x)(0.86)(x)(0.81) = 1.0 \times 10^{-14} \Rightarrow x = [H^+] = 1.2 \times 10^{-7}$ M. pH $= -\log(1.2 \times 10^{-7})(0.86) = 6.99$.

G. (a) moles of $I^- = 2$(moles of Hg_2^{2+})

$$(V_e)(0.100 \text{ M}) = 2(40.0 \text{ mL})(0.040\,0 \text{ M})$$

$$\Rightarrow V_e = 32.0 \text{ mL}$$

(b) Virtually all the Hg_2^{2+} has precipitated, along with 3.20 mmol of I^-. The ions remaining in solution are

$$[NO_3^-] = \frac{3.20 \text{ mmol}}{100.0 \text{ mL}} = 0.032\,0 \text{ M}$$

$$[I^-] = \frac{2.80 \text{ mmol}}{100.0 \text{ mL}} = 0.028\,0 \text{ M}$$

$$[K^+] = \frac{6.00 \text{ mmol}}{100.0 \text{ mL}} = 0.060\,0 \text{ M}$$

$$\mu = \tfrac{1}{2}\sum c_i z_i^2 = 0.060\,0 \text{ M}$$

(c) $$\mathcal{A}_{Hg_2^{2+}} = K_{sp}/\mathcal{A}_{I^-}^2$$

$$= 4.5 \times 10^{-29}/(0.028\,0)^2(0.795)^2$$

$$= 9.1 \times 10^{-26} \Rightarrow pHg_2^{2+} = 25.04$$

Chapter 9

A. $[H^+] + 2[Ca^{2+}] + [CaF^+] = [OH^-] + [F^-]$

Remember that H^+ and OH^- are present in every aqueous solution.

B. (a) $[Cl^-] = 2[Ca^{2+}]$

(b) $\underbrace{[Cl^-] + [CaCl^+]}_{\text{Moles of Cl}} = \underbrace{2([Ca^{2+}] + [CaCl^+])}_{\text{Moles of Ca}}$

C. (a) $[F^-] + [HF] = 2[Ca^{2+}]$

(b) $\underbrace{[F^-] + [HF] + 2[HF_2^-]}_{\text{Moles of F}} = 2[Ca^{2+}]$

(One mole of HF_2^- contains two moles of fluorine.)

D. $2[Ca^{2+}] = 3\{[PO_4^{3-}] + [HPO_4^{2-}]$
$\qquad\qquad + [H_2PO_4^-] + [H_3PO_4]\}$

E. (a) charge balance: invalid because pH is fixed

mass balance: $[Ag^+] = [CN^-] + [HCN]$ (1)

equilibria: $K_b = \dfrac{[HCN][OH^-]}{[CN^-]}$ (2)

$$K_{sp} = [Ag^+][CN^-] \qquad (3)$$

$$K_w = [H^+][OH^-] \qquad (4)$$

Because $[H^+] = 10^{-9.00}$ M, $[OH^-] = 10^{-5.00}$ M. Putting this value $[OH^-]$ into Equation 2 gives

$$[HCN] = \frac{K_b}{[OH^-]}[CN^-] = 1.6[CN^-]$$

Substituting into Equation 1 gives

$$[Ag^+] = [CN^-] + 1.6[CN^-] = 2.6[CN^-]$$

Substituting into Equation 3 gives

$$[Ag^+]\left(\frac{[Ag^+]}{2.6}\right) = K_{sp}$$

$$\Rightarrow [Ag^+] = 2.4 \times 10^{-8} \text{ M}$$

$$[CN^-] = [Ag^+]/2.6 = 9.2 \times 10^{-9} \text{ M}$$

$$[HCN] = 1.6[CN^-] = 1.5 \times 10^{-8} \text{ M}$$

(b) With activities:

$$[Ag^+] = [CN^-] + [HCN] \qquad (1')$$

$$K_b = \frac{[HCN]\gamma_{HCN}[OH^-]\gamma_{OH^-}}{[CN^-]\gamma_{CN^-}} \qquad (2')$$

$$K_{sp} = [Ag^+]\gamma_{Ag^+}[CN^-]\gamma_{CN^-} \qquad (3')$$

$$K_w = [H^+]\gamma_{H^+}[OH^-]\gamma_{OH^-} \qquad (4')$$

Because pH = 9.00, $[OH^-]\gamma_{OH^-} = K_w/[H^+]\gamma_{H^+} = 10^{-5.00}$. Putting this value into Equation 2' gives

$$[HCN] = \frac{K_b\gamma_{CN^-}[CN^-]}{\gamma_{HCN}[OH^-]\gamma_{OH^-}}$$

$$= \frac{(1.6 \times 10^{-5})(0.755)[CN^-]}{1 \cdot 10^{-5.00}}$$

$$= 1.208[CN^-]$$

Here we have assumed $\gamma_{HCN} = 1$. Using the above relation in the mass balance (Equation 1') gives

$$[Ag^+] = 2.208[CN^-]$$

Substituting into Equation 3' gives

$$K_{sp} = [Ag^+](0.75)\left(\frac{Ag^+}{2.208}\right)(0.755)$$

$$\Rightarrow [Ag^+] = 2.9 \times 10^{-8}\ M$$

$$[CN^-] = \frac{[Ag^+]}{2.208} = 1.3 \times 10^{-8}\ M$$

$$[HCN] = 1.208[CN^-] = 1.6 \times 10^{-8}\ M$$

F. charge balance: invalid because pH is fixed

mass balance:

$$[Zn^{2+}] = [C_2O_4^{2-}] + [HC_2O_4^-] \qquad (1)$$
$$+ [H_2C_2O_4]$$

equilibria:

$$K_{sp} = [Zn^{2+}][C_2O_4^{2-}] \qquad (2)$$

$$K_{b1} = \frac{[HC_2O_4^-][OH^-]}{[C_2O_4^{2-}]} \qquad (3)$$

$$K_{b2} = \frac{[H_2C_2O_4][OH^-]}{[HC_2O_4^-]} \qquad (4)$$

$$K_w = [H^+][OH^-] \qquad (5)$$

If pH = 3.0, $[OH^-] = 1.0 \times 10^{-11}$. Putting this into Equation 3 gives

$$[HC_2O_4^-] = \frac{K_{b1}}{[OH^-]}[C_2O_4^{2-}] = 18[C_2O_4^{2-}]$$

Using this result in Equation 4 gives

$$[H_2C_2O_4] = \frac{K_{b2}}{[OH^-]}[HC_2O_4^-]$$

$$= \frac{K_{b2}}{[OH^-]}\frac{K_{b1}}{[OH^-]}[C_2O_4^{2-}]$$

$$= 0.324[C_2O_4^{2-}]$$

Using these values of $[H_2C_2O_4]$ and $[HC_2O_4^-]$ in Equation 1 gives

$$[Zn^{2+}] = [C_2O_4^{2-}](1 + 18 + 0.324)$$

$$= [C_2O_4^{2-}](19.324)$$

Putting this last equality into Equation 2 produces the result

$$[Zn^{2+}]\left(\frac{[Zn^{2+}]}{19.324}\right) = K_{sp}$$

$$\Rightarrow [Zn^{2+}] = 3.8 \times 10^{-4}\ M$$

If $[Zn^{2+}] = 3.8 \times 10^{-4}$ M, this many moles of ZnC_2O_4 (= 0.058 g) must be dissolved in each liter.

Chapter 10

A. pH = $-\log \mathscr{A}_{H^+}$. But $\mathscr{A}_{H^+}\mathscr{A}_{OH^-} = K_w \Rightarrow \mathscr{A}_{H^+} = K_w/\mathscr{A}_{OH^-}$. For 1.0×10^{-2} M NaOH, $[OH^-] = 1.0 \times 10^{-2}$ M and $\gamma_{OH^-} = 0.900$ (using Table 8-1, with ionic strength = 0.010 M).

$$\mathscr{A}_{H^+} = \frac{K_w}{[OH^-]\gamma_{OH^-}}$$

$$= \frac{1.0 \times 10^{-14}}{(1.0 \times 10^{-2})(0.900)}$$

$$= 1.11 \times 10^{-12}$$

$$\Rightarrow pH = -\log \mathscr{A}_{H^+} = 11.95$$

B. (a) charge balance: $[H^+] = [OH^-] + [Br^-]$

mass balance: $[Br^-] = 1.0 \times 10^{-8}$ M

equilibrium: $[H^+][OH^-] = K_w$

Setting $[H^+] = x$ and $[Br^-] = 1.0 \times 10^{-8}$ M, the charge balance tells us that $[OH^-] = x - 1.0 \times 10^{-8}$. Putting this into the K_w equilibrium gives

$$(x)(x - 1.0 \times 10^{-8}) = 1.0 \times 10^{-14}$$

$$\Rightarrow x = 1.0_5 \times 10^{-7}\ M$$

$$\Rightarrow pH = 6.98$$

(b) charge balance: $[H^+] = [OH^-] + 2[SO_4^{2-}]$

mass balance: $[SO_4^{2-}] = 1.0 \times 10^{-8}$ M

equilibrium: $[H^+][OH^-] = K_w$

As above, writing $[H^+] = x$ and $[SO_4^{2-}] = 1.0 \times 10^{-8}$ M gives $[OH^-] = x - 2.0 \times 10^{-8}$ and $[H^+][OH^-] = (x)[x - (2.0 \times 10^{-8})] = 1.0 \times 10^{-14}$ $\Rightarrow x = 1.10 \times 10^{-7}$ M $\Rightarrow$ pH = 6.96.

C.

2-Nitrophenol FW = 139.110

$C_6H_5NO_3$ $K_a = 6.2 \times 10^{-8}$

F_{HA} (formal concentration)

$$= \frac{1.23 \text{ g}/(139.110 \text{ g/mol})}{0.250 \text{ L}}$$

$$= 0.035\,4 \text{ M}$$

$$HA \rightleftharpoons H^+ + A^-$$
$$F - x \quad\quad x \quad\quad x$$

$$\frac{x^2}{0.035\,4 - x} = 6.2 \times 10^{-8}$$

$$\Rightarrow x = 4.7 \times 10^{-5} \text{ M}$$

$$\Rightarrow pH = -\log x = 4.33$$

D.

$$F - x \quad\quad x \quad\quad x$$

But $[H^+] = 10^{-pH} = 8.9 \times 10^{-7}$ M $\Rightarrow [A^-] = 8.9 \times 10^{-7}$ M and $[HA] = 0.010 - [H^+] = 0.010$.

$$K_a = \frac{[H^+][A^-]}{[HA]} = \frac{(8.9 \times 10^{-7})^2}{0.010}$$

$$= 7.9 \times 10^{-11} \Rightarrow pK_a = 10.10$$

E. As $[HA] \to 0$, pH $\to$ 7. If pH = 7,

$$\frac{[H^+][A^-]}{[HA]} = K_a \Rightarrow [A^-] = \frac{K_a}{[H^+]} [HA]$$

$$= \frac{10^{-5.00}}{10^{-7.00}} [HA] = 100[HA]$$

$$\alpha = \frac{[A^-]}{[HA] + [A^-]} = \frac{100[HA]}{[HA] + 100[HA]}$$

$$= \frac{100}{101} = 99\%$$

If $pK_a = 9.00$, we find $\alpha = 0.99\%$.

F. $CH_3CH_2CH_2CO_2^- + H_2O \rightleftharpoons$
$F - x$

$$CH_3CH_2CH_2CO_2H + OH^-$$
$$x \quad\quad\quad x$$

$$K_b = \frac{K_w}{K_a} = 6.58 \times 10^{-10}$$

$$\frac{x^2}{F - x} = K_b \Rightarrow x = 5.7_4 \times 10^{-6} \text{ M}$$

$$pH = -\log\left(\frac{K_w}{x}\right) = 8.76$$

G. (a) $CH_3CH_2NH_2 + H_2O \rightleftharpoons$
$F - x$

$$CH_3CH_2NH_3^+ + OH^-$$
$$x \quad\quad\quad x$$

Because pH = 11.80, $[OH^-] = K_w/10^{-pH} = 6.3 \times 10^{-3}$ M $= [BH^+]$. $[B] = F - x = 0.094$ M.

$$K_b = \frac{[BH^+][OH^-]}{[B]} = \frac{(6.3 \times 10^{-3})^2}{0.094}$$

$$= 4.2 \times 10^{-4}$$

(b) $CH_3CH_2NH_3^+ \rightleftharpoons CH_3CH_2NH_2 + H^+$
$F - x \quad\quad\quad x \quad\quad x$

$$K_a = \frac{K_w}{K_b} = 2.4 \times 10^{-11}$$

$$\frac{x^2}{F - x} = K_a \Rightarrow x = 1.5_5 \times 10^{-6} \text{ M}$$

$$\Rightarrow pH = 5.81$$

H.

Compound	pK_a (for conjugate acid)	
Ammonia	9.24	← Most suitable, because pK_a
Aniline	4.60	is closest to pH
Hydrazine	8.48	
Pyridine	5.23	

I. pH = 4.25 + log 0.75 = 4.13

J. (a) $pH = pK_a + \log\dfrac{[B]}{[BH^+]}$

$$= 8.20 + \log\frac{[(1.00 \text{ g})/(74.083 \text{ g/mol})]}{[(1.00 \text{ g})/(110.543 \text{ g/mol})]} = 8.37$$

(b) $pH = pK_a + \log\dfrac{\text{mol B}}{\text{mol BH}^+}$

$$8.00 = 8.20 + \log\frac{\text{mol B}}{(1.00 \text{ g})/(110.543 \text{ g/mol})}$$

$$\Rightarrow \text{mol B} = 0.005\,708 = 0.423 \text{ g of glycine amide}$$

(c)

	B	+	H$^+$	→	BH$^+$
Initial moles:	0.013 498		0.000 500		0.009 046
Final moles:	0.012 998		—		0.009 546

$$pH = 8.20 + \log\left(\frac{0.012\,998}{0.009\,546}\right) = 8.33$$

(d)

	BH$^+$	+	OH$^-$	$\rightarrow$	B
Initial moles:	0.009\,546		0.001\,000		0.012\,998
Final moles:	0.008\,546		—		0.013\,998

$$pH = 8.20 + \log\left(\frac{0.013\,998}{0.008\,546}\right) = 8.41$$

K. The reaction of phenylhydrazine with water is

$$B + H_2O \rightleftharpoons BH^+ + OH^- \quad K_b$$

We know that pH = 8.13, so we can find [OH$^-$].

$$[OH^-] = \frac{\mathcal{A}_{OH^-}}{\gamma_{OH^-}} = \frac{K_w/10^{-pH}}{\gamma_{OH^-}} = 1.78 \times 10^{-6}\ M$$

(using $\gamma_{OH^-} = 0.76$ for $\mu = 0.10$ M)

$$K_b = \frac{[BH^+]\gamma_{BH^+}[OH^-]\gamma_{OH^-}}{[B]\gamma_B}$$

$$= \frac{(1.78 \times 10^{-6})(0.80)(1.78 \times 10^{-6})(0.76)}{[0.010 - (1.78 \times 10^{-6})](1.00)}$$

$$= 1.93 \times 10^{-10}$$

$$K_a = \frac{K_w}{K_b} = 5.19 \times 10^{-5} \Rightarrow pK_a = 4.28$$

To find K_b, we made use of the equality [BH$^+$] = [OH$^-$].

Chapter 11

A. (a) $H_2SO_3 \rightleftharpoons HSO_3^- + H^+$
  0.050 − x   x   x

$$\frac{x^2}{0.050 - x} = K_1 = 1.23 \times 10^{-2}$$

$$\Rightarrow x = 1.94 \times 10^{-2}$$

$$[HSO_3^-] = [H^+] = 1.94 \times 10^{-2}\ M$$

$$\Rightarrow pH = 1.71$$

$$[H_2SO_3] = 0.050 - x = 0.031\ M$$

$$[SO_3^{2-}] = \frac{K_2[HSO_3^-]}{[H^+]} = K_2$$

$$= 6.6 \times 10^{-8}\ M$$

(b) $$[H^+] = \sqrt{\frac{K_1 K_2 (0.050) + K_1 K_w}{K_1 + (0.050)}}$$

$$= 2.55 \times 10^{-5} \Rightarrow pH = 4.59$$

$$[H_2SO_3] = \frac{[H^+][HSO_3^-]}{K_1}$$

$$= \frac{(2.55 \times 10^{-5})(0.050)}{1.23 \times 10^{-2}}$$

$$= 1.0 \times 10^{-4}\ M$$

$$[SO_3^{2-}] = \frac{K_2[HSO_3^-]}{[H^+]} = 1.3 \times 10^{-4}\ M$$

$$[HSO_3^-] = 0.050\ M$$

(c) $SO_3^{2-} + H_2O \rightleftharpoons HSO_3^- + OH^-$
  0.050 − x   x   x

$$\frac{x^2}{0.050 - x} = K_{b1} = \frac{K_w}{K_{a2}} = 1.52 \times 10^{-7}$$

$$[HSO_3^-] = x = 8.7 \times 10^{-5}\ M$$

$$[H^+] = \frac{K_w}{x} = 1.15 \times 10^{-10}\ M$$

$$\Rightarrow pH = 9.94$$

$$[SO_3^{2-}] = 0.050 - x = 0.050\ M$$

$$[H_2SO_3] = \frac{[H^+][HSO_3^-]}{K_1} = 8.1 \times 10^{-13}$$

B. **(a)** $$pH = pK_2 \text{ (for } H_2CO_3) + \log\frac{[CO_3^{2-}]}{[HCO_3^-]}$$

$$10.80 = 10.329 + \log\frac{(4.00/138.206)}{(x/84.007)}$$

$$\Rightarrow x = 0.822\ g$$

(b)

	CO$_3^{2-}$	+	H$^+$	$\rightarrow$	HCO$_3^-$
Initial moles:	0.0289$_4$		0.0100		0.00978
Final moles:	0.0189$_4$		—		0.0197$_8$

$$pH = 10.329 + \log\frac{0.0189_4}{0.0197_8} = 10.31$$

(c)

	CO$_3^{2-}$	+	H$^+$	$\rightarrow$	HCO$_3^-$
Initial moles:	0.0289$_4$		x		—
Final moles:	0.0289$_4$ − x		—		x

$$10.00 = 10.329 + \log\frac{0.0289_4 - x}{x}$$

$$\Rightarrow x = 0.0197\ mol$$

$$\Rightarrow \text{volume} = \frac{0.0197\ mol}{0.320\ M} = 61.6\ mL$$

C. $H_2C_2O_4$: FW = 90.035, so 3.38 g = 0.037 5 mol.

$$H_2Ox \xrightarrow{OH^-} HOx^- \xrightarrow{OH^-} Ox^{2-}$$

The pH of HOx^- is approximately $\frac{1}{2}(pK_1 + pK_2)$ = 2.76. At pH = 2.40, the predominant species will be H_2Ox and HOx^-.

	H_2Ox	+ OH^-	→ HOx^-	+ H_2O
Initial				
moles:	0.037 5	x	—	—
Final				
moles:	0.037 5 − x	—	x	—

$$pH = pK_1 + \log \frac{[HOx^-]}{[H_2Ox]}$$

$$2.40 = 1.252 + \log \frac{x}{0.037 5 - x}$$

$$x = 0.035\,0 \text{ mol}$$

$$\Rightarrow \text{volume} = \frac{0.035\,0 \text{ mol}}{0.800 \text{ M}} = 43.8 \text{ mL}$$

(Note that this problem involves a rather low pH, at which Equations 10-20 and 10-21 would be more appropriate to use.)

D. (a) Calling the three forms of glutamine H_2G^+, HG, and G^-, the form shown is HG.

$$[H^+] = \sqrt{\frac{K_1K_2(0.010) + K_1K_w}{K_1 + 0.010}}$$

$$= 1.9_9 \times 10^{-6} \Rightarrow pH = 5.70$$

(b) Calling the four forms of cysteine H_3C^+, H_2C, HC^-, and C^{2-}, the form shown is HC^-.

$$[H^+] = \sqrt{\frac{K_2K_3(0.010) + K_2K_w}{K_2 + 0.010}}$$

$$= 2.8_1 \times 10^{-10} \Rightarrow pH = 9.55$$

(c) Calling the four forms of arginine H_3A^{2+}, H_2A^+, HA, and A^-, the form shown is HA.

$$[H^+] = \sqrt{\frac{K_2K_3(0.010) + K_2K_w}{K_2 + 0.010}}$$

$$= 3.6_8 \times 10^{-11} \Rightarrow pH = 10.43$$

E.

	pH 9.00	pH 11.00
Principal species:		
Secondary species:		
Percentage in major form:	66.5%	52.9%

The percentage in the major form was calculated with the formulas for α_{H_2A} (Equation 11-19 at pH 9.00) and α_{HA^-} (Equation 11-20 at pH 11.00).

F.

	pH 9.0	pH 10.0
Predominant form	NH_3^+ CHCH$_2$CH$_2$CO$_2^-$ CO$_2^-$	NH_2 CHCH$_2$CH$_2$CO$_2^-$ CO$_2^-$
Secondary form	NH_2 CHCH$_2$CH$_2$CO$_2^-$ CO$_2^-$	NH_3^+ CHCH$_2$CH$_2$CO$_2^-$ CO$_2^-$

G. The isoionic pH is the pH of a solution of pure neutral lysine, which is

$$
\begin{array}{c}
NH_2 \\
| \\
CHCH_2CH_2CH_2CH_2NH_3^+ \\
| \\
CO_2^-
\end{array}
$$

$$[H^+] = \sqrt{\frac{K_2 K_3 F + K_2 K_w}{K_2 + F}} \Rightarrow pH = 9.88$$

H. We know that the isoelectric point will be near $\frac{1}{2}(pK_2 + pK_3) \approx 9.88$. At this pH, the fraction of lysine in the form $H_3 L^{2+}$ is negligible. Therefore, the electroneutrality condition reduces to $[H_2 L^+] = [L^-]$, for which the expression, isoelectric pH = $\frac{1}{2}(pK_2 + pK_3) = 9.88$, applies.

Chapter 12

A. The titration reaction is $H^+ + OH^- \rightarrow H_2O$ and $V_e = 5.00$ mL. Three representative calculations are given:

At 1.00 mL: $[OH^-] = \left(\frac{4.00}{5.00}\right)(0.0100)\left(\frac{50.00}{51.00}\right)$
$$= 0.00784 \text{ M}$$

$$pH = -\log\left(\frac{K_w}{[OH^-]}\right) = 11.89$$

At 5.00 mL: $H_2O \rightleftharpoons \underset{x}{H^+} + \underset{x}{OH^-}$

$$x^2 = K_w \Rightarrow x = 1.0 \times 10^{-7} \text{ M}$$

$$pH = -\log x = 7.00$$

At 5.01 mL: $[H^+] = \left(\frac{0.01}{55.01}\right)(0.100)$

$$= 1.82 \times 10^{-5} \text{ M} \Rightarrow pH = 4.74$$

V_a (mL)	pH	V_a	pH	V_a	pH
0.00	12.00	4.50	10.96	5.10	3.74
1.00	11.89	4.90	10.26	5.50	3.05
2.00	11.76	4.99	9.26	6.00	2.75
3.00	11.58	5.00	7.00	8.00	2.29
4.00	11.27	5.01	4.74	10.00	2.08

B. The titration reaction is $HCO_2H + OH^- \rightarrow HCO_2^- + H_2O$ and $V_e = 50.0$ mL. For formic acid, $K_a = 1.80 \times 10^{-4}$. Four representative calculations are given:

At 0.0 mL: $\underset{0.0500 - x}{HA} \rightleftharpoons \underset{x}{H^+} + \underset{x}{A^-}$

$$\frac{x^2}{0.0500 - x} = K_a \Rightarrow x = 2.91 \times 10^{-3}$$

$$\Rightarrow pH = 2.54$$

At 48.0 mL: $HA + OH^- \rightarrow A^- + H_2O$

Initial:	50	48	—	—
Final:	2	—	48	48

$$pH = pK_a + \log\frac{[A^-]}{[HA]} = 3.745 + \log\frac{48.0}{2.0} = 5.13$$

At 50.0 mL: $\underset{F - x}{A^-} + H_2O \overset{K_b}{\rightleftharpoons} \underset{x}{HA} + \underset{x}{OH^-}$

$$K_b = \frac{K_w}{K_a} \quad \text{and} \quad F = \left(\frac{50}{100}\right)(0.05)$$

$$\frac{x^2}{0.0250 - x} = 5.56 \times 10^{-11} \Rightarrow x = 1.18 \times 10^{-6} \text{ M}$$

$$pH = -\log\left(\frac{K_w}{x}\right) = 8.07$$

At 60.0 mL: $[OH^-] = \left(\frac{10.0}{110.0}\right)(0.0500)$
$$= 4.55 \times 10^{-3} \text{ M} \Rightarrow pH = 11.66$$

V_b (mL)	pH	V_b	pH	V_b	pH
0.0	2.54	45.0	4.70	50.5	10.40
10.0	3.14	48.0	5.13	51.0	10.69
20.0	3.57	49.0	5.44	52.0	10.99
25.0	3.74	49.5	5.74	55.0	11.38
30.0	3.92	50.0	8.07	60.0	11.66
40.0	4.35				

C. The titration reaction is $B + H^+ \rightarrow BH^+$ and $V_e = 50.00$ mL. Representative calculations:

At $V_a = 0.0$ mL: $\underset{0.100 - x}{B} + H_2O \rightleftharpoons \underset{x}{BH^+} + \underset{x}{OH^-}$

$$\frac{x^2}{0.100 - x} = 2.6 \times 10^{-6} \Rightarrow x = 5.09 \times 10^{-4}$$

$$pH = -\log\left(\frac{K_w}{x}\right) = 10.71$$

At $V_a = 20.0$ mL: $B + H^+ \rightarrow BH^+$

Initial:	50.00	20.0	—
Final:	30.0	—	20.0

$$pH = pK_a \text{ (for } BH^+) + \log\frac{[B]}{[BH^+]}$$

$$= 8.41 + \log\frac{30.0}{20.0} = 8.59$$

At $V_a = V_e = 50.0$ mL: All B has been converted to the conjugate acid, BH^+. The formal concentration of BH^+ is $(\frac{100}{150})(0.100) = 0.0667$ M. The pH is determined by the reaction

$$\underset{0.0667 - x}{BH^+} \rightleftharpoons \underset{x}{B} + \underset{x}{H^+}$$

$$\frac{x^2}{0.0667 - x} = K_a = \frac{K_w}{K_b} \Rightarrow x$$

$$= 1.60 \times 10^{-5} \Rightarrow pH = 4.80$$

At $V_a = 51.0$ mL: There is excess H^+:

$$[H^+] = \left(\frac{1.0}{151.0}\right)(0.200) = 1.32 \times 10^{-3}$$

$$\Rightarrow pH = 2.88$$

V_a (mL)	pH	V_a	pH	V_a	pH
0.0	10.71	30.0	8.23	50.0	4.80
10.0	9.01	40.0	7.81	50.1	3.88
20.0	8.59	49.0	6.72	51.0	2.88
25.0	8.41	49.9	5.71	60.0	1.90

D. The titration reactions are

$$HO_2CCH_2CO_2H + OH^- \rightarrow$$
$$^-O_2CCH_2CO_2H + H_2O$$

$$^-O_2CCH_2CO_2H + OH^- \rightarrow$$
$$^-O_2CCH_2CO_2^- + H_2O$$

and the equivalence points occur at 25.0 and 50.0 mL. We will designate malonic acid as H_2M.

At 0.0 mL: $H_2M \rightleftharpoons H^+ + HM^-$
 $0.0500 - x \quad\ x \quad\ \ x$

$$\frac{x^2}{0.0500 - x} = K_1 \Rightarrow x = 7.75 \times 10^{-3}$$

$$\Rightarrow pH = 2.11$$

At 8.0 mL: $H_2M + OH^- \rightarrow HM^- + H_2O$

Initial:	25	8	—	—
Final:	20	—	8	—

$$pH = pK_1 + \log\frac{[HM^-]}{[H_2M]} = 2.847 + \log\frac{8}{17}$$

$$= 2.52$$

At 12.5 mL: $V_b = \frac{1}{2}V_e \Rightarrow pH = pK_1 = 2.85$

At 19.3 mL: $H_2M + OH^- \rightarrow HM^- + H_2O$

Initial:	25	19.3	—	—
Final:	5.7	—	19.3	—

$$pH = pK_1 + \log\frac{19.3}{5.7} = 3.38$$

At 25.0 mL: At the first equivalence point, H_2M has been converted to HM^-.

$$[H^+] = \sqrt{\frac{K_1K_2F + K_1K_w}{K_1 + F}}$$

where $F = \left(\frac{50}{75}\right)(0.0500)$

$$[H^+] = 5.23 \times 10^{-5} \text{ M} \Rightarrow pH = 4.28$$

At 37.5 mL: $V_b = \frac{3}{2}V_e \Rightarrow pH = pK_2 = 5.70$

At 50.0 mL: At the second equivalence point, H_2M has been converted to M^{2-}:

$$M^{2-} + H_2O \rightleftharpoons HM^- + OH^-$$
$$\left(\frac{50}{100}\right)(0.0500) - x \qquad x \qquad x$$

$$\frac{x^2}{0.0250 - x} = K_{b1} = \frac{K_w}{K_{a2}}$$

$$\Rightarrow x = 1.12 \times 10^{-5} \text{ M}$$

$$\Rightarrow pH = -\log\left(\frac{K_w}{x}\right) = 9.05$$

At 56.3 mL: There is 6.3 mL of excess NaOH.

$$[OH^-] = \left(\frac{6.3}{106.3}\right)(0.100) = 5.93 \times 10^{-3} \text{ M}$$

$$\Rightarrow pH = 11.77$$

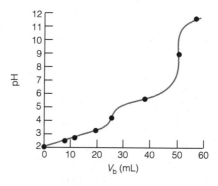

E.

HHis

H_2His^+

H_3His^{2+}

The equivalence points occur at 25.0 and 50.0 mL.

At 0 mL: HHis is the second intermediate form derived from the triprotic acid, H_3His^{2+}.

$$[H^+] = \sqrt{\frac{K_2K_3(0.0500) + K_2K_w}{K_2 + (0.0500)}}$$

$$= 2.81 \times 10^{-8} \text{ M} \Rightarrow \text{pH} = 7.55$$

At 4.0 mL:

HHis	+	H^+	$\rightarrow$	H_2His^+
Initial:	25	4		—
Final:	21	—		4

$$\text{pH} = \text{p}K_2 + \log\frac{21}{4} = 6.74$$

At 12.5 mL: $\text{pH} = \text{p}K_2 = 6.02$

At 25.0 mL: The histidine has been converted to H_2His^+ at the formal concentration $F = (\frac{25}{50}) \times (0.0500) = 0.0250$ M.

$$[H^+] = \sqrt{\frac{K_1K_2F + K_1K_w}{K_1 + F}}$$

$$= 1.03 \times 10^{-4} \Rightarrow \text{pH} = 3.99$$

At 26.0 mL:

H_2His^+	+	H^+	$\rightarrow$	H_3His^{2+}
Initial:	25	1		—
Final:	24	—		1

$$\text{pH} = \text{p}K_1 + \log\frac{24}{1} = 3.08$$

At 50.0 mL: The histidine has been converted to H_3His^{2+} at the formal concentration $F = (\frac{25}{75})(0.0500) = 0.0167$ M.

$$H_3His^{2+} \rightleftharpoons H_2His^+ + H^+$$
$$0.0167 - x \qquad x \qquad x$$

$$\frac{x^2}{0.0167 - x} = K_1 \Rightarrow x = 0.0108 \text{ M}$$

$$\Rightarrow \text{pH} = 1.97$$

F. Figure 12-1: bromothymol blue: blue $\rightarrow$ yellow

Figure 12-2: thymol blue: yellow $\rightarrow$ blue

Figure 12-3: thymolphthalein: colorless $\rightarrow$ blue

G. (a) $A = 2080[\text{HIn}] + 14\,200[\text{In}^-]$

(b) $[\text{HIn}] = x$; $[\text{In}^-] = 1.84 \times 10^{-4} - x$

$A = 0.868$

$= 2080x + 14\,200(1.84 \times 10^{-4} - x)$

$\Rightarrow x = 1.44 \times 10^{-4}$ M

$$\text{p}K_a = \text{pH} - \log\frac{[\text{In}^-]}{[\text{HIn}]}$$

$$= 6.23 - \log\frac{(1.84 \times 10^{-4}) - (1.44 \times 10^{-4})}{1.44 \times 10^{-4}}$$

$$= 6.79$$

H. The titration reaction is $HA + OH^- \rightarrow A^- + H_2O$. It requires one mole of NaOH to react with one mole of HA. Therefore, the formal concentration of A^- at the equivalent point is

$$\underbrace{\left(\frac{27.63}{127.63}\right)}_{\substack{\text{Dilution factor} \\ \text{for NaOH}}} \times \underbrace{(0.09381)}_{\substack{\text{Initial concentration} \\ \text{of NaOH}}} = 0.02031 \text{ M}$$

Because the pH is 10.99, $[\text{OH}^-] = 9.77 \times 10^{-4}$ and we can write

$$A^- + H_2O \rightleftharpoons HA + OH^-$$

$$K_b = \frac{[\text{HA}][\text{OH}^-]}{[\text{A}^-]} = \frac{(9.77 \times 10^{-4})^2}{0.02031 - (9.77 \times 10^{-4})}$$

$$= 4.94 \times 10^{-5}$$

$$K_a = \frac{K_w}{K_b} = 2.03 \times 10^{-10} \Rightarrow \text{p}K_a = 9.69$$

For the 19.47-mL point, we have

HA	+	OH^-	$\rightarrow$	A^-	+	H_2O
Initial:	27.63	19.47		—		—
Final:	8.16	—		19.47		—

$$\text{pH} = \text{p}K_a + \log\frac{[\text{A}^-]}{[\text{HA}]} = 9.69 + \log\frac{19.47}{8.16}$$

$$= 10.07$$

I. When $V_b = \frac{1}{2}V_e$, $[\text{HA}] = [\text{A}^-] = 0.0333$ M (using a correction for dilution by NaOH). $[\text{Na}^+] = 0.0333$ M, as well. Ionic strength $= 0.0333$ M.

$$\text{p}K_a = \text{pH} - \log\frac{[\text{A}^-]\gamma_{A^-}}{[\text{HA}]\gamma_{HA}} \quad \left(\substack{\text{from} \\ \text{Equation 10-18}}\right)$$

$$= 4.62 - \log\frac{(0.0333)(0.854)}{(0.0333)(1.00)} = 4.69$$

The activity coefficient of A^- was found by interpolation in Table 8-1.

J. (a) The derivatives are shown in the spreadsheet on the next page.

	A	B	C	D	E	F	G
1			First derivative		Second derivative		
2	µL NaOH	pH	µL	Derivative	µL	Derivative	Vb*10^–pH
3	107	6.921					
4	110	7.117	108.5	6.533E–02			
5	113	7.359	111.5	8.067E–02	110	5.11E–03	4.94E–06
6	114	7.457	113.5	9.800E–02	112.5	8.67E–03	3.98E–06
7	115	7.569	114.5	1.120E–01	114	1.40E–02	3.10E–06
8	116	7.705	115.5	1.360E–01	115	2.40E–02	2.29E–06
9	117	7.878	116.5	1.730E–01	116	3.70E–02	1.55E–06
10	118	8.09	117.5	2.120E–01	117	3.90E–02	9.59E–07
11	119	8.343	118.5	2.530E–01	118	4.10E–02	5.40E–07
12	120	8.591	119.5	2.480E–01	119	–5.00E–03	3.08E–07
13	121	8.794	120.5	2.030E–01	120	–4.50E–02	1.94E–07
14	122	8.952	121.5	1.580E–01	121	–4.50E–02	1.36E–07
15							
16	C4 = (A4+A3)/2			E5 = (C5+C4)/2		G5 = A5*10^–B5	
17	D4 = (B4–B3)/(A4–A3)			F5 = (D5–D4)/(C5–C4)			

In the first derivative graph, the maximum is near 119 mL. In Figure 12-7, the second derivative graph gives an end point of 118.9 µL.

(b) Column G in the spreadsheet gives $V_b(10^{-pH})$. In a graph of $V_b(10^{-pH})$ vs. V_b, the points from 113 to 117 µL give a straight line whose slope is -1.178×10^6 and whose intercept (end point) is 118.7 µL.

K. (a) pH 9.6 is past the equivalence point, so excess volume (V) is given by

$$[OH^-] = 10^{-4.4} = (0.1000 \text{ M}) \frac{V}{50.00 + 10.00 + V}$$

$$\Rightarrow V = 0.024 \text{ mL}$$

(b) pH 8.8 is before the equivalence point:

$$8.8 = 6.15 + \log \frac{[A^-]}{[HA]} \Rightarrow \frac{[A^-]}{[HA]} = 446.7$$

Titration reaction:	HA	+ OH⁻ → A⁻	+ H₂O	
Relative initial quantities:	10	V	—	—
Relative final quantities:	10 – V	—	V	—

To attain a ratio $[A^-]/[HA] = 446.7$, we need $V/(10 - V) = 446.7 \Rightarrow V = 9.978$ mL. The indicator error is $10 - 9.978 = 0.022$ mL.

Chapter 13

A. For every mole of K^+ entering the first reaction, four moles of EDTA are produced in the second reaction.

moles of EDTA = moles of Zn^{2+} used in titration

$$[K^+] = \frac{(\frac{1}{4})(\text{moles of } Zn^{2+})}{\text{volume of original sample}}$$

$$= \frac{(\frac{1}{4})[28.73 \ (\pm 0.03)][0.043 \ 7 \ (\pm 0.000 \ 1)]}{250.0 \ (\pm 0.1)}$$

$$= \frac{[\frac{1}{4}(\pm 0\%)][28.73 \ (\pm 0.104\%)][0.043 \ 7 \ (\pm 0.229\%)]}{250.0 \ (\pm 0.040 \ 0\%)}$$

$$= 1.256 \ (\pm 0.255\%) \times 10^{-3} \text{ M}$$

$$= 1.256 \ (\pm 0.003) \text{ mM}$$

B. Total $Fe^{3+} + Cu^{2+}$ in 25.00 mL = (16.06 mL) × (0.050 83 M) = 0.816 3 mmol.

Second titration:

millimoles EDTA used:	(25.00)(0.050 83) =	1.270 8
millimoles Pb^{2+} needed:	(19.77)(0.018 83) =	0.372 3
millimoles Fe^{3+} present:	(difference)	0.898 5

Because 50.00 mL of unknown was used in the second titration, the millimoles of Fe^{3+} in 25.00 mL are 0.449 2. The millimoles of Cu^{2+} in 25.00 mL are $0.816 \ 3 - 0.449 \ 2 = 0.367 \ 1$ mmol/25.00 mL = 0.014 68 M.

C. Designating the total concentration of free EDTA as [EDTA], we can write

$$\frac{[GaY^-]}{[Ga^{3+}][EDTA]} = \alpha_{Y^{4-}}K_f = 7.58 \times 10^{11}$$

Some representative calculations are shown below:

At 0.1 mL:

$$[EDTA] = \left(\frac{25.0 - 0.1}{25.0}\right)(0.040\,0)\left(\frac{50.0}{50.1}\right)$$

$$= 0.039\,8 \text{ M}$$

$$[GaY^-] = \left(\frac{0.1}{50.1}\right)(0.080\,0)$$

$$= 1.60 \times 10^{-4} \text{ M}$$

$$[Ga^{3+}] = \frac{(1.32 \times 10^{-12})[GaY^-]}{[EDTA]}$$

$$= 5.31 \times 10^{-15}$$

$$\Rightarrow pGa^{3+} = 14.28$$

At 25.0 mL: formal concentration of GaY^-

$$= \left(\frac{25.0}{75.0}\right)(0.080\,0) = 0.026\,7 \text{ M}$$

	Ga^{3+} +	EDTA ⇌	GaY^-
Initial concentration:	—	—	0.026 7
Final concentration:	x	x	$0.026\,7 - x$

$$\frac{0.026\,7 - x}{x^2} = 7.58 \times 10^{11}$$

$$\Rightarrow [Ga^{2+}] = 1.88 \times 10^{-7} \text{ M}$$

$$\Rightarrow pGa^{3+} = 6.73$$

At 26.0 mL: $Ga^{3+} = \left(\frac{1.0}{76.0}\right)(0.080\,0)$

$$= 1.05 \times 10^{-3} \text{ M}$$

$$\Rightarrow pGa^{3+} = 2.98$$

Summary:

Volume (mL)	pGa^{3+}	Volume	pGa^{3+}
0.1	14.28	24.0	10.50
5.0	12.48	25.0	6.73
10.0	12.06	26.0	2.98
15.0	11.70	30.0	2.30
20.0	11.28		

D.
$$\begin{array}{ll} HY^{3-} \rightleftharpoons H^+ + Y^{4-} & K_6 \\ H_2Y^{2-} \rightleftharpoons H^+ + HY^{3-} & K_5 \\ \hline H_2Y^{2-} \rightleftharpoons 2H^+ + Y^{4-} & K = K_5K_6 \end{array}$$

$$= \frac{[H^+]^2[Y^{4-}]}{[H_2Y^{2-}]}$$

$$[H_2Y^{2-}] = \frac{[H^+]^2[Y^{4-}]}{K_5K_6} = \frac{[H^+]^2\alpha_{Y^{4-}}[EDTA]}{K_5K_6}$$

Using the values $[H^+] = 10^{-4.00}$, $\alpha_{Y^{4-}} = 3.8 \times 10^{-9}$, and $[EDTA] = 1.88 \times 10^{-7}$ gives $[H_2Y^{2-}] = 1.8 \times 10^{-7}$ M.

E. (a) One volume of Fe^{3+} will require two volumes of EDTA to reach the equivalence point. The formal concentration of FeY^- at the equivalence point is $(\frac{1}{3})(0.010\,0) = 0.003\,33$ M.

$$\begin{array}{ccc} Fe^{3+} + & EDTA \rightleftharpoons & FeY^- \\ x & x & 0.003\,33 - x \end{array}$$

$$\frac{0.003\,33 - x}{x^2} = \alpha_{Y^{4-}}K_f$$

$$\Rightarrow x = [Fe^{3+}] = 8.8 \times 10^{-8} \text{ M}$$

(b) Because the pH is 2.00, the *ratio* $[H_3Y^-]/[H_2Y^{2-}]$ is constant throughout the *entire* titration.

$$\frac{[H_2Y^{2-}][H^+]}{[H_3Y^-]} = K_4 \Rightarrow \frac{[H_3Y^-]}{[H_2Y^{2-}]} = \frac{[H^+]}{K_4} = 4.6$$

F. K_f for $CoY^{2-} = 10^{16.31} = 2.0 \times 10^{16}$.

$$\alpha_{Y^{4-}} = 0.054 \text{ at pH } 9.00$$

$$\alpha_{Co^{2+}} = \frac{1}{1 + \beta_1[C_2O_4^{2-}] + \beta_2[C_2O_4^{2-}]^2}$$

$$= 6.8 \times 10^{-6}$$

(using $\beta_1 = K_1 = 10^{4.69}$ and $\beta_2 = K_1K_2 = 10^{4.69}\,10^{2.46}$)

$$K_f' = \alpha_{Y^{4-}}K_f = 1.1 \times 10^{15}$$

$$K_f'' = \alpha_{Co^{2+}}\alpha_{Y^{4-}}K_f = 7.5 \times 10^9$$

At 0 mL: $[Co^{2+}] = \alpha_{Co^{2+}}(1.00 \times 10^{-3})$

$$= 6.8 \times 10^{-9} \text{ M} \Rightarrow pCo^{2+} = 8.17$$

At 1.00 mL: $C_{Co^{2+}} = \left(\frac{1.00}{2.00}\right)(1.00 \times 10^{-3})\left(\frac{20.00}{21.00}\right)$

Fraction remaining · Initial concentration · Dilution factor

$$= 4.76 \times 10^{-4} \text{ M}$$

$$[Co^{2+}] = \alpha_{Co^{2+}}C_{Co^{2+}}$$

$$= 3.2 \times 10^{-9} \text{ M} \Rightarrow pCo^{2+} = 8.49$$

At 2.00 mL: This is the equivalence point.

$$C_{Co^{2+}} + EDTA \overset{K_f''}{\rightleftharpoons} CoY^{2-}$$

$$x \qquad x \qquad \left(\frac{20.00}{22.00}\right)(1.00 \times 10^{-3}) - x$$

$$K_f'' = \frac{9.09 \times 10^{-4} - x}{x^2}$$

$$\Rightarrow x = 3.5 \times 10^{-7} \text{ M} = C_{Co^{2+}}$$

$$[Co^{2+}] = \alpha_{Co^{2+}}C_{Co^{2+}}$$

$$= 2.4 \times 10^{-12} \text{ M} \Rightarrow pCo^{2+} = 11.63$$

At 3.00 mL:

$$\text{concentration of excess EDTA} = \frac{1.00}{23.00}(1.00 \times 10^{-2})$$

$$= 4.35 \times 10^{-4} \text{ M}$$

$$\text{concentration of CoY}^{2-} = \frac{20.00}{23.00}(1.00 \times 10^{-3})$$

$$= 8.70 \times 10^{-4} \text{ M}$$

Knowing [EDTA] and [CoY^{2-}], we can use the K_f' equilibrium to find [Co^{2+}]:

$$K_f' = \frac{[CoY^{2-}]}{[Co^{2+}][EDTA]} = \frac{[8.70 \times 10^{-4}]}{[Co^{2+}][4.35 \times 10^{-4}]}$$

$$\Rightarrow [Co^{2+}] = 1.8 \times 10^{-15} \text{ M} \Rightarrow pCo^{2+} = 14.74$$

G. 25.0 mL of 0.120 M iminodiacetic acid = 3.00 mmol

25.0 mL of 0.0500 M Cu^{2+} = 1.25 mmol

	Cu^{2+} +	2 iminodiacetic acid	$\rightleftharpoons$ CuX$_2^{2-}$
Initial millimoles:	1.25	3.00	—
Final millimoles:	—	0.50	1.25

$$\frac{[CuX_2^{2-}]}{[Cu^{2+}][X^{2-}]^2} = K_f$$

$$\frac{[1.25/50.0]}{[Cu^{2+}][(0.50/50.0)(4.6 \times 10^{-3})]^2} = 3.5 \times 10^{16}$$

$$\Rightarrow [Cu^{2+}] = 3.4 \times 10^{-10} \text{ M}$$

Chapter 14

A. The cell voltage will be 1.35 V because all activities are unity.

$$I = P/E = 0.0100 \text{ W}/1.35 \text{ V} = 7.41 \times 10^{-3} \text{ C/s}$$

$$= 7.68 \times 10^{-8} \text{ mol e}^-/\text{s} = 2.42 \text{ mol e}^-/365 \text{ days}$$

$$= 1.21 \text{ mol HgO}/365 \text{ days} = 0.262 \text{ kg HgO}$$

$$= 0.578 \text{ lb}$$

B. (a) $5Br_2(aq) + 10e^- \rightleftharpoons 10Br^-$

$$2IO_3^- + 12H^+ + 10e^- \rightleftharpoons I_2(s) + 6H_2O$$

$$I_2(s) + 5Br_2(aq) + 6H_2O \rightleftharpoons 2IO_3^- + 10Br^- + 12H^+$$

$$E_+^\circ = 1.098 \text{ V}$$

$$E_-^\circ = 1.210 \text{ V}$$

$$E^\circ = 1.098 - 1.210 = -0.112 \text{ V}$$

$$K = 10^{10(-0.112)/0.05916} = 1 \times 10^{-19}$$

(b) $Cr^{2+} + 2e^- \rightleftharpoons Cr(s)$

$$Fe^{2+} + 2e^- \rightleftharpoons Fe(s)$$

$$Cr^{2+} + Fe(s) \rightleftharpoons Cr(s) + Fe^{2+}$$

$$E_+^\circ = -0.89 \text{ V}$$

$$E_-^\circ = -0.44 \text{ V}$$

$$E^\circ = -0.89 - (-0.44) = -0.45 \text{ V}$$

$$K = 10^{2(-0.45)/0.05916} = 10^{-15}$$

(c) $Cl_2(g) + 2e^- \rightleftharpoons 2Cl^-$

$$Mg^{2+} + 2e^- \rightleftharpoons Mg(s)$$

$$Mg(s) + Cl_2(g) \rightleftharpoons Mg^{2+} + 2Cl^-$$

$$E_+^\circ = 1.360 \text{ V}$$

$$E_-^\circ = -2.360 \text{ V}$$

$$E^\circ = 1.360 - (-2.360) = 3.720 \text{ V}$$

$$K = 10^{2(3.720)/0.05916} = 6 \times 10^{125}$$

(d) $3[MnO_2(s) + 4H^+ + 2e^- \rightleftharpoons Mn^{2+} + 2H_2O]$

$$2[MnO_4^- + 4H^+ + 3e^- \rightleftharpoons MnO_2(s) + 2H_2O]$$

$$5MnO_2(s) + 4H^+ \rightleftharpoons 2MnO_4^- + 3Mn^{2+} + 2H_2O$$

$$E_+^\circ = 1.230 \text{ V}$$

$$E_-^\circ = 1.692 \text{ V}$$

$$E_+^\circ = 1.230 - 1.692 = -0.462 \text{ V}$$

$$K = 10^{6(-0.462)/0.05916} = 1 \times 10^{-47}$$

An alternate way to answer (d) is the following:

$$5[MnO_2(s) + 4H^+ + 2e^- \rightleftharpoons Mn^{2+} + 2H_2O]$$

$$2[MnO_4^- + 8H^+ + 5e^- \rightleftharpoons Mn^{2+} + 4H_2O]$$

$$5MnO_2(s) + 4H^+ \rightleftharpoons 2MnO_4^- + 3Mn^{2+} + 2H_2O$$

$$E_+^\circ = 1.230 \text{ V}$$

$$E_-^\circ = 1.507 \text{ V}$$

$$E^\circ = 1.230 - 1.507 = -0.277 \text{ V}$$

$$K = 10^{10(-0.277)/0.05916} = 2 \times 10^{-47}$$

(e) $Ag^+ + e^- \rightleftharpoons Ag(s)$

$$\underline{-\quad Ag(S_2O_3)_2^{3-} + e^- \rightleftharpoons Ag(s) + 2S_2O_3^{2-}}$$

$$Ag^+ + 2S_2O_3^{2-} \rightleftharpoons Ag(S_2O_3)_2^{3-}$$

$$E_+^\circ = 0.799 \text{ V}$$

$$E_-^\circ = 0.017 \text{ V}$$

$$E^\circ = 0.799 - 0.017 = 0.782 \text{ V}$$

$$K = 10^{0.782/0.05916} = 2 \times 10^{13}$$

(f) $CuI(s) + e^- \rightleftharpoons Cu(s) + I^-$

$$\underline{-\quad Cu^+ + e^- \rightleftharpoons Cu(s)}$$

$$CuI(s) \rightleftharpoons Cu^+ + I^-$$

$$E_+^\circ = -0.185 \text{ V}$$

$$E_-^\circ = 0.518 \text{ V}$$

$$E^\circ = -0.185 - 0.518 = -0.703 \text{ V}$$

$$K = 10^{-0.703/0.05916} = 1 \times 10^{-12}$$

C. (a) $Br_2(l) + 2e^- \rightleftharpoons 2Br^- \quad E_+^\circ = 1.078 \text{ V}$

$\underline{-\quad Fe^{2+} + 2e^- \rightleftharpoons Fe(s) \qquad E_-^\circ = -0.44 \text{ V}}$

$$Br_2(l) + Fe(s) \rightleftharpoons 2Br^- + Fe^{2+}$$

$$E = \left\{1.078 - \frac{0.05916}{2} \log (0.050)^2\right\} -$$

$$\left\{-0.44 - \frac{0.05916}{2} \log \frac{1}{0.010}\right\} = 1.65 \text{ V}$$

(b) $Fe^{2+} + 2e^- \rightleftharpoons Fe(s) \qquad E_+^\circ = -0.44 \text{ V}$

$\underline{-\quad Cu^{2+} + 2e^- \rightleftharpoons Cu(s) \qquad E_-^\circ = 0.339 \text{ V}}$

$$Fe^{2+} + Cu(s) \rightleftharpoons Fe(s) + Cu^{2+}$$

$$E = \left\{-0.44 - \frac{0.05916}{2} \log \frac{1}{0.050}\right\} -$$

$$\left\{0.339 - \frac{0.05916}{2} \log \frac{1}{0.020}\right\} = -0.77 \text{ V}$$

(c) $Cl_2(g) + 2e^- \rightleftharpoons 2Cl^- \qquad E_+^\circ = 1.360 \text{ V}$

$\underline{-\quad Hg_2Cl_2(s) + 2e^- \rightleftharpoons 2Hg(l) + 2Cl^- \quad E_-^\circ = 0.268 \text{ V}}$

$$Cl_2(g) + 2Hg(l) \rightleftharpoons Hg_2Cl_2(s)$$

$$E = \left\{1.360 - \frac{0.05916}{2} \log \frac{(0.040)^2}{0.50}\right\} -$$

$$\left\{0.268 - \frac{0.05916}{2} \log (0.060)^2\right\} = 1.094 \text{ V}$$

D. (a) $H^+ + e^- \rightleftharpoons \frac{1}{2}H_2(g) \qquad E_+^\circ = 0 \text{ V}$

$\qquad Ag^+ + e^- \rightleftharpoons Ag(s) \qquad E_-^\circ = 0.799 \text{ V}$

$$E^\circ = E_+^\circ - E_-^\circ = -0.799 \text{ V}$$

$$E = \left\{0 - 0.05916 \log \frac{P_{H_2}^{1/2}}{[H^+]}\right\} -$$

$$\left\{0.799 - 0.05916 \log \frac{1}{[Ag^+]}\right\}$$

(b) $[Ag^+] = \dfrac{K_{sp}}{[I^-]} = \dfrac{8.3 \times 10^{-17}}{0.10} = 8.3 \times 10^{-16} \text{ M}$

$$E = \left\{0 - 0.05916 \log \frac{\sqrt{0.20}}{0.10}\right\} -$$

$$\left\{0.799 - 0.05916 \log \frac{1}{8.3 \times 10^{-16}}\right\} = 0.055 \text{ V}$$

(c) $H^+ + e^- \rightleftharpoons \frac{1}{2}H_2(g) \qquad E_+^\circ = 0 \text{ V}$

$\qquad AgI(s) + e^- \rightleftharpoons Ag(s) + I^- \qquad E_-^\circ = \text{?}$

$$0.055 = \left\{0 - 0.05916 \log \frac{\sqrt{0.20}}{0.10}\right\} -$$

$$\left\{E_-^\circ - 0.05916 \log (0.10)\right\}$$

$$\Rightarrow E_-^\circ = -0.153 \text{ V}$$
(The appendix gives $E_-^\circ = -0.152 \text{ V}$.)

E. (a) $Ag(CN)_2^- + e^- \rightleftharpoons Ag(s) + 2CN^-$

$\qquad Cu^{2+} + 2e^- \rightleftharpoons Cu(s)$

$$E_+^\circ = -0.310 \text{ V}$$

$$E_-^\circ = 0.339 \text{ V}$$

$$E = \left\{-0.310 - 0.05916 \log \frac{[CN^-]^2}{[Ag(CN)_2^-]}\right\} -$$

$$\left\{0.339 - \frac{0.05916}{2} \log \frac{1}{[Cu^{2+}]}\right\}$$

We know that $[Ag(CN)_2^-] = 0.010 \text{ M}$ and $[Cu^{2+}] = 0.030 \text{ M}$. To find $[CN^-]$ at pH 8.21, we write

$$\frac{[CN^-]}{[HCN]} = \frac{K_a}{[H^+]} \Rightarrow [CN^-] = 0.10 \, [HCN]$$

But because $[CN^-] + [HCN] = 0.10$ M, $[CN^-] = 0.0091$ M.

Putting this concentration into the Nernst equation gives $E = -0.481$ V.

F. **(a)** $PuO_2^+ + e^- + 4H^+ \rightleftharpoons Pu^{4+} + 2H_2O$

$$
\begin{array}{ll}
PuO_2^{2+} \rightarrow PuO_2^+ & \Delta G = -1F\,(0.966) \\
PuO_2^+ \rightarrow Pu^{4+} & \Delta G = -1F\,E° \\
Pu^{4+} \rightarrow Pu^{3+} & \Delta G = -1F\,(1.006) \\
\hline
PuO_2^{2+} \rightarrow Pu^{3+} & \Delta G = -3F\,(1.021)
\end{array}
$$

$$-3F\,(1.021) = -1F\,(0.966) - 1F\,E°$$

$$-1F\,(1.006) \Rightarrow E° = 1.091 \text{ V}$$

(b)

$$2PuO_2^{2+} + 2e^- \rightleftharpoons 2PuO_2^+ \qquad E_+° = 0.966 \text{ V}$$

$$-\tfrac{1}{2}O_2(g) + 2H^+ + 2e^- \rightleftharpoons H_2O \qquad E_-° = 1.229 \text{ V}$$

$$\overline{2PuO_2^{2+} + H_2O \rightleftharpoons 2PuO_2^+ + \tfrac{1}{2}O_2(g) + 2H^+}$$

$$E = \left\{ 0.966 - \frac{0.059\,16}{2} \log \frac{[PuO_2^+]^2}{[PuO_2^{2+}]^2} \right\} -$$

$$\left\{ 1.229 - \frac{0.059\,16}{2} \log \frac{1}{P_{O_2}^{1/2}\,[H^+]^2} \right\}$$

$[PuO_2^+]$ cancels $[PuO_2^{2+}]$ because they are equal. At pH 2.00, we insert $[H^+] = 10^{-2.00}$ and $P_{O_2} = 0.20$ to find $E = -0.134$ V. Because $E < 0$, the reaction is not spontaneous and water is not oxidized. At pH 7.00, we find $E = +0.161$ V, so water will be oxidized.

G. $2H^+ + 2e^- \rightleftharpoons H_2(g) \qquad E_+° = 0$ V

$Hg_2Cl_2(s) + 2e^- \rightleftharpoons 2Hg(l) + 2Cl^- \quad E_-° = 0.268$ V

$$E = \left\{ -\frac{0.059\,16}{2} \log \frac{P_{H_2}}{[H^+]^2} \right\} -$$

$$\left\{ 0.268 - \frac{0.059\,16}{2} \log [Cl^-]^2 \right\}$$

We find $[H^+]$ in the right half-cell by considering the acid-base chemistry of KHP, the intermediate form of a diprotic acid:

$$[H^+] = \sqrt{\frac{K_1 K_2 (0.50) + K_1 K_w}{K_1 + 0.050}} = 6.5 \times 10^{-5} \text{ M}$$

$$E = \left\{ \frac{-0.059\,16}{2} \log \frac{1}{(6.5 \times 10^{-5})^2} \right\} -$$

$$\left\{ 0.268 - \frac{0.059\,16}{2} \log (0.10)^2 \right\} = -0.575 \text{ V}$$

H. $2H^+ + 2e^- \rightleftharpoons H_2(g) \qquad E_+° = 0$ V

$Hg^{2+} + 2e^- \rightleftharpoons Hg(l) \qquad E_-° = 0.852$ V

$$E = \left\{ \frac{-0.059\,16}{2} \log \frac{P_{H_2}}{[H^+]^2} \right\} -$$

$$\left\{ 0.852 - \frac{0.059\,16}{2} \log \frac{1}{[Hg^{2+}]} \right\}$$

$$-0.321 = \left\{ \frac{-0.059\,16}{2} \log \frac{1}{1^2} \right\} -$$

$$\left\{ 0.852 - \frac{0.059\,16}{2} \log \frac{1}{[Hg^{2+}]} \right\}$$

$$\Rightarrow [Hg^{2+}] = 1.1 \times 10^{-18} \text{ M}$$

Because $[Hg^{2+}]$ is so small, $[HgI_4^{2-}] = 0.0010$ M. To make this much HgI_4^{2-}, the concentration of I^- must have been reduced from 0.010 M to 0.0060 M, since one Hg^{2+} ion reacts with four I^- ions.

$$K = \frac{[HgI_4^{2-}]}{[Hg^{2+}][I^-]^4} = \frac{(0.0010)}{(1.1 \times 10^{-18})(0.0060)^4}$$

$$= 7 \times 10^{23}$$

I. $CuY^{2-} + 2e^- \rightleftharpoons Cu(s) + Y^{4-} \qquad E_+° = ?$

$Cu^{2+} + 2e^- \rightleftharpoons Cu(s) \qquad\qquad E_-° = 0.339$ V

$$\overline{CuY^{2-} \overset{1/K_f}{\rightleftharpoons} Cu^{2+} + Y^{4-} \qquad\qquad E°}$$

$$E° = \frac{0.059\,16}{2} \log \frac{1}{K_f} = -0.556 \text{ V}$$

$$E_+° = E° + E_-° = -0.217 \text{ V}$$

J. To compare glucose and H_2 at pH = 0, we need to know $E°$ for each. For H_2, $E° = 0$ V. For glucose, we find $E°$ from $E°'$:

$$\underset{\text{Gluconic acid}}{HA} + 2H^+ + 2e^- \rightleftharpoons \underset{\text{Glucose}}{G} + H_2O$$

$$E = E° - \frac{0.059\,16}{2} \log \frac{[G]}{[HA][H^+]^2}$$

But $F_G = [G]$ and $[HA] = \dfrac{[H^+]F_{HA}}{[H^+] + K_a}$. Putting these into the Nernst equation gives

$$E = E° - \frac{0.059\,16}{2} \log \frac{F_G}{\left(\dfrac{[H^+]F_{HA}}{[H^+] + K_a} \right)[H^+]^2}$$

$$= E° - \frac{0.059\,16}{2} \log \underbrace{\frac{[H^+] + K_a}{[H^+]^3}}$$

This is $E°' = -0.45$ V

$$-\frac{0.059\,16}{2} \log \frac{F_G}{F_{HA}}$$

$$-0.45\text{ V} = E° - \frac{0.059\,16}{2} \log \frac{10^{-7.00} + 10^{-3.56}}{(10^{-7.00})^3}$$

$$\Rightarrow E° = +0.066\text{ V for glucose.}$$

Because $E°$ for H_2 is more negative than $E°$ for glucose, H_2 is the stronger reducing agent at pH 0.00.

K. **(a)** Each H^+ must provide $\frac{1}{2}(34.5$ kJ$)$ when it passes from outside to inside.

$$\Delta G = -(\tfrac{1}{2})(34.5 \times 10^3\text{ J}) = -RT \ln \frac{A_{high}}{A_{low}}$$

$$\frac{A_{high}}{A_{low}} = 1.05 \times 10^3$$

$$\Rightarrow \Delta\text{pH} = \log(1.05 \times 10^3)$$

$$= 3.02\text{ pH units}$$

(b) $\Delta G = -nFE$ (where n = charge of $H^+ = 1$)
$-(\tfrac{1}{2})(34.5 \times 10^3\text{ J}) = -1FE \Rightarrow E = 0.179$ V

(c) If $\Delta\text{pH} = 1.00$, $A_{high}/A_{low} = 10$.

$$\Delta G(\text{pH}) = -RT \ln 10 = -5.7 \times 10^3\text{ J}$$

$$\Delta G(\text{electric}) = [\tfrac{1}{2}(34.5) - 5.7]\text{ kJ} = 11.5\text{ kJ}$$

$$E = \frac{\Delta G(\text{electric})}{F} = 0.120\text{ V}$$

Chapter 15

A. The reaction at the silver electrode (written as a reduction) is $Ag^+ + e^- \rightleftharpoons Ag(s)$, and the cell voltage is written as

$$E = E_+ - E_-$$
$$= (+0.200) - \left(0.799 - 0.059\,16 \log \frac{1}{[Ag^+]}\right)$$
$$= -0.599 - 0.059\,16 \log[Ag^+]$$

Titration reactions:

$$Br^- + Ag^+ \rightarrow AgBr(s) \quad (K_{sp} = 5.0 \times 10^{-13})$$

$$Br^- + Tl^+ \rightarrow TlBr(s) \quad (K_{sp} = 3.6 \times 10^{-6})$$

The two equivalence points are at 25.0 and 50.0 mL. Between 0 and 25 mL, there is unreacted Ag^+ in the solution.

At 1.0 mL: $[Ag^+] = \left(\dfrac{24.0}{25.0}\right)(0.050\,0)\left(\dfrac{100.0}{101.0}\right)$

Initial concentration of Ag^+

$$= 0.047\,5\text{ M} \Rightarrow E = -0.521\text{ V}$$

At 15.0 mL: $[Ag^+] = \left(\dfrac{10.0}{25.0}\right)(0.050\,0)\left(\dfrac{100.0}{115.0}\right)$

$$= 0.017\,4\text{ M} \Rightarrow E = -0.495\text{ V}$$

At 24.0 mL: $[Ag^+] = \left(\dfrac{1.0}{25.0}\right)(0.050\,0)\left(\dfrac{100.0}{124.0}\right)$

$$= 0.001\,61\text{ M} \Rightarrow E = -0.434\text{ V}$$

At 24.9 mL: $[Ag^+] = \left(\dfrac{0.10}{25.0}\right)(0.050\,0)\left(\dfrac{100.0}{124.9}\right)$

$$= 1.60 \times 10^{-4}\text{ M}$$

$$\Rightarrow E = -0.374\text{ V}$$

Between 25 mL and 50 mL, all AgBr has precipitated and TlBr is in the process of precipitating. There is some unreacted Tl^+ left in solution in this region.

At 25.2 mL:

$$[Tl^+] = \left(\dfrac{24.8}{25.0}\right)(0.050\,0)\left(\dfrac{100.0}{125.2}\right)$$

$$= 3.96 \times 10^{-2}\text{ M}$$

$$[Br^-] = K_{sp} \text{ (for TlBr)}/[Tl^+] = 9.0_9 \times 10^{-5}\text{ M}$$

$$[Ag^+] = K_{sp} \text{ (for AgBr)}/[Br^-] = 5.5 \times 10^{-9}\text{ M}$$

$$E = -0.599 - 0.059\,16 \log[Ag^+]$$

$$= -0.110\text{ V}$$

At 35.0 mL: $[Tl^+] = \left(\dfrac{15.0}{25.0}\right)(0.050\,0)\left(\dfrac{100.0}{135.0}\right)$

$$= 0.022\,2\text{ M} \Rightarrow$$

$$[Br^-] = 1.62 \times 10^{-4}\text{ M} \Rightarrow$$

$$[Ag^+] = 3.08 \times 10^{-9}\text{ M} \Rightarrow$$

$$E = -0.095\text{ V}$$

50.0 mL is the second equivalence point, at which $[Tl^+] = [Br^-]$.

At 50.0 mL: $[Tl^+][Br^-] = K_{sp}$ (for TlBr)

$$\Rightarrow [Tl^+] = \sqrt{K_{sp}}$$

$$= 1.90 \times 10^{-3}\text{ M}$$

$$\Rightarrow [Br^-] = 1.90 \times 10^{-3} \text{ M}$$

$$\Rightarrow [Ag^+] = 2.64 \times 10^{-10} \text{ M} \Rightarrow E = -0.032 \text{ V}$$

At 60.0 mL, there is excess Br^- in the solution:

$$[Br^-] = \left(\frac{10.0}{160.0}\right)(0.200) = 0.0125 \text{ M}$$

$$\Rightarrow [Ag^+] = 4.00 \times 10^{-11} \text{ M} \Rightarrow E = +0.016 \text{ V}$$

B. The cell voltage is given by Equation C, in which K_f is the formation constant for $Hg(EDTA)^{2-}$ ($= 5.0 \times 10^{21}$). To find the voltage, we must calculate $[HgY^{2-}]$ and $[Y^{4-}]$ at each point. The concentration of HgY^{2-} is 1.0×10^{-4} M when $V = 0$, and is thereafter affected only by dilution because $K_f(HgY^{2-}) >> K_f(MgY^{2-})$. The concentration of Y^{4-} is found from the Mg-EDTA equilibrium at all but the first point. At $V = 0$ mL, the Hg-EDTA equilibrium determines $[Y^{4-}]$.

At 0 mL:

$$\frac{[HgY^{2-}]}{[Hg^{2+}][EDTA]} = \alpha_{Y^{4-}}K_f \text{ (for } HgY^{4-})$$

$$\frac{1.0 \times 10^{-4} - x}{(x)(x)} = 1.8 \times 10^{21}$$

$$\Rightarrow x = [EDTA] = 2.36 \times 10^{-13} \text{ M}$$

$$[Y^{4-}] = \alpha_{Y^{4-}} [EDTA]$$

$$= 8.49 \times 10^{-14} \text{ M}$$

Using Equation C, we write

$$E = 0.852 - 0.241$$

$$- \frac{0.05916}{2} \log \frac{5.0 \times 10^{21}}{1.0 \times 10^{-4}}$$

$$- \frac{0.05916}{2} \log(8.49 \times 10^{-14})$$

$$= 0.237 \text{ V}$$

At 10.0 mL: Because $V_e = 25.0$ mL, $\frac{10}{25}$ of the Mg^{2+} is in the form MgY^{2-}, and $\frac{15}{25}$ is in the form Mg^{2+}.

$$[Y^{4-}] = \frac{[MgY^{2-}]}{[Mg^{2+}]} \bigg/ K_f \text{ (for } MgY^{2-})$$

$$= \left(\frac{10}{15}\right) \bigg/ 6.2 \times 10^8 = 1.08 \times 10^{-9} \text{ M}$$

$$[HgY^{2-}] = \left(\frac{50.0}{60.0}\right)(1.0 \times 10^{-4}) = 8.33 \times 10^{-5} \text{ M}$$

Dilution
factor

$$E = 0.852 - 0.241 - \frac{0.05916}{2} \log \frac{5.0 \times 10^{21}}{8.33 \times 10^{-5}}$$

$$- \frac{0.05916}{2} \log(1.08 \times 10^{-9})$$

$$= 0.114 \text{ V}$$

At 20.0 mL:

$$[Y^{4-}] = \left(\frac{20}{5}\right) \bigg/ 6.2 \times 10^8 = 6.45 \times 10^{-9} \text{ M}$$

$$[HgY^{2-}] = \left(\frac{50.0}{70.0}\right)(1.0 \times 10^{-4}) = 7.14 \times 10^{-5} \text{ M}$$

$$\Rightarrow E = 0.089 \text{ V}$$

At 24.9 mL: $[Y^{4-}] = \left(\frac{24.9}{0.1}\right) \bigg/ 6.2 \times 10^8$

$$= 4.02 \times 10^{-7} \text{ M}$$

$$[HgY^{2-}] = \left(\frac{50.0}{74.9}\right)(1.0 \times 10^{-4})$$

$$= 6.68 \times 10^{-5} \text{ M}$$

$$\Rightarrow E = 0.035 \text{ V}$$

At 25.0 mL: This is the equivalence point, at which $[Mg^{2+}] = [EDTA]$.

$$\frac{[MgY^{2-}]}{[Mg^{2+}][EDTA]} = \alpha_{Y^{4-}}K_f \text{ (for } MgY^{2-})$$

$$\frac{\left(\frac{50.0}{75.0}\right)(0.0100) - x}{x^2} = 2.22 \times 10^8$$

$$\Rightarrow x = 5.48 \times 10^{-6} \text{ M}$$

$$[Y^{4-}] = \alpha_{Y^{4-}}(5.48 \times 10^{-6})$$

$$= 1.97 \times 10^{-6} \text{ M}$$

$$[HgY^{2-}] = \left(\frac{50.0}{75.0}\right)(1.0 \times 10^{-4})$$

$$= 6.67 \times 10^{-5} \text{ M}$$

$$\Rightarrow E = 0.014 \text{ V}$$

At 26.0 mL: Now there is excess EDTA in the solution:

$$[Y^{4-}] = \alpha_{Y^{4-}}[EDTA] = (0.36)\left[\left(\frac{1.0}{76.0}\right)(0.0200)\right]$$

$$= 9.47 \times 10^{-5} \text{ M}$$

$$[HgY^{2-}] = \left(\frac{50.0}{76.0}\right)(1.0 \times 10^{-4}) = 6.58 \times 10^{-5} \text{ M}$$

$$\Rightarrow E = -0.036 \text{ V}$$

C. At intermediate pH, the voltage will be constant at 100 mV. When $[OH^-] \approx [F^-]/10 = 10^{-6}$ M (pH = 8), the electrode begins to respond to OH^- and the voltage will decrease (i.e., the electrode potential will change in the same direction as if more F^- were being added). Near pH = 3.17 (= pK_a for HF), F^- reacts with H^+ and the concentration of free F^- decreases. At pH = 1.17, $[F^-] \approx 1\%$ of 10^{-5} M = 10^{-7} M, and $E \approx 100 + 2(59) = 218$ mV. A qualitative sketch of this behavior is shown below. The slope at high pH is less than 59 mV/pH unit, because the response of the electrode to OH^- is less than the response to F^-.

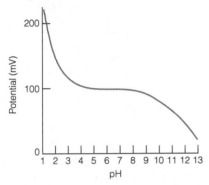

D. (a) For 1.00 mM Na^+ at pH 8.00, we can write

$$E = \text{constant} + 0.059\,16 \log([Na^+] + 36[H^+])$$

$$-0.038 = \text{constant} + 0.059\,16$$
$$\times \log[(1.00 \times 10^{-3}) + (36 \times 10^{-8})]$$

$$\Rightarrow \text{constant} = +0.139 \text{ V}$$

For 5.00 mM Na^+ at pH 8.00, we have

$$E = +0.139 + 0.059\,16$$
$$\times \log[(5.00 \times 10^{-3}) + (36 \times 10^{-8})]$$

$$= 0.003 \text{ V}$$

(b) For 1.00 mM Na^+ at pH 3.87, we have

$$E = +0.139 + 0.059\,16$$
$$\times \log[(1.00 \times 10^{-3}) + (36 \times 10^{-3.87})]$$

$$= 0.007 \text{ V}$$

E. A graph of E(mV) versus $\log [NH_3(M)]$ gives a straight line whose equation is $E = 563.4 + 59.05 \times \log [NH_3]$. For $E = 339.3$ mV, $[NH_3] = 1.60 \times 10^{-4}$ M. The sample analyzed contains (100 mL) $\times$ $(1.60 \times 10^{-4}$ M) = 0.016 0 mmol of nitrogen. But

this sample represents just 2.00% (20.0 mL/1.00 L) of the food sample. Therefore, the food contains $0.016/0.020\,0 = 0.800$ mmol of nitrogen = 11.2 mg of N = 3.59 wt% nitrogen.

F. (a) For the original solution, we can write

$$[Ag^+] = \frac{[Ag(CN)_2^-]}{[CN^-]^2(7.1 \times 10^{19})}$$

$$= \frac{(1.00 \times 10^{-5})}{(8.0 \times 10^{-6})^2(7.1 \times 10^{19})}$$

$$= 2.2 \times 10^{-15} \text{ M}$$

$$E = \text{constant} + 0.059\,16 \log (2.2 \times 10^{-15})$$

$$= 206.3 \text{ mV}$$

$$\Rightarrow \text{constant} = 1.073\,4 \text{ V}$$

After addition of CN^-, we can say that

$$[Ag^+] = \frac{(1.00 \times 10^{-5})}{(12.0 \times 10^{-6})^2(7.1 \times 10^{19})}$$

$$= 9.8 \times 10^{-16} \text{ M}$$

$$E = 1.073\,4 + 0.059\,16 \log (9.8 \times 10^{-16})$$

$$= 185.4 \text{ mV}$$

(b) Let there be x mol of CN^- in 50.0 mL of unknown. After the standard addition, the unknown contains

$$x + (1.00 \times 10^{-3} \text{ L})(2.50 \times 10^{-4} \text{ M})$$

$$= (x + 2.50 \times 10^{-7}) \text{ mol } CN^-$$

$$\Rightarrow [CN^-] = \frac{x + 2.50 \times 10^{-7} \text{ mol}}{0.051\,0 \text{ L}}$$

$$[Ag^+]_2 = \frac{(1.00 \times 10^{-5})}{\left(\dfrac{x + 2.50 \times 10^{-7}}{0.051\,0}\right)^2 (7.1 \times 10^{19})}$$

Before the addition, we have

$$E = \text{constant} + 0.059\,16 \log [Ag^+]$$

$$0.134\,8 = \text{constant} + 0.059\,16 \log [Ag^+]_1$$

After the addition, we can write

$$0.118\,6 = \text{constant} + 0.059\,16 \log [Ag^+]_2$$

Subtracting the second equation from the first gives

$$0.134\,8 - 0.118\,6 = 0.059\,16 \log \frac{[Ag^+]_1}{[Ag^+]_2}$$

$$\Rightarrow \frac{[Ag^+]_1}{[Ag^+]_2} = 1.879$$

But we can also write that

$$1.879 = \frac{[Ag^+]_1}{[Ag^+]_2}$$

$$= \frac{\left(\dfrac{x}{0.0500}\right)^2 (7.1 \times 10^{19})}{\dfrac{(1.00 \times 10^{-5})(50.0/51.0)}{\left(\dfrac{x + 2.50 \times 10^{-7}}{0.0510}\right)^2} (7.1 \times 10^{19})}$$

with the numerator shown as $\dfrac{1.00 \times 10^{-5}}{\left(\dfrac{x}{0.0500}\right)^2 (7.1 \times 10^{19})}$

$$\Rightarrow x = 6.50 \times 10^{-7} \text{ mol}$$

$$[CN^-] = 6.50 \times 10^{-7} \text{ mol}/0.0500 \text{ L}$$

$$= 1.30 \times 10^{-5} \text{ M}$$

Chapter 16

A. Titration reaction: $Sn^{2+} + 2Ce^{4+} \rightarrow Sn^{4+} + 2Ce^{3+}$
$$V_e = 10.0 \text{ mL}$$

Representative calculations:

At 0.100 mL:

$$E_+ = 0.139 - \frac{0.059\,16}{2} \log \frac{[Sn^{2+}]}{[Sn^{4+}]}$$

$$= 0.139 - \frac{0.059\,16}{2} \log \frac{9.90}{0.100} = 0.080 \text{ V}$$

$$E = E_+ - E_- = 0.080 - 0.241 = -0.161 \text{ V}$$

At 10.00 mL:

$$2E_+ = 2(0.139) - 0.059\,16 \log \frac{[Sn^{2+}]}{[Sn^{4+}]}$$

$$E_+ = 1.47 - 0.059\,16 \log \frac{[Ce^{3+}]}{[Ce^{4+}]}$$

$$\rule{7cm}{0.4pt}$$

$$3E_+ = 1.748 - 0.059\,16 \log \frac{[Sn^{2+}][Ce^{3+}]}{[Sn^{4+}][Ce^{4+}]}$$

At the equivalence point, $[Sn^{4+}] = \frac{1}{2}[Ce^{3+}]$ and $[Sn^{2+}] = \frac{1}{2}[Ce^{4+}]$, which makes the log term zero. Therefore $3E_+ = 1.748$ and $E_+ = 0.583$ V.

$$E = E_+ - E_- = 0.583 - 0.241 = 0.342 \text{ V}$$

At 10.10 mL:

$$E_+ = 1.47 - 0.059\,16 \log \frac{[Ce^{3+}]}{[Ce^{4+}]}$$

$$= 1.47 - 0.059\,16 \log \frac{10.0}{0.10} = 1.35_2 \text{ V}$$

$$E = E_+ - E_- = 1.35_2 - 0.241 = 1.11 \text{ V}$$

mL	E (V)	mL	E (V)
0.100	-0.161	10.00	0.342
1.00	-0.130	10.10	1.11
5.00	-0.102	12.00	1.19
9.50	-0.064		

B. Titration reaction: $Br_2(aq) + 2I^- \rightarrow 2Br^- + I_2(aq)$
$$V_e = 5.00 \text{ mL}$$

Representative calculations:

At 0.100 mL: $E_+ = 0.620 - \dfrac{0.059\,16}{2} \log \dfrac{[I^-]^2}{[I_2]}$

$$[I_2] = \underset{\text{Stoichi-}}{\frac{1}{2}} \underset{\substack{\text{Fraction} \\ \text{converted}}}{\left(\frac{0.100}{5.00}\right)} \underset{\substack{\text{Initial } [I^-] \\ \text{concentration}}}{(1.00 \text{ mM})} \underset{\substack{\text{Dilution} \\ \text{factor}}}{\left(\frac{50.0}{50.1}\right)}$$

$$= 9.98 \times 10^{-6} \text{ M}$$

$$[I^-] = \underset{\substack{\text{Fraction} \\ \text{remaining}}}{\left(\frac{4.90}{5.00}\right)} \underset{\substack{\text{Initial} \\ \text{concentration}}}{(1.00 \text{ mM})} \underset{\substack{\text{Dilution} \\ \text{factor}}}{\left(\frac{50.0}{50.1}\right)}$$

$$= 9.78 \times 10^{-4} \text{ M}$$

$$E_+ = 0.620 - \frac{0.059\,16}{2} \log \frac{(9.78 \times 10^{-4})^2}{9.98 \times 10^{-6}} = 0.650 \text{ V}$$

$$E = E_+ - E_- = 0.650 - 0.241 = 0.409 \text{ V}$$

At 5.01 mL: $E_+ = 1.098 - \dfrac{0.059\,16}{2} \log \dfrac{[Br^-]^2}{[Br_2]}$

$$[Br^-] = \underset{\substack{\text{Stoichi-} \\ \text{ometry}}}{2} \underset{\substack{\text{Fraction} \\ \text{converted}}}{\left(\frac{5.00}{5.01}\right)} \underset{\substack{\text{Initial } [Br_2] \\ \text{concentration}}}{(5.00 \text{ mM})} \underset{\substack{\text{Dilution} \\ \text{factor}}}{\left(\frac{5.01}{55.01}\right)}$$

$$= 9.09 \times 10^{-4} \text{ M}$$

$$[Br_2] = \underset{\substack{\text{Fraction} \\ \text{unreacted}}}{\left(\frac{0.01}{5.01}\right)} \underset{\substack{\text{Initial} \\ \text{concentration}}}{(5.00 \text{ mM})} \underset{\substack{\text{Dilution} \\ \text{factor}}}{\left(\frac{5.01}{55.01}\right)}$$

$$= 9.09 \times 10^{-7} \text{ M}$$

$$E_+ = 1.098 - \frac{0.059\,16}{2} \log \frac{(9.09 \times 10^{-4})^2}{9.09 \times 10^{-7}} = 1.099 \text{ V}$$

$$E = E_+ - E_- = 1.099 - 0.241 = 0.858 \text{ V}$$

mL	E (V)	mL	E (V)
0.100	0.409	5.01	0.858
2.50	0.468	6.00	0.918
4.99	0.620		

C. Titration reactions:

$$V^{2+} + Ce^{4+} \rightarrow V^{3+} + Ce^{3+}$$

$$V^{3+} + Ce^{4+} + H_2O \rightarrow VO^{2+} + Ce^{3+} + 2H^+$$

$$VO^{2+} + Ce^{4+} + H_2O \rightarrow VO_2^+ + Ce^{3+} + 2H^+$$

At 5.0 mL:

$$E_+ = -0.255 - 0.05916 \log \frac{[V^{2+}]}{[V^{3+}]}$$

$$\Rightarrow E = -0.496 \text{ V}$$

At 15.0 mL:

$$E_+ = 0.337 - 0.05916 \log \frac{[V^{3+}]}{[VO^{2+}](1.00)^2}$$

$$\Rightarrow E = 0.096 \text{ V}$$

At 25.0 mL:

$$E_+ = 1.001 - 0.05916 \log \frac{[VO^{2+}]}{[VO_2^+](1.00)^2}$$

$$\Rightarrow E = 0.760 \text{ V}$$

At 35.0 mL:

$$E_+ = 1.70 - 0.05916 \log \frac{[Ce^{3+}]}{[Ce^{4+}]}$$

$$= 1.70 - 0.05916 \log \left(\frac{30.0}{5.0}\right)$$

$$\Rightarrow E = 1.41 \text{ V}$$

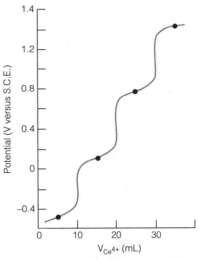

D. (a) The second end point is halfway between 0.096 and 0.760 = 0.43 V versus S.C.E. (= 0.67 V versus S.H.E.). Diphenylamine would change from colorless to violet at this end point.

(b) The third end point is approximately midway between 1.41 and 0.760 = 1.08 V versus S.C.E. (= 1.32 V versus S.H.E.). Tris(2,2'-bipyridine)ruthenium would change from yellow to pale blue at this end point.

E. Titration reaction: $5Fe^{2+} + MnO_4^- + 8H^+ \rightarrow 5Fe^{3+} + Mn^{2+} + 4H_2O$

$$V_e = 15.0 \text{ mL}$$

$$Fe^{3+} + e^- \rightleftharpoons Fe^{2+}$$

$$E = 0.68 - 0.05916 \log \frac{[Fe^{2+}]}{[Fe^{3+}]}$$

	A	B	C	D	E	F	G
1	E°(T) =	E (vs S.H.E.)	Tau	Alpha	Phi	E(vs S.C.E.)	Volume (mL)
2	1.507	0.500	1.28E+85	1.10E+03	0.00091	0.259	0.014
3	E°(A) =	0.600	4.53E+76	2.25E+01	0.04255	0.359	0.638
4	0.68	0.680	7.86E+69	1.00E+00	0.50000	0.439	7.500
5	Nernst =	0.800	5.67E+59	9.37E–03	0.99072	0.559	14.861
6	0.05916	1.369	4.61E+11	2.26E–12	1.00000	1.128	15.000
7	Ve =	1.480	1.91E+02	3.00E–14	1.00522	1.239	15.078
8	15	1.490	2.73E+01	2.03E–14	1.03658	1.249	15.549
9	pH =	1.5	3.90E+00	1.38E–14	1.25608	1.259	18.841
10	0	1.51	5.58E–01	9.34E–15	2.79287	1.269	41.893
11							
12	C2 = 10^((5*($A$2–B2)/$A$6)–8*$A$10)					F2 = B2–0.241	
13	D2 = 10^(($A$4–B2)/$A$6)					G2 = $A$8*E2	
14	E2 = (1+C2)/(C2*(1+D2))						

Spreadsheet for Exercise 16-E

$$\Rightarrow \alpha = \frac{[Fe^{2+}]}{[Fe^{3+}]} = 10^{(0.68 - E)/0.059\,16}$$

$$MnO_4^- + 8H^+ + 5e^- \rightarrow Mn^{2+} + 4H_2O$$

$$E = 1.507 - \frac{0.059\,16}{5} \log \frac{[Mn^{2+}]}{[MnO_4^-][H^+]^8}$$

$$\Rightarrow \tau = \frac{[Mn^{2+}]}{[MnO_4^-]} = 10^{\{[5(1.507 - E)/0.059\,16] - 8\,pH\}}$$

$$\phi = \frac{5Mn_{total}}{Fe_{total}} = \frac{(1 + \tau)}{\tau(1 + \alpha)}$$

F. (a) $I_3^- + 2S_2O_3^{2-} \rightarrow 3I^- + S_4O_6^{2-}$

823 μL of 0.098 8 M $S_2O_3^{2-}$ = 81.3_1 μmol of $S_2O_3^{2-}$ = 40.6_6 μmol of I_3^-. But one mole of unreacted IO_4^- gives one mole of I_3^-. Therefore, 40.6_6 μmol of IO_4^- was left from the periodate oxidation of the amino acids. The original amount of IO_4^- was 2.000 mL of 0.048 7 M IO_4^- = 97.4_0 μmol. The difference (97.40 − 40.66 = 56.74) is the number of micromoles of serine + threonine in 128.6 mg of protein. But with MW = 58 600, 128.6 mg of protein = 2.195 μmol. (Serine + threonine)/protein = 56.74 μmol/2.195 μmol = 25.85 ≈ 26 residues/molecule.

(b) 40.66 μmol of I_3^- will react with 40.66 μmol of H_3AsO_3 (Section 16-9), which is produced by $\frac{1}{4}$(40.66) = 10.16 μmol of As_4O_6 = 4.02 mg.

Chapter 17

A. The anode reaction is $Zn(s) \rightarrow Zn^{2+} + 2e^-$

$$5.0 \text{ g Zn} = 7.65 \times 10^{-2} \text{ mol Zn}$$

$$= 1.52 \times 10^{-1} \text{ mol } e^-$$

$$(0.152 \text{ mol } e^-)(96\,485 \text{ C/mol}) = 1.48 \times 10^4 \text{ C}$$

The current flowing through the circuit is $I = E/R$ = 1.02 V/2.8Ω = 0.364 A = 0.364 C/s.

$$1.48 \times 10^4 \text{ C}/(0.364 \text{ C/s}) = 4.06 \times 10^4 \text{ s}$$
$$= 11.3 \text{ h}$$

B. Anode:

$$H_2O \rightleftharpoons \tfrac{1}{2}O_2(g) + 2H^+ + 2e^- \quad E° = -1.229 \text{ V}$$

Cathode:

$$\underline{2H^+ + 2e^- \rightleftharpoons H_2(g) \qquad E° = 0.00 \text{ V}}$$

$$H_2O \rightleftharpoons \tfrac{1}{2}O_2(g) + H_2(g) \qquad E° = -1.229 \text{ V}$$

$$E_{eq} = -1.229 - \frac{0.059\,16}{2} \log P_{O_2}^{1/2} P_{H_2}$$

$$= -1.229 \text{ V}$$

$$E_{applied} = E_{eq} - IR - \text{overpotential}$$

$$= -1.229 - (0.100 \text{ A})(2.00 \text{ Ω})$$

$$\underbrace{-0.85 \text{ V}}_{\substack{\text{Anode} \\ \text{overpotential}}} \; \underbrace{- 0.068 \text{ V}}_{\substack{\text{Cathode} \\ \text{overpotential}}} = -2.35 \text{ V}$$

From Table 17-1

For Au electrodes, $E_{applied} = -2.78$ V.

C. (a) To electrolyze 0.010 M SbO^+ requires a potential of

E(cathode)

$$= 0.208 - \frac{0.059\,16}{3} \log \frac{1}{[SbO^+][H^+]^2}$$

$$= 0.208 - \frac{0.059\,16}{3} \log \frac{1}{(0.010)(1.0)^2}$$

$$= 0.169 \text{ V}$$

The concentration of $[Cu^{2+}]$ that would be in equilibrium with $Cu(s)$ at this potential is found as follows:

$$Cu^{2+} + 2e^- \rightleftharpoons Cu(s) \qquad E° = 0.339$$

$$E\text{(cathode)} = 0.339 - \frac{0.059\,16}{2} \log \frac{1}{[Cu^{2+}]}$$

$$0.169 = 0.339 - \frac{0.059\,16}{2} \log \frac{1}{[Cu^{2+}]}$$

$$\Rightarrow [Cu^{2+}] = 1.8 \times 10^{-6} \text{ M}$$

Percentage of Cu^{2+} not reduced

$$= \frac{1.8 \times 10^{-6}}{0.10} \times 100 = 1.8 \times 10^{-3}\%$$

Percentage of Cu^{2+} reduced = 99.998%

(b) In (a), E(cathode versus S.H.E.) = 0.169 V.

E(cathode versus Ag|AgCl)

$$= E\text{(versus S.H.E.)} - E(Ag|AgCl)$$

$$= 0.169 - 0.197 = -0.028 \text{ V}$$

D. (a) $Co^{2+} + 2e^- \rightleftharpoons Co(s)$ $E° = -0.282$ V

E(cathode versus S.H.E.)

$$= -0.282 - \frac{0.059\,16}{2} \log \frac{1}{[Co^{2+}]}$$

Putting in $[Co^{2+}] = 1.0 \times 10^{-6}$ M gives $E = -0.459$ V and

E(cathode versus S.C.E.)

$$= -0.459 - \underbrace{0.241}_{E(\text{S.C.E.})} = -0.700 \text{ V}$$

(b) $Co(C_2O_4)_2^{2-} + 2e^- \rightleftharpoons$

$\qquad Co(s) + 2C_2O_4^{2-}$ $E° = -0.474$ V

E(cathode versus S.C.E.)

$$= -0.474 - \frac{0.059\,16}{2} \log \frac{[C_2O_4^{2-}]^2}{[Co(C_2O_4)_2^{2-}]} - 0.241$$

Putting in $[C_2O_4^{2-}] = 0.10$ M and $[Co(C_2O_4)_2^{2-}] = 1.0 \times 10^{-6}$ M gives $E = -0.833$ V.

(c) We can think of the reduction as $Co^{2+} + 2e^- \rightleftharpoons$ $Co(s)$, for which $E° = -0.282$ V. But the concentration of Co^{2+} is that tiny amount in equilibrium with 0.10 M EDTA plus 1.0×10^{-6} M $Co(EDTA)^{2-}$. In Table 13-2, we find that the formation constant for $Co(EDTA)^{2-}$ is $10^{16.31} = 2.0 \times 10^{16}$.

$$K_f = \frac{[Co(EDTA)^{2-}]}{[Co^{2+}][EDTA^{4-}]} = \frac{[Co(EDTA)^{2-}]}{[Co^{2+}]\alpha_{Y4-}F}$$

where F is the formal concentration of EDTA (= 0.10 M) and $\alpha_{Y4-} = 5.0 \times 10^{-4}$ at pH 7.00 (Table 13-1). Putting in $[Co(EDTA)^{2-}] = 1.0 \times 10^{-6}$ M and solving for $[Co^{2+}]$ gives $[Co^{2+}] = 1.0 \times 10^{-18}$ M.

$$E = -0.282 - \frac{0.059\,16}{2} \log \frac{1}{1.0 \times 10^{-18}}$$

$$- 0.241 = -1.055 \text{ V}$$

E. (a) 75.00 mL of 0.023 80 M KSCN = 1.785 mmol of SCN^-, which gives 1.785 mmol of AgSCN, containing 0.103 7 g of SCN. Final mass = 12.463 8 + 0.103 7 = 12.567 5 g.

(b) The concentration of Ag^+ in equilibrium with 0.10 M Br^- is $[Ag^+] = K_{sp}/[Br^-] = (5.0 \times 10^{-13})/(0.10) = 5.0 \times 10^{-12}$ M. Writing both cell reactions as reductions

Anode: $Ag^+ + e^- \rightleftharpoons Ag(s)$ $E° = 0.799$ V

Cathode: $Hg_2Cl_2(s) + 2e^- \rightleftharpoons 2Hg(l) + 2Cl^-$

$\qquad\qquad\qquad\qquad\qquad E(\text{S.C.E.}) = 0.241$ V

we can say

$$E_{eq} = E(\text{cathode}) - E(\text{anode})$$

$$= 0.241 - \left(0.799 - 0.059\,16 \log \frac{1}{[Ag^+]}\right)$$

Putting in $[Ag^+] = 5.0 \times 10^{-12}$ M gives $E_{eq} = 0.111$ V.

(c) To remove 99.99% of 0.10 M KI will leave $[I^-] = 1.0 \times 10^{-5}$ M. The concentration of Ag^+ in equilibrium with this much I^- is $[Ag^+] = K_{sp}/[I^-] = 8.3 \times 10^{-17}/1.0 \times 10^{-5} = 8.3 \times 10^{-12}$ M. In **(b)**, we found that it takes only 5.0×10^{-12} M Ag^+ to precipitate 0.10 M Br^-. Therefore, the separation is not possible.

F. 1.00 ppt corresponds to $30.0/1\,000 = 0.030\,0$ mL of O_2/min $= 5.00 \times 10^{-4}$ mL of O_2/s. The moles of oxygen in this volume are

$$n = \frac{PV}{RT} = \frac{(1.00 \text{ atm})(5.00 \times 10^{-7} \text{ L})}{(0.082\,06 \text{ L atm K}^{-1} \text{ mol}^{-1})(293 \text{ K})}$$

$$= 2.080 \times 10^{-8} \text{ mol}$$

For each mole of O_2, four moles of e^- flow through the circuit, so $e^- = 8.320 \times 10^{-8}$ mol/s $= 8.03 \times 10^{-3}$ C/s $= 8.03$ mA. An oxygen content of 1.00 ppm would give a current of 8.03 μA instead.

G. The Zn^{2+} reacts first with PDTA freed by the reduction of $Hg(PDTA)^{2-}$ in the region BC. Then additional Zn^{2+} goes on to liberate Hg^{2+} from $Hg(PDTA)^{2-}$. This additional Hg^{2+} is reduced in the region DE. The total Hg^{2+}, equivalent to the added Zn^{2+}, equals one-half the coulombs measured in regions BC and DE (because $2e^-$ reacts with 1 Hg^{2+}). Coulombs $= 3.89 + 14.47 = 18.36$. Moles of Hg^{2+} reduced $= 0.5(18.36 \text{C})/(96\,485$ C/mol$) = 9.514 \times 10^{-5}$ mol. $[Zn^{2+}] = 9.514 \times 10^{-5}$ mol$/2.00 \times 10^{-3}$ L $= 0.047\,57$ M.

Chapter 18

A. The standard curve is moderately linear with slope = 0.004 19 μA/ppb and intercept of 0.019 8. The concentration of Ni^{2+} when 54.0 μL of 10.0 ppm solution is added to 5.00 mL is

$$\left(\frac{0.054\,0 \text{ mL}}{5.054\,0 \text{ mL}}\right) (10.0 \text{ ppm}) = 0.107 \text{ ppm}$$

$$= 107 \text{ ppb}$$

The expected current is

$$I = m[Ni^{2+}] + b$$

$$= (0.004\,19)(107) + 0.019\,8 = 0.468 \text{ μA}$$

A careful examination of the standard curve shows that a better fit might be obtained if the slope and intercept of just the first seven points are calculated. The curve appears to be starting to level off at the higher concentrations in this experiment.

B. (a) $[Cd^{2+}]$ standard added to unknown $= \left(\dfrac{10.00}{50.00}\right)$

$\times (3.23 \times 10^{-4}$ M$) = 6.46 \times 10^{-5}$ M

$\dfrac{(Cd^{2+}\text{ signal}/Pb^{2+}\text{ signal}) \text{ in known}}{(Cd^{2+}\text{ signal}/Pb^{2+}\text{ signal}) \text{ in unknown}}$

$\qquad = \dfrac{([Cd^{2+}]/[Pb^{2+}]) \text{ in known}}{([Cd^{2+}]/[Pb^{2+}]) \text{ in unknown}}$

$\dfrac{1.64/1.58}{2.00/3.00} = \dfrac{3.23/4.18}{6.46 \times 10^{-5}/x}$

$\Rightarrow x = 1.30 \times 10^{-4}$ M

The concentration of Pb^{2+} in the diluted unknown is 1.30×10^{-4} M. In the undiluted sample, the concentration is

$\left(\dfrac{50.00}{25.00}\right)(1.30 \times 10^{-4}) = 2.60 \times 10^{-4}$ M

(b) $\dfrac{(1.64 \pm 0.03)/(1.58 \pm 0.03)}{(2.00 \pm 0.03)/(3.00 \pm 0.03)}$

$= \dfrac{(3.23 \pm 0.01)/(4.18 \pm 0.01)}{\left[\dfrac{(10.00 \pm 0.05)}{(50.00 \pm 0.05)}(3.23 \pm 0.01) \times 10^{-4}\right]/x}$

$\Rightarrow x = 1.30_2(\pm 3._{28}\%) \times 10^{-4}$ M

$[Pb^{2+}]$

$= \left(\dfrac{50.00 \pm 0.05}{25.00 \pm 0.05}\right)(1.30_2 \pm 3._{28}\%) \times 10^{-4}$ M

$= 2.60(\pm 3._{29}\%) \times 10^{-4}$ M

$= 2.60(\pm 0.09) \times 10^{-4}$ M

C. Sample height (mm) $= 26.8 - 2.4 = 24.4$
Sample $+$ 1 ppm Cu $= 42.2 - 5.6 = 36.6$
Sample $+$ 2 ppm Cu $= 57.8 - 8.7 = 49.1$.

The average response to added Cu is

$\dfrac{(36.6 - 24.4) + (49.1 - 24.4)}{3} = 12.3 \dfrac{\text{mm}}{\text{ppm Cu}}$

The initial sample must have contained

$\dfrac{24.4}{12.3} = 1.98$ ppm Cu

D. A graph of $E_{1/2}$ versus $\log [OH^-]$ is shown below. All but the lowest two points appear to lie on a line whose equation is

$$E_{1/2} = -0.0806 \log [OH^-] - 0.763$$

According to Equation 18-10, the slope of the graph is $-0.05916\,p/n$. Assuming that $n = 2$, we calculate p as follows:

$$p = \dfrac{(n)(\text{slope})}{-0.05916} = 2.72 \approx 3$$

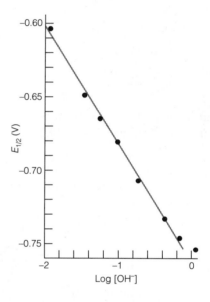

The intercept of Equation 18-10 is given by

$$\text{intercept} = E_{1/2}(\text{for free Pb}^{2+}) - \dfrac{0.05916}{n} \log \beta_3$$

$$-0.763 = -0.41 - \dfrac{0.05916}{2} \log \beta_3$$

$$\Rightarrow \beta_3 = 9 \times 10^{11}$$

E. We see two consecutive reductions. From the value of $E_{pa} - E_{pc}$, we find that one electron is involved in each reduction (using Equation 18-11). A possible sequence of reaction is

$$Co(III)(B_9C_2H_{11})_2^- \rightarrow Co(II)(B_9C_2H_{11})_2^{2-}$$
$$\rightarrow Co(I)(B_9C_2H_{11})_2^{3-}$$

The equality of the anodic and cathodic peak heights suggests that the reactions are reversible. The expected DC (a) and DPP (b) polarograms are sketched on the next page.

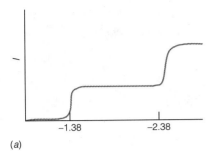

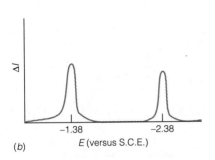

(a)

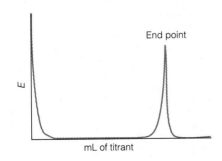

(b)

E (versus S.C.E.)

F. In curve a, the current decreases prior to the end point because Pb^{2+} is reduced to $Pb(Hg)$ at -0.8 V. Beyond the end point, excess $Cr_2O_7^{2-}$ can be reduced and the current increases again. Curve b is level near zero current prior to the equivalence point because Pb^{2+} is not reduced at 0 V (versus S.C.E.). Beyond the end point, excess $Cr_2O_7^{2-}$ is reduced, even at 0 V (versus S.C.E.).

G. Initially there is no redox couple to carry current, so the potential will be high. As Ce^{4+} is added, Fe^{2+} is converted to Fe^{3+}, the mixture of which can support current flow by the reactions

$$\text{anode: } Fe^{2+} \rightleftharpoons Fe^{3+} + e^-$$

$$\text{cathode: } Fe^{3+} + e^- \rightleftharpoons Fe^{2+}$$

The potential will therefore decrease. At the equivalence point, all of the Fe^{2+} and all of the Ce^{4+} are consumed, so the potential is very high. Beyond the equivalence point, the redox couple $Ce^{4+}|Ce^{3+}$ can support a current and the potential will be low again. The expected curve is shown below.

End point

E

mL of titrant

H. Electricity required for H_2O in 0.848 g of polymer
$= (63.16 - 4.23) = 58.93$ C.

$$\frac{58.93 \text{ C}}{964\,85 \text{ C/mol}} = 0.610\,8 \text{ mmol of } e^-$$

which corresponds to $\frac{1}{2}(0.610\,8) = 0.305\,4$ mmol of $I_2 = 0.305\,4$ mmol of $H_2O = 5.502$ mg H_2O.

$$\text{water content} = 100 \times \frac{5.502 \text{ mg } H_2O}{848 \text{ mg polymer}}$$
$$= 0.649 \text{ wt}\%$$

Chapter 19

A. (a) $c = A/\epsilon b = 0.463/(4170)(1.00) = 1.110 \times 10^{-4}$ M $= 8.99$ g/L $= 8.99$ mg of transferrin/mL. The Fe concentration is 2.220×10^{-4} M $= 0.0124$ g/L $= 12.4$ μg/mL.

(b)
$$A_\lambda = \sum \epsilon bc$$

$$0.424 = 4\,170[T] + 2\,290[D]$$

$$0.401 = 3\,540[T] + 2\,730[D]$$

where $[T]$ and $[D]$ are the concentrations of transferrin and desferrioxamine, respectively. Solving for $[T]$ and $[D]$ gives $[T] = 7.30 \times 10^{-5}$ M and $[D] = 5.22 \times 10^{-5}$ M. The fraction of iron in transferrin (which binds two ferric ions) is $2[T]/(2[T] + [D]) = 73.7\%$.

B. The appropriate Scatchard plot is a graph of $\Delta A/[X]$ versus ΔA (Equation 19-16).

Experiment	ΔA	$\Delta A/[X]$
1	0.090	20 360
2	0.181	19 890
3	0.271	16 940
4	0.361	14 620
5	0.450	12 610
6	0.539	9 764
7	0.627	7 646
8	0.713	5 021
9	0.793	2 948
10	0.853	1 453
11	0.904	93.6

Points 2–10 lie on a reasonably straight line whose slope is -2.72×10^4 M^{-1}, giving $K = 2.72 \times 10^4$ M^{-1}.

C.

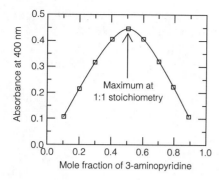

Maximum at
1:1 stoichiometry

D. If self-absorption can be neglected, Equation 19-27 reduces to

$$I = k'P_0(1 - 10^{-\epsilon_{ex}b_2c}) \quad (a)$$

At low concentrations, this expression reduces to

$$I = k'P_0(\epsilon_{ex}b_2c \ln 10) \quad (b)$$

(using the first term of a power series expansion). As the concentration is increased, Expression b becomes greater than a. When Expression a is 5% below b, we can say

$$k'P_0(1 - 10^{-\epsilon_{ex}b_2c}) = 0.95k'P_0\epsilon_{ex}b_2c \ln 10$$

$$1 - 10^{-A} = 0.95A \ln 10$$

By trial and error, this equation can be solved to find that when $A = 0.045$, $1 - 10^{-A} = 0.95A \ln 10$. (Alternatively, you could make a graph of $1 - 10^{-A}$ versus A and $0.95A \ln 10$ versus A. The solution is the intersection of the two curves.)

Chapter 20

A. $T = 0.820 = \dfrac{1 - R}{1 + R} \Rightarrow R = 0.0989$

Now put this value of R into Equation 20-7 and use $n_1 = 1$ for air:

$$0.0989 = \left(\frac{1 - n_2}{1 + n_2}\right)^2$$

$$\pm 0.3145 = \frac{1 - n_2}{1 + n_2} \Rightarrow n_2 = 1.92 \text{ or } 0.522$$

B. (a) $R = \left(\dfrac{1 - 2.17}{1 + 2.17}\right)^2 = 0.1362$

$$T = \frac{(1 - R)^2 e^{-\alpha b}}{1 - R^2 e^{-2\alpha b}}$$

$$= \frac{(1 - 0.1362)^2 e^{-(0.47 \text{ cm}^{-1})(1.20 \text{ cm})}}{1 - (0.1362)^2 e^{-2(0.47 \text{ cm}^{-1})(1.20 \text{ cm})}}$$

$$= 0.427$$

(b) $T = \dfrac{(1 - 0.1362)^2 e^{-(0.47 \text{ cm}^{-1})(0.120 \text{ cm})}}{1 - (0.1362)^2 e^{-2(0.47 \text{ cm}^{-1})(0.120 \text{ cm})}}$

$$= 0.717$$

(c) $T = \dfrac{(1 - 0.1362)^2 e^{-[(0.47 + 0.20) \text{ cm}^{-1}](1.20 \text{ cm})}}{1 - (0.1362)^2 e^{-2[(0.47 + 0.20) \text{ cm}^{-1}](1.20 \text{ cm})}}$

$$= 0.335$$

C. (a) $\dfrac{dM_\lambda}{d\lambda} = 0$

$$= 2\pi hc^2\left[-5\lambda^{-6}e^{-hc/\lambda kT}\right.$$

$$+ \left.\lambda^{-5}e^{-hc/\lambda kT}\left(\frac{hc}{kT}\right)\lambda^{-2}\right]$$

$$\Rightarrow \lambda_{max} \cdot T = \frac{hc}{5k} = 2.878 \times 10^{-3} \text{ m·K}$$

(b)

T (K)	λ_{max} (μm)
100	28.8
500	5.76
5000	0.576

D. (a) $\lambda/\Delta\lambda = 10^4$. If $\lambda = 10.00$ μm, $\Delta\lambda = 10^{-3}$ μm. Therefore, 10.01 μm could be resolved from 10.00 μm.

(b) $\lambda = \dfrac{1}{\tilde{\nu}} = \dfrac{1}{(1000 \text{ cm}^{-1})(10^{-4} \text{ cm}/\mu m)} = 10 \ \mu m$

$$\Delta\lambda = \frac{\lambda}{10^4} = 10^{-3} \ \mu m$$

$\Rightarrow 10.001$ μm could be resolved from 10.000 μm.

$$\left.\begin{array}{l}10.000 \ \mu m = 1000.0 \text{ cm}^{-1}\\ 10.001 \ \mu m = 999.9 \text{ cm}^{-1}\end{array}\right\} \text{ difference} = 0.1 \text{ cm}^{-1}$$

(c) 5.0 cm × 2500 lines/cm = 12500 lines

resolution = $1 \cdot 12500 = 12500$ for $n = 1$

$= 10 \cdot 12500 = 125000$ for $n = 10$

(d) $\dfrac{\Delta\phi}{\Delta\lambda} = \dfrac{n}{d \cos \phi} = \dfrac{2}{\left(\dfrac{1 \text{ mm}}{250}\right) \cos 30°}$

$$= 577 \text{ radians/mm} = 0.577 \text{ radian}/\mu m$$

$$\text{degrees} = \frac{\text{radians}}{\pi} \times 180$$

$$\Rightarrow \frac{\Delta\phi}{\Delta\lambda} = 33.1 \text{ degrees}/\mu m$$

The two wavelengths are $1\,000$ cm^{-1} = 10.00 µm and $1\,001$ cm^{-1} = 9.99 µm $\Rightarrow \Delta\lambda$ = 0.01 µm.

$$\Delta\phi = 0.577\,\frac{\text{radian}}{\text{µm}} \times 0.01\,\text{µm}$$
$$= 6 \times 10^{-3}\,\text{radian} = 0.3°$$

E. true transmittance = $10^{-1.000}$ = 0.100

apparent transmittance = 0.100 + 0.010 = 0.110

apparent absorbance = $-\log 0.110$ = 0.959

relative error in concentration

$$= \text{relative error in absorbance} = \frac{0.041}{1.000} = 4.1\%$$

F. (a) $\Delta\tilde{\nu} = 1/2\delta = 1/(2 \cdot 1.266\,0 \times 10^{-4}\,\text{cm})$
$$= 3\,949\,\text{cm}^{-1}$$

(b) Each interval is $1.266\,0 \times 10^{-4}$ cm. $4\,096$ intervals = $(4\,096)(1.266\,0 \times 10^{-4}\,\text{cm})$ = 0.518\,6 cm. This is a range of $\pm\Delta$, so Δ = 0.259\,3 cm.

(c) resolution $\approx 1/\Delta = 1/(0.259\,3\,\text{cm})$
$$= 3.86\,\text{cm}^{-1}$$

(d) mirror velocity = 0.693 cm/s
$$\text{interval} = \frac{1.266\,0 \times 10^{-4}\,\text{cm}}{0.693\,\text{cm/s}} = 183\,\text{µs}$$

(e) $(4\,096\,\text{points})(183\,\text{µs/point})$ = 0.748 s

(f) The beamsplitter is germanium on KBr. The KBr absorbs light below 400 cm^{-1}, which the background transform shows clearly.

G. The graphs show that signal-to-noise ratio is proportional to $\sqrt{n}$.

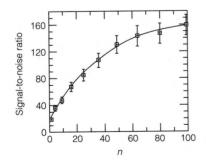

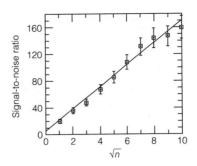

Chapter 21

A.

Emission intensity	Concentration of added standard (µg/mL)
309	0
452	0.081
600	0.162
765	0.243
906	0.324

A graph of intensity versus concentration of added standard intercepts the x axis at -0.164 µg/mL. Because the sample was diluted by a factor of 10, the original sample concentration is 1.64 µg/mL.

B. The concentration of Mn in the unknown mixture is $(13.5)(1.00/6.00)$ = 2.25 µg/mL.

$$\frac{\text{concentration ratio in unknown}}{\text{concentration ratio in standard}}$$
$$= \frac{\text{signal ratio in unknown}}{\text{signal ratio in standard}}$$
$$\frac{[\text{Fe}]/2.25}{2.50/2.00} = \frac{0.185/0.128}{1.05}$$
$$[\text{Fe}] = 3.87\,\text{µg/mL}$$

The original concentration of Fe must have been

$$\frac{6.00}{5.00}(3.87) = 4.65\,\text{µg/mL} = 8.33 \times 10^{-5}\,\text{M}$$

C. The ratio of signal to peak-to-peak noise level is measured to be 17 in the figure. The concentration of Fe needed to give a signal-to-noise ratio of 2 is $(\frac{2}{17})(0.048\,5\,\text{µg/mL})$ = 0.005\,7 µg/mL (= 5.7 ppb).

D. In the excitation spectrum, we are looking at emission over a band of wavelengths 1.6 nm wide, while exciting the sample with different narrow bands (0.03 nm) of laser light. The sample can absorb the laser light only when the laser frequency coincides with the atomic frequency. Therefore, emission is observed only when the narrow laser line is in resonance with the atomic levels. In the emission spectrum, the sample is excited by a fixed laser frequency and then emits radiation. The monochromator bandwidth is not narrow enough to discriminate between emission at different wavelengths, so a broad envelope is observed.

Chapter 22

A. (a) $k_1' = \dfrac{t_{r1} - t_m}{t_m}$

$$\Rightarrow t_m = \frac{t_{r1}}{k_1' + 1} = \frac{10.0}{5.00} = 2.00 \text{ min}$$

$$t_{r2} = t_m(k_2' + 1) = 2.00\,(5.00 + 1) = 12.0 \text{ min}$$

$$\sigma_1 = \frac{t_{r1}}{\sqrt{N}} = \frac{10.0}{\sqrt{1\,000}} = 0.316 \text{ min}$$

$$\Rightarrow w_{1/2} \text{ (peak 1)} = 2.35\sigma_1 = 0.74 \text{ min}$$

$$w_1 = 4\sigma_1 = 1.26 \text{ min}$$

$$\sigma_2 = \frac{t_{r2}}{\sqrt{N}} = \frac{12.0}{\sqrt{1\,000}} = 0.379 \text{ min}$$

$$\Rightarrow w_{1/2} \text{ (peak 2)} = 2.35\sigma_2 = 0.89 \text{ min}$$

$$w_2 = 4\sigma_2 = 1.52 \text{ min}$$

(b)

(c) $\text{resolution} = \dfrac{\Delta t_r}{w_{av}} = \dfrac{2}{[(1.26 + 1.52)/2]} = 1.44$

B. (a) $\text{fraction remaining} = q = \dfrac{V_1}{V_1 + KV_2}$

$$0.01 = \frac{10}{10 + 4.0V_2} \Rightarrow V_2 = 248 \text{ mL}$$

(b) $q^3 = 0.01 = \left(\dfrac{10}{10 + 4.0V_2}\right)^3 \Rightarrow V_2 = 9.1 \text{ mL},$

and total volume $= 27.3 \text{ mL}$.

C. (a) Relative distances measured from Figure 22-7:

$$t_m = 10._4$$

$$t_r' = 39._8 \text{ for octane}$$

$$t_r' = 76._0 \text{ for nonane}$$

$k' = t_r'/t_m = 3.8_3$ for octane and 7.3_1 for nonane.

(b) Let t_s = time in stationary phase, t_m = time in mobile phase, and t be total time on column. We know that $k' = t_s/t_m$. But

$$t = t_s + t_m = t_s + \frac{t_s}{k'}$$

$$= t_s\left(1 + \frac{1}{k'}\right) = t_s\left(\frac{k' + 1}{k'}\right)$$

Therefore,

$$t_s/t = \frac{k'}{k' + 1} = 3.8_3/4.8_3 = 0.79_3.$$

(c) $\alpha = t_r'$ (nonane)$/t_r'$(octane) $= 76._0/39._8 = 1.9_1.$

(d) $K = k'V_m/V_s = 3.8_3\,(V_m/\tfrac{1}{2}V_m) = 7.6_6.$

D. (a) For ethyl acetate, we measure $t_r = 11.3$ and $w = 1.5$ mm. Therefore, $N = 16\,t_r^2/w^2 = 910$ plates. For toluene, the figures are $t_r = 36.2$, $w = 4.2$, and $N = 1\,200$ plates.

(b) We expect $w_{1/2} = (2.35/4)w$. The measured value of $w_{1/2}$ is in good agreement with the calculated value.

E. The column is overloaded, causing a gradual rise and an abrupt fall of the peak. As the sample size is decreased, the overloading decreases and the peak becomes more symmetric.

F. (a) $k_2' = \alpha k_1' = (1.068)(5.16) = 5.51_1$

$$k_{av} = \tfrac{1}{2}(k_1' + k_2') = 5.33_5$$

$$R = \frac{\sqrt{N}}{4}\left(\frac{\alpha - 1}{\alpha}\right)\left(\frac{k_2'}{1 + k_{av}}\right)$$

$$1.00 = \frac{\sqrt{N}}{4}\left(\frac{0.068}{1.068}\right)\left(\frac{5.51_1}{1 + 5.33_5}\right) \Rightarrow N = 5\,215$$

$$L = (5\,215 \text{ plates})(0.520 \text{ mm/plate}) = 2.71 \text{ m}$$

(b) $k_1' = \dfrac{t_{r1} - t_m}{t_m}$

$$5.16 = \frac{t_{r1} - 2.00}{2.00} \Rightarrow t_{r1} = 12.32 \text{ min}$$

$$k_2' = 5.51_1 = \frac{t_{r2} - 2.00}{2.00} \Rightarrow t_{r2} = 13.02 \text{ min}$$

$$\Delta t = 13.02 - 12.32 = 0.70 \text{ min}$$

$$w_1 = \frac{4t_{r1}}{\sqrt{N}} = \frac{4(12.32)}{\sqrt{5215}} = 0.68 \text{ min}$$

$$w_2 = \frac{4t_{r2}}{\sqrt{N}} = \frac{4(13.02)}{\sqrt{5215}} = 0.72 \text{ min}$$

(c)　　$k' = KV_s/V_m$

　　　$5.16 = K(0.30) \Rightarrow K = 17._2$

Chapter 23

A.　$\dfrac{\text{concentration ratio (X/S) in unknown}}{\text{concentration ratio in standard mixture}}$

　　　$= \dfrac{\text{area ratio (X/S) in unknown}}{\text{area ratio in standard mixture}}$

　　　X = hexanol and S = internal standard = pentanol

$$\frac{x \text{ mmol}/0.57 \text{ mmol}}{1.53 \text{ mmol}/1.06 \text{ mmol}} = \frac{816/843}{1570/922} \Rightarrow$$

$$x = 0.47 \text{ mmol}$$

B.　(a) A plot of $\log t_r'$ versus (number of carbon atoms) should be a fairly straight line for a homologous series of compounds.

Peak	t_r'	$\log t_r'$
$n = 7$	2.9	0.46
$n = 8$	5.4	0.73
$n = 14$	85.8	1.93
Unknown	41.4	1.62

From a graph of $\log t_r'$ versus n, it appears that $n = 12$ for the unknown.

(b) $k' = t_r'/t_m = 41.4/1.1 = 38$

C.　In a, the low ionic strength reservoir drains more rapidly than the high ionic strength reservoir, because the levels must remain equal. In b, the high ionic strength reservoir drains faster than the low ionic strength reservoir.

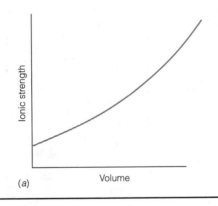

(a)

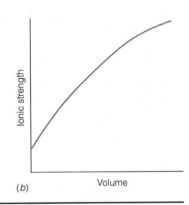

(b)

Chapter 24

A.　13.03 mL of 0.022 74 M NaOH = 0.296 3 mmol of OH⁻, which must equal the total cation charge (= $2[VO^{2+}] + 2[H_2SO_4]$) in the 5.00-mL aliquot. 50.0 mL therefore contains 2.963 mmol of cation charge. The VO^{2+} content is (50.0 mL) (0.024 3 M) = 1.215 mmol = 2.43 mmol of charge. The H_2SO_4 must therefore be $(2.963 - 2.43)/2 = 0.267$ mmol.

1.215 mmol $VOSO_4$ = 0.198 g $VOSO_4$ in
　　　　　　0.244 7-g sample = 80.9%

0.267 mmol H_2SO_4 = 0.026 2 g H_2SO_4 in
　　　　　　0.244 7-g sample = 10.7%

H_2O (by difference) = 8.4%

B.　(a) Because the fractionation range of Sephadex G-50 is 1 500–30 000, hemoglobin should not be retained and ought to be eluted in a volume of 36.4 mL.

(b) ²²NaCl ought to require one column volume to pass through. Neglecting the volume occupied by gel, we expect NaCl to require $\pi r^2 \times \text{length} = \pi(1.0 \text{ cm})^2(40 \text{ cm}) = 126 \text{ mL}$ of solvent.

(c) $K_{av} = \dfrac{V_r - V_0}{V_t - V_0} \Rightarrow V_r = K_{av}(V_t - V_0) + V_0$

　　　$= 0.65(126 - 36.4) + 36.4 = 95 \text{ mL}$

C. (a)

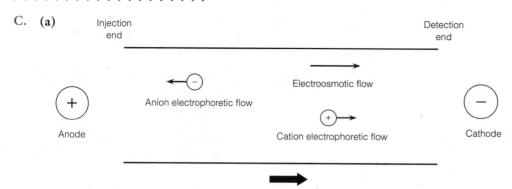

Net flow of cations and anions is to the right,
because electroosmotic flow is stronger than
electrophoretic flow at high pH

(b) I^- has a greater mobility than Cl^-. Therefore, I^- swims upstream faster than Cl^- (because electrophoresis opposes electroosmosis) and is eluted later than Cl^-. The mobility of Br^- is greater than that of I^- in Table 15-1. Therefore, Br^- will have a longer migration time than I^-.

(c) Bare I^- is a larger ion than bare Cl^-, so the charge density in I^- is lower than the charge density in Cl^-. Therefore, I^- should have a smaller hydrated radius than Cl^-. This means that I^- has less friction than Cl^- and a greater mobility than Cl^-.

(d) $^{37}Cl^-$ has a smaller electrophoretic mobility than the lighter $^{35}Cl^-$, because the same electrostatic force will accelerate the lighter ion faster. $^{35}Cl^-$ therefore swims upstream against electroosmosis faster than $^{37}Cl^-$ and is eluted later.

Chapter 25

A. One mole of ethoxyl groups produces one mole of AgI. 29.03 mg of AgI = 0.123 65 mmol. The amount of compound analyzed is 25.42 mg/(417 g/mol) = 0.060 96 mmol. There are

$$\frac{0.123\,65 \text{ mmol ethoxyl groups}}{0.060\,96 \text{ mmol compound}}$$

$$= 2.03\ (=2) \text{ ethoxyl groups/molecule.}$$

B. There is one mole of SO_4^{2-} in each mole of each reactant and of the product. Let x = g of K_2SO_4 and y = g of $(NH_4)_2SO_4$.

$$x + y = 0.649 \text{ g} \qquad (1)$$

$$\underbrace{\frac{x}{174.27}}_{\substack{\text{Moles of} \\ K_2SO_4}} + \underbrace{\frac{y}{132.14}}_{\substack{\text{Moles of} \\ (NH_4)_2SO_4}} = \underbrace{\frac{0.977}{233.40}}_{\substack{\text{Moles of} \\ BaSO_4}} \qquad (2)$$

Making the substitution $y = 0.649 - x$ in Equation 2 gives $x = 0.397$ g $= 61.1\%$ of the sample.

C. Formula and atomic weights: Ba(137.327), Cl(35.453), K(39.098), H_2O(18.015), KCl(74.551), $BaCl_2 \cdot 2H_2O$(244.26). H_2O lost $= 1.783\,9 - 1.562\,3 = 0.221\,6$ g $= 1.230\,1 \times 10^{-2}$ mol of H_2O. For every two moles of H_2O lost, one mole of $BaCl_2 \cdot 2H_2O$ must have been present. $1.230\,1 \times 10^{-2}$ mol of H_2O implies that $6.150\,4 \times 10^{-3}$ mol of $BaCl_2 \cdot 2H_2O$ must have been present. This much $BaCl_2 \cdot 2H_2O$ equals $1.502\,3$ g. The Ba and Cl contents of the $BaCl_2 \cdot 2H_2O$ are

$$Ba = \left(\frac{137.33}{244.26}\right)(1.502\,3 \text{ g}) = 0.844\,64 \text{ g}$$

$$Cl = \left(\frac{2(35.453)}{244.26}\right)(1.502\,3 \text{ g}) = 0.436\,10 \text{ g}$$

Because the total sample weighs $1.783\,9$ g and contains $1.502\,3$ g of $BaCl_2 \cdot 2H_2O$, the sample must contain $1.783\,9 - 1.502\,3 = 0.281\,6$ g of KCl, which contains

$$K = \left(\frac{39.098}{74.551}\right)(0.281\,6) = 0.147\,68 \text{ g}$$

$$Cl = \left(\frac{35.453}{74.551}\right)(0.281\,6) = 0.133\,92 \text{ g}$$

Weight percent of each element:

$$Ba = \frac{0.844\,64}{1.783\,9} = 47.35\%$$

$$K = \frac{0.147\,68}{1.783\,9} = 8.276\%$$

$$Cl = \frac{0.436\,10 + 0.133\,92}{1.783\,9} = 31.95\%$$

D. Let x = mass of $Al(BF_4)_3$ and y = mass of $Mg(NO_3)_3$. We can say that $x + y = 0.2828$ g. We also know that

$$\text{moles of nitron tetrafluoroborate} = 3(\text{moles of } Al(BF_4)_3) = \frac{3x}{287.39}$$

$$\text{moles of nitron nitrate} = 2(\text{moles of } Mg(NO_3)_2) = \frac{2y}{148.31}$$

Equating the mass of product to the mass of nitron tetrafluoroborate plus the mass of nitron nitrate, we can write

$$1.322 = \underbrace{\left(\frac{3x}{287.39}\right)(400.18)}_{\substack{\text{Mass of nitron} \\ \text{tetrafluoroborate}}} + \underbrace{\left(\frac{2y}{148.31}\right)(375.39)}_{\substack{\text{Mass of nitron} \\ \text{nitrate}}}$$

$\underbrace{}_{\substack{\text{Mass of} \\ \text{product}}}$

Making the substitution $x = 0.2828 - y$ in the above equation allows us to find $y = 0.1589$ g of $Mg(NO_3)_2 = 1.072$ mmol of Mg = 0.02605 g of Mg = 9.210% of the original solid sample.

Chapter 26

A. (a) Expected number of red marbles = np_{red} = $(1\,000)(0.12) = 120$. Expected number of yellow = $nq_{yellow} = (1\,000)(0.88) = 880$.

(b) absolute: $\sigma_{red} = \sigma_{yellow} = \sqrt{npq}$

$$= \sqrt{(1\,000)(0.12)(0.88)} = 10.28$$

relative: $\sigma_{red}/n_{red} = 10.28/120 = 8.56\%$

$$\sigma_{yellow}/n_{yellow} = 10.28/880 = 1.17\%$$

(c) For $4\,000$ marbles, $n_{red} = 480$ and $n_{yellow} = 3\,520$.

$$\sigma_{red} = \sigma_{yellow} = \sqrt{npq} = \sqrt{(4\,000)(0.12)(0.88)} =$$

20.55. $\sigma_{red}/n_{red} = 4.28\%$. $\sigma_{yellow}/n_{yellow} = 0.58\%$

(d) $2, \sqrt{n}$

(e) $\dfrac{\sigma_{red}}{n_{red}} = 0.02 = \dfrac{\sqrt{n(0.12)(0.88)}}{(0.12)n}$

$$\Rightarrow n = 1.83 \times 10^4$$

B. (a) $mR^2 = K_s \Rightarrow m(10)^2 = 36 \Rightarrow m = 0.36$ g

(b) An uncertainty of ± 20 counts per second per gram is $100 \times 20/237 = 8.4\%$.

$$n = \frac{t^2 s_s^2}{e^2} \approx \frac{(1.96)^2(0.10)^2}{(0.084)^2} = 5.4 \approx 5$$

$$\Rightarrow t = 2.776 \quad (4 \text{ degrees of freedom})$$

$$n \approx \frac{(2.776)^2(0.10)^2}{(0.084)^2} = 10.9 \approx 11 \Rightarrow t = 2.228$$

$$n \approx \frac{(2.228)^2(0.10)^2}{(0.084)^2} = 7.0 \approx 7 \Rightarrow t = 2.447$$

$$n \approx \frac{(2.447)^2(0.10)^2}{(0.084)^2} = 8.5 \approx 8 \Rightarrow t = 2.365$$

$$n \approx \frac{(2.365)^2(0.10)^2}{(0.084)^2} = 7.9 \approx 8$$

C. 1. The figure shows that coprecipitation occurs in distilled water, so it should first be shown that coprecipitation occurs in seawater, which has a high salt concentration. This can be done by adding standard Pb^{2+} to real seawater and to artificial seawater (NaCl solution) and repeating the experiment in the figure.

2. To demonstrate that coprecipitation of Pb^{2+} does not decrease the concentration of alkyl lead compounds, artificial seawater samples containing known quantities of alkyl lead compounds (but no Pb^{2+}) should be prepared and stripping analysis performed. Then a desired excess of Pb^{2+} would be added to a similar solution containing an alkyl lead compound and coprecipitation with $BaSO_4$ would be carried out. The remaining solution would then be analyzed by stripping voltammetry. If the signal is the same as it was without Pb^{2+} coprecipitation, then it is safe to say that coprecipitation does not reduce the concentration of alkyl lead in the seawater.

D. The acid-soluble inorganic matter and the organic material can probably be dissolved (and oxidized) together by wet ashing with $HNO_3 + H_2SO_4$ in a Teflon-lined bomb in a microwave oven. The insoluble residue should be washed well with water and the washings combined with the acid solution. After the residue has been dried, it can be fused with one of the fluxes in Table 26-5, dissolved in dilute acid, and combined with the previous solution.

E. Nonstoichiometry means that the actual composition of a compound is not exactly what the label says. For example, $CaCO_3$ might contain some $Ca(HCO_3)_2$.

Answers to Problems

Chapter 1

5. **(a)** milliwatt = 10^{-3} watt **(b)** picometer = 10^{-12} meter **(c)** kiloohm = 10^3 ohm **(d)** microfarad = 10^{-6} farad **(e)** terajoule = 10^{12} joule **(f)** nanosecond = 10^{-9} second **(g)** femtogram = 10^{-15} gram **(h)** decipascal = 10^{-1} pascal

6. **(a)** 100 fJ or 0.1 pJ **(b)** 43.1728 nF **(c)** 299.79 THz **(d)** 0.1 nm or 100 pm **(e)** 21 TW **(f)** 0.483 amol or 483 zmol

7. **(a)** 6.0 amol/vesicle **(b)** 3.6×10^6 molecules

8. 7.457×10^4 J/s, 6.416×10^7 cal/h

9. **(a)** 2.0 W/kg **(b)** The person consumes 107 W.

10. **(a)** $\dfrac{0.0254 \text{ m}}{1 \text{ inch}}$, 39.37 inches **(b)** 0.214 $\dfrac{\text{mile}}{\text{s}}$, 770 mile/h **(c)** 1.04×10^3 m, 1.04 km, 0.643 mile

11. 1.47×10^3 J/s, 1.47×10^3 W

13. 1.51 m

16. 1.10 M

17. 5.48 g

18. 10^{-3} g/L, 10^3 µg/L, 1 µg/mL, 1 mg/L

19. 7×10^{-10} M

20. 26.5 g $HClO_4$, 11.1 g H_2O

21. **(a)** 1670 g solution **(b)** 1.18×10^3 g $HClO_4$ **(c)** 11.7 mol

22. **(a)** 3.35×10^{-20} m^3, 3.35×10^{-17} L **(b)** 0.30 M

23. 4.4×10^{-3} M, 6.7×10^{-3} M

24. **(a)** 1046 g, 377 $\dfrac{\text{g}}{\text{L}}$ **(b)** 9.07 m

25. Cal/g, Cal/ounce: shredded wheat (3.6, 102); doughnut (3.9, 111); hamburger (2.8, 79); apple (0.48, 14)

26. 2540 kg F$^-$, 5600 kg NaF

27. **(a)** 2.14×10^{-7} M **(b)** Ar: 3.82×10^{-4} M, Kr: 4.66×10^{-8} M, Xe: 3.6×10^{-9} M

28. 6.18 g in a 2-L volumetric flask

29. Dissolve 6.18 g $B(OH)_3$ in 2.00 kg H_2O.

30. 3.2 L

31. 8.0 g

32. **(a)** 55.6 mL **(b)** 1.80 g/mL

33. 1.51 g/mL

Chapter 2

3. $PbSiO_3$ is insoluble and will not leach into groundwater.

4. The lab notebook must (1) state what was done; (2) state what was observed; and (3) be understandable to a stranger.

5. See Section 2-3.

6. The buoyancy correction is zero when the substance being weighed has the same density as the weight used to calibrate the balance.

7. 14.85 g

8. smallest: lead dioxide; largest: lithium

9. 190 Hz

10. (a) 0.000166 g/mL (b) 0.823 g

11. (a) 7.4 torr (b) 0.00011 g/mL (c) 1.0010 g

20. 9.97909 mL

21. 0.2%; 0.4990 M

22. 49.947 g in vacuum; 49.892 g in air

23. true mass = 50.506 g; mass in air = 50.484 g

Chapter 3

1. (a) 5 (b) 4 (c) 3

2. (a) 1.237 (b) 1.238 (c) 0.135 (d) 2.1
   (e) 2.00

3. (a) 0.217 (b) 0.216 (c) 0.217

4. (b) 1.18 (3 significant figures) (c) 0.71 (2 significant figures)

5. (a) 3.71 (b) 10.7 (c) 4.0×10^1 (d) 2.85×10^{-6} (e) 12.6251 (f) 6.0×10^{-4} (g) 242

6. (a) 208.232 (b) 560.604

7. (a) 12.3 (b) 75.5 (c) 5.520×10^3 (d) 3.04
   (e) 3.04×10^{-10} (f) 11.9 (g) 4.600
   (h) 4.9×10^{-7}

11. (a) Carmen (b) Cynthia (c) Chastity
    (d) Cheryl

12. 3.124 (±0.005), 3.124 (±0.2%)

13. (a) 2.1 (±0.2 or ±11%) (b) 0.151 (±0.009 or ±6%) (c) $0.22_3 \pm 0.02_4$ (±11%) (d) $0.097._1 \pm 0.002_2$ ($\pm 2._2\%$)

14. (a) 10.18 (±0.07 or ±0.7%) (b) 174 (±3 or ±2%) (c) 0.147 (±0.003 or ±2%)
    (d) 7.86 (±0.01 or ±0.1%) (e) 2185.8 (±0.8 or ±0.04%) (f) 1.464_3 ($\pm 0.007_8$ or $\pm 0.5_3\%$)
    (g) 0.496_9 ($\pm 0.006_9$ or $\pm 1.3_9\%$)

16. (b) 0.4507 (±0.0005) M

17. 1.05457267 (±0.00000064) × 10^{-34} J·s

18. 1.0357 (±0.0002) g

Chapter 4

2. (a) 0.6826 (b) 0.9546 (c) 0.3413
   (d) 0.1915 (e) 0.1498

3. (a) 1.52767 (b) 0.00126 (c) 1.59×10^{-6}

4. (a) 0.05004 (b) 0.4037

5. 104.7

6. 0.14111

12. 90%: $0.14_8 \pm 0.02_8$; 99%: $0.14_8 \pm 0.05_6$

13. (a) $\bar{x} \pm 0.00010$ (1.52783 to 1.52803) (b) all, real variation

14. no

15. yes

16. yes

17. 1-2 difference *is* significant; 2-3 difference is not significant.

18. no

19. 1.7_4 ppm

21. Retain 216.

25. slope = −1.299 (±0.001) × 10^4, intercept = 3(±3) × 10^2

26. 30.82 ± 0.05; 30.82 ± 0.16

27. (a) $2.0_0 \pm 0.5_0$ (b) 0.3_8 (c) 0.2_6

28. 5.6 (±0.8) × 10^3

29. (a) $y(\pm 0.17) = -0.159_{75}(\pm 0.019_{15})x + 308._{95}(\pm 38._{00})$

    (b) temperature = 7×10^{-13} K. Extrapolations like this are nonsense.

Chapter 5

4. (a) $K = 1/[Ag^+]^3[PO_4^{3-}]$ (b) $K = P_{CO_2}^6/P_{O_2}^{15/2}$

5. 1.2×10^{10}

6. 2.0×10^{-9}

7. (a) decrease (b) give off (c) negative

8. 5×10^{-11}

9. (a) right (b) right (c) neither (d) right
   (e) smaller

10. (a) 4.7×10^{-4} atm (b) 153°C

11. (a) 7.82 kJ/mol (b) A graph of ln K vs $1/T$ will have a slope of $-\Delta H°/R$.

12. (a) right (b) $P_{H_2} = 1366$ Pa, $P_{Br_2} = 3306$ Pa, $P_{HBr} = 57.0$ Pa (c) neither (d) formed

13. (a) $7._1 \times 10^{-5}$ M (b) $1._0 \times 10^{-3}$ g/100 mL

14. (a) 5.1×10^{-10} M (b) 3.3×10^{-8} g/100 mL
    (c) 0.22 ppb

15. AgCl: 1 400 ppb; AgBr: 76 ppb; AgI: 0.98 ppb

16. 5.2×10^{-7} M

17. true

18. (a) 0.67 g/L   (b) 6.5×10^{-3} g/L

19. 0.018

20. (a) 1.3 mg/L   (b) 1.3×10^{-13} M
    (c) 8.4×10^{-4} M

24. 1.0×10^{-6} M

25. $I^- < Br^- < Cl^- < CrO_4^{2-}$

26. no; 0.001 4 M

28. (a) BF_3 (b) AsF_5

29. 0.096 M

30. $[Zn^{2+}] = 2.9 \times 10^{-3}$ M, $[ZnOH^+] = 2.3 \times 10^{-5}$ M, $[Zn(OH)_3^-] = 6.9 \times 10^{-7}$ M, $[Zn(OH)_4^{2-}] = 8.6 \times 10^{-14}$ M

31. 15%

38. (a) HI   (b) H_2O

39. $2H_2SO_4 \rightleftharpoons HSO_4^- + H_3SO_4^+$

40. (a) (H_3O^+, H_2O) $(H_3\overset{+}{N}CH_2CH_2\overset{+}{N}H_3, H_3\overset{+}{N}CH_2CH_2NH_2)$

    (b) $(C_6H_5CO_2H, C_6H_5CO_2^-)$ $(C_5H_5NH^+, C_5H_5N)$

41. (a) 2.00   (b) 12.54   (c) 1.52   (d) -0.48
    (e) 12.00

42. 7.472, 6.998, 6.132

43. 1.0×10^{-56}

44. 7.8

45. (a) endothermic   (b) endothermic
    (c) exothermic

48. $Cl_3CCO_2H \rightleftharpoons Cl_3CCO_2^- + H^+$

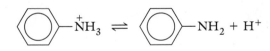

49.

$HOCH_2CH_2S^- + H_2O \rightleftharpoons HOCH_2CH_2SH + OH^-$

50. K_a: $HCO_3^- \rightleftharpoons H^+ + CO_3^{2-}$

    K_b: $HCO_3^- + H_2O \rightleftharpoons H_2CO_3 + OH^-$

51. (a) $H_3\overset{+}{N}CH_2CH_2\overset{+}{N}H_3$

    $\overset{K_{a1}}{\rightleftharpoons} H_2NCH_2CH_2\overset{+}{N}H_3 + H^+$

    $H_2NCH_2CH_2\overset{+}{N}H_3$

    $\overset{K_{a2}}{\rightleftharpoons} H_2NCH_2CH_2NH_2 + H^+$

    (b) $^-O_2CCH_2CO_2^- + H_2O$

    $\overset{K_{b1}}{\rightleftharpoons} HO_2CCH_2CO_2^- + OH^-$

    $HO_2CCH_2CO_2^- + H_2O$

    $\overset{K_{b2}}{\rightleftharpoons} HO_2CCH_2CO_2H + OH^-$

52. (a),   (c)

53. $CN^- + H_2O \rightleftharpoons HCN + OH^-$
    $K_b = 1.6 \times 10^{-5}$

54. $H_2PO_4^- \overset{K_{a2}}{\rightleftharpoons} HPO_4^{2-} + H^+$
    $HC_2O_4^- + H_2O \overset{K_{b2}}{\rightleftharpoons} H_2C_2O_4 + OH^-$

55. $K_{a1} = 7.04 \times 10^{-3}$, $K_{a2} = 6.29 \times 10^{-8}$,
    $K_{a3} = 7.1 \times 10^{-13}$

56. 3.0×10^{-6}

57. (a) 1.2×10^{-2} M   (b) Solubility will be greater.

58. 0.22 g

Chapter 6

1. (a) double   (b) halve   (c) double

2. (a) 184 kJ/mol   (b) 299 kJ/mol

3. 5.33×10^{14} Hz, 1.78×10^4 cm^{-1},
   3.53×10^{-19} J/photon, 213 kJ/mol

5. $\nu = 5.088\,491\,0$ and $5.083\,335\,8 \times 10^{14}$ Hz, $\lambda = 588.985\,54$ and $589.582\,86$ nm, $\tilde{\nu} = 1.697\,834\,5$ and $1.696\,114\,4 \times 10^4$ cm^{-1}

12. 3.56×10^4 M^{-1} cm^{-1}

13. violet blue

14. 2.19×10^{-4} M

15. (a) 325 nm: $T = 0.90$ $A = 0.045$, 300 nm: $T = 0.061$ $A = 1.22$

    (b) 2.0%   (c) $T_{winter} = 0.142$; $T_{summer} = 0.095$; 49%

17. (a) 6.97×10^{-5} M   (b) 6.97×10^{-4} M
    (c) 1.02 mg

18. Yes.

19. (a) $4.97 \times 10^4 \ M^{-1} \ cm^{-1}$ (b) $4.69 \ \mu g$

20. (a) $k = 0.485, b = 0.084$ (b) $4.6_2 \ ng/mL = 82._7 \ nM$

22. $10.1 \ \mu g$

23. (a) absorbance $= 0.266_0 \pm 0.006_1$; blank $= 0.097_0 \pm 0.004_3$ (b) $0.169_0 \pm 0.007_5$ (c) $10.1 \pm 0.5 \ \mu g$ (d) $10.1 \pm 0.4 \ \mu g$

24. (a) $m = 869 \pm 11, b = -22.1 \pm 8.9$ (b) unknown $= 154.0 \pm 1.4 \ mV$, blank $= 9.0 \pm 1.3$ (c) 145.0 ± 1.9 (d) $0.192 \ (\pm 0.011)$

25. (b) $m = 15.12 \ (\pm 0.10), b = 18.9 \ (\pm 3.8), s_y = 5.7$ (c) $31.9 \ (\pm 0.5) \ \mu M$

26. $21.9 \ \mu g$

31. (a) $[Cu^{2+}]_f = [Cu^{2+}]_i \dfrac{V_i}{V_f}$ (b) $1.00 \ ppm$ (c) $1.04 \ ppm$

32. (a) $0.847 \ mM$ (b) $6.16 \ mM$ (c) $12.3 \ mM$

33. $11.3_7 \ (\pm 0.3_9) \ ppm$; no

34. $1.56_2 \ (\pm 0.05_9) \ mM$

35. $9.09 \ mM$

Chapter 7

5. $32.0 \ mL$

6. $43.2 \ mL \ KMnO_4$, $270.0 \ mL \ H_2C_2O_4$

7. $0.149 \ M$

8. 15.1%

9. (a) $0.020\,34 \ M$ (b) $0.125\,7 \ g$ (c) $0.019\,83 \ M$

10. 56.28%

11. 8.17%

12. $0.092\,54 \ M$

15. (a) $2.33 \times 10^{-7} \ mol \ Fe(III)$ (b) $5.83 \times 10^{-5} \ M$

16. Theoretical equivalence point $= 13.3 \ \mu L$. Observed end point $= 12.2 \ \mu L = 1.83$ Fe/transferrin. Ga does not appear to bind in the absence of oxalate.

17. Transmittance decreases and scattering increases up to the end point.

19. 13.08, 8.04, 2.53

20. (a) $20.17 \ mL$ (b) 9.57 (c) 5.11 (d) 2.61

21. (a) $25.0 \ mL$ (b) 24.07

22. 13.91

23. (a) 14.45 (b) 13.80 (c) 8.07 (d) 4.87 (e) 2.61

24. (a) $19.00, 18.85, 18.65, 17.76, 14.17, 13.81, 7.83, 1.95$ (b) no

25. $V_{e1} = 18.76 \ mL, V_{e2} = 37.52 \ mL$

28. $V_X = V_M \ (C_M^0 - [M^+] + [X^-])/(C_X^0 + [M^+] - [X^-])$

30. 99.723% of Br^- and 0.277% of Cl^- are precipitated at the first equivalence point.

37. $947 \ mg$

38. upper $= 9.6 \times 10^3 \ M$ (which means there is no upper limit), lower $= 6.2 \times 10^{-4} \ M$

Chapter 8

2. (a) true   (b) true   (c) true

3. (a) $0.008\,7 \ M$ (b) $0.001\,2 \ M$ (c) $0.08 \ M$ (d) $0.30 \ M$

5. (a) 0.660 (b) 0.54 (c) 0.18 (d) 0.83

6. 0.88_7

7. (a) 0.42_2 (b) 0.43_2

8. 0.929

9. increase

10. 0.33

11. (a) $0.80, 0.56, 0.29, 0.19, 0.20, 0.22, 0.26, 0.37$;   (b) $0.28_6 \pm 0.04_3$

13. (a) $2.6 \times 10^{-8} \ M$ (b) $3.0 \times 10^{-8} \ M$ (c) $4.2 \times 10^{-8} \ M$

14. $1.5 \times 10^{-6} \ M$

15. $2.5 \times 10^{-3} \ M$

16. (a) $9.2 \ mM$ (b) Remainder is ion-paired $Ca^{2+}SO_4^{2-}(aq)$.

17. $[Li^+] = [F^-] = 0.050\,15 \ M$

18. $[Ca^{2+}] = 0.017\,3 \ M$

19. 13.61

20. $\gamma_{H^+} = 0.86$, pH $= 2.07$

21. $11.94, 12.00$

Chapter 9

2. $[H^+] + 2[Ca^{2+}] + [Ca(HCO_3)^+] + [Ca(OH)^+] + [K^+] = [OH^-] + [HCO_3^-] + 2[CO_3^{2-}] + [ClO_4^-]$

3. $[H^+] = [OH^-] + [HSO_4^-] + 2[SO_4^{2-}]$

4. $[H^+] = [OH^-] + [H_2AsO_4^-] + 2[HAsO_4^{2-}] + 3[AsO_4^{3-}]$

5. (a) $2[Mg^{2+}] + [H^+] = [Br^-] + [OH^-]$

    (b) $2[Mg^{2+}] + [H^+] + [MgBr^+] = [Br^-] + [OH^-]$

6. 5.2×10^5 pounds

8. (a) $0.20\ M = [Mg^{2+}]$ (b) $0.40\ M = [Br^-]$
   (c) $0.20\ M = [Mg^{2+}] + [MgBr^+]$
   (d) $0.40\ M = [Br^-] + [MgBr^+]$

9. $[CH_3CO_2^-] + [CH_3CO_2H] = 0.1\ M$

10. $[Y^{2-}] = [X_2Y_2^{2+}] + 2[X_2Y^{4+}]$

12. 7.5×10^{-6}

13. $[OH^-] = 1.5 \times 10^{-7}\ M$

14. (a) $12._4$ (b) $[Sr^{2+}] = 1.6 \times 10^{-4}\ M$, $[F^-] = 4.3 \times 10^{-3}\ M$, $[Pb^{2+}] = 2.0 \times 10^{-3}\ M$

15. (a) $0.10 = [M^{2+}] + \dfrac{2K_1[M^{2+}]^2}{1-K_1[M^{2+}]} + \dfrac{4K_1K_2[M^{2+}]^3}{(1-K_1[M^{2+}])^2}$

16. (a) $[Zn^{2+}] = 5.00 \times 10^{-3}\ M$
    (b) $[Zn^{2+}] = 6.64 \times 10^{-3}\ M$
    (c) $[Zn^{2+}] = 6.85 \times 10^{-3}\ M$, 32% ion-paired
    (d) $\gamma_\pm = 0.579$

18. $[Ca^{2+}] = 1.4 \times 10^{-3}\ M$, $[F^-] = 1.7 \times 10^{-4}\ M$, $[HF] = 2.5 \times 10^{-3}\ M$

19. $4.0 \times 10^{-5}\ M$

20. $5.8 \times 10^{-4}\ M$

21. (a) $5.0 \times 10^{-9}\ M$ (b) $5 \times 10^{-9}\ M$
    (c) $1.1 \times 10^{-8}\ M$

22. $4.3 \times 10^{-3}\ M$

23. (a) $[NH_4^+] + [H^+] = 2[SO_4^{2-}] + [HSO_4^-] + [OH^-]$

    (b) $[NH_3] = [NH_4^+] = 2\{[SO_4^{2-}] + [HSO_4^-]\}$

    (c) $6.59\ M$

24. (a) $[FeG^+] + [H^+] = [G^-] + [OH^-]$ (b) $[FeG_2] + [FeG^+] = 0.0500\ M$, $[FeG^+] + 2[FeG_2] + [G^-] + [HG] = 0.100\ M$ (c) $0.016_4\ M$

27. 7.307

28. 9.876

Chapter 10

2. (a) 3.00 (b) 12.0

3. $6.89, 0.61$

4. (a) 0.809 (b) 0.791 (c) Activity coefficient depends slightly on counterion.

5. (a)

   (b)

   (c)

   (d)

6. $pH = 3.00$, $\alpha = 0.995\%$

7. 5.50

8. 5.51, $3.1 \times 10^{-6}\ M$, $0.060\ M$

9. 99% dissociation when $F = (0.010\,2)K_a$

10. 4.19

11. 5.79

12. (a) $3.03, 9.4\%$ (b) $7.00, 99.9\%$

13. $0.005\,4\%$

14. $2.93, 0.118$

15. (a) endothermic (b) exothermic

16. For 10^{-5} F HA, $[H^+]$ and $[A^-]$ differ by more than 10% when $pK_a \geq 8$; for 10^{-4} F HA, $pK_a \geq 9$; for 10^{-3} F HA, $pK_a \geq 10$.

17. $11.00, 0.995\%$

18. 11.28, $[B] = 0.058\ M$, $[BH^+] = 1.9 \times 10^{-3}\ M$

19. 10.95

20. $0.007\,56\%$, $0.023\,9\%$, 0.568%

21. 3.6×10^{-9}

22. 4.0×10^{-5}

24. pH = 9.978, 7.150

30. 4-aminobenzenesulfonic acid

31. 4.70

32. 0.180, 1.00, 1.80

33. **(a)** 14 **(b)** 1.4×10^{-7}

34. **1.** Weigh out $(0.250)(0.0500) = 0.0125$ mol of HEPES and dissolve in ~200 mL. **2.** Adjust the pH to 7.45 with NaOH (HEPES is an acid). **3.** Dilute to 250 mL.

35. 3.27 mL

36. **(b)** 7.18 **(c)** 7.00 **(d)** 6.86 mL

37. **(a)** 2.56 **(b)** 2.61 **(c)** 2.86

38. 13.7 mL

39. [HA] = 0.00199 M, $[A^-]$ = 0.00401 M

40. 11.70, 11.48

41. **(a)** 6.001 **(b)** 6.371 **(c)** 2.074 **(d)** 11.617

Chapter 11

2.
$$\overset{\underset{\displaystyle |}{R}}{H_3\overset{+}{N}-CH-CO_2^-} \quad pK \text{ values apply to } -\overset{+}{N}H_3,$$
$-CO_2H$, and, in some cases, R.

3. 4.37×10^{-4}, 8.93×10^{-13}

4. **(a)** pH = 2.51, $[H_2A]$ = 0.0969 M, $[HA^-]$ = 3.11×10^{-3}, $[A^{2-}]$ = 1.00×10^{-8} M

   (b) 6.00, 1.00×10^{-3} M, 1.00×10^{-1} M, 1.00×10^{-3} M

   (c) 10.50, 1.00×10^{-10} M, 3.16×10^{-4} M, 9.97×10^{-2} M

5. **(a)** pH = 1.95, $[M^{2-}]$ = 2.01×10^{-6} M

   (b) pH = 4.28, $[H_2M]$ = 3.7×10^{-3} M

   (c) pH = 9.35, $[H_2M]$ = 7.04×10^{-12} M

6. pH = 11.60. [B] = 0.296 M, $[BH^+]$ = 3.99×10^{-3} M, $[BH_2^{2+}]$ = 2.15×10^{-9} M

7. pH = 3.69, $[H_2A]$ = 2.9×10^{-6} M, $[HA^-]$ = 7.9×10^{-4} M, $[A^{2-}]$ = 2.1×10^{-4} M

8. 3.96

9. 5.72

10. **(a)** $\Delta H°$ = -12 (± 1) kJ/mol, $\Delta S°$ = -27 (± 4) J/(mol·K)

    (b) 5.0 (± 3.6)

11.

	(a)	(b)	(c)
pH	6.002	4.504	4.266
$[HA^-]$	0.0098 M	0.0061 M	0.0048 M

12. 2.96 g

13. 2.23 mL

14. Procedure: Dissolve 10.0 mmol (1.23 g) picolinic acid in ~ 75 mL H_2O in a beaker. Add NaOH (~ 5.63 mL) until the measured pH is 5.50. Transfer to a 100-mL volumetric flask and use small portions of H_2O to rinse the beaker into the flask. Dilute to 100.0 mL and mix well.

15. 22.20 g Na_2SO_4 + 4.29 g H_2SO_4

16. no

17.

glutamic acid

tyrosine

18. 2.8×10^{-3}, 3.4×10^{-8}

19. pH = 5.70
    $[H_2L^+] = 0.010\,0$ M,
    $[H_3L^{2+}] = 2.19 \times 10^{-6}$ M,
    $[HL] = 4.17 \times 10^{-6}$ M,
    $[L^-] = 4.19 \times 10^{-11}$ M

20. (a) 5.88 (b) 5.59

21. (a) HA (b) A^- (c) 1.0, 0.10

22. (a) 4.00 (b) 8.00 (c) H_2A (d) HA^-
    (e) A^{2-}

23. (a) 9.00 (b) 9.00 (c) BH^+ (d) 1.0×10^3

24.

25. $\alpha_{HA} = 0.091$, $\alpha_{A^-} = 0.909$, $[A^-]/[HA] = 10$

26. 0.91

27. $\alpha_{HA^-} = 0.123$, 0.694

28. $\alpha_{HA^-} = 0.110$, 0.500, 0.682, 0.500,
    2.15×10^{-4}

29. (b) 8.6×10^{-6}, 0.61, 0.39, 2.7×10^{-6}

30. 0.37

31. 18%

33. At pH 10: $\alpha_{H_3A} = 1.53 \times 10^{-9}$ $\alpha_{H_2A^-} = 0.104$
    $\alpha_{HA^{2-}} = 0.669$ $\alpha_{A^{3-}} = 0.227$

34. (b) $[Cr(OH)_3(aq)] = 10^{-6.84}$ M $[Cr(OH)_2^+] = 10^{-4.44}$ M $[Cr(OH)^{2+}] = 10^{-2.04}$ M

36. The *average* charge is zero. There is no pH at which *all* molecules have zero charge.

37. isoelectric pH 5.59, isoionic pH 5.72

Chapter 12

2. 13.00, 12.95, 12.68, 11.96, 10.96, 7.00, 3.04, 1.75

4. 3.00, 4.05, 5.00, 5.95, 7.00, 8.98, 10.96, 12.25

5. $V_e/11$; $10\,V_e/11$; $V_e = 0$: pH = 2.80; $V_e/11$: pH = 3.60; $V_e/2$: pH = 4.60; $10V_e/11$: pH = 5.60; V_e: pH = 8.65; $1.2\,V_e$: pH = 11.96

6. 8.18

7. 7.1×10^7

8. 0.092 M

9. 9.72

11. 11.00, 9.95, 9.00, 8.05, 7.00, 5.02, 3.04, 1.75

12. $\frac{1}{2}V_e$

13. 2.2×10^9

14. 10.92, 9.57, 9.35, 8.15, 5.53, 2.74

15. (a) 9.45 (b) 2.55 (c) 5.15

19. 11.49, 10.95, 10.00, 9.05, 8.00, 6.95, 6.00, 5.05, 3.54, 1.79

20. 2.51, 3.05, 4.00, 4.95, 6.00, 7.05, 8.00, 8.95, 10.46, 12.21

21. 11.36, 10.21, 9.73, 9.25, 7.53, 5.81, 5.33, 4.85, 3.41, 2.11, 1.85

22. 5.09

23. (a) 1.99

24. (b) 7.18

25. 2.66

26. (a) 9.56 (b) 7.4×10^{-10}

27. 6.28 g

28. $pK_2 = 9.84$

30. 23.40 mL

31. end point = 10.727 mL

34. yellow, green, blue

35. red, orange, yellow

36. red, orange, yellow, red

37. no

38. (a) 2.47

39. violet, blue, yellow

40. 5.62

41. 2.859%

42. 0.139 M

43. 0.815

44. 4.00

48. 0.079 34 mol/kg

49. 0.100 0 M

50. 0.30 g

51. (a) 20.254% (b) 17.985 g

60. (b) 4.16

Chapter 13

2. (a) 3.4×10^{-10} (b) 0.64

3. (a) 3.3×10^7 (b) 3.9×10^{-5} M

4. 5.00 g

5. (a) 100.0 mL (b) 0.01667 M (c) 0.054
   (d) 5.4×10^{10} (e) 6.8×10^{-7}
   (f) 1.9×10^{-10} M

6. (a) 2.93 (b) 6.77 (c) 10.49

7. 49.9: 4.87; 50.0: 6.90; 50.1: 8.92

8. 49.9: 7.55; 50.0: 6.21; 50.1: 4.88

9. 2.7×10^{-11} M

14. (a) 4.3×10^3 (b) 0.017

15. 0 mL: pCu^{2+} = 11.08; 1.00 mL: pCu^{2+} =
    11.09; 45.00 mL: pCu^{2+} = 12.35; 50.00 mL:
    pCu^{2+} = 15.06; 55.00 mL: pCu^{2+} = 17.73

16. (b) α_{ML} = 0.28, α_{ML_2} = 0.70

17. (d) [T] = 0.27773, [Fe_aT] = 0.55532, [Fe_bT] =
    0.09222, [Fe_2T] = 0.07773

23. 1. With metal ion indicators. 2. With a mer-
    cury electrode. 3. With a glass electrode.
    4. By polarography.

24. HIn^{2-}, wine-red, blue

25. Buffer A

30. 10.0 mL, 10.0 mL

31. 0.02000 M

32. 0.995 mg

33. 21.45 mL

34. [Ni^{2+}] = 0.01224 M, [Zn^{2+}] = 0.007 18 M

35. 0.024 30 M

36. 0.092 28 M

Chapter 14

2. (a) 6.2415063×10^{18} (b) 96485.309

3. (a) $71._5$ A (b) 4.35 A (c) 79 W

4. (a) 1.87×10^{16} e^-/s (b) 9.63×10^{-19} J/e^-
   (c) 5.60×10^{-5} mol (d) 447 V

5. (a) I_2 (b) $S_2O_3^{2-}$ (c) 861 C (e) 14.3 A

7. (a) $Fe(s) | FeO(s) | KOH(aq) | Ag_2O(s) | Ag(s)$;
   $Fe(s) + 2OH^- \rightleftharpoons FeO(s) + H_2O + 2e^-$; $Ag_2O(s)$
   $+ H_2O + 2e^- \rightleftharpoons 2Ag(s) + 2OH^-$

   (b) $Pb(s) | PbSO_4(s) | K_2SO_4(aq) \| H_2SO_4(aq) |$
   $PbSO_4(s) | PbO_2(s) | Pb(s)$; $Pb(s) + SO_4^{2-} \rightleftharpoons$
   $PbSO_4(s) + 2e^-$; $PbO_2 (s) + 4H^+ + SO_4^{2-} + 2e^-$
   $\rightleftharpoons PbSO_4(s) + 2H_2O$

10. Cl_2

12. (a) Fe^{3+} (b) Fe^{2+}

14. (b) -0.359 V

15. (b) -2.854 V (c) Br_2 (d) 1.31 kJ
    (e) 2.69×10^{-8} g/s

16. (a) 0.572 V (b) 0.568 V

17. 0.054 20, 0.061 54

18. 0.799 3 V

19. $HOBr + 2e^- + H^+ \rightleftharpoons Br^- + H_2O$; 1.341 V

20. $3X^+ \rightleftharpoons X^{3+} + 2X(s)$ $E_2^\circ > E_1^\circ$

21. 0.580 V

22. (a) 1.33 V (b) 9×10^{44}

23. (a) $K = 10^{47}$ (b) $K = 1.9 \times 10^{-6}$

24. (b) $K = 2 \times 10^{16}$ (c) -0.02_0 V (d) 10 kJ
    (e) 0.21

25. $K = 1.0 \times 10^{-9}$

26. 0.101 V

27. 34 g/L

28. 0.117 V

29. -1.664 V

30. -0.447 V

31. $K = 3.2 \times 10^5$

33. (d) 0.14_3 M

34. (b) A = -0.414 V, B = 0.059 16 V
    (c) $Hg \rightarrow Pt$

35. 9.6×10^{-7}

36. 5.4×10^{14}

37. 0.76

38. 7.5×10^{-8}

40. (c) 0.317 V

41. -0.041 V

42. -0.158 V

43. -0.036 V

44. 7.2×10^{-4}

45. (a) $[Ox] = 3.82 \times 10^{-5}$ M, $[Red] = 1.88 \times 10^{-5}$ M

    (b) $[S^-] = [Ox]$, $[S] = [Red]$ (c) -0.092 V

Chapter 15

1. (b) 0.044 V

3. 0.684 V

4. 0.243 V

5. 0.627

6. (c) 0.068 V

8. 0.481 V; 0.445 V; 0.194 V; -0.039 V

9. (a) $K_I/0.033\,3$ (b) $K_{Cl}/0.020\,0$
   (d) 2.2×10^6

10. $5._2 \times 10^{21}$

11. (b) $1._2 \times 10^{11}$

12. 0.29_6 M

15. left

16. H^+ : 42.4 s; NO_3^- : 208 s

17. (a) $3._2 \times 10^{13}$ (b) 8% (c) 49.0, 8%.

19. (a) 26.9 mV

22. 10.67

25. 0.10 pH unit

26. (a) 274 mV (b) 285 mV

32. (a) -0.407 V (b) $1.5_5 \times 10^{-2}$ M
    (c) $1.5_2 \times 10^{-2}$ M

33. $+0.0296$ V

34. (a) K^+ (b) Group I > Group II

35. 3.8×10^{-9} M

36. (a) 2.4×10^{-3} M (b) $0.951\,3$

37. (b) $2.43\ (\pm 0.09) \times 10^{-3}$ M

38. -0.332 V

39. (a) 2.32×10^{-4} M (c) Divide each ΔE by 2.

41. $E = 120.2 + 28.80 \log ([Ca^{2+}]) + 6.0 \times 10^{-4} [Mg^{2+}]$

42. 1.4×10^{-4} M

43. (a) 8.9×10^{-8} M (b) 1.90 mmol

44. (a) 1.13×10^{-4} (b) 4.8×10^4

Chapter 16

2. (d) 0.490, 0.526, 0.626, 0.99, 1.36, 1.42, 1.46 V

3. (d) 1.58, 1.50, 1.40, 0.733, 0.065, 0.005, -0.036 V

4. (d) -0.120, -0.102, -0.052, 0.21, 0.48, 0.53 V

6.

	$0.01\ V_e$	$0.5\ V_e$	$0.99\ V_e$	$0.999\ V_e$
a	-0.779	-0.661	-0.543	-0.483
b	0.321	0.439	0.557	0.617
c	0.609	0.491	0.373	0.314
d	0.091	0.060	-0.025	-0.034

	V_e	$1.01\ V_e$	$1.1\ V_e$	$2\ V_e$
a	-0.067	0.408	0.467	0.526
b	1.049	1.148	1.159	1.171
c	0.223	0.149	0.119	0.090
d	-0.035	-0.046	-0.072	-0.102

9. -0.107, -0.102, 0.213, 0.511, 0.529, 0.839, 0.984 V

10. (e) $E = 0.086$ V at V_{e1}; $E = 0.797$ V at V_{e2}

11. (d) 0.273, 0.434, 0.767, 1.458 V

12. diphenylamine sulfonic acid: colorless → red-violet; diphenylbenzidine sulfonic acid: colorless → violet; *tris*-(2,2'-bipyridine)iron: red → pale blue; ferroin: red → pale blue

13. no

14. no

15. 0.631, 0.691, 0.767, 0.843, 0.903, 1.055, 1.147, 1.326, 1.487, 1.505, 1.561, 1.623 V

16. -0.207 V

22. 0.011 29 M

23. 3.826 mM

24. 41.9%

25. 78.67%

26. oxidation number = 3.761; 217 µg/g

29. iodometry

30. (a) 0.029 14 M (b) no

31. (a) 7×10^2 (b) 1.0 (c) 0.34 g/L

32. mol NH_3 = 2 (initial mol H_2SO_4) − mol thiosulfate

33. 5.730 mg

34. **(a)** 0.125   **(b)** 6.875 ± 0.038

35. Bi oxidation state = +3.200 0 (±0.003 3)
    Cu oxidation state = +2.200 1 (±0.004 6)
    formula = $Bi_2Sr_2CaCu_2O_{8.4001}$ (±0.0057)

Chapter 17

5. 2.68 h

6. −1.23 V

7. **(a)** −1.906 V   **(b)** 0.20 V   **(c)** −2.71 V
   (d) −2.82 V

8. **(a)** 0.84 V   **(b)** −1.06 V

9. 0.53 V

10. **(a)** 6.64×10^3 J   **(b)** 0.0124 g/h

13. V_2

16. 54.77%

17. −0.619 V, negative

18. **(a)** anode   **(b)** 52.0 g/mol   **(c)** 0.0396 M

19. −2.178 V

20. −0.744 V

21. yes

24. **(b)** 0.000 26 mL

25. **(a)** 5.32×10^{-5} mol   **(b)** 2.66×10^{-5} mol
    (c) 5.32×10^{-3} M

26. **(a)** 1.946 mmol   **(b)** 0.041 09 M   **(c)** 4.75 h

27. 26.3% trichloroacetic acid, 49.5% dichloroacetic
    acid

28. 1.51×10^2 μg/mL

29. **(a)** current density = 1.00×10^2 A/m^2,
    overpotential = 0.85 V

    (b) −0.036 V   **(c)** 1.160 V   **(d)** −2.57 V

30. $96\,486.6_7$ ±0.2_8 C/mol

Chapter 18

6. 1.03×10^{-9} m^2/s

7. **(a)** $0.000\,34_8$ min^{-1} **(b)** 3.4 min   **(c)** 0.118%

8. $E_{3/4} - E_{1/4} = 0.0565/n$ V

9. 1.7 mM

10. 1.21 mM

11. 31.2 μg NO_2^-/g; 67.6 μg NO_3^-/g

12. 2.37 (±0.02) mM

13. 0.096 mM

14. 0.000 35 wt%

15. −0.01 V

17. 0.90 μg/L = 1.4×10^{-8} M

18. 0.53 μg/mL

19. Peak B: RNHOH → RNO + 2H$^+$ + 2e$^-$
    Peak C: RNO + 2H$^+$ + 2e$^-$ → RNHOH.
    There was no RNO present before the initial
    scan.

21. Fe^{2+}

22. 7.8×10^{-10} m^2/s

25. **(c)** $E' = 0.853$ V (vs. S.H.E.)   **(d)** $n \approx 2$

30. **(a)** Similar to curve (b) in Exercise 18-F.

    (b) Similar to the curve in Demonstration 18-1.

31. **(c)** 4.3 (±0.2) μM

Chapter 19

1. [X] = 8.03×10^{-5} M, [Y] = 2.62×10^{-4} M

2. [X] = 1.78×10^{-4} M, [Y] = 8.42×10^{-5} M

5. [A] = 9.11×10^{-3} M, [B] = 4.68×10^{-3} M

7. [X] = 6.47×10^{-5} M, [Y] = 1.42×10^{-5} M,
   [Z] = 3.98×10^{-5} M

8. **(b)** $K = 88.2$

9. **(b)** $K = 0.464$, $\epsilon = 1.07 \times 10^4$ M^{-1} cm^{-1}

10. **(a)** 1:1

12. $\Delta E(T_1 - S_1) = 36$ kJ/mol

14. Wavelength: absorption < fluorescence <
    phosphorescence

21. 78 (±3) μL

Chapter 20

3. 0.78

4. 0.970

5. **(a)** 34°   **(b)** 0°

6. 77 K: 1.99 W/m^2; 298 K: 447 W/m^2

7. 71% (independent of thickness)

9. **(a)** 80.7°   **(b)** 0.955

10. **(a)** 4.2×10^2 cm^{-1} **(b)** 0.036   **(c)** 0.86
    (d) 0.80

11. 0.051 cm^{-1}

12. 2×10^{-4}

13. **(b)** blue

14. **(a)** $61.04°$ **(b)** $51.06°$

15. $n_{prism} > \sqrt{2}$

16. quartz or MgF$_2$

17. **(a)** $M_\lambda = 8.789 \times 10^9$ W/m^3 at 2.00 μm; $M_\lambda = 1.164 \times 10^9$ W/m^3 at 10.00 μm

    (b) 1.8×10^2 W/m^2 **(c)** 2.3×10^1 W/m^2

    (d) $\dfrac{M_{2.00\ \mu m}}{M_{10.00\ \mu m}} = 7.551$ at 1 000 K

    $\dfrac{M_{2.00\ \mu m}}{M_{10.00\ \mu m}} = 3.163 \times 10^{-22}$ at 100 K

19. D$_2$

25. **(a)** 2.38×10^3 **(b)** 143

27. **(a)** 1.7×10^4 **(b)** 0.05 nm **(c)** 5.9×10^4
    (d) $0.000\,43°$, $0.013°$

28. $T = 0.036\,6$, $A = 1.436$

29. $0.124\,2$ mm

30. **(a)** ± 2 cm **(c)** 0.5 cm^{-1} **(d)** 2.5 μm

32. 7

Chapter 21

7. 589.3 nm

8. 0.025

9. Na: 3.3 GHz; Hg: 2.6 GHz

10. **(a)** 283.0 kJ/mol **(b)** 3.67×10^{-6} **(c)** 8.4%
    (d) 1.03×10^{-2}

12. 0.079 μg/mL

16. **(b)** 204 μg/mL

17. **(a)** 7.49 μg/mL **(b)** 25.6 μg/mL

18. 17.4 μg/mL

19. 4.54 μM

Chapter 22

2. 3

7. **(a)** 0.080 M **(b)** 0.50

8. 0.088

9. **(c)** 4.5 **(d)** greater

11. **(a)** 0.16 M in benzene **(b)** 2×10^{-6} M in benzene

12. 2 pH units

13. **(a)** 2.6×10^4 at pH 1 and 2.6×10^{10} at pH 4
    (b) 3.8×10^{-4}

15. 1-C, 2-D, 3-A, 4-E, 5-B

18. **(a)** 17.4 cm/min **(b)** 0.592 min
    (c) 6.51 min

19. **(a)** 13.9 m/min, 3.00 mL/min **(b)** $k' = 7.02$, fraction of time = 0.875 **(c)** 295

20. **(a)** 40 cm long × 4.25 cm diameter
    (b) 5.5 mL/min **(c)** 1.11 cm/min for both

21. **(a)** 2.0 **(b)** 0.33 **(c)** 20

22. 19 cm/min

23. 0.6, 6

24. $K = 4.69$, $k' = 3.59$

25. 603, 0.854

27. 0.1 mm

30. 33 mL/min

34. 2.65 mm

35. **(a)** 1.1×10^2 **(b)** 0.89 mm

36. 138

37. resolution = 0.83

38. 10.4 mL

39. 110 s^2, 43 s^2, 26.9 s

40. 4.0×10^5

41. **(a)** 9.8×10^3 **(b)** 2.6×10^3 **(c)** 8.2×10^3

42. **(a)** $k' = 11.25, 11.45$ **(b)** 1.018 **(c)** C$_6$HF$_5$: 60 800 plates, height = 0.493 mm; C$_6$H$_6$: 66 000 plates, height = 0.455 mm **(d)** C$_6$HF$_5$: 55 700 plates, C$_6$H$_6$: 48 800 **(e)** 0.96 **(f)** 0.94

Chapter 23

6. 38–75 μm; 200/400 mesh

7. **(a)** hexane < butanol < benzene < 2-pentanone < heptane < octane

    (b) hexane < heptane < butanol < benzene < 2-pentanone < octane

    (c) hexane < heptane < octane < benzene < 2-pentanone < butanol

8. **(a)** 3,1,2,4,5,6 **(b)** 3,4,1,2,5,6 **(c)** 3,4,5,6,2,1

9. (a) 4.7 min, 1.3 (b) 1.8

10. 836

11. 27.1 min

12. 0.41 μM

13. 932

14. (a) 0.58, 1.9 (b) 0.058 mm, 0.19 mm
     (c) 3.0×10^5 (d) 4

15. 3.2

22. (a) shorter (b) amine

23. 126 mm

24. 0.418 mg/mL

26. (a) 30% tetrahydrofuran (b) 30%
     tetrahydrofuran (c) 14%

27. $t_1 = 23.5$ min, $t_2 = 71.2$ min, $w_{av} = 6.2$ min

28. 0.27 m^2

Chapter 24

6. Exchange position of buffers.

8. (a) 30 (b) 3.3 (c) decrease

9. 38.0%

11. (b) 29 ng/mL

14. (a) 40.2 mL (b) 0.42

15. transferrin: 0.127; ferric citrate: 0.789

16. (a) 2000 (b) 300

17. (a) 5.7 mL (b) 11.5 mL (c) solutes must be
     adsorbed

18. 320 000

23. (a) 0.167 mm (b) 0.016 s (c) 0.000 40 s
     (d) 0.048 s

26. (a) 1.15×10^4 Pa (b) 1.17 m

27. (a) 29.5 fmol (b) 3.00×10^3 V

28. 9.2×10^4 plates, 4.1×10^3 plates (My measure-
     ments are about $1/3$ lower than the values la-
     beled in the figure from the original source.)

29. (a) maleate (b) fumarate is eluted first
     (c) maleate is eluted first

30. (a) pH 2: 920 s; pH 12: 150 s (b) pH 2:
     never; pH 12: 180 s

32. 4.7×10^4 plates, 13 μm

33. 20.5 min

34. 2.0×10^5 plates

35. Thiamine < (niacinamide + riboflavin) < niacin.
     Thiamine is most soluble.

36. 5.55

Chapter 25

9. 0.022 86 M

10. 1.94 wt%

11. 0.086 57 g

12. 50.80 wt%

13. 0.191 4 g, 0.107 3 g

14. 104.1 ppm

15. 7.22 mL

16. 0.339 g

17. 14.5 wt% K_2CO_3, 14.6 wt% NH_4Cl

18. 40.4 wt%

19. 22.65 wt%

20. (a) 40.05 wt% (b) 39%

21. (a) 1.82 (b) $Y_2O_2(OH)Cl$ or $Y_2O(OH)_4$

22. (b) 0.204 ($\pm$0.004)

23. (a) 5.5 mg/100 mL (b) 5.834 mg; yes

28. 11.69 mg CO_2, 2.051 mg H_2O

30. $C_4H_9NO_2$

31. 10.5 wt%

32. $C_8H_{9.06 \pm 0.17}N_{0.997 \pm 0.010}$

33. 12.4 wt%

Chapter 26

3. (a) 5% (b) 2.6%

5. 1.0 g

6. 120/170 mesh

7. $10^4 \pm 0.99\%$

8. (a) 15.8 (b) 1.647 (c) 474–526

9. 95%: 8, 90%: 6

10. (a) 5.0 g (b) 7

11. (a) Na_2CO_3: 4.47 μg, 8.94×10^5 particles
     K_2CO_3: 4.29 μg, 2.24×10^7 particles
     (b) 2.33×10^4
     (c) Na_2CO_3: 3.28%, K_2CO_3: 0.131%

12. Zn, Fe, Co, Al

13. Avoids possible explosion

18. (a) 53

19. 64.90 wt%

20. 81.1 mL

Index

Abbreviations

b = box   f = footnote   p = problem
d = demonstration   i = illustration   r = reference
e = experiment   m = marginal note   t = table

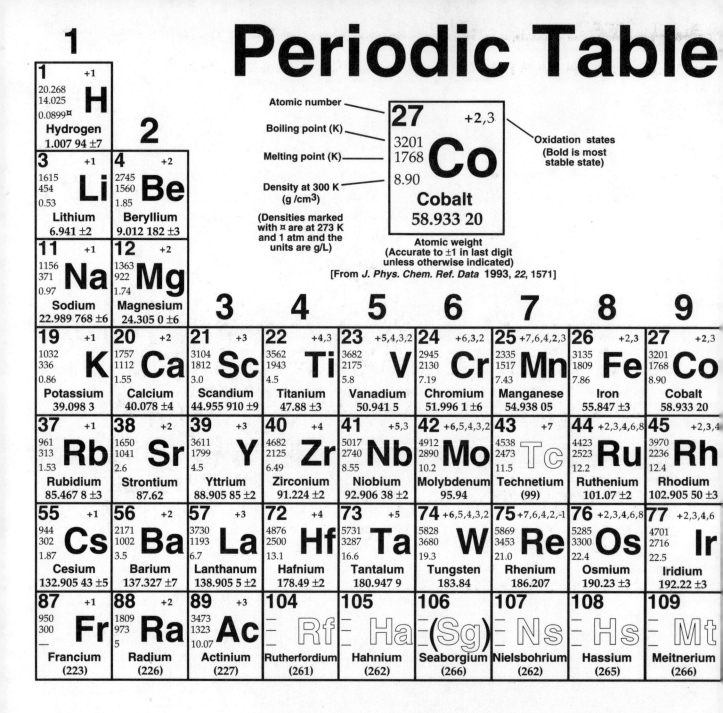
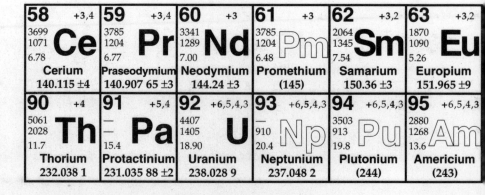